Introduction to Plant Physiology

Introduction to Plant Physiology

Second Edition

William G. Hopkins

The University of Western Ontario

John Wiley & Sons, Inc.

New York • Chichester • Weinheim • Brisbane • Singapore • Toronto

Acquisitions Editor	*David Harris*
Developmental Editor	*Marian Provenzano*
Marketing Manager	*Catherine Beckham*
Production Editor	*Sandra Russell*
Cover and Text Designer	*Karin Gerdes Kincheloe*
Illustration Editor	*Edward Starr*
Cover Photo	*Charles Krebs/Tony Stone Images, N.Y.*

This book was set in 10/12 Janson text by UG and printed and bound by Courier Companies, Inc. (Westford). The cover was printed by Phoenix Color Corporation. The insert was printed by Phoenix Color Corporation.

This book is printed on acid-free paper. 

The paper in this book was manufactured by a mill whose forest management programs include sustained yield harvesting of its timberlands. Sustained yield harvesting principles ensure that the numbers of trees cut each year does not exceed the amount of new growth.

Library of Congress Cataloging in Publication Data:
Hopkins, William G.
Introduction to plant physiology, 2e/William G. Hopkins

ISBN 0-471-19281-3

Printed in the United States of America

10 9 8 7 6 5 4 3

Preface

This second edition of *INTRODUCTION TO PLANT PHYSIOLOGY* is, like its predecessor, intended as a text for undergraduate students encountering plant physiology for the first time. Its purpose is to introduce the student, using a narrative format, to fundamental concepts of plant physiology within a framework of historical origins and modern approaches. The subject is "whole plant" physiology, in that it blends modern molecular approaches with traditional physiological and biochemical methods and environmental physiology in order to understand how plants work.

The text assumes that the student has completed a first course in botany (or biology with a strong botanical component) and chemistry. It is appropriate for a one-semester course in plant physiology for general students, and as an introduction for those interested in advanced study in plant physiology, environmental plant physiology, or physiological plant ecology.

The organization of the first edition has been retained. Chapter 1 is primarily an introduction and review, touching on the terminology of botany and cell biology, that provides a foundation for the discussions that follow. Part *1*, beginning with Chapter 2, covers the water relations of plants and plant cells and the acquisition of inorganic nutrients. The significance of roots and root-soil interactions is a prominent theme in this section. A separate chapter (Chap. 6) is devoted to the assimilation of nitrogen. Part *2* is built around the common theme of energy transduction and carbon metabolism. I have presented individual photosynthetic and respiratory pathways with an appropriate level of detail while maintaining a holistic overview of carbon metabolism. This section includes a discussion of the leaf as a photosynthetic machine, carbon translocation in the phloem, and a chapter that summarizes the various roles of photosynthesis, respiration, and translocation patterns in determining plant productivity. The focus of Part *3* is plant development and its regulation. This part opens with a general review of patterns in plant development and fundamental aspects of cell enlargement. Plant hormones are covered in two chapters: The first develops the hormone concept in plants and principles of hormone action, while the second discusses the biochemistry and mechanism of hormone action within the context of those principles. The chapter on photomorphogenesis includes a discussion of how phytochrome might function in the natural environment. Part *4* deals with the physiology of plants under stress and concludes with a discussion of plant physiology and biotechnology.

For this edition, many arguments have been rewritten in order to clarify their presentation. Some figures have been redrawn for the same reason and figure legends rewritten to better stand alone. Several topics have been expanded:

- A discussion of aquaporins has been added to Chapter 5.
- The role of carotenoids in Chapter 9 has been expanded to take into account the contribution of the xanthophyll cycle to photoprotection.
- The significance of the role of molecular genetic approaches to the study of hormone action, photoperiodism, and other aspects of development has been amplified in Part *3*.

- A new section on insects and disease stress has been added to Chapter 22.
- The chapter on biotechnology has been expanded and updated to show how rapidly this aspect of plant physiology is moving and its potential for the future.
- A new chapter (Chapter 14, Molecules and Metabolism) has been added, introducing relevant aspects of secondary metabolism that were not covered in the previous edition.

Even with these additions, the philosophy of the previous edition remains. The focus on ideas and experimental approaches is deliberate. The rapid expansion of the plant physiology literature in recent years makes it literally impossible for a book such as this to be current in many areas. Thus it is even more important to focus on a conceptual framework. For the student who takes but one course in plant physiology, this balanced approach will provide a general appreciation of the field. For those who go on in the field, this book provides a foundation in physiological principles that will enable the student to grow with the discipline as it moves forward and new frontiers are expanded.

My criterion for selecting what has been included in this book is simple: What are the fundamental problems facing a plant and has this piece of work significantly advanced our understanding of those problems? Space alone dictates that not all relevant work can be cited. I have therefore chosen citations to reflect the broad range of research literature, general reviews, and opinion on the subject of plant physiology. The number of references cited in the text has also been limited in order to avoid disrupting the narrative and interfering with the flow of ideas that is essential to developing an understanding of a subject. Each chapter also concludes with a list of suggested readings that provides the interested student with an entry into the relevant primary literature. Many topics have also been presented within a historical perspective. I believe students benefit when they have a sense of where concepts originated and how they developed over time. I can only hope that I have made appropriate choices.

In the preparation of this second edition, I acknowledge my continuing debt of gratitude to those reviewers whose suggestions and criticisms contributed to the success of the first edition: M. S. Brower *Armstrong State College*, R. E. Cordero *St. Joseph's University*, D. Cosgrove *Pennsylvania State University*, R. P. Donaldson *The George Washington University*, J. D. DuBois *Middle Tennessee State University*, R. C. Evans *The State University of New Jersey*, N. Grant *William Patterson College*, J. G. Harris *Utah Valley Community College*, R. M. Klein *University of Vermont*, M. Rincón *Midwestern State University*, H. Sweet *University of Central Florida*, and D. E. Wivagg *Baylor University*. Numerous others have responded with constructive criticisms based on their use of the previous edition. To this list I add those who have graciously reviewed the manuscript for the second edition: A. Scott Holaday *Texas Tech University*, Thomas Vogelmann *University of Wyoming*, C. James Lovelace *Humboldt State University*, David Orcutt *Virginia Polytechnic Institute & State University*, and Leslie Towill *Arizona State University*.

A special thank-you goes to Marguerite Kane for kindly providing the drawing of *Arabidopsis* that accompanies Box 15.1.

I am also indebted to the many colleagues who have generously provided photographs or who have consented to have their work included in this book. Finally, I continue to be indebted to the students who have contributed far more than they might ever know. In the end, of course, responsibility for any omissions or errors is mine alone. I can only hope that, whatever its faults, this book opens new roads to students encountering the excitement, mystery, and challenge of plant physiology for the first time.

William G. Hopkins
London, Ontario
December 1997

To the Student

This book is about how plants work. It is not an encyclopedia and will not, I hope, overwhelm you with detail. This book is mainly about ideas. It is about the questions that plant physiologists ask and how they go about seeking answers to those questions. Most of all, it is about how plants do the things they do in their everyday life.

The book contains several pedagogic features that are intended to assist your learning. New terms and concepts are identified with boldfaced type. Some of these terms may be boldfaced when encountered a second time, for emphasis. You should attempt to understand each boldfaced term—what it means and its significance to the problem under discussion.

Each chapter concludes with a Summary, which attempts to highlight the principal topics discussed in that chapter. There is also a series of review questions (Chapter Review). For many of these questions, there is no single or simple answer. The questions are intended as a guide to your review of the chapter and, perhaps, as a stimulus to help you integrate apparently diverse aspects or to extend what you have learned to new situations.

A word about literature citations: I have purposely kept the number of literature citations in the text to a minimum. These are compiled at the end of each chapter along with a selection of further readings. Some are reports of research results and some are review articles. If you find a particular topic interesting and wish to learn more about it, the listed publications are your gateway into the relevant literature. Plant physiology is a very active field of study and new revelations about how plants work are reported in the literature almost daily. To learn what has happened since this book was written, seek out recent publications in the same journals cited in the reference lists. Many of the journals listed publish review articles that summarize the status of a topic up to that point.

I hope that you enjoy reading this book as much as I have enjoyed writing it.

William G. Hopkins

Contents

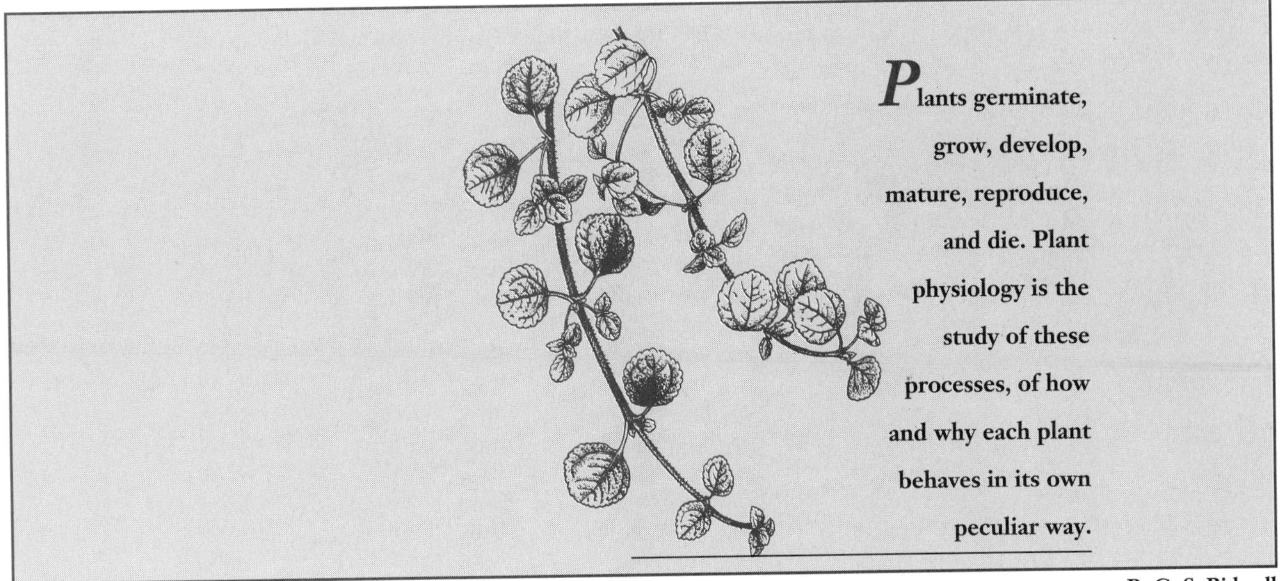

R. G. S. Bidwell
(*Plant Physiology*, Macmillan, 1979)

1

Introduction: The Organization of Plants and Plant Cells

THE SCOPE OF PLANT PHYSIOLOGY

What is plant physiology? The term *physiology* is constructed from the Greek words *physis*, meaning nature, and *logos*, meaning discourse. Taken literally, then, plant physiology is a discourse about the nature of plants. This is a rather broad mandate indeed, encompassing virtually everything of a botanical nature. For good reason, physiologists have traditionally taken a more limited, mechanistic approach to plants. From the physiological perspective, plants are viewed primarily as biochemical machines; machines that take in energy and simple inorganic molecules from the physical environment and use that energy and those molecules to assemble complex chemical structures. The processes that enable plants to carry out these activities are themselves the sum of a multitude of chemical reactions. In other words, everything that plants are and everything that they do are based on chemical and physical processes.

Plant physiology is about how plants use the energy of the sun to assimilate carbon, and how they convert that carbon to the stuff of which they are made. It is about how plants obtain and distribute nutrients and water. It is about how they grow and develop, how they respond to their environment, how they react to stress, and how they reproduce. In short, plant physiology is about how plants function. The task of the plant physiologist is to explain how plants function in terms of known chemical and physical laws.

THE PLANT CELL

The vital processes of the individual cells form the first indispensible and fundamental basis for . . . vegetable physiology. . . .

M. J. Schleiden (1838)

The basic functional unit of plants, as with all living organisms, is the cell. Indeed, the study of plant physiology is very much a study of the physiology of plant cells and how their coordinated activities are reflected in the physiology of the whole organism. In a similar fashion, the architecture of a plant reflects the number, morphology, and arrangement of its individual cells. Architecture and function are inseparable. It is therefore appropriate that we begin our study of plant physiology with a general review of the basic structure and organization of plant cells and tissues.

A representative plant cell is shown in Figure 1.1. A **cell** is an aqueous solution of chemicals called **protoplasm** surrounded by a **plasma membrane.** The membrane and the protoplasm it contains are collectively referred to as a **protoplast.** Of course, the protoplasm and

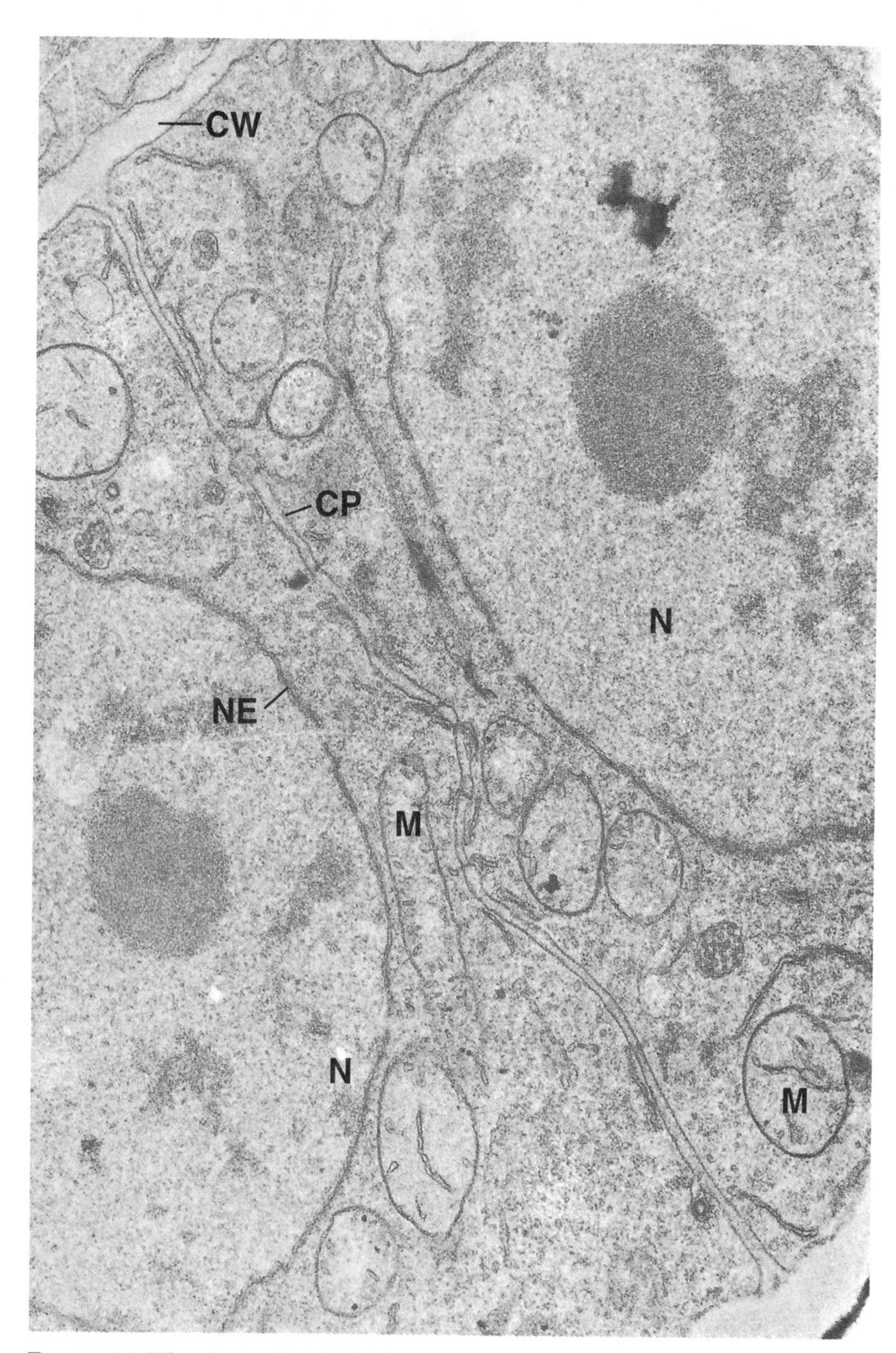

FIGURE 1.1 The plant cell. (*A*) Electron micrograph of two adjacent meristematic cells in the shoot apex of *Chenopodium album*. The original meristematic cell has recently divided. A cell plate, an early stage in the new cell wall, can be seen running diagonally from the upper left to the lower right. Note the large nucleii and the absence of vacuoles. M, mitochondrion; N, nucleus; NE, nuclear envelope; CP, cell plate; CW, cell wall. (*B*) A mature mesophyll cell from a *Coleus* leaf. Note the large central vacuole and chloroplasts. (*A*; courtesy of E. M. Gifford and K. Stewart. *B*; Electron micrograph by Wm. P. Wergin, courtesy of E. H. Newcomb, University of Wisconsin—Madison.)

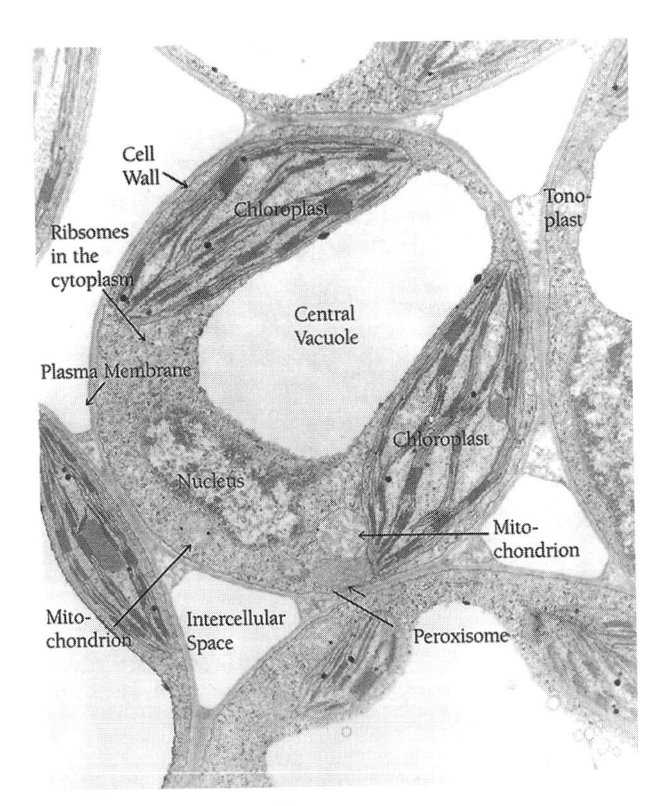

FIGURE 1.1 **(continued)**

all of the components that make up protoplasm have important roles to play in the life of a cell, but the plasma membrane is particularly significant because it represents the boundary between the living and nonliving worlds. The plasma membrane is **selectively permeable,** which means that it allows some materials to pass through but not others. Thus, the plasma membrane not only physically limits the cell, it also controls the exchange of material and serves to maintain essential differences between the cell and its environment.

When we examine a cross-section of a cell in an electron micrograph, that is, a picture of a cell as seen through the electron microscope, we see that membranes are a singularly prominent feature (Fig. 1.1). Additional selectively permeable membranes are found throughout the protoplast where they form a variety of subcellular structures called **organelles.** These cellular organelles serve to compartmentalize major metabolic activities, much in the same way that an automobile factory is set up for metal fabrication, paint shop, assembly line, and so forth. One of these organelles, the **nucleus,** contains the genetic information and is the control center of the cell. The balance of the protoplasm, excluding the nucleus but including other organelles, is called **cytoplasm.** Different organelles in the cytoplasm are the sites of cellular respiration, photosynthesis, protein synthesis, secretion, and so forth.

In spite of the wide variation in cellular morphology and function, cells are remarkably alike in that they are all built according to the same basic plan and contain certain fundamental structures. Perhaps, because they define the boundary between the living and nonliving world, the most significant of those structures are the membranes.

BIOLOGICAL MEMBRANES

Biological membranes are composed primarily of lipids and proteins, with smaller amounts of carbohydrate. Although the relative proportions of lipid and protein, as well as the specific lipid and protein composition, vary widely from one membrane to another, there are fundamental structural and functional concepts that apply to all membranes.

THE MEMBRANE BILAYER

The most abundant lipids in cell membranes are the phospholipids. **Phospholipids** are amphipathic molecules; they have two *hydrophobic* ("water-fearing") nonpolar hydrocarbon tails and a *hydrophilic* ("water-loving") polar head group (see Box 1.1). In an aqueous environment, phopholipids spontaneously form a **bilayer,** with their hydrophobic tails buried in the interior and the hydrophillic heads exposed to the water (Fig. 1.2). Bilayers form because the nonpolar hydrophobic groups tend to avoid contact with water and associate only with other nonpolar groups, a tendency known as **hydrophobic bonding.** The bilayer structure gives membranes two properties that are particularly important to our understanding of how membranes perform their many diverse functions: They are (1) highly fluid and (2) impermeable to most polar molecules.

The phospholipid bilayer of biological membranes contains a high proportion of unsaturated fatty acids, giving the bilayer the consistency of light machine oil. At the same time, the bilayer forms a very stable structural basis for biological membranes. These apparently contradictory properties arise largely because the hydrocarbon tails of the phospholipid molecules exclude water and develop strong hydrophobic interactions. The diffusion of a phospholipid from one side of the bilayer to the other, called *flip-flop*, is a very rare event. Flip-flop is prevented because the hydrophilic head group does not readily dissolve in the hydrophobic core of the bilayer. On the other hand, the individual lipids and other hydrophobic molecules may diffuse very rapidly—as much as 1 $\mu m\ s^{-1}$—*in the plane of the membrane.* This movement is called **lateral diffusion.** The membrane thus behaves as a very stable two-dimensional fluid.

The membrane bilayer is also impermeable to most polar or charged solutes, which includes virtually all ions

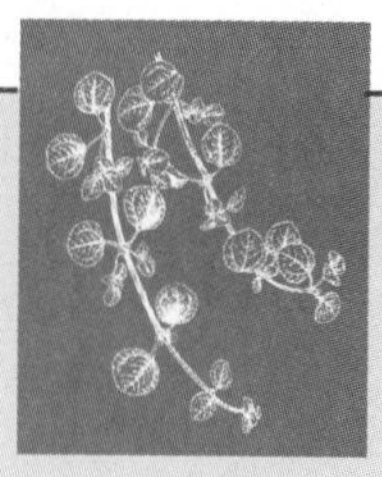

Box 1.1
Lipids

Lipids (Gr. *lipos;* fat) are a chemically diverse group of molecules that dissolve readily in organic solvents, but only sparingly in water. These include fats, oils, and the phospho- and glycolipids that make up the membrane bilayer. Sterols (steroid alcohols) and molecules containing long chain hydrocarbons such as the pigments chlorophyll and carotene, although chemically distinct, are also considered lipids on the basis of solubility.

Fats and oils are composed of long chain hydrocarbons called **fatty acids**

$$HOOC{-}(CH_2)_n{-}CH_3$$

and the 3-carbon alcohol **glycerol.**

$$\begin{array}{c} H_2COH \\ | \\ HCOH \\ | \\ H_2COH \end{array}$$

Fatty acids are attached to the glycerol through an ester link formed between the carboxyl group of the fatty acid and a hydroxyl group on the glycerol molecule. Because a fatty acid is esterified to each of the three glycerol carbons, a molecule of fat or oil is known as **triacylglycerol,** or a **triglyceride.**

$$\begin{array}{l} H_2{-}C{-}O{-}\overset{\overset{\displaystyle O}{\|}}{C}{-}(CH_2)_n{-}CH_3 \\ \quad\;\; | \\ H{-}C{-}O{-}\overset{\overset{\displaystyle O}{\|}}{C}{-}(CH_2)_n{-}CH_3 \\ \quad\;\; | \\ H_2{-}C{-}O{-}\overset{\overset{\displaystyle O}{\|}}{C}{-}(CH_2)_n{-}CH_3 \end{array}$$

Fatty acids vary according to the length of the hydrocarbon chain (that is, the value of *n*) and the number and position of carbon-carbon double bonds in the chain. Fatty acids that have no double bonds are known as saturated fatty acids, while those with double bonds are unsaturated. Examples of several fatty acids commonly encountered in plants are listed in the following. The number before the colon indicates the number of carbon atoms in the chain. The number following the colon indicates the number of double bonds.

Saturated fatty acids	
Lauric	12:0
Myristic	14:0
Palmitic	16:0
Stearic	18:0

Unsaturated fatty acids	
Oleic	18:1
Linoleic	18:2
Linolenic	18:3

Unsaturated fatty acids are identified not only by the number of double bonds, but also by their location in the molecule. For example, the formula for linoleic acid is:

$$CH_3{-}(CH_2)_4{-}CH{=}CH{-}CH_2{-}CH{=}CH{-}(CH_2)_7{-}COOH$$

The difference between fats and oils is a matter of melting point. The melting point of fatty acids increases with chain length and the extent of saturation. Saturated fatty acids tend to be solid at room temperature and unsaturated fatty acids tend to be liquid at room temperature. Thus fats are triglycerides com-

---

and molecules normally encountered by cells. Again, in order to diffuse across the membrane, a molecule must first dissolve in the fatty acid matrix, and polar or charged molecules do not readily dissolve in a strongly hydrophobic environment. Three notable exceptions to this rule are water, carbon dioxide, and oxygen, which permeate the membrane more rapidly than expected on the basis of their solubility properties. In particular, the anomalous behavior of water, which appears to cross membranes freely, is only now beginning to be understood (Chap. 5).

MEMBRANE PROTEIN

Most membranes contain as much as 50 percent protein by weight, although some, such as the inner membranes posed of predominantly saturated fatty acids and oils with predominantly unsaturated fatty acids. Plants, especially seeds, contain predominantly oils that are stored in lipid bodies called **oleosomes.**

Phospholipids are the principal class of lipids found in most membranes. Phospholipids are **diglycerides,** with only two fatty acids esterified to the glycerol molecule; the third position is occupied by a phosphate group.

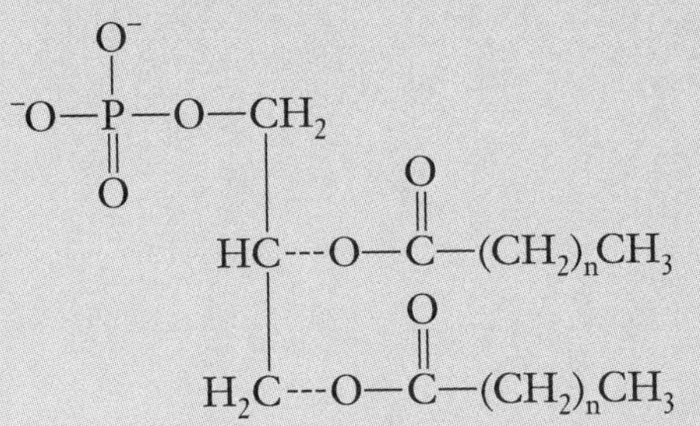

```
     O⁻
     |
⁻O—P—O—CH2
     ||      |          O
     O       |          ||
           HC---O—C—(CH2)nCH3
             |          O
             |          ||
           H2C---O—C—(CH2)nCH3
```

Phospholipids are often described as having a charged, polar phosphate "head" and a long hydrocarbon "tail" represented by the two fatty acids. This gives the molecule a dual character in that the phosphate head is **hydrophilic** ("water-loving") and the fatty acid tail is **hydrophobic** ("water-fearing"). A molecule with both hydrophilic and hydrophobic properties is known as **amphipathic.**

Variation in phospholipids is introduced by the nature of the two fatty acids; in plant membranes, the most abundant are 16:0, 16:1, 18:0, 18:1, 18:2, and 18:3. A high proportion of unsaturated fatty acids in the membrane lipids contributes to the fluidity of membranes. Further variation is introduced by the addition of other small, polar molecules to the phosphate head group. Two of the more commonly encountered polar molecules in plant membrane phospholipids are **choline** and **ethanolamine.** Phospholipids containing choline or ethanolamine are known as **phosphatidylcholine (PC)** and **phosphatidylethanolamine (PE),** respectively.

Some plant membranes contain a large amount of **glycolipid.** The internal membranes of the chloroplast are particularly distinctive in this regard. They contain only about 10 percent phospholipid and about 80 percent **monogalactosyl diglyceride (MGDG)** and **digalactosyl diglyceride (DGDG).** MGDG and DGDG are similar to phospholipids except that the phosphate group is *replaced* with one or two molecules of the sugar galactose. The remaining 10 percent of the chloroplast membrane lipid is accounted for by **sulfolipid;** in this case, a sulfur group is attached to the galactose in MGDG.

Where are the membrane lipids synthesized in plant cells? This question has been answered by feeding isolating organelles radioactive precursors and seeing which fatty acids or lipids can be formed. These experiments have shown that the principal sites of fatty acids and lipid biosynthesis are the endoplasmic reticulum, the chloroplast, and the mitochondria. Chloroplasts appear to have the enzymatic machinery for synthesizing all the C_{16} and C_{18} fatty acids and the addition of galactosyl residues to the diglycerides in order to make MGDG and DGDG. Chloroplasts do not, however, appear to make phospholipids; this is accomplished in the endoplasmic reticulum, which contains all the enzymes necessary for the synthesis of PC and PE. Mitochondria do not appear to synthesize either PC or PE, even though these are the principal phospholipids of the mitochondrial membranes. Thus, it is necessary to conclude that there is significant exchange of fatty acids and finished lipids between the chloroplast, endoplasmic reticulum, and other cellular organelles.

of mitochondria, may contain even more. Membrane protein may be categorized as either **integral protein** or **peripheral protein,** according to whether it is integrated into the bilayer or bound to the hydrophillic surface, respectively (Fig. 1.2). Integral proteins, also known as **intrinsic** protein, may consist of a single protein or large complexes made up of several proteins with additional nonprotein components. Some integral proteins have access to only one side of the membrane, while others, called **transmembrane** proteins, span the membrane and have access to both sides. Integral proteins are able to integrate into the bilayer because they have large *hydrophobic domains* that interact with the hydrocarbon tails of the membrane lipids. Hydrophobic domains are regions of the protein made up predominantly of amino acids with nonpolar side chains (see Box

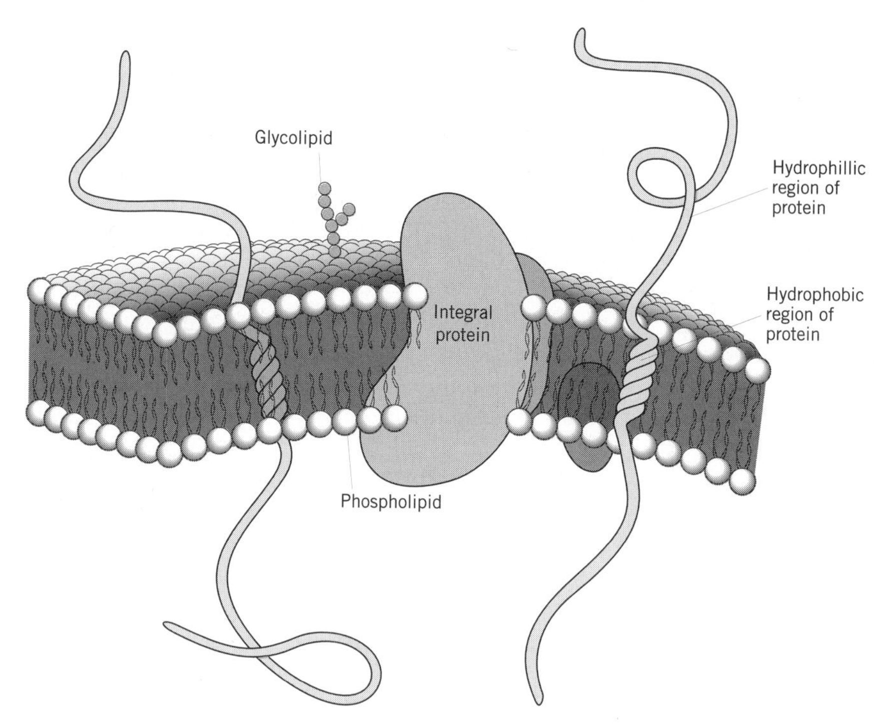

FIGURE 1.2 A general model for membrane structure. The bilayer is shown composed of phospholipid molecules. The circles indicate the polar phosphate head groups and the wiggly lines the fatty acid tails. This model is known as the Fluid-Mosaic, referring to a mosaic of protein "icebergs" floating in a lipid "sea." (From D. Voet, J. G. Voet. 1990, *Biochemistry*. New York, Wiley. Reprinted by permission.)

1.2). Integral proteins can thus be removed from the membrane only by use of detergents or other agents that disrupt the membrane by countering the strong hydrophobic interactions. Integral proteins also contain *hydrophilic domains*, consisting of predominantly polar amino acids, that project into the aqueous environment on either side of the membrane. Peripheral proteins, on the other hand, are predominantly hydrophilic and bind loosely with the polar phospholipid heads. Peripheral proteins can be dislodged with relatively mild procedures using salt solutions or buffers that leave the membrane intact.

Integral membrane proteins are to some extent structural in that they occupy space in the plane of the membrane and interact strongly with the membrane bilayer. The principal role of membrane proteins, however, is functional rather than structural. Membrane proteins are responsible for all metabolic activities associated with membranes. For example, some membrane proteins function as enzymes, others assist in the selective transport of solute molecules across the membrane, and still others participate in energy transduction. The fluidity of the lipid bilayer is very important in this respect since it allows proteins to diffuse laterally in order to achieve the necessary interactions. The functions of different membranes in the cell are very diverse, so it should not be too surprising that membrane composition is equally diverse, yet very characteristic of particular membranes.

CELLULAR ORGANELLES

The cytoplasmic interior of a cell is organized into a number of discrete, membrane-limited compartments called **organelles.** The nonparticulate portion of the cytoplasm, that is, the liquid portion that is not included as one of the organelles, is called the **cytosol.** The cytosol contains large quantities of protein and other solutes, often assuming the physical properties of a gel.

Vacuole. A conspicuous feature of most mature higher plant cells is a large central **vacuole** (Fig. 1.1). In a mature cell, the vacuole may occupy as much as 80 to 90 percent of the cell volume. In such cells, the cytoplasm with all the other organelles will appear as a thin film pressed against the cell wall. Vacuoles are surrounded by a membrane called the **vacuolar membrane** or **tonoplast.** The vacuole contains a variety of inorganic ions, organic acids, sugars, enzymes, and secondary metabolic products including pigments. The high solute content of the vacuole is instrumental in the uptake of water that is required for cell enlargement. Young, actively dividing cells normally contain a number of small vacuoles. As the cell matures, these smaller vacuoles coalesce and expand to form the single large vacuole characteristic of a mature cell.

Nucleus. The nucleus is the information center of the cell. It contains the bulk of the cell's genetic material,

Box 1.2 Proteins

Proteins play a central role in the biochemistry of cells and are responsible for virtually all the properties of life as we know it. Of the several classes of biological macromolecules, proteins are the most abundant. Excluding cell walls, proteins and their monomers, amino acids, comprise some 60 to 75 percent of the dry weight of most cells. The role of proteins may be structural, such as in the microtubules and microfilaments of the cytoskeleton, or storage, as in protein bodies found in seeds. Proteins serve as ion channels in cell membranes, as hormone receptors, as pigments, and a variety of other important functions in the cell. But clearly, the predominant role of proteins is as biological catalysts, or enzymes (see Chap. 12).

Proteins are formed of long chains of amino acids linked by **peptide bonds.** All amino acids have the same basic structure, in which a single carbon atom (the α-carbon) carries both an *amino group* ($-NH_2$) and a *carboxyl group* ($-COOH$):

$$\begin{array}{c} R \\ | \\ H_2N-C-COOH \\ | \\ H \end{array}$$

The structure of individual amino acids varies according to the nature of the R group, or side chain. The side chain may consist of nonpolar groups, giving the amino acid *hydrophobic* properties, or either neutral or charged polar groups, in which case the amino acid has *hydrophilic* properties.

The peptide bond forms when the amino group of one amino acid forms a *covalent* link with the carboxyl group of a second amino acid

$$\begin{array}{ccccccc} & R_1 & & H & & & \\ & | & & | & & & \\ & H-C & & N & & COOH & \\ H_2N & & C & & CH & & \\ & & \| & & | & & \\ & & O & & R_2 & & \end{array}$$

The structure formed when two amino acids are linked by a single peptide bond is called a *dipeptide;* three amino acids, a *tripeptide;* and so forth. A large number of amino acids in a chain is called a **polypeptide.** Most proteins are made up of 20 standard amino acids that can be assembled in various combinations. Some proteins contain additional unusual amino acids, such as the hydroxyproline found in cell wall proteins. In virtually all cases, however, these unusual amino acids are the result of specific modifications to a standard amino acid after the peptide chain has been synthesized.

Proteins are complex molecules with several levels of organization, as shown in Figure 1.3. The sequence of amino acids in a polypeptide is referred to as the *primary structure.* Primary structure is determined by the sequence of nucleotides in the deoxyribonucleic acid (DNA) that makes up the gene for that protein. This sequence is *transcribed* into a messenger ribonucleic acid (mRNA), which moves from the nucleus into the cytoplasm, where it attaches to ribosomes. The amino acids are assembled on the ribosome according to the sequence of nucleotides, or message, in the mRNA. This process is known as *translation.*

When the polypeptide chains are released from the ribosome, they fold spontaneously in various ways to form three-dimensional shapes. The backbone of peptide bonds may form a coiled, or α-helical struc-

deoxyribonucleic acid (DNA). The DNA in turn comprises the genes, which encode information for the synthesis of **ribonucleic acid (RNA).** The RNA is exported to the cytoplasm where it directs the synthesis of specific proteins. In a nondividing cell, the DNA is dispersed along with associated proteins as an indistinct tangle called **chromatin.** During cell division, the DNA condenses into short, threadlike structures called **chromosomes.** Nondividing cells also contain small, densely staining bodies called the nucleoli (sing. *nucleolus*). Nucleoli are the site of ribosome synthesis; they disappear during cell division.

Nuclei are in the order of 5 to 30 μm in diameter and are surrounded by a double membrane called the **nuclear envelope.** The inner and outer nuclear membranes are fused at sites called **nuclear pores,** which interrupt the integrity of the envelope and provide a channel for the export of RNA from the nucleus to the cytoplasm. In most eukaryote cells, the nuclear envelope disappears during cell division and reforms around the DNA in the daughter cells.

Endoplasmic Reticulum and Golgi Complex. The **endoplasmic reticulum (ER)** and the **Golgi Complex**

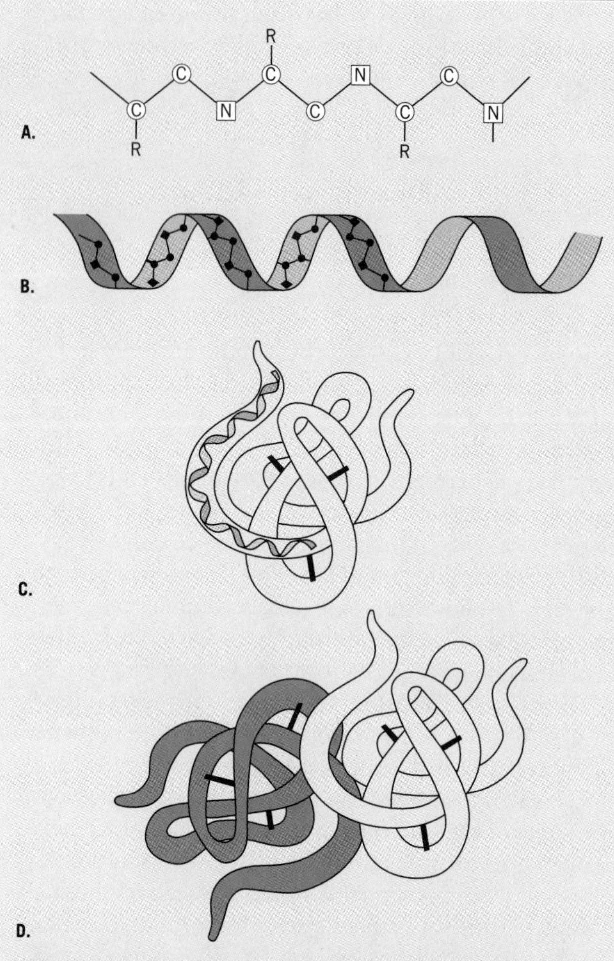

FIGURE 1.3 The structural hierarchy in proteins. (*A*) Primary structure—the amino acid sequence in a peptide chain. (*B*) Secondary structure—an α-helix. (*C*) Tertiary structure—the 3-dimensional shape of a complete peptide chain. (*D*) Quaternary structure—an assembly of multiple peptide chains.

ture, or may fold such that segments of the chain lie side by side to form a pleated sheet, or may form random coils. These specific spatial arrangements are referred to as *secondary structure.* The number and distribution of helices, pleated sheets, and random coils in the polypeptide determines the three-dimensional configuration of the entire polypeptide, called *tertiary structure.* Tertiary structure of the protein is commonly referred to as **conformation.** Some proteins are composed of a single polypeptide, but many are composed of multiple polypeptide chains, called *subunits.* The arrangement of subunits in the protein molecule is referred to as *quaternary structure.* The three-dimensional structure of a protein is stabilized by a variety of noncovalent bonds, including hydrogen bonds (Chap. 2) and electrostatic interactions between ionized amino and carboxyl groups on the amino acid side chains. At the tertiary level, covalent disulfide bonds (—S—S—) may form between nearby amino acids with sulfur groups in the side chains. Alternatively, neighboring amino acids with nonpolar side groups may form strong hydrophobic interactions.

The extreme diversity of proteins can be explained by the nearly infinite number of variations possible in the sequence of 20 different amino acids in polypeptide chains containing hundreds and even thousands of amino acid residues. The sequence of amino acids, with the numerous possible interactions between side chains, determines the three-dimensional shape of the protein. The three-dimensional shape, in turn, determines the biological properties of protein molecules. Subtle differences in conformation give each protein its unique characteristics and its ability to discriminate between other molecules with which it interacts.

FURTHER READING

Voet, D., J. G. Voet. 1990. *Biochemistry.* New York: Wiley.

together form an elaborate system of membranes involved in lipid and protein biosynthesis and secretion. The ER is believed to be a single, highly convoluted membrane continuous with the outer membrane of the nuclear envelope (Fig. 1.4). In cross-section the ER forms a double membrane enclosing a **lumen.** Much of the ER has ribosomes associated with its cytosolic surface; this is called **rough ER** because of the granular appearance of the ribosomes. Proteins synthesized on rough ER are passed through the membrane into the lumen, where they move into a region of **smooth ER,** that is, an area of the ER devoid of ribosomes. Once in the ER, the protein is modified and sugars are added to form glycoproteins. The glycoproteins are then packaged in spherical **transport vesicles** that bud off the smooth ER.

Smooth ER is also a major site for lipid biosynthesis and membrane formation. Of interest in this regard is the unique character of the membrane surrounding lipid storage bodies or **oleosomes** found in seeds. Oleosomes appear to be enclosed by a membrane that consists of a lipid **monolayer,** or half-membrane, rather than a bilayer. The polar head groups of the monolayer face the aqueous cytoplasm, and the hydrophobic fatty acid tails face the stored lipids. The origin of the half-membrane can be traced to the formation of oleosomes in the ER. Cytological studies of oleosome formation have shown that the newly synthesized lipid accumulates in the *mid-*

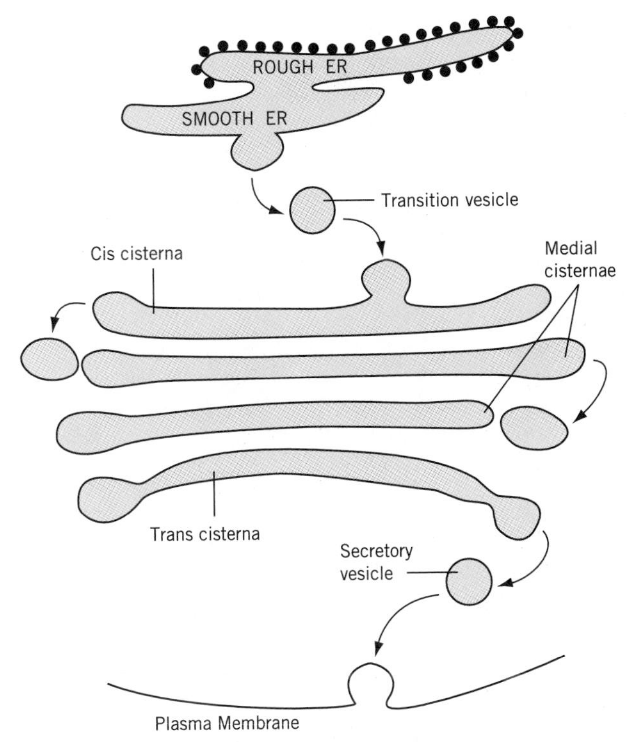

FIGURE 1.4 Endoplasmic reticulum (ER) and Golgi complex. A schematic representation illustrating the role of transition vesicles and secretory vesicles. Glycoproteins synthesized in the ER are transported to the Golgi complex in transition vesicles. Transition vesicles fuse with the Golgi and the glycoproteins are released to the lumen of the Golgi cisterna. As the glycoproteins are passed from the cis cisternae to the trans cisternae, they are modified and additional sugar groups are added. Processed glycoproteins are then packaged in secretory vesicles, which bud off the Golgi and move to the plasma membrane.

dle of the bilayer, thus driving the two halves of the bilayer apart until they pinch off to form the oleosome.

The Golgi complex is a stack of flattened, membranous sacs, called **cisternae** (sing. *cisterna*), that is separate from the ER (Fig. 1.4). The Golgi serves to assemble and process carbohydrate (oligosaccharide) chains of glycoproteins that are transferred to it, via transport vesicles, from the ER. The transport vesicle membranes fuse with the Golgi membranes, delivering their contents to the Golgi cisternae. Here the sugar chains are modified and enlarged and other sugars are added. The modified glycoproteins leave the Golgi in **secretory vesicles,** which deliver their contents to sites inside the cell, such as sites for protein storage (called **protein bodies**), or to the plasma membrane for discharge outside the cell. Another principal function of the Golgi complex in plant cells is to synthesize the complex polysaccharides that make up the cell wall matrix (described below) and deliver them to the site of wall formation in dividing and growing cells.

Mitochondria. The **mitochondria** and chloroplasts, described below, are the two energy-transducing organelles in plant cells. Mitochondria are found in all but a few living cells and are central to the energy metabolism of cells. In plant cells, mitochondria tend to be spherical in cross-section and approximately 1 μm in diameter (Fig. 1.5). Mitochondria are limited by a smooth outer membrane and contain a highly invaginated inner membrane.

The inner mitochondrial membrane is about 75 percent protein, which reflects its special role in energy transduction. Mitochondria are the site of cellular respiration, the process in which the energy of sugar oxidation is used to drive the synthesis of adenosine triphosphate (ATP). The ATP is then exported to other regions of the cell where the energy is used for various cellular activities. The structure and function of mitochondria will be discussed in detail in Chapter 12.

Plastids. **Plastids** are a family of double membrane–bound organelles common to plant cells (Fig. 1.6). Most prominent are the **chloroplasts,** which contain the photosynthetic pigments, carry out photosynthesis, and are responsible for the prominent green color of leaves. Chloroplasts will be discussed further in Chapter 8.

Plastids arise from **proplastids,** small vesicular bodies produced in dividing cells. Colorless plastids, that is, plastids without pigments, are called **leucoplasts.** Leucoplasts are often the site of starch accumulation, in which case they are called **amyloplasts.** Leucoplasts and chloroplasts are often interconvertible, depending on the physiological state of the cell and light conditions. **Chromoplasts** contain pigments other than chlorophyll. The characteristic colors of tomato fruit and carrot roots are due to the presence of chromoplasts containing the orange carotene pigments. The pigments in chromoplasts are often so concentrated that they form a crystalline deposit.

Microbodies. Plant cells contain **microbodies,** which are small organelles about the size of mitochondria (ca. 1 μm) and bound by a *single* membrane. Microbodies have specialized roles in specific metabolic pathways and are usually characterized by a particular enzyme or enzymes. **Peroxisomes** contain large amounts of the enzyme catalase, which is used to remove often harmful metabolic byproducts. Another microbody, the **glyoxysome,** is prominent in seeds, especially those that store a significant proportion of their carbon reserves as lipids. Glyoxysomes contain enzymes of the glyoxylate cycle, which is used to convert fatty acids to glucose through a process known as **gluconeogenesis.** The glucose is then transported to the growing embryo to be used as a carbon and energy source.

Ribosomes. **Ribosomes** are the site of protein synthesis. Strictly speaking, ribosomes should not be con-

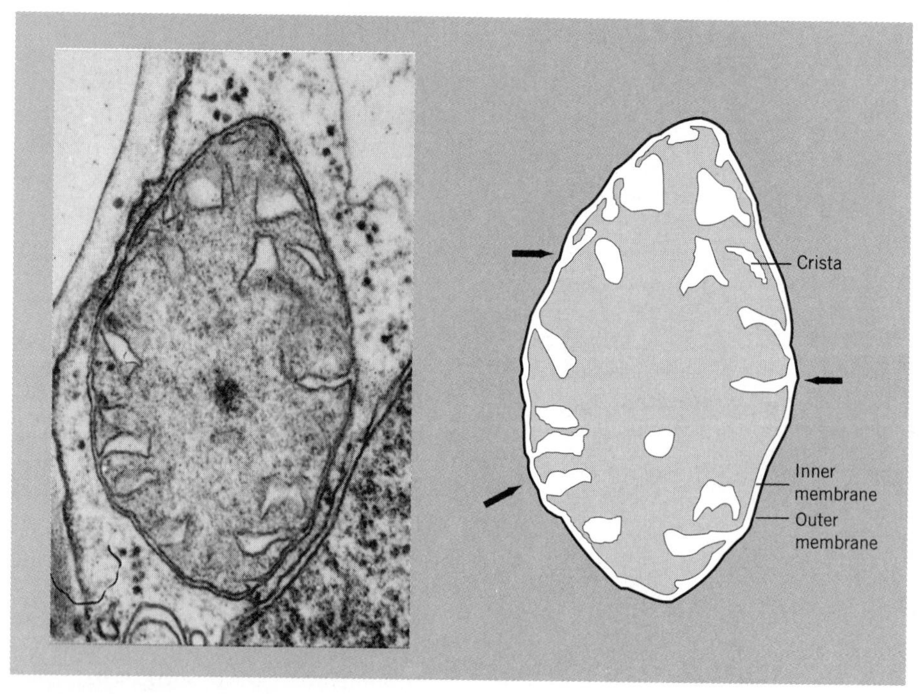

FIGURE 1.5 The Mitochondrion. (*A*) An electron micrograph of a mitochondrion from a maize (*Zea mays* L.) leaf cell (× 10,000). (*B*) A diagram to illustrate the essential features. Note that the cristae are continuous with the inner membrane (arrows).

sidered organelles since they are not membrane limited. However, because of their discrete nature and their importance in protein synthesis, ribosomes will be considered in this discussion along with other cellular organelles.

Ribosomes are complex aggregates of RNA and protein, which appear as small granules (approximately 25 nm diameter) in electron micrographs. The complete ribosome is assembled in the cytosol from a large and a small subunit synthesized in the nucleolus. Some ribosomes occur free in the cytosol while others appear to be attached to membranes such as the endoplasmic reticulum. A cell may contain up to several hundred thousand ribosomes, depending on the amount of protein required.

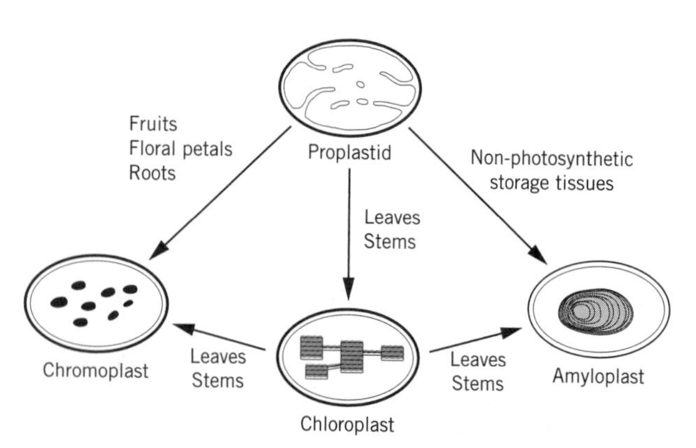

FIGURE 1.6 Plastids. A diagram illustrating the interrelationships between various types of plastids. The structure of the chloroplast is illustrated in Chapter 8.

CYTOSKELETON

Virtually all eukaryotic cells, both animal and plant, contain a three-dimensional, interconnected network of fibrous protein called the **cytoskeleton.** The cytoskeleton plays vital roles in determining the organization of cytoplasm and cell shape, and in cell division, growth, and differentiation.

The cytoskeleton of plant cells is composed of two different elements: **microtubules** and **microfilaments.** Microtubules are long rods approximately 24 nm in diameter and a hollow core about 12 nm in diameter. They are assembled from subunits of a globular protein called **tubulin,** which has a molecular mass of approximately 100,000 daltons[1] (100 kD) and is made up of two

[1]The **molecular mass** of a molecule or particle is expressed in units of daltons, defined as 1/12 the mass of a carbon atom. Molecular mass should not be confused with **molecular weight,** which is a dimensionless quantity expressing the ratio of particle mass to 1/12 the mass of a carbon atom. Molecular weight is symbolized M_r for *relative* molecular mass.

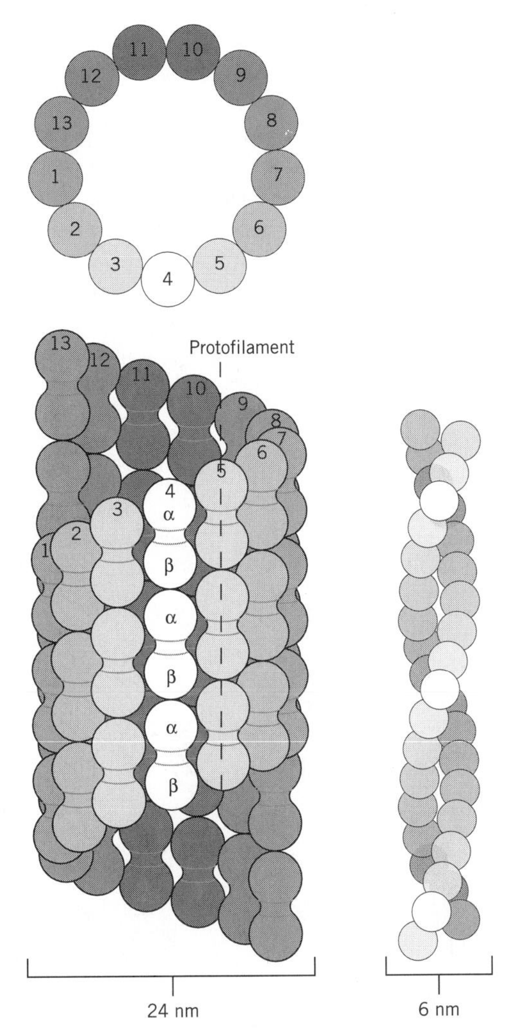

FIGURE 1.7 Microtubules and microfilaments. Left: Diagram of a microtubule in cross-section and longitudinal view. The number of vertical protofilaments varies from 11 to 15, but is usually 13, as shown here. Protofilaments are offset to form a helix. Right: A microfilament is composed of two parallel strands of globular subunits twisted to form a helix.

polypeptides (α-tubulin and β-tubulin). Microtubules are formed when tubulin subunits spontaneously self-assemble into long chains called **protofilaments.** Protofilaments then line up laterally to form the microtubule wall (Fig. 1.7). Microfilaments are solid threads composed of **actin,** also a globular protein. Two chains of actin subunits assemble in a helical fashion to form a microfilament approximately 6 nm in diameter.

The cytoskeleton has been studied in animal cells since the early 1960s, but its presence in plant cells was uncertain for many years. Because the cell wall appeared to provide support for the protoplast and determine the shape of a plant cell, the necessity for a cytoskeleton was questioned. It was almost 20 years before microtubules and microfilaments were clearly demonstrated in nondividing plant cells. Microtubules and microfilaments can be seen under the electron microscope or revealed by digesting away the cell wall and removing the plasma membrane and other cytoplasmic components with detergent. Only recently has a third cytoskeletal element, intermediate filaments, been detected in plant cells. Intermediate filaments are so named because they have diameters of 10 to 15 nm, intermediate to the diameters of microtubules and microfilaments.

The term **cytoskeleton** is an unfortunate choice because it implies a rigid structure with a static function. Instead, the cytoskeleton is very dynamic. Microtubules in particular are constantly being assembled, disassembled, and rearranged as the cell divides, enlarges, and differentiates. Microtubules form the mitotic spindle, which plays a significant role in the movement of chromosomes during cell division. Microtubules also determine the orientation and location of the new cell wall between daughter cells, and the deposition of cellulose in growing cell walls (see next section).

Microfilaments appear to control the direction of **cytoplasmic streaming,** the continuous flow of cytoplasmic particles and organelles around the periphery of the cell. The microfilaments form aggregates or bundles oriented parallel to the direction of cytoplasmic flow. Microfilaments are also involved in the growth of pollen tubes. When a pollen grain germinates, it develops a tubular extension that grows down the stigma of the flower and serves to deliver the male nucleus to the egg. Growth of the tube is only at the tip, and vesicles that contain cell wall precursors are guided through the cytoplasm to the growing tip by a network of microtubules.

The importance of the cytoskeleton in organizing and coordinating the dynamic properties of growing cells is only beginning to be appreciated. Techniques for the study of plant cell cytoskeleton are rapidly improving, and we can expect exciting advances in the future.

THE EXTRACELLULAR MATRIX

Cells are characterized not only by their cytoplasmic contents and *intracellular* organization, but also by a complex mixture of *extracellular* materials, collectively called **extracellular matrix (ECM).** Components of the ECM, predominantly carbohydrate and protein, are synthesized inside the cell and transported through the plasma membrane where they are assembled. In the case of plant cells, the ECM is dominated by the **cell wall,** which provides rigidity and protection for the underlying protoplast and is ultimately responsible for maintaining cell shape. Two types of cell walls are recognized: **primary walls,** which surround young, actively growing cells, and **secondary walls** that are laid down as the cells mature and are no longer growing.

THE PRIMARY CELL WALL

The primary wall is thin, measuring only a few micrometers in thickness, and consists of randomly arranged threadlike polymers of glucose, called **cellulose.** Cellulose is a long, unbranched β-1 → 4-glucan; that is, a chain of glucose molecules in which the number 1 carbon of one glucose molecule is linked to the number 4 carbon of the next glucose molecule in the chain (Fig. 1.8). A single molecule of cellulose may contain as many as 3,000 or more glucose units. In the cell wall, the cellulose molecules are grouped together in long parallel arrays called **microfibrils** (Fig. 1.8). Each microfibril is approximately 5 to 12 nm in diameter and contains approximately 50 to 60 cellulose molecules held together by hydrogen bonds between hydroxyl (—OH) groups on adjacent glucose units. (Hydrogen bonds are discussed in Chapter 2.) This arrangement gives the microfibrils a crystalline character and a great measure of strength. The tensile strength (the ability to withstand tension without breaking) of a microfibril is similar to that of a steel wire of the same size. The orientation of microfibrils in the primary wall is more-or-less random, although in elongating cells they tend to orient parallel to the direction of growth.

The cellulose microfibrils are embedded in a matrix of noncellulosic polysaccharides, principally **hemicellulose** and **pectic substances.** Hemicellulose is a complex, heterogeneous mixture of sugars and sugar derivatives that forms a highly branched network. Hemicellulose is characterized primarily by its extractability into dilute alkali, such as 7.5 percent sodium hydroxide. The pectic substances, which may make up as much as 35 percent of the primary wall, are also a heterogeneous group of polysaccharides especially rich in galacturonic acid—the acid form of the sugar galactose. The pectic substances are also the principal constituent of the **middle lamella,** the cement that holds together the primary walls of adjacent cells. The softening of fruit as it ripens, for example, is due in part to the enzymatic degradation of pectic substances in the middle lamella.

The cell wall matrix has traditionally been considered an amorphous gel with little or no structure. This property is reflected in the age-old practice of using pectic substances extracted from unripe fruit when making jellies. More recently, however, it appears that both the hemicelluloses and pectic substances may coat the cellulose microfibrils and are oriented more-or-less parallel to them (Fig. 1.9).

Primary cell walls also contain approximately 10 percent glycoprotein with an unusually high content of the amino acid hydroxyproline. These hydroxyproline-rich glycoproteins are called **extensin.** Although the precise function of extensin is unknown, it is thought to form a structural network that adds strength to the wall and is involved in cell growth. Several other families of cell wall proteins rich in proline, glycine, or threonine have been described more recently, adding to the apparent complexity of cell wall structure.

One of the final acts of cell division (cytokinesis) is the formation of a **cell plate,** the initial partition between daughter cells. The cell plate is formed by the fusion of Golgi vesicles that deliver the cell wall precursors. The fused membrane vesicles eventually form

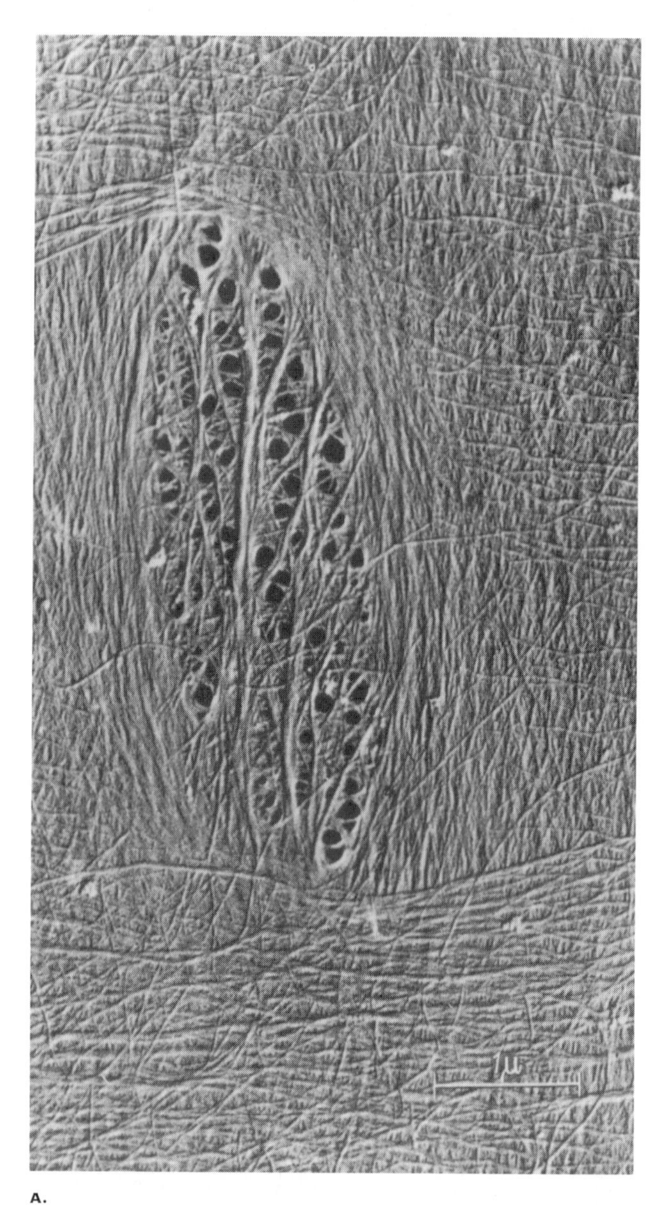

A.

FIGURE 1.8 Cell walls. (*A*) Electron micrograph of the primary wall of a parenchyma cell from the coleoptile of an oat (*Avena*) seedling. Note the pores through which plasmodesmata pass. (*B*) The principal structural components of cell walls are cellulose microfibrils constructed of β-1→4-linked glucose chains (cellulose). Adjacent chains in the microfibril are joined by intermolecular hydrogen bonds. (*A* from H. Böhmer, Untersuchungen über das Wachstum und den Feinbau der Zellwände in der *Avena*-Koleoptile, *Planta* 50:461–497, Fig. 20, 1958. Copyright Springer-Verlag, Heidelberg.)

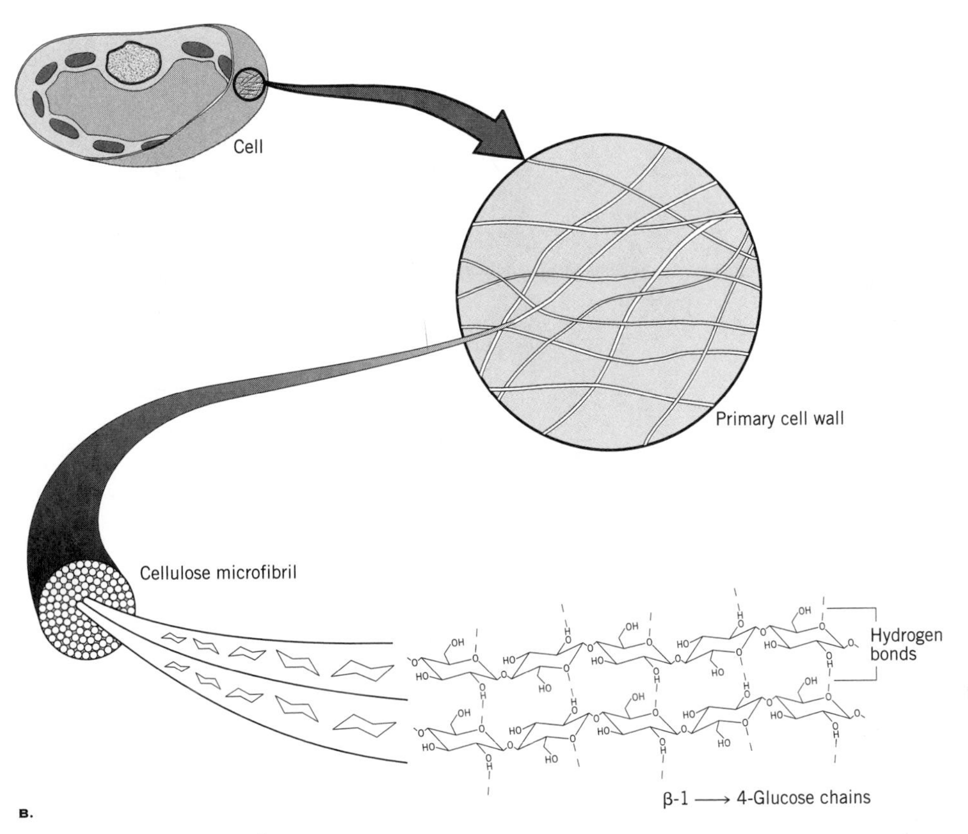

FIGURE 1.8 (continued)

the plasma membranes of the daughter cells, and their contents become incorporated into the newly formed middle lamella. The details of cellulose biosynthesis are not understood, but it is believed that a complex of enzymes (called a **rosette**) in the plasma membrane is responsible for the simultaneous synthesis of cellulose, its assembly into a microfibril, and extrusion of the microfibril into the extracellular space (Fig. 1.10).

The hemicelluloses and pectic substances are highly hydrated and, in spite of extensive cross-linking between

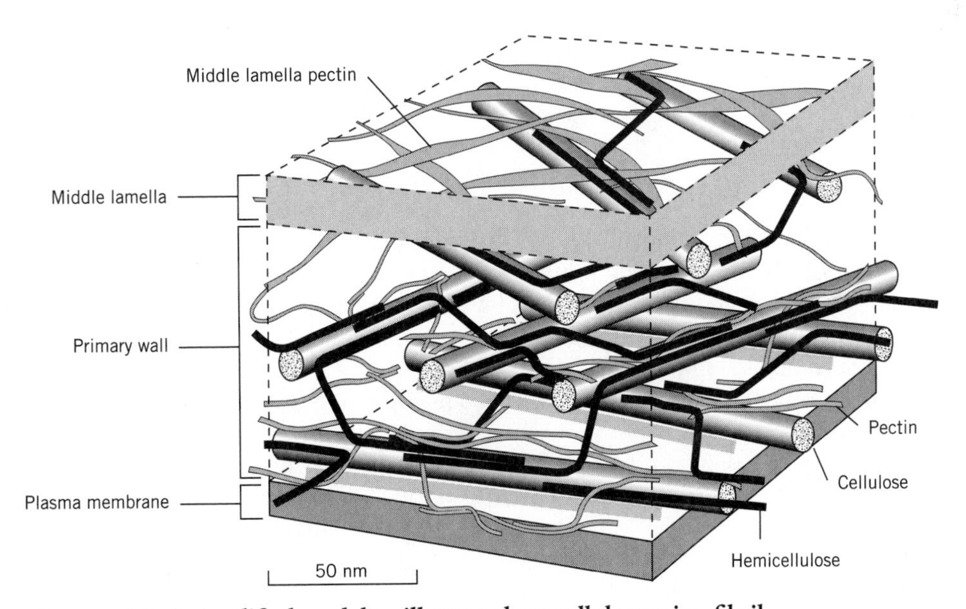

FIGURE 1.9 A simplified model to illustrate how cellulose microfibrils, hemicellulose, and pectic substances might be arranged in the cell wall. (From M. C. McCann, K. Roberts, Architecture of the primary cell wall, in C. W. Lloyd (ed.), *The Cytoskeletal Basis of Plant Growth and Form*, Academic Press, 1991. Original figure courtesy of Dr. M. C. McCann.)

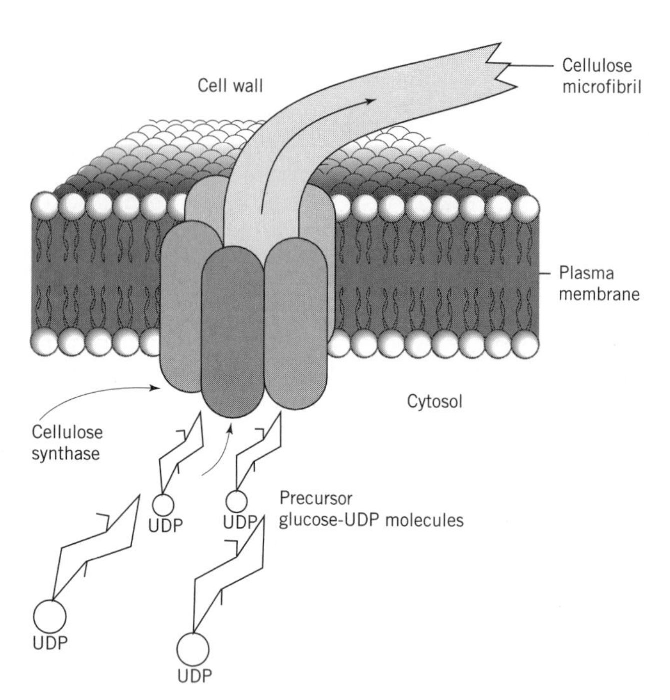

FIGURE 1.10 A model for cellulose synthesis from precursor molecules, uridine diphospho-glucose (UDP-glucose). The enzyme, formally known as *UDP-glucose:(1→4)-β-glucan glucosyltransferase*, is believed to be located in the plasma membrane where it simultaneously synthesizes the cellulose chains, assembles them into microfilaments, and extrudes the microfilaments into the cell wall.

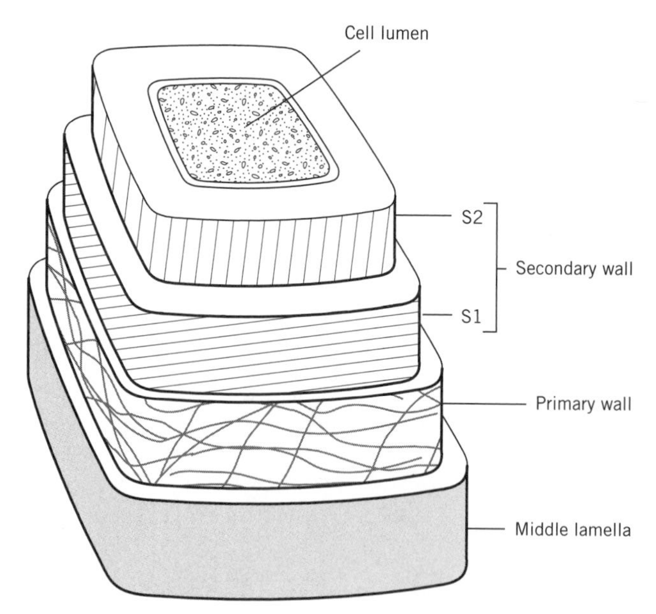

FIGURE 1.11 Differing orientations of cellulose microfibrils in primary and secondary cell walls.

the cellulose microfibrils, the wall remains sufficiently porous to allow the free passage of water and most dissolved solutes. While the wall maintains the structural rigidity and shape of the cell, it exerts no specific control over the exchange of material between the cell and its environment. That, as noted earlier, is the responsibility of the plasma membrane.

THE SECONDARY CELL WALL

As the cell stops enlarging and begins to mature, a secondary wall is laid down on the inside of the primary wall. Secondary walls are thicker and more rigid than primary walls. They contain up to 45 percent cellulose, correspondingly less hemicellulose, and relatively little pectic substance. In thick-walled, woody cells, the secondary wall frequently consists of two distinct zones, depending on the orientation of the microfibrils (Fig. 1.11). In both zones the microfibrils are oriented helically around the cell. In the outermost layer, adjacent to the primary wall, the microfibrils are oriented at a large angle to the long axis of the cell. In the inner layer, the microfibrils are almost parallel to the long axis.

Most secondary walls also contain **lignin,** which may account for as much as 35 percent of the dry weight of woody tissues. Lignin has a high degree of strength, stronger even than cellulose microfibrils. It is extremely resistant to extraction without chemical degradation, which makes its chemistry difficult to study. It is known to consist of a complex system of interlocking bonds between several relatively simple phenolic alcohols (Chap. 14). Next to cellulose, lignin is probably one of the most important biological substances in terms of structural importance. The combination of cellulose embedded in lignin is responsible for the exceptional strength of wood.

PLASMODESMATA

When the cell plate forms, membrane fusion is not complete, thus leaving locations where cytoplasm continuity is maintained between daughter cells. As the cellulose is laid down and the wall increases in thickness, these connections form membrane-lined channels called **plasmodesmata** (sing. *plasmodesma*) (Fig. 1.12). The membrane that lines the channel is a continuation of the plasma membranes from adjacent cells. Running through the center of the plasmodesma is a second membranous tube—a tube within a tube—called the **desmotubule.** The desmotubule is believed to be an extension of the endoplasmic reticulum that was entrapped during the formation of the cell plate. A sleeve of cytosol fills the space between the desmotubule and plasmodesmata itself.

Plasmodesmata are not large—approximately 60 nm in diameter—but there are often large numbers of them. Estimated frequencies are in the range of 0.1 to 10.0 μm^{-2} of cell wall (Robards and Lucas, 1990), although there is a tendency for plasmodesmata to be

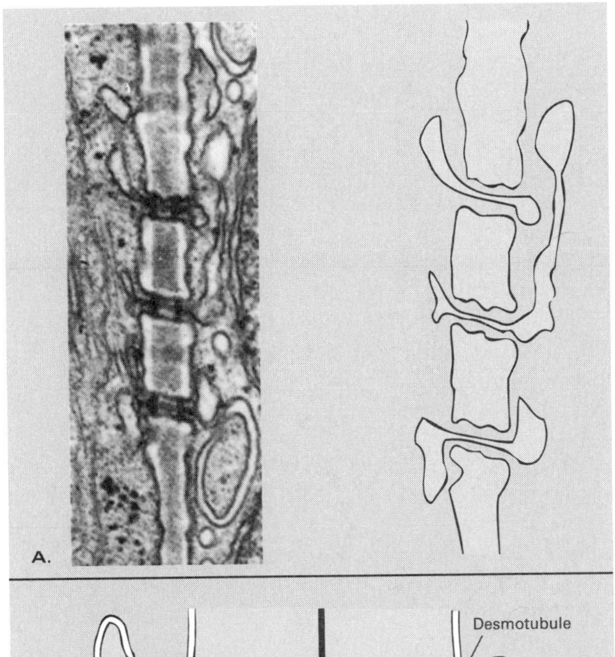

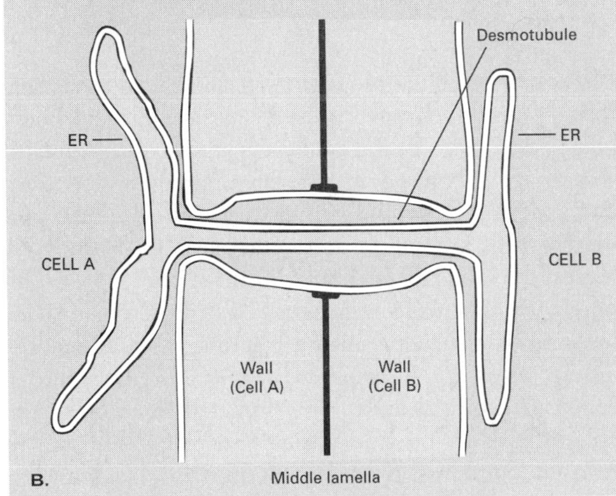

FIGURE 1.12 Plasmodesmata. (*A*) Electron micrograph showing plasmodesmata connecting adjacent cells. (*B*) Diagram of a plasmodesma showing the relationship between plasma membranes, ER, and desmotubules.

grouped in roughly oval areas called **primary pit fields** (Fig. 1.8A). Plasmodesmata are small enough to preclude the exchange of organelles between cells but large enough to permit the diffusion of small solute molecules through the cytosolic sleeve. Plasmodesmata thus provide a measure of membrane and cytosolic continuity between cells. Their function is undoubtedly similar to the protein-lined gap junctions of animal cells, which also provide a means of cytoplasmic communication between cells.

When the secondary wall is laid down, it is excluded over the area of primary pit fields. This leaves a **pit** in the secondary wall. The nature of the primary pit field is such that pits normally form opposite each other (called a **pit-pair**). The primary wall and middle lamella remain intact between the two cells to form a **pit membrane.** In cells that have heavy secondary wall and no protoplast at maturity, such as the water-conducting elements of the tissue, the pit-pairs provide a route for the flow of water and dissolved solute between adjacent conduits. The role of pit-pairs in water-conducting elements is discussed in Chapter 3.

The connection of neighboring protoplasts through plasmodesmata creates a continuous cytoplasmic network, referred to as the **symplast,** throughout the plant. In a similar manner, the **apoplast** consists of continous non-cytoplasmic space. The apoplast is comprised of interconnected cell walls, intercellular spaces, and non-living vascular tissue. The concept of symplast and apoplast is especially useful when considering the movement of water and dissolved solutes throughout the plant.

CELLS AND TISSUES

Not only is the cell the basic unit of life, it is also the fundamental morphological unit of the plant body. Cells are in turn organized into groups whose structure or function, or both, are distinct from others. These groups of cells are called **tissues.** Some tissues are relatively simple, containing only one cell type. Other tissues are more complex and are made up of several different types of cells.

In spite of the diversity of plant structure and morphology, there is a consistency in the arrangement of tissues that reflects a fundamental structural and functional organization of the plant as a whole. Thus, the epidermis provides a continuous layer of protective cells over the surface of a young plant. Tissues concerned with the distribution of nutrients and water form a coherent system that interconnects roots and photosynthetic tissues with regions of active growth and storage.

A plant begins as a fertilized egg, or **zygote,** in the ovary of a flower. The zygote undergoes cell division and differentiation, giving rise to an **embryo** (Fig. 1.13). The embryo is a rudimentary plant; it has an embryonic root (the **radicle**) and shoot (the **plumule**) and contains the first leaves and primordia for perhaps several more. As the embryo develops into first a young seedling and then an adult plant, the embryonic character persists in regions of continued cell division called **meristems** (Fig. 1.14). Through continued cell division, meristems contribute new cells to the plant body as the older cells begin to differentiate and mature. Unlike animals, where the size and form of the adult are determined in the early stages of embryo development, meristems give plants their unique character of indeterminant or continuous open growth.

The two principal meristems are the **apical meristems** located at the apex of the shoot and the root. The apical meristems are responsible for adding to the length of the shoot and root axis, called **primary growth,** and

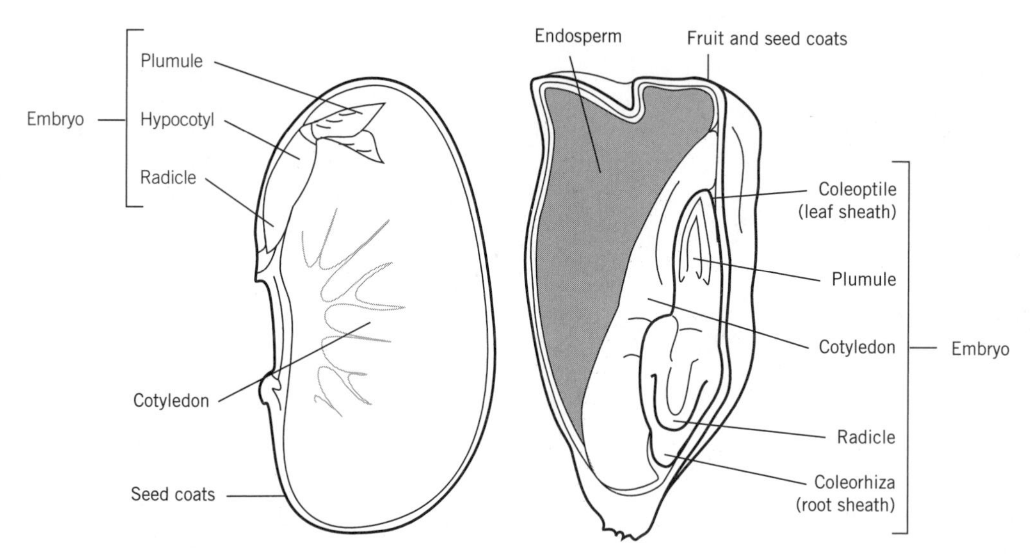

FIGURE 1.13 Seeds of common bean (*Phaseolus vulgaris*) (left), and maize (*Zea mays*) (right), a monocot, a dicot, showing the embryo.

the cells and tissue derived from these meristems form the **primary plant body.** Many plants undergo an increase in girth or thickening of the stems and roots by adding new vascular tissue to the primary body. This lateral growth is called **secondary growth** and is produced by a **lateral meristem** called the **vascular cambium.**

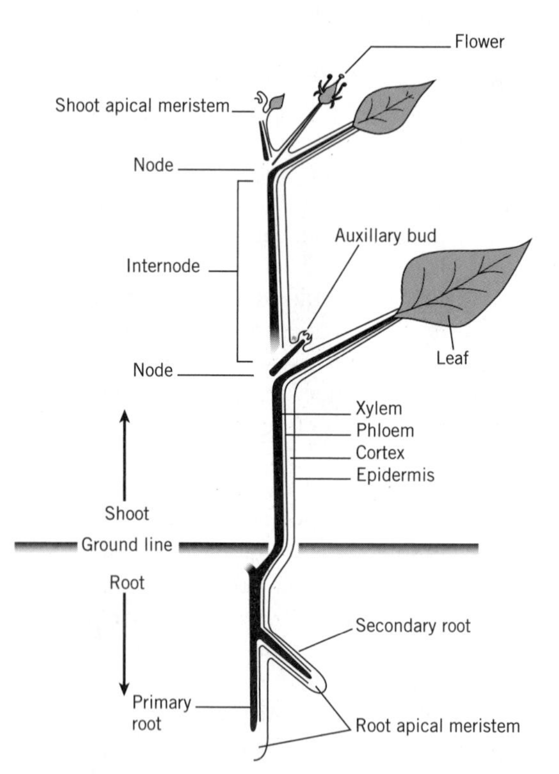

FIGURE 1.14 The general organization of a primary plant body, showing the principal tissues and organs. The relative positions of apical meristems, nodes, and internodes are also shown.

Epidermis. The **epidermis** is a superficial tissue that forms a continuous layer over the surface of the primary plant body. Cells of the epidermis are usually regular in shape, are appressed very tightly together, and their outer walls are covered with a waxy cuticle. Some epidermal cells are specialized as hairs or **trichomes.** In leaves, the integrity of the epidermis is interrupted with pores, which allow for exchange of carbon dioxide, water, and oxygen between the leaf and the ambient air. These pores are surrounded by specialized epidermal cells called **guard cells.** The principal function of the epidermis is to provide mechanical protection and to restrict water loss. Some stems undergoing secondary growth produce a cork or **phellem** on the outer surface, which is derived from a cork cambium or **phellogen.**

Parenchyma. **Parenchyma** cells are found throughout the plant body in the cortical regions of stems and roots and in the mesophyll of leaves, and are scattered throughout the vascular tissues. Parenchyma cells tend to be more-or-less isodiametric, although they may be irregular in shape or, in some cases, elongated. Parenchyma cells are usually surrounded only by a primary wall, although secondary wall thickenings are sometimes encountered. Parenchyma cells serve primarily in photosynthesis (in which case they may be called *chlorenchyma*), storage, and wound healing.

Supporting Tissues. Although most plants depend on hydrostatic pressures acting against the cell wall or on lignified vascular tissues for structural support, there are some tissues that serve principally as strengthening and supporting elements. The two principal supporting tissues are **collenchyma** and **sclerenchyma.** Collenchyma cells may be considered parenchyma cells that are specialized for support in young tissues. They are commonly found in the cortex of stems and petioles or along the veins in leaves and are characterized by thickened primary walls. Sclerenchyma cells, on the other hand, are scattered throughout the plant, in both primary and secondary tissues. There are two types of sclerenchyma: **sclerids** and **fibers.** Sclerids tend to be polyhedral or branched, while fibers are generally very long and slender. Both sclerids and fibers have thick secondary walls that may be heavily lignified. The lumen of the cell is very reduced due to the thick walls and usually lacks a protoplast at maturity.

Sclerenchyma fibers, known commercially as *bast fibers*, have been used since antiquity to make cloth. Fibers from the stem of flax plants (*Linum sps.*), which may reach lengths of two inches, have little or no lignin and are used to produce fine linen thread. Fibers of the hemp plant (*Cannabis sativa*) contain more lignin and produce a coarser cloth. Because of their exceptional strength, heavily lignified fibers from leaves of *Musa textilis* and *Agave sps.* are used primarily for cordage (rope and twine).

Vascular Tissues. The vascular tissues are concerned primarily with the distribution of nutrients, water, and the products of photosynthesis. The origin of vascular tissues represented a significant evolutionary advance, since they enable the spatial separation of the photosynthetic organs (the leaves) from organs for the absorption of water and nutrients (the roots), which must remain fixed in the soil. The evolution of vascular tissue is one of the most significant factors in the colonization of terrestrial habitats by plants.

There are two types of vascular tissue: **xylem** and **phloem.** Xylem is a structurally and functionally complex tissue concerned primarily with water conduction, storage, and support. The cells most characteristic of xylem are the principal water conducting elements, **tracheids** and **vessel members.** Tracheids and vessel members have heavy, often lignified, secondary walls and are devoid of protoplasm at maturity. The structure and function of tracheids and vessel members are discussed further in Chapter 3. Xylem tissue also contains parenchyma cells, which function primarily as storage, and mechanical cells (sclerids and fibers), which provide support.

Phloem tissue is also both structurally and functionally complex. Phloem is concerned with the distribution of primarily organic molecules between "sources," that is, photosynthetic or storage tissues, and "sinks," or regions of active growth and metabolism. The principal conducting elements are **sieve cells** or **sieve tubes**—long vertical arrays of **sieve tube members** joined end-to-end. Sieve cells and sieve tube members are enucleate at maturity and contain highly modified protoplasts. Phloem also contains parenchyma cells, some of which are specialized **companion cells** or **transfer cells,** as well as sclerids and fibers. The structure and function of phloem tissue are discussed further in Chapter 11.

Vascular tissues may be of either primary or secondary origin; that is, derived from either the apical meristem (primary vascular tissues) or from the vascular cambium (secondary vascular tissues). Increase in the diameter of stems and roots is due to the vascular cambium, which produces secondary xylem to the inside and secondary phloem to the outside. The vascular cambium is the layer that separates when bark is peeled from a tree.

PLANT ORGANS

Just as cells are grouped into tissues with distinct structures and functions, tissues are also grouped into organs. The principal organs of a plant are roots, stems, leaves, flowers, and fruits. Brief sketches of the three principal vegetative organs—roots, stems, and leaves—are given here. Additional details will be presented in later chapters when functional aspects are discussed.

Roots. A schematic diagram of a root is shown in Fig. 1.15. The principal regions of a root are a central core of vascular tissues, called the **stele**; a surrounding cortex composed of parenchyma; and a protective epidermis. The primary functions of a root are anchorage, storage, and absorption of water and mineral nutrients.

Stems. The specific arrangement of stem tissues may vary considerably with age or depending on whether it is a **monocotyledonous** or **dicotyledonous** plant (Fig.

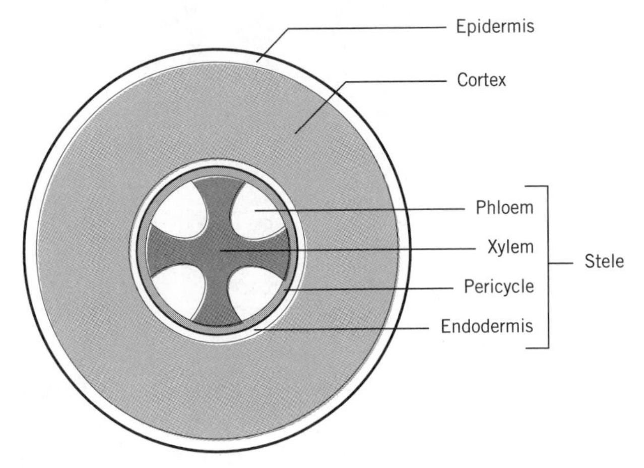

FIGURE 1.15 Schematic diagram of a cross-section of a typical root.

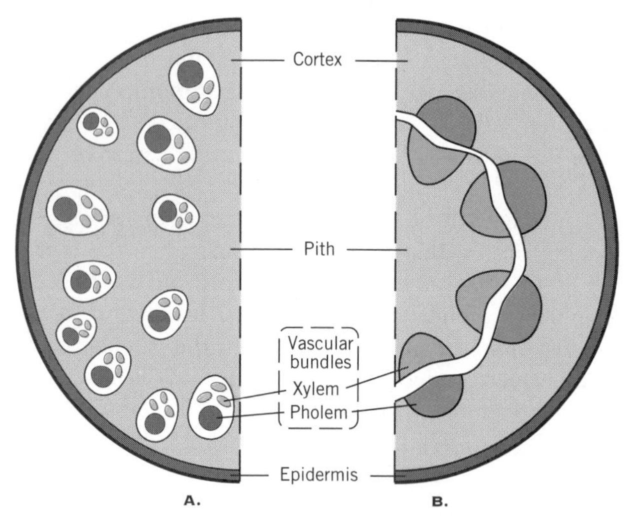

FIGURE 1.16 Schematic diagram of a cross-section of a monocotyledonous stem (*A*) and a dicotyledonous stem (*B*).

1.16). In monocotyledonous stems, the vascular tissues are arranged in bundles more-or-less scattered throughout a core of parenchyma. The bundle usually includes a cap of sclerenchyma fibers that support the stem. Young dicotyledonous stems are more orderly, with the vascular tissues (xylem and phloem) forming concentric rings enclosing a central pith of parenchyma. Additional parenchyma forms a cortex around the vascular tissues, all of which is covered with one or more epidermal layers.

The apical meristem not only adds to the stem tissues but also gives rise to leaves, branches, and other appendages. The points along the stem at which these appendages arise are referred to as **nodes** and the intervening stem regions as **internodes**. Increase in stem length is due to elongation of the internode cells. Failure of internode elongation gives rise to bulbs and the rosette form, in which all the leaves appear to arise from a common point.

Leaves. Leaves are typically laminar or bladelike structures attached to the stem by a stalk or **petiole.** The blade surfaces are composed of an upper and lower epidermis, which enclose photosynthetic parenchyma tissue known as the **mesophyll** (meso, *middle;* phyll, *leaf*) (Fig. 1.17). Most dicotyledonous leaves contain two types of mesophyll; the **palisade mesophyll,** a layer of rather tightly packed, elongated cells toward the upper leaf surface, and the **spongy mesophyll,** a zone of more loosely packed, irregular cells with an extensive network of air spaces.

Leaves also contain a network of vascular bundles that are continuous, through the petiole, with the vascular tissues of the stem. In dicotyledonous leaves, the bundles form an anastomosing system of veins of diminishing size, ensuring efficient delivery of water and nutrients to each photosynthetic cell and rapid removal of photosynthetic product. In monocotyledonous leaves, a system of interconnected, closely parallel veins serves the same purpose.

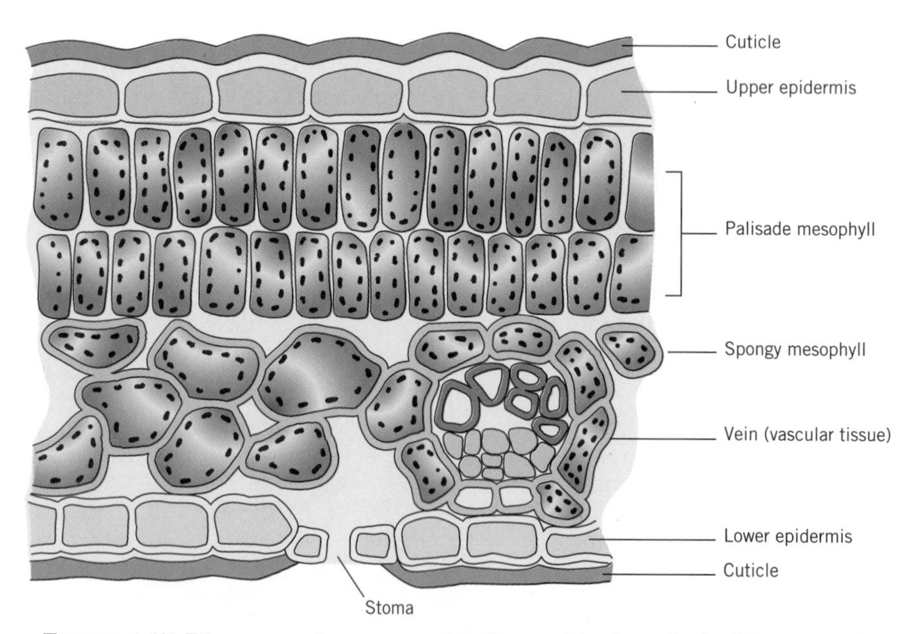

FIGURE 1.17 Diagrammatic cross-section illustrating the principal features of a typical dicotyledonous leaf. Chloroplasts are located in the photosynthetic palisade and spongy mesophyll cells.

CHAPTER REVIEW

1. Describe the composition and structure of plant membranes. What forces are responsible for the stability of the bilayer and how is the protein organized in the membrane?
2. List the principal organelles in a plant cell and describe the distinctive features of each.
3. Describe the cytoskeleton and identify some of its roles in the cell.
4. Describe the extracellular matrix of a plant cell. What are plasmodesmata, and what is their relationship to the extracellular matrix?
5. Understand the relationships between cells, tissues, and organs.

FURTHER READING

Alberts, B., D. Bray, J. Lewis, M. Raff, K. Roberts, J. D. Watson. 1989. *Molecular Biology of the Cell.* 2nd. ed. New York: Garland.

Esau, K. 1977. *Anatomy of Seed Plants.* 2nd ed. New York: Wiley.

Gunning, B. E. S., M. W. Steer. 1975. *Ultrastructure and the Biology of Plant Cells.* London: Arnold Press.

Lloyd, C. W. 1991. *The Cytoskeletal Basis of Plant Growth and Form.* New York: Academic Press.

O'Brien, T. P., M. E. McCully. 1969. *Plant Structure and Development.* London: Collier-Macmillan.

Robards, A. W., W. J. Lucas. 1990. Plasmodesmata. *Annual Review of Plant Physiology and Plant Molecular Biology* 41:369–419.

Roberts, K. 1990. Structures at the plant cell surface. *Current Opinion in Cell Biology* 2:920–928.

Voet, D., J. G. Voet. 1990. *Biochemistry.* New York: Wiley.

Zhang, G. F., L. A. Staehelin. 1992. Functional compartmentation of the Golgi aparatus of plant cells. *Plant Physiology* 99:1070–1083.

PART ONE

Plants, Water, and Minerals

PREVIEW

Plants live in an inorganic world consisting largely of water, mineral nutrients, carbon dioxide, and oxygen. Carbon dioxide enters through the shoot, oxygen enters through both shoot and root, and water and mineral nutrients are taken up primarily through the root system. A healthy, active root system together with an adequate supply of water and nutrients are critical to the physiology and development of the plant as a whole.

Root systems of most plants are very extensive, often accounting for more than 50 percent of total plant body weight. Roots form an intimate and dynamic association with soils in which they grow and from which they extract mineral nutrients and water. Water is important to a plant because it forms the milieu in which vital biochemical reactions occur and its hydraulic properties drive cell expansion and provide structural support. Because the shoot is exposed to a relatively dry atmosphere and the roots are surrounded by a relatively moist soil, there is a continuous flow of water through the plant, in the direction of decreasing water potential. The flow of water through a plant requires a continuous column of water maintained in specific transport tissues. From a practical perspective, water deficiency is a principal limiting factor in crop production worldwide.

Mineral nutrients are selectively taken into the root from the soil solution. The concentration and availability of minerals in the soil solution are determined by properties of the soil and metabolic activities of the roots. A limited number of mineral elements are actually required to support the normal growth of plants. Some minerals are required in large quantity, while others are required in only trace amounts; the specific requirement depends on the biochemical function of the element. Control over which mineral elements are taken up by the roots is exercised at the plasma membrane of individual root cells.

Plants are assisted in the uptake of mineral nutrients by a variety of microorganisms that form close associations with the roots. Fungi that infect roots expand the volume of soil accessible to the roots, and bacteria play an important role in converting nitrogen into a form that can be taken up and utilized by plants.

In the war of sun and dryness against living things, life has its secrets of survival. Life, no matter on what level, must be moist or it will disappear.

John Steinbeck
(Travels with Charley)

2

Plant Cells and Water

Without water, life as we know it could not exist. Water is the most abundant constituent of most organisms. The actual water content will vary according to tissue and cell type and it is dependent to some extent upon environmental and physiological conditions, but water typically accounts for more than 70 percent by weight of nonwoody plant parts. But the water content of plants is in a continual state of flux, depending on the level of metabolic activity, the water status of the surrounding air and soil, and a host of other factors. Although certain desiccation-tolerant plants may experience water contents of only 20 percent and dry seeds may contain as little as 5 percent water, both are metabolically inactive and resumption of significant metabolic activity is possible only after the water content has been restored to normal levels.

Water fills a number of important roles in the physiology of plants; roles for which it is uniquely suited because of its physical and chemical properties. The **thermal properties** of water ensure that it is in the liquid state over the range of temperatures at which most biological reactions occur. This is important because most of these reactions can only occur in an aqueous medium. The thermal properties of water also contribute to temperature regulation, helping to ensure that plants do not cool down or heat up too rapidly. Water also has excellent **solvent properties,** making it a suitable medium for the uptake and distribution of mineral nutrients and other solutes required for growth. Many of the **biochemical reactions** that characterize life occur in water and water is itself a participant in a large number of those reactions. The **transparency** of water to visible light enables sunlight to penetrate the aqueous medium of cells where it can be used to power photosynthesis or control development.

Water in land plants is part of a very dynamic system. Plants that are actively carrying out photosynthesis experience substantial water loss, largely through evaporation from the leaf surfaces. Equally large quantities of water must therefore be taken up from the soil and moved through the plant in order to satisfy deficiencies that develop in the leaves.

This constant flow of water through plants is a matter of considerable significance to their growth and survival. The uptake of water by cells generates a pressure known as *turgor;* in the absence of any skeletal system, plants must maintain cell turgor in order to remain erect. As will be shown in later chapters, the uptake of water by cells is also the driving force for cell enlargement. Few plants can survive desiccation. There is no doubt that the water relations of plants and plant cells are fundamental to an understanding of their physiology.

This chapter is concerned with the water relations of cells. Topics to be addressed include the following:

- A review of the unique physical and chemical properties of water which make it particularly suitable as a medium for life;
- Physical processes that underlie water movement in plants, including diffusion, osmosis, and bulk flow as mechanisms for water movement; and
- The chemical potential of water and the concept of water potential.

These concepts provide the basis for understanding water movement within the plant and between the plant and its environment, to be discussed in Chapter 3.

WATER AND HYDROGEN BONDS

Water has a number of unique physical and chemical properties when compared with other molecules of similar molecular size. The key to many of these properties is found in the structure of the water molecule and the strong intermolecular attractions that result from that structure.

Water consists of an oxygen atom covalently bonded to two hydrogen atoms (Fig. 2.1). The oxygen atom is strongly **electronegative,** which means that it has a tendency to attract electrons. One consequence of this strong electronegativity is that, in the water molecule, the oxygen tends to draw electrons away from the hydrogen. The shared electrons that make up the O—H bond are, on the average, closer to the oxygen nucleus than to hydrogen. As a consequence, the oxygen atom carries a *partial negative charge* and a corresponding *partial positive charge* is shared between the two hydrogen atoms. This asymmetric electron distribution makes water a **polar** molecule. Overall, water remains a neutral molecule, but the separation of negative and positive charges generates a strong mutual (electrical) attraction between adjacent water molecules or between water and other polar molecules. This attraction is called **hydrogen bonding** (Fig 2.1). The energy of the hydrogen bond is about 20 kJ mol^{-1}. The hydrogen bond is thus weaker than either covalent or ionic bonds, which typically measure several hundred kJ mol^{-1}, but stronger than the short-range, transient attractions known as Van der Waals forces (about 4 kJ mol^{-1}). Hydrogen bonding is largely responsible for the many unique properties of water, compared with other molecules of similar molecular size (Table 2.1).

In addition to interactions between water molecules, hydrogen bonding also accounts for attractions between water and other molecules or surfaces. Hydrogen bonding, for example, is the basis for hydration

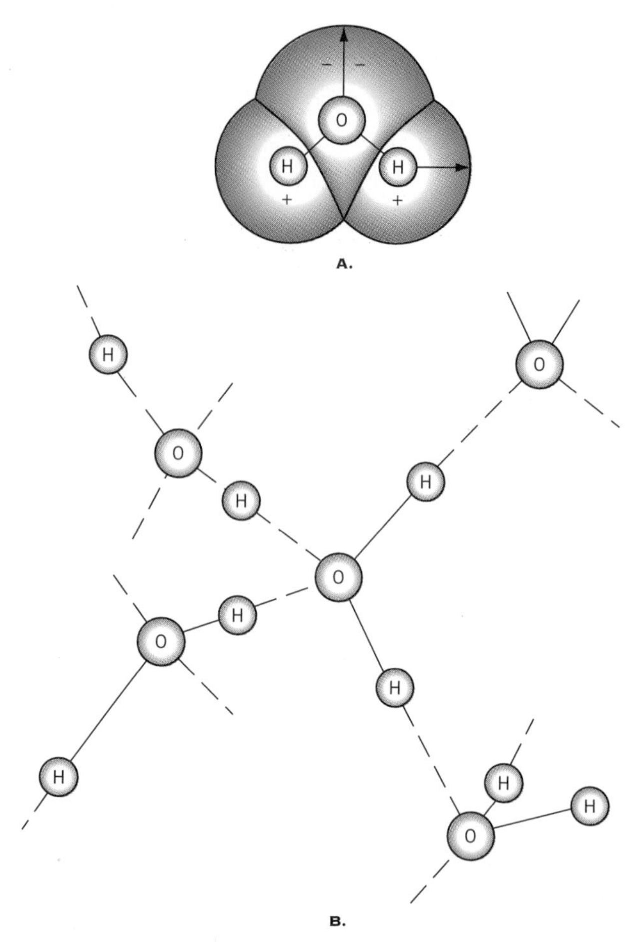

FIGURE 2.1 **(*A*) Schematic structure of two water molecules. (*B*) The hydrogen bond (dashed line) results from the electrostatic attraction between the partial positive charge on one molecule and the partial negative charge on the next.**

shells that form around biologically important macromolecules such as proteins, nucleic acids and carbohydrates. These layers of tightly bound and highly oriented water molecules are often referred to as *bound* water. It has been estimated that bound water may account for as much as 30 percent by weight of hydrated protein molecules. Bound water is important to the stability of protein molecules. Bound water "cushions" protein, preventing the molecules from approaching close enough to form aggregates large enough to precipitate.

Hydrogen bonding, although characteristic of water, is not limited to water. It arises wherever hydrogen is found between electronegative centers. This includes alcohols, which can form hydrogen bonds because of the —OH group, and macromolecules such as proteins and nucleic acids where hydrogen bonds between amino (—NH_2) and carbonyl (>C=O) groups help to stabilize structure.

TABLE 2.1 **Some physical properties of water compared with other molecules of similar molecular size. Because thermal properties are defined on an energy-per-unit mass basis, values are given in units of joules per gram.**

	Molecular mass (Da)	Specific heat (J/g/°C)	Melting point (°C)	Heat of fusion (J/g)	Boiling point (°C)	Heat of vaporization (J/g)
Water	18	4.2	0	335	100	2452
Hydrogen sulphide	34	—	−86	70	−61	—
Ammonia	17	5.0	−77	452	−33	1234
Carbon dioxide	44	—	−57	180	−78	301
Methane	16	—	−182	58	−164	556
Ethane	30	—	−183	96	−88	523
Methanol	32	2.6	−94	100	65	1226
Ethanol	46	2.4	−117	109	78	878

PHYSICAL AND CHEMICAL PROPERTIES OF WATER

THERMAL PROPERTIES

The thermal properties of water that result from hydrogen bonding are among the most biologically important. These properties can generally be attributed to the larger quantity of energy (that is, higher temperatures) required to overcome the relatively strong intermolecular attraction of the hydrogen bond.

Temperature and Physical State Perhaps the single most important property of water is that it is a liquid over the range of temperatures most compatible with life. Boiling and melting points are generally related to molecular size, such that state transitions for smaller molecules occur at lower temperatures than for larger molecules. On the basis of size alone, water might be expected to exist primarily in the vapor state at temperatures encountered over most of the earth. However, both the melting and boiling points of water are higher than expected when compared with other molecules of similar size, especially ammonia (NH_3) and methane (CH_4) (Table 2.1). Molecules such as ammonia and the hydrocarbons (methane and ethane) are associated only through weak Van der Waals forces and relatively little energy is required to change their state. Note, however, that the introduction of oxygen raises the boiling points of both methanol (CH_3OH) and ethanol (CH_3CH_2OH) to temperatures much closer to that of water. This is because the presence of oxygen introduces polarity and the opportunity to form hydrogen bonds.

Absorption and Dissipation of Heat The term *specific heat*[1] is used to describe the thermal capacity of a substance or the amount of energy that can be absorbed for a given temperature rise. The specific heat of water is 4.184 J g^{-1} $°C^{-1}$, higher than that of any other substance except liquid ammonia (Table 2.1). Because of its highly ordered structure, liquid water also has a high **thermal conductivity.** This means that it rapidly conducts heat away from the point of application. The combination of high specific heat and thermal conductivity enables water to absorb and redistribute large amounts of heat energy without correspondingly large increases in temperature. For plant tissues that consist largely of water, this property provides for an exceptionally high degree of temperature stability. Localized overheating in a cell due to the heat of biochemical reactions is largely prevented because the heat may be quickly dissipated throughout the cell. In addition, large amounts of heat can be exchanged between cells and their environment without extreme variation in the internal temperature of the cell.

Melting and Vaporizing Water Energy is required to cause changes in the state of any substance, such as from solid to liquid or liquid to gas, without a change in temperature. The energy required to convert a sub-

[1]Specific heat is defined as the amount of energy required to raise the temperature of one gram of substance by 1 °C (usually at 20 °C). The specific heat of water is the basis for the definition of a quantity of energy called the *calorie.* The specific heat of water was therefore assigned the value of 1.0 calorie. In accordance with the International System of Units (Système Internationale d'Unites, or SI), the preferred unit for energy is the *joule* (J). 1 calorie = 4.184 joule.

stance from the solid to the liquid state is known as the **heat of fusion.** The heat of fusion for water is 335 J g^{-1}, which means that 335 J of energy are required to convert 1 gram of ice to 1 gram of liquid water at 0 °C (Table 2.1). Expressed on a molar basis, the heat of fusion of water is 6.0 kJ mol^{-1} (18 g of water per mole × 335 J g^{-1}). The heat of fusion of water is one of the highest known, second only to ammonia. The high heat of fusion of water is attributable to the large amount of energy necessary to overcome the strong intermolecular forces associated with hydrogen bonding.

The density of ice is another important property. At 0 °C, the density of ice is less than that of liquid water. Thus water, unlike other substances, reaches its maximum density in the liquid state (near 4 °C), rather than as a solid. This occurs because molecules in the liquid state are able to pack more tightly than in the highly ordered crystalline state of ice. Consequently, ice floats on the surface of lakes and ponds rather than sinking to the bottom where it might remain year-round. This is extremely important to the survival of aquatic organisms of all kinds.

Just as hydrogen bonding increases the amount of energy required to melt ice, it also increases the energy required to evaporate water. The **heat of vaporization** of water, or the energy required to convert one mole of liquid water to one mole of water vapor, is about 44 kJ mol^{-1} at 25 °C. Because this energy must be absorbed from its surroundings, the heat of vaporization accounts for the pronounced cooling effect associated with evaporation. Evaporation from the moist surface cools the surface because the most energetic molecules escape the surface, leaving behind the lower-energy (hence, cooler) molecules. As a result, plants may undergo substantial heat loss as water evaporates from the surfaces of leaf cells. Such heat loss is an important mechanism for temperature regulation in the leaves of terrestrial plants that are often exposed to intense sunlight.

WATER AS A SOLVENT

Water comes close to being the "universal" solvent—more substances will dissolve in water than in any other common liquid.

The excellent solvent properties of water are due to the highly polar character of the water molecule. Water has the ability to partially neutralize electrical attractions between charged solute molecules or ions by surrounding the ion or molecule with one or more layers of oriented water molecules, called a **hydration shell.** Hydration shells encourage solvation by reducing the probability that ions can recombine and form crystal structures (Fig. 2.2).

The polarity of molecules can be measured by a quantity known as the **dielectric constant.** Water has one of the highest known dielectric constants (Table 2.2). The dielectric constants of alcohols are somewhat lower and those of nonpolar organic liquids such as benzene and hexane are very low. Water is thus an excellent solvent for charged ions or molecules, which dissolve very poorly in nonpolar organic liquids. Many of the solutes of importance to plants are charged. On the other hand, the low dielectric constants of nonpolar molecules helps to explain why charged solutes do not

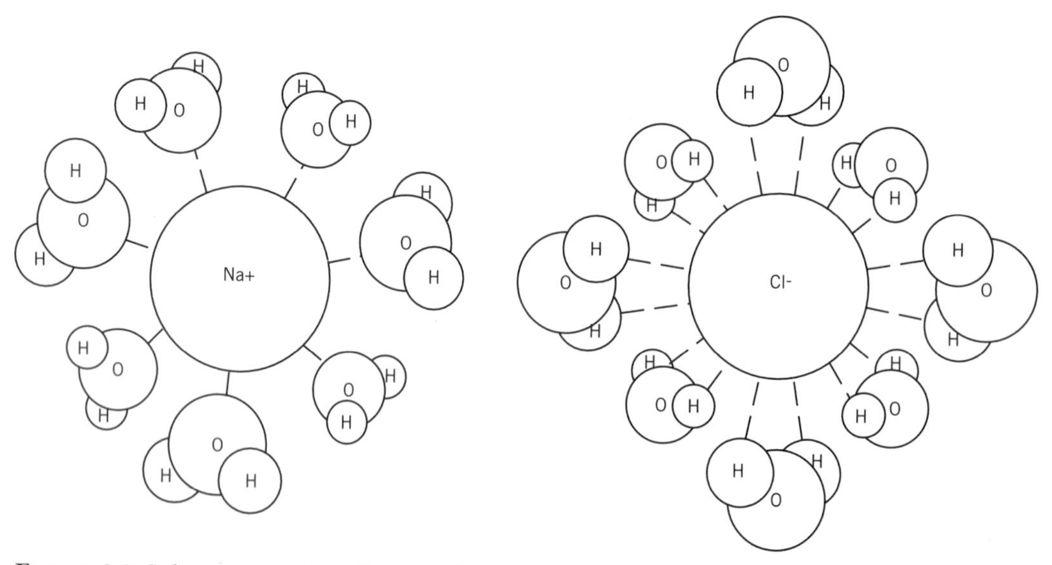

FIGURE 2.2 Solvent properties of water. The orientation of water molecules around the sodium and chloride ions screens the local electrical fields around each ion. The screening effect reduces the probability of the ions reuniting to form a crystalline structure.

TABLE 2.2 Dielectric constants for some common solvents at 25 °C.

Water	78.4
Methanol	33.6
Ethanol	24.3
Benzene	2.3
Hexane	1.9

readily cross the predominantly nonpolar, hydrophobic lipid regions of cellular membranes (Chap. 1).

COHESION AND ADHESION

The strong mutual attraction between water molecules gives rise to several other closely related, important properties.

The strong mutual attraction between water molecules resulting from hydrogen bonding is also known as **cohesion.** One consequence of cohesion is that water has an exceptionally high **surface tension,** which is most evident at interfaces between water and air. Surface tension arises because the cohesive force between water molecules is much stronger than interactions between water and air. The result is that water molecules at the surface are constantly being pulled into the bulk water (Fig. 2.3). The surface thus tends to contract and behaves much in the manner of an elastic membrane. A high surface tension is the reason water drops tend to be spherical or that a water surface will support the weight of small insects.

Cohesion is directly responsible for the unusually high **tensile strength** of water. Tensile strength is the maximum tension that an uninterrupted column of any material can withstand without breaking. High tensile strength is normally associated with metals but, under the appropriate conditions, water columns are also capable of withstanding extraordinarily high tensions—on the order of 30 megapascal (MPa).[2]

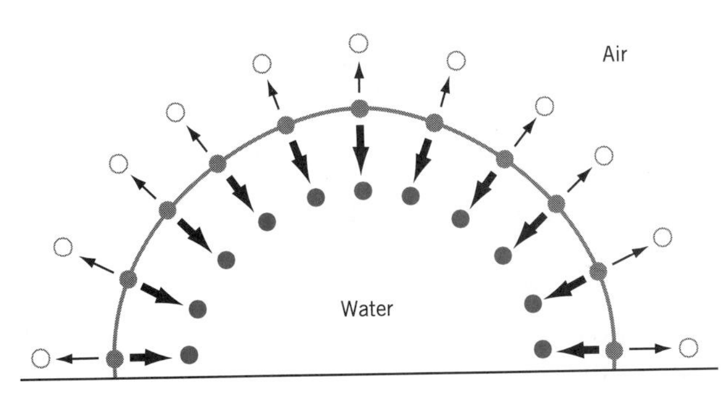

FIGURE 2.3 Schematic demonstration of surface tension in a water drop. Intermolecular attractions between neighboring water molecules (heavy arrows) are greater than attractions between water and air (light arrows), thus tending to pull water molecules at the surface into the bulk water.

The same forces that attract water molecules to each other will also attract water to solid surfaces, a process known as adhesion. Adhesion is an important factor in the capillary rise of water in small-diameter conduits.

The combined properties of cohesion, adhesion, and tensile strength help to explain why water rises in capillary tubes and are exceptionally important in maintaining the continuity of water columns in plants. Cohesion, adhesion, and tensile strength will be discussed in greater detail in Chapter 3, when evaporative water loss from plants and water movement in the xylem are examined.

TRANSLOCATION OF WATER

A principal focus of studies on the water economy of plants and plant cells is the factors that govern water movement from cell to cell or between cells and their environment. The movement of liquid water may be pressure-driven or may occur by diffusion.

One objective of plant physiology is to understand the dynamics of water as it flows into and out of cells or from the soil, through the plant, into the atmosphere. Movement of substances from one region to another is commonly referred to as **translocation.** Mechanisms for translocation may be classified as either *active* or *passive*, depending on whether or not metabolic energy is expended in the process. It is sometimes difficult to distinguish between active and passive transport, but the translocation of water is clearly a passive process. Although in the past many scientists argued for an active component, there is simply no evidence that water movement in plants requires a direct expenditure of metabolic energy. Passive movement of most substances can be accounted for by one of two physical processes: either **bulk flow** or **diffusion.** In the case of water, a special case of diffusion known as **osmosis** must also be taken into account.

BULK FLOW

Movement of materials by **bulk flow** (or mass flow) is pressure-driven. Bulk flow occurs when an external force, such as gravity or pressure, is applied. As a result, all of the molecules of the substance move in a mass. Movement of water by bulk flow is a part of our everyday experience. Water in a stream flows in response to the hydrostatic pressure established by gravity. It flows from

[2]The pascal (Pa), equal to a force of 1 newton per square meter, is the standard SI unit for pressure. MPa = Pa × 10^6.

the faucet in the home or workplace because of pressure generated by gravity acting on standing columns of water in the municipal water tower. Bulk flow also accounts for some water movement in plants, such as through the conducting cells of xylem tissue or the movement of water into roots. Later, in Chapter 11, it will be shown that bulk flow is a major component of the most widely accepted hypothesis for transport of solutes through the vascular tissue.

DIFFUSION

Like bulk flow, **diffusion** is also a part of our everyday experience. When a small amount of sugar is placed in a cup of hot drink, the sweetness soon becomes dispersed throughout the cup. The scent of perfume from a bottle opened in the corner of a room will soon become uniformly distributed throughout the air. If the drink were not stirred and there were no mass movements of air in the room, the distribution of these substances has occurred by diffusion. Diffusion can be interpreted as a *directed movement* from a region of a high concentration to a region of lower concentration, but it is accomplished through the *random thermal motion* of individual molecules (Fig. 2.4). Thus, while bulk flow is pressure-driven, diffusion is driven principally by concentration differences. Diffusion is a significant factor in the uptake and distribution of water, gasses and solutes throughout the plant. In particular, diffusion is an important factor in the supply of carbon dioxide for photosynthesis as well as the loss of water vapor from leaves.

The process of diffusion was first examined quantitatively by A. Fick. Fick's *first law*, formulated in 1855, forms the basis for the modern-day quantitative description of the process.

$$J = -\frac{DA\Delta C}{\ell} \quad (2.1)$$

J is the flux or the amount of material crossing a unit area per unit time (for example, mol m^{-2} s^{-1}). D is the **diffusion coefficient,** a proportionality constant that is a function of the diffusing molecule and the medium through which it travels. A and ℓ are the cross-sectional area and the length of the diffusion path, respectively. The term ΔC represents the difference in concentration between the two regions, also known as the concentration gradient. ΔC is the driving force for simple diffusion. In the particular case of gaseous diffusion, it is more convenient to use the difference in density (g m^{-3}) or vapor pressure (KPa, kilopascal) in place of concentration. The negative sign in Fick's law accounts for the fact that diffusion is toward the *lower* concentration or vapor pressure. In summary, Fick's law tells us that the rate of diffusion is directly proportional to the cross-sectional area of the diffusion path and to the concentration or vapor pressure gradient, and it is inversely proportional to the length of the diffusion path.

OSMOSIS—THE DIFFUSION OF WATER

Diffusion of water occurs when two regions are separated by a membrane that prevents the exchange of solute but allows unrestricted passage of water.

Fick's law is most readily applicable to the diffusion of solutes and gasses. In the general model illustrated in Figure 2.4, for example, the diffusing molecules could be glucose dissolved in water, carbon dioxide dissolved in water, or carbon dioxide in air. While, theoretically, Fick's law applies to the diffusion of *solvent* molecules as well, it can at first be difficult to imagine a situation in which diffusion of solvent molecules could occur. Consider what would happen if, for example, water were added to one of the chambers in Figure 2.4. As soon as

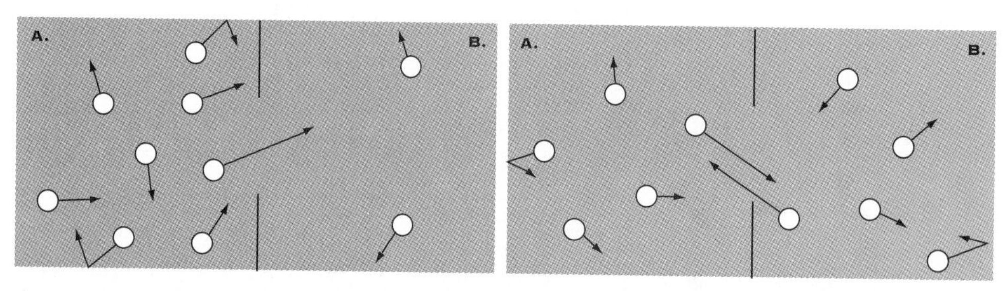

FIGURE 2.4 Diffusion is the directed movement of a substance from a region of high concentration to a region of lower concentration due to the random thermal motion of molecules. Initially (left) there is a much higher probability that a molecule in chamber A will pass through the open window into chamber B. With time (right), the number of molecules in chamber B will increase and the number in chamber A will decrease. This will continue until the molecules are uniformly distributed between the two chambers. At that point, the probability of molecules passing between the chambers in either direction will be equal and net diffusion will cease.

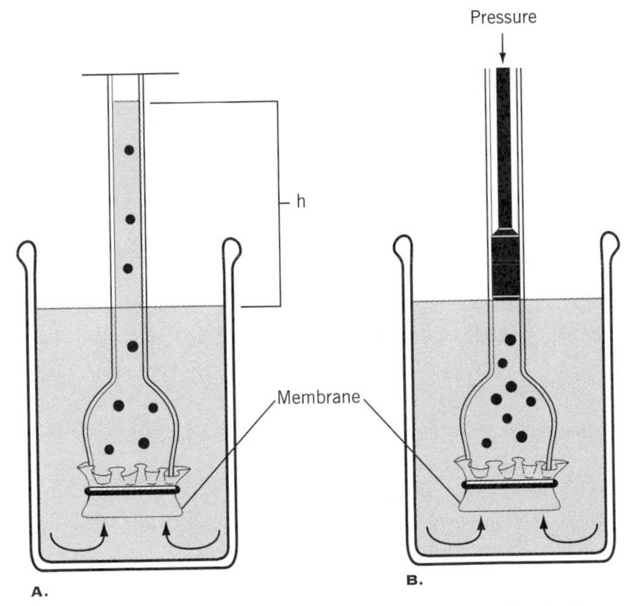

FIGURE 2.5 A demonstration of osmosis. A selectively permeable membrane is stretched across the end of a thistle tube containing a sucrose solution and the tube is inverted in a container of pure water. Initially, water will diffuse across the membrane in response to a chemical potential gradient. Diffusion will continue until the force tending to drive water into the tube is balanced by (*A*) the force generated by the hydrostatic head (h) in the tube or (*B*) the pressure applied by the piston. When the two forces are balanced, the system has achieved equilibrium and no further net movement of water will occur.

the water level in the first chamber reached the *open* window, it would flow over into the second chamber—an example of bulk flow.

Diffusion of water, a process known as *osmosis*, will occur only when the two chambers are separated from one another by a **selectively permeable** membrane. Recall that a selectively permeable membrane allows virtually free passage of water and certain small molecules, but restricts the movement of large solute molecules, and that all cellular membranes are selectively permeable (Chap. 1). Osmosis, then, is simply a special case of diffusion through a selectively permeable membrane.

Osmosis can easily be demonstrated using a device known as an **osmometer,** constructed by closing off the open end of a thistle tube with a selectively permeable membrane (Fig. 2.5). If the tube is then filled with a sugar solution and inverted in a volume of pure water, the volume of solution in the tube will increase over time. The increase in volume is due to a net diffusion of water across the membrane into the solution. This occurs because the **chemical potential** of water in the solution is lower than the chemical potential in the surrounding pure water. We will explore the concept of chemical potential further in the next section. For now, simply note that the increase in the volume of the solution will continue until the hydrostatic pressure developed in the tube is sufficient to balance the force driving the water into the solution (Fig. 2.5A). Alternatively, the tube could be fitted with a piston that would allow us to measure the amount of force required to *just prevent* any increase in the volume of solution (Fig. 2.5B). This force, measured in units of pressure (force per unit area), is known as **osmotic pressure** (symbol = π; pi).

The magnitude of the osmotic pressure that develops is a function of solute concentration,[3] at least in dilute solutions. It is useful to note that an isolated solution cannot have an osmotic pressure. It has only the *potential* to manifest a pressure when placed in an osmometer. For this reason, we say that the solution has an **osmotic potential** (symbol = Ψ_S). It is convention to define osmotic potential as the negative of the osmotic pressure, since they are equal but opposite forces ($\Psi_S = -\pi$). The significance of this distinction will become evident later when we introduce the concept of water potential.

Whereas simple diffusion is driven entirely by concentration differences, it is apparent from Figure 2.5 that pressure is also a factor in determining both the direction and rate of water movement in an osmometer. Sufficient pressure applied to the piston (Fig. 2.5B) will prevent further net movement of water into the solution. If additional pressure were applied, we might expect the net movement of water to reverse its direction and instead flow out of the solution. Thus, osmosis is driven not only by the concentration of dissolved solute but by pressure differences as well. Both of these factors influence the overall chemical potential of water, which is the ultimate driving force for water movement in plants. In order to fully grasp the significance of osmotic potential in plant water relations, we must first examine the concept of chemical potential and its role in determining water movement.

OSMOSIS AND CHEMICAL POTENTIAL

Osmosis requires energy. Water, like any other substance, will only move down an energy gradient—that is, when there is a difference in the energy of water in two parts of a system. In the case of the osmometer, water initially moves into the osmometer because the

[3]Osmotic pressure (potential) is one of the four *colligative properties* of a solution. The other three are freezing point depression, boiling point elevation and the lowering of the vapor pressure of the solvent. These four properties depend on the concentration of solutes, irrespective of their chemical nature.

energy of pure water in the beaker is greater than the energy of the water in the sugar solution. Net movement of water stops when there is no longer an energy gradient across the membrane.

The energy content of water, like any substance, is most easily described in terms of its chemical potential. Chemical potential (μ) is defined as the free energy per mole of that substance. The concept of free energy is described more fully in Chapter 9. For now, it is sufficient to note that chemical potential is a measure of the capacity of a substance to react or move. The rule is that *osmosis occurs only when there is a difference in the chemical potential ($\Delta\mu$) of water on two sides of a selectively permeable membrane.*

The chemical potential of water (or any other chemical species) is influenced by several factors. Two of those factors—concentration and pressure—are illustrated in the example of an osmometer (Fig. 2.5).

Consider first the concentration factor. In chemical systems, concentration is normally replaced by chemical activity (a). Concentration and activity are related by means of a proportionality constant called an activity coefficient ($a = \gamma C$). Except for ideal solutions, γ is less than unity; that is, chemical activity is invariably less than the actual concentration. Chemical activity may be converted to units of energy per mole simply by multiplying the logarithm of a by R (the universal gas constant) and T (the absolute temperature).

For reasons of simplicity and convenience, concentration (C) is often substituted for activity when discussing plant water relations. Biological situations often involve relatively dilute solutions and so—at least with respect to neutral molecules—do tend to approach this ideal state. Thus, the chemical potential of water in a solution may be expressed as follows:

$$\mu = \mu^* + \mathrm{RT} \ln C \qquad (2.2)$$

Equation 2.2 says that the chemical potential of water in a solution (μ) is equal to the chemical potential of pure water at atmospheric pressure and at the same temperature as the solution under consideration (μ^*) plus the (negative) contribution of concentration or chemical activity of water in that solution.

The effect of pressure on chemical potential is represented by the value $\overline{V}$P. $\overline{V}$ is the partial molal volume, or the volume occupied by one mole of the chemical species. Since one liter ($= 10^{-3}$ m^3) of water contains 55.5 moles, $\overline{V}$ = (1000/55.5) or 18 ml per mole ($=18\,000$ mm$^3 \cdot$ mol^{-1}) of water. P is the pressure. Measurements in plant physiology are commonly made at atmospheric pressure but the presence of relatively rigid cell walls allows plant cells to develop significant hydrostatic pressures. Hence it is convention to express P as the difference between actual pressure and atmospheric pressure. The influence of pressure on chemical potential can now be added to Equation 2.2:

$$\mu = \mu^* + \mathrm{RT} \ln C + \overline{V}\mathrm{P} \qquad (2.3)$$

The chemical potential of water may also be influenced by electrical potential and gravitational field. In spite of its strong dipole nature, the net electrical charge for water is zero and so the electrical term can be ignored. This is why we talk about the *chemical* potential of water rather than its *electrochemical* potential.

While the gravitational term may be large and must be considered where water movement in tall trees is concerned, it is not significant at the cellular level. Where water movement involves heights of five to ten meters or less, the gravitational term is commonly omitted. The chemical potential of water may be considered a function primarily of its chemical activity and pressure.

In what way is the diffusion of water by osmosis related to its chemical potential? We noted earlier that addition of solute generates an osmotic pressure in a solution and that the magnitude of the osmotic pressure is related to the concentration of solute. This relationship arises because the addition of solute decreases the chemical activity of water (a_w). To a very limited extent this may be viewed as a dilution effect, although the concentration of water in a solution may not always decrease. It may remain unchanged or even increase depending on the characteristics of the particular solute being added. The addition of solute, however, does decrease the number of water molecules as a fraction of the total number of water plus solute molecules. In other words, the presence of solute decreases the **mole fraction** of water. As the mole fraction of water decreases, the osmotic pressure of the solution increases. This relationship between osmotic pressure, water activity, and the mole fraction of water can be expressed in the following way:

$$\pi = -\frac{\mathrm{RT}\ln a_w}{\overline{V}_w} \qquad (2.4)$$

and

$$\mathrm{RT} \ln a_w = -\overline{V}_w\pi \qquad (2.5)$$

Equation 2.5 is the formal definition of π. It tells us that addition of solute decreases the value of a_w. As a matter of convenience, the value of a_w for pure water is arbitrarily designated as unity (=1). Consequently, *a_w for a solution is always less than one.*

By substituting Equation 2.5 in Equation 2.3, we can now arrive at an expression for the chemical potential of water:

$$\mu_w = \mu_w^* - \overline{V}_w\pi + \overline{V}_w\mathrm{P} \qquad (2.6)$$

According to Equation 2.6, the extent to which the chemical potential of water in a solution differs from that of pure water is a function of an osmotic component and a pressure component.

The chemical potential of water (μ_w) is particularly useful in the study of cellular and plant water relations because it defines the amount of work that can be done by water in one location, for example, a cell or vacuole, compared with pure water at atmospheric pressure and the same temperature. More importantly, a difference in μ_w between two locations means that water is not in equilibrium and there will be a tendency for water to flow toward the location with the lower value. At this point, then, it can be said that the driving force for water movement is a gradient in chemical potential.

THE CONCEPT OF WATER POTENTIAL

The chemical potential of water in a system can be expressed as water potential, defined as the sum of two easily measured quantities: hydrostatic pressure and osmotic pressure. Water always moves from a region of high water potential to a region of low water potential.

It is usually more convenient to measure relative values than it is to measure absolute values. The absolute chemical potential of water in solutions is one of those quantities that is not conveniently measured. However, Equation 2.6 can be rearranged as:

$$\frac{\mu_w - \mu_w^*}{\overline{V}_w} = P - \pi \qquad (2.7)$$

where $\mu_w - \mu_w^*$ is the difference between the chemical potential of water in solution and that of pure water. Although the value of $\mu_w - \mu_w^*$ is more easily measured, the task of plant physiologists was simplified even further when, in 1960, R. O. Slatyer and S. A. Taylor introduced the concept of **water potential** (symbolized by the Greek uppercase PSI, Ψ) (Slatyer and Taylor, 1960). Water potential is proportional to $\mu_w - \mu_w^*$ and can be defined as:

$$\Psi = \frac{\mu_w - \mu_w^*}{\overline{V}_w} = P - \pi \qquad (2.8)$$

or simply:

$$\Psi = P - \pi \qquad (2.9)$$

where P is the hydrostatic pressure and π is the osmotic pressure.

The concept of water potential has been widely accepted by plant physiologists because it avoids the difficulties inherent in measuring chemical activity. Instead, it enables experimenters to predict the behavior of water on the basis of two easily measured quantities, P and π. It also makes it possible to express water potential in units of pressure (pascals), which is more relevant to soil-plant-atmosphere systems than units of energy (joules). This distinction is not trivial. In practice it is far easier to measure pressure changes than it is to measure the energy required to effect water movement. Finally, we can restate the driving force for water movement as the *water potential gradient;* that is, water will move from a region of high water potential to a region of lower water potential. As we shall see, however, water potentials are usually negative. This means that water moves from a region of less negative water potential to a region where the water potential is more negative.

In the same way that the chemical potential of water in a solution is measured against that of pure water, water potentials of solutions are also measured against a reference. For water potential, the reference state is arbitrarily taken as pure water at atmospheric pressure. Under these conditions there is neither hydrostatic pressure nor dissolved solutes; that is, both P and π are zero. According to Equation 2.9, the value of Ψ for pure water is therefore also zero. This is not to say that the chemical activity of pure water is zero. The value of μ^* is in fact very high but Ψ for pure water is zero by definition.

THE COMPONENTS OF WATER POTENTIAL

Water potential may be also be defined as the sum of its component potentials:[4]

$$\Psi = \Psi_P + \Psi_S \qquad (2.10)$$

The symbol Ψ_P represents the **pressure potential.** It is identical to P and represents the hydrostatic pressure in excess of ambient atmospheric pressure. The term Ψ_S represents the *osmotic potential.* Note the change in sign. As pointed out earlier, osmotic potential is equal to osmotic pressure but carries a negative sign. Osmotic potential is also called **solute potential** (hence the designation S) because it is the contribution due to dissolved solute. The term *osmotic* (or solute) *potential* is preferred over *osmotic pressure* because it is more properly a property of the solution.

[4]Students reading further in the literature will find that a variety of conventions, names, and symbols have been used to describe the components of water potential. These differences are for the most part superficial, but careful reading is required to avoid confusion.

We can see from Equation 2.10 that hydrostatic pressure and osmotic potential are the principal factors contributing to water potential. A third component, the **matric potential** (M), is often included in the equation for water potential. Matric potential is a result of the adsorption of water to solid surfaces. It is particularly important in the early stages of water uptake by dry seeds (called **imbibition**) and when considering water held in soils (Chap. 3). There is also a matric component in cells, but its contribution to water potential is relatively little compared with solute component. It is also difficult to distinguish the matric component from osmotic potential. Consequently, matric potential may be excluded for purposes of the present discussion. We will return to matric potential when we discuss soil water in Chapter 3.

Returning to Equation 2.10, we can see that an increase in hydrostatic pressure or osmotic potential will increase water potential while a decrease in the pressure or osmotic potential (more negative) lowers it. We can use these changes to explain what happened earlier in our example of the osmometer (Fig. 2.5). The dissolved sucrose generated an osmotic potential in the thistle tube, thereby lowering the water potential of the solution ($\Psi < 0$) compared with the pure water ($\Psi = 0$) on the other side of the membrane. Water thus diffused across the membrane into the solution. As the volume of the solution increased, a hydrostatic pressure developed in the thistle tube. When the positive hydrostatic pressure was sufficient to offset the negative osmotic potential, the water potential of the solution was reduced to zero. At that point $\Psi = 0$ on both sides of the membrane and there was no further *net* movement of water. It is also interesting to note that where volume in the osmometer is permitted to increase, the osmotic potential will decrease. This is because along with the volume change accompanying diffusion of water into the solution, the mole fraction of water would also increase. In effect, the solute concentration decreases due to dilution. At equilibrium, then, the osmotic potential of the solution would be higher (i.e., less negative) than at the beginning. In the end the pressure required to balance osmotic potential is less than what would have been required had the volume increase not occurred. Because the contribution of S to water potential is always negative, water will, at constant pressure, always move from the solution with the higher (less negative) osmotic potential to the solution with the lower (more negative) osmotic potential.

We can now ask what contributes to the osmotic and pressure potentials, and thus water potential, in plant cells. The osmotic potential of most plant cells is due primarily to the contents of the large central vacuole. With the exception of meristematic and certain other highly specialized cells, cell vacuoles contain on the order of 50 to 80 percent of the cellular water and a variety of dissolved solutes. These may include sugars, inorganic salts, organic acids, and anthocyanin pigments. Most of the remaining cellular water is located in the cell wall spaces, while the cytoplasm accounts for as little as 5 to 10 percent. Methods for determining the osmotic potential of cells and tissues do not generally discriminate between the cytoplasmic and vacuolar contributions—the result is an average of the two. The osmotic potential of a parenchyma cell is typically in the range of -0.1 to -0.3 MPa, the largest part of which is due to dissolved salts in the vacuole.

In a laboratory osmometer, pressure (Ψ_P) can be estimated as the difference between atmospheric pressure (0.1 MPa) and the hydrostatic pressure generated by the height of the water column. In cells, the pressure component arises from the force exerted outwardly against the cell walls by the expanding protoplast. This is known as **turgor pressure.** An equal but opposite inward pressure, called **wall pressure,** is exerted by the cell wall. A cell experiencing turgor pressure is said to be **turgid.** A cell that experiences water loss to the point where turgor pressure is reduced to zero is said to be **flaccid.** Instruments are available for measuring P directly in large algal cells (see Appendix), but in higher plants it is usually calculated as the difference between water potential and osmotic potential. In nonwoody herbaceous plants, turgor pressure is almost solely responsible for maintaining an erect habit. Indeed, one of the first outward signs of water deficit in plants is the wilting of leaves due to loss of turgor in the leaf cells.

WATER MOVEMENT IN CELLS AND TISSUES

The water status of plant cells is constantly changing as the cells adjust to fluctuations in the water content of the environment or to changes in metabolic state. This dynamic flux of water will inevitably be accompanied by changes in protoplast volume as well as changes in the water potential, osmotic potential, and especially, turgor pressure.

The relationship between the components of water potential as a function of protoplast volume in a cell may be described by the diagram in Figure 2.6. For purposes of this discussion, protoplast volume is assigned a relative value of 1.0 at **incipient plasmolysis,** the condition in which the protoplast just fills the cell volume. At incipient plasmolysis the protoplast exerts no pressure against the wall but neither is it withdrawn from the wall. Consequently, turgor pressure (Ψ_P) is zero and the

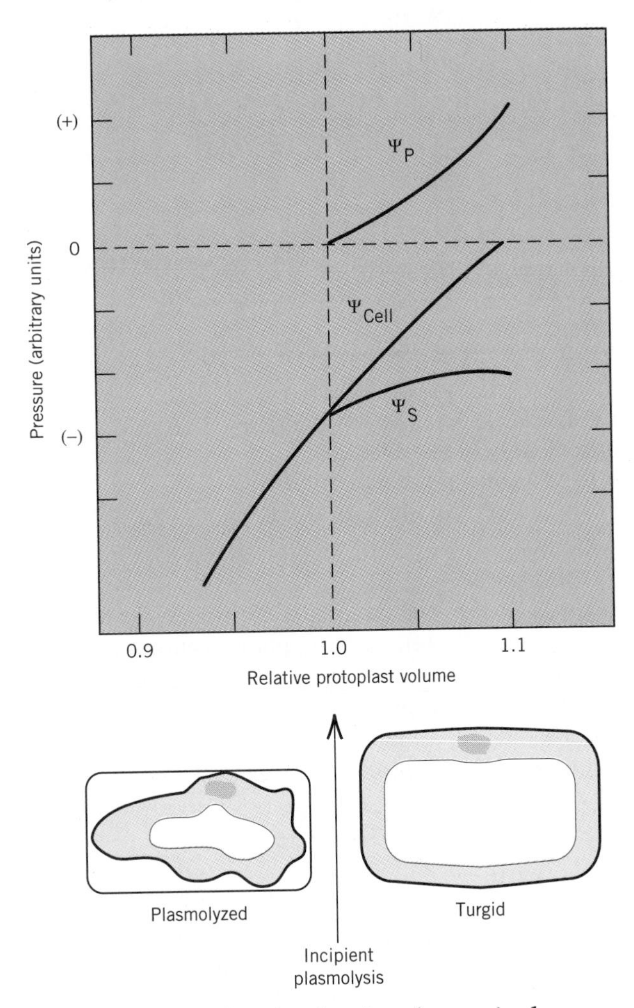

FIGURE 2.6 A Höfler plot showing changes in the relationship between turgor pressure (Ψ_P), solute potential (Ψ_S), and water potential (Ψ) with changing protoplast volume. It is assumed that the cell is bathed with pure water. The cell is assigned a relative volume of 1.0 at incipient plasmolysis. At incipient plasmolysis, turgor pressure is zero. A 10 percent increase in cell volume will give rise to a fully turgid cell with a small change in osmotic potential (Ψ_S) but a large increase in turgor pressure (Ψ_P). When fully turgid, the osmotic potential is balanced by the turgor pressure and the water potential of the cell (the algebraic sum of Ψ_S and Ψ_P) is zero.

water potential of the cell (Ψ_{cell}) is equal to its osmotic potential (Ψ_S).

When the cell is bathed by a *hypotonic*[5] solution such as pure water ($\Psi = 0$), water will enter the cell down the water potential gradient. This causes simultaneously a small dilution of the vacuolar contents (with a corresponding increase in osmotic potential) and the generation of a turgor pressure. Net movement of water into the cell will cease when the osmotic potential of the cell is balanced by its turgor pressure and, by Equation 2.10, the water potential of the cell is therefore also zero. When the cell is bathed by a *hypertonic* solution, which has a more negative osmotic potential than the cell, the water potential gradient favors loss of water from the cell. The protoplast then shrinks away from the cell wall, a condition known as **plasmolysis.** Continued removal of water concentrates the vacuolar contents, further lowering the osmotic potential. Turgor pressure remains at zero and the water potential of the cell is determined solely by its osmotic potential. In either situation described above, the water potential of the cell is determined as the algebraic sum of the turgor pressure and osmotic potential (compare Eq. 2.10 with Fig. 2.6).

In addition to water movement between cells and their environment, diffusion down a water potential gradient can also account for water movement between cells (Fig. 2.7). Individual cells in a series may experience different values for Ψ_S and Ψ_P, depending on the specific circumstances of each cell. Nonetheless, water will flow through the series of cells so long as a continuous gradient in water potential is maintained.

The phenomena of plasmolysis and wilting are superficially the same, but there are some important differences. Plasmolysis can be studied in the laboratory simply by subjecting tissues to hypertonic solutions and observing protoplast volume changes under the microscope. As plasmolysis progresses, protoplast volume progressively decreases, plasmodesmata (Chap. 1) are broken and the protoplast pulls away from the cell wall. The void between the outer protoplast surface (the plasma membrane) and the cell wall will become filled with external solution, which readily penetrates the cell wall. For this reason, plasmolysis does not normally give rise to a significant negative pressure (or tension) on the protoplast. Plasmolysis remains essentially a laboratory phenomenon and, with the possible exception of conditions of extreme water stress or saline environments, seldom occurs in nature.

Wilting, on the other hand, is the typical response to dehydration in air under natural conditions. Because of its extreme surface tension, water in the small pores of the cell wall resists the entry of air and the collapsing protoplast maintains contact with the cell wall. This tends to pull the wall inward and substantial negative pressures may develop. The water potential of wilted cells becomes even more negative as it is the sum of the *negative* osmotic potential plus the *negative* pressure potential.

[5]A solution with a lower solute content than a cell or another solution and, hence, less negative osmotic potential, is referred to as *hypotonic*. A *hypertonic* solution has a higher solute content and more negative osmotic potential. A solution with an equivalent osmotic potential is known as *isotonic*.

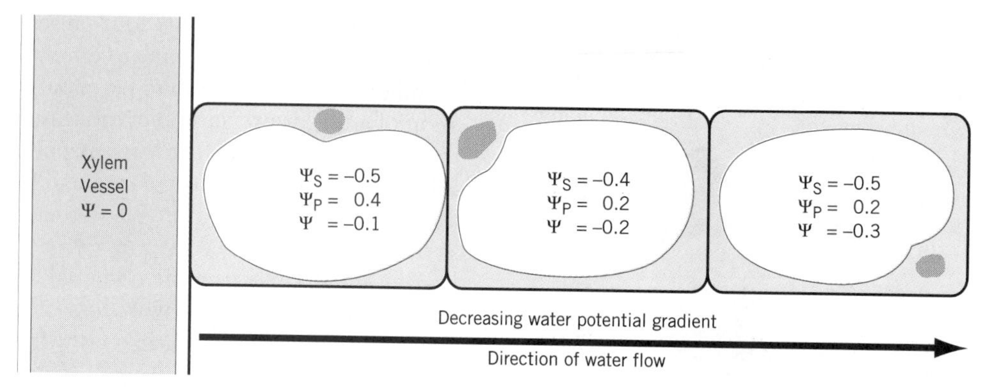

FIGURE 2.7 Diagram illustrating the contributions of osmotic potential (Ψ_S), turgor pressure (Ψ_P) and water potential (Ψ) to water movement between cells. The direction of water movement is determined solely by the value of the water potential in adjacent cells.

HOW ELASTIC ARE CELL WALLS?

Cell walls have a high elastic modulus, which means that they resist deformation. It is this property that enables cells to develop high turgor pressures while maintaining structural integrity.

In an ideal osmometer there is no volume change and, consequently, no change in solute concentration or osmotic potential of the solution. Cells do shrink and swell, but the volume changes they undergo are relatively small. Most cells undergo at most a difference in cell volume of about 10 percent between incipient plasmolysis and full turgor. This is reflected in Figure 2.6 where it is shown that a large change in water potential between incipient plasmolysis and full turgor is comprised of a minimal change in osmotic potential (Ψ_S) and a relatively large change in turgor pressure (Ψ_P).

The basis for this pattern is found in the elastic properties of cell walls. The elasticity of walls will vary, of course, from one cell type or tissue to another. The walls of young, actively growing cells, for example, will be relatively more elastic than those of older, mature cells with appreciable secondary thickenings. In general, however, the walls of plant cells are quite resistant to mechanical deformation. That is to say they are relatively rigid or inelastic. As cells take up and lose water, they experience minimal volume and osmotic potential changes. On the other hand, because of the very strong and relatively unyielding cell wall, plant cells are able to develop substantial turgor pressures. This important property of cell walls should be taken into account when considering water movement at the cellular level.

In physical terms, elasticity is described as the amount of stress, in units of force per unit area, or pressure, which must be applied in order to create a particular level of strain. Strain is normally expressed as a change in dimension such as length or volume. In the particular case of cell walls, applied stress is the pressure exerted by the expanding protoplast, or turgor pressure. The strain that results is measured as a change in cell volume. The relationship between applied pressure (ΔP) and the relative volume change ($\Delta V/V$), known as the **volumetric elastic modulus** ($\in$), may be quantified as:

$$\in = \frac{\Delta P}{\Delta V/V} \tag{2.11}$$

According to Equation 2.11, a high value for $\in$ means that a large pressure must be applied to produce a given increment in cell volume. Put another way, the higher the value for $\in$, the more resistant the wall is to deformation. By rearranging Equation 2.11 as

$$\in(\Delta V/V) = \Delta P \tag{2.12}$$

it is evident that cells with a higher value of $\in$ will experience a larger change in turgor pressure for a given increment of change in volume.

A large number of measurements of $\in$ have been made, particularly in giant algal cells in which pressure and volume changes can be measured directly (see Appendix). In higher plant cells, which are not as amenable to direct measurements, the value of $\in$ is more commonly estimated from measurements of water potential and known osmotic potentials (Equation 2.10). Values as high as 100 MPa have been reported for the giant alga *Nitella*. In higher plant cells the range is generally from 1 to 12 MPa (Dainty, 1976). Assuming a nominal value of $\in$ = 10 MPa, a 1 percent change in volume ($\Delta V/V = 0.01$) will give rise to a change in turgor pressure ($\Delta\Psi_P$) of 0.1 MPa. For a typical cell with a turgor pressure of 0.3 to 0.5 MPa, this represents a change of 20 to 33 percent in turgor pressure. At the same time osmotic pressure, which is proportional to volume, would be changed by only 1 percent (refer to Fig. 2.6). Thus, the osmotic potential of a cell remains relatively constant, a fact that has been verified by experimental measurement. Changes in turgor pressure are then left as the principal determinant of water potential as the water content of cells changes.

One consequence of a high elastic modulus is that plants, unlike animals, should not have to **osmoregulate** in order to maintain the structural integrity of their cells. In order to prevent continued uptake of water and eventual lysis of the cell membrane, animal cells must continually expend metabolic energy to excrete solutes and maintain a suitable osmotic potential. In plant cells, that function is taken over by the cell wall. The turgor pressure developed due to a very strong and relatively inelastic cell wall is normally sufficient to maintain an appropriate water potential and prevent excess water uptake. Interestingly, however, many plant cells do appear to maintain a particular level of pressure that is below the maximum possible. This suggests that cells must have pressure sensors that help to regulate solute content and thus maintain turgor in an appropriate range. Although the existence of pressure sensors has not been established, it does suggest an interesting line of future study.

SUMMARY

Water has numerous chemical and physical properties that make it particularly suitable as a medium in which life can occur. Most of these properties are the result of the tendency of water molecules to form hydrogen bonds. At the cellular level, water moves primarily by osmosis, in response to a chemical potential gradient across a selectively permeable membrane. The movement of water can be predicted on the basis of water potential. Water potential is a particularly useful concept because it can be calculated from two readily measured quantities: pressure and osmotic potential. Plants derive mechanical support from the turgidity of cells, due at least in part to the high structural strength of cell walls and their ability to resist deformation.

CHAPTER REVIEW

1. What is a hydrogen bond and how does it help to explain many of the unique physical and chemical properties of water?
2. Describe osmosis as a special case of diffusion. Distinguish between osmotic *pressure* and osmotic *potential.*
3. Understand the concept of water potential and how it is related to the chemical potential of water. In what way does the concept of water potential help the plant physiologist explain water movement?
4. Can you suggest an important role for turgor in the plant?
5. Is it osmotic potential or turgor pressure that has the more significant role in regulating water potential of plant cells?
6. Estimate the values of Ψ_{cell}, S, and P for a tissue that neither gains nor loses weight when equilibrated with a 0.4 molal mannitol solution and that, when placed in a 0.6 molal mannitol solution, 50 percent of the cells are plasmolysed. (Hint: use the methods described in the Appendix.)

FURTHER READING

Dainty, J. 1976. Water relations of plant cells. In: U. Luttge, M. G. Pitman (eds.), Transport in Plants. *Encyclopedia of Plant Physiology,* New Series. Vol. 2, Part A, pp. 12–32.

Milburn, J. A. 1979. *Water Flow in Plants.* White Plains, NY: Longman.

Nobel, P. S. 1991. *Physicochemical and Environmental Plant Physiology.* New York: Academic Press.

REFERENCES

Boyer, J. S., E. B. Knipling. 1965. Isopiestic technique for measuring leaf water potentials with a thermocouple psychrometer. *Proceedings of the National Academy of Sciences USA* 54:1044–1051.

Green, P. B., R. W. Stanton. 1967. Turgor pressure: Direct manometric measurement in cells of *Nitella. Science* 155:1675–1676.

Scholander, P. F., H. T. Hammel, E. D. Bradstreet. 1965. Sap pressure in vascular plants. *Science* 148:339–346.

Slatyer, R. O., S. A. Taylor. 1960. Terminology in plant-soil-water relations. *Nature* 187:922–924.

Zimmermann, U., E. Steudle. 1974. The pressure-dependence of the hydraulic conductivity, the membrane resistance and membrane potential during turgor regulation in *Valonia utricularis. Journal of Membrane Biology* 16:331–352.

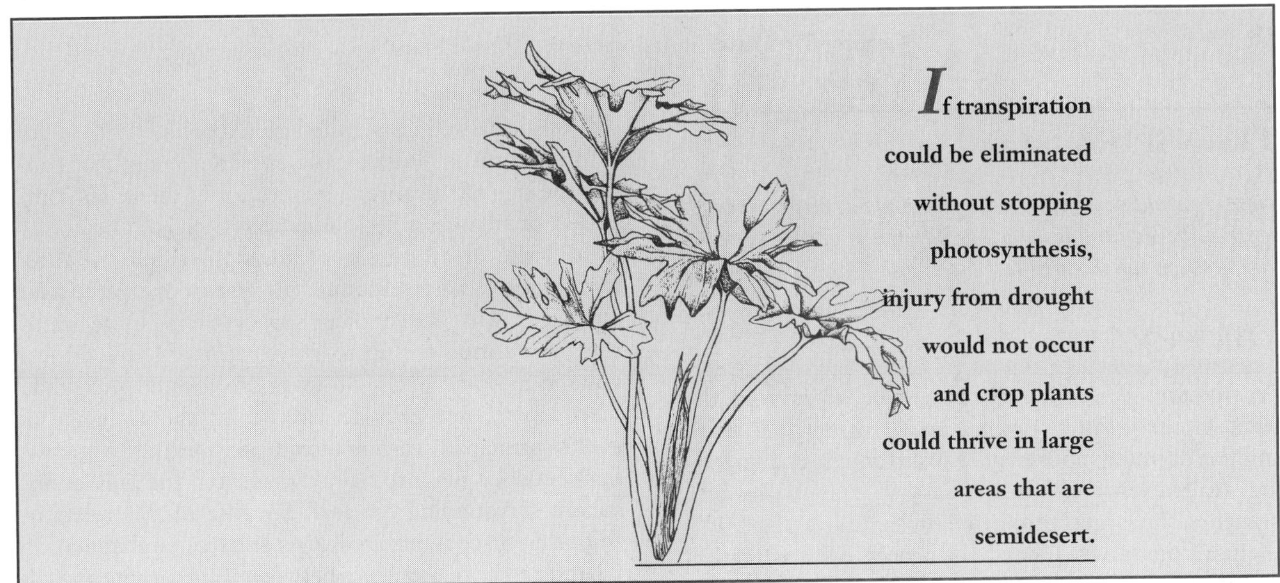

P. J. Kramer
(1983)

3

Water Relations of the Whole Plant

The dominant process in water relations of the whole plant is the absorption of large quantities of water from the soil, its translocation through the plant and eventual loss to the surrounding atmosphere as water vapor. Of all the water absorbed by plants, less than 5 percent is actually retained for growth and even less is used biochemically. The balance passes through the plant to be lost as water vapor, a phenomenon known as **transpiration.** Nowhere is transpiration more evident than in crop plants, where several hundred kilograms of water may be required to produce each kilogram of dry matter and excessive transpiration can lead to significant reductions in productivity.

The quantitative importance of transpiration has been indicated by a variety of studies over the years. In his classic 1938 physiology textbook, E. C. Miller reported that a single maize plant might transpire as much as 200 liters of water over its lifetime—approximately 100 times its own body weight. Extrapolated to a field of maize plants, this volume of water is sufficient to cover the field to a depth of 38 cm over the course of a growing season. A single, 14.5 m open-grown silver maple tree may lose as much as 225 liters of water per hour. In a deciduous forest, such as that found in the southern Appalachians of the United States, one-third of the annual precipitation will be absorbed by plants only to be returned to the atmosphere as vapor.

Whether there is any positive advantage to be gained by transpiration is a point for discussion, but the potential for such massive amounts of water loss clearly has profound implications for the growth, productivity, and even survival of plants. Were it not for transpiration, for example, a single rainfall might well provide sufficient water to grow a crop. As it is, the failure of plants to grow because of water deficits produced by transpiration is a principal cause of economic loss and crop failure across the world. Thus on both theoretical and practical grounds, transpiration is without doubt a process of considerable importance.

This chapter will examine the phenomena of transpiration and water movement through plants. The principal topics to be addressed include

- The process of transpiration and the role of vapor pressure differences in directing the exchange of water between leaves and the atmosphere;
- The role of environmental factors, in particular temperature and humidity, in regulating the rate of transpirational water loss;
- The anatomy of the water-conducting system in plants and how plants are able to maintain standing columns of water to the height of the tallest trees; and
- Water in the soil and how water is taken up by roots to meet the demands of water loss at the other end.

TRANSPIRATION

The principal loss of water vapor from plants occurs through pores in the leaf and is driven by differences in vapor pressure between the internal leaf spaces and the ambient air.

THE PROCESS

Transpiration is defined as the loss of water from the plant in the form of water vapor. Although a small amount of water vapor may be lost through small openings (called **lenticels**) in the bark of young twigs and branches, the largest proportion by far (more than 90%) escapes from leaves. Indeed, the process of transpiration is strongly tied to leaf anatomy (Fig. 3.1). The outer surfaces of a typical vascular plant leaf are covered with a multilayered waxy deposit called the **cuticle.** The principal component of the cuticle is **cutin,** a heterogeneous polymer of long-chain—typically 16 or 18 carbons—hydroxylated fatty acids. Ester formation between the hydroxyl and carboxyl groups of neighboring fatty acids forms cross-links, establishing an extensive polymeric network.

The cutin network is embedded in a matrix of cuticular **waxes** which are complex mixtures of long-chain (up to 37 carbon atoms) saturated hydrocarbons, alcohols, aldehydes, and ketones. Because cuticular waxes are very hydrophobic, they offer extremely high resistance to diffusion of both liquid water and water vapor from the underlying cells. The cuticle thus serves to restrict evaporation of water directly from the outer surfaces of leaf epidermal cells and protects both the epidermal and underlying mesophyll cells from potentially lethal desiccation.

The integrity of the epidermis and the overlying cuticle is occasionally interrupted by small pores called **stomata** (sing. = *stoma*). Each pore is surrounded by a pair of specialized cells, called **guard cells.** These guard cells function as hydraulically operated valves that control the size of the pore. The interior of the leaf is comprised of photosynthetic **mesophyll** cells. The somewhat loose arrangement of mesophyll cells in most leaves creates an interconnected system of intercellular air spaces. This system of air spaces may be quite extensive, accounting for up to 70 percent of the total leaf volume in some cases. Stomata are located such that, when open, they provide a route for the exchange of gasses (principally carbon dioxide, oxygen, and water vapor) between the internal air space and the bulk atmosphere surrounding the leaf. Because of this relationship, this space is referred to as **substomatal space.** In Chapter 8 the relationship between leaf anatomy and gas exchange, and especially the structure and role of stomata, will be discussed in greater detail. For the moment it is sufficient to note that the cuticle is generally impermeable to water and open stomata provide the primary route for escape of water vapor from the plant.

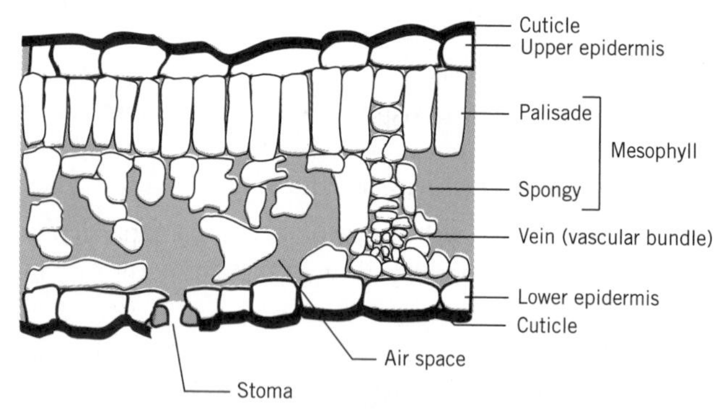

FIGURE 3.1 Diagrammatic representation of a typical mesomorphic leaf (*Acer* sp.) shown in cross-section. Note especially the presence of a cuticle covering the outer surfaces of both the upper and lower epidermis. Note also the extensive intercellular spaces with access to the ambient air through the open stomata.

Transpiration may be considered a two-stage process: (1) the evaporation of water from the moist cell walls into the substomatal air space and (2) the diffusion of water vapor from the substomatal space into the atmosphere. It is commonly assumed that evaporation occurs primarily at the surfaces of those mesophyll cells that border the substomatal air spaces. However, several investigators have proposed a more restricted view, suggesting instead that most of the water evaporates from the inner surfaces of epidermal cells in the immediate vicinity of the stomata (Kramer, 1983). Known as **peristomal evaporation,** this view is based on numerous reports indicating the presence of cuticle layers *on mesophyll cell walls.* In addition, mathematical modeling of diffusion in substomatal cavities has predicted that as much as 75 percent of all evaporation occurs in the immediate vicinity of the stomata (Tyree and Yianoulis, 1980). The importance of peristomal evaporation versus evaporation from mesophyll surfaces generally remains to be established by direct experiment. Whether the evaporation occurs principally at the mesophyll or epidermal cell surfaces is an interesting problem, reminding us that physiological processes are often not as straightforward as they may first appear.

The diffusion of water vapor from the substomatal space into the atmosphere is relatively straightforward. Once the water vapor has left the cell surfaces, it diffuses through the substomatal space and exits the leaf through the stomatal pore. Diffusion of water vapor through the stomatal pores, known as **stomatal transpiration**, accounts for 90 to 95 percent of the water loss from leaves. The remaining 5 to 10 percent is accounted for by **cuticular transpiration.** Although the cuticle is composed of waxes and other hydrophobic substances and is generally impermeable to water, small quantities of water vapor can pass through. The contribution of cuticular transpiration to leaf water loss varies considerably between species. It is to some extent dependent on the thickness of the cuticle. Thicker cuticles are characteristic of plants growing in full sun or dry habitats, while it is generally thinner on the leaves of plants growing in shaded or moist habitats. Cuticular transpiration may become more significant, particularly for leaves with thin cuticles, under dry conditions when stomatal transpiration is prevented by closure of the stomata.

MEASURING TRANSPIRATION

At the beginning of this chapter the magnitude of transpiration was indicated with some representative values. But how is transpiration measured? Before proceeding with our discussion of transpiration, it would be instructive to discuss briefly how these values are obtained. Two principal methods are commonly used: weight loss and gas exchange.

The weight loss method can be demonstrated by sealing a well-watered potted plant to prevent evaporation through the pot or from the soil surface. The potted plant can then be weighed at intervals and any weight loss attributed to loss of water via transpiration through the shoot. The weight loss method, also known as the **lysimeter** method, has been scaled up for agricultural field studies by constructing large containers filled with soil (perhaps several cubic meters), mounted on weighing devices usually buried in the ground (Kramer, 1983). In such cases, records must be kept of water input (rainfall, irrigation) and evaporation from the soil as well. The lysimeter method is generally considered most reliable and accurate for field studies, but lysimeters are expensive to construct and are not usually considered portable.

The gas exchange method, often used in conjunction with experiments on photosynthesis, involves sealing a leaf or branch in a transparent chamber with a flowing air stream. Transpiration can be estimated as the difference in water content of the air entering the chamber and the air leaving the chamber. Temperature, carbon dioxide content and other parameters can also be measured, coupling measurements of transpiration with stomatal opening and the rate of photosynthesis. This method has also been scaled up for field studies by enclosing entire trees or other large plants within a sealed plastic canopy. Gas exchange methods, whether on a small scale in the laboratory or in large-scale field measurements, are usually limited to short-term studies. The act of enclosing the plant may, over the long term, significantly alter the microclimate surrounding the leaves. Conditions such as temperature, humidity of the incoming air stream, and air velocity must be carefully monitored and controlled. On the other hand, chambers and measuring systems can be made quite portable and a number of commercial instruments are now available for field studies.

Transpiration in large-scale natural ecosystems is difficult to measure and is most commonly estimated indirectly. Here the investigator essentially calculates a water balance sheet, taking into account inputs (rainfall) and outputs such as soil storage, drainage, runoff, and so forth. The difference between measured inputs and outputs is taken as a measure of transpiration.

THE DRIVING FORCE OF TRANSPIRATION

In the previous chapter, it was shown that water movement is determined by differences in water potential. It can then be assumed that the driving force for transpiration is the difference in water potential between the substomatal air space and the external atmosphere. However, because the problem is now concerned with the diffusion of water vapor rather than liquid water, it will be more convenient to think in terms of vapor systems. Consider what happens, for example, when a volume of pure water is introduced into a closed chamber (Fig. 3.2). Initially the more energetic water molecules will escape into the air space, filling that space with water vapor. Some of those water molecules will then begin to condense into the liquid phase. Eventually water in the chamber will reach a dynamic equilibrium; the rate of evaporation will be balanced by the rate of condensation. The air space will then contain the maximum amount of water vapor that it can hold at that temperature. In other words, *at equilibrium the gas phase will be saturated with water vapor.*

The concentration of water molecules in a vapor phase may be expressed as the vapor mass per unit volume (g m^{-3}), called **vapor density.** Alternatively, the concentration may be expressed in terms of the pressure exerted by the water vapor molecules against the fluid surface and walls of the chamber. This is called **vapor pressure** (symbol = e). With an appropriate equation, vapor density and vapor pressure are interconvertible. However, because we are now accustomed to dealing with the components of water potential in pressure units, it will be more consistent for us to use vapor pressure (expressed as kilopascals, kPa) in our discussion. We can then say that when a gas phase has reached equilibrium and is saturated with water vapor, the system will have achieved its **saturation vapor pressure.**

The vapor pressure over a solution at atmospheric pressure is influenced by both solute concentration and temperature. As was previously discussed with respect to water potential (Chap. 2), the effect of solute concentration on vapor pressure may be expressed in terms of the mole fraction of water molecules. This relationship is given by a form of Raoult's law, which states:

$$e = X_i e^{o} \tag{3.1}$$

where e is vapor pressure of the solution, X_i is the mole fraction of water (= number of water molecules/number

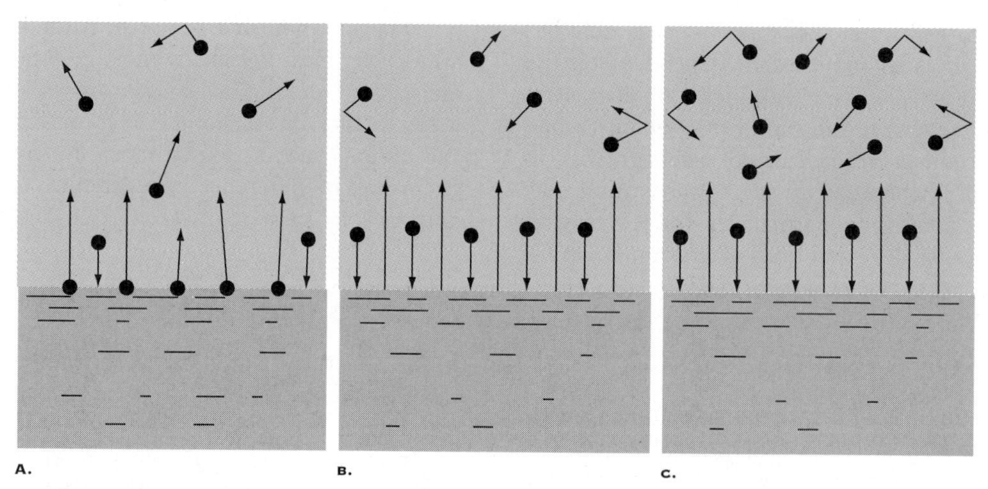

FIGURE 3.2 Vapor pressure in a closed container. Initially (*A*), more molecules escape from the water surface than condense, filling the air space with water vapor molecules. The vaporous water molecules exert pressure—vapor pressure—against the walls of the chamber and the water surface. At equilibrium (*B*) the rate of condensation equals evaporation and the air is saturated with water vapor. The vapor pressure when the air is saturated is known as the saturation vapor pressure. At higher temperature (*C*), a higher proportion of water molecules have sufficient energy to escape. Both the concentration of water molecules in the vapor phase and the saturation vapor pressure are correspondingly higher.

of water molecules + number of solute molecules), and e^o is the saturation vapor pressure over pure solvent.

The actual reduction in vapor pressure due to solute turns out to be quite small. This is because even in relatively concentrated solutions the mole fraction of solvent remains large. Consider, for example, a 0.5 molal solution, which is approximately the concentration of vacuolar sap in a typical plant cell. A 0.5 molal solution contains 1/2 mole of solute dissolved in 1,000 grams (55.5 mol) of water. The mole fraction of water in a 0.5 molal solution is therefore $55.5/(55.5 + 0.5) = 0.991$. According to Equation 3.1, the saturation vapor pressure of a half-molal solution would be reduced by less than one percent compared with pure water.

Temperature, on the other hand, has a significant effect on vapor pressure. This is due to the effect of temperature on the average kinetic energy of the water molecules. As the temperature of a volume of water or an aqueous solution increases, the proportion of molecules with sufficient energy to escape the fluid surface also increases. This in turn will increase the concentration of water molecules in the vapor phase and, consequently, the equilibrium vapor pressure. An increase in temperature of about 12°C will nearly double the saturation vapor pressure.

According to Ficks' law of diffusion (Chap. 2), molecules will diffuse from a region of high concentration to a region of low concentration, or, down a concentration gradient. Because vapor pressure is proportional to vapor concentration, water vapor will also diffuse down a vapor pressure gradient; that is, from a region of high vapor pressure to a region of lower vapor pressure. In principle, we can assume that the substomatal air space of a leaf is normally saturated or very nearly saturated with water vapor. This is because the mesophyll cells which border the air space present a large, exposed surface area for evaporation of water. On the other hand, the atmosphere which surrounds the leaf is usually unsaturated and may often have a very low water content. These circumstances create a gradient between the high water vapor pressure in the interior of the leaf and the lower water vapor pressure of the external atmosphere. This difference in water vapor pressure between the internal air spaces of the leaf and the surrounding air is the driving force for transpiration.

THE INFLUENCE OF HUMIDITY, TEMPERATURE, AND WIND SPEED ON TRANSPIRATION RATE

The rate of transpiration will naturally be influenced by factors such as *humidity* and *temperature*, and *wind speed*, which influence the rate of water vapor diffusion between the substomatal air chamber and the ambient atmosphere. Fick's law of diffusion tells us that the rate of diffusion is proportional to the difference in concentration of the diffusing substance. It therefore follows that the rate of transpiration will be governed in large measure by the *magnitude* of the vapor pressure gradient between the leaf and the surrounding air. In other words,

$$T \propto e_{leaf} - e_{air} \qquad (3.2)$$

At the same time, the escape of water vapor from the leaf is controlled to a considerable extent by resistances

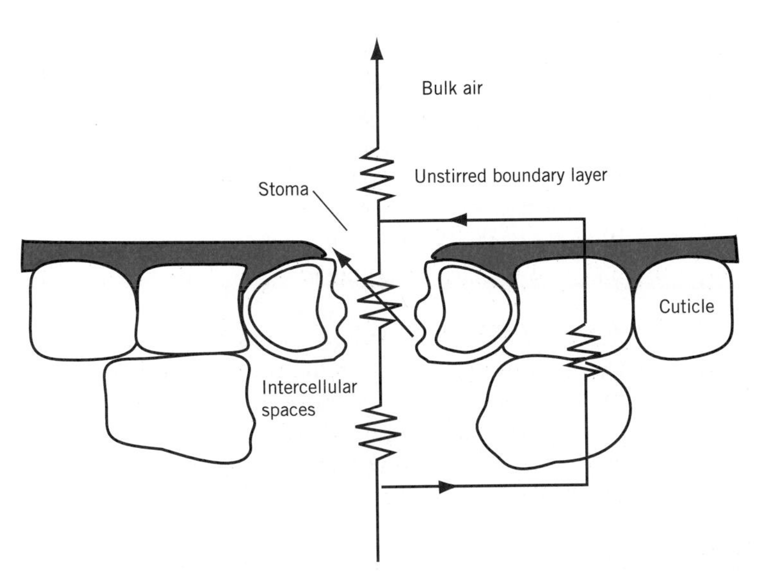

FIGURE 3.3 A schematic representation of the principal resistances encountered by water vapor diffusing out of a leaf. Symbols for electrical resistance are used because resistance to water diffusion is analogous to resistance in an electrical circuit. Note the symbol for stomatal resistance indicates it is variable—taking into account the capacity of the stomata to open and close.

encountered by the diffusing water molecules both within the leaf and in the surrounding atmosphere (Fig. 3.3). Resistance is encountered by the vapor molecules as they pass through the intercellular spaces, which are already saturated with water vapor, and the stomatal pores. Note that the symbol for stomatal resistance indicates that it is variable, to account for the fact that stomata may at any time be fully open, partially open, or closed. Additional resistance is encountered by the **boundary layer,** a layer of undisturbed air on the surface of the leaf. The boundary layer and its effect on tranpiration are described more fully later in this chapter. The transpiration equation (Equation 3.2) now requires additional terms to account for these resistances.

$$T \propto \frac{e_{leaf} - e_{air}}{r_{air} + r_{leaf}} \quad (3.3)$$

Equation 3.3 tells us that the rate of transpiration is proportional to the difference in vapor pressure between the leaf and the atmosphere divided by the sum of resistances encountered in the air and the leaf.

Effect of Humidity **Humidity** is the actual water content of air, which, as noted earlier, may be expressed either as vapor density (g m^{-3}) or vapor pressure (kPa). In practice, however, it is more useful to express water content as the **relative humidity (RH).** Relative humidity is the ratio of the actual water content of air to the maximum amount of water that can be held by air at that temperature. Expressed another way, relative humidity is the ratio of the actual vapor pressure to the saturation vapor pressure. Relative humidity is most commonly expressed as RH × 100, or **percent relative humidity.** The effects of humidity and temperature on the vapor pressure of air are illustrated in Table 3.1. Air at 50 percent RH by definition contains one-half the amount of water possible at saturation. Its vapor pressure is therefore one-half the saturation vapor pressure. Note also that a 10 °C rise in temperature nearly doubles the saturation vapor pressure. Relative humidity and temperature also have a significant effect on the water potential of air (Table 3.2).

As indicated earlier, the vapor pressure of the substomatal leaf spaces is probably close to saturation most of the time. Even in a rapidly transpiring leaf the relative humidity would probably be greater than 95 percent and the resulting water potential would be close to zero (Table 3.2). Under these conditions, the vapor pressure in the substomatal space will be the saturation vapor pressure at the leaf temperature. The vapor pressure of atmospheric air, on the other hand, depends on both the relative humidity of the air and its temperature. Humidity and temperature thus have the potential to modify the magnitude of the vapor pressure gradient ($e_{leaf} - e_{air}$), which, in turn, will influence the rate of transpiration.

Effect of Temperature Temperature modulates transpiration rate through its effect on vapor pressure, which, in turn, affects the vapor pressure gradient as illustrated by the three examples in Table 3.3. In the first example (A), assuming an ambient temperature of 10°C and a relative humidity of 50 percent, the leaf-

TABLE 3.1 Water vapor pressure (kPa) in air as a function of temperature and varying degrees of saturation. Air is saturated with water vapor at 100% relative humidity (RH).

	Relative Humidity				
Temperature (C)	100%	80%	50%	20%	10%
30	4.24	3.40	2.12	0.85	0.42
20	2.34	1.87	1.17	0.47	0.23
10	1.23	0.98	0.61	0.24	0.12

TABLE 3.2 Some values for water potential (Ψ) as a function of relative humidity (RH) at 20 °C.

RH(%)	Ψ (MPa)*
100	0
95	−6.9
90	−14.2
50	−93.5
20	−217.1

*Water potential is calculated from the following relationship:

$$\Psi = 1.06\ T \log (RH/100)$$

to-air vapor pressure gradient is 0.61 kPa. This might be a typical situation in the early morning hours. As the sun comes up, the air temperature will increase. A 10°C increase in temperature (Table 3.3B), assuming the water content of the atmosphere remains constant, will increase the leaf-to-air vapor pressure gradient and, consequently, the potential for transpiration, by a factor of almost three. Note that in this example it is assumed that leaf temperature is in equilibrium with the atmosphere. This is not always the case. A leaf exposed to full sun may actually reach temperatures 5°C to 10°C higher than that of the ambient air. Under these circumstances, the vapor pressure gradient may increase as much as sixfold (Table 3.3C).

As long as the stomata remain open and a vapor pressure gradient exists between the leaf and the atmosphere, water vapor will diffuse out of the leaf. This means transpiration may occur even when the relative humidity of the atmosphere is 100 percent. This is often the case in tropical jungles where leaf temperature and, consequently, saturation vapor pressure is higher than the surrounding atmosphere. Because the atmosphere is already saturated, the water vapor condenses upon exiting the leaf, thereby giving substance to the popular image of the steaming jungle.

TABLE 3.3 The effect of temperature and relative humidity on leaf-to-air vapor pressure gradient. In this example it is assumed that the water content of the atmosphere remains constant.

Leaf	Atmosphere	$e_{leaf} - e_{air}$
(A)		
T = 10 °C	T = 10 °C	
e = 1.23 kPa	e = 0.61 kPa	0.61 kPa
RH = 100%	RH = 50%	
(B)		
T = 20 °C	T = 20 °C	
e = 2.34 kPa	e = 0.61 kPa	1.73 kPa
RH = 100%	RH = 26%	
(C)		
T = 30 °C	T = 20 °C	
e = 4.24 kPa	e = 0.61 kPa	3.63 kPa
RH = 100%	RH = 26%	

Effect of Wind Wind speed has a marked effect on transpiration because it modifies the effective length of the diffusion path for exiting water molecules. This is due to the existence of the boundary layer introduced earlier (Fig. 3.3). Before reaching the bulk air, water vapor molecules exiting the leaf must diffuse not only through the thickness of the epidermal layer (i.e., the guard cells), but also through the boundary layer. The thickness of the boundary layer thus adds to the length of the diffusion path. According to Ficks' law, this added length will decrease the rate of diffusion and, hence, the rate of transpiration.

The thickness of the boundary layer is primarily a function of leaf size and shape, the presence of leaf hairs (trichomes), and wind speed (Nobel, 1991). The calculated thickness of the boundary layer as a function of wind speed over a typical small leaf is illustrated in Figure 3.4. With increasing wind speed, the thickness of the boundary layer and, consequently, the length of the diffusion path decreases. In accordance with Ficks' law, the vapor pressure gradient steepens and, all other factors being equal, the rate of transpiration increases. This relationship holds truest at lower wind speeds, however. As wind speed increases it tends to cool the leaf and may cause sufficient desiccation to close the stomata. Either one of these factors tends to lower the rate of transpiration. High wind speeds will therefore have less of an

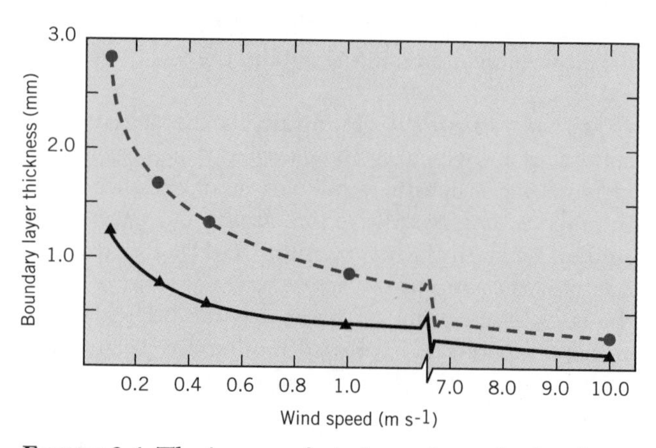

FIGURE 3.4 The impact of wind speed on calculated boundary layer thickness for leaves 1.0 cm (triangles) or 5.0 cm (circles) wide. A wind speed of 0.28 m s^{-1} = 1 km hr^{-1}. (Plotted from the data of Nobel, 1991.)

Box 3.1
Why Transpiration?

Our discussion of transpiration in this chapter has focused on the mechanism of water loss and the role of transpiration in the ascent of sap; a sort of operational approach to the problem. We cannot fail to be impressed by the amount of water that must be made available to a plant in order to support transpiration and the possible consequences of such water loss to plant survival. Transpiration often results in water deficits and desiccation injury, especially when high temperature and low humidity favor transpiration but the soil is deficient of water. This raises an interesting and often controversial question: Is there any positive advantage in transpiration to be gained by the plant? It has been argued that transpiration is required to bring about the ascent of sap, that it increases nutrient absorption and that it assists in the cooling of leaves. It has also been argued that transpiration is little more than a necessary "evil."

Transpiration does speed up the movement of xylem sap, but it seems unlikely that this is an essential requirement. The growth of cells alone would cause a slow ascent of xylem sap, even in the absence of transpiration. Transpiration serves only to increase the rate and quantity of water moved and there is no evidence that the higher rates are beneficial. Another argument is that, because mineral nutrients absorbed by the roots move largely in the xylem sap, transpiration may benefit nutrient distribution. It is true that minerals in the xylem sap will be carried along with a rapidly moving transpiration stream, but the rate-limiting step in nutrient supply is more likely to be the rate at which the nutrients are absorbed by the roots and delivered to the xylem. Moreover, experiments with radioactive tracers have shown that minerals continue to circulate within the plant in the absence of transpiration.

Because transpiration involves the evaporation of water, it can assume a significant role in the cooling of leaves. The is illustrated by the energy budget for a typical mesophyte leaf shown in the accompanying table. Because leaves are heavily pigmented, they absorb large amounts of direct solar radiation. Some of this absorbed solar radiation will not be utilized in photochemical reactions, such as photosynthesis, but will instead account for a significant heat gain by the leaves. Leaves also exchange infra-red energy with their surroundings, both absorbing and radiating infrared. Overall a leaf will radiate more infrared energy than it gains, leaving a negative net infrared exchange. This leaves a net radiation gain in this example of 370 W m^{-2}, which must be dissipated by other means. One way of dissipating the heat load is by evaporation of water from the leaf surface, or transpiration. The latent heat of vaporization of water is 44 kJ mol^{-1} and a typical mesophyte leaf might transpire at the rate of about 4 mmol of water m^{-2} per second. The heat energy consumed by transpiration may be calculated as: $(4 \times 10^{-3}\ \text{mol m}^{-2}\ \text{s}^{-1})\ (44 \times 10^{3}\ \text{J mol}^{-1}) = 176\ \text{J m}^{-2}\ \text{s}^{-1} = 176\ \text{W m}^{-2}$. In this example, transpiration can thus account for dissipation of approximately one-half of the net radiation balance. Dissipation of the remaining heat is probably accounted for by convection from the leaf to the surrounding air.

The energy budget for a typical mesophyte leaf.
(Data from Nobel, 1991).

Energy gain	W m^{-2}
a. Absorbed solar radiation	+ 605
b. Net infrared exchange	− 235
c. Net radiation balance (a + b)	+ 370
Energy loss	
d. Loss by transpiration	− 176
e. Loss by convection	− 194
	− 370

One argument raised against a significant role for transpiration is that there is seldom any clear correlation between transpiration and plant growth. While some plants may develop more slowly at high humidity, many are able to complete their life cycle without apparent harm under conditions such as 100 percent relative humidity, where transpiration is minimal. Under such circumstances, the supply of water and nutrients is clearly adequate. If the leaf were not cooled by evaporative water loss, other processes such as convection might remove sufficient energy to prevent the leaf reaching lethal temperatures.

In Chapter 8 it will be argued that the evolutionary function of stomata is to ensure an adequate supply of carbon dioxide for photosynthesis. It has been suggested that transpiration is simply an unfortunate consequence of this structure; that is, a structure that is efficient for the diffusive uptake of carbon dioxide is equally efficient for the outward diffusion of water vapor. According to this view, leaf structure represents a compromise between the need to restrict desiccation of leaf cells while at the same time maintaining access to atmospheric carbon dioxide.

effect on transpiration rate than expected on the basis of their effect on boundary layer thickness alone.

Boundary layer thickness can also be influenced by a variety of plant factors. Boundary layers are thicker over larger leaves and leaf shape may influence the wind pattern. Leaf pubescence, or surface hairs, helps to maintain the boundary layer, and thus reduce transpiration, by breaking up the air movement over the leaf.

EFFICIENCY OF TRANSPIRATION

Before leaving our discussion of transpiration, we raise two particularly interesting questions. First, how can plants lose such large quantities of water vapor through pores so small as the stomata? It has been estimated, for example, that diffusion through stomata is 50 to 70 times more efficient than can be accounted for on the basis of stomatal area alone. It turns out that small pores are remarkably efficient for the diffusion of gasses, including water vapor. However, we will leave a full discussion of this question to Chapter 8, when we can examine it within the context of carbon dioxide supply for photosynthesis. The second question is whether transpiration offers any positive advantage to plants. This question is addressed in Box 3.1.

THE ANATOMY OF WATER CONDUCTION

The principal pathway for conducting water through plants is a series of tracheary elements that form an interconnected system of open conduits.

The distinguishing feature of vascular plants is the presence of **vascular tissues,** the **xylem** and **phloem,** which conduct water and nutrients between the various organs. Vascular tissues begin differentiating a few mm from the root and shoot apical meristems and extend as a continuous system into other organs such as branches, leaves, flowers, and fruits. In organs such as leaves, the larger veins subdivide into smaller and smaller veins such that no photosynthetic leaf cell is more than a few cells removed from a small vein ending. Xylem tissue is responsible for the transport of water, dissolved minerals, and, on occasion, small organic molecules upward through the plant; from the root through the stem to the aerial organs. Phloem, on the other hand, is responsible primarily for the translocation of organic materials from sites of synthesis to storage sites or sites of metabolic demand. The role of phloem tissue in the translocation of photosynthetic products and organic substances in general will be discussed more fully in Chapter 11.

Xylem consists of **fibers, parenchyma cells,** and **tracheary elements** (Fig. 3.5). Fibers are very elongated cells with thickened secondary walls. Their principal function is to provide structural support for the plant. Parenchyma cells provide for storage as well as the lateral translocation of solutes. The tracheary elements include both **tracheids** and **vessel elements** (Fig.

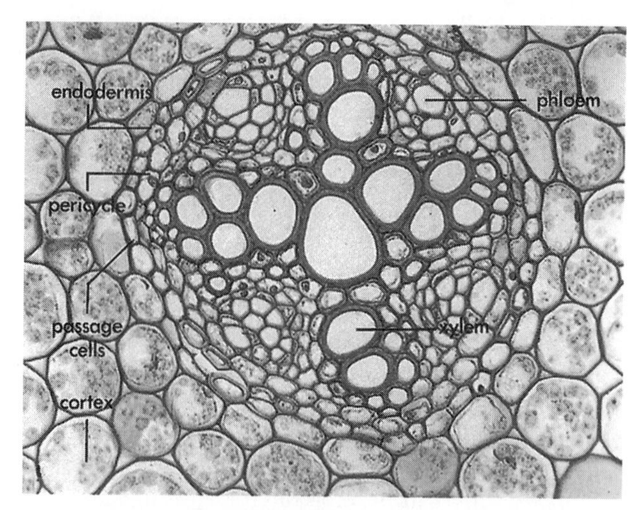

FIGURE 3.5 Vascular tissue from a young root of buttercup (*Ranunculus* sp.), showing the large xylem vessels in a cross-section. (From T. E. Weier et al. *Botany*. 6th edition. 1982. New York, Wiley, Fig. 9.16. Used by permission of the authors.)

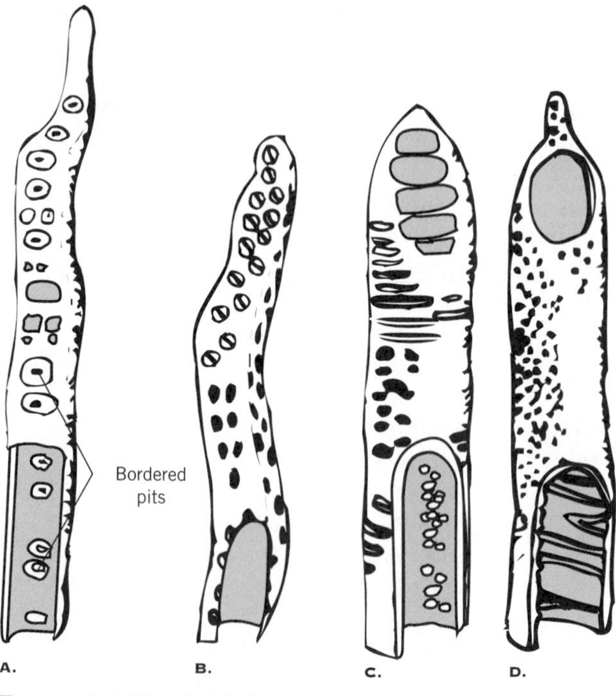

FIGURE 3.6 Tracheids from (*A*) spring wood of white pine (*Pinus*) and (*B*) oak (*Quercus*). Vessel elements from (C) *Magnolia* and (*D*) basswood (*Tilia*). Only short tip sections are shown. (From T. E. Weier et al. *Botany*. 6th edition. 1982. New York, Wiley, Fig. 7.18. Used by permission of the authors.)

3.6). Tracheary elements are the most highly specialized of the xylem cells and are the principal water-conducting cells. Tracheids and vessels are both elongated cells with heavy, often sculptured, secondary cell walls. Their most distinctive feature, however, is that when mature and functioning, both tracheids and vessels form an interconnected network of nonliving cells, devoid of all protoplasm. The hollow, tubular nature of these cells together with their extensive interconnections facilitates the rapid and efficient transport of large volumes of water throughout the plant.

Tracheids are single cells with diameters in the range of 10 to 50 μm. They are typically less than 1 cm in length, although in some species they may reach lengths up to 3 cm. Tracheids also have thickened secondary walls composed mainly of cellulose, hemicellulose and lignin. Because of the high lignin content, secondary walls are less permeable to water than are the primary walls of growing cells. On the other hand, the additional strength of secondary walls helps to prevent the cells from collapsing under the extreme negative pressure that may develop in the actively transpiring plants. Although their principal function is to conduct water, the thickened secondary walls of tracheids also contribute to the structural support of the plant.

The movement of water between tracheids is facilitated by interruptions, known as **pit pairs,** in the secondary wall (Fig. 3.7). During the development of tracheids, regions that are to become pit pairs avoid the deposition of secondary mall material. This leaves only the middle lamella and primary walls to separate the hollow core, or **lumen,** of one cell from that of the adjacent cell. The combined middle lamella and primary wall is known as the pit **membrane.** Pit membranes are not solid but have openings (about 0.3 μm diameter) that permit the relatively free passage of water and solutes. The origin of these openings is not clear, but they are thought to represent regions where cytoplasmic strands (**plasmodesmata**) once penetrated the cell walls when the cells were still alive. **Bordered pit pairs** have secondary wall projections over the pit area and a swollen central region of the pit membrane called the **torus.** When pressure is unequal in adjacent vascular elements, such as when one contains an air bubble, the torus is drawn toward the element with the lower pressure. Pressure of the torus against the borders seals off the pit as shown in Figure 3.7.

Successive tracheids commonly overlap at their tapered ends. As a result, tracheids line up in files running longitudinally through the plant. Water moves between adjacent tracheids (either vertically or laterally) through the pit pairs in those regions where overlap occurs. The movement of water is no doubt facilitated by the openings in the pit membranes.

Vessels are very long tracheary elements made up of individual units, known as **vessel members,** which are arranged end-to-end in longitudinal series. At maturity, the end walls of the vessel members have dissolved away, leaving openings called **perforation plates.** As the cell that is to become a vessel member develops, the future perforation area becomes thickened due to swelling of the middle lamella (Fig. 3.8). The thickened area contains little cellulose, consisting almost entirely of noncellulosic polysaccharides. As the cell matures and the cytoplasm begins to break down, the unprotected site of the perforation is attacked by hydrolytic enzymes and dissolved away. The rest of the wall area has been covered with lignified secondary wall materials and is protected from degradation. In some cases, the resulting perforation will encompass virtually the entire end wall, leaving only a ring of secondary wall to mark the junction between two successive vessel members (Fig. 3.8). In other cases, the plate may be multiperforate. If the perforations are elongate and parallel, the pattern is called **scalariform.** An irregular, net-like pattern is called **reticulate.** The perforations generally allow for a relatively free flow of water between successive vessel members. There are no perforation plates at the ends of vessels (that is, the last vessel member in a sequence), but water is able to move laterally from one

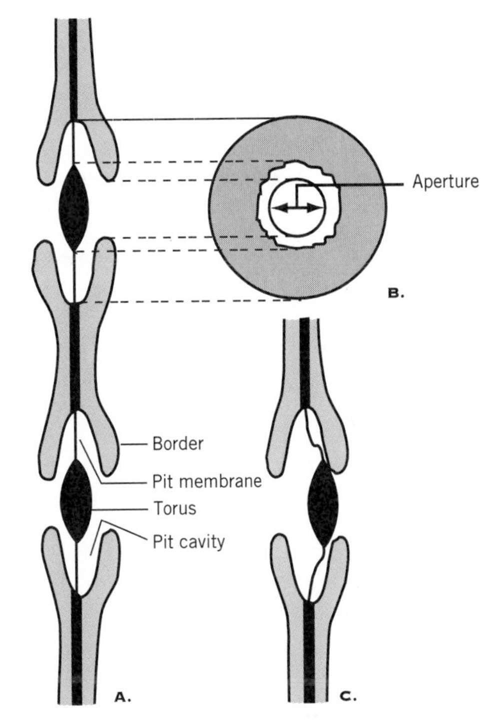

FIGURE 3.7 Diagram of bordered pit pairs. (*A*) Two bordered pit pairs in the wall between two xylem tracheids, in side view. (*B*) Surface view. (*C*) A pressure differential—lower to the right—pushes the torus against the border, thereby sealing the pit and preventing water movement. (From K. Esau, *Anatomy of Seed Plants*, New York, Wiley, 1977. Reprinted by permission.)

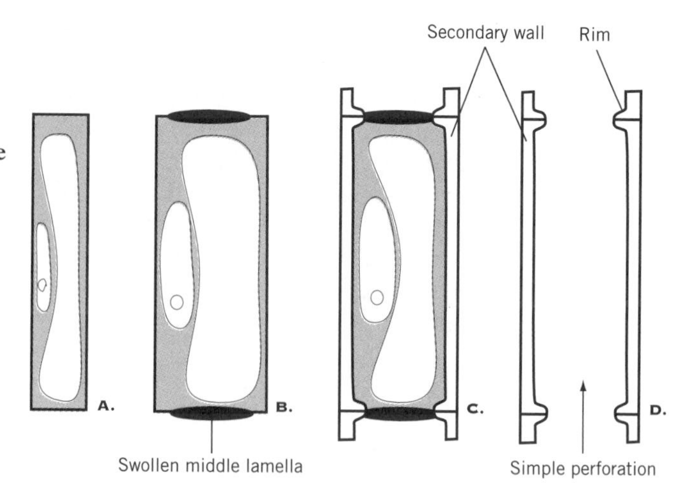

FIGURE 3.8 The development of a vessel member. (*A*) meristematic cell. (*B*) Swelling of the middle lamella in the region of a future perforation plate. (*C*) Secondary wall deposition except over area of future perforation. (*D*) Mature vessel member. The primary wall and middle lamella have dissolved away and the protoplast has disappeared. (From K. Esau, *Anatomy of Seed Plants,* 1st ed., New York, Wiley, 1960. Reprinted by permission.)

vessel to the next due to the presence of pit pairs similar to those found in tracheids.

The size of vessels is highly variable, although they are generally larger than tracheids. In maples (*Acer* sps.), for example, vessels range from 40 μm to 60 μm in diameter, while in some species of oak (*Quercus* sps.) diameters may range up to 300 μm to 500 μm. (The large diameter vessels account for the ring porus character of spring wood in woody species.) The length of vessels in maple are generally 4 cm or less, but some may reach lengths of 30 cm. In oak, on the other hand, vessel lengths up to 10 m have been recorded (Zimmermann, 1983). However, because of extensive branching of the vascular system and the large number of lateral connections between overlapping tracheary elements, the xylem constitutes a single continuous, interconnected system of water-conducting conduits between the extremes of the plant—from the tip of the longest root to the outermost margins of the highest leaf.

Vessels are considered evolutionarily more advanced than tracheids. For example, xylem tissue in the gymnosperms, considered evolutionarily more primitive than the angiosperms, consists entirely of tracheids. Although tracheids do occur in angiosperms, the bulk of the water is conducted in vessels. Also because of their larger size, vessels are considerably more efficient than tracheids when it comes to conducting water. An empirical equation relating flow rate to the size of conduits was developed in the nineteenth century by the French scientist Jean L. M. Poiseuille. Poiseuille showed that when a fluid is pressure-driven, the volume flow rate (J_v) is a function of the viscosity of the liquid (η), the difference in pressure or pressure drop (ΔP), and the radius of the conduit:

$$\frac{\Delta P \pi r^4}{8\eta} = J_v \qquad (3.4)$$

Equation 3.4 applies to water movement in the xylem tracheary elements because, as will be shown below, it is driven by a difference in pressure between the soil and the leaves. The important point to note, then, is that the volume flow rate is *directly proportional to the fourth power of the radius.* The impact of this relationship can be seen by comparing the relative volume flow rates for a 40 μm diameter (r = 20 μm) tracheid and a 200 μm diameter (r = 100 μm) vessel. Although the relative diameter of the vessel is 5 times that of the tracheid, its relative volume flow rate will be 625 (that is, 5^4) times that of the tracheid. The high rate of flow in the larger vessels occurs because the flow rate of water is not uniform across the conduit. The flow rate of molecules near the conduit wall is reduced by friction, due to adhesive forces between the water and the conduit wall. As the diameter of the conduit increases, the proportion of molecules near the wall and consequently subject to these frictional forces will decrease. Put another way, the faster moving molecules in the center of the conduit constitute a larger proportion of the population and the overall rate of flow increases accordingly.

THE ASCENT OF XYLEM WATER

A problem of some interest to plant physiologists for many years has been how the integrity of the xylem water column is maintained and how it moves to the tops of the tallest trees. Several mechanisms have been proposed, but the only one to have stood the test of time combines transpiration with the strong cohesive forces of water.

The tallest-standing trees are generally found growing in the rainforests along the Pacific coast of northwestern United States and southwestern British Columbia. The best known are the redwoods (*Sequoia*

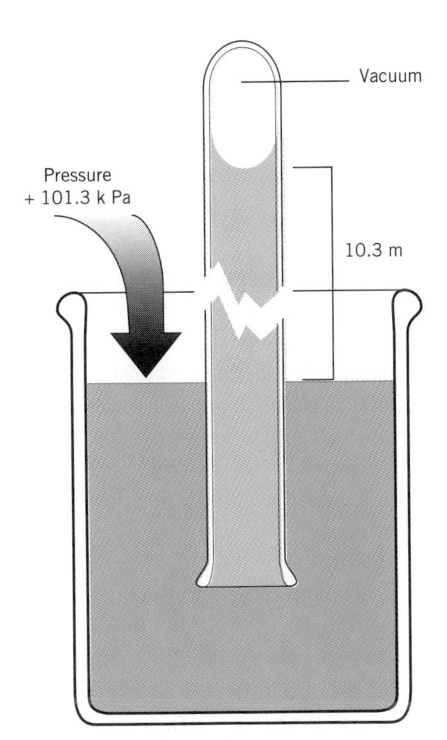

FIGURE 3.9 Atmospheric pressure can support a water column to a maximum height of 10.3 m.

sempervirens) of northern California, some of which exceed 110 m in height. Individual specimens of Douglas fir (*Pseudotsuga menziesii*) have been reported in excess of 100 m and a sitka spruce (*Picea sitchensis*) measuring 95 m has been located in the Carmanah Valley of Vancouver Island. In Australia, there have been reports of *Eucalyptus* trees measuring more than 130 m in height.

The forces required to move water to such heights are substantial. Were we able to devise a sufficiently long tube closed at one end, fill it with water, and invert it as shown in Figure 3.9, we would find that atmospheric pressure (ca. 101 kPa at sea level) would support a column of water approximately 10.3 m in height. To push the water column any higher would require a correspondingly greater pressure acting on the open surface. Clearly, elevating water to the height of the tallest trees would require a force 10 to 15 times greater than atmospheric, or 1.0 to 1.5 MPa. This force would be equivalent to the pressure at the base of a standing column of water 100 m to 150 m high.

But even a force of this magnitude would not be sufficient. In addition to the force of gravity, water moving through the plant will encounter a certain amount of resistance inherent in the structure of the conducting tissues—irregular wall surfaces, perforation plates, and so forth. We can assume that a force at least equal to that required to support the column would be necessary to overcome these resistances. In that case, a force on the order of 2.0 to 3.0 MPa would be required to move water from ground level to the top of the tallest known trees.

How can such a force be generated? This is a question that has long held the interest of plant physiologists and over the years a number of theories have been advanced. The three most prominent are **root pressure, capillarity** and the **cohesion theory.**

ROOT PRESSURE

If the stem of a well-watered herbaceous plant is cut off above the soil line, xylem sap will exude from the cut surface (Fig. 3.10). Exudation of sap, which may persist for several hours, indicates the presence of a positive pressure in the xylem. The magnitude of this pressure can be measured by attaching a manometer to the cut surface (Fig. 3.11). This pressure is known as **root pres-**

FIGURE 3.10 A demonstration of xylem sap exudation due to root pressure in tomato. Photograph was taken 10 minutes after excising the stem of a well-watered plant.

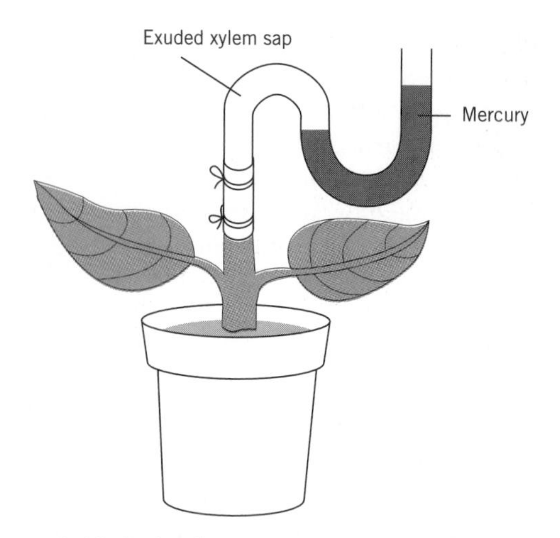

FIGURE 3.11 A simple manometer for measuring root pressure. Root pressure can be calculated from the height of the mercury in the glass tube.

sure because the forces which give rise to the exudation originate in the root.

Root pressure has its basis in the structure of roots and the active uptake of mineral salts from the soil. The xylem vessels are located in the central core of a root, the region known as the **stele.** Surrounding the stele is a layer of cells known as the **endodermis.** In most roots, the radial and transverse walls of the endodermal cells develop characteristic thickenings called the **Casparian band** (Fig. 3.12). The Casparian band is principally composed of **suberin,** a complex mixture of *hydrophobic,* long-chain fatty acids and alcohols. These hydrophobic molecules impregnate the cell wall, filling in the spaces between the cellulose microfibrils as well as the intercellular spaces between the cells. Because it is both space-filling and hydrophobic, the Casparian band presents an effective barrier to the movement of water through the apoplastic space of the endodermis. The result is that water can move into or out of the stele only by first passing through the membranes of the endodermal cells and then through the plasmodesmatal connections.

As roots take up mineral ions from the soil, the ions are transported into the stele where they are actively deposited in the xylem vessels. The accumulation of ions in the xylem lowers the osmotic potential and, consequently, the water potential of the xylem sap. In response to the lowered water potential, water follows, also passing from the cortical cells into the stele through the membranes of the endodermal cells. Since the Casparian band prevents the free return of water to the cortex, a positive hydrostatic pressure is established in the xylem vessels. In a sense, the root may be thought of as a simple osmometer (refer to Fig. 2.5) in which the endodermis constitutes the differentially permeable membrane, the ions accumulated in the xylem represent the dissolved solute, and the xylem vessels are the vertical tube. So long as the root continues to accumulate ions in the xylem, water will continue to rise in the vessels or exude from the surface when the xylem vessels are severed.

The question to be answered at this point is whether root pressure can account for the rise of sap in a tree. The answer is probably no, for several reasons. To begin with, xylem sap is not as a rule very concentrated and measured root pressures are relatively low. Values in the range of 0.1 to 0.5 MPa are common, which are no more than 16 percent of that required to move water to the top of the tallest trees. In addition, root pressure has not been detected in all species and is not always detectable

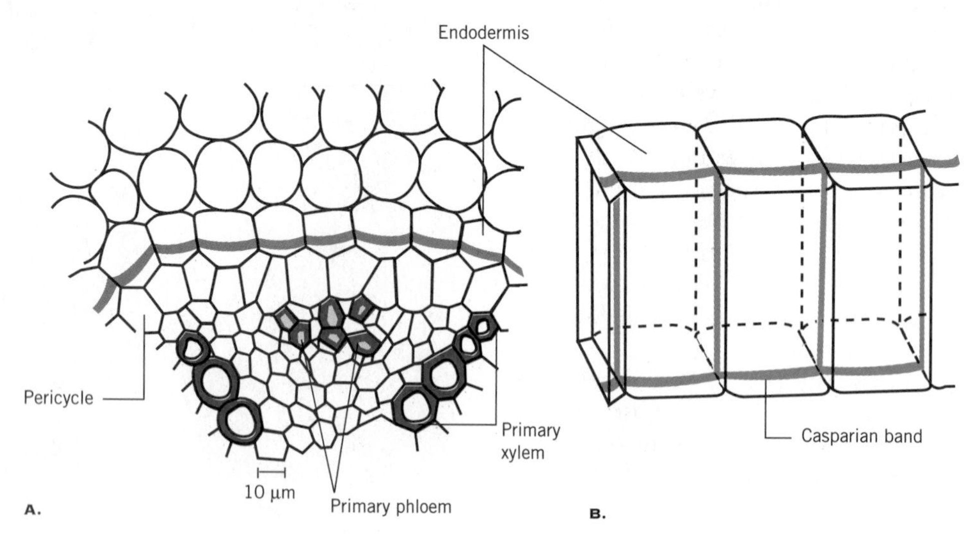

FIGURE 3.12 Suberin deposits (Casparian strip) in the walls of root endodermal cells. (*A*), cross section; (*B*), 3-dimensional view. Suberin deposits in the radial walls establish a barrier to movement of water and salts in the apoplast of the endodermis. (From K. Esau, *Anatomy of Seed Plants,* New York, Wiley, 1977. Reprinted by permission.)

even in those species which do exhibit it. Finally, it has been clearly established that during periods of active transpiration, when water movement through the xylem would be expected to be most rapid, the xylem is under **tension** (i.e., *negative* pressure). Root pressure clearly cannot serve as the mechanism for the ascent of sap in all cases. However, root pressure could serve to fill vessels in small, herbaceous plants and in some woody species in the spring when sap moves up to the developing buds (Nobel, 1991).

WATER RISE BY CAPILLARITY

If a glass capillary tube (that is, a tube of small diameter) is inserted into a volume of water, water will rise in the tube to some level above the surface of the surrounding bulk water. This phenomenon is called **capillary rise,** or simply **capillarity.** Capillary rise is due to the interaction of several forces. These include **adhesion** between water and polar groups along the capillary wall, **surface tension** (due to cohesive forces between water molecules), and the force of gravity acting on the water column (see Box 3.2). Adhesive forces attract water molecules to polar groups along the surface of the tube. When these water-to-wall forces are strong, as they are between water and glass tubes or the inner surfaces of tracheary elements, the walls are said to be *wettable.* As water flows upward along the wall, strong cohesive forces between the water molecules act to pull the bulk water up the lumen of the tube. This will continue until these lifting forces are balanced by the downward force of gravity acting on the water column.

As shown in Box 3.2, the calculated rise of water in a capillary tube is inversely proportional to the radius of the tube. In a large tracheid or small vessel, with a diameter of 50 μm (r = 25 μm), water will rise to a height of about 0.6 m. For a large vessel (r = 200 μm), capillarity would account for a rise of only 0.08 m. On the basis of these numbers, capillarity in tracheids and small vessels might account for the rise of xylem sap in small plants, say less than 0.75 m in height. However, to reach the height of a 100 m tree by capillarity, the diameter of the capillary would have to be about 0.15 μm—much smaller than the smallest tracheids. Clearly capillarity is inadequate as a *general* mechanism for the ascent of xylem sap.

THE COHESION THEORY

The most widely accepted theory for movement of water through plants is known as the **cohesion theory.** This theory depends on there being a continuous column of water from the tips of the roots, through the stem and into the mesophyll cells of the leaf. The theory is generally credited to H. H. Dixon, who gave the first detailed account of it in 1914.

The Driving Force According to the cohesion-tension theory, the driving force for water movement in the xylem is provided by evaporation of water from the leaf and the tension or negative pressure that results. Water covers the surfaces of the mesophyll cells as a thin film, adhering to cellulose and other hydrophillic surfaces. As water evaporates from this film, the air–liquid interface retreats into the small spaces between cellulose microfibrils and the angular junctions between adjacent cells. This creates very small curved surfaces or microscopic menisci (Fig. 3.13). As the radii of these menisci progressively decrease, surface tension at the air–water interface generates an increasingly negative pressure, which in turn tends to draw more liquid water toward the surface. Because the water column is continuous, this negative pressure, or tension,[1] is transmitted through the column all the way to the soil. As a result, water is literally pulled up through the plant from the roots to the surface of the mesophyll cells in the leaf.

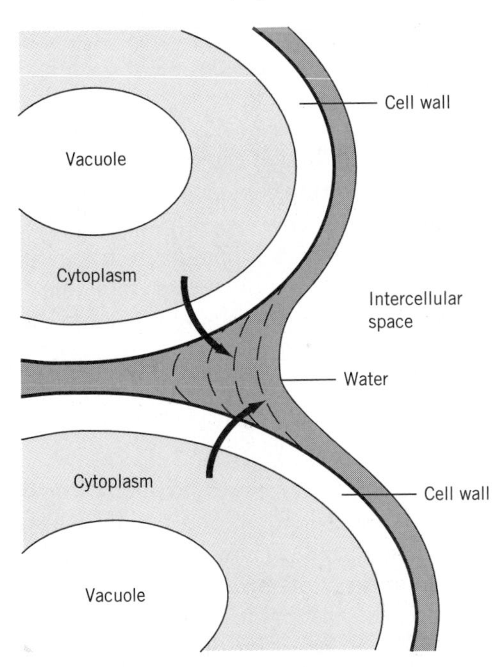

FIGURE 3.13 Tension (negative pressure) in the water column. Evaporation into the leaf spaces causes the water–air interface (dashed lines) to retreat into the spaces between and at the junctions of leaf mesophyll cells. As the water retreats, the resulting surface tension pulls water from the adjacent cells. Because the water column is continuous, this tension is transmitted through the column, ultimately to the roots and soil water.

[1]Recall that in plant water relations, pressure is always given as the difference between actual pressure and atmospheric pressure (actual minus atmospheric). Tension is less than atmospheric pressure and therefore is always negative.

Box 3.2 Forces Involved in Capillary Rise

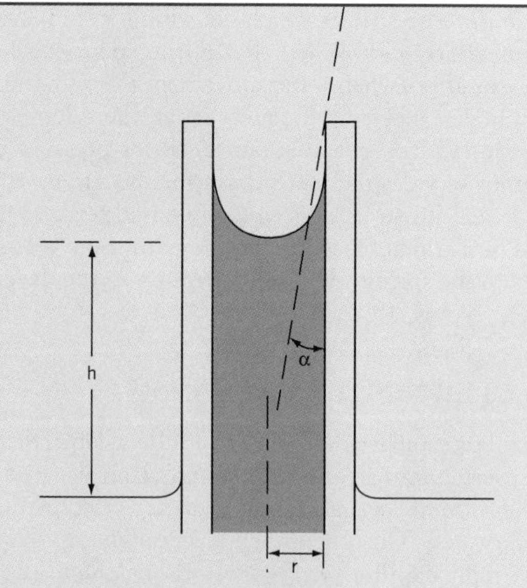

FIGURE 3.14 Diagram illustrating the forces involved in capillary rise.

The height to which a fluid may rise in a capillary tube may be calculated by taking into account two balancing forces. The lifting force, which tends to pull the fluid up the tube, is determined by the adhesive forces between the fluid and the capillary wall and surface tension (see Chap. 2). This lifting force is balanced by the force of gravity acting on the fluid column, which tends to pull the column downward.

Adhesion results from the attraction between fluid molecules and a surface. In the case of water, this attraction results from hydrogen bonding of the water molecules and polar groups on the capillary wall. The attraction of water molecules to the capillary wall causes the water to creep up the wall. Normally this would "stretch" the surface of the water column, but the high surface tension of water resists stretching and pulls the column up. The result is a curved surface called a **meniscus**. Adhesion can be measured indirectly as the cosine of the contact angle (*cos* α) between the meniscus and the capillary wall (Figure 3.14). The extent to which a fluid creeps up the wall and thus both the radius of the meniscus and its contact angle with the capillary wall are related to the strength of the adhesive forces between the fluid and the wall. This relationship is shown in the accompanying illustration.

Surface tension (γ) is measured as a force per unit length, usually N m^{-1}. (The newton, N, is a quantity of force calculated as mass times gravitational acceleration.) In this case, the force acts along the inside circumference of the capillary tube ($2r\pi$). The **total lifting force** for a column of water in a capillary tube is

The cohesion theory raises two very important questions: (1) Is the xylem sap of a rapidly transpiring plant under tension, and (2) how is the integrity of very tall water columns maintained? In the absence of any direct evidence, the answers to these questions provide substantial indirect support for the theory.

Except in certain circumstances, such as when root pressure is active, pressures in the xylem are rarely positive. On the other hand, several lines of evidence support the conclusion that xylem water is instead under significant tension. First, if one listens carefully when the xylem of a rapidly transpiring plant is severed, it is sometimes possible to hear the sound of air being drawn rapidly into the wound. If severed beneath the surface of a dye solution, the dye will be very rapidly taken up into tracheary elements in the immediate vicinity of the wound. A second line of evidence involves sensitive measuring devices, called dendrographs, which can be used to measure small changes in the diameter of woody stems. Diameters decrease significantly during periods of active transpiration (Dobbs and Scott, 1971). This will happen because the stem is slightly elastic and the tension in the water column pulls the tracheary walls inward. In the evening, when transpiration declines, the tension is released and stem diameter recovers. Moreover, the shrinkage occurs first in the upper part of the tree, closest to the transpiring leaves, when transpiration begins in the morning. Only later does it show up in the lower part. This observation has been confirmed by experiments in which a localized pulse of heat is generated in the xylem and the flow of heat away from that point is monitored. Flow rates indicate that tensions are greater near the top of the tree. A variety of other experiments have demonstrated that rapidly transpiring shoots are able to pull a column of mercury to heights greater than can be accounted for by atmospheric pressure alone.

Direct measurement of tension in xylem vessels was made possible by the introduction of the *pressure bomb* technique by P. F. Scholander (Scholander et al., 1965) (see Appendix). If the xylem solution is under tension, it will withdraw from the cut surface but can be forced therefore given by the product of surface tension and adhesion, or:

$$2r\pi\gamma \cos \alpha$$

At the same time as adhesion and surface tension are pulling water up the capillary, the force of gravity is acting to pull the water column downward. This downward force is a function of the volume of the water column ($\pi r^2 h$), the density of water (ρ), and the acceleration due to gravity (g). Water will rise in the capillary tube until the lifting forced is balanced by the downward force, or:

$$2r\pi\gamma \cos \alpha = \pi\rho^2 h\rho g$$

Solving for h,

$$h = \frac{2\pi r^2 \gamma \cos \alpha}{\pi r^2 \rho g} = \frac{2\gamma \cos \alpha}{r\rho g}$$

Known values for water at 20°C are:

$$\gamma = 0.073 \text{ N m}^{-1} \text{ (where N = kg m s}^{-2}\text{)}$$

$$\rho = 998 \text{ kg m}^{-3}$$
$$g = 9.8 \text{ m s}^{-2}$$

For water in capillaries with exposed polar groups, such as glass or tracheary elements, the contact angle is zero and $\cos 0° = 1$. This is because the adhesive forces between water and the capillary walls are at least as strong, if not stronger, than the intermolecular cohesive forces which rise to surface tension.

With the above values for water and once again solving for h:

$$h = \frac{149 \times 10^{-5} \text{ m}^2}{\text{r m}}$$

Note that both the capillary radius and the capillary rise are given in meters.

The above equation can now be used to calculate the height to which water will rise in a capillary tube with a diameter of 50 μm ($r = 25 \times 10^{-6}$m):

$$h = \frac{1.49 \times 10^{-5} \text{ m}^2}{25 \times 10^{-6} \text{ m}} = 0.596 \text{ m}$$

or about 0.6 m.

It is interesting to note that not all fluids will *rise* in a capillary tube. When a capillary tube is immersed in mercury, for example, the mercury level inside the tube will fall *below* the level of the bulk fluid and the meniscus will appear inverted. This happens because the cohesive forces of mercury (and, consequently, its surface tension) are much stronger than the adhesive forces between mercury and the glass wall. The result is a contact angle greater than 90° (for mercury $\alpha = 150°$) and a negative value for $\cos \alpha$. Thus the lifting force for mercury is actually negative, contributing to a depression of the mercury level in the tube.

back to the surface by increasing the pressure in the chamber. With such a device, Scholander and others have measured tensions on the order of −0.5 to −2.5MPa in rapidly transpiring, temperate zone trees.

Finally, it has been observed that water potentials near the bottom of the tree are less negative than water potentials higher in the crown. Such a pressure drop between the bottom and the top of the crown is consistent with a tension resulting from forces originating at the top of the tree. The weight of evidence clearly supports the hypothesis that the xylem water column is literally pulled up the tree in response to transpiration.

Maintaining the Water Column Turning now to the second question—how is the integrity of the water column maintained?—the ability to resist breakage is a function of the **tensile strength** of the water column. Tensile strength is a measure of the maximum tension a material can withstand before breaking. Tensile strength is expressed as force per unit area, where the area for the purpose of our discussion is the cross-sectional area of the water column. Tensile strength is yet another property of water attributable to the strong intermolecular cohesive forces, or hydrogen bonding, between the water molecules.

The tensile strength of water (or any fluid, for that matter) is not easily measured—a column of water does not lend itself to testing in the same way as a steel bar or a copper wire. The tensile strength of water will also depend on the diameter of the conduit, the properties of the conduit wall, and the presence of any dissolved gasses or solute. Still, a number of ingenious approaches have been developed to measure the tensile strength of water with fairly consistent results (Hayward, 1971). It is now generally accepted that pure water, free of dissolved gas, is able to withstand tensions as low as −25 to −30 MPa at 20°C. This is approximately 10 percent of the tensile strength of copper, and 10 times greater than the −2.5 to −3.0 MPa required to pull an uninterrupted water column to the top of the tallest trees. As noted above, tensions in the xylem are more typically in the range of −0.5 to −2.5 MPa for temperate deciduous trees such as maple (*Acer* sps.), but may sometimes be as low as −10 MPa (Tyree and Sperry, 1989)

Because xylem water is under tension, it must remain in the liquid state well below its vapor pressure—recall that the vapor pressure of water at 20°C is 2.3 kPa or 0.0023 MPa (Table 3.1). A water column under tension is therefore physically unstable. Physicists call this condition a *metastable state*, a state in which change is ready to occur but does not occur in the absence of an external stimulus. Stability can be achieved in a water column under tension by introducing a vapor phase. Water molecules in the vapor phase have very low cohesion, which allows the vapor to expand rapidly, thus causing the column to rupture and relieve the tension.

How might a vapor phase be introduced to the xylem column? Xylem water contains several dissolved gasses, including carbon dioxide, oxygen, and nitrogen. When the water column is under tension, there is a tendency for these gasses to come out of solution. Submicroscopic bubbles first form at the interface between the water and the walls of the tracheid or vessel, probably in small, hydrophobic crevices or pores in the walls. These small bubbles may redissolve or they may coalesce and expand rapidly to fill the conduit. This process of rapid formation of bubbles in the xylem is called **cavitation** (L. *cavus* = hollow). The resulting large gas bubble forms an obstruction, called an **embolism** (Gr. *embolus* = stopper), in the conduit. The implications of embolisms with respect to the cohesion theory are quite serious, because a conduit containing an embolism is no longer available to conduct water. Indeed, the potential for frequent cavitation in the xylem was raised as a principal objection to the cohesion theory when it was initially proposed. In order to satisfy these objections, it was necessary to determine just how vulnerable the xylem was to cavitation.

Early attempts to relate cavitation to tensions developed in the xylem were largely inconclusive. There were no satisfactory methods for observing cavitation in the xylem itself and model systems, employing glass tubes, did not necessarily duplicate the interface conditions present in plant tissues. This all changed in 1966 when J. A. Milburn and R. P. C. Johnson introduced an acoustic method for detecting cavitation in plants (Milburn and Johnson, 1966). In laboratory experiments with glass tubes, the rapid relaxation of tension which follows cavitation produces a shockwave which can be heard as an audible click. Milburn and Johnson found that similar clicks could be "heard" in plant tissue by using sensitive microphones and amplifiers. Each click is believed to represent formation of an embolism in a single vessel element.

Milburn and Johnson studied cavitation in water-stressed leaves of castor bean (*Ricinus communis*). Water stress was introduced by detaching the leaf from the plant and permitting it to wilt. As the leaf wilted, the number of clicks occurring in the petiole was recorded. A total of 3,000 clicks were detected, which is approximately equal to the number of vessels that might be expected in such a petiole. Cavitation could be prevented by adding water to the severed end of the petiole. Various methods that either increased or decreased transpiration from the leaf resulted in a corresponding increase or decrease in the number of clicks. These results indicate a reasonably straightforward relationship between cavitation and tension in the xylem, which appears to support the cohesion theory. They further suggest that cavitation is readily induced by water stress, a condition that herbaceous plants might be expected to encounter on a daily basis.

A long-term study of cavitation in a stand of sugar maple (*Acer saccharum*) has been conducted by J. S. Sperry and his colleagues (Sperry et al., 1988). Rather than using the acoustic method, Sperry employed a method that measures changes in **hydraulic conductance** (Sperry et al., 1987). In its simplest form, conductance is the inverse of resistance. Hydraulic conductance is therefore a measure of the total capacity of the tissue to conduct water. The acoustic method is limited to counting the number and frequency of cavitations. The hydraulic method, on the other hand, assesses the impact of the resulting embolisms on the capacity of the tissue to transport water.

During the summer growing season, embolisms appeared to be confined to the main trunk and reduced hydraulic conductance by 31 percent. During the winter, loss of conductance in the main trunk increased to 60 percent, while some twigs suffered a 100 percent loss! A decline in conductance during the summer months is no doubt attributable to water stress, as it is in herbaceous plants. The rise in embolisms during the winter is probably related to freeze-thaw cycles. The solubility of gasses is very low in ice; when tissue freezes, gas is forced out of solution. During a thaw, these small bubbles will expand and nucleate cavitation.

Problems related to cavitation are not limited to mature trees. Newly planted seedlings often experience water stress due to poor root-soil contact and may be vulnerable to cavitation. In a recent study, Kavanagh and Zaerr (1997) found that seedlings of western hemlock (*Tsuga heterophylla*) experienced water stress and declining xylem pressure potential when planted out. The resulting cavitation and embolism formation in the tracheids caused a decline in hydraulic conductance in the seedling. If the decline in hydraulic conductance is severe enough, it can lead to defoliation or death of the seedling.

Clearly, the effect of cavitation and embolisms on long-term survival of plants would be disastrous if there were not means for their removal or for minimizing their effects. The principal mechanism for minimizing

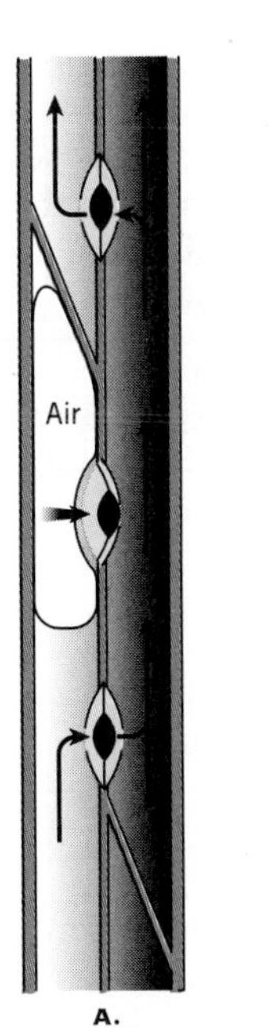

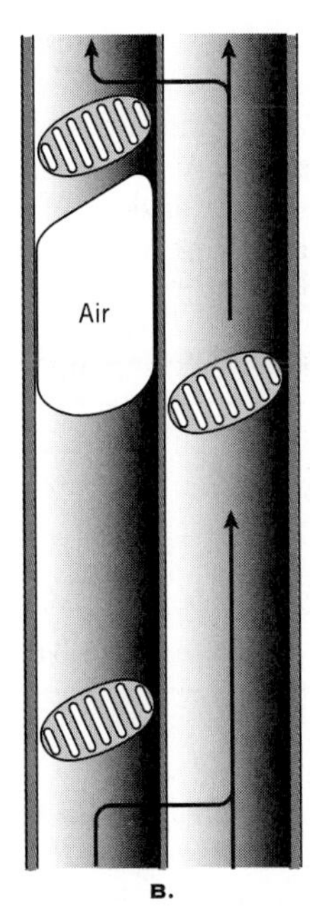

FIGURE 3.15 Diagram to illustrate how water flow bypasses embolisms in tracheids and vessels. In tracheids (*A*), the pressure differential resulting from an embolism causes the torus to seal off bordered pits lining the affected tracheary element. In vessels (*B*), the bubble may expand through perforation plates, but will eventually be stopped by an imperforate end wall. In both tracheids and vessels, surface tension prevents the air bubbles from squeezing through small pits or capillary pores in the side walls. Water, however, continues to move around the blockage by flowing laterally into adjacent conducting elements.

the effect of embolisms is a structural one. The embolism is simply contained within a single tracheid or vessel member. In those tracheary elements with bordered pit pairs, the embolism is contained by the structure of the pit membrane (Fig. 3.15A). A difference in pressure between the vessel member containing the embolism and the adjacent water-filled vessel causes the torus to press against the pit border, thus preventing the bubble from being pulled through. At the same time, surface tension prevents the bubble from squeezing through the small openings in the perforation plates between successive vessel members (Fig. 3.15B). Water, however, will continue to flow laterally through available pits, thus detouring around the blocked element by moving into adjacent conduits. In addition to bypassing embolisms, plants may also avoid long-term damage by repairing the embolism. This can happen at night, for example, when transpiration is low or absent. Reduced tension in the xylem water permits the gas to simply redissolve in the xylem solution. An alternative explanation, particularly in herbaceous species, is that air may be forced back into solution on a nightly basis by positive root pressure.

The repair of embolisms in taller, woody species is not so easily explained. As noted above, sugar maples do recover from freezing-induced embolisms in the spring. Maples and many other tree species also exhibit positive xylem pressures in the spring. These pressures account for the springtime sap flow and could also serve to reestablish the continuity of the water column. Springtime recovery has also been documented for grapevines, although this is apparently due to relatively high root pressures at this time of the year (Sperry et al., 1987). It might be possible that woody plants in general develop higher than normal root pressures in the spring in order to overcome winter damage. Finally, most woody species produce new secondary xylem each spring. This new xylem tissue is laid down before the buds break and may meet the hydraulic conductance needs of the plant, replacing older, nonfunctional xylem.

It is clear, even on the basis of the relatively few studies that have been completed, that xylem is particularly vulnerable to cavitation and embolism. If herbaceous plants were unable to recover on a nightly basis or woody species were unable to recover in the spring, their growth and ultimate survival would be severely compromised. This might lead one to question why plants have not evolved mechanisms to lessen their effects. Interestingly, M. H. Zimmermann turned this question around, proposing instead that cell walls might actually be designed to cavitate (Zimmermann, 1983). By rewriting the equation for capillary rise (see Box 3.2) it is possible to calculate the force required to pull an air bubble through a small pore. This statement may appear to contradict an earlier observation that surface tension prevents the passage of embolisms through perforations in xylem conduits, but it does not. Where a water-filled conduit under tension is separated from an air volume (at atmospheric pressure) by a porous wall, a concave meniscus (the "lifting force") will form in the pore to balance the negative pressure. As the pressure differential increases, the radius of the meniscus will decrease. At a "suitable" tension in the xylem, the meniscus will reach a radius less than the diameter of the pore (a fraction of a μm in diameter) and will be pulled through the pore into the water-filled conduit. The resulting bubble nucleates a cavitation, which relieves the tension, reduces the pressure differential, and prevents the further entry of air. Results consistent with Zimmermann's hypothesis have been obtained by Tyree and Sperry (1988), who compared the pressure differences required to embolize sugar maple stems with diameters of the pores in the pit membranes.

Accordingly, Zimmermann's **designed leakage** hypothesis represents a kind of safety valve. Plants appear to be constructed in such a way as to allow cavitation when water potentials reach critically low levels, yet still allow the "damage" to be repaired when conditions improve. Water stress and, consequently, cavitations would be expected to appear first in leaves and smaller branches. These localized cavitations would serve the additional advantage of cutting off peripheral structures while preserving the integrity of the main stem or trunk during extended dry periods.

Roots are even more vulnerable to cavitation than shoots, which could benefit the whole plant during periods of drought (Sperry and Ikeda, 1997). Complete cavitation of the xylem in the smallest roots, for example, would isolate those roots from drying soil, reduce hydraulic conductance, and ultimately reduce transpiration rates. This would help buffer the water status of the stem until the drought eased and the cavitated conduits were refilled or new growth replaced the damaged roots.

TABLE 3.4 **Classification of soil particles and some of their properties. A mixture of 40 percent sand, 40 percent silt, and 20 percent clay is known as a *loam soil.* A sandy soil contains less than 15 percent silt and clay, while a clay soil contains more than 40 percent clay particles.**

Particle Class	Particle Size (mm)	Water Retention	Aeration
Coarse sand	2.00 – 0.2		
		poor	excellent
Sand	0.20 – 0.02		
Silt	0.02 – 0.002	good	good
Clay	less than 0.002	excellent	poor

ROOTS, SOIL, AND THE UPTAKE OF WATER

In order to maintain the turgidity of leaves and shoots as well as to support biochemical activities at a level that ensures survival, water lost to a plant by transpiration must be continuously replenished. Many plants can absorb atmospheric moisture in the form of mist or dew, but foliar absorption of water is negligible compared with absorption through the roots.

Water lost by transpiration must be replenished by the absorption of an equivilent amount of water from the soil through the root system. This establishes an integrated flow of water from the soil, through the plant, and into the atmosphere, referred to as the **soil-plant-atmosphere continuum.** The concept of a soil-plant-atmosphere continuum reinforces the observation that plants do not exist in isolation, but are very interdependent with their environment.

SOIL—A COMPLEX MEDIUM

In order to understand interactions between roots and soil water, a review of the nature of soils would be helpful. Soil is a very complex medium, consisting of a solid phase comprised of inorganic rock particles and organic material, a soil solution containing dissolved solutes, and a gas phase generally in equilibrium with the atmosphere. The inorganic solid phase of soils is derived from parent rock that is degraded by weathering processes to produce particles of varying size (Table 3.4). In addition to the solid, liquid and gas phases, soils also contain organic material in varying stages of decomposition as well as algae, bacteria, fungi, earthworms and various other organisms.

The clay particles in a soil combine to form complex aggregates that, in combination with sand and silt, determine the structure of a soil. Soil structure in turn affects the porosity of a soil and, ultimately, its water retention and aeration. *Porosity*, or pore space, refers to the interconnected channels between irregularly shaped soil particles. Pore space typically occupies approximately 40 percent to 60 percent of a soil by volume. Two major categories of pores—**large pores** and **capillary pores**—are recognized. Although there is no sharp line of demarcation between large pores and capillary pores—the shape of the pore is also a determining factor—water is not readily held in pores larger than 10 to 60 μm diameter. When a soil is freshly watered, such as by rain or irrigation, the water will percolate down through the pore space until it has displaced most, if not all, of the air. The soil is then **saturated** with water. Water will drain freely from the large pore space due to gravity. The water that remains after free (gravity) drainage is completed is held in the capillary pores. At this point, the water in the soil is said to be at **field capacity.** Under natural conditions, it might require two to three days for a loam soil to come to field capacity following a heavy rainfall. The relative proportions of large and capillary pore space in a soil can be estimated by determining the water contents of the soil when freshly watered and at field capacity. Water content, expressed as the weight of water per unit weight of dry soil, may be determined by drying the soil at 105°C.

It should not be surprising that a sandy soil, with its coarse particles, will have a relatively high proportion of large pores. A sandy soil will therefore drain rapidly, has a relatively low field capacity, and is well-aerated (Table 3.4). The pore space of a clay soil, on the other hand, consists largely of capillary pores. Clay soils hold correspondingly larger quantities of water and are poorly aerated. A loam soil represents a compromise, balancing water retention against aeration for optimal plant growth.

The water held by soil at or below field capacity is found in capillary channels and the interstitial spaces between contacting soil particles, much as it is in the cell walls and intercellular spaces of mesophyll cells in a leaf. Soil water is therefore also subject to the same forces of surface tension found in the capillary spaces of mesophyll tissues, as discussed earlier. Consequently, soil water at or below field capacity will be under tension and its water potential will be negative. As the water content of the soil decreases, either by evaporation from the soil surface or because it is taken up by the roots, the air-water interface will retreat into the capillary spaces between the soil particles. Because water adheres strongly to the soil particles, the radius of the meniscus decreases and pressure becomes increasingly negative. In principle, water movement in the soil is primarily pressure-driven in the same manner that capillary tension in the mesophyll cells of a leaf draw water from the xylem column. As water is removed from the soil by a root, tensions in the soil water will draw more bulk water toward the root. If there is an abundance of water in the soil, these pressure differences may draw water from some distance.

Except in highly saline soils, the solute concentration of soil water is relatively low—on the order of 10^{-3} M—and soil water potential is determined principally by the negative pressure potential. As might be expected, the uptake of water by roots occurs because of a water potential gradient between the soil and the root. Thus as the soil dries and its water potential declines, plants may experience difficulty extracting water from the soil rapidly enough to balance losses by transpiration. Under such conditions, plants will lose turgor and wilt. If transpiration is reduced or prevented for a period of time (such as at night, or by covering the plant with a plastic bag), water uptake may catch up, turgor will be restored, and the plants will recover. Eventually, however, a point can be reached where the water content of the soil is so low that, even should all water loss by transpiration be prevented, the plant is unable to extract sufficient water from the soil and the loss of turgor is permanent. The soil water content at this point, measured as a percentage of soil dry weight, is known as the **permanent wilting percentage.** The actual value of the permanent wilting percentage varies between soil types; it is relatively low (in the range of 1 to 2 percent) for sand and high (20 to 30 percent) for clay. Loam soils fall between these two extremes, depending on the relative proportion of sand and clay. Regardless of soil type, however, the water potential of the soil at the permanent wilting percentage is relatively uniform at about -1.5 MPa. Although there are some exceptions to the rule, most plants are unable to extract significant amounts of water when the soil water potential falls below -1.5 MPa. In a sense, field capacity may be considered a property of the soil, while the permanent wilting percentage is a property of the plant.

The water content of the soil between field capacity and the permanent wilting percentage is considered **available water,** or water that is available for uptake by plants. The range of available water is relatively high in silty loam soils, somewhat less in clay, and relatively low in sand. Not all water in this range is uniformly available, however. In a drying soil, plants will begin to show signs of water stress and reduced growth long before the soil water potential reaches the permanent wilting percentage.

ABSORPTION OF WATER BY ROOTS

Roots have four important functions. Roots (1) anchor the plant in the soil; (2) provide a place for storage of carbohydrates and other organic molecules; (3) are a site of synthesis for important molecules such as alkaloids and some hormones; and (4) absorb and transport upward to the stem virtually all the water and minerals taken up by plants.

ROOTS, THE HIDDEN HALF

The effectiveness of roots as absorbing organs is related to the extent of the root system. Over the years, a number of efforts have attempted to establish the true extent of root systems. Such studies involve careful and often tedious procedures for excavating the plant and removing the soil without damaging the roots. More elaborate efforts involved construction of subterranean walkways, called "rhizotrons," equipped with windows through which the growth and development of roots could be observed. J. E. Weaver and R. W. Darland (1949) examined the root systems of several prairie grasses in different types of soils. By carefully excavating the grasses and washing the roots free of soil, they were able to show massive networks of roots penetrating to a depth of 1.5 m. Using a radioactive tracer technique, N. S. Hall and his colleagues (1953) later found the roots of a single fourteen-week-old corn plant had penetrated to a depth of more than 6 m and extended horizontally as much as 5 m in all directions. In one of the more impressive efforts, H. J. Dittmer (1937) has reported the measurements of the roots of a *single* mature rye plant grown in a box of soil measuring 30 cm × 30 cm by 56 cm deep. The combined length of all roots was 623 km with an estimated total surface area of 639 m^2. Many species invest substantially more than 50 percent of their body weight in roots. Roots clearly comprise a large proportion of the plant body, although their importance to plants is less obvious to the casual observer than the more visible shoot system. It is no wonder that a recent monograph summarizing our current understanding of root development and function was entitled "Plant Roots. The Hidden Half" (Waisel et al., 1996).

THE ABSORBING REGIONS OF A ROOT

A large number of anatomical and physiological studies have established that the region of most active water uptake lies near the root tip. Beyond this generalization, the permeability of roots to water varies widely with age, physiological condition, and the water status of the plant. Studies with young roots have shown that the region most active with respect to water uptake starts about 0.5 cm from the tip and may extend down the root as far as 10 cm (Fig. 3.16). Little water is absorbed in the meristematic zone itself, presumably because the protoplasm in this zone is dense and there are no differentiated vascular elements to carry the water away. The region over which water appears to be taken up most rapidly corresponds generally with the zone of cell maturation. This is the region where vascular tissues, in particular the xylem, has begun to differentiate. Also in this region the deposition of suberin and lignin in the walls of endodermal cells is only beginning and has not yet reached the point of offering significant resistance to water movement.

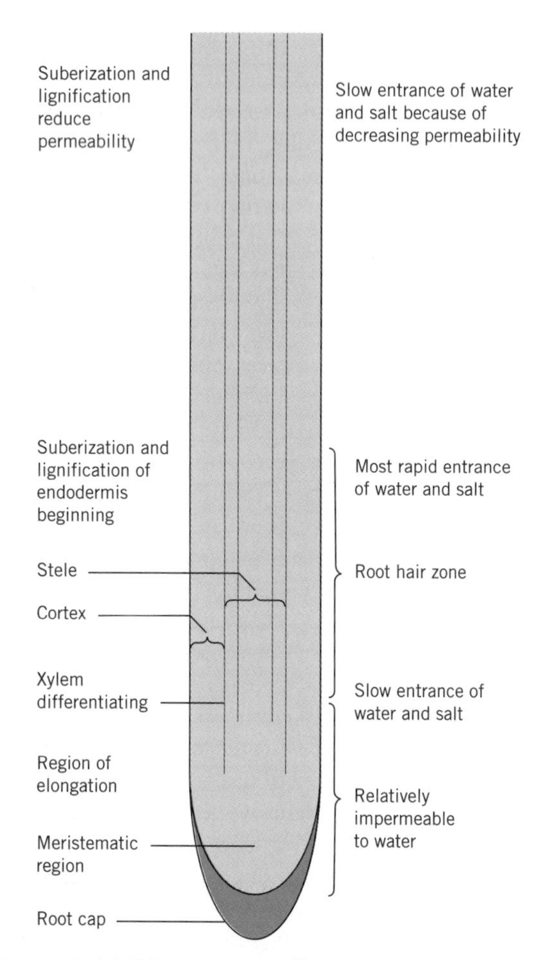

FIGURE 3.16 Diagrammatic illustration of the relationship between differentiation of root tissues and water uptake.

The region of most rapid water uptake also coincides with the region of active root hair development. Root hairs are thin-walled outgrowths of epidermal cells which increase the absorptive surface area and extend the absorptive capacity into larger volumes of soil (Fig. 3.17). Depending on the species and environmental conditions, root hairs may reach lengths of 0.1 mm to 10 mm and an average diameter of 10 μm. In some species such as peanut, pecan, and certain conifers, root hairs are rare or absent. More commonly a single root tip may contain as many as 2500 hairs cm^{-2} and may increase the absorbing surface of the root 1.5 to twentyfold. For two reasons, root hairs greatly increase the contact of the root with soil water. First their small diameter permits root hairs to penetrate capillary spaces not accessible to the root itself (Fig. 3.17B). Second, root hairs extend contact into a cylinder of soil whose diameter is twice that of the length of the hair (Fig. 3.17C).

RADIAL MOVEMENT OF WATER THROUGH THE ROOT

Once water has been absorbed into the root hairs or epidermal cells, it must traverse the cortex in order to reach the xylem elements in the central stele. In principle, the path of water through the cortex is relatively straightforward: There appear to be two options. Water may either flow past the cells through the apoplast of the cortex or from cell to cell through the plasmodesmata of the symplast. In practice, however, the two pathways are not separate. Apoplastic water is in constant equilibration with water in the symplast and cell vacuoles. This means that water is constantly being exchanged across both the cell and vacuolar membranes. In effect, then, water flow through the cortex involves both pathways. The cortex consists of loosely packed cells with numerous intercellular spaces. The apoplast would thus appear to offer the least resistance and probably accounts for a larger proportion of the flow.

In the less mature region near the tip of the root, water will flow directly from the cortex into the developing xylem elements, meeting relatively little resistance along the way. Moving away from the tip toward the more mature regions, water will encounter the endodermis (see earlier discussion of root pressure). Suberization of the endodermal cell walls imposes a permeability barrier, forcing water taken up in these regions to pass through the cell membranes. While the endodermis does increase resistance to water flow, it is far from being an absolute barrier. Indeed, under conditions of rapid transpiration the region of most rapid water uptake will shift toward the basal part of the root (Table 3.5).

The resistance offered by the root to water uptake is reflected in the absorption lag commonly observed

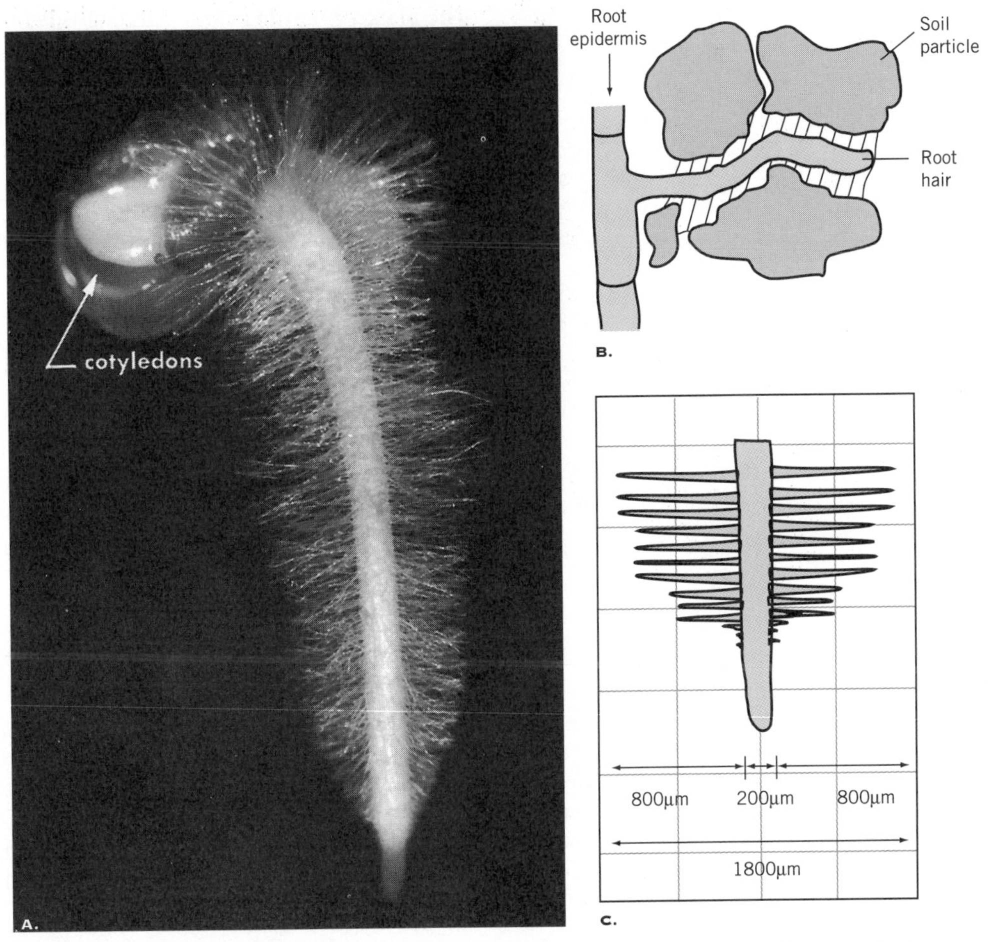

FIGURE 3.17 Root hairs and water uptake. (*A*) Root hairs on radish (*Raphanus sativus*). (*B*) Root hairs enhance water uptake by their ability to penetrate water-containing capillary spaces between soil particles. (*C*) Root hairs increase by several times the volume of soil that can be extracted of water by a root. (*A* from T. E. Weier et al. *Botany*. 6th edition. 1982. New York, Wiley. Fig. 9.7A. Used by permission of the authors.)

TABLE 3.5 Relative water uptake for different zones along a *Vicia faba* root as a function of transpiration rate. The measured xylem tension at the low transpiration rate was −0.13 MPa and −10.25 MPa at the high transpiration rate. When transpiration is low, most of the water uptake occurs near the root tip. When transpiration is high, the resulting increase in tension shifts the region of uptake toward more basal regions.

Water Uptake Zone (cm from apex)	Transpiration Rate	
	Low	High
0–2.5	100	132
2.5–5.0	103	216
5.0–7.5	54	216
7.5–10.0	27	270
10.5–12.5	19	283

From data of Brouwer, 1965.

when water loss by transpiration is compared with absorption by the roots (Fig. 3.18). That this lag is due to resistance in the root can be demonstrated experimentally. If the roots of an actively transpiring plant are cut off (under water, of course), there is an immediate increase in the rate of absorption into the xylem. In some species, absorption lag may cause a water deficit in the leaves sufficient to stimulate a temporary closure of the stomata (Fig. 3.18). This phenomenon is known as **midday closure.** Closure of the stomata reduces transpiration, allowing absorption of water to catch up and the stomata to then reopen.

Finally, although the absorption of water by roots is believed to be a passive, pressure-driven process, it is nonetheless dependent on respiration in the root cells. Respiratory inhibitors (such as cyanide or dinitrophenol), high carbon dioxide levels and low oxygen (anaerobiosis) all stimulate a decrease in the hydraulic conductance of most roots. Anaerobiosis is most commonly

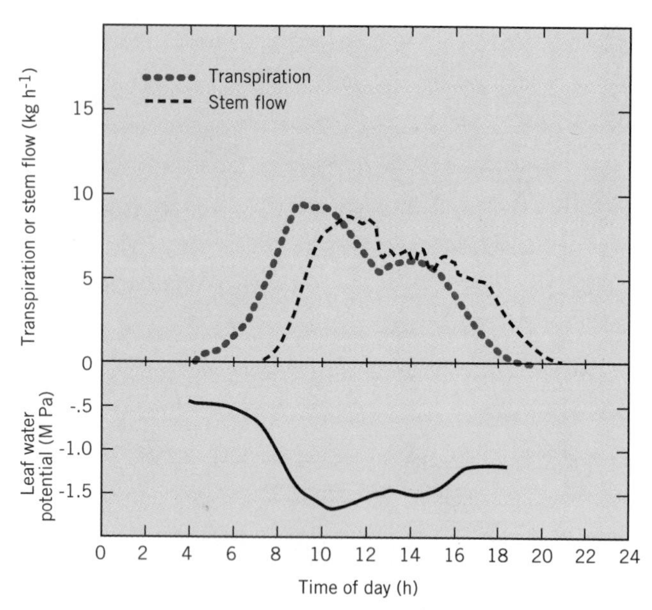

FIGURE 3.18 The absorption lag in a *Larix* (larch) tree. Upper: Absorption of water, measured as the flow of water through the stem, lags about two hours behind transpiration. A transient decline in transpiration rate near midday may occur if the lag is sufficient to create a water deficit and stimulate temporary stomatal closure. Lower: Rapid transpiration causes a decrease in the leaf water potential, which slowly recovers as water moves in to satisfy the deficit. (Adapted from E.-D. Schulze et al., Canopy transpiration and water fluxes in the xylem of *Larix* and *Picea* trees—A comparison of xylem flow, porometer and cuvette measurements, *Oecologia* 66:475–483, 1985. Reprinted by permission of Springer-Verlag.)

encountered by plants in water-logged soils and may lead to extreme wilting in species not specifically adapted to such situations. The exact role of respiration is not clear. The requirement is probably indirect, such as maintaining the cellular integrity and continued elongation of roots. Active uptake of nutrient ions (Chap. 5) may also be a factor. On the other hand, killing roots outright—for example, by immersion in boiling water—dramatically reduces resistance and allows water to be absorbed more rapidly than when the roots were alive.

SUMMARY

Large amounts of water are lost by plants through evaporation from leaf surfaces, a process known as transpiration. Transpiration is driven by differences in water vapor pressure between internal leaf spaces and the ambient air. A variety of factors influence transpiration rate, including temperature, humidity, wind, and leaf structure. Water is conducted upward through the plant primarily in the xylem, a tubelike system of tracheary elements including tracheids and vessels. The principal driving force for water movement in the xylem is transpiration and the resulting tension in the water column. The water column is maintained because of the high tensile strength of water. Water lost by transpiration is replenished by the absorption of water from the soil through the root system.

CHAPTER REVIEW

1. Explain why transpiration rate tends to be greatest under conditions of low humidity, bright sunlight, and moderate winds.
2. Describe the anatomy of xylem tissue and explain why it is an efficient system for the transport of water through the plant.
3. Trace the path of water from the soil, through the root, stem, and leaf of a plant, and into the atmosphere.
4. Explain how water can be moved to the top of a 100 m tree, but a mechanical pump can lift water no higher than about 10.3 m. What prevents the water column in a tree from breaking? Under what conditions might the water column break, and, if it does break, how is it reestablished?
5. Many farmers have found that fertilizing their fields during excessively dry periods can be counter-productive, as it may significantly damage their crops. Based on your knowledge of the water economy of plants and soils, explain how this could happen.
6. Does transpiration serve any useful function in the plant?
7. Explain the relationships between field capacity, permanent wilting percentage, and available water. Even though permanent wilting percentage is based on soil weight, it is often said to be a property of the plants. Explain why this might be so.

FURTHER READING

Boyer, J. S. 1985. Water transport. *Annual Review of Plant Physiology* 36:473–516.

Hinckley, T. M., H. Richter, P. J. Schulte. 1991. Water relations. In: Raghavendra, A. S. *Physiology of Trees.* New York: Wiley, pp. 137–162.

Kramer, P. J. 1983. *Water Relations of Plants.* New York: Academic Press.

Milburn, J. A. 1991. Cavitation and embolisms in xylem conduits. In: Raghavendra, A. S. *Physiology of Trees.* New York: Wiley, pp. 162–174.

Nobel, P. S. 1991. *Physicochemical and Environmental Plant Physiology.* New York: Academic Press.

Waisel, Y., A. Eshel, U. Kafkafi. 1996. 2nd ed. *Plant Roots: The Hidden Half.* New York: M. Dekker.

Zimmermann, M. H. 1983. *Xylem Structure and the Ascent of Sap.* Berlin: Springer-Verlag.

REFERENCES

Brouwer, R. 1965. Water movement across the root. *Symposium of the Society for Experimental Biology.* 19:131–149.

Dittmer, H. J. 1937. A quantitative study of the roots and root hairs of a winter rye plant (Secale cereale). *American Journal of Botany* 24:417–420.

Dobbs, R. C., D. R. M. Scott. 1971. Distribution of diurnal fluctuations in stem circumference of Douglas fir. *Canadian Journal of Forest Research* 1:80–83.

Hall, N. S., W. F. Chandler, C. H. M. van Bravel, P. H. Ried, J. H. Anderson. 1953. A tracer technique to measure growth and activity of plant root systems. *N. C. Agriculture Experiment Station Technical Bulletin No. 101.* 40 pp.

Hayward, A. T. J. 1971. Negative pressure in liquids: can it be harnessed to serve man? *American Scientist* 59:434–443

Kavanagh, K. L., J. B. Zaer. 1997. Xylem cavitation and loss of hydraulic conductance in western hemlock following planting. *Tree Physiology* 17:59–63.

Kramer, P. J. 1983. *Water Relations of Plants.* New York: Academic Press.

Milburn, J. A., R. P. C. Johnson. 1966. The conduction of sap. II. Detection of vibrations produced by sap cavitation in Ricinus xylem. *Planta* 69:43–52.

Nobel, P. S. 1991. *Physicochemical and Environmental Plant Physiology.* New York: Academic Press.

Scholander, P. F., H. T. Hammel, E.D. Bradstreet, E. A. Hemmingsen. 1965. Sap pressure in vascular plants. *Science* 148:339–346.

Schulze, E.-D., J. Cermák, R. Matyssek, W. Penka, R. Zimmermann, F. Vasicek, W. Gries, J. Kučera. 1985. Canapoy transpiration and water fluxes in the xylem of Larix and Picea trees—A comparison of xylem flow, porometer and cuvette measurements. *Oecologia* 66:475–483.

Sperry, J. S., J. R. Donnelly, M. T. Tyree. 1988. Seasonal occurrence of xylem embolisms in sugar maple (*Acer saccharum*). *American Journal of Botany* 75:1212–1218.

Sperry, J. S., N. M. Holbrook, M. H. Zimmermann, M. T. Tyree. 1987. Spring filling of xylem vessels in wild grapevine. *Plant Physiology* 83:414–417.

Sperry, J. S., T. Ikeda. 1997. Xylem cavitation in roots and stems of Douglas-fir and white fir. *Tree Physiology* 17:275–280.

Tyree, M. T., J. S. Sperry. 1988. Do woody plants operate near the point of catastrophic xylem disfunction caused by dynamic water stress? *Plant Physiology* 88:574–580.

Tyree, M. T., J. S. Sperry. 1989. Vulnerability of xylem to cavitation and embolism. *Annual Review of Plant Physiology* 40:19–38.

Tyree, M. T., P. Yianoulis. 1980. The site of water evaporation from sub-stomatal cavities, liquid path resistances and hydroactive stomatal closure. *Annals of Botany* 46:175–193.

Weaver, J. E., R. W. Darland. 1949. Soil-root relationships of certain native grasses in various soil types. *Ecological Monographs* 19:303–338.

Zimmermann, M. H. 1983. *Xylem Structure and the Ascent of Sap.* Berlin: Springer-Verlag.

E. Epstein
(1972)

4

Plants and Inorganic Nutrients

Plants are **autotrophic** organisms. They live in an entirely inorganic environment, taking in CO_2 from the atmosphere and water and mineral nutrients from the soil. Animals, on the other hand, are **heterotrophic** organisms. They depend for their existence on energy-rich organic molecules previously synthesized by other organisms. Since plants stand at the bottom of the food chain, mineral nutrients assimilated by plants eventually find their way into the matter that makes up animals, including man. Whether autotrophic or heterotrophic, all organisms must constantly draw material substance from their environment in order to maintain their metabolism, growth, and development. The means for making these materials available to the organism is the subject of **nutrition.**

The nutritional needs of plants are traditionally discussed as two separate topics: organic nutrition and inorganic nutrition. The organic nutrition of plants focuses on the production of carbon compounds, specifically the incorporation of carbon, hydrogen, and oxygen via photosynthesis. Inorganic nutrition, on the other hand, is concerned with the acquisition of mineral elements from the soil. This distinction between organic and inorganic nutrition is more a matter of convenience than real. The subject of this and the next two chapters is the inorganic nutrition of plants, with the focus on acquisition of mineral elements and the role of those elements in plant metabolism. The acquisition and assimilation of carbon will be addressed later in Chapters 9–13.

Interest in plant nutrition is inextricably bound up with our interest in agriculture and crop productivity. Much of the groundwork for modern nutritional studies was laid in Europe in the early to mid-nineteenth century, in response to a combination of political and social factors (Steward, 1968). The Napoleonic wars had devastated Europe and the industrial revolution was gaining momentum. The traditional agricultural economy, relying heavily on organic manures, could no longer meet the demands of rising populations and massive migration to the cities. Greater efficiency in agriculture was required and this was not possible without a more thorough understanding of plant nutrition.

One of the first to make significant progress in the field of plant nutrition was N. T. de Saussure (1767–1845), who studied both photosynthesis and the absorption of nutrient elements with the same careful, quantitative methods. De Saussure conducted some of the first elemental analyses of plant material and introduced the concept that some, but not necessarily all, of the elements found might be indispensable (i.e., *essential*) to plant growth. De Saussure's ideas concerning the importance of elements derived from the soil generated considerable debate at the time, but received support from the work of C. S. Sprengel (1787–1859), working in Germany, and Jean-Baptiste Boussingault in

France (Epstein, 1972). Sprengel introduced the idea that soils might be unproductive if deficient in but one single element necessary for plant growth and Boussingault stressed quantitative relationships between the effects of fertilizer and nutrient uptake on crop yields. Boussingault is also credited with providing the first evidence that legumes had the unique capacity to assimilate atmospheric nitrogen, a finding that was later confirmed by the discovery of the nitrogen-fixing role of bacteria in root nodules.

By the middle of the nineteenth century, many pieces of the nutritional puzzle were beginning to fall into place. In 1860, Julius Sachs, a prominent German botanist, demonstrated for the first time that plants could be grown to maturity in defined nutrient solutions in the complete absence of soil. Finally, by 1850, J. B. Lawes and J. H. Gilbert, working at Rothamsted in England, had successfully converted insoluble rock phosphate to soluble phosphate (called *superphosphate*) and by the end of the century the use of nitrogen, phosphorous, and potassium (N–P–K) fertilizers in agriculture was well established in Europe.

This chapter will examine the nutritional requirements of plants that are satisfied by mineral elements. This will include

- methods employed in the study of mineral nutrition;
- the concept of essential and beneficial elements and the distinction between macronutrients and micronutrients;
- a general discussion of the metabolic roles of the fourteen essential mineral elements, the concept of critical and deficient concentration, and symptoms that are associated with deficiencies of the mineral elements; and
- a brief discussion of micronutrient toxicity.

In Chapter 5, we will discuss soils as a nutrient reservoir and examine how minerals are taken up and distributed throughout the plant. The unique situation with respect to uptake and metabolism of nitrogen will be addressed in Chapter 6.

STUDYING THE MINERAL REQUIREMENTS OF PLANTS

Methods for studying the nutritional requirements of plants usually involve the culture of plants in a soil-free, defined mineral solution. These methods require careful purification of water and nutrient salts and other procedures to avoid contamination.

In the mid-nineteenth century, J. Sachs, with several of his contemporaries, was interested in determining the minimal nutrient requirements of plants. Recognizing that it would be difficult to pursue such studies in a medium as complex as soil, Sachs devised an experimental system such that the roots grew not in soil but in an aqueous solution of mineral salts. With this simplified system, Sachs was able to demonstrate the growth of plants to maturity on a relatively simple nutrient solution containing six inorganic salts (Table 4.1). Variations on Sachs' system, known as **solution culture** or **hydroponics** (growing plants in a defined nutrient solution), have remained to this day the principal experimental system for study of plant nutrient requirements. Hydroponics has also enjoyed some success in the commercial production of vegetables such as lettuce, tomato, and seedless cucumber.

TABLE 4.1 The composition of Sachs' nutrient solution (1860) used for solution culture of plants.

Salt	Formula	Approximate Concentration (mM)
Potassium nitrate	KNO_3	9.9
Calcium phosphate	$Ca_3(PO_4)_2$	1.6
Magnesium sulfate	$MgSO_4{\cdot}7H_2O$	2.0
Calcium sulfate	$CaSO_4$	3.7
Sodium chloride	NaCl	4.3
Iron sulfate	$FeSO_4$	trace

In the simplest form of solution culture, a seedling is supported in the lid of a container, with its roots free to grow in the nutrient solution (Fig. 4.1). Note that the solution must be aerated in order to obtain optimal growth. A solution that is not aerated becomes depleted of oxygen, a condition known as **anoxia.** Anoxia inhibits

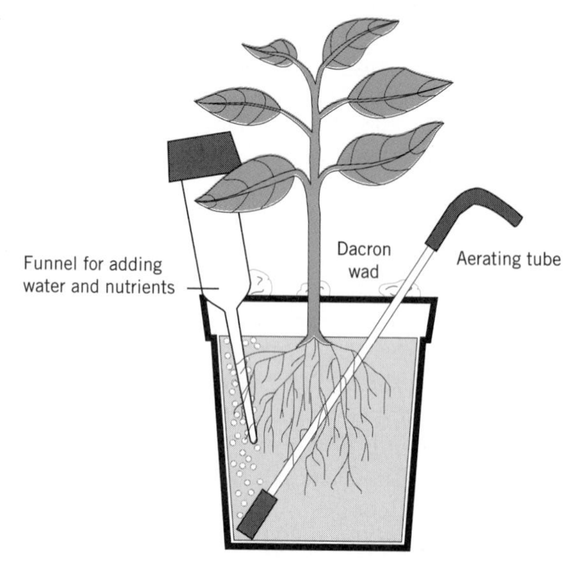

FIGURE 4.1 Diagram of a typical setup for nutrient solution culture. (From Epstein, 1972. Reprinted by permission.)

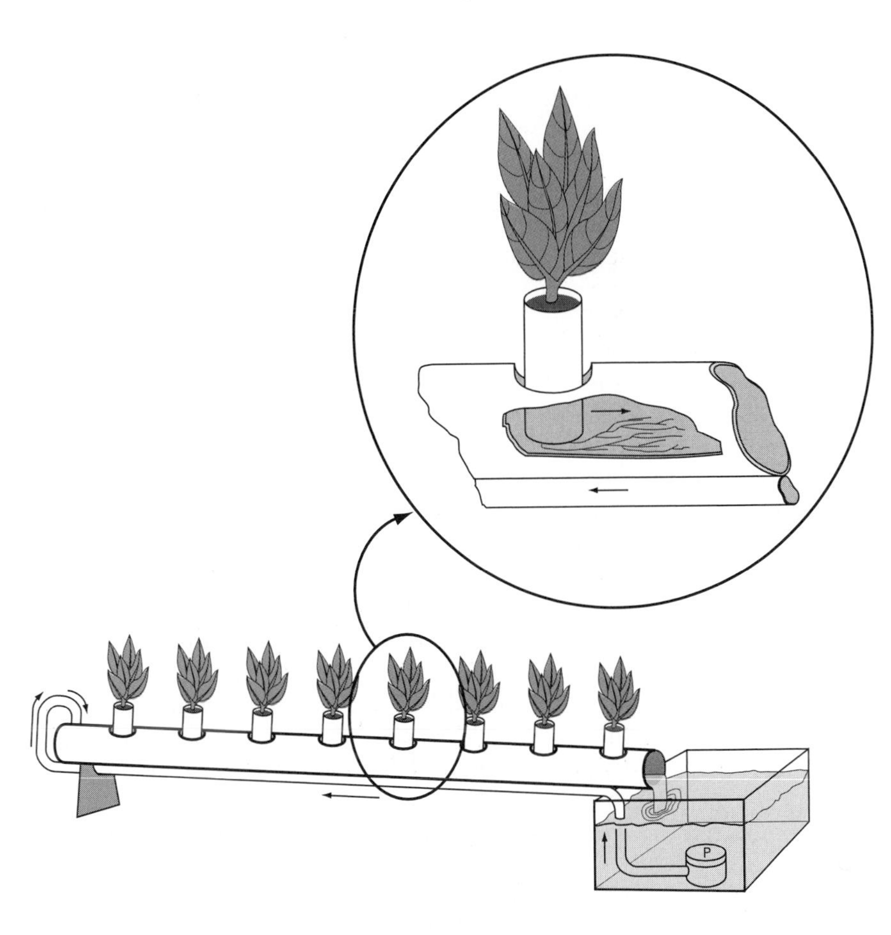

FIGURE 4.2 The nutrient film technique for hydroponic plant production. Plants are grown in a tube or trough placed on a slight incline. A pump (P) circulates nutrient solution from a reservoir to the elevated end of the tube. The solution then flows down the tube by gravity, returning to the reservoir. Inset: the roots grow along the bottom of the tube, bathed continuously in a thin film of aerated nutrient solution. Arrows indicate the direction of nutrient flow.

the respiration of root cells and reduces nutrient uptake. The container in which the plants are grown is usually painted black or wrapped with an opaque material in order to keep out light. The purpose of excluding light is to reduce the growth of algae that would compete with the plants for nutrients.

There are disadvantages to the solution culture technique. Selective ion depletion and associated changes in the pH of the solution as the roots take up nutrients are the major problems. Plants maintained in pure solution culture will continue to grow vigorously only if the nutrient solution is replenished on a regular basis. In order to avoid such problems, some investigators grow the plants in a *nonnutritive* medium such as acid-washed quartz sand, *perlite*, or *vermiculite*.[1] Plants can then be watered by daily application of fresh nutrient solution from the top of the medium (a technique called *slop culture*) or by slowly dripping onto the culture from a reservoir (*drip culture*). Alternatively, the nutrient solution can be *subirrigated*. Here the nutrient solution is pumped into the culture from below and then allowed to drain out. This fill-and-empty process is repeated on a regular basis and serves both to replenish the nutrient solution and to **aerate** the roots. Most commercial hydroponic operations now utilize some variation of the **nutrient film** technique in which the roots are continuously bathed with a thin film of recirculating nutrient solution (Fig 4.2).

These methods overcome some of the problems inherent in pure solution culture, but may not be suitable for many laboratory experiments. This is because no medium is truly nonnutritive. Any medium, even the glass, plastic, or ceramic containers used in pure solution culture, may provide some nutrients at very low levels. For example, soft (sodium silicate) glass provides sodium, hard (borosilicate) glasses provide boron, and plastics might provide chloride or fluoride, and so forth. Water used to prepare nutrient solutions must be carefully distilled, avoiding, wherever possible, metallic components in the distillation apparatus.

The nutrient solution devised by Sachs contributed a total of ten mineral nutrients.[2] It was at least another

[1]Vermiculite is a silicate mineral of the mica family. It expands on heating to produce a lightweight product that has high water retention and is commonly used as a mulch in seed beds. Perlite is a coarsely ground glassy volcanic rock. Both vermiculite and perlite provide virtually no nutrients for plants.

[2]Carbon, oxygen, and hydrogen are excluded from this total. They are provided in the form of carbon dioxide and water and are not considered mineral elements.

half century before others demonstrated the need for additional mineral nutrients. There was no magic to the success of Sachs' experiments. Many of the mineral nutrients used by plants are required in very low amounts and Sachs unknowingly provided these nutrients as impurities in the salts and water he used to make up his nutrient solution. Analytical techniques have now improved to the point where it is possible to detect mineral contents several orders of magnitude lower than was possible in Sachs' time. Most mineral elements are now measured by either **atomic absorption spectrometry** or **atomic emission spectrometry.** These techniques involve vaporization of the elements at temperatures in excess of several thousand degrees. In the vaporous state, the element will (depending on the temperature) either absorb or emit light at very narrow wavelength bands (see Chap. 7). The wavelength of light absorbed or emitted is characteristic of a particular element and the quantity of absorbed or emitted energy is proportional to the concentration of the element in the sample. In this way, concentrations as low as 10^{-8} g ml^{-1} for some elements can be measured in samples of plant tissue, soil, or nutrient solutions within a few minutes.

THE ESSENTIAL NUTRIENT ELEMENTS

Most plants require a relatively small number of nutrient elements in order to successfully complete their life cycle.

Nutrient elements that are required for the growth and development of plants are deemed to be **essential.** Essentiality is based primarily on two criteria formulated by E. Epstein in 1972. According to Epstein, an element is considered essential if (a) *in its absence the plant is unable to complete a normal life cycle*, or (b) *that element is part of some essential plant constituent or metabolite.* By the first criterion, if a plant is unable to produce viable seed when deprived of that element, the element is deemed essential. By the second criterion, an element such as magnesium would be considered essential because it is a constituent of the chlorophyll molecule and chlorophyll is essential for photosynthesis. Similarly, chlorine is essential because it is a necessary factor in the photosynthetic oxidation of water. Most elements satisfy both of the two criteria, although either one alone is usually considered sufficient. It is generally agreed, based on these criteria, that only 17 elements are essential for the growth of all higher plants (Table 4.2).

The essential elements are traditionally segregated into two categories: (a) the so-called **macronutrients** and (b) the **trace elements** or **micronutrients.** Such a distinction is more convenient than real, since it simply reflects the relative concentrations found in tissue or required in nutrient solutions (Table 4.2). Any distinction between macro- and micronutrients should not be interpreted as assigning different levels of importance relative to the nutritional needs of the plant. The first nine elements in the list are called macronutrients because they are required in large amounts (in excess of 10 mmole kg^{-1} of dry weight). The macronutrients are largely, but not exclusively, involved in the structure of molecules, which to some extent accounts for their need in large quantities. The remaining eight essential elements are the micronutrients. Micronutrients are required in relatively small quantities (less than 30 mmole kg^{-1} of dry weight) and serve catalytic and regulatory roles such as enzyme activators. Some macronutrients, calcium and magnesium, for example, serve as regulators in addition to their structural role.

Although the criteria for essentiality are quite clear, it is not always easy to demonstrate that an element is or is not essential. D. Arnon and P. Stout had earlier suggested a third criterion: In order to be considered essential the element must act directly in the metabolism of the plant and not simply to correct an unfavorable microbial or chemical condition in the nutrient medium (Arnon and Stout, 1939). The use of solution cultures from which the element in question has been omitted has largely circumvented the need to apply this third criterium. On the other hand, some plants may form viable seeds even though a particular element has been excluded from the nutrient solution and other symptoms of deficiency are evident. In such cases there may be present in the seed, or contaminating the nutrient solution, a quantity of the element sufficient to moderate the deficiency and allow seed formation. It is assumed that in the complete absence, the deficiency symptoms would be severe enough to kill the plant before viable seed could be formed. If required, this can be confirmed by careful purification of nutrient salts and exclusion of atmospheric contaminants. Where a sufficient quantity of the element may be carried within the seed, essentiality can be confirmed by growing several successive generations from seed which was itself produced in the absence of that element. This is usually sufficient to reduce the concentration of that element in the seed to the deficient range.

The essentiality of micronutrients is particularly difficult to establish because they are required in such small quantities. Most micronutrient requirements are fully satisfied by concentrations in the range of 0.1 to 1.0 $\mu g\ L^{-1}$—amounts readily obtained from impurities in water or macronutrient salts, the containers in which the plants are grown, and contamination by atmospheric dust. Since elements may be required at concentrations below detectable limits, it is far easier to establish that a micronutrient *is* essential than that it *is not.* In 1866, for example, H. Birner and B. Lucanus were able to grow oats successfully in solution culture containing only calcium, nitrogen, potassium, phosphorous, mag-

TABLE 4.2 The essential nutrient elements of higher plants and their concentrations considered adequate for normal growth.

Element	Chemical Symbol	Available Form	Concentration in Dry Matter (mmol/kg)
Macronutrients			
Hydrogen	H	H_2O	60,000
Carbon	C	CO_2	40,000
Oxygen	O	O_2, CO_2	30,000
Nitrogen	N	NO_3^-, NH_4^+	1,000
Potassium	K	K^+	250
Calcium	Ca	Ca^{2+}	125
Magnesium	Mg	Mg^{2+}	80
Phosphorous	P	HPO_4^-, HPO_4^{2-}	60
Sulfur	S	SO_4^{2-}	30
Micronutrients			
Chlorine	Cl	Cl^-	3.0
Boron	B	BO_3^{3-}	2.0
Iron	Fe	Fe^{2+}, Fe^{3+}	2.0
Manganese	Mn	Mn^{2+}	1.0
Zinc	Zn	Zn^{2+}	0.3
Copper	Cu	Cu^{2+}	0.1
Nickel	Ni	Ni^{2+}	0.05
Molybdenum	Mo	Mo_4^{2-}	0.001

nesium, sulfur, and trace amounts of iron (plus the carbon, hydrogen and oxygen supplied by CO_2 and water). The plants grew normally because sufficient quantities of micronutrients were available as impurities in the four salts used to make up their nutrient solution.

As micronutrients go, iron is usually supplied at relatively high concentrations. This is necessary because availability of iron is very sensitive to pH and other soil conditions. At high pH, iron forms insoluble iron hydroxides and calcium complexes and in acidic solution it reacts with aluminum. In both cases iron readily precipitates out of solution and, consequently, is frequently deficient in natural situations. For these reasons, the need for iron as an essential plant nutrient was thus established early. The need for other micronutrients, however, was not recognized until salts of sufficient purity became available in the early part of this century. In 1922, J. S. McHargue demonstrated that the disorder known as *grey speck of oats* was caused by a **manganese** deficiency and the following year Katherine Warington showed that **boron** was required for several legume species. By 1939, the need for **zinc, copper,** and **molybdenum** had also been clearly established. In each case, the nutrient deficiency was found to cause a well-known disorder previously thought to be a disease. **Chlorine** was not added to the list until 1954, although its essential nature was suggested nearly one hundred years earlier (Hewitt, 1963). The need for chlorine became evident in the course of experiments to determine whether cobalt was required for tomato (*Lycopersicum esculentum*). T. C. Broyer and his coworkers had purified their nutrient salts by methods which removed not only cobalt but halides as well (Broyer et al., 1954). Plants grown in solutions prepared from these purified salts developed browning and necrosis of the leaves. The symptoms could be avoided by supplementing the nutrient solution with *cobalt chloride.* Subsequent investigation, however, established that it was the deficiency of chloride rather than cobalt that gave rise to the symptoms.

There is now mounting evidence that **nickel** should be added to the list of essential elements (Dalton et al., 1988). Nickel is an essential component of urease, an enzyme widely distributed in plants, microorganisms, and some marine invertebrates. Urease is an enzyme that catalyses the hydrolysis of urea into NH_3 and CO_2 and is thought to play an important role in mobilization of nitrogenous compound in plants. In 1987, P. H. Brown and colleagues showed that nickel depletion led to the formation of nonviable seed in barley (*Hordeum vulgare*). The addition of nickel would bring to 17 the total number of nutrient elements essential for higher plants.

TABLE 4.3 The composition of a typical one-half strength "modified" Hoagland's nutrient solution, showing the nutrient salts used and their approximate millimolar (mM) concentrations.

		Concentration (mM)
Calcium nitrate	$Ca(NO_3)_2$	2.5
Potassium phosphate	KH_2PO_4	0.5
Potassium nitrate	KNO_3	2.5
Magnesium sulfate	$MgSO_4$	1.0
Zinc sulfate	$ZnSO_4$	0.00039
Manganous sulfate	$MnSO_4$	0.0046
Copper sulfate	$CuSO_4$	0.00016
Boric acid	H_3BO_3	0.0234
Molybdic acid	MoO_3	0.000051
Iron sequestrene	Fe	0.179

From Downs and Hellmers, 1975.

Aside from the commercial applications of hydroponic plant culture, a great deal of plant physiology and other botanical research is conducted with plants grown under controlled environments. This may include relatively simple greenhouses or complex growth rooms in which temperature, lighting, and other environmental conditions are carefully regulated (Downs and Hellmers, 1975). Plant nutrient supply is one of those factors that must be regulated and over the years a large number of nutrient solutions have been formulated for this purpose. Most modern formulations are based on a solution originally developed by D. R. Hoagland, a pioneer in the study of plant mineral nutrition. Individual investigators may introduce minor modifications to the composition of the nutrient solution in order to accommodate specific needs. Such formulations are commonly referred to as *modified Hoagland's solutions* (Table 4.3, 4.4).

TABLE 4.4 The quantity of each nutrient element in modified Hoagland's nutrient solution.

Element	mg/L
Calcium	103
Nitrogen	105
Potassium	118
Sulfur	33
Magnesium	25
Phosphorous	15
Iron	10
Boron	0.25
Manganese	0.25
Zinc	0.025
Copper	0.01
Molybdenum	0.0052

The concentration of minerals in nutrient solutions is many times greater than that normally found in soils (Downs and Hellmers, 1975). An excess is necessary in order to maintain a continual supply of nutrients as they are taken up by the roots. The nutrient concentration of the soil solution, on the other hand, is relatively low but is continually replenished by nutrients adsorbed on the soil **particles** (Chap. 5).

BENEFICIAL ELEMENTS

Beneficial elements satisfy specific additional nutrient requirements for particular plants.

In addition to the 17 essential elements listed in Table 4.1, some plants appear to have additional requirements. However, because these have not been shown to be a requirement of higher plants generally, they are excluded from the list of essential elements. They are referred to instead as **beneficial elements.** If these elements are essential to all plants, they are required by most at concentrations well below what can be reliably detected by present analytical techniques. The definition of beneficial currently applies primarily to sodium, silicon, selenium, and cobalt. With time, and as experimental methods improve, one or more of these beneficial elements may be added to the list of essential elements.

SODIUM

A **sodium** requirement was first demonstrated for the bladder salt-bush (*Atriplex vesicaria*), a perennial pasture species of arid inland areas of Australia (Brownell and Wood, 1957). By carefully purifying the water, recrystallizing the nutrient salts, and using sodium-free vessels, Brownell and Wood were able to reduce the sodium content of the final culture medium to less than 1.6 $\mu g\ L^{-1}$. Plants grown in the depleted solution showed reduced growth, **chlorosis** (yellowing), and **necrosis** (dead tissue) of the leaves. Brownell and Crossland (1972) have since then surveyed 32 species of plants for their sodium requirement and have concluded that sodium is generally essential as a micronutrient for plants having one particular photosynthetic pathway (called the C4 pathway), but not for most plants which use what is known as the C3 pathway (see Chap. 10). The metabolic basis for the sodium requirement has not yet been resolved, but it may be related to the transport of pyruvate, a critical intermediate in the C4 pathway, between the bundle-sheath and mesophyll cells (Chap. 1, 10).

SILICON

Given the high content of silicon dioxide in normal soils, it should not be surprising that many plants take up appreciable quantities of silicon. Silicon may comprise 1 to 2 percent of the dry matter of maize (*Zea mays*) and other grasses and as much as 16 percent of the scouring rush (*Equisetum arvense*), yet experiments have generally failed to demonstrate that silicon is essential for most other plants. The ubiquitous presence of silicon in glass, nutrient salts, and atmospheric dust makes it especially difficult to exclude silicon from nutrient experiments. However, there are numerous reports of beneficial effects of silicon in a variety of species. Silicon seems to be particularly beneficial to grasses, where it accumulates in the cell walls, especially of epidermal cells, and may play a role in fending off fungal infections or preventing **lodging** (the condition in which stems are bent over by heavy winds or rain).[3]

COBALT

Cobalt is essential for the growth of legumes which are host to symbiotic nitrogen-fixing bacteria (Chap. 6). In this case, the requirement can be traced to the needs of the nitrogen-fixing bacterium rather than the host plant. A similar cobalt requirement has been demonstrated for the free-living, nitrogen-fixing bacteria, including the cyanobacteria. In addition, when legumes are provided with fixed nitrogen such as nitrate, a cobalt requirement cannot be demonstrated.

SELENIUM

Selenium salts tend to accumulate in poorly drained, arid regions of the western plains of North America. Although selenium is generally toxic to most plants, certain members of the legume genus *Astragalus* (milk-vetch or poison-vetch) are known to tolerate high concentrations of selenium (up to 0.5 percent dry weight) and are found only on seleniferous soils. Such concentrations of selenium would be toxic to most other plants. At one time it was thought that selenium might be essential to these "accumulator species," but there is no definitive supporting evidence. Selenium accumulators are of considerable importance to ranchers, however, as they are among a diverse group of plants known as "loco weeds." The high selenium content in these plants causes a sickness known as alkali poisoning or blind-staggers in grazing animals.

[3]Grazing animals appear to have adapted to the high silicon content of grasses. The teeth of grazing animals (such as cows and horses) grow continuously, compensating for the wear caused largely by silicon. On the other hand, the teeth of browsing animals such as deer, whose diets contain little grass, do not continue to grow.

NUTRIENT ROLES AND DEFICIENCY SYMPTOMS

The essential elements have specific metabolic roles in plants. When they are absent, plants will exhibit characteristic deficiency symptoms that, in most cases, are related to one or more of those roles.

Some students of plant mineral nutrition prefer to classify the macro- and micronutrients along functional lines. For example, elements such as carbon, hydrogen, and oxygen have a predominantly structural role—they are the stuff of which molecules are made—while others appear to be predominantly involved in regulatory roles, such as maintaining ion balance and activating enzymes. Other investigators have proposed more complicated schemes with up to four categories of biochemical function. Unfortunately, any attempt to categorize the nutrient elements in this way runs into difficulty because the same element often fills both structural and nonstructural roles. Magnesium, for example, is an essential component of the chlorophyll molecule but also serves as a cofactor for many enzymes, including those involved in critical energy transfer reactions (Chaps. 9, 10, 12). Calcium is an important constituent of cell walls where its role is largely structural, but it is also implicated as a second messenger in hormone (Chap. 16) and phytochrome responses (Chap. 18). Regardless of how they are classified, it is clear that these elements are essential because they satisfy specific metabolic requirements of the plant. When those requirements are not met or are only partially met, the plant will exhibit characteristic deficiency symptoms that, if severe enough, result in death.

The need of individual plants for any particular element is normally defined in terms of **critical concentration.** This is the concentration of that nutrient, *measured in the tissue*, just below the level which gives maximum growth (Epstein, 1972). This concept is illustrated in Figure 4.3. At concentrations above the critical concentration, additional increments in nutrient content have no particular effect on growth and the nutrient content is said to be **adequate.** Below the critical concentration, growth falls off sharply as the nutrient content becomes **deficient.** In other words, at tissue nutrient levels below the critical concentration, that nutrient is *limiting to growth.*

When nutrient levels exceed the critical concentration, that nutrient is, with one qualification, no longer limiting. The qualification is that at sufficiently high tissue levels, virtually all nutrients become toxic. Toxic levels are seldom achieved with the macronutrients, but are common in the case of the micronutrients. Normal con-

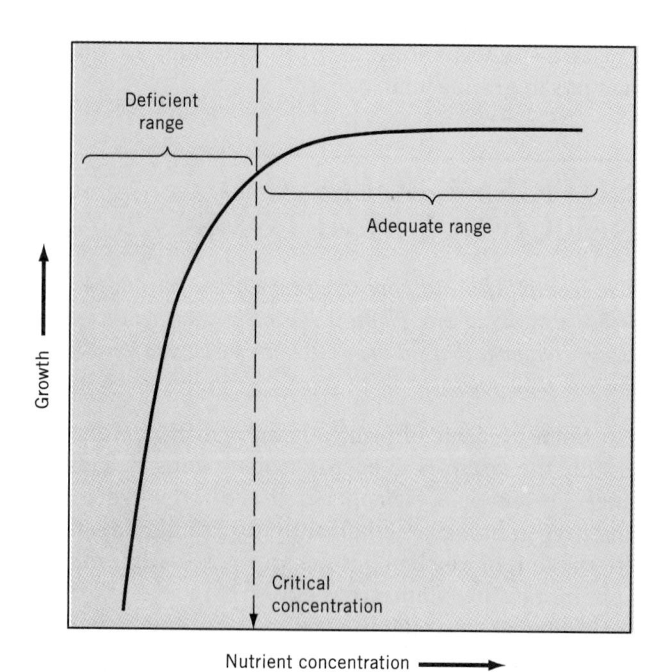

FIGURE 4.3 Generalized plot of growth as a function of nutrient concentration in tissue. The critical concentration is that concentration giving a 10 percent reduction in growth.

FIGURE 4.4 Nutrient deficiency symptoms in tomato (*Lysopersicum esculentum*). From left: Complete nutrient solution (control); minus P; minus Ca; minus Fe; minus N. (From a student experiment.) (See color plate 1.)

centrations of copper, for example, are in the range of 4 to 15 $\mu g\ g^{-1}$ of tissue dry weight (dwt). Deficiency occurs at concentrations below 4 μg but most plants are severely damaged by concentrations in excess of 20 μ g^{-1} dwt. Boron is toxic above 75 $\mu g\ g^{-1}$ dwt and zinc above 200 $\mu g\ g^{-1}$ dwt.

Since each element has one or more specific structural or functional roles in the plant, in the absence of that element the plant will be expected to exhibit certain morphological or biochemical symptoms of that deficiency (Fig 4.4). In some cases the deficiency symptoms will clearly reflect the functional role of that element. One example is yellowing, or, chlorosis, in the absence of magnesium. Chlorosis occurs because, in the absence of magnesium, the plant is unable to synthesize the green pigment chlorophyll. In other cases, the relationship between deficiency symptoms and functional role of the element may not always be so straightforward. Moreover, deficiency symptoms for some elements are not always consistent between one plant and the next. Nonetheless, for each element there are certain generalizations that can be made with respect to deficiency symptoms. Deficiency symptoms also depend in part on the mobility of the element in the plant. Where elements are mobilized within the plant and exported to young developing tissues, deficiency symptoms tend to appear first in older tissues. Other elements are immobile—once located in a tissue they are not readily mobilized for use elsewhere. In this case, the symptoms tend to appear first in the younger tissues.

In this section we will review the functional roles of the essential elements and, in general terms, morphological and biochemical abnormalities that result from their deficiencies. Carbon, hydrogen, and oxygen will be excluded from this discussion as they are required for the structural backbone of all organic molecules. A deficiency of carbon, consequently, leads quickly to starvation of the plant, while a deficiency of water leads to desiccation. Instead, our discussion will be limited to deficiencies of those essential elements taken in from the soil solution. Remember that deficiency means only that the concentration is less than the critical concentration; it does not necessarily mean total absence of the element.

Although we will discuss each element individually, the reader is reminded that elements are capable of complex interactions with each other and with soil conditions. These interactions can significantly influence symptoms of deficiency under natural conditions.

NITROGEN

Although the atmosphere is approximately 80 percent nitrogen, only certain prokaryote species—bacteria and cyanobacteria—can utilize gaseous nitrogen directly. The special problems of nitrogen availability and metabolism will be addressed in Chapter 6. For the present discussion, it is sufficient to note that most plants absorb nitrogen from the soil solution primarily as inorganic nitrate ion (NO_3^-) and, in a few cases, as ammonium (NH_4^+) ion. Once in the plant, NO_3^- is reduced to NH_4^+ before incorporation into amino acids, proteins, and other nitrogenous organic molecules (see Chap. 6). Nitrogen is most often limiting in agricultural situations—many plants, such as maize (*Zea mays*) are known as "heavy feeders" and require heavy applications of nitrogen fertilizer. The manufacture and distribution of nitrogen fertilizers for agriculture is, in both energy and financial terms, an extremely costly process (Chap. 6).

Nitrogen is a constituent of many important molecules, including proteins, nucleic acids, certain hormones (e.g., indole-3-acetic acid; cytokinin), and chlorophyll. It should not be surprising, then, that the most overt symptoms of nitrogen deficiency are a slow, stunted growth and a general chlorosis of the leaves. Nitrogen is very mobile in the plant. As the older leaves yellow and die, the nitrogen, largely in the form of soluble amines and amides, is mobilized from the older leaves and exported to the younger, more rapidly developing leaves. Thus the symptoms of nitrogen deficiency generally appear first in the older leaves and do not occur in the younger leaves until the deficiency becomes severe. At this point, the older leaves will turn completely yellow or tan and fall off the plant. Conditions of nitrogen stress also lead to an accumulation of anthocyanin pigments in many species, contributing a purplish color to the stems, petioles, and the underside of leaves. The precise cause of anthocyanin accumulation in nitrogen-starved plants is not known. It may be related to an overproduction of carbon structures that, in the absence of nitrogen, cannot be utilized to make amino acids and other nitrogen-containing compounds.

Excess nitrogen stimulates abundant growth of the shoot system, favoring a high shoot/root ratio, and will often delay the onset of flowering in agricultural crops. Similarly, a deficiency of nitrogen stimulates early flowering.

PHOSPHOROUS

Phosphorous is available in the soil solution primarily as forms of the polyprotic phosphoric acid (H_3PO_4). A polyprotic acid contains more than one proton, each with a different dissociation constant. Soil pH thus assumes a major role in the availability of phosphorous. At a soil pH less than 6.8, the predominant form of phosphorous is the monovalent orthophosphate anion ($H_2PO_4^-$). Orthophosphate is readily absorbed by plant roots. Between pH 6.8 and pH 7.2, the predominant form is HPO_4^{2-}, which is less readily absorbed. In alkaline soils (pH greater than 7.2), the predominant form is the trivalent HPO_4^{3-}, which is virtually unavailable for uptake by plants. The actual concentration of soluble phosphorous in most soils is relatively low, due to several factors. At neutral pH, phosphorous tends to form insoluble complexes with aluminum and iron, while in basic soils calcium and magnesium complexes will precipitate the phosphorous. Because insoluble phosphates are only very slowly released into the soil solution, phosphorous is always limited in highly calcarious soils.

Substantial amounts of phosphorous may also be bound up in organic forms, which are not available for uptake by plants. Organic phosphorous must first be converted to an inorganic form by the action of soil microorganisms before it is available for uptake by plants. In addition, plants must compete with the soil microflora for the small amounts of phosphorous which are available. For these sorts of reasons, phosphorous rather than nitrogen is often the limiting element in natural ecosystems. One of the apparent advantages of mycorrhizal associations, discussed in Chapter 5, is that the fungus enhances the uptake of phosphorous.

In the plant, phosphorous is found largely as phosphate esters—including the sugar-phosphates, which play such an important role in photosynthesis and intermediary metabolism. Other important phosphate esters are the nucleotides that make up DNA and RNA as well as the phospholipids present in membranes. Phosphorous in the form of ATP, ADP and P_i, phosphorylated sugars, and phosphorylated organic acids also plays an integral role in the energy metabolism of cells.

The most characteristic manifestation of phosphorous deficiency is an intense green coloration of the leaves. In the extreme, the leaves may become malformed and exhibit necrotic spots. In some cases, anthocyanins also accumulate, giving the leaves a dark greenish-purple color. Like nitrogen, phosphorous is readily mobilized and redistributed in the plant, leading to the rapid senescence and death of the older leaves. The stems are usually shortened and slender and the yield of fruits and seeds is markedly reduced.

An excess of phosphorous has the opposite effect of nitrogen in that it preferentially stimulates growth of roots over shoots, thus reducing the shoot/root ratio. Fertilizers with a high phosphorous content, such as bone meal, are often applied when transplanting perennial plants in order to encourage establishment of a strong root system.

POTASSIUM

Potassium is available as the monovalent cation, K^+. In agricultural practice, potassium is usually provided as potash (potassium carbonate, K_2CO_3). Potassium is required in large amounts by most plants and is frequently deficient in sandy soils because of its high solubility and the ease with which K^+ leaches out of sandy soils (Chap. 5). Potassium ion serves to activate a number of enzymes, notably those involved in photosynthesis and respiration. Starch and protein synthesis are also affected by potassium deficiency. Potassium, like sodium in animal tissues, also serves an important function in osmoregulation. Movement of potassium ion is a principal factor in plant movements, such as opening and closure of stomatal guard cells (Chap. 8) and the sleep movements, or daily changes in the orientation of leaves (Chap. 18).

Unlike other macronutrients, potassium does not appear to be structurally bound in the plant, but like nitrogen and phosphorous, is highly mobile. Deficiency symptoms first appear in older leaves, which characteristically develop mottling or chlorosis, followed by necrotic lesions (spots of dead tissue), at the leaf margins. In monocotyledonous plants, especially maize and other cereals, the necrotic lesions begin at the older tips of the leaves and gradually progress along the margins to the younger cells near the leaf base. Stems are shortened and weakened and susceptibility to root-rotting fungi is increased. The result is that potassium-deficient plants are easily lodged.

SULFUR

There are several forms of sulfur in most soils, including iron sulfides and elemental sulfur. It is taken up by plants, however, as the divalent sulfate anion (SO_4^{2-}). Sulfur deficiency is not a common problem, since there are numerous microorganisms capable of oxidizing sulfides or decomposing organic sulfur compounds. In addition, heavy consumption of fossil fuels in industry as well as natural phenomena such as geysers, hot sulfur springs, and volcanos together contribute large amounts of sulfur oxides (SO_2 and SO_3) to the atmosphere. Indeed it is often difficult to demonstrate sulfur deficiencies in greenhouses in industrial areas because of the high concentrations of airborne sulfur.

Sulfur is particularly important in the structure of proteins where disulphide bonds between neighboring cysteine and methionine residues contribute to the tertiary structure. Sulfur is also a constituent of the vitamins thiamine and biotin and of coenzyme A, an important component in respiration and fatty acid metabolism. In the form of iron-sulfur proteins it is important in electron transfer reactions of photosynthesis and nitrogen fixation. The sulfur-containing **thiocyanates** and **isothiocyanates** (mustard oils) are responsible for the pungent flavors of mustards, cabbages, turnips, horseradish and other plants of the family Brassicaceae (formerly Cruciferae, or "crucifers") (Chap. 14). Because of the presence of mustard oils, many species of crucifers prove fatal to livestock that graze on them. Mustard oils also appear to serve as a defense against insect herbivory of crucifers.

Sulfur deficiency, like nitrogen, results in a generalized chlorosis of the leaf, including the tissues surrounding the vascular bundles. This is due to reduced protein synthesis rather than a direct impairment of chlorophyll synthesis. However, chlorophyll is stabilized by binding to protein in the chloroplast membranes. With impaired protein synthesis, the ability to form stable chlorophyll-protein complexes is also impaired. Unlike nitrogen, however, sulfur is not readily mobilized in most species and the symptoms tend to occur initially in the younger leaves.

CALCIUM

Calcium is taken up as the divalent cation (Ca^{2+}). It is abundant in most soils and seldom deficient under natural conditions. Calcium is important to dividing cells for two reasons. It plays a role in the mitotic spindle during cell division and it forms calcium pectates in the middle lamella of the cell plate that forms between daughter cells. It is also required for the physical integrity and normal functioning of membranes and, more recently, has been implicated as a second messenger in certain hormonal and environmental responses. As a second messenger involved in protein phosphorylation, Ca^{2+} may be an important factor in regulating the activities of a number of enzymes.

Because of its role in dividing cells, calcium deficiency symptoms characteristically appear in the meristematic regions where cell division is occurring and new cell walls are being laid down. Young leaves are typically deformed and necrotic and, in extreme cases, death of the meristem ensues. In solution cultures, calcium deficiency results in poor root growth. The roots are discolored and may feel "slippery" to the touch because of the deterioration of the middle lamella. Calcium is relatively immobile and the symptoms typically appear in the youngest tissues first.

MAGNESIUM

Like calcium, magnesium is also taken up as the divalent cation (Mg^{2+}). Magnesium is generally less abundant in soils than calcium but is required by plants in relatively large amounts. Magnesium deficiencies are most likely in strongly acid, sandy soils. Magnesium has several important functions in the plant. By far the largest proportion is found in the porphyrin moiety of the chlorophyll molecule but it is also required to stabilize ribosome structure and is involved as an activator for numerous critical enzymes. It is critical to reactions involving ATP, where it serves to link the ATP molecule to the active site of the enzyme. Mg^{2+} is also an activator for both ribulosebisphosphate carboxylase and phosphoenolpyruvate carboxylase, two critical enzymes in photosynthetic carbon fixation (Chap. 10).

The first and most pronounced symptom of magnesium deficiency is a chlorosis due to a breakdown of chlorophyll in the regions of the leaf that lie between the veins (the interveinal regions). Chloroplasts in the region of the veins are for some reason less susceptible to magnesium deficiency and retain their chlorophyll much longer. Magnesium is also quite mobile and readily withdrawn from the older leaves. Consequently, chlorosis due to Mg^{2+} deficiency is most pronounced in the older leaves.

IRON

Of all the micronutrients, iron is required by plants in the largest amounts (it is considered a macronutrient by

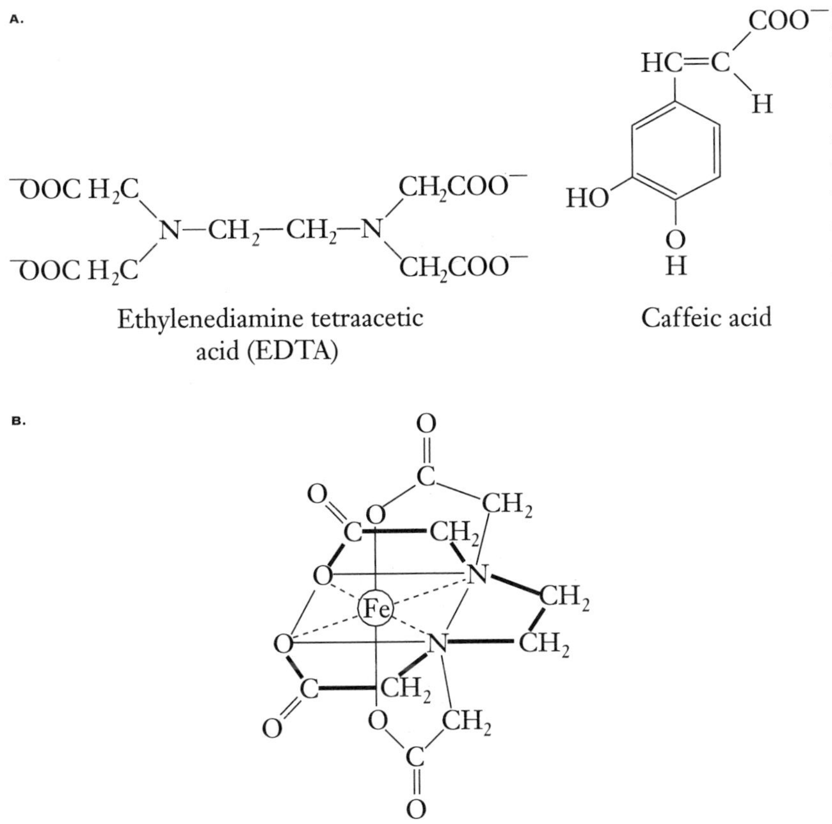

A.

$^{-}OOC\,H_2C$, CH_2COO^{-}, $N{-}CH_2{-}CH_2{-}N$, $^{-}OOC\,H_2C$, CH_2COO^{-}

Ethylenediamine tetraacetic acid (EDTA)

COO^{-}, $HC{=}C$, H, HO, O, H

Caffeic acid

B.

O, C, CH_2, O, O, C, CH_2, O, N, CH_2, Fe, O, N, CH_2, C, CH_2, O, O, CH_2, C, O

FIGURE 4.5 Chelating agents. (*A*) Examples of organic acids that function as chelating agents. Ethylenediamine-tetraacetic acid (EDTA) is a synthetic acid in common commercial use. Complexed with iron, it is sold under the tradename Versenate. Caffeic acid is one of several naturally occurring phenolic acids that may be secreted by roots. (*B*) Diagram of an ethylenediaminetetraacetic acid-Fe^{3+} complex. The coordinate bonds are indicated by the dashed lines.

some). Iron may be taken up as the ferric[4] (Fe^{3+}) or ferrous (Fe^{2+}) ion, although the latter is more common due to its greater solubility. The importance of iron is related to two important functions in the plant. It is part of the catalytic group for many redox enzymes and it is required for the synthesis of chlorophyll. Important redox enzymes include the heme-containing cytochromes and non-heme iron-sulfur proteins (for example, Rieske proteins; ferredoxin) involved in photosynthesis, nitrogen fixation, and respiration. During the course of electron transfer the iron is reversibly reduced from the Fe^{3+} to the Fe^{2+} state. Iron is also a constituent of several oxidase enzymes, such as catalase and peroxidase. Iron is not a constituent of the chlorophyll molecule itself and its precise role in chlorophyll synthesis remains somewhat of a mystery. There is, for example, no definitive evidence that any of the enzymes involved in chlorophyll synthesis are iron-dependent. Instead, the iron requirement may be related to a more general need for iron in the synthesis of the chloroplast constituents, especially the electron transport proteins. Iron deficiencies invariably lead to a simultaneous loss of chlorophyll and degeneration of chloroplast structure. Chlorosis appears first in the interveinal regions of the youngest leaves, because the mobility of iron in the plant is very low and cannot be withdrawn from the older leaves. Chlorosis may progress to the veins and, if the deficiency is severe enough, the leaves may actually turn white.

Iron deficiencies are common because of the propensity of Fe^{3+} to form insoluble hydrous oxides ($Fe_2O_3 \cdot 3H_2O$) at biological pH. This problem is particularly severe in neutral or alkaline calcareous soils. On the other hand, iron is very soluble in strongly acidic soils and iron toxicity due to excess iron uptake can result. The problem of iron deficiency can usually be overcome by providing chelated iron, either directly to the soil or as a foliar spray. A **chelate** (from the Greek, *chele* or "claw") is a stable complex formed between a metal ion and an organic molecule, called the **ligand** or **chelating agent.** The ligand and the metal ion share electron pairs, forming what is known as a *coordinate bond* (Freifelder, 1985). Because chelating agents have a rather high affinity for most metal ions, formation of the complex reduces the possibility for formation of insoluble precipitates. At the same time, the metal can easily be withdrawn from the chelate for uptake by the plant. One of the more common synthetic chelating agents is the sodium salt of ethylenediaminetetraacetic acid, or EDTA (Fig. 4.5), known commercially as *versene* or *sequestrene*. EDTA and similar commercially available chelating agents, however, are not as a rule highly specific and will bind a range of cations, including iron, copper, zinc, manganese, and calcium. Natural chelating agents, including heme groups (as in haemoglobin, cy-

[4]In the scientific literature, particularly that body of literature dealing with iron uptake by organisms, the ferric form of iron, Fe^{3+}, is commonly referred to as Fe(III) (*iron-three*). By the same convention, ferrous iron, Fe^{2+}, is referred to as Fe(II) (*iron-two*).

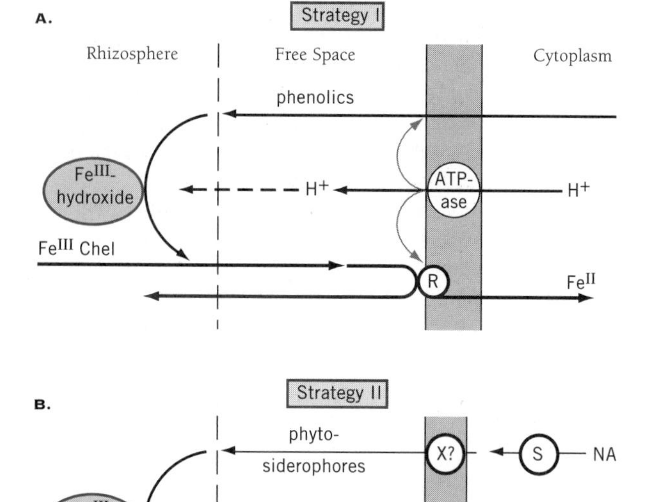

FIGURE 4.6 Two models for the solubilization and uptake of sparingly soluble inorganic iron by higher plants. *Model I:* Fe^{3+} is solubilized by one of several phenolic acids secreted into the rhizosphere by the roots. The iron is reduced at the root surface by an inducible reductase enzyme and the resulting Fe^{2+} immediately transported across the plasma membrane of the root cell. *Model II:* Fe^{3+} is solubilized by a phytosiderophore secreted into the rhizosphere by the root. The entire ferrisiderophore complex is then taken into the root cell where the iron is subsequently released. (From V. Römheld, H. Marschner. *Plant Physiology* 80:175–180, 1986. Copyright American Society of Plant Physiologists.)

tochromes and chlorophyll) and a variety of organic and phenolic acids, are far more specific for iron.

The importance of iron in plant nutrition is highlighted by the strategies plants have developed for uptake under conditions of iron stress (Marschner, 1986; Römheld, 1987). Iron deficiency induces several morphological and biochemical changes in the roots of dicots and nongraminaceous monocots. These include the formation of specialized transfer cells in the root epidermis, enhanced proton secretion into the soil surrounding the roots, and the release of strong ligands, such as **caffeic acid** (Fig. 4.5), by the roots. Simultaneously, there is an induction of reducing enzymes in the plasma membrane of the root epidermal cells. Acidification of the rhizosphere encourages chelation of the Fe^{3+} with caffeic acid, which then moves to the root surface where the iron is reduced to Fe^{2+} at the plasma membrane (Fig. 4.6A). Reduction to Fe^{2+} causes the ligand to release the iron, which is immediately taken up by the plant before it has the opportunity to form insoluble precipitates.

A second strategy for iron uptake by organisms involves the synthesis and release *by the organism* of low molecular weight, iron-binding ligands called **siderophores** (Gr. *iron-bearers*). Most of our knowledge of siderophores comes from studies with aerobic microorganisms (bacteria, fungi, and algae), where they were first discovered and have been studied most extensively. More recently, however, it has been discovered that siderophores are also released by the roots of higher plants (Fig. 4.7). Known as **phytosiderophores,** to distinguish them from ligands of microbial origin, these highly specific iron-binding ligands have thus far been found only in members of the family Gramineae, including the cereal grains (Römheld and Marschner, 1986). Phytosiderophores are synthesized and released by the plant only under conditions of iron stress, have a high affinity for Fe^{3+}, and very effectively scavenge iron from the rhizosphere. The distinctive feature of the siderophore system is that *the entire iron-phytosiderophore complex, or ferrisiderophore, is then reabsorbed into the roots* (Fig. 4.6B). Once inside the root, the iron is presumably reduced to Fe^{2+} and released for use by the cell. The fate of the phytosiderophore is unknown. In microorganisms it may either be chemically degraded and metabolized or the same molecule may again be secreted by the cell in order to pick up more iron.

The study of phytosiderophores is a relatively young field and, although substantial progress has been made in recent years, there is still much to be learned. It is not yet known, for example, how widespread the use of phytosiderophores is and the nature of the ferrisiderophore transport system has not been demonstrated in plants. One thing is clear: In those plants that use them, phytosiderophores are an important and effective strategy for supplying iron to the plant under conditions of iron stress.

Avenic acid (AA)

Mugineic acid (MA)

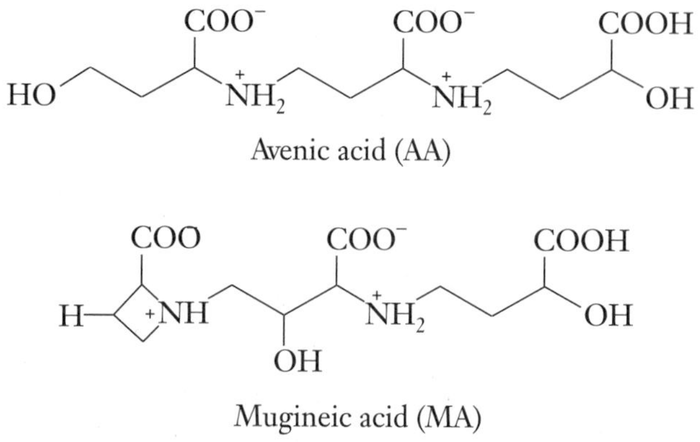

FIGURE 4.7 Phytosiderophores. The structures of two phytosiderophores released by the roots of higher plants. Ferric iron, Fe^{3+}, forms coordinate bonds with the nitrogen and carboxyl groups.

BORON

In aqueous solution, boron is present as the weak acid $B(OH)_3$ (boric acid, or H_3BO_3). At physiological pH (< 8), it is found predominantly in the undissociated form, which is preferred for uptake by roots. With respect to its biochemical and physiological role, boron is perhaps the least understood of all the micronutrients. Indeed, much of what we know is based entirely on studies of what happens to plants when boron is withheld. There is no solid evidence for involvement of boron with specific enzymes, either structurally or as an activator.

A substantial proportion of the total borate content of cells is found in the cell wall. This is apparently because borate has a propensity to form stable esters with cell wall saccharides that have adjacent hydroxyl groups. This so-called *cis*-diol configuration is characteristic of mannose and its derivatives, common cell wall polysaccharides. Glucose, fructose, and galactose, on the other hand, do not have this configuration and so do not bind boron. The primary walls of boron-deficient cells exhibit marked structural abnormalities, suggesting that boron is required for the structural integrity of the cell wall.

Other responses to boron deficiency point toward a role in cell division and elongation. One of the most rapid responses to boron deficiency, for example, is an inhibition of both cell division and elongation in primary and secondary roots. This gives the roots a stubby and bushy appearance. Cell division in the shoot apex and young leaves is also inhibited, followed by necrosis of the meristem. In addition, boron is known to stimulate pollen tube germination and elongation. It is not known how boron is involved in cell growth, but both hormone and nucleic acid metabolism have been implicated. Inhibition of cell division and elongation is accompanied by an increased activity of enzymes that oxidize the hormone indole-3-acetic acid and a decrease in RNA content (possibly through impaired synthesis of uracil, a RNA precursor).

In addition to the effects on shoot meristems noted above, common symptoms of boron deficiency include shortened internodes, giving the plant a bushy or rosette appearance, and enlarged stems, leading to the disorder known as "stem crack" in celery. In storage roots such as sugar beets, the disorder known as heart rot is due to the death of dividing cells in the growing region.

COPPER

In well-aerated soils, copper is generally available to the plant as the divalent cupric ion, Cu^{2+}. Cu^{2+} readily forms a chelate with humic acids in the organic fraction of the soil and may be involved in providing copper to the surface of the root. In wet soils with little oxygen, Cu^{2+} is readily reduced to the cuprous form, Cu^{+}, which is unstable. As a plant nutrient, copper seems to function primarily as a cofactor for a variety of oxidative enzymes. These include the photosynthetic electron carrier plastocyanin, cytochrome oxidase, which is the final oxidase enzyme in mitochondrial respiration, and ascorbic acid oxidase. The browning of freshly cut apple and potato surfaces is due to the activity of copper-containing **polyphenoloxidases** (or phenolase). **Superoxide dismutase** (SOD), which detoxifies superoxide radicals (O_2^-), is another important copper enzyme. Common disorders due to copper deficiency are generally stunted growth, distortion of young leaves and, particularly in citrus trees, a loss of young leaves referred to as "summer dieback."

ZINC

Zinc is absorbed as the divalent cation Zn^{2+}. It is an activator of a large number of enzymes, including alcohol dehydrogenase (ADH), which catalyses the reduction of acetaldehyde to ethanol, carbonic anhydrase (CA), which catalyses the hydration of carbon dioxide to bicarbonate, and, along with copper, superoxide dismutase. However, there is general agreement that disorders associated with zinc deficiency reflect disturbances in the metabolism of the auxin hormone indole-3-acetic acid. Typically, zinc-deficient plants have shortened internodes and smaller leaves (for example, "little leaf" of fruit trees). The precise role of zinc in auxin metabolism remains obscure, but it has been observed that auxin levels in zinc-deficient plants decline *before* the overt symptoms appear. Furthermore, restoration of the zinc supply is followed first by a rapid increase in hormone level and then resumption of growth. Available evidence supports the view that zinc is required for synthesis of the hormone precursor tryptophan (Marschner, 1986).

MANGANESE

Manganese is required as a cofactor for a number of enzymes—particularly decarboxylase and dehydrogenase enzymes, which play a critical role in the respiratory carbon cycle. Interestingly, manganese can often substitute for magnesium in reactions involving, for example, ATP. However, its best known and most studied function is in photosynthetic oxygen evolution. Manganese, in the form of a *manganoprotein*, is part of the oxygen evolving complex (OEC) where it appears to be involved in the accumulation of charges during the oxidation of water. It is absorbed and transported within the plant mainly as the divalent cation Mn^{2+}. Manganese deficiency can be widespread in some areas, depending on soil conditions, weather and crop species. It is aggravated by low soil pH (< 6) and high organic content. Manganese deficiency is responsible for "grey speck" of cereal grains, a disorder characterized by the appearance of greenish-grey, oval-shaped spots on the basal regions of young leaves. It may cause extreme chlorosis between the leaf veins as well as discoloration and deformities in legume seeds.

MOLYBDENUM

Although molybdenum is a metal, its properties more closely resemble those of nonmetals. In aqueous solution it occurs mainly as the molybdate ion MoO_4^{2-}. Molybdenum requirements are among the lowest of all known micronutrients and appear to be primarily related to nitrogen metabolism. Among the several enzymes found to contain molybdenum are **dinitrogenase** and **nitrate reductase.** The molybdenum requirement thus depends to some extent on the mode of nitrogen supply (Chap. 6). Dinitrogenase is the enzyme used by prokaryotes, including those in symbiotic association with higher plants, to reduce atmospheric nitrogen. Nitrate reductase is found in roots and leaves where it catalyses the reduction of nitrate to nitrite, a necessary first step in the incorporation of nitrogen into amino acids and other metabolites. In plants such as legumes, which depend on nitrogen fixation, molybdenum deficiency gives rise to symptoms of nitrogen deficiency. When nitrogen supplies are adequate, a deficiency of molybdenum shows up as a classic disorder known as "whiptail" in which the young leaves are twisted and deformed. The same plants may exhibit interveinal chlorosis and necrosis along the veins of older leaves. Like many of the micronutrients, molybdenum deficiency is highly species dependent—it is particularly widespread for legumes, members of the family Brassicaceae, and for maize. It is aggravated in acid soils with a high content of iron precipitates, which strongly absorb the molybdate ion.

CHLORINE

Chloride ion (Cl^-) is ubiquitous in nature and highly soluble. It is thus rarely, if ever, deficient. Deficiencies can normally be shown only in very carefully controlled solution culture experiments. Along with manganese, chloride is required for the oxygen evolving reactions of photosynthesis. Cl^- is a highly mobile anion with two principal functions: it is both a major counterion, maintaining electrical neutrality across membranes, and one of the principal osmotically active solutes in the vacuole. Chloride ion also appears to be required for cell division in both leaves and shoots. Chloride is readily taken up and most plants accumulate chloride ion far in excess of their minimal requirements. Plants deprived of chloride tend to exhibit reduced growth, wilting of the leaf tips, and a general chlorosis.

NICKEL

Nickel has only recently been added to the list of essential nutrient elements (Dalton et al., 1988). Nickel is an abundant metallic element and is readily absorbed by roots. It is ubiquitous in plant tissues, usually in the range of 0.05 to 5.0 mg Kg^{-1} dry weight (Mishra and Kar, 1974). One of the principal difficulties encountered in attempting to establish a role for nickel is its extremely low requirement. It has been estimated that the quantity of nickel needed by a plant to complete one life cycle is approximately 200 ng; a quantity that can be met by the initial nickel content of the seed in most cases. In order to establish a nickel deficiency it is therefore necessary to undertake extensive purification of the nutrient salts and then grow several generations in nickel-deficient solutions. The strongest evidence in favor of essential status for nickel is based on studies with legumes and cereal grains. In one such study of barley (*Hordeum vulgare*), the critical nickel concentration for seed germination was found to be 90 ng g^{-1} of seed dry weight (Brown et al., 1987). By growing plants for three generations in the absence of nickel, the nickel content of the seed could be reduced to 7.0 ng g^{-1} dry weight. Germination of these seeds was less than 12 percent. When the plants were grown for the same number of generations in nutrient solution supplemented with 0.6 μM or 1.0 μM nickel, seed germination was 57 and 95 percent, respectively. In other studies, nickel deficiencies have led to depressed seedling vigor, chlorosis, and necrotic lesions in leaves (Dalton et al., 1988).

The basis for a nickel requirement by plants is not clear, but it may be related to mobilization of nitrogen during seed germination. Nickel is known to be a component of two enzymes; urease and hydrogenase. Urease catalyses the hydrolysis of urea into NH_3 and CO_2 and is found widely through the plant kingdom. It is of interest to note that urease from jack bean seeds (*Canavalia ensiformis*) was the first protein to be crystallized by J. B. Sumner in 1926. One of the principal effects of Ni deficiency in soybean (*Glycine max*) is decreased urease activity in the leaves although the metabolic significance of urease is not yet clear. Free urea is rarely, if ever, detected in plant tissue, but it is formed by the action of the enzyme arginase on arginine and its structural analog, canavanine (Chap. 6). Canavanine, a nonprotein amino acid, is abundant in the seeds of some plant groups, such as jack bean, but its concentration diminishes rapidly upon germination. Arginine is also abundant in seeds and both amino acids could function as stored nitrogen that is readily mobilized during seed germination. If this view should be proven valid, then urease, and thus nickel as well, would play an important role in the mobilization of nitrogen during germination and early seedling growth.

A common form of mobile nitrogen in some legumes is a family of urea-based compounds known as **ureides,** such as allantoic acid or citrulline (Chap. 6). Ureides are formed in root nodules during nitrogen fixation and transported via the xylem throughout the host plant. Ureides are also formed in senescing leaves and transported out to the developing seeds for storage. The breakdown of ureides produces urea, which accumulates to toxic levels in Ni-deficient plants. Furthermore, the metabolism of purine bases (adenine and guanine) in all plants also produces ureides. It seems reasonable to assume that most, if not all, plants have a requirement for urease and nickel.

Hydrogenase is another important enzyme in some nitrogen-fixing plants. Hydrogenase is responsible for recovering hydrogen for use in the nitrogen-fixing process (Chap. 6). A deficiency of nickel leads to depressed levels of hydrogenase activity in the nodules of soybean, which in turn would be expected to depress the efficiency of nitrogen fixation.

TOXICITY OF MICRONUTRIENTS

As a group, the micronutrient elements are an excellent example of the dangers of excess. Most have a rather narrow adequate range and become toxic at relatively low concentrations.

Critical toxicity levels, defined as the tissue concentration that gives a 10 percent reduction in dry matter, vary widely between the several micronutrients as well as between plant species. As noted earlier, critical concentrations for copper, boron, and zinc are on the order of 20, 75, and 200 $\mu g\ g^{-1}$ dry weight, respectively. On the other hand, critical toxicity levels for manganese vary from 200 $\mu g\ g^{-1}$ dry weight for corn to 600 $\mu g\ g^{-1}$ for soybean and 5300 $\mu g\ g^{-1}$ for sunflower. Toxicity symptoms are often difficult to decipher because an excess of one nutrient may induce deficiencies of other nutrients. For example, the classic symptom of manganese toxicity, which often occurs in waterlogged soils, is the appearance of brown spots due to deposition of MnO_2 surrounded by chlorotic veins. But excess manganese may also induce deficiencies of iron, magnesium, and calcium. Manganese competes with both iron and magnesium for uptake and with magnesium for binding to enzymes. Manganese also inhibits calcium translocation into the shoot apex, causing a disorder known as "crinkle leaf." Thus the dominant symptoms of manganese toxicity may actually be the symptoms of iron, magnesium, and/or calcium deficiency.

Excess micronutrients typically inhibit root growth, not because the roots are more sensitive than shoots but because roots are the first organ to accumulate the nutrient. This is particularly true of both copper and zinc. Copper toxicity is of increasing concern in vineyards and orchards due to long-term use of copper-containing fungicides as well as urban and industrial pollution. Zinc toxicity can be a problem in acid soils or when sewage sludge is used to fertilize crops.

In spite of the apparent toxicity of micronutrients, many plant species have developed the capacity to tolerate extraordinarily high concentrations. For example, most plants are severely injured by nickel concentrations in excess of 5 $\mu g\ g^{-1}$ dry weight, but species of the genus *Alyssum* can tolerate levels in excess of 10 000 $\mu g\ g^{-1}$ dry weight.

SUMMARY

Plants are autotrophic organisms, taking their entire nutritional needs from the inorganic environment. Plants require carbon, hydrogen, and oxygen, plus 14 other naturally occurring elements that are taken from the soil. These 17 elements are considered essential because it has been demonstrated that in their absence all plants are unable to complete a normal life cycle.

Essential elements may be considered either macronutrients or micronutrients, depending on the quantity normally required. Micronutrients are normally required in concentrations less than 10 mmole/kg of dry weight.

Each essential element has a role to play in the biochemistry and physiology of the plant and its absence is characterized by one or more deficiency symptoms, commonly related to that role. Additional elements may be considered beneficial because they satisfy special requirements for particular plants. Essential elements, especially micronutrients, may be toxic when present in excess amounts.

CHAPTER REVIEW

1. Explain the difference between autotrophic and heterotrophic nutrition.
2. What is meant by essentiality? What is the difference between an essential element and a beneficial element? Describe the steps you would go through in order to determine the essentiality or non-essentiality of an element for a higher plant.
3. List the 17 elements that are essential for the growth of all higher plants. Be able to identify one or more principal structural or metabolic roles for each essential element.
4. Deficiencies of iron, magnesium, and nitrogen all cause chlorosis. Iron chlorosis develops only between the veins of young leaves while chlorosis due to both magnesium and nitrogen deficiencies

develop more generally in older leaves. Explain these differences. Why does each deficiency lead to chlorosis and why are the patterns different?

5. For what reasons might a soil rich in calcium supply too little phosphorous for plant growth?
6. What is a chelating agent? Explain how chelating agents help to maintain iron availability in nutrient cultures and soils. Explain how phytosiderophores help to supply iron to plants under conditions of iron stress.
7. What is meant by critical toxicity level? Which elements are most likely to be both essential and toxic to plants?
8. There are currently 17 elements know to be essential for higher plants. Is it possible that other elements might be added to this list in the future? Explain your answer.

FURTHER READING

Epstein, E. 1972. *Mineral Nutrition of Plants: Principles and Perspectives.* New York: Wiley.

Marschner, H. 1986. *Mineral Nutrition of Higher Plants.* London: Academic Press.

Steward, F. C. 1963. *Plant Physiology. A Treatise. III. Inorganic Nutrition of Plants.* New York: Academic Press.

REFERENCES

Arnon, D. I., P. R. Stout. 1939. The essentiality of certain elements in minute quantity for plants with special reference to copper. *Plant Physiology* 14:371–375.

Brown, P. H., R. M. Welch, E. E. Cary. 1987 Nickel: a micronutrient essential for higher plants. *Plant Physiology* 85:801–803.

Brownell, P. F., C. J. Crossland. 1972. The requirement for sodium as a micronutrient by species having the C_4 dicarboxylic photosynthetic pathway. *Plant Physiology* 49:794–797.

Brownell, P. F., J. G. Wood. 1957. Sodium as an essential micronutrient element for *Atriplex vesicaria* Heward. *Nature* 179:635–636.

Broyer, T. C., A. B. Carlton, C. M. Johnson, P. R. Stout. 1954. Chlorine—A micronutrient element for higher plants. *Plant Physiology* 29:526–532.

Dalton, D. A., S. A. Russell, H. J. Evans. 1988. Nickel as a micronutrient element for plants. *BioFactors* 1:11–16.

Downs, R. J., H. Hellmers. 1975. *Environment and the Experimental Control of Plant Growth.* New York: Academic Press.

Freifelder, D. 1985. *Principles of Physical Chemistry with Applications to the Biological Sciences.* 2nd ed. Boston: Jones and Bartlett.

Hewitt, E. J. 1963. Mineral nutrition of plants in culture media. In: Steward, F. C. (ed.), *Plant Physiology. A Treatise. III. Inorganic Nutrition of Plants.* New York: Academic Press. pp. 97–134.

MacDonald, I. R., P. C. DeKock, A. H. Knight. 1960. Variations in the mineral content of storage tissue disks maintained in tap water. *Physiologia Plantarum* 13:76–89.

Marschner, H. 1986. *Mineral Nutrition of Higher Plants.* London: Academic Press.

Mishra, D. and M. Kar. 1974. Nickel in plant growth and metabolism. *Botanical Review* 40:395–452.

Römheld, V. 1987. Different strategies for iron acquisition in higher plants. *Physiologia Plantarum* 70:231–234.

Römheld, V., H. Marschner. 1986. Evidence for a specific uptake system for iron phytosiderophores in roots of grasses. *Plant Physiology* 80:175–180.

Steward, F. C. 1968. *Growth and Organization in Plants.* Reading, Mass.: Addison-Wesley.

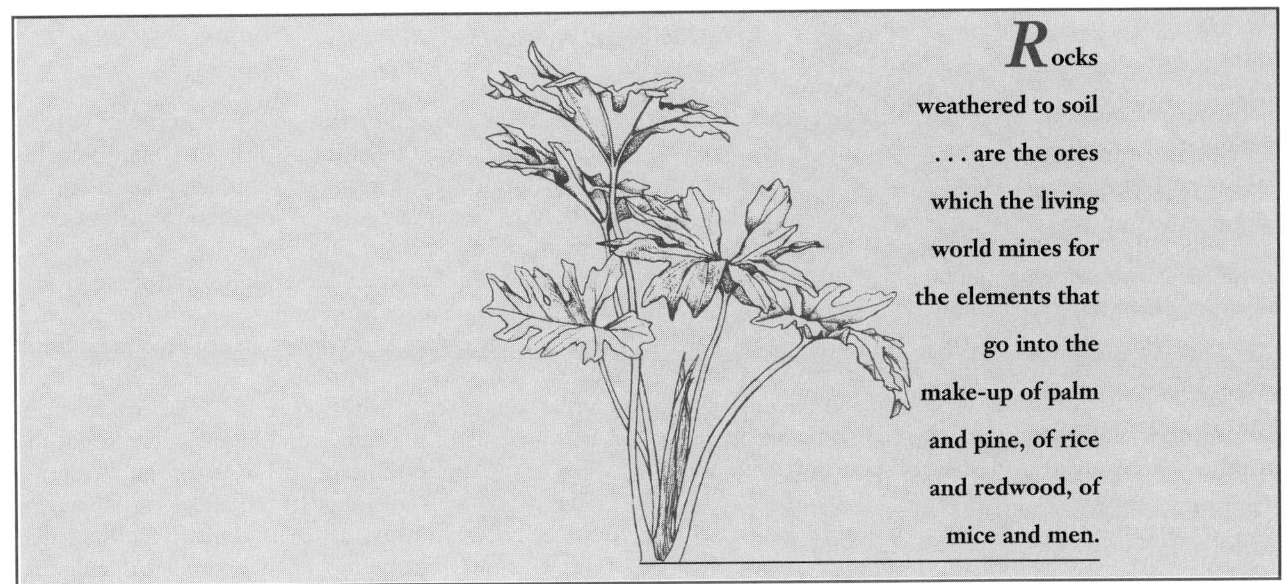

E. Epstein
(1972)

5

Roots, Soils, and Nutrient Uptake

With the exception of carbon and oxygen, which are supplied as carbon dioxide from the air, terrestrial plants generally take up nutrient elements from the soil solution through the root system. As we saw in Chapter 3, root systems are surprisingly extensive. Through a combination of primary roots, secondary and tertiary branches, and root hairs, root systems penetrate massive volumes of soil in order to, in the words of E. Epstein, "mine the soil" for required nutrients and water (Epstein, 1972).

As discussed earlier in Chapter 3, soil is a complex medium. It consists of a solid phase comprised of mineral particles derived from parent rock plus organic material in various stages of decomposition, a liquid phase comprised of water or the soil solution, gasses in equilibrium with the atmosphere, and a variety of microorganisms. The solid phase, in particular the mineral particles, is the primary source of nutrient elements. In the process of weathering, various elements are released into the soil solution, which then becomes the immediate source of nutrients for uptake by the plant. The soil solution, however, is very dilute (total mineral content is on the order of 10^{-3} M) and would quickly become depleted by the roots were it not continually replenished by the release of elements from the solid phase. In one study of phosphorous uptake, for example, it was calculated that the phosphate content of the soil was renewed on the average of ten times each day (Stout and Overstreet, 1950). Availability of nutrient elements is not, however, limited to the properties of the soil itself. Access to elements is further enhanced by the continual growth of the very dynamic root system into new regions of the soil.

In this chapter we will examine the availability of nutrients in the soil and their uptake by roots. This will include

- soil as a source of nutrient elements, the colloidal nature of soil, and ion exchange properties that determine the availability of nutrient elements in a form that can be taken up by roots;
- mechanisms of solute transport across membranes, including simple and facilitated diffusion and active transport, the function of membrane proteins as ion channels and carriers, and the role of electrochemical gradients;
- ion traffic into and through the root tissues and the concept of apparent free space; and
- the beneficial role of microorganisms, especially fungi, with respect to nutrient uptake by roots.

THE SOIL AS A NUTRIENT RESERVOIR

Soils vary widely with respect to composition, structure, and nutrient supply. Especially important from the nutritional perspective are inorganic and organic soil particles called colloids. *Soil colloids retain nutrients for release into the soil solution where they are available for uptake by roots.*

Some soil characteristics were described previously in the discussion of water uptake by roots (Chap. 3). There it was noted that the mineral component of soils consisted predominantly of **sand, silt,** and **clay,** which are differentiated on the basis of particle size (see Chap. 3, Table 3.4). A further discussion of soil chemistry is now necessary in order to understand how nutrients are made available to the soil solution and how they are taken up by plants.

When a small quantity of soil is stirred into water, the larger particles of *sand* settle out almost immediately, leaving a turbid suspension. Over the course of hours or perhaps days, if left undisturbed, the finer particles of *silt* gradually settle to the bottom as well and the turbidity will in all likelihood disappear. The very small *clay* particles, however, remain in stable suspension and will not settle out, at least not within a reasonable time frame. Clay particles in suspension are not normally visible to the naked eye—they are simply too small—but they are there. They can be detected, however, by directing a beam of light through the suspension. The suspended clay particles will scatter the light, causing the path traversed by the light beam to become visible. Known as the **Tyndall effect,** this light-scattering phenomenon is one characteristic of colloidal suspensions. **Colloids** are particles small enough to remain in suspension but too large to go into true solution. Solutions, sodium chloride or sucrose in water, for example, will not scatter light because the solute and solvent constitute a single phase. A colloidal suspension, on the other hand, is a two-phase system. It consists of a solid phase, the colloidal **micelle,** suspended in a liquid phase. Light scattered by the solid phase is responsible for the Tyndall effect.

But why this interest in colloidal clay? It has to do with two factors: (1) the large specific surface area of colloids and (2) numerous negative charges on the colloid surface that are able to bind **cations** (positively charged ions). Indeed, with respect to the nutrition of plants, the ability to retain and exchange cations on colloidal surfaces is the single most important property of soils.

Because of their small size, one of the distinguishing features of colloids is a high surface area per unit mass, also known as **specific surface area.** Consider, for example, a cube with a mass of one gram that measures 10 mm on a side. The specific surface area of this cube is $10 \text{ mm} \times 10 \text{ mm} \times 6 \text{ sides} = 600 \text{ mm}^2 \text{ g}^{-1}$. If the cube is then subdivided into particles of colloidal dimensions, say 0.001 mm on a side, the surface area of each particle decreases to $10^{-3} \text{ mm} \times 10^{-3} \text{ mm} \times 6 \text{ sides} = 6 \times 10^{-6} \text{ mm}^2$. However, for each gram there are now 10^{12} particles, so the specific surface area has increased to $10^{12} \times 6 \times 10^{-6} = 6{,}000{,}000 \text{ mm}^2 \text{ g}^{-1}$, a 10,000-fold increase. On a mass basis, then, colloids provide an incredibly large surface area for interaction with mineral elements in the soil solution.

Interaction between soil colloids and mineral elements is further enhanced by charges on the colloidal surfaces. Colloidal clays consist primarily of aluminum silicates (the chemical formula for kaolinite, one of the simplest clays, is $Al_2Si_2O_5{\cdot}(OH)_4$). Negative charges arise by virtue of ionization of alumina and silica at the edges of the mineral particle (Flegmann and George, 1975). Many soils also contain colloidal carbonaceous residue, called **humus,** which is organic material that has been slowly but incompletely degraded to colloidal dimension through the action of weathering and microorganisms. Because colloidal carbon is derived largely from lignin and carbohydrates, it carries negative charges arising from exposed carboxyl and hydroxyl groups.

Colloids, whether mineral or organic, are highly hydrated. This is because the positive pole of adjacent water molecules is attracted to the negatively charged surface, thus forming a hydration shell. Since the forces are of an electrostatic nature, this attraction is not an all-or-none phenomenon. It is strongest at the colloid surface and decreases as the distance between the surface and the water molecule increases. This combination of negative charge and hydration shell is a major factor in the stability of a colloidal suspension, as it prevents the micelles from approaching close enough to form aggregates large enough to settle out. As might be expected, colloids also attract, to varying degrees, cations from the surrounding soil solution. If sufficient cations are available, their adsorption will have the effect of neutralizing the surface charge.

Colloidal suspensions can be destabilized if the hydration shells are removed and the surface charges neutralized. The micelles are then able to approach, aggregate, and settle out. This approach, for example, is commonly employed in the isolation and purification of protein. Although it is common practice to speak of protein "solutions," proteins do not form a true solution but are in fact suspended colloids. Proteins may be precipitated from suspension by adding organic solvents such as alcohol or acetone, which serve to dehydrate the colloid by decreasing the water potential of the surrounding solution. Alternatively, the protein may be precipitated by adding an excess of salt, a process known as *salting out.* The salt neutralizes the surface charges and, at the same time, removes the hydration shell. Ammonium sulphate is commonly used for this purpose because of its extremely high solubility. Salting out is a popular method because it removes protein from suspension and concentrates it without denaturing them. The same principles influence the stability of soil col-

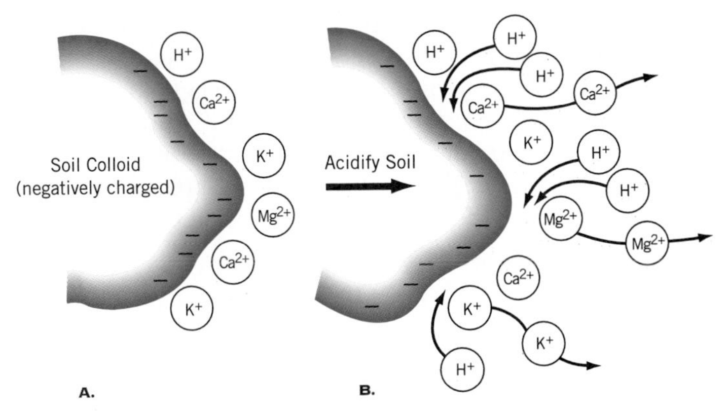

FIGURE 5.1 Ion exchange in the soil. (*A*) Cations are adsorbed to the negatively charged soil particles by electrostatic attraction. (*B*) Acidifying the soil increases the concentration of hydrogen ions in the soil. The additional hydrogen ions have a stronger attraction for the colloidal surface charges and so displace other cations into the soil solution.

loids and the association of mineral ions with the colloidal surface.

The association of cations with negative surfaces depends on electrostatic interactions; hence, binding affinity varies according to the **lyotropic series:**

$$Al^{3+} > H^{+} > Ca^{2+} > Mg^{2+} > K^{+} = NH_4^{+} > Na^{+}$$

In this series, aluminum ions have the highest binding affinity and sodium ions the least, reflecting the general rule that trivalent ions (3+) are retained more strongly than divalent (2+) and divalent more than monovalent (1+) (Fig. 5.1). Ions, however, are also hydrated and electrostatic rules are modulated by the *relative hydrated size* of the ion. Since ions of smaller hydrated size can approach the colloidal surfaces more closely, they tend to be more tightly bound. As in the case of hydration, described above, cation adsorption is not an all-or-none phenomenon. Both the degree of association and ion concentration decline in a continuous gradient with increasing distance from the surface of the colloid.

Cation adsorption is also *reversible* and consequently any ion with a higher affinity (e.g., H^{+}) is capable of displacing an ion lower in the series (e.g., Ca^{2+}). Alternatively, an ion with a lower affinity can, if provided in sufficient quantity, displace an ion with higher affinity by mass action (Fig. 5.1). This process of exchange between adsorbed ions and ions in solution is known as **ion exchange.** The ease of removal, or **exchangeability,** is indicated by the reverse of the lyotropic series shown above. Thus, sodium ion is the most readily exchanged in the series and aluminum the least.

Although the immediate source of mineral nutrients for the plant are the ions in the soil solution, the colloidal fraction with its absorbed ions represents the principal nutrient reservoir. The soil must be viewed as a very dynamic system, with cations in solution freely exchangeable with cations adsorbed to colloidal surfaces. As the soluble nutrients are taken up by the roots from the dilute soil solution, they are continually replaced by exchangeable ions held in the colloidal reservoir. The reservoir is then replenished by ions derived from the weathering of rock particles. In this way, ion exchange plays a major role in providing a controlled release of nutrients to the plant. It may not always work to the advantage of the plant, however. One effect of acid rain, for example, appears to be the displacement of cations from the colloidal reservoir due to the high concentration of hydrogen ions. Those solubilized nutrient ions not immediately assimilated by the roots are readily leached out of the soil by the rain and percolating ground water. Both the soil solution and the reservoir of nutrients are depleted more rapidly than they can be replenished and the plants are deprived of an adequate nutrient supply. Under normal circumstances, roots also secrete hydrogen ions, which assist in the uptake of nutrients.

What about **anions** (negatively charged ions)? Since the soil colloids are predominantly negatively charged, they do not tend to attract anions. Although some of the clay minerals do contain cations such as Mg^{2+}, the anion exchange capacity of most soils is generally low. Consequently, anions are not held in the soil but tend to be readily leached out by percolating ground water. This situation has important consequences for agricultural practice. Nutrients supplied in the form of anions, in particular nitrogen (NO_3^{-}), must be provided in large quantity to ensure sufficient uptake by the plants. As a rule, farmers must apply at least twice—sometimes more—the amount of nitrogen required to produce a crop. Unfortunately, much of the excess nitrate is leached into the ground water and eventually finds its way into streams and lakes, where it contributes to problems of eutrophication by stimulating the growth of algae.

MEMBRANE TRANSPORT

In order for mineral nutrients to be taken up by the plant, they must, at some point, be taken across the plasma membranes of root cells. Nutrient uptake by roots is therefore fundamentally a cellular problem, governed by the rules of membrane transport.

The study of solute transport across membranes is extensive and a detailed discussion is beyond the scope of this book. However, in this section, some of the more important principles of transport across membranes—in particular as they apply to the uptake of nutrient elements by roots—will be reviewed. This is not to suggest that membrane transport is of peripheral interest to the plant physiologist. Indeed the opposite is the case, since it is the plasma membrane that defines the cell and selectively regulates the exchange of material and energy between the cell and its environment, or between organelles within the cell. Moreover, membrane transport is not limited to the exchange of nutrients and waste products. It also assumes a critical role in hormonal action, plant movements, and signal transduction.

Over the past decade, our understanding of membrane transport in plants has advanced significantly, primarily because of advances in technology for isolating protoplasts (cells from which the cell wall has been removed) and a technique, known as patch-clamp recording (see Box 5.1: Electrophysiology—Exploring Ion Channels). This revolutionary electrophysiological technique enables investigators to measure cross-membrane currents as small as 10^{-12} amperes. The reader who wishes a more detailed treatment of transport by cells is referred to the several excellent cell biology texts and reviews listed at the end of this chapter.

Membrane transport is inherently an abstract subject (Harold, 1986). That is to say, investigators measure the kinetics of solute movement across various natural and artificial membranes under a variety of circumstances. Models are then constructed that attempt to explain these kinetic patterns in terms of what is currently understood about the composition and architecture of membranes. As our understanding of membrane structure has changed over the years, so have the models that attempt to interpret how solutes cross these membranes. There are, however, three fundamental concepts—**simple diffusion, facilitated diffusion,** and **active transport**—that have persevered, largely because they have proven particularly useful in categorizing and interpreting experimental observations. These three concepts now make up the basic language of transport across all membranes of all organisms. These three basic modes of transport are interpreted schematically in Figure 5.2.

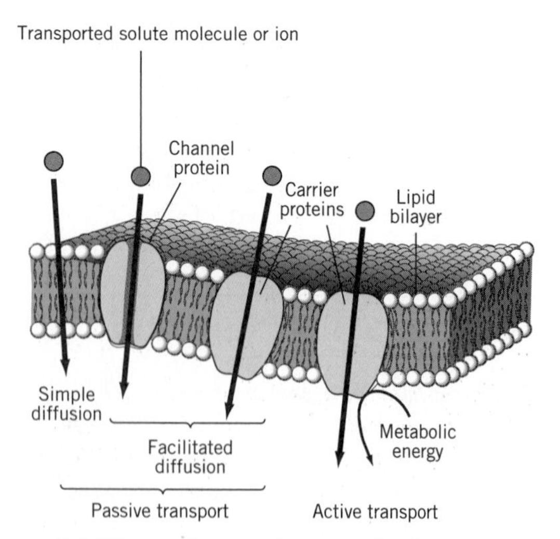

FIGURE 5.2 **The exchange of ions and solutes across membranes may involve simple diffusion, facilitated diffusion, or active transport.**

SIMPLE DIFFUSION

According to Fick's law (Chap. 2, Eq. 2.1), the rate at which molecules in solution diffuse from one region to another is a function of their concentration difference. For a membrane-bound cell, Fick's law may be restated as:

$$J = PA(C^o - C^i) \qquad (5.1)$$

where J is the flux, or amount of solute crossing the membrane per unit time. A is the cross-sectional area of the diffusion path, which, in this case, is the area of the cell membrane (in cm^2). P is the permeability coefficient; it measures the velocity (in $cm\ s^{-1}$) with which the solute crosses that membrane and is specific for a particular membrane-solute combination. Since the membrane barrier is primarily lipid in character, nonpolar solute molecules tend to pass through more rapidly. Thus, the permeability coefficient generally reflects the lipid solubility of diffusing molecules. Few solutes of biological importance are nonpolar and only three (O_2, CO_2, NH_3) appear to traverse membranes by simple diffusion through the lipid bilayer. Water, in spite of its high polarity, also diffuses rapidly through lipid bilayers; this is apparently because water passes through water-selective channels called aquaporins. Aquaporins are discussed later in this chapter.

FACILITATED DIFFUSION

Lipid bilayers are particularly impermeable to charged solutes or ions. Their charge and high degree of hydration renders ions insoluble in lipids and thus prevents them from entering the hydrocarbon phase of membranes. Synthetic lipid bilayers, or artificial membranes, for example, are some nine orders of magnitude less permeable to smaller ions such as K^+ or Na^+ than to water. In the 1930s, students of membrane transport recognized that certain ions entered cells far more quickly than would be expected on the basis of their diffusion through a lipid bilayer. We now know that this is because natural membranes contain a large number of proteins, many of which function as **transport proteins.** Some of these transport proteins *facilitate* the diffusion of solutes, especially charged solutes or ions, into the cell by effectively overcoming the solubility problem. The term facilitated diffusion was coined to describe this rapid, assisted diffusion of solutes across the membrane.

In facilitated diffusion, the direction of transport is still determined by the concentration gradient (for uncharged solute) or electrochemical gradient (for ions). The difference between concentration and electrochemical gradients will be described more fully below. Facilitated diffusion is also bidirectional and, like simple diffusion, net movement ceases when the rate of movement across the membrane is the same in both directions.

Transport by diffusion, whether simple or facilitated, is considered a **passive** process. Passive means that the transport process does not require a direct input of metabolic energy. The energy for transport by diffusion comes from the concentration or electrochemical gradient of the solute being transported. As a consequence, transport by diffusion does not lead to an accumulation of solute against an electrochemical gradient.

TRANSPORT PROTEINS—CARRIERS AND CHANNELS

Two major classes of transport proteins are known. **Carrier proteins** (also known as carriers, transporters, or simply, *porters*) bind the particular solute to be transported, much along the lines of an enzyme—substrate interaction. Binding of the solute induces a conformational change in the carrier protein, which delivers the solute to the other side of the membrane. Release of the solute at the other surface of the membrane completes the transport and the protein then reverts to its original conformation, ready to pick up another solute.

Channel proteins form a charged-lined, water-filled channel that extends across the membrane. Channels are normally identified by the ion species that is able to permeate the channel, which is in turn dependent upon the *hydrated* size of the ion and its charge. Diffusion through a channel is dependent upon the hydrated size because the associated water molecules must diffuse along with the ion. The number of ion channels discovered in the membranes of plant cells is increasing. Currently there is solid evidence for K^+, Cl^-, and Ca^{2+} channels, while additional channels for organic ions are strongly suggested.

Channel proteins are frequently **gated,** which means they may be open or closed (Box 5.1). Solutes of an appropriate size and charge may diffuse through only when the channel "gate" is open. Two types of gates are known. An electrically gated channel opens in response to membrane potentials of a particular magnitude. Other channels may open only in the presence of the ion that is to be transported and may be modulated by light, hormones, or other stimuli. The precise mechanism of gated channels is not known, although it is presumed to involve a change in the three-dimensional shape, or conformation, of the protein. *All* channel proteins and many carriers allow solute entry by facilitated diffusion.

The importance of carriers lies in the selectivity they impart with respect to which solutes are permitted to enter or exit the cell. Channels, on the other hand, appear to be involved wherever large quantities of solute, particularly charged solutes or ions, must cross the membrane rapidly. Whereas a carrier may transport between 10^4 to 10^5 solute molecules per second, a channel may pass on the order of 10^8 ions per second. It should also be stressed that large numbers of channels are not required to satisfy the needs of most cells. The rate of efflux through guard cell K^+ channels during stomatal closure, for example, has been estimated at 10^7 K^+ ions sec^{-1}—a rate that could conceivably be accommodated by a single channel!

ACTIVE TRANSPORT

Many transport processes, in addition to being rapid and specific, will lead to an **accumulation** of solute inside the cell. In other words, the transport process will establish significant concentration or electrochemical gradients and will continue to transport solute against those gradients. The transport process involved is known as **active transport.** By definition, active transport is tightly coupled to a metabolic energy source—usually, although not always, hydrolysis of adenosine triphosphate (ATP) (see Chap. 9). In other words, active transport requires an input of energy and does not occur spontaneously. Unlike facilitated diffusion, active transport is unidirectional—either into *or* out of the cell—and is always mediated by carrier proteins. Active transport serves to accumulate solutes in the cell when solute concentration in the environment is very low. When used to transport solute out of the cell, active transport serves to maintain a low internal solute concentration. Because active transport systems move solutes against a concentration or electrochemical gradient, they are frequently referred to as **pumps.**

Armed with this basic introduction to membrane transport, we can now proceed with our discussion of ion uptake by roots.

SELECTIVE ACCUMULATION OF IONS

There are two important characteristics of solute transport across the membranes of all cells, including ion uptake by roots: **accumulation** and **selectivity.** These are illustrated by a typical set of ion uptake data shown in Table 5.1. Accumulation refers to the observation noted above that the concentrations of some ions inside the cell may reach levels much higher than in the surrounding medium. This phenomenon is expressed quantitatively by the **accumulation ratio,** which can be defined as the ratio of the concentration inside the cell (C^i) to

Box 5.1
Electrophysiology—Exploring Ion Channels

The exchange of ions across cellular membranes is facilitated by the presence of transmembrane proteins referred to as ion channels. Most ion channels are highly specific for one or a limited number of ion species, which can diffuse through an open channel at rates as high as 10^8 s^{-1}.

Channel proteins may exist in two different conformations, referred to as *open* and *closed.* In the open conformation, the core of the protein forms a pathway for diffusion of ions through the membrane (Fig. 5.3). A channel that can open and close is said to be *gated*—in the open conformation, the "gate" is open and ions are free to diffuse the channel. When the gate is "closed," the channel is not available for ion diffusion. A number of stimuli, including voltage, light, hormones, and ions themselves, are known to influence the frequency or duration of channel opening. The channel protein is believed to contain a *sensor* that responds to the appropriate stimulus by changing the conformation of the protein and opening the gate.

Because ions are mobile and carry a charge, their movement across membranes establishes an electrical current. These currents, typically on the order of picoamperes (pA = 10^{-12} ampere), can be measured using microelectrodes constructed from finely drawn-out

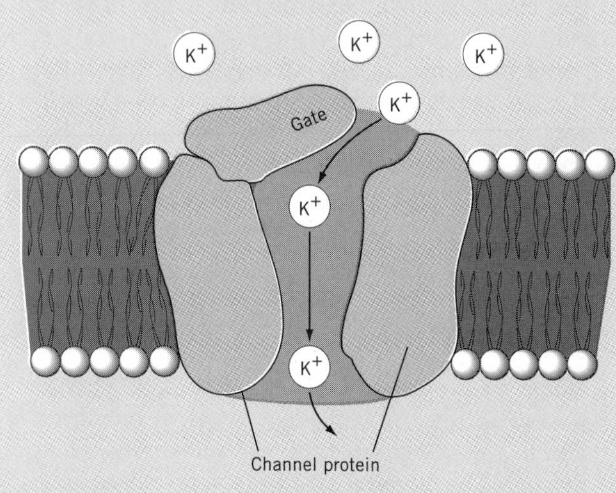

FIGURE 5.3 A gated membrane channel. Gated channels may be open, in which case ions are permitted to pass through the channel, or closed to ion flow. Opening may be stimulated by changes in membrane potential, the presence of hormones, or the ion itself.

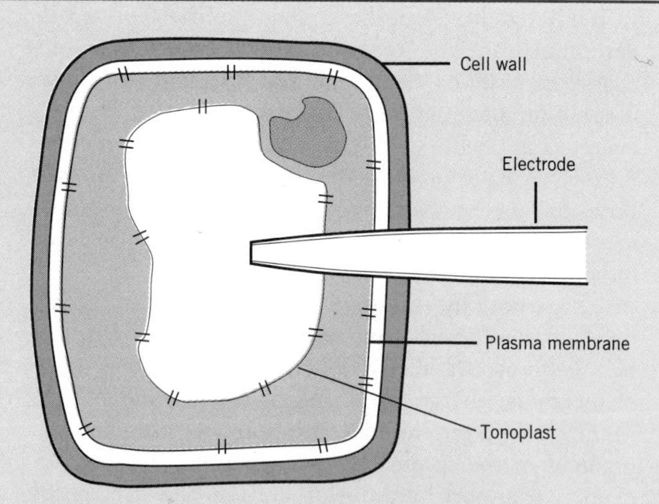

FIGURE 5.4 Changes in electrical properties of the cell related to ion flow can be measured by inserting a microelectrode directly into the cell.

glass tubing. The first evidence for gated channels was based on experiments in which the electrode was inserted directly into the cell (Fig. 5.4). This method has several limitations. It requires relatively large cells and the results reflect the activities of many different channels of various types at the same time. Moreover, when applied to plant cells, the electrode usually penetrates the vacuolar membrane (tonoplast) as well as the plasma membrane, thus summing the behavior of channels in both membranes.

These problems were largely circumvented by development of the patch clamp method that permits the study of single ion channels in selected membranes. In this technique, the tip of the microelectrode is placed in contact with the membrane surrounding an isolated protoplast (a cell from which the cell wall has been removed) (Fig. 5.5). A tight seal between the electrode and the membrane is formed by applying a slight suction. The small region of membrane in contact with the electrode is referred to as the "patch." Measurements may be made in this configuration, with the whole cell attached, or, alternatively, the electrode can be pulled away from the cell. In that case the patch then remains attached to the electrode tip and can be bathed in solutions of known composition. Note that the exterior surface of the membrane is in contact with the microelectrode solution (called the "inside-out" configuration). Variations in the technique permit the orientation of the patch to be reversed, with the internal surface facing the electrode solution (the "outside-out" configuration). With a sufficiently small electrode tip (ca. 1.0 μm diameter), the patch may contain a single ion channel.

With an appropriate electrical circuit, the experimenter can hold, or "clamp," the potential difference (voltage) across the patch at some predetermined value. At the same time, the circuit will monitor any current that flows through the membrane patch. Typical experimental results are shown in Figure 5.6. Figure 5.6A illustrates a patch-clamp recording for a single K^+ channel, using an inside-out patch of plasma membrane isolated from a stomatal guard cell protoplast. For the lower trace, the voltage was clamped at 0 mV. Note the appearance of one small, transient current pulse. For the upper trace, the voltage was stepped up to +90 mV. The current trace indicates that the channel opened immediately, in response to the voltage stimulus, and remained open for about 50 milliseconds (ms). It then spontaneously closed and reopened again for about 12 ms. This is an example of a voltage-gated ion channel. Figure 5.6B is a trace obtained with guard cell protoplasts, using a whole cell configuration. Note the difference in time scale. In this case, a pulse of blue light initiates current flow by activating a blue-light activated proton pump.

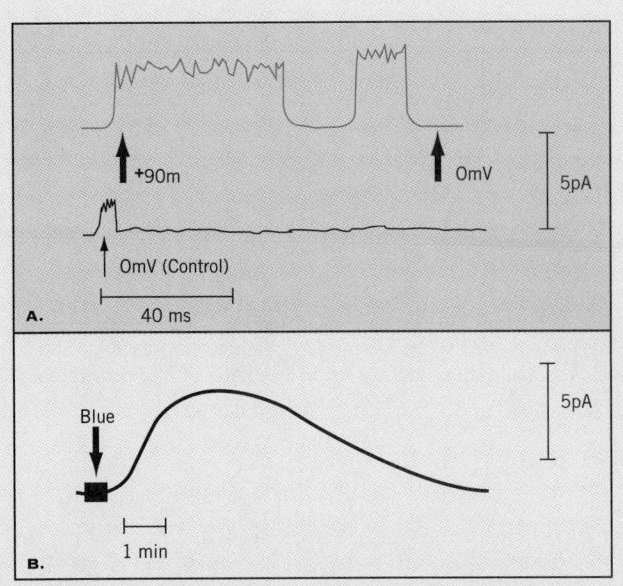

FIGURE 5.6 **(*A*) Current generated by a K^+-selective channel in the plasma membrane of a stomatal guard cell protoplast. Channel opening was stimulated by applying a 90 mV pulse across the membrane. Note that the channel exhibited a transient closure and opening before the voltage pulse was terminated. A 0 mV control is shown for comparison. (Reprinted with permission from J. I. Schroeder et al., *Nature* 312:361–362. Copyright 1984 Macmillan Magazines Ltd.) (*B*) Blue light stimulates a current in whole-cell membranes, an indication that membrane ion pumps are activated by blue light. (Reprinted with permission from S. M. Assmann et al., *Nature* 318:285–287. Copyright 1985 Macmillan Magazines Ltd.)**

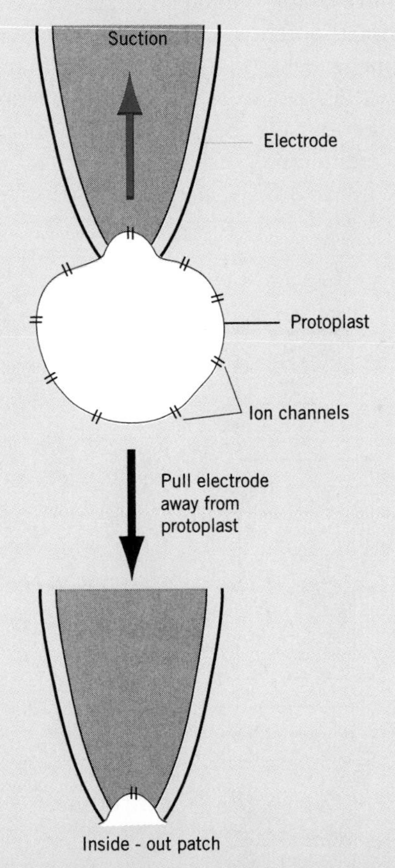

FIGURE 5.5 **Current flow through individual channels can be measured by the patch-clamp technique. A small piece of membrane containing a single channel can be isolated at the tip of a microelectrode. Ions flowing through the channel carry the current, which can be measured with sensitive amplifiers.**

The patch-clamp technique represents a major advance in electrophysiology that is only just beginning to provide insights into the mechanisms of plant membrane transport.

FURTHER READING

Assmann, S. M., L. Simoncini, J. I. Schroeder. 1985. Blue light activates electrogenic ion pumping in guard cell protoplasts of *Vicia faba. Nature* 318:285–287.

Hedrich, R., J. I. Schroeder. 1989. The physiology of ion channels and electrogenic pumps in higher plants. *Annual Review of Plant Physiology and Plant Molecular Biology* 40:539–569.

Satter, R. L., N. Moran. 1988. Ionic channels in plant cell membranes. *Physiologia Plantarum* 72:816–820.

Schroeder, J. I., R. Hedrich, J. M. Fernandez. 1984. Potassium-selective single channels in guard cell protoplasts of *Vicia faba. Nature* 312:361–362.

TABLE 5.1 The uptake of selected ions by maize roots.

	Accumulation Ratio		
Ion	C^o (m)	C^i (m)	$[C^i/C^o]$
K^+	0.14	160	1142
Na^+	0.51	0.6	1.18
NO_3^-	0.13	38	292
SO_4^{2-}	0.61	14	23

Maize roots were bathed in nutrient solutions for four days. C^o and C^i are the ion concentrations of the medium and root tissue, respectively. C^i was measured as the concentration of ions in the sap expressed from the roots.

From data of H. Marschner, 1986.

the concentration outside the cell (C^o). Note that in Table 5.1 the internal concentration of K^+ is more than 1,000 times greater than it is in the bathing medium. In the past, an accumulation ratio greater than one has been considered compelling evidence in favor of active transport, since that solute has evidently moved in against a concentration gradient. Conversely, an accumulation ratio less than one implies that the solute has been actively *excluded* or *extruded* from the cell. As will be shown below, this is not necessarily true, especially where charged solutes are involved. When assessing solute uptake by cells, it is especially important to distinguish between uncharged and charged solutes.

Ion uptake is highly selective. Note there is virtually no accumulation of Na^+ by maize roots (Table 5.1) and accumulation ratios for K^+ and NO_3^- are substantially higher than for SO_4^{2-}. The low concentrations of Na^+ in plant cells (unlike animal cells) may result from limited uptake of Na^+ in the first place, but also because Na^+ (along with certain other ions) is actively expelled from plant cells.

ELECTROCHEMICAL GRADIENTS AND ION MOVEMENT

Accumulation ratios for uncharged solutes, such as sugars, are relatively straightforward. It can be assumed that uptake is fundamentally dependent on the difference in concentration on the two sides of the membrane. In other words, for uncharged solutes it is the concentration gradient alone that determines the gradient in *chemical potential*. It is a relatively simple matter to measure experimentally internal and external concentrations of the solute and thus calculate the accumulation ratio.

With charged solutes, or ions, the situation is more complex and the accumulation ratio is not always a valid indication of passive or active transport. Because ions carry an electrical charge, they will diffuse in response to a gradient in *electrical potential* as well as chemical potential. Positively charged potassium ions, for example, will naturally be attracted to a region with a preponderance of negative charges. Consequently, the movement of ions is determined by a gradient that has two components: one concentration and one electrical. In other words, ions will move in response to an **electrochemical gradient** and the electrical properties of the cell, or its **transmembrane potential,** must be taken into account.

A voltage or potential difference will develop across a membrane quite simply because of an unequal distribution of anionic and cationic charges. The cytosol, for example, contains a large number of fixed or nondiffusible charges such as the carboxyl ($R{\cdot}COO^-$) and amino ($R{\cdot}NH_4^+$) groups of proteins. At the same time, cells use energy to actively pump cations, in particular H^+, Ca^{2+}, and Na^+, into the exterior space. The resulting unequal distribution of cations establishes a potential difference, or voltage, across the membrane. The cytosol remains negative relative to the cell wall space, which accumulates the positively charged cations.

A simple example should suffice to illustrate how a transmembrane potential can influence ion movement into and out of cells (Fig. 5.7). In this example, it is assumed that (a) the internal K^+ concentration is high relative to that outside the cell; (b) K^+ can move freely across the membrane, perhaps through K^+ channels; and (c) the internal K^+ concentration is balanced by a

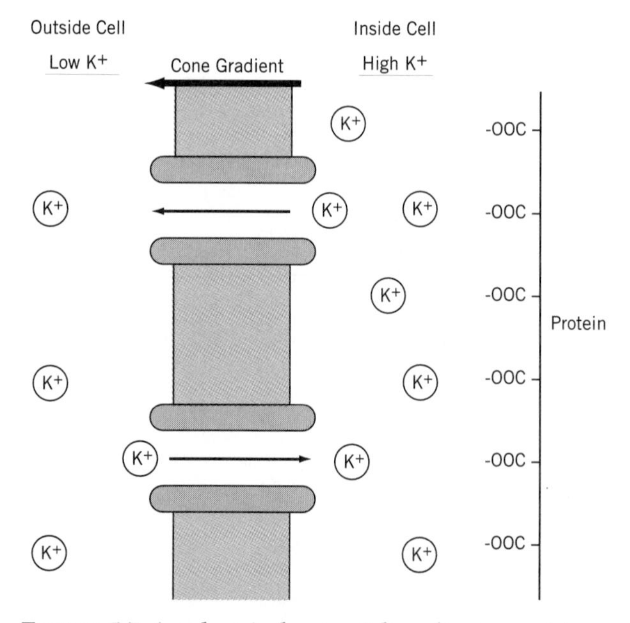

FIGURE 5.7 An electrical potential gradient may drive an apparent accumulation of cations (e.g., K^+) against a concentration gradient. The combination of concentration differences and charge differences constitutes an *electrochemical* gradient. Thus, potassium ions tend to accumulate in cells in response to the large number of fixed charges on proteins and other macromolecules.

number of organic anions restrained within the cell. Under these conditions, it might be expected that K^+ will diffuse out of the cell, driven by its concentration gradient, until the concentrations of K^+ outside and inside the cell are equal. However, as K^+ diffuses out of the cell it leaves behind the nondiffusible anionic charges, thus creating a charge imbalance or voltage (potential) difference across the membrane. The negative charges tend to pull the positively charged potassium ions back onto the cell. As a result, equilibrium is achieved not when the *concentrations* of K^+ are equal on both sides of the membrane, but when the membrane potential difference reaches a value such that the force of the concentration gradient pulling K^+ out of the cell is balanced by the force of the electrical gradient pulling K^+ back into the cell. Under these circumstances, the cell will maintain a high internal K^+ concentration and the accumulation ratio will be greater than unity, yet the movement of potassium ion is solely by passive diffusion.

Anion movement would also be influenced by the membrane potential, but in the opposite direction. Anions would be repelled by the preponderance of internal negative charges and attracted by the preponderance of external positive charges, thus leading to an accumulation ratio less than unity. It is clear from these examples that an accumulation ratio other than unity does not necessarily mean that active transport is involved.

Transmembrane potentials can be measured with a microelectrode made from finely drawn-out glass tubing. With the aid of a microscope, the electrode is inserted into the vacuole of a cell. A reference electrode is placed in the medium surrounding the cell. The difference in potential between the two electrodes can be measured with the aid of a sensitive voltmeter. It is by no means an easy technique, but many experimenters have become quite proficient with it. Potentials measured in this way are commonly in the range of −100 to −130 mV for young roots and stems, although potentials as high as —200 mV have been recorded for some algal cells. The cytosol is always negative with respect to the surrounding medium. Such potentials do not require a large charge imbalance. As few as one unbalanced charge in a population of a million ions is sufficient to generate a potential of 100 mV (Clarkson, 1991).

The relationship between transmembrane potential gradient and ion distribution across the membrane can be expressed quantitatively by the **Nernst equation:**

$$\Delta E_{nj} = \frac{2.3RT}{zF} \times \log \frac{C_j^i}{C_j^o} \qquad (5.2)$$

where ΔE_n is the electrical potential difference for the ion j (also known as the **Nernst potential**), and C_j^i/CC_j^o is the ratio of the molar concentrations inside and outside the cell. R is the gas constant, F is the Faraday constant (96 500 J V^{-1} $mole^{-1}$), and z is the valency or charge for ion j. The value of z for a univalent cation, for example, would be 1 while for calcium or magnesium it would be 2. For chloride or nitrate it would be −1 and for sulphate it would be −2. By substituting numerical values for the constants and rearranging, this relationship can be simplified to:

$$\log \frac{C_j^i}{C_j^o} = -\frac{z\Delta E_{nj}}{59} \qquad (5.3)$$

for an ion at 25°. Finally,

$$-z\Delta E_{nj} = 59 \log \frac{C_j^i}{C_j^o} \qquad (5.4)$$

In its final version, the Nernst equation allows us to make certain predictions about the equilibrium concentrations for ions inside the cell (C^i) when ion transport is due to facilitated diffusion. In order to apply the equation, it is necessary to first measure the transmembrane potential and the concentrations of ions both inside and outside the cell. Deviations from the concentrations predicted by the Nernst equation are considered evidence that either active uptake or expulsion of the ions is involved. In other words, the Nernst equation can be used to determine whether an ion is actively or passively distributed across the membrane. If the measured internal concentrations are approximately equal to the calculated Nernst value, it can be assumed that the ion has been distributed passively. If the measured concentration is greater than predicted, active uptake is probably involved, and, if lower, the ion is being actively expelled from the cell.

Application of the Nernst equation is illustrated by the experiment of Higinbotham and colleagues shown in Table 5.2. At equilibrium the measured membrane potentials were −110 mV for pea roots and −84 mV for oat roots. C^o, the concentration of ions in the external solution, was known and the predicted C^i was calculated from the Nernst equation. The ratio of the predicted value to the actual value is a measure of how well the Nernst relationship applies to that particular ion. The accumulation ratios for almost all of the ions are greater than 1.0, indicating some degree of accumulation in the cell. Only in the case of K^+ was the ratio of predicted concentration to actual concentration near 1.0. This indicates that K^+ is near electrochemical equilibrium and was probably accumulated passively, at least in pea roots. There appears to be a possibility of some active accumulation of K^+ by oat roots. Cellular concentrations of both Na^+ and Ca^{2+} are lower than predicted. Since other evidence supports the existence of Na^+ and Ca^{2+} pumps in the membrane, these ions probably entered passively down their electrochemical gradients but were then actively expelled. Internal concen-

TABLE 5.2 The uptake of selected ions by roots of pea (*Pisum sativum*) and oat (*Avena sativa*). The Nernst equation was used to predict the internal concentration (C_i) assuming cell membrane potentials of −110 mV and −84 mV for pea and oat roots, respectively. The accumulation ratio was calculated on the basis of the measured (or actual) C_i. The symbol E under probable uptake mechanism refers to active exclusion from the root. The symbol U denotes active uptake by the roots.

Ion	C^o	Predicted C^i	Actual C^i	Accumulation Ratio	Actual / Predicted	Probable Uptake Mechanism
			Pea Root			
K^+	1.0	74	75	75	1.01	Diffusion
Na^+	1.0	74	8	8	0.108	E
Ca^{2+}	1.0	5400	1.0	1.0	0.00018	E
NO_3^-	2.0	0.027	28	14	1037	U
$H_2PO_4^-$	1.0	0.014	21	21	1500	U
SO_4^{2-}	0.25	0.000047	9.5	38	202,127	U
			Oat Root			
K^+	1.0	27	66	66	2.4	Diffusion (?)
Na^+	1.0	27	3	3	0.11	E
Ca^{2+}	1.0	700	1.5	1.5	0.0021	E
NO_3^-	2.0	0.076	56	28	741	U
$H_2PO_4^-$	1.0	0.038	17	17	447	U
SO_4^{2-}	0.25	0.00036	2	8	5,555	U

From data of Higinbotham et al., 1967.

trations for all three anions are much higher than predicted, indicating they are actively taken up by the cells. This is understandable, since energy would be required to move *negatively* charged ions against a *negative* electrochemical gradient.

Although these results are indicative of active and passive transport, further tests would help to confirm the conclusion in each case. Since active transport requires a direct input of metabolic energy, it is sensitive to oxygen and respiratory poisons. Thus reduced uptake of a particular ion in the absence of oxygen or in the presence of cyanide or 2,4-dinitrophenol would be evidence in support of active transport. Even this evidence is not always compelling, however, since effects of inhibitors may indirectly influence nutrient uptake. For example, even transport by diffusion ultimately requires an expenditure of metabolic energy, if only to establish and maintain the organization of membranes and other aspects of the cell that make transport possible. It is not always a simple matter to distinguish between direct and indirect involvement of energy. The criteria for active transport usually require that the solute distribution not be in electrochemical equilibrium and that there be a quantitative relationship between energy expended and the amount of solute transported.

Finally, it should be noted that when ion concentrations are known, the Nernst equation can also be used to estimate the transmembrane potential, or Nernst potential, *contributed by that ion*. At steady state, however, calculation of the membrane potential is complicated by the fact that many different ion species, each with a different permeability, are simultaneously crossing the membrane in both directions. The individual contributions of all ion gradients must consequently be taken into account and summed in order to arrive at the overall potential for the cell. In practice, however, K^+, Na^+, and Cl^- are the dominant ions with the highest permeabilities and concentrations in plant cells and a reasonable estimation of transmembrane potential can be based on these three ions alone.

ACTIVE TRANSPORT AND ELECTROGENIC PUMPS

As noted above, energy to drive active transport comes chiefly from the hydrolysis of ATP. The energetics and operation of **ATPase-proton pumps** found in the energy-transducing membranes of chloroplasts and mitochondria will be described in Chapter 9. These large multiprotein complexes are able to utilize the energy associated with an electrochemical proton gradient across a membrane to drive the *synthesis* of ATP. An important characteristic of these pumps is that they are reversible. In other words, they can use the energy of

ATP *hydrolysis* to establish a proton gradient across the membrane. This proton gradient, together with the normal membrane potential, establishes a **proton motive force** (see Chap. 9) that tends to move protons back across the membrane. It is now well established that similar ATPase-proton pumps are found in the plasma and vacuolar membranes and possibly other membranes as well. It is generally conceded that the proton motive force established by pumping protons across membranes is the primary source of energy for a variety of plant activities. Included are activities such as active transport of solutes (cations, anions, amino acids and sugars), regulation of cytoplasmic pH, stomatal opening and closure (Chap. 8), sucrose transport during phloem loading (Chap. 11), and auxin-mediated cell elongation (Chap. 17) (Sze, 1984).

A schematic model relating ATPase-proton pumps to solute exchange across membranes is shown in Figure 5.8. Energy to drive the pump is provided in the form of ATP, produced by oxidative phosphorylation in the mitochondria (Chap. 12). The proton-transporting protein is shown extending across the plasma membrane, with its ATP binding site on the cytosolic side. Hydrolysis of the ATP results in the translocation of a proton from the cytosol to the surrounding apoplastic cell wall space. The ATPase thus serves as a proton-translocating carrier protein.

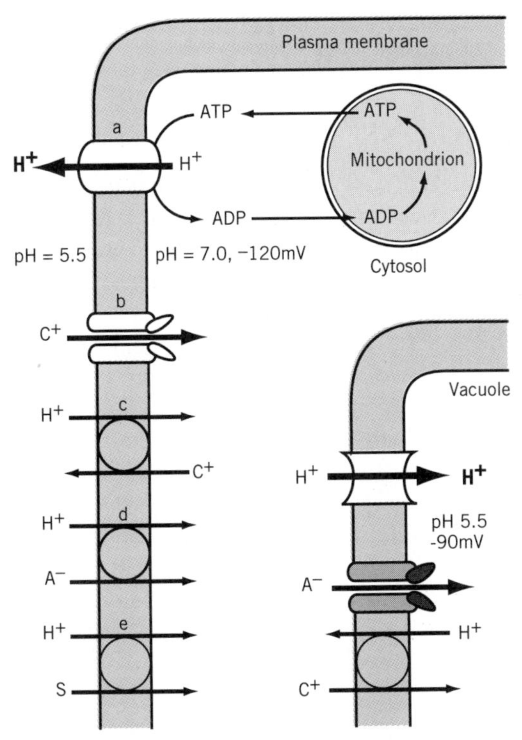

FIGURE 5.8 Schematic diagram relating the activity of a membrane ATPase proton pump to solute exchange. The proton pump (*a*) uses the energy of ATP to establish both a proton gradient and a potential difference (negative inside) across the membrane. The energy of the proton gradient may (*b*) activate an ion channel, or (*c*) drive the removal of ions from the cell by an antiport carrier, or drive the uptake of ions (*d*) or uncharged solute (*e*) by a symport carrier. Similar pumps and carriers operate across the tonoplast (vacuolar membrane). C^+, cation; A^-, anion; S, uncharged solute.

There are several particularly interesting consequences of the ATPase-proton pump. First, a single ion species is translocated in one direction. This form of transport is consequently known as a **uniport** system. Second, because the ion transported carries a charge, an electrochemical gradient is established across the membrane. In other words, the ATPase-proton pump is **electrogenic;** it contributes directly to the negative potential difference across the plasma membrane. In fact, the electrogenic proton pump is a major factor in the membrane potential of most plant cells. Recall that a tenfold difference in proton concentration (one pH unit) at 25° contributes 59 mV to the potential (Eq. 5.3). Since the proton gradient across the plasma membrane is normally on the order of 1.5 to 2 pH units, it can account for approximately 90 to 120 mV of the total membrane potential. Third, since the species translocated is protons, the ATPase-proton pump establishes a proton gradient as well as an electrical gradient across the membrane. Energy stored in the resulting electrochemical proton gradient (also known as the *proton motive force*) can then be coupled to cellular work in accordance with Mitchell's *chemiosmotic hypothesis* (Chap. 9). Indeed this is an excellent demonstration of how chemiosmotic coupling is not restricted to ATP synthesis in chloroplasts and mitochondria but can be used to perform other kinds of work elsewhere in the cell.

Four different ways of coupling the electrochemical proton gradient to solute movement across the membrane are illustrated in Figure 5.8. First, since the pump is electrogenic, it contributes to the charge-dependent, passive uptake of cations (e.g., K^+) through ion-specific channels. Second, the return of protons to the cytosol can be coupled with the **cotransport** of other solute molecules. In each case, transport of both ions is mediated by the same carrier protein and the movement of the second solute is obligatorily coupled to the inward flux of protons down their electrochemical gradient. In the first example, proton flux into the cell is shown coupled to the efflux of other cations (e.g., Ca^{2+}) out of the cell. Here the energy of the proton electrochemical gradient is used to maintain low internal concentrations of specific cations. Any cations that do chance to "leak" into the cell, no doubt passively through ion channels, are thus pumped out against their electrochemical gradient. Since in this case the two ions are moving in opposite directions, this form of cotransport is known as **antiport.** Finally, there are two examples of **symport,** by which the two solutes move in the same direction at the same time. In the first example, proton flux into the cell is coupled with the uptake of anions (A^-) against their electrochemical gradient. In the second example of symport, the proton gradient can be used to power the uptake of uncharged solutes (S), such as sugars. All three examples of cotransport are forms of active transport mediated by specific carrier proteins.

Much of the pioneering experimental work on membrane ATPases has been conducted with small, spherical vesicles derived from isolated cellular membrane (Sze, 1984). When membranes are disrupted, the pieces naturally seal off to form vesicles because of their strongly hydrophobic nature. While the technique is relatively straightforward, the preparation of vesicles from a single membrane source, and thus containing a single type of ATPase, presents some difficulties. Contamination by chloroplast ATPase can be avoided by isolating the membranes from dark-grown, etiolated tissue while mitochondria can usually be separated from other membranes by differential centrifugation. It is more difficult, however, to separate plasma membranes from other cellular membranes such as the tonoplast and, consequently, many of the early studies were characterized by inconsistent results from different laboratories. These inconsistencies were resolved when it became clear that the membrane preparations often contained two types of electrogenic ATPase proton pumps: one associated with the plasma membrane and one with the tonoplast. Improved techniques have enabled at least partial separation of the two membranes by density gradient centrifugation and it is now possible to characterize their respective ATPases.

The plasma membrane-type proton-pumping ATPase is characteristically inhibited by vanadate ion (VO_3^-) but is generally insensitive to other anions such as NO_3^-. Interestingly, activity of the plasma membrane ATPase is stimulated by K^+, although there is no evidence that it is directly involved in K^+ transport. Just how the ATPase transports protons across the membrane is not known. We can surmise on the basis of its vanadate sensitivity—vanadate competes with phosphate for binding sites—that ATP transfers a phosphate group to the protein. The resulting energy-rich *phosphoenzyme* is presumed to undergo a conformational change that effects the proton transfer. Evidence thus far indicates that a single proton is translocated for each ATP hydrolyzed.

The tonoplast-type proton-pumping ATPase differs from the plasma membrane-type in several ways. It is, for example, insensitive to vanadate but strongly inhibited by nitrate (NO_3^-). In this respect it is similar to mitochondrial ATPase, which is also insensitive to vanadate. Should there be any errant mitochondrial ATPase present, however, its activity can be blocked by including oligomycin or azide in the assay medium. Both inhibit mitochondrial ATPase without affecting the activity of the tonoplast-type. Insensitivity to vanadate suggests that the tonoplast ATPase, unlike its plasma membrane counterpart, does not form a phosphorylated intermediate. The tonoplast version also appears to differ in that it transports two protons for each ATP hydrolyzed.

The function of the tonoplast ATPase is to pump protons from the cytosol into the vacuole, thus accounting for the fact that the potential of the vacuole is more positive than the cytosol by some 20 to 30 mV (Fig. 5.8). In extreme cases, large pH gradients can be maintained across the tonoplast and the vacuolar sap may become quite acidic. For example, the pH of lemon juice (which is predominantly vacuolar sap), is normally about 2.5. Like the plasma membrane pump the tonoplast pump is also electrogenic, except that the accumulated protons serve to reduce the potential of the vacuole relative to the cytosol. The resulting potential difference serves to drive anions (e.g., Cl^- or malate) into the vacuole, which is less negative than the cytosol. The electrochemical proton gradient can also be used to drive cations (e.g., K^+ or Ca^{2+}) into the vacuole by an antiport carrier. Both of these activities make important contributions to the turgor changes that drive stomatal guard cell movement (Chap. 8) and the specialized motor cells that control nyctinastic responses (Chap. 19).

AQUAPORINS

In Chapter 1, it was noted that water permeated biological membranes more rapidly than would be expected on the basis of its solubility properties. Water is a polar molecule. It is not highly soluble in lipid, yet it freely passes through most membranes.

The free movement of water across membranes has traditionally been attributed to the simple diffusion of water molecules through the bilayer. However, the high hydraulic conductivity of both animal and plant membranes has been studied extensively over the past several decades and the results have generally supported the existence of channel-mediated water transport. More recently, molecular genetic techniques have led to the isolation of integral membrane proteins that appear to function as water-selective channels (Chrispeels and Maurel, 1994). These proteins are now called **aquaporins**.

Aquaporins are channel-forming proteins that facilitate the diffusion of water through membranes. It is important to remember that aquaporins do not act as pumps. As discussed in Chapter 2, the driving force behind water movement is the difference in water potential.

Aquaporins have been identified in numerous mammalian membranes as well as the tonoplast and the plasma membrane of *Arabidopsis thaliana*. The aquaporin γ-TIP is located in the tonoplast while the one designated RD28 appears to be a plasma membrane protein.

It is interesting that the gene coding for the tonoplast aquaporin, γ-TIP, is most highly expressed during cell elongation. RD28, on the other hand, is induced by desiccation. Thus both aquaporins appear to be produced when there is a high demand for water—to support cell growth or to avoid water stress—and the hydraulic conductivity of the cells is also high. Just how aquaporins achieve a high selectivity for water remains unclear and how they actually modulate intra- and intercellular water flow is a major challenge for future research.

ION UPTAKE BY ROOTS

Even though the uptake of ions by roots is essentially a cellular problem, the organization of roots at the tissue level cannot be totally ignored. The organization and architecture of roots (and other plant organs) are such that they can absorb mineral salts without them ever entering a cell.

DIFFUSION AND APPARENT FREE SPACE

Most nutrient uptake studies are carried out with "intact" tissues, such as excised barley roots or tissue slices of storage tissue from potato, carrot, or beet. Tissue slices normally require a period of aging before they are capable of rapid accumulation of ions. During this period a number of metabolic changes occur and it is not clear what relationship exists between aging and nutrient uptake. The most popular organ for study of ion uptake is excised roots, using a technique pioneered by Hoagland and Broyer (1936) and refined by Epstein (1961). Roots of grass seedlings, usually barley, are produced under conditions that encourage depletion of nutrients (Fig. 5.9). These so-called "low-salt" roots are excised and placed in cheesecloth "teabags" that can quickly and easily be transferred between experimental solutions. With this technique, a significant uptake of mineral elements can be demonstrated within as short a time as one minute.

A typical pattern for the uptake of a cation (such as Ca^{2+}) by roots is illustrated by the kinetic diagram in Figure 5.10. Note that initially, usually within the first few minutes, uptake of Ca^{2+} is very rapid. It then settles into a slow but steady accumulation over time. If at some point the roots are transferred to a large volume of solution lacking calcium, Ca^{2+} will be lost from the root into the bathing solution as shown by the dashed lines. When the bathing solution is distilled water, the quantity of ions lost is usually less than the quantity taken up during the initial rapid phase. If the roots are then trans-

FIGURE 5.9 Growing low-salt barley roots for ion uptake studies. Seeds are germinated on a stainless steel screen, allowing the roots to grow into a dilute sulfate solution. The solution is constantly aerated to provide healthy root growth.

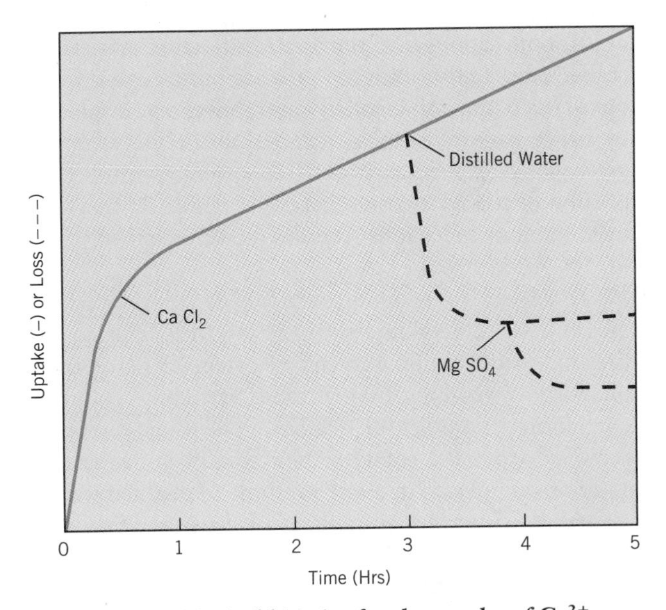

FIGURE 5.10 Typical kinetics for the uptake of Ca^{2+} into roots prepared as shown in Fig 5.9. When low-salt roots are placed in a solution of calcium chloride, an initial rapid uptake is followed by a slower but steady accumulation of calcium ion. If the roots are then transferred to a large volume of distilled water, some of the calcium diffuses out of the roots. Transfer to a strong magnesium solution releases additional calcium into the medium. The total amount of calcium released is equivalent to the amount taken up by free diffusion during the initial rapid phase.

ferred from distilled water to a bathing solution containing another cation, say Mg^{2+}, an additional quantity of Ca^{2+} will be lost from the tissue. If volumes of the bathing solutions are sufficiently large, the total quantity of Ca^{2+} lost from the tissue will approximately equal the quantity taken up during the initial rapid phase.

The kinetics of Ca^{2+} uptake and release can be interpreted as follows. Assume that there is a fraction of the root tissue volume that is not separated from the environment by a membrane or other diffusion barrier. This volume, called **apparent free space** (AFS), would therefore be accessible by simple diffusion. When root tissue is immersed in the calcium solution, Ca^{2+} will rapidly diffuse into the AFS until the Ca^{2+} concentration in the AFS reaches equilibrium with the bathing solution. Thereafter, calcium ions are more slowly but steadily transported across the cell membrane and accumulated by the tissue. When the roots are transferred to distilled water, only some of those cations present in the free space are free to diffuse back into the surrounding solution—and they will do so until equilibrium is again reached. Those cations already transported across the cell membrane remain in the cells.

Loss of additional calcium when the roots are transferred to the solution containing magnesium ions is taken as evidence that the tissue behaves as a cation exchange material. That is to say, the AFS matrix, primarily cell wall components, is negatively charged and holds some cations by electrostatic attraction just as soil colloids do. These absorbed ions are not free to diffuse out of the tissue into distilled water, but can be displaced by other cations, such as magnesium in the example given above. Thus, apparent free space describes that portion of the root tissue that is accessible by free diffusion and includes ions restrained electrostatically due to charges within the space.

A variety of techniques have been developed to measure the root volume given over to AFS. In principle they follow the example given by Epstein (1972). A sample of roots weighing 1.0 g were immersed in a solution containing 20 μmoles ml^{-1} K_2SO_4. The roots were then removed from the solution, blotted to remove excess solution and placed in a large volume of distilled water. A total of 4.5 μmoles of sulphate were released from the roots. If it is assumed that sulphate in the AFS is in equilibrium with the external solution, that is, 20 μmoles ml^{-1}, then the volume occupied by the sulphate in the tissue can be calculated: 4.5 μmole/20 μmole ml^{-1} = 0.22 ml. Thus the volume of root tissue freely accessible by diffusion is 0.22 ml. By further assuming that 1 g of root tissue occupies approximately 1 ml of volume, the proportion of tissue freely accessible by diffusion is approximately 22 percent by volume. Estimates for the volume occupied by AFS do vary, depending on the species, conditions under which the roots were grown, whether the measurements are corrected for surface films, and so forth. Still, most measured values for AFS tend to fall in the 10-to-25 percent range.

Exactly what constitutes AFS in a root? The answer to this question almost certainly is: the cell walls and intercellular spaces (equivalent to the apoplastic space) of the epidermis and cortex. That is to say, AFS comprises that volume of the root that can be entered *without crossing a membrane*. In other words, it is the volume of the root that is accessible by free diffusion. In most cases there is a strong correlation between the calculated volume of AFS and the calculated volume of cell walls in the cortex of the root. The cation exchange capacity of the AFS can be traced to the carboxyl groups (R·COO$^-$) associated with the galacturonic acid residues in the cell wall pectic compounds.

Almost as certainly, AFS stops at the endodermis where, in most roots, the radial and transverse walls develop characteristic thickenings called the **Casparian band** (see Chap. 3, Fig. 3.12). The Casparian band is principally composed of **suberin,** a complex mixture of *hydrophobic,* long-chain fatty acids and alcohols. These hydrophobic substances impregnate the cell wall, filling in the spaces between the cellulose microfibrils. They are, in addition, strongly attached to the plasma membrane of the endodermal cells. The hydrophobic, space-filling nature of the Casparian band along with its attachment to the membrane greatly reduces the possibility that ions or small hydrophilic molecules can pass between the cortex and stele without first passing through the endodermal protoplast. This means, of course, that they must pass through the plasma membrane of the endodermal cells and are, consequently, subject to all of the control and selectivity that implies.

THE RADIAL PATH OF ION MOVEMENT THROUGH ROOTS

Rapid distribution of nutrient ions throughout the plant is accomplished in the xylem vessels. In order to reach these conducting tissues, which are found in the central core (**stele**) of the root, the ions must move in a radial path through the root. The path these ions must follow is diagrammed in Figure 5.11. For these purposes we may consider the root as consisting of three principal regions. The outermost region consists of the root epidermis (often referred to as the **rhizodermis**) and the cortical cells. The innermost region consists of vascular tissues—the vessel elements and associated parenchyma cells are of particular interest for our discussion. Separating the two is the endodermis with its suberized Casparian band.

Ion movement begins with free diffusion into the apparent free space. As we noted above, the apparent free space is equivalent to the apoplast outside the endodermis but the Casparian band effectively prevents further apoplastic diffusion through the endodermis

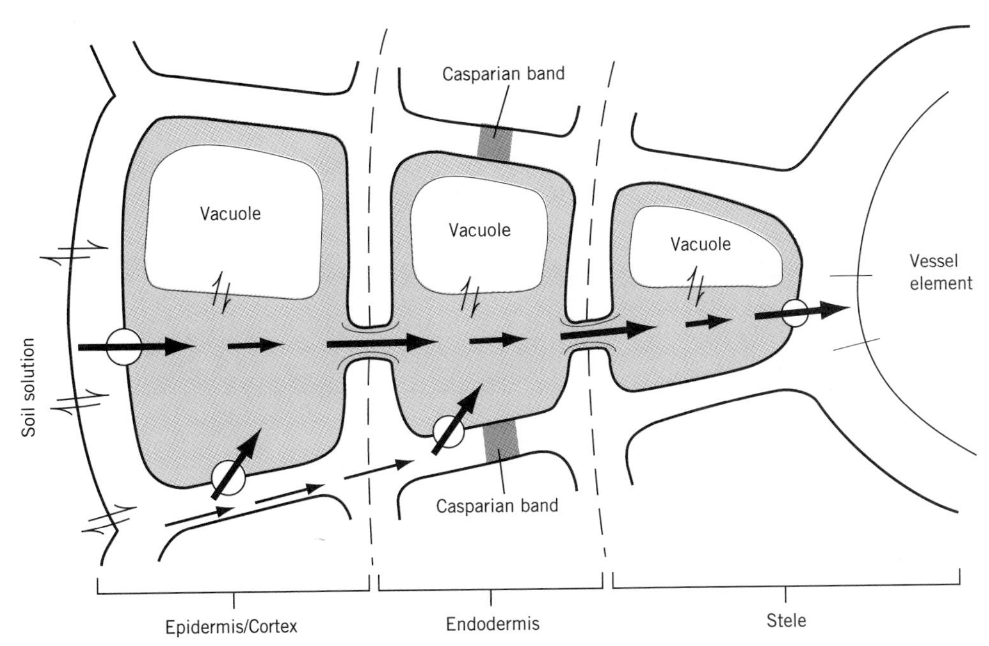

FIGURE 5.11 The radial path of ion movement through a root. Arrows indicate the alternate paths that may be taken by nutrient ions as they move from the soil solution into the vascular elements in the stele. Arrows with circles indicate active transport of ions across plasma membranes.

into the stele. Hence the only possible route for ions to pass through the endodermis is to enter the symplast by some carrier- or channel-mediated transport at the cell membrane. This may occur either on the outer tangential wall of the endodermal cell itself or through any of the epidermal or cortical cells. Regardless of which cell takes up the ions, symplastic connections (i.e., plasmodesmata) facilitate their passive movement from cell to cell until they arrive at a xylem parenchyma cell in the stele. At this point the ions are unloaded into the xylem vessels.

Except at the very tip of the root where the young xylem vessel elements are still maturing, functional xylem is part of the apoplast. The interconnected vessel elements are devoid of cytoplasm and consist only of nonliving, water-filled tubes. Release of ions into the xylem thus requires a transfer from the symplast into the apoplast. At one time, it was thought that this transfer was simply a passive leakage, but the evidence now suggests that ions are actively secreted from the xylem parenchyma. Although there is some conflicting evidence, ion concentration in the stelar apoplast is generally much higher than in the surrounding cortex. This suggests that ions are being accumulated against a concentration gradient, presumably by some carrier-mediated process. It is also interesting to speculate, in this regard, that another and perhaps principal function of the Casparian band is to prevent loss of ions from the stele by diffusion down their concentration gradient.

In addition to working uphill against a concentration gradient, delivery of ions into the vessels is sensitive to metabolic inhibitors such as carbonyl cyanide m-chlorophenylhydrazone (CCCP), which uncouples ATP formation. It is interesting that ion transport into the xylem is also sensitive to cycloheximide, an inhibitor of protein synthesis, but uptake into the root, at least initially, is not affected. Two plant hormones (abscisic acid and cytokinin) have a similar effect. Whether inhibitors of protein synthesis and hormones are affecting symplastic transport through the endodermis or unloading of ions from the endodermis into the xylem is not certain, but these results at least raise the possibility that ion release into the vessels is a different kind of process than ion uptake by the roots.

The possibility remains that a limited portion of ion uptake may be accomplished entirely through the apoplast, at least in some roots. More basal endodermal cells—the distance from the tip is variable, but measured in centimeters—are characterized by additional suberin deposits that cover the entire radial and inner tangential wall surfaces. This would seem to present an additional barrier to apoplastic flow. However, in *some* plants, a small number of endodermal cells, called **passage cells,** remain unsuberized. Passage cells might represent a major point of entry for solutes into the stele (Clarkson, 1991).

Apoplastic continuity between the cortex and stele may also be established at the point of lateral root formation. In one series of experiments, Carol Peterson and her coworkers traced the uptake of fluorescent dyes, which are normally confined to the apoplast, into the vascular tissues and shoots of corn (*Zea mays*) and broad

TABLE 5.3 Uptake and translocation of potassium and calcium as a function of position along a corn root.

Zone of application[1]	Ion	Total Uptake[2]	Percent Retained	Percent translocated to	
				Root Tip	Shoot
0–3	K^+	15.3	75	—	25
	Ca^{2+}	6.3	63	—	37
6–9	K^+	22.7	17	19	64
	Ca^2	3.8	42	—	58
12–15	K^+	19.5	10	10	80
	Ca^{2+}	2.8	14	—	86

[1]Distance from root tip, cm.

[2]Uptake expressed as microequivalents per 24 hours. Based on data of H. Marschner and C. Richter, 1973, Z. Pflenzenernaehr, Bodenkd, 135:1–15.

bean (*Vicia faba*) seedlings (Peterson et al., 1981). The point of dye entry was traced to recently emerged secondary roots. These branch roots arise in the **pericycle,** a layer of cells immediately to the inside of the endodermis. The emergence of the root primordia through the endodermis disrupts the continuity of the Casparian band and establishes, at least temporarily, the apoplastic continuity required to allow diffusion of the dye into the vascular tissue. Continuity of the apoplast through passage cells and secondary roots has been cited to explain increased calcium uptake in certain regions of corn roots. It may also help to account for the fact that a plant appears to contain virtually every element in its environment, even those not known to be essential or not accumulated by plant cells.

Finally, it should be noted the uptake of ions is not uniform along the length of the root. As shown in Table 5.3, uptake of calcium is highest in the apical 3 cm of the root while potassium is taken up in roughly equivalent amounts along the first 15 cm. Moreover, most of what is taken up in the tip (almost two-thirds of the calcium and three-fourths of the potassium) remains in the root. The proportion of ions translocated to the shoot increases with increasing distance from the tip. It is also interesting that when calcium is taken up further along the root (12–15 cm from the tip), it is translocated to the shoot but not to the tip. Clearly, although substantial progress has been made in several laboratories, the transport of ions through roots and into the xylem remains a complex and challenging field of study.

ROOT–MICROBE INTERACTIONS

The influence of living roots extends well beyond the immediate root surface into a region of the soil defined as the rhizosphere. *A principal manifestation of this influence is the numerous associations that develop between roots and soil microorganisms, especially bacteria and fungi.*

Root–microbe associations can at times be quite complex and may involve invasion of the host root by the microorganism. Alternatively the microorganism may remain free-living in the soil. In either case, the association may prove beneficial to the plant or it may be pathogenic and cause injury. Root–microbe associations have interested microbiologists for years but, with the exception of nitrogen-fixing organisms, have only recently attracted the attention of plant physiologists. Awakening interest is largely due to the realization that many of these microorganisms may play a significant role in the inorganic nutrition of plants.

BACTERIA

Plant roots generally support large populations of bacteria, principally because of the large supply of energy-rich nutrients provided by the growing root system. The immediate environment of the roots is so favorable that the bacterial population in the rhizosphere may exceed that in the surrounding bulk soil by as much as 50 percent. Nutrients provided by the roots are comprised largely of amino acids and soluble amides, reducing sugars, and other low molecular weight compounds. These compounds may either leak from the cells (a nonmetabolic process) or be actively secreted into the apoplastic space from whence they readily diffuse into the surrounding rhizosphere.

Dominant among the root secretions are the mucilages: polysaccharides secreted by Golgi vesicles in cells near the growing tip. Secretion of mucilage appears to be restricted to cells such as root cap cells, young epidermal cells, and root hairs where secondary walls have yet to form. Secretion of mucilage in the more basal regions of the root appears to be restricted by development of secondary walls.

In nonsterile soils, the mucilage is rapidly invaded by soil bacteria that contribute their own metabolic products, including mucopolysaccharides of the capsule. In addition, mucilage also attracts colloidal mineral and organic matter from the soil. The resulting mixture of root secretions, living and dead bacteria, and colloidal soil particles is commonly referred to as **mucigel.**

To what extent do the bacteria influence plant nutrition? There is no doubt that bacteria are intimately involved in the nitrogen nutrition of plants. Both invasive and free-living nitrogen-fixing bacteria, known since the late nineteenth century, are the primary source of nitrogen for plants. In addition, other soil bacteria convert ammonium nitrogen to nitrate. The role of these bacteria in nitrogen nutrition will be discussed more fully in the next chapter.

In the previous chapter we pointed out that phosphorous is sparingly soluble in most soils and, in natural ecosystems, is often the limiting nutrient. There is some evidence that soil bacteria can assist in making phosphorous available by solubilizing water-insoluble forms. It is considered unlikely, however, that this represents a major source of phosphorous for plants, especially in light of the extensive fungal associations described below. Bacteria can, however, enhance nutrient uptake other than by simply making nutrients more available. One way is to influence the growth and morphology of roots. One of the more striking examples is the formation of "proteoid" roots. This is a phenomenon of localized, intense lateral root production observed originally in the Proteaceae, a family of tropical trees and shrubs. (The Proteaceae includes the genus *Macadamea*, the source of the popular macadamia nut). Proteoid roots have now been found in several other families. Their induction has been traced to localized aggregations of bacteria in the mucigel. The larger number of lateral roots allows a more intensive mining of the soils for poorly mobile nutrients, such as phosphorous. In addition, proteoid roots are generally found near the soil surface where they can take advantage of nutrients leached out of the litter. The mechanism for proteoid root induction has not been determined, but could be related to IAA production by the bacteria.

MYCORRHIZA

Perhaps the most widespread—and from the nutritional perspective, more significant—associations between plants and microorganisms are those formed between roots and a wide variety of soil fungi (Marschner, 1986). A root infected with a fungus is called a **mycorrhiza** (literally, *fungus roots*). Mycorrhizae are a form of *mutualism*, in that both partners derive benefit from the intimate association. The significance of mycorrhizae is reflected in the observation that more than 80 percent of plants studied, including virtually all plant species of economic importance to man, form mycorrhizal associations.

Two major forms of mycorrhizae are known: ectotrophic and endotrophic (Fig. 5.12). The ectotrophic form, also known as **ectomycorrhizae,** is restricted to a few families consisting largely of temperate trees and shrubs, such as pines (Pinaceae) and beech (Fagaceae). Ectomycorrhizae are typically short, highly branched, and ensheathed by a tightly interwoven mantle of fungal hyphae (Fig. 5.12a, b). The fungus also penetrates the intercellular or apoplastic space of the root cortex, forming an *intercellular* network called a **Hartig net** (Fig. 5.12c). Endotrophic mycorrhizae, or **endomycorrhizae,** are found in some species of virtually every angiosperm family and most gymnosperms as well (except the Pinaceae). Unlike the ectomycorrhizae, the hyphae of endomycorrhizae develop extensively within cortical cells of the host roots (Fig. 5.12d). By far the most common type of endomycorrhiza, found in the majority of the world's vegetation, is the vesicular-arbuscular mycorrhiza (VAM). The hyphae of VAM grow between and into root cortical cells, where they form highly branched "treelike" structures called **arbuscules** (meaning dwarf tree) (Fig. 5.12e). Each branch of the arbuscule is surrounded by the plasma membrane of the host cell. Thus, while the hyphae do penetrate the host cell wall, they do not actually invade the protoplast. The arbuscule serves to increase contact surface area between the hypha and the cell by two to three times. At the same time, it apparently influences the host cell, which may increase its cytoplasmic volume by as much as 20 to 25 percent. Less frequently, VAMs form large ellipsoid **vesicles** either between or within the host cells. The presence of arbuscules and vesicles provides a large surface for the exchange of nutrients between the host plant and the invading fungus. Although VAMs do not form a well-defined sheath around the root, the hyphae, like those of the ectomycorrhizae, do effectively extend the rhizosphere by growing outward into the surrounding soil.

What can be said of the physiological role of mycorrhizae? When originally discovered by the nineteenth-century German botanist Frank, he concluded, on the basis of experiments conducted with beech seedlings, that mycorrhizal inoculation stimulated seedling growth. Although not universally accepted in the beginning, these results have been amply confirmed by more modern studies. Numerous studies with pine and other tree seedlings in the United States, Australia, and the former Soviet Union have demonstrated 30 to 150 percent increases in dry weight of tree seedlings infected with mycorrhizae when compared with noninfected controls. Similar results have been obtained in studies with agricultural plants such as maize (Fig. 5.13). In one

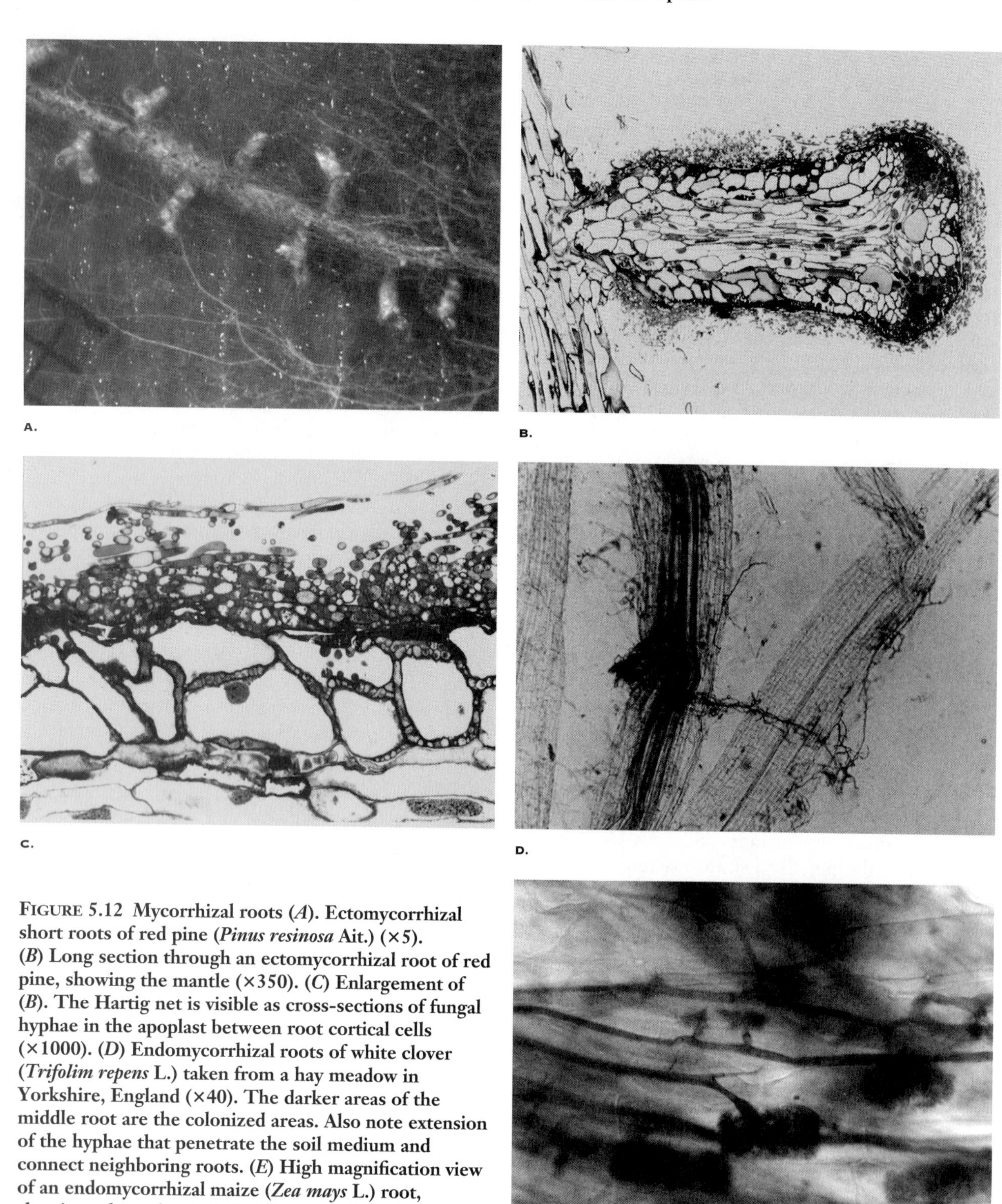

FIGURE 5.12 Mycorrhizal roots (*A*). Ectomycorrhizal short roots of red pine (*Pinus resinosa* Ait.) (×5). (*B*) Long section through an ectomycorrhizal root of red pine, showing the mantle (×350). (*C*) Enlargement of (*B*). The Hartig net is visible as cross-sections of fungal hyphae in the apoplast between root cortical cells (×1000). (*D*) Endomycorrhizal roots of white clover (*Trifolim repens* L.) taken from a hay meadow in Yorkshire, England (×40). The darker areas of the middle root are the colonized areas. Also note extension of the hyphae that penetrate the soil medium and connect neighboring roots. (*E*) High magnification view of an endomycorrhizal maize (*Zea mays* L.) root, showing arbuscules (×200). (Photographs *A*, *B*, and *C* courtesy of R. L. Peterson, University of Guelph; *D* and *E* courtesy of T. McGonigle, University of Guelph.)

experiment, for example, the dry weight of VAM-infected *Lavendula* plants increased 8.5 times over noninfected controls (Azecon et al., 1976).

The primary cause of mycorrhizal-enhanced growth appears to be enhanced uptake of nutrients, especially phosphorous. In a classic experiment, Hatch demonstrated in 1937 that infected pine seedlings absorbed 2 to 3 times more nitrogen, potassium, and phosphorous (Jackson and Mason, 1984). Coupled with enhanced nutrient uptake is the observation that

FIGURE 5.13 The response of maize (*Zea mays* L.) seedlings to mycorrhizal colonization (left) compared with uncolonized plants (right). (Photograph courtesy of T. McGonigle, University of Guelph.)

mycorrhiza-induced growth responses are more pronounced in nutrient-deficient soils. VAM infection, for example, can be effectively eliminated by supplying the plant with readily available phosphorous. With a surplus of phosphorous fertilizer, uninoculated plants will grow as well as those inoculated with mycorrhizal fungi.

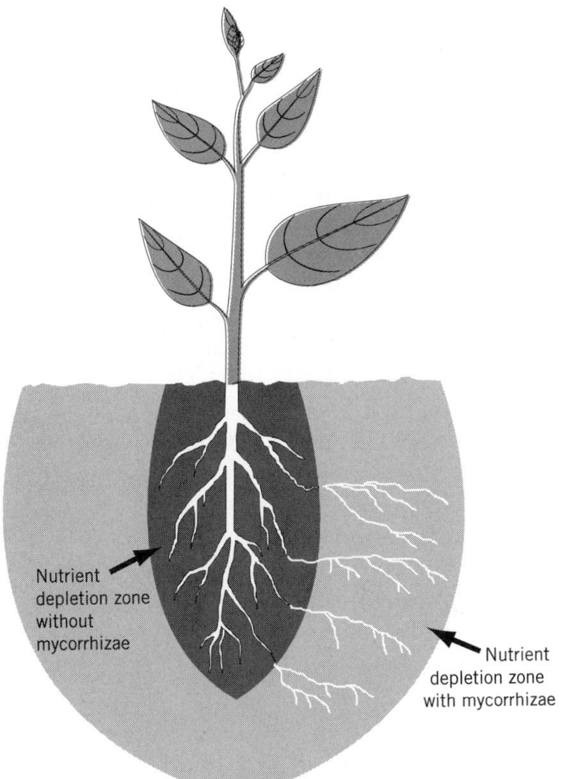

FIGURE 5.14 Infection of roots with mycorrhizal fungi extends the nutrient depletion zone for a plant.

The beneficial role of mycorrhizae, particularly with respect to the uptake of phosphorous, appears to be related to the **nutrient depletion zone** that surrounds the root (Fig. 5.14). This zone defines the limits of the soil from which the root is able to readily extract nutrient elements. Additional nutrients can be made available only by extension of the root into new regions of the soil or by diffusion of nutrients from the bulk soil into the depletion zone. The extent of the depletion zone varies from one nutrient element to another, depending on the solubility and mobility of the element in the soil solution. The depletion zone for nitrogen, for example, extends some distance from the root because nitrate is readily soluble and highly mobile. Phosphorous, on the other hand, is less soluble and relatively immobile in soils and, consequently, the depletion zone for phosphorous is correspondingly smaller. Mycorrhizal fungi assist in the uptake of phosphorous by extending their mycelia beyond the phosphorous depletion zone (Fig. 5.14). Apparently mycorrhizal plants find it advantageous to expend their carbon resources supporting mycorrhizal growth as opposed to more extensive growth of the root system itself.

As we continue to learn about mycorrhizae, their nutritional role becomes increasingly evident. Many mycorrhizae are host species–specific. Attempts to establish a plant species in a new environment may be unsuccessful if the appropriate mycorrhizal fungus is not present. Inoculation of fields with mycorrhizal fungi is already an additional factor taken into account by the forest and agricultural industries when attempting to resolve problems of soil infertility.

SUMMARY

The uptake of nutrient salts by plants involves a complex interaction between plant roots and the soil. The colloidal component of soil, comprised of clay particles and humus, presents a highly specific surface area carrying numerous, primarily negative, charges. Ions adsorbed to the charged colloidal surfaces represent the principal reservoir of nutrients for the plant. As ions are taken up from the dilute soil solution by the roots, they are replaced by exchangeable ions from the colloidal reservoir.

For nutrients to be taken up by a plant, they must be transported across the cell membrane into the root cell—thus making nutrient uptake fundamentally a cellular problem. Solutes may cross a membrane by simple diffusion, facilitated diffusion, or active transport. Facilitated diffusion and active transport are mediated by channel and carrier proteins—proteins which span the lipid bilayer. Only active transport achieves accumulation of ions against an electrochemical gradient. Active transport requires a source of metabolic energy, normally in the form of ATP. The free movement of water across membranes has long been an enigma, although it now appears to be a special case of facilitated diffusion through water-selective channels called aquaporins.

The uptake of nutrients by most plants is enhanced by association of the roots with soil microorganisms, especially fungi. Fungal-root associations (mycorrhizae) benefit the plant by significantly increasing the volume of soil accessible to the roots.

CHAPTER REVIEW

1. Distinguish between simple diffusion, facilitated diffusion, and active transport. Which of these three mechanisms would most probably account for:
   a) entry of a small lipid-soluble solute;
   b) extrusion of sodium ions leaked into a cell;
   c) rapid entry of a neutral hydrophilic sugar;
   d) accumulation of potassium ions?
2. The Casparian band was encountered earlier with regard to root pressure and again in this chapter with regard to ion uptake. What is the Casparian band and how does it produce these effects?
3. Trace the pathway taken by a potassium ion from the point where it enters the root to a leaf epidermal cell.
4. What are channel proteins and what role do they play in nutrient uptake?
5. Describe the concept of apparent free space. What role does apparent free space play in the uptake of nutrient ions?
6. Why is an accumulation ratio greater than 1.0 not necessarily an indication that active transport is involved?
7. How do mycorrhizae assist a plant in the uptake of nutrient elements?
8. Describe the colloidal properties of soil. How do the properties of colloids help to ensure the availability of nutrient elements in the soil?

FURTHER READING

Alberts, B., D. Bray, J. Lewis, M. Raff, K. Roberts, J. D. Watson. 1989. *Molecular Biology of the Cell.* 2nd ed. New York: Garland.

Clarkson, D. T. 1985. Factors affecting mineral nutrient acquisition by plants. *Annual Review of Plant Physiology* 36:77–115.

Epstein, E. 1972. *Mineral Nutrition of Plants: Principles and Perspectives.* New York: Wiley.

Marschner, H. 1986. *Mineral Nutrition of Higher Plants.* London: Academic Press.

Maurel, C. 1997. Aquaporins and water permeability of plant membranes. *Annual Review of Plant Physiology and Plant Molecular Biology* 48:399–429.

REFERENCES

Azecon, R., J. M. Barea, D. S. Hayman. 1976. Utilization of rock phosphate in alkaline soils by plants inoculated with mycorrhizal fungi and phosphate solubilizing bacteria. *Soil Biology and Biochemistry* 8:135–138.

Chrispeels, M. J., C. Maurel. 1994. Aquaporins: The molecular basis of facilitated water movement through living plant cells. *Plant Physiology* 105:9–13.

Clarkson, D. T. 1991. Root structure and sites of ion uptake. In: Y. Waisel, A. Eshel, U. Kafkafi, 1991, *Plant Roots: The Hidden Half.* New York: M. Dekker, pp. 417–453.

Epstein, E. 1961. The essential role of calcium in selective cation transport by plant cells. *Plant Physiology* 36:437–444.

Epstein, E. 1972. *Mineral Nutrition of Plants: Principles and Perspectives.* New York: Wiley.

Flegmann, A. W., R. A. T. George. 1975. *Soils and Other Growth Media.* London: Macmillan.

Harold, F. M. 1986. *The Vital Force: A Study of Bioenergetics.* New York: Freeman.

Higinbotham, N., B. Etherton, R. J. Foster. 1967. Mineral ion contents and cell transmembrane electropotentials of pea and oat seedling tissue. *Plant Physiology* 42:37–46.

Hoagland, D. R., T. C. Broyer. 1936. General nature of the process of salt accumulation by roots with description of experimental methods. *Plant Physiology* 11:471–507.

Jackson, R. M., P. A. Mason. 1984. *Mycorrhiza.* London: E. Arnold.

Marschner, H. 1986. *Mineral Nutrition of Higher Plants.* London: Academic Press.

Peterson, C. A., M. E. Emanual, G. B. Humphreys. 1981. Pathway of movement of apoplastic fluorescent dye tracers through the endodermis at the site of secondary root formation in corn (*Zea mays* cultivar Seneca Chief) and broad bean (*Vicia faba* cultivar Windsor). *Canadian Journal of Botany* 59:618–625.

Stout, P. R., R. Overstreet. 1950. Soil chemistry in relation to inorganic nutrition of plants. *Annual Review of Plant Physiology* 1:305–342.

Sze, H. 1984. H^+-translocating ATPases of the plasma membrane and tonoplast of plant cells. *Physiologia Plantarum* 61:683–691.

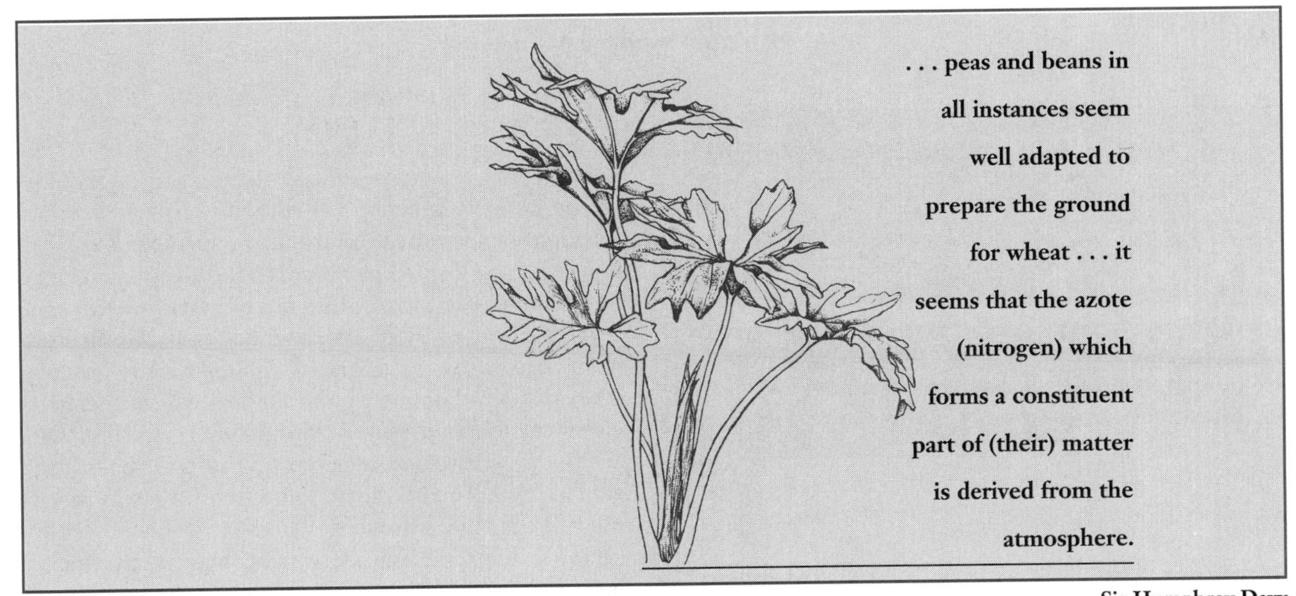

Sir Humphrey Davy
Agricultural Chemistry (1818)

6

Plants and Nitrogen

On a dry weight basis, nitrogen is the fourth most abundant nutrient element in plants. It is an essential constituent of proteins, nucleic acids, hormones, chlorophyll, and a variety of other important primary and secondary plant constituents. Most plants obtain the bulk of their nitrogen from the soil in the form of either nitrate (NO_3^-) or ammonium (NH_4^+), but the supply of nitrogen in the soil pool is limited and plants must compete with a variety of soil microorganisms for what nitrogen is available. As a result, nitrogen is often a limiting nutrient for plants, in both natural and agricultural ecosystems.

The bulk of the atmosphere, 78 percent by volume, consists of molecular nitrogen (N_2, or **dinitrogen**), an odorless, colorless gas. In spite of its abundance, however, higher plants are unable to convert dinitrogen into a biologically useful form. The two nitrogen atoms in dinitrogen are joined by an exceptionally stable bond ($N{\equiv}N$) and plants do not have the enzyme that will break this bond. Only certain prokaryote species are able to carry out this important reaction. This situation presents plants with a unique problem in respect to the uptake and assimilation of nitrogen; plants must depend on prokaryote organisms to convert atmospheric dinitrogen into a usable form. The nature of this problem and the solutions that have evolved are the subject of this chapter.

The principal topics discussed in this chapter include:

- a review of the nitrogen cycle: the flow of nitrogen between three major global nitrogen pools;
- the biology and biochemistry of biological nitrogen-fixing systems; and
- pathways for assimilation of ammonium and nitrate nitrogen by plants.

THE NITROGEN CYCLE

The global nitrogen supply is generally distributed between three major pools: the atmospheric pool, the soil (and associated groundwater) pool, and nitrogen contained within the biomass. The complex pattern of nitrogen exchange between these three pools is known as the **nitrogen cycle.**

Central to the idea of a nitrogen cycle (Fig. 6.1) is the pool of nitrogen found in the soil. Nitrogen from the soil pool enters the biomass principally in the form of nitrate (NO_3^-) taken up by plants and microorganisms. Once assimilated, nitrate nitrogen is converted to organic nitrogen in the form of amino acids and other nitrogenous building blocks of proteins and other macromolecules. Nitrogen moves further up the food chain

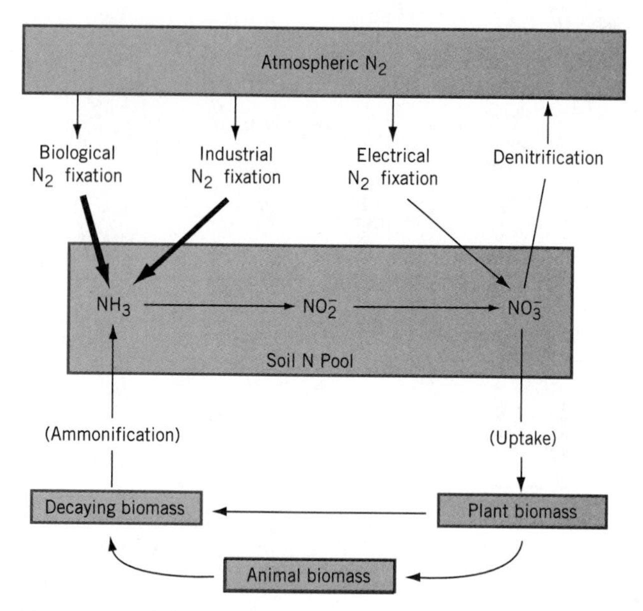

FIGURE 6.1 The nitrogen cycle, illustrating relationships between the three principal nitrogen pools: atmospheric, soil, and biomass.

when animals consume plants. Nitrogen is returned to the soil through animal wastes or the death and subsequent decomposition of all organisms.

AMMONIFICATION, NITRIFICATION, AND DENITRIFICATION

In the process of decomposition, organic nitrogen is converted to ammonia by a variety of microorganisms. This process is known as **ammonification** (Fig. 6.1). Some of the ammonia may volatilize and reenter the atmosphere, but most of it is recycled to nitrate by soil bacteria. The first step in the formation of nitrate is the oxidization of ammonia to nitrite (NO_2^-) by bacteria of the genera *Nitrosomonas* or *Nitrococcus.* Nitrite is further oxidized to nitrate by members of the genus *Nitrobacter.* These two groups are known as nitrifying bacteria and the result of their activities is called **nitrification.** Nitrifying bacteria are **chemoautotrophs;** that is, the energy obtained by oxidizing inorganic substances such as ammonium or nitrite is used to convert carbon dioxide to organic carbon.

In taking up nitrate from the soil, plants must compete with bacteria known as **denitrifiers** (e.g., *Thiobacillus denitrificans*). By the process of **denitrification,** these bacteria reduce nitrate to dinitrogen, which is then returned to the atmosphere. Estimates for the amount of nitrogen lost to the atmosphere by denitrification range from 93 million to 190 million metric tons annually.

NITROGEN FIXATION

The loss of nitrogen from the soil pool through denitrification is largely offset by additions to the pool through conversion of atmospheric dinitrogen to a combined or *fixed* form. The process of reducing dinitrogen to ammonia is known as **nitrogen fixation.**[1] Just how much dinitrogen is fixed on a global basis is difficult to determine with any accuracy. Figures on the order of 200 to 250 million metric tons annually have been suggested (Sprent and Sprent, 1990; Vitousek et al., 1997).

Approximately 10 percent of the dinitrogen fixed annually is accounted for by nitrogen oxides in the atmosphere. Lightning strikes and ultraviolet radiation each provide sufficient energy to convert dinitrogen to nitrogen oxides (NO, N_2O). Other sources of atmospheric nitrogen oxides (NO_x) are industrial combustion, forest fires, automotive exhausts, and power-generating stations. Although the chemistry of atmospheric nitrogen oxides is complicated and poorly understood, it is believed that both NO and N_2O are quickly oxidized to HNO_3 (nitric acid). Nitric acid in turn may be carried to earth in rain droplets or adsorbed to particles in the atmosphere and settle out as "dry deposits."

Another 30 percent of the total dinitrogen fixed, 80 million metric tons, is accounted for by industrial nitrogen fixation. The industrial process (known as the Haber-Bosch process) combines dinitrogen with hydrogen at elevated temperature (300 to 400°C) and pressure (35 to 100 MPa). Industrial nitrogen fixation is a costly process as it relies heavily on fossil fuels, both for the hydrogen source (natural gas) and the energy to achieve the necessary temperature and pressure. Most industrially fixed nitrogen is destined for use as agricultural fertilizers and about 1.5 kg of fuel oil is required to produce and deliver each kg of nitrogen to the farm (Nutman, 1976).

The balance of the nitrogen fixed on a global scale, about 60 percent or 150 to 190 million metric tons annually, is accounted for by the reduction of dinitrogen to ammonia by living organisms. This process is known as **biological nitrogen fixation.** Agriculture dating back to the ancient Greeks and Romans has taken advantage of biological nitrogen fixation through the cultivation of leguminous crops, although it was not until 1888 that the role of microorganisms in nitrogen fixation was discovered by H. Hellriegel and H. Wilfarth (Evans and Burris, 1992). Anticipating the day when fossil fuels required for industrial nitrogen fixation are in short sup-

[1]The process is commonly referred to simply as **nitrogen fixation,** but the term **dinitrogen fixation** has recently been introduced into the literature to reflect specifically the involvement of molecular nitrogen—N_2 or **dinitrogen.**

ply, an intensive research effort is now being directed toward understanding biological nitrogen fixation with a view toward improving on its contribution to agriculture.

BIOLOGICAL NITROGEN FIXATION

Biological nitrogen fixation is exclusively a prokaryote domain.

Plants are **eukaryotic** organisms, distinguished by the presence of a membrane-limited nuclear compartment. Eukaryotic organisms are unable to fix dinitrogen because they do not have the appropriate biochemical machinery. Bacteria and cyanobacteria are **prokaryotic** organisms; the genetic material is not contained within a membrane-limited organelle. Nitrogen fixation is a prokaryote domain, because only prokaryote organisms have the enzyme complex, called **dinitrogenase,** that catalyses the reduction of dinitrogen to ammonia. Simple as this may seem, biological nitrogen fixation turns out to be a complex biochemical and physiological process.

Prokaryotes that fix nitrogen, called **nitrogen-fixers,** include both free-living organisms and those that form symbiotic associations with other organisms.

FREE-LIVING NITROGEN FIXERS

Free-living, nitrogen-fixing bacteria are widespread. Their habitats include marine and freshwater sediments, soils, leaf and bark surfaces, and the intestinal tracts of various animals. Although some species are aerobic (e.g., *Azotobacter, Beijerinckia*), most will fix dinitrogen only under anaerobic conditions or in the presence of very low oxygen partial pressures (a condition known as **microaerobic**). These include both nonphotosynthetic genera (*Clostridium, Bacillus, Klebsiella*) and photosynthetic genera (*Chromatium, Rhodospirillum*) of bacteria. In addition to the bacteria, several genera of cyanobacteria (principally *Anabaena, Nostoc, Lyngbia*, and *Calothrix*) are represented by nitrogen-fixing species.

Although free-living nitrogen-fixing organisms are widespread, most grow slowly and, except for the photosynthetic species, tend to be confined to habitats rich in organic carbon. Because a high proportion of their respiratory energy is required to fix dinitrogen, less energy is therefore available for growth.

SYMBIOTIC NITROGEN FIXERS

Several types of symbiotic nitrogen-fixing associations are known, including the well-known association between various species of bacteria and leguminous plants. Some of the more important associations are listed in Table 6.1. In symbiotic associations the plant is identified as the **host** and the microbial partner is known as the **microsymbiont.** The most common form of symbiotic association results in the formation of enlarged, multicellular structures, called **nodules,** on the root (or occasionally the stem) of the host plant (Fig 6.2). In the case of legumes,[2] the microsymbiont is a bacterium of one of three genera: *Rhizobium, Bradyrhizobium*, or *Azorhizobium*. Collectively, these organisms are referred to as **rhizobia.** Curiously, only one nonleguminous genus, *Parasponia* (of the family Ulmaceae), is known to form root nodules with a rhizobia symbiont.

TABLE 6.1 **Some examples of specificity in rhizobia-legume symbiosis.**

Bacterium	Host
Azorhizobium	*Sesbania*
Bradyrhizobium japonicum	*Glycine* (soybean)
Rhizobium meliloti	*Medicago* (alfalfa), *Melilotus* (sweet clover)
Rhizobium leguminosarum	
biovar viciae	*Lathyrus* (sweet pea), *Lens* (lentil), *Pisum* (garden pea), *Vicia* (vetch, broad bean)
biovar trifolii	*Trifolium* (clover)
biovar phaseoli	*Phaseolus* (bean)
Rhizobium loti	*Lotus* (bird's-foot trefoil)

The rhizobia are further divided into species and subgroups called biovars (a biological variety) according to their host range (Table 6.1). Most rhizobia are restricted to nodulation with a limited number of host plants while others are highly specific, infecting only one host species.

Nodules are also found in certain nonleguminous plants such as alder (*Alnus*), bayberry (*Myrica*), Australian pine (*Casuarina*), some members of the family Rosaceae, and certain tropical grasses. However, the microsymbiont in these nonleguminous nodules is a

[2]The legumes are a heterogeneous group traditionally assigned to the family Leguminosae. Modern treatments split the group into three families: Mimosaceae, Caesalpiniaceae, and Fabaceae (S. B. Jones, A. E. Luchsinger, *Plant Systematics*, New York, McGraw-Hill, 1986). Most of the economically important, nitrogen-fixing legumes are assigned to the Fabaceae.

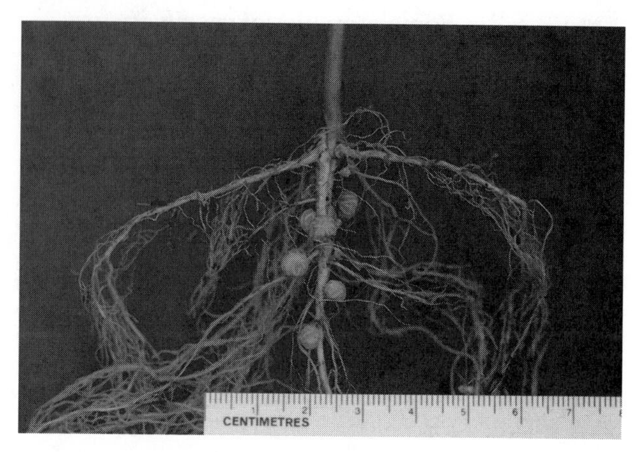

FIGURE 6.2 Nitrogen-fixing nodules on roots of soybean (*Glycine max*). (See color plate 2.)

filamentous bacterium (*Frankia*) of the group actinomycetes. Both *Rhizobium* and *Frankia* live freely in the soil, but fix dinitrogen only when in symbiotic association with an appropriate host plant.

A limited number of non-nodule-forming associations have been studied, such as that between *Azolla* and the cyanobacterium *Anabaena*. *Azolla* is a small aquatic fern that harbors *Anabaena* in pockets within its leaves. In southeast Asia, *Azolla* has proven useful as green manure in the rice paddy fields where it is either applied as a manure or co-cultivated along with the rice plants. Because more than 75 percent of the rice acreage consists of flooded fields, free-living cyanobacteria and anaerobic bacteria may also make a significant contribution. These practices have allowed Asian rice farmers to maintain high productivity for centuries without resorting to added chemical fertilizers.

SYMBIOTIC NITROGEN FIXATION IN LEGUMES

Symbiotic nitrogen fixation in legumes involves complex anatomical, morphological, and biochemical interactions between the host plant and the invading microorganisms.

It is generally agreed that symbiotic nitrogen fixers, particularly legumes, contribute substantially more nitrogen to the soil pool than do free-living bacteria. Typically a hectare of legume-*Rhizobium* association will fix 25 to 60 kg of dinitrogen annually, while nonsymbiotic organisms fix less than 5 kg ha^{-1} (Sprent and Sprent, 1990). There are over 17,000 species of legumes. Even though only 20 percent have been examined for nodulation, 90 percent of those examined do form nodules. Given the obvious significance of the legume symbiosis to the nitrogen cycle, it is worth examining in some detail.

INFECTION AND NODULE DEVELOPMENT

The sequence of events beginning with bacterial infection of the root and ending with formation of mature, nitrogen-fixing nodules has been studied extensively in the legumes, historically from the morphological perspective and more recently from the biochemical/molecular genetic perspective. Overall the process involves a sequence of multiple interactions between the bacteria and the host roots. In effect, the rhizobia and the roots of the prospective host plant establish a dialogue in the form of chemical messages passed between the two partners. Based on studies carried out primarily with *Glycine*, *Trifolium*, and *Pisum*, as many as nine or ten separate developmental stages have been recognized (Sprent and Sprent, 1990). In order to simplify our discussion, however, we will consider the events in four principal stages:

1. Multiplication of the rhizobia, colonization of the rhizosphere, and attachment to epidermal and root hair cells.
2. Characteristic curling of the root hairs and invasion of the bacteria to form an infection thread.
3. Nodule initiation and development in the root cortex. This stage is concurrent with stage 1.
4. Release of the bacteria from the infection thread and their differentiation as specialized nitrogen-fixing cells.

The four principal stages are illustrated in Figure 6.3. In this section, we will concentrate on the physiology and morphology of infection and nodule development. Genetics aspects will be addressed in a subsequent section.

The Early Stage—Colonization and Nodule Initiation Rhizobia are free-living, saprophytic soil bacteria. Their numbers in the soil are highly variable, from as few as zero or 10 to as many as 10^7 $gram^{-1}$ of soil, depending upon the structure of the soil, water content, and a variety of other factors. In the presence of host roots, the bacteria are encouraged to multiply and colonize the rhizosphere. The initial attraction of rhizobia to host roots appears to involve **positive chemotaxis,** or movement toward a chemical stimulant. Chemotaxis is an important adaptive feature in microorganisms generally. It allows the organism to detect nutrients and other chemicals that are either beneficial or required for their growth and reproduction. Roots are known to exude a variety of amino acids, sugars, and organic acids that may function as nutrients for rhizobia and other soil microorganisms. Pea roots, for example, exude an unusual amino acid, *homoserine*, which is a preferred source of carbon and nitrogen for the pea root symbiont, *Rhizobium leguminosarum* (biovar *viciae*). A large increase in

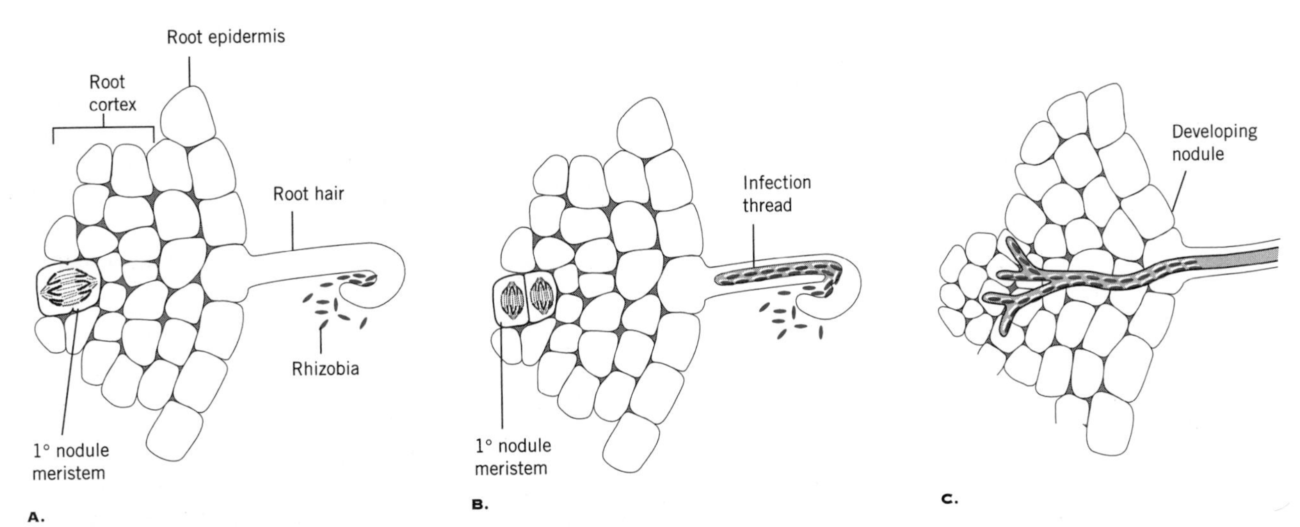

FIGURE 6.3 Schematic diagram of the infection process leading to nodule formation. (*A*) Rhizobia colonize the soil in the vicinity of the root hair in response to signals sent out from the host root. The rhizobia in turn stimulate the root hair to curl while, at the same time, sending mitogenic signals that stimulate cell division in the root cortex. (*B*) Rhizobia invade the root by digesting the root hair cell wall and forming an infection thread. The rhizobia continue to multiply as the infection thread elongates toward the root cortex. (*C*) The infection thread branches to penetrate numerous cortical cells as a visibly evident nodule develops on the root. The final stage (not shown) is the release of rhizobia into the host cells and the activation of the nitrogen-fixing machinery.

the number of rhizobia on the pea root surfaces was found to correlate with the liberation of significant amounts of homoserine (Egeraat, 1975).

Another group of chemicals that have been implicated in attraction of rhizobia are the **flavonoids** (Fig. 6.4). Flavonoids, which include the red, blue, and purple pigments of leaves and floral petals as well as a number of other important plant products, are described more fully in Chapter 7. A wide variety of flavonoids have now been characterized in root exudates, many of which stimulate nodulation but some of which actually inhibit the process (Rolfe, 1988). It must be noted, however, that flavonoids are found in root exudates of many non-leguminous species as well and there is no direct evidence that flavonoids serve as chemotactic agents.

Once rhizobia have colonized the rhizosphere, they begin to synthesize morphogenic signal molecules called nodulation factors, or **nod factors** (Dénarié and Cullimore, 1993). Nod factors are derivatives of chitin, a β-1→4-linked polymer of N-acetyl-D-glucosamine found in the cell walls of fungi and exoskeletons of insects. Nod factors are similar polymers except that a

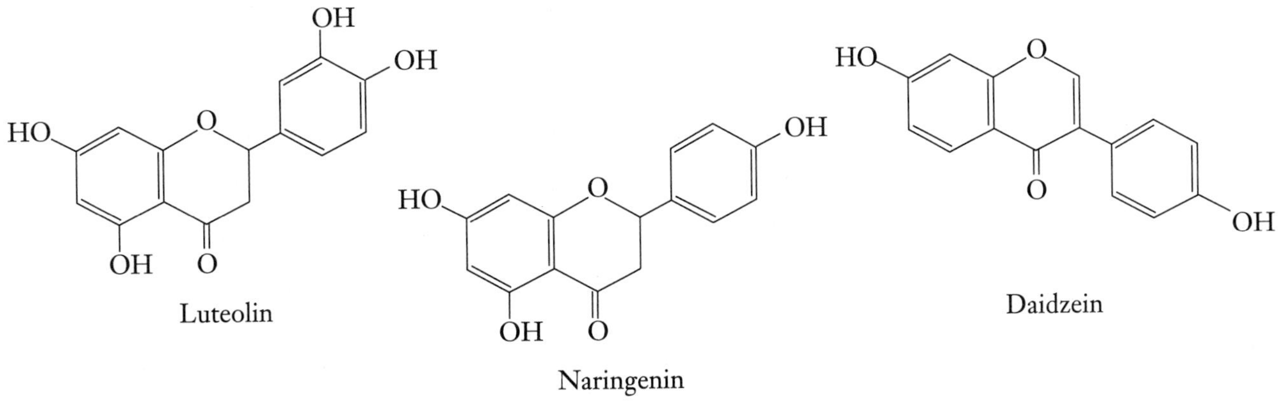

Luteolin

Naringenin

Daidzein

FIGURE 6.4 Structures of three common flavonoids implicated in rhizobia-host interactions. Leuteolin (flavone), narigenin (flavanone), and daidzein (isoflavone) are released by the host root. The flavonoid interacts with the product of the bacterial nodD gene, leading to the induction of other nodulation genes.

Box 6.1
Lectins—Proteins with a Sweet Tooth

Castor oil, pressed from the seeds of castor bean (*Ricinus communis*), has long been used both as a medicinal purgative and as a lubricating oil for fine machinery. For just as long, castor bean seed has also been recognized as one of the most potent animal toxins known. The toxic principle, a protein called **ricin,** is found in the residue, or press cake, which remains following the extraction of the oil.

Ricin was discovered by a nineteenth-century Estonian scientist in the course of his studies toward a degree in medicine. H. Stillmark began his studies as a pharmacological investigation of the toxic principle in *Ricinus* seed. In his dissertation, published in 1888, Stillmark reported the fortuitous observation that aqueous extracts of castor bean caused agglutination of mammalian erythrocytes. With time, other **hemagglutinins** were found in a variety of plants. They were named **lectins** (L. *legere* = to choose), as it was learned that their hemagglutinin properties resulted from the presence of specific carbohydrate binding sites on the protein. It is now well established that lectins are not restricted to plants but are found in bacteria, algae, fungi, invertebrates and vertebrates as well. In plants, lectins have been described in every major taxonomic group of flowering plants and many nonflowering plants as well.

In the broadest sense, lectins are carbohydrate-binding proteins capable of recognizing and reversibly binding with specific complex carbohydrates. As proteins go, lectins are relatively small; molecular masses range from 50 kD to 120 kD. They usually consist of two (dimer) or four (tetramer) identical subunits. Two lectins studied in greatest detail are **concanavalin A** (or Con A) isolated from seeds of the tropical legume Jackbean (*Conavalina ensiformis*) and **wheat germ agglutinin** isolated from embryos of wheat (*Triticum* sps.). Con A, a tetrameric protein with a molecular mass of 104 kD, specifically recognizes α-D-mannosyl residues. Wheat germ agglutinin is a dimeric protein (MM = 36 kD) that recognizes oligomers of β-(1 → 4)-N-acetyl-D-glucosamine. Interestingly, lectins from at least 90 species of closely related cereals all have properties very similar to wheat germ agglutinin.

Lectins, because of their ability to bind with specific carbohydrates, have proven very useful as an analytical and preparatory tool in the laboratory. One example is the use of lectins in mapping the surface architecture of mammalian cell membranes. Cell

fatty acid replaces the acetyl group at one end of the molecule. Nod factors are consequently considered *lipo-chitooligosaccharides*. Nod factors secreted into the soil solution by the rhizobia induce several significant changes in the growth and metabolism of the host roots as a prelude to rhizobial invasion of the root hair and subsequent nodule development. These changes (Fig. 6.3A) include increased root hair production and the development of shorter, thicker roots. Stimulated by the nod factors to renew their growth, the root hairs develop branching and curl at the tip.

Before actually invading the host, rhizobia also release mitogenic signals that stimulate localized cell divisions in the root cortex. These cell divisions form the **primary nodule meristem,** defining the region in which the nodule will eventually develop (Fig. 6.3A). A second center of cell division arises in the pericycle. Eventually these two masses of dividing cells will fuse to form the complete nodule.

The nature of the mitogenic signal is unknown, although there is some evidence that the plant hormone ethylene could be involved. Ethylene promotes root hair development in some plants and responses similar to those induced by nod factors can be mimicked by ethephon, a chemical that releases ethylene. More recent experiments have shown that ethylene has no role in root hair deformation induced by nod factors, but enzymes involved in the synethesis of ethylene are expressed in the region of the cortex where the primary nodule meristem arises (Heidstra et al., 1997). It has been suggested that, although ethylene normally inhibits cortical cell division, it could be involved in determining the position of the primary nodule meristem in the cortex.

Rhizobia-host specificity is probably determined when the rhizobia attach to the root hairs and must involve some form of recognition between symbiont and host. As a general principle, recognition between cells involves chemical linkages that form between unique molecules on cell surfaces. In the case of rhizobia-host interactions, recognition appears to involve two classes of molecules: lectins and complex polysaccharides. Lectins are small, nonenzymatic proteins synthesized by the host and have the particular ability to recognize and bind to specific complex carbohydrates (see Box 6.1: Lectins—Proteins with a Sweet Tooth).

membranes contain both glycolipids and glycoproteins, with their sugar residues located on the external membrane surface. If a lectin is first conjugated with an electron-dense marker (such as the Fe-protein, ferritin) and then bound to membrane glycoproteins or glycolipids, the locations of those membrane components can be visualized in the electron microscope. Indeed, the ability of lectins to cause agglutination of red blood cells can be explained by their capacity to form crosslinks between sugar residues on several red blood cells. Hemagglutination is still the principal test for detecting lectins. Lectins have also been used to isolate and analyze complex carbohydrates and for separation of isolated cells based on their specific surface carbohydrates.

Unfortunately, in spite of extensive information on the chemistry of lectins—the amino acid sequences of several lectins are well known—their biological role in plants remains obscure. The abundance of lectins in seeds has directed attention to possible roles in seed maturation and germination or in enforcing seed dormancy. Lectins are found primarily in seed protein bodies, leading to the suggestion that they function in organizing or mobilizing seed storage protein. Because of their ability to inhibit fungal growth and germination, lectins have been implicated in plant defense mechanisms, particularly of seeds. A related role, postulating recognition between host roots and bacteria in symbiosis between *Rhizobia* and legumes is described more fully in this chapter. It has been suggested that a similar recognition between lectin and microorganism may be a factor in determining pathogenicity. While many of these postulated roles may seem particularly appealing, the evidence in support of any one of them is far from unequivocal. Still, it is difficult to accept that molecules of such abundance and with such evident functional properties would not have some significant biological function. We can only anticipate that continuing research in "lectinology" will soon resolve this question.

FURTHER READING

Etzler, M. E. 1985. Plant lectins: Molecular and biological aspects. *Annual Review of Plant Physiology* 36:209–234.

Franz, H. 1988. *Advances in Lectin Research*. Berlin: Springer-Verlag, 187 pp.

Goldstein, I. J., R. C. Hughes, M. Monsigny, T. Osawa, N. Sharon. 1980. What should be called a lectin? *Nature* (London) 285:66.

Sharon, N., H. Lis. 1989. Lectins as cell recognition molecules. *Science* 246:227–246.

Individual legume species each produce different lectins with different sugar-binding specificities. Lectins appear to recognize complex polysaccharides found on the surface of the potential symbiont. Although bacterial surfaces normally contain an array of complex extracellular polysaccharides, the synthesis of additional nodulation-specific extracellular polysaccharides is directed by bacterial genes that are activated in the presence of flavonoids in the host root exudate. Host range specificity would thus result from attachment of the rhizobium to the host root hair because of specific lectin-surface polysaccharide interactions. Support for this hypothesis comes from experiments in which the gene for pea lectin was introduced into roots of white clover (Diaz et al., 1989). The result was that clover roots could be nodulated by strains of *Rhizobium leguminosarum*, biovar *viciae*, which are usually specific for peas.

Lectin may, however, be only part of the story. Other experiments have indicated the involvement of a calcium-binding protein, called **rhicadhesin,** located on the surface of the *rhizobial* cell (Smit et al., 1989). Rhicadhesin appears to be common to all rhizobia and is required for attachment, at least in pea. In addition to lectin and rhicadhesin, other physicochemical factors may also have a role in attachment.

Invasion of the Root Hair and the Infection Thread In the second stage of nodulation, the bacterium must penetrate the host cell wall in order to enter the space between the wall and the plasma membrane. In pea, the preferred attachment site is the tip of the growing root hair. The root hairs of pea grow by **tip growth;** that is, new wall material is laid down only at the tip of the elongating hair cell. Colonies of attached rhizobia become entrapped by the tip of the root hair as it curls around. How rhizobia actually breach the cell wall is not known, but the process almost certainly includes some degree of wall degradation. There is some evidence that rhizobia release enzymes such as pectinase, hemicellulase, and cellulase, which degrade cell wall materials. These enzymes could result in localized interference with the assembly of the growing wall at the root tip and allow the bacteria to breach the cell wall and gain access to the underlying plasma membrane.

Once the rhizobia reach the outer surface of the plasma membrane, tip growth of the root hair ceases and the cell membrane begins to invaginate. The result is a tubular intrusion into the cell called an **infection thread,** which contains the invading rhizobia (Fig. 6.3B). The infection thread elongates by adding new membrane material by fusion with vesicles derived from the Golgi apparatus. As the thread moves through the root hair cell, a thin layer of cellulosic material is deposited on the inner surface of its membrane. Because this new wall material is continuous with the original cell wall, the invading bacteria never actually enter the host cell but remain technically outside the cell.

The infection thread continues to elongate until it reaches the base of the root hair cell. Here it must again breach the cell wall in order for the bacteria to gain access to the next cell in their path. This is apparently accomplished by fusing the infection thread membrane with the plasmamembrane (Bauer, 1981). In the process, some bacteria are released into the apoplastic space. These bacteria apparently degrade the walls of the next cell in line, thus allowing the infection process to continue into successive cells in the cortex. As the infection thread moves through the root hair into the cortex, the bacteria continue to multiply. When the thread reaches the developing nodule, it branches so that many individual cells in the young nodule become infected (Fig. 6.3C).

The Release of Bacteria The final step in the infection process occurs when the bacteria are "released" into the host cells. Actually the membrane of the infection thread buds off to form small vesicles, each containing one or more individual bacteria. Shortly after release, the bacteria cease dividing, enlarge, and differentiate into specialized nitrogen-fixing cells called **bacteroids.** The bacteroids remain surrounded by a membrane, now called the **peribacteroid membrane.** Differentiation into a bacteroid is marked by a number of metabolic changes, including the synthesis of the enzymes and other factors that the organism requires for the principal task of nitrogen fixation.

The infection process continues throughout the life of the nodule. As the nodule increases in size due to the activity of the nodule meristem, bacteria continue to invade the new cells. Also as the nodule enlarges and matures, vascular connections are established with the main vascular system of the root (Fig 6.5). These vascular connections serve to import photosynthetic carbon into the nodule and export fixed nitrogen from the nodule to the plant.

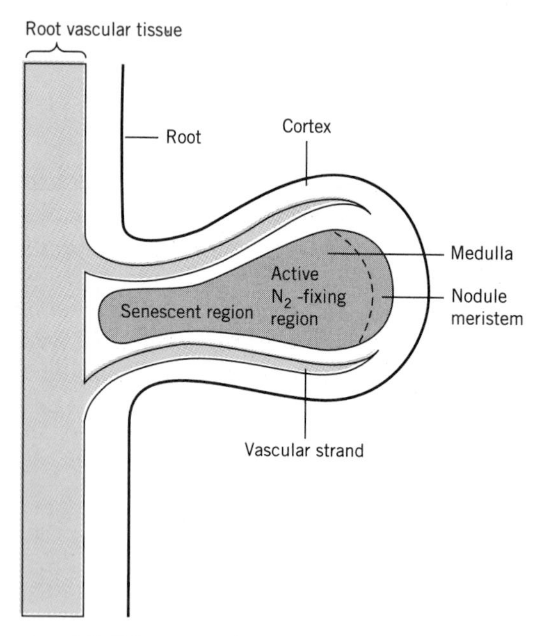

FIGURE 6.5 Schematic diagram of a cross-section through a mature nodule. Vascular connections with the host plant provide for the exchange of carbon and nitrogen between the host and the microsymbiont.

THE BIOCHEMISTRY OF NITROGEN FIXATION

Dinitrogen is not easily reduced because the interatomic nitrogen bond (N≡N) is very stable. In the industrial process, reduction of the dinitrogen triple bond with hydrogen can be achieved only at high temperature and pressure and at the cost of considerable energy. Biological reduction of dinitrogen is equally costly, consuming a large proportion of the photoassimilate provided by the host plant.

DINITROGENASE

Nitrogen fixation is catalyzed by an enzyme complex known as **dinitrogenase** (Box 6.2). Only prokaryote cells are able to fix dinitrogen principally because only they have the gene coding for this enzyme. Dinitrogenase has been purified from virtually all known nitrogen-fixing prokaryotes. It is a multimeric protein complex made up of two proteins of different size (Fig. 6.6). The smaller protein is a **dimer** consisting of two identical subunit polypeptides. The molecular mass of each

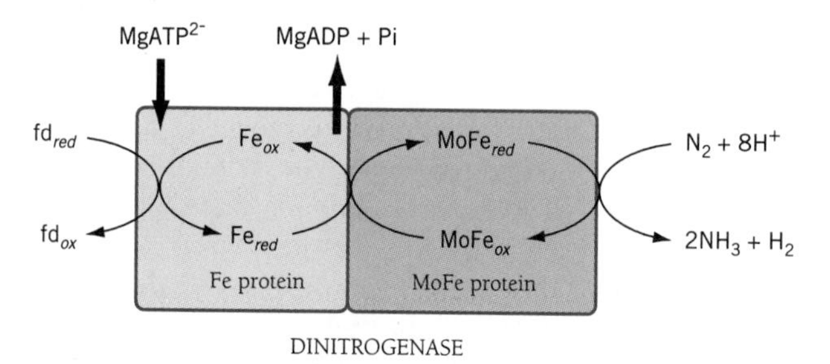

FIGURE 6.6 Schematic diagram of the nitrogenase reaction in bacteroids. Electron flow is from left to right. The principal electron donor is ferredoxin (fd), which receives its electron from respiratory substrate.

subunit ranges from 24 to 36 kD, depending on the bacterial species. It is called the **Fe protein** because the dimer contains a single cluster of four iron ions bound to four sulphur groups (Fe_4S_4). The larger protein in the dinitrogenase complex is called the **MoFe protein.** It is a **tetramer** consisting of two pairs of identical subunits with a total molecular mass of approximately 220 kD. Each MoFe protein contains two molybdenum ions in the form of an iron-molybdenum- sulphur cofactor. The MoFe protein also contains Fe_4S_4 clusters, although the exact number is uncertain. It varies as a function of the species or its physiological condition.

The overall reaction for reduction of dinitrogen to ammonia by dinitrogenase is shown in the following equation:

$$8H^+ + 8e^- + N_2 + 16\ ATP \longrightarrow 2NH_3 + H_2 + 16\ ADP + 16\ P_i \quad (6.1)$$

Note that the principal product of biological nitrogen fixation is ammonia, but that for every dinitrogen molecule reduced one molecule of hydrogen is generated. We will return to the problem of hydrogen evolution later. Also note that reduction of dinitrogen is a two-step process. In the first step, the Fe protein is reduced by a primary electron donor; usually **ferredoxin.** Ferredoxin is a small (14 to 24 kD) protein containing an iron-sulphur group. Electrons are carried by the iron moiety, which can exist in either the reduced ferrous (Fe^{2+}) or the oxidized ferric (Fe^{3+}) states. It is of interest to note that ferredoxin not only participates in nitrogen fixation, but is an important electron carrier in photosynthesis as well (see Chap. 9).

In the second step, the reduced Fe protein passes electrons to the MoFe protein, which catalyzes the reduction of both dinitrogen gas and hydrogen. The precise role of ATP in the reaction is not yet clear, but it is known to react with reduced Fe protein and is believed to play a role in the transfer of electrons between the Fe protein and the MoFe protein.

THE ENERGY COST OF NITROGEN FIXATION

Biological reduction of dinitrogen, as is industrial nitrogen fixation, is very costly in terms of energy. One measure of energy cost is the number of **adenosine triphosphate (ATP)** required. (ATP is the energy coinage of the cell. Its role in biological energetics is developed more fully in Chapter 9). At least 16 **ATP** are required for each molecule of dinitrogen reduced; two for each electron transferred (Eq. 6.1). By comparison, only 3 ATP are required to fix a molecule of carbon dioxide in photosynthesis (see Chap. 10). The total energy cost of biological nitrogen fixation, however, must take into account the requirement for reduced ferredoxin as well. Were this reducing potential not required for the reduction of dinitrogen, it could have been made available for the production of additional ATP or other uses by the plant. It has been estimated that the reducing potential used in nitrogen fixation is equivalent to at least a further 9 ATP, bringing the total investment to a minimum of 25 ATP for each molecule of dinitrogen fixed. A similar calculation for CO_2 fixation brings the total to 9 ATP, about one-third the cost for nitrogen.

Another and perhaps better way to assess the cost of nitrogen fixation is by measuring the amount of carbon utilized in the process. The ultimate source of energy for symbiotic nitrogen fixation is carbohydrate produced by photosynthesis in the host plant. A portion of that carbohydrate is diverted from the plant to the bacteroid where it is metabolized to produce the required reducing potential and ATP. P. G. Heytler et al. (1984) have calculated that, in soybean, aproximately 12 grams of carbon are required to fix a gram of dinitrogen. It is clear that nitrogen fixation represents a considerable drain on the carbon resources of the host plant. A diagram summarizing the integration of photosynthesis, respiration, and nitrogen fixation is presented in Figure 6.7.

DINITROGENASE AND OXYGEN

One of the more critical problems facing nitrogen fixing organisms is the sensitivity of dinitrogenase to molecular oxygen. Both the Fe protein and the MoFe protein are rapidly and irreversibly inactivated by molecular oxygen. The **half-life,** or time to reduce activity by one-half, of isolated Fe protein in air is 30 to 45 seconds; the half-life of MoFe protein is 10 minutes. This extreme sensitivity of dinitrogenase to oxygen raises a problem for nitrogen-fixing organisms. The large amounts of energy required (in the form of ATP and reductant) are produced through a cellular respiratory pathway that can operate efficiently only when molecular oxygen is present (see Chap. 12). How then does the organism reconcile the conflicting demands of the respiratory pathway for oxygen and the sensitivity of dinitrogenase to oxygen?

Several strategies for regulating oxygen level have developed to resolve this conflict. Many free-living bacterial nitrogen fixers have retained an anaerobic lifestyle or, if facultative, fix dinitrogen only under anaerobic conditions. Production of ATP and reductant is markedly less efficient under anaerobic conditions, which may offer a partial explanation for why, in spite of their numbers, free-living nitrogen fixers contribute a relatively small proportion of the total nitrogen fixed biologically. Other organisms, such as cyanobacteria, have structurally isolated the nitrogen fixing apparatus (Fig. 6.8). The nitrogen-fixing cells of the cyanobacteria are specialized cells called **heterocysts.** Heterocysts have

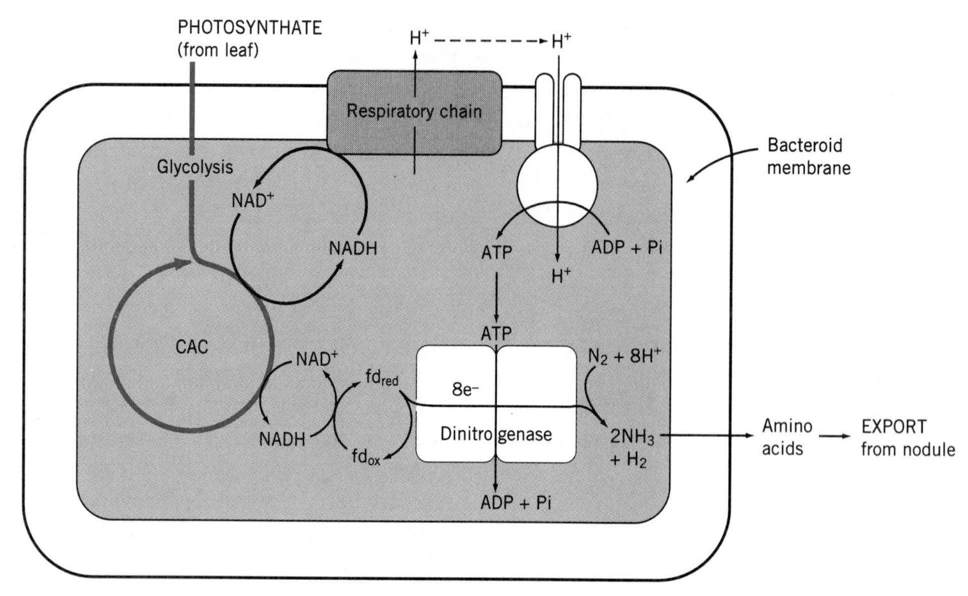

FIGURE 6.7 Summary diagram illustrating the interactions between photosynthesis, respiration, and nitrogen fixation in bacteroids.

thickened, multilayered cell walls that restrict the diffusion of oxygen. They are also characterized by a high respiratory activity that maintains a low intracellular oxygen concentration. Finally, although heterocysts are photosynthetic cells, they have eliminated the oxygen-evolving portion of photosynthesis and reserved only that portion that generates ATP.

In legume nodules, the oxygen supply is regulated to a large extent by an oxygen-binding protein called **leghemoglobin.** Leghemoglobin is synthesized by the host plant and is located within the bacteroid-infected host cell. Leghemoglobin may comprise as much as 30 percent of the host cell protein and gives the nodule a distinctive pink color when a cut surface is exposed to air. Leghemoglobin is similar in structure to the hemoglobin of mammalian blood. Its function is also similar, since it apparently binds oxygen and controls the release of oxygen in the region of the bacteroid. The equilibrium concentration of oxygen in the bacteroid zone is thus kept at a level (about 10 nM) sufficient to support bacteroid respiration—and the production of ATP and reducing potential—while at the same time preventing excess oxygen from inactivating dinitrogenase. Oxygen levels must be carefully balanced, because too low an oxygen concentration can also limit dinitrogenase activity in nodules (Layzell et al., 1990). This could be a result of limiting ATP availability.

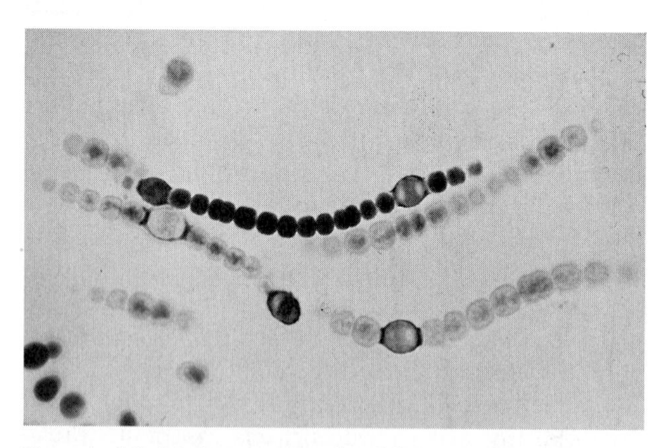

FIGURE 6.8 Light micrograph of the cyanobacterium *Anabaena* showing heterocysts. Nitrogen fixation is carried out in the heterocyst, whose structure and metabolism limits the concentration of free oxygen. (Copyright E. Reschke. Peter Arnold, Inc. Reprinted by permission.) (See color plate 3.)

DINITROGENASE AND HYDROGEN PRODUCTION

A final problem facing nitrogen-fixing organisms is the evolution of hydrogen. Although the mechanism is not well understood, it appears that hydrogen production by dinitrogenase is an inescapable byproduct of the nitrogen fixation reaction. As noted earlier in Equation 6.1, at least one molecule of hydrogen (H_2) is evolved for every molecule of N_2 reduced. This is actually a minimum value. When dinitrogenase is not operating optimally, as might be the case when the supply of reductant to dinitrogenase is suboptimal, even more electrons may be diverted to the production of hydrogen. As much as 25 to 30 percent of the ATP and electrons supplied to dinitrogenase may be consumed by hydrogen production. In 1980, it was estimated that more than one million tons of hydrogen were released to the atmosphere annually from nitrogen-fixing root nodules (Conrad and Seiler, 1980).

Box 6.2
Enzymes

Enzymes are wonderful substances—they consist of wonderful molecules.

L. Pauling (1956)

Living cells must carry out an enormous variety of biochemical reactions, yet cells are able to rapidly construct very large and complicated molecules, or regulate the flow of materials through complex metabolic pathways, with unerring precision and accuracy. All this is made possible by **enzymes.** Enzymes are biological catalysts; they facilitate the conversion of *substrate* molecules to product, but are not themselves permanently altered by the reaction. Cells contain thousands of enzymes, each catalyzing a particular reaction.

Enzyme-catalyzed reactions differ from ordinary chemical reactions in four important ways:

1. *High specificity.* Enzymes are capable of recognizing subtle and highly specific differences in substrate and product molecules, to the extent of discriminating between mirror images of the same molecules (called stereoisomers or *enantiomers*) in the same way you do not fit your right hand into your left glove.
2. *High reaction rates.* The rates of enzyme-catalyzed reactions are typically 10^6 to 10^{12} greater than rates of uncatalyzed reactions. Many enzymes are capable of converting thousands of substrate molecules every second.
3. *Mild reaction conditions.* Enzyme reactions typically occur at atmospheric pressure, relatively low temperature, and within a narrow range of pH near neutrality. There are exceptions, such as certain protein-degrading enzymes that operate in vacuoles with a pH near 4.0, or enzymes of thermophilic bacteria that thrive in hot sulfur springs, where temperatures are close to 100°C. Most enzymes, however, enable biological reactions to occur under conditions far milder than those required for most chemical reactions.
4. *Opportunity for regulation.* The presence of a particular enzyme and its amount is regulated by controlled gene expression and protein turnover. In addition, enzyme activity is subject to regulatory control by a variety of activators and inhibitors.

These opportunities for regulation are instrumental in keeping complex and often competing metabolic reactions in balance.

The first step in an enzyme-catalyzed reaction is the reversible binding of a substrate molecule (S) with the enzyme (E) to form an enzyme-substrate complex (ES):

$$E + S \rightleftharpoons ES \longrightarrow E + P$$

The enzyme-substrate complex then dissociates to release the product molecule (P). The free enzyme is regenerated and is then available to react with another molecule of substrate.

Enzymes are proteins and the site on the protein where the substrate binds and the reaction occurs is called the **active site.** Active sites are usually located in a cleft or pocket in the folded protein, and contain reactive amino acid side chains, such as carboxyl ($—COO^-$), amino ($—NH_3^+$), or sulfur ($—S^-$) groups that position the substrate and participate in the catalysis. The shape and polarity of the active site is largely responsible for the specificity of an enzyme, since the shape and polarity of the substrate molecule must complement or "fit" the geometry of the active site in order for the substrate to gain access and bind to the catalytic groups. Where two or more substrates participate in a common reaction, binding of the first substrate may induce a change in the conformation of the protein, which then allows the second substrate access to the active site.

Enzymes increase the rate of a reaction because they lower the amount of energy, known as the *activation barrier*, required to initiate the reaction. This effect is illustrated by the ball and hill analogy (Fig. 6.9A). In order for the ball to roll down the hill, it must first be pushed over the lip of the depression in which it sits. This act increases the potential energy of the ball. When the ball is poised at the very top of the lip, it is in a **transition state;** that is, there is an equal probability that it will fall back into the depression or roll forward and down the hill.

Chemical reactions go through a similar transition state (Fig. 6.9B). As reacting molecules come together, they increasingly repel each other and the potential energy of the system increases. If the reactants approach with sufficient kinetic energy, however, they will achieve a transition state where there is an equal

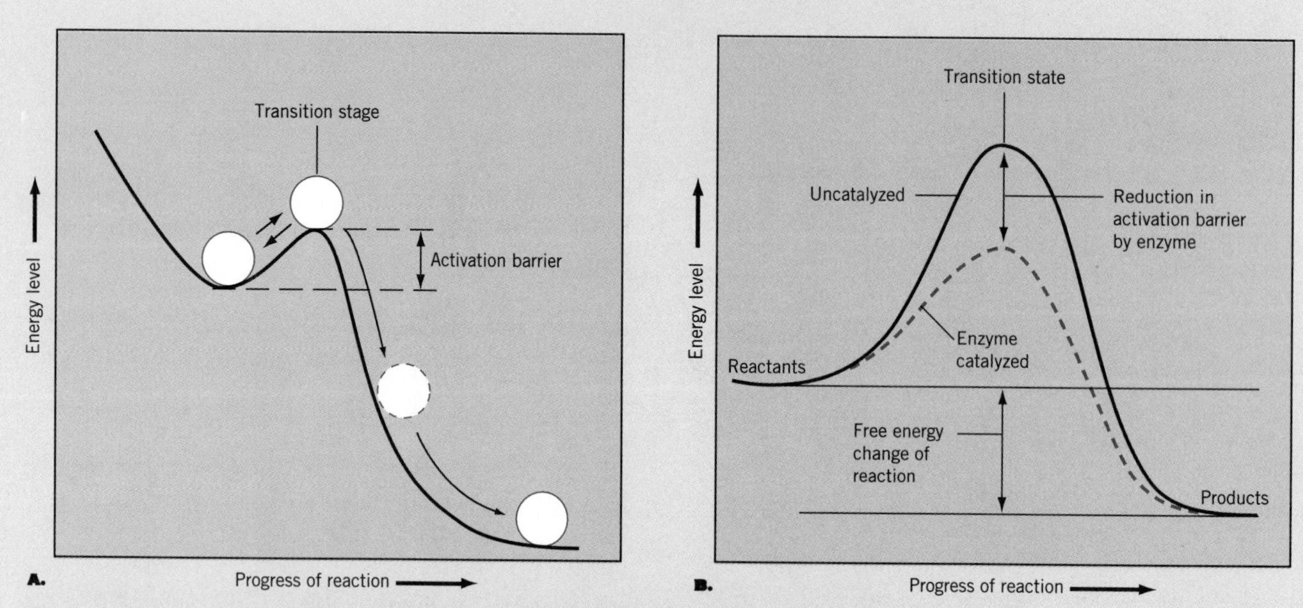

FIGURE 6.9 Enzymes. (*A*) The "ball and hill" analogy of chemical reactions. (*B*) Enzymes reduce the activation barrier, measured as transition state energy, for a reaction.

probability that they will decompose back to reactants or proceed to products. In the case of an enzyme-catalyzed reaction, the enzyme-substrate complex takes a different reaction pathway; a pathway that has a transition state energy level substantially lower than that of the uncatalyzed reaction (Fig. 6.9B).

It is important to note that enzymes do not alter the course of a reaction. They do not change the equilibrium between reactants and products, nor do they alter the free energy change (ΔG) for the reaction. (See Chap. 9 for a discussion of free energy changes.) Enzymes change only the rate of a reaction.

Most enzymes are identified by adding the suffix *-ase* to the name of the substrate, often with some indication of the nature of the reaction. For example, α-amyl*ase* digests amylose (starch), malate dehydroge*nase* oxidizes (that is, removes hydrogen) from malic acid, and phosphoenolpyruvate carboxyl*ase* adds carbon dioxide (a carboxyl group) to a molecule of phosphoenolpyruvate.

Many enzymes do not work alone, but require the presence of nonprotein cofactors. Some cofactors, called **coenzymes,** are transiently associated with the protein and are themselves changed in the reaction. Many electron carriers, such as NAD^+ or FAD, for example, serve as coenzymes for many dehydrogenase enzymes (see Chap. 9). They are, in fact, cosubstrates and are reduced to NADH or $FADH_2$ in the reaction. **Prosthetic groups** are nonprotein cofactors more or less permanently associated with the enzyme protein. The heme group of hemoglobin is an example of a tightly bound prosthetic group. Many plant enzymes utilize ions such as iron or calcium as prosthetic groups.

Enzymes and enzyme reactions are sensitive to both temperature and pH. Like most chemical reactions, enzyme reactions have a Q_{10} of about 2, which means that the rate of the reaction doubles for each 10°C rise in temperature. The rate increases with temperature until an optimum is reached, beyond which the rate usually declines sharply. The decline is normally caused by **thermal denaturation,** or unfolding of the enzyme protein. With most enzymes, thermal denaturation occurs in the range of 40 to 45°C, although many enzymes exhibit temperatures closer to 25 or 30°C. Some enzymes exhibit instability at lower temperatures as well. One example is pyruvate, pyrophosphate dikinase (PPDK) (Chap. 10). PPDK is unstable and loses activity at temperatures below about 12 to 15°C. Enzyme reactions are also sensitive to pH, since pH influences the ionization of catalytic groups at the active site. The conformation of the protein may also be modified by pH.

A variety of ions or molecules may combine with an enzyme in such a way that it reduces the catalytic activity of the enzyme. These are known as **inhibitors.** Inhibition of an enzyme may be either **irreversible** or **reversible.** Irreversible inhibitors act by chemically modifying the active site so that the substrate can no longer bind, or by permanently altering the protein in some other way. Reversible inhibitors often have chemical structures that closely resemble the natural substrate. They bind at the active site, but either do not react or react very slowly. For example, the oxidation of succinate to fumarate by the enzyme succinic dehydrogenase is competitively inhibited by malonate, an analog of succinate (Fig. 6.10).

Because substrate and inhibitor *compete* with one another for attachment to the active site, this form of inhibitor is known as **competitive inhibition.** Another form of reversible inhibitor, the **noncompetitive inhibitor,** does not compete with the substrate for the active site, but binds elsewhere on the enzyme and, in doing so, restricts access of the substrate to the active site. Alternatively, noncompetitive inhibitors may bind directly to the enzyme-substrate complex, thereby rendering the enzyme catalytically inactive.

Enzymes play a key role in **feedback inhibition,** one of the most common modes for metabolic regulation. Feedback inhibition occurs when the end product of a metabolic pathway controls the activity of an enzyme near the beginning of the pathway. When demand for the product is low, excess product inhibits the activity of a key enzyme in the pathway, thereby reducing the synthesis of product. Once cellular activities have depleted the supply of product, the enzyme is deinhibited and the rate of product formation increases. The enzyme subject to feedback regulation is usually the first one past a metabolic branch point. This is known as the **committed step.** In the example shown in Figure 6.11, reactions A → B. C → D, and C → F all represent committed steps. In this example, an excess of product G would reduce the flow of precursor through the reaction of C → F, thereby diverting more precursor, C, to product E. Alternatively, an excess of both E and G would regulate the conversion of A to B. Feedback regulation is an effective way of coordinating product formation within complex pathways. Many of the enzymes of respiratory metabolism, for example, are subject to feedback regulation, thereby balancing the flow of carbon against the constantly changing energy demands of the cell.

Enzymes are remarkable biological catalysts that both enable and control the enormous variety of biochemical reactions that comprise life.

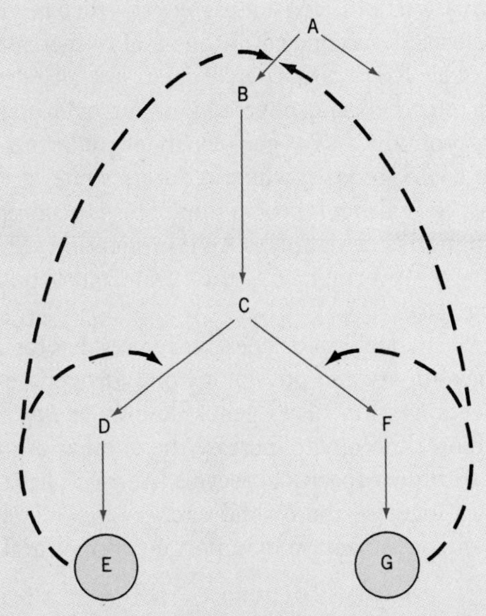

FIGURE 6.11 Feedback inhibition. Excess product inhibits the enzyme that catalyzes a first committed step leading to product formation.

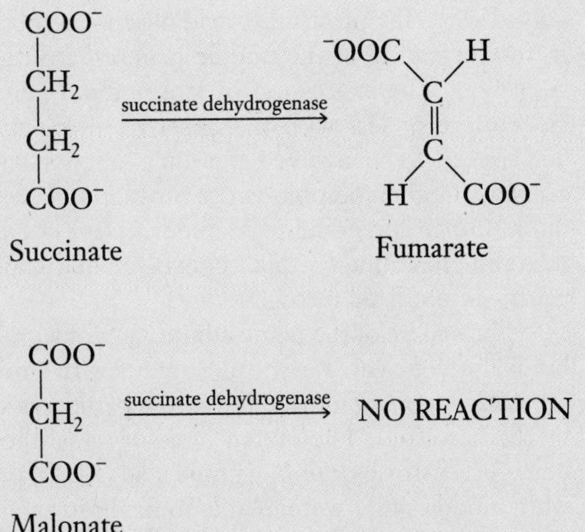

$^-OOC-CH_2-CH_2-COO^-$ (Succinate) $\xrightarrow{\text{succinate dehydrogenase}}$ $^-OOC-CH=CH-COO^-$ (Fumarate)

$^-OOC-CH_2-COO^-$ (Malonate) $\xrightarrow{\text{succinate dehydrogenase}}$ NO REACTION

FIGURE 6.10 Malonate, a structural analog of succinate, inhibits the enzyme succinate dehydrogenase. Malonate binds to the enzyme in place of succinate, but does not enter into a reaction.

FURTHER READING

Voet, D., J. G. Voet. 1990. *Biochemistry.* New York: Wiley.

Needless to say, H_2 production along with nitrogen fixation is wasteful, consuming energy that might otherwise be used to reduce dinitrogen. However, although all nitrogen fixers *produce* hydrogen, not all *release* hydrogen into the atmosphere. Many nitrogen-fixing organisms contain an oxygen-dependent enzyme, called **uptake hydrogenase,** which recovers some of the energy lost to hydrogen production. This is accomplished by coupling H_2 oxidation to ATP production. The electrons are returned to the reductant pool for dinitrogenase.

Understandably, considerable interest has focused on the biochemistry and physiology of hydrogenase action and the expression of its genes (known as *hup* genes). Using biotechnology to increase the number of rhizobia strains with the capacity to recycle hydrogen has the potential to increase the overall energy efficiency of biological nitrogen fixation in important agricultural crops.

THE GENETICS OF NITROGEN FIXATION

Several sets of genes, both in the microsymbiont and host plant roots, contribute to nodulation and biological nitrogen fixation.

It should be evident from the discussion up to this point that nitrogen fixation involves very complex relationships between organisms and their environment or between rhizobia and host in the case of symbiotic fixation. The switch to nitrogen-fixing metabolism in anaerobic environments or infection and subsequent nodule development in symbiotic relationships requires major changes in the genetic programs of the organisms involved. The genetics of infection, nodulation, and the nitrogen-fixing machinery is currently one of the more exciting and rapidly advancing areas in the study of nitrogen fixation.

In free-living nitrogen fixers, the principal requirement is for the synthesis of the enzyme dinitrogenase. Dinitrogenase synthesis is directed by a set of genes known as *nif* genes. Best characterized is *Klebsiella pneumonieae*, where at least 17 *nif* genes have been described. The *nif* genes include structural genes that encode for dinitrogenase protein as well as a number of regulatory genes. Two genes, the *nif*D and *nif*K genes, for example, encode the two different subunits of the MoFe protein. The Fe protein and ferredoxin are encoded by *nif*H and *nif*F, respectively. Other *nif* genes are involved in insertion of the FeMo cofactor and the activation and processing of the enzyme complex.

At least three different sets of genes, including *nif* genes, are involved in the symbiotic process. In the early stages of nodulation, prior to infection of the root, a set of rhizobial *nod* genes is switched on by flavonoids in the host root exudate. The *nod* genes are located in a large circular piece of rhizobial DNA (or plasmid) known as the **Sym** (for symbiosis) **plasmid.** Three *nod* genes (*nod*A, *nod*B, *nod*C) are basic nodulation genes common to all rhizobia. They code for the chitooligosaccharide core of the nod factors.

The role of the *nod*D gene appears to be pivotal. Its expression is differentially affected by root exudates and its product in turn activates transcription of both the *nod*ABC group and a series of host-specific genes (*nod*EFGH). These host-specific genes code for modifications to the nod factors that are important in determining host specificity. The pivotal role of *nod*D has been demonstrated by transferring the *nod*D gene from a strain of *Rhizobium* that infects *Parasponia* to a strain that normally infects only clover. The clover strain was then able to nodulate *Parasponia.*

During the latter stages of nodule development, rhizobial *nif* and *fix* genes are switched on. The discrimination between *nif* and *fix* genes is not always clear, except that, as in the free-living forms, *nif* genes are involved in the synthesis and regulation of dinitrogenase. The *fix* genes are restricted to symbiotic nitrogen fixers. At least one (*fix*X) encodes a ferredoxin and others may be involved in the transport of electrons to dinitrogenase.

Development of an active nodule requires a number of nodule-specific proteins contributed by the host cells. These proteins, called **nodulins,** are encoded by *NOD* genes located in the host cell genome. Early nodulins are expressed during the infection process and nodule development. Although several have been identified, their role is not clear. Early nodulins appear to be involved with the infection thread plasma membrane and in the formation of the nodule primordia (Franssen et al., 1992). The expression of late nodulin genes coincides more or less with the onset of nitrogen fixation and appears to be involved in nodule function and maintenance. Leghemoglobin is the most abundant late nodulin. Other late nodulins include enzymes such as uricase and glutamine synthetase involved in the metabolic processing of fixed nitrogen.

The source of the heme component of leghemoglobin is not yet clear. Early experiments with a mutant of *Rhizobium melilotii* indicated that the heme was supplied by the bacteroid. The mutant, unable to synthesize the heme precursor δ-aminolevulinic acid (ALA), produced white nodules that were unable to fix dinitrogen. These results suggest the host plant is unable to provide sufficient heme to build adequate levels of leghemoglobin. However, in later experiments with soybean, plants infected with *Bradyrhizobium japonicum* carrying the same mutation produced fully competent nodules (Werner, 1992).

Symbiotic nitrogen fixation clearly requires the coordinated expression of many genes of both the host and microsymbiont. Understanding how these genes are regulated and how the complex processes of infection and nodule development are coordinated constitutes one of the more challenging problems facing plant physiologists today. Armed with sufficient understanding of the process and the tools of modern molecular genetics, plant scientists may one day be able to extend the range of biological nitrogen fixation to other important crop species—thus extending the benefits of nitrogen fixation to nitrogen-poor soils and reducing the economic and environmental costs of chemical nitrogen fertilizers (Triplett, 1996).

NITROGEN ASSIMILATION

The ammonia produced by nitrogen fixation must be converted to organic nitrogen before it can be exported from the nodule for utilization by the host plant.

The first stable product of nitrogen fixation is ammonia (NH_3), although at physiological pH ammonia is almost certainly protonated to form ammonium ion:

$$NH_3 + H^+ \rightleftharpoons NH_4^+ \quad (6.2)$$

Plants that cannot fix dinitrogen meet their nutritional needs by taking in nitrogen from the soil. While there are exceptions, most plants are able to assimilate either NH_4^+ or NO_3^-, depending on their relative availability in the soil. In most soils, ammonia is rapidly converted to nitrate by the nitrifying bacteria described earlier in this chapter. Nitrifying bacteria do not grow well under anaerobic conditions and consequently ammonia will accumulate in soils that are poorly drained. Nitrification itself is also inhibited in strongly acidic soils. Some members of the family Ericaceae, typically found on acidic soils, have adapted by preferentially utilizing ammonium as their nitrogen source. One extreme example is the cranberry (*Vaccinium macrocarpon*), native to swamps and bogs of eastern North America, which cannot exploit NO_3^- as a nitrogen source and must take up nitrogen in the form of ammonium ion.

Regardless of the route taken, assimilation of mineral (inorganic) nitrogen into organic molecules is a complex process that can be very energy-intensive. It has been estimated, for example, that assimilation of ammonium nitrogen consumes from 2 to 5 percent of the plant's total energy production (Oaks and Hirel, 1985). Nitrate, on the other hand, must first be reduced to ammonium before it can be assimilated, at a cost of nearly 15 percent of total energy production. In this section, we will review the assimilation first of ammonium nitrogen and then of nitrate nitrogen.

ASSIMILATION OF AMMONIUM

Although NH_4^+ is readily available to many plants, either as the product of nitrogen fixation or by uptake from the soil, it is also quite toxic to plants. In nitrogen-fixing systems, NH_4^+ will inhibit the action of dinitrogenase. Ammonium also interferes with the energy metabolism of cells, especially ATP production. Even at low concentrations, NH_4^+ will dissociate ATP formation from electron transport in both mitochondria and chloroplasts (see Chap. 9). Consequently, plants can ill afford to accumulate excess free NH_4^+. Most plants avoid any toxicity problem by rapidly incorporating the NH_4^+ into amino acids.

The general pathway for NH_4^+ assimilation in nitrogen-fixing symbionts has been worked out largely by supplying nodules with labeled dinitrogen ($^{13}N_2$ or $^{15}N_2$). These studies have indicated that the initial organic product is the amino acid **glutamine.** Assimilation of NH_4^+ into glutamine by legume nodules is accomplished by the **glutamate synthase cycle,** a pathway involving the sequential action of two enzymes: **glutamine synthetase (GS)** and **glutamate synthase (GOGAT)**[3] (Fig. 6.12). Both GS and GOGAT are nodulin proteins that are expressed at high levels in the host cytoplasm of infected cells, outside the peribacteroid membrane. The NH_4^+ formed in the bacteroid must therefore diffuse across the peribacteroid membrane before it can be assimilated.

In the first reaction of the glutamate synthase cycle, catalyzed by GS, the addition of an NH_4^+ group to **glutamate** forms the corresponding amide, **glutamine:**

$$\text{glutamate} + NH_4^+ + \text{ATP} \longrightarrow \text{glutamine} + \text{ADP} + P_i \quad (6.3)$$

Energy to drive the ammination of glutamate is provided by ATP, yet an additional cost of nitrogen fixation. Glutamine is then converted back to glutamate by the transfer of the amide group to a molecule of **α-ketoglutarate.**

$$\text{glutamine} + \alpha\text{-ketoglutarate} + \text{NADH} \longrightarrow 2\ \text{glutamate} + \text{NAD}^+ \quad (6.4)$$

Reaction 6.4 is catalyzed by GOGAT and requires reducing potential in the form of NADH. The α-ketoglutarate is probably derived from photosynthetic carbon through respiration in the host cell. α-Ketoglutarate is an intermediate in the respiratory pathway for the oxidation of glucose. Note that Reaction 6.4 gives rise to two molecules of glutamate, each of which gives rise to a molecule of glutamine. Since only one molecule of

[3]The acronym *GOGAT* refers to glutamine-2-oxoglutarate-amino-transferase. 2-Oxoglutarate is an alternative name for α-ketoglutarate.

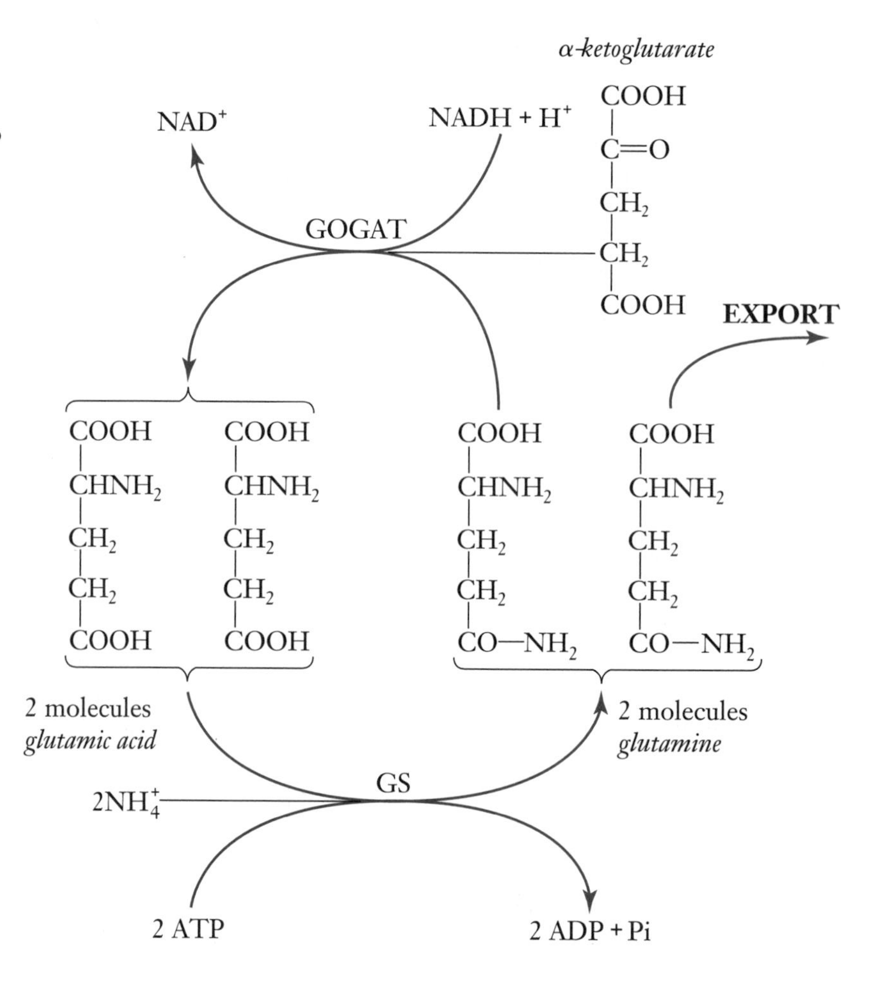

FIGURE 6.12 Assimilation of ammonium by the glutamate synthase cycle. Two molecules of glutamine are formed by the addition of ammonium to two molecules of glutamate catalyzed by the enzyme glutamine synthetase (GS). One molecule of glutamine is available for export to the host plant while the second molecule undergoes a transamination reaction with α-ketoglutarate, regenerating two molecules of glutamate. This second reaction is catalyzed by glutamate synthase (GOGAT).

glutamine is required to keep the cycle going, the other is available for export to the host plant (Fig. 6.12). Overall, then, carbon skeletons originating with photosynthesis and nitrogen fixed by the microsymbiont are combined to form organic nitrogen that is exported out of the nodule for use by the host.

There is a possible alternative pathway for nitrogen assimilation, involving the direct reductive amination of α-ketoglutarate by the enzyme **glutamate dehydrogenase (GDH).** However, although GDH activity has been detected in nodules, there is no convincing evidence that it plays a significant role in NH_4^+ assimilation under normal circumstances. Both the quantities and activities of GS and GOGAT are much higher than GDH. GS alone may account for as much as 2 percent of the total soluble protein outside the bacteroid. In addition, GDH has a much lower affinity for NH_4^+ than does GS and could hardly be expected to compete with GS for available NH_4^+. The cost of the glutamate synthase cycle is one ATP for each NH_4^+ assimilated, but the benefit is rapid assimilation. The high affinity of GS for NH_4^+, together with the high concentration of the enzyme, ensures that the free NH_4^+ concentration is kept below toxic levels.

Before leaving the glutamate synthase cycle, it is important to note that GS and GOGAT are not restricted to nodules. These enzymes are located in the roots and leaves of non-nitrogen-fixing plants where they also catalyze the assimilation of NH_4^+ nitrogen.

EXPORT OF FIXED NITROGEN FROM NODULES

The final step in nitrogen fixation is the export of the fixed nitrogen from the nodule to other regions of the host plant. Export of the organic nitrogen products from nodules is primarily through the xylem. Consequently, the form in which the nitrogen is exported has been identified primarily by analysis of xylem sap. There are some pitfalls to such analyses, however, as there is no guarantee that all of the nitrogen present in sap represents current nodule production. Some of the better analyses have been conducted directly on detached nodules or by monitoring the flow of organic nitrogen following fixation of $^{15}N_2$. These studies have shown that although glutamine is the principal organic product of nitrogen fixation, it rarely accounts for a significant fraction of the nitrogen exported, at least in legumes (Atkins, 1991). In some groups of legumes, largely those of temperate origins, such as pea and clover, the amino acid **asparagine** is the predominant form translocated. Legumes of tropical origins, for example soybean and cow-

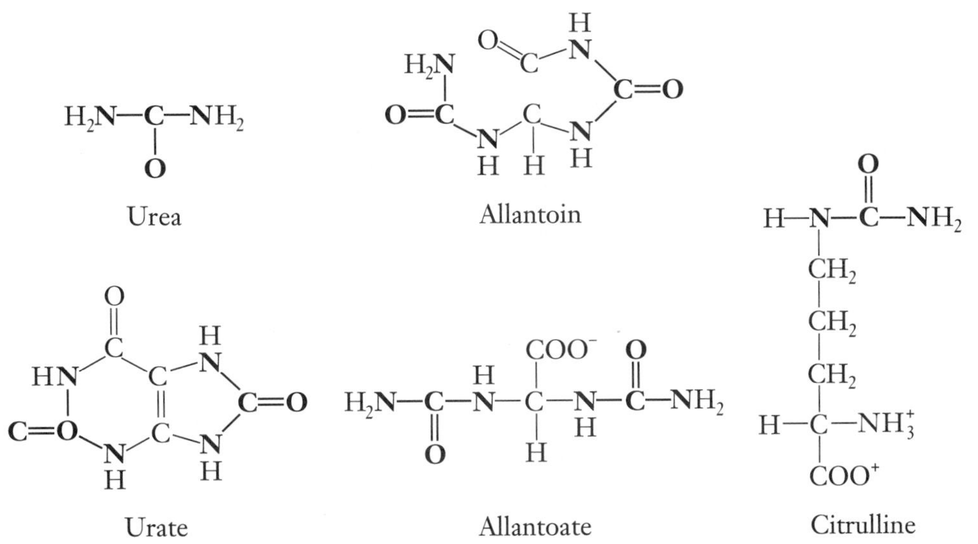

FIGURE 6.13 **Structures of the principal ureides used in the transport of assimilated nitrogen in some nitrogen-fixing species. Ureides are considered derivatives of urea and are formed principally from uric acid (urate). The N-C-N urea backbone is shown in bold print.**

pea, appear to export predominantly derivatives of urea, known as **ureides** (Fig. 6.13).

The biosynthetic pathway for asparagine in nodules involves two **transamination** reactions. A transamination reaction is the transfer of an amino group from an amino acid to the carboxyl group of a keto acid. Transamination reactions, catalyzed by a class of enzymes known as **aminotranferases,** enable nitrogen initially fixed in glutamate to be incorporated into other amino acids and, ultimately, into protein. Aminotransferases are found throughout the plant—in the cytosol, chloroplasts and in microbodies—wherever protein synthesis activity is high.

The enzymes involved in asparagine biosynthesis in nodules appear to be similar to those found elsewhere in the plant. The first step is the transfer of an amino group from glutamate to **oxaloacetate,** catalyzed by the enzyme **aspartate aminotransferase.**

$$\text{glutamate} + \text{oxaloacetate} \longrightarrow \alpha\text{-ketoglutarate} + \text{aspartate} \quad (6.5)$$

The glutamate used in this reaction is derived from the GS-GOGAT reactions in the nodule. In order to continue the synthesis and export of asparagine, the nodule requires a continued supply of the 4-carbon acid oxaloacetate. This could be provided through the oxidation of carbon in the nodule; oxaloacetate is another intermediate in the respiratory oxidation of glucose. However, nodules from a number of species exhibit high activities of the enzyme **p**hospho**e**nol**p**yruvate carboxylase (**PEP** carboxylase). PEP carboxylase catalyzes the addition of a carbon dioxide (a **carboxylation** reaction) to the 3-carbon **phosphoenolpyruvate (PEP).** PEP carboxylase is involved in a number of important metabolic pathways in plants and animals, including respiration and photosynthesis.

In the second step of asparagine biosynthesis, the amide nitrogen is transferred from glutamine to aspartate.

$$\text{glutamine} + \text{aspartate} + \text{ATP} \longrightarrow \text{glutamate} + \text{asparagine} + \text{ADP} + \text{P}_i \quad (6.6)$$

The enzyme for this reaction is **asparagine synthetase** and the reaction is driven by the energy of one molecule of ATP for each asparagine synthesized.

The synthesis of ureides is more complex, both biochemically and with respect to the division of labor between the microsymbiont and tissues of the host plant (Sprent and Sprent, 1990). **Allantoin** and **allantoic acid** (Fig. 6.13) are formed by the oxidation of purine nucleotides, which apparently requires an active symbiosis. It has been observed that *de novo* synthesis of purines increases sharply as nodules develop but can be prevented by replacing dinitrogen with argon. It has also been demonstrated that cell-free extracts of cowpea nodules contain high activities of the enzymes necessary to support the conversion of a range of purines to allantoin. Evidence to date suggests that purine synthesis occurs in the cytosol of the bacteroid-infected cell, utilizing NH_4^+ exported from the bacteroid (Fig 6.14). The purine is then oxidized to uric acid, which is translocated to adjacent uninfected nodule cells. In the peroxisome, the enzyme urate oxidase converts uric acid to allantoin, which is in turn converted to allantoic acid by the enzyme allantoinase in the endoplasmic reticulum (Atkins, 1991).

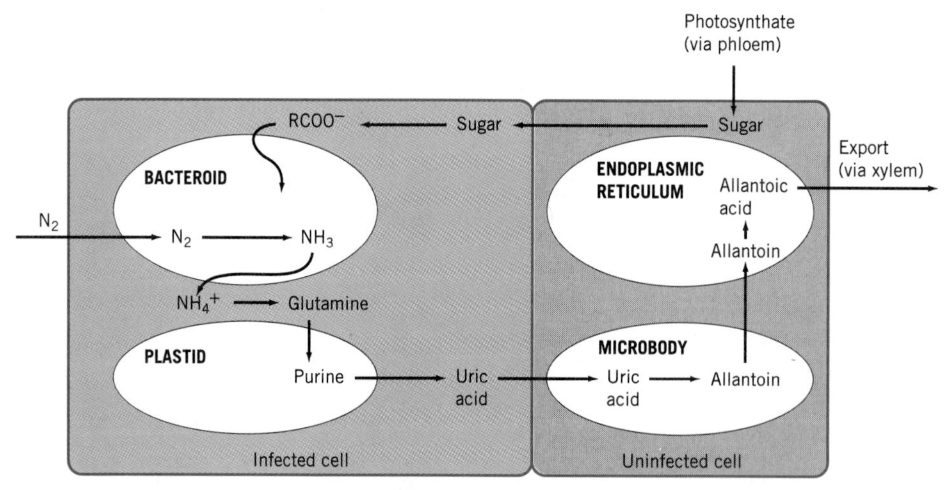

FIGURE 6.14 A generalized outline for the assimilation of fixed nitrogen in ureide-forming nodules. (Based on Atkins, 1991.)

Ureides apparently serve specifically for the transport of nitrogen. They are translocated through the xylem to other regions of the plant, where they are rapidly metabolized. In the process, NH_4^+ is released which is then reassimilated via GS and GOGAT in the target tissue.

Although the synthesis of asparagine and especially the ureides both appear to be complex processes, there are some advantages relating to the energy costs and efficiencies of nitrogen export. J. S. Pate and his coworkers have estimated that the carbon metabolism associated with nitrogen export may consume as much as 20 percent of the photosynthate diverted to nitrogen fixation (Atkins, 1984). One way to judge efficiency is to consider the amount of carbon required for each nitrogen exported. The ureides, with a carbon to nitrogen ratio of 1, are the most economic in the use of carbon. Both asparagine and citrulline (C:N = 2) require more carbon in their transport and glutamine (C:N = 2.5) would be the least economic. The energy costs of ureides, asparagine, and citrulline, in terms of ATP consumed, are about the same, so the principal advantage to be gained by the ureide-formers appears to be a favorable carbon economy.

ASSIMILATION OF NITRATE

Plants that do not form nitrogen-fixing associations generally take up nitrogen in the form of nitrate from the soil. Nitrate must first be reduced to ammonium before it can be incorporated into amino acids and other organic molecules.

Except in extreme situations noted earlier, NO_3^- is the more abundant form of nitrogen in soils and is most available to plants that do not form nitrogen-fixing associations. However, in spite of numerous studies describing the physiology of NO_3^- uptake, there is a great amount of uncertainty surrounding the mechanism of NO_3^- transport into roots. It has been shown in various studies that uptake of NO_3^- is sensitive to (1) low temperature, (2) inhibitors of both respiration and protein synthesis, and (3) anaerobic conditions. All of these results support the hypothesis that NO_3^- transport across the root cell membrane is an energy-dependent process mediated by a carrier protein (see Chap. 5).

In root cells that have never been exposed to nitrate, there appears to be a limited capacity for NO_3^- uptake. This suggests a small amount of carrier is present in the membrane at all times (that is, a **constitutive** protein). On exposure to external nitrate, the rate of uptake increases from 2- to 5-fold but addition of inhibitors of protein synthesis, causes the rate to fall rapidly back to the constitutive level. This pronounced sensitivity of NO_3^- uptake to inhibitors of protein synthesis suggests that the bulk of the carrier protein is **inducible,** that is, the presence of NO_3^- in the soil stimulates the synthesis of new carrier protein. Once inside the root, NO_3^- may be stored in the vacuole, assimilated directly in the root cells, or translocated in the xylem to the leaves for assimilation.

Nitrate cannot be assimilated directly but must first be reduced to NH_4^+ in order to be assimilated into organic compounds. This is a two-step process, the first being the reduction of NO_3^- to nitrite (NO_2^-) by the enzyme **nitrate reductase (NR).**

$$2H^+ + NO_3^- + 2e^- \longrightarrow NO_2^- + H_2O \quad (6.8)$$

NR is generally assumed to be a cytosolic enzyme. The product NO_2^- then moves into plastids (in roots) or chloroplasts (in leaves) where it is quickly reduced to NH_4^+ by the enzyme **nitrite reductase (NiR).**

$$8H^+ + NO_2^- + 6e^- \longrightarrow NH_4^+ + 2H_2O \quad (6.9)$$

Nitrite is toxic and is rarely found at high concentrations in plants. This is no doubt because the activity of NiR (per gram dry weight of tissue) is normally several times higher than the activity of NR. The resulting ammonia is then rapidly assimilated into organic compounds via the GS/GOGAT system already described. In non-nitrogen-fixing systems, both GS and GOGAT are commonly found in root and leaf cells. GS is found in the cytosol of root cells and in both the cytosol and chloroplasts of leaf cells. GOGAT is a plastid enzyme, localized in the chloroplasts of leaves and in plastids in roots. Depending on its location, GOGAT may use ferredoxin, NADH or NADPH as electron donors.

Nitrate reductase is a ubiquitous enzyme found in both prokaryote and eukaryote cells. In prokaryotes, the principal electron donor is ferredoxin, while in higher plants electrons are donated by the reduced forms of one of the pyrimidine nucleotides, **n**icotinamide **a**denine **d**inucleotide (**NAD**) or **n**icotinamide **a**denine **d**inucleotide **p**hosphate (**NADP**) (Chap. 9). The enzyme isolated from a variety of higher plants is composed of two identical subunits with a molecular mass of approximately 115 kD. A key constituent of NR is molybdenum; NR is the principal Mo-protein in non-nitrogen-fixing plants. One of the results of Mo deficiency is markedly reduced levels of nitrate reductase activity and consequent nitrogen starvation (see Chap. 4).

NR is a highly regulated, inducible enzyme. It has long been recognized that both substrate (NO_3^-) and light are required for maximum activity and that induction involves an increase in the level of NR messenger RNA followed by *de novo* synthesis of NR protein (Hageman and Flesher, 1960; Bowsher et al., 1991). Treatment of cereal seedlings such as barley (*Hordeum vulgare*) or maize (*Zea mays*) with nitrate in the dark induces relatively low levels of NR activity but activity is strongly promoted if seedlings are also exposed to light. Induction by light is eliminated by various treatments that interfere with chloroplast development or photosynthetic energy transformations, implying a requirement for photosynthetic energy. NR activity can also be reversibly regulated by red and far-red light, indicating control by the phytochrome system (Chap. 18).

More recent work has established that NR activity is also subject to post-translational regulation by a specific NR **protein kinase** (Huber et al., 1996). Protein kinases, first characterized by Edwin Krebs and Edmund Fischer in the 1950s, are a ubiquitous class of enzymes that phosphorylate other proteins by transferring a phosphate group from adenosine triphosphate (ATP). The phosphate group can then be removed by a second enzyme called **protein phosphatase**. It is increasingly evident that, by switching enzymes and other proteins on and off, reversible protein phosphorylation plays a central role in regulating metabolism. The fundamental role of protein kinases was recognized by the award of the Nobel Prize to Krebs and Fischer in 1992.

In the case of NR, the enzyme appears to be active in the both the phosphorylated and non-phosphorylated states. When NR is phosphorylated and the leaf is transferred from the light to dark, however, the enzyme is rapidly inactivated by binding with a small inhibitor protein. NR activity is slowly restored on return to light by a release of the inhibitor protein and subsequent phosphatase action. The question of why there exists such a complex system for regulation of NR activity has yet to be answered. The overall effect, however, is to coordinate nitrate reduction with photosynthetic activity. It ensures that nitrate reduction is engaged only after photosynthesis is fully active and able to provide both the energy required and the carbon skeletons necessary for incorporation of ammonia.

As indicated earlier, nitrate assimilation can be carried out in either the root or shoot tissues in most plants. Several studies have shown that the proportion of NO_3^- reduced in the root or shoot depends to a large extent on the external NO_3^- concentration. At low concentrations, most of the NO_3^- can be reduced within the root tissues and translocated to the shoot as amino acids or amides. At higher concentrations of NO_3^-, assimilation in the roots becomes limiting and a higher proportion of the NO_3^- finds its way into the translocation stream. Thus, at higher concentrations, a higher proportion of the nitrogen is assimilated in the leaves.

Not all plants have the same capacity to metabolize NO_3^- in their roots. In the extreme, NO_3^- is virtually the sole nitrogen source in the xylem sap of cocklebur (*Xanthium strumarium*). This is because cocklebur has no detectable NR in its roots. On the other hand, plants such as barley (*Hordeum vulgare*) and sunflower (*Helianthus annus*) translocate roughly equal proportions of NO_3^- and amino acid/amide nitrogen and radish (*Raphanus sativus*) translocates only about 15 percent of its nitrogen as NO_3^- (Pate, 1973).

NITROGEN CYCLING

Nitrogen uptake by most plants is highest during its early rapid growth phase and declines as reproductive growth begins and the plant ages. Cereals, for example, take up as much as 90 percent of their total nitrogen requirement before the onset of reproductive growth. Most of this nitrogen is directed toward young, expanding leaves, which reach their maximum nitrogen content just prior to full expansion. The leaf then begins to *export* nitrogen. Several studies have shown that mature leaves continue to import nitrogen, even though they have become net nitrogen exporters and the total nitrogen content of the leaf is in decline. This simultaneous import and export of nitrogen is known as **nitrogen cycling** (Simpson, 1986).

The export of nitrogen from leaves becomes particularly significant as the seed begins to develop. The nitrogen requirement of developing seeds is sufficiently

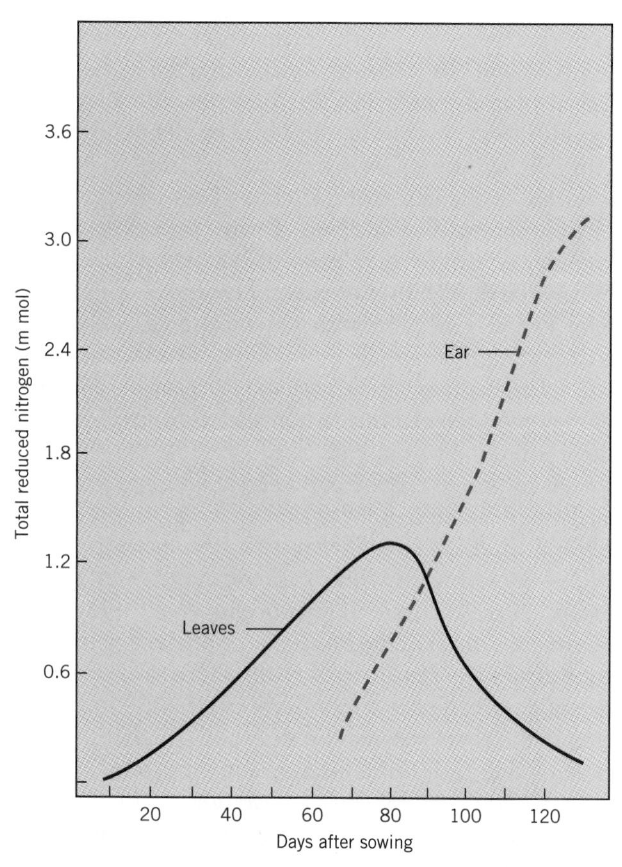

FIGURE 6.15 Nitrogen redistribution in wheat. (From Abrol et al., in H. Lambers, J. J. Neeteson, I. Stuhlen (eds.), *Fundamental, Ecological and Agricultural Aspects of Nitrogen Metabolism in Higher Plants*, Dordrecht, Martinus Nijhoff, 1986. Reprinted by permission of Kluwer Academic Publishers.)

great that it cannot be met by uptake from the soil (in the cases of cereals, for example) or by nitrogen fixed in nodules. The additional nitrogen must come from vegetative parts, principally leaves (Fig. 6.15). This may have significant implications for the photosynthetic capacity of leaves. The major leaf protein is Rubisco, the enzyme that catalyzes photosynthetic incorporation of carbon dioxide (see Chap. 10). Rubisco may comprise from 40 to 80 percent of the total soluble protein in leaves of soybean and cereal grains. Perhaps because of its abundance, Rubisco also functions as a storage protein; it may be degraded when nitrogen is required elsewhere in the plant, such as developing seeds. With the loss of Rubisco there is a concomitant decline in photosynthetic carbon fixation. In the case of soybean and other symbiotic legumes, this means less energy available to the nodule to support nitrogen fixation. This competition between nitrogen and carbon supply may be a major factor limiting seed development in legumes. In the case of cereals or plants growing in nitrogen-poor soils, mobilization of accumulated nitrogen from the leaves represents the principal source of nitrogen for developing fruits and seeds. In perennial plants, nitrogen from senescing leaves is mobilized and translocated to the roots for storage over the winter. In this way, the nitrogen is conserved and made available to support the first flush of renewed growth the following spring.

NITROGEN NUTRITION: AGRICULTURAL AND ECOLOGICAL ASPECTS

In terms of quantity, nitrogen is the fourth most abundant element in plants and is the most abundant mineral element (see Table 4.2). On a dry weight basis, herbaceous plant material typically contains between 1 and 4 percent nitrogen, mostly in the form of protein. At the same time, the availability of nitrogen in the soil may be limited by a number of environmental factors, such as temperature, oxygen, water status, and pH, which influence the activity of microorganisms responsible for nitrogen-fixation, nitrification, and ammonification. Moreover, a substantial quantity of nitrogen is removed each year with the harvested crop. It is not too surprising, then, that crop growth is most often limited by nitrogen supply.

In agricultural situations, the application of nitrogen fertilizers overcomes environmentally imposed nitrogen limitation. Most crops respond to applied nitrogen with increases in yield (Fig. 6.16). At sufficiently high application rates, factors other than nitrogen become limiting and there is no further gain. At even higher application rates, yield may decline slightly, but this is probably due to excess salt in the soil rather than some form of nitrogen toxicity. Data such as that shown in Figure 6.16 is of considerable practical value to farmers, who want to maximize the ratio of yield to input costs. Throughout North America, corn is the leading consumer of nitrogen fertilizer and farmers typically apply 100 to 150 kg N ha^{-1} each growing season. During its early rapid growth phase, a well-irrigated stand of corn will take up as much as 4 kg of nitrogen ha^{-1} day^{-1}. The use of such large amounts of nitrogen fertilizers is also costly in terms of energy. It has been estimated, for example, that fully one-third of the energy cost of a corn crop is accounted for by the production and distribution of nitrogen fertilizers.

Without continued application of fertilizers, yields of nonleguminous crops traditionally decline over a period of years. In a few situations where records have been kept, yields on plots from which nitrogen fertilizers have been withheld will eventually stabilize at a lower level that can be sustained indefinitely. Sustained low yields are possible because the extraction of nitrogen (and other nutrients) from the soil is balanced against replenishment from all sources, including rainfall, irrigation water, dust, and weathering of parent rock.

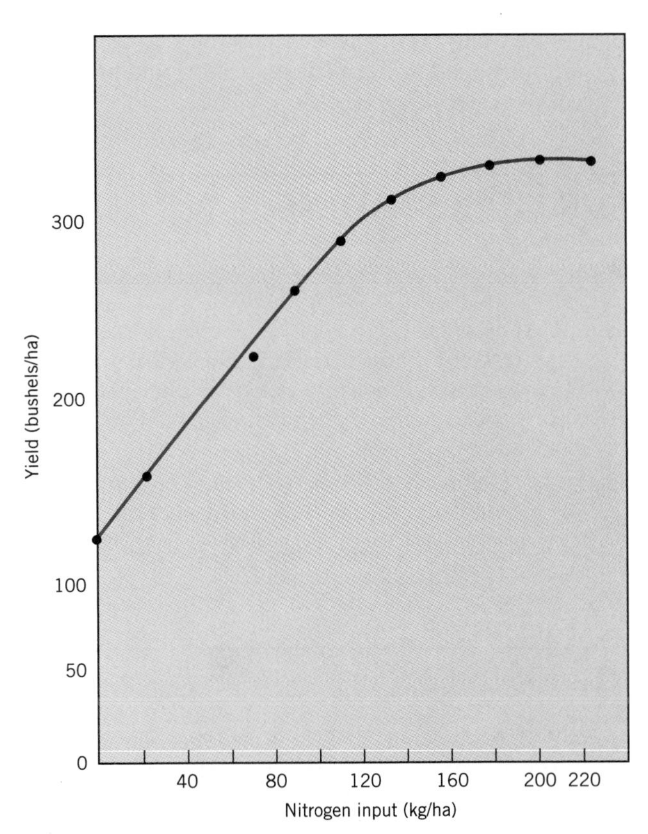

FIGURE 6.16 The effect of applied nitrogen fertilizer on corn yield. (Recalculated and plotted from data cited in Doering, 1982.)

The role of nitrogen in natural ecosystems is much more difficult to define, in part because the level of inputs is very low relative to the total nitrogen pool and much of the nitrogen is recycled. Still, it is generally agreed that nitrogen is limiting in most natural ecosystems just as it is in agriculture. In forest ecosystems, approximately two-thirds of the annual nitrogen input is contributed by nitrogen fixation, while the other third is believed derived primarily from atmospheric sources: either through rainfall or dry deposition of nitrogen oxides. A recent study has shown that, on average, close to half of the incoming fixed nitrogen is retained in the canopy. This implies that foliar absorption of nitrogen could play a significant role in nitrogen uptake by forest species (van Miegroet et al., 1992). This seems particularly true of trees growing at high elevations, which are frequently bathed in cloud cover, or trees that grow near urban industrialized areas.

Except in mature, slowly growing forests, most of the nitrogen is taken up and either retained in the canopy or held in long-term storage in the litter on the forest floor where it is slowly recycled. The nitrogen content of the litter is slowly leached into the soil by rain and surface water or is broken down into simpler compounds by a variety of soil bacteria, fungi, earthworms, and other decomposing organisms. The final step in the breakdown is **mineralization,** or the formation of inorganic nitrogen from organic nitrogen. Mineralization is largely due to the process of ammonification described earlier. Mineralization is invariably accompanied by **immobilization,** or the retention and use of nitrogen by the decomposing organisms. Availability of litter nitrogen to plants depends above all on **net mineralization,** or the extent to which mineralization exceeds immobilization (Haynes, 1986).

The balance between mineralization and immobilization and oxidation of the mineralized NH_4^+-nitrogen by nitrifying bacteria is regulated by environmental parameters. Principal among these are temperature, pH, soil moisture, and oxygen supply. The optimum temperature for nitrification generally falls between 25°C and 35°C, although climatic adaptations of indigenous nitrifying bacteria to more extreme temperatures have been demonstrated (Haynes, 1986). In one study, for example, the optimum temperature for nitrification in soils of northern Australia was 35°C, in Iowa (USA) it was 30°C, and in Alberta (Canada), 20°C (Malhi and McGill, 1982). As noted earlier, soil pH is a major limiting factor in the growth of nitrifying bacteria. The growth of *Nitrobacter* is probably inhibited by ammonium toxicity at high pH (7.5 and above), while aluminum toxicity is suspected as the cause of limited nitrification in acid soils. Soil moisture and oxygen supply go hand-in-hand—little nitrification occurs in water-saturated soils because of the limited O_2 supply. At the other extreme, the rate of nitrification declines with decreasing soil water potential (Ψ_{soil}) below about -0.03 to -0.04 MPa.

The relative significance of nitrification in the nitrogen cycle of natural ecosystems is not altogether clear. Most studies indicate a relatively minor role, since little if any surplus NO_3^- is found in the soil or streams of most undisturbed ecosystems. Experimental deforestation, however, leads to a rapid rise (as much as 50-fold) in the levels of NO_3^- in stream water (Bormann and Likens, 1979). NO_3^- levels gradually returned to normal only as the vegetation began to regrow. These results suggest that nitrification is a significant source of nitrogen, but rapid uptake of NO_3^- by plants is an important factor in maintaining low levels of NO_3^- in the soil solution.

Trees and other plants also tend to conserve a large proportion of their nitrogen, withdrawing nitrogen from the leaves and flowers before they are shed and placing it in storage in the roots and stem tissues. Between one-third and two-thirds of a plant's nitrogen may be conserved by such internal cycling. In the case of deciduous trees, for example, this stored nitrogen offers a degree of nutritional independence from the often nitrogen-poor soil during the flush of growth in early spring.

SUMMARY

Nitrogen is often a limiting nutrient for plants, even though molecular nitrogen is readily available in the atmosphere. Plants do not have the gene coding for dinitrogenase but must depend instead on the nitrogen-fixing activities of certain prokaryote organisms to produce nitrogen in a combined form.

Nitrogen-fixing organisms may be free-living or form symbiotic associations with plants. Symbiotic nitrogen fixation involves complex genetic and biochemical interactions between host plant roots and bacteria. The invading rhizobia induce the formation of root nodules, where the protein leghemoglobin helps to ensure a low-oxygen environment in which the enzyme dinitrogenase can function. The host plant provides energy in the form of photosynthate and, in turn, receives a supply of combined nitrogen for its own growth and development.

The product of nitrogen fixation is ammonium, which is rapidly incorporated into amino acids before it is exported from the nodule. Plants that do not form nitrogen-fixing associations generally take up nitrogen in the form of nitrate. Nitrate must first be reduced to ammonium before it can be incorporated into organic molecules.

CHAPTER REVIEW

1. What are ammonification, nitrification, and denitrification? What are their respective contributions to the nitrogen cycle?
2. What is meant by the statement that biological nitrogen fixation is exclusively a prokaryote domain?
3. Describe the process of rhizobial infection and nodule development in a legume root.
4. Review the biochemistry of nitrogen fixation. How does a bacteroid differ from a bacterium?
5. What is the function of leghemoglobin in symbiotic nitrogen fixation?
6. The product of nitrogen fixation is ammonia. Trace the path of nitrogen as the ammonia is converted to organic nitrogen and translocated to a leaf cell.
7. While most plants take up nitrogen in the form of nitrate ion, there are some that seem to prefer ammonium. Can you suggest a possible biochemical basis for this difference?
8. Heavy fertilization of agricultural crops with nitrogen is a costly process, both economically and energetically. Is it feasible to produce crops without nitrogen fertilizers? If so, what would be the consequences with respect to yields?

FURTHER READING

Haynes, R. J. 1986. *Mineral Nitrogen in the Plant-Soil System.* New York: Academic Press, 482 pp.

Lam, H.-M., K. T. Coschigano, I. C. Oliveira, R. Melo-Oliveira, G. M. Coruzzi. 1996. The molecular genetics of nitrogen assimilation into amino acids in higher plants. *Annual Review of Plant Physiology and Plant Molecular Biology* 47:569–593.

Sprent, J.I., P. Sprent. 1990. *Nitrogen Fixing Organisms. Pure and Applied Aspects.* London: Chapman and Hall.

Stacy, G., R. H. Burris, H. J. Evans. 1992. *Biological Nitrogen Fixation.* London: Chapman and Hall.

REFERENCES

Abrol, Y. P., M. S. Kaim, T. V. R. Nair. 1986. Nitrogen redistribution and its loss in wheat. In: H. Lambers, J. J. Neeteson, I. Stuhlen (eds.), *Fundamental, Ecological and Agricultural Aspects of Nitrogen Metabolism in Higher Plants.* Dordrecht: Martinus Nijhoff.

Atkins, C. A., 1984. Leghaemoglobin and *Rhizobium* respiration. *Annual Review of Plant Physiology* 35:443–478.

Atkins, C. A. 1991. Ammonia assimilation and export of nitrogen from the legume nodule. In: M. J. Dilworth, A. R. Glenn, (eds.), *Biology and Biochemistry of Nitrogen Fixation.* Amsterdam: Elsevier. pp. 293–319.

Bauer, W. D. 1981. Infection of legumes by rhizobia. *Annual Review of Plant Physiology.* 32:407–449.

Bormann, F. H., G. E. Likens. 1979. *Patterns and Process in a Forested Ecosystem.* Berlin: Springer-Verlag.

Bowsher, C. G., D. M. Long, A. Oaks, S. J. Rothstein. 1991. The effect of light/dark cycles on expression of nitrate assimilatory genes in maize shoots and roots. *Plant Physiology* 95:281–285.

Conrad, R., W. Seiler. 1980. Contribution of hydrogen production by biological nitrogen fixation to the global hydrogen budget. *Journal of Geophysical Research* 85:5493–5498.

Dénarié, J., J. Cullimore. 1993. Lipo-oligosaccharide nodulation factors: A new class of signalling molecules mediating recognition and morphogenesis. *Cell* 74:951–954.

Diaz, C. L., L. S. Melchers, P. J. J. Hooykaass, G. J. J.. Lugtenberg, J. W. Kijne. 1989. Root lectin as a determinant of host-plant specificity in the *Rhizobium*-legume symbiosis. *Nature* 338:579–581.

Egeraat, A. W. S. M. van. 1975. The possible role of homoserine in the development of *Rhizobium leguminosarum* in the rhizosphere of pea seedlings. *Plant and Soil* 42:381–386.

Evans, H. J., R. H. Burris. 1992. Highlights in biological nitrogen fixation during the last 50 years. In: G. Stacey, R. H. Burris, H. J. Evans (eds.), *Biological Nitrogen Fixation.* New York: Chapman and Hall, pp. 1–42.

Franssen, H. J., J.-P. Nap, T. Bisseling. 1992. Nodulins in root nodule development. In: G. Stacey, R. H. Burris, H. J. Evans (eds.), *Biological Nitrogen Fixation.* New York: Chapman and Hall, pp. 598–624.

Hageman, R. H. and D. Flesher. 1960. Nitrate reductase activity in corn seedlings as affected by light and nitrate content of nutrient media. *Plant Physiology* 35:700–708.

Haynes, R. J. 1986. *Mineral Nitrogen in the Plant-Soil System.* New York: Academic Press.

Heidstra, R., W. C. Wang, Y, Yalcin, S. Peck, A. M. Emons, A. Vankammen, T. Bisseling. 1997. Ethylene provides positional information on cortical cell division but is not involved in nod factor-induced root hair tip growth in rhizobium-legume interaction. *Development* 124:1781–1787.

Heytler, P. G., G. S. Reddy, R. W. F. Hardy. 1984. In vivo energetics of symbiotic nitrogen fixation in soybeans. In: P. W. Ludden, P. J. Burris (eds.), *Nitrogen Fixation and CO_2 Metabolism.* New York: Elsevier, pp. 283–292.

Huber, S. C., M. Bachmann, J. L. Huber. Post-translational regulation of nitrate reductase activity: a role for Ca^{2+} and 14-3-3 proteins. *Trends in Plant Science* 1:432–438.

Layzell, D. B., S. Hunt, G. R. Palmer. 1990. Mechanism of nitrogenase inhibition in soybean nodules. *Plant Physiology* 92:1101–1107.

Malhi, S. S., W. B. McGill. 1982. Nitrification in three Alberta soils: effect of temperature, moisture and substrate concentration. *Soil Biology and Biochemistry* 14:393–399.

Nutman, P. S. 1976. Alternative sources of nitrogen for crops. *Journal of the Agricultural Society*, pp. 86–94.

Oaks, A., B. Hirel. 1985. Nitrogen metabolism in roots. *Annual Review of Plant Physiology* 36:345–365.

Pate, J. S. 1973. Uptake, assimilation and transport of nitrogen compounds by plants. *Soil Biology and Biochemistry* 5:109–119.

Pate, J. S., C. A. Atkins, R. M. Rainbird. 1981. Theoretical and experimental costing of nitrogen fixation and related processes in the nodules of legumes. In: A. H. Gibson and W. E. Newton, (eds.), *Current Perspectives in Nitrogen Fixation.* Canberra: Australian Academy of Sciences. pp. 105–116.

Rolfe, B. G. 1988. Flavones and isoflavones as inducing substances of legume nodulation. *BioFactors* 1:3–10.

Simpson, R. J. 1986. Translocation and metabolism of nitrogen: whole plant aspects. In: H. Lambers, J. J. Neeteson, I. Stulen (eds.), *Fundamental, ecological and agricultural aspects of nitrogen metabolism in higher plants.* Dordrecht: Martinus Nijhoff, pp. 71–96.

Smit, G., T. J. J. Logman, M. E. T. I. Boerrigter, J. W. Kijne, B. J. J. Lugtenberg. 1989. Purification and partial characterization of the *Rhizobium leguminosarum* bv. viciae Ca^{2+}-dependent adhesin, which mediates the first step in attachment of cells of the family Rhizobiaceae to plant root hair tips. *Journal of Bacteriology* 171:4054–4062.

Sprent, J. I., P. Sprent. 1990. *Nitrogen Fixing Organisms. Pure and Applied Aspects.* London: Chapman and Hall.

Triplett, E. W. 1996. Diazotrophic endophytes: Progress and prospects for nitrogen fixation in monocots. *Plant & Soil* 186:29–38.

van Miegroet, H., G. M. Lovett, D. W. Cole. 1992. Nitrogen chemistry, deposition, and cycling in forests. Summary and conclusions. In: D. W. Johnson, S. E. Linberg (eds.), *Atmospheric Deposition and Forest Nutrient Cycling.* New York: Springer-Verlag, pp. 202–207.

Vitousek, P. M., H. A. Mooney, J. Lubchenco, J. M. Melillo. 1997. Human domination of earth's ecosystems. *Science* 277:494–499.

Werner, D. 1992. Physiology of nitrogen-fixing legume nodules: Compartments and functions. In: G. Stacey, R. H. Burris, H. J. Evans (eds.), *Biological Nitrogen Fixation.* New York: Chapman and Hall, pp. 399–431.

Part Two

Plants, Energy, and Carbon

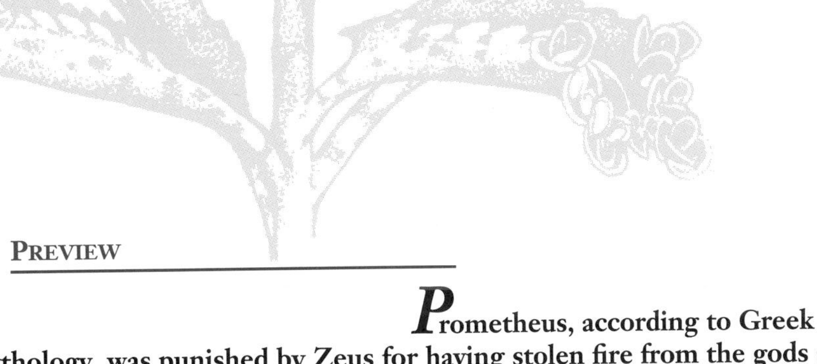

Preview

Prometheus, according to Greek mythology, was punished by Zeus for having stolen fire from the gods and given it to man. It is fortunate that Zeus did not feel similarly disposed toward plants, which have been "stealing" sunlight and converting it to their own use almost since the beginning of life on earth. Energy is required by all living organisms in order to build and maintain their complex structures. The flow of energy into the biosphere begins with green plants, which live by capturing the energy of sunlight and converting it to a usable chemical form. This process, known as photosynthesis, occurs in discrete subcellular organelles called *chloroplasts,* which contain the pigment chlorophyll and all the enzymes and other factors required to assemble inorganic carbon and other elements into the complex carbon compounds of which living organisms are made. Leaves, the site of photosynthesis in most higher plants, are uniquely designed for the efficient absorption of light and CO_2 while restricting the potentially disastrous loss of water that can occur in a hostile terrestrial environment.

Photosynthesis occurs in two stages. The first involves a series of electron transport reactions in which light energy is converted to stable chemical forms: NADPH and ATP. In the second stage, the NADPH and ATP are used to reduce carbon dioxide to sugars. Ultimately, the goal of photosynthesis is to provide organic carbon for increasing the biomass of the organism and the energy to build and maintain that biomass.

The energy orginally stored in sugars by photosynthesis is retrieved by cellular respiration, a complex metabolic pathway that also converts simple sugars into the building blocks used to build the more complex molecules required by cells.

Plant productivity, or carbon gain, is determined largely by the balance beween photosynthesis and respiration, taking into consideration the allocation of photosynthetic product between different matabolic pathways, partitioning of product to different plant organs, and the proportion of carbon expended in maintaining the organism. Variations in the basic photosynthetic pathway ensure that certain organisms are able to maintain high productivity under specific environmental conditions.

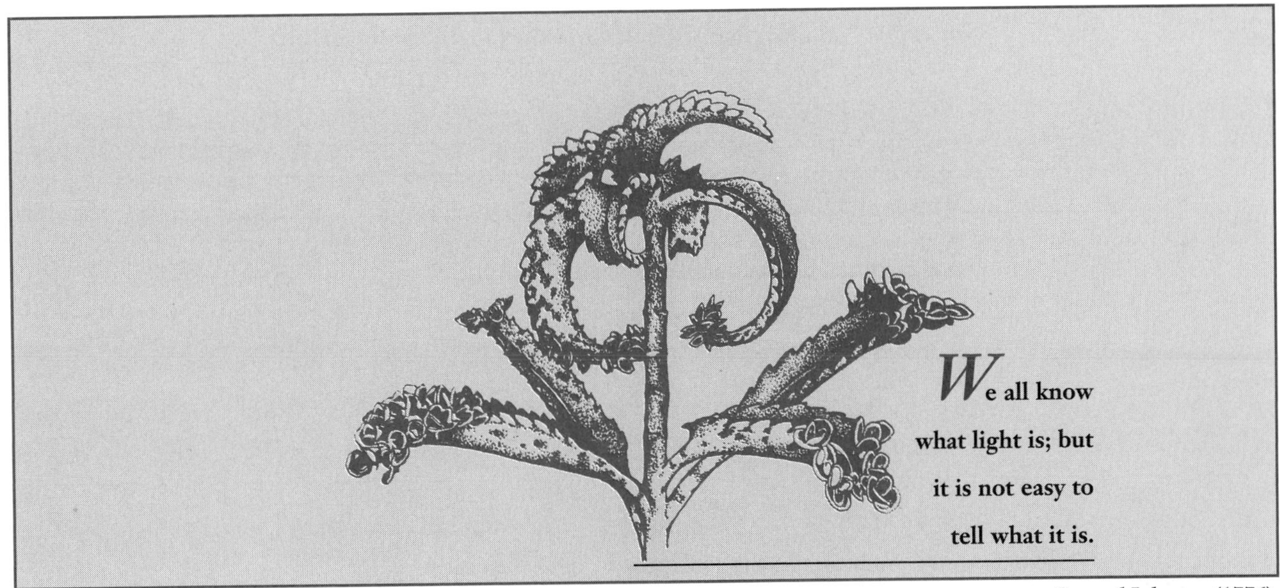

Samuel Johnson (1776),
according to Boswell

7

Light and Pigments: An Introduction to Photobiology

Sunlight satisfies two very important needs of biological organisms: energy and information. On the one hand, radiant energy from the sun maintains the planet's surface temperature in a range suitable for life and, through the process of photosynthesis, is the ultimate source of energy that sustains life. Radiation, primarily in the form of light, also provides critical information about the environment—information that is used by plants to regulate movement, trigger developmental events, and mark the passage of time. The importance of light in the life of green plants is reflected in the study of *photo*biology, which encompasses phenomena with names such as *photo*morphogenesis, *photo*periodism, and *photo*synthesis.

In order to fully appreciate the pervasive importance of light to plants, it is necessary to understand something of the nature of light and the molecules with which light interacts in plants. This chapter will

- explore the physical nature of light and how light interacts with matter;
- discuss some of the terminology used in describing light and methods for measuring it;
- discuss briefly the characteristics of light in the natural environment of plants; and
- review the principal pigments and pigment systems found in plants.

The various ways in which light is used by plants to power photosynthesis and regulate development will be discussed throughout many of the subsequent chapters.

THE PHYSICAL NATURE OF LIGHT

Light is a form of electromagnetic energy that has dual attributes of continuous waves and discrete particles. When transmitted through space, reflected by an object, or refracted by a lens, electromagnetic radiation behaves as a wave phenomenon. Light being emitted or absorbed behaves as though it were a stream of particles.

What is light? As Johnson recognized more than 200 years ago, this question is not easily answered. The simplest answer is that light is a form of radiant energy, a narrow band of energy within the continuous **electromagnetic spectrum** of radiation emitted by the sun (Fig. 7.1). The term "light" describes that portion of the electromagnetic spectrum that causes the physiological sensation of vision in humans. In other words, light is defined by the range of wavelengths—between 400 and approximately 700 nanometers—capable of stimulating the receptors located in the retina of the human eye. Strictly speaking, those regions of the spectrum we per-

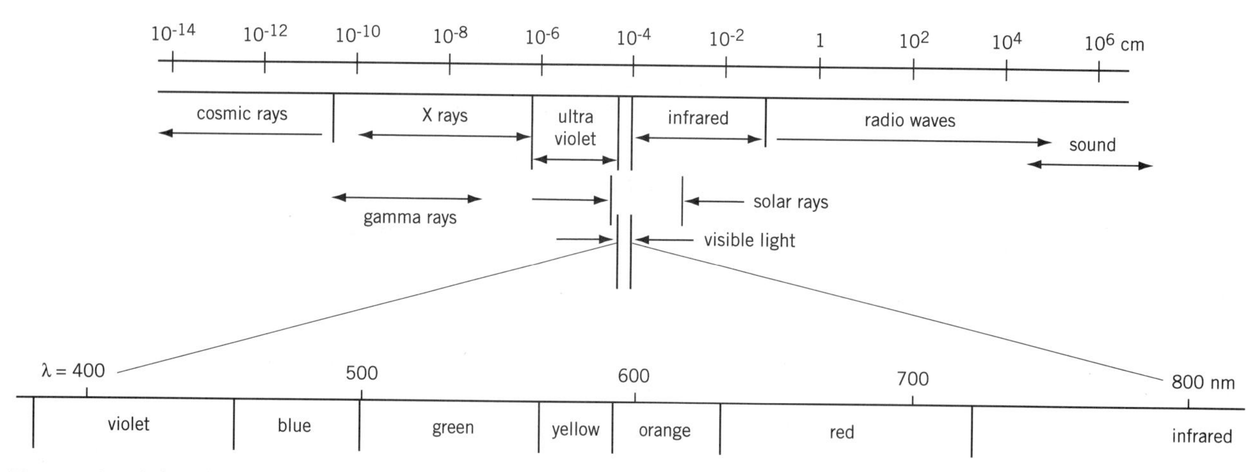

FIGURE 7.1 The electromagnetic spectrum. Visible radiation, or light, represents a very small portion of the total electromagnetic spectrum.

ceive as red, green, or blue are called light, whereas the ultraviolet and infrared regions of the spectrum, which our eyes cannot detect (although they may have significant biological effects), are referred to as ultraviolet or infrared *radiation*, respectively. While the following discussion will focus on light, it is understood that the principles involved apply to radiant energy in the broader sense.

Like other forms of energy, light is a bit of an enigma and is difficult to define. It is more easily described not by what it is but by how it interacts with matter. Physicists of the late nineteenth and early twentieth centuries resolved that light has attributes of both continuous waves and discrete particles. Both of these attributes are important in understanding the biological role of light.

LIGHT AS A WAVE PHENOMENON

The propagation of light through space is characterized by regular and repetitive changes, or waves, in its electrical and magnetic properties. Electromagnetic radiation actually consists of two waves—one electrical and one magnetic—which oscillate at 90° to each other and to the direction of propagation (Fig. 7.2). The wave properties of light may be characterized by either **wavelength** or **frequency.** The distance in space between wave crests is known as the wavelength and is represented by the Greek letter lambda (λ). Biologists commonly express wavelengths in units of *nanometers* (nm), where 1 nm = 10^{-9}m. Frequency, represented by the Greek letter nu (ν), is the number of wave crests, or cycles, passing a point in space in one second. Frequency is thus related to wavelength in the following way:

$$\nu = c / \lambda \tag{7.1}$$

where c is the speed of light (3×10^8 m s^{-1}). Biologists most commonly use wavelength to describe light and other forms of radiation, although frequency is useful in

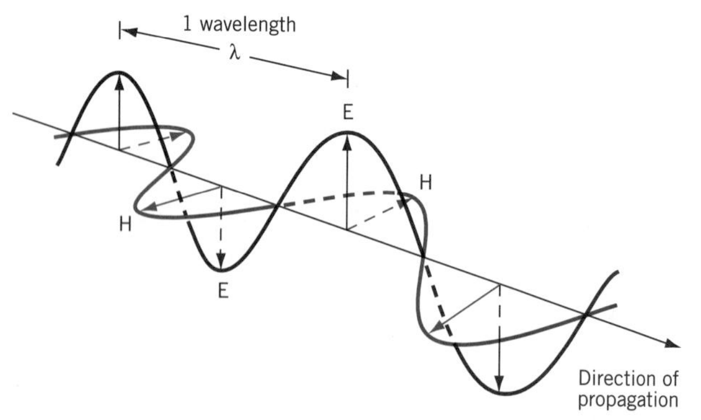

FIGURE 7.2 Wave nature of light. Electrical vectors (E) and magnetic vectors (H) oscillate at 90° to each other.

TABLE 7.1 Radiation of principal interest to biologists.

Color	Wavelength Range (nm)	Average Energy (kJ mol^{-1} photons)
Ultraviolet	*100–400*	
UV-C	100–280	471
UV-B	280–320	399
UV-A	320–400	332
Visible	*400–740*	
Violet	400–425	290
Blue	425–490	274
Green	490–550	230
Yellow	550–585	212
Orange	585–640	196
Red	640–700	181
Far-red	700–740	166
Infrared	*longer than 740*	85

certain situations. Wavelengths of primary interest to photobiologists fall into three distinct ranges: ultraviolet, visible, and infrared (Table 7.1).

LIGHT AS A STREAM OF PARTICLES

When light is emitted from a source or interacts with matter, it behaves as though its energy is divided into discrete units or particles called **photons.** The energy carried by a photon is called a **quantum** (pl. = *quanta*), to reflect the fact that the energy can be *quantized*, or divided into multiple units.

The energy carried by a photon is related to wavelength and frequency in accordance with the following relationship:

$$E_q = hc/\lambda = h\nu \quad (7.2)$$

where h is a proportionality constant, called Planck's constant. The value of h is 6.62×10^{-34} J s photon^{-1}. Accordingly, *the quantum energy of radiation is inversely proportional to its wavelength or directly proportional to its frequency*. The symbol $h\nu$ (pronounced "h nu") is commonly used to represent a photon in figures and diagrams.

Since both h and c are constants, the energy of a photon is easily calculated for any wavelength of interest. The following example illustrates a calculation of the energy content of red light, with a representative wavelength of 660 nm (6.6×10^{-7} m).

$$E_q = \frac{(6.62 \times 10^{-34} \text{ J s photon}^{-1})(3 \times 10^{8} \text{ m s}^{-1})}{6.6 \times 10^{-7} \text{ m}} \quad (7.3)$$

Solving for E_q:

$$E_q = 3.01 \times 10^{-19} \text{ J photon}^{-1} \quad (7.4)$$

For blue light, with a representative wavelength of 435 nm (4.35×10^{-7} m),

$$E_q = \frac{(6.62 \times 10^{-34} \text{ J s photon}^{-1})(3 \times 10^{8} \text{ m s}^{-1})}{4.35 \times 10^{-7} \text{ m}} \quad (7.5)$$

Again, solving for E_q:

$$E_q = 4.56 \times 10^{-19} \text{ J photon}^{-1} \quad (7.6)$$

As the above numbers indicate, the energy content of a single photon is a very small number. For practical purposes it is convenient to multiply the energy of a single photon by Avogadro's number ($N = 6.023 \times 10^{23}$ photons mol^{-1}). This gives the value of energy for a mole of photons, which is a more workable quantity. The energy carried by a mole of photons of red light, for example, is 181 292 J mol^{-1}, or 181 kJ mol^{-1} (Table 7.1). The energy carried by a mole of photons of blue light is correspondingly 274 kJ mol^{-1}. The concept of a mole of photons is more useful than dealing with individual photons. For example, as will become apparent in the following section, a mole of photons of a particular wavelength would be required to excite a mole of pigment molecules.

THE ABSORPTION AND FATE OF LIGHT ENERGY

The energy of light can be absorbed by molecules called pigments. *The resulting excited pigment molecules are unstable and the energy must be dissipated as heat, reemitted as light, or utilized through participation in a photochemical reaction.*

For light to be used by plants, it must first be absorbed. This initial action of light in photochemistry and photobiology is based on two fundamental principles. The first principle, known as the *Gotthaus-Draper principle*, tells us that only light that is absorbed can be active in a photochemical process. Because the initial event in any light-driven physiological process is photochemical in nature, this principle can be extended to include photobiological phenomena as well. Therefore, any photobiological phenomenon requires the participation of a molecule that absorbs light. Such a molecule is known as a **pigment.** Plants contain a variety of pigments that are prominent visual features and important physiological components of virtually all plants. The characteristic green color of leaves, for example, is due to a family of pigments known as the chlorophylls. Chlorophyll absorbs the light energy used in photosynthesis. The pleasing colors of floral petals are due to the anthocyanin pigments that serve to attract insects as pollen vectors. Other pigments, such as phytochrome, are present in quantities too small to be visible but nonetheless serve important roles in plant morphogenesis. These and other important plant pigments will be described later in this chapter.

What actually happens when a pigment molecule absorbs light? Absorption of light by pigment molecule is a rapid event, occurring within a femtosecond ($= 10^{-15}$ s). During that time, the energy of the absorbed photon is transferred to an electron in the pigment molecule. The energy of the electron is thus elevated from a low energy level, the **ground state,** to a higher energy level known as the **excited,** or **singlet,** state. This change in energy level is illustrated graphically in Figure 7.3. Like photons, the energy states of electrons are also *quantized*, that is, an electron can exist in only one of a series of discrete energy levels. A photon can be absorbed only if its energy content matches the energy required to raise the energy of the electron to one of the higher, allowable energy states.

In the same way that quanta cannot be subdivided,

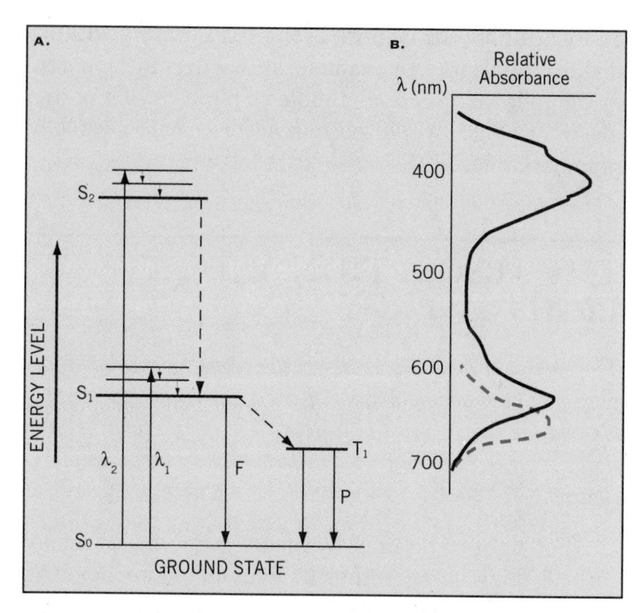

FIGURE 7.3 The absorption of light by a molecule. (*A*) An energy level diagram depicting the various possible transitions when light is absorbed. A nonexcited molecule is said to be in the ground state. S_2 and S_1 are excited singlet states achieved by absorption (solid arrows) of λ_2 and λ_1, respectively. Dashed arrows represent radiationless decay through which energy is given up primarily as heat. Fluorescence (F) is the emission of light from the lowest excited singlet state. T_1 represents the metastable excited triplet state. Energy from the triplet state may be lost by radiationless decay or by delayed emission of light known as phosphorescence (P). The triplet state is sufficiently longlived to allow for photochemical reactions to occur. (*B*) An absorption spectrum is a graph of absorption as a function of wavelength. Peaks, or absorption bands, correspond to principle excitation levels. Also shown is a fluorescence emission spectrum (dashed line), which corresponds to the emission of absorbed energy from the lowest excited singlet state.

electrons cannot be partially excited. This is the second photochemical principle (known as the *Einstein–Stark law of photochemical equivalence*)—a single photon can excite only one electron. On the other hand, complex pigment molecules, such as chlorophyll, will have many different electrons, each of which may absorb a photon of a different energy level and, consequently, different wavelength. Moreover, each electron may exist in a variety of excitation states (called vibrational and rotational levels), which broadens even further the number of photons that may be absorbed (Fig. 7.3). Pigment molecules such as chlorophyll, when exposed to white light, will thus exhibit many different excited states at one time.

An excited molecule has a very short lifetime (on the order of a nanosecond, or 10^{-9} s) and, in the absence of any chemical interaction with other molecules in its environment, it must rid itself of any excess energy and return to the ground state. Dissipation of excess energy may be accomplished in several ways.

1. **Thermal deactivation** occurs when a molecule loses excitation energy as heat (Fig 7.3A). The electron will very quickly drop to the lowest excited singlet state. The excess energy is given off as heat to its environment. If the electron then returns to the ground state, that energy will also be dissipated as heat.
2. **Fluorescence** is the emission of a photon of light by the excited molecule. However, fluorescence emission can occur only after the electron has first undergone thermal deactivation to the lowest excited singlet state (Fig 7.3A). Consequently the emitted photon has a lower energy content, and, in accordance with Equation 7.1, a longer wavelength, than the exciting photon. In the case of the photosynthetic pigment chlorophyll, for example, peak fluorescent emission falls to the long wavelength side of the red absorption band (Fig. 7.3B). This is true regardless of whether the pigment was excited with blue light (450 nm, 262 KJ mol^{-1}) or red light (660 nm, 181 kJ mol^{-1}). For pigments such as chlorophyll in solution, a return to the ground state by emission of light is often the only option available (Fig. 7.4).

FIGURE 7.4 An example of fluorescence. An extract of leaf pigments, containing predominantly chlorophyll, is irradiated with blue light from above. The absorbed energy is reemitted as red fluorescent light. (See color plate 4.)

3. Energy may be transferred between pigment molecules by what is known as **inductive resonance** or **radiationless transfer.** Such transfers will occur with high efficiency, but require that the pigment molecules are very close together and that the fluorescent emission band of the donor molecule overlaps the absorption band of the recipient. Inductive resonance accounts for much of the transfer of energy between pigment molecules in the chloroplast (Chap. 9).
4. The molecule may revert to another type of excited state, called the **triplet** state (Fig. 7.3A). The difference between singlet and triplet states, related to the spin of the valence electrons, is not so important here. It is sufficient to know that the triplet state is more stable than the singlet state—it is considered a *metastable* state. The longer lifetime of the metastable triplet state (on the order of 10^{-3} seconds) is sufficient to allow for photochemical reactions to occur.
5. The energetic electron may be given up to an acceptor molecule. When this occurs, the pigment is said to be **photooxidized** and the acceptor molecule becomes reduced. Photooxidation of chlorophyll, for example, is the primary photochemical act in photosynthesis (Chap. 9). As in the case of fluorescence, pigments will first undergo rapid thermal deactivation to the lowest excited state before participating in a photochemical act. This means that the energy available to carry out photosynthesis is no greater for a photon of blue light than it is for a photon of red light. In addition to photooxidations, the energy may be used to induce a conformational change, as is believed to occur in the case of the phytochrome molecule (Chap. 18).

ABSORPTION AND ACTION SPECTRA

The efficiency of light absorption and resulting physiological action can be displayed as a function of wavelength. The resulting graphs are called absorption spectra and action spectra, respectively.

Figure 7.3B illustrates absorption of light by a pigment (in this case, chlorophyll) as a function of wavelength. The resulting graph, in which the efficiency of absorption is plotted against wavelength, is known as an **absorption spectrum.** In this figure, the absorption spectrum has actually been turned 90° from its normal orientation. This has been done in order to emphasize the correspondence between possible excitation states of the molecule and the principal bands in the absorption spectrum. The energy level diagram in Figure 7.3A represents, in a broad sense, absorption of light by chlorophyll. In this case, λ_1 would represent red light and λ_2 would represent blue light.

An absorption spectrum is in effect a probability statement. The height of the absorption curve (or width, as presented in Fig. 7.3B) at any given wavelength reflects the probability by which light of that energy level will be absorbed. More importantly, an absorption spectrum is like a fingerprint of the molecule. Every light-absorbing molecule has a unique absorption spectrum that is often a key to its identification. For example, there are several different variations of the green pigment chlorophyll. The pattern of the absorption spectrum for each variation generally resembles that shown in Figure 7.3B, yet each variation of chlorophyll differs with respect to the precise wavelengths at which maximum absorbance occurs.

Because light must first be absorbed in order to be effective in a physiological process, it follows that there must be a pigment that absorbs the effective light. One of the first tasks facing a photobiologist when studying a light-dependent response is to identify the responsible pigment. One important piece of information is called an **action spectrum.** An action spectrum is a graph that shows the effectiveness of light in a particular process plotted as a function of wavelength. The underlying assumption is that light most efficiently absorbed by the

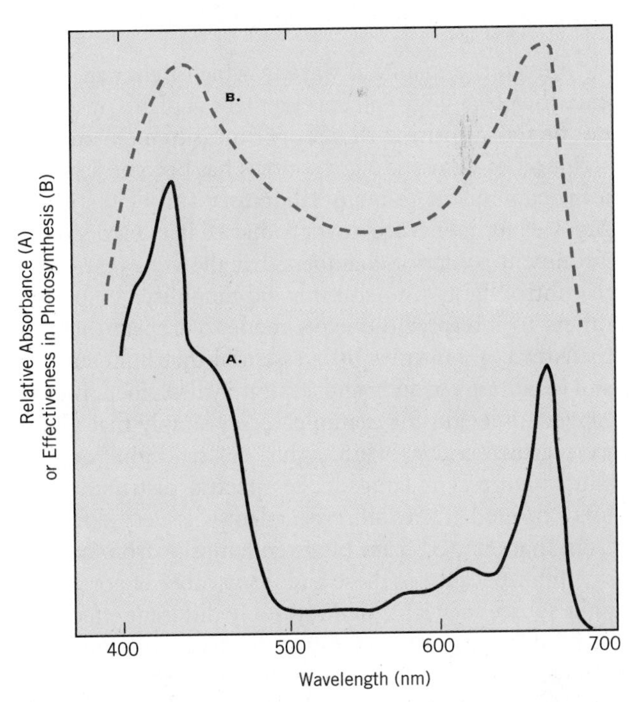

FIGURE 7.5 A typical action spectrum for photosynthesis (*B*) compared with the absorption spectrum of leaf extract (*A*), that contains predominantly chlorophyll. The action spectrum has peaks in the blue and red regions of the spectrum, which correspond with the principal absorption peaks for the pigment in solution.

responsible pigment will also be most effective in driving the response. In other words, the action spectrum for a light-dependent response should closely resemble the absorption spectrum of the pigment that absorbs the effective light. A comparison of an action spectrum with the absorption spectra of suspected pigments can therefore provide useful clues to the identity of the pigment responsible for a photosensitive process. As an example, a typical action spectrum for photosynthesis in a green plant is shown in Figure 7.5. It is compared with the absorption spectrum for a leaf extract that contains primarily chlorophyll and some carotenoid. Note that the action spectrum has pronounced peaks in the red and blue regions of the spectrum and that these action maxima correspond to the absorption maxima for chlorophyll. This is part of the evidence that identifies a role for chlorophyll in photosynthesis. Similar comparisons between action spectra and absorption spectra have been used to identify the responsible pigment as a first step in unravelling a number of plant photoresponses (see Chap. 18).

MEASURING LIGHT

The careful description and accurate measurement of light are important parameters in the study of photobiology. A variety of methods are available for measuring the quantity and wavelength composition of light.

Given the manifold ways in which light can influence the physiology and development of plants, it should not be too surprising that proper measurement and description of light and light sources has become a significant component of many laboratory and field studies. Many experiments are now conducted in controlled environment-rooms or chambers that allow the researcher to control light, temperature, and humidity. To permit others to interpret the experiments or repeat them in their own laboratories, it is essential that light sources and conditions be fully and accurately described. It is no longer sufficient, for example, to say simply that plants were grown under "white light" or "red light" or cool white fluorescent lamps. The spectral distribution of light emitted from fluorescent lamps is very different from that emitted from tungsten lamps or from natural sky light. Because of these and many other factors, each light source may have demonstrably different effects on plant development and behavior. Even natural light changes in quality from dawn through midday to dusk, or between shady and sunny habitats or cloudy and open skies. Understanding photobiology thus requires an understanding of how light is measured and what those measurements mean. Also required is a consistent terminology that is understood by everyone working in the field.

There are three parameters of primary concern when describing light. The first is **light quantity**—how much light has the plant received? The second is the composition of light with respect to wavelength, known as **light quality, spectral composition,** or **spectral energy distribution (SED).** The third factor is **timing.** What are the duration and periodicity of the light treatment?

The measure of light quantity most widely accepted by plant photobiologists is based on the concept of *fluence.* Fluence is defined as the quantity of radiant energy falling on a small sphere, divided by the cross-section of the sphere. Since light is a form of energy that can be emitted or absorbed as discrete packets or photons, fluence can be expressed in terms of either the number of photons or quanta (in moles, mol) or the amount of energy (in joules, J). **Photon fluence** (units = mol m^{-2}) refers to the total number of photons incident on the sphere while **energy fluence** (units = J m^{-2}) refers to the total amount of energy incident on the sphere. The corresponding rate terms are **photon fluence rate** (units = mol m^{-2} s^{-1}) and **energy fluence rate** (units = J m^{-2} s^{-1}, or W m^{-2}). The term **irradiance** is frequently used interchangeably with energy fluence rate although in principle the two are not equivalent. Irradiance refers to the flux of energy on a flat surface rather than a sphere.

Many instruments for measuring radiation actually measure total energy, including energy outside the visible portion of the spectrum, such as infrared, which is not directly relevant to photobiological processes. In order to avoid such complications, instruments are now commercially available that are limited to that portion of the spectrum between 400 nm and 700 nm. This range of light is broadly defined as **photosynthetically active radiation (PAR).** Thus photon fluence rates, expressed as mol photons m^{-2} s^{-1} PAR, or energy fluence rates, expressed as W m^{-2} PAR, are widely accepted for routine laboratory work. The only serious limitation to PAR measurements is that they exclude light in the 700 to 750 nm range—light which, although inactive in higher plant photosynthesis, plays a significant role in regulating plant development (Chap. 18).

The term light quality refers to spectral composition and is usually defined by an emission or incidence spectrum. SED is measured with a spectroradiometer, an instrument capable of measuring fluence rate over narrow wavelength bands. Depending upon the instrument, either **spectral photon fluence rate** (units = mol photons m^{-2} s^{-1} nm^{-1}) or **spectral energy fluence rate** (units = W m^{-2} nm^{-1}) is plotted against wavelength. In practice, spectroradiometers are also equipped with flat surface detectors that measure **spectral irradiance** (W m^{-2} nm^{-1}).

SED can vary depending on the nature of the light source and a number of other factors (Fig. 7.6). The SED of natural sunlight, for example, can vary depend-

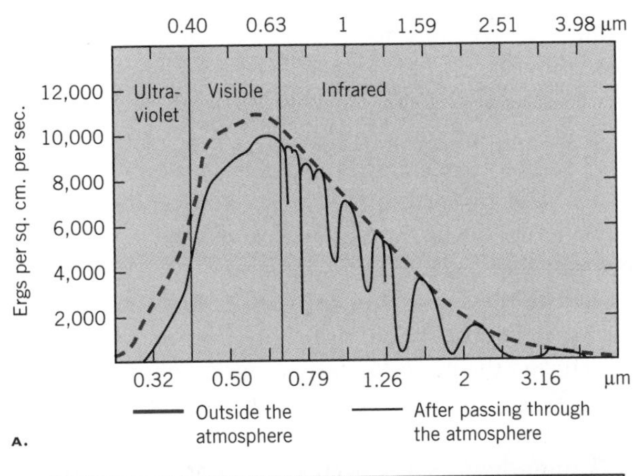

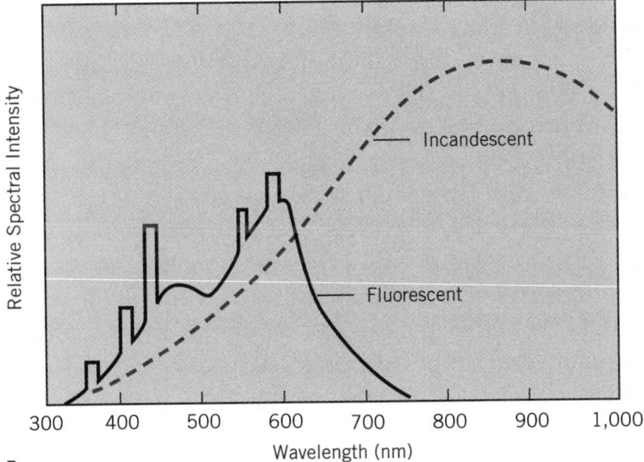

FIGURE 7.6 Spectral energy distribution of sunlight (*A*) compared with fluorescent and incandescent light (*B*). Note the difference in wavelength scale between (*A*) and (*B*). About 20% of the incoming energy from the sun, particularly in the infrared region, is absorbed by atmospheric gasses (primarily CO_2, H_2O). The solar spectrum was drawn from measurements made in clear weather from an observatory in Australia by C. W. Allen. (From H. H. Lamb, *Climate: Present, Past and Future*, Methuen & Co., 1972.

ing on the quality of the atmosphere, cloud cover, and the time of day. The SED of artificial sources, such as incandescent and fluorescent lamps, are significantly different from natural light. Fluorescent light has a relatively high emission in the blue but drops off sharply in the red. Incandescent light, on the other hand, contains relatively little blue light but high emissions in the far-red and infrared.

Most biologists do not have ready access to spectroradiometers—they are usually expensive, require frequent calibration, and are often too cumbersome for routine laboratory work. Simpler, less expensive instruments fitted with filters are available. These enable investigators to measure average spectral irradiance over broad regions of the spectrum; comparing, for example, the proportions of red, blue, and far-red light.

THE NATURAL RADIATION ENVIRONMENT

Plants are exposed to an ever-changing radiation environment. Regular and predictable changes convey to the plant useful information about its environment.

A relatively small proportion of the radiation originating in the sun reaches the earth's atmosphere and even less actually reaches the surface (Fig. 7.6A). However, both the quantity and spectral distribution of radiant energy that reaches (or fails to reach) earth may have a significant impact on the physiology of the plant. As well, radiant energy is central to several problems of more immediate and profound consequences for man.

Significant amounts of infrared radiation are absorbed by the water vapor and carbon dioxide and other gasses present in the earth's atmosphere (Fig. 7.6A), giving rise to a phenomenon known as the **greenhouse effect** (Fig. 7.7). Although public awareness of the greenhouse effect has increased markedly in recent years, it is not a phenomenon restricted to the late twentieth century. Indeed the greenhouse effect has been with us since the beginnings of life on earth. Without it, life as we know it would not be possible. Infrared radiation is of low frequency (or long wavelength) and therefore low energy. Its principle effect is to increase vibrational activity in molecules—that is, heat. Absorption of infrared by atmospheric water vapor and carbon dioxide creates a "thermal blanket" that helps to prevent extreme variations in temperature such as occur on the

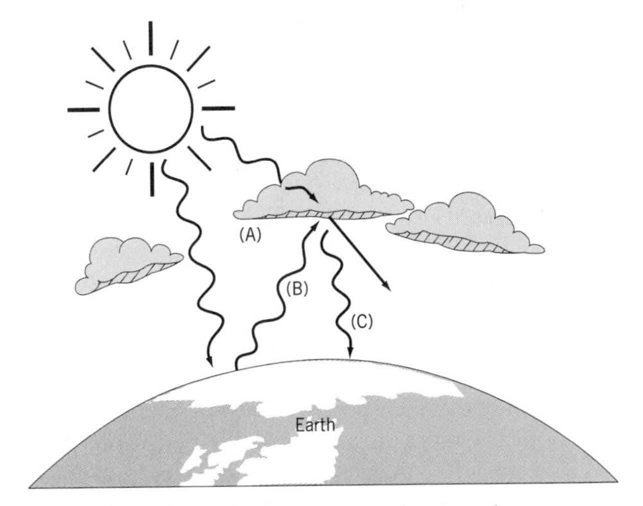

FIGURE 7.7 Diagram representing the greenhouse effect. Radiation from the sun warms both the atmosphere and the earth (*A*). The earth then reradiates infrared back into the atmosphere (*B*). Here it is either reflected back to earth or absorbed by atmospheric gasses (CO_2, H_2O vapor, methanol), thus preventing its escape (*C*). Some of the trapped infrared is reradiated back to earth giving rise to increased temperatures.

lunar surface where these gasses are absent. Similar, although less extreme, temperature variations are characteristic of dry, desert regions on earth where high daytime temperatures alternate with very cool nights.

Public concern about the greenhouse effect arises from evidence that, since the beginnings of the industrial revolution, our prodigious consumption of fossil fuels has contributed to a steady increase in atmospheric carbon dioxide and other so-called "greenhouse gasses." Many believe that continued release of carbon dioxide will lead to greater heat retention in the atmosphere and global warming. This could result in partial melting of polar ice caps with extensive flooding of low-lying land areas and major shifts in agricultural productivity. Another scenario predicts that higher carbon dioxide levels will stimulate photosynthesis and increase the amount of plant material on earth. These scenarios are balanced by the argument that the world's oceans, by exchanging large quantities of carbon dioxide with the atmosphere, are capable of buffering the carbon dioxide concentration and counteracting any warming trend. This is a controversial area and we clearly know far less than we need to about such problems.

At the other end of the spectrum, ultraviolet radiation is characterized by short wavelength, high frequency and high energy levels (Table 7.1). Absorption of ultraviolet radiation creates highly reactive molecules, often causing the ejection of an electron, or ionization of the molecule. Such ionizations usually have deleterious effects on organisms. A principal action of UV-C (about 254 nm), for example, is to induce thiamine dimers (hence, mutations) in deoxyribonucleic acid. In the natural environment, UV-induced mutation is not normally a major problem because little far-ultraviolet radiation reaches the surface. Virtually all of the UV-C and most of the UV-B is absorbed by ozone (O_3) and aerosols (dispersed particles of solids or liquids) in the stratosphere. If, however, the atmospheric ozone concentration were to be lowered, there would be an increased potential for harmful effects to all organisms. In recent years, just such a depletion of the stratospheric ozone layer, leading to increases in UV-B radiation reaching the earth's surface, has become a matter of some concern. Data compiled over the past two decades has revealed that approximately one-half of the plant species studied are adversely affected by elevated UV-B radiation. It is perhaps not surprising that plants most sensitive to UV-B radiation are those native to lower elevations where UV-B fluxes are normally low.

With respect to both frequency and energy level, visible light falls between UV and infrared radiation. Absorption of visible light raises the energy level of outer electrons of the absorbing molecule and thus has the potential for initiating useful photochemical reactions. Moreover, the fluence rate and spectral quality of visible light are constantly changing, often predictably, throughout the day or season. These variations convey information about the environment; information that the plant can use to advantage.

The two most significant changes in visible light on a daily basis are seen in the fluence rate and in spectral distribution. Typically at midday under full sun, the fluence rate approaches 2000 μmole m^{-2} s^{-1}. At twilight, just before the sun sets below the horizon, fluence rate will have dropped to the order of 10 μmol m^{-2} s^{-1} or less. During the period known as dusk, fluence rate falls rapidly—by as much as one order of magnitude every 10 minutes.

Falling light levels at end of day are accompanied by shifts in spectral quality (see Chap. 18). Normal daylight consists of direct sunlight and diffuse skylight. Diffuse skylight is enriched with blue wavelengths because the shorter wavelengths are preferentially scattered by moisture droplets, dust, and other components of the atmosphere. Consequently, normal daylight is enriched with blue (hence, blue skies!). At twilight, often defined as a solar elevation of 10° or less from the horizon, a combination of scattering and refraction of the sun's rays as they enter the earth's atmosphere at a low angle, enriches the light with longer red and far-red wavelengths. This is because, at twilight, the path traversed by sunlight through the atmosphere to an observer on earth may be up to 50 times longer than it is when the sun is directly overhead. Much of the violet and blue light is thus scattered out of the line of sight, leaving predominantly the longer red and orange to reach the observer.

Atmospheric factors, such as clouds and air pollution, also influence the spectral distribution of sunlight. Cloud cover reduces irradiance and increases the proportion of scattered (i.e., blue) light. Airborne pollutants will cause scattering, but will also absorb certain wavelengths.

Plants growing under a canopy must cope with severe reduction in red and blue light as it is filtered through the chlorophyll-containing leaves above or with **sunflecks,** spots of direct sunlight that suddenly appear through an opening in the canopy. These sudden changes in irradiance may have a significant impact on the photosynthetic capacity of a plant (Chap. 9).

It is clear that plants are exposed to an ever-changing light environment. Many of these changes, such as cloud cover, are unpredictable but others such as daily changes in fluence rate and spectral energy distribution occur with great regularity. The more regular changes convey precise information about the momentary status of the environment as well as impending changes (see Chaps. 18, 20). It is perhaps not surprising that plants have evolved sophisticated means for interpreting this information as a matter of survival.

PHOTORECEPTORS

Plants contain a variety of pigment molecules that absorb physiologically useful radiation.

Pigments that absorb light for eventual use in a physiological process are called **photoreceptors.** These molecules process the energy and informational content of light into a form that can be used by the plant. The principal photoreceptors found in plants are described here. Their roles in various physiological processes will be discussed in detail in later chapters.

CHLOROPHYLLS

As noted earlier, chlorophyll is the pigment primarily responsible for harvesting light energy used in photosynthesis. The chlorophyll molecule consists of two parts, a **porphyrin** head and a long hydrocarbon, or **phytol** tail (Fig. 7.8). A porphyrin is a cyclic tetrapyrrole, made up of four nitrogen-containing **pyrrole rings** arranged in a cyclic fashion. Porphyrins are ubiquitous in living organisms and include the heme group found in mammalian hemoglobin and the photosynthetic and respiratory pigments, cytochromes (Chaps. 9, 12). Esterified to ring IV of the porphyrin in chlorophyll is a 20 carbon alcohol, **phytol.** This long, lipid-soluble hydrocarbon tail is a derivative of the 5-carbon isoprene. Isoprene is the precursor to a variety of important molecules, including other pigments (the carotenes), hormones (the gibberellins), and steroids (Chap. 14).

Completing the chlorophyll molecule is a magnesium ion (Mg^{2+}) chelated to the four nitrogen atoms in the center of the ring. Loss of the magnesium ion from chlorophyll results in the formation of a nongreen product **pheophytin.** Pheophytin is readily formed during extraction under acidic conditions, but small amounts are also found naturally in the chloroplast where it serves as an early electron acceptor (Chap. 9).

Four species of chlorophyll, designated chlorophyll *a*, *b*, *c*, and *d*, are known. The chemical structure of chlorophyll *a*, the primary photosynthetic pigment in all higher plants, algae, and the cyanobacteria, is shown in Figure 7.8. Chlorophyll *b* is similar except that a formyl group (—CHO) substitutes for the methyl group on ring II. Chlorophyll *b* is found in virtually all higher plants and green algae, although viable mutants deficient of chlorophyll *b* are known. The principal difference between chlorophyll *a* and chlorophyll *c* (found in the diatoms, dinoflagellates, and brown algae) is that chlorophyll *c* lacks the phytol tail. Finally, chlorophyll *d*, found only in the red algae, is similar to chlorophyll *a* except that a (—O—CHO) group replaces the (—CH=CH_2) group on ring I.

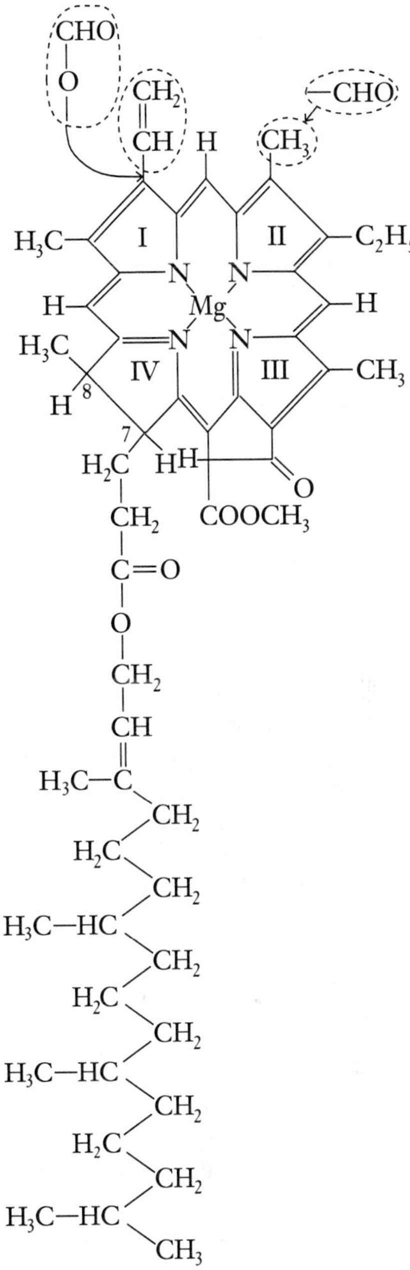

FIGURE 7.8 Chemical structure of chlorophyll a. Chlorophyll b is similar except that a formyl group replaces the methyl group on ring II. Chlorophyll c is similar to chlorophyll a except that it lacks the long hydrocarbon tail. Chlorophyll d is similar to chlorophyll a except that a —O—CHO group is substituted on ring I as shown.

When grown in the dark, angiosperm seedlings do not accumulate chlorophyll. Their yellow color is due primarily to the presence of carotenoids. Dark-grown seedlings do, however, accumulate significant amounts of **protochlorophyll a,** the immediate precursor to chlorophyll a. The chemical structure of protochloro-

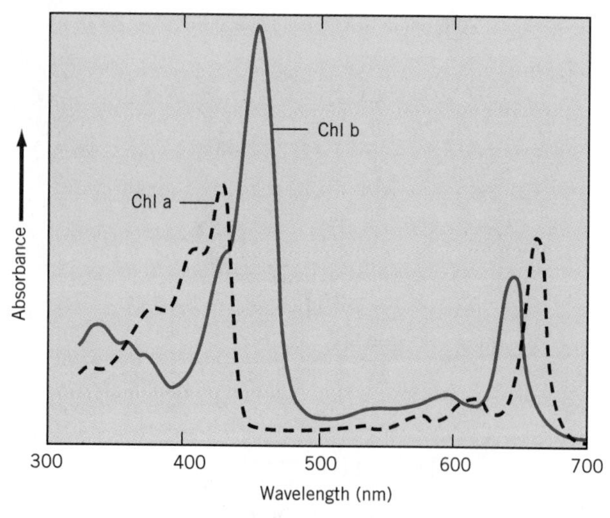

FIGURE 7.9 Absorption spectra of chlorophyll a (solid line) and chlorophyll b (broken line) in acetone.

phyll differs from chlorophyll only by the presence of a double bond between carbons 7 and 8 in ring IV (Fig. 7.8). The reduction of this bond is catalyzed by the enzyme **NADPH:protochlorophyll oxidoreductase.** In angiosperms this reaction requires light, but in gymnosperms and most algae chlorophyll can be synthesized in the dark. There is a general consensus among investigators that chlorophyll *b* is synthesized from chlorophyll *a*.

Note that the respective chlorophylls exhibit generally similar absorption spectra in organic solvents, but with absorption maxima at distinctly different wavelengths, in both the blue and the red regions of the spectrum (Fig. 7.9). Note also that chlorophyll does not absorb strongly in the green (490–550 nm). The strong absorbance in the blue and red and transmittance in the green is what gives chlorophyll its characteristic green color.

The presence of the long hydrocarbon phytol exerts a dominant effect on the solubility of chlorophyll, rendering it virtually insoluble in water. In the plant, chlorophyll is found exclusively in the lipid domain of the chloroplast membranes, where it forms noncovalent associations with hydrophobic proteins. Only a very small percentage of the chlorophyll found *in vivo* is free chlorophyll and that is probably nascent chlorophyll that has yet to be inserted into the membrane. The absorption spectra of these chlorophyll-protein complexes are markedly different from that of free pigment in solution. For example, chlorophyll a-protein absorbs primarily in the region of 675 nm as opposed to 663 nm for chlorophyll a in acetone. Conjugation of chlorophylls with protein in the membrane is important for two reasons. One is that it helps to maintain the pigment molecules in the precise relationship required for efficient energy transfer. A second reason is that it provides each pigment with a unique environment that in turn gives each molecule a slightly different absorption maximum. These slight absorbance differences are an important factor in the orderly transfer of energy through the pigment bed toward the reactive center where photochemical conversion actually occurs (Chap. 9).

PHYCOBILINS

Phycobilins serve as accessory light-harvesting pigments in the red algae and cyanobacteria or as a critical regulatory system in green plants.

Mammalian liver possesses enzymes that degrade porphyrins (e.g., hemoglobin) by oxidation of the double bond between rings III and IV. The result is an open, or straight chain, tetrapyrrole that is subsequently excreted in the bile (Fig. 7.10). These straight-chain tetrapyrroles are thus known as bile pigments or bilins. Although plants certainly do not have livers, they do have pigments with properties similar to the bile pigments. These pigments are called **phycobilins,** the prefix *phyco-* designating pigments of plant origin. Four phycobilins are known. Three of these are involved in photosynthesis and the fourth, phytochromobilin, is an important photoreceptor that regulates various aspects of growth and development (Chap. 18).

The three photosynthetic phycobilins are *phycoerythrin* (also known as phycoerythrobilin), *phycocyanin* (phycocyanobilin), and *allophycocyanin* (allophycocyanobilin). In addition to the open chain tetrapyrrole, the phycobilin pigments differ from chlorophyll in that the

FIGURE 7.10 The open chain tetrapyrrole chromophore of phycocyanin. Compare with the cyclic tetrapyrrole group in the chlorophyll molecule (Fig. 7.8).

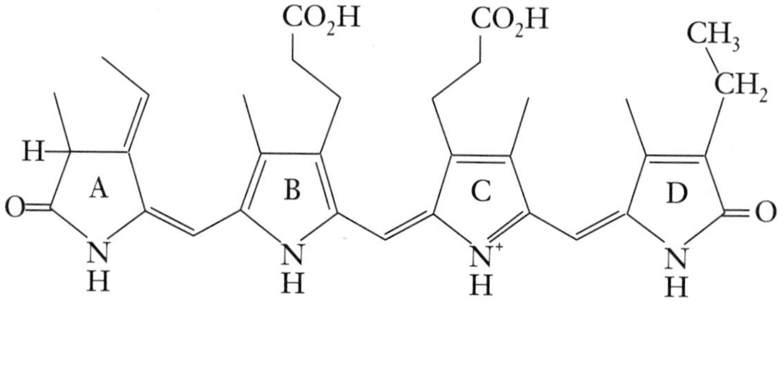

tetrapyrrole group is covalently linked with a protein that forms a part of the molecule. A pigment that contains protein as an integral part of the molecule is known as a **chromoprotein.** The tetrapyrrole portion of the molecule is a **chromophore** (Gr. *phoros* = bearing). A chromophore is the portion of the molecule responsible for absorbing light and, hence, color. The protein portion of a chromoprotein molecule is called the **apoprotein.** The complete molecule, or **holochrome,** consists of the chromophore plus the protein. In the cell, phycobiliproteins are organized into large macromolecular complexes called **phycobilisomes.**

With the exception of phytochromobilin, phycobilin pigments are not found in higher plants but occur exclusively in the cyanobacteria and the red algae (Rhodophyta) where they assume a light harvesting function in photosynthesis. Phycobilins, and in particular phycoerythrin, are useful as light harvesters for photosynthesis because they absorb light energy in the green window where chlorophyll does not absorb (Fig. 7.11). The red algae, for example, appear almost black because the chlorophyll and phycoerythrin together absorb almost all of the visible radiation for use in photosynthesis (compare Fig. 7.11 with Fig. 7.9).

The fourth phycobiliprotein, of particular significance to higher plants, is **phytochrome,** a receptor that plays an important role in many photomorphogenic phenomena. Its chromophore structure and absorption spectrum are similar to that of allophycocyanin. Phytochrome (literally, plant pigment) is unique because it exists in two forms that are photoreversible. The form

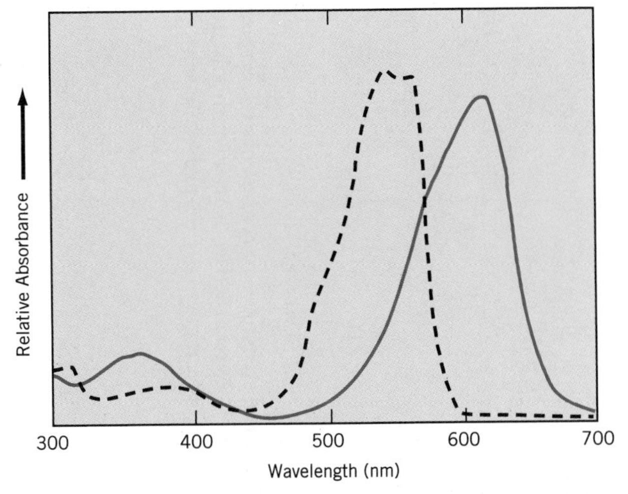

FIGURE 7.11 Absorption spectra of phycocyanin (solid line) and phycoerythrin (broken line) in dilute buffer. Compare with the absorption spectra of chlorophyll (Fig. 7.9). Note that the phycobilins, phycoerythrin in particular, absorb strongly in the 500–600 nm range where chlorophyll absorption is minimal.

P660 (or Pr) absorbs maximally at 660 nm. However, absorption of 660 nm light converts the pigment to a second, far-red-absorbing form P735 (or Pfr). Absorption of far-red light by Pfr converts it back to the red-absorbing form. Pfr is believed to be an active form of the pigment that is capable of initiating a wide range of morphogenetic responses. Phytochrome will be discussed in more detail in Chapter 18.

CAROTENOIDS

Carotenoids comprise a family of orange and yellow pigments present in most photosynthetic organisms. Found in large quantity in roots of carrot and tomato fruit, carotenoid pigments are also prominent in green leaves. In the fall of the year, the chlorophyll pigments are degraded and the more stable carotenoid pigments account for the brilliant orange and yellow colors so characteristic of autumn foliage.

Carotenoid pigments are C_{40} terpenoids biosynthetically derived from the isoprenoid pathway described in Chapter 14. Because the carotenoids are predominantly hydrocarbons, they are lipid soluble and found either in the chloroplast membranes or in specialized plastids called **chromoplasts.** The concentration of pigment in chromoplasts may reach very high levels, to the extent that the pigment actually forms crystals. The carotenoid family of pigments includes **carotenes** and **xanthophylls** (Fig. 7.12). Carotenes are predominantly orange or red-orange pigments. β-carotene is the major carotenoid in algae and higher plants. Note that in β-carotene and α-carotene (a minor form), both ends of the molecule are cyclized. Other forms, such as γ-carotene found in the green photosynthetic bacteria, have only one end cyclized. Lycopene, the principal pigment of tomato fruit, has both ends open. The yellow carotenoids, xanthophylls, are oxygenated carotenes. Lutein and zeaxanthin, for example, are hydroxylated forms of α-carotene and β-carotene, respectively.

The functions of carotenoids in the plant are not clearly understood. Like chlorophyll, β-carotene in the chloroplast is complexed with protein. β-Carotene absorbs blue light (Fig. 7.13) and, to some extent, is able to pass the energy on to chlorophyll for ultimate use in photosynthesis. Because carotenoids absorb so strongly in the blue, they may also serve to protect the chlorophyll from photooxidation by absorbing excess blue light. Some albino mutants are lacking both carotenoids and chlorophyll under normal light conditions but will accumulate chlorophyll under very dim light. Carotenes appear to prevent chlorophyll destruction by reversibly combining with oxygen radicles (a highly reactive form of oxygen) to form xanthophylls (eg., violaxanthin). This phenomenon is discussed further in Chapter 9.

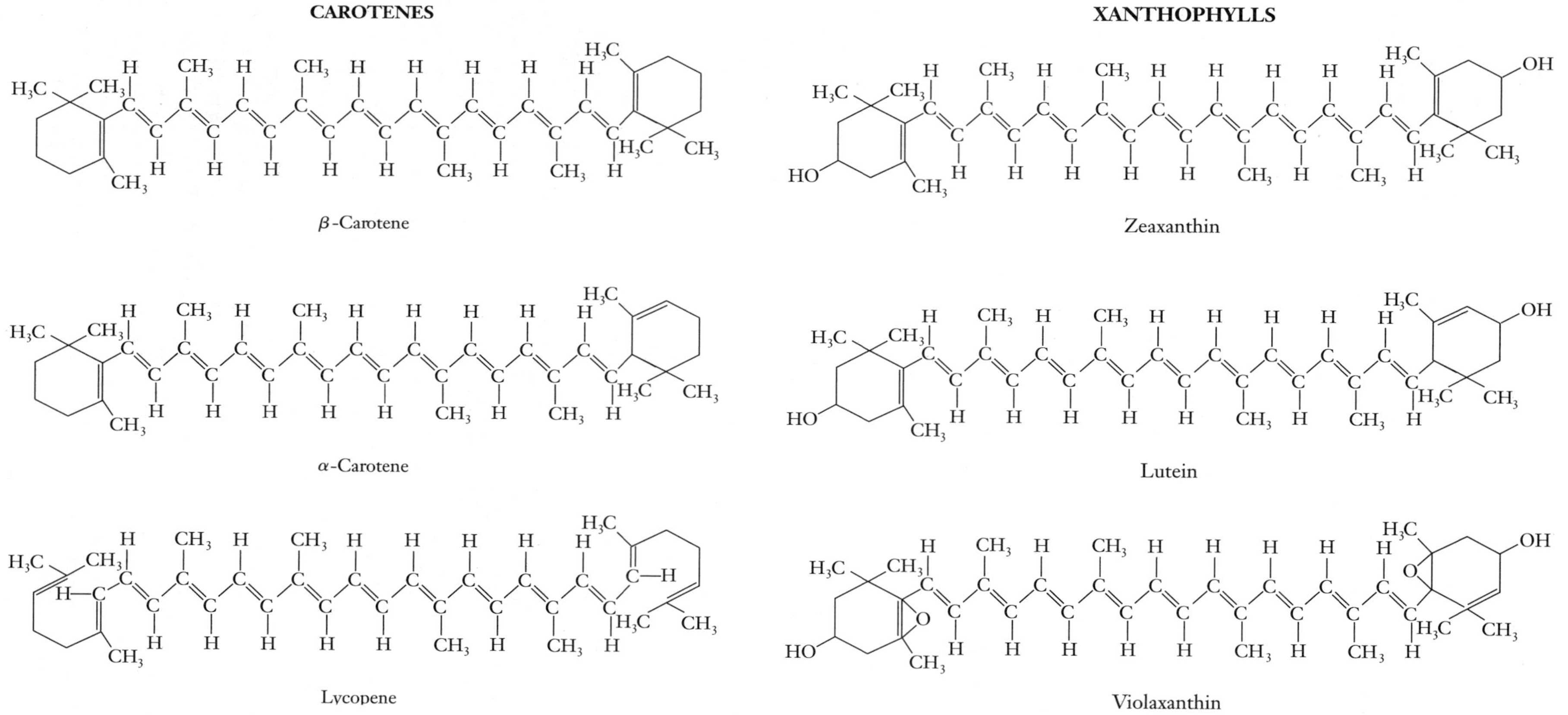

CAROTENES

β-Carotene

α-Carotene

Lycopene

XANTHOPHYLLS

Zeaxanthin

Lutein

Violaxanthin

FIGURE 7.12 The chemical structures of representative carotenes and xanthophylls. The principal distinction between the two is that xanthophylls contain oxygen and carotenes do not. Carotenes are generally orange while xanthophylls are yellow.

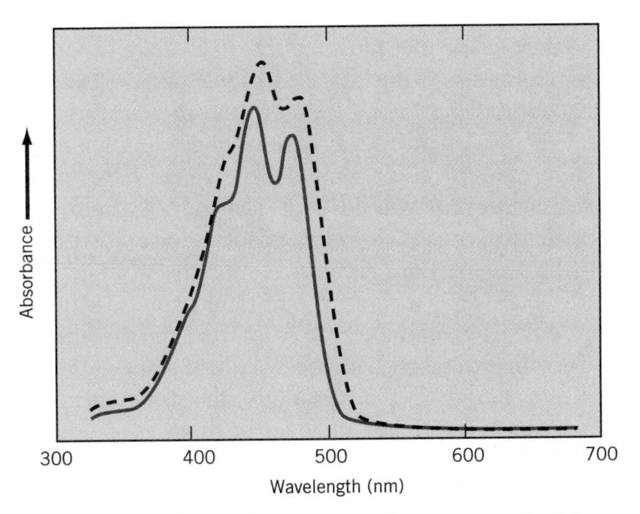

FIGURE 7.13 Absorption spectra of α-carotene (solid line) and β-carotene (broken line).

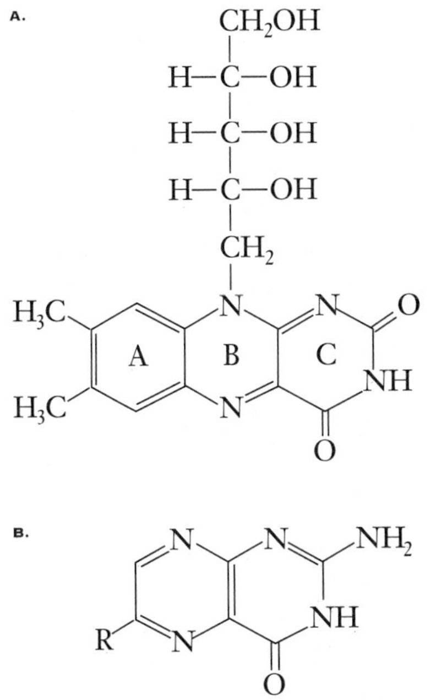

CRYPTOCHROME

A wide range of plant responses to blue and UV-A radiation have been known or suspected for a long time. Although blue/UV-A responses appear to be especially prevalent in lower plants (ferns, mosses, and fungi), these responses all share remarkably similar action spectra with higher plant responses such as phototropism and hypocotyl elongation. Typically, these action spectra exhibit prominent activity in the blue region at about 450 nm with shoulders in the region of 480 nm and in the UV at 380 nm.

The identity of the blue/UV-A photoreceptor has proven difficult to unravel—hence the name **cryptochrome** (literally, hidden pigment). Candidates for cryptochrome include carotenoids and flavins, or both. Like carotenoids, flavins are ubiquitous in living organisms. The three most common are riboflavin (Fig. 7.14) and its two nucleotide derivatives, flavin mononucleotide (FMN) and flavin adenine dinucleotide (FAD). The flavins may occur free or complexed with protein, in which case they are called flavoprotein. The flavoproteins are currently the most favored candidates for cryptochrome (Galston and Baker, 1949; Galland and Senger, 1991; Warpeha et al., 1992). However, those flavins that function as photoreceptors probably constitute a very small portion of a much larger pool. Both FMN and FAD, for example, are important cofactors in cellular oxidation-reduction reactions (Chaps. 9, 12). This makes it difficult to isolate that small portion of the population that might serve as cryptochrome and establish unequivocally its physiological role.

The search for the elusive cryptochrome may now be over, however. Using a combination of genetic, photobiological, and biochemical approaches, the *Arabidopsis HY4* gene has been shown to encode the blue photoreceptor that mediates inhibition of hypocotyl elongation (Ahmad and Cashmore, 1993, 1996). Named **CRY1** (for cryptochrome 1), the *HY4* gene product is a cytoplasmic protein with a mass of about 75 kDa. Interestingly, the sequence of the CRY1 protein is similar to **photolyase,** a unique class of flavoproteins that use blue light to stimulate repair of UV-induced damage to microbial DNA. Photolyases contain two chromophores; one a flavin (FAD) and one a **pterin** (Fig. 7.14). Although the precise nature of the CRY1 chromophores remains to be determined, it appears that one is FAD and the second is likely to be a pterin (Cashmore, 1997). Thus CRY1 appears to qualify as cryptochrome, at least with respect to inhibition of hypocotyl elongation in *Arabidopsis.* Whether the *Arabidopsis* CRY1 is representative of a more general class of molecules remains to be established.

FIGURE 7.14 The structure of riboflavin (*A*) and pterin (*B*). Note the similarity between the pterin structure and the B and C rings of riboflavin. See Chap. 9 for the structures of riboflavin derivatives, FMN and FAD.

THE UV-B RECEPTORS

More recently, a small number of responses, such as anthocyanin synthesis in young milo seedlings (*Sorghum vulgare*) and suspension cultures of parsley or carrot cells have been described with an action spectrum peak near 290 nm and no action at wavelengths longer than about 350 nm (Caldwell et al., 1989; Takeda and Abe, 1992). These findings would seem to indicate the presence of one or more UV-B (280–320 nm) receptors in plants, although the nature of the photoreceptors has yet to be identified with certainty.

The impact of ultraviolet radiation, especially UV-B, on plants is receiving increasing attention because of concerns about the thinning of the atmospheric ozone layer. A reduction in the ozone layer results in an increase in UV-B radiation, specifically between 290 and 314 nm, which can cause damage to nucleic acids, proteins, and the photosynthetic apparatus and lead to shorter plants and reduced biomass. The UV-B receptor also appears to modulate responses to phytochrome in some systems. It has yet to be identified.

FLAVONOIDS

Although the plant world is predominantly green, it is the brilliant colors of floral petals, fruits, bracts, and occasionally leaves that most attracts man and a variety of other animals to plants. These various shades of scarlet, pink, purple, and blue are due to the presence of pigments known as **anthocyanins.** Anthocyanins belong to a larger group of compounds known as **flavonoids.** Other classes of flavonoids (e.g., chalcones and aurones) contribute to the yellow colors of some flowers. Yet others (the flavones) are responsible for the whiteness of floral petals that, without them, might appear translucent. The flavonoids are readily isolated and, because of their brilliant colors, have been known since antiquity as a source of dyes. Consequently, the flavonoids have been extensively studied since the beginnings of modern

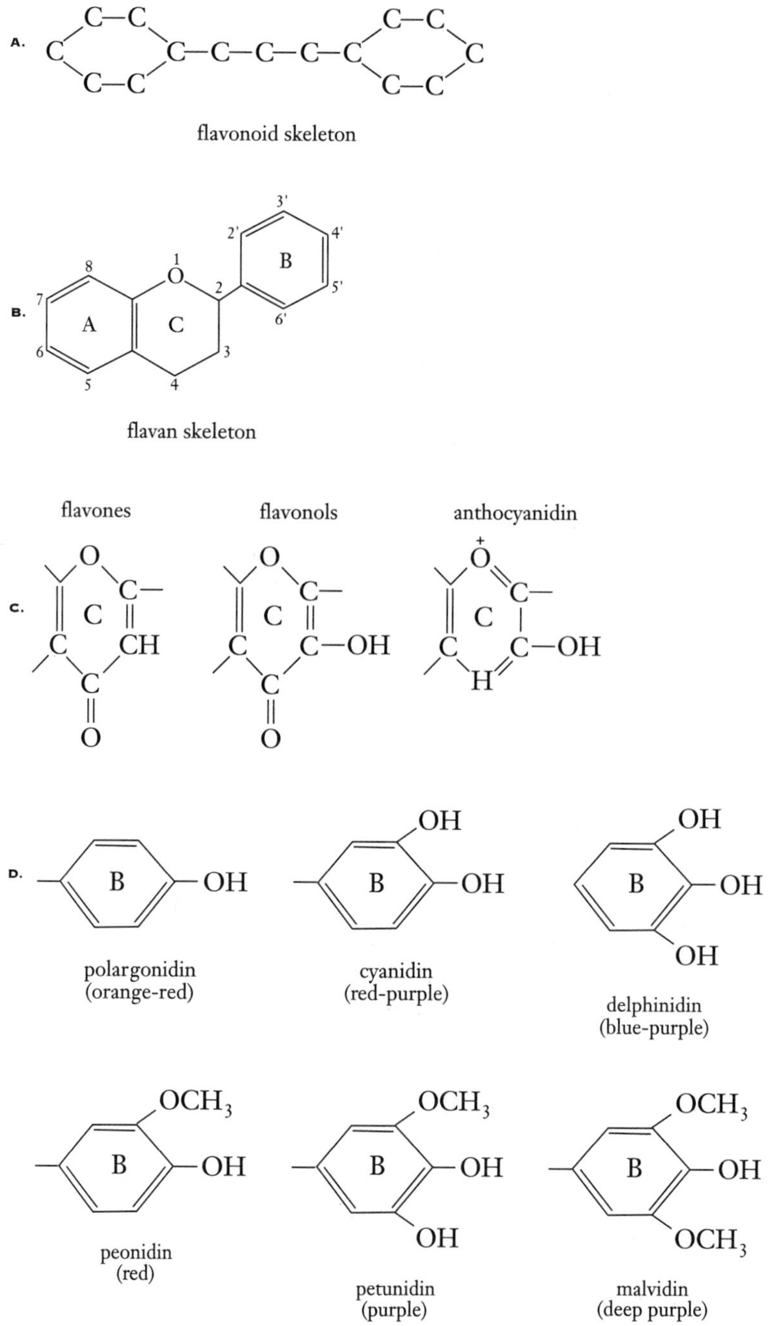

A. flavonoid skeleton

B. flavan skeleton

C. flavones flavonols anthocyanidin

D. polargonidin (orange-red) cyanidin (red-purple) delphinidin (blue-purple) peonidin (red) petunidin (purple) malvidin (deep purple)

FIGURE 7.15 Flavonoids and anthocyanidins. (*A*) Flavonoids are phenylpropane derivatives with a basic C_6—C_3—C_6 composition. (*B*) The parent skeleton for the flavonoid group is flavan, in which C_3-link has formed a heterocyclic pyrone ring. (*C*) The difference between flavones, flavonols, and anthocyanidins is based on substitutions on the C ring. (*D*) The six most common anthocyanidins.

TABLE 7.2 **The composition of some common anthocyanins.**

Anthocyanidin	+ Glycoside	=	Anthocyanin	Source
pelargonidin	3,5-diglucoside		pelargonin	geranium (*Pelargonium*) petals
cyanidin	3,5-diglucoside		cyanin	red rose petals
cyanidin	3-galactoside		idaein	cranberry fruit
delphinidin	3-rhamnoglucoside		violanin	violet (*Viola*) flowers
malvidin	3-glucoside		oenin	blue grapes

organic chemistry and their chemistry is well known. The biosynthesis of flavonoids is discussed in Chapter 14.

Flavonoids are phenylpropane derivatives with a basic C_6—C_3—C_6 composition (Fig. 7.15A). The parent skeleton for the group is flavan, in which C_3-link has formed a heterocyclic pyrone ring (Fig 7.15B). In flavan the heterocyclic ring is fully reduced. There are about twelve recognized groups of flavonoids that differ from one another only by the oxidation state of this heterocyclic ring. Identifying features of the three major groups of flavonoids: **flavones, flavonols,** and **anthocyanidins,** are shown in Figure 7.15C. In addition, all have 5,7-hydroxylation or, in some cases, methoxylation on the A ring.

The most strongly colored of the flavonoids are the anthocyanidins and anthocyanins. Twelve anthocyanidins are known; structures of six most common anthocyanidins are represented in Figure 7.15D. They are distinguished by the pattern of hydroxylation or methoxylation on the B ring, which ultimately determines their color. As a general rule, hydroxylation increases the blueness of the pigment while addition of methoxyl groups increases redness. Anthocyanins differ from anthocyanidins by the addition of various sugars, primarily at the 3-OH position. Some anthocyanins together with their anthocyanidin counterparts (the aglycone) are listed in Table 7.2. The aglycone is rarely encountered in nature. Glycosides of cyanidin, pelargonidin, and delphinidin occur most frequently and contribute to the color of all pink, scarlet, orange-red, crimson, and blue flowers. Glycosides of cyanidin alone are found in more than 80 percent of pigmented leaves and 50 percent of flowers (Swain, 1965).

Unlike chlorophyll, the anthocyanins are water-soluble pigments and are found predominantly in the vacuolar sap. They are readily extracted into weakly acidic solution. The color of anthocyanins is sensitive to pH: both anthocyanidins and anthocyanins are natural indicator dyes. For example, the color of cyanidin changes from red (acid) to violet (neutral) to blue (alkaline). The deep violet extract of boiled red cabbage will turn a definitely unappetizing blue-green if boiled in alkaline water!

Anthocyanins in leaves such as *Coleus* and red-leaved cultivars of maple (*Acer* sps.) are found in the vacuoles of the epidermal cells where they appear to mask the chlorophylls. However, the anthocyanins absorb strongly between 475 nm to 560 nm while transmitting both blue and red light. Consequently, the presence of anthocyanins does not interfere with photosynthesis in the chloroplasts of the underlying mesophyll cells.

Virtually all flavonoids, especially the flavones and flavonols, absorb strongly in the UV-B region of the spectrum (Fig. 7.16). Since these compounds also occur in leaves, one possible function of the flavonoids is thought to be protection of the underlying leaf tissues from damage due to ultraviolet radiation. As flower pigments, the flavonoids attract insect pollinators. Many insects can detect ultraviolet light and thus can perceive patterns contributed by the colorless flavonoids as well as the colored patterns visible to humans (McCrea and Levey, 1983). The synthesis of anthocyanins is stimulated by light, both UV and visible, as well as by nutrient stress (especially nitrogen and phosphorous deficiencies) and low temperature.

A physiological role for flavonoids in plants has been the subject of much speculation over the years. Several investigators in the early 1960s reported that

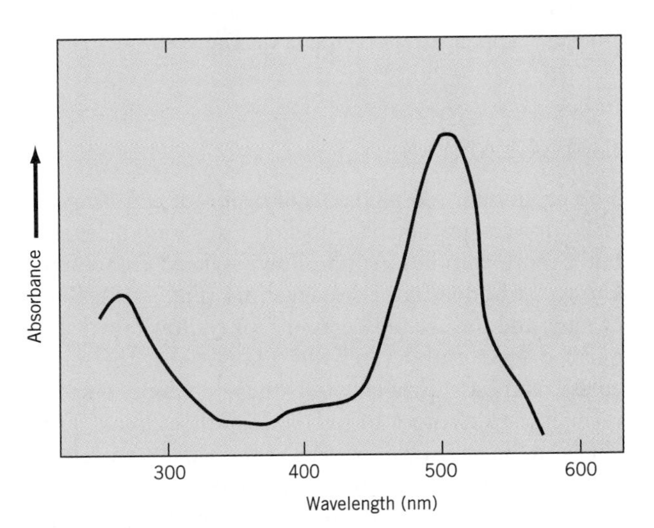

FIGURE 7.16 **Absorption spectrum of pelargonin (perlargonidin-3,5-diglucoside).**

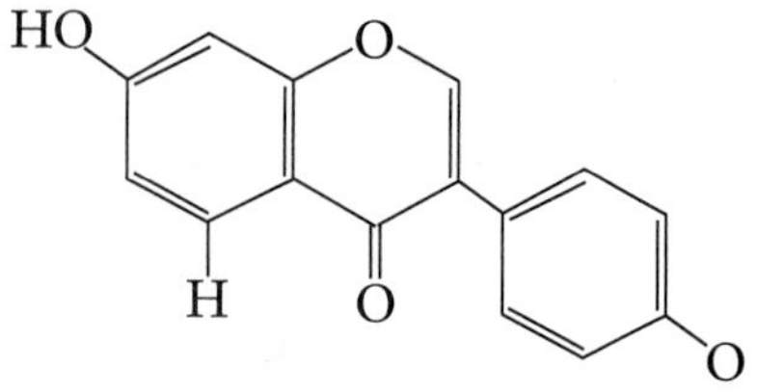

FIGURE 7.17 Structure of the isoflavonoid formononetin, isolated from red clover (*Trifolium pratense*).

two flavonols, kaempferol and quercitin, were involved in light-regulated stem growth of pea seedlings. Kaempferol inhibited growth by stimulating oxidation of IAA and this effect was counteracted by quercitin. Other studies have implicated flavonoid in root growth and dormancy, but as yet there is no definitive evidence that flavonoids play a significant role in plant growth.

At least one group of flavonoids, the **isoflavonoids,** have become known for their antimicrobial activities (Fig. 7.17). Isoflavonoids are one of several classes of chemicals of differing chemical structures, known as **phytoalexins,** which help to limit the spread of bacterial and fungal infections in plants. Phytoalexins are generally absent or present in very low concentrations, but are rapidly synthesized following invasion by bacterial and fungal pathogens. The details of phytoalexin metabolism are not yet clear. Apparently a variety of small polysaccharides, glycoproteins and proteins of fungal or bacterial origin, serve as **elicitors** that stimulate the plant to begin synthesis of phytoalexins. Studies with soybean cells infected with the fungus *Phytophthora* indicate that the fungal elicitors trigger transcription of mRNA for enzymes involved in the synthesis of isoflavonoids (Ebel and Greisebach, 1988). The production of phytoalexins appears to be a common defense mechanism. Isoflavonoids are the predominant phytoalexin in the family Leguminoseae, but other families, such as Solanaceae, appear to use terpene derivatives.

BETACYANINS

The prominent red pigments of beet root and *Bougainvillea* flowers are not flavonoids (as was long believed), but a more complex group of glycosylated compounds known as **betalains** or **betacyanins** (Fig. 7.18). Betacyanins and the related betaxanthins (yellow) are distinguished from anthocyanins by the fact that the molecules contain nitrogen and they do not respond reversibly to changes in pH as do anthocyanins. They appear to be restricted to a small group of closely related families in the order Chenopodiales, including the goosefoot, cactus, and portulaca families, which are not known to produce anthocyanins.

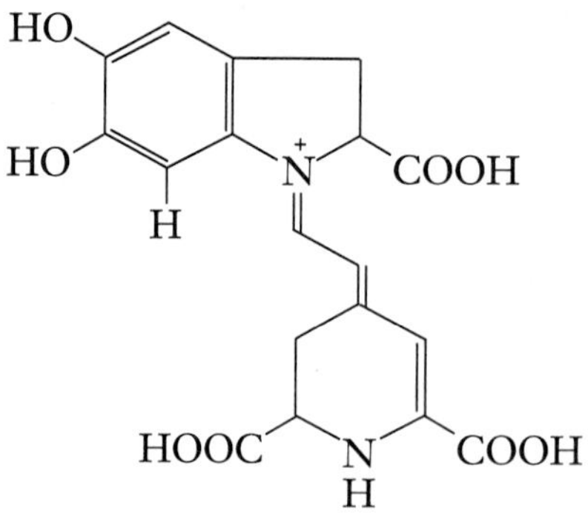

FIGURE 7.18 Betanidin is the basic structure of the betacyanins, which differ by the nature of sugars attached to the hydroxyl groups.

SUMMARY

Sunlight provides plants with energy to drive photosynthesis and critical information about the environment. Light is a form of electromagnetic energy that has attributes of continuous waves and discrete particles. The energy of a particle of light (a quantum) is inversely proportional to its wavelength.

Light is absorbed by pigments and pigments that absorb physiologically useful light are called photoreceptors. All pigments have a characteristic absorption spectrum that describes the efficiency of light absorption as a function of wavelength. Because only light that is absorbed by pigments can be effective in a physiological or biochemical process, a comparison of absorption spectra with the action spectrum for a process helps to identify the responsible pigment. When light is absorbed, the pigment becomes excited, or unstable. The excess energy must be dissipated as heat, re-emitted as light, or used in a photochemical reaction, thus allowing the pigment to return to its stable, ground state.

Regular and predictable changes in fluence rate and spectral energy distribution provide plants with information about the momentary status of their environment as well as impending changes.

The biochemical characteristics of the principal plant pigments of physiological interest are described.

CHAPTER REVIEW

1. Although, as Samuel Johnson said, it is not easy to tell what light is, what is it? Describe the various parameters of light and how it can be measured.
2. Describe the relationship between an absorption spectrum and an action spectrum. Of what signifi-

cance is an action spectrum to the plant physiologist?

3. When is a pigment a photoreceptor? Make a list of the major plant pigments and identify one or more principal functions of each.
4. Chlorophylls and carotenoids are found predominantly in cellular membranes while anthocyanins are located in vacuoles. What does this distribution tell you about the chemistry of these pigments?
5. Assume you are writing a paper in which you report the effects of artificial light on the growth and photosynthesis of plants. How would you describe the light environment so that a reader could attempt to repeat your experiments in their own laboratory?
6. Describe how light energy is absorbed and dissipated by a pigment.

FURTHER READING

Goodwin, T. W. (ed.) 1965. *Chemistry and Biochemistry of Plant Pigments.* London and New York: Academic Press.

Nobel, P. S. 1991. *Physiochemical and Environmental Plant Physiology.* New York: Academic Press.

Smith, H. (ed.). 1981. *Plants and the Daylight Spectrum.* London: Academic Press.

Smith, H., M. G. Holmes. 1984. *Techniques in Photomorphogenesis.* London: Academic Press.

REFERENCES

Ahmad, M., A. R. Cashmore. 1993. *HY4* gen of *A. thaliana* encodes a protein with characteristics of a blue-light photoreceptor. *Nature* 366:162–166.

Ahmad, M., A. R. Cashmore. 1996. Seeing blue: The discovery of cryptochrome. *Plant Molecular Biology* 30:851–861.

Caldwell, M. M., A. H. Teramura, M. Tevini. 1989. The changing solar ultraviolet climate and the ecological consequences for higher plants. *Trends in Ecology and Evolution* 4:363–367.

Cashmore, A. R. 1997. The cryptochrome family of photoreceptors. *Plant, Cell and Environment* 20:764–767.

Ebel, J., H. Greisebach. 1988. Defense strategies of soybean against the fungus *Phytophthora megasperma* f.sp. *glycinea:* A molecular analysis. *Trends in Biochemical Sciences* 13:23–27.

Galland, P., H. Senger. 1991. Flavins as possible blue light photoreceptors. In: M. G. Holmes (ed.), *Photoreceptor Evolution and Function.* New York: Academic Press, pp. 65–124.

Galston, A. W., R. S. Baker. 1949. Studies on the physiology of light action. II. The photodynamic action of riboflavin. *American Journal of Botany* 36:773–780.

Holmes, M. G. 1984. In: Smith, H., M. G. Holmes. *Techniques in Photomorphogenesis.* London: Academic Press.

McCrea, K. D., M. Levey. 1983. Photographic visualization of floral colors as perceived by honeybee pollinators. *American Journal of Botany* 70:369–375.

Swain, T. 1965. Nature and properties of flavonoids. In: Goodwin, T. W. (ed.), 1965, *Chemistry and Biochemistry of Plant Pigments.* London and New York: Academic Press.

Takeda, J., S. Abe. 1992. Light-induced synthesis of anthocyanin in carrot cells in suspension. IV. the action spectrum. *Photochemistry and Photobiology* 56:69–74.

Warpeha, K. M. F., L. S. Kaufman, W. R. Briggs. 1992. A flavoprotein may mediate the blue light-activated binding of GTP to isolated plasma membranes of *Pisum sativum* L. *Photochemistry and Photobiology* 55:595–603.

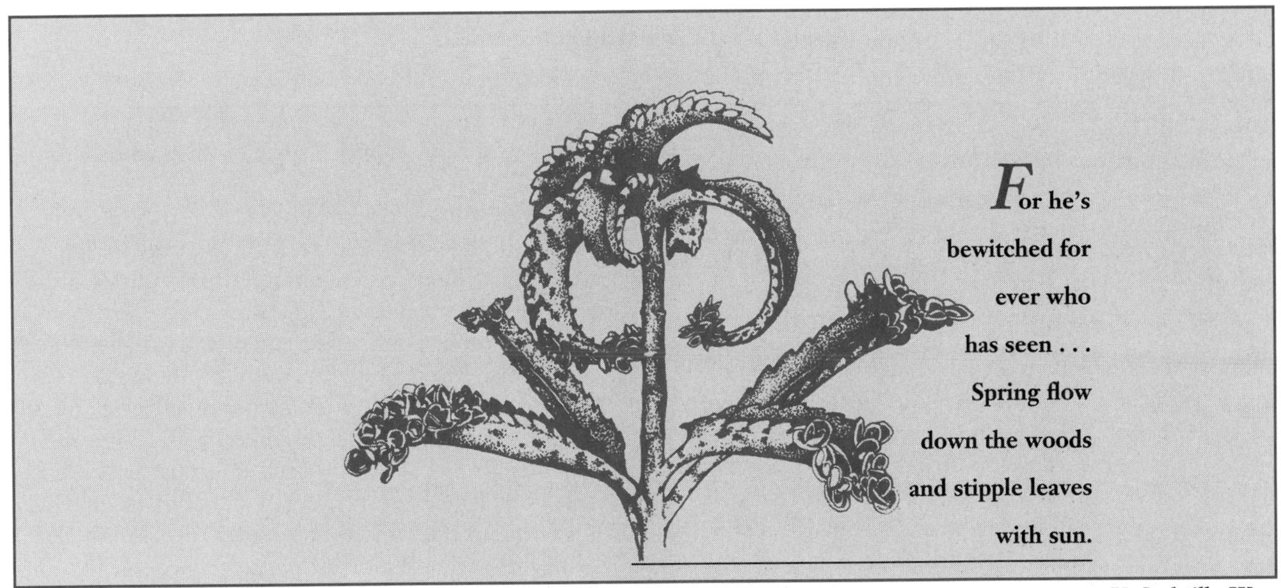

V. Sackville-West
(1926)

8

Leaves and Photosynthesis

Photosynthesis is the fundamental basis of competitive success in green plants and the principal organ of photosynthesis in higher plants is the leaf. From the delicate, pastel hues of early spring through the brilliant reds and oranges of autumn, foliage leaves are certainly one of the dominant features of higher plants. The biologist's interest in leaves, however, goes far beyond their aesthetic quality. Biologists are interested in the structure of organs and how those structures are adapted to effectively carry out certain physiological and biochemical functions. Leaves provide an excellent demonstration of this structure-function relationship. While some leaves may be modified for special purposes (for example, tendrils, spines, and floral parts), the *primary* function of leaves remains photosynthesis. From this perspective, the leaf may be viewed as a photosynthetic machine—superbly engineered to carry out photosynthesis efficiently in an extremely hostile environment.

The evolution of leaves as photosynthetic organs has revolved around three major themes: (1) the exploitation of light, (2) efficient gas exchange, particularly with respect to CO_2 uptake, and (3) a well-organized transport system for the rapid export of product. In order to absorb light efficiently, a typical leaf presents a large surface area at right angles to the incoming sunlight. The high surface area per unit volume contributed by the thin laminar structure enhances CO_2 uptake as well, but there is a price to be paid. In the process of moving into a terrestrial habitat, photosynthetic organisms (like all others) have remained essentially aquatic. Their cells are comprised largely of water and most of their biochemistry is carried out in the aqueous milieu. Unfortunately, an efficient CO_2 exchanger is also an efficient evaporator. Leaves have evolved some unique structural and metabolic solutions to this problem of balancing CO_2 supply against excessive water loss. Finally, leaves are highly vascularized. An extensive network of interconnected veins serves to supply nutrients and other raw materials to the photosynthetic cells as well as to collect and export the products of photosynthesis to nonphotosynthetic tissues and organs elsewhere in the plant.

The focus of this chapter is the first two evolutionary themes; that is, the organization of leaves with respect to the exploitation of light and efficient gas exchange. Following a brief review of structure/function relationships in algae and liverworts, we will examine

- the structure of higher plant leaves with respect to interception of light;
- gas exchange by leaves, including the efficiency of diffusion through pores, the structure of stomata, and mechanisms for stomatal opening and closure;
- the effects of environmental factors, such as CO_2, light, temperature, and water stress on stomatal movements; and

Box 8.1
Historical Perspective—The Discovery of Photosynthesis

Photosynthesis assumes a role of such dominant proportions in the organization and development of plants, not to mention feeding world populations, it is somewhat surprising that so little was known about the process before the final decades of the eighteenth century. The practice of agriculture was already several thousand years old and practical discussions of crop production had been written at least 2,000 years before. The origins of plant nutrition as a science can be traced as far back as Aristotle and other Greek philosophers who taught that plants absorbed organic material directly from the soil. This theory, known as the humus theory, prevailed in agricultural circles until the late nineteenth century, long after the principles of photosynthesis had been established.

The first suggestions of photosynthesis appear in the writings of Stephen Hales, an English clergyman and naturalist who is considered "the father of plant physiology." In 1727 Hales surmised that plants obtain a portion of their nutrition from the air and wondered, as well, whether light might also be involved. Hales' insights were remarkably prescient, contrary as they were to the long-established humus theory. However, chemistry had yet to come of age as a science and Hales' ideas were not provable by experiment or by reference to any well-established chemical laws.

Rabinowitch and Govindjee (1969) date the "discovery" of photosynthesis as 1776, the year Joseph Priestly published his two-volume work entitled *Experiments and Observations on Different Kinds of Air*. But as with many other phenomena, there was no one moment of discovery. The story gradually fell into place through the cumulative efforts of several clergy, physicians, and chemists over a period of nearly 75 years. J. Priestly (1733–1804) was an English minister whose nonconformist views on religion and politics led to his emigration to the United States in 1794. He was also a scientist engaged in pioneering experiments with gasses and is perhaps best known for his discovery of oxygen. Priestly's experiments, begun in 1771 and first published in 1772, led him to observe that air "contaminated" by burning a candle could not support the life of a mouse. He then found that the air could be restored by plants—a sprig of mint was introduced into the contaminated air and "after eight or nine days I found that a mouse lived perfectly in that part of the air in which the sprig of mint had grown" (Priestly, 1772).

Priestly failed to recognize the role of light in his experiments and it was perhaps serendipitous that his laboratory was well enough lighted for the experiments to have succeeded at all. In 1773, Priestly's experiments came to the attention of Jan Ingen-Housz (1730–1799), a physician to the court of Austrian Empress Maria Theresa. During a visit to London, Ingen-Housz heard Priestly's experiments described by the

- the ultrastructure and biochemical compartmentation of the chloroplast.

The third major theme—a well-organized transport system—has been addressed in part in Chapter 3 with a discussion of water transport in the xylem. The export of photosynthetic product through the phloem tissue will be addressed later in Chapter 11. Details of the energy-conversion reactions of photosynthesis and carbon metabolism will be explored in Chapters 9 and 10, respectively.

PHOTOSYNTHESIS IN ALGAE AND LIVERWORTS

The algae and liverworts illustrate the basic structural requirements of photosynthetic organisms.

The three principal raw materials for photosynthesis are water, carbon dioxide, and light energy (see Box 8.1: Historical Perspective—The Discovery of Photosynthesis). The relative abundance of these three factors, but especially water, may vary widely with environment. In order to fully appreciate the significance of leaf structure in photosynthesis, it would help to first review, in a very general way, structure-function relationships in two "lower" plants—the unicellular algae and liverworts—especially as these relationships relate to the availability of photosynthetic raw materials and their access to the photosynthetic machinery.

Unicellular algae are aquatic organisms also known as phytoplankton. Because these organisms are bathed in water and because water freely exchanges across the cell membrane, water is readily available to the unicellular algae.

As a rule, light should not normally be in short supply either. Other aquatic microorganisms and suspended microscopic debris do attenuate light to some degree, depending upon the quantities present. Even the phytoplankton themselves filter out some photosyntheti-

President of the Royal Society. He was intrigued by these experiments and six years later returned to England to conduct experiments. In the course of a single summer, Ingen-Housz performed and had published some 500 experiments on the purification of air! He observed that plants could purify air within hours, not days as observed by Priestly, but only when the *green* parts of plants were exposed to *sunlight*. Together, Priestly and Ingen-Housz had confirmed Hale's guesses made some 52 years earlier.

Although Priestly continued his experiments—in 1781 he agreed with Ingen-Housz on the value of light and green plant parts—neither Priestly nor Ingen-Housz recognized the role of "fixed air," as CO_2 was known at the time. This was left to the Swiss pastor and librarian Jean Senebier (1742–1809). In 1782 Senebier published a three-volume treatise in which he demonstrated that the purification of air by green plants in the light was dependent upon the presence of "fixed" air. It is interesting to note that all three scientists had emphasized the purification of air in relation to its capacity to support animal life—plant nutrition was not a central theme. At the same time chemists across Europe, including Priestly in England, Scheele in Germany, and Lavoisier in France, were actively investigating the chemical and physical properties of gasses. By 1785 Lavoisier had identified "fixed" air as CO_2 and by 1796 Ingen-Housz had correctly deduced that CO_2 was the source of carbon for plants.

Another important component in the equation of photosynthesis was added by the work of a Geneva chemist, N. T. de Saussure (1767–1845). It was de Saussure who first approached photosynthesis in a sound, quantitative fashion. In his book *Recherches Chimiques sur la Végétation* (1804) he showed that the weight of organic matter plus oxygen formed by photosynthesis was substantially larger than the weight of CO_2 consumed. He thus concluded that the additional weight was provided by water as a reactant. The equation for photosynthesis, using the new language of chemistry founded by Lavoisier, could now be written:

$$CO_2 + H_2O \longrightarrow O_2 + \text{organic matter} \quad (8.1)$$

Finally, it remained for a German surgeon, Julius Mayer (1814–1878), to clarify the energy relationships of photosynthesis. In 1845, he correctly deduced, for the first time, that the energy used by plants and animals in their metabolism is derived from the energy of the sun and that it was transformed by photosynthesis from the radiant to the chemical form. Thus by the middle of the nineteenth century the general outline of photosynthesis was complete. Despite the importance of the process, however, it would be almost another century before the structural and chemical details of photosynthesis would yield to modern methods of microscopic and radiochemical analysis.

REFERENCES

Priestly, J. 1772. Observations on different kinds of air. *Philosophical Transactions of the Royal Society of London* 62:166–170.

Rabinowitch, E., Govindjee. 1969. *Photosynthesis*. New York: Wiley.

cally active radiation. However, because the phytoplankton generally float close to the surface, they are essentially unshaded. Thus light, at least for the unicellular algae, is effectively limited only by latitude, season, and atmospheric conditions.

The third substrate for photosynthesis is carbon dioxide and at first glance its availability would not appear to be a problem for aquatic organisms. CO_2 is soluble in water and readily reacts to form bicarbonate (HCO_3^-) and carbonate (CO_3^{2-}). But the solubility of CO_2 in water is relatively low—about 15 μM at 15° compared with 320 μM for O_2—and its solubility declines with increasing temperature. Moreover, the rate of diffusion of CO_2 in water is approximately 10^{-4} that of diffusion in air. In fact, the diffusion rate for all three forms of inorganic carbon (CO_2, carbonate, and bicarbonate) is rather low. This combination of low solubility and low diffusion rate tends to restrict the supply of CO_2 and may limit photosynthesis in aqueous environments. Most algae, however, do not depend on diffusion for their CO_2 supply but have instead evolved CO_2-concentrating mechanisms—mechanisms for actively transporting inorganic carbon (CO_2 or HCO_3^-) into the cell (Canvin, 1990). The ability to concentrate CO_2 for the most part overcomes the problems of low solubility and diffusion rates.

Liverworts are primitive land forms assigned to the division Bryophyta. This group includes the hornworts, liverworts, and mosses. With regard to the present discussion, the important thing about liverworts is that they thrive only in very moist habitats. Although most liverworts closely resemble the mosses in basic body plan, about 20 percent of the known species are characterized by a flattened, dichotomously lobed body called a **thallus** (Fig. 8.1). In these forms, the photosynthetic cells are found along the upper region of the thallus where they have access to air. Cells on the lower surface of the thallus contain no chlorophyll and function primarily as storage or absorptive cells.

A close look at the upper surface of a liverwort thallus reveals it is divided into polygonal sections. These

FIGURE 8.1 The liverwort *Marchantia*.

polygonal sections mark a hollow air chamber below. In the center of each polygonal section is a small pore that opens into the chamber (Fig. 8.2). Within each chamber are irregular, branching columns of photosynthetic cells resembling a garden of prickly pear cacti. This arrangement ensures that all photosynthetic cells have direct access to ambient air through the pore in the roof of the chamber.

The liverworts are particularly interesting because their basic body plan anticipates the design of a higher plant leaf. Spreading out in two dimensions, the thallus presents a broad but relatively thin laminar surface for effective interception of light. CO_2 availability is effectively increased by direct exposure of the photosynthetic cells to air spaces inside the thallus, thus reducing the length of the relatively slow aqueous diffusion path. Because each cell contains a large central vacuole, the chloroplasts are located in the thin zone of cytosol pressed against the plasma membrane. Thus in most cases the aqueous diffusion path is effectively limited to little more than the thickness of the cell wall and the short distance between the wall and the chloroplast.

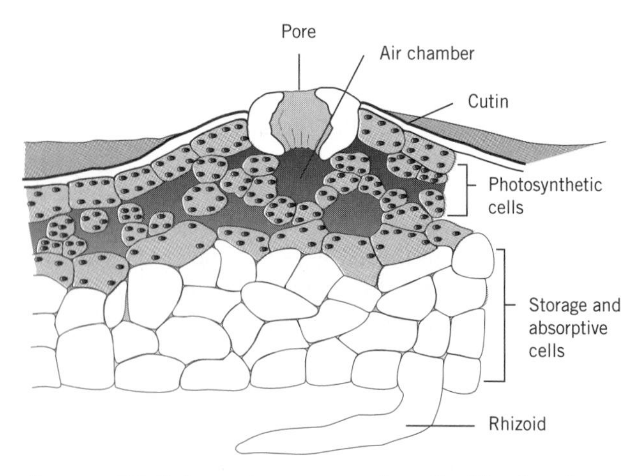

FIGURE 8.2 Schematic cross-section of a *Marchantia* thallus.

Exposure of the photosynthetic cells to air facilitates the uptake of CO_2, but it creates serious problems with regard to water loss. Although terrestrial and aquatic organisms are distinguished on the basis of habitat, all organisms remain fundamentally aquatic. Water is the principal component of all cells and most of their biochemistry takes place in an aqueous milieu. The cell wall spaces of the surface cells are filled with liquid water and their water potential is correspondingly high. On the other hand, the water potential of the atmosphere is often very low (at 22° and 50% relative humidity, $\Psi = -100$ MPa). The resulting water potential gradient would encourage rapid water loss by evaporation from the surface cells were they not coated with a water-impermeable barrier, the **cuticle** (Chap. 3). The cuticle is not limited to the liverworts but is found on all terrestrial plants with surfaces exposed to the atmosphere. The presence of a cuticle thus serves to reduce water loss in those organs with surfaces exposed directly to air.[1] In the case of liverworts, rhizoids located on the underside of the thallus serve not only to anchor the thallus to the moist soil, but also to absorb water and stabilize the water potential of the cells.

Unfortunately, in addition to restricting the flow of water vapor, the cuticle also restricts the exchange of other gasses such as CO_2 and O_2. An uninterrupted cuticle would therefore prevent photosynthesis by effectively cutting off the supply of carbon dioxide.[2] This situation creates a conundrum for the plant—how to control evaporative water loss while at the same time maintaining an adequate supply of CO_2 for photosynthesis. In the liverworts the solution was provided by the combination of air chamber and pore. This arrangement provides that virtually every photosynthetic cell—including those in the "prickly pear" columns, the chamber walls and roof—has at least one surface exposed to the air chamber. The central pore provides a route for diffusion of CO_2 into the chamber while moderating the flow of water vapor from the humid chamber. In liverworts this pore is fixed and permanently open. As a result, water loss cannot be completely prevented and the plants can thrive only in moist habitats.

[1]As a secondary function, the cuticle also appears to reduce the opportunity for infection by bacterial and fungal pathogens.

[2]Many nonphotosynthetic organs such as fruits are also covered with a cuticle but contain no pores to permit exchange of CO_2 or O_2. Although these organs do not carry out photosynthesis, they must respire. Apparently the diffusion gradient for O_2 is large enough to ensure that sufficient O_2 crosses the cuticle to support respiratory needs in such organs.

PHOTOSYNTHESIS IN LEAVES

Leaves are aerial structures that must carry out photosynthesis in a hostile environment.

Since the liverwort thallus lies more or less flattened on the substratum, these organisms inhabit what is essentially a two-dimensional world. Any advantage gained by extensive surface area is to some extent ameliorated by the tendency of the thalli to overlap. The condition of overlapping thalli introduces mutual shading, which in turn tends to decrease photosynthetic efficiency. Leaves of higher terrestrial plants, on the other hand, are generally aerial structures, borne on a branching system of stems. This configuration provides some compensation for the tendency to overlap by permitting the organism to occupy a three-dimensional space. At the same time it moves the photosynthetic structures into an extremely hostile environment with respect to water loss and away from intimate contact with the moist substratum. Problems of water supply in this situation are at least partially resolved by the introduction of stems with their complement of vascular tissues. The stem with its interconnected system of conduits permits aerial leaves to remain in contact with the adsorptive root tissues which, in turn, remain anchored in the moist soil. Vascular tissue provides for the transport of water and dissolved nutrients from the roots to the leaves and the transport of energy and foodstuffs from the leaves back to the roots. The structure of most leaves is such that no photosynthetic cell is more than one or two cells removed from a vascular strand. This proximity of vascular tissue means that both the import of nutrients and the export of photosynthetic product can be accomplished quickly and efficiently.

Like the thallus of liverworts, the leaves of higher terrestrial plants are also covered with a cuticle and contain pores. Unlike the fixed pores of liverworts, however, the leaf pores, called **stomata** (sing. = *stoma*), are hydraulically operated valves that can be adjusted to control the size of the opening depending upon the circumstances (Ziegler, 1987). Together, stomata and the vascular system represent the two most important structural innovations that permitted successful colonization of the terrestrial environment.

ABSORPTION OF LIGHT BY LEAVES

The morphology, anatomy, and optical properties of leaves are designed to efficiently intercept light and channel it to the chloroplasts where photosynthesis occurs.

The architecture of a typical higher plant leaf is particularly well suited to absorb light. As with the liverworts, a broad, laminar surface serves to maximize interception of light. In addition, the bifacial nature of the leaf allows it to collect incident light on the upper surface and diffuse (both scattered and reflected) light on the lower surface. Gross morphology is not, however, the only factor enhancing interception of light—internal cellular arrangements also play an important role.

The anatomy of a typical dicotyledonous mesomorphic leaf is shown in Figure 8.3A,C. The leaf is sheathed with an upper and lower **epidermis.** The exposed surfaces of the epidermal cells are coated with a cuticle. The photosynthetic tissues are located between the two epidermal layers and are consequently identified as **mesophyll** (*meso* = middle; *phyll* = leaf) tissues. The upper photosynthetic tissue generally consists of one to three layers of **palisade** mesophyll cells. Palisade cells are elongated, cylindrical cells with the long axis perpendicular to the surface of the leaf. Below is the **spongy** mesophyll, so named because of the prominent air spaces between the cells. The shape of spongy mesophyll cells is somewhat irregular but tends toward isodiametric. The plan of a monocotyledonous leaf is similar except that it lacks the distinction between palisade and spongy mesophyll (Fig. 8.3B,D).

Palisade cells generally have larger numbers of chloroplasts than spongy mesophyll cells. In leaves of *Camellia*, for example, the chlorophyll concentration of the palisade cells is 1.5 to 2.5 times that of the spongy mesophyll cells. The higher number of chloroplasts in the palisade cells no doubt reflects an adaptation to the higher fluence rates for photosynthetically active light generally incident on the upper surfaces of the leaf.

In spite of the relatively large number of chloroplasts in the palisade layers of a dicotyledonous leaf, there is a significant proportion of the cell volume that does not contain chloroplasts. Because the absorbing pigments are confined to the chloroplast, a substantial amount of light may thus pass through the first cell layer without being absorbed. This has been called the **sieve effect** (Terashima and Saeki, 1983). Multiple layers of photosynthetic cells is one way of increasing the probability that photons passing through the first layer of cells will be intercepted by successive layers (Fig. 8.4).

The impact of the sieve effect on the efficiency of light absorption is to some extent balanced by factors that change the direction of the light path within the leaf (Vogelmann et al., 1996). Light may first of all be *reflected* off the many surfaces associated with leaf cells. Second, light that is not reflected but passes between the aqueous volume of mesophyll cells and the air spaces that surround them (especially in the spongy mesophyll) will be bent by *refraction.* Third, light may be *scattered* when it strikes particles or structures with diameters comparable to its wavelengths. In the leaf cell, for example, both mitochondria and the grana structures within chloroplasts have dimensions (500 to 1000 nm)

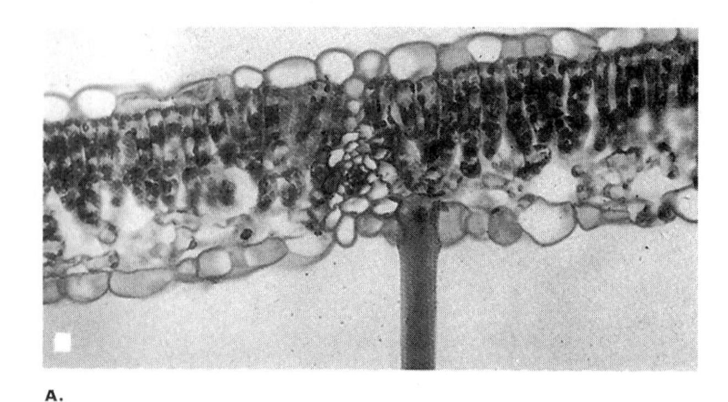
A.

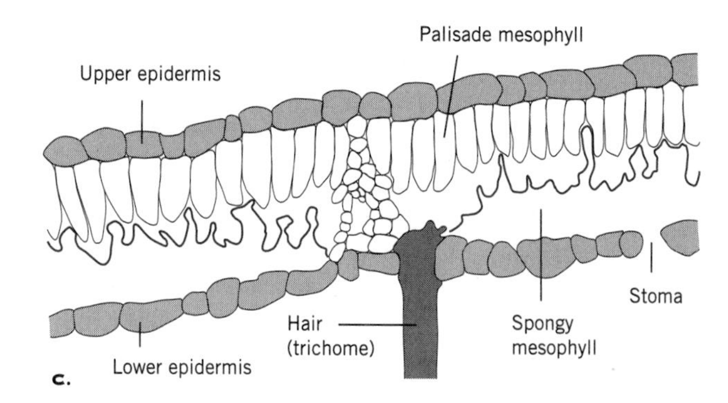

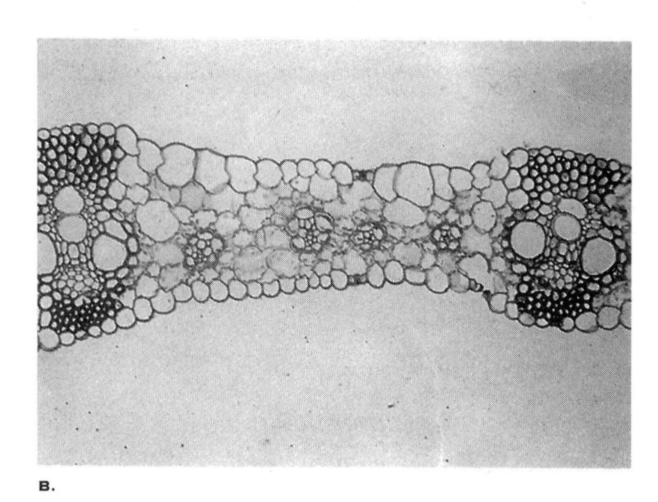
B.

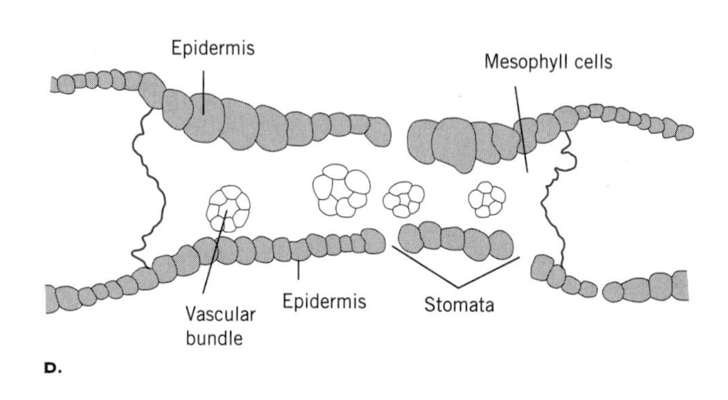

FIGURE 8.3 The structure of leaves shown in cross-section. (*A, C*) A dicotyledonous leaf (*Acer* sp.). (*B, D*) A monocotyledonous leaf (*Zea mays*), showing a section between two major veins. (A,B: From T. E. Weier et al. *Botany*. 6th edition. New York, Wiley. 1982. Figs. 10.22, 10.17. Used by permission of the authors.)

similar to the wavelengths active in photosynthesis. Both organelles will scatter light. These three factors—reflection, refraction, and scattering—combine to increase the effective path length as light passes through the leaf. The longer light path increases the probability that any given photon will be absorbed by a chlorophyll molecule before it can escape from the leaf (Fig. 8.4).

Careful studies of the optical properties of leaves have shown that, in spite of their scattering properties, palisade cells do not appear to absorb as much light as might be expected. That is to say that the palisade cells have a lower than expected *efficiency* of light attenuation. This is apparently because they also act to some extent as a **light guide.** Some of the incident light is channelled through the intercellular spaces between the palisade cells in much the same way that light is transmitted by an optical fiber (Fig. 8.4). It is probable that photosynthesis in the uppermost palisade layer is frequently light saturated. Any excess light would be wasteful and could, in fact, give rise to photoinhibition and other harmful effects (Chap. 9). Thus the increased transmission of light to the lower cell layers resulting from both scattering and the light guide effect would no doubt be advantageous by contributing to a more efficient allocation of photosynthetic energy throughout the leaf.

Not all leaves are designed like the "typical" dicotyledonous mesomorphic leaf described above. Leaves may be modified in many ways to fit particular environmental situations. Pine leaves (or needles), for example, are more circular in cross-section. Their capacity for light interception has been compromised in favor of a reduced surface-to-volume ratio, a modification that helps to combat desiccation when exposed to dry winter air. In other cases, such as dry land or desert species, the leaves are much thicker in order to provide for storage of water. In extreme cases, such as the cacti, the leaves have been reduced to thorns and the stem has taken over the dual functions of water storage and photosynthesis. These and other modifications to leaf morphology will be discussed more fully in later chapters.

LEAVES AND GAS EXCHANGE

Pores in the leaf have a high diffusive capacity, thus providing an efficient mechanism for the absorption of carbon dioxide along a shallow concentration gradient, while at the same time protecting against excessive water loss.

As noted earlier, the epidermis of leaves contains pores that provide for the exchange of gasses between

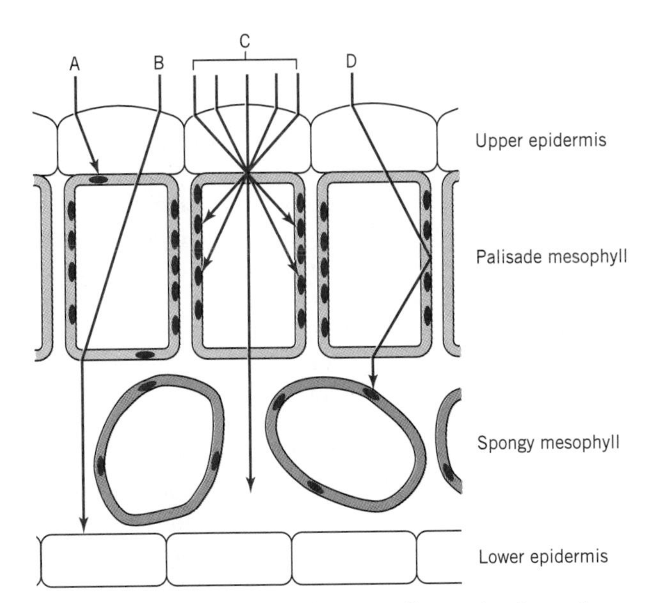

FIGURE 8.4 A simplified diagram illustrating how the optical properties of leaves help to redistribute incoming light and maximize interception by chlorophyll. (*A*) Photon strikes a chloroplast and is absorbed by chlorophyll. (*B*) The sieve effect—a photon passes through the first layer of mesophyll cells without being absorbed. It may be absorbed in the next layer of cells or pass through the leaf, to be absorbed by another leaf below. (*C*) The planoconvex nature of epidermal cells creates a lens effect, redirecting incoming light to chloroplasts along the lateral walls of the palisade cells. (*D*) The light-guide effect. Because the refractive index of cells is greater than that of air, light reflected at the cell-air interfaces may be channeled through the palisade layer(s) to the spongy mesophyll below.

the internal air spaces and the ambient environment. The opening, or stoma, is bordered by a pair of unique cells called **guard cells** (Fig. 8.5). In most cases the guard cells are in turn surrounded by specialized, differentiated epidermal cells called **subsidiary cells.** The stoma together with its bordering guard cells and subsidiary cells is referred to as the **stomatal complex,** or **stomatal apparatus.**

The distinguishing feature of the stomatal complex is the pair of guard cells that functions as a hydraulically operated valve. Guard cells take up water and swell to open the pore when CO_2 is required for photosynthesis, and lose water to close the pore when CO_2 is not required or when water stress overrides the photosynthetic needs of the plant. The mechanical, physiological, and biochemical properties of the guard cells have attracted scholars almost since their occurrence was first reported by M. Malpighi in the late seventeenth century. A continuing interest in stomatal movement is understandable, given the foremost importance of stomata in regulating gas exchange and consequent effects on photosynthesis and productivity. More than 90 percent of the CO_2 and water vapor exchanged between a plant and its environment passes through the stomata. Stomata are therefore involved in controlling two very important but competing processes: uptake of CO_2 for photosynthesis and, as discussed in Chapter 3, transpirational water loss. It is important, therefore, to take into account stomatal function when considering photosynthetic productivity and crop yields. More recently, additional interest in stomatal function has been prompted by recognition

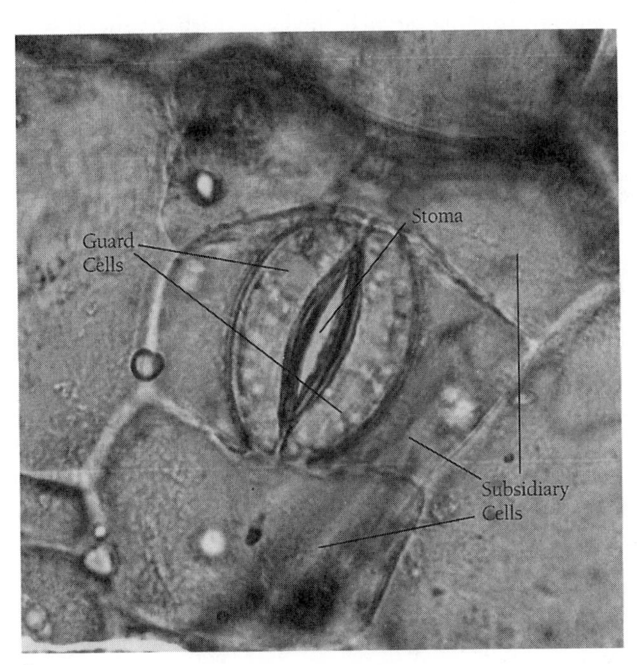

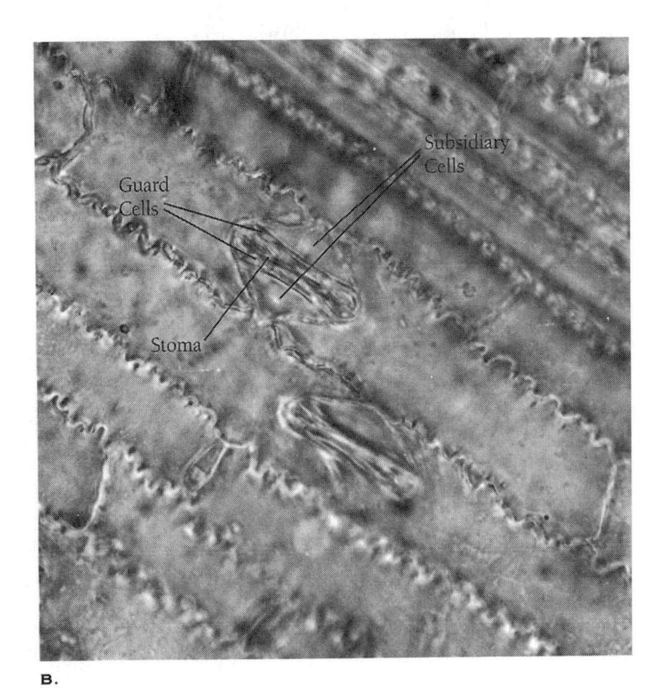

FIGURE 8.5 Stomata. (*A*) Elliptic type in the lower epidermis of *Zebrina*. In this picture the stoma is open. (× 250). (*B*) Graminaceous type from the adaxial surface of maize (*Zea mays*) leaf. These stomata are closed. (× 250)

TABLE 8.1 Stomatal frequencies on the upper and lower surfaces of leaves.

Genus	Number of stomata mm^{-2} upper surface	lower surface
Monocotyledonae		
Allium (onion)	175	175
Hordeum (barley)	70	85
Triticum (wheat)	50	40
Dicotyledonae		
Herbaceous species		
Helianthus (sunflower)	120	175
Medicago (alfalfa)	169	188
Pelargonium (geranium)	29	179
Woody species		
Aesculus (horse chestnut)	—	210
Quercus (oak)	—	340
Tilia (linden)	—	370

Data from Meidner and Mansfield, 1968.

that airborne pollutants such as ozone (O_3) and sulphur dioxide (SO_2) also enter the leaf through open stomata.

Stomata are found in the leaves of virtually all higher plants (angiosperms and gymnosperms) and most lower plants (mosses and ferns) with the exception of submerged aquatic plants and the liverworts. In angiosperms and gymnosperms they are found on most aerial parts including nonleafy structures such as floral parts and stems, although they may be nonfunctional in some cases. The frequency and distribution of stomata is quite variable and depends on a number of factors including species, leaf position, ploidy level (the number of chromosome sets), and growth conditions. A frequency in the range of 20 to 400 stomata mm^{-2} of leaf surface is representative, although frequencies of 1000 mm^{-2} or more have been reported. Although there are exceptions to every rule, the leaves of herbaceous monocots such as grasses usually contain stomata on both the adaxial ("upper") and abaxial ("lower") surfaces with roughly equal frequencies (Table 8.1). Stomata occur on both the upper and lower surfaces of herbaceous dicots leaves, but the frequency is usually lower on the upper surface. Most woody dicots and tree species have stomata only on the lower leaf surface while floating leaves of aquatic plants (e.g., water lily) have stomata only on the upper surface. In most cases the stomata are randomly scattered across the leaf surface, although in monocots with parallel-veined leaves the stomata are arranged in linear arrays between the veins.

The most striking feature of the stomatal complex is the pair of guard cells that border the pore. These specialized epidermal cells have the capacity to undergo reversible turgor changes that in turn regulate the size of the aperture between them. When the guard cells are fully turgid the aperture is open, and when flaccid, the aperture is closed. While there are many variations on the theme, anatomically we recognize two basic types of guard cells: the graminaceous type and the elliptic type (Fig. 8.5).

Elliptic or kidney-shaped guard cells are so called because of the elliptic shape of the opening. In surface view, these guard cells resemble a pair of kidney beans with their concave sides opposed. In cross-section the cells are roughly circular in shape, with a **ventral wall**

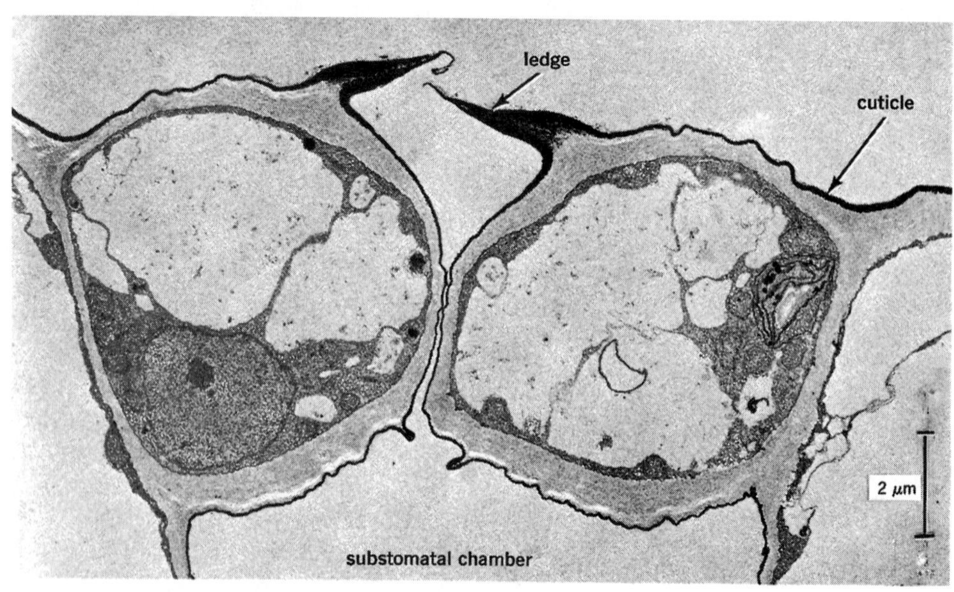

FIGURE 8.6 Guard cells seen in cross-section. (From K. Esau. *Anatomy of Seed Plants*. 1977. New York, Wiley. Reprinted by permission.)

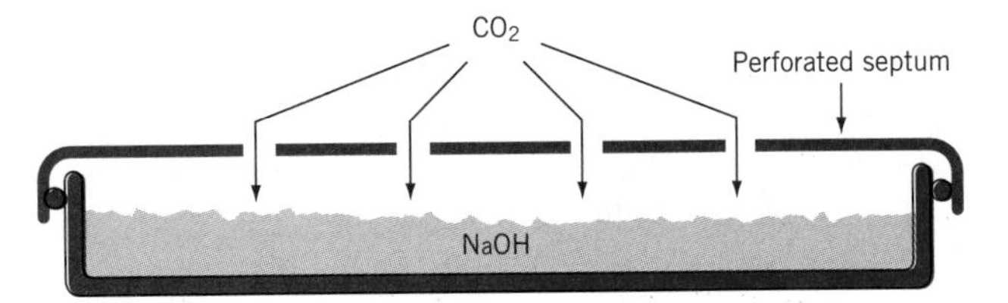

FIGURE 8.7 An experimental setup for studying diffusion through a perforated septum. The amount of CO_2 absorbed by the sodium hydroxide (NaOH) can be measured after an interval of time.

bordering the pit and a **dorsal wall** adjacent to the surrounding epidermal cells (Fig. 8.6). The mature guard cell has characteristic wall thickenings, mainly along the outer and inner margins of the ventral wall. These thickenings extend into one or two **ledges** that protect the **throat** of the stoma. In some plants, particularly the gymnosperms and aquatic species, the inner ledge may be small or absent. The outer ledge appears to be an architectural adaptation that helps to prevent the penetration of liquid water from the outside into the substomatal air space (Ziegler, 1987), which would otherwise have disastrous consequences for gas exchange.

The graminaceous type of guard cell is largely restricted to members of the Gramineae and certain other monocots (e.g., palms). Often described as dumbbell-shaped, the graminaceous-type guard cells have thin-walled, bulbous ends that contain most of the cell organelles (Fig. 8.5). The "handle" of the dumbbell is characterized by walls thickened toward the lumen. The pore in this case is typically an elongated slit. The guard cells are flanked by two prominent subsidiary cells.

DIFFUSION THROUGH PORES

Diffusion of CO_2 into the leaf through the stoma is more efficient than would be predicted on the basis of stomatal area alone. A fully open stomatal pore typically measures 5 to 15 μm wide and about 20 μm long (Nobel, 1991). The combined pore area of open stomata thus amounts to no more than 0.5 to 2 percent of the total area of the leaf (Meidner and Mansfield, 1968). Since leaves contain no active pumps, all of the CO_2 taken into the leaf for photosynthesis must enter by diffusion through these extremely small pores. One might think that diffusion through such a limited area would be extremely restricted, yet it has been calculated that the rate of CO_2 uptake by an actively photosynthesizing leaf may approach 70 percent of the rate over an absorbing surface with an area equivalent to that of the entire leaf (Bidwell, 1979)! This extraordinarily high diffusive efficiency appears to be related to the special geometry of gaseous diffusion through small pores.

The high efficiency of gaseous diffusion through stomata can be demonstrated experimentally by measuring CO_2 diffusion into a container of CO_2-absorbing agent such as sodium hydroxide (Fig. 8.7). The container is covered with a thin membrane perforated with pores of known dimensions. Diffusion of CO_2 through the membrane can be measured as the amount of carbonate present in the sodium hydroxide solution after, for example, one hour. The results of such an experiment, shown in Table 8.2, are clear: The rate of CO_2 diffusion through a perforated membrane varies *in proportion to the diameter of the pores, not the area.* Indeed, as the pore size decreases, the *efficiency* of CO_2 diffusion per unit area increases severalfold! How can these results be reconciled with Fick's law, which states that diffusion is a function of area?

The physical explanation for this paradox lies in the pattern of diffusive flow as the gasses *enter* and *exit* the stomatal pore. This is illustrated schematically for a stoma in Figure 8.8. Note that in the aperture itself (i.e., in the throat of the stoma), CO_2 molecules can flow only straight through and diffusion is proportional to the cross-sectional area of the throat as predicted by Fick's law of diffusion (Chap. 2). But when the gas molecules pass through the aperture into the substomatal cavity, they can "spill over" the edge of the pore. The additional diffusive capacity contributed by spillover is proportional to the amount of edge, or the perimeter of the pore. Because the area of a pore decreases by the square of the radius (r) while perimeter varies directly with the diameter (2r), the relative contribution of the perimeter effect increases as the pore size decreases. Thus in very

TABLE 8.2 Diffusion of CO_2 through small apertures.

Diameter (mm)	Relative CO_2 Diffusion*	Relative Diameter	Relative Area	Relative Efficiency: Diffusion per Unit Area
22.7	1.00	1.00	1.00	1.00
12.1	0.42	0.53	0.28	1.51
6.0	0.26	0.26	0.07	3.72
3.2	0.17	0.14	0.02	8.26

*$\mu g\ CO_2\ hr^{-1}$.

Data from Bidwell, *Plant Physiology*, New York, Macmillan, 1979.

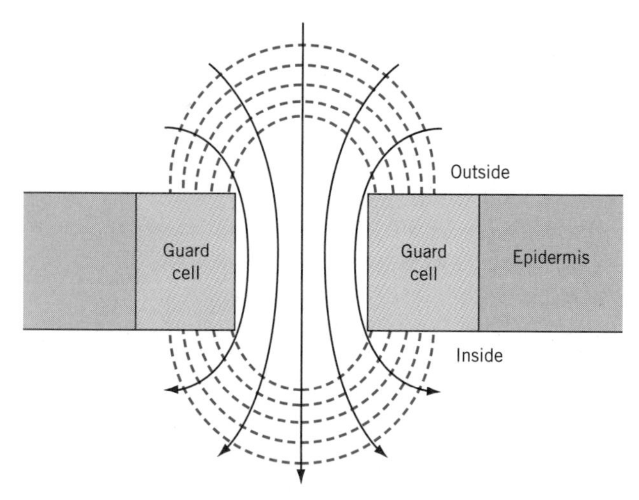

FIGURE 8.8 The spillover effect for diffusion of CO_2 through a stomata pore. The dashed lines are *isobars*, representing regions of equivalent CO_2 partial pressure.

small pores (for example, the size of stomata), the bulk of gas movement is accounted for by diffusion over the perimeter. Even this effect is exaggerated with respect to stomata. Because of their elliptical shape, the ratio of perimeter to area is greater than for circular pores.

The above argument, of course, represents an ideal situation. In reality, the stomatal pore itself is not the only barrier to gaseous diffusion between the leaf and its environment. A number of other factors—such as unstirred air layers on the leaf surface and the aqueous path between the air space and the chloroplast—offer resistance to the uptake of CO_2 into the leaf and complicate the actual situation (Nobel, 1991). Nonetheless, stomata are remarkably efficient structures. They permit very high rates of CO_2 absorption, without which photosynthesis would be severely limited. This creates a paradox. A system that is efficient for the uptake of CO_2 is also efficient for the loss of water vapor from the internal surfaces of the leaf (Chap. 3). Thus the principal functional advantage offered by the stomatal apparatus is an ability to conserve water by closing the pore when CO_2 is not required for photosynthesis or when water stress overrides the leaf's photosynthetic needs.

THE MECHANISM OF GUARD CELL MOVEMENT

How do stomata open and close? This question may be answered by first asking what mechanical forces are involved in guard cell movement. The driving force for stomatal opening is known to be the osmotic uptake of water by the guard cells and the consequent increase in hydrostatic pressure. The result is a deformation of the opposing cells that increases the size of the opening between them. In the case of elliptic guard cells the thickened walls become concave while in the dumbbell-shaped cells the handles separate but remain parallel. Stomatal closure follows a loss of water, and the consequent decrease in hydrostatic pressure and relaxation of the guard cell walls.

Deformation of elliptic guard cells during opening is due to the unique structural arrangement of the guard cell walls (Sharpe et al., 1987). In normal cells, bands of cellulose microfibrils encircle the cell at right angles to the long axis of the cell. Studies with polarized light and electron microscopy have demonstrated that the microfibrils in the guard cell walls are oriented in radial fashion, fanning out from the central region of the ventral wall (Fig. 8.9). Additional microfibrils are arranged longitudinally within the ventral wall thickenings, cross-linking with the radial bands and restricting expansion along the ventral wall. When the guard cells take up water, expansion follows the path of least resistance—which is to push the relatively thin dorsal walls outward into the neighboring epidermal cells. This causes the cells to arch along the ventral surface and form the stomatal opening. The dumbbell-shaped guard cells of the grasses also depend on the osmotic uptake of water, but operate in a slightly different way. In this case the bulbous ends of the cells push against each other as they swell, driving the central handles apart in parallel and widening the pore between them.

What controls stomatal opening and closure? To answer this question it is necessary instead to ask what regulates the osmotic properties of the guard cells. This question has proven difficult to answer, partly because so many factors seem to be involved and partly because it has been difficult to study guard cell metabolism free of complications introduced by the surrounding epidermal and mesophyll cells. This problem has been par-

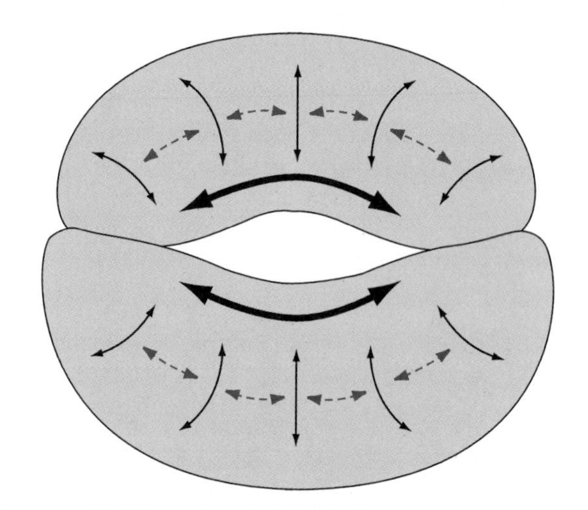

FIGURE 8.9 The role of microfibrils in guard cell movement. The orientation of microfibrils (solid arrows) in elliptic guard cells allows expansion of the cells only in the direction shown by the dashed arrows. This causes the cells to buckle and thereby increase the size of the opening between their adjacent walls.

TABLE 8.3 **Potassium content of open and closed guard cells.**

Species	K^+ Content			
	pmol/guard cell		mM	
	open	closed	open	closed
Vicia faba	2.72	0.55	552	112
Commelina communis	3.1	0.4	448	95

Data from MacRobbie, 1987.

tially resolved by studying guard cell behavior in peeled strips of epidermal cells. More recently, techniques for preparation of guard cell protoplasts have become available, making it possible to study guard cell metabolism and ion movement in isolation.

Over the years a variety of mechanisms have been offered to explain changing osmotic concentrations of guard cells.[3] Most have centered on the observation that guard cells normally contain chloroplasts and were assumed to be photosynthetically competent. One way or another it was proposed that an accumulation of photosynthetic product—sugars and other small molecules—contributed directly to the observed osmotic changes in the guard cells. While it is true that most guard cells do have chloroplasts, the number of chloroplasts varies considerably. As well, the guard cells of some species (e.g., some orchids and variegated regions of *Pelargonium*) have no chloroplasts but remain fully functional. Furthermore, investigators have been unable to detect significant levels of Rubisco (the principal carbon-fixing enzyme; see Chap. 10) in the guard cells of at least 20 species, leading to the conclusion that the carbon-fixing portion of photosynthesis does not operate in guard cells (Outlaw, 1987). The conclusion is inescapable: photosynthetic carbon metabolism cannot be invoked as a general mechanism to explain guard cell movement.

In the late 1960s it became evident that K^+ levels are very high in open guard cells and very low in closed guard cells (Table 8.3).[4] A variety of techniques, including electron microprobes and histochemical methods specific for K^+, have confirmed that the K^+ content of closed guard cells is low compared with that of the surrounding subsidiary and epidermal cells. Upon opening, large amounts of K^+ move from the subsidiary and epidermal cells into the guard cells. Consequently, an accumulation of K^+ in guard cells is now accepted as a universal process in stomatal opening. This work gave rise to the current hypothesis that the osmotic potential of guard cells and, consequently, the size of the stomatal opening, is determined by the extent of K^+ accumulation in the guard cells.

Although we lack a thorough understanding of the mechanisms involved, available information about guard cell metabolism and stomatal movements is summarized in the general model shown in Figure 8.10. It is widely accepted that accumulation of ions by most plant cells is driven by an ATP-powered proton pump located on the plasma membrane (Chap. 5). Two lines of evidence indicate that K^+ uptake by stomatal guard cells fits this general mechanism. First, the fungal toxin **fusicoccin,** which is known to stimulate active proton extrusion by the pump, stimulates stomatal opening. Second, vanadate (VO_3^-), which inhibits the proton pump, also inhibits stomatal opening. This constitutes reasonably good evidence that proton extrusion is one of the initial events in stomatal opening. By removing positively charged ions, proton extrusion would tend to hyperpolarize the plasma membrane (i.e., lower the electrical potential inside the cell relative to the outside) as well as establish a pH gradient. Hyperpolarization is thought to open K^+ channels in the membrane, which then al-

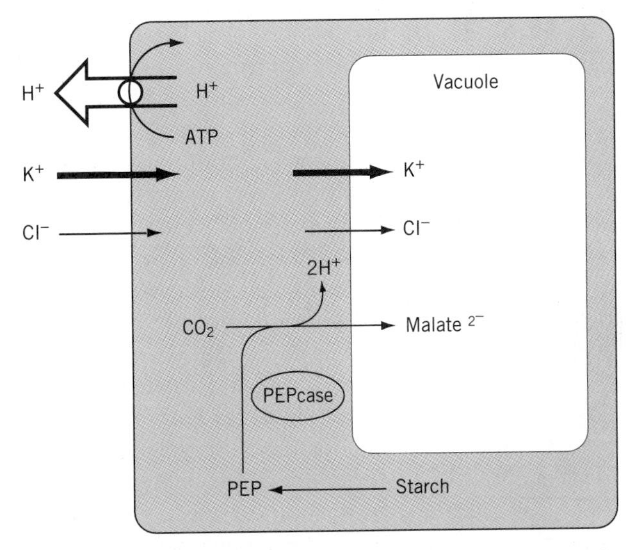

FIGURE 8.10 **A simplified model for ion flow associated with the guard cells during stomatal opening. Potassium uptake is driven by an ATPase-proton pump located in the plasma membrane. The accumulation of ions in the vacuole lowers the water potential of the guard cell, thereby stimulating the osmotic uptake of water and increased turgor.**

[3]For the student interested in historical aspects, the monograph by C. M. Willmer provides a summary of the early hypotheses.

[4]Accumulation of K^+ in open guard cells was actually studied by Japanese scientists in the 1940s and 1950s but their work was not brought to the attention of western scientists until the publication of a report by M. Fujino in 1967. The Japanese findings were quickly confirmed by U.S. and British plant physiologists in 1968 and 1969 (Willmer, 1983).

lows the passive uptake of K^+ in response to the potential difference or charge gradient across the membrane.

In order to maintain electrical neutrality, excess K^+ ion accumulated in the cells must be balanced by a counterion carrying a negative charge. According to the model shown in Figure 8.10, charge balance is achieved partly by balancing K^+ uptake against proton extrusion, partly by an influx of chloride ion (Cl^-), and partly by production within the cell of organic anions such as malate. In most species, malate production probably accounts for the bulk of the required counterion while in others, such as corn (*Zea mays*), as much as 40 percent of the K^+ moving into the cell is accompanied by Cl^-. In those few species whose guard cells lack chloroplast or starch, Cl^- is probably the predominant counterion (Nobel, 1991).

In addition to its role in maintaining charge balance, the accumulation of malate also helps to maintain cellular pH during solute accumulation. Proton extrusion would tend to deplete the intracellular proton concentration and increase cellular pH. However, because malate is an organic anion, each carboxyl group (COO^-) accumulated releases one proton into the cytosol. The synthesis of malate therefore tends to replenish the supply of protons lost by extrusion and maintain cellular pH at normal levels.

The evidence for malate as a counterion is quite strong. To begin with, malate levels in guard cells of open stomata are five to six times that of closed stomata. Second, guard cells contain high levels of the enzyme phosphoenolpyruvate carboxylase (PEPcase), which catalyzes the formation of malate (see Chap. 12). Third, there is a decrease in the starch content of open stomata that correlates with the amount of malate formed. Finally, factors that influence stomatal opening and closure also influence the activity of PEPcase. For example, fusicoccin, which induces stomatal opening, also causes an increase in both malate concentration and the activity of PEPcase (Zhirong et al., 1997). Conversely, the plant hormone abscisic acid, which normally induces stomatal closure, antagonizes the effect of fusicoccin. The effect of fusicoccin is to stimulate the phosphorylation of PEPcase, a process well known to activate a variety of enzymes and other proteins in the cell.

The accumulation of K^+, Cl^- and malate in the vacuoles of the guard cells would lower both the osmotic potential and the water potential of the guard cells. The consequent uptake of water would increase the turgor and cause the stomata to open. At present, this remains a working model for stomata opening since many of the details have yet to be experimentally verified.

Stomatal closure has not received the same attention that opening has, but it is generally assumed that closure is effected by a simple reversal of the events leading to opening. On the other hand the rate of closure is often too rapid to be accounted for simply by a passive leakage of ions from the guard cells, leading to the suggestion that other specific metabolic pumps are responsible for actively extruding ions upon closure (MacRobbie, 1987; Schroeder and Keller, 1992). One possibility is that signals for stomatal closure stimulate the uptake of Ca^{2+} into the cytosol. Ca^{2+} uptake would depolarize the membrane, thus initiating a chain of events that includes opening anion channels to allow the release of Cl^- and malate. According to this scenario, a loss of anions would further depolarize the membrane, opening K^+ channels and allowing the passive diffusion of K^+ into the adjacent subsidiary and epidermal cells.

What is the source of ATP that powers the guard cell proton pumps? The two most logical sources would be either photosynthesis in the guard cell chloroplasts or cellular respiration. Although most guard cells do contain chloroplasts, they are generally smaller, less abundant, and with fewer thylakoids than those of underlying mesophyll cells. As noted above, guard cell chloroplasts apparently lack the enzymatic machinery for photosynthetic carbon fixation. On the other hand, although ATP production has not been measured directly, indirect evidence indicates that they are capable of using light energy to produce ATP, a process known as **photophosphorylation** (see Chap. 9). Photosynthesis is probably not the only immediate source of energy, however, since stomatal movement can occur in the dark. An alternative source of energy is cellular respiration. Guard cells do have large numbers of mitochondria and high levels of respiratory enzymes. They may well be able to derive sufficient ATP from the oxidation of carbon through normal respiratory pathways. Assmann and Zeiger have conducted an extensive analysis of energy production by guard cells and have concluded that guard cells have more than adequate capacity to produce, through either respiration or photosynthesis, all the energy necessary to drive stomatal opening (Assmann and Zeiger, 1987).

CONTROL OF STOMATAL MOVEMENTS

Stomatal opening and closure is controlled by a number of internal and environmental factors. The major role of stomata is to allow entry of CO_2 into the leaf for photosynthesis while at the same time preventing excessive water loss. In this sense they evidently serve a **homeostatic** function; they operate to maintain a constancy of the internal environment of the leaf. It should come as no surprise, then, to find that stomatal movement is regulated by a variety of environmental and internal factors such as light, CO_2 levels, water status of the plant, and temperature. It might be expected, for example, that stomata will open in the light in order to admit CO_2 for photosynthesis or partially close when CO_2 levels are high in order to conserve water while allowing photosynthesis to continue. On the other hand, conditions of extreme water stress should override the plant's immediate photosynthetic needs and lead to closure, protecting the leaf against the potentially more damaging effects of desiccation. In general, these expectations have been verified by direct observation. Each of these factors can theoretically be studied independently under the controlled conditions of the laboratory, but the extent to which they interact under natural conditions makes it far more difficult to study the effects of one relative to another. Moreover, it must be kept in mind that stomatal opening is not an all-or-none phenomenon. At any given time, the extent of stomatal opening and its impact on both photosynthesis and water loss will be determined by the sum of all of these factors and not by any one alone.

Light and Carbon Dioxide Both light and CO_2 appear to make a substantial contribution to the daily cycle of stomatal movements. Their effects are also tightly coupled, which makes it very difficult to distinguish their relative contributions. In general, low CO_2 concentrations and light stimulate opening while high CO_2 concentrations cause rapid closure even in the light. The response of the stomata is to the *intracellular concentration of CO_2 in the guard cells.* Recall that the outer surfaces of the epidermis, *including the guard cells,* are covered with the CO_2-impermeable cuticle. Once induced to close by high CO_2 treatment, stomata are not easily forced to open by treatment with CO_2-free air. This is because the closed guard cells remain in equilibrium with the high CO_2 content of the air trapped in the substomatal chamber. Consequently, it is the CO_2 content of the substomatal chamber rather than the ambient atmosphere that is most important in regulating stomatal opening. The actual mechanism by which CO_2 regulates stomatal opening is not understood.

Stomata normally open at dawn. As well, stomata closed by exposure to high CO_2 can be induced to open slowly if placed in the light. Both responses appear to result from two separate effects of light; one indirect and one direct. The indirect effect requires relatively high fluence rates and is usually attributed to a reduction in intercellular CO_2 levels due to photosynthesis in the mesophyll cells. By the same argument, closure of the guard cells in the dark can be attributed to the accumulation of respiratory CO_2 inside the leaf. This interpretation is reinforced by the observation that the action spectrum for moderate to high fluence rates resembles that for photosynthesis with peaks in both the red and blue. Thus it appears that CO_2 is a primary trigger and that, at least in intact leaves, the indirect effect of light may operate through regulation of intercellular CO_2 levels. A significant difficulty with this interpretation, however, is that similar action spectra have been obtained for isolated epidermal peels. Such a result in the absence of an intact leaf argues strongly for an important but yet undefined role of the guard cell chloroplasts (Zeiger, 1983).

Perhaps one of the more significant advances to emerge in recent years is the unequivocal demonstration of a direct effect of low-fluence blue light on stomatal opening. If the stomata depended solely on photosynthetically active light, it would likely suffer from two limitations. First, the guard cells would be unable to respond to light levels below the photosynthetic compensation point (i.e., the minimum fluence rate at which photosynthesis exceeds respiration). Second, the system would be prone to extreme oscillations as the rate of photosynthesis fluctuated with rapid changes in PAR (Zeiger, 1983). A direct effect of blue light on stomatal opening would seem to circumvent these limitations.

The blue light effect has been demonstrated in a variety of ways. Although stomatal opening is promoted by both red and blue light, it is generally more sensitive to blue light than to red. At low fluence rates, below 15 μmol m^{-2} s^{-1}, blue light will cause stomatal opening but red light is ineffective. At higher fluence rates stomatal opening under blue light (which presumably activates both systems) is consistently higher than under red at the same fluence rate. The two responses can be further distinguished by providing a dual-beam experimental protocol, in which a short (30 s) pulse of blue light is superimposed over a background of continuous red light. Under such conditions, the pulse of blue light stimulates a significant increase in **stomatal conductance**[5] (Fig. 8.11). It is assumed that the fluence rate of the red light is sufficient to saturate photosynthesis in both the guard cells and the surrounding mesophyll cells. The slight rise in photosynthesis shown in Figure 8.11 therefore probably results from the enhanced intercellular CO_2 levels made possible by the increased stomatal opening.

The response of stomata to red light is probably indirect, mediated by the guard cell chloroplasts and involving photosynthetic ATP production. The action spectrum of the blue light response, on the other hand, is typical of other blue light responses and is probably mediated by cryptochrome, a putative blue light receptor (Chap. 7). The mode of action of blue light is not certain, but blue light does cause swelling of isolated guard cell protoplasts. This result indicates that blue light acts directly on the guard cells. Several investigators have reported that blue light activates proton extrusion by the guard cells and stimulates malate biosynthesis; both are prerequisites to stomatal opening.

But what function does the blue light response serve under natural conditions? One interesting and plausible

[5]*Conductance* is the reciprocal of *resistance*. It is a measure of the ease with which a substance (in this case, a gas) passes through a piece of material (in this case, a leaf). Stomatal conductance is proportional to stomatal opening.

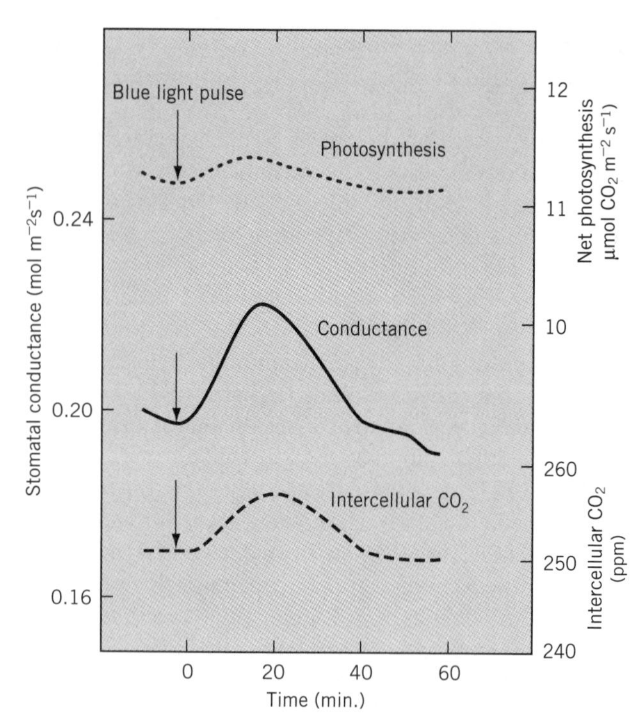

FIGURE 8.11 Blue light induces stomatal opening. Leaves of *Commelina communis* were presented with a brief pulse (30s) of blue light over a background of continuous red light. The pulse of blue light stimulates an increase in stomatal opening (conductance), intercellular CO_2 concentration, and photosynthesis. (Reprinted from *Photochemistry and Photobiology*, Vol. 42, E. Zeiger et al., The blue light response of stomata: Pulse kinetics and some mechanical implications, pp. 759–763, 1985. With kind permission from Pergamon Press, Headington Hill Hall, Oxford OX3 OBW, UK.)

suggestion is that it may have a role in the early morning opening of stomata (Zeiger et al., 1981). Opening can often be observed before sunrise, when fluence rates are much lower than that required to drive photosynthesis. They may also remain open after sunset. The high sensitivity of the blue light response to low fluence rates together with the relatively high proportion of blue light in sunlight at dawn and dusk (Hughes et al., 1984) suggests that the blue light response could function as an effective "light-on" signal. From an ecophysiological standpoint, the blue light response anticipates the need for atmospheric CO_2 and drives stomatal opening in preparation for active photosynthesis. Another possible role is to stimulate rapid stomatal opening in response to sunflecks—the sunfleck itself would be analogous to a blue light pulse—in order to maximize the opportunity for photosynthesis under this particular condition.

Water Status and Temperature Since the degree of stomatal opening is subject to the water status of the guard cells, virtually any change in the water status of the plant can be expected to elicit some effect on stomatal opening and closure. Guard cells are themselves exposed directly to the atmosphere and can lose water by evaporation. At times the rate of water loss from the guard cells might exceed the rate of movement into the guard cells from the surrounding epidermal cells. When this happens, the decrease in turgidity of the guard cells will override all other controls, the cells become flaccid and the stomata close. Such closure is called **hydropassive closure.**

Another response to water stress, known as **hydroactive closure,** occurs when the plant senses a water deficit and initiates a specific mechanism to induce closure. The mechanism for hydroactive closure involves the same ion fluxes normally associated with closure, but is triggered by water deficits in the leaf and is mediated by the hormone abscisic acid (ABA) (Chap. 22). ABA is a normal constituent of leaves; it is synthesized at low rates in mesophyll cells and accumulates in the chloroplasts. In response to moderate water deficit in the leaf, stored ABA is released into the apoplast where it is carried in the transpiration stream to the guard cells and initiates closure. Within as little as seven minutes of the first signs of closure, an increase in the level of ABA synthesis can be detected. Apparently the small amount of stored ABA serves as the first line of defense against water loss; subsequently, an increased synthesis serves to prolong closure. Because of its ability to stimulate stomatal closure and thus reduce transpirational loss, ABA has been referred to as an "anti-transpirant."

It is not clear how the cell can sense water deficits or how ABA initiates closure. The distribution of ABA, which changes during water stress, may be related to the weakly-acid properties of the ABA molecule and changing pH gradients in the cell (Hartung et al., 1988).[6] ABA appears to accumulate in the chloroplast in the first place because the pH of the chloroplast in an actively photosynthesizing cell is normally higher than the cytosol. Moderate water stress causes a decrease in the pH of the chloroplast and an increase in the cytoplasmic and apoplastic pH. Such changes in pH could lead to the release of ABA from the chloroplast into the apoplast. Alternatively, ABA may be released from the cytosol or even imported into the leaves from the roots during periods of water stress (Zeevaart and Creelman, 1988). The role of ABA in stomatal closure is discussed further in Chapter 22.

Temperature can influence stomatal opening in a variety of ways, most of them indirect. An increase in temperature generally increases metabolic activity in the guard cells and the leaf as a whole, reaching an optimum, and then declining. Stomatal opening follows a parallel

[6]ABA is a weak acid (pK 4.7) that dissociates to differing degrees in different cellular compartments. The protonated form diffuses freely across membranes where the dissociated form does not. ABA thus tends to accumulate in those compartments having a higher pH.

pattern. Temperature also affects the balance of respiration and photosynthesis and, consequently, the intercellular CO_2 content. Higher temperatures will generally stimulate respiration but may impair photosynthesis. The resulting higher CO_2 levels will bring about stomatal closure.

The stomata of some species often undergo a distinct "midday" closure, resulting in a transient reduction in photosynthesis (Tenhunen et al., 1980). Traditionally this closure has been attributed to water stress brought about by high temperatures and excessive transpiration. Subsequent reopening occurs by midafternoon when the temporary water deficit has been satisfied by a redistribution of water within the plant.

Circadian Rhythms Many biological processes undergo periodic fluctuations that persist under constant enviromental conditions. This phenomenon, known as endogenous rhythm, is discussed further in Chapter 20. In 1971, Martin and Meidner demonstrated that stomatal opening and closure in *Tradescantia* leaves persisted for at least three days, even though the plants were maintained under continuous light. A periodicity of approximately 24 hours was maintained, although the timing of opening or closure could be shifted by a six-hour dark period. Results such as these clearly indicate an involvement of an endogenous circadian rhythm in control of stomatal opening, although it is not clear how the rhythm interacts with other stimuli.

THE CHLOROPLAST

Photosynthesis, along with several other metabolic activities, takes place in discrete, highly structured cellular organelles called chloroplasts.

It is generally believed that photosynthesis originated in the oceans of a primordial world. The first photosynthetic cells were no doubt prokaryotes, similar perhaps to some of our present-day photosynthetic bacteria. The environment was highly reducing—there was virtually no uncombined oxygen—and these organisms probably relied on reduced substrate such as H_2S for the supply of necessary electrons. At some point in history, photosynthetic prokaryotes probably invaded primitive, nucleated eukaryotic cells and established a symbiotic relationship. According to this scenario, the prokaryote gradually gave up its independence and, although it retained a functional genome, eventually came under the genetic and biochemical control of the host cell. Out of this association there eventually evolved what we now recognize as a **chloroplast.**

Chloroplasts are one of a family of organelles bounded by a double membrane and known generally as *plastids* (Chap. 1). As the name implies, *chloro*plasts are identified by the fact that they contain the *chlorophyll* pigments responsible for the green color of leaves. In addition to chlorophyll, chloroplasts contain large amounts of carotenes and xanthophylls (Chap. 7).

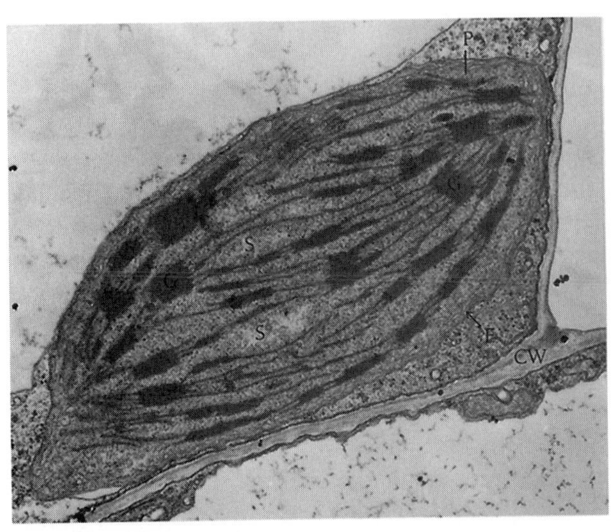

FIGURE 8.12 Electron micrograph of a mesophyll chloroplast of maize (*Zea mays*). S, stroma, G, granum; P, peripheral reticulum; E, envelope; CW, cell wall (× 5000).

A typical higher plant chloroplast is generally described as discoid with a diameter of 3 to 10 μm (Fig. 8.12). In cross-section the chloroplast appears lens-shaped. Chloroplasts are located in the cytosol of the cell and, consequently, are normally seen pressed between the cell wall and the prominent central vacuole. Although there is a remarkable consistency in the size and shape of chloroplasts in higher plants, the algae are noted for the wide variety in shape and size of their chloroplasts.

Chloroplasts are most often limited to the inner, or mesophyll, leaf cells and stomatal guard cells, although Butterfass lists a number of species for which chloroplasts may be found in epidermal cells as well (Butterfass, 1979). In those species that have epidermal chloroplasts, both the size of the epidermal chloroplasts and their number per cell are generally smaller than in the mesophyll cells. The number of chloroplasts in a mesophyll cell is typically in the range of 20 to 60, although values of several hundred have been reported for some species. As well, the chloroplasts in palisade mesophyll cells are generally larger and more numerous than in the spongy mesophyll cells (Butterfass, 1979).

ULTRASTRUCTURE AND BIOCHEMICAL COMPARTMENTATION OF CHLOROPLASTS

The structure and development of chloroplasts has been studied extensively with the electron microscope (Fig.

8.12). As a result, we recognize four major structural regions or compartments: (1) a pair of outer limiting membranes, collectively known as the **envelope,** (2) an unstructured background matrix or **stroma,** (3) a highly structured internal system of membranes, called **thylakoids,** and (4) the intrathylakoid space, or **lumen.** The envelope defines the outer limits of the organelle. These membranes are 5.0 to 7.5 nm thick and are separated by a 10 nm **intermembrane space.** Because the inner envelope membrane is selectively permeable, the envelope also serves to isolate the chloroplast and regulate the exchange of metabolites between the chloroplast and the cytosol that surrounds it. Experiments with spinach chloroplasts have shown that the intermembrane space is freely accessible to metabolites in the cytoplasm (Heldt, 1976). Thus it appears that the outer envelope membrane offers little by way of a permeability barrier. It is left to the inner envelope membrane to regulate the flow of molecular traffic between the chloroplast and cytoplasm.

The envelope encloses the unstructured background matrix of the chloroplast, or stroma. The composition of the stroma is predominantly protein. The stroma contains all of the enzymes responsible for photosynthetic carbon reduction, including *ribulose-1,5-bisphosphate carboxylase/oxygenase*, generally referred to by the acronym **rubisco** (Chap. 10). Rubisco, which accounts for fully half of the total chloroplast protein, is no doubt the world's single most abundant protein. In addition to rubisco and other enzymes involved in carbon reduction, the stroma contains enzymes for a variety of other metabolic pathways as well as DNA, RNA, and the necessary machinery for transcription and translation of protein.

Embedded within the stroma is a complex system of membranes often referred to as **lamellae** (Fig. 8.12). This system is composed of individual pairs of parallel membranes that appear to be joined at the end, a configuration that in cross-section gives the membranes the appearance of a flattened sack, or **thylakoid** (Gr. = sack-like; Menke, 1962). In some regions adjacent thylakoids appear to be closely appressed, giving rise to membrane stacks known as **grana.** The thylakoids found within a region of membrane sacking are called **grana thylakoids.** Some thylakoids, quite often every second one, extend beyond the grana stacks into the stroma as single, nonappressed thylakoids. These **stroma thylakoids** most often continue into another grana stack, thus providing a network of interconnections between grana. While the organization of thylakoids into stacked and unstacked regions is typical, it is by no means universal. One particularly striking example is the chloroplast in cells that surround the vascular bundles in C4 photosynthetic plants and that have no grana stacks (Fig. 8.13). Here the thylakoids form long, unpaired arrays extending almost the entire diameter of the chloroplast.

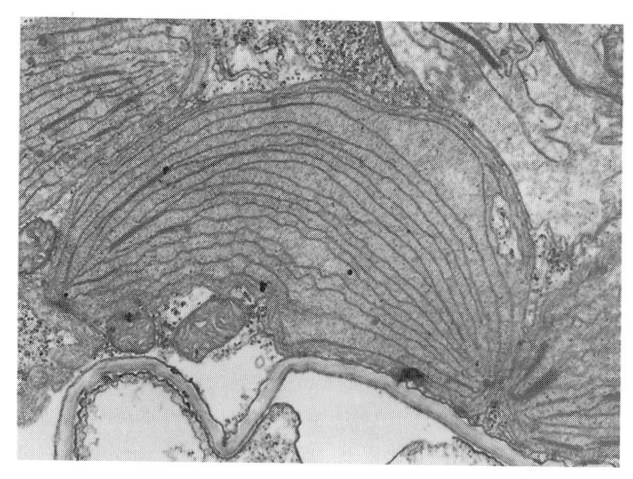

FIGURE 8.13 Electron micrograph of a bundle-sheath chloroplast of maize (*Zea mays*). Note the absence of grana stacks.

The thylakoid membranes contain the chlorophyll and carotenoid pigments and are the site of the light-dependent, energy-conserving reactions of photosynthesis.

The interior space of the thylakoid is known as the **lumen.** The lumen is the site of water oxidation and, consequently, the source of oxygen evolved in photosynthesis. Otherwise it functions primarily as a reservoir for protons that are pumped across the thylakoid membrane during electron transport and that are used to drive ATP synthesis (Chap. 9). Although thylakoids arise during development of the chloroplast as invaginations of the inner envelope membrane, at maturity they are no longer continuous with the envelope and the lumen is no longer connected with intermembrane space of the envelope. Although a number of models have been proposed to account for the interconnections between grana and stroma thylakoids (Coombs and Greenwood, 1976), most of our information about chloroplast structure is derived from thin sections viewed in the electron microscope and relatively little is actually known about the three-dimensional structure of the thylakoids. However, it is generally assumed that the thylakoids represent a single, continuous network of membranes. This means, of course, that the lumens of each thylakoid seen in cross-section also represent but a small part of a single, continuous system.

Chloroplasts may also contain **starch grains,** which represent stored photosynthate, and lipid droplets, called **plastoglobuli.** Since starch grains do not take up the electron-opaque material used to stain electron microscope preparations, they usually appear as electron-transparent areas in the photographs. Plastoglobuli, on the other hand, are darkly staining electron-dense bodies. Plastoglobuli appear to function primarily as lipid-storage bodies and may contain particularly large amounts of the electron carrier plastoquinone. Some chloroplasts, particularly in aging or senescing leaves, may contain deposits of **phytoferritin,** an iron-binding protein thought to prevent accumulation of excess free iron.

NONPHOTOSYNTHETIC METABOLISM OF CHLOROPLASTS

The focus of chloroplast biochemistry is naturally on various aspects of energy transformation and carbon metabolism. This is as it should be, for photosynthesis is a principal activity of the chloroplast and is the only significant route of entry for both energy and carbon into the biosphere. But in addition to the synthesis of carbohydrate, energy generated in the chloroplast is used for other metabolic needs, including the assimilation of nitrogen and sulphur as well as the biosynthesis of protein and fatty acids.

As discussed earlier (Chap. 6), nitrogen is normally taken up by plants in the form of nitrate and must be chemically reduced to ammonia before it can be incorporated in amino acids and other nitrogenous compounds in the cell. The initial enzyme in the process, nitrate reductase, is located in the cytoplasm of leaves and roots. The second step is the reduction of nitrite to ammonia which, in leaves, is associated with the photosynthetic electron transport reactions. Some of the reactions of amino acid biosynthesis appear to be located in the chloroplast as well, although they may duplicate reactions located in the mitochondria and elsewhere. The principal reaction by which ammonia is incorporated into amino acids involves two enzymes, **glutamine synthetase (GS)** and **glutamate synthase (GOGAT)** (Chap. 6). GS is located in the chloroplast as well as the cytosol of leaf cells and in the cytosol of root cells. GOGAT is located both in the chloroplasts of leaf cells and in root cell plastids. Glutamine appears to play a central role as an organic nitrogen donor in amino acid biosynthesis. Several enzymes that catalyze the synthesis of other amino acids are also located in the chloroplast.

Sulphate, like nitrate, must also be reduced, usually to cysteine, before it can be incorporated into amino acids and other organic compounds. The complete sulphate-reducing cycle appears to be localized within the chloroplasts of leaf cells and in proplastids of root cortical cells. It is interesting to note that the reactions for synthesis of the sulphur-containing amino acid methionine are also localized entirely in the chloroplast.

Fatty acid biosynthesis is another activity localized at least in part in chloroplasts and proplastids. The two major products are both saturated fatty acids: **palmitate** (C_{16}) and **stearate** (C_{18}). These fatty acids are then transferred from the plastid to the endoplasmic reticulum where they undergo further modification before being assembled with glycerol to form lipids.

Chloroplasts contain a complete genome, including DNA, RNA, ribosomes, and the enzymes necessary for protein biosynthesis. Chloroplast DNA is circular, like prokaryote DNA, and is capable of encoding about 120 to 125 polypeptides. The chloroplast genome encodes a number of polypeptides with significant roles in both energy conversion and carbon metabolism. These include the large subunit of the enzyme Rubisco (the carbon-fixing enzyme), several subunits of the ATP-synthesizing complex, and important components of the electron transport chain. The identities of many other chloroplast gene products, however, have yet to be described. In spite of having its own genome, the chloroplast must depend on the nucleus and cytosol for additional polypeptides. Indeed, a majority of chloroplast polypeptides are encoded in the nucleus, translated on cytosolic ribosomes and imported into the chloroplast.

SUMMARY

The leaf is a remarkable example of the relationship between structure and function in plants. In order to achieve its primary function, which is to carry out photosynthesis, a leaf requires adequate supplies of light, carbon dioxide, and water. To this end, most leaves present a broad surface area for efficient interception of light. As well, the internal optical properties of a leaf help to ensure that incoming light is distributed throughout the leaf to maximize absorption by chlorophyll. An extensive vascular system serves to import water and nutrients to the photosynthetic cells and export photosynthate to nonphotosynthetic tissues and organs elsewhere in the plant.

The presence of stomata in the leaf epidermis helps to balance loss of water vapor against the need to take up carbon dioxide from the surrounding air. Although stomata are microscopic pores which, when fully open, account for less than 2 percent of the total leaf area, the efficiency of gaseous diffusion through the stomata is very high. The carbon dioxide gradient between the air and the interior of the leaf is relatively shallow, so the high efficiency of diffusion thorough stomata ensures an adequate supply of carbon dioxide to support photosynthesis. For the same reasons, stomata support the efficient diffusion of water vapor from the interior of the leaf. In order to prevent excessive water loss, the stomata can be closed by the action of the surrounding guard cells.

Regulation of the stomatal opening is an osmotic phenomenon, driven by the exchange of potassium ions between the guard cells and the surrounding epidermal cells. Potassium uptake is followed by the osmotic uptake of water. The consequent swelling and distortion of the guard cells opens the pore, or stoma. Stomatal opening and closure is controlled by several environmental factors, including light, internal leaf carbon dioxide concentration, leaf water status, and temperature.

Chloroplasts, the site of photosynthesis in the leaf, are one of a family of organelles bound by a double membrane. The chloroplast contains four structural compartments: an outer membrane envelope, an unstructured stroma, an internal thylakoid membrane system, and an intrathylakoid lumen. The thylakoid membranes are the site of the light-dependent, energy-conserving reactions of photosynthesis, while carbon reduction takes place in the stroma. The stroma is also the site of other metabolic activities, including the assimilation of nitrogen and sulfur, fatty acid biosynthesis, maintenance and transcription of the chloroplast genome, and translation of protein.

CHAPTER REVIEW

1. The evolution of leaves of photosynthetic organs has revolved around what three major themes? What is important about these three themes?
2. Early in the chapter it is stated that leaves must carry out photosynthesis in a "hostile" environment. What is hostile about the environment of a leaf? In what way does a leaf cope with these problems?
3. How does the anatomy of a leaf lend itself to the efficient uptake of CO_2 for photosynthesis by leaf mesophyll cells?
4. In what way(s) does the design of a leaf lend itself to the efficient capture of light for photosynthesis?
5. Explain how guard cells regulate the size of the stomatal aperture.
6. What are the four compartments in a chloroplast and what principal functions are associated with each?
7. The leaf is often described as a photosynthetic "machine." Do you agree or disagree with this analogy? What arguments can you advance for or against this analogy?

FURTHER READING

Meidner, H., D. Mansfield. 1968. *Physiology of Stomata.* New York: McGraw-Hill.

Vogelmann, T. C. 1994. Light within the plant. In: R. E. Kendrick, G. H. M. Kronenberg, eds. *Photomorphogenesis in Plants.* Dordrecht: M. Nijhoff Publishers. pp. 491–535.

Zeiger, E., G. O. Farquhar, I. R. Cowan (eds.). 1987. Stomatal Function. Stanford, CA: Stanford University Press.

REFERENCES

Assmann, S. M., E. Zeiger. 1987. Guard cell bioenergetics. In: E. Zeiger, G. D. Farquhar, I. R. Cowas (eds.) *Stomatal Function.* Stanford, CA: Stanford University Press. pp. 163–193.

Bidwell, R. G. S. 1979. *Plant Physiology.* New York: Macmillan.

Butterfass, T. 1979. *Patterns of Chloroplast Reproduction.* Vienna: Springer-Verlag.

Canvin, D. T. 1990. Photorespiration and CO_2-concentrating mechanisms. In: D. T. Dennis and D. H. Turpin (eds.), *Plant Physiology, Biochemistry and Molecular Biology.* England: Longman Scientific & Technical, pp. 253–273.

Coombs, J., A. D. Greenwood. 1976. Compartmentation of the photosynthetic apparatus. In: J. Barber (ed.), *Topics in Photosynthesis. Vol. 1. The Intact Chloroplast.* Amsterdam: Elsevier/North-Holland Biomedical Press, pp. 1–51.

Hartung, W., J. W. Radin, D. L. Hendrix. 1988. Abscisic acid movement into the apoplastic solution of water-stressed cotton leaves. Role of apoplastic pH. *Plant Physiology* 86:908–913.

Heldt, H. W. 1976. Metabolite transport in intact spinach chloroplasts. In: J. Barber (ed.), *Topics in Photosynthesis. Vol. 1. The Intact Chloroplast.* Amsterdam: Elsevier/North-Holland Biomedical Press, pp. 171–214.

Hughes, J. E., D. C. Morgan, P. A. Lambton, C. R. Black, H. Smith. 1984. Photoperiodic time signals during twilight. *Plant, Cell and Environment* 7:269–277.

MacRobbie, E. A. C. 1987. Ionic relations of guard cells. In: E. Zeiger, G. D. Farquhar, I. R. Cohen (eds.), *Stomatal Function.* Stanford, CA: Stanford University Press, pp. 125–162.

Meidner, H., D. Mansfield. 1968. *Physiology of Stomata.* New York: McGraw-Hill.

Menke, W. 1962. Structure and chemistry of plastids. *Annual Review of Plant Physiology* 13:27–44.

Nobel, P. S. 1991. *Physicochemical and Environmental Plant Physiology.* New York: Academic Press.

Outlaw, W. H., Jr. 1987. An introduction to carbon metabolism in guard cells. In: E. Zeiger, G. D. Farquhar, I. R. Cohen (eds.), *Stomatal Function.* Stanford, CA: Stanford University Press, pp. 115–123.

Schroeder, J. I., B. U. Keller. 1992. Two types of anion channel currents in guard-cells with distinct voltage regulation. *Proceedings of the National Academy of Sciences, US* 89:5025–5029.

Sharpe, P. J. H., H. Wu, R. D. Spence, 1987. Stomatal mechanics. In: E. Zeiger, G. D. Farquhar, I. R. Cohen (eds.), *Stomatal Function.* Stanford, CA: Stanford University Press, pp. 91–114.

Tenhunen, J. D., O. L. Lange, M. Braun, A. Meyer, R. Losch, J. S. Pereira. 1980. Midday stomatal closure in *Arbutus unedo* leaves in a natural macchia and under simulated habitat conditions in an environmental chamber. *Oecologia* 47:365–367.

Terashima, I., T. Saeki. 1983. Light environment within a leaf. I. Optical properties of paradermal sections of Camellia leaves with special reference to differences in the optical properties of palisade and spongy tissues. *Plant and Cell Physiology* 24:1493–1501.

Vogelmann, T. C., J. N. Nishio, W. K. Smith. 1996. Leaves and light capture: light propagation and gradients of carbon fixation within leaves. *Trends in Plant Science* 1:65–70.

Willmer, C. M. 1983. *Stomata.* Essex: Longman Group, 166 pp.

Zeevaart, J. A. D., R. A. Creelman. 1988. Metabolism and physiology of abscissic acid. *Annual Review of Plant Physiology* 39:439–473.

Zeiger, E. 1983. The biology of stomatal guard cells. *Annual Review of Plant Physiology* 34:441–475.

Zeiger, E., C. Field, H. A. Mooney. 1981. Stomatal opening at dawn: Possible roles of the blue light response in nature. In: H. Smith (ed.), *Plants and the Daylight Spectrum.* London: Academic Press, pp. 391–407.

Zeiger, E., M. Iino, T. Ogawa. 1985. The blue light response of stomata: Pulse kinetics and some mechanical implications. *Photochemistry and Photobiology* 42:759–763.

Zhirong, D., K. Aghoram, W. H. Outlaw. 1997. *In vivo* phosphorylation of phospho*enol*pyruvate carboxylase in guard cells of *Vicia faba* L. is enhanced by fusicoccin and suppressed by abscisic acid. *Archives of Biochemistry and Biophysics* 337:345–350.

Ziegler, H. 1987. The evolution of stomata. In: E. Zeiger, G. D. Farquhar, I. R. Cohen (eds.), *Stomatal Function.* Stanford, CA: Stanford University Press, pp. 29–55.

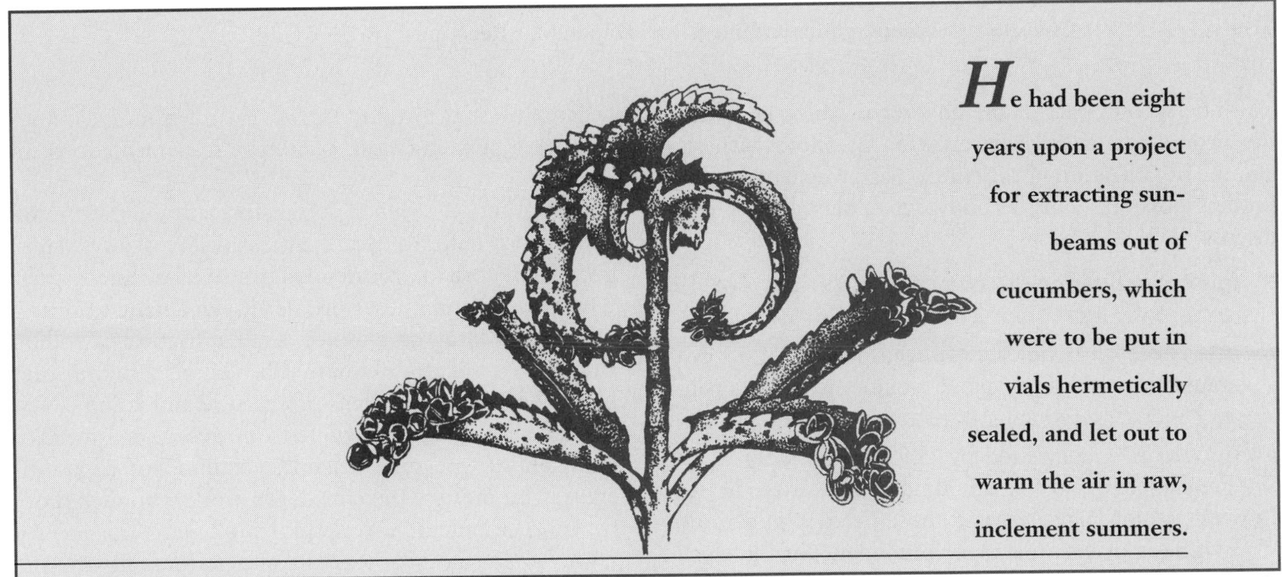

Jonathon Swift,
A Voyage to Laputa

9

Bioenergetics and the Light-Dependent Reactions of Photosynthesis

At every level—cell, tissue, organ, or even community—biological systems are complex and highly organized. Organization is the very essence of biology and yet it is constantly under attack. Proteins, nucleic acids, and other molecules that make up the cell are continually subject to breakdown by hydrolysis. Membranes leak solutes to the environment. Everything on earth, cells and environment alike, is subject to persistent oxidation. Still, all around us we see biological organisms extracting materials from their environment and using them to maintain their organization or to build new, complex structures.

To build and maintain structure requires energy. Energy to build and preserve order in the face of a constantly deteriorating environment is a fundamental need of all organisms. Two strategies have developed to satisfy this need. One is photosynthesis—the **photoautotropic** lifestyle—which traps energy from the sun to build complex structures out of simple inorganic substances. By contrast, organisms that live by the alternative, **chemoheterotropic,** lifestyle require a constant intake of energy-rich, organic substances from their environment. But even these substances trace their origins back to photosynthesis. In the end, all life on earth is powered by energy from the sun through **photosynthesis.**

Photosynthesis occurs in green plants, the cyanobacteria, and certain groups of bacteria. In higher plants the reactions of photosynthesis occur in the chloroplast, which is, put quite simply, a thermodynamic machine. The chloroplast traps the radiant energy of sunlight and conserves some of it in a stable chemical form. The reactions that accomplish these energy transductions are identified as the **light-dependent reactions** of photosynthesis. Energy generated by the light-dependent reactions is subsequently used to reduce inorganic carbon dioxide to organic carbon in the form of sugars. Both the carbon and the energy conserved in those sugars are then used to build the order and structure that distinguishes living organisms from their inorganic surroundings.

This chapter is divided into two parts. The first part is concerned with **bioenergetics**—the study of energy transformations in living organisms—and will discuss

- thermodynamic laws and the concepts of free energy and entropy;
- free energy and its relationship with chemical equilibria, illustrating how displacement of a reaction from equilibrium can be used to drive vital reactions;
- oxidation-reduction reactions, showing how they also are involved in biological energy transformations; and
- the chemiosmotic model for synthesis of adenosine triphosphate (ATP), a key mediator of biological energy metabolism.

The second part of the chapter is concerned with the process of energy conservation in photosynthesis and the organization of the chloroplast, which make this process possible. This second part of the chapter will discuss

- photosynthesis as the reduction of carbon dioxide to carbohydrate;
- the photosynthetic electron transport chain, its organization in the thylakoid membrane, and its role in generating reducing potential and ATP;
- problems encountered by chloroplasts when they are subject to varying amounts of light, often in excess, which may decrease the efficiency of photosynthesis or even damage components of the electron transport chain;
- the dynamic nature of the thylakoid membrane, showing how changes in the organization of light-harvesting apparatus influence the absorption and distribution of light energy;
- the role of carotenoids as accessory pigments and in photoprotection of chlorophyll; and
- the use of herbicides that specifically interact with photosynthetic electron transport.

BIOENERGETICS—ENERGY TRANSFORMATIONS IN LIVING ORGANISMS

Bioenergetics is the application of thermodynamic laws to the study of energy transformations in biological systems. The energetics of cellular processes can be related to chemical equilibrium and oxidation-reduction potentials of chemical reactions.

Whether at the level of molecules, cells or ecosystems, the flow of energy is central to the course of biology. A basic understanding of energy flow is therefore essential to grasping the true beauty and significance of biology. The field of study concerned with the flow of energy through living organisms is called **bioenergetics.** For the past several decades, however, the central focus of bioenergetics has been to unravel the complexities of energy transductions in photosynthesis and respiration and understand how that energy is used to drive energy-requiring reactions such as ATP synthesis and accumulation of ions across membranes (Nicholls, 1982). Energy flow is dictated by certain fundamental thermodynamic rules. The science of **thermodynamics** arose from nineteenth-century interests in the workings of steam engines or why heat was evolved when boring cannon barrels. Although the study of thermodynamics is now concerned with energy flow in a more general sense, the name *thermo*dynamics and much of its language reflect this historical interest in heat. A general understanding of thermodynamic principles is necessary because these principles provide the quantitative framework for understanding energy transformations in biology.

In addition to energy transformations, thermodynamics also helps to describe the capacity of a system to do work. Work may be defined in several different ways. The physicist defines work as displacement against a force: the sliding of an object against friction or rolling a boulder uphill, for example. The chemist, on the other hand, views work in terms of pressure and volume. For example, work must be done to overcome the force of atmospheric pressure when the volume of a gas increases. In biology the concept of work is applied more broadly, embracing a variety of work functions against a wide spectrum of forces encountered in cells and organisms. In addition to mechanical work such as muscular activity, the biologist is concerned with such diverse activities as chemical syntheses, the movement of solute against electrochemical gradients, osmosis, and ecosystem dynamics. These and a host of other essential activities of living things can all be described in thermodynamic terms. The purpose of this section is to facilitate an understanding of biological energy transductions by assembling, in the simplest form possible, some basic thermodynamic principles. The interested student will find a more comprehensive treatment of thermodynamics and bioenergetics in the excellent monographs by D. G. Nicholls (1982, 1992) and F. Harold (1986). These publications have provided the basis for the discussion that follows.

ENERGY CONSERVATION, ORDER AND DISORDER

Biological energy transductions are based on two thermodynamic laws. The first law, commonly known as the *law of conservation of energy,* states that the energy of the universe is constant. This is not a difficult concept to comprehend—it means simply that there is a fixed amount of energy and, while it may be moved about or changed in form, it can all be accounted for somewhere. More to the point, energy is never "lost" in a reaction—an apparent decrease in one form of energy will be balanced by an increase in some other form of energy. In one of the examples mentioned earlier, for example, some of the energy expended in displacing an object appears as work while some appears as heat generated due to friction. In the same way, some of the chemical energy released in the combustion of glucose will also be found as heat in the environment, while some will be found as bond energy in the product molecules, CO_2 and water.

As biologists we are concerned above all with how much work can be done. But, as suggested earlier, not all energy is available to do work. This brings us to the second law of thermodynamics and the concept of **entropy.** Because it involves the concept of entropy (S), the second law is a bit more difficult to comprehend.

What is entropy? Entropy has been variously described as a measure of randomness, disorder, or chaos. However, since entropy is a thermodynamic concept, it is probably most useful to describe it in terms of thermal energy.

Temperature is defined as the mean molecular kinetic energy of matter. Thus any molecular system not at absolute zero (−273 °C, or 0 K) contains a certain amount of thermal energy—energy in the form of the vibration and rotation of its constituent molecules and their movement through space. This quantity of thermal energy and temperature go hand–in–hand: as the quantity of energy increases or decreases, so does temperature. Because temperature cannot be held constant when this energy is given up, it is said to be "isothermally unavailable" (Gr. *isos* = equal). Quantitatively, isothermally unavailable energy is given by the term TS, where T is the absolute temperature and S is entropy.

Since isothermally unavailable energy, and consequently, entropy, are related to the energy of molecular motion, it follows that the more molecules are free to move about—that is, the more random or less ordered or chaotic the system—the greater will be their entropy. (By this same argument it also follows that at absolute zero, a state in which all molecular motion ceases, entropy is also zero.) Consider a familiar example: the combustion of glucose. The highly ordered structure of a glucose molecule imposes certain constraints on the movement of the constituent carbon atoms. In the form of six individual carbon dioxide molecules, however, those same atoms are far less constrained. They are individually free to rotate and tumble through space. The glucose molecules and carbon dioxide molecules each have entropy, but the product carbon dioxide molecules are less ordered, their freedom of movement is greater, and so is their entropy. The natural tendency is for entropy to increase; that is, for systems to become increasingly chaotic. This tendency was summarized by R. J. Clausius: *The entropy of the universe tends toward a maximum.*

Clausius' dictum is one way of stating the second law of thermodynamics. As biologists, however, our concern with entropy is primarily that it represents energy that is *not* available to do work. In this context, the second law can then be restated as: *The capacity of an isolated system to do work continually decreases.* In other words, it is never possible to utilize all of the energy of a system to do work.

From the above discussion, it is apparent that some energy will be available under isothermal conditions and is, consequently, available to do work. This energy is called **Gibbs free energy** in honor of J. W. Gibbs, the nineteenth-century physical chemist who introduced the concept. Free energy (G) is related to TS in the following way:

$$H = G + TS \tag{9.1}$$

H is the total heat energy (also called **enthalpy**), including any work that might be done. H is comprised of isothermally unavailable TS *plus* G. Equation 9.1 thus identifies two kinds of energy: free energy, which is available to do work, and entropy, which is not. Except in a limited number of situations, it is free energy—the energy available to do work—that is of greatest interest to the biologist. Equation 9.1 also suggests a corollary of the second law—*the free energy of the universe tends toward a minimum.*

It is neither convenient nor relevant to measure absolute energies (either G or S), but *changes* (designated by the symbol Δ) in energy during the course of a reaction can usually be measured with little difficulty—as, for example, heat gain or loss, or work. Equation 9.1 can be restated:

$$\Delta G = \Delta H - T\Delta S \tag{9.2}$$

Changes in free energy can tell us much about a reaction. It can tell us, for example, the feasibility of a reaction actually taking place and the quantity of work that might be done if it does take place. Feasibility is indicated by the sign of ΔG. If the sign of ΔG is negative, the reaction is considered **spontaneous**—meaning that it will proceed without an input of energy. Since the free energy of the products is less than the reactants, reactions with a negative ΔG are sometimes known as **exergonic,** or energy yielding. If, on the other hand, ΔG is positive, an input of energy is required for the reaction to occur. The oxidation of glucose is an example of a reaction with a negative ΔG. Once an activation barrier is overcome, glucose will spontaneously oxidize to form CO_2 and water. Despite the existence of large quantities of CO_2 and water in the atmosphere, however, they are not known to spontaneously recombine to form glucose! This is because the equilibrium constant favors the formation of CO_2 and H_2O and the ΔG for glucose formation is positive. Reactions with a positive ΔG are known as **endergonic,** or energy consuming.

The magnitude of free energy changes is very much a function of the particular set conditions for that reaction. For that reason it is convenient to compare the free energy changes of reactions under standard reaction conditions. In biochemistry the **standard free energy change,** $\Delta G^{\circ\prime}$, defines the free energy change of a reaction that occurs at physiological pH (pH = 7.0) under conditions where both reactants and products are at unit concentration (1M).

FREE ENERGY AND CHEMICAL EQUILIBRIA

Under appropriate conditions, all chemical reactions will achieve a state of equilibrium, at which there will be no further *net* change in the concentrations of reactants and products. There is a fairly straightforward re-

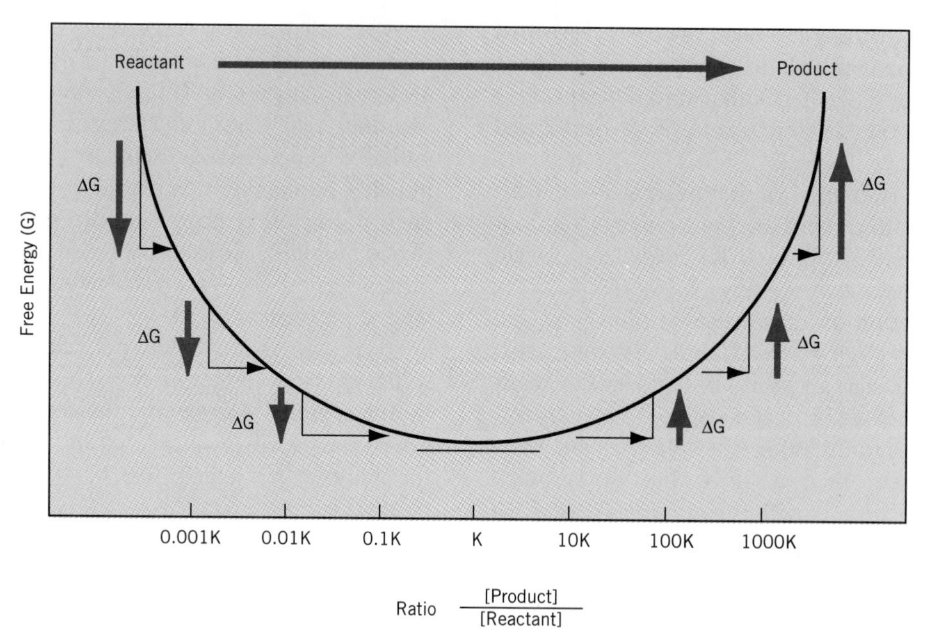

FIGURE 9.1 The free energy of a reaction is a function of its displacement from equilibrium. K is the mass-action ratio when the reaction is at equilibrium. Vertical arrows indicate the slope of the free energy curve, or change in free energy, as reactant is converted to product. Note that the free energy change at equilibrium is zero and that the magnitude of free energy change, indicated by the length of the arrow, increases as the reaction moves away from equilibrium toward pure reactant or pure product. A downward arrow indicates a negative free energy change; an upward arrow indicates a positive free energy change. (Redrawn from D. G. Nicholls, *Bioenergetics*, Academic Press, 1982. Reprinted by permission.)

lationship between free energy and chemical equilibria. This relationship, which is central to an understanding of bioenergetics, is illustrated diagrammatically in Figure 9.1. In a reaction where the reactant is converted to product, K is the equilibrium mass-action ratio—the ratio of concentration of products to the concentration of reactants when the reaction has come to equilibrium. Thus

$$K = \frac{[B]_{eq}}{[A]_{eq}}$$

In Figure 9.1 the slope of the line represents the change in free energy (ΔG) when a small amount of the reactant A is converted to the product B. Several useful points can be drawn from this diagram.

1. At equilibrium, the slope of the line is zero. Consequently, *the free energy change of a reaction at equilibrium is zero.*
2. The further the mass-action ratio is displaced from equilibrium (K), the greater the free energy change for conversion of the same small amount of A to B. *The free energy change for a reaction is a function of its displacement from equilibrium. The further a reaction is poised away from equilibrium, the more free energy is available as the reaction proceeds toward equilibrium.*
3. As A approaches equilibrium, ΔG is *negative* and free energy is available to do work. As the reaction proceeds past equilibrium toward B, ΔG is *positive* and energy must be supplied. *A system can do work as it moves toward equilibrium.* Note that if the reaction were initiated with pure B, the direction of the arrows would be reversed and work could be done as B approached equilibrium.

The relationship between the standard free energy change ($\Delta G^{\circ\prime}$) and equilibrium can be expressed quantitatively as:

$$\Delta G^{\circ\prime} = -RT \ln K = -2.3RT \log K \quad (9.3)$$

and the actual free energy change (ΔG) of a reaction not at equilibrium is given by:

$$\Delta G = \Delta G^{\circ\prime} + 2.3RT \log \Gamma \quad (9.4)$$

where Γ equals the observed (i.e., nonequilibrium) mass-action ratio. Equation 9.3 can then be substituted in Equation 9.4 and rearranged to give:

$$\Delta G = -2.3RT \log (K/\Gamma) \quad (9.5)$$

Equation 9.5 reinforces the observation that the value of ΔG is a function of the degree to which a reaction is displaced from equilibrium. When Γ = K, the reaction is at equilibrium. When Γ is less than K the value of ΔG is negative; it is positive when Γ is greater than K. Finally, a reaction with a negative ΔG can drive a reaction with a positive ΔG, if they are biochemically coupled.

To illustrate the above relationships, consider the **adenosine triphosphate (ATP)** system. When ATP was first isolated from extracts of muscle in 1929, F. Lipmann was one of the first to recognize its significant role in the energy metabolism of the cell. He called it a *high-energy molecule* and designated the terminal phosphate bonds by the squiggle symbol (~) (Fig. 9.2). Actually, the designation of ATP as a high-energy molecule is somewhat misleading because it implies that ATP is in some way a unique molecule. It is not. The two terminal phosphate bonds of the ATP molecule are normal, covalent anhydride bonds. Hydrolysis of an anhydride is accompanied by a favorable increase in entropy, largely due to resonance stabilization of the product molecules: the electrons have one additional bond through which they can resonate. In addition, both the ADP and P_i products are acidic anions and the two negatively charged products, because of mutual charge repulsion, do not readily recombine. Consequently, the equilibrium constant for ATP hydrolysis is rather large—on the order of 10^5.

A large equilibrium constant helps to explain why ATP is so important in cellular metabolism. Under conditions that define standard free energy change (unit concentrations of products and reactants), the mass-action ratio is five orders of magnitude less than equilibrium. ATP hydrolysis has a correspondingly large standard free energy change ($\Delta G^{\circ\prime}$ = 31 kJ/mol). In cells, where the free nucleotide pool is relatively small and hydrolysis is influenced by the concentration of Mg^{2+}, the actual free energy change can be even higher! Photosynthesis and respiration serve to maintain a large pool of ATP and the observed mass-action ratio (Γ) can be maintained as low as 10^{-4}. This is nine orders of magnitude away from equilibrium (ΔG ≈ 56 kJ/mol). It is the cell's capacity to maintain the mass-action ratio so far from equilibrium that enables ATP to function as an energy store. Indeed, the capacity to avoid equilibrium is an essential characteristic of living organisms.

It is important to emphasize that it is the extent of displacement of G from equilibrium that defines the capacity of a reactant to do work, rather than any intrinsic property of the molecule. In the words of D. G. Nicholls, if the glucose-6-phosphate reaction ($\Delta G^{\circ\prime}$ = 13.8 kJ/mol) "were maintained ten orders of magnitude away from equilibrium, then glucose-6-phosphate would be just as capable of doing work in the cell as is ATP. Conversely, the Pacific ocean could be filled with an equilibrium mixture of ATP, ADP and P_i but the ATP would have no capacity to do work" (Nicholls, 1982). Of course, it goes without saying that the biochemistry of the cell is structured so as to use ATP, not glucose-6-phosphate, in this capacity.

The relationship between free energy and the ca-

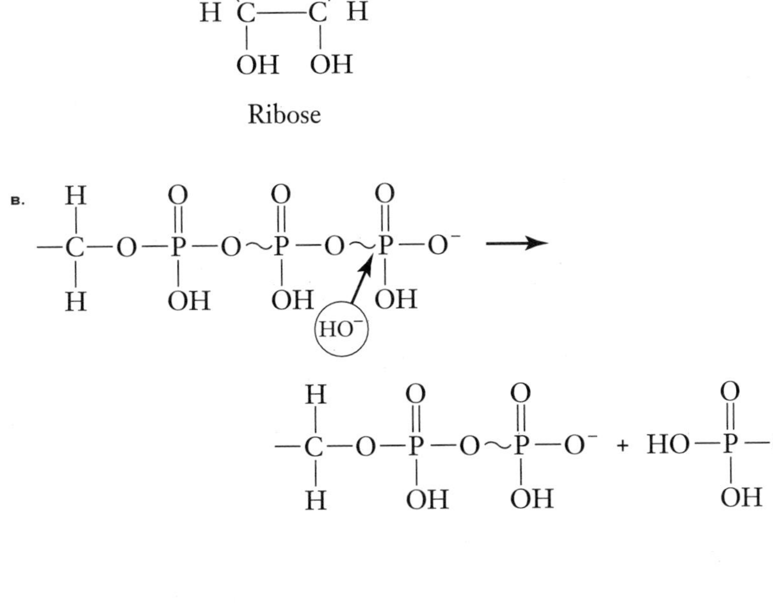

A.

Adenine

Ribose

Phosphate [×3]

B.

FIGURE 9.2 **Adenosine triphosphate (ATP). (*A*) The ATP molecule consists of adenine (a nitrogenous base), ribose (a sugar), and three terminal phosphate groups. (*B*) Hydrolysis of ATP yields ADP (adenosine diphosphate) plus an inorganic phosphate molecule. The terminal O-P bond in ADP is energetically equivalent to the terminal bond in ATP.**

pacity to do work is not restricted to chemical reactions. Any system not at equilibrium has a capacity to do work, such as an unequal distribution of solute molecules across a membrane. Solute gradients provide the energy for diffusion, for example—as was discussed earlier in Chapters 2 and 5.

FREE ENERGY OF OXIDATION-REDUCTION REACTIONS

Photosynthesis and respiration are electrochemical phenomena. Each operates as a sequence of oxidation-reduction reactions in which electrons are transferred from one component to another. Such reactions are known as **red**uction-**ox**idation or **redox** reactions. As an example, consider the reduction of 3-phosphoglyceric acid (PGA) to glyceraldehyde-3-P (GAP):

$$\text{PGA} + \text{NADPH} + \text{H}^+ \rightleftharpoons \text{GAP} + \text{NADP}^+ \quad (9.6)$$

Redox reactions may be conveniently dissected into two half-reactions involving the donation and acceptance of electrons. Thus the reduction of PGA to GAP may be considered as the two half-reactions:

$$\text{NADPH} \rightleftharpoons \text{NADP}^+ + \text{H}^+ + 2e^- \quad (9.7)$$

$$\text{PGA} + 2e^- + 2\text{H}^+ \rightleftharpoons \text{GAP} \quad (9.8)$$

A reduced/oxidized pair such as NADPH/NADP$^+$ is known as a **redox couple.** Note that although oxidation/reduction reactions are often shown as a transfer of protons, it is actually electrons that are exchanged. The structures of some principal redox compounds involved in biological electron transport are shown in Figure 9.3.

Since each of the half-reactions described above, as well as the overall reaction, is reversible, their free energies could be described on the basis of chemical equilibria. However, it is not clear how to treat the electrons, which have no independent existence. Moreover, our interest in redox couples is more in their tendency to accept electrons from or donate electrons to another couple, a tendency known as **redox potential.** Redox potentials allow the feasibility and direction of electron transfers between components in a complex system to be predicted. Indeed, in order to understand electron flow in photosynthesis and respiration it is necessary to have a working understanding of redox potential and how it is applied.

Like most reactions, redox potentials are defined against some arbitrary standard. The primary standard for redox potentials is the hydrogen half-cell

$$\text{H}_2 \rightleftharpoons 2\text{H}^+ + 2e^- \quad (9.9)$$

in which hydrogen gas is bubbled over a platinum electrode immersed in 1 M H$^+$ (pH = 0) (Fig. 9.4). All other redox couples are described on the basis of their capacity to either receive electrons from or donate electrons to the standard hydrogen half-cell. Redox potentials, measured in **volts,** can be determined by connecting a second half-cell, containing the couple of interest, to the reference (hydrogen) half-cell by a voltmeter. Since there would be no net exchange of electrons when one hydrogen half-cell is compared with another, the standard redox potential (E_0) of the hydrogen half-cell is zero.

The standard hydrogen half-cell is fine for chemists, but a pH of zero is not frequently encountered in biological systems. Here it is more convenient, and more relevant, to define redox potentials at pH 7.0. Redox potentials are sensitive to pH. For the generalized redox couple

$$\text{ox} + ne^- + m\text{H}^+ = \text{red} \quad (9.10)$$

the redox potential, at 25°C, becomes more negative at the rate of 2.3RT/F(m/n) mV per pH unit. R, the universal gas constant has a value of 8.31 J mol^{-1} °K^{-1}, T is absolute temperature (298°C), and F is the faraday constant (96 500 coulombs mol^{-1}). The volt, a measure of electrical potential, is defined as 1 J/coulomb. Thus when the ratio m/n = 1, the redox potential becomes more negative by 60 mV per pH unit.

FIGURE 9.3 The chemical structures of some common biological redox agents in the oxidized and reduced states. (*A*) Nicotinamide adenine dinucleotide (NAD) and nicotinamide adenine dinucleotide phosphate (NADP). Note that only the nicotinamide ring is changed in the reaction. The nicotinamide ring accepts two electrons but only one proton. Arrow indicates where the electrons are added to the nicotinamide ring. (*B*) Flavin adenine dinucleotide (FAD) consists of adenosine (adenine plus ribose) and riboflavin (ribitol plus isoalloxazine). Flavin mononucleotide (FMN) consists of riboflavin alone. Reduction occurs on the isoalloxazine moiety, which accepts two electrons and two protons. (*C*) Quinones. A quinone ring is attached to a hydrocarbon chain composed of 5-carbon isoprene units. The value of *n* is usually 9 for plastoquinone, found in chloroplasts, and 10 for ubiquinone, found in mitochondria. Reduction of the quinone ring is a two-step reaction. The transfer of one electron produces the partially reduced, negatively charged semiquinone (not shown). Addition of a second electron plus two protons yields the fully reduced hydroquinone form.

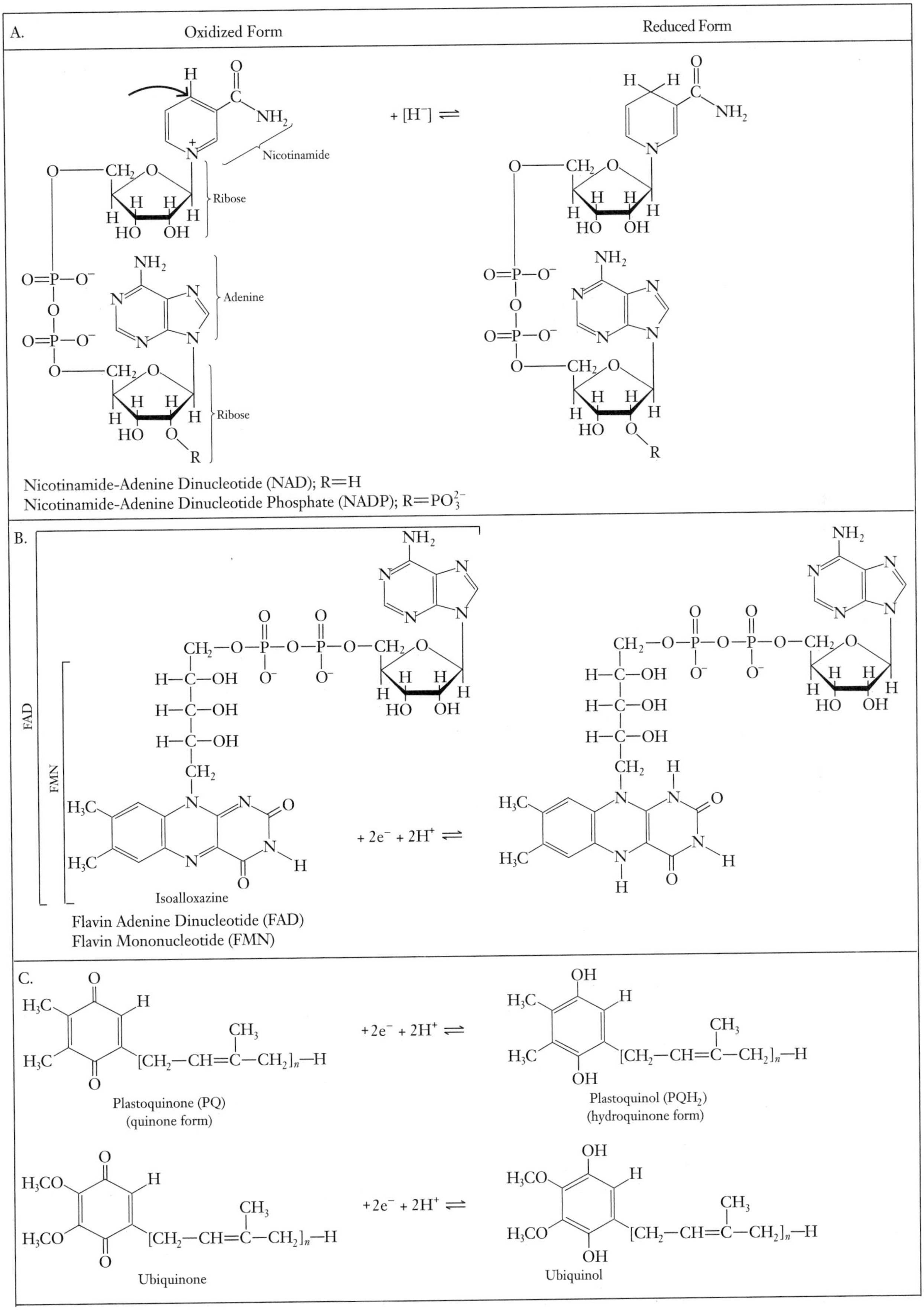

A. Oxidized Form | Reduced Form

Nicotinamide

Ribose

Adenine

Ribose

$+ [H^-] \rightleftharpoons$

Nicotinamide-Adenine Dinucleotide (NAD); R=H

Nicotinamide-Adenine Dinucleotide Phosphate (NADP); $R=PO_3^{2-}$

B.

FAD

FMN

Isoalloxazine

$+ 2e^- + 2H^+ \rightleftharpoons$

Flavin Adenine Dinucleotide (FAD)

Flavin Mononucleotide (FMN)

C.

$+2e^- + 2H^+ \rightleftharpoons$

Plastoquinone (PQ)
(quinone form)

Plastoquinol (PQH_2)
(hydroquinone form)

$+2e^- + 2H^+ \rightleftharpoons$

Ubiquinone

Ubiquinol

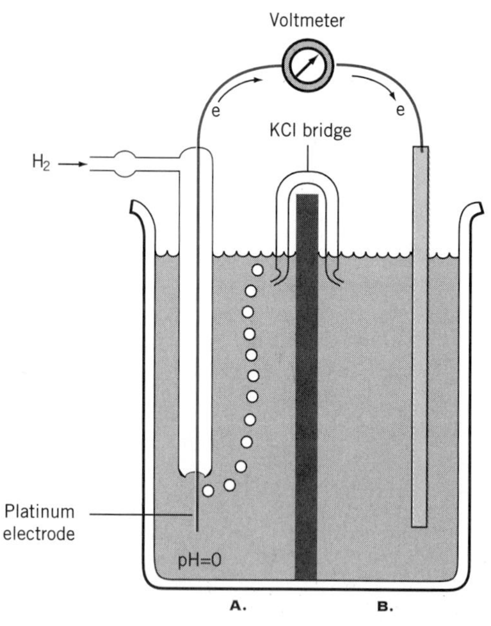

FIGURE 9.4 The potential (voltage) of an unknown redox couple (*B*) can be measured by its tendency to accept or donate electrons from a standard hydrogen half-cell (*A*).

From the above discussion, it is apparent that redox potentials are, like Gibbs free energy changes, dependent on the relative concentrations of products and reactants (as indicated in Fig. 9.1). The convention in biological energetics is to define standard redox potentials for a pH of 7.0 and a ratio of the reduced form to the oxidized form equal to one. Since under these conditions the redox couple is 50 percent reduced, the redox potential is known as the **midpoint potential** and is designated E_m.[1] Thus, although the standard redox potential (E_0) for the hydrogen half-cell is zero, the midpoint potential (E_m) for the $H_2/2H^+$ couple is $7 \times (-60) = -420$ mV. E_m values for selected redox couples are listed in Table 9.1. It is also convention in biological energetics to assign the more negative potentials to couples that have a tendency to *donate* electrons. Thus reduced ferredoxin is a strong electron donor, reducing agent or **reductant.** Ferredoxin is, conversely, a weak oxidant. By comparison, the water/oxygen couple is a strong electron acceptor or **oxidant** and a weak reductant.

The direction of electron transfer between redox couples can be predicted by comparing their midpoint potentials. Thermodynamically spontaneous electron transfer will proceed from couples with the more negative (less positive) redox potential to those with the less negative (more positive) redox potential. The energy-transducing membranes of bacteria, mitochondria, and chloroplasts all contain electron transport systems involving a number of electron carriers with different midpoint redox potentials (Fig. 9.2).

TABLE 9.1 Midpoint redox potentials for a selection of redox couples involved in photosynthesis and respiration.

	E_m (mV)
Ferredoxin red/ox	−430
$H_2/2H^+$	−420
$NADH + H^+/NAD^+$	−320
$NADPH + H^+/NADP^+$	−320
Succinate/fumarate	+30
Ubiquinone red/ox	+40
$Cyt\ c^{2+}/Cyt\ c^{3+}$	+220
$2H_2O/O_2$	+820

In addition to allowing us to predict direction of electron transfer, redox potentials also permit the calculation of Gibbs free energy changes for electron transfer reactions. This can be done using the following relationship:

$$\Delta G^{\circ\prime} = -nF\Delta E_m \tag{9.11}$$

where n is the number of electrons transferred. Biological electron transfers may involve either single electrons or pairs, but energy calculations are almost always based on $n = 2$. ΔE_m is the redox interval through which the electrons are transferred and is determined as

$$\Delta E_m = E_m\ (\text{acceptor}) - E_m\ (\text{donor}) \tag{9.12}$$

Thus for a transfer of electrons from NADH (the donor) to oxygen (the acceptor), as occurs in mitochondrial electron transport, $\Delta E_m = 0.82 - (-0.32) = 1.14$ V. Substituting this value in Equation 9.11, the value of $\Delta G^{\circ\prime}$ for a two-electron transfer from NADH to oxygen is -218 kJ mol^{-1}. Note that the sign of $\Delta G^{\circ\prime}$ is negative, indicating that this is a spontaneous electron transfer. For the reverse transfer, from water to $NADP^+$, the value of ΔE_m, and consequently $\Delta G^{\circ\prime}$, will be the same but the sign will be positive. In photosynthesis, light energy is used to drive this endergonic reaction.

A MODEL FOR ATP SYNTHESIS

The synthesis of ATP is coupled to electron transport reactions through an electrochemical proton gradient across the energy-transducing membranes.

It has been known for many years that the three principal energy-transducing membrane systems (in bacteria, chloroplasts, and mitochondria) were able to link electron transport with the synthesis of ATP. The mechanism, however, was not understood until Peter Mitchell proposed his **chemiosmotic hypothesis** in

[1]Strictly speaking the symbol $E_{m,7}$ is used to designate a midpoint potential at pH 7.0. However, in biological energetics a pH of 7.0 is understood and the symbol may be simplified to E_m.

1961 (Mitchell, 1961; Nicholls, 1982). Although not readily accepted by many biochemists in the beginning, Mitchell's hypothesis is now firmly supported by experimental results. In honor of his pioneering work, Mitchell was awarded the Nobel prize for chemistry in 1978.

Mitchell's hypothesis is based upon two fundamental requirements. First, energy-transducing membranes are impermeable to H^+. Second, electron carriers are organized asymmetrically in the membrane. The result is that, in addition to transporting electrons, some carriers also serve to translocate protons across the membrane against a proton gradient. The effect of these **proton pumps** is to conserve some of the free energy of electron transport as an unequal or *nonequilibrium* distribution of protons, or ΔpH, across the membrane.

In chloroplasts the protons are pumped across the thylakoid membrane, from the stroma into the lumen. The difference in proton concentration (ΔpH) across the membrane may be quite large—as much as three or four orders of magnitude. Since protons carry a positive charge, a ΔpH also contributes to an electrical potential gradient across the membrane. A trans-thylakoid ΔpH of 3.5, for example, establishes a potential difference of $3.5 \times 60\ \text{mV pH unit}^{-1} = 210$ mV at 25°. Together the membrane potential plus the proton gradient constitute a **proton motive force.** In order to pump protons into the lumen against a proton motive force of this magnitude, a large amount of energy is required. This energy is provided by the $-\Delta G$ of electron transport.

The direction of the proton motive force also favors the return of protons to the stroma, but the low proton conductance of the thylakoid membrane will not allow the protons to simply diffuse back. In fact, the return of protons to the stroma is restricted to highly specific, protein-lined channels that extend through the membrane and that are a part of the ATP synthesizing enzyme, **ATP synthase** (Fig. 9.5). This large (400 kDa) multisubunit complex, also known as **coupling factor** or CF_0-CF_1, consists of two multipeptide complexes. A hydrophobic complex called CF_0 is largely embedded in the membrane. Attached to the CF_0 on the stroma side is a hydrophilic complex called CF_1. The CF_1 complex contains the active site for ATP synthesis while the CF_0 forms a H^+ channel across the membrane, channelling the energy of the electrochemical proton gradient toward the active site of the enzyme. The interactions of the several subunits within each complex and the mechanism for harnessing the energy of the proton gradient is an active focus of current research.

When the electron transport complexes and the ATP synthesizing complex are both operating, a proton circuit is established (Fig. 9.6). The electron transport complex pumps the protons from the stroma into the lumen and thus establishes the proton gradient. At the same time, the ATP synthase allows the protons to return to the stroma. Some of the free energy of electron transport is initially conserved in the proton gradient.

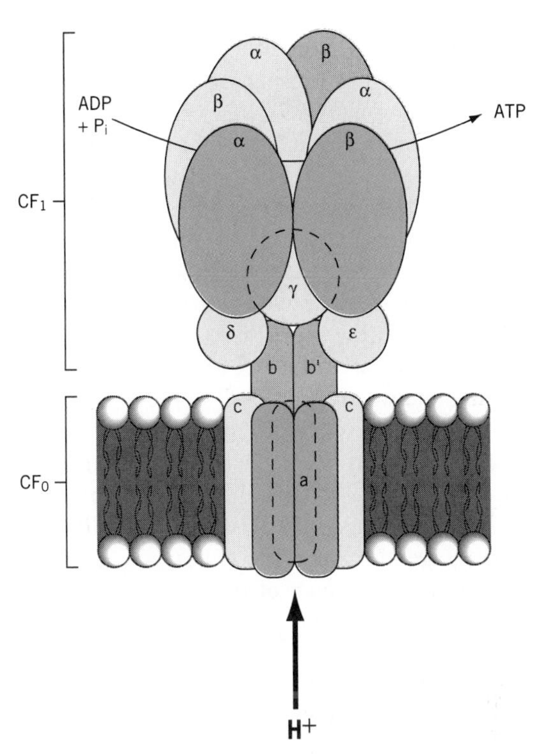

FIGURE 9.5 A model of the chloroplast ATP synthase. CF_0 is an integral membrane protein that forms a proton channel through the membrane. CF_1 is attached to the stromal side of CF_0 and contains the active site for ATP synthesis. CF_1 consists of five different subunits with a stoichiometry of α_3, β_3, γ, δ, ϵ. CF_0 consists of four different subunits with a probable stoichiometry of a, b, b′, c_{10}.

As the energy-rich, proton gradient collapses through the CF_0-CF_1 complex, that conserved energy is available to drive the synthesis of ATP.

From the beginning Mitchell had proposed that both oxidative phosphorylation and photophosphorylation were linked to a proton gradient across the appropriate membrane. Indeed some of the earliest evidence supporting chemiosmotic ATP synthesis was provided by A. Jagendorf and his colleagues, who were studying ATP synthesis in illuminated, washed thylakoids. Hind and

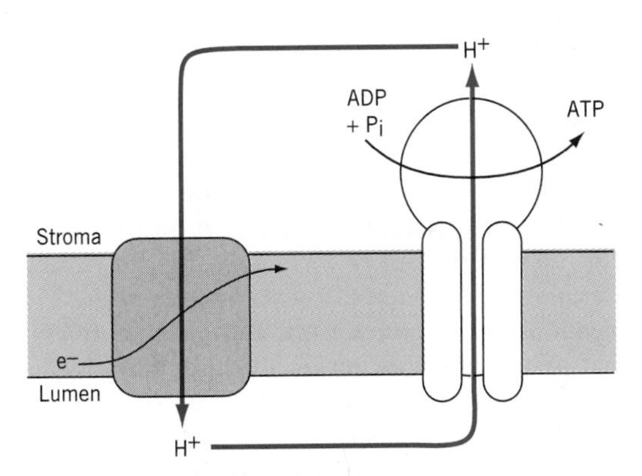

FIGURE 9.6 The coupling of electron transport system with ATP synthesis establishes a proton circuit.

Jagendorf (1963) had noted that illuminated thylakoids took up protons, thus causing an increase in the pH of the bathing solution (Fig. 9.7A). Furthermore, there appeared to be a correlation between the extent of proton uptake and the amount of ATP synthesized. Could a proton gradient be driving ATP synthesis as Mitchell had proposed? To test this idea, Jagendorf and Uribe (1966) devised a simple but very elegant experiment (Fig. 9.7B). Washed thylakoids were allowed to equilibrate with an acid solution (pH 4) in the dark. At equilibrium, the concentration of protons in the lumen would be the same as in the surrounding medium. Still in the dark, the thylakoids were rapidly transferred to a basic solution (pH 8) containing ADP and inorganic phosphate, immediately establishing a ΔpH of four units across the membrane. The result was a burst of ATP synthesis! With this experiment, Jagendorf and his colleagues had demonstrated that, in the absence of any light energy, a proton gradient could drive ATP synthesis in accordance with Mitchell's predictions.

An essential element of Mitchell's chemiosmotic hypothesis is the reversibility of the ATP synthase reaction. This means that under appropriate conditions CF_0-CF_1 and other similar complexes can use the negative free energy of ATP hydrolysis to *establish* a proton gradient. For example, both the plasma membrane and tonoplast contain **ATPase-proton pumps,** which pump protons out of the cell or the vacuole, as the case may be. The energy of ATP is thus conserved in the form of a proton gradient that may then be coupled to various forms of cellular work. ATPase proton pumps are a principal means of utilizing ATP to provide energy for the transport of other ions and small solute molecules across cellular membranes. Plasma membrane ATPase proton pumps are probably involved in auxin-induced cell expansion according to the acid growth hypothesis discussed in Chapter 17.

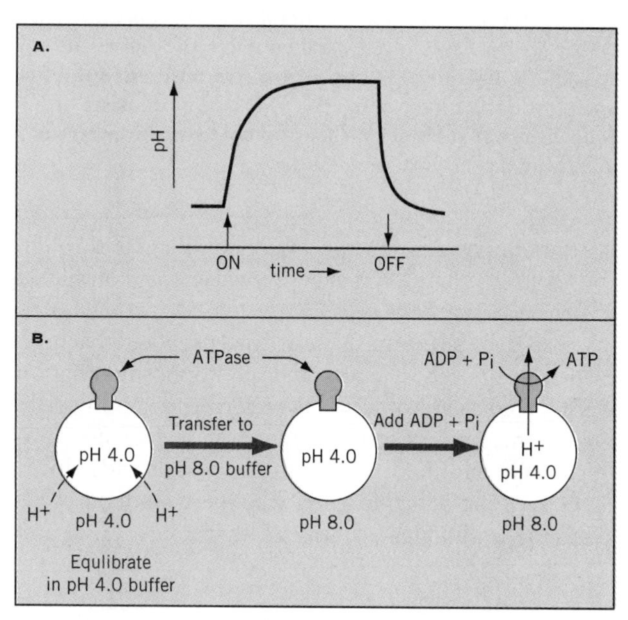

FIGURE 9.7 **Jagendorf's experiment demonstrating chemiosmotic synthesis of ATP by isolated chloroplast thylakoids. (*A*) Graph showing light-driven pH changes. When the light is turned on, pH on the stroma side of the thylakoid increases, indicating that protons are being transferred into the lumen. When the light is turned off, protons slowly leak back through the membrane. (*B*) Jagendorf's acid bath experiment. The proton content of the lumen is increased by equilibrating washed thylakoids in a low pH buffer in the dark. The thylakoids are then transferred to a high pH buffer, which artificially establishes a proton gradient. ADP and inorganic phosphate are quickly added to the bathing medium and the thylakoids generate ATP *in the dark*.**

Bioenergetics is a fundamental science. Its study is challenging but this discussion has, of necessity, been restricted to general principles. With this brief background in mind, however, we can now proceed to a discussion of energy conservation through the light reactions of photosynthesis.

ENERGY CONSERVATION IN PHOTOSYNTHESIS

Photosynthesis uses the energy of sunlight to photochemically reduce carbon dioxide, thus conserving energy in the sugars that are produced. Sugars are a form of portable energy that can be mobilized for use elsewhere in the plant.

PHOTOSYNTHESIS AS AN OXIDATION-REDUCTION REACTION

Although it may not be obvious at first glance, photosynthesis is fundamentally an oxidation-reduction reaction. This can be seen by examining the summary equation for photosynthesis introduced earlier in Chapter 8:

$$6CO_2 + 12H_2O \longleftrightarrow C_6H_{12}O_6 + 6O_2 + 6H_2O \quad (9.13)$$

Here photosynthesis is shown as a reaction between CO_2 and water to produce glucose, a six-carbon carbohydrate or hexose. Although glucose is not the first product of photosynthesis, it is a common form of accumulated carbohydrate and provides a convenient basis for discussion. Note that equal molar quantities of CO_2 and O_2 are consumed and evolved, respectively. This is convenient for the experimenter since it means that photosynthesis can be measured in the laboratory as either the uptake of CO_2 or the evolution of O_2.

For simplicity, we can reduce Equation 9.13 to

$$CO_2 + 2H_2O \longleftrightarrow (CH_2O) + O_2 + H_2O \quad (9.14)$$

where the term (CH_2O) represents the basic building block of carbohydrate. Equation 9.14 can be interpreted as a simple redox reaction, that is, a reduction of CO_2 to carbohydrate, where H_2O is the reductant and CO_2 is the oxidant. But it might also be interpreted as a hydration of carbon (e.g., carbo*hydrate*), as it was in early studies of photosynthesis. How do we know that it is one and not the other? And why is it necessary to write the equation with two molecules of water as reactant (and one as product) when one would appear to suffice? These questions can best be answered by reviewing some of the early studies on photosynthesis.

One of the earliest clues to the redox nature of photosynthesis was provided by studies of C. B. van Niel in the 1920s. As a microbiologist, van Niel was interested in the photosynthetic sulphur bacteria that use hydrogen sulphide (H_2S) as a reductant in place of water. Consequently, unlike algae and higher plants, the photosynthetic sulphur bacteria do not evolve oxygen. Instead, they deposit elemental sulphur according to the following equation:

$$CO_2 + 2H_2S \longleftrightarrow (CH_2O) + 2S + H_2O \quad (9.15)$$

The reaction in Equation 9.15 can also be written as two partial reactions:

$$2H_2S \longleftrightarrow 4e^- + 4H^+ + 2S \quad (9.16)$$

$$CO_2 + 4e^- + 4H^+ \longleftrightarrow (CH_2O) + H_2O \quad (9.17)$$

Equations 9.16 and 9.17 describe photosynthesis in the purple sulphur bacteria as a straightforward oxidation/reduction reaction. van Niel adopted a comparative biochemistry approach and argued that the mechanisms for **oxygenic** (i.e., oxygen-evolving) photosynthesis in green plants and **nonoxygenic** photosynthesis in the sulphur bacteria both followed the general plan:

$$2H_2A + CO_2 \longleftrightarrow 2A + (CH_2O) + H_2O \quad (9.18)$$

In this equation, A can represent either oxygen or sulphur, depending on the type of photosynthetic organism. According to Equation 9.18, the O_2 released in oxygenic photosynthesis would be derived from the reductant, water. Correct stoichiometry would therefore require the participation of four electrons and hence two molecules of water.

A second important clue was provided by R. Hill who, in 1939, was first to demonstrate the partial reactions of photosynthesis in *isolated* chloroplasts (Hill, 1939). In Hill's experiments with chloroplasts, artificial electron acceptors, such as ferricyanide, were used. Under these conditions, no CO_2 was consumed and no carbohydrate was produced, but light-driven reduction of the electron acceptors was accompanied by O_2 evolution:

$$4Fe^{3+} + 2H_2O \longleftrightarrow 4Fe^{2+} + O_2 + 4H^+ \quad (9.19)$$

Hill's experiments confirmed the redox nature of green plant photosynthesis and added further support for the argument that water was the source of evolved oxygen. Direct evidence for the latter point was finally provided by S. Ruben and M. Kamen in the early 1940s. Using either CO_2 or H_2O labeled with ^{18}O, a heavy isotope of oxygen, they showed that the label appeared in the evolved oxygen only when supplied as water ($H_2{}^{18}O$), not when supplied as $C^{18}O_2$. If the evolved O_2 is derived from water, then two molecules of water must participate in the reduction of each molecule of CO_2.

Based on these results, *photosynthesis can be viewed as a photochemical reduction of CO_2*. The energy of light is used to generate strong reducing equivalents from H_2O—strong enough to reduce CO_2 to carbohydrate. These reducing equivalents are in the form of reduced $NADP^+$ (or, NADPH + H^+). Additional energy for carbon reduction is required in the form of ATP, which is also generated at the expense of light. *The principal function of the light-dependent reactions of photosynthesis is therefore to generate the NADPH and ATP required for carbon reduction.* This is accomplished through a series of reactions that constitute the **photosynthetic electron transport chain.**

PHOTOSYNTHETIC ELECTRON TRANSPORT

The photosynthetic electron transport chain consists of a series of chlorophyll molecules and electron carriers arranged in multimolecular aggregates in the thylakoid membranes of the chloroplast.

PHOTOSYSTEMS AND REACTION CENTERS

The key to the photosynthetic electron transport chain is the presence of two large, multimolecular complexes known as **photosystem I (PSI)** and **photosystem II (PSII)** (Fig. 9.8). These two photosystems operate in

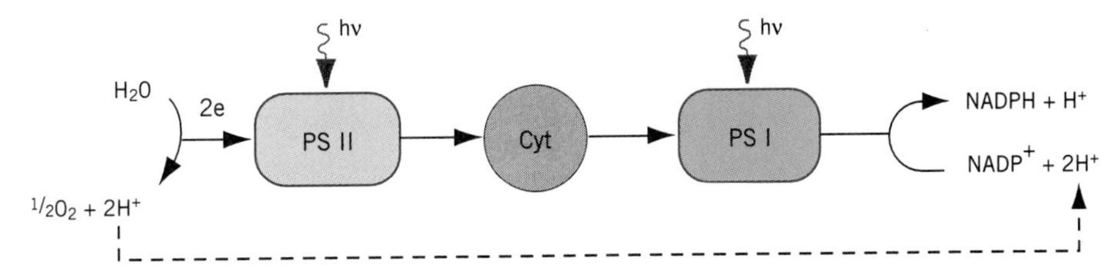

FIGURE 9.8 A linear representation of the photosynthetic electron transport chain. A sequential arrangement of three multimolecular membrane complexes extracts low-energy electrons from water and, using light energy, produces a strong reductant, NADPH + H^+.

Box 9.1
The Case for Two Photosystems

The photosynthetic unit of oxygenic photosynthetic organisms is organized as two separate **photosystems** that operate in series. While the two-step series formulation, or "Z-scheme," for photosynthesis is general knowledge today, the idea generated considerable excitement when it was first proposed in the early 1960s. The two-step idea was based on a series of experiments conducted during the 1950s, which laid the foundation for significant advances in our understanding of photosynthetic electron transport. The first of these experiments was centered around the concept of quantum efficiency.

Information about quantum efficiency is very useful when attempting to understand photochemical processes. Quantum efficiency can be expressed in two ways—either as **quantum yield** or as **quantum requirement.** Quantum yield (Φ) expresses the efficiency of a process as a ratio of the yield of product to the number of photons absorbed. In photosynthesis, for example, product yield would be measured as the amount of CO_2 taken up or O_2 evolved. Alternatively, the quantum requirement ($1/\Phi$) (sometimes referred to as quantum number) tells how many photons are required for every molecule of CO_2 reduced or oxygen evolved. Equation 9.17 identifies that a minimum of four electrons are required for every molecule of CO_2 reduced. In Chapter 7 it was established that one photon is required for each electron excited. Therefore, the minimum *theoretical* quantum requirement for photosynthesis is four. However, it has been well established *experimentally* that the minimum quantum requirement for photosynthesis is eight to ten photons for every CO_2 reduced. If eight photons are required (it is usual to assume the minimum) for four electrons, then *each electron must be excited twice!*

A second line of evidence, again from the laboratory of R. Emerson, was based on attempts to deter-

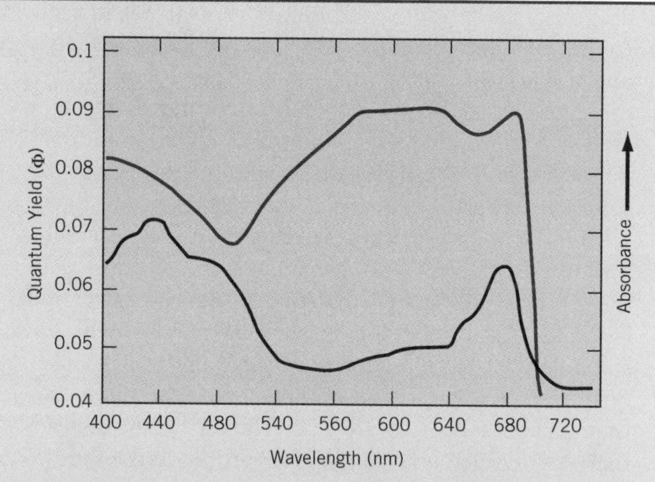

FIGURE 9.9 The Emerson "red drop" in the green alga *Chlorella*. Lower curve: Absorption spectrum of photosynthetic pigments. Upper curve: Action spectrum for quantum yield of photosynthesis. (Redrawn from the data of R. Emerson, C. M. Lewis, *American Journal of Botany* 30:165–178, 1943.)

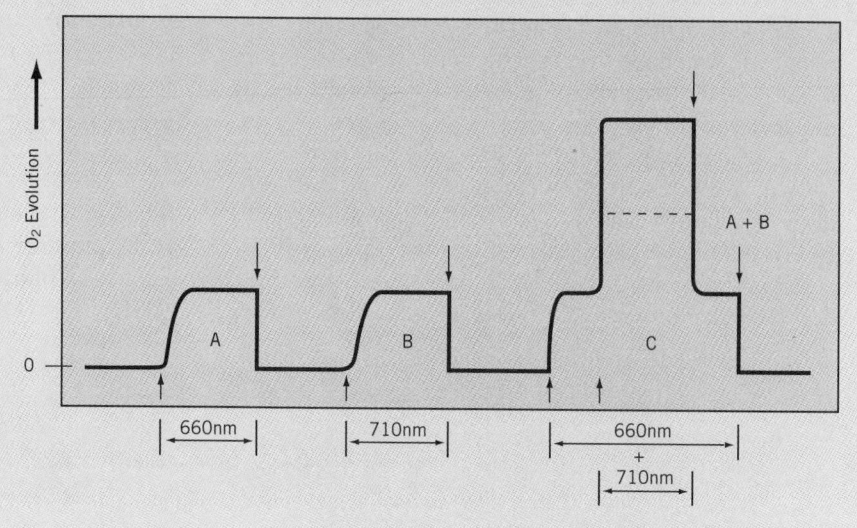

FIGURE 9.10 Schematic to illustrate the Emerson "enhancement effect." Two beams of light (660 nm and 710 nm) were presented either singly or in combination. Beam energies were adjusted to give equal rates of oxygen evolution. When presented simultaneously (*C*), the rate of oxygen evolution exceeded the sum of the rates when each beam was presented singly (*A* + *B*). Up arrows indicate light on. Down arrows indicate light off. (Reproduced with permission from the *Annual Review of Plant Physiology*, Vol. 22, copyright 1971 by Annual Reviews, Inc.)

mine the action spectra for photosynthesis in *Chlorella*. Emerson and his colleague C. M. Lewis reported in 1943 that the value of Φ was remarkably constant over most of the spectrum (Emerson and Lewis, 1943). This would indicate that any photon absorbed by chlorophyll was more or less equally effective in driving photosynthesis. However, there was an unexpected drop in the quantum yield at wavelengths greater than 680 nm, even though chlorophyll still absorbed in that range (Fig. 9.9). This puzzling drop in quantum efficiency in the long red portion of the spectrum was called the *red drop*.

In another experiment, Emerson and his colleagues set up two beams of light—one in the region of 650 to 680 nm and the other in the region of 700 to 720 nm. The fluence rates of both beams were adjusted to give equal rates of photosynthesis. Emerson discovered that when the two beams were applied simultaneously, the rate of photosynthesis was *two to three times greater than the sum of the rates* obtained with each beam separately! This phenomenon has become known as the **Emerson enhancement effect** (Fig. 9.10) (Emerson et al., 1957). The enhancement effect suggests that photosynthesis involves two photochemical events or systems, one driven by short wavelength light (≤680 nm) and one driven by long wavelength light (>680 nm). For optimal photosynthesis to occur, both systems must be driven simultaneously or in rapid succession.

In an attempt to explain conflicting information about the role of cytochromes and redox potential values, R. Hill and Fay Bendall, in 1960, proposed a new model for electron transport (Hill and Bendall, 1960). The Hill and Bendall model involved two photochemical acts operating in series; one serving to oxidize the cytochromes and one serving to reduce them (Fig. 9.11). The following year, L. Duysens confirmed the Hill and Bendall model, showing that cytochromes were oxidized in the presence of long wavelength light. The effect could be reversed by short wavelength light (Duysens et al., 1961).

Although the scheme has been significantly modified and considerable detail has been added since it was originally proposed, the Hill and Bendall scheme provided the catalyst that has led to our present understanding of photosynthetic electron transport and oxygen evolution.

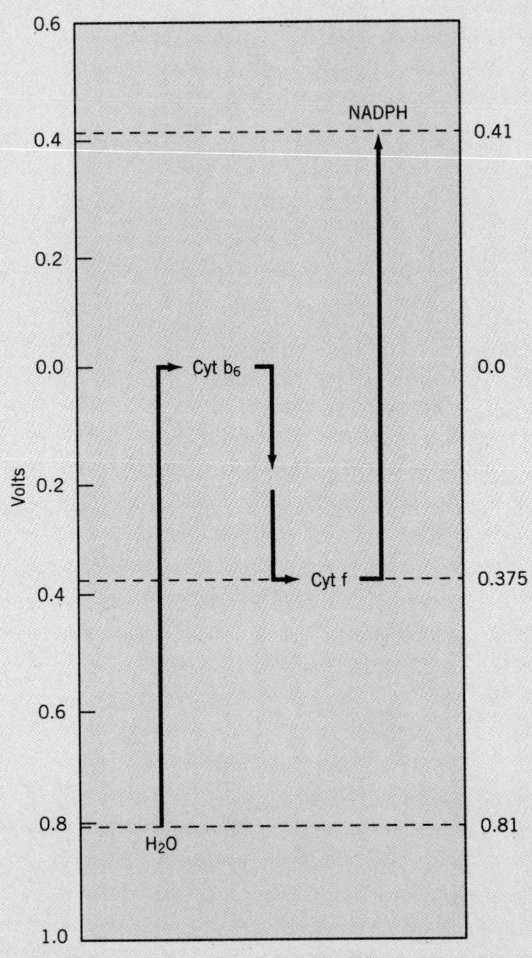

FIGURE 9.11 The Z scheme as originally proposed by Hill and Bendall. For a current version, see Figure 9.14. (Redrawn from Hill and Bendall, 1960.)

REFERENCES

Emerson, R., C. M. Lewis. 1943. The dependence of the quantum yield of *Chlorella* photosynthesis on wavelength of light. *American Journal of Botany* 30:165–178.

Emerson, R., R. Chalmers, C. Cederstrand. 1957. Some factors influencing the long-wave length limit of photosynthesis. *Proceedings of the National Academy of Science USA* 43:133–143.

Duysens, L. N. M., J. Amesz, B. M. Kamp. 1961. Two photochemical systems in photosynthesis. *Nature* 190:510–511.

Hill, R., F. Bendall. 1960. Function of the two cytochrome components in chloroplasts: A working hypothesis. *Nature* 186:136–137.

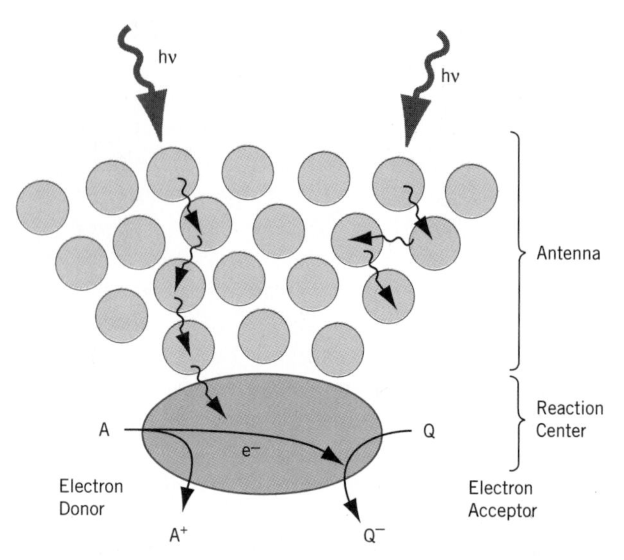

FIGURE 9.12 A photosystem consists of an antenna and a reaction center. Antenna chlorophyll molecules absorb incoming photons and funnel the excitation energy to the reaction center where the photochemical oxidation-reduction reaction occurs.

series linked by a third multiprotein aggregate called the **cytochrome complex.** Overall, the effect of the chain is to extract low-energy electrons from water and, using light energy trapped by chlorophyll, raise the energy level of those electrons to produce a strong reductant NADPH (see Box 9.1: The Case for Two Photosystems).

The composition, organization, and function of the photosynthetic electron transport chain has been an area of active study and rapid progress in recent years. This interest has led to the development of a variety of experimental methods for the study of PSI, PSII, and other large membrane protein aggregates. Most significant among these are techniques for the removal of the complexes from the thylakoid membranes by first solubilizing the membrane with a range of detergents (Leegood and Malkin, 1986). The different photosystems or classes of molecular aggregates can then be separated from each other by centrifugation. If the detergents and the conditions under which the treatments are carried out are carefully selected, not only can complexes be isolated but individual complexes can be further subdivided into smaller aggregates that retain varying parts of the overall activity. These purified complexes or subunits may then be analyzed for their composition with respect to pigments, protein, or other components (Markwell, 1986), or assayed for their capacity to carry out specific photochemical or electron transport reactions.

Such fractionation studies have revealed that PSI and PSII each contain several different proteins together with a collection of chlorophyll and carotenoid molecules that absorb photons (Äkerlund, 1993; Andersen and Scheller, 1993). The bulk of the chlorophyll in the photosystem functions as **antenna chlorophyll** (Fig. 9.12). The association of chlorophyll with specific proteins forms a number of different **chlorophyll-protein (CP) complexes,** which can be separated by gel electrophoresis and identified on the basis of their molecular mass. The antenna for photosystem II, for example, consists of two chlorophyll-proteins known as CP43 and CP47 (Fig. 9.13). These two CP complexes each contain 20 to 25 molecules of chlorophyll *a*. Antenna pigments absorb light but do not participate directly in photochemical re-

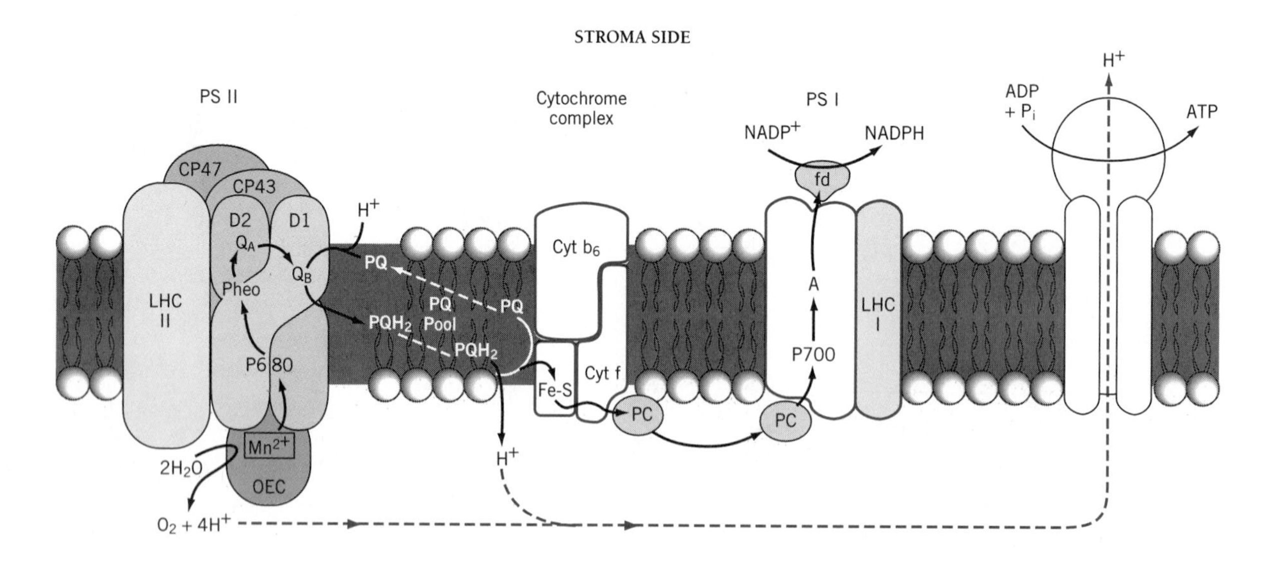

FIGURE 9.13 The organization of the photosynthetic electron transport system in the thylakoid membrane. See text for details.

actions. However, antenna chlorophylls lie very close together such that excitation energy can easily pass between adjacent pigment molecules by a radiationless transfer process (Chap. 7). The energy of absorbed photons thus migrates through the antenna complex, passing from one chlorophyll molecule to another until it eventually arrives at the **reaction center** (Fig. 9.12).

The reaction center consists of four to six molecules of chlorophyll *a*, called the **reaction center chlorophyll,** plus associated proteins and cofactors (Green and Durnford, 1996). The reaction center chlorophyll is, in effect, an energy sink—it is the longest wavelength, lowest energy-absorbing chlorophyll in the complex. Because the reaction center is the site of the primary photochemical redox reaction, it is here that light energy is actually converted to chemical energy. The reaction centers of PSI and PSII are designated as P700 and P680, respectively. These designations identify the reaction center as a species of chlorophyll *a*, or pigment (P), with an absorbance maximum at either 700 nm (PSI) or 680 nm (PSII). The efficiency of energy transfer through the antenna chlorophyll to the reaction is very high—only about 10 percent of the energy is lost.

The principal advantage of associating a single reaction center with a large number of antenna chlorophyll molecules is to increase efficiency in the collection and utilization of light energy. Even in bright sunlight it is unlikely that an individual chlorophyll molecule would be struck by a photon more than a few times every second. Since events at the reaction center occur within a millisecond time scale, any reaction center that depended on a single molecule of chlorophyll for its light energy would no doubt lie idle much of the time. Thus, the advantage of a photosystem is that while the reaction center is busy processing one photon, other photons are being intercepted by the antenna molecules and funnelled its way. This increases the probability that as soon as the reaction center is free, more excitation energy is immediately available. Plants growing under low light augment their antenna systems with even more chlorophyll in order to ensure efficient light harvesting. A principal function of this expanded antenna system is to facilitate harvesting available photons and keep the reaction center operating at near optimal rates. The augmented antenna complexes are known as **light-harvesting complexes.** The organization and function of light-harvesting complexes will be explored later in this chapter.

A schematic of the photosynthetic electron transport chain depicting the arrangement of PSI, PSII, and the cytochrome *b/f* complex in the thylakoid membrane is presented in Figure 9.13. A fourth complex—the CF_0-CF_1 coupling factor or ATP synthase—is also shown. All four complexes are membrane-spanning, integral proteins with a substantial portion of their structure buried in the hydrophobic lipid bilayer. Note that the orientation of the complex and their individual constituents is not random—specific polypeptide regions will be oriented toward the stroma or lumen respectively. Such a **vectorial** arrangement of proteins is characteristic of all energy transducing membranes, if not all membrane proteins, and is an essential element of their capacity to conserve energy. One particularly significant consequence of this arrangement is the directed movement of protons between the stroma and the thylakoid lumen as shown in Fig. 9.13. It is this arrangement that gives rise to the proton gradient necessary for ATP synthesis. This aspect of the electron transport chain will be revisited later. Another consequence of the vectorial arrangement is that the oxidation of water and reduction of $NADP^+$ occur on opposite sides of the thylakoid membrane. Water is oxidized and protons accumulate on the lumen side of the membrane where they contribute to the gradient which drives ATP synthesis. However, both NADPH and ATP are produced in the stroma where they are used in the carbon reduction cycle or other chloroplast activities.

Also shown in Figure 9.13 are two additional chlorophyll-protein complexes, depicted in close association with PSII and PSI. These are the light-harvesting complexes referred to earlier and are known as light-harvesting complex II (LHCII) and light-harvesting complex I (LHCI), respectively. LHCII is associated with PSII and LHCI is associated with PSI. As their names imply, the light-harvesting complexes function as extended antenna systems for harvesting additional light energy. LHCI and LHCII together contain as much as 70 percent of the total chloroplast pigment, including virtually all of the chlorophyll b. LHCI is relatively small, has a chlorophyll *a/b* ratio of about 4/1, and appears rather tightly bound to the core photosystem. LHCII, on the other hand, contains 50 to 60 percent of the total chlorophyll and, with a chlorophyll *a/b* ratio of about 1.2, most of the chlorophyll b. LHCII also contains most of the xanthophyll. The light-harvesting complexes have an important role in the dynamic regulation of energy distribution and electron transport (discussed later).

PHOTOSYSTEM II AND THE OXIDATION OF WATER

Electron transport actually begins with the arrival of excitation energy at the photosystem II reaction center chlorophyll, P680. The excited form of P680, designated P680*, is rapidly (within picoseconds, 10^{-12}s) photooxidized as it passes an electron to **pheophytin (Pheo)**. Pheophytin, considered the primary electron acceptor in PSII, is a form of chlorophyll *a* in which the magnesium ion has been replaced by two hydrogens. This initial photochemical act results in the formation of $P680^+$ and $Pheo^-$, a *charge separation.* This charge separation effectively stores light energy as redox energy and represents the actual conversion of light energy to chemical energy. It is essential that this charge separation be stabilized. If the electron were permitted to recombine with $P680^+$, there would be no forward movement of electrons, the energy would be wasted and, ultimately, carbon could not be reduced.

Charge stabilization in PSII is achieved in two ways. First, pheophytin quickly passes the electron on to another electron acceptor called Q_A and then to **plastoquinone (PQ)**. PQ is a quinone (see Fig. 9.3C) that binds transiently to a binding site (Q_B) on the D1 reaction center protein. The reduction of PQ to **plastoquinol** (PQH_2) decreases its affinity for the binding site. The plastoquinol is thus released from the reaction center, to be replaced by another molecule of PQ. The initial charge separation is further stabilized because $P680^+$ is a very strong oxidant (perhaps the strongest known in biological systems) and is able to "extract" electrons from water. Thus $P680^+$ is rapidly reduced, again within picoseconds, to P680, which is then ready for another excitation.

The electrons that reduce $P680^+$ are most immediately supplied by a cluster of four manganese ions associated with a small complex of proteins called the **oxygen-evolving complex (OEC).** As the name implies, the OEC is responsible for the splitting (oxidation) of water and the consequent evolution of molecular oxygen. The OEC is located on the lumen side of the thylakoid membrane. The OEC is bound to the D1 and D2 proteins of the PSII reaction center and functions to stabilize the manganese cluster. It also binds Cl^-, which is necessary for the water-splitting function (Hankamer et al., 1997).

It has been determined that only one PSII reaction center and OEC is involved in the release of a single oxygen molecule. Thus, in order to complete the oxidation of two water molecules, expel four protons, and produce a single molecule of O_2, the reaction center must be excited four times. Experiments in which electron transport was driven by extremely short flashes of light—short enough to excite essentially one electron at a time—have demonstrated that the OEC has the capacity to store charges. Each excitation of P680 is followed by withdrawal of one electron from the manganese cluster, which stores the residual positive charge. When four positive charges have accumulated, the complex oxidizes two molecules of water and releases the product oxygen molecule.

THE CYTOCHROME COMPLEX AND PHOTOSYSTEM I

Following its release from PSII, plastoquinol diffuses laterally through the membrane until it encounters a cytochrome b_6f complex. This is another multiprotein, membrane-spanning complex whose principal constituents are cytochrome b_6 (Cyt b_6) and cytochrome f (Cyt f). The cytochrome complex also contains **Rieske iron-sulphur (FeS) proteins**—iron-binding proteins in which the iron complexes with sulphur residues rather than a heme group as in the case of the cytochromes. Electrons from plastoquinol are passed first to the FeS protein and then to Cyt f. From Cyt f, the electrons are picked up by a copper-binding protein, **plastocyanin (PC).** PC is a small peripheral protein that is able to diffuse freely along the luminal surface of the thylakoid membrane.

In the meantime, a light-driven charge separation similar to that involving P680, has also occurred in the reaction center in PSI. In PSI, the reaction center chlorophyll, P700, is first excited to P 700*, then oxidized to $P700^+$. The primary electron electron acceptor (A) in PSI is a molecule of chlorophyll *a*; the electron is then passed through a quinone and additional FeS centers and finally, on the stroma side of the membrane, to **ferredoxin.** Ferredoxin is another FeS-protein that is soluble in the stroma.

Ferredoxin in turn is used to reduce $NADP^+$, a reaction mediated by the enzyme **ferredoxin-$NADP^+$-oxidoreductase.** Finally, the electron deficiency in $P700^+$ is satisfied by withdrawing an electron from reduced PC.

The overall effect of the complete electron transport scheme is to establish a continuous flow of electrons between water and $NADP^+$, passing through the two separate photosystems and the intervening cytochrome complex (Fig. 9.13). The energetics of this process are illustrated in Figure 9.14. In the process, electrons are removed from water, a very weak reductant ($E_m = 0.82$ V), and elevated to the energy level of ferredoxin, a very strong reductant ($E_m = -0.42$ V). Ferredoxin in turn reduces $NADP^+$ to NADPH ($E_m = -0.32$ V). NADPH, also a strong reductant, is a water-soluble, mobile electron carrier that diffuses freely through the stroma where it is used to reduce CO_2 in the carbon reduction cycle (Chap. 10). Since two excitations—at PSII and PSI—are required for each electron moved through the entire chain, a substantial amount of energy is put into the system. Based on one 680 nm photon (175 kJ per mol quanta) and one 700 nm photon (171 kJ per mol quanta), 692 kJ are used to excite each mole pair of electrons ($2 \times (175 + 171)$). Only a portion (about 32 percent) of that energy is conserved in NADPH (218 kJ mol^{-1}).

An additional portion of electron transport energy is conserved as ATP. This occurs in part because transfer of electrons between PSII and PSI is energetically downhill—that is, it is accompanied by a negative ΔG. In the process of moving electrons between plastoquinone and the cytochrome complex, some of that energy is used to move protons from the stroma side of the membrane to the lumen side. These protons contribute

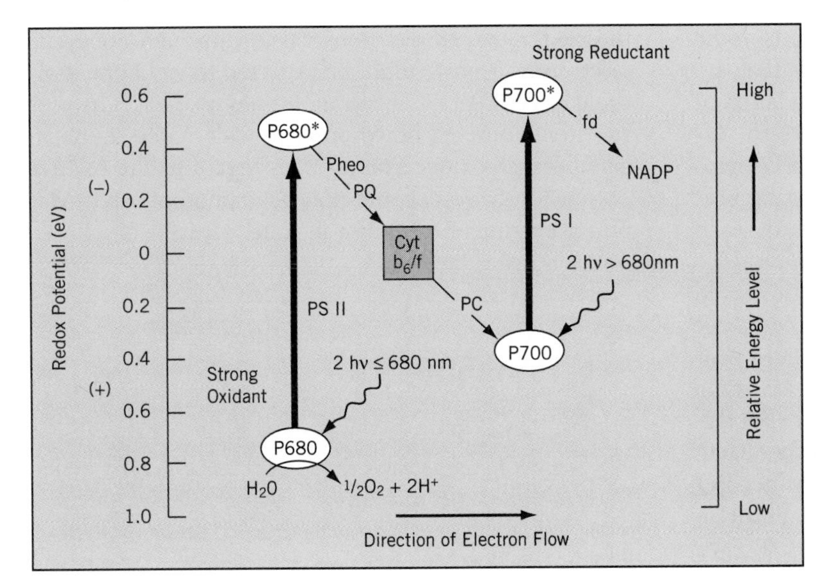

FIGURE 9.14 The Z scheme for photosynthetic electron transport. The redox components are placed at their approximate midpoint redox potential. The vertical direction indicates a change in energy level. The horizontal direction indicates electron flow. The net effect of the process is to use the energy of light to generate a strong reductant from low-energy electrons of water. The downhill transfer of electrons between P680* and P700 represents a negative free energy change. Some of this energy is used to establish a proton gradient which in turn drives ATP synthesis. Indicated redox potentials are only approximate.

to a proton gradient that can be used to drive ATP synthesis.

PHOTOPHOSPHORYLATION

The ATP required for carbon reduction and other metabolic activities of the chloroplast is synthesized by photophosphorylation in accordance with Mitchell's chemiosmotic mechanism.

Light-driven production of ATP by chloroplasts is known as **photophosphorylation.** Photophosphorylation is very important because, in addition to using ATP (along with NADPH) for the reduction of CO_2, a continual supply of ATP is required to support a variety of other metabolic activities in the chloroplast. These activities include synthesis of protein in the stroma and the transport of proteins and metabolites across the envelope membranes.

When electron transport is operating according to the scheme shown in Figures 9.13 and 9.14, electrons are continuously supplied from water and withdrawn as NADPH. This flow-through form of electron transport is consequently known as **noncyclic electron transport.** Formation of ATP in association with noncyclic electron transport is known as **noncyclic photophosphorylation.** However, as will be shown later, PSI units and PSII units in the membrane are not physically linked as implied by the Z scheme, but are even segregated into different regions of the thylakoid. One consequence of this heterologous distribution in the membranes is that PSI units may transport electrons independently of PSII, a process known as **cyclic electron transport** (Fig. 9.15). In this case ferredoxin transfers the electron back to PQ rather than to $NADP^+$. The electron then returns to $P700^+$, passing through the cytochrome complex and plastocyanin. Since these electrons also pass through PQ and the cytochrome complex, cyclic electron transport will also support ATP synthesis, a process known as **cyclic photophosphorylation.** Cyclic photophosphorylation is a source of ATP required for chloroplast activities over and above that required in the carbon reduction cycle.

A key to energy conservation in photosynthetic electron transport and the accompanying production of

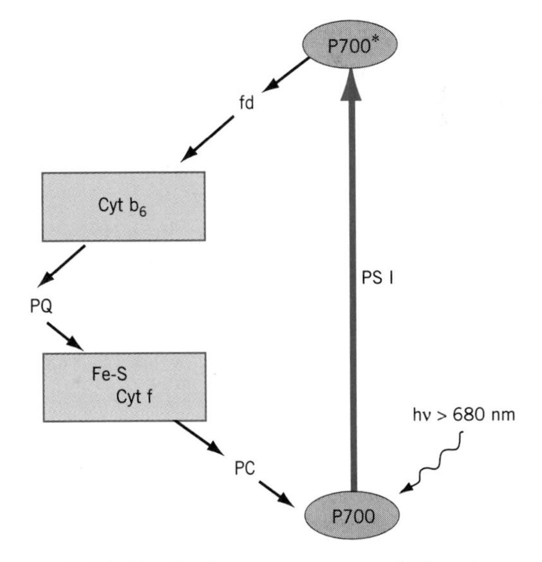

FIGURE 9.15 Cyclic electron transport. PSI units operating independently of PSII may return electrons from P700* to P700 through ferredoxin (fd), plastoquinone (PQ), and the cytochrome complex. In cyclic electron transport the cytochrome complex is involved both before and after the PQ, as the electrons are passed from fd through Cyt b_6 to PQ, and then back to P700 through the Rieske FeS center and Cyt *f*. No NADPH is produced by cyclic electron transport, but the energy can be used to produce ATP.

ATP is the light-driven accumulation of protons in the lumen. There are two principal mechanisms that account for this accumulation of protons: the oxidation of water, in which two protons are deposited into the lumen for each water molecule oxidized, and a PQ-cytochrome proton pump. The energy of the resulting proton gradient is then used to drive ATP synthesis in accordance with Mitchell's chemiosmotic hypothesis.

The precise mechanism by which protons are moved across the membrane by the cytochrome complex is not yet understood, although several models have been proposed. The most widely accepted model is known as the Q-cycle, based on an original proposal by Mitchell (Harold, 1986). A simplified version of the Q-cycle during steady-state operation is shown in Figure 9.16. When PQ is reduced by PSII, it binds temporarily to the D1 protein as a semiquinone and accepts an electron from Q_A. Two protons are picked up from the surrounding medium to produce a fully reduced plastoquinol, PQH_2. PQH_2 dissociates from the PSII complex and diffuses laterally through the membrane until it encounters a cytochrome *b/f* complex. There one electron is transferred to a Rieske FeS-protein and the two protons are expelled into the lumen. The electron then passes from the Rieske protein to cytochrome — to plastocyanin, which carries the electron to PSI. The second electron from PQH_2 reduces cytochrome *b*, which ultimately passes the electron back to a plastoquinone, reducing it to the semiquinone. Thus for *each* electron passing from plastoquinone to plastocyanin, *two* protons are translocated from the stroma into the lumen of the thylakoid. If this scheme is correct, then each pair of electrons passing through noncyclic electron transport contributes six protons to the gradient—four from the Q-cycle plus two from water oxidation. For cyclic electron transport, the number of protons would be four.

Since it is generally agreed that three protons must be transported through the CF_0-CF_1 for each ATP synthesized, a pair of electrons passing through noncyclic electron transport would be expected to yield one molecule of NADPH and two molecules of ATP. The precise stoichiometry, however, is difficult to determine, in part because of uncertainty with regard to the relative proportions of cyclic and noncyclic photophosphorylation occurring at any point in time.

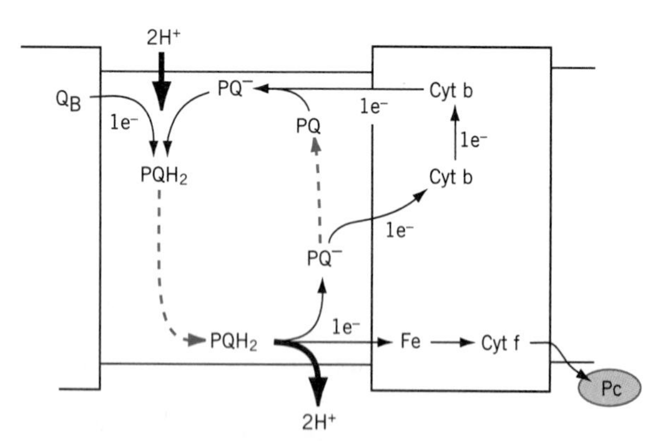

FIGURE 9.16 The Q-cycle, a model for coupling electron transport from plastoquinol to cytochrome with the translocation of protons across the thylakoid membrane. Two protons are translocated for each electron that passes through the electron transport chain.

LATERAL HETEROGENEITY OF THE ELECTRON TRANSPORT CHAIN

Photosystem I and photosystem II units and cytochrome complexes are unequally distributed within the fluid membranes of the grana and stroma thylakoids. Electrons are carried between complexes by mobile electron carriers.

In addition to the vectorial arrangement of electron transport components across the membrane, there is also a distinct **lateral heterogeneity** with respect to their distribution within the thylakoids (Fig. 9.17). The result is that PSI and PSII, for example, are spatially segregated, rather than arranged as some kind of supercomplex that might be suggested by the static representation in the previous figures. The PSI/LHCI complexes and the CF_0-CF_1 ATPase are located exclusively in nonappressed regions of the thylakoid; that is, those regions where the membranes are not paired to form grana. These regions include the stroma thylakoids, the margins of the grana stacks and membranes at either end of the grana stacks, all of which are in direct contact with the stroma (Fig. 9.17A). Virtually all of the PSII complexes and LHCII, on the other hand, are located in the appressed regions of the grana membranes (Fig. 9.17B). The cytochrome *b/f* complexes are uniformly distributed throughout both regions.

Spatial segregation also requires that the electron transport complexes be linked with each other through one or more mobile carriers that can deliver electrons between complexes. These carriers are plastoquinone (PQ), plastocyanin (PC), and ferredoxin. All three are mobile carriers that are not permanently part of any electron transport complex. Plastoquinone is a hydrophobic molecule and is consequently free to diffuse laterally within the lipid matrix of the thylakoid membrane. Its estimated diffusion coefficient is 10^6 cm^{-2} s^{-1}, which means that it could travel more than the diameter of a typical granum in less than one millisecond. The lateral mobility of PQ allows it to carry electrons between PSII and the cytochrome complex. Plastocyanin is a small (10.5 kDa) peripheral copper-protein found on the lumenal side of the membrane. It readily diffuses along the lumenal surface of the membrane and carries electrons between the cytochrome complex and PSI. Ferredoxin, a small (9 kDa) iron-sulphur protein, is found on the

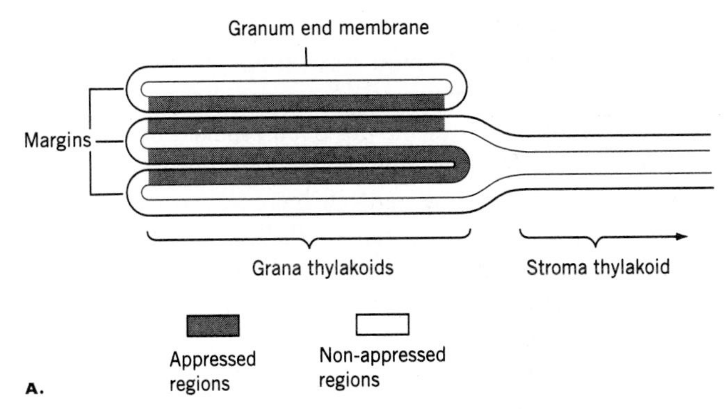

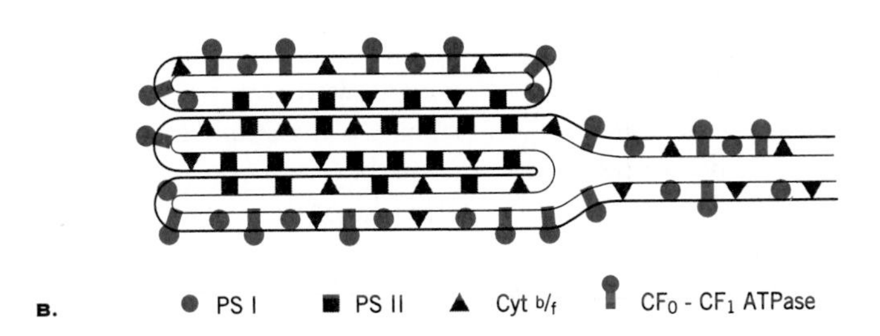

FIGURE 9.17 Lateral heterogeneity in the thylakoid membrane. (*A*) Nonappressed membranes of the stroma thylakoids, grana end membranes, and grana margins are exposed to the stroma. Appressed membranes in the interior of grana stacks are not exposed to the stroma. (*B*) PSII units are located almost exclusively in the appressed regions while PSI and ATP synthase units are located in nonappressed regions. The cytochrome complex, plastoquinone, and plastocyanin are uniformly distributed throughout the membrane system.

stroma side of the membrane. It receives electrons from PSI and, with the assistance of the ferredoxin-NADP oxidoreductase, reduces NADP to NADPH.

The extreme spatial segregation of PSI and PSII has several consequences. For example there is no *a priori* requirement for an equal number of PSI and PSII complexes. Indeed the ratio of PSI to PSII may vary widely between plant species. This ratio may also vary within individual plants as they adjust to changing light conditions in their environment (Anderson and Andersson, 1988). Furthermore, existing components often undergo rapid redistribution in response to sudden, short-term fluctuations in light and temperature.

An unequal number and spatial segregation of PSI and PSII also means that both cyclic and noncyclic photophosphorylation can occur more or less simultaneously. Thus the output of ATP and NADPH can be adjusted to meet the demands not only of photosynthesis but of other biosynthetic energy requirements within the chloroplast (see Chap. 8).

LIGHT-HARVESTING COMPLEXES AND DYNAMIC REGULATION OF PHOTOSYNTHESIS

Light-harvesting complexes are super antenna complexes that regulate the input of energy into photosystem II and photosystem I.

The large proportion of total chlorophyll associated with LHCII—over half the chlorophyll *a* and almost all the chlorophyll *b*—suggests a major role for LHCII. Yet LHCII is not photochemically active in the sense that it participates directly in a photochemical reduction as do PSII and PSI. LHCII has no reaction center and is not necessary for the survival of the plant. Mutants such as the *chlorina* mutant of barley (*Hordeum vulgare*), for example, have no detectable LHCII. These and similar chlorophyll *b*-less mutants do have the balance of the electron transport system, do carry out photosynthesis, and are fully viable plants. What, then, is the role of LHCII?

One role of LHCII is illustrated by comparing plants grown in full sunlight with plants grown in the shade. There are many differences between shade-grown and sun-grown leaves: sun leaves are thicker, with longer palisade cells or additional palisade layers, and generally smaller surface areas. Chloroplasts of shade-grown plants (or shaded leaves in large canopies) have more thylakoids with large grana and thus a higher proportion of appressed thylakoids (Anderson and Andersson, 1988). One of the more striking differences, however, is the high proportion of LHCII and, consequently, chlorophyll *b* in shade leaves. The larger antenna system provided by additional LHCII helps to intercept more light and thus increase the activity of PSII under conditions of low irradiance—much in the same way a large antenna improves radio or television reception in isolated areas where the signal is weak. Sun plants, on the other hand, have less LHCII but take advantage of the high irradiance by increasing the levels of the cytochrome *b/f* complex, plastoquinone, plastocyanin, ferredoxin, and CF_0-CF_1 per unit chlorophyll.

The differences between sun and shade plants are,

of course, long-term adaptations and require synthesis and turnover of chlorophyll-proteins and electron transport complexes. In addition to daily and seasonal variations in light, plants are often subject to extreme short-term fluctuations in fluence rate and light quality, which require rapid, short-term adaptations. Under natural conditions, the amount of light driving PSII and PSI is not necessarily balanced or consistent. Moreover, because of differences in the number of antenna molecules, their absorption coefficients, and a variety of other factors that influence absorption, the capacities of PSII and PSI to absorb light (often referred to as their *absorption cross-section*) are not equal. This might be expected to place unique constraints on the overall efficiency of photosynthesis. Optimal CO_2 reduction requires an efficient supply of NADPH and ATP that would, in turn, require a steady, balanced electron flow through PSII and PSI. But if light is not saturating and the delivery of excitation energy to PSII and PSI is not balanced, then the rate of electron transport and, consequently, photosynthesis will be limited by the photosystem receiving the least energy. Some mechanism for short-term regulation of excitation energy distribution between the two photosystems is required in order to maintain an efficient flow of electrons to $NADP^+$.

In addition to the inherent inequities of spectral distribution, plants often face other situations requiring rapid adjustments in the amount of energy being fed into PSII and PSI. Plants growing in the shade of a canopy, for example, are frequently subject to sudden transient fluctuations in fluence rate, known as **sunflecks.** A sunfleck is a spot of direct sunlight impinging on the leaf through an open gap in the canopy. Sunfleck lifetimes are variable; from a few seconds' duration due to wind flutter up to 20 minutes or longer under woodland canopies. Similar fluctuations occur when the sun suddenly reappears after having been blocked by extensive cloud cover. In situations such as these the leaf may be subject to as much as tenfold increases in energy that, particularly when directed to PSII, could have severe damaging effects.

One of the best understood mechanisms for short-term regulation of energy distribution is based on reversible phosphorylation of LHCII protein (Fig. 9.18). The phosphorylation of proteins is a ubiquitous mechanism for regulating many aspects of gene regulation and response to environmental stimuli in all eukaryote organisms (Bennett, 1991). The phosphorylation of proteins is catalyzed by a class of enzymes known as **protein kinases.** Chloroplasts contain a thylakoid membrane-bound protein kinase capable of phosphorylating LHCII. The activity of this kinase is sensitive to the redox state of the membrane and is activated when excess energy drives PSII, resulting in a buildup of reduced plastoquinone (PQH_2) (Larsson et al., 1987). The resulting phosphorylation of LHCII increases the negative charge of the protein, causing LHCII to dissociate from PSII. The same negative charge also loosens the appression of the thylakoid membranes in the grana stacks, freeing the LHCII to migrate into the PSI-rich stroma thylakoids. This shifts the balance of energy away from the PSII complexes, which remain behind in the appressed region, in favor of PSI. As the plastoquinone pool becomes reoxidized, due to increased PSI activity, the protein kinase is deactivated and a phosphatase enzyme dephosphorylates the LHCII, causing it to

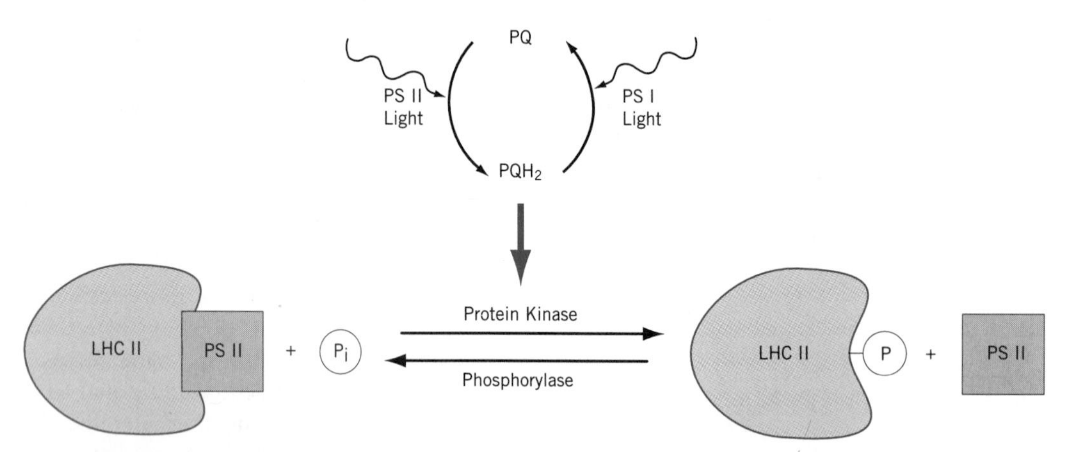

FIGURE 9.18 Reversible phosphorylation of LHCII. When PSII is overexcited relative to PSI, reduced plastoquinone (PQH_2) accumulates. A high level of PQH_2 activates a protein kinase that phosphorylates LHCII. Addition of a phosphate group weakens the interaction between LHCII and PSII, causing LHCII to dissociate from PSII. The input of light to PSII is diminished and PSII slows down, thus allowing PSI to remove excess PQH_2, which, in turn, deactivates the protein kinase. LHCII is dephosphorylated by the enzyme phosphorylase, allowing LHCII to reform an association with PSII.

migrate back into the appressed region and recombine with PSII. The net result is a very dynamic, continuous adjustment of excitation energy distribution between PSII and PSI. This continual adjustment of energy input in turn maintains an optimal flow of electrons through the two photosystems.

THE ROLE OF CAROTENOIDS IN PHOTOSYNTHESIS

Carotenoid pigments serve a dual function: collecting energy that can be used in photosynthesis and protecting chlorophyll against photodestruction in times of excess light.

The pigment-protein complexes of thylakoid membranes contain not only the chlorophyll pigments, but carotenes and xanthopylls as well (Chap. 7). The principal carotene in most higher plant species is β-carotene, although smaller amounts of α-carotene may be present in some species. The principal xanthophylls are lutein, violaxanthin, and zeaxanthin (see Fig. 7.12). The distribution of carotenoids throughout the pigment-protein complexes is not uniform; β-carotene appears to be concentrated in the antenna closely surrounding the PSI and PSII reaction centers, while the xanthophylls tend to predominate in the light-harvesting complexes. As well, the relative proportions of individual pigments, especially the xanthophylls, can vary significantly with environmental conditions.

It appears that carotenoids may serve two principal functions in photosynthesis: light harvesting and photoprotection (Demmig-Adams et al., 1996). Primarily on the basis of action spectra, it has long been believed that the principal function of carotenoids is to transfer absorbed light energy to chlorophyll. In this sense, the carotenoids serve a light-harvesting function and the energy they absorb is eventually used in photosynthesis.[2] Experimental techniques involving analysis of chlorophyll fluorescence have clearly established a light-harvesting role for a variety of carotenoids in the algae. Unfortunately, because technical problems limit the applicability of these techniques to leaves, the situation in higher plants is not as clear. Based on the limited information available, however, it has been suggested that either β-carotene or lutein *may be* capable of serving a light-harvesting role in higher plants.

In contrast to the putative light-harvesting role of carotenoids, there is now substantial evidence that carotenoids play an important role in protecting the photosynthetic system from photooxidative damage. Some of the earliest indications came from studies of albino (pigment-deficient) mutants of maize and other species. Many albino seedlings are, of course, mutant in the chlorophyll biosynthetic pathway. Others, however, are carotenoid-deficient mutants, that will bleach under normal conditions but green up when grown under low irradiance or in an oxygen-free atmosphere. Similar results are observed when seedlings are treated with norflurazon, which chemically inhibits carotenoid biosynthesis. In either case, the absence of carotenoids leads to the photooxidation of chlorophyll and other chloroplast constituents.

Photooxidation is a potential problem in genetically normal plants as well. During periods of peak irradiance, plants typically absorb more energy than they can utilize in the reduction of carbon dioxide. Rapidly growing crops, for example, may utilize no more than 50 percent of absorbed radiation, while other species, such as evergreens, may utilize as little as 10 percent. Any excess absorbed energy must be dissipated. If not, the reduced products of PSI, especially ferredoxin, may instead react with molecular oxygen to produce a toxic form of oxygen known as a superoxide radical (O_2^-). (A radical is a molecule with an unpaired electron.) The superoxide radical is highly reactive; it will oxidize, and thus destroy, not only chlorophyll but most organic molecules in the cell.

There are at least two ways to prevent photooxidation—either remove superoxide as it is formed or prevent its formation. An effective system for the removal of superoxide is the ubiquitous enzyme, **superoxide dismutase (SOD)**. SOD is found in several cellular compartments including the chloroplast. It is able to scavenge and inactivate superoxide radicals by forming hydrogen peroxide and molecular oxygen:

$$O_2^- + O_2^- + 2H^+ \longrightarrow H_2O_2 + O_2 \quad (9.20)$$

The H_2O_2 is in turn reduced to water by sequential reduction with ascorbate, glutathione, and NADPH. Reduction of H_2O_2 is necessary in order to prevent its reaction with O_2^- to form $OH^{\cdot}$, a highly toxic radical that can rapidly damage proteins.

Alternatively, formation of superoxide radical in chloroplasts can be prevented by trapping and dissipating excess excitation energy before it reaches the reaction center and enters the electron transport chain. Recent studies have established an important link between energy dissipation and the presence of the xanthophyll zeaxanthin (Demmig-Adams and Adams, 1996). Zeaxanthin is formed from violaxanthin by a process known as the **xanthophyll cycle** (Fig. 9.19). Violaxanthin is a diepoxide; it contains two epoxy groups, one on each ring. Under conditions of excess light, violaxanthin is converted to zeaxanthin through the removal of those two oxygens (de-epoxidation). De-epoxidation is step-

[2]In this role, the carotenoids have been referred to as *accessory pigments*. However, as it is now understood that the bulk of chlorophylls also serve in a light-harvesting capacity, the distinction of carotenoids as accessory pigments serves no useful purpose.

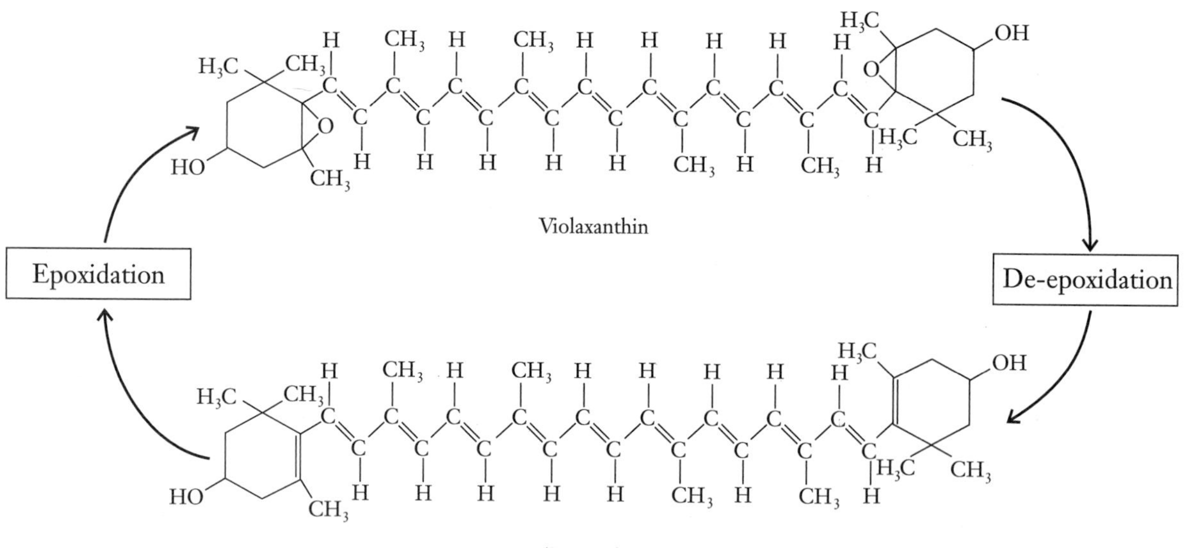

Violaxanthin

Epoxidation

De-epoxidation

Zeaxanthin

FIGURE 9.19 The xanthophyll cycle. De-epoxidation removes the two oxygen (epoxy groups) from the rings of violaxanthin. This is a two-step reaction with the intermediate antheraxanthin (not shown) containing only one epoxy group. De-epoxidation is induced by light, low pH, and reduced ascorbate. Note that in zeaxanthin, the number of carbon–carbon double bonds is increased by two. Zeaxanthin and antheraxanthin, but not violaxanthin, are able to accept a downhill energy transfer from excited chlorophyll. The reverse reaction, epoxidation, occurs in low light or darkness.

wise—removal of the first oxygen generates an intermediate monoepoxide (antheraxanthin). De-epoxidation is also induced by a low pH in the lumen, which is a normal consequence of electron transport under high light conditions. The reaction reverse in the dark as zeaxanthin is again epoxidated to form violaxanthin.

Although it has been established that the xanthophyll cycle plays a key role in photoprotection of the chloroplast, the precise molecular mechanism is not yet clear. It is known that carotenoids readily lose energy in the form of heat. It has also been shown that zeaxanthin with its increased number of conjugated carbon-carbon double bonds can accept a downhill transfer of energy from excited chlorophyll. These observations form the basis for the hypothesis that zeaxanthin in the antenna complexes receives excess energy directly from chlorophyll, dissipating it harmlessly as heat (Demmig-Adams and Adams, 1996). The xanthophyll cycle thus operates as an effective switch, generating zeaxanthin whenever dissipation of excess energy is required but removing the zeaxanthin under conditions of low irradiance when more of the energy is required for photosynthesis (Fig. 9.20).

The potential value of the xanthophyll cycle is evident in the way plants alter their carotenoid content in response to a variety of environmental conditions. For example, the carotenoid/chlorophyll ratio is typically higher in leaves exposed to full sun compared with shade-grown leaves (Demmig-Adams et al., 1996). Particularly striking is the relative increase in the propor-

FIGURE 9.20 Energy dissipation by zeaxanthin. Under low light, the energy of excited chlorophyll (Chl*) is preferentially allocated to photosynthesis (left). As irradiance increases, zeaxanthin (Z) is formed by de-epoxidation of violaxanthin (V) and an increasing proportion of excitation energy is transferred to zeaxanthin to be dissipated as heat (right).

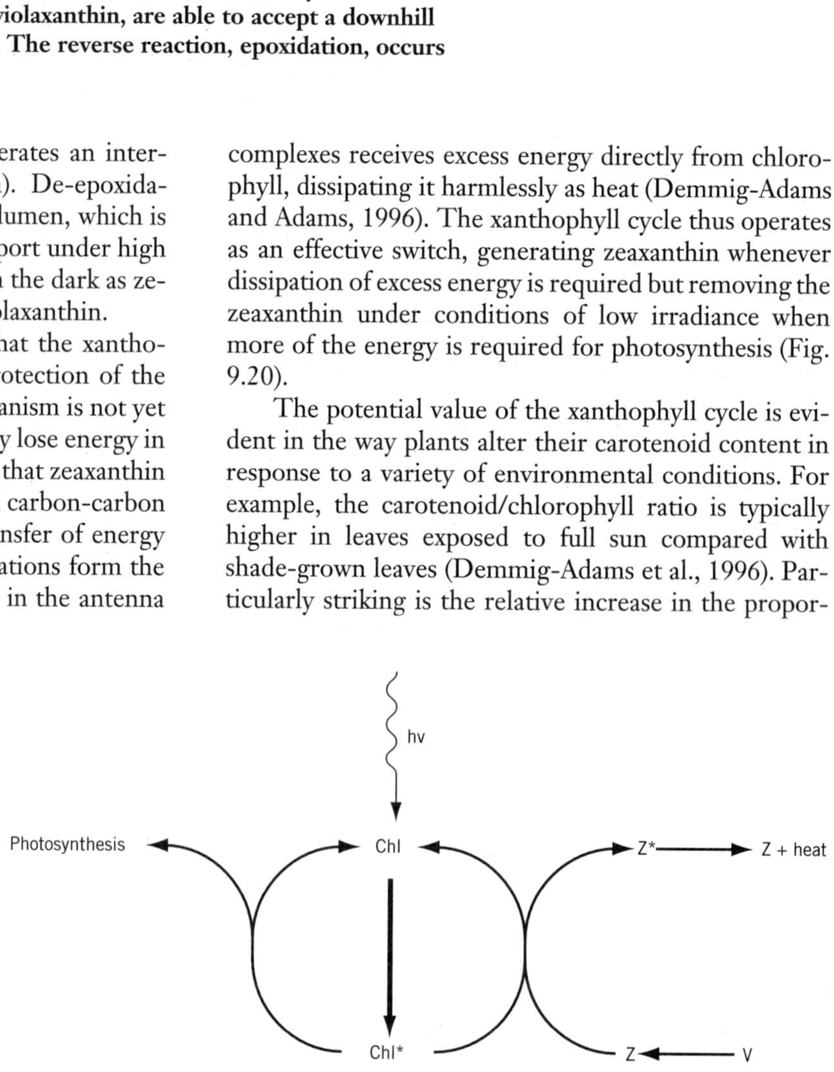

hv
Photosynthesis
Chl
Z* → Z + heat
Chl*
Z ← V

tion of xanthophyll cycle carotenoids. In one case, the proportion of violaxanthin, antheraxanthin, and zeaxanthin combined increased from 13 percent of total carotenoids in shade leaves to 32 percent in sun leaves. At the same time, the proportion of absorbed light allocated to photosynthesis *decreased* from 91 percent to 12 percent while the proportion allocated to dissipation as heat *increased* from 6 percent to 79 percent. Similar trends are evident in other environmental situations that lead to greater light stress. It appears that the xanthophyll cycle is a ubiquitous process for protecting the chloroplast against potentially damaging effects of excess light.

ELECTRON TRANSPORT AND WEED CONTROL

Inhibitors of photosynthetic electron transport are effective herbicides.

Since the dawn of agriculture, man has waged war against weeds. Weeds compete with crop species for water, nutrients, and light and ultimately reduce crop yields. Traditional methods of weed control, such as crop rotation, manual hoeing, or tractor-drawn cultivators were largely replaced in the 1940s by labor-saving chemical weed control. Modern agriculture now depends exclusively on the intensive use of **herbicides.**

A wide spectrum of herbicides is now available that interfere with a variety of cell functions. Many of the commercially more important herbicides, however, act by interfering with photosynthetic electron transport. Two major classes of such herbicides are **derivatives of urea,** such as monuron and diuron, and the **triazine herbicides,** triazine and simazine (Fig. 9.21). Both the urea and triazine herbicides are taken up by the roots and transported to the leaves. There they bind to the Q_B binding site of the D1 protein in PSII (also known as the herbicide-binding protein). The herbicide interferes with the binding of plastoquinone to the same site and thus blocks the transfer of electrons to plastoquinone. Because of its action in blocking electron transport at this point, DCMU is commonly used in laboratory experiments where the investigator wishes to block electron transport between PSII and PSI.

The triazine herbicides are used extensively to control weeds in cornfields, since corn roots contain an en-

Substituted Urea Herbicides

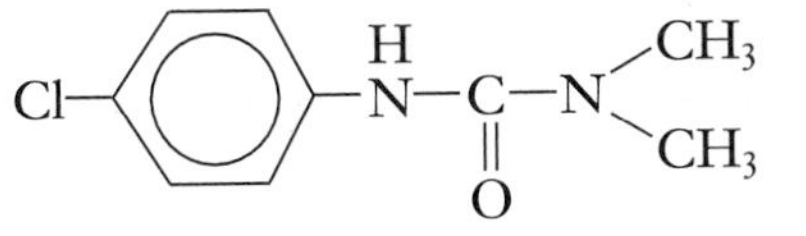

3-(p-Chlorophenyl)-1,1-dimethylurea
(common name: Monuron)

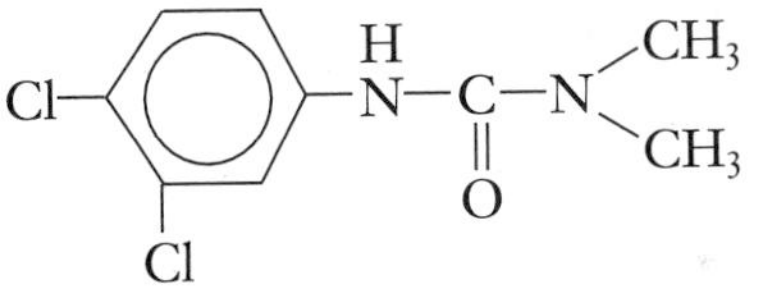

3-(3,4-Dichlorophenyl)-1,1-dimethylurea
(common names: Diuron, DCMU)

s-Triazine Herbicides

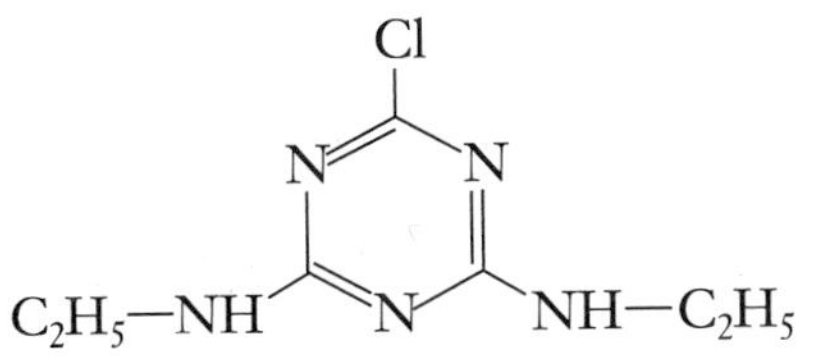

2-Chloro-4,6-bis(ethylamino)-s-triazine
(common name: Simazine)

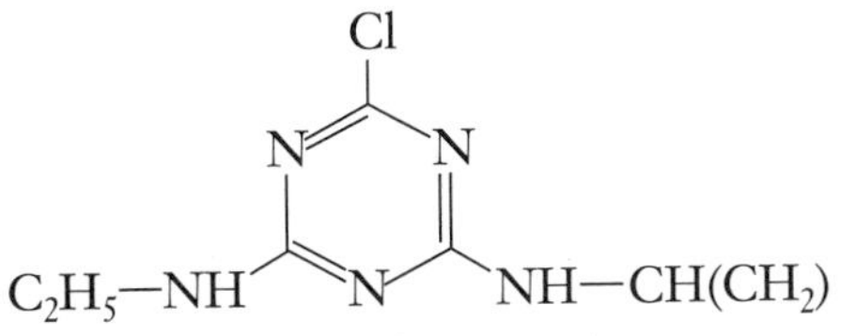

2-Chloro-4-ethylamino-6-isopropylamino-s-triazine
(common name: Atrazine)

Bipyridylium Herbicides

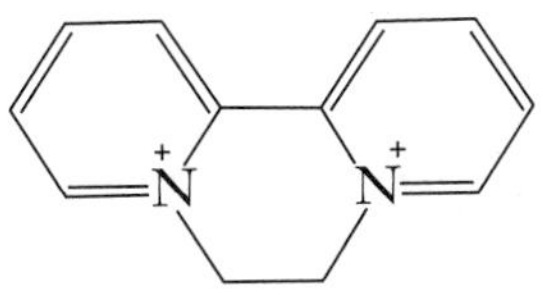

Diquat

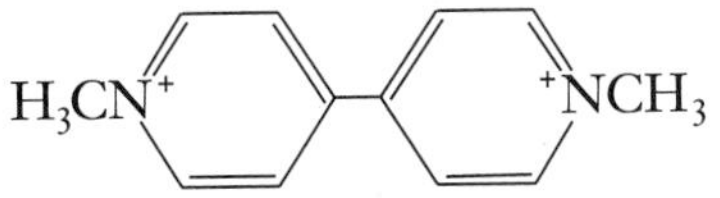

Paraquat
(methyl viologen)

FIGURE 9.21 The chemical structures of some common herbicides that act by interfering with photosynthesis.

zyme that degrades the herbicide to an inactive form. Other plants are also resistant. Some, such as cotton, sequester the herbicide in special glands while others avoid taking it up by way of root systems that penetrate deep below the application zones. In many cases, however, weeds have developed triazine-resistant races, or biotypes. In several cases, the resistance has been traced to a single amino acid substitution in the D1 protein. The change in amino acid reduces the affinity of the protein for the herbicide but does not interfere with plastoquinone binding and, consequently, electron transport.

The availability of herbicide-resistant genes together with recombinant DNA technology has stimulated considerable interest in the prospects for developing additional herbicide-resistant crop plants. It is possible, for example, to transfer the gene for the altered D1 protein into crop species and confer resistance to triazine herbicides. This approach will be successful, however, only if weed species do not continue to acquire resistance to the same herbicides through natural evolutionary change (see Chap. 23).

Another class of herbicides are the bipyridylium **viologen dyes**—diquat and paraquat (Fig. 9.21)—which act by intercepting electrons on the reducing side of PSI. The viologen dyes are autooxidizable, immediately reducing oxygen to superoxide. Thus the viologen dyes not only interfere with photosynthetic electron transport, but the superoxide they produce causes additional damage by rapidly inactivating chlorophyll and oxidizing chloroplast membrane lipids. Because viologen herbicides are also highly toxic to animals, their use is banned or tightly regulated in many jurisdictions.

Chemical herbicides have become an important management tool for modern agriculture, but their value as a labor-saving device must be carefully weighed against potentially harmful side effects.

SUMMARY

The application of thermodynamic laws to the study of energy flow through living organisms is called bioenergetics. There are two forms of energy, one that is available to do work (free energy) and one that is not (entropy). In a biochemical system such as living organisms, free energy is related to chemical equilibrium. The further a reaction is held away from equilibrium, the more work can be done. Cells utilize this principal to link energy-yielding reactions with energy-consuming reactions. In fact, life exists because cells are able to avoid equilibrium.

Most energy-exchange reactions to the cell are mediated by phosphorylated intermediates, especially ATP and related molecules. ATP is useful in this regard because it has a large equilibrium constant and is highly mobile within the cell. Because ATP is turned over rapidly, it is maintained far from equilibrium. In chloroplasts (and mitochondria), ATP synthesis is linked to an energy-rich proton gradient (a nonequilibrium proton distribution) across the energy-transducing thylakoid (or inner) membrane. The free energy of electron transport is used to establish the proton gradient and ATP is synthesized as the protons return through transmembrane ATP synthesizing complex.

The function of the light-dependent reactions of photosynthesis is to generate the ATP and reducing potential (as NADPH) required for subsequent carbon reduction. The electron transport chain in the thylakoid membrane is composed of two photosystems (PSI, PSII) and a cytochrome complex. The three complexes are linked by plastoquinone and plastocyanin, mobile carriers that freely diffuse within the plane of the membrane. Each photosystem consists of a reaction center and associated antenna and light-harvesting (LHC) complexes. Light energy gathered by the antenna and LHC is passed to the reaction center. In the reaction center, electron flow is initiated by a charge separation (photooxidation). As a result, electrons obtained from the oxidation of water are passed through PSII, the cytochrome complex, and PSI to $NADP^+$. Protons pumped across the membrane between PSII and PSI drive photophosphorylation. Several classes of economically important herbicides act by interfering with photosynthetic electron transport.

Lateral heterogeneity within the thylakoid membrane allows for adjustments to the relative numbers of PSI and PSII and for the output of ATP and NADPH. The presence of light-harvesting complexes allows chloroplasts to adjust to short-term changes in irradiance by regulating energy input into PSI and PSII.

Carotenoids may function in a light-harvesting mode, but their best-understood function is photoprotection. The capacity of zeaxanthin to absorb excess light energy and dissipate it as heat helps to protect chlorophyll against photooxidative damage.

CHAPTER REVIEW

1. The second law of thermodynamics states that the free energy of the universe tends toward a minimum or that entropy tends toward a maximum. This idea is sometimes referred to as as **entropic doom**. Explain what is meant by entropic doom.
2. British writer C. P. Snow has written that understanding the second law of thermodynamics is as

much a mark of the literate individual as having read a work of Shakespeare. Can you offer an explanation of what he means by this?

3. Explain the relationship between free energy and chemical equilibria.
4. ATP formation in chloroplasts is based on the step-wise *conservation* of energy. Trace the conservation of energy from the initial absorption of light by an antenna chlorophyll molecule to the final formation of a molecule of ATP.
5. Describe the concept of a photosystem and how it is involved in converting light energy to chemical energy.
6. Explain the difference between cyclic and non-cyclic electron transport. How can non-cyclic photosynthetic electron transport function if the PSII and PSI units are located in different regions of the thylakoid membrane?
7. What is LHCII and why is there more in the thylakoids of shade-grown leaves than in sun-grown leaves? How does LHCII work?
8. The herbicide DCMU is commonly used in laboratory investigations of electron transport reactions in isolated chloroplasts. Can you suggest why DCMU might be useful for such studies?

FURTHER READING

Andersson, B., S. Styring. 1991. Photosystem II: Molecular organization, function, and acclimation. *Current Topics in Bioenergetics* 16:1–81.

Cramer, W. A., P. N. Furbacher, A. Szczepaniak, G.-S. Tae. 1991. Electron transport between photosystem II and photosystem I. *Current Topics in Bioenergetics.* 16:179–217.

Golbeck, J. H., D.A. Bryant. 1991. Photosystem I. *Current Topics in Bioenergetics* 16:83–177.

Gregory, R. P. F. 1989. *Biochemistry of Photosynthesis.* New York: Wiley.

Harold, F. M. 1986. *The Vital Force. A Study of Bioenergetics.* New York: W. H. Freeman.

Nicholls, D. G., S. J. Ferguson. 1992. *Bioenergetics 2.* New York: Academic Press.

Ort, D. T., C. F. Yokum, eds. 1996. *Oxygenic Photosynthesis: The Light Reactions.* Dordrecht: Kluwer.

REFERENCES

Åkerlund, H-E. 1993. Function and organization of photosystem II. In: C. Sundqvist, M. Ryberg (eds.). *Pigment-Protein Complexes in Plastids: Synthesis and Assembly.* New York: Academic Press pp. 419–447.

Andersen, B., H. V. Scheller. 1993. Structure, function, and assembly of photosystem I. In: C. Sundqvist, M. Ryberg (eds.). *Pigment-Protein Complexes in Plastids: Synthesis and Assembly.* New York: Academic Press, pp. 383–418.

Anderson, J. M., B. Andersson. 1988. The dynamic photosynthetic membrane and regulation of solar energy conversion. *Trends in Biochemical Sciences* 13:351–355.

Bennett, J. 1991. Protein phosphorylation in green plant chloroplasts. *Annual Review of Plant Physiology and Plant Molecular Biology* 42:281–311.

Demmig-Adams, B., W. W. Adams. 1996. The role of xanthophyll cycle carotenoids in the protection of photosynthesis. *Trends in Plant Science* 1:21–26.

Demmig-Adams, B., A. M. Gilmore, W. W. Adams. 1996. *In vivo* functions of carotenoids in higher plants. *FASEB Journal* 10:403–412.

Green, B. R., D. G. Durnford. 1996. The chlorophyll-carotenoid proteins of oxygenic photosynthesis. *Annual Review of Plant Physiology and Plant Molecular Biology* 47:685–714.

Hankamer, Bz., J. Barber, E. J. Boekema. 1997. Structure and membrane organization of photosystem II from green plants. *Annual Review of Plant Physiology and Plant Molecular Biology* 48:641–671.

Hill, R. 1939. Oxygen produced by isolated chloroplasts. *Proceedings of the Royal Society of London, Series B* 127:192–210.

Hind, G., A. T. Jagendorf. 1963. Separation of light and dark stages in photophosphorylation. *Proceedings of the National Academy of Sciences USA* 49:715–722.

Jagendorf, A. T., E. Uribe. 1966. ATP formation caused by acid-base transition of spinach chloroplasts. *Proceedings of the National Academy of Sciences USA* 55:170–177.

Larsson, U. K., C. Sundby, B. Andersson. 1987. Characterization of two different subpopulations of spinach light-harvesting chlorophyll a/b-protein complex (LHCII): Polypeptide composition, phosphorylation pattern and association with photosystem II. *Biochimica Biophysica Acta* 894:56–68.

Leegood, R. C., R. Malkin. 1986. Isolation of subcellular photosynthetic systems. In: M. F. Hipkins, N. R. Baker (eds.), *Photosynthesis Energy Transduction: A Practical Approach.* Oxford: IRL Press, pp. 9–26.

Markwell, J. 1986. Electrophoretic analysis of photosynthetic pigment-protein complexes. In: M. F. Hipkins, N. R. Baker (eds.), *Photosynthesis Energy Transduction: A Practical Approach.* Oxford: IRL Press, pp. 27–49.

Mitchell, P. 1961. Coupling of phosphorylation to electron and hydrogen transfer by a chemiosmotic type of mechanism. *Nature* (London) 191:144–148.

Nicholls, D. G. 1982. *Bioenergetics.* New York: Academic Press.

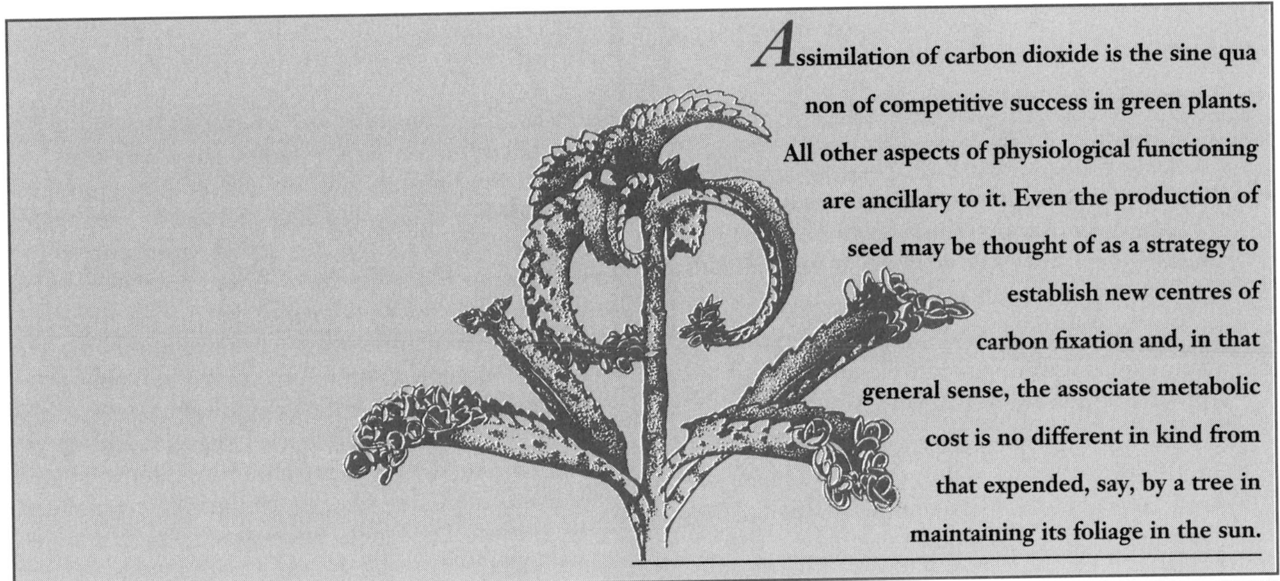

I. R. Cowan (1978)

10

Photosynthesis: Carbon Metabolism

The previous chapter showed how chloroplasts conserve light energy by converting it to reducing potential (NADPH) and ATP. In this chapter, attention is focused on the second half of photosynthesis—the metabolic pathways in which the NADPH and ATP produced by the light reactions are used to reduce inorganic CO_2 to organic carbon. These reactions have traditionally been designated the "dark reactions" of photosynthesis, but this designation is quite misleading, since it implies they can proceed in the absence of light. However, several critical enzymes in the carbon reduction cycle are light activated; in the dark they are either inactive or exhibit low activity. Consequently, carbon reduction cannot occur in the dark, even if energy could be made available from some source other than the photochemical reactions.

Only a few decades ago, knowledge of carbon metabolism was in its infancy and, as is to be expected when opening up new areas of study, understanding of the process was somewhat unsophisticated. Photosynthesis and respiration, long recognized as the two major divisions of carbon metabolism, were thought to be separate and independent metabolic processes, neatly compartmentalized in the cell. Photosynthesis was localized in the chloroplast while respiration appeared restricted to the cytoplasm and the mitochondrion. The task of photosynthesis was to reduce carbon and store it as sugars or starch. When required, these storage products could be mobilized, and exported to the mitochondrion where they were oxidized to satisfy the energy and carbon needs of the cell through respiration. Thus, the relationships between photosynthesis and respiration appeared simple and uncomplicated.

Over the past 40 years, however, knowledge and understanding of carbon metabolism has improved considerably and, with that, so has the apparent complexity. Photosynthetic carbon metabolism can no longer be explained by a single, invariable cycle. It is no longer restricted to just the chloroplast or even to a single cell. In addition to carbon reduction, photosynthetic energy is used to drive nitrogen assimilation, sulphate reduction, and other aspects of intermediary metabolism. The rate of photosynthesis is influenced and even controlled by events occurring outside the chloroplast and elsewhere in the plant. These and other complexities of metabolic integration within the cell and between different parts of the plant are only beginning to become obvious. One thing is certain: A more holistic approach to carbon metabolism is called for. The traditional, compartmentalized vision of independent processes no longer adequately explains carbon metabolism in plants.

This chapter will describe several interrelated variations of photosynthetic carbon metabolism in higher plants, including

- the path of carbon, energetics, and regulation of the photosynthetic carbon reduction cycle, or C3 metabolism—the pathway that all organisms ultimately use to assimilate carbon;
- photorespiration and how limitations are imposed on carbon assimilation in C3 plants by the photosynthetic carbon oxidation cycle;
- the biochemistry and ecological implications of a specialized carbon dioxide concentrating mechanism, known as the C4 syndrome, exhibited by a large number of tropical and subtropical plants;
- the biology of crassulacean acid metabolism (CAM), in which carbon dioxide uptake and carbon reduction are separated in time as means for improving water economy; and
- the synthesis of starch (for storage) in the chloroplast and sucrose (for export) in the cytosol, and some of the factors involved in regulating allocation of carbon between these two forms of carbohydrate.

Later chapters will address respiratory carbon metabolism and factors influencing the distribution of carbon throughout the plant.

THE PHOTOSYNTHETIC CARBON REDUCTION (PCR) CYCLE

The PCR cycle condenses a molecule of CO_2 with a 5-carbon sugar to produce two molecules of a 3-carbon sugar. Energy in the form of ATP and NADPH, from the light-dependent reactions, is required to reduce the product molecules and regenerate the 5-carbon acceptor molecule.

The pathway by which all photosynthetic eukaryotic organisms ultimately incorporate CO_2 into carbohydrate is known as carbon fixation or the **photosynthetic carbon reduction (PCR) cycle.** It is also referred to as the **Calvin cycle,** in honor of Melvin Calvin, who directed the research effort that elucidated the pathway. Mapping the complex sequence of reactions involving the formation of organic carbon and its conversion to complex carbohydrates represented a major advance in plant biochemistry. For his efforts and those of his associates, Calvin was awarded the Nobel Prize for chemistry in 1961.

The solution to the puzzle of carbon fixation was made possible largely because of two technological advances in the late 1940s and early 1950s: the discovery of radioactive carbon-14 (^{14}C) and the development of paper chromatography as an analytical tool. In order to trace the path of carbon it was necessary to provide CO_2 that was labeled in some way so that its fate in the cell could be followed. Early experiments were conducted with ^{11}C but its short half-life[1] (about 22 minutes) required that complex organic analyses be completed rather quickly before the radioactivity disappeared. The availability of ^{14}C with its much longer half-life (5730 *years*) clearly solved the time problem!

The technique of paper chromatography made it possible to separate complex mixtures of sugars and other small molecules. Following chromatographic separation, the radioactively labeled sugars could be detected by placing the paper chromatogram in contact with a sheet of X-ray film, a technique known as **autoradiography.** The radioactivity exposed the film which, when developed, would show the location of the labeled compounds as dark spots on the autoradiogram. The corresponding regions on the chromatogram could be cut out, the substance chemically identified, and the amount of radioactivity measured. The sugars could also be systematically degraded so that the distribution of radioactivity in the various carbon atoms could be determined. This was an important step in helping to understand how simple sugars were assembled to build more complex molecules.

THE CARBOXYLATION REACTION

Calvin's strategy for unraveling the path of carbon in photosynthesis was conceptually very straightforward: Identify the first stable organic product formed following uptake of radiolabeled CO_2. In order to achieve this, cultures of the photosynthetic green alga *Chlorella* were first allowed to establish a steady rate of photosynthesis. $^{14}CO_2$ was then introduced and photosynthesis continued for various periods of times before the cells were dropped rapidly into boiling methanol. The hot methanol served two functions: It denatured the enzymes, thus preventing any further metabolism, while at the same time extracting the sugars for subsequent chromatographic analysis. When the time of photosynthesis in the presence of $^{14}CO_2$ was reduced to as little as two seconds, most of the radioactivity was found in a three-carbon acid, **3-phosphoglycerate** (3-PGA). Thus 3-PGA appeared to be the first stable product of photosynthesis. Other sugars that accumulated the label later in time were probably derived from 3-PGA. Because Calvin's group determined that the first product was a three-carbon molecule, the PCR cycle is commonly referred to as the **C3 cycle.**

The next step was to determine what molecule served as the acceptor—the molecule to which CO_2 was added in order to make the three-carbon product. Systematic degradation of 3-PGA demonstrated that the ^{14}C label was predominantly in the carboxyl carbon. A

[1] A half-life is the time required for one-half of the material to decay.

Box 10.1
Carbohydrates—A Primer

Carbohydrates are a group of organic compounds containing carbon, hydrogen, and oxygen in the general proportions of 1:2:1. In other words, O and H are found in the same proportions found in water. The basic composition of carbohydrates is thus $(CH_2O)_n$. The simplest carbohydrates, which illustrate most of the basic features of this class of molecules, are the three-carbon **(triose)** sugars **glyceraldehyde** and **dihydroxyacetone.**

	Glyceraldehyde	Dihydroxyacetone
	H	
①	C=O	CH_2OH
②	CHOH	C=O
③	CH_2OH	CH_2OH

Note that each molecule contains a *carbonyl* oxygen (=O). In glyceraldehyde, the carbonyl oxygen forms an **aldehyde group** (—CHO) while in dihydroxyacetone it forms a **ketone** group (—C—CO—C—). Glyceraldehyde is also known as an *aldo sugar*, or **aldose,** and dihydroxyacetone is known as a *keto sugar*, or **ketose.** Carbohydrates with free aldehyde or ketone groups are capable of *reducing* cupric ion (Cu^{3+}) to cuprous ion (Cu^{2+}) under alkaline conditions. For this reason, they are known as **reducing sugars.** Because the other carbons all carry hydroxyl groups, carbohydrates may be broadly classified as *polyhydroxyaldehydes* or *polyhydroxyketones.*

Glyceraldehyde and dihydroxyacetone are also known as **monosaccharides** (L. *saccharum* = sugar), because the products of hydrolysis of these molecules are not carbohydrates. Examples of other common monosaccharides are the four-carbon (**tetrose**) *erythrose*, five-carbon (**pentose**) *ribose*, six-carbon (**hexose**) *glucose*, and seven-carbon (**heptose**) *sedoheptulose.*

	D-erythrose	D-ribose	D-glucose	D-sedoheptulose
①	H–C=O	H–C=O	H–C=O	
②				H_2C—OH
②				C=O
③	H—C—OH	H—C—OH	H—C—OH	
④	H—C—OH	H—C—OH	HO—C—H	HO—C—H
⑤	H_2C—OH	H—C—OH	H—C—OH	H—C—OH
⑥		H_2C—OH	H—C—OH	H—C—OH
⑦			H_2C—OH	H—C—OH
⑧				H_2C—OH

Triose sugars serve as important intermediates in both photosynthesis and respiration. Erythrose, the most common tetrose, is another intermediate in photosynthesis and some forms of respiration and serves as a precursor in several biosynthetic pathways. Pentoses, also important intermediates in photosynthesis and respiration, serve as structural elements in nucleic acids (both RNA and DNA), ATP, and key electron transport components such as the nicotinamide and flavin nucleotides. The hexoses include glucose and fructose and several other naturally occurring sugars. They are commonly considered the starting material and end products of respiration and photosynthesis, respectively, and are involved in the synthesis of more complex carbohydrates.

Sedoheptulose is the only common heptose, principally as an intermediate in photosynthesis.

The four most common hexoses found in plants are D-glucose, D-fructose, D-mannose, and D-galactose:

	D-glucose	D-fructose	D-mannose	D-galactose
①	H–C=O	CH_2OH	H–C=O	H–C=O
②	H—C—OH	C=O	HO—C—H	H—C—OH
③	HO—C—H	HO—C—H	HO—C—H	HO—C—H
④	H—C—OH	H—C—OH	H—C—OH	HO—C—H
⑤	H—C—OH	H—C—OH	H—C—OH	H—C—OH
⑥	CH_2OH	CH_2OH	CH_2OH	CH_2OH

Note the rather subtle differences between these molecules. Fructose differs from glucose because the former is a ketose while the latter is an aldose. Mannose is similar to glucose except for the orientation of the hydroxyl group on the third carbon. Similarly, galactose differs from glucose only in the orientation of the hydroxyl group on carbon four. These are examples of **stereoisomers.** Stereoisomerism arises because the four bonds of a carbon atom do not lie in a single plane as they are usually represented on paper, but form a three-dimensional tetrahedron:

As shown in the above illustration, when the carbon atom is attached to four *different* substituents, there are two possible configurations that are mirror images of each other, known as **enantiomers.** The carbons capable of forming enantiomers are known as **asymmetric** carbons. Glucose, mannose, and galactose each contain four asymmetric carbon atoms (carbons 2, 3, 4, 5). This allows for several different combinations that have differing chemical, physical, and biological properties. Note that in each of these sugars, the first and sixth carbons each have only three *different* kinds of substituents and are thus not asymmetric.

Note that the names of the four sugars above contain the prefix D-. This identifies the structure of the D-enantiomer, which, by convention, is drawn with the hydroxyl group on the *highest numbered* asymmetric carbon placed to the right of the carbon chain. If this hydroxyl is written to the left (and all other hydroxyl groups are similarly inverted), the designation is L. Most naturally occurring sugars are in the D configuration. One exception is L-galactose, a constituent of agar.

Asymmetric carbon atoms are also optically active; that is, in solution they can cause rotation of the plane of polarized light. Rotation to the right is known as *dextrorotary* (designated d, or +) and to the left, *levorotary* (l, or −). Unfortunately, there is no simple relationship between the D and L configurations and optical activity. Thus D-glucose is dextrorotary and D-fructose is levorotary.

Although the structures of sugars are often written as linear molecules, as they are above, sugars of five or more carbons normally exist as a five-carbon (*pyran*) or four-carbon (*furan*) **lactone ring:**

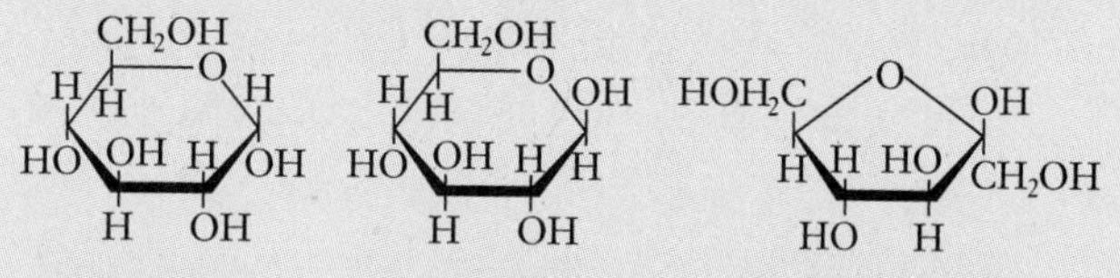

α-D-glucose
α-D-glucopyranose

β-D-glucose
α-D-glucopyranose

β-D-fructose
β-D-fructofuranose

Note that ring closure generates a hydroxyl group on carbon 1, thus creating an additional asymmetric center at carbon 1. Depending on the position of this hydroxyl group, the configuration is known as either α or β. When glucose molecules are linked together in a chain, the distinction between α and β is not trivial. For example, starch is comprised of α-D-glucose while cellulose is comprised of β-D-glucose. The physical and chemical properties of starch and glucose are very distinct.

Individual monosaccharides may be linked together to form **oligosaccharides.** These are identified by the number of monosaccharides found in the structure: disaccharides (2), trisaccharides (3), tetrasaccharides (4), and pentasaccharides (5). Oligosaccharides may be composed of either identical or nonidentical monosaccharides. The principal disaccharide found in higher plants is sucrose, formed by a condensation of glucose with fructose:

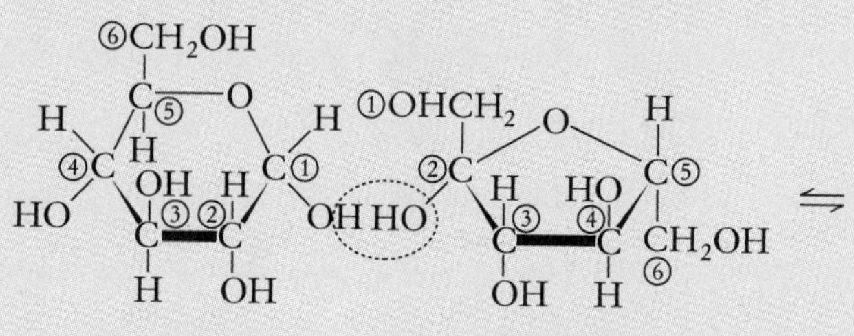

Glucose

Fructose

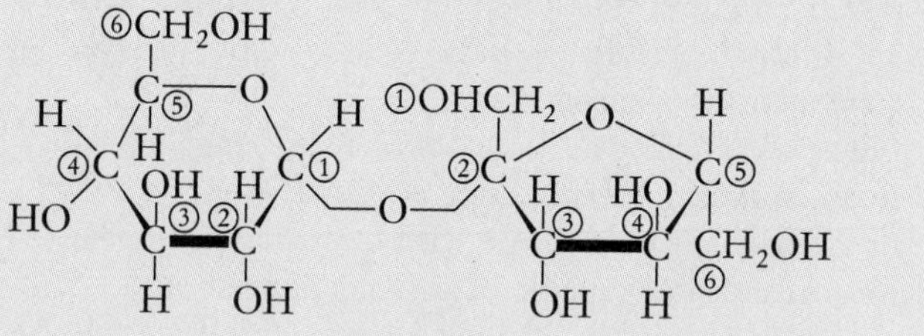

Sucrose

two-carbon acceptor molecule would be logical, but the search was long and futile. No two-carbon molecule could be found. Instead, Calvin recognized that the acceptor was the five-carbon keto sugar **ribulose-1,5-bisphosphate (RuBP)** (see Box 10.1). This turned out to be the key to the entire puzzle. The reaction is a carboxylation in which CO_2 is added to RuBP, forming a six-carbon intermediate (Fig. 10.1). The intermediate, which is transient and unstable, remains bound to the enzyme and is quickly hydrolyzed to *two* molecules of

FIGURE 10.1 The carboxylation reaction of the Photosynthetic Carbon Reduction Cycle.

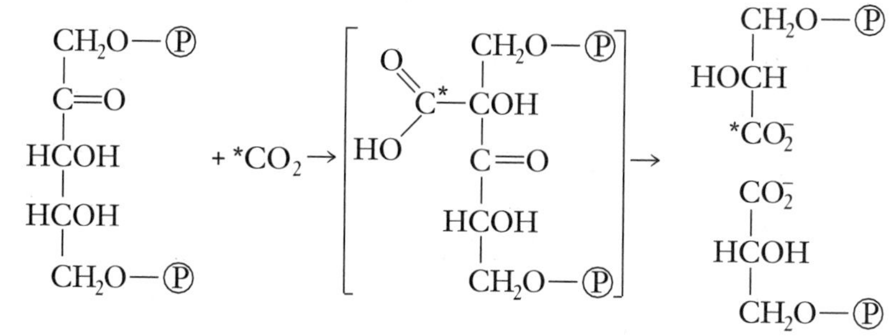

Ribulose-1,5-bisphosphate (RuBP)

3-phosphoglycerate (3-PGA)

Note that both reducing groups are involved in formation of a bond between the number 1 carbon of glucose and the number 2 carbon of fructose. This is designated as an α(1 → 2) link. Therefore, sucrose is a nonreducing sugar. Other disaccharides such as maltose (two glucose *residues* joined by an α(1 → 4) linkage) are reducing sugars.

Polysaccharides are large, high molecular weight polymers of monosaccharides. Polysaccharides usually consist of a single sugar, although polysaccharides consisting of more than one sugar are known. The two predominant polysaccharides in higher plants are starch, made up entirely of α(1→4) linked and α(1→6) branched D-glucose residues (see Fig. 10.17), and cellulose, made entirely of β-D-glucose residues. Starch and cellulose are both examples of **glucans.** Starch forms a helically coiled structure used primarily for storage of energy and carbon. Cellulose forms cross-linked arrays of linear molecules with a high degree of structural strength. Cellulose is probably the most abundant organic substance in the world and perhaps one of the most economically important.

Another important class of polysaccharides are the **fructans.** Fructans are polymers with different numbers of fructose molecules added to the fructose end of sucrose and are particularly common in grasses. Fructans built up of β(2→6)-links are known as **levans,** while those with β(1→6)-links are known as **inulins.** Inulin is an important sotage carbohydrate in Jerusalem artichoke (*Helianthus tuberosum*).

The aldehyde or ketone group of sugars may be reduced to an alcohol or oxidized to a carboxyl group. Reduction results in a polyhydric alcohol, or **sugar alcohol.** Reduction of glucose yields **sorbitol** and mannose yields **mannitol.** The acid form of glucose is **glucuronic acid** and of galactose, **galacturonic acid.**

The sugar alcohols occur frequently as storage carbohydrate, particularly in lower plants. Galacturonic acid is the monomer that makes up **pectic acid,**

Sorbitol	Mannitol	Glucuronic acid	Galacturonic acid
CH_2OH	CH_2OH	CHO	CHO
HCOH	HCOH	HCOH	HCOH
HOCH	HOCH	HOCH	HOCH
HCOH	HCOH	HCOH	HCOH
HCOH	HCOH	HCOH	HCOH
CH_2OH	CH_2OH	COOH	COOH

found in abundance in the middle lamella of cell walls.

3-PGA. The carboxylation reaction is catalyzed by the enzyme **ribulose-1,5-bisphosphate carboxylase-oxygenase**, or **Rubisco.** Rubisco is without doubt the most abundant protein in the world, accounting for approximately 50 percent of the soluble protein in most leaves. The enzyme also has a high affinity for CO_2 that, together with its high concentration in the chloroplast stroma, ensures rapid carboxylation at the normally low atmospheric concentrations of CO_2.

ENERGY INPUT IN THE PCR CYCLE

The carboxylation reaction, with a $\Delta G^{\circ\prime}$ of -35 kJ mol^{-1}, is energetically very favorable (Leegood et al., 1985). This poses an interesting question. If the equilibrium constant of the reaction favors carboxylation with such a high negative free energy change, where is the need for an input of energy from the light reactions of photosynthesis? Energy is required at two points: first for the reduction of 3-PGA and second for regeneration of the RuBP acceptor molecule. Each of these requirements will be discussed in turn.

Reduction of 3-PGA In order for the chloroplast to continue to take up CO_2, two conditions must be met. First, the product molecules (3-PGA) must be continually removed and, second, provisions must be made to maintain an adequate supply of the acceptor molecule (RuBP). Both require energy in the form of ATP and NADPH.

The 3-PGA is removed by *reduction* to the triose phosphate, **glyceraldehyde-3-phosphate.** This is a two-step reaction (Fig. 10.2) in which the 3-PGA is first phosphorylated to 1,3-bisphosphoglycerate, which is then reduced to glyceraldehyde-3-phosphate (G3P). Both the ATP and the NADPH required in these two

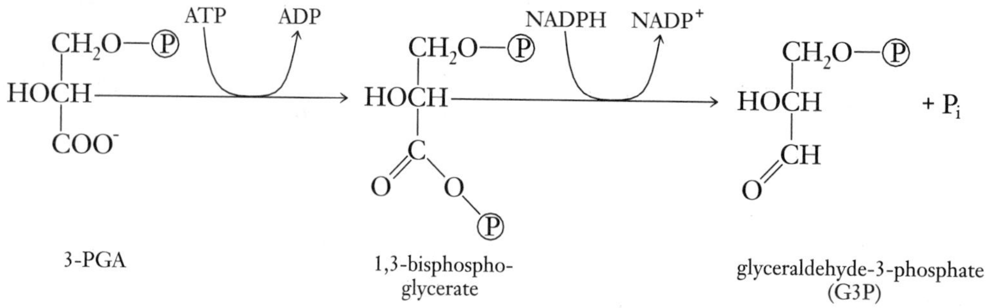

$$\text{3-PGA} \xrightarrow{\text{ATP} \rightarrow \text{ADP}} \text{1,3-bisphosphoglycerate} \xrightarrow{\text{NADPH} \rightarrow \text{NADP}^+} \text{glyceraldehyde-3-phosphate (G3P)} + P_i$$

FIGURE 10.2 Reduction of phosphoglyceric acid (PGA) to glyceraldehyde-3-phosphate (G3P).

steps are products of the light reactions and together represent one of two sites of energy input. The resulting triose sugar-phosphate, G3P, is available for export to the cytoplasm, probably after conversion to dihydroxyacetone phosphate (DHAP). Once in the cytoplasm, the triose molecules can easily be joined to synthesize hexoses, fructose-phosphate and glucose-phosphate. These two hexose-phosphates then combine to form sucrose-phosphate. These reactions will be described in more detail later in the chapter.

Regeneration of the Acceptor Molecule In order to maintain the process of CO_2 reduction, it is necessary to ensure a continuing supply of the acceptor molecule, RuBP. This is accomplished by a series of reactions involving 4-, 5-, 6-, and 7-carbon sugars (Figs. 10.3, 10.4). These reactions include the condensation of a 6-carbon fructose-phosphate with a triose-phosphate to form a 5-carbon sugar and a 4-carbon sugar. Another triose joins with the 4-carbon sugar to produce a 7-carbon sugar. When the 7-carbon sugar is combined with a third triose-phosphate, the result is two more 5-carbon sugars. All of the five-carbon sugar can be isomerized to form ribulose-5-phosphate (Ru5P). Ru5P can, in turn, be phosphorylated to regenerate the required ribulose-1,5-bisphosphate.

The net effect of these reactions is to recycle the carbon from five out of every six G3P molecules, thus

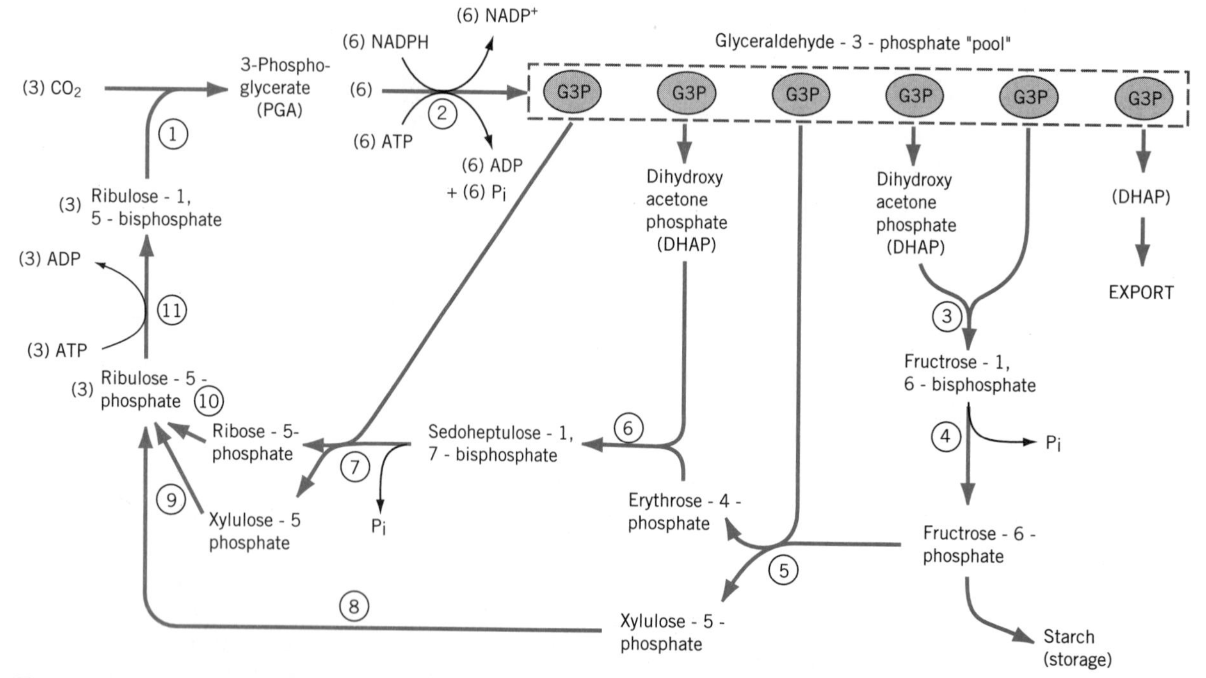

FIGURE 10.3 The photosynthetic carbon reduction (PCR) cycle. Numbers in brackets indicate stoichiometry. Enzymes, indicated by circled numbers are: (1) ribulose-1,5-bisphosphate carboxylase/oxygenase (Rubisco); (2) 3-phosphoglycerate kinase and glyceraldehyde-3-phosphate dehydrogenase; (3) aldolase; (4) fructose-1,6-bisphosphatase; (5) transketolase; (6) aldolase; (7) sedoheptulose-1,7-bisphosphatase; (8, 9) ribulose-5-phosphate epimerase; (10) ribose-5-phosphate isomerase; (11) ribulose-5-phosphate kinase.

$$
\begin{array}{lrcl}
(1) & 3\,\text{RuBP} + 3\,CO_2 & \longrightarrow & 6\,\cancel{\text{PGA}} \\
(2) & 6\,\cancel{\text{PGA}} & \longrightarrow & 5\,\cancel{\text{G3P}} + \text{G3P} \\
(3) & 2\,\cancel{\text{G3P}} & \longrightarrow & \cancel{\text{FBP}} \\
(4) & \cancel{\text{FBP}} & \longrightarrow & \cancel{\text{F6P}} \\
(5) & \cancel{\text{G3P}} + \cancel{\text{F6P}} & \longrightarrow & \cancel{\text{E4P}} + \cancel{\text{Xu5P}} \\
(6) & \cancel{\text{G3P}} + \cancel{\text{E4P}} & \longrightarrow & \cancel{\text{SBP}} \\
(7) & \cancel{\text{G3P}} + \cancel{\text{SBP}} & \longrightarrow & \cancel{\text{R5P}} + \cancel{\text{Xu5P}} \\
(8,9) & 2\,\cancel{\text{Xu5P}} & \longrightarrow & 2\,\cancel{\text{R5P}} \\
(10) & \cancel{\text{R5P}} & \longrightarrow & \text{RuBP} \\
(11) & 2\,\cancel{\text{R5P}} & \longrightarrow & 2\,\text{RuBP}
\end{array}
$$

$$\text{SUM} = 3\,\text{RuBP} + 3\,CO_2 \rightarrow 3\,\text{RuBP} + \text{G3P}$$

FIGURE 10.4 Summary reactions of the PCR cycle. Three turns of the cycle result in the regeneration of 3 molecules of the acceptor ribulose-1,5-bisphosphate (RuBP) plus an additional molecule of glyceraldehyde-3-phosphate (G3P). Additional abbreviations are: PGA, 3-phosphoglyceric acid; FBP, fructose-1,6-bisphosphate; F6P, fructose-6-phosphate; E4P, erythrose-4-phosphate; XuP, xyulose-5-phosphate; SBP, sedoheptulose-1,7-bisphosphate; R5P, ribulose-5-phosphate.

regenerating three RuBP molecules to replace those used in the earlier carboxylation reactions. The summary reactions shown in Figures 10.3 and 10.4 include three molecules of RuBP on each side of the equation. This is to emphasize that the cycle serves to regenerate the original number of acceptor molecules and maintain a steady state carbon reduction. Figures 10.3 and 10.4 show that for every three turns of the cycle (i.e., the uptake of three CO_2) there is sufficient carbon to regenerate the required number of acceptor molecules *plus one additional triose phosphate*, which is available for export from the chloroplast. The stoichiometry in Figures 10.3 and 10.4 was chosen to illustrate this point. Six turns of the cycle would regenerate 6 molecules of RuBP, leaving the equivalent of one additional hexose sugar as net product. Twelve turns would generate the equivalent of a sucrose molecule, and so on.

As a general rule it is necessary to show that the required enzymes are present and active before a complex metabolic scheme can be accepted as fact. Calvin's PCR cycle has met this criterion since all of the enzymes required by the scheme in Figure 10.3 have now been demonstrated in the stroma. Moreover, all of the reactions have been demonstrated *in vitro*, at rates that would support maximal rates of photosynthesis.

Energetics of the PCR Cycle Figure 10.3 shows that for three turns of the cycle, that is, the uptake of 3 molecules of CO_2, a total of 6 molecules of NADPH and 9 molecules of ATP are required. Therefore, the reduction of each molecule of CO_2 requires 2 molecules of NADPH and 3 molecules of ATP. This total represents an energy input of 529 kJ mol^{-11}. Oxidation of one mole of hexose would yield about 2817 kJ, or 469 kJ mol^{-1} of CO_2. This represents an energy storage efficiency of about 88 percent.

ACTIVATION AND REGULATION OF THE PCR CYCLE

As with most metabolic pathways, the PCR cycle is closely regulated in order to make the most efficient use of energy and carbon resources.

It was originally believed that the PCR cycle did not require a significant level of regulation, in part because early *in vitro* studies of Rubisco suggested a low, and probably rate-limiting, reactivity for this critical enzyme. (Its *in vivo* reactivity is now known to be much higher, although it may still be rate limiting.) In addition, plants were widely believed to be opportunistic and would use available light, water, and CO_2 to conduct photosynthesis at maximum rates (Bidwell, 1983). However, it is now recognized that photosynthesis does not operate in isolation and an unregulated photosynthetic machinery is incompatible with an orderly and integrated metabolism. Changing levels of intermediates between light and dark periods and competing demands for light energy and carbon with other cellular needs (nitrate reduction, for example) demand some degree of regulation. The most effective control is, of course, at the level of enzyme activities. Although relatively little information is yet available, the sophisticated nature of photosynthetic enzyme regulation is becoming increasingly apparent and a principal factor in regulation is, perhaps not surprisingly, light.

AUTOCATALYSIS

The rate of carbon reduction is partly dependent upon the availability of an adequate pool of acceptor molecules. The PCR cycle can utilize newly fixed carbon to increase the size of this pool, when necessary, through the **autocatalytic** regeneration of RuBP (Bidwell, 1983). During the night when photosynthesis is shut down and carbon is required for other metabolic activities, the concentrations of intermediates in the cycle (including RuBP) will fall to low levels. When photosynthesis starts up again, the rate could be severely limited by the availability of CO_2 acceptor molecules. Normally the extra carbon taken in through the PCR cycle is accumulated as starch or exported from the chloroplast. However, the PCR cycle has the potential to augment supplies of acceptor by retaining that extra carbon and diverting it toward generating increasing amounts of RuBP instead (Fig. 10.5). In this way the amount of acceptor can be

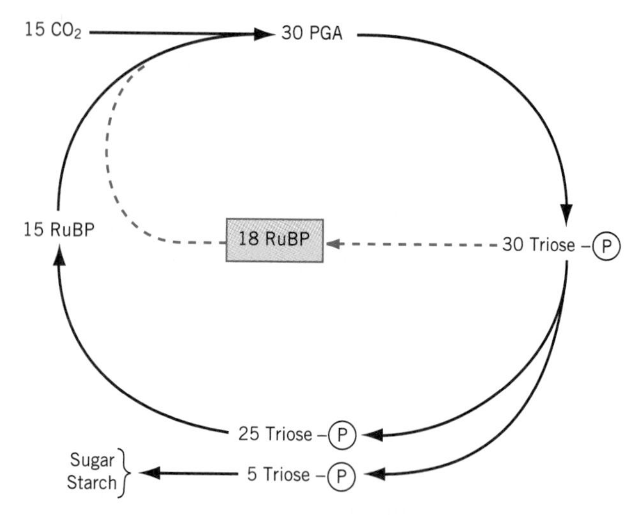

FIGURE 10.5 Autocatalytic properties of PCR cycle. When required, carbon can be retained within the PCR cycle (dashed arrows) to build up the amount of receptor molecules and increase the rate of photosynthesis.

quickly built up within the chloroplast to the level needed to support rapid photosynthesis. Only after the level of RuBP has been built up to adequate levels will carbon be withdrawn for storage or export. No other sequence of photosynthetic reactions has this capacity, which may help to explain why *all* photosynthetic organisms ultimately rely on the C3 cycle for carbon reduction. How autocatalysis is regulated is not clear. The most effective control would be to enhance the activities of enzymes favoring recycling over those leading to starch synthesis or export of product, but little is known about the relative activities of these enzymes.

REGULATION OF RUBISCO ACTIVITY

Rubisco activity is light regulated. Its activity declines rapidly to zero when the light is turned off and is regained only slowly when the light is once again turned on. Light activation is apparently indirect and involves complex interactions between Mg^{2+} fluxes across the thylakoid, CO_2 activation, chloroplast pH changes, and an activating protein.

As noted in the previous chapter, light-driven electron transport leads to a net movement of protons into the lumen of the thylakoids. The movement of protons across the thylakoid membrane generates a proton gradient equivalent to 2.5 to 3.5 pH units and an increase in the pH of the stroma from around pH 7 to near pH 8.0. *In vitro*, Rubisco is generally more active at pH 8 than at pH 7. The Mg^{2+} requirement for Rubisco activity was noted some years ago. Light also brings about an increase in the free Mg^{2+} of the stroma as it moves out of the lumen to compensate for the proton flux in the opposite direction.

Work in the laboratory of G. H. Lorimer, again using isolated Rubisco *in vitro*, has shown that Rubisco uses CO_2 not only as a substrate but also as an activator. The activating CO_2 must bind to an activating site that is separate and distinct from the substrate-binding site. Based on these *in vitro* studies, Lorimer and Miziorko have proposed a model for *in vivo* activation that takes into account all three factors: CO_2, Mg^{2+}, and pH (Lorimer and Miziorko, 1980). According to this model, the CO_2 first reacts with an ε-amino group of a lysine residue, forming what is known as a **carbamate** (Fig 10.6). Carbamate formation requires the release of two protons and, consequently, would be favored by increasing pH. The Mg^{2+} then becomes coordinated to the carbamate to form a carbamate-Mg^{2+} complex, which is the active form of the enzyme.

Further experiments, however, indicated that the *in vitro* model could not fully account for the activation of Rubisco in leaves (Portis, 1990). In particular, measured values for *in vivo* Mg^{2+} and CO_2 concentrations and pH differences were not sufficient to account for more than half the expected activation level. This paradox was resolved by the discovery of an *Arabidopsis* mutant that failed to activate Rubisco in the light, even though the enzyme isolated from the mutant was apparently identical to that isolated from the wildtype. Electrophoretic analysis revealed that the *rca* mutant, as it was called, was missing a soluble chloroplast protein. Subsequent experiments demonstrated that full activation of Rubisco could be restored *in vitro* simply by adding the missing protein to a reaction mixture containing Rubisco, RuBP, and physiological levels of CO_2. This protein has been named **Rubisco activase** to signify its role in promoting light-dependent activation of Rubisco.

The details of Rubisco activase and how it operates are still being worked out, but it is known to require energy in the form of ATP. The protein has been indentified in at least 10 genera of higher plants as well as

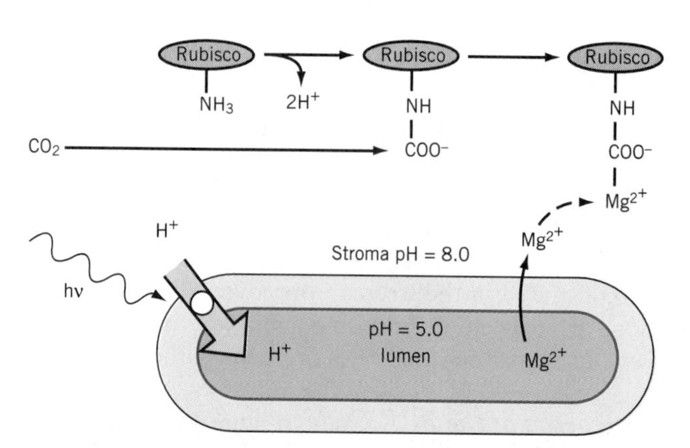

FIGURE 10.6 Light-driven ion fluxes and activation of Rubisco. Activation of Rubisco is facilitated by the increase in stromal pH and Mg^{2+} concentration that accompanies light-driven electron transport.

the green alga *Chlamydomonas*. It is clear that Rubisco activase has a significant and probably ubiquitous role to play in regulating eukaryotic photosynthesis.

REGULATION OF OTHER PCR ENZYMES

Rubisco is not the only PCR cycle enzyme requiring light activation. Studies with algal cells, leaves and isolated chloroplasts have shown that the activities of at least four other PCR cycle enzymes are also stimulated by light. These include glyceraldehyde-3-phosphate dehydrogenase (reaction 2, Fig. 10.3), fructose-1,6-bisphosphatase (FBPase) (reaction 4, Fig. 10.3), sedoheptulose-1,7-bisphosphatase (reaction 7, Fig. 10.3) and ribulose-5-phosphate kinase (reaction 11, Fig. 10.3).

The mechanism for light-activation is different from that of Rubisco and is best demonstrated in the case of FBPase. Light activation of FBPase can be blocked by the electron transport inhibitor DCMU and agents that selectively modify sulfhydryl groups. On the other hand, the enzyme can be activated in the dark by the reducing agent, dithiothreitol (DTT). It gradually emerged that activation requires the participation of both chloroplast ferredoxin, a product of the light-dependent reactions, and **thioredoxin** (Fig. 10.7). Like ferredoxin, thioredoxin is a small (12 kDa) iron-sulphur protein, known to biochemists for its role in the reduction of ribonucleotides to deoxyribonucleotides. It contains two cysteine residues in close proximity that undergo reversible reduction-oxidation from the disulphide (—S—S—) state to the sulfhydryl (—SH HS—) state. In the chloroplast, PSI drives the reduction of ferredoxin, which in turn reduces thioredoxin. The reaction is mediated by the enzyme ferredoxin-thioredoxin reductase (Buchanan, 1980). Thioredoxin subsequently reduces the appropriate disulphide bond on the target enzyme, resulting in its activation. Subsequent deactivation of the enzymes in the dark is not well understood, but clearly the sulfhydryl groups are in some way reoxidized and the enzymes rendered inactive.

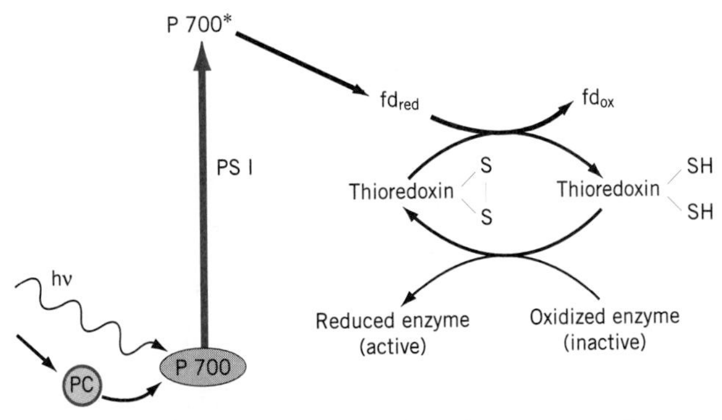

FIGURE 10.7 The ferredoxin/thioredoxin system for light-driven enzyme activation.

The traditional view of the PCR cycle was that it did not require the direct input of light. These reactions were consequently referred to as the "dark reactions" of photosynthesis. In view of the fact that at least five critical enzymes in the cycle require light activation, such a designation is clearly not appropriate.

PHOTORESPIRATION AND THE PHOTOSYNTHETIC CARBON OXIDATION CYCLE

Plants utilizing the PCR cycle exclusively for carbon fixation also exhibit a competing process of light- and oxygen-dependent CO_2 evolution.

The most widely used method for assessing the rate of photosynthesis in whole cells (e.g., algae) or intact plants is to measure gas exchange—either CO_2 uptake or O_2 evolution. This is, at best, a complicated process since there are several different and competing metabolic reactions that contribute to the gas exchange of an algal cell or a higher plant leaf. Cellular (or mitochondrial) respiration is an example of opposite gas exchange, since it results in an evolution of CO_2 and uptake of O_2. Here the term "respiration" is used solely to indicate the pattern of gas exchange, without reference to the specific metabolic origins of those gases. Historically, it was assumed that mitochondrial-based respiration and chloroplast-based photosynthesis were effectively independent and that their respective contributions to gas exchange could also be assessed independently. (One argument held that photosynthesis could supply the entire energy need of the leaf directly and the mitochondria would consequently "shut down" in the light!) The measured CO_2 uptake in the light was termed *apparent* or *net* photosynthesis, since it represented photosynthetic CO_2 uptake *less* the CO_2 evolved from mitochondria. *True* or *gross* photosynthesis was thus calculated by adding the amount of mitochondrial-respired CO_2 (measured during a dark period) to that taken up in the light.

We now know that measuring gas exchange is a far less certain process, complicated in part by oxidative metabolism and the consequent *evolution* of CO_2 directly associated with photosynthetic metabolism. Called **photorespiration,** this process involves the reoxidation of products just previously assimilated in photosynthesis (Bidwell, 1983). The photorespiratory pathway involves the activities of at least three different cellular organelles (the chloroplast, the peroxisome, and the mitochondrion) and, because CO_2 is evolved, results in a net loss of carbon from the cell.

Early experiments based on discrimination between carbon isotopes suggested there were both qualitative and quantitative differences between the process of respiration (i.e., CO_2 *evolution*) as it occurred in the dark and in the light. On this basis, CO_2 evolution in the light was called photorespiration. The concept that light would alter either the rate or path of respiration was, to say the least, controversial (Decker, 1970). However, in 1955, J. P Decker reported significantly higher rates of CO_2 evolution immediately after the lights were turned off (Decker, 1955) (Fig. 10.8). For this experiment, tobacco leaves were placed in a closed chamber—that is, a chamber with a fixed supply of CO_2. The leaves were illuminated until photosynthesis reduced the CO_2 concentration in the chamber to the **CO_2 compensation concentration.** The CO_2 compensation concentration is the concentration of CO_2 at which uptake for photosynthesis is balanced by the evolution of CO_2 by respiration or other processes. Once this point was reached, the light was turned off and the change in CO_2 concentration during the ensuing dark period was monitored. For the first two minutes, the rate of CO_2 evolution was high, after which it slowly declined to the normal rate in darkness. This *postillumination burst* was interpreted as a light-dependent CO_2 evolution (hence, *photo*respiration), which came to a halt more slowly than photosynthetic CO_2 uptake when the light was extinguished. Decker's observation stimulated more studies that linked the magnitude of the postillumination CO_2 burst to factors that affected the rate of photosynthesis during the previous light period. A variety of experiments from different laboratories have confirmed that a different pathway for CO_2 evolution comes into play in illuminated leaves.

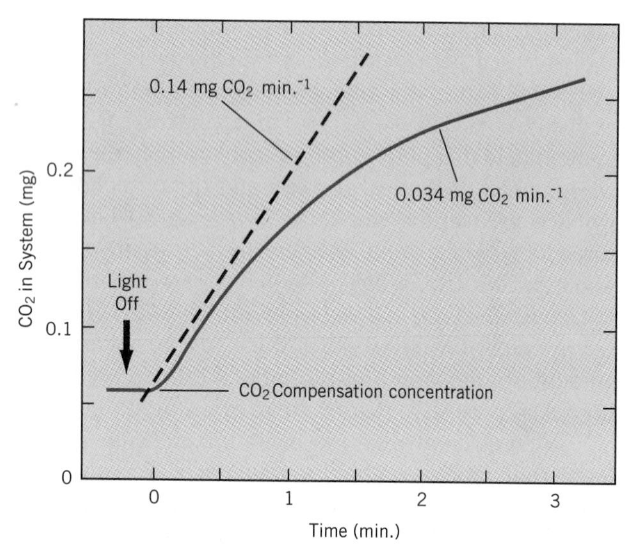

FIGURE 10.8 Photorespiration. A *postillumination burst* experiment illustrating a rapid, light-dependent evolution of CO_2. The tracing from a CO_2 monitor shows changes in the CO_2 content of a closed system in darkness immediately following a period of photosynthesis. Note the initial high rate of CO_2 evolution when the lights are turned off (dashed line). Within about 2 minutes, the rate of CO_2 evolution has declined to a lower level, which is maintained through a prolonged dark period. (Redrawn from J. P. Decker, *Plant Physiology* 30:82–84, 1955. Copyright American Society of Plant Physiologists.)

RUBP OXYGENASE AND THE C2 GLYCOLATE PATHWAY

While the legitimacy of photorespiration was being established during the 1960s, the attention of several investigators was attracted to the synthesis and metabolism of a two-carbon compound, **glycolate.** It gradually emerged that glycolate metabolism was related to photorespiration and that the enzymes involved were located in peroxisomes and mitochondria as well as the chloroplast. The key to photorespiratory CO_2 evolution and glycolate metabolism is the bifunctional nature of Rubisco, first demonstrated by Bowes et al. in 1971. In addition to the **carboxylation** reaction, Rubisco also catalyzes an **oxygenase** reaction, hence the name ribulose-1,5-bisphosphate carboxylase-oxygenase.[2] With the addition of a molecule of oxygen, RuBP is converted into one molecule of 3-PGA and one molecule of **phosphoglycolate** (Fig. 10.9). The phosphoglycolate is subsequently metabolized in a series of reactions that result in the release of a molecule of CO_2 and recovery of the remaining carbon by the PCR cycle (Fig. 10.10).

The **C2 glycolate cycle,** also known as the **photosynthetic carbon oxidation (PCO) cycle,** begins with the oxidation of RuBP to 3-PGA and P-glycolate. The 3-PGA is available for further metabolism by the PCR cycle, but the P-glycolate is rapidly dephosphorylated to glycolate in the chloroplast. The glycolate is exported from the chloroplast and diffuses to a **peroxisome.** Taken up by the peroxisome, the glycolate is oxidized to glyoxylate and hydrogen peroxide. The peroxide is broken down by catalase and the glyoxylate undergoes a transamination reaction to form the amino acid glycine. Glycine is then transferred to a mitochondrion where

CH_2O—Ⓟ
|
C=O
|
HCOH
|
HCOH
|
CH_2O—Ⓟ

RuBP

+ O_2 ⟶

CH_2O—Ⓟ
|
COO^-

2-phosphoglycolate

COO^-
|
HCOH
|
CH_2O—Ⓟ

3-phosphoglycerate

FIGURE 10.9 The RuBP oxygenase reaction.

[2]In the current literature, Rubisco is sometimes referred to as RuBPCO.

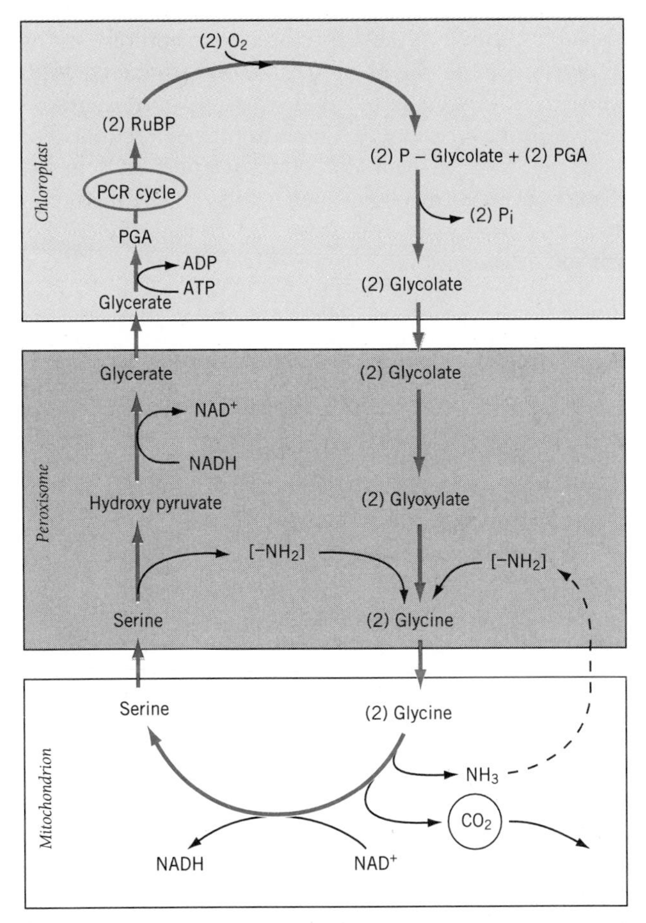

FIGURE 10.10 The photorespiratory glycolate pathway.

two molecules of glycine (4 carbons) are converted to one molecule of serine (3 carbons) plus one CO_2. *Glycine is thus the immediate source of photorespired CO_2.* The serine then leaves the mitochondrion, returning to a peroxisome where the amino group is given up in a transamination reaction and the product, hydroxypyruvate, is reduced to glycerate. Finally, glycerate is returned to the chloroplast where it is phosphorylated to 3-PGA.

The release of carbon as CO_2 during the conversion of glycine to serine is accompanied by the release of an equivalent amount of nitrogen in the form of ammonia. During active photorespiration, the rate of ammonia release may be substantially greater than the rate of nitrogen assimilation (Lea et al., 1992). This nitrogen is not lost, however, as the ammonia is rapidly reassimilated in the chloroplast, using the enzymes of the glutamate synthase cycle (Chap. 6).

The C2 glycolate pathway involves complex interactions between photosynthesis, photorespiration, and various aspects of nitrogen metabolism in at least three different cellular organelles. Nonetheless, there is a substantial amount of supporting evidence. Much of the support comes from labeling studies employing either $^{14}CO_2$ or specific intermediates, or $^{18}O_2$, in which the fate of the label is followed through the various suspected chemical transformations. As with the PCR cycle, all of the enzymes necessary to carry out the C2 glycolate cycle have been demonstrated. The distribution of intermediates between the three organelles, however, is not conclusively established. It is largely inferred from the location of the enzymes. All of the subcellular organelles involved have been isolated and shown to contain the appropriate enzymes.

WHY PHOTORESPIRATION?

In normal air (21% O_2), the rate of photorespiration in sunflower leaves is about 17 percent of gross photosynthesis (Fock et al., 1974). Every photorespired CO_2, however, requires an input of two molecules of O_2 (Fig. 10.10). The true rate of oxygenation is therefore about 34 percent and the ratio of carboxylation to oxygenation is about 3 to 1 (100/0.34). This experimental value agrees with similar values calculated for several species based on the known characteristic of purified Rubisco. The ratio of carboxylation to oxygenation depends, however, on the relative levels of O_2 and CO_2 since both gases compete for binding at the active site on Rubisco. As the concentration of O_2 declines, the relative level of carboxylation increases until, at zero O_2, photorespiration is also zero. On the other hand, increases in the relative level of O_2 (or decrease in CO_2) shifts the balance in favor of oxygenation. An increase in temperature will also favor oxygenation, since as the temperature increases the solubility of gasses in water declines, but O_2 solubility is less affected than CO_2. Thus O_2 will inhibit photosynthesis, measured by net CO_2 reduction, in plants that photorespire. The inhibition of photosynthesis by O_2 was first recognized by Otto Warburg in the 1920s, but fifty years were to pass before the bifunctional nature of Rubisco offered the first satisfactory explanation for this phenomenon.

There is also an energy cost associated with photorespiration and the glycolate pathway. The amount of ATP and NAD(P)H expended in the glycolate pathway following oxygenation is roughly equal to that expended for the reduction of one CO_2 in the PCR cycle, *yet there is a net loss of carbon.* On the surface, then, photorespiration appears to be a costly and inefficient process with respect to both energy and carbon. It is logical to ask, as many have, why the plant should indulge in such an apparently wasteful process.

This question is not easily answered, although several ideas have been put forward. One has it that the oxygenase function of Rubisco is inescapable. Rubisco evolved at a time when the atmosphere contained large amounts of CO_2 but little oxygen. Under these conditions, an inability to discriminate between the two gasses would have had little significance to the survival of the organism. Both CO_2 and O_2 react with the enzyme at the same active site and oxygenation requires activation by CO_2 just as carboxylation does. It is believed that oxygen began to accumulate in the atmosphere primarily due to photosynthetic activity, but by the time the atmospheric content of O_2 had increased to significant proportions, the bifunctional nature of the enzyme had been established without recourse. In a sense, C3 plants were the architect of their own problem—generating the oxygen that functions as a competitive inhibitor of carbon reduction. By this view, then, the oxygenase function is an evolutionary "hangover" that has no useful role. On the other hand, any inefficiencies resulting from photorespiration are apparently not severe. There is no evidence that selection pressures have caused evolution of a form of Rubisco with lower affinity for O_2.

While most agree that oxygenation is an unavoidable consequence of evolution, many have argued that plants have capitalized on this evolutionary deficiency by turning it into a useful, if not essential, metabolic sequence. The glycolate pathway, for example, undoubtedly serves a scavenger function. For each two turns of the cycle, two molecules of phosphoglycolate are formed by oxygenation. Of these four carbon atoms, one is lost as CO_2 and three are returned to the chloroplast. The glycolate pathway thus recovers 75 percent of the carbon that would otherwise be lost as glycolate. The salvage role alone may be sufficient justification for the complex glycolate cycle (Canvin, 1990). There is also the possibility that some of the intermediates, serine and glycine, for example, are of use in other biosynthetic pathways, although this possibility is still subject to some debate.

Some investigators feel that photorespiration could also function as a sort of safety valve in situations that require dissipation of excess excitation energy. Osmond and Bjorkman (1972), for example, have shown a significant decline in the photosynthetic capacity of leaves irradiated in the absence of CO_2 and O_2. Injury is prevented, however, if sufficient O_2 is present to permit photorespiration to occur. Apparently the CO_2 generated by photorespiration is sufficient to protect the plant from photooxidative damage by permitting continued operation of the electron transport system. This could be of considerable ecological value under conditions of high light and limited CO_2 supply, for example, when the stomata are closed due to moisture stress.

A claim made frequently in the literature is that crop productivity might be significantly enhanced by inhibiting or genetically eliminating photorespiration. As a result, substantial effort has been expended in the search for chemicals that inhibit the glycolate pathway or selective breeding for low photorespiratory strains. Others have surveyed large numbers of species in an effort to find a Rubisco with a significantly lower affinity for oxygen. All of these efforts have been unsuccessful, presumably because the basic premise that photorespiration is detrimental to the plant and counterproductive is incorrect. Clearly, success in increasing photosynthesis and improving productivity lies in other directions. For example, a mechanism for concentrating CO_2 in the photosynthetic cells could be one way to suppress photorespiratory loss and improve the overall efficiency of carbon assimilation. That is exactly what has been achieved by C4 and CAM plants, described in the next two sections.

THE C4 SYNDROME

The initial incorporation of CO_2 by some plants utilizes an additional pathway involving 4-carbon molecules.

Plants that incorporate carbon solely through the PCR or Calvin cycle are generally known as C3 plants because the first product to incorporate $^{14}CO_2$ is the 3-carbon acid PGA. Certain other groups, known as C4 plants, are distinguished by the fact that the first product is a 4-carbon acid **oxaloacetate (OAA)**. These C4 plants also exhibit a number of specific anatomical, physiological and biochemical characteristics that constitute the *C4 syndrome.* One particular anatomical feature characteristic of most C4 leaves is the presence of two distinct photosynthetic tissues (Fig. 10.11). In C4 leaves the vascular bundles are quite close together and each bundle is surrounded by a tightly fitted layer of cells called the **bundle sheath.** Between the vascular bundles and adjacent to the air spaces of the leaf are the more loosely arranged **mesophyll** cells. This distinction between mesophyll and bundle sheath photosynthetic cells, called Kranz anatomy (see below), plays a major role in the C4 syndrome.

C4 plants are generally of tropical or subtropical origin representing nearly 1,500 species spread through at least 18 different angiosperm families (3 monocots, 15 dicots) (Raghavendra and Das, 1978). Interestingly, no one family has been found to express the C4 syndrome exclusively—all 18 families contain both C3 and C4 representatives. This suggests that the C4 cycle has arisen rather recently in evolution of angiosperms and in a number of diverse taxa at different times. Under conditions of high fluence rates and high temperature (30° to 40°C) the photosynthetic rate of C4 species may be two to three times greater than that of C3 species. They appear to be better equipped to withstand drought and are able to maintain active photosynthesis under conditions of water stress that would lead to stomatal closure and consequent reduction of CO_2 uptake by C3 species. All of these features appear to be a consequence of the CO_2-concentrating capacity of C4 plants and the resulting suppression of photorespiratory CO_2 loss.

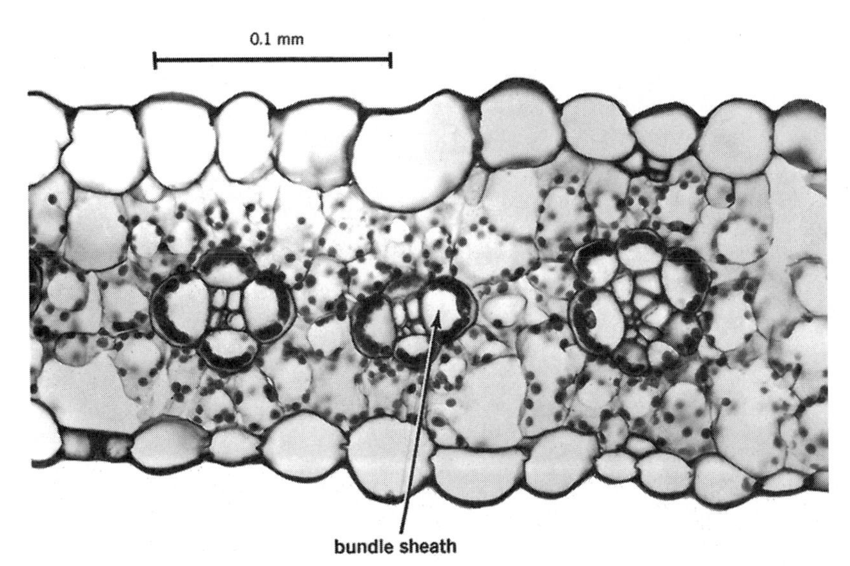

FIGURE 10.11 Leaf of a C4 grass. Cross-section of a leaf of maize (*Zea mays*), showing the arrangement of mesophyll and bundle-sheath cells. Note high concentration of chloroplasts in bundle sheath cells. (From K. Esau. *Anatomy of Seed Plants*. 1977. New York, Wiley. Reprinted by permission.)

DISCOVERY AND GENERAL PRINCIPLES OF THE C4 SYNDROME

The discovery of the C4 syndrome can be traced to the efforts of H. P. Kortschack and his coworkers to demonstrate the presence of the PCR cycle in sugarcane (Kortschack et al., 1965). Previously it had been shown that 4-carbon acids do accumulate in C3 plants, but not until photosynthesis had been ongoing for some time. By contrast, Kortschack found that certain C4 acids were very early products of photosynthesis in sugarcane and that the label did not appear in sugar phosphates until later. These observations were confirmed and extended by M. D. Hatch and C. R. Slack, who proposed a cyclic mechanism in which the carbon was first incorporated into a C4 acid. The β-carboxyl carbon was subsequently transferred, as CO_2, to the PCR cycle (Hatch and Slack, 1966). The cycle has since been refined in a number of ways, resulting in the general outline shown in Fig. 10.12.

The key to the C4 cycle is the enzyme **phospho*enol* pyruvate carboxylase (PEPcase)** which catalyses the β-carboxylation of **phospho*enol* pyruvate (PEP)** using

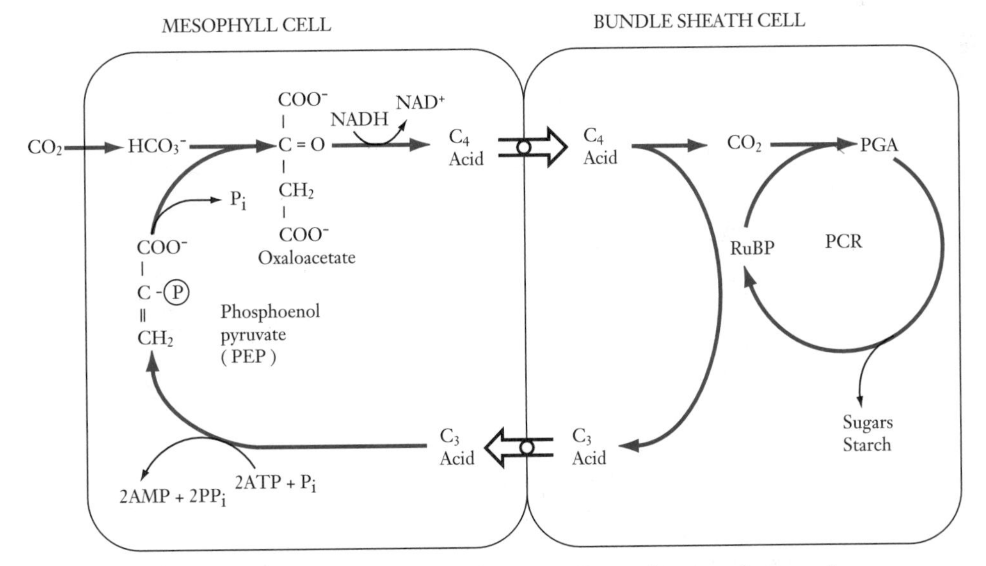

FIGURE 10.12 Schematic of the general plan of the C4 photosynthetic carbon assimilation cycle. Initial carboxylation of phosphoenol pyruvate (PEP) to oxaloacetate (OAA) in the mesophyll cell is followed by translocation of a C4 acid to the bundle-sheath cell where it undergoes a decarboxylation. The resulting C3 acid is returned to the mesophyll cell to complete the cycle. The CO_2 released in the decarboxylation step is assimilated by the PCR cycle in the bundle-sheath chloroplast. The C4 cycle serves to concentrate CO_2 in the bundle-sheath cell and suppress photorespiration.

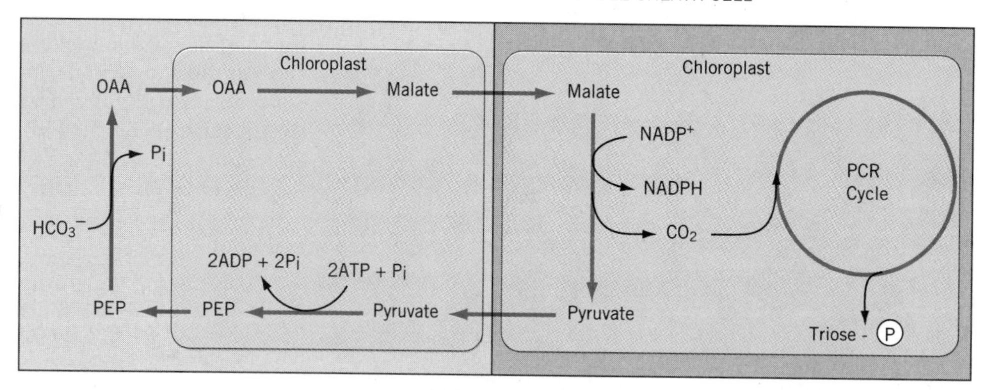

A. NADP - Malic enzyme type

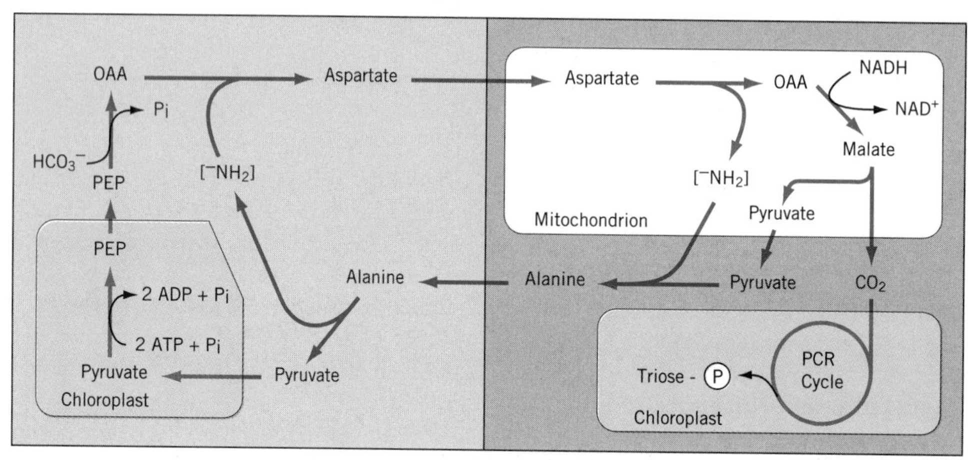

B. NAD - Malic enzyme type

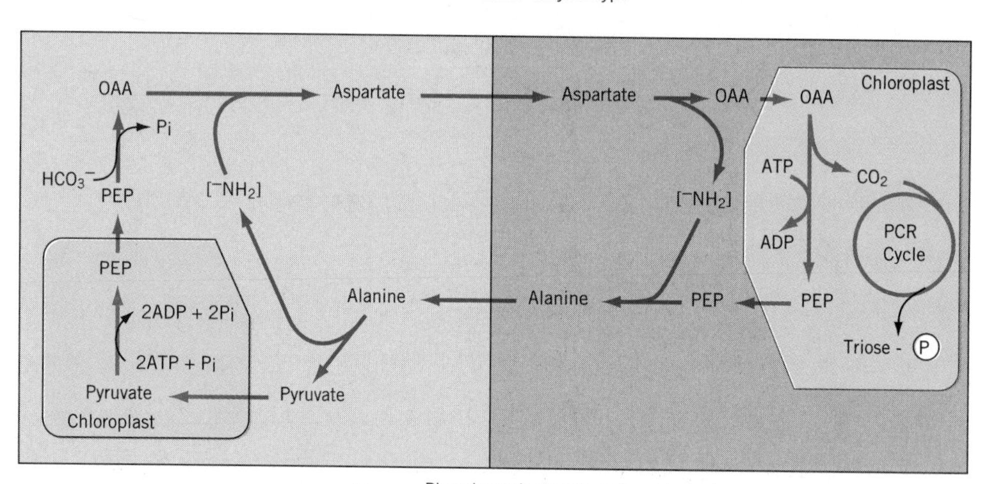

C. Phosphoenolpyruvate carboxykinase type

FIGURE 10.13 Three variations of the C4 cycle. The principal differences are: (1) the form of the C4 acid translocated into the bundle-sheath cell; (2) the nature and location (cytosol, mitochondrion, or chloroplast) of the decarboxylating enzyme; and (3) the form of the C3 acid returned to the mesophyll cell.

the bicarbonate ion HCO_3^- as the substrate (rather than CO_2). C4 plants also have a specialized leaf anatomy comprised of **mesophyll** cells and **bundle-sheath** cells. The anatomical arrangements are explored further in the next section. The product of the PEPcase reaction, **oxaloacetate (OAA)**, is moderately unstable. It is quickly either reduced to malate or transaminated to aspartate—both of which are more stable—and transported out of the mesophyll cell into an adjacent bundle-sheath cell. Once in the bundle-sheath cell, the acid undergoes a decarboxylation and the resulting CO_2 is available for reduction to triose sugars via the PCR cycle in the bundle-sheath chloroplast. The C3 acid (either pyruvate or alanine) that remains after decarboxylation is then transported back into the mesophyll cell. Here the alanine is converted to pyruvate. The pyruvate is then phosphorylated to regenerate the original acceptor molecule, PEP.

There are certain similarities between the PCR cycle and C4 metabolism. Like Rubisco, the PEPcase carboxylation reaction is virtually irreversible and, consequently, energetically very favorable. Reducing potential is required at some point to remove the product and ATP is required to regenerate the acceptor molecule, PEP, and thus keep the reaction going. A very significant difference, however, is that once in the bundle-sheath cell, the C4 acid is decarboxylated, giving up the CO_2 originally assimilated in the mesophyll cell. This decarboxylation means that, unlike the C3 cycle, the C4 cycle does not *of itself* result in any net carbon reduction. The plant relies ultimately on the operation of the PCR cycle in the bundle-sheath chloroplast for the synthesis of triose phosphates.

Within the general pattern of the C4 cycle described above there are three variations (Fig. 10.13). The primary carboxylation reaction occurs in the cytosol of the mesophyll cell and is the same for all three types—they are differentiated primarily by the nature and intracellular location of the enzyme that catalyzes decarboxylation in the bundle-sheath cells. In the *NADP-malic enzyme (NADP-ME) type* (Fig. 10.13A), the oxaloacetate is first converted to malate. The malate is then transported to the chloroplast of bundle-sheath cell where it is decarboxylated to yield pyruvate and CO_2. The pyruvate is then transported back to the mesophyll cell where it is phosphorylated to form PEP. The phosphorylation reaction, catalyzed by the enzyme **pyruvate, phosphate dikinase,** requires two ATP and is also common to all three types. In the other two variants—known as the *NAD-malic enzyme (NAD-ME) type* (Fig. 10.13B) and the *PEP-carboxykinase (PCK) type* (Fig. 10.13C), oxaloacetate undergoes a transamination reaction and the resulting aspartate is transported to the bundle-sheath cell. Decarboxylation in the PCK-type variant also occurs in the chloroplast, while in the NAD-ME variant the decarboxylating malic enzyme is located in the mitochondrion. In both of these variants, the C3 acid transported back into the mesophyll cell is alanine. There is no great taxonomic distinction between the variants, as even within a single genus (e.g., *Panicum*) species expressing any one of three types may be found.

Regardless of the variant employed, the principal effect of the C4 cycle remains to concentrate CO_2 in the bundle-sheath cells where the enzymes of the PCR cycle are located. By shuttling the CO_2 in the form of organic acids it is possible to build much higher CO_2 concentrations in the bundle-sheath cells than would be possible relying on the diffusion of CO_2 alone. Results of studies employing radiolabeled $^{14}CO_2$ have indicated the concentration of CO_2 in bundle-sheath cells may reach 60 μM; about 10-fold higher than that in C3 plants (Edwards and Walker, 1983). Higher CO_2 concentrations would suppress photorespiration and support higher rates of photosynthesis.

Under optimal conditions, C4 crop species can assimilate CO_2 at rates two to three times that of C3 species. All this productivity does not, however, come "free." There is an energy cost to building the CO_2 concentration in the bundle-sheath cells. For every CO_2 assimilated, *two ATP* must be expended in the regeneration of PEP. This is in addition to the ATP and NADPH required in the PCR cycle. Thus the net energy requirement for assimilation of CO_2 by the C4 cycle is five ATP and two NADPH.

KRANZ ANATOMY

The anatomy of a typical C4 leaf is shown in Figure 10.11. Note that the photosynthetic parenchyma cells in a typical C3 leaf are organized into two distinct tissues—an upper region of tightly packed palisade cells and the more loosely arranged spongy mesophyll cells bordering large air spaces. The C4 leaf, on the other hand, is generally thinner than the C3 leaf, the vascular bundles are closer together, and the air spaces are smaller. Moreover, there is only one type of mesophyll cell, loosely arranged in the fashion of the spongy mesophyll in the C3 leaf. Surrounding each vascular bundle is a sheath of tightly packed, thick-walled cells containing large numbers of chloroplasts. Indeed, C4 plants can often be recognized by the prominent, dark-green veins. Because of the wreathlike configuration of these bundle-sheath cells (originally observed by the German anatomist G. Haberlandt) this arrangement is known as *Kranz anatomy* (*Kranz*-wreath).

The characteristic anatomical arrangement of the C4 leaf ensures a short diffusion path for CO_2 to the initial carboxylation site in the mesophyll cell. A short diffusion path together with the uniform distribution of the initial carboxylating enzyme throughout the cytosol makes for a more efficient CO_2 trapping. Moreover, no mesophyll cell is more than two or three cells distant from a bundle-sheath cell, which no doubt facilitates the transfer of the C4 and C3 acids. The high chloroplast density in the bundle-sheath cell is apparently necessary to process the high concentrations of CO_2 generated by the C4 system. Finally, the close juxtaposition of the PCR cycle cells to the vascular tissue means that product (i.e., sugars) can be quickly exported from the leaf when required. Overall it is safe to conclude that the effectiveness of the C4 system is enhanced by these anatomical adaptations. It should be pointed out, however, that some C3 dicots do have well-developed bundle sheaths, although they generally have few or no chloroplasts. Thus the presence of Kranz anatomy cannot in itself be taken as evidence of the C4 syndrome.

ECOLOGICAL SIGNIFICANCE OF THE C4 SYNDROME

C4 plants exhibit a number of physiological attributes that appear to be immediate consequences of their unique carbon metabolism. It is generally perceived that these physiological attributes may lead to higher photosynthetic productivity under certain conditions and have significant ecological consequences.

Unlike C3 plants, photosynthesis of C4 plants is not inhibited by O_2, and they exhibit no postillumination CO_2 burst and have a very low **CO_2 compensation concentration** (Table 10.1). The CO_2 compensation concentration is the ambient carbon dioxide concentration at which the rate of CO_2 uptake (for photosynthesis) is balanced by the rate of CO_2 evolution (by respiration). In a closed environment, such as that shown in Figure 10.8, the CO_2 compensation concentration would be the stable CO_2 concentration in the air when CO_2 uptake and evolution have come into equilibrium. For C3 plants, values fall into the range of 20 to 100 μl CO_2 per liter. Comparable values for C4 plants are in the range of 0 to 5 $\mu l\ l^{-1}$.

On the basis of the above observations, it is reasonable to conclude that photorespiration is either absent from C4 plants or that the process is suppressed. However, although the activity of the glycolate pathway is very low in some C4 plants (*Sorghum bicolor*, for example), most C4 plants appear to have both peroxisomes and the metabolic machinery to support photorespiration. The weight of evidence thus favors the conclusion that C4 plants do photorespire, but at much reduced rates (Bidwell, 1983). The high level of CO_2 developed in the bundle-sheath cells would tend to suppress photorespiration by out-competing O_2 for binding to Rubisco. In adddition, the anatomical and biochemical adaptations of C4 leaves ensures that any CO_2 that might escape the bundle-sheath cell is trapped and reassimilated by PEPcase in the mesophyll cells, before it has the opportunity to escape from the leaf. Thus C4 leaves are not only efficient CO_2 absorbers, but also effectively trap and recirculate any CO_2 that might be produced in the leaf.

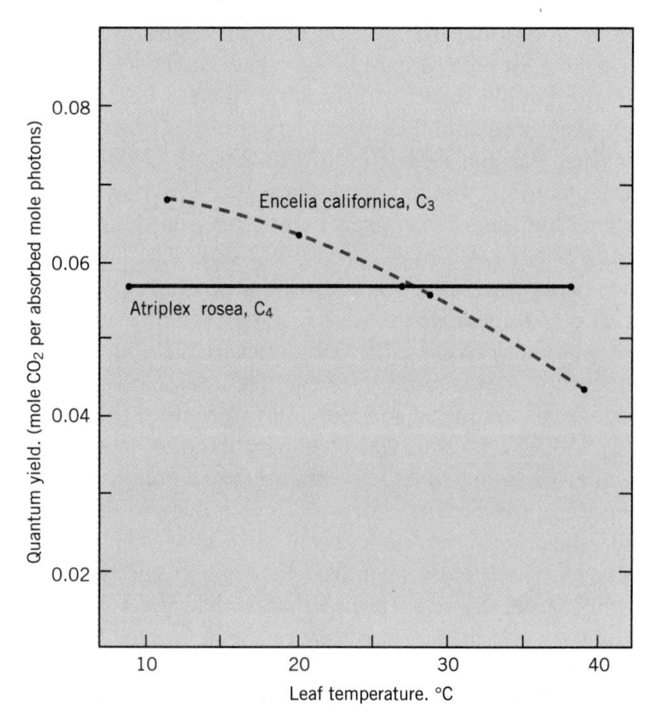

FIGURE 10.14 Quantum yield for uptake of CO_2 in a C4 and C3 plant as a function of leaf temperature. (From J. Ehleringer, O. Björkman, *Plant Physiology* 59:86–90, 1977. Copyright American Society of Plant Physiologists.)

TABLE 10.1 A comparison of significant features of C3 and C4 plants.

	C3	C4
Photorespiration	yes	no*
CO_2 compensation ($\mu l\ CO_2\ l^{-1}$)	20–100	0–5
Temperature optimum (°C)		
photosynthesis	20–25	30–45
Rubisco	20–25	
PEPcase		30–35
Quantum yield as a function of temperature	declining	steady
Transpiration Ratio	500–1000	200–350
Light saturation (μmole photons $m^{-2}\ s^{-1}$)	400–500	does not saturate

*Not detectable.

In addition to the virtual absence of photorespiration, most C4 plants tend to have a higher temperature optimum (30–45 °C) than C3 plants (20–25 °C). This difference is due at least in part to the higher temperature stability of some of the C4 cycle enzymes. Maximal activity of PEPcase, for example, is in the range of 30 to 35 °C compared with 20 to 25 °C for Rubisco. As a result, *the quantum yield* of photosynthesis in C3 plants tends to decline with increasing leaf temperature while *the quantum yield* of C4 plants remains essentially constant (Fig. 10.14). The decline of quantum yield in C3 plants is due to decreased carboxylation activity of Rubisco at the higher temperatures. This trend is exacerbated by changes in the relative solubility of CO_2 and O_2. The solubility of gasses generally declines with increasing temperature, but the solubility of CO_2 is affected more than the solubility of O_2. Consequently, higher temperatures increasingly favor oxygenation by Rubisco.

Another interesting feature of C4 plants is their general low-temperature sensitivity. While there are some cold-tolerant C4 species, most perform poorly, if at all, at low temperature. *Zea mays*, for example, will not grow at temperatures below 12 to 15 °C. This lower limit for growth is probably set by the enzyme pyruvate, phosphate dikinase, which is cold labile and experiences a substantial loss of activity below 12 °C.

Photosynthesis in most situations is limited by available CO_2 and water. In C3 plants, even moderate water stress will initiate closure of the stomata and reduce the available supply of CO_2. The low CO_2 compensation concentration of C4 plants means that they can maintain higher rates of photosynthesis at lower CO_2 levels. Thus C4 plants gain an advantage over C3 plants when the stomata are partially closed to conserve water during a period of water stress. An effective measure of this advantage is the value of the **transpiration ratio.** The transpiration ratio (TR) relates the uptake of CO_2 to the loss of water by evaporation (transpiration) from the leaf:

$$TR = \text{moles } H_2O \text{ transpired/moles } CO_2 \text{ assimilated}$$

The inverse of transpiration ratio, called **water use efficiency,** is often cited in ecological literature (although occasionally the numbers are actually transpiration ratios!). Transpiration ratios for C4 plants are typically in the range of 200 to 350, while for C3 plants values in the range of 500 to 1000 are often cited. The low transpiration ratio for C4 plants reflects their capacity to maintain high rates of photosynthesis while effectively conserving water.

Even under ideal conditions, CO_2 supply limits photosynthesis in C3 plants to the extent that light saturation occurs at fluence rates about 25 percent of full sunlight. C4 plants, on the other hand, never really saturate, even at full sunlight (Fig. 10.15). Even so, C4 photosynthesis is not necessarily more efficient than C3 photosynthesis. At leaf temperatures below 30 °C the quantum yield for C4 plants is actually lower than for C3 plants—that is, C4 photosynthesis is *less efficient* (Table 10.2, Fig. 10.14). The low quantum efficiency of C4 plants reflects an additional light requirement that can be accounted for by the ATP required by the pyruvate, phosphate dikinase reaction (Fig. 10.12). How can the lower photosynthetic efficiency of C4 plants be reconciled with their apparent higher productivity? Recall that C4 plants are native to tropical or subtropical habitats where there is usually an abundance of light. They can take advantage of some of this excess available light to generate the ATP needed to run the C4 cycle, concentrate the CO_2, and increase net carbon assimilation.

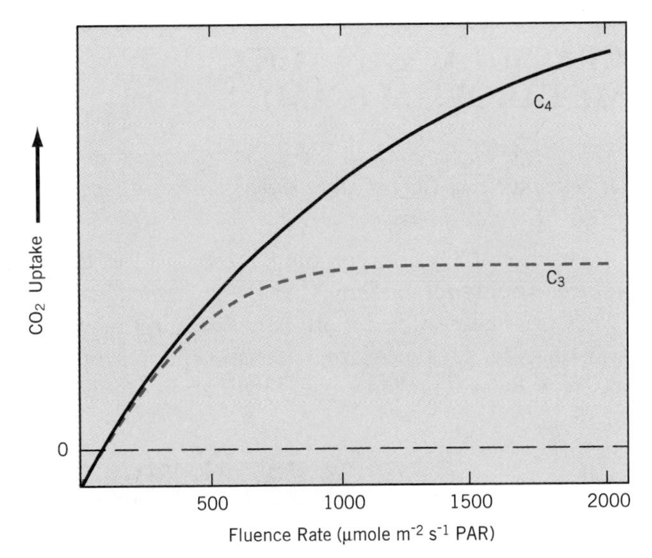

FIGURE 10.15 Fluence response curves for typical C3 and C4 plants.

Although C4 plants are not competitive in all situations—some C3 plants may even equal or exceed C4 plants in productivity—given the right combination of high temperature, high light and low water, the C4 syndrome confers a definite advantage. This advantage is reflected in the observation that, to the frustration of homeowner and farmer alike, many of our more aggressive weeds are C4 species. These include crabgrass (*Digitaria sanguinalis*), Russian thistle (*Salsola kali*), and several species of pigweed (*Amaranthus*) that often take over during the hot, dry months in the middle of summer. Many of the more highly productive crop species also fall within the C4 group, including sugarcane (*Saccharum officinarum*), sorghum (*Sorghum bicolor*), maize (*Zea mays*), and millet (*Panicum miliaceum*).

TABLE 10.2 Effect of temperature on quantum yield of photosynthesis in C3 and C4 plants.

	Quantum yield (mol CO_2/mol quanta)		
	15 °C	25 °C	35°
C3 plants			
Triticum aestivum	0.055	0.050	0.041
Encelia californica	0.067	0.059	0.046
C4 plants			
Zea mays	0.057	0.059	0.058
Atriplex rosea	0.054	0.054	0.054

CRASSULACEAN ACID METABOLISM (CAM)

A number of plants growing in dry, desert habitats store large quantities of CO_2 at night, later releasing it for use by the PCR cycle during the day.

Another CO_2 concentrating mechanism is **Crassulacean acid metabolism (CAM)**—so named because it was originally studied most extensively in the family Crassulaceae. This specialized pattern of photosynthesis has now been found in some 23 different families of flowering plants (including the Cactaceae and Euphorbiaceae), one family of ferns (the Polypodiaceae), and in the primitive plant *Welwitschia*. Like C4 plants, however, most families, with the exception of Crassulaceae and Cactaceae, are not exclusively CAM. Most families will have C3 representatives as well and some are known to contain all three photosynthetic patterns; C3, C4, and CAM.

The unique features of CAM permit a remarkable degree of water conservation. Individual species utilizing this pathway are thus especially adapted to survival in extremely dry, or xerophytic, habitats. They are also,

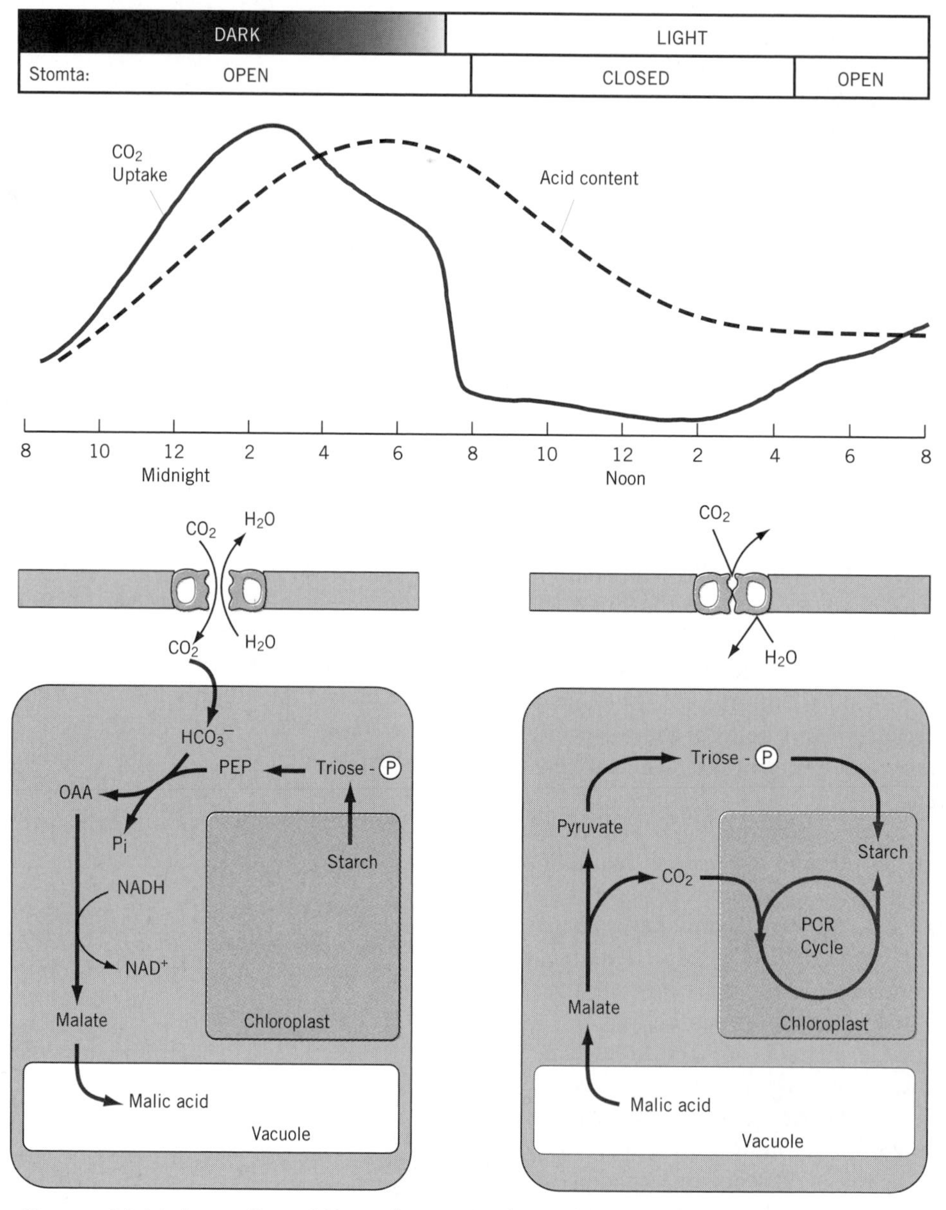

FIGURE 10.16 An outline of Crassulacean Acid Metabolism. Above: Curves illustrating stomatal opening, CO_2 uptake, and changing acid content of cell vacuoles over a 24 hour period. Stomata open in the dark to admit CO_2 and close during the day to conserve water. Below, right: While the stomata are closed during the day, stored CO_2 is released to be assimilated via the PCR cycle.

without exception, succulent plants—characterized by thick, fleshy leaves (or, as in the cacti, photosynthetic stems) whose cells contain large, water-filled vacuoles. Other than succulence no particular anatomical modifications appear to be required for CAM. However, although succulence appears prerequisite for CAM, not all succulent plants exhibit CAM.

One of the most striking features of CAM plants is an inverted stomatal cycle—the stomata open mainly during the nighttime hours and are usually closed during the day. This means that *CO_2 uptake also occurs mainly at night* (Fig. 10.16). In addition, CAM plants are characterized by an accumulation of malate at night and its subsequent depletion during daylight hours and storage carbohydrate levels that fluctuate inversely with malate levels (Kluge and Ting, 1978). Nocturnal stomatal opening supports a β-carboxylation reaction, producing C4 acids that are stored in the large, watery vacuole (Fig 10.16). Accumulation of the organic acids leads to a marked acidification of these cells at night. The acids are subsequently decarboxylated during daylight hours and the resulting CO_2 is fixed by the PCR cycle.

As in C4 plants, the enzyme PEP carboxylase is central to CAM operation. The immediate product, OAA, is rapidly reduced by NAD-dependent malate-dehydrogenase to malate, which is then stored in the vacuole. During the daylight hours, malate is retrieved from the vacuole, decarboxylated (by NAD-malic enzyme in the Crassulaceae), and the CO_2 diffuses into the chloroplast where it is converted to triose phosphates by the C3 PCR cycle. The large amounts of PEP required to support the carboxylation reaction appear to be derived from the breakdown of starch and other storage glucans by the enzymes of the glycolytic pathway (Chap. 12). The fate of C3 acid (pyruvate or PEP) resulting from decarboxylation is uncertain, but the weight of evidence is that it is reduced to triose phosphate, which in turn can be converted back to glucose or starch (Fig. 10.16).

IS CAM A VARIATION OF THE C4 SYNDROME?

CAM and C4 plants share certain similarities, but there are significant differences.

A comparison between CAM and the C4 cycle is unavoidable since they both use cytoplasmic PEPcase to form C4 acids from PEP and bicarbonate, and in both cases the acids are subsequently decarboxylated to provide CO_2 for the PCR cycle. However, there are two significant differences. The first is that the C4 cycle requires a specialized anatomy by which C4 carboxylation is spatially separated from the C3 PCR cycle—*in CAM both occur in the same cells but are separated in time.* Secondly, in CAM there is no closed cycle of carbon intermediates as there is in C4 plants. Instead, the PEP required as substrate for the carboxylation reaction is derived from stored carbohydrate. The C3 product of decarboxylation is disposed of by a variety of metabolic paths, which no doubt includes the resynthesis of storage carbohydrate. Thus the reaction is cyclic in time only (Bidwell, 1983). Since CAM occurs in the more primitive ferns and *Welwitschia* where the C4 cycle is found only in angiosperms, it appears that CAM preceded C4 photosynthesis in evolutionary time.

ECOLOGICAL SIGNIFICANCE OF CAM

CAM plants are particularly suited to dry habitats because they use water very efficiently.

As mentioned above, CAM represents a particularly significant adaptation to exceptionally dry habitats. Many CAM plants are true desert plants, growing in shallow, sandy soils with little available water. Nocturnal opening of the stomata allows for CO_2 uptake during periods when conditions leading to evaporative water loss are at a minimum. Then, during the daylight hours when the stomata are closed to reduce water loss, photosynthesis can proceed by using the reservoir of stored CO_2. This interpretation is supported by the transpiration ratio for CAM plants, in the range of 50 to 100, which is substantially lower than that for either C3 or C4 plants. There is a price to be paid, however. Rates for daily carbon assimilation by CAM plants are only about one-half that of C3 plants and one-third that of C4. CAM plants can be expected to grow more slowly under conditions of adequate moisture. On the other hand, CO_2 uptake by CAM plants will continue under conditions of water stress that would cause complete cessation of photosynthesis in C3 plants and severely restrict carbon uptake by C4 plants. CAM plants enjoy the further advantage of being able to retain and reassimilate respired CO_2, thus preventing loss of carbon and helping to maintain a favorable dry weight through extended periods of severe drought.

While some species, in particular the cacti, are obligatory CAM plants, many other succulents exhibit a facultative approach to CAM. One well-studied example is the ice plant, *Mesembryanthemum crystallinum*, a fleshy annual of the family Aizoaceae. Under conditions of abundant water supply, *Mesembryanthemum* assimilates carbon as a typical C3 plant—there is no significant uptake of CO_2 at night and no diurnal variation in leaf cell acidity. Under conditions of limited water availability or high salt concentration in the soil, CAM metabolism is switched on (Winter and Troughton, 1978). Although carbon assimilation by CAM is slower than with conventional C3 photosynthesis, its higher water use efficiency permits photosynthesis to continue in times of water stress and the plant is better able to complete its reproductive development.

REGULATION OF C4 PHOTOSYNTHESIS AND CAM

One intriguing aspect of both C4 photosynthesis and CAM is the requirement for precise regulation and temporal integration of enzyme activities and other potentially competitive metabolic processes.

In addition to regulation of PCR cycle enzymes discussed earlier, successful operation of C4 photosynthesis and CAM also requires regulation of starch-PEP interconversions, storage and retrieval of malate, PEPcase and Rubisco competition for CO_2, and the temporal operation of PEP carboxylation. Regulation of C4 photosynthesis and CAM in general is poorly understood, with the exception of the primary carboxylation catalyzed by PEPcase, which has attracted increasing interest in recent years.

PEPcase is a cytoplasmic enzyme found in virtually all cells of higher plants where it serves a variety of important metabolic functions. However, plants with the C4 and CAM modes of photosynthesis contain specific isozymes[3] with higher activity levels than associated with C3 or nonphotosynthetic cells. Based on a variety of physiological and biochemical considerations, it appears that PEPcase activity in C4 and CAM plants must be regulated by light–dark transitions. In C4 plants, for example, PEPcase activity should be high in the light in order to maximize availability of CO_2 for the PCR cycle in the bundle-sheath cells. Although the substrate PEP consumed in C4 photosynthesis is derived from the bundle-sheath chloroplast, PEP is also a critical intermediate in glycolysis. Glycolysis is also a cytoplasmic metabolic sequence that represents the first stage in respiration. Continued high PEPcase activity at night could result in uncontrolled utilization of PEP and seriously impair respiratory metabolism. In the case of CAM it is clear that efficient operation requires the carboxylation and decarboxylation reactions—which occur within the same cell—not be allowed to compete with each other at the same time. In CAM, PEP carboxylase activity must be high at night when atmospheric CO_2 is available, but should be switched off during daylight hours in order to avoid a futile recycling of CO_2 derived from malate. Competition for CO_2 is not a problem at night since Rubisco and the PCR cycle are inoperative in the dark.

PEPcase regulation was initially studied in CAM plants but it is now evident that PEPcase activity in C4 plants as well as CAM is subject to reversible activation by light—dark conditions and inhibition by malate—a form of feedback inhibition in which accumulated product reduces the activity of the enzyme (Jiao and Chollet, 1991). In C4 plants, light induces an increase in the catalytic activity of the enzyme while at the same time reducing its sensitivity to inhibition by the product molecule, malate. This results in a fivefold increase in activity in the light when CO_2 assimilation and malate production is required to boost CO_2 levels in the bundle-sheath cells. In the case of CAM, the situation is reversed (Kluge et al., 1980). Enzyme extracted during the night part of a diurnal cycle exhibits a high affinity for PEP and is relatively insensitive to inhibition by malate. Enzyme extracted during the day has a low affinity for PEP and is more sensitive to inhibition by malate.

Recent evidence from studies of *Bryophyllum* and *Kalanchoe*, both CAM plants, and the C4 plants *Zea mays* and *Sorghum* have shown that PEPcase exists in two states; the biochemically more active form of the enzyme is phosphorylated but the less active form is not (Jiao and Chollet, 1991). From this it can be concluded that PEPcase activity is regulated by a light-sensitive protein kinase. Evidence for reversible activation of a protein kinase has been obtained from both *in vivo* and *in vitro* experiments. The reaction requires ATP and phosphorylation occurs at a serine residue near the N-terminal end of the enzyme molecule. Just how light activates the protein kinase is not known at this stage, although there is some indication that it is related to photosynthetic electron transport. It may be that photophosphorylation supplies the ATP required for phosphorylation of the enzyme.

It is interesting to note that light appears to have the opposite effect in the two systems—the protein kinase is activated by light in C4 plants but inactivated during the day in CAM. The effect of light and darkness in CAM plants, however, may be indirect. Studies with *Bryophyllum* have indicated that CAM physiology and the sensitivity of PEPcase to inhibition by malate in CAM plants may be controlled by an endogenous circadian rhythm (Nimmo et al., 1987). Whatever the mechanism, spatial and temporal coordination of C4 and C3 metabolism in C4 and CAM plants is an important area where we can expect to see exciting advances in the future.

EXPORT AND STORAGE OF PHOTOSYNTHETIC PRODUCT

The products of the PCR cycle may be stored as starch in the chloroplast or exported from the chloroplast to the cytosol where they are converted to sucrose.

The primary function of photosynthesis is to provide energy and carbon sufficient to support maintenance and growth not only of the photosynthetic tissues

[3]Isozymes are different species of enzyme that catalyze the same reaction. Isozymes are frequently coded by different genes and consequently have different protein structures. Although they act on the same substrate, isozymes may be tissue- or organ-specific and subject to regulation by different chemical or environmental factors.

but of the plant as a whole. During daylight hours, some photosynthetic product, or **photoassimilate,** is exported as sucrose from the leaf to other tissues of the plant and a portion is temporarily accumulated in the leaf. Many plants, such as soybean, spinach, and tobacco, store excess photoassimilate as starch in the chloroplast while others, such as wheat, barley and oats, accumulate little starch but temporarily hold large amounts of sucrose in the vacuole. The starch and sucrose will later be mobilized to support respiration and other metabolic needs at night or during periods of limited photosynthetic output. Sucrose exported from the leaf cell to nonphotosynthetic tissues may be metabolized immediately, stored temporarily as sucrose in the vacuoles, or converted to starch for longer-term storage in the chloroplasts.

STARCH SYNTHESIS IN THE CHLOROPLAST

The dominant storage carbohydrate in higher plants is the polysaccharide starch, which exists in two forms (Fig. 10.17). **Amylose** is a linear polymer of glucose created by linking adjacent glucose residues between the first and fourth carbons. Amylose is consequently known as an α-(1,4)-glucan. **Amylopectin** is similar to amylose except that occasional α-(1,6) linkages, about every 24 to 30 glucose residues, create a branched molecule. Amylopectin is very similar to glycogen, the principal storage carbohydrate in animals. Glycogen is more highly branched, with one α-(1,6) linkage for every 10 glucose residues compared with one in 30 for amylopectin.

The site of starch synthesis in leaves is the chloroplast. Large deposits of starch are clearly evident in electron micrographs of chloroplasts from C3 plants and the bundle-sheath chloroplasts of C4 plants. In addition, the two principal enzymes involved—*ADPglucose phosphorylase* and *starch synthase* are found localized in the chloroplast (Preiss, 1988). Starch synthesis in the chloroplast begins with the PCR cycle intermediate fructose-6-phosphate (F6P). Two chloroplast enzymes, hexose-phosphate isomerase (10.1) and phosphoglucomutase (10.2), convert the F6P to glucose-1-phosphate, which then reacts with ATP to form ADP-glucose (10.3). Reaction 10.3 is catalyzed by the enzyme ADPglucose phosphorylase. ADPglucose is an activated form of glucose and serves as the immediate precursor for starch synthesis.

$$\text{fructose-6-P} \rightleftharpoons \text{glucose-6-P} \qquad (10.1)$$

$$\text{glucose-6-P} \rightleftharpoons \text{glucose-1-P} \qquad (10.2)$$

$$\text{ATP} + \text{glucose-1-P} \rightleftharpoons \text{ADP-glucose} + PP_i \qquad (10.3)$$

$$PP_i + H_2O \rightleftharpoons 2P_i \qquad (10.4)$$

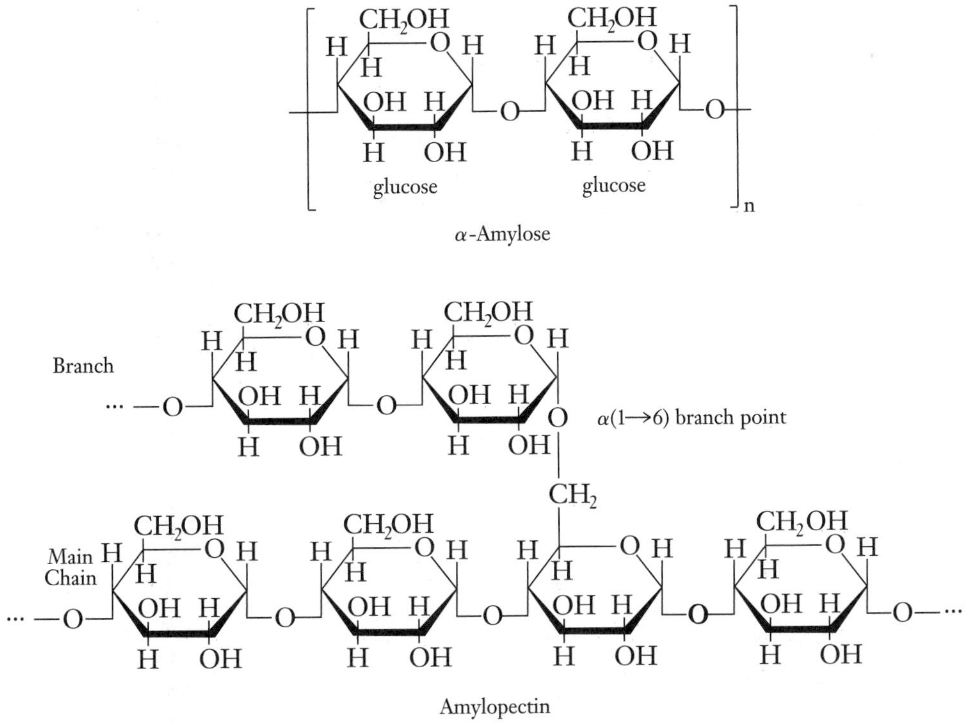

CH_2OH H O H H OH H H OH glucose — O — CH_2OH H O H H OH H H OH glucose — O — n

α-Amylose

Branch ··· —O— CH_2OH — O — CH_2OH — O α(1→6) branch point

CH_2

Main Chain ··· —O— CH_2OH — O — CH_2OH — O — — O — CH_2OH — O—···

Amylopectin

FIGURE 10.17 The chemical structures of the two forms of starch: amylose and amylopectin. Amylose is a long chain of α(1 → 4) linked glucose residues. Amylopectin is a multibranched polymer of α(1 → 4) linked glucose containing α(1 → 6) branch points.

Finally, the enzyme starch synthase catalyzes formation of a new α-(1,4) link, adding one more glucose to the elongating chain (10.5).

$$\text{ADPglucose} + \alpha\text{-}(1\rightarrow4)\text{-glucan} \rightleftharpoons \text{ADP} + \alpha\text{-}(1\rightarrow4)\text{-glucosyl-glucan} \quad (10.5)$$

Formation of the α-(1,6) branching linkages, giving rise to amylopectin, is catalyzed by the *branching enzyme*, also known as the Q enzyme.

SYNTHESIS OF SUCROSE

Sucrose is a soluble disaccharide containing a glucose and a fructose residue. It is one of the more abundant natural products that not only plays a vital role in plant life but is also a leading commercial commodity. Sucrose may function as a storage product as it does in sugarbeets or sugarcane where it is stored in the vacuoles of specialized storage cells. Alternatively sucrose may be translocated to other, nonphotosynthetic tissues in the plant for direct metabolic use or for conversion to starch. Sucrose is by far the most common form of sugar found in the translocation stream (Chap. 11).

The site of sucrose synthesis in the cell was the subject of debate for some time. On the basis of cell fractionation and enzyme localization studies it has now been clearly established that sucrose synthesis occurs exclusively in the cytosol of photosynthetic cells. Earlier reports of sucrose synthesis in isolated chloroplasts appear attributable to contamination of the chloroplast preparation with cytosolic enzymes. Moreover the inner membrane of the chloroplast envelope is impermeable to sucrose, so that if sucrose were synthesized inside the chloroplast it would be unable to exit the chloroplast and enter the translocation stream.

Two routes of sucrose synthesis are possible. The principal pathway for sucrose synthesis in photosynthetic cells is provided by the enzymes **sucrose phosphate synthase** (10.6) and **sucrose phosphate phosphatase** (10.7).[4]

$$\text{UDPglucose} + \text{fructose-6-phosphate} \rightleftharpoons \text{UDP} + \text{sucrose-}6^{F}\text{-phosphate} \quad (10.6)$$

$$\text{Sucrose-}6^{F}\text{-phosphate} + H_2O \rightleftharpoons \text{sucrose} + P_i \quad (10.7)$$

Energy provided by the hydrolysis of sucrose-6-phosphate (about 12.5 kJ mol^{-1}) may play a role in the accumulation of high sucrose concentrations typical of sugarcane and other sucrose-storing plants (see Chap. 11).

[4]The superscript F indicates that the phosphate is on the number 6 carbon of the fructose moiety.

Another cytoplasmic enzyme capable of synthesizing sucrose is sucrose synthase (SS) (10.8):

$$\text{UDPglucose} + \text{fructose} \rightleftharpoons \text{UDP} + \text{sucrose} \quad (10.8)$$

With a free energy change of approximately +4 kJ mol^{-1}, this reaction is not spontaneous. Most of the evidence indicates that under normal conditions SS operates in the reverse direction to break down sucrose (see Eq. 10.11).

Note that, in contrast with starch biosynthesis, sucrose biosynthesis by either pathway requires activation of glucose with the nucleotide *uridine triphosphate* (UTP) rather than ATP (10.9).

$$\text{UTP} + \text{glucose-1-phosphate} \rightleftharpoons \text{UDPglucose} + PP_i \quad (10.9)$$

$$PP_i + \text{H2O} \rightleftharpoons 2P_i \quad (10.10)$$

Although sucrose phosphate synthase in some tissues can use ADPglucose, UDPglucose is clearly predominant.

Carbon for cytoplasmic sucrose biosynthesis is exported from the chloroplast through a special orthophosphate (P_i)-dependent transporter located in the chloroplast envelope membranes. This **P_i/triose phosphate transporter** exchanges P_i and triose phosphate—probably as dihydroxyacetone phosphate (DHAP)—on a one-for-one basis. Once in the cytoplasm, two molecules of triose phosphates (glyceraldehyde-3-phosphate and DHAP) are condensed to form fructose-1,6-bisphosphate. Subsequently, the fructose-1,6-bisphosphate is converted to glucose-1-phosphate as it is in the chloroplast, employing cytoplasmic counterparts of the chloroplastic enzymes. Some of the orthophosphate generated in sucrose synthesis is used to regenerate UTP while the rest can reenter the chloroplast in exchange for triose-P.

Sucrose translocated from the leaf to storage organs such as roots, tuber tissue, and developing seeds is most commonly stored as starch. The conversion of sucrose to starch is generally thought to involve a reversal of the sucrose synthase reaction:

$$\text{Sucrose} + \text{UDP} \rightleftharpoons \text{fructose} + \text{UDPglucose} \quad (10.11)$$

Because ADPglucose is preferred for starch biosynthesis, UDPglucose is converted to ADPglucose as shown in 10.12 and 10.13:

$$\text{UDPglucose} + PP_i \rightleftharpoons \text{UTP} + \text{glucose-1-phosphate} \quad (10.12)$$

$$\text{glucose-1-phosphate} + \text{ATP} \rightleftharpoons \text{ADPglucose} + PP_i \quad (10.13)$$

The resulting ADPglucose is then converted to starch by starch synthase.

SUCROSE OR STARCH—THE FATE OF TRIOSE PHOSPHATE

Starch and sucrose synthesis are competitive processes—each relies on the same pool of triose phosphate produced by the PCR cycle. One of the more interesting problems associated with starch and sucrose synthesis is the way in which that triose phosphate is allocated between chloroplastic starch synthesis on the one hand and cytoplasmic sucrose synthesis on the other.

It has traditionally been held that carbohydrate metabolism is to a large extent governed by source-sink relationships. The photosynthetically active leaf, for example, would be a "source," providing assimilated carbon that is available for transport to the "sink," a storage organ or developing flower or fruit, for example, which utilizes that assimilate. This concept of source and sink will be developed further in Chapter 11. With respect to relationships between sucrose and starch, it was often observed that removal of a sink, thus reducing demand for photoassimilate, resulted in accumulation of starch in the leaves. This led to the assumption that starch represented little more than excess carbon. There is now good evidence that this assumption is false.

In soybean plants (*Glycine max*), starch accumulation is not related to the length of the photosynthetic period (Chatterton and Silvius, 1979). Plants maintained on a 7-hour light period put a larger proportion of their daily photoassimilate into starch than those maintained on a 14-hour light period, even though the assimilation period is only half as long. Thus it appears that foliar starch accumulation is more closely related to the energy needs of the daily dark period than photosynthetic input. Just how these needs are anticipated by the plant is unknown. However, many species are now known to distribute different proportions of carbon between starch and sucrose in ways apparently unrelated to sink capacity or inherent capacities of isolated chloroplasts to form starch. Carbon distribution thus appears to be a programmed process, implying some measure of control beyond a simple source-sink relationship. Moreover, it is essential that sucrose synthesis be controlled in order to maintain an efficient operation of photosynthesis itself. If the rate of sucrose synthesis should exceed the rate of carbon assimilation, demand for triose-P in the cytoplasm could deplete the pool of PCR cycle intermediates, thereby decreasing the capacity of Calvin cycle enzymes for regeneration of RuBP and seriously inhibiting photosynthesis.

While the enzyme sucrose phosphate synthase determines the maximum capacity for sucrose synthesis, it appears that cytosolic fructose-1,6-bisphosphate phosphatase (FBPase) plays the more important role in balancing sucrose and starch synthesis. The highly exergonic reaction (fructose-1,6-bisphosphate → fructose-6-phosphate + P_i) occupies a strategic site in the sucrose synthetic pathway—it is the first *irreversible* reaction in the conversion of triose-P to sucrose. Consequently the flow of carbon into sucrose can easily be controlled by regulating the activity of FBPase—similar to regulating the flow of water by opening or closing a valve.

The activity of cytosolic FBPase, like its chloroplastic counterpart, appears to be regulated by light; in sugarbeet leaves the activity is highest at the end of a light period and lowest at the end of a dark period (Khayat et al., 1993). Unlike the chloroplast enzyme, activation of the cytosolic enzyme appears to be indirect, by some yet-unknown mechanism. Cytosolic FBPase is, however, quite sensitive to inhibition by fructose-2,6-bisphosphate (F26BP). F26BP, an analog of the natural substrate fructose-1,6-bisphosphate, is considered a *regulator metabolite* because it functions as a regulator rather than a substrate. F26BP levels are, in turn, sensitive to a number of interacting factors including the concentration of F6P and the cytosolic triose-P/P_i ratio (Fig 10.18). Control of sucrose synthesis by F26BP appears to be particularly relevant when the rate of CO_2 assimilation lags behind the rate of sucrose synthesis, when light is limiting for example. As the rate of photosynthesis declines, the concentration of triose-P also declines. A low triose-P/P_i favors formation of F26BP, which inhibits FBPase and slows down sucrose synthesis. This action reduces the withdrawal of triose-P from the chloroplast and allows the pool of PCR cycle intermediates to be maintained at an appropriate level. Accumulation of triose-P in the chloroplast also favors the formation of starch by stimulating the activity of ADPglucose pyrophosphorylase (10.3).

Typically, sucrose export from the cell slows in the light, leading to an accumulation of intermediates such as F6P and triose-P and a shift in allocation in favor of starch. When the demand for sucrose decreases, sucrose and its precursors (e.g., F6P) will accumulate in the leaf.

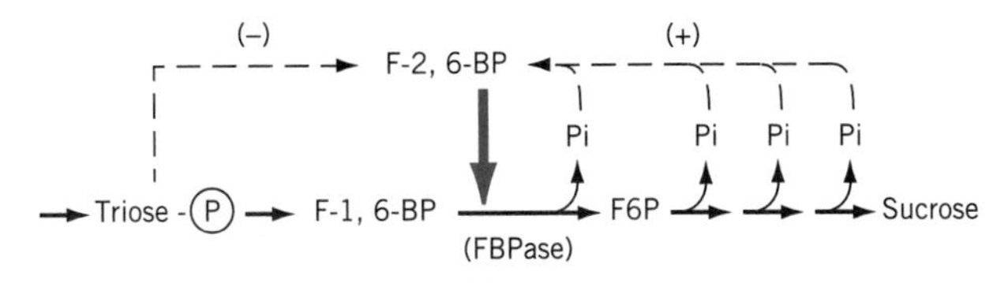

FIGURE 10.18 The synthesis of sucrose is regulated by the triose-P/orthophosphate ratio. A low ratio favors the synthesis of fructose-2,6-bisphosphate (F26BP), which in turn inhibits the activity of the enzyme fructose-1,6-bisphosphatase (FBPase) and decreases the rate of sucrose synthesis. A high proportion of triose-P inhibits the synthesis of F26BP, releases the enzyme from inhibition, and favors sucrose synthesis.

Since F6P is also the precursor for F26BP, levels of the inhibitor will increase as well—leading to an inhibition of FBPase and an accumulation of triose-P. The accumulation of phosphorylated intermediates probably also leads to a decrease in the concentration of P_i. The combined accumulation of triose-P and decrease of P_i will in turn decrease the rate at which triose-P can be exported from the chloroplast through the transporter. The consequent accumulation of triose-P and decrease of orthophosphate in the chloroplast in turn stimulate the synthesis of starch.

Many of the details remain to be worked out, but it is clear that starch synthesis in the chloroplast, triose-P export, and sucrose synthesis in the cytosol are in delicate balance. The balance is modulated by very subtle changes in the level of triose-P and P_i as well as precise regulation of a number of enzymes and requires intimate communication between the two cellular compartments.

SUMMARY

The photosynthetic carbon reduction (PCR) cycle occurs in the chloroplast stroma. It is the sequence of reactions all plants use to reduce carbon dioxide to organic carbon. The key enzyme is ribulose-1,5-bisphosphate carboxylase-oxygenase, which catalyzes the addition of a carbon dioxide molecule to an acceptor molecule, ribulose-1,5-bisphosphate (RuBP). The product is two molecules of 3-phosphoglycerate (3-PGA). Energy from the light-dependent reactions is required at two stages: ATP and NADPH for the reduction of 3-PGA and ATP for the regeneration of the acceptor molecule RuBP. The bulk of the cycle involves a series of sugar rearrangements that (1) regenerate RuBP and (2) accumulate excess carbon as 3-carbon sugars. This excess carbon can be stored in the chloroplast in the form of starch or exported from the chloroplast for transport to other parts of the plant.

Photosynthesis, like all other complex metabolic reactions, is subject to regulation. In this case, the primary activator is light. Several key PCR cycle enzymes, including Rubisco, are light activated. This is one way of integrating photosynthesis with other aspects of metabolism, regulating changing levels of intermediates between light and dark periods and competing demands for carbon with other cellular needs.

Plants that utilize the PCR cycle exclusively for carbon fixation also exhibit a competing process of light- and oxygen-dependent carbon dioxide evolution, called photorespiration. The source of carbon dioxide is the photosynthetic carbon oxidation (PCO) cycle. The PCO cycle also begins with Rubisco, which, in the presence of oxygen, catalyzes the oxidation, as well as carboxylation, of RuBP. The product of RuBP oxidation is one molecule of 3-PGA plus one 2-carbon molecule, phosphoglycolate. Phosphoglycolate is subsequently metabolized in a series of reactions that result in the release of carbon dioxide and recovery of the remaining carbon by the PCR cycle. The role of the PCO cycle is not yet clear, although it has been suggested that it helps protect the chloroplast from photooxidative damage during periods of moisture stress, when the stomata are closed and the carbon dioxide supply is cut off.

Plants with the C4 photosynthetic pathway or CAM have evolved a mechanism for avoiding the impact of photorespiratory carbon dioxide loss by concentrating carbon dioxide in the carbon-fixing cells. C4 plants exhibit a division of labor between mesophyll cells, which pick up carbon dioxide from the ambient air, and bundle-sheath cells, which contain the PCR cycle and actually fix carbon. Mesophyll cells contain the enzyme PEP carboxylase, which catalyzes the carboxylation of phospho*enol*pyruvate. The product C4 acid is transported into the bundle-sheath cell. There it is decarboxylated and the resulting carbon dioxide is fixed via the PCR cycle. C4 plants have a very low carbon dioxide compensation concentration and low transpiration ratio. This means C4 plants are able to maintain higher rates of photosynthesis at lower carbon dioxide levels, even when the stomata are partially closed to conserve water during periods of water stress.

Crassulacean acid metabolism (CAM) is also a means of maintaining higher rates of photosynthesis in habitats with little available water. CAM plants exhibit an inverted stomatal cycle, opening for carbon dioxide uptake at night (when water stress is lower) and closing during the day (when water stress is high). The carbon dioxide is stored as organic acids, again through the action of PEP carboxylase. Decarboxylation during the day furnishes the necessary carbon dioxide for photosynthesis.

The products of photosynthesis may be stored as starch in the chloroplast or exported from the chloroplast to the cytosol where they are converted to sucrose. Storage as starch or export to the cytoplasm are competing processes subject to regulation by subtle changes in the level of triose phosphate and inorganic phosphate and the activity of a cytosolic enzyme, fructose-2,6-bisphosphatase.

CHAPTER REVIEW

1. Review the reactions of the photosynthetic carbon reduction cycle and show how:
   a) product is generated;
   b) the carbon is recycled to regenerate the acceptor molecule.

2. In what chemical form(s) and where is energy put into the photosynthetic carbon reduction (PCR) cycle? What is the source of this energy?
3. The photosynthetic carbon reduction (PCR) cycle is said to be autocatalytic. What does this mean and of what advantage is it?
4. Describe the photorespiratory C2 glycolate pathway. What is the relationship between photorespiration and photosynthesis?
5. Debate the position that the oxygenase function of rubisco is an evolutionary "hangover."
6. Trace the path of carbon in a typical C4 leaf, from entry into the leaf through the stomata to its export in the vascular tissue.
7. A distinctive feature of the bundle sheath cells in a typical C4 leaf is a high density of chloroplasts. What advantage does this offer? Why is it advantageous to have the PCR cycle located in the bundle sheath cells in a C4 leaf?
8. Review the ecological significance of the C4 cycle. In what situations can a C3 species be more productive than a C4 species?
9. In what significant way does Crassulacean acid metabolism (CAM) differ from C4 metabolism?
10. What factors determine whether the product of the PCR cycle (triose phosphate) will be converted to starch in the chloroplast or sucrose in the cytosol?

FURTHER READING

Briggs, W. R. (ed.). 1989. *Photosynthesis.* New York: Alan R. Liss.

Dennis, D. T., D. H. Turpin (eds.). 1990. *Plant Physiology, Biochemistry and Molecular Biology.* London: Longman Scientific & Technical.

Edwards, G., D. Walker. 1983. *C_3, C_4: Mechanisms and Cellular and Environmental Regulation of Photosynthesis.* Oxford: Blackwell Scientific Publications.

Gregory, R. P. F. 1989. *Biochemistry of Photosynthesis.* New York: Wiley.

Huber, S. C., J. L. Huber. 1992. Role of sucrose-phosphate synthase in sucrose metabolism in leaves. *Plant Physiology* 99:1275–1278.

Ogren, W. L. 1984. Photorespiration: pathways, regulation and modification. *Annual Review of Plant Physiology* 35:415–422.

REFERENCES

Bidwell, R. G. S. 1983. Carbon metabolism of plants: photosynthesis and photorespiration. In: Steward, F. C. (ed.), *Plant Physiology. A Treatise. Vol VII: Energy and Carbon Metabolism.* New York: Academic Press, pp. 287–458.

Bowes, G., W. L. Ogren, R. H. Hageman. 1971. Phosphoglycolate production catalyzed by ribulose diphosphate carboxylase. *Biochemical Biophysical Research Communications* 45:716–722.

Buchanan, B. B. 1980. Role of light in the regulation of chloroplast enzymes. *Annual Review of Plant Physiology* 31:341–374.

Canvin, D. T. 1990. Photorespiration and CO_2-concentrating mechanisms. In: D. T. Dennis and D. H. Turpin (eds.), *Plant Physiology, Biochemistry and Molecular Biology.* London: Longman Scientific & Technical. pp. 253–273.

Chatterton, N. J., J. E. Silvius. 1979. Photosynthate partitioning into starch in soybean leaves. I. Effects of photoperiod versus photosynthetic period duration. *Plant Physiology* 64:749–753.

Decker, J. P. 1955. A rapid, postillumination deceleration of respiration in green leaves. *Plant Physiology* 30:82–84.

Decker, J. P. 1970. Early history of photorespiration. *Bioengineering Bulletin* 10.

Edwards, G. E., D. A. Walker. 1983. *C3, C4 Mechanisms, Cellular and Environmental Regulation of Photosynthesis.* Oxford: Blackwell Scientific Publications.

Fock, H., J. Mahon, D. T. Canvin, B. R. Grant. 1974. Estimation of carbon fluxes through the photosynthetic and photorespiratory pathways. In: R. L. Bieleski, A. R. Furguson, M. M. Cresswell (eds.), *Mechanisms of Regulation of Plant Growth.* Bulletin 12, pp. 235–242. Wellington: The Royal Society of New Zealand.

Hatch, M. D., C. R. Slack. 1966. Photosynthesis by sugarcane leaves. A new carboxylation reaction and the pathway of sugar formation. *Biochemical Journal* 101:103–111.

Jiao, J., R. Chollet. 1991. Posttranslational regulation of photsphoenolpyruvate carboxylase in C4 and crassulacean acid metabolism in plants. *Plant Physiology* 95:981–985.

Khayat, E., C. Harn, J. Daie. 1993. Purification and light-dependent modulation of the cytosolic fructose-1,6-bisphosphatase in sugarbeet leaves. *Plant Physiology* 101:57–64.

Kluge, M., M. Böcher, G. Jungnickel. 1980. Metabolic control of Crassulacean acid metabolism: evidence for diurnally changing sensitivity against inhibition by malate of PEP carboxylase in *Kalenchoe tubiflora* Hamet. *Zeitschrift für Pflanzenphysiologie* 97:197–204.

Kluge, M., I. P. Ting. 1978. *Crassulacean Acid Metabolism: Analysis of an Ecological Adaptation.* Berlin: Springer-Verlag.

Kortschack, H. P., C. C. E. Hartt, G. O. Burr. 1965. Carbon dioxide fixation in sugarcane leaves. *Plant Physiology* 40:209–213.

Lea, P. J., R. D. Blackwell, K. W. Joy. 1992. Ammonia assimilation in higher plants. In: K. Mengel, D. J. Pilbeam (eds.), *Nitrogen Metabolism of Plants.* Oxford: Clarendon Press, pp. 153–186.

Leegood, R. C., D. A. Walker, C. H. Foyer. 1985. Regulation of the Benson-Calvin cycle. In: J. Barber, N. R. Baker (eds.), *Photosynthetic Mechanisms and the Environment.* Amsterdam: Elsevier, pp. 189–258.

Lorimer, G. H., H. M. Miziorko. 1980. Carbamate formation in the ϵ-amino group of a lysyl residue as the basis for the activation of ribulosebisphosphate carboxylase by CO_2 and Mg^{2+}. *Biochemistry* 19:5321–28.

Nimmo, G. A., M. B. Wilkins, C. A. Fewson, H. G. Nimmo. 1987. Persistent circadian rhythms in the phosphorylation state of phosphoenolpyruvate carboxylase from *Bryophyllum fedtschenkoi* leaves and its sensitivity to inhibition by malate. *Planta* 170:408–415.

Osmond, C. B., O. Björkman. 1972. Simultaneous measurement of oxygen effects on net photosynthesis and glycolate metabolism in C3 and C4 species of *Atriplex. Carnegie Institution of Washington Yearbook* 71:141–148.

Portis, A. R. 1990. Rubisco activase. *Biochimica et Biophysica Acta* 1015:15–28.

Preiss, J. 1988. Biosynthesis of starch and its regulation. In: J. Preiss (ed), *Biochemistry of Plants. Vol. 10. Carbohydrates.* New York: Academic Press. pp. 181–254.

Raghavendra, A. S., V. S. R. Das. 1978. The occurrence of C4-photosynthesis: A supplementary list of C4 plants reported during late 1974—mid-1977. *Photosynthetica* 12:200–208.

Winter, K., J. H. Troughton. 1978. Carbon assimilation pathways in *Mesembryanthemum nodiflorum. Zeitschrift für Pflanzenphysiologie* 88:153–162.

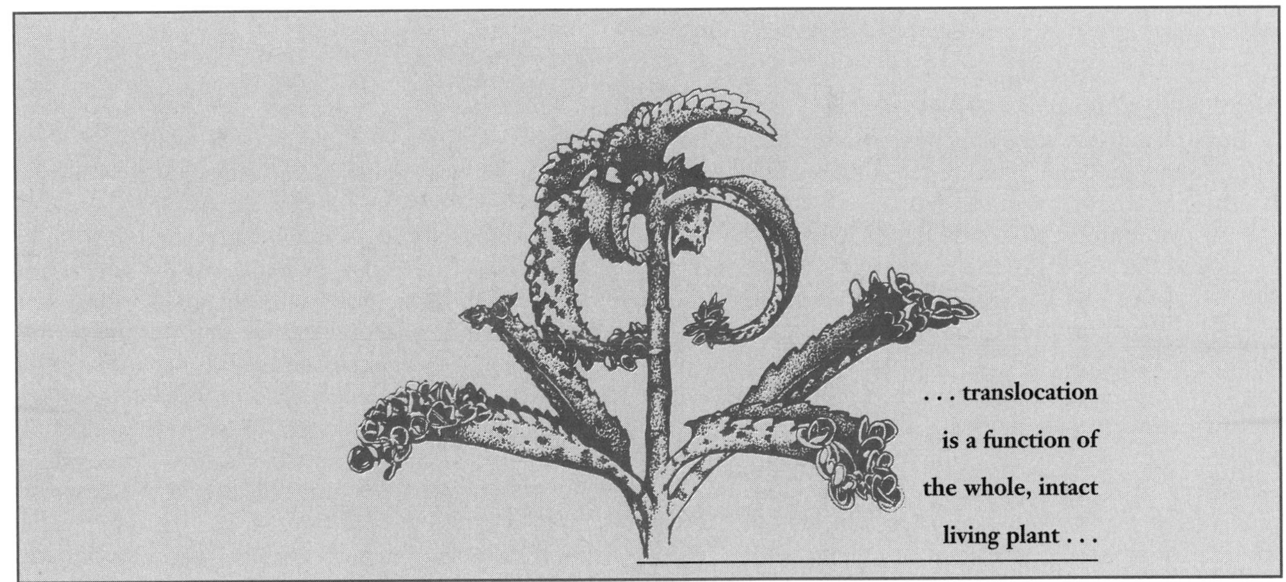

A. S. Crafts (1961)

11

Translocation and Distribution of Photoassimilates

The previous two chapters showed how energy was conserved in the form of carbon compounds, or **photoassimilate.** Although a portion of the carbon assimilated on a daily basis is retained by the leaf to support its continued growth and metabolism, the majority is exported out of the leaf to nonphotosynthetic organs and tissues. There, it is either metabolized directly or placed in storage for retrieval and metabolism at a later time. The transport of photoassimilates over long distances is known as **translocation.** Translocation occurs in the vascular tissue called **phloem.** Phloem translocation is a highly significant process that functions to ensure an efficient distribution of photosynthetic energy and carbon throughout the organism. Phloem translocation is also important from an agricultural perspective because it plays a significant role in determining productivity, crop yield, and the effectiveness of applied herbicides and other xenobiotic chemicals.

The subject of this chapter is the structure of the phloem and its function in the translocation and distribution of photoassimilates. The principal topics to be covered include

- the basis for identifying phloem as the route for translocation of photoassimilate and the nature of substances translocated in the phloem;
- the structure of the phloem tissue, especially the several unique aspects of sieve tube structure and composition;
- the source-sink concept and the significance of sources and sinks to the translocation process;
- the pressure-flow hypothesis for phloem translocation and the processes by which photoassimilates gain entry into the phloem in the leaf and are subsequently removed from the translocation stream at the target organ;
- factors that regulate the distribution of photoassimilate between competing sinks; and
- the loading and translocation of xenobiotic agrochemicals.

TRANSLOCATION OF PHOTOASSIMILATES

Girdling experiments that interrupt the phloem, analyses of phloem exudates, and experiments with radioisotopes have confirmed that photoassimilates and other small organic substances are translocated over long distances through the phloem tissue.

Attempts to distinguish between the translocation of inorganic and organic substances in plants can be traced back to the seventeenth-century plant anatomist M. Malpighi. In his experiments, Malpighi removed a ring of bark (containing phloem) from the wood (containing xylem) of young stems by separating the two at the vascular cambium, a technique known as **girdling.** Because the woody xylem tissue remained intact, water and inorganic nutrients continued to move up to the leaves and the plant was able to survive for some time. Girdled plants, however, developed characteristic swellings of the bark in the region immediately above the girdle (Fig. 11.1).

Over the years this experiment has been repeated and refined to include nonsurgical girdling such as by localized steam-killing or chilling. The characteristic swelling is attributed, in part, to an accumulation of photoassimilate flowing downward, which is blocked from moving further by removal or otherwise interfering with the activity of the phloem. As we now know, the downward stream also contains nitrogenous material and probably hormones that help to stimulate proliferation and enlargement of cells above the blockage. Eventually, of course, the root system will starve from the lack of nutrients and the girdled plant will die.

An analysis of phloem exudate provides more direct evidence in support of the conclusion that photoassimilates are translocated through the phloem. Unfortunately, phloem tissue does not lend itself to analysis as easily as xylem tissue does (described in Chap. 3). This is because the translocating elements in the phloem are, unlike xylem vessels and tracheids, living cells when functional. These cells contain a dense, metabolically active cytoplasm and, because of an inherent sealing action of its cytoplasm, do not exude their contents as readily as

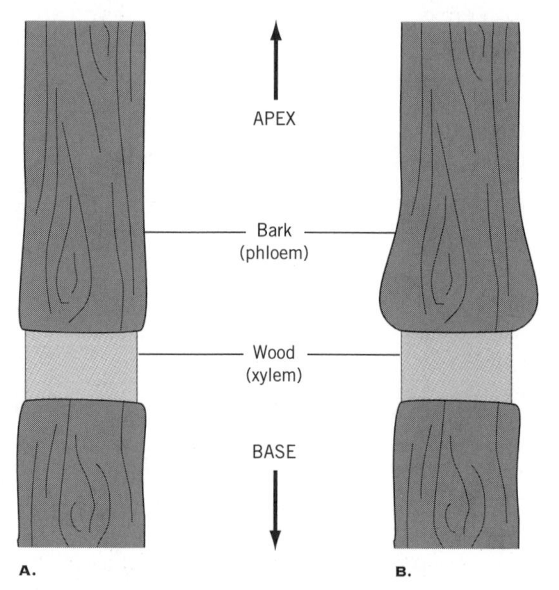

FIGURE 11.1 The results of girdling on woody stems. (*A*) The phloem tissue can be removed by separating the phloem (the bark) from the xylem (the wood) at the vascular cambium. (*B*) The girdle interrupts the downward flow of nutrients and hormones, resulting in a proliferation of tissue immediately above the girdle.

do xylem vessels. Moreover, phloem contains numerous parenchyma cells that, while not directly involved in the transport process, do provide contaminating cytoplasm. Cutting the stems of some herbaceous plants will produce an exudate of largely phloem origin, but in some plants, such as some representatives of the family Cucurbitaceae, the exudate may quickly gel on contact with oxygen, making collection and subsequent analyses difficult. The gelling of phloem exudate is due to the properties of a particular phloem protein, which is described more fully later in this chapter. In spite of these difficulties, however, numerous investigators have successfully completed analyses of phloem exudates obtained by making incisions into the phloem tissue, assisted in part by the development of modern analytical techniques applicable to very small samples.

One intriguing solution to the problem of obtaining the contents of sieve tubes uncontaminated by other cells was provided by insect physiologists studying the nutrition of aphids. Aphids are one of several groups of small insects that feed on plants by inserting a long mouthpart (the stylus) directly into individual sieve tubes. When feeding aphids are anaesthetized with a stream of carbon dioxide and the stylus carefully severed with a razor blade, phloem sap continues to exude from the cut stylus for several days. The aphid technique works well for a number of herbaceous plants and some woody shrubs, but it is restricted to those plants on which the aphids naturally feed. The principal advantage of this technique is that the severed aphid stylet delivers an uncontaminated sieve tube sap. Although the volumes delivered are relatively low, this technique has proven extremely useful in studies of phloem transport. The continued exudation, incidentally, demonstrates that phloem sap is under pressure, an important observation with respect to the proposed mechanism for phloem transport to be discussed later.

The third line of evidence involves the use of radioactive tracers, predominantly ^{14}C and usually fed to a leaf. A typical example is the translocation of photoassimilate in petioles of sugarbeet (*Beta vulgaris*) leaves (Mortimer, 1965). In these experiments, attached leaves were allowed to photosynthesize in a closed chamber containing a radioactive carbon source ($^{14}CO_2$). After 10 minutes, the radiolabeled photoassimilate being transported out of the leaf was immobilized by freezing the petiole in liquid nitrogen. Cross-sections of the frozen petiole were prepared and placed in contact with X-ray film. The resulting image on the X-ray film, or radioautograph, indicated that the radioactive photoassimilate being translocated out of the leaf was localized exclusively in the phloem (Fig 11.2). Similar experiments have been conducted on a variety of herbaceous and woody plants and with other radioactive nuclides such as phosphorous and sulphur with the same conclusion—*the translocation of photoassimilates and other organic*

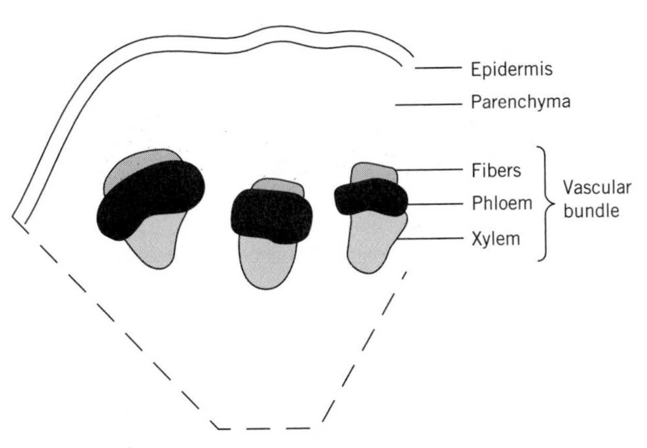

FIGURE 11.2 Location of radioactivity (black area) in the phloem of sugarbeet petioles after 10 minutes of photosynthesis in the presence of $^{14}CO_2$. (Diagram is based on the work of Mortimer, 1965.)

TABLE 11.1 The chemical composition of phloem exudate from stems of actively growing castor bean (*Ricinus communis*).

Organic	$mg\ l^{-1}$
Sucrose	80–106
Protein	1.45–2.20
Amino acids	5.2
Malic acid	2.0–3.2
Inorganic	$meq\ l^{-1}$
Anions (Inorganic)	20–30
Cations (Inorganic)	74–138
Total dry matter	100–125 $mg\ l^{-1}$

Data from Hall and Baker, 1972.

compounds over long distances occurs through the phloem tissue. There are, as already noted in Chapter 3, exceptions to this rule, such as when stored sugars are mobilized in the spring of the year and translocated through the xylem to the developing buds.

THE COMPOSITION OF PHLOEM EXUDATE

In what form is photoassimilate translocated? This question may be answered by analyzing the chemical composition of phloem exudate. Phloem sap can be collected from aphid stylets or, alternatively, from some plants by simply making an incision into the bark. If done carefully, to avoid cutting into the underlying xylem, the incision opens the sieve tubes and a relatively pure exudate can be collected in very small microcapillary tubes for subsequent analysis. As might be expected, the chemical composition of phloem exudate is highly variable. It depends on the species, age, and physiological condition of the tissue sampled. Even for a particular sample under uniform conditions, there may be wide variations in the concentrations of particular components between subsequent samples. For example, an analysis of phloem exudate from stems of actively growing castor bean (*Ricinus communis*) (Table 11.1) shows that the exudate contains sugars, protein, amino acids, the organic acid malate, and a variety of inorganic anions and cations. The predominant amino acids are glutamic acid and aspartic acid, which, as shown earlier, are common forms for the translocation of assimilated nitrogen (Chap. 6). The inorganic anions include phosphate, sulphate, and chloride—nitrate is conspicuously absent—while the predominant cation is potassium. Although not shown in Table 11.1, some plant hormones (auxin, cytokinin, and gibberellin) were also detected, but at very low concentrations. Of course, many of the components identified in phloem exudate—inorganic ions, for example—are cytoplasmic constituents of the translocating cells and do not necessarily represent translocated photoassimilate. Protein found in phloem exudates includes a wide variety of enzymes as well as one predominant protein (called **P-protein**) that is unique to the translocating cells. We will return to a discussion of P-protein later in this chapter.

The principal constituent of phloem exudate in most species is sugar. In castor bean it is sucrose, which comprises approximately 80 percent of the dry matter (Table 11.1). Such a preponderance of sucrose in the translocation stream strongly suggests that it is the predominant form of translocatable photoassimilate. This suggestion has been amply confirmed by labeling experiments. In the example of translocation in sugarbeet petioles described earlier, more than 90 percent of the radioactivity, following 10 minutes of labeling with $^{14}CO_2$, was recovered as sucrose. There are exceptions to this rule—one is the squash family (Cucurbitaceae) where nitrogenous compounds (principally amino acids) are quantitatively more important—but overall sugar, particularly sucrose, accounts for the bulk of the translocated carbon. Zimmermann and Ziegler (1975) have compiled a survey of over 500 species representing approximately 100 dicotyledonous families. Their survey confirms that sucrose is almost universal as the dominant sugar in the phloem stream.

A small number of families translocate, in addition to sucrose, oligosaccharides of the raffinose series (raffinose, stachyose or verbascose) (Fig. 11.3). Stachyose, for example, accounts for about 46 percent of the sugars in stem internodes of *Cucurbita maxima* (Richardson et al., 1982). Yet other families (Oleaceae, Rosaceae) translocate some of their photoassimilates as the sugar alcohols mannitol or sorbitol (Zimmermann and Ziegler, 1975).

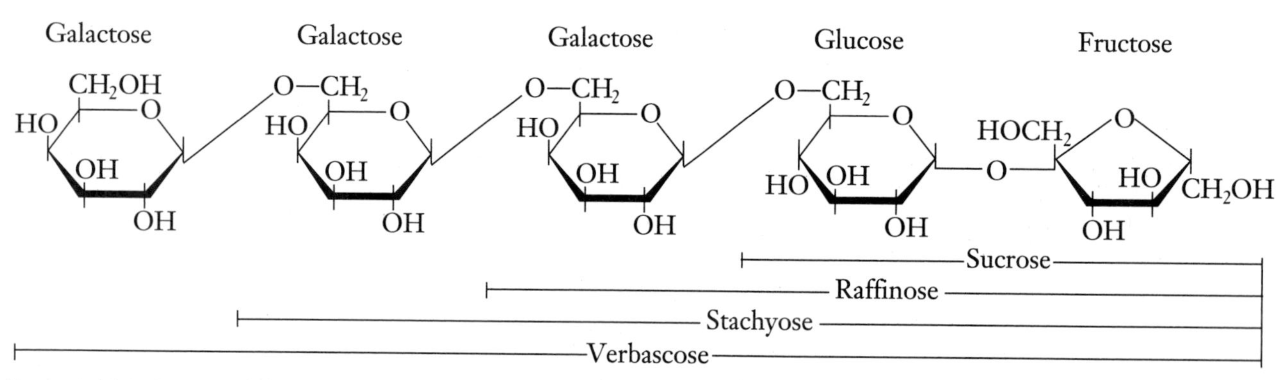

FIGURE 11.3 Sugars of the raffinose series. Raffinose, stachyose, and verbascose consist of sucrose with 1, 2, or 3 galactose units, respectively. All sugars in the raffinose series, including sucrose, are nonreducing sugars.

Why sucrose? It is interesting to speculate on why sucrose is the preferred vehicle for long-distance translocation of photoassimilate. One possibility is that sucrose, a disaccharide, and its related oligosaccharides are nonreducing sugars. On the other hand, all monosaccharides, including glucose and fructose, are *reducing sugars.* Reducing sugars have a free aldehyde or ketone group that is capable of *reducing* cupric ion (Cu^{3+}) to cuprous ion (Cu^{2+}) under alkaline conditions. Some oligosaccharides, such as sucrose, are *nonreducing* sugars because the acetal link between the subunits is stable and nonreactive in alkaline solution. The exclusive use of nonreducing sugars in the translocation of photoassimilate may be related to this greater chemical stability. Nonreducing sugars are less likely to react with other substances along the way. Indeed, free glucose and fructose, both reducing sugars, are rarely found in phloem exudates. The occasional report of reducing sugars in phloem exudate probably indicates contamination by nonconducting phloem cells, where reducing sugars are readily formed by hydrolysis of sucrose or other oligosaccharides.

A second possible factor is that the β-fructoside linkage between glucose and fructose, a feature of sucrose and other members of the raffinose series, has a relatively high negative free energy of hydrolysis—about -27 kJ mol^{-1} compared with about -31 kJ mol^{-1} for ATP. Sucrose is thus a small and highly mobile but relatively stable packet of energy, which may account for its "selection" as the principal form of assimilate to be translocated in most plants (Giaquinta, 1983).

THE STRUCTURE OF PHLOEM TISSUE

The principal conducting elements in phloem tissue are elongated ranks of cells that contain a living, but highly modified, protoplast.

The distinguishing feature of phloem tissue is the conducting cell called the **sieve element.** Also known as a **sieve tube,** the sieve element is an elongated rank of individual cells, called **sieve-tube members,** arranged end-to-end (Fig. 11.4). Unlike xylem tracheary elements, phloem sieve elements lack rigid walls and contain living protoplasts when mature and functional. The protoplasts of contiguous sieve elements are inter-

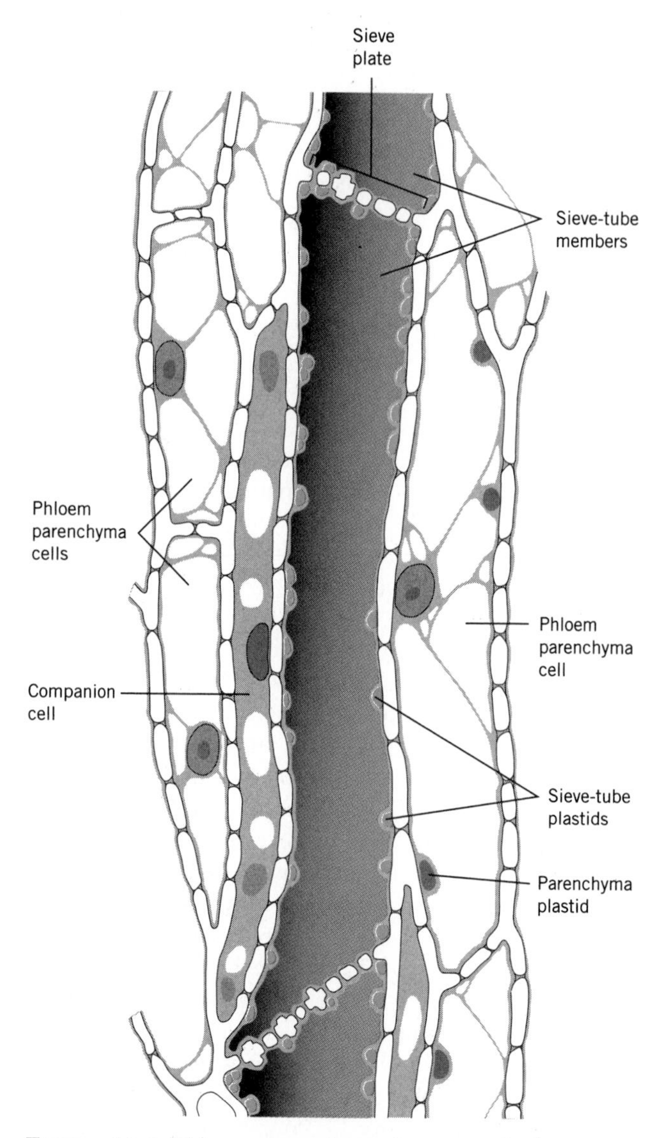

FIGURE 11.4 Phloem tissue from the stem of tobacco.

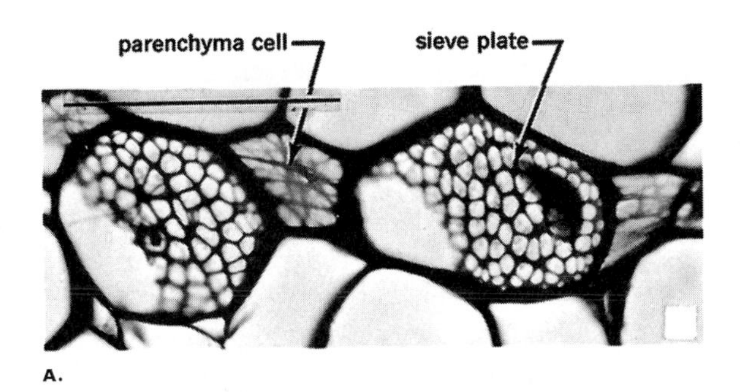

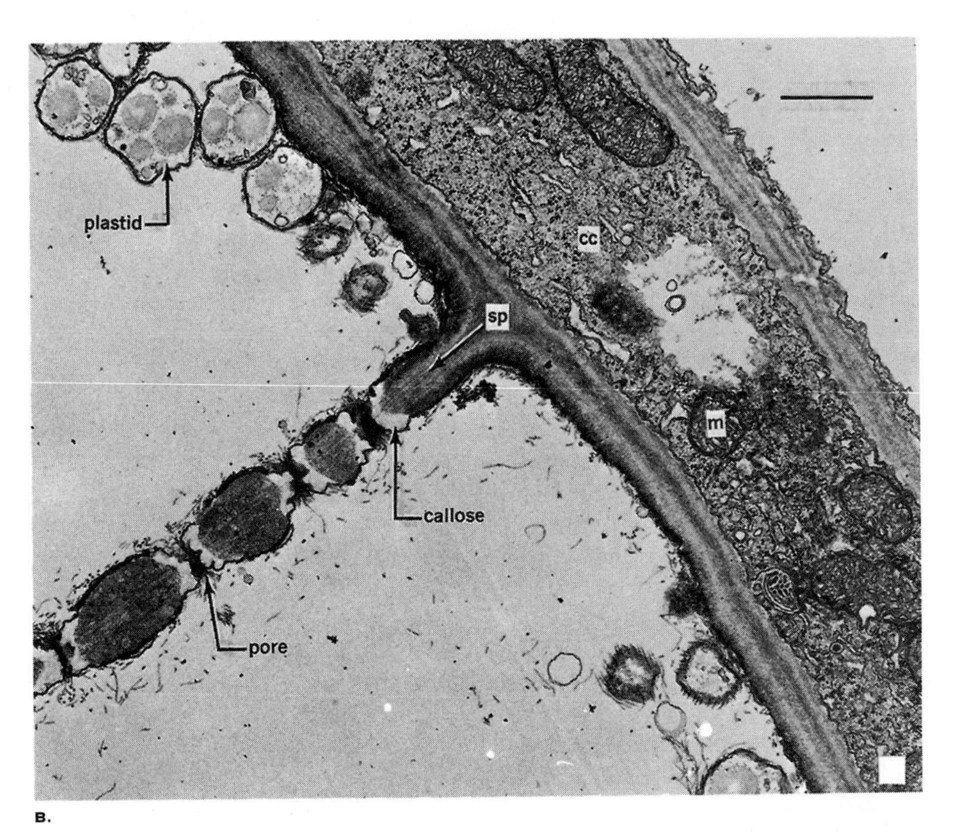

FIGURE 11.5 A phloem sieve plate as seen in the electron microscope. (*A*) Surface view. (*B*) Long section. (From K. Esau. *Anatomy of Seed Plants*. 1977. New York, Wiley. Reprinted by permission.)

connected through specialized **sieve areas** in adjacent walls. Where the pores of the sieve area are relatively large and are found grouped in a specific area, they are known as **sieve plates** (Fig. 11.5). Sieve plates are typically found in the end walls of sieve-tube members and provide a high degree of protoplasmic continuity between consecutive sieve-tube members. Additional pores are found in sieve areas located in lateral walls. These are generally smaller and are not, as a rule, grouped in distinct areas. These sieve areas nonetheless provide cytoplasmic continuity through the lateral walls of adjacent sieve elements.

As noted earlier, mature sieve elements contain active cytoplasm. However, as the sieve element matures it undergoes a series of progressive changes that result in the breakdown and loss of the nucleus, the vacuolar membrane (or tonoplast), ribosomes, the Golgi apparatus (or dictyosomes), as well as microtubules and filaments. At maturity, the cells retain the plasmalemma, and endoplasmic reticulum (although it is somewhat modified), and mitochondria. Even though there is no central vacuole as such, the cytoplasmic components appear to assume a parietal position in the cell; that is, along the inner wall of the cell.

In addition to sieve elements, phloem tissue also contains a variety of parenchyma cells. Some of these cells are intimately associated with the sieve-tube members and for this reason are called **companion cells.** Companion cells contain a full complement of cytoplasm and cellular organelles. A companion cell is derived from the same mother cell as its associated sieve-tube member and shares numerous cytoplasmic connections with it. The interdependence of the sieve-tube member and companion cell is reflected in their lifetimes—the companion cell remains alive only so long as the sieve-tube member continues to function. When the sieve-tube member dies, its associated companion cell also dies. Companion cells are believed to provide metabolic support for the sieve-tube member and, perhaps, are involved in the transport of sucrose or other sugars into the sieve tube (Gunning, 1976).

The rest of the phloem parenchyma cells are not always readily distinguishable from companion cells, even at the ultrastructural level. The single exception is found in the minor leaf veins of some plants, typically of herbaceous dicotyledonous plants. Here certain phloem parenchyma cells develop extensive ingrowths of the cell wall. The result is a significant increase in the surface area of the plasma membrane. These cells are called **transfer cells.** The precise role of transfer cells is not understood but, as the name implies, they are thought to be involved in collecting and passing on photoassimilates produced in nearby mesophyll cells. They may also be involved in recycling solutes that enter the apoplast from the transpiration stream (Evert, 1990). These proposed functions are speculative, based largely on the assumption that the high protoplasmic surface area would be expected to facilitate solute exchange between the transfer cell and the surrounding apoplast.

P PROTEIN AND CALLOSE

It was noted above that phloem exudate contains a significant amount of protein. One, now called **P-protein** (for *p*hloem *protein*), characteristically accumulates in the sieve elements in such large quantities that it can be observed under the light microscope. When P-protein was first observed by early anatomists, it was assumed to be a carbohydrate and was called slime. The accumulations observed in the region of the sieve plates were called slime plugs. It was not until the 1960s that cytochemical tests revealed the proteinaceous nature of this material and the name P-protein was introduced (Cronshaw, 1975).

In the early stages of sieve element differentiation, P-protein appears in the form of discrete protein bodies. As the sieve elements mature, the P-protein bodies continue to enlarge. At the time the nucleus, vacuole, and other cellular organelles disappear, the P-protein bodies disperse in the cytoplasm. In some species, such as maple (*Acer rubrum*), the P-protein takes the form of a loose network of filaments, ranging from 2 to 20 nm in width. In others, such as tobacco (*Nicotiana* sps.), the filaments appear tubular in cross-section. In yet others, such as some leguminous plants, P-protein takes the form of crystalline inclusions.

Biochemical investigations of phloem proteins began in the early 1970s, principally in exudates of *Cucurbita.* Some caution must be exercised when interpreting these results, however, since phloem exudates contain proteins in addition to P-protein. Using the technique of **s**odium **d**odecyl **s**ulphate **p**oly**a**crylamide **g**el **e**lectrophoresis (**SDS-PAGE**), a variety of polypeptide subunits with molecular mass values ranging from 15 to 220 kD have been reported. Apparently phloem protein varies widely between species, with respect to both its subunit composition and its chemical properties. One particularly interesting property of phloem protein is its capacity to form a gel. Walker and Thaine (1971) showed that gelation could be prevented by 2-mercaptoethanol, a reducing agent that prevents formation of intermolecular disulfide (—S—S—) bonds. The effect of reducing agent is fully reversible—removal of the 2-mercaptoethanol allows gelling to proceed. This effect was traced to a single basic protein in the phloem exudate. This protein probably accounts for the propensity of certain phloem exudates, such as from *Cucurbita,* to gel rapidly on exposure to air. Another interesting property of some phloem proteins is their capacity to function as carbohydrate-binding proteins, or lectins (see Chap. 6).

P-protein has been the subject of considerable attention over the years because of its prominence in sieve elements and its propensity to plug the pores in the sieve plates. Still, its role and that of other phloem-specific proteins is not yet clear. P-protein has been implicated in various ways in the transport function of sieve elements. According to some theories, P-protein is considered an active participant in the transport process. At the same time, the presence of P-protein in sieve elements is invoked as an argument against other theories. It is now generally accepted that, *in intact, functioning sieve elements,* P-protein is located principally along the inner wall of the sieve element and does not plug the sieve plate.

The formation of plugs in the sieve plates occurs only when the sieve element is injured. This occurs because the sieve element is normally under *positive* hydrostatic pressure, as evidenced by the continued flow of exudate from aphid stylets. When the pressure is released through injury to the sieve element, the contents, including P-protein, surge toward the site of injury. This results in the accumulation of P-protein, possibly assisted by its gelling properties, as "slime" plugs on the side of the sieve plate away from the pressure release. Thus, it appears that at least one function of P-protein is protective. By sealing off sieve plates in areas where the integrity of the phloem has been breached, P-protein helps to maintain the positive hydrostatic pressure in the phloem and reduce unnecessary loss of translocated photoassimilate.

Another prominent and somewhat controversial feature of sieve elements is the presence of **callose.** Callose, a $\beta 1 \rightarrow 3$-glucan, is related to starch and cellulose.

Small amounts of callose are deposited on the surface of the sieve plate or lining the pores through which the interconnecting strand of cytoplasm pass between contiguous cells (Fig. 11.5). Controversy over the role of callose arises from the frequent observation that callose appears to accumulate in the pores to the extent that it would appear to interfere with translocation. However, it is now known that callose can be synthesized very rapidly (within a matter of seconds) and, similar to P-protein, will accumulate in the sieve area in response to injury (Eschrich, 1975). Large amounts of callose also appear to be deposited on the sieve plates of older, nonfunctional sieve elements. In both cases, the function of callose appears to be one of sealing off sieve elements that have been injured or are no longer functional, thus preserving the integrity of the translocating system.

SOURCES AND SINKS

The direction of long-distance translocation in the phloem is determined largely by relationships between sources and sinks.

Source-sink relationships were briefly introduced in the previous chapter. Because the source-sink concept is central to our understanding of how photoassimilate is translocated and distributed through the phloem, we will now examine this concept more thoroughly.

Identification of an organ or tissue as a source or sink depends on the direction of its *net* assimilate transport (Ho, 1988; Dickson, 1991). An organ or tissue that produces more assimilate than it requires for its own metabolism and growth is a **source.** A source is thus a net exporter or producer of photoassimilate; that is, it exports more assimilate than it imports. Mature leaves and other actively photosynthesizing tissues are the predominant sources in most plants. A **sink,** on the other hand, is a net importer or consumer of photoassimilate. Roots, stem tissues and developing fruits are examples of organs and tissues that normally function as sinks. *The underlying principle of phloem translocation is that photoassimilates are translocated from a source to a sink.* Sink organs may respire the photoassimilate, use it to build cytoplasm and cellular structure, or place it into storage as starch or other carbohydrate.

Any organ, at one time or another in its development, will function as a sink and may undergo a conversion from sink to source. Leaves are an excellent example. In its early stages of development a leaf will function as a sink, drawing photoassimilates from older leaves to support its active metabolism and rapid enlargement. However, as a leaf approaches maximum size and its growth rate slows, its own metabolic demands diminish and it will gradually switch over to a net exporter. The mature leaf then serves as a source of photoassimilate for sinks elsewhere in the plant. The conversion of a leaf from sink to source is a gradual process, paralleling the progressive maturation of leaf tissue. In simple leaves, for example, the export of photoassimilate from mature regions of the leaf may begin while other regions are still developing and functioning as sinks. In compound leaves, such as ash (*Fraxinus pennsylvanica*) and honeylocust (*Gleditsia triacanthos*), the early-maturing basal leaflets may export photoassimilate to the still-developing distal leaflets as well as out of the leaf (Dickson, 1991).

MECHANISM OF TRANSLOCATION IN THE PHLOEM

Sugars are translocated in the phloem by mass transfer along a hydrostatic pressure gradient between the source and sink.

What is the mechanism for assimilate translocation over long distances through the phloem? Any comprehensive theory must take into account a number of factors. These include: (a) the structure of sieve elements, including the presence of active cytoplasm, P-protein, and resistances imposed by sieve plates; (b) observed rapid rates of translocation (50 to 250 cm hr^{-1}) over long distances; (c) translocation in different directions at the same time; (d) the initial transfer of assimilate from leaf mesophyll cells into sieve elements of the leaf minor veins (called "**phloem loading**"); and (e) final transfer of assimilate out of the sieve elements into target cells (called "**phloem unloading**)." Phloem loading and unloading will be discussed in a following section.

At various times assimilate transport has been explained in terms of simple diffusion, cytoplasmic streaming, ion pumps operating across the sieve plate, and contractile elements in the transcellular protoplasmic strands. All of these proposals have been largely rejected on both theoretical and experimental grounds. (For a review of these various proposals and the problems associated with each, the reader is referred to the excellent review by R. F. Evert, 1982.)

The most credible and generally accepted model for phloem translocation is one of the earliest. Originally proposed by E. Münch in 1930 but modified by a series of investigators since, the **pressure-flow** hypothesis remains the simplest model and continues to earn widespread support amongst plant physiologists. The pressure-flow mechanism is based on the mass transfer of solute from source to sink along a hydrostatic (turgor) pressure gradient (Fig. 11.6). Translocation of solute in the phloem is closely linked to the flow of water in the transpiration stream and a continuous recirculation of water in the plant.

Assimilate translocation begins with the loading of

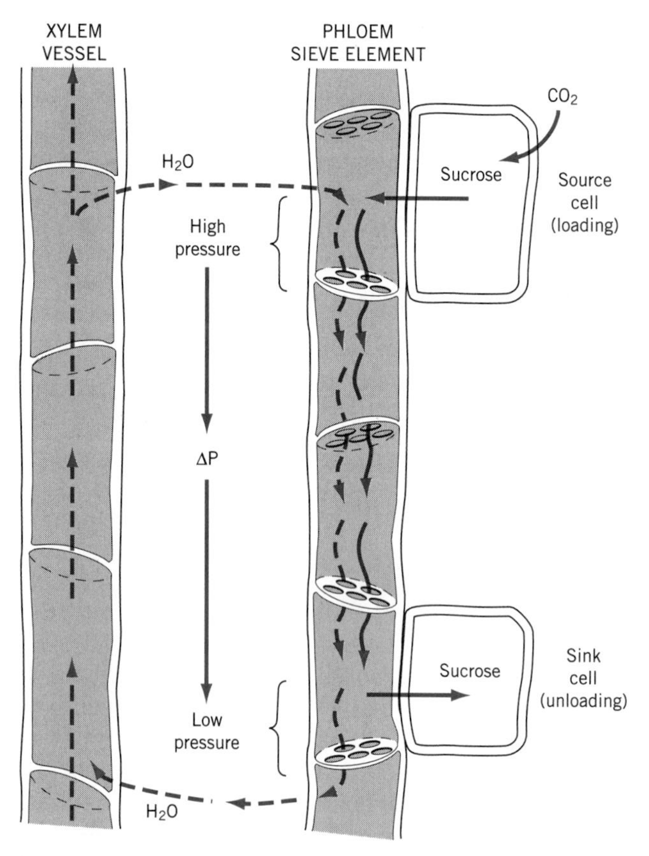

FIGURE 11.6 A diagram of pressure flow. The loading of sugar into the sieve element adjacent to a source cell causes the osmotic uptake of water from nearby xylem elements. The uptake of water increases the hydrostatic (turgor) pressure in the sieve element. The pressure is lowered at the sink end when sugar is unloaded into the receiver cell and the water returns to the xylem. This pressure differential causes a flow of water from the source region to the sink. Sugar is carried passively along.

sugars into sieve elements at the source. Typically, loading would occur in the minor veins of a leaf, close to a photosynthetic mesophyll or bundle-sheath cell. The increased solute concentration in the sieve element lowers its water potential and, consequently, is accompanied by the osmotic uptake of water from the nearby xylem. This establishes a higher turgor or hydrostatic pressure in the sieve element at the source end. At the same time, sugar is unloaded at the sink end—a root or stem storage cell, for example. The hydrostatic pressure at the sink end is lowered as water leaves the sieve elements and returns to the xylem. So long as assimilates continue to be loaded at the source and unloaded at the sink, this pressure differential will be maintained, water will continue to move in at the source and out at the sink, and assimilate will be carried passively along.

According to the pressure-flow hypothesis, solute translocation in the phloem is fundamentally a passive process; that is, translocation requires no direct input of metabolic energy to make it function. Yet for years it has been observed that translocation of assimilates was sensitive to metabolic inhibitors, temperature, and other conditions suggesting that metabolic energy was required. More recent experiments, however, have been designed to discriminate between energy requirements for the actual movement of assimilate within the sieve elements and the global energy requirements for translocation from source to sink. The results are clear. The effects of low temperature and metabolic inhibitors are either transient or cause disruption of the P-protein and plug the sieve plates. Energy requirements for translocation within the sieve elements are therefore minimal and compatible with the passive character of the pressure-flow hypothesis. The energy requirement indicated in earlier translocation experiments no doubt reflected the needs for loading and unloading the sieve elements.

The principle of pressure flow can easily be demonstrated in the laboratory by connecting two osmometers (Fig. 11.7), but a simple physical demonstration does not in itself prove the hypothesis. A number of questions must be answered. First, is the sieve tube under pressure? The prolonged exudation of phloem sap

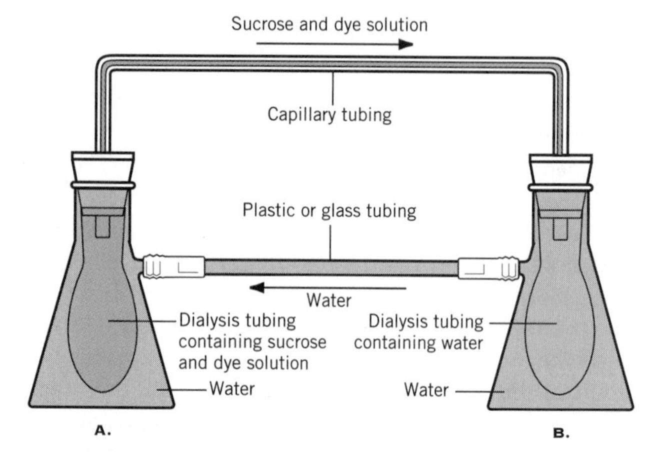

FIGURE 11.7 A physical model of the pressure-flow hypothesis for translocation in the phloem. Two osmometers are constructed from side-arm flasks and dialysis tubing. Osmometer A (the source) initially contains a concentrated sucrose solution and a dye. Osmometer B (the sink) contains only water. The two osmometers are connected by capillary tubing (the phloem). Water moves into osmometer A by osmosis, generating a hydrostatic pressure that forces water out of osmometer B. Water returns via the tubing (the xylem) that connects the side-arm flasks. As a consequence of the flow of water between the two ends of the system, the sucrose-dye solution flows through the capillary form osmometer A to osmometer B. In the model, the system will come to equilibrium and flow will cease when the sucrose concentration is equal in the two osmometers. In the plant, flow is maintained because sucrose is continually added to the source (*A*) and withdrawn at the sink (*B*).

from excised aphid stylets clearly demonstrates that it is. The total volume of exudate may exceed the volume of an individual sieve tube by several thousand times. It is difficult to measure the turgor pressure of individual sieve elements, although a number of attempts have been made over the years. Turgor pressure can be calculated as the difference between sieve tube water potential (Ψ) and osmotic potential (Ψs) (Chap. 2, Eq. 2.10), or it may be measured directly by inserting a small pressure-sensing device, or micromanometer, into the phloem tissue. J. Wright and D. B. Fisher, for example, measured the turgor pressure in willow saplings by sealing a closed glass capillary over a severed aphid stylet (Wright and Fisher, 1980). Pressure was calculated from the ratio of the compressed and uncompressed air columns in the capillary. As might be expected, values reported in the literature range widely, depending on the method chosen, plant material, the time of day, and physiological status of the subject plant. Whether calculated or measured directly, values of 0.1 MPa to 2.5 MPa are typical.

A second question to be addressed is whether differences in sugar concentration and the turgor pressure drop in the sieve tube is sufficient to account for the measured rates of transport. Sugar concentration is, of course, highly variable, depending on the rate of photosynthesis and the general physiological condition of the plant. However, most studies have confirmed that the sugar content of phloem exudate taken near the source is higher than in exudates taken near sinks. Using the Hagen-Poiseuille equation (Chap. 3, Eq. 3.3), it has been calculated that a pressure drop of about 0.06 MPa m^{-1} would be required for a 10 percent sucrose solution to flow at 100 cm hr^{-1} through a sieve tube with a radius of 12 μm (Weatherly and Johnson, 1968). In these calculations, the resistance offered by sieve plates was taken into account by assuming that (a) the area of the pores in the sieve plate was equal to one-half the area of the sieve tube, (b) there were 60 sieve plates per cm of sieve tube, and (c) the sieve plate pores were not blocked. Assuming that the turgor pressure of sieve tubes in the source regions is typically in the range 1.0 to 1.5 MPa, and that it is zero in the sink (which may not be true), a pressure drop of 0.06 MPa m^{-1} would be sufficient to push a solution through the sieve tubes over a distance of 15 to 25 m. Flow over longer distances could be accomplished if the source sucrose concentration were higher and/or the flow rate were reduced (Tyree et al., 1974). D. B. Fisher compared pressure drop with flow rate in soybean (Fisher, 1978). He found that assimilates moved from the petiolule[1] (the source) to the root (the sink), a distance of 160 cm, at a velocity of 48 cm hr^{-1}. Fisher calculated, again using the Hagen-Poiseuille equation, that a pressure drop of 0.2 MPa would be required to achieve this velocity if the sieve plate pores were completely open. He then measured the sucrose concentration in the source and sink and calculated that the actual pressure drop was 0.44 MPa, twice that required. A pressure drop of 0.44 MPa would be sufficient to accomplish a velocity of 48 cm hr^{-1} even if the pores were only 70 to 75 percent open.

Another question that is frequently raised in discussions of the pressure-flow hypothesis is that of bidirectional transport. The translocation of assimilates simultaneously in opposite directions would at first seem incompatible with the pressure-flow hypothesis, but it does occur. Bidirectional transport is first of all a logical necessity. At any one time, plants will likely have more than one sink being served by the same source—roots for metabolism and storage and developing apical meristems or flowers, for example. It is also easy to demonstrate experimentally the movement of radiolabeled carbon and phosphorous in opposite directions through the same internode or petiole at the same time. This observation might easily be explained by movement through two separate vascular bundles or even through different sieve tubes in the same bundle. As long as the sieve elements are connected to different sinks, the pressure-flow hypothesis does not require that translocation occur in the same direction or even at the same velocity at any one time. A few experiments, however, have purported to show bidirectional translocation in the same sieve tube which, if true, would be far more difficult to reconcile with the pressure-flow hypothesis. It must be pointed out that bidirectional transport in the same sieve element has not been convincingly demonstrated. Most of these experiments suffer technical difficulties that allow the results to be interpreted in other ways.

Finally, it has often been argued that sieve elements, because of their structure and composition, offer a substantial resistance to flow and that pressure-flow might not provide sufficient force to overcome this resistance. In this regard it is important to note once again that the sieve plates *in functioning sieve tubes* are not occluded by either P-protein or callose. The presence of viscous cytoplasm and sieve plates undoubtedly imposes some resistance, but a variety of experiments have indicated that the capacity of the phloem to translocate assimilates is not normally a limiting factor in the growth of sinks. The phloem is a flexible system for translocation. It is easily capable of bypassing localized regions of high resistance and the hydrostatic pressure can be adjusted in response to demand at either the source or sink. There are even developmental controls, apparently to ensure that the phloem is adequately "sized" to meet anticipated demand. In wheat, for example, both the number and size of the vascular bundles serving a floral head

[1]Soybean has a trifoliate leaf, that is, a leaf comprised of three leaflets. The **petiolule** is the stemlike structure supporting the center leaflet. The **petiole** is the structure that attaches the entire trifoliate leaf to the stem.

correlates with the number of flowers. Thus, although there is little, if any, direct proof for the pressure-flow hypothesis, on the balance of evidence it is strongly favored.

PHLOEM LOADING AND UNLOADING

Phloem loading and unloading play a major role in regulating both the rate of translocation and the partitioning of assimilates between competing sinks. The loading and unloading processes thus offer considerable attraction as potential sites for manipulating assimilate distribution and, ultimately, crop yields.

PHLOEM LOADING

A discussion of phloem translocation is not complete without considering how assimilates are translocated from the photosynthetic mesophyll cells into the sieve elements at the source end (**phloem loading**) or from the sieve elements into the target cells at the sink end (**phloem unloading**). The path traversed by assimilate from the site of photosynthesis to the sieve element is not long. Most mesophyll cells are within a few tenths of a mm, at most three or four cells, distance from a minor vein ending where loading of assimilate into the **sieve element-companion cell complex** (se-cc) actually occurs.[2] It is generally agreed that sucrose moves from the mesophyll cells to the phloem, probably phloem parenchyma cells, principally by diffusion through the plasmodesmata (i.e., the symplasm). At this point, the pathway becomes less certain and the subject of some debate. From the phloem parenchyma there are two possible routes into the se-cc complex (Fig. 11.8). Sucrose may continue through the symplasm—that is, through plasmodesmata—directly into the se-cc complex. This route is known as the **symplastic pathway.** Alternatively, the sugar may be transported across the mesophyll cell membrane and released into the cell wall (apoplastic) solution. From there it would be taken up across the membrane of the se-cc complex where it enters the long-distance transport stream. This route is known as the **apoplastic pathway.**

The apoplastic model for phloem loading gained favor in the mid-1970s, based largely on studies of translocation in sugarbeet leaves. D. R. Geiger and his coworkers used leaves that had been abraded with carbo-

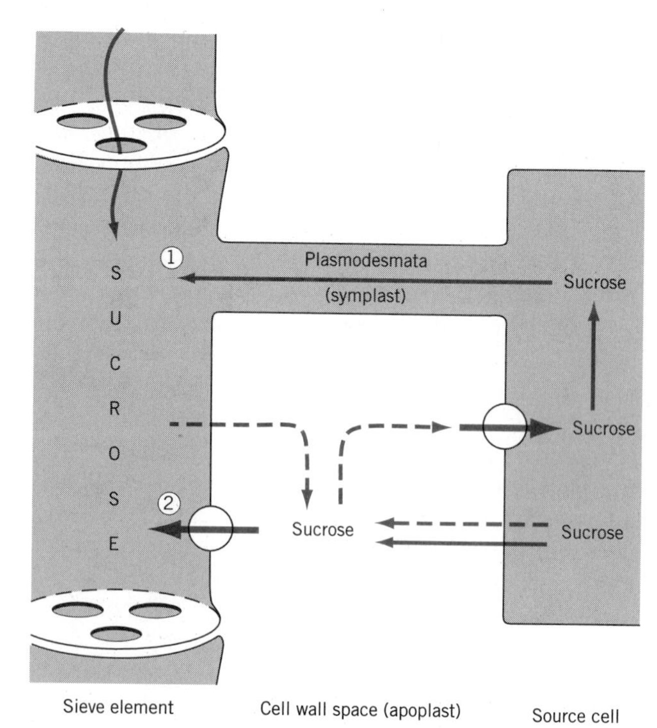

FIGURE 11.8 Loading and retrieval of sugars at the source. Sucrose may be loaded into sieve elements at minor vein endings via one of two paths. In path 1, the symplastic path, sugar moves through plasmodesmata that connect the protoplasts of the source cell and the sieve element. In path 2, the apoplastic path, sugar is released into the cell wall (apoplastic) space, from which it is actively transported across the plasma membrane of the sieve element by sugar-H^+ cotransport. Alternatively, sugar may leak into the apoplast (dashed lines) and be actively retrieved by either the source cell or the sieve element.

rundum to remove the cuticle, thus improving access to the leaf apoplast (Geiger et al., 1974). They found that radioactive sucrose appeared in the apoplast following a period of photosynthesis in the presence of $^{14}CO_2$. They also found that exogenously supplied sugar was readily absorbed into the se-cc complex when abraded leaves were bathed in a solution containing ^{14}C-sucrose. These results indicate that sucrose is normally found in the apoplast and can be taken into the sieve elements from the apoplast. Phloem loading in some plants is also inhibited by chemicals such as **p-chloromercuribenzene sulfonic acid (PCMBS)** when applied to abraded leaves or leaf disks. PCMBS and certain other sulfhydryl-specific reagents presumably interfere with carrier proteins involved in the transport of sucrose across the plasma membranes. (Carrier systems for the transport of solutes across membranes were discussed earlier in Chap. 5.) Because these reagents do not penetrate the cell membrane, any effect they have must be localized on the apoplastic surface of the membrane.

[2]It is not technically possible to discriminate between the respective roles of the companion cell and the sieve element in phloem loading and unloading experiments. For this reason, the sieve element and companion cells are considered as a single se-cc complex.

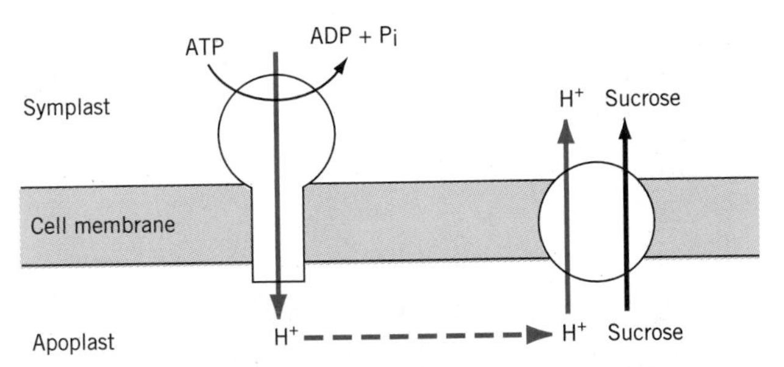

FIGURE 11.9 An illustration of sugar-H^+ cotransport. The energy for sugar uptake may be provided by a plasma membrane ATPase proton pump.

Sucrose and other sugars are selectively loaded into the se-cc complex against a concentration gradient, which usually infers active transport. In addition, there is an increasingly large body of evidence supporting the existence, in plant cells generally and phloem loading in particular, of a sucrose uptake mechanism that is both ATP-dependent and linked to the uptake of protons; that is, a sugar-H^+ cotransport (Fig. 11.9) (Giaquinta, 1983). This conclusion is supported by the observation that sugar uptake is accompanied by an increase in pH (i.e., a depletion of protons) or polarity changes in the apoplast. Conversely, if the pH is experimentally increased—that is, protons are removed from the apoplast by infiltrating the apoplast of abraded leaves with a basic buffer—uptake into the phloem cells will be inhibited. Finally, in most cases it can be shown that only sugars taken up into the se-cc complex and translocated in the sieve elements will elicit pH responses. Other sugars, which are not normally taken up by the phloem, elicit no pH changes. In summary, much of the evidence is consistent with an apoplastic pathway for phloem loading, by which sugars first pass from mesophyll or phloem parenchyma cells into the leaf apoplast. The sugar is subsequently taken up into the se-cc complex by sugar-H^+ symport, to be translocated out of the source region.

Although the exact nature of the sucrose-H^+ symport carrier is not yet worked out, genes (*SUT1*, *SUC2*) coding for the carrier have been identified and cloned from several species. These include spinach (*Spinacea oleraceae*), potato (*Solanum tuberosum*), *Plantago major*, and *Arabidopsis thaliana* (Stadler and Sauer, 1996). In several species, expression of the *SUC2* sucrose carrier gene appears limited to the companion cells, while in potato the *SUT1* gene product was located in the plasma membrane of the sieve elements but could not be detected within the companion cells. Such conflicting results might suggest that apoplastic loading occurs differently in different species or, alternatively, that different carriers are active in companion cells and sieve elements.

In spite of the strong evidence in support of an apoplastic pathway for phloem loading, there are other data that support transport through the symplast. Much of the data from sucrose feeding experiments, for example, indicate that a significant proportion—up to 60 percent—of the radioactive sucrose can be detected in *leaf mesophyll cells*. This leaves the experiments open to an alternate interpretation: that sucrose-H^+ cotransport exists as a mechanism for retrieving sucrose that has leaked from the photosynthetic cells into the apoplast. Such a retrieval mechanism would not discriminate between sugar that leaked from mesophyll, or possibly the se-cc complex, and sugar that was supplied exogenously by the experimenter. If such leakage does occur, a mechanism for retrieval would serve to prevent unnecessary loss of sugar from the transport stream. Indeed it has been postulated that a leakage-retrieval cycle occurs normally along the entire length of the translocation path.

If the uptake data do reflect retrieval by mesophyll cells rather than phloem loading, one is left with the conclusion that phloem loading occurs via the symplastic pathway. Several laboratories have presented evidence that appears to offer further support for this hypothesis. When mesophyll cells of *Ipomea tricolor* are injected with a fluorescent dye, the dye moves readily into neighboring mesophyll cells and appears in the minor veins within 25 minutes (Madore and Lucas, 1986). Since the dye is water-soluble and unable to cross membranes, it is assumed that the dye traveled into the minor veins via the symplastic connection between cells. It is also of interest that much of the data supporting apoplastic loading have come from experiments with one species: sugarbeet. In a recent survey, van Bel and others selected plants on the basis of whether they had abundant symplastic connections between the se-cc complex and adjacent cells of the minor veins, or whether these cells were **symplastically isolated,** that is, had no symplastic connections (van Bell et al., 1992). Those plants whose se-cc complexes were symplastically isolated exhibited characteristics of apoplastic loading, while those with abundant symplastic connections exhibited characteristics of symplastic loading.

The concept of symplastic loading does, however, raise some questions. For example, if sugars diffuse freely from the mesophyll into the se-cc complex, they should be equally free to diffuse back into the mesophyll cells. How, then, is it possible for the se-cc complex to accumulate sugars by simple diffusion through the plasmodesmata? Based on studies of phloem loading in *Cucurbita* sps., R. Turgeon has proposed a **polymer trap** model to account for symplastic loading (Turgeon, 1996). Species, such as the cucurbits, which have abundant plasmodesmata connections with the se-cc complex and appear to load sympastically, also translocate oligosaccharides in the raffinose series. According to the polymer trap model, sucrose diffuses from the mesophyll or bundle-sheath cells into the companion cells through the connecting plasmodesmata. The route is not certain, but the sucrose is probably able to diffuse through the sleeve of cytoplasm that lies between the plasma membrane and the desmotubule (see Fig. 1.12). The core of the desmotubule is believed to be too small to carry sucrose. In the companion cell, the sucrose is converted to an oligosaccharide, such as the tetrasaccharide stachyose, which is too large to diffuse back through the plasmodesmata. The polymer (i.e., stachyose) thus remains "trapped" in the se-cc complex, to be carried away by mass flow.

The symplastic model assumes that the plasmodesmata limit the passage of large molecules, but this may not be the case. Several recent studies of the sucrose transporter gene have indicated that both the transporter protein and its mRNA are able to pass through plasmodesmata between companion cells and sieve elements (Sauer, 1997). If macromolecules can pass through plasmodesmata, it is difficult to imagine why small oligosaccharides cannot. Perhaps plasmodesmata are more than simple tubes allowing solute flux between cells. This is an exciting issue that will no doubt receive considerable attention in the future.

Why there is more than one pathway for phloem loading is not clear. The symplastic pathway appears to have an energetic advantage by avoiding two carrier-dependent membrane transport steps. The observed energy-dependence of loading and translocation, however, is more readily explained by the apoplastic model. In the sympastic model, on the other hand, energy is required for the synthesis of oligosaccharides in the companion cells. It has also been suggested that species employing the symplastic pathway are more ancestral or that the apoplastic pathway is an evolutionary adaptation that arose as plants spread from tropical climates into more temperate regions. The new molecular approaches now available will be no doubt allow investigators to discriminate between available options. It may be that there is no universal pathway but that the path of phloem loading is family or species-specific. Given the theoretical and potential practical significance of phloem loading in determining yields, we can expect the investigation and debate to continue.

PHLOEM UNLOADING

Once assimilate has reached its target sink, it must be unloaded from the se-cc complex into the cells of the sink tissue. In principle, the problem is similar to loading; only the direction varies. In detail there are some significant differences. As with phloem loading, phloem unloading may occur via symplastic or apoplastic routes (Fig. 11.10) (Thorne, 1986). The symplastic route (pathway 1, Fig. 11.10) has been described predominantly in young, developing leaves and root tips. Sucrose

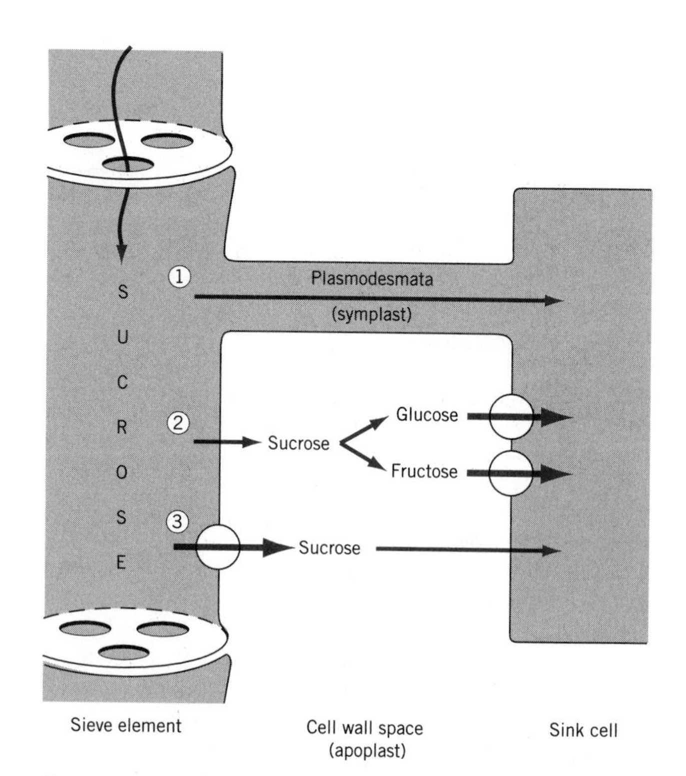

FIGURE 11.10 Three possible routes for sugar unloading into sink cells. In all three possible routes, a favorable diffusion gradient is maintained by metabolizing the sugar once it enters the sink cell.

flows, via interconnecting plasmodesmata, down a concentration gradient from the se-cc complex to sites of metabolism in the sink. The gradient and, consequently, flow into the sink cell is maintained by hydrolyzing the sucrose to glucose and fructose.

There are two possible apoplastic routes, shown as paths 2 and 3 in Figure 11.10. Pathway 2, which has been studied most extensively in the storage parenchyma cells of sugarcane, involves the release of sucrose from the se-cc complex into the apoplast. Release is insensitive to metabolic inhibitors or PCMBS and therefore does not involve an energy-dependent carrier. Once in the apoplast, sucrose is hydrolyzed by the enzyme **acid invertase,** which is tightly bound to the cell wall and catalyzes the reaction:

$$\text{sucrose} + H_2O \rightleftharpoons \text{glucose} + \text{fructose} \quad (11.1)$$

This reaction is essentially irreversible and the hydrolysis products, glucose and fructose, are actively taken up by the sink cell. Once in the cell, they are again combined as sucrose and actively transported into the vacuole for storage. Hydrolysis of sucrose in the apoplast, perhaps combined with the irreversibility of the acid invertase reaction, serves to maintain the gradient and allows the unloading to continue. This pathway seems to be prominent in seeds of maize, sorghum, and pearl millet.

The third pathway for phloem unloading has been studied extensively in legume seeds, employing the "empty-ovule" technique (Fig. 11.11). The key to the empty ovule technique is that there are no plasmodesmata connecting the embryonic tissue with the maternal tissues in the seeds of many legumes; that is, the embryo is symplastically isolated from the **endothelium,** or inner lining of the seed coat. This "apoplastic boundary" has two consequences: (1) transfer of solute from the maternal tissues to the developing embryo (the sink) must occur via an apoplastic route and (2) the boundary provides a natural point for surgical separation of the se-cc complex from the sink tissue. The empty seed-coat "cup" that remains can then be filled with agar or buffer solutions that trap any solutes released.

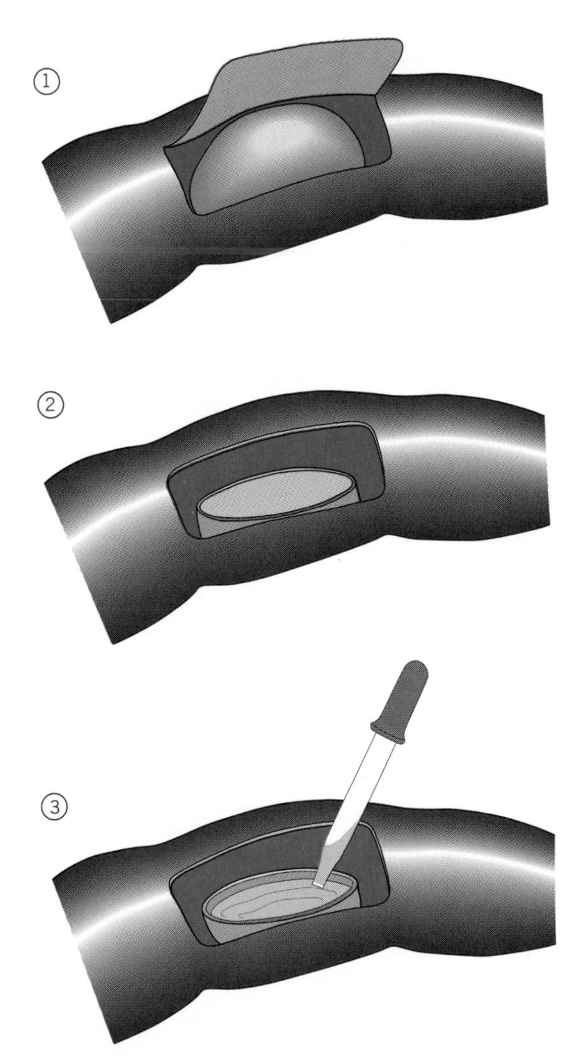

FIGURE 11.11 The empty ovule technique for studying phloem unloading in legumes. A flap is cut into the wall of a young bean pod (1), allowing access to the developing seed. The distal (unattached) half of the seed is surgically removed and discarded (2). The embryo tissues of the remaining half-seed are scooped out (3), leaving a cup-shaped structure (the empty ovule) comprised only of maternal seed coat tissue. The cup can then be filled with a buffer or agar, which traps substances from vascular tissue that, in the intact seed, would supply the embryo.

Following a brief period of adjustment, phloem unloading into the empty seed-coat "cup" will continue for up to several hours. This allows kinetic studies to be conducted by periodically replacing the trapping solution, and the unloading process can be manipulated by changing the pH or by adding other solutes, inhibitors, and so forth. It is interesting to note that this imaginative and highly useful technique was developed by three research groups working simultaneously, but independently, in the United States, the Netherlands, and Australia!

The empty-ovule technique has since been used to study phloem unloading in several legumes in addition to soybean and developing maize kernels. Experiments with the empty-ovule technique have shown that unloading into the embryo is sensitive to anoxia, low temperature, metabolic poisons, and PCMBS. These results indicate that, at least in legumes, sucrose is unloaded into the apoplast by an energy-dependent carrier (Thorne, 1986). The nature of the carrier has not been conclusively identified, but evidence to date suggests it is probably the same sucrose-H^+ cotransporter described earlier. As with phloem loading, there does not appear to be a universal path for phloem unloading into the developing embryo.

It was established through experiments using the empty-ovule technique that unloading in maize ovule is unaffected by PCMBS, metabolic poisons, and anoxia. These data were discussed earlier. Thus it appears that legume and maize have adapted slightly different mechanisms for unloading; and active sucrose pump is involved in legume ovules while in maize ovules sucrose is passively unloaded into the apoplast.

ASSIMILATE DISTRIBUTION

Regulating the distribution of photoassimilate between various metabolic pathways and plant organs is an important but complex and poorly understood problem.

Some of the newly fixed carbon or photoassimilate in a source leaf is retained within the leaf, and the rest is distributed to various nonphotosynthetic tissues and organs. This raises several interesting questions. What, for example, determines how much carbon is retained and in what form? What determines how much is exported and to where? What determines how much assimilate, for example, is exported to the roots of a wheat or corn plant and how much is translocated to fill the developing grain? Questions of this sort have been receiving increasing attention of late, because the patterns of distribution, or more to the point, *regulation* of the distribution patterns is highly significant with respect to productivity and yield. One maize farmer may wish to maximize grain yield while another may require more of the carbon be put into production of vegetative (i.e., leafy) material. Each farmer will assess the **harvest index** (the ratio of usable plant material to total biomass) for the crop in a different way.

The traditional route to improving harvest index has been through breeding and selection. The uncultivated progenitors of modern-day wheat and maize, for example, produced sparse heads with small seeds. Centuries of agricultural selection and, in the last century, careful breeding, have been required to produce the high-yielding wheat and maize varieties in use today. However, the more we learn about the factors regulating carbon distribution and utilization, the greater the prospects for using modern genetic methods to manipulate the harvest index. The distribution of photoassimilate occurs at two levels: **allocation** and **partitioning.** Each of these will be discussed in turn.

ALLOCATION

Newly assimilated carbon may be allocated to a variety of metabolic functions in the source organ or the sink.

Allocation refers to the *metabolic* fate of carbon either newly assimilated in the source leaf or delivered to a sink. At the source, there are three principal uses for photoassimilate: leaf metabolism and maintenance of leaf biomass, short-term storage, or export to other parts of the plant (Fig. 11.12).

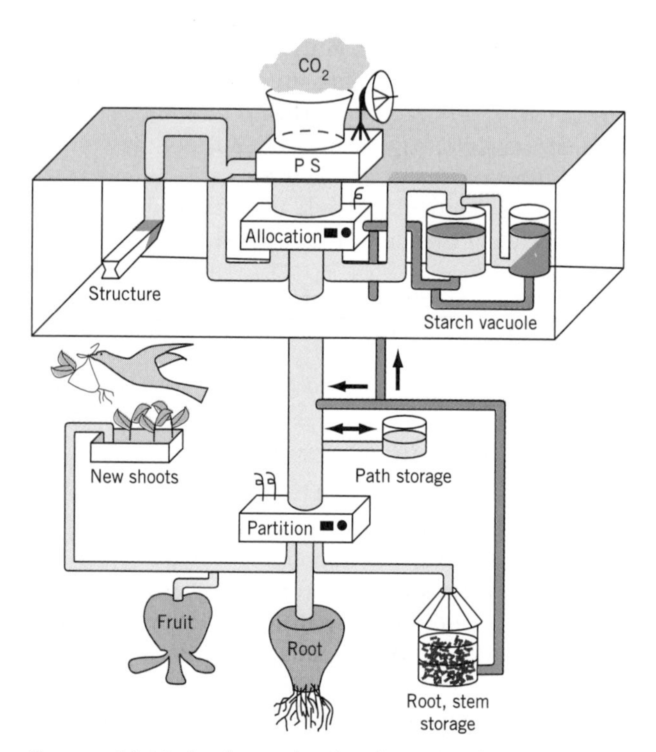

FIGURE 11.12 A schematic of pathways and processes involved in the allocation and partitioning of photoassimilates. (Redrawn from D. R. Geiger, in J. Cronshaw et al. (eds.), *Phloem Transport*, A. R. Liss, 1986. Reprinted by permission.)

Leaf Metabolism and Biomass Some of the carbon will be allocated to the immediate metabolic needs of the leaf itself. These needs include the maintenance of cell structure, synthesis of additional leaf biomass, and the maintenance of the photosynthetic system itself. Most of this carbon is metabolized through respiration, which provides both the energy and carbon skeletons necessary to support ongoing synthetic activities.

Storage Under normal light-dark regimes, plants face a dilemma—photosynthesis is restricted to the daylight hours, but a supply of photoassimilate for growth must be maintained over the entire 24 hours. A partial solution to this dilemma is to allocate a portion of the newly fixed carbon for storage in the leaves. Most plants, especially dicots, store the bulk of their carbon as starch, with a smaller amount stored as sucrose. Some, such as barley (*Hordeum vulgare*), sugarcane (*Saccharum spontaneum*), and sugarbeet (*Beta vulgaris*) accumulate little if any starch, but store carbon primarily as sucrose in the vacuoles of leaf, stem, or root cells, respectively. Many grasses accumulate fructose polymers called **fructans** (see Box 10.1, Chap. 10). Carbon stored in the leaves serves primarily as a buffer against fluctuations in metabolite levels and is available for reallocation to metabolism when required.

Alternatively, most plants appear to be programmed to maintain a fairly constant rate of translocation and supply to sink tissues. Leaf reserves are therefore available for reallocation to export at night or during periods of stress when photosynthesis is very low. In plants that store both starch and sucrose, there are generally two pools of sucrose, one in the cytoplasm and one in the vacuole. The vacuolar pool, which is larger and turns over more slowly than the cytoplasmic pool, is the first source of sucrose for export at night. Only when the vacuolar pool is depleted will the starch, stored in the chloroplast, be mobilized for export.

Export from the Leaf Normally about half the newly assimilated carbon is allocated for immediate export from the leaf via the phloem. In many plants, a portion of this exported carbon may be stored along the translocation path. As in the leaf, this stored carbohydrate helps to buffer the carbon supply at times when the rate of translocation through the phloem might otherwise be reduced.

Regulating the allocation of photoassimilate is a complex and poorly understood process, involving the interactions of a number of metabolic pathways. Allocation within a source leaf is to a large extent genetically programmed but there is a strong developmental component. Young leaves, for example, retain a large proportion of their newly fixed carbon for growth, but as leaves mature the proportion allocated for export increases. In soybean leaves there are corresponding changes in the activities of enzymes such as acid invertase (Eq. 11.1) and sucrose synthase (see Chap. 10, Eq. 10.11). The activities of these two degradative enzymes are highest in young, rapidly expanding leaves, which no doubt reflects the need to metabolize sucrose in the early stages of leaf development when the leaf is functioning primarily as a sink.

As a leaf matures and becomes photosynthetically self-sufficient, both its need and capacity to import assimilate decline and the metabolism of the leaf switches over to the synthesis of sucrose for export. There is a corresponding decline in the activities of acid invertase and sucrose synthase and a steady increase in the activity of sucrose phosphate synthase (SPS), a key enzyme in the synthesis of sucrose (see Chap. 10, Eq. 10.6). Because sucrose is the predominant form of translocated carbohydrate and SPS activity is closely correlated with sucrose production, the increase in SPS activity may be a critical factor in determining the transition of the leaf from a sink to a source.

The allocation of photoassimilate between storage and export has been extensively described (see Cronshaw et al., 1986), but there are few answers to the question of how this allocation is regulated. In most plants, the level of starch fluctuates on a daily basis—increasing during the light period and declining at night. The rate of sucrose export exhibits similar, but less extreme, diurnal fluctuation. The distribution of carbon between starch and sucrose depends primarily on the allocation of triose phosphate between starch synthesis in the chloroplast and sucrose synthesis in the cytoplasm.

Metabolic regulation of starch and sucrose biosynthesis has been discussed in the previous chapter. The two key enzymes are fructose-1,6-bisphosphatase (FBPase) and SPS, and it is reasonable to expect that factors that influence allocation do so at least in part by influencing the activities of these two enzymes. There is some data to bear out these expectations, as illustrated by the experiments of Hendrix and Huber (1986). In cotton leaves, for example, there is a strong correlation between SPS activity, sucrose content, and export of carbon. All three increase more or less in concert during the photoperiod and drop precipitously at the beginning of the dark period. During the dark period, sucrose content and SPS activity remain low, but the drop in export activity is only transient. The pattern of export recovery during the dark period corresponds very closely with the pattern of starch mobilization. Although there is considerable variation in timing and magnitude, similar diurnal fluctuations in carbon metabolites and enzymes have been found in other species. It thus appears that during periods of active photosynthesis, carbon allocation is largely determined by the activity of SPS. At night, the determining factor appears to be the breakdown of reserve starch.

The single most consistent aspect of source leaf allocation, however, is the generally steady rate of export. Except for transient increases at "dawn" or "dusk," diurnal fluctuations in export are small or nonexistent. Apparently plants are programmed to maintain a steady rate of assimilate translocation over the entire 24-hour period. Whether this program is imposed by photoperiod or some other factor is not known. An understanding of how allocation is regulated in source leaves awaits further investigation.

PARTITIONING OF ASSIMILATE AMONG SINKS

The distribution of newly assimilated carbon between competing sinks is determined by sink strength.

The distribution of assimilate between sinks is referred to as **partitioning**. In a vegetative plant, the principal sinks are the meristem and developing leaves at the shoot apex, roots, and nonphotosynthetic stem tissues. With the onset of reproductive growth, the development of flowers, fruits, and seeds creates additional sinks. In general, sinks are competitive and the photoassimilate is partitioned to all active sinks. If the number of sinks is reduced, a correspondingly higher proportion of the photoassimilate is directed to each of the remaining sinks. This is the basis for the common practice of pruning fruit trees to ensure a smaller number of fruit per tree (see Chap. 16). Partitioning the assimilate among a smaller number of fruit encourages the development of larger, more marketable fruit.

Partitioning of assimilate between competing sinks depends primarily on three factors: the nature of vascular connections between source and sinks, the proximity of the sink to the source, and sink strength. Translocation is clearly facilitated by direct vascular connections between the source leaf and the sink. Each leaf is connected to the main vascular system of the stem by a vascular trace, which diverts from the vascular tissue of the stem into the petiole. Experiments have shown that photoassimilate will move preferentially toward sink leaves above and *in line* (that is, in the same rank) with the source leaf. These sink leaves are most directly connected with the source leaf. Sink leaves not in the same rank, such as those on the opposite side of the stem, are less directly linked; the assimilate must make its way through extensive radial connections between sieve elements.

One of the more significant factors in determining the direction of translocation is **sink strength.** Sink strength is a measure of the capacity of a sink to accumulate metabolites (Waring and Patrick, 1975). It is given as the product of sink size and sink activity:

$$\text{sink strength} = \text{sink size} \times \text{sink activity}$$

Sink size is the total mass of the sink (usually as dry weight). Sink activity is the rate of uptake, or, assimilate intake per unit dry weight of sink per unit time. Differences in sink strength can be measured experimentally, although it is not known exactly what determines sink strength or what causes sink strength to change with time. The rate of phloem unloading is surely a factor, as well as the rate of assimilate uptake by the sink and allocation to metabolism and storage within the sink. Environmental factors (e.g., temperature) and hormones will also have an impact to the extent that they influence the growth and differentiation of the sink tissue.

Photoassimilate from most source leaves is readily translocated in either vertical direction; upward toward the apex or downward toward the roots. All else being equal, however, there is a marked bias in favor of translocation toward the closest sink. In the vegetative plant, photoassimilate from young source leaves near the top of the plant is preferentially translocated toward the stem apex, while older, nonsenescent leaves near the base of the plant preferentially supply the roots. Intermediate leaves may translocate photoassimilate equally in both directions. The direction of translocation is probably related to the magnitude of the hydrostatic pressure gradient in the sieve elements. Given two equivalent sinks at different distances, the sink closest to the source will be served by the steeper pressure gradient. The bias in favor of the shorter translocation distance is sufficient to overcome even sink size.

Because sink strength is closely related to productivity and yield, most studies have been conducted with crop species—in particular the filling of grain in cereals such as wheat (*Triticum aestivum*) and maize (*Zea mays*). Developing grain is a particularly active sink and has a major impact on translocation patterns. From the time of anthesis, when the floral parts open to receive pollen, the developing grain becomes the dominant sink. The influence of developing grain on translocation patterns is illustrated by the results shown in Table 11.2. In this experiment the supply of photoassimilate was altered by reducing the supply of carbon dioxide and the dry weight increase of various plant parts was monitored over the grain filling period. Reducing photoassimilate supply had virtually no effect on grain weight, which means that a higher proportion of the carbon was translocated to the grain. The difference was made up by an equivalent decrease in the proportion of carbon directed to the roots. Roots and the developing grain are competing sinks. When the supply of photoassimilate is limited, it is preferentially directed toward the sink with the greater strength. The dominant role of developing grain as a sink is also shown by experiments with wheat. When photosynthesis was limited by lowering the light level, the proportion of ^{14}C-photoassimilate from the flag leaf (the leaf directly below the floral head) increased from 49 percent to 71 percent. In this case, however, the difference was made up by an equivalent reduction in the proportion translocated in the lower stem.

TABLE 11.2 Patterns of photoassimilate distribution in *Sorghum* plants subjected to high (400 $\mu l\ l^{-1}$) and low (250 $\mu l\ l^{-1}$) concentrations of carbon dioxide. Values are percentage of total dry weight gain during the grain filling period. Final grain weight was the same under the two conditions.

	Carbon Dioxide Level	
	High	Low
Grain	71.5	87
Roots	18	4
Other	10.5	9

Based on the data of K. Fischer and G. Wilson, 1975, *Australian Journal of Agricultural Research* 26:11–23.

The above discussion indicates that sink strength is a significant factor in determining the pattern of translocation, but to suggest that sink strength alone is responsible for the partitioning of assimilate would be to grossly oversimplify the problem. At the very least, assimilate partitioning is a highly integrated system, depending upon interactions between the source leaf, the actively growing sinks, and the translocation path itself. We intuitively expect that such an integrated system will be subject to regulation at one or more points. However, beyond the observation that transport rate generally responds to sink demand—sudden changes in sink activity will cause corresponding changes in transport rate to that sink—relatively little is known about regulation of sink strength and interactions between sink strength and translocation rate.

Two factors that have been implicated in influencing sink strength are cell turgor and hormones. While investigating phloem exudate of castor bean (*Ricinus communis*), J. A. C. Smith and J. A. Milburn noted that the act of collecting exudate by making bark incisions, which causes a sudden reduction in the turgor pressure in the sieve elements, gave rise to a marked increase in sucrose loading at the source (Smith and Milburn, 1980). Subsequently, a series of experiments involving artificial manipulation of turgor led Smith and Milburn to conclude that phloem loading is dependent on turgor pressure in the sieve elements. Turgor-dependent phloem loading now forms the basis for a relatively simple hypothesis to explain the regulation of transport rate by sink demand. When the se-cc complex is rapidly unloaded at the sink, the reduction in solute concentration causes a corresponding reduction in the hydrostatic pressure, or turgor, at the sink end of the sieve elements (refer to Fig. 11.6). This reduced hydrostatic pressure will be transmitted throughout the interconnected system of sieve elements, quickly stimulating increased phloem loading at the source. The resulting increase in solute concentration at the source end of the system would serve to counter the drop in hydrostatic pressure, thus maintaining the pressure gradient and, in accordance with the pressure-flow mechanism, stimulating the flow of assimilate toward the sink. A reduction in sink demand would have the opposite result, leading to a lower rate of solute withdrawal and a higher turgor in the sieve elements. Loading at the source and the hydrostatic pressure gradient would be reduced, thereby lowering the rate of translocation. According to this model, changes in sieve element turgor would be an important message in the long-distance communication between sinks and sources.

It is not known how the se-cc complex or the mesophyll cells sense changes in turgor. The mechanism by which pressure changes can be translated into changes in sucrose loading is also unknown. However, some experiments have demonstrated that sucrose transport across cell membranes of beetroot tissue is turgor regulated, possibly by controlling the activity of an ATPase proton pump located in the plasma membrane (Wyse et al., 1986).

Plant hormones (see Chaps. 16, 17) have been implicated in directing long-distance translocation, particularly with regard to redirection of assimilates to new sinks. Hormone-directed transport, however, may be simply an indirect consequence of hormone action. We know that hormones are one of several intrinsic factors involved in regulating the growth and development of organs. Through their influence on the size and metabolic activity of sink organs, hormones will undoubtedly influence sink strength and, as a result, translocation rates. The role of hormones is complicated by the fact that they may, at least in part, be delivered to new sink organs by the phloem. As well, new sinks often themselves become sources of hormones that may act locally or be translocated to other regions of the plant.

While a role for hormone-directed transport over long distance may be uncertain, there is an accumulating body of evidence that seems to indicate a more direct involvement of hormones in the transfer of solute over short distances. For example, there are a number of reported correlations between the concentration of abscisic acid (ABA; see Chap. 16) and the growth rate of developing fruits. ABA also stimulates the translocation of sugar into the roots of intact bean plants, the uptake of sucrose by sugarbeet root tissue, and the unloading of sucrose into the apoplast of soybean seed coats and its subsequent uptake into the embryo. There have been conflicting reports on whether ABA stimulates the translocation of ^{14}C-photoassimilate into filling wheat ears. The hormone auxin (IAA; see Chap. 16), on the other hand, inhibits sucrose uptake by sugarbeet roots but stimulates loading in bean leaves. These and other results suggest that loading and unloading may be susceptible to control by hormones.

Although it appears that sink strength is a major factor in determining assimilate distribution, the process of assimilate partitioning remains a complex, highly integrated, and poorly understood phenomenon. Investigators have only begun to address the respective roles of turgor and hormones, while genetic questions and other potential means of regulation have yet to be addressed in any serious way. The regulation of loading, unloading, and source-sink communication should continue to be active and productive areas of research in the future.

TRANSLOCATION OF XENOBIOTIC CHEMICALS

Overall, the composition of phloem exudate is rather limited, which raises an interesting question—what properties make a molecule mobile in the phloem?

Phloem mobility is of particular interest to the agrochemical industry producing **xenobiotic** chemicals (Lichtner, 1984). (The term xenobiotic refers to biologically active molecules that are foreign to an organism). The rate of absorption and translocation of xenobiotic chemicals often determines their effectiveness as herbicides, growth regulators, fungicides, or insecticides. One excellent example is the broad-spectrum herbicide N-(phosphonomethyl)glycine, or **glyphosate.** Glyphosate acts by preventing the synthesis of aromatic amino acids, which in turn blocks the synthesis of protein, auxin hormones, and other important metabolites. Because it is highly mobile in the phloem, glyphosate applied to leaves is rapidly translocated to meristematic areas or to underground rhizomes for effective control of perennial weeds.

The principal problem with xenobiotics appears to be in gaining entry into the phloem at the minor vein endings in the leaf, that is, phloem loading. Although a few theories have been advanced to explain phloem mobility of xenobiotics, there are relatively few consistent chemical and physical characteristics that describe these molecules. Because xenobiotics are not normally encountered by plants, there are no carriers to mediate their uptake by the cell. Entry is probably by passive diffusion. One consistent characteristic of mobile xenobiotics is their relative level of lipid solubility, or *lipophilicity*, a factor that helps to predict their ability to diffuse through cell membranes.

Efforts to further understand factors controlling the entry of xenobiotic chemicals into plants and their systemic mobility may ultimately lead to advances in our understanding of phloem translocation generally.

SUMMARY

The long-distance translocation of photoassimilate and other small organic molecules occurs in the phloem tissue. The distinguishing feature of phloem tissue is the conducting tissue called the sieve element or sieve tube. Filled with modified, but active, protoplasm at maturity, sieve tubes are interconnected through perforated end walls called sieve plates.

The direction of long-distance translocation in the phloem is determined largely by source–sink relationships. An organ or tissue that produces more assimilate than it requires for its own metabolism is a source, while a sink is a net importer of assimilate. Sinks include meristems and developing leaves at the apex, nonphotosynthetic stem tissues, roots, and storage organs. Organs such as leaves are commonly sinks in their early stages, but become sources as they mature.

Sugars are translocated in the phloem by mass transfer along a hydrostatic pressure gradient between the source and sink. Loading of sugars into the sieve element–companion cell complex (se-cc) in minor veins of the source is followed by the osmotic uptake of water. The resulting hydrostatic pressure is transmitted throughout the system of sieve elements. Unloading of sugars from the minor veins in the sink maintains the pressure differential that causes mass flow. Phloem loading and unloading may occur through the symplast (plasmodesmata) directly into the se-cc complex. Alternatively, sucrose may be transported across the mesophyll cell membrane into the apoplastic space. From there it would be taken across the membrane of the se-cc complex and enter the long-distance transport stream. There is evidence to support both pathways, but there are a number of issues yet to be resolved.

The distribution of photoassimilate between metabolic pathways and plant organs occurs at two levels: allocation and partitioning. Allocation refers to the immediate metabolic fate of assimilate. It may be allocated to the immediate metabolic needs of the leaf itself and maintenance of leaf biomass, it may be stored for use during nonphotosynthetic periods, or it may be exported from the leaf. Once exported, assimilate will be partitioned between competing sinks. Partitioning is determined by sink strength, which is a combination of sink size and metabolic activity.

CHAPTER REVIEW

1. What tissues are removed when a tree is girdled? What causes hypertrophic growth above a girdle wound?
2. Describe the structure of mature phloem tissue. What are its unique features? What kinds of problems do these features raise with respect to phloem translocation?
3. Describe the source-sink concept. To what extent are source-sink relationships involved in determining the direction and rate of translocation in the phloem?
4. Describe the Münch pressure-flow hypothesis and show how it operates to drive translocation in the phloem.
5. How are sugars loaded into the phloem sieve tubes at the source and removed at the sink?
6. Distinguish between allocation and partitioning. What factors determine allocation of carbon within a source leaf? What factors determine partitioning between more than one potential sink?

FURTHER READING

Baker, D. A., J. A. Milburn (eds.). 1989. *Transport of Photoassimilates.* Essex: Longman Scientific and Technical.

Cronshaw, J., W. J. Lucas, R. T. Giaquinta. 1986. *Phloem Transport.* New York: Alan R. Liss, pp. 211–224.

Evert, R. F. 1982. Sieve-tube structure in relation to function. *BioScience* 32:789–795.

Pollock, C. J., J. F. Farrar, A. J. Gordon. 1992. *Carbon Partitioning Within and Between Organisms.* Oxford: BIOS Scientific Publishers.

Turgeon, R. 1996. Phloem loading and plasmodesmata. *Trends in Plant Science* 1:418–422.

Van Bel, A. J. E. 1993. Strategies of phloem loading. *Annual Review of Plant Physiology and Plant Molecular Biology* 44:253–281.

REFERENCES

Cronshaw, J. 1975. P-proteins. In: S. Aronoff, J. Dainty, P. R. Gorham, S. M. Srivastava, C. A. Swanson (eds.), *Phloem Transport.* New York: Plenum. pp. 79–115.

Dickson, R. E. 1991. Assimilate distribution and storage. In: A. S. Raghavendra (ed.), *Physiology of Trees.* New York: Wiley, pp. 51–86.

Eschrich, W. 1975. Sealing systems in phloem. In: M. H. Zimmermann, J. A. Milburn (eds.), *Encyclopedia of Plant Physiology. New Series.* Berlin: Springer-Verlag, Vol. 1, pp. 39–56.

Evert, R. F. 1982. Sieve-tube structure in relation to function. *BioScience* 32:789–795.

Evert, R. F. 1990. Dicotyledons. In: H.-D. Behnke, R. D. Sjolund (eds.), *Sieve Elements. Comparative Structure, Induction and Development.* Berlin: Springer-Verlag, pp. 103–137.

Fisher, D. B. 1978. An evaluation of the Münch hypothesis for phloem transport in soybean. *Planta* 139:25–28.

Giaquinta, R. T. 1983. Phloem loading of sucrose. *Annual Review of Plant Physiology* 34:347–387.

Geiger, D. R. 1986. Processes affecting carbon allocation and partitioning among sinks. In: J. Cronshaw, W. J. Lucas, R. T. Giaquinta, 1986, *Phloem Transport.* New York: Alan R. Liss. pp. 375–388.

Geiger, D. R., S. A. Sovonick, T. L. Shock, R. J. Fellows. 1974. Role of free space in translocation in sugar beets. *Plant Physiology* 54:892–898.

Gunning, B. E. S. 1976. The role of plasmodesmata in short distance transport to and from the phloem. In: B. E. S. Gunning, A. W. Robards (eds.), *Intercellular Communication in Plants: Studies on Plasmodesmata.* Berlin: Springer-Verlag. pp. 203–227.

Hall, S. M., D. A Baker. 1972. The chemical composition of *Ricinus* phloem exudate. *Planta* 106:131–140.

Hendrix, D. L., S. C. Huber. 1986. Diurnal fluctuations in cotton leaf carbon exchange rate, sucrose synthesizing enzymes, leaf carbohydrate content and carbon export. In: J. Cronshaw, W. J. Lucas, R. T. Giaquinta, (eds.) *Phloem Transport.* New York: Alan R. Liss, pp. 369–373.

Ho, L. C. 1988. Metabolism and compartmentation of imported sugars in sink organs in relation to sink strength. *Annual Review of Plant Physiology and Plant Molecular Biology* 39:355–378.

Lichtner, F. T. 1984. Phloem transport of xenobiotic chemicals. *What's New in Plant Physiology* 15:29–32.

Madore, M. A., W. J. Lucas. 1986 Characterization of the source leaf symplast by means of lucifer yellow CH. In: J. Cronshaw, W. J. Lucas, R. T. Giaquinta (eds.) *Phloem Transport.* New York: Alan R. Liss, pp. 129–133.

Mortimer, D. C. 1965. Translocation of the products of photosynthesis in sugar beet petioles. *Canadian Journal of Botany* 43:269–280.

Richardson, P. T., D. A. Baker, L. C. Ho. 1982. The chemical composition of cucurbit vascular exudates. *Journal of Experimental Botany* 33:1239–1247.

Sauer, N. 1997. Sieve elements and companion cells—extreme division of labour. *Trends in Plant Science* 2:285–286.

Smith, J. A. C., J. A. Milburn. 1980. Phloem turgor and regulation of sucrose loading in *Ricinus communis* L. *Planta* 148:42–48.

Stadler, R., N. Sauer. 1996. The *Arabidopsis thaliana AtSUC2* gene is specifically expressed in companion cells. *Botanica Acta* 109:299–306.

Thorne, J. H. 1986. Sieve tube unloading. In: J. Cronshaw, W. J. Lucas, R. T. Giaquinta, 1986, *Phloem Transport.* New York: Alan R. Liss. pp. 211–224.

Turgeon, R. 1996. Phloem loading and plasmodesmata. *Trends in Plant Science* 1:418–422.

Tyree, M. T., A. L. Christy, J. M. Ferrier. 1974. A simpler iterative steady state solution of Münch pressure-flow systems applied to long and short translocation paths. *Plant Physiology* 54:589–600.

Van Bel, A. J. E., Y. V. Gamalei, A. Ammerlaan, L. P. M. Bik. 1992. Dissimilar phloem loading in leaves with symplasmic or apoplasmic minor-vein configurations. *Planta* 186:518–525.

Walker, J. T. S., R. Thaine. 1971. Proteins and fine structure components in exudate from sieve tubes in *Curcurbita pepo* stems. *Annals of Botany* 35:773–790.

Waring, P. F., J. Patrick. 1975. In: J. P. Cooper (ed.) *Photosynthesis and Productivity in Different Environments.* Cambridge: Cambridge University Press.

Weatherly, P. E., R. P. C. Johnson. 1968. The form and function of the sieve tube: A problem in reconciliation. *International Review of Cytology* 24:149–192.

Wright, J. P., D. B. Fisher. 1980. Direct measurement of sieve tube turgor pressure using severed aphid stylets. *Plant Physiology* 65:1133–1135.

Wyse, R. E., E. Zamski, A. D. Tomos. 1986. Turgor regulation of sucrose transport in sugar beet taproot tissue. *Plant Physiology* 81:478–481.

Zimmerman, M. H., H. Ziegler. 1975. Appendix III. List of sugars and sugar alcohols in sieve-tube exudates. In: M. H. Zimmermann, J. A. Milburn (eds.) *Transport in Plants I. Phloem Transport. Encyclopedia of Plant Physiology*, New Series. Vol. 1. New York: Springer Verlag. pp. 480–503.

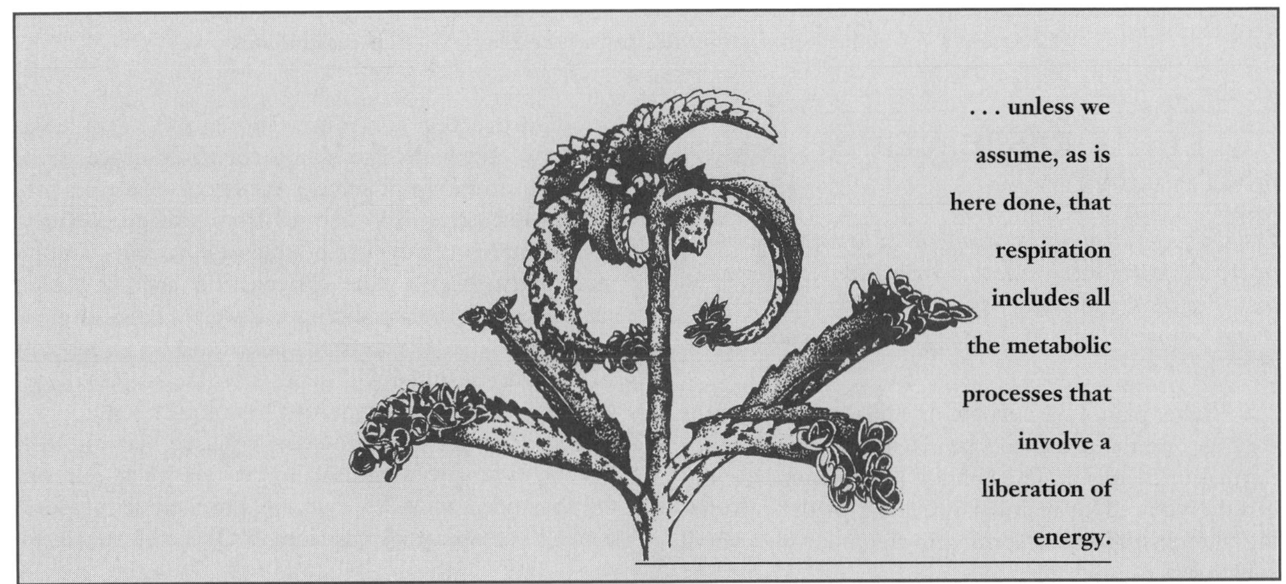

W. Pfeffer (1900)

12

Cellular Respiration: Retrieving the Energy in Photoassimilates

The previous four chapters have been devoted to the conservation of light energy as compounds of carbon, or photoassimilates, and factors directing the distribution of those carbon compounds into different plant organs and tissues. Sugars and other photoassimilates represent two important acquisitions by the plant. They represent, first of all, a highly mobile form of stored photosynthetic energy, and second, a source of carbon skeletons. Through respiration, the plant is able to retrieve the energy in a more useful form, and, in the process, the sugars are modified to form the carbon skeletons that make up the basic building blocks of cell structure.

This chapter is divided into three principal parts. The first part is devoted to the biochemistry and physiology of cellular respiration. After presenting a brief overview of respiration, the following topics will be discussed:

- pathways and enzymes involved in the degradation of sucrose and starch to hexose sugars;
- the conversion of hexose to pyruvate via the glycolytic pathway and the alternate oxidative pentose phosphate pathways;
- the structure and organization of the mitochondrion, which is the site of oxidative respiratory metabolism;
- the pathway for the complete oxidation of pyruvate to CO_2, known as the citric acid cycle, the passage of electrons to molecular oxygen via the terminal electron transport chain, and the conservation of energy as reducing potential and ATP;
- several alternative pathways for electron transport that are unique to plants and their possible physiological consequences; and
- the respiration of oils in seeds by first converting fatty acids to hexose sugars via a process known as gluconeogenesis.

In the second part of this chapter, respiration in intact plants and tissues is discussed, showing how environmental factors such as light, temperature, and oxygen availability influence respiration.

Finally, the role of respiration in the accumulation of biomass and plant productivity will be examined briefly.

Certain principles introduced in the earlier discussion of photosynthesis (Chap. 9) apply equally well to the present discussion. These include bioenergetics, oxidation and reduction reactions, proton gradients, and the synthesis of adenosine triphosphate (ATP). You may find it helpful at this time to review the appropriate sections of Chapter 9.

CELLULAR RESPIRATION: AN OVERVIEW

Cellular respiration consists of a series of pathways by which carbohydrate and other molecules are oxidized for the purpose of retrieving the energy stored in photosynthesis and to obtain carbon skeletons used in the growth and maintenance of the cell.

Higher plants are **aerobic** organisms, which means they require the presence of molecular oxygen (O_2) for normal metabolism. They obtain both the energy and carbon required for maintenance and growth by oxidizing photoassimilates according to the following overall equation:

$$C_6H_{12}O_6 + 6O_2 + 6H_2O \rightleftharpoons 6CO_2 + 12H_2O \quad \Delta G^{\circ\prime} = -2872 \text{ kJ mol}^{-1} \quad (12.1)$$

Note that this equation is written as a reversal of the equation for photosynthesis (Chap. 9, Eq. 9.13). The photosynthetic equation is written as the *reduction* of carbon dioxide to hexose sugar, with water as the source of electrons. The equation for respiration, on the other hand, is written as the *oxidation* of hexose to carbon dioxide, with water as a product. Respiration is accompanied by the release of an amount of free energy equivalent to that consumed in the synthesis of the same carbon compounds by photosynthesis. Here the similarity basically ends. Although the two processes share the same reactants and products and their energetics are similar, the complex of enzymes involved and the metabolic routes taken are fundamentally different, and they occur in different locations in the cell. Moreover, respiration is a process shared by all living cells in the plant, while photosynthesis is restricted to those cells containing chloroplasts.

Equation 12.1 is written as the direct oxidation of hexose by molecular oxygen, with the consequent release of all of the free energy as heat. Cells do not, of course, oxidize sugars in this way. The release of such a large quantity of energy all at once would literally consume the cells. Instead, the overall process of respiration occurs in three separate but interdependent stages—called **glycolysis,** the **citric acid cycle,** and the **electron transport chain**—comprised of some fifty or more individual reactions in total. The transfer of electrons to oxygen is but the final step in this long and complex process. From the energetic perspective, the function of such a complex process is clear: By breaking the oxidation of hexose down into a series of small, discrete steps, the release of free energy is also controlled so that it can be conserved in metabolically useful forms. Equally important to the cell, as we noted earlier, is the fact that respiration also serves to produce a variety of carbon skeletons that are then used to build other molecules required by the cell. We will return to this point later in the chapter.

The equation for respiration (Eq. 12.1) is commonly written with hexose (in particular, glucose) as the initial substrate. In practice, a variety of substrates may serve. Glucose is itself derived from storage polymers such as starch (a polymer of glucose), fructans (a polymer of fructose), or the disaccharide, sucrose. Other sugars may also be metabolized, as well as lipids, organic acids, and to a lesser extent, protein. The actual substrate being respired will depend on the species or organ, stage of development, or physiological state.

The type of substrate being respired may on occasion be indicated by measuring the relative amounts of O_2 consumed and CO_2 evolved. From these measurements the **respiratory quotient (RQ)** can be calculated:

$$RQ = \frac{\text{moles } CO_2 \text{ evolved}}{\text{moles } O_2 \text{ consumed}} \quad (12.2)$$

The value of the respiratory quotient is a function of the oxidation state of the substrate being respired. Note that when carbohydrate is being respired (Eq. 12.1), the theoretical value of RQ is $6CO_2/6O_2 = 1.0$. Experimental values actually tend to vary in the range 0.97 to 1.17. Because lipids and proteins are more highly reduced than carbohydrate, more oxygen is required to complete their oxidation and the RQ value may be as low as 0.7. On the other hand, organic acids, such as citrate or malate, are more highly oxidized than carbohydrate, less oxygen is required for complete oxidation, and RQ values when organic acids are being respired are typically about 1.3.

While RQ values may provide some useful information, care must be taken when interpreting them. For example, should more than one type of substrate be respired at any one time, the measured RQ will be an average value. Should fermentation be occurring (see below), little or no oxygen will be consumed and an abnormally high RQ may result. Or should either CO_2 or O_2 be trapped in the tissue for any reason, results will be correspondingly misleading. Still, respiratory quotients less than one are typical of plants under starvation conditions as lipids and possibly proteins replace carbohydrate as the principal respiratory substrate. Another example of the use of RQ is in germinating seeds. During germination, seeds that store large quantities of lipids will initially exhibit RQ values less than one. Values will gradually approach one as the seedlings consume the lipid reserves and switch over to carbohydrate as the principal respiratory substrate.

BREAKDOWN OF SUCROSE AND STARCH

Photoassimilate is commonly stored as starch, which must be broken down to its component glucose units before it can be used in the respiratory pathways.

Because most plants store their carbohydrate as starch or sucrose, the breakdown of these carbohydrates is an appropriate point at which to begin the path of respiratory carbon.

Starch normally consists of a mixture of two polysaccharides: amylose and amylopectin. **Amylose,** which probably represents no more than one-third of the starch present in most higher plants, consists of very long, straight chains of (1→4)-linked α-D-glucose units. **Amylopectin,** on the other hand, is a highly branched molecule in which relatively short (1→4)-linked α-D-glucose chains are connected by (1→6) links (Fig. 12.1). Starch is normally deposited in plastids as water-insoluble granules or grains. The complete breakdown of starch to its component glucose residues requires the participation of several hydrolytic enzymes:

α-AMYLASE

α-Amylase randomly cleaves α-(1→4) glucosyl bonds in both amylose and amylopectin (Fig. 12.1). α-Amylase, however, does not readily attack terminal α-(1→4) bonds. In the case of amylopectin, α-amylase will not cleave the α-(1→6) glucosyl bonds, nor those α-(1→4) bonds in the immediate vicinity of the branch points. Consequently, about 90 percent of the sugar released on hydrolysis of amylose and amylopectin by α-amylase consists of the disaccharide **maltose** ((1→4)-α-D-glucosylglucose). The balance consists of a small amount of glucose and, in the case of amylopectin, **limit dextrins.** Limit dextrins are comprised of a small number of glucose residues, perhaps 4 to 10, and contain the original branch points. α-Amylase is not restricted to plants but can be found widely in nature, including bacteria and mammals (including human saliva). Indeed this enzyme can be expected in any tissue that rapidly metabolizes starch. A unique and important property of α-amylase is its ability to use starch grains as a substrate. α-Amylase plays an important role in the early stages of seed germination, where it is regulated by the plant hormones gibberellin and abscisic acid (see later chapter on hormones).

β-AMYLASE

β-Amylase degrades amylose by selectively hydrolyzing every second bond, beginning at the nonreducing end of the chain. β-Amylase thus produces exclusively maltose. β-Amylase will degrade the short chains in amylopectin molecules as well. However, because the enzyme can work only from the nonreducing end and cannot cleave the (1→6) branch points, β-amylase will degrade only the short, outer chains and will leave the interior of the branched molecule intact (Fig. 12.1).

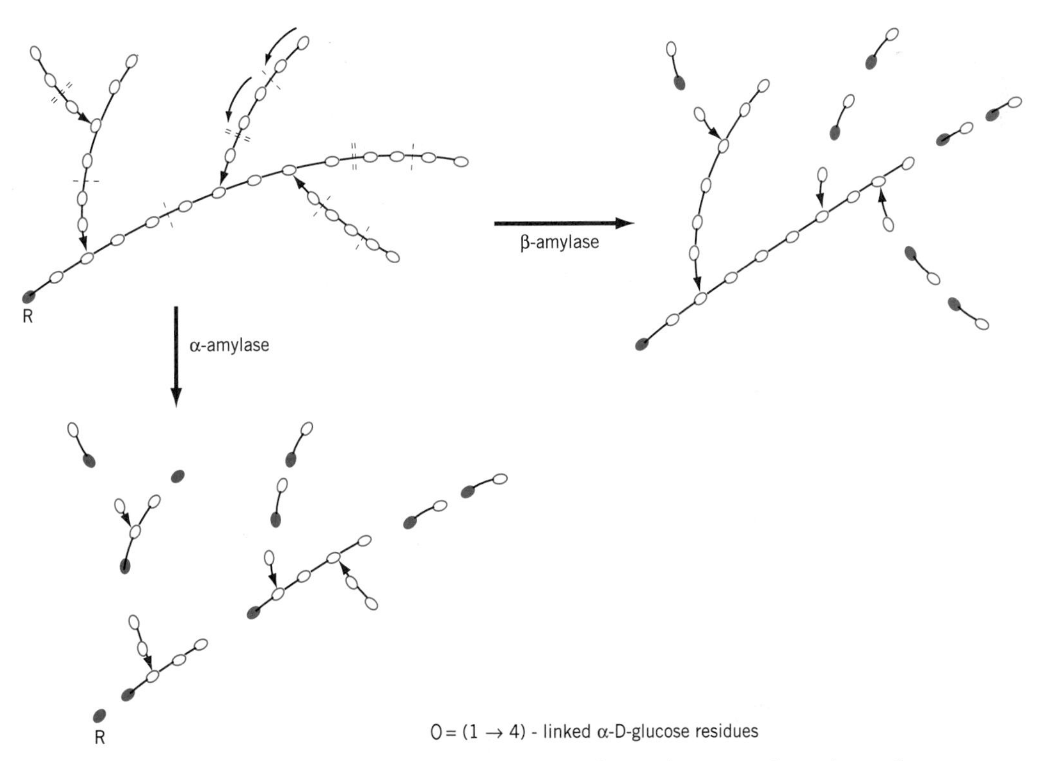

FIGURE 12.1 A schematic representation of starch (amylopectin) degradation by α- and β-amylases. Circles indicate (1→4)-linked α-D-glucose residues. Filled circles indicate the reducing end of the chain.

LIMIT DEXTRINASE

Limit dextrinase is a *debranching enzyme.* Its action is to cleave the (1→6) branching bond, thus allowing both α- and β-amylase to continue degrading the starch to maltose.

α-GLUCOSIDASE

The final step is the hydrolysis of maltose to two molecules of glucose by the enzyme **α-glucosidase.**

STARCH PHOSPHORYLASE

All of the above enzymes mediate the **hydrolytic** breakdown of starch to free sugar; that is, the molecule is cleaved essentially by the addition of water across the bond. However, when the inorganic phosphate level is high (greater than 1 mM), the breakdown of starch is accompanied by an accumulation of *phosphorylated* sugars. This is due to the action of the enzyme **phosphorylase,** which catalyzes the **phosphorolytic** degradation of starch:

$$\text{starch} + nP_i \rightleftharpoons (\text{glucose-1-phosphate})_n \quad (12.3)$$

We will return to this point later, but because the end product is glucose-1-phosphate rather than free glucose, the action of phosphorylase offers a slight energetic advantage. Phosphorylase cannot operate alone—it is unable to degrade starch grains and, like β-amylase, its action is confined to the outer chains of amylopectin molecules. Phosphorylase thus can work only in conjunction with α-amylase, which initiates degradation of the insoluble grains, and debranching enzymes, which render the interior glucose chains accessible to the phosphorylase enzyme. The relative importance of phosphorylase *in vivo* is not known, but in laboratory experiments phosphorylase accounts for less than half of the degradation of potato starch. The balance is degraded via α- and β-amylase.

Starch is stored and degraded inside plastids (either chloroplasts or amyloplasts—see Chap. 1), but the initial stages of cellular respiration occur in the cytosol. The products of starch degradation must therefore make their way across the plastid envelope in order to gain access to the respiratory machinery. This is accomplished by two transporter systems located in the membranes of the plastid envelope (Fig. 12.2). The product of phosphorolytic breakdown, glucose-1-phosphate, is first converted to triose-P in the chloroplast and then exits the plastid via the P_i-triose phosphate transporter described earlier in Chapter 10. Free glucose is able to exit the plastid via a separate hexose transporter (Beck, 1985).

Sucrose synthesis and breakdown has been described in Chapters 10 and 11. Two enzymes are responsible for breakdown: sucrose synthase (Chap. 10, Eq. 10.11) and invertase (Chap. 11, Eq. 11.1). Invertase occurs in two forms, alkaline invertase, with a pH optimum near 7.5, and acid invertase, with a pH optimum near 5. Sucrose synthase and alkaline invertase appear to be localized in the cytosol, while acid invertase is found associated with cell walls and vacuoles. Clearly the relative contributions of these three enzymes will depend to some extent on the cellular location of the sucrose being metabolized. Acid invertase, for example, would be important to the mobilization of sucrose in sugarcane (*Saccharum spontaneum*), which stores excess carbohydrate primarily as sucrose in the vacuoles of stem cells.

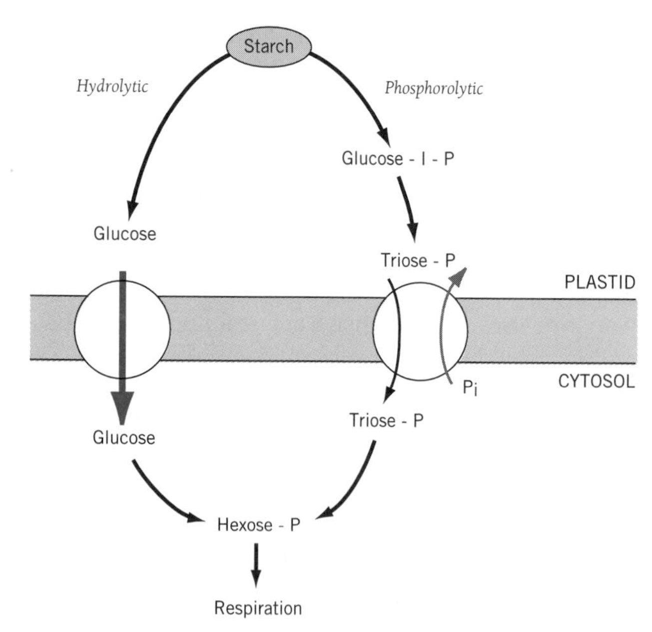

FIGURE 12.2 Mobilization of starch in the stroma of the chloroplast, showing the breakdown of starch by both hydrolytic and phosphorolytic enzymes. Triose-P and glucose are transported out of the chloroplast by specific translocator proteins in the envelop membranes.

GLYCOLYSIS

The reactions of glycolysis convert hexose sugars to pyruvic acid.

The first stage of respiratory carbon metabolism is a group of reactions by which hexose sugars undergo a partial oxidation to the 3-carbon acid pyruvic acid or **pyruvate.** These reactions, collectively known as **glycolysis,** are catalyzed by enzymes located in the cytosol of the cell. Parallel reactions occur independently in plastids, in particular amyloplasts and some chloroplasts (Plaxton, 1996). The reactions of glycolysis, which literally means the lysis or breakdown of sugar, was originally worked out by Meyerhof and others in Germany during the early part of this century, in order to explain fermentation in yeasts and the breakdown of glycogen in animal muscle tissue. Like much of respiratory metabolism, glycolysis is now known to occur universally in all organisms. It is also believed to represent the most primitive form of carbon catabolism since it can lead to fermentation products such as alcohol and lactic acid in the absence of molecular oxygen. Although the energy yield of glycolysis is low, it can be used to support growth in anaerobic organisms or in some aerobic organisms or tissues under anaerobic conditions. Under normal aerobic conditions, however, the pyruvate formed by glycolysis will be further metabolized by the mitochondria to extract yet more energy.

Glycolysis is conveniently considered in two parts. The first is a set of reactions by which the several forms of glucose and fructose derived from storage carbohydrate are converted to the common intermediate fructose-1,6-bisphosphate. The fructose-1,6-bisphosphate is then converted to pyruvate, the end product of glycolysis.

ENTRY OF HEXOSE INTO GLYCOLYSIS

In order for carbon from storage carbohydrate to enter glycolysis, the glucose and fructose derived from hydrolysis of starch, sucrose, or fructans must first be converted to fructose-1,6-bisphosphate (FBP) (Fig. 12.3). Note that the conversion of glucose and fructose to FBP requires an initial expenditure of energy in the form of ATP. Two molecules of ATP are consumed for each molecule of hexose that enters glycolysis. This is analogous to priming a pump; glucose is a relatively stable molecule and the initial phosphorylations, first to glucose-6-P and then to fructose-1,6-P, are a form of activation energy. These two ATP molecules will be recovered during glycolysis. It is here the phosphorolytic breakdown of starch offers a slight energetic advantage over hydrolytic degradation. Because the product is glucose-1-P, for each molecule of hexose entering via the phosphorolytic route the initial expenditure of ATP is reduced by one.

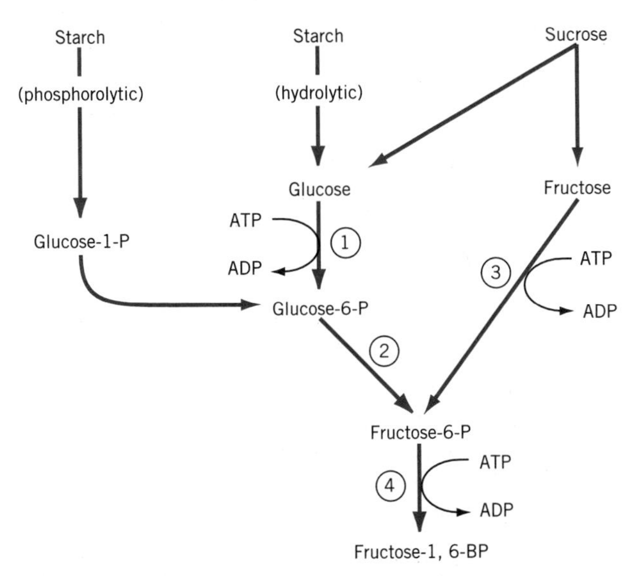

FIGURE 12.3 The conversion of storage carbohydrate to fructose-1,6-bisphosphate (FBP). Enzymes are (1) hexokinase, (2) hexosephosphate isomerase, (3) fructokinase, and (4) phosphofructokinase. FBP is the principal substrate for glycolysis. All reactions shown are reversible, except for the conversion of F6P to FBP.

CONVERSION OF FRUCTOSE-1, 6-BISPHOSPHATE TO PYRUVATE

The reactions for the conversion of FBP to pyruvate are shown summarized in Figure 12.4. The initial step in the sequence (reaction 1) is the cleavage of FBP into two 3-carbon molecules: dihydroxyacetone phosphate and glyceraldehyde-3-phosphate. These two triose phosphates are readily interconvertible (reaction 2), which

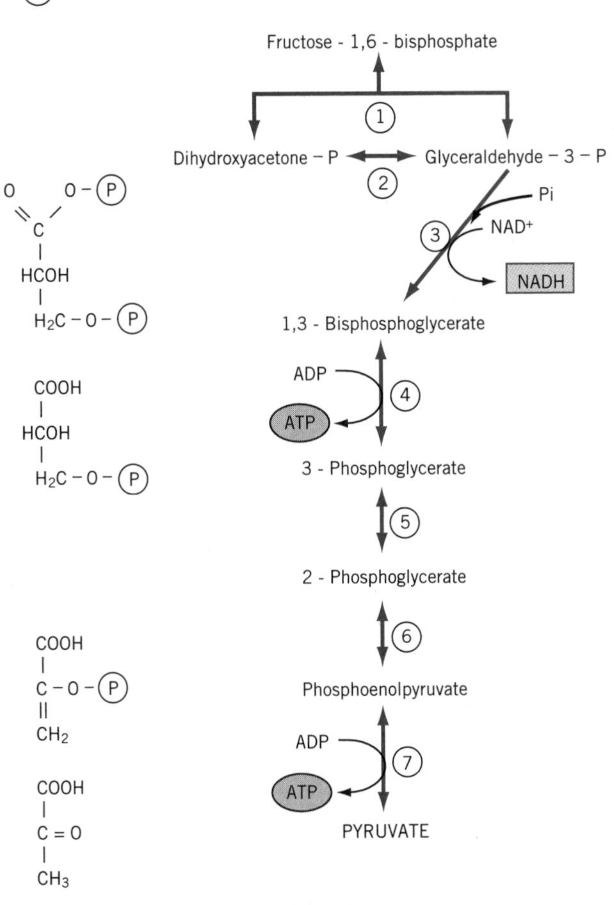

FIGURE 12.4 The conversion of hexose-P to pyruvate via glycolysis. Enzymes are (1) aldolase, (2) triosephosphate isomerase, (3) glyceraldehydephosphate dehydrogenase, (4) phosphoglycerate kinase, (5) phosphoglycerate mutase, (6) enolase, (7) pyruvate kinase.

means that all of the carbon in the original hexose molecule will eventually be converted to pyruvate. In other words, one molecule of FBP will yield two molecules of pyruvate. Thus, in order to account for the hexose molecule originally entering the pathway, everything from this point on must be multiplied by two.

A principal function of glycolysis is energy conservation, which occurs in two ways. The first is **reducing potential** in the form of NADH. In reaction 3, two molecules of NADH (one for each triose phosphate) are produced as glyceraldehyde is oxidized to 1,3-bisphosphoglycerate. This partial oxidation does not require molecular oxygen and does not result in the release of any CO_2. The NADH may be used as reducing potential by the cell for synthesis of other molecules or, if oxygen is present, can be metabolized by the mitochondria to produce ATP.

Energy contained in the original hexose molecule is also conserved as ATP produced via reactions 4 and 7 (Fig. 12.4). For each molecule of hexose entering into glycolysis, four ATP are formed (two for each triose phosphate). Note that formation of ATP at this point does not involve a proton gradient and cannot be explained by Mitchell's chemiosmotic hypothesis. It is instead linked directly to conversion of substrate in the pathway. This form of ATP production is called a **substrate-level phosphorylation.** Depending on whether the storage carbohydrates were initially degraded by the hydrolytic or the phosphorolytic pathways, this represents a net gain of either two or three ATP.

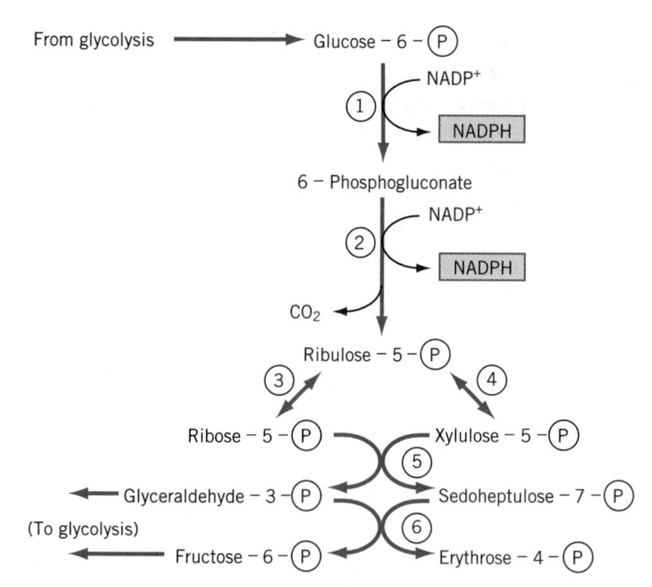

FIGURE 12.5 The oxidative pentose phosphate pathway. A principal function of this alternate pathway is to generate reducing potential in the form of NADPH and pentose sugars for nucleic acid biosynthesis. For the origin of glucose-6-P, see Fig. 12.3. Glyceraldehyde-3-P and fructose-6-P may be returned to the glycolytic pathway for further metabolism. Enzymes are: (1) glucose-6-phosphate dehydrogenase, (2) 6-phosphogluconate dehydrogenase, (3) phosphoriboisomerase, (4) phosphopentoepimerase, (5) transketolase, (6) transaldolase.

THE OXIDATIVE PENTOSE PHOSPHATE PATHWAY

The oxidative pentose phosphate pathway is an alternate route for glucose metabolism that produces tetrose and pentose sugars and NADPH.

Most organisms, including both plants and animals contain an alternate route for glucose metabolism called the **oxidative pentose phosphate pathway** (Fig. 12.5). Also located in the cytosol, the oxidative pentose phosphate pathway shares several intermediates with glycolysis and is closely integrated with it. The first step in the oxidative pentose phosphate pathway is the oxidation of glucose-6-P (Fig. 12.3) to 6-phosphogluconate. This initial step, which is sensitive to the level of $NADP^+$, is apparently the rate determining step for the oxidative pentose phosphate pathway. This is the reaction that determines the balance between glycolysis and the oxidative pentose phosphate pathway. The second step is another oxidation accompanied by the removal of a CO_2 group to form ribulose-5-P. The electron acceptor in both reactions is $NADP^+$, rather than NAD^+. Subsequent reactions in the pathway result in the formation of glyceraldehyde-3-P and fructose-6-P, both of which are then further metabolized via glycolysis.

The role of the oxidative pentose phosphate pathway and its contribution to carbon metabolism overall is difficult to assess because the pathway is not easily studied in green plants. This is largely because many of the intermediates and enzymes of this respiratory cycle are shared by the more dominant *reductive* pentose phosphate pathway, or PCR cycle, in the chloroplasts (Chap. 10). From studies of animal metabolism, however, it can be concluded that the oxidative pentose phosphate pathway has two significant functions. The first is to generate reducing potential in the form of NADPH. $NADP^+$ is distinguished from NAD^+ by an extra phosphoryl group. NADPH serves primarily as an electron donor when required to drive normally reductive biosynthetic reactions whereas NADH is used predominantly to generate ATP through oxidative phosphorylation (see below). This distinction allows the cell to maintain separate pools of NADPH and NAD^+ in the same compartment: a high NADPH/$NADP^+$ ratio to support reductive biosynthesis and a high NAD^+/NADH ratio to support glycolysis. The oxidative pentose phosphate pathway is therefore thought to be a means to generate NADPH required to drive biosynthetic reactions in the cytosol. In animals, for example, the oxidative pentose phosphate pathway is extremely active in fatty tissues where NADPH is required for active fatty acid synthesis.

The second function for the oxidative pentose phosphate pathway is the production of pentose phosphate, which serves as a precursor for the ribose and deoxyribose required in the synthesis of nucleic acids. Another intermediate of the oxidative pentose phosphate pathway with potential significance to plants is the 4-carbon erythrose-4-P, a precursor for the biosynthesis of aromatic amino acids, lignin, and flavonoids.

THE FATE OF PYRUVATE

The fate of pyruvate depends on the availability of molecular oxygen.

The fate of pyruvate produced by glycolysis depends primarily on whether or not oxygen is present (Fig. 12.6). Under normal aerobic conditions, pyruvate is transported into the mitochondrion where it is further oxidized to CO_2 and water, transferring its electrons ultimately to molecular oxygen. We will address mitochondrial respiration further in the following section.

Although higher plants are obligate aerobes and are able to tolerate anoxia for only short periods, tissues or organs are occasionally subjected to anaerobic conditions. A typical situation is that of roots when the soil is saturated with water. When there is no oxygen to serve as the terminal electron acceptor, mitochondrial respiration will shut down and metabolism will shift over to **fermentation.** Fermentation converts pyruvate to either ethanol through the action of the enzyme alcohol dehydrogenase (ADH) or lactate via lactate dehydrogenase (LDH). In most plants, the principal products of fermentation are CO_2 and ethanol (reactions 1, 2, Fig. 12.6). Some lactate may be formed, primarily in the early stages of anoxia. However, lactate lowers the pH of the cytosol, which in turn activates pyruvate decarboxylase and initiates the production of ethanol.

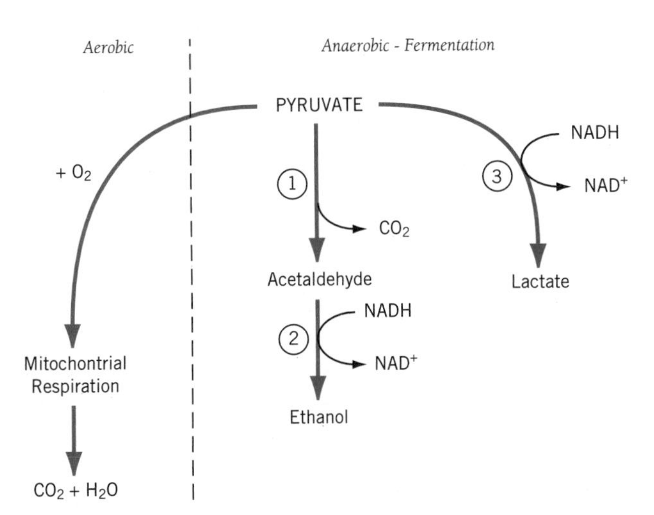

FIGURE 12.6 The fate of pyruvate depends largely on available oxygen. Enzymes are: (1) pyruvate decarboxylase, (2) alcohol dehydrogenase (ADH), (3) lactate dehydrogenase (LDH).

Note that either one of the fermentation reactions consumes the NADH produced earlier in glycolysis by the oxidation of glyceraldehyde-3-P (reaction 3, Fig. 12.4). Although this means there is no net gain of reducing potential in fermentation, this recycling of NADH is still important to the cell. The pool of NADH plus NAD^+ in the cell is relatively small and if the NADH is not recycled, there will be no supply of NAD^+ to support the continued oxidation of glyceraldehyde-3-P. If this were the case, glycolysis and the production of even the small quantities of ATP necessary to maintain the cells under anaerobic conditions would then grind to a halt.

OXIDATIVE RESPIRATION

THE MITOCHONDRION

The next two stages of respiration—the citric acid cycle and the oxidation of NADH along with its accompanying production of ATP—are carried out within an organelle called the ***mitochondrion.***

Plant mitochondria are typically spherical or short rods approximately 0.5 μm in diameter and up to 2 μm in length (Fig. 12.7). The number of mitochondria per cell is variable but generally relates to the overall metabolic activity of the cell. In one study of rapidly growing sycamore cells in culture, R. Douce reported 250 mitochondria per cell (Douce, 1985). The mitochondria accounted for about 0.7 percent of total cell volume and contained 6 to 7 percent of the total cell protein. In other metabolically more active cells, such as secretory

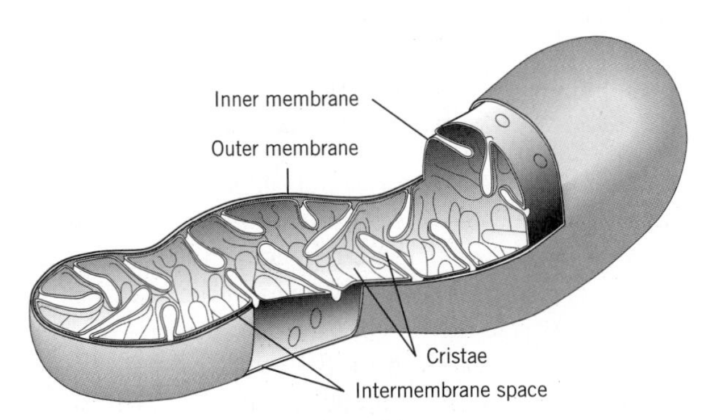

FIGURE 12.7 The mitochondrion. Diagram of the mitochondrion, showing the principal metabolic compartments.

and transfer cells, the number is even higher and may exceed a thousand per cell.

Mitochondria, like the other energy transducing organelle, the chloroplast, are organized into several ultrastructural compartments with distinct metabolic functions: an **outer membrane,** an **inner membrane,** the **intermembrane space,** and the **matrix** (Fig. 12.7; see also Fig. 1.5). The composition and properties of the two membranes are very different. The outer membrane, which has few enzymatic functions, is rich in lipids and contains relatively little protein. The inner membrane, on the other hand, contains over 70 percent protein on a dry weight basis. The outer membrane is also highly permeable to most metabolites. It contains large channel forming proteins, known as **porins,** which allow essentially free passage of molecules and ions with a molecular mass of 10,000D or less. The permeability of the inner membrane, like that of the chloroplast, is far more selective—it is freely permeable only to a few small molecules such as water, O_2, and CO_2. Like the chloroplast thylakoid membrane, the permeability of the mitochondrial inner membrane to protons is particularly low, a significant factor with respect to its role in ATP synthesis.

The inner membrane of the mitochondrion is extensively infolded. These invaginations form a dense network of internal membranes called **cristae.** The typical biology textbook picture of cristae is based on animal mitochondria where the infoldings are essentially lamellar and form platelike extensions into the matrix. The common pattern in plant mitochondria is less regular, forming a system of tubes and sacs. The unstructured interior of the mitochondrion, or matrix, is an aqueous phase consisting of 40 to 50 percent protein by weight. Much of this protein is comprised of enzymes involved in carbon metabolism, but the matrix also contains a mitochondrial genome, including DNA, RNA, ribosomes, and the necessary machinery for transcribing genes and synthesizing protein.

THE CITRIC ACID CYCLE

In the presence of molecular oxygen, pyruvate is completely oxidized to CO_2 and water by the citric acid cycle.

The second stage of respiration is the complete oxidation of pyruvate to CO_2 and water through a series of reactions known as the **citric acid cycle** (Fig. 12.8). The citric acid cycle is also known as the tricarboxylic acid (TCA) cycle or the Krebs cycle, in honor of Hans Krebs, whose research in the 1930s was responsible for elucidating this central metabolic process. Krebs was awarded the Nobel Prize in medicine in 1954 for his outstanding contribution.

Schemes for the citric acid cycle such as that shown in Figure 12.8 invariably begin with pyruvate. Although pyruvate is technically not a part of the cycle, it does provide the major link between glycolysis and subsequent carbon metabolism. Note that pyruvate is produced in the cytosol while the enzymes of the citric acid cycle are located in the matrix space of the mitochondrion. Thus in order for pyruvate to be metabolized by the citric acid cycle, it must first be translocated through the inner membrane. This is accomplished by a pyruvate-OH^- antiport carrier—that is, pyruvate is taken up by the mitochondrion in exchange for a hydroxyl ion carried into the intermembrane space.

Once inside the matrix, pyruvate is oxidatively decarboxylated by a large multienzyme complex **pyruvate dehydrogenase.** Pyruvate dehydrogenase catalyzes a series of five linked reactions, the overall effect of which is to oxidize one molecule of pyruvate to a 2-carbon acetate group:

$$\text{pyruvate} + NAD^+ + \text{CoA} \longrightarrow \text{acetyl-CoA} + NADH + H^+ + CO_2 \quad (12.4)$$

Pyruvate is first decarboxylated, then oxidized. The resulting 2-carbon acetyl group is finally linked via a thioester bond to a sulphur-protein **coenzyme A (CoA).** In the process, NAD^+ is reduced to NADH. The CO_2 given off represents the first of three carbon atoms in the degradation of pyruvate.

The citric acid cycle proper begins with the enzyme **citrate synthase,** which transfers the acetyl group from acetyl-CoA to the 4-carbon **oxaloacetate** to form the 6-carbon, tricarboxylic **citric acid** (hence the designation *citric acid cycle*). The next step is an isomerization of citrate to isocitrate (reaction 3, Fig. 12.8), followed by two successive oxidative decarboxylations (reactions 4, 5, Fig. 12.8). The two CO_2 molecules given off effectively complete the degradation of pyruvate and the oxidations add two more molecules of NADH to the pool of reductant in the matrix.

The balance of the citric acid cycle serves two functions. First, additional energy is conserved at three more locations. One molecule of ATP is formed from ADP and inorganic phosphate when succinate is formed from succinyl-CoA (reaction 6, Fig. 12.8). Because the ATP formation is linked directly to conversion of substrate, this is another example of **substrate-level** phosphorylation. Additional energy is conserved with the oxidation of succinate to fumarate (reaction 7, Fig. 12.8) and the oxidation of malate to oxaloacetate (reaction 9). Second, the cycle serves to regenerate a molecule of oxaloacetate and so prepare the cycle to accept another molecule of acetyl-CoA. Regeneration of oxaloacetate is critical to the catalytic nature of the cycle in that it allows a single oxalacetate molecule to mediate the oxidation of an endless number of acetyl groups.

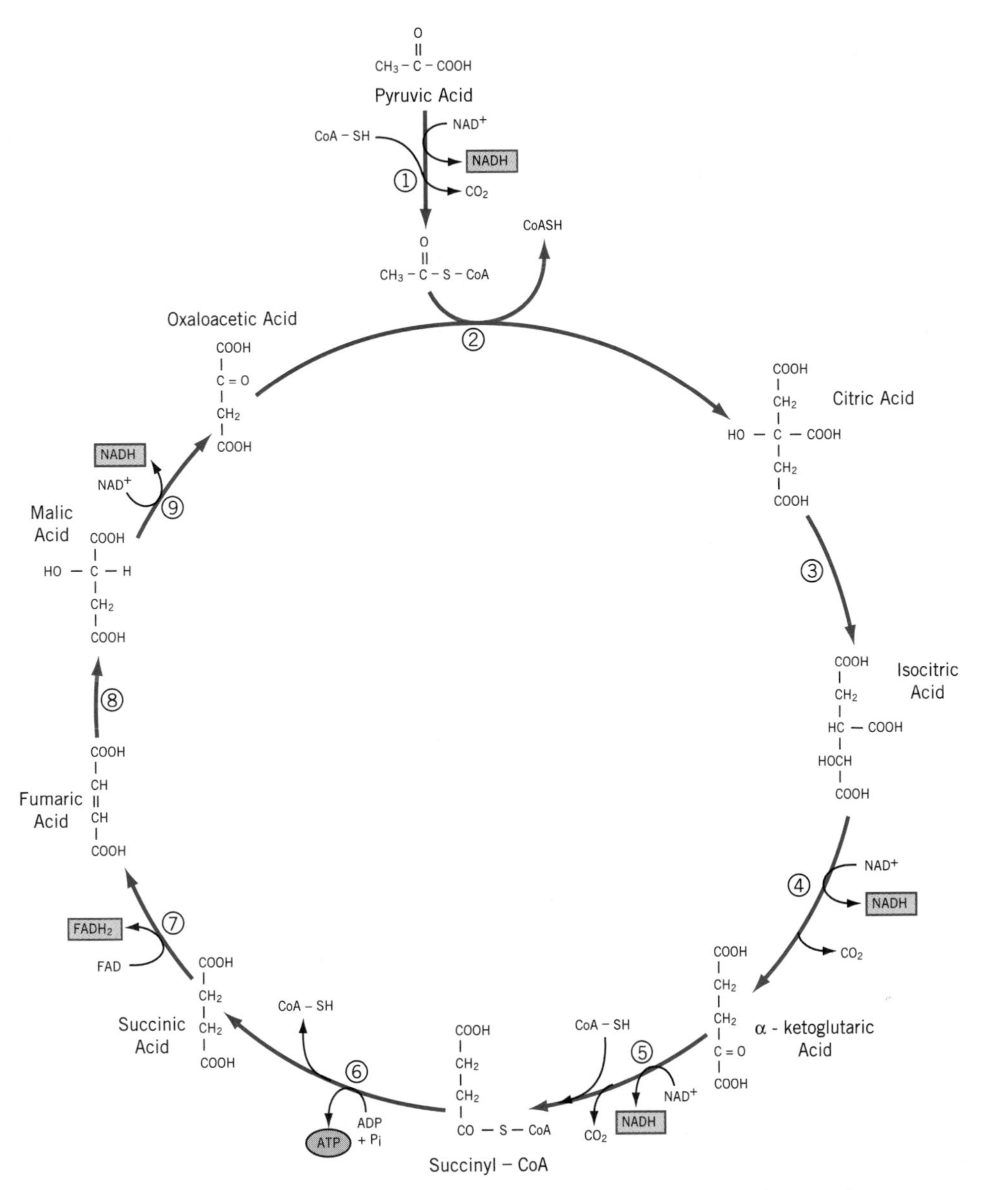

FIGURE 12.8 The reactions of the citric acid cycle. The citric acid cycle completes the oxidation of pyruvate to carbon dioxide. Reducing potential is stored as NADH and $FADH_2$.

In summary, the citric acid cycle consists of eight enzyme-catalyzed steps, beginning with the condensation of a 2-carbon acetyl group with the 4-carbon oxaloacetate to form a molecule of the 6-carbon citrate. The acetyl group is then degraded to two molecules of CO_2. The cycle includes four oxidations, which yield NADH at three steps and $FADH_2$ at one step. One molecule of ATP is formed by substrate-level phosphorylation. Finally, the oxaloacetate is regenerated, which allows the cycle to continue.

Before leaving the citric acid cycle for the moment, it is useful to point out that the cycle must turn twice to metabolize the equivalent of a hexose sugar.

OXIDATION OF NADH AND $FADH_2$

Electrons removed from substrate in the citric acid cycle are passed to molecular oxygen through the mitochondrial electron transport chain.

We noted earlier that one of the principal functions of the respiration is to retrieve, in useful form, some of the energy initially stored in assimilates. Our traditional measure of useful energy in most processes is the number of ATP molecules gained or consumed. By this measure alone, the yield from both glycolysis and the citric acid cycle is quite low. After two complete turns of the cycle, one molecule of glucose has been completely oxidized to 6 molecules of CO_2, but only four molecules of ATP have been produced (a net of 2 ATP from glycolysis plus one for each turn of the cycle). At this point, most of the energy associated with the glucose molecule has been conserved in the form of electron pairs generated by the oxidation of glycolytic and citric acid cycle intermediates. For each molecule of glucose, a total of 12 electron pairs were generated; 10 as NADH and 2 as $FADH_2$. In this section we will discuss the third stage of cellular respiration—the transfer of electrons from NADH and $FADH_2$ to oxygen and the accompanying conversion of redox energy to ATP.

The transfer of electrons from NADH and $FADH_2$ to oxygen involves a sequence of electron carriers arranged in an **electron transport chain.** Membrane fractionation studies have shown that the enzymes and electron carriers making up the electron transport chain are organized predominantly into four large multimolecular complexes (complexes I—IV) and two mobile carriers located in the inner mitochondrial membrane (Fig. 12.9). In this sense, there are a great number of similarities between the mitochondrial inner membrane and the thylakoid membranes of the chloroplast (compare Fig. 12.9 with Fig. 9.13). This is not unexpected, since the principal function of each membrane is energy transduction and many of the same or similar components are involved.

The path of electrons from NADH to oxygen can be summarized as follows. Electrons from NADH enter the electron transport chain through Complex I, known as **NADH-ubiquinone oxidoreductase.** In addition to several proteins, this complex also contains a tightly bound molecule of flavin mononucleotide (FMN) and several nonheme iron-sulphur centers. Complex I conveys the electrons from NADH to ubiquinone. Ubiquinone is a benzoquinone—its structure and function are similar to the plastoquinone found in the thylakoid membranes of chloroplasts (see Fig. 9.3). Like plastoquinone, ubiquinone is highly lipid soluble and diffuses freely in the plane of the membrane. It is not permanently associated with Complex I, but forms a pool of mobile electron acceptor that conveys electrons between Complex I and Complex III. Ubiquinol (the fully reduced form of ubiquinone) is oxidized by Complex III, or **cytochrome *c* reductase.** Complex III contains cytochromes *b* and c_1 and an iron-sulphur center. Complex III in turn reduces a molecule of cytochrome *c*. Cytochrome *c* is a peripheral protein, located on the side of the membrane facing the intermembrane space. Like plastoquinone, cytochrome *c* is a mobile carrier and conveys electrons between Complex III and the terminal complex in the chain, Complex IV. Also known as **cytochrome *c* oxidase,** Complex IV contains cytochromes *a* and a_3 and copper. Electrons are passed first from cytochrome *c* to cytochrome *a*, then to cytochrome a_3, and finally to molecular oxygen.

All of the oxidative enzymes of the citric acid cycle, with one exception, are located in the matrix. The one exception is succinic dehydrogenase (reaction 7, Fig. 12.8). This enzyme is an integral protein complex (Complex II) that is tightly bound to the inner mitochondrial membrane. In fact, succinic dehydrogenase is the preferred marker enzyme for inner membranes when doing mitochondrial fractionations. Complex II,

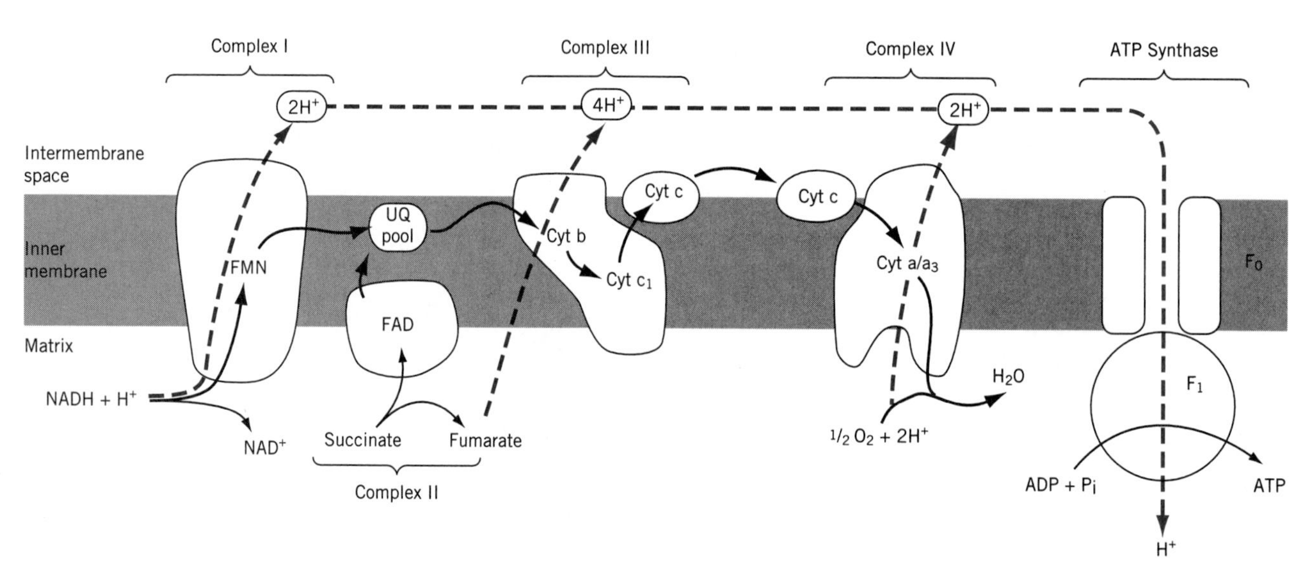

FIGURE 12.9 A schematic representation of the electron transport chain and proton "pumping" sites in the inner membrane of a plant mitochondrion. Solid arrow indicates the path of electrons from NADH or succinate to molecular oxygen. Energy conserved in the proton gradient is used to drive ATP synthesis through the F_0—F_1 ATPase coupling factor elsewhere in the membrane.

known as **succinate-ubiquinone oxido-reductase,** contains flavin adenine di-nucleotide (FAD), several nonheme iron proteins, and iron-sulphur centers. Like Complex I, succinic dehydrogenase transfers electrons from succinate to a molecule of ubiquinone from the membrane pool. From there the electrons pass through Complexes III and IV to molecular oxygen.

In our earlier discussion of glycolysis, it was mentioned that NADH generated in the oxidation of glyceraldehyde-3-P could be metabolized by the mitochondria to eventually form ATP. Electrons from this so-called *cytosolic or external NADH* can enter the electron transport chain through another membrane-bound dehydrogenase located at the membrane surface facing the intermembrane space. Like Complex II, the "external" dehydrogenase contains a molecule of FAD and transfers electrons directly into the ubiquinone pool.

It is important to note that Figure 12.9 presents a static, essentially linear representation of electron flow in mitochondria. *In vivo*, the organization is far more dynamic. In Chapter 9 we showed that the electron transport complexes of the photosynthetic membranes were independently distributed, rather than organized in one supermolecular complex. Similar arguments apply here. The several complexes are not found in equal stoichiometry and are free to diffuse independently within the plane of the inner membrane. The large complexes are functionally linked through the pools of ubiquinone and cytochrome *c*, which function as mobile carriers and convey electrons from one complex to the other largely on the basis of random collision.

PROTON GRADIENTS AND SYNTHESIS OF ATP

Energy released, as electrons are passed down the electron transport chain, is conserved in an electrochemical proton gradient across the inner mitochondrial membrane. The energy stored in this gradient is then used to drive ATP synthesis in accordance with Mitchell's chemiosmotic mechanism.

As electrons are passed from NADH (or $FADH_2$) to oxygen through the electron transport chain, there is a substantial drop in free energy. The actual free energy change is quantitatively the same, but of opposite sign, to the amount consumed when electrons are moved from water to NADPH in photosynthesis (Chap. 9). This energy is conserved first in the form of a proton gradient and ultimately as ATP. The mechanism for ATP synthesis is explained by Mitchell's chemiosmotic hypothesis, described earlier in Chapter 9. Here the focus is on specific details of the proton gradient and ATP synthesis as it applies to mitochondria.

Studies of P/O ratios (the atoms of phosphorous esterified as ATP relative to the atoms of oxygen reduced) and various inhibitors have established that there are three transitions in the electron transport chain that are associated with ATP synthesis. Put another way, when internal or matrix NADH is oxidized, the P/O ratio is approximately 3. According to Mitchell's hypothesis, then, these transitions represent locations, generally described as proton pumps, where contributions are made to a proton gradient across the mitochondrial inner membrane. The three locations, associated with complexes I, III, and IV, respectively, are identified in Figure 12.9. The resulting proton gradient then drives ATP synthesis via a F_0—F_1 ATP synthase complex located in the same membrane (see below). Because mitochondrial ATP synthesis is closely tied to oxygen consumption, it is referred to as **oxidative phosphorylation.**

In the course of mitochondrial electron transport, protons are extruded from the matrix into the intermembrane space. Proton extrusion associated with Complex I (site 1) can be explained by the vectorial arrangement of the complex across the membrane. When a pair of electrons is donated to the complex by NADH, a pair of protons are picked up from the matrix. When the electrons are subsequently passed on to ubiquinone, the protons are released into the intermembrane space. Proton extrusion associated with Complex III (site 2) is probably due to the operation of a "Q-cycle" described previously in Chapter 9 (see Fig. 9.16). The contribution of cytochrome *c* oxidase (site 3) to the proton gradient has been the subject of some discussion for many years. Experiments with isolated enzyme incorporated into lipid vesicles indicate that cytochrome *c* oxidase was capable of transferring protons across membranes. These results are difficult to explain, because cytochromes exchange only electrons, not protons, when reduced and oxidized. However, the H^+/electron pair ratio for site 3 is about 2, which can readily be explained by the two protons consumed from the matrix when oxygen is reduced to water. This is similar in principle to the production of protons in the intrathylakoid space as water is oxidized early in photosynthetic electron transport (see Fig. 9.13). The stoichiometry of proton "extrusion" (the term is applied whether or not the protons are physically carried across the membrane) has been studied extensively. It appears that approximately nine protons are extruded for each pair of electrons conveyed from internal (or matrix) NADH to oxygen.

The link between a proton gradient and ATP synthesis in the mitochondrion embodies the same principles previously described for ATP synthesis in the chloroplast: (1) the inner membrane is virtually impermeable to protons; (2) a proton motive force is established across the membrane by a combination of membrane potential and proton disequilibrium; and (3) ATP synthesis is driven by the return of protons to the matrix

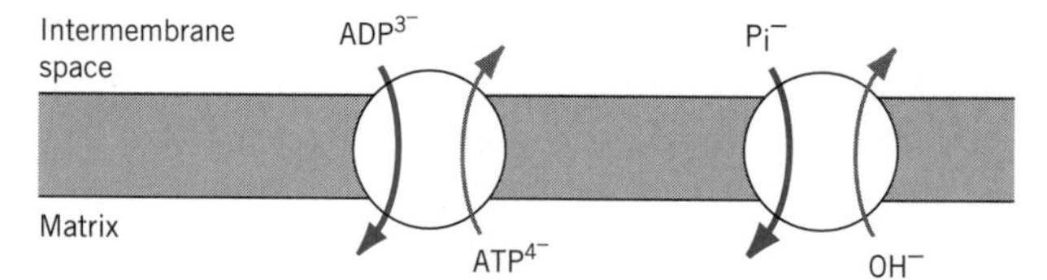

FIGURE 12.10 The adenine nucleotide transporter. The one-for-one exchange of mitochondrial ATP and cytosolic ADP across the inner membrane is driven by the membrane potential. Inorganic phosphate is returned to the matrix in exchange for hydroxyl ions.

through an integral membrane protein complex known variously as **ATP synthase,** coupling factor, or F_0—F_1-ATPase. Mitochondrial ATP synthase is structurally and functionally similar to the chloroplast enzyme. It consists of a hydrophobic, channel-forming portion (F_0) that spans the membrane plus a multimeric, matrix-facing peripheral protein (F_1) that couples proton translocation to ATP synthesis. As in the case of chloroplasts, the H^+/ATP ratio is approximately 3. Because 9 protons are extruded for each pair of electrons moving through the entire chain, this means that a total of 3 ATP molecules could be formed. For electrons entering the chain from extramitochondrial NADH, succinate, or via the rotenone-insensitive dehydrogenase (see below), all three of which bypass site 1, a maximum of two ATP could be formed.

Unlike the chloroplast, most of the ATP synthesized in the mitochondrion is utilized elsewhere in the cell. This requires that the ATP be readily translocated out of the organelle. As well, a supply of ADP and inorganic phosphate is required in order to maintain maximum rates of electron transport and ATP synthesis. This is accomplished by two separate translocator proteins located in the inner membrane. An **adenine nucleotide transporter** located in the inner membrane (Fig. 12.10) exchanges ATP and ADP on a one-for-one basis. An inorganic phosphate translocator exchanges P_i for hydroxyl ions.

ALTERNATE ELECTRON PATHWAYS IN PLANTS

In addition to the common electron transport pathway shared with other organisms, plants contain several alternate electron transport pathways.

The electron transport chain described above is shared in essentially the same form by virtually all organisms: plants, animals, and microorganisms. Plant mitochondria contain, in addition, several other redox enzymes, at least two of which are unique to plants (Fig. 12.11). These enzymes have been discovered largely by virtue of their insensitivity to certain classic inhibitors of electron transport. A comprehensive list of these inhibitors with their sites of action is given by Lambers (1990).

EXTERNAL NADPH DEHYDROGENASE

Plant mitochondria contain an "external" dehydrogenase that faces the intermembrane space and is capable of oxidizing cytosolic NADPH (Fig. 12.11). It is not clear yet whether this is a separate enzyme complex or whether the "external" NADH dehydrogenase described above can oxidize NADPH as well. In either case, electrons from the oxidation of cytosolic NADPH

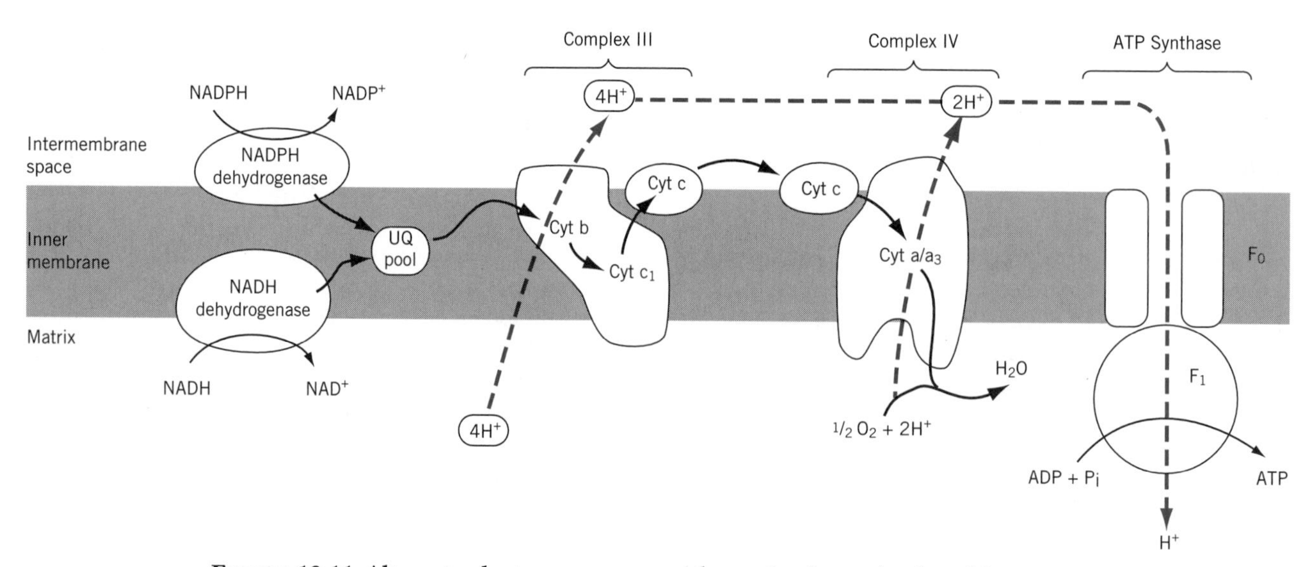

FIGURE 12.11 Alternate electron transport pathways in plant mitochondria. Electrons entering the chain through the alternate dehydrogenases will pass through two phosphorylating sites.

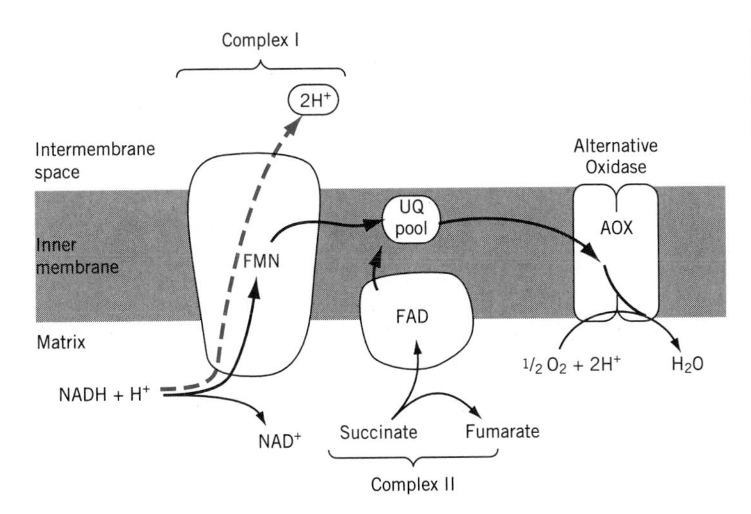

FIGURE 12.12 The alternative respiratory pathway. Electrons intercepted by the alternative oxidase (AOX) pass through one or no phosphorylating sites.

are donated directly to the ubiquinone pool. Because the external NADPH dehydrogenase enzyme does not span the membrane, it will not translocate protons as Complex I does. Consequently, only two ATP can be formed from the transfer of each pair of electrons to oxygen.

ROTENONE-INSENSITIVE NADH DEHYDROGENASE

The reduction of ubiquinone by Complex I is sensitive to inhibition by **rotenone**[1] and **amytal.** Plants, however, appear to have another NADH dehydrogenase that is insensitive to both of these electron transport inhibitors. Called the **rotenone-insensitive dehydrogenase,** this enzyme will oxidize only internal, or matrix, NADH (Fig. 12.11). The enzyme must therefore be located on the inner surface of the membrane, facing the matrix. As with the external NADH and NADPH dehydrogenase enzymes, electrons entering the chain via the rotenone-insensitive dehydrogenase can generate only two ATP per electron pair.

CYANIDE-RESISTANT RESPIRATION

Cytochrome *c* oxidase (Complex IV) is inhibited by cyanide (CN^-), carbon monoxide (CO) and azide (NH_3^-). In animals, all three of these inhibitors completely inhibit respiratory O_2 uptake. By contrast, many plants or plant tissues show considerable resistance to these inhibitors. In tissues such as roots and leaves of spinach (*Spinacea oleraceae*) or pea (*Pisum sativum*), for example, cyanide-resistant respiration may account for as much as 40 percent of total respiration (Lambers, 1990). Cyanide-resistant respiration is, however, sensitive to inhibition by hydroxamic acid derivatives such as **salicylhydroxamic acid (SHAM).** This cyanide-resistant, SHAM-sensitive respiration is attributed to a so-called **alternative oxidase** (Fig. 12.12). The pathway is commonly referred to as the **alternative respiratory pathway,** or, simply, the alternative pathway. Although the existence of a cyanide-resistant alternative respiratory pathway has been widely accepted for more than a decade, relatively little is known of the nature of the oxidase enzyme itself (Siedow and Berthold, 1986; Moore and Siedow, 1991). The enzyme has been difficult to study by conventional biochemical techniques; the protein appears to be relatively unstable and loses its activity rapidly upon isolation from the membrane. However, a gene encoding for the oxidase protein has been cloned, first from *Sauromatum guttatum*, the voodoo lily (Rhoads and McIntosh, 1991), and since from tobacco, soybean, and other plants. This has led to significant advances in the molecular biology of the enzyme and will no doubt lead to better understanding of its biochemistry and physiology in the future (see Vanlerberghe and McIntosh, 1997).

The alternative oxidase is composed of two identical subunits (a homodimer) that span the inner mitochondrial membrane, with the active site facing the matrix side of the membrane. It functions as a ubiquinone: O_2 oxidoreductase; that is, it accepts electrons from the ubiquinone pool and transfers them directly to oxygen. This is an important characteristic of the alternative oxidase because it means that electrons processed by this enzyme bypass at least two sites for proton-extrusion. Consequently, energy that would otherwise be conserved as ATP is, in the case of the alternative oxidase, converted to heat instead. Depending upon whether electrons are initially donated to Complex I, the rotenone-insensitive NADH-dehydrogenase, or succinic dehydrogenase, electrons passing through the alterna-

[1]Rotenone is an isoflavonoid isolated from the roots of *Derris* and commonly used as an insecticide. It is also highly toxic to fish, leading to the practice in primitive native communities of using *Derris* root to harvest fish for human consumption.

tive oxidase will contribute to the synthesis of one ATP or none (Fig. 12.12).

The physiological role of alternate pathway respiration is still uncertain. One possible role is **thermogenesis,** a hypothesis based largely on events in the floral development in certain members of the family Araceae. Just prior to pollination in species such as skunk cabbage (*Symplocarpus foetidus*), the tissues of the spadix (the structure that bears both male and female flowers) undergoes a surge in oxygen consumption, called a *respiratory crisis.* The respiratory crisis is attributed almost entirely to an increase in alternative pathway respiration and can elevate the temperature of the spadix by as much as 10 °C above ambient. The high temperature volatilizes certain odoriferous amines (hence, *skunk* cabbage) which attract insect pollinators. Thermogenesis does not, however, appear to be the function of the alternative pathway in roots and leaves. In one study of an arctic herb, for example, the alternative pathway accounts for up to 75 percent of total respiration but, in part because the heat is rapidly dissipated, accounts for no more than a 0.02 °C rise in leaf temperature (McNulty and Cummins, 1987).

A second hypothesis to explain the alternative pathway is referred to as the **energy overflow hypothesis.** This hypothesis is based on two general observations (Lambers, 1990). First, in most tissues the alternative pathway is inoperative until the normal cytochrome pathway has become saturated. Second, the rate of the alternative pathway can be increased by increasing the supply of carbohydrate to cells. In spinach (*Spinacea oleraceae*), for example, the alternative pathway is engaged only after photosynthesis has been in operation for several hours and has built up a supply of carbohydrate. In other words, the alternative pathway is generally engaged when there is an excess supply of carbohydrate, over and above what is required for metabolism or processed for storage. The function of the alternative pathway, according to this hypothesis, would be to burn off temporary accumulations of excess carbon that might otherwise interfere with source-sink relationships and inhibit translocation.

PLANT OILS, THE GYOXYLATE CYCLE, AND GLUCONEOGENESIS

Many seeds store carbon as fats and oils, which must first be converted to sugar in order to be respired.

Although lipids are a principal constituent of membranes and are stored by many tissues, they are not frequently used as a source of respiratory carbon. A major exception to this rule is found in germinating seeds, many of which store large quantities of lipids, principally triglycerides, as reserve carbon (Table 12.1). Storage lipids are deposited as **oil droplets** (also called oil bodies, oleosomes, or spherosomes), which are normally found in storage cells of cotyledons or endosperm.

TABLE 12.1 **Approximate lipid content of selected seeds.**

Species		Oil Content (% dry weight)
Macadamia nut	*Macadamia ternifolia*	75
Hazel nut	*Coryllus avellana*	65
Safflower	*Carthmus tinctoris*	50
Oil palm	*Elaeis guineensis*	50
Canola	*Brassica napus*	45
Castor bean	*Ricinus communis*	45
Sunflower	*Helianthus annum*	40
Maize	*Zea mays*	5

Plants are unable to translocate fats and oils from seed storage tissues to the elongating roots and shoots where the energy and carbon are required to support growth. The fatty acids must first be converted to a form that is more readily translocated. Usually this is sucrose (or sometimes stachyose), which is readily translocated from the storage cells containing the oil droplets to the embryo where the sucrose is metabolized. Complete conversion of triglycerides to sucrose is a complex process, involving the interaction of the oil bodies, glyoxysomes, mitochondria, and the cytosol (Fig. 12.13).

We can summarize the conversion of triglycerides to sucrose as follows. The first step is the hydrolysis of triglycerides to free fatty acids and glycerol. This is accomplished through the action of **lipase** enzymes, which probably act at the surface of the oil droplet. The fatty acid then enters the **glyoxysome,** an organelle similar in structure to the peroxisome found in leaves but with many different enzymes. In the glyoxysome, the fatty acid undergoes **β-oxidation;** the fatty acid chain is cleaved at every second carbon, resulting in the formation of acetyl-CoA.

Some of the acetyl-CoA combines with oxaloacetate (originating in the mitochondrion) to form citrate (6 carbons) in what is known as the **glyoxylate cycle.** The citrate in turn is converted to isocitrate, which then breaks down into one molecule of succinate (4 carbons) and one molecule of glyoxylate (2 carbons). The succinate returns to the mitochondrion where it enters the citric acid cycle, regenerating oxaloacetate, which is necessary to keep the glyoxylate cycle turning. Glyoxylate combines with another acetyl-CoA to produce malate. The malate then enters the cytosol where it is first oxidized to oxaloacetate and decarboxylated to phosphoenolpyruvate (PEP).

The glyoxylate cycle thus involves enzymes of both the gloxysome and the mitochondrion. Two enzymes of the cycle are unique to plants: **isocitrate lyase,** which

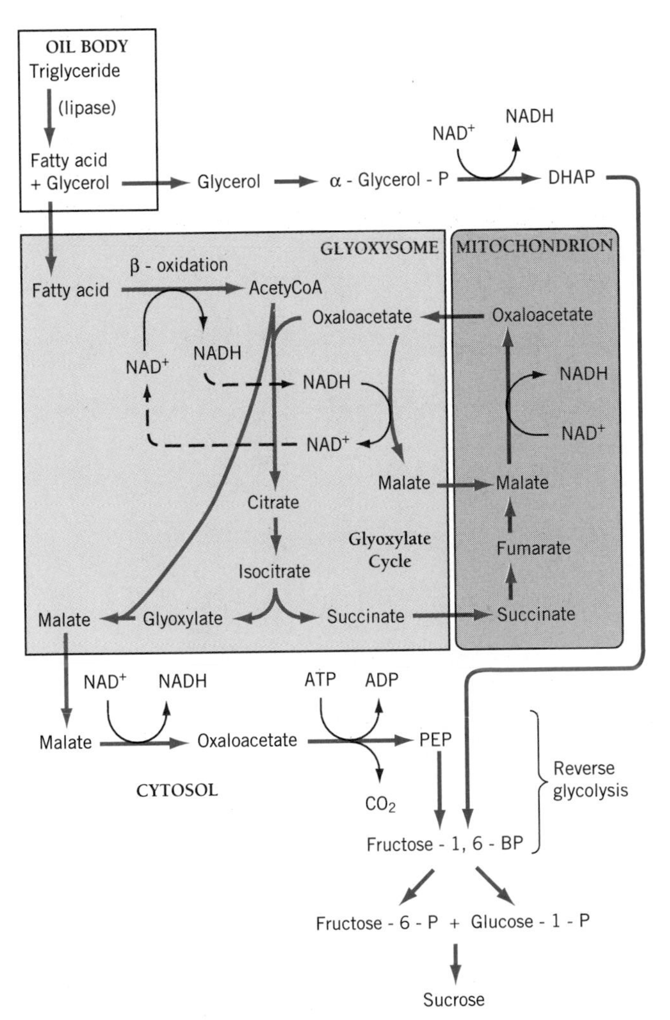

FIGURE 12.13 Lipid catabolism, the glyoxylate cycle and gluconeogenesis. (Based on ap Rees, 1990; Bewley and Black, 1985.

converts isocitrate to succinate plus glyoxylate, and **malate synthase,** which condenses an acetyl group with glyoxylate to form malate. The malate is then translocated from the glyoxysome into the cytosol where it is quickly oxidized to oxaloacetate by the enzyme malate dehydrogenase. The overall effect of the glyoxylate cycle is to catalyze the formation of oxaloacetate from two molecules of acetyl-CoA.

In the cytosol, oxaloacetate derived from the glyxoylate cycle is decarboxylated via the enzyme phosphoenolpyruvate carboxykinase to form phosphoenolpyruvate (PEP). Through a sequence of reactions that is essentially a reversal of glycolysis, PEP is converted to glucose. The conversion of PEP to glucose by a reversal of glycolysis is known as **gluconeogenesis.** Gluconeogenesis utilizes the enzymes of glycolysis, with significant differences. The glycolytic phosphofructokinase and hexokinase reactions (Fig. 12.3) are effectively irreversible—their free energy changes are highly unfavorable in direction of glucose synthesis. During gluconeogenesis, these reactions are replaced by reactions that make glucose synthesis more thermodynamically favorable. The conversion of fructose-1,6-bisphosphate to fructose-6-P is catalyzed by cytosolic fructose-1,6-bisphosphatase and the conversion of glucose-6-P to glucose is catalyzed by glucose-6-phosphatase. These differences are significant because they allow both directions to be thermodynamically favorable, yet be independently regulated. One direction can be activated while the other is inhibited, thus avoiding what might otherwise end up as a futile cycle.

The glycerol resulting from lipase action in the oil droplet also enters the cytosol where it is first phosphorylated with ATP to form α-glycerolphosphate and then oxidized to dihydroxyacetone phosphate (DHAP). The DHAP can also be converted to sucrose by reversal of glycolysis.

Some of the energy stored in triglycerides is conserved in the sucrose formed by gluconeogenesis, but β-oxidation of fatty acids in the glyoxysome also produces a large amount of NADH. The glyoxysome is unable to reoxidize NADH directly, but it can be used to reduce oxaloacetate to malate (Fig. 12.13). The malate then moves into the mitochondrion where it is reoxidized by malate dehydrogenase. Malate thus serves as a shuttle, carrying reducing equivalents between the glyoxysome and the mitochondrion. Reoxidation of malate inside the mitochondrion yields NADH, which can then enter the electron transport chain and drive ATP synthesis.

RESPIRATION AND BUILDING BLOCKS

In addition to energy in the form of ATP, cellular respiration also provides the carbon skeletons required for biosynthesis of other molecules required by the cell.

Before leaving the subject of cellular respiration, it is important to note that production of reducing potential and ATP is not the sole purpose of the respiratory pathways. In addition to energy, the synthesis of nucleic acids, protein, cellulose, and all other cellular molecules requires carbon skeletons as well. As noted at the beginning of this chapter, respiration also serves to modify the carbon skeletons of storage compounds to form these basic building blocks of cell structure. A few of the more important building blocks that can be formed from intermediates in glycolysis and the citric acid cycle are represented in Figure 12.14.

The withdrawal of glycolytic and citric acid cycle intermediates for the synthesis of other molecules means, of course, that not all of the respiratory substrate will be fully oxidized to CO_2 and water. The flow of carbon through respiration no doubt represents a balance between the metabolic demands of the cell for ATP to drive various energy-consuming functions on the one hand and demands for the reducing equivalents and car-

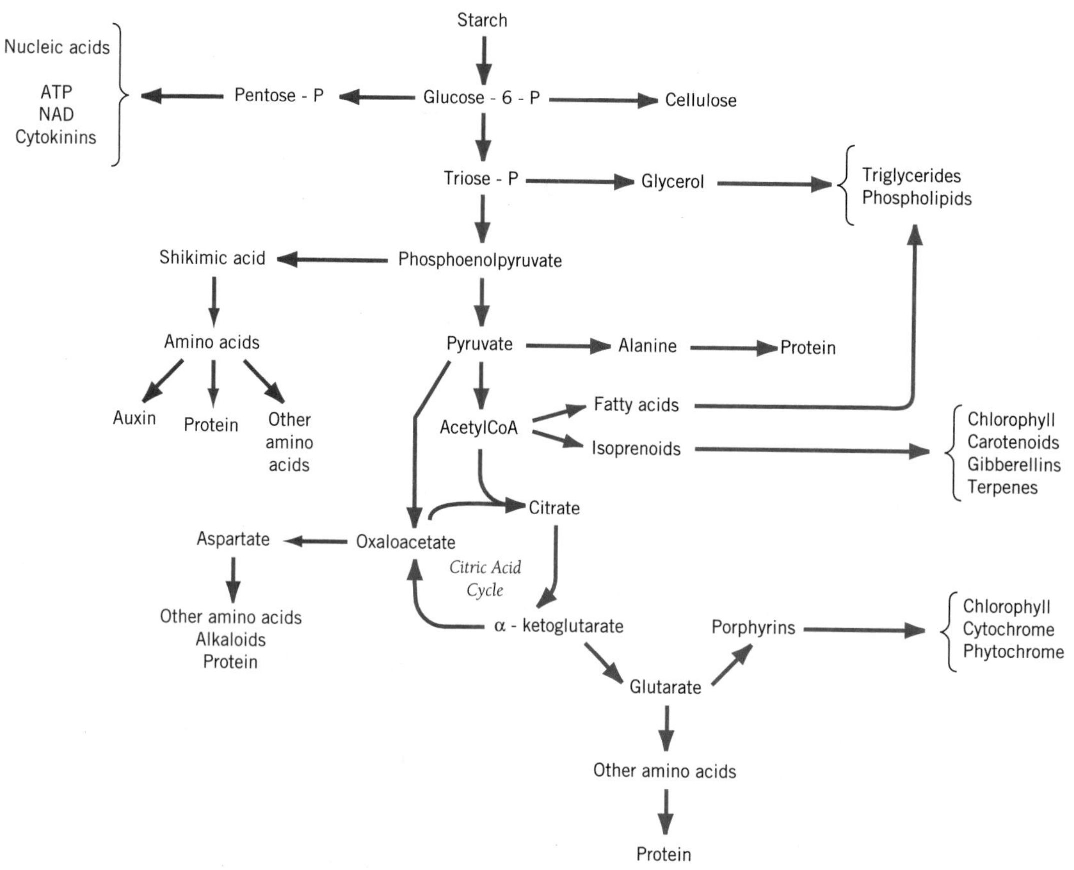

FIGURE 12.14 The role of respiration in biosynthesis. Intermediates in glycolysis and the citric acid cycle are drawn off to serve as building blocks for the synthesis of cellular molecules. Carbon in the cycle is maintained by the synthesis of oxaloacetate through anaplerotic reactions. This scheme is incomplete and is intended only to give some indication of the importance of these two schemes in biosynthesis.

bon skeletons required to build cell structure on the other.

It is also important to note that during periods of active synthesis, diversion of carbon from the citric acid cycle for synthetic reactions will lead to a significant reduction in the level of oxaloacetate. These synthetic reactions require not only carbon, but energy in the form of reducing potential and ATP as well. Without some means of compensating for this loss of oxaloacetate, the cycle will slow down or, in the extreme case, come to a complete halt and energy production will be impaired. This eventuality is precluded by the action of two cytosolic enzymes: phosphoenolpyruvate (PEP) carboxylase (see Chap. 10) and malate dehydrogenase. All plants, not just those with C_4 photosynthetic activity (see Chap. 10), have some level of PEP carboxylase activity that converts phosphoenolpyruvate (PEP) into oxaloacetate:

$$\text{PEP} + \text{HCO}_3^- \rightleftharpoons \text{oxaloacetate} \qquad (12.5)$$

In this case the PEP is derived from glycolysis.

Although there is some evidence that oxaloacetate may be translocated directly into the mitochondrion, it is more likely that oxaloacetate is quickly reduced to malate by the action of cytosolic malate dehydrogenase:

$$\text{oxaloacetate} + \text{NADH} \rightleftharpoons \text{malate} + \text{NAD}^+ \qquad (12.6)$$

The malate would then pass into the mitochondrion, via a malate (or dicarboxylate) translocator, where it is reoxidized to oxaloacetate by the action of a mitochondrial malate dehydrogenase:

$$\text{malate} + \text{NAD}^+ \rightleftharpoons \text{oxaloacetate} + \text{NADH} \qquad (12.7)$$

The replenishment of oxaloacetate in this way is an example of a "filling-up" mechanism or **anaplerotic** pathway. Thus carbon from glycolysis is delivered to the citric acid cycle through two separate but equally important streams: (1) to citrate via pyruvate and acetyl-CoA and (2) from PEP via oxaloacetate and malate to compensate for carbon "lost" to synthesis. Anaplerotic reactions such as the latter help to ensure that diversion of carbon for synthesis does not adversely influence the overall carbon balance between energy-generating catabolic reactions and biosynthetic anabolic reactions.

In addition to the normal enzymes of the citric acid cycle, plant mitochondria tend to have significant levels of **NAD⁺-malic enzyme,** which catalyzes the oxidative decarboxylation of malate:

$$\text{malate} + NAD^+ \rightleftharpoons \text{pyruvate} + CO_2 + NADH \quad (12.8)$$

The pyruvate may be further metabolized by pyruvate dehydrogenase to acetyl-CoA and from there enter the citric acid cycle. Thus the mitochondrial pool of malate may replenish the citric acid intermediates through either oxaloacetate or pyruvate.

In addition to serving an anaplerotic role, the uptake and oxidation of malate by mitochondria via either malic enzyme or malate dehydrogenase also provides an alternate pathway for metabolizing malate. This alternate pathway may be particularly significant in plants such as those of the family Crassulaceae (see Chap. 10) and others that store significant levels of malate in their vacuoles. Finally, it should be noted that diversion of pyruvate through oxaloacetate and malate bypasses the pyruvate kinase step in glycolysis (reaction 7, Fig. 12.4) and thus reduces the yield of ATP by one. This reduction is, however, offset by gains achieved by reduction of malate in the cytosol and its subsequent reoxidation in the mitochondrion. This sequence of reactions effectively shuttles extramitochondrial NADH (generated during glycolysis) into the mitochondrion, where it can be used to generate 3 molecules of ATP. This a gain of 1 ATP over the 2 ATP generated via the NADH-reductase route described earlier for extramitochondrial NADH.

RESPIRATION IN INTACT PLANTS AND ORGANS

The respiratory rate of whole plants and organs varies widely with age, metabolic state, and environmental conditions.

The study of respiration at the level of individual organs or the whole plant becomes much more difficult than it is for the study of individual cells. Whole plant respiration is normally studied by measuring the uptake of oxygen or the evolution of CO_2, but respiration rates obtained this way are highly variable. The balance of O_2 and CO_2 exchange is dependent on the substrate being respired and the balance of fermentation, citric acid cycle, and alternative pathway activities at any point in time. In addition, respiration rates differ between organs, change with age and developmental state, and are markedly influenced by temperature, oxygen, salts, and other environmental factors. Nevertheless, the study of respiration at the organ and plant level is a field of active study. Understanding respiration at this level has important implications for the plant physiologist interested in growth and development, for the physiological ecologist interested in plant biomass production, and the agricultural scientist because of its impact on productivity and yield.

As a general rule, respiratory rate is a reflection of metabolic demands. Younger plants, organs, or tissues respire more rapidly than older plants, organs or tissues (Fig. 12.15). The rapid rate of respiration during early stages of growth is presumably related to synthetic requirements of rapidly dividing and enlarging cells. As the plant or organ ages and approaches maturity, growth and its associated metabolic demands decline. Many organs, especially leaves and some fruits, experience a transient rise in respiration, called a **climacteric,** that marks the onset of senescence and the degenerative changes that precede death. Typically the climacteric rise in O_2 consumption is accompanied by a decline in oxidative phosphorylation, indicating that ATP production is no longer tightly coupled to electron transport. The respiration rate of woody stems and branches, expressed on a weight or mass basis, also declines as they grow. This is because as the diameter increases, the relative proportion of nonrespiring woody tissue also increases.

Carbon lost to the plant due to respiration can represent a significant proportion of the available carbon. Actual respiration rates for plant tissues range in the extreme between barely detectable (0.005 μmol CO_2 $gW_d^{-1}\ h^{-1}$) in dormant seeds to 1000 μmol CO_2 $gW_d^{-1}\ h^{-1}$ or more in the spadix of skunk cabbage during the respiratory crisis. More typically rates for vegetative tissues range from 10 to 200 μmol CO_2 $gW_d^{-1}\ h^{-1}$ (Table 12.2). This may represent a considerable fraction of the carbon assimilated by photosyn-

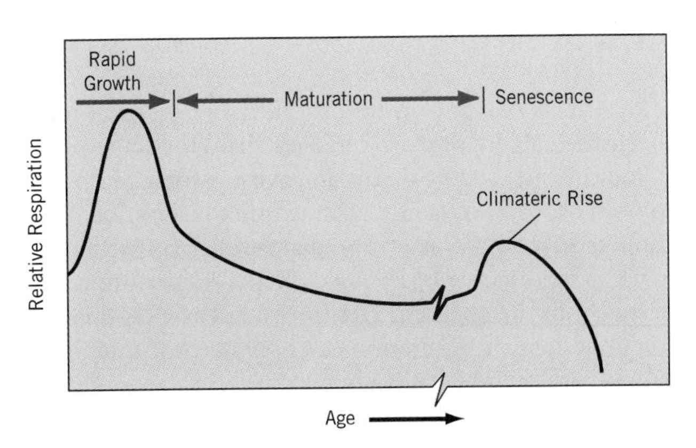

FIGURE 12.15 Respiratory rate as a function of age. This type of curve applies generally to most herbaceous plants, tissues, and organs. The magnitude of the climacteric will vary—some organs exhibit little or no climacteric rise.

TABLE 12.2 **Approximate specific dark respiration rates at 20 °C for crop species, deciduous foliage, and conifers.**

	Specific Respiration Rate μmol CO_2 evolved · g^{-1} dry mass · h^{-1}
Crops	70–180
Deciduous foliage	
(sun leaves)	70–90
(shade leaves)	20–45
Conifers	4–25

Data recalculated from Larcher, 1980.

thesis during a 24-hour period. (Recall that photosynthesis occurs only during daylight hours, but respiration, especially of roots and similar tissues, is ongoing 24 hours a day.) This is evident in the list of calculated respiratory carbon loss compiled by H. Lambers (1985). On the average, 30 to 60 percent of daily photoassimilate is lost as respiratory CO_2. In tropical rainforest species, probably because of accelerated enzymic activities at higher temperatures, this loss may exceed 70 percent. Of the total daily carbon loss in any given plant, some 30 to 70 percent is accounted for by respiration in the roots alone, that "hidden half" of the plant.

FACTORS INFLUENCING RESPIRATION RATE

In addition to the effects of age and tissue type described briefly above, respiration, as any other metabolic process, responds to a variety of external conditions. These include light, temperature, gaseous composition of the atmosphere, salts, and mechanical stimuli.

LIGHT

The effects of light on mitochondrial respiration have been the subject of considerable debate for some time. Traditionally, photosynthesis investigators and crop physiologists have tacitly assumed that respiration continues in the light at a rate comparable to that in the dark. The true rate of photosynthesis is therefore taken as equal to the apparent rate (measured as CO_2 uptake) plus the rate of respiration (CO_2 evolved) in the dark. However, attempts to study respiration in green leaves have led to alternate and conflicting conclusions (Graham, 1980). These range from complete inhibition of mitochondrial activities, to partial operation of the citric acid cycle, or to stimulation of respiration by light. The problem lies in the difficulty of measuring respiration during a period when gas exchange is dominated by the overwhelming flux of CO_2 and O_2 due to photosynthesis, the recycling of CO_2 within the leaf, and the exchange of metabolites by chloroplasts and mitochondria. One view holds that production of ATP over and above that required by chloroplasts could make oxidative phosphorylation by the mitochondria superfluous while others argue that at least a partial operation of the citric acid cycle is necessary in order to provide carbon skeletons for synthesis and growth in the light. It is unlikely that the true answer to this problem will be known until it is possible to obtain accurate estimates of carbon flux through mitochondria in the light.

Light effects on respiration during a subsequent dark period have been demonstrated. For example, dark respiratory rates in leaves adapted to full sun (sun leaves) are generally higher than those of leaves of the same species adapted to shade (shade leaves) (Table 12.2). As well, the rate is consistently higher in mature leaves of shade-intolerant species than in shade-tolerant species. Indeed, a reduced respiratory rate appears to be a fairly consistent response to low irradiance. This is probably related to lower growth rates also observed in shade-grown plants, but it is not known which is cause and which is effect. It is projected that low respiratory and growth rates may confer a survival advantage under conditions of deep shade.

The basis for light regulation of respiratory rate is unknown, although some have suggested that low respiratory rates under conditions of low irradiance may reflect availability of substrate. For example, the respiration rate 1 to 2 hours following a period of active photosynthesis is higher than after a long dark period. Other experiments, however, have shown that dark respiratory rates do not correlate with CO_2 supply during the previous light period. This seems to suggest a more direct effect on respiration, the nature of which remains unknown.

TEMPERATURE

One of the most commonly applied quantitative measures used to describe the effect of temperature on a process is the **temperature coefficient** or (**Q_{10}**), given by the expression:

$$Q_{10} = \frac{\text{rate at } (t + 10)°\text{C}}{\text{rate at } t°\text{C}} \quad (12.8)$$

At temperatures between 5 °C and about 25 °C or 30 °C, respiration rises exponentially with temperature and the Q_{10} value is approximately 2.0. Within this temperature range, a doubling of rate for every 10 °C rise in temperature is typical of enzymic reactions. At temperatures above 30 °C, the Q_{10} in most plants begins to fall off as substrate availability becomes limiting. In particular, the solubility of O_2 declines as temperatures increase and the diffusion rate (with a Q_{10} close to 1) does not increase sufficiently to compensate. As temperatures approach 50 °C to 60 °C, thermal denaturation of respiratory enzymes and damage to membranes bring respiration to a halt.

Some investigators have observed differences in the rate of respiration in tropical, temperate, and arctic species at different temperatures (see Lambers, 1985). For example, the respiration rate of leaves of tropical plants at 30 °C is about the same as that of arctic species at 10 °C. The temperature coefficient (Q_{10}) for respiration is the same in both cases and there is no evidence of intrinsic differences in the biochemistry of respiration. It is likely that the differences reflect differences in temperature optima for growth of arctic and tropical species—optima determined by factors other than respiration—and the consequent metabolic demand for ATP.

OXYGEN AVAILABILITY

As the terminal electron acceptor, oxygen availability is obviously an important factor in determining respiration rate. The oxygen content of the atmosphere is relatively stable at about 21 percent O_2. The equilibrium concentration of oxygen in air-saturated water, including the cytosol, is approximately 250 μM. However, cytochrome *c* oxidase has a very high affinity for oxygen with a K_m less than 1 μM. Under normal circumstances, oxygen is rarely a limiting factor.

There are some situations, however, where oxygen availability may become a significant factor. One is in bulky tissues with low surface to volume ratios, such as potato tubers and similar storage tissues, where the diffusion of oxygen may be slow enough to restrict respiration. This may not be a serious problem, however. A significant volume—as much as 40 percent—of roots and similar tissues may be occupied by intercellular air spaces that aid in the rapid distribution of O_2 absorbed from the soil or, in some cases, from the aerial portions of the plant. Plants are most likely to experience oxygen deficits during periods of flooding, when air in the large pore spaces of the soil is displaced by water, thereby decreasing the oxygen supply to the roots. For similar reasons, plants grown in hydroponic culture must be aerated to maintain adequate oxygen levels in the vicinity of the roots (Chap. 4).

SUMMARY

Cellular respiration consists of a series of interdependent pathways by which carbohydrate and other molecules are oxidized for the purpose of retrieving the energy stored in photosynthesis and to obtain the carbon skeletons that serve as precursors for other molecules used in the growth and maintenance of the cell. Plants store excess photosynthate as starch, a long linear or branched polymer of glucose. Starch and other storage carbohydrates are enzymatically degraded to glucose or fructose, which are then converted to fructose-1,6-bisphosphate (FBP). FBP is the starting point for glycolysis, a series of reactions that ultimately produce pyruvate. In the process, a small amount of ATP and reducing potential is generated. The intermediates in glycolysis are 3-carbon sugars, many of which are precursors to triglycerides and amino acids. Precursors with four and five carbons are produced by an alternate route for glucose metabolism called the oxidative pentose phosphate pathway. The oxidative pentose phosphate pathway also produces NADPH (as opposed to NADH), which provides reducing potential when required for biosynthetic reactions in plants.

The fate of pyruvate depends on the availability of oxygen. In an anaerobic environment, pyruvate is reduced (usually to ethanol), while in the presence of oxygen pyruvate is first oxidized to acetyl-CoA and carbon dioxide. The acetate group is then further oxidized to carbon dioxide and water through the citric acid cycle (CAC). The CAC enzymes are located predominantly in the matrix of the mitochondrion. Altogether, eight enzyme-catalyzed steps degrade the acetate group to carbon dioxide and water. The cycle includes four oxidations that yield NADH at three steps and $FADH_2$ at another. One molecule of ATP is generated in a substrate-level phosphorylation and the original acetate acceptor, oxaloacetate, is regenerated, which allows the cycle to continue.

The NADH and $FADH_2$ produced in the CAC are oxidized via an electron transport chain found in the mitochondrial inner membrane. The chain consists of four multiprotein complexes linked by mobile electron carriers. The final complex, cytochrome oxidase, transfers the electrons to molecular oxygen, forming water. At three points in the chain, the free energy drop associated with electron transport is used to establish a proton gradient across the membrane. As in the chloroplasts, this proton gradient is used to drive ATP synthesis.

The CAC and electron transport chain are virtually identical in all organisms. Plants, however, have an alternative oxidase that intercepts electrons early in the chain, thus bypassing two of the three proton-pumping sites. When using this route, at least two-thirds less ATP is formed and much of the electron energy is converted to heat. The alternative oxidase, at least in some plants, has been associated with thermogenesis, particularly in certain members of the Araceae where the higher temperature volatilize amines that appear to attract insect pollinators. The alternative oxidase may also serve to "burn off" excess carbohydrate.

Many seeds store carbon as fats and oils, which

must first be converted to sugar in order to be respired. After the fatty acids are broken down into acetyl-CoA units, a complex series of reactions involving enzymes of the mitochondrion, the glyoxysome, and the cytosol convert the acetate units to phospho*enol*pyruvate. The pyruvate is then converted to glucose by gluconeogenesis, a process that is essentially a reversal of glycolysis.

The respiratory rate of whole plants and organs varies widely with age, metabolic state, and environmental conditions. The latter include light, temperature, gaseous composition of the atmosphere, salts, and mechanical stimuli.

CHAPTER REVIEW

1. Compare respiration and fermentation. Aerobic organisms are generally much larger than anaerobic organisms. Can you suggest how this may be related to respiration?
2. Phosphorous plays an important role in respiration, photosynthesis, and metabolism generally. What is this role?
3. Oils are a common storage form in seeds, particularly small seeds. What advantage does this offer the seed?
4. There are a number of similarities between chloroplasts and mitochondria. Compare these two organelles form the perspective of:
   1) ultrastructure and biochemical compartmentation;
   2) organization of the electron transport chain;
   3) proton gradients and ATP synthesis.
5. How does cyanide-resistant respiration differ from normal respiration? Of what value might cyanide-resistant respiration be to plants?
6. Discussions of respiration most often emphasize energy retrieval and ATP production. What other very important metabolic role(s) does respiration fulfill?

FURTHER READING

Amthor, J. S. 1989. *Respiration and Crop Productivity.* New York: Springer-Verlag.

Dennis, D. T., D. H. Turpin. 1990. *Plant Physiology, Biochemistry and Molecular Biology.* Essex: Longman Scientific and Technical.

Douce, R. 1985. *Mitochondria in Higher Plants. Structure, Function, and Biogenesis.* New York: Academic Press.

Manners, D. J. 1985. Starch. In: P. M. Dey, R. A. Dixon (eds.), *Biochemistry of Storage Carbohydrates in Green Plants.* London: Academic Press, pp. 149–203.

Siedow, J. N., D. A Berthold. 1986. The alternative oxidase: A cyanide-resistant respiratory pathway in higher plants. *Physiologia Plantarum* 66:569–573.

Vanlerberghe, G. C., L. McIntosh. 1997. Alternative oxidase: from gene to function. *Annual Review of Plant Physiology and Plant Molecular Biology* 48:703–734.

REFERENCES

ap Rees, T. 1990. Carbon metabolism in mitochondria. In: D. T. Dennis, D. H. Turpin (eds.), *Plant Physiology, Biochemistry and Molecular Biology.* Essex: Longman Scientific and Technical, pp. 106–123.

Beck, E. 1985. The degradation of transitory starch in chloroplasts. In: R. L. Heath, J. Preiss (eds.), *Regulation of Carbohydrate Partitioning in Photosynthetic Tissue.* Rockville, MD: American Society of Plant Physiologists, pp. 27–40.

Bewley, J. D., M. Black 1985. Seeds. *Physiology of Development and Germination.* New York: Plenum Press.

Graham, D. 1980. Effects of light on "dark" respiration. In: D. D. Davies (ed.), *The Biochemistry of Plants. Vol 2. Metabolism and Respiration.* New York: Academic Press, pp. 525–579.

Lambers, H. 1985. Respiration in intact plants and tissues. Its regulation and dependence on environmental factors, metabolism and invaded organisms. In: R. Douce, D. A. Day (eds.), *Higher Plant Cell Respiration. Encyclopedia of Plant Physiology, New Series* 18:418–473.

Lambers, H. 1990. Oxidation of mitochondrial NADH and the synthesis of ATP. In: D. T. Dennis, D. H. Turpin (eds.), *Plant Physiology, Biochemistry and Molecular Biology.* Essex: Longman Scientific and Technical, pp. 124–143.

Larcher, W. 1980. *Physiological Plant Ecology.* Berlin: Springer-Verlag.

McNulty, A. K., W. R. Cummins. 1987. The relationship between respiration and temperature in leaves of the arctic plant *Saxifraga cernua. Plant, Cell and Environment* 10:319–325.

Moore, A. L., J. N. Siedow. 1991. The regulation and nature of the cyanide-resistant alternative oxidase of plant mitochondria. *Biochimica et Biophysica Acta* 1059:121–140.

Plaxton, W. C. 1996. The organization and regulation of plant glycolysis. *Annual Review of Plant Physiology and Plant Molecular Biology* 47:185–214.

Rhoads, D. M., L. McIntosh. 1991. Isolation and characterization of a cDNA clone encoding an alternative oxidase protein of *Sauromatum guttatum* (Schott). *Proceedings of the National Academy of Science USA* 88:2122–2126.

Siedow, J. N., D. A Berthold. 1986. The alternative oxidase: A cyanide-resistant respiratory pathway in higher plants. *Physiologia Plantarum* 66:569–573.

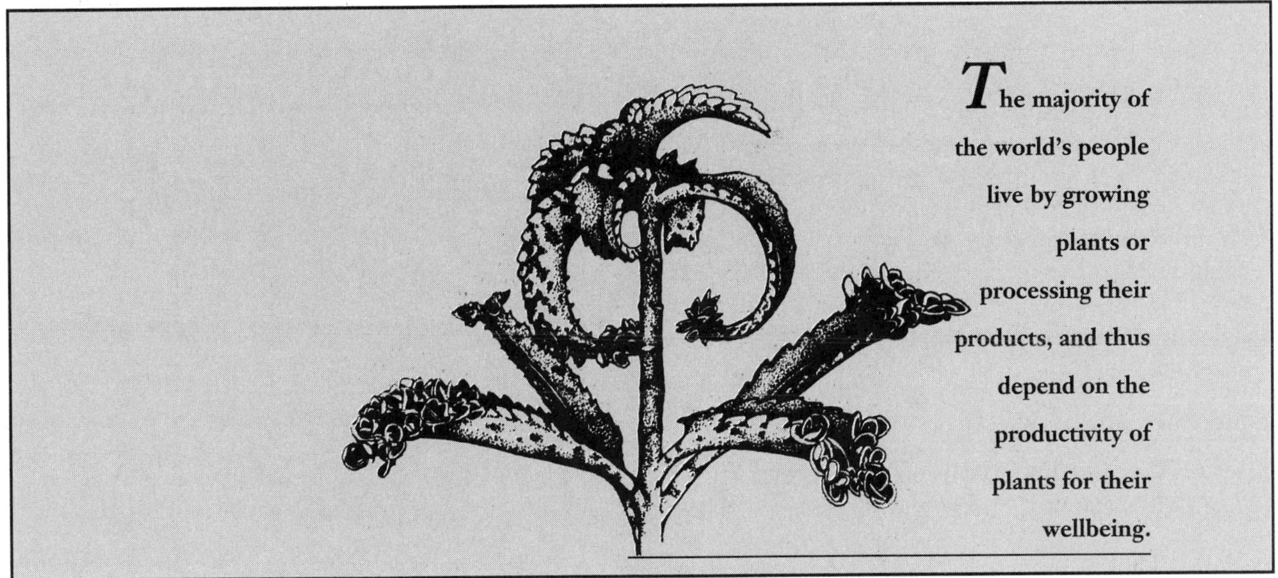

C. L. Beadle et al. (1985)

13

Carbon Assimilation and Productivity

In an earlier chapter, the theme was introduced that photosynthesis is the fundamental basis of competitive success in green plants (Chap. 10). The significance of carbon assimilation, however, extends well beyond the performance of green plants. Carbon assimilation creates plant biomass, or dry matter, which in turn supports humans and virtually all other heterotrophic organisms in the biosphere. Since the beginnings of the industrial revolution, growth of the human population and industries has been putting increasing pressures on the biosphere. These pressures have in turn stimulated interest in studies of plant productivity, which is directly related to carbon gain through photosynthesis. Productivity and its relationship to yield is of obvious concern to agriculture, but it also has relevance in broader ecological terms because it provides the energetic and material basis for other organisms. Knowledge of plant productivity provides a basis level for ecosystem research, helping to elucidate problems of energy and nutrient flow and their relationships to the structures of communities. Plant productivity also carries implications for the upper limit of the earth's sustainable human population. Knowing how plants function in a natural environment, their potential for harvest, and how they might respond to potentially stressful environmental change is essential to learning how to manage world resources in a time of burgeoning world population.

The physiology of photosynthesis—light capture, energy conversion, and partitioning of carbon—are at the root of productivity. This chapter will address

- the concepts of carbon gain and productivity;
- interactions between respiration and photosynthesis and how these relationships determine the overall carbon gain of a plant;
- environmental factors such as light, available carbon dioxide, temperature, availability of soil water and nutrients, which influence photosynthesis and productivity and various aspects of leaf and canopy structure and their impact on productivity.

The chapter will finish with a brief discussion of primary productivity on a global scale.

PRODUCTIVITY

*The term **productivity** refers to an increase in biomass—the dry matter content of an organ, organism, or population.*

Although inorganic nutrients are a part of this dry matter, by far the bulk of dry matter for any organism consists of carbon. The basic input into the biosphere is the conversion of solar energy into organic matter by photosynthetic plants and microorganisms, known as **primary productivity** (**PP**). Total carbon assimilation is known as **gross primary productivity (GPP)**. Not all of the GPP is available for increased biomass, however, since there is a respiratory cost that must be taken into account. The principal focus of most productivity studies is therefore **net primary productivity (NPP)**. NPP is determined by correcting GPP for energy and carbon loss due to respiration. NPP is a measure of the net increase in carbon, or **carbon gain,** and reflects the additional biomass that is available for harvest by animals.

RESPIRATION AND CARBON ECONOMY

Overall carbon gain by a plant depends on net primary productivity, which is in turn dependent on the balance between carbon uptake by photosynthesis and carbon loss to respiration.

Because photosynthesis occupies such a prominent position in the metabolism of higher plants, its rate is often regarded as the primary factor regulating biomass production and crop productivity. Yet it has often been observed that plants with similar photosynthetic rates may differ markedly with respect to growth rate and biomass accumulation. Clearly other factors such as partitioning and allocation of carbon, translocation rates, and respiration rates must be considered when attempting to understand the overall carbon budget or **carbon economy** of a plant. Carbon economy is the term used to describe the balance between carbon acquisition and its utilization. Respiration is the principal counterbalance to photosynthesis. Respiration consumes assimilated carbon in order to obtain the energy required to increase and maintain biomass. Respiratory loss of carbon constitutes one of the most significant intrinsic limitations on plant productivity.

In an effort to better understand the impact of respiration on the carbon economy of plants, some physiologists have sought to distinguish experimentally between the carbon and energy costs of *growth* on the one hand and *maintenance* on the other (Lambers, 1985, Pearcy et al., 1989). The term **growth respiration** has been coined to account for the carbon cost of growth. Growth respiration includes the carbon actually incorporated plus the carbon respired to produce the energy (in the form of reducing potential and ATP) required for biosynthesis and growth. **Maintenance respiration,** on the other hand, provides the energy for processes that do not result in a net increase in dry matter, such as turnover of organic molecules, maintenance of membrane structure, turgor, and solute exchange. These distinctions are not easily made, but may be estimated by relating respiration rate to the relative growth rate (Fig. 13.1). When respiration rate is extrapolated back to zero growth, it can be assumed that the residual respiration represents the carbon and energy requirements for maintenance of the nongrowing cells.

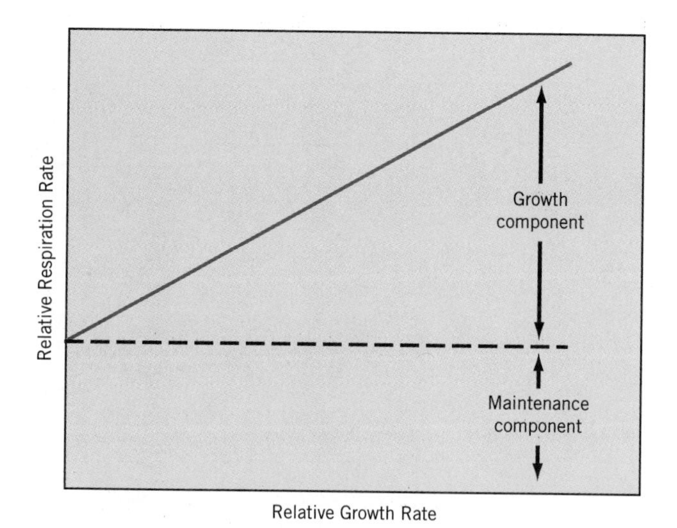

FIGURE 13.1 Growth and maintenance respiration. The proportion of respiration devoted to maintenance can be estimated by extrapolation to zero growth rate.

From Figure 13.1 it can be seen that maintenance respiration relative to total respiration will vary with the growth rate. The proportion devoted to maintenance will be least in a young, rapidly growing plant or organ while it will account for the bulk of respiration in a nongrowing organ such as a mature leaf. Indeed, measuring maintenance respiration by one commonly used experimental approach assumes that the respiration of a mature leaf is essentially 100 percent maintenance, although a small (but unknown) amount must be used for the translocation of solutes into and out of the leaf. Maintenance respiration also tends to be higher in roots than in shoots and other above-ground organs. This may be related to the expenditure of maintenance energy by roots for ion uptake to satisfy not only their own needs but the needs of the shoot as well. It may also reflect the observation that the cyanide-resistant alternative pathway tends to be higher in roots.

Respiration produces the metabolic energy that is required for various growth processes that increase biomass and agricultural yield, but it can also consume carbon with little or no apparent yield of useful energy. Since the latter situation represents a loss of carbon to the plant, it has been assumed that lower respiration rates would establish a more positive carbon economy that might, in turn, result in more rapid growth and increased productivity. Is it possible to manipulate respiration rates in favor of higher productivity? In a series of studies, genotypes of perennial rye grass (*Lolium perenne*) were selected for respiratory rates ranging from "slow" (2.0 mg CO_2 g^{-1} h^{-1}) to "rapid" (3.5 mg CO_2 g^{-1} h^{-1}) (Wilson, 1982). Selection was based on the specific respiration rate of mature leaves at 25 °C (i.e., maintenance respiration). The results (Fig. 13.2) establish a

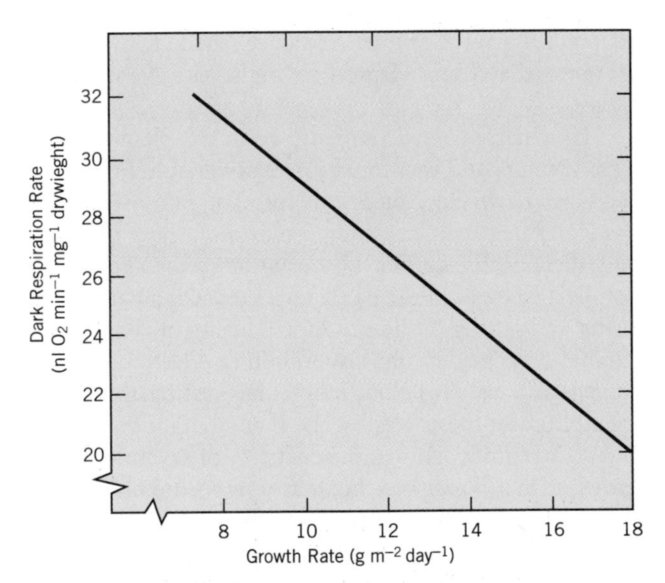

FIGURE 13.2 The inverse correlation between respiratory rate and growth rate in genotypes of perennial rye grass (*Lolium pratense*). (From D. Wilson, *Annals of Botany* (London) 49:303–312, 1982. With permission of the Annals of Botany Company.)

negative correlation between respiration and growth rate; that is, the highest growth rates were recorded for genotypes having the lowest rates of respiration. Wilson concluded that the higher growth rates resulted from a more efficient use of carbon "as a consequence of reduced respiratory evolution of CO_2 from fully grown tissues." In other words, with less of the carbon consumed in maintenance respiration, a higher proportion of the carbon was available for allocation to growth. Other investigators have found evidence of similar negative correlations between respiration and growth with wheat, barley, and oats.

Another method for improving respiratory efficiency might be to reduce the contribution, if any, of cyanide-resistant alternative pathway activity (Chap. 12). The alternative pathway oxidizes carbon with most of the energy released as heat. The overall impact of the alternative pathway remains to be firmly established, but there are some promising indications. Certain cultivars have been identified that differ with respect to the engagement of alternative pathway respiration at two different CO_2 levels (Musgrave et al., 1986). The cultivars were crossed to produce progeny that either strongly expressed the alternative pathway or showed no expression. The accumulation of biomass was greater in the progeny from which the alternative pathway was absent.

Studies such as these suggest that respiration has a significant impact on biomass accumulation and yield. They also indicate that improving yield by manipulating respiration might be feasible. Given that approximately half of the carbon assimilated in photosynthesis is eventually lost by respiration, reducing either the level of maintenance respiration or the engagement of the alternative pathway should shift the carbon balance in favor of growth respiration and increased biomass production. It is assumed, of course, that maintenance respiration and/or the alternative pathway can be reduced without detrimental effects on the plant, but this remains to be clearly established. Certainly some level of maintenance respiration is essential to the health of the plant—this level is yet to be determined.

One approach to improving the efficiency of net primary production is through breeding and selection programs. It is quite possible that this has already been achieved to some extent in existing breeding programs, without consciously evaluating the role of respiration. Another approach would be to manipulate respiration through genetic engineering. However, respiration is a complex process, both biochemically and physiologically. It is central to the metabolism of the cell, involves many different enzymes, and ultimately the coordinated activities of cellular organelles plus the cytosol. It is difficult to know which enzymes or reactions—that is, which genes—might be profitably manipulated or how an altered respiratory balance will affect other physiological processes. Although the prospects for improving productivity by manipulating respiration are encouraging, there is clearly a great deal of fundamental research yet to be conducted before significant progress can be expected.

FACTORS INFLUENCING PHOTOSYNTHESIS AND PRODUCTIVITY

Photosynthesis and, consequently, productivity are influenced by a variety of genetic and environmental factors, including light, available carbon dioxide, temperature, soil water, nutrients, and canopy structure.

The rate of photosynthesis may be limited by a host of variables. Which of these variables have an influence and the extent to which the influence is felt depend on whether one is concerned with a single leaf, a whole plant, or a population of plants that form a canopy. Included in these variables are both environmental factors and genetic factors. Environmental factors include light, availability of CO_2, temperature, soil water, nutrient supply, pathological conditions, and pollutants. Genetic factors include carbon assimilation pathway (C3 versus C4 metabolism, CAM), which was discussed in Chapter 10, leaf age and morphology, leaf area index, leaf angle, and leaf orientation.

LIGHT

Typical responses of photosynthesis to fluence rate are illustrated in Fig. 13.3. At very low fluence rates the rate

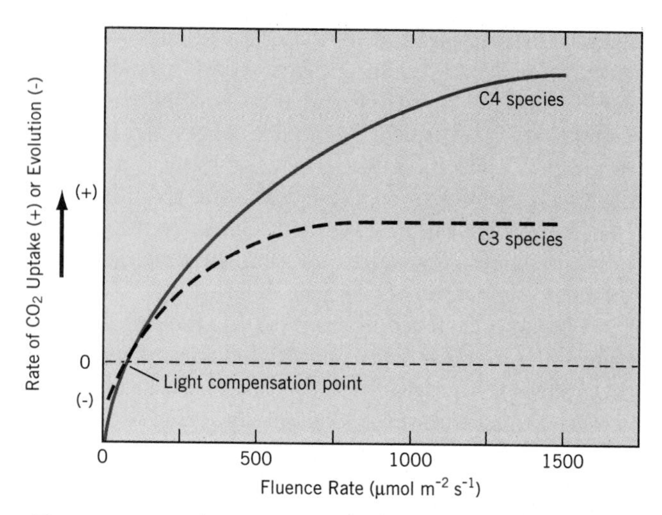

FIGURE 13.3 A graph showing the typical response of photosynthesis in C3 and C4 species to fluence rate.

of CO_2 evolution due to dark respiration exceeds the rate of photosynthetic CO_2 uptake. This results in a negative CO_2 uptake, or net CO_2 evolution. As fluence rate increases, photosynthesis also increases and so does CO_2 uptake until the rate of CO_2 exchange equals zero. This is the fluence rate, known as the **light compensation point,** at which the competing processes of photosynthesis and respiration are balanced. The light compensation point for most plants falls somewhere in the range of 10 to 40 $\mu mol\ m^{-2}\ s^{-1}$, roughly equivalent to the light level found in a well-lighted office, laboratory, or classroom.

At fluence rates above the compensation point, the rate of photosynthesis continues to increase until, at least in C3 plants, it reaches light saturation. In most C3 plants at normal atmospheric CO_2 levels, photosynthesis saturates with light levels of about 500 to 1000 $\mu mol\ m^{-2}\ s^{-1}$; that is, about one-quarter to one-half of full sunlight. Light saturation occurs because some other factor, usually CO_2 levels, becomes limiting. In most cases, both the saturation rate of photosynthesis and the fluence rate at which saturation occurs can be increased by increasing the CO_2 level above ambient. The light compensation point for C4 plants generally falls within the same range as for C3 plants but, as noted earlier (Chap. 10), C4 plants never really achieve light saturation (Fig. 13.3). The failure of C4 plants to saturate is probably related to their capacity to use the additional energy to drive phosphoenolpyruvate carboxylation and their capacity to continue photosynthesis at very low CO_2 levels (Chap. 10). There are also a small number of C3 plants, such as peanut (*Arachis hypogea*), which do not light saturate. It is not clear why this is the case, but these are exceptions to the rule. Individual leaves and plants will also adapt to the light environment in which they are grown. The light saturated rate of photosynthesis, for example, is lower in leaves that have adapted to growth at low irradiance (shade leaves) than in those that have adapted to higher irradiance (sun leaves).

In a natural environment, even C3 plants rarely light saturate and then only for relatively brief periods. Between dawn and dusk, the rate of photosynthesis gradually increases, reaching a maximum near midday, and then declines. The photosynthetic rate generally parallels changes in the irradiance that accompanies the rising and setting of the sun. Even during midday, measurable decreases in photosynthetic rate have been observed with passing cloud cover, suggesting that even then photosynthesis was barely, if at all, light saturated. In another study, annual productivity of several species growing in a European hedgerow was limited to less than half their potential maximum (Küppers, 1984). The failure of carbon gain to match leaf photosynthetic capacity was attributed to reduced average irradiance due to effects of dawn and dusk, short photoperiods in the spring and fall, and cloud cover. Long-term carbon gain is clearly dependent on cumulative irradiance over the growing season.

AVAILABLE CARBON DIOXIDE

The carbon dioxide concentration of the atmosphere is relatively low and reasonably stable, at least over the short term, at about 0.035 percent by volume or 350 $\mu l\ l^{-1}$. This is well below the CO_2 saturation level for most C3 plants at normal fluence rates (Fig. 13.4), which means that availability of CO_2 is often a limiting factor in photosynthesis. In C3 plants, increased photosynthetic rates with higher CO_2 levels results from two factors: increased substrate for the carboxylation reaction and, through competition with oxygen, reduced photorespiration. Note the interaction between ambient CO_2 levels and light. At higher fluence rates, both the maximum rate of

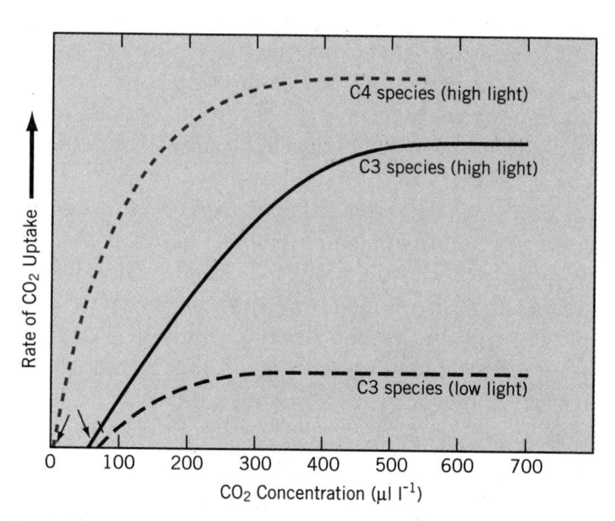

FIGURE 13.4 A graph showing the typical response of C3 and C4 species to ambient CO_2 concentration. Arrows indicate CO_2 compensation concentration.

photosynthesis and the CO_2 saturation level increase. Interestingly, most C4 plants appear to saturate at CO_2 levels at or just above normal atmospheric concentrations, regardless of fluence rate (Fig. 13.4).

Assessing the impact of CO_2 levels on photosynthesis is not quite so straightforward as it might at first appear. The rate of photosynthesis is actually determined not by the ambient CO_2 concentration, as much as by the intracellular CO_2 concentration, that is, the supply of CO_2 at the carboxylation site in the chloroplast (Farquhar and Sharkey, 1982). It is assumed that the intracellular CO_2 concentration is in equilibrium with the intercellular spaces. Since diffusion rates depend in part on concentration gradients, the primary effect of increasing ambient CO_2 levels would be to increase the intercellular CO_2 concentration by increasing the rate of diffusion into the leaf. Here it is assumed that water supply is adequate and, consequently, stomatal CO_2 conductance is not limiting.

Although it was once thought that stomatal CO_2 conductance was the principal factor limiting photosynthesis, more recent studies suggest it may be the other way around—stomatal conductance varies in response to photosynthetic capacity (Farquhar and Sharkey, 1982). Photosynthetic capacity is determined by the balance between carboxylation capacity and electron transport capacity (Fig. 13.5). At low CO_2 concentrations, the rate of photosynthesis is limited by available CO_2 and, hence, the carboxylation capacity of the system, but is saturated with respect to availability of the acceptor molecule, ribulose-1,5-bisphosphate (RuBP) (see Chap. 10). However, any excess generation of RuBP, which is in turn dependent upon the electron transport reactions, over that required to support carboxylation would represent an inefficient use of resources. Conversely, at high CO_2 concentrations or in low light, the limiting factor would be the energy-limited capacity to regenerate the acceptor molecule, ribulose-1,5-bisphosphate. In this case, an excess of carboxylating capacity—that is, an excess of Rubisco—would be an inefficient use of resources.

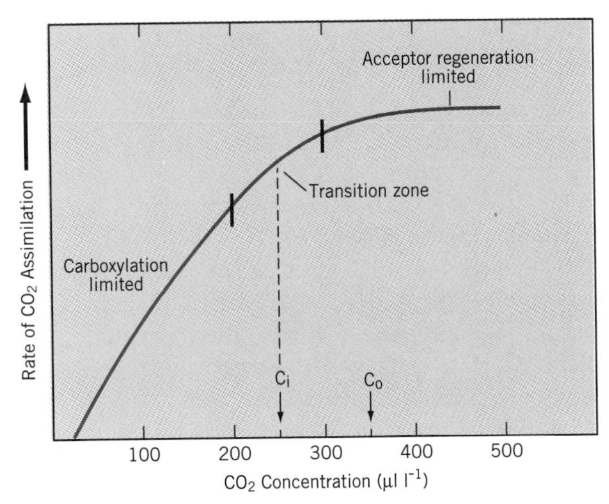

FIGURE 13.5 A model to describe limitation of photosynthetic rate as a function of CO_2 concentration. At low CO_2 concentrations, photosynthesis is limited by the carboxylation capacity of the enzyme Rubisco. At high concentrations of CO_2, the rate is limited by the rate or regeneration of the acceptor molecule, ribulose-1,5-bisphosphate. Stomata probably operate to keep the intercellular CO_2 concentration within the transition zone where there is neither an excess of carboxylating capacity nor an excess of electron transport. C_i and C_0 indicate the intercellular and ambient CO_2 concentrations, respectively. (Redrawn from G. Farquhar and T. Sharkey, 1982.)

The most efficient use of resources for the plant would be to maintain intercellular CO_2 levels in the transition zone, where there is neither an excess of electron transport capacity nor an excess of carboxylating capacity (Farquhar and Sharkey, 1982, Pearcy et al., 1987). Because intercellular CO_2 levels are at least partly determined by stomatal conductance, it appears that the principal function of the stomata might be to regulate CO_2 uptake in order to keep intercellular CO_2 levels as much as possible within the transition range. Note that this is not the traditional view of stomatal function, which says that stomata operate principally to regulate water loss. Note also that CO_2 enrichment at high fluence rates leads to both higher photosynthetic maxima and higher CO_2 saturation levels (Fig. 13.4). These observations suggest that plants are also able to compensate for higher light levels by increasing their carboxylating capacity. Such an increase could be achieved by regulating the amount or catalytic activities of photosynthetic enzymes, principally Rubisco.

CO_2 limitation is a particular problem in greenhouses, especially in winter when the greenhouses are closed and CO_2 levels are reduced due to photosynthesis. Even under more normal conditions, most plants will grow significantly faster and increase yields when the atmosphere is enriched with CO_2. For these reasons, CO_2 enrichment has become common practice for commercial growers of vegetable crops such as lettuce, tomato, and cucumbers. In practice, the CO_2 content of the greenhouse atmosphere is increased up to twice present atmospheric levels. Much beyond 700μl l^{-1} there is an increasing risk of stomatal closure and other toxic effects that will cause a reduction in photosynthesis. CO_2 enrichment might be expected to improve growth and productivity of field crops as well, but there are obvious technical and economic problems related to controlling the supply of gas in an open environment.

TEMPERATURE

Photosynthesis, like most other biological processes, is sensitive to temperature. The temperature response for most biological processes reflects the temperature dependence of the enzymic and other chemical reactions involved. The temperature response curve can be characterized by three **cardinal points:** the **minimum** and **maximum** temperatures (**T_{min}** and **T_{max}**, respectively) at which the reaction can proceed and the **optimum** temperature (**T_{opt}**) (Fig. 13.6). Thus there is a range of temperatures below the optimum over which the rate of the reaction or process is stimulated with increasing temperature and a range, beyond the optimum, over which the rate declines. These points are determined largely by biochemical factors such as the binding of substrate with active sites and protein (enzyme) stability.

The temperature response of chemical and biological reactions can generally be characterized by comparing the rate of the reaction at two temperatures 10 °C apart, a value known as the **Q_{10}**:

$$Q_{10} = \frac{R_{T+10}}{R_T} \qquad (13.1)$$

The value of Q_{10} for enzyme-catalyzed reactions is about 2, meaning that the rate of the reaction will approximately double for each 10 °C rise in temperature. This value for Q_{10} applies primarily to stimulation of the reaction by temperatures between T_{min} and T_{opt}. Once the optimum is reached, the reaction rate may decline sharply due to enzyme inactivation (Fig. 13.6).

The photochemical reactions in the thylakoid membranes are largely independent of temperature—their Q_{10} is close to 1.0. Consequently, the temperature response of photosynthesis largely reflects the effect of temperature on the reactions of carbon metabolism. Because photosynthesis is a complex, multienzyme process, its temperature response will tend to reflect the average temperature characteristics for all of the enzymes. In spite of that, however, the temperature characteristics of C3 and C4 photosynthesis seem to be dominated by the temperature response curves for Rubisco and PEPcarboxylase, respectively. As noted earlier in Chapter 10, the low temperature sensitivity of C4 photosynthesis probably reflects the low temperature inactivation of the enzyme pyruvate, phosphate di-kinase.

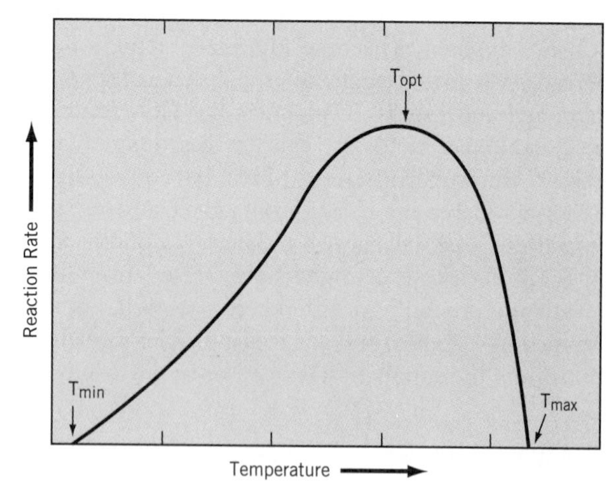

FIGURE 13.6 Temperature dependence and cardinal points for a typical biological reaction.

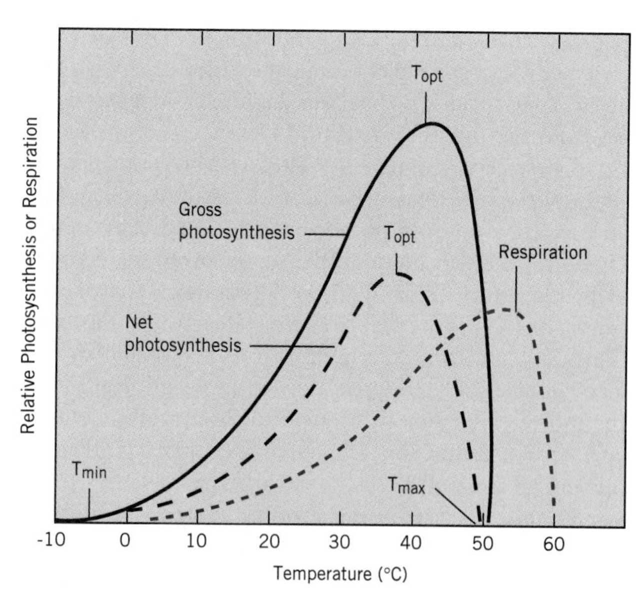

FIGURE 13.7 Diagram illustrating the temperature dependence of gross photosynthesis, respiration, and net photosynthesis. Gross photosynthesis increases with thermal activation of the participating enzymes until inhibitory factors (enzyme inactivation, stomatal closure) take effect. Respiration increases more slowly with temperature and has a higher temperature optimum, but declines more rapidly at high temperature. Net photosynthesis (dashed curve) is determined as the difference between gross photosynthesis and respiration. The resulting cardinal points for net photosynthesis are indicated.

Measurements of photosynthetic activity in leaves and whole plants are normally based on net gas exchange—the difference between the rate of photosynthetic CO_2 uptake and CO_2 evolution due to respiration. Because gross photosynthesis and respiration respond very differently to temperature, the optimum temperature for net photosynthesis is not the same as for gross photosynthesis (Fig. 13.7). Note that the rate of respiration continues to increase with temperature, reaching a maximum near 50 °C, where it drops off sharply due to inactivation of enzymes.

SOIL WATER POTENTIAL

The importance of available water in determining productivity cannot be underestimated. The rate of photosynthesis declines under conditions of water stress and, in cases of severe water stress, may cease completely. Stomatal closure and the resultant decrease in CO_2 supply due to water stress imposes a major limitation on photosynthesis. Low water potentials reduce turgor pressure in leaf cells, which in turn reduces leaf expansion (see Chap. 22). Under prolonged water stress this results in a reduced photosynthetic surface area. There may be some compensation as stored reserves are mobilized to offset the loss of new assimilate, but overall even mild water stress causes a reduction in net productivity. C4 plants enjoy some advantage over C3 plants with respect to photosynthesis and water stress because of their higher water use efficiency (Chap. 10).

NUTRIENT SUPPLY, PATHOLOGY, AND POLLUTANTS

The maximum possible photosynthetic capacity of a leaf, known as the **leaf photosynthetic capacity,** is determined as the rate of photosynthesis per unit leaf area under conditions of saturating incident light, normal CO_2 and O_2 concentrations, optimum temperature, and high relative humidity. Although leaf photosynthetic capacity may vary as much as a hundredfold, it is generally highest in plants adapted to resource-rich environments; that is, where light, water, and nutrients are abundant (Pearcy et al., 1987). Reduced photosynthesis is a consequence of deficiencies of virtually all essential elements, but photosynthetic capacity is particularly sensitive to nitrogen supply. As a basic constituent of chlorophyll, redox carriers in the photosynthetic electron transport chain, and all of the enzymes involved in carbon metabolism, nitrogen plays a critical role in primary productivity. In a C3 species, Rubisco alone will account for more than half of the total leaf nitrogen. In one study of C3 and C4 grasses, net photosynthesis increased linearly with nitrogen content (Fig. 13.8). In barley seedlings, a 5-fold increase in nitrate supply stimulated a 25-fold increase in net photosynthesis. One impact of nitrogen deficiency is to reduce the amount and activity of photosynthetic enzymes, but leaf expansion and other factors no doubt contribute to reductions in photosynthetic capacity as well.

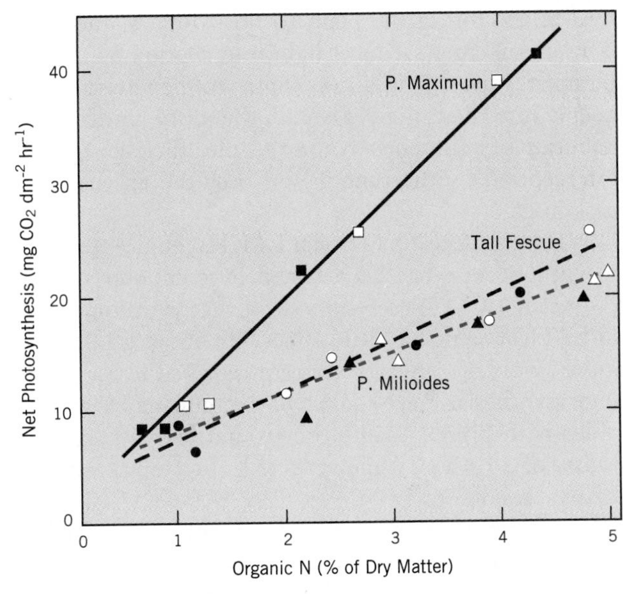

FIGURE 13.8 The relationship between leaf organic nitrogen content and net photosynthesis. Organic nitrogen is measured as a percent of dry matter. *P. maximum* is a C4 species, tall fescue is a C3 species, and *P. melioides* has characteristics intermediate between C3 and C4. (From J. K. Bolton, R. H. Brown, *Plant Physiology* 66:97–100, 1980. Copyright American Society of Plant Physiologists.)

In addition to nutrient supply, photosynthetic capacity is also reduced in leaves subjected to stress caused by invading pathogens and environmental pollutants such as sulphur dioxide, oxides of nitrogen, ozone, and heavy metals. The impact of these factors will be discussed further in Chapter 22.

LEAF FACTORS

The net carbon gain of an individual leaf depends on its photosynthetic capacity limited by environmental parameters and balanced against its construction and maintenance costs. For example, the net carbon gain of a leaf varies markedly during leaf development and aging. During initial development and the rapid growth phase, the photosynthetic capacity of a leaf also increases. However, the developing leaf functions as a sink, utilizing carbon assimilated locally as well as importing carbon to support its expansion. Leaf photosynthetic capacity then declines as the aging leaf undergoes senescence, a progressive deterioration of the leaf characterized in part by the loss of chlorophyll and photosynthetic enzymes. Only in the period between full expansion and the onset of senescence does a leaf produce a profit in terms of carbon gain.

Different types of leaves may also have different photosynthetic capacities. Evergreen leaves, for example, have a lower photosynthetic capacity than deciduous leaves and, because they take longer to develop, their construction costs are also higher (Chabot and Hicks, 1982). Still, the evergreen leaf may be favored if the cost of maintaining the leaf over winter together with the cost of a lower photosynthetic capacity are less than the cost of producing a new leaf.

Regardless of leaf photosynthetic capacity, unfavorable environmental conditions will cause a reduction in long-term carbon gain. In any natural environment, available water, the quantity of light, and temperature will all vary widely and to some extent independently, often keeping the photosynthetic rate well below full capacity. In addition, a prominent feature in the environment of any leaf is the presence of other leaves; that is, leaves are normally part of a **canopy.** Net primary productivity of a stand of plants is markedly influenced by canopy structure. Canopy structure is in turn determined by the age, morphology, angle, and spacing of individual leaves.

A herbaceous C3 annual plant is characterized by a gradient in leaf age and development along the stem axis. The young, growing leaves at the top are exposed to full sunlight while older leaves further down may be heavily shaded. Irradiance reaching shaded leaves may be reduced to 10 percent or less, thus producing a very low net photosynthesis. Very often the fluence rate reaching leaves lowermost in a canopy may fall below the light compensation point for a large part of the day. Those leaves would not only no longer contribute to net photosynthesis, but would incur a negative carbon gain through respiratory loss. Many herbaceous annuals avoid the costs of maintaining such nonproductive leaves by undergoing **sequential senescence;** that is, the older leaves lower in the canopy senesce as new leaves are being formed at the top of the canopy. Senescing leaves may lose as much as 50 percent of their dry weight, largely in the form of soluble organic nitrogen compounds, before dying and falling to the ground. These compounds are exported to developing leaves and other sinks where they are reused. In this way, limited resources are redistributed amongst leaves of varying ages in order to maximize whole plant carbon gain.

Canopy architecture is important when considering agricultural crops and natural ecosystems because it determines how efficiently light is absorbed. High productivity is in part dependent upon the extent to which ground area is covered with photosynthetic surface. Because sunlight striking exposed soil does not contribute to productivity, most agricultural systems are designed so that the young plants fill in the canopy rapidly in order to maximize interception of available light. On the other hand, planting at too high a density will introduce mutual shading by leaves in the canopy. Shading reduces the overall efficiency of light interception and, consequently, reduces long-term carbon gain (Fig. 13.9). The ratio of photosynthetic leaf area to covered ground area is known as the **leaf area index (LAI)**. Leaf area is usually taken as the area of a single surface (or projected area, where leaves are not planar). Because both leaf surface and the covered ground are measured as areas (m^2), LAI is dimensionless. Values of LAI in productive agricultural ecosystems typically fall in the range of 3 to 5.

The optimum LAI for a given stand of plants depends on the angle between the leaf and the stem. Horizontal leaves, typical of beans (*Phaseolus*) and similar crops, are efficient light absorbers because of the broad surface presented to the sun, but they also more effectively shade leaves lower down in the canopy. Erect leaves, typical of grasses like wheat (*Triticum*) and maize (*Zea mays*), produce less shading but, because of their steeper angle, are not as efficient at intercepting light.

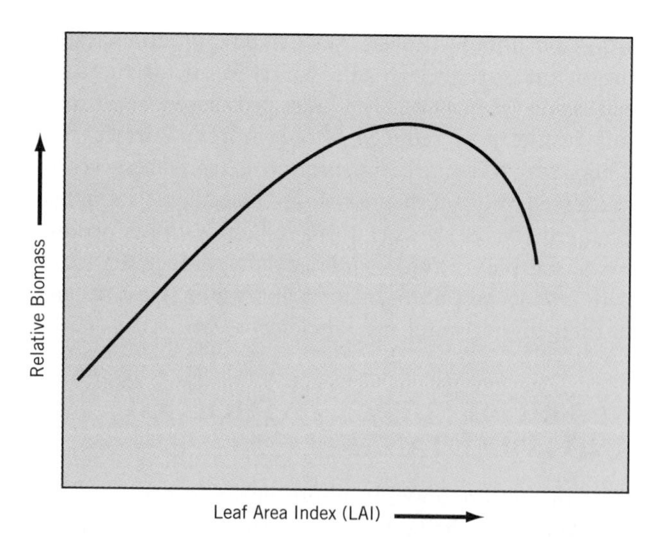

FIGURE 13.9 The relationship between biomass and leaf area index (LAI) in a crop. LAI is varied by varying the density of plants in the stand. Higher planting densities lead to a decline in biomass because of mutual shading of leaves and loss of carbon by respiration in the shaded, nonproductive leaves.

Experiments with closely spaced rice plants have shown that carbon gain dropped by a third when leaves were weighted in the horizontal position. With maize, tying upper leaves into a more erect posture increased yield by as much as 15 percent. In some crop plants, such as sugarbeet (*Beta vulgaris*), leaf angle varies from near vertical at the top of the plant to horizontal at the base. This arrangement reduces light interception by the uppermost leaves, but allows light to penetrate more deeply into the canopy. Overall, the more uniform distribution of light tends to improve the efficiency of light interception by the canopy and thus the efficiency of carbon gain.

The relationship between LAI, leaf angle, and photosynthetic rate has been tested by computer models (Duncan, 1971). The results show that with low values of LAI leaf angle has little effect, but above a LAI of 3, more layers of vertical leaves are required to maximize photosynthesis. Field studies have confirmed that canopies with predominantly horizontal leaves have LAI values of 2 or less, while vertical leaf canopies support LAI values of 3 to 7.

In cases where the leaves are more or less fixed in space, the efficiency of light interception will change with the angle of the sun. Many desert and agricultural species, however, have the capacity to alter the orientation of their leaves, allowing them to track the sun as it moves across the sky through the day (Ehleringer and Forseth, 1980). Called **heliotropism,** or **solar tracking,** the leaf blades move in such a way that their surfaces remain perpendicular to the sun's direct rays. Solar tracking would help to maximize daily carbon gain in those plants that must complete their life cycle in a brief period before the onset of unfavorable conditions such as drought or high temperatures.

PRIMARY PRODUCTIVITY ON A GLOBAL SCALE

Understanding primary productivity on an ecosystem and global basis helps us to better understand agricultural potential and human impact on the biosphere.

The quantitative evaluation of primary productivity on a global basis is an expensive and time-consuming exercise that is, at best, imprecise. However, a large quantity of information on global productivity was obtained during the International Biological Programme, which ran between 1964 and 1974 under the sponsorship of the International Council of Scientific Unions. Drawing on the collaborative efforts of biologists from around the world, the program coordinated a major study of photosynthesis, nitrogen fixation, and biological productivity.

A number of models for estimating global productivity have been produced and the estimates arising from application of these models are reasonably consistent (Lieth and Whittaker, 1975). Total global productivity is estimated to be approximately 172 billion metric tons of dry organic matter per year (Table 13.1). Approximately 68 percent (117.5×10^9 t yr^{-1}) of that total is accounted for by terrestrial ecosystems. The remaining 32 percent is accounted for by marine ecosystems. Productivity on land is thus somewhat more than twice that of the oceans on an area less than half as large. On an area basis, land-based production is about five times greater than the oceans. This difference is at least partly explained by differences in nutrient supply. Over the major portion of the oceans, dead organisms and sinking particles carry nutrients out of the lighted zone near the surface where photosynthesis occurs. Terrestrial ecosystems, on the other hand, retain a much larger nutrient capital in the soil and litter where they can continue to support growth and productivity.

By far the highest-producing terrestrial ecosystem, through a combination of high rate of net primary production and large area, is comprised of the tropical forests. This includes both the rainforests, which alone contribute more than 37×10^9 t yr^{-1}, and tropical seasonal forests. The high net primary productivity of the tropical forests is largely related to the long growing season that extends through a full twelve months of the year. By contrast, temperate and boreal forest productivity on an annual basis is limited by the length of the growing season.

Agriculture accounts for only 5.3 percent of global productivity. This is in part because the amount of land

TABLE 13.1 Estimates for mean net primary production (NPP), total production (P), and total biomass for various worldwide ecosystems. Bracketed numbers indicate the percentage of world total.

Ecosystem	Area (10^6 km^2)		NPP (g(m^2 yr)$^{-1}$)	Total P (10° t yr^{-1})		Total Biomass (10^9 t)	
Forest							
Tropical	24.5	(4.8)	3800	49.4	(28.6)	1025	(55.7)
Temperate	12.0	(2.4)	2500	14.9	(8.6)	385	(20.9)
Boreal	12.0	(2.4)	800	9.6	(5.6)	240	(13.0)
Woodlands	8.5	(1.7)	700	6.0	(3.5)	50	(2.7)
Grasslands	24.0	(4.7)	1500	18.9	(11.0)	74	(4.0)
Tundra and alpine	8.0	(1.6)	140	1.1	(0.6)	5	(0.3)
Desert	42.0	(8.2)	93	1.7	(1.0)	14	(0.8)
Freshwater	4.0	(0.8)	3400	6.9	(3.9)	30	(1.6)
Cultivated land	14.0	(2.7)	650	9.1	(5.3)	14	(0.8)
Total terrestrial	149.0	(29.2)		117.5	(68.1)	1837	(99.7)
Marine							
Open oceans	332.0	(65.1)	125	41.5	(24.0)	1	(0.05)
Continental shelf	26.6	(5.2)	360	9.6	(5.6)	0.3	(0.016)
Other marine	2.4	(0.5)	4500	3.9	(2.3)	2.6	(0.14)
Total marine	361.0	(70.8)		55.0	(31.9)	3.9	(0.2)
World total	510	(100.0)		172.5	(100.0)	1841.0	(100.0)

Total productivity and biomass figures are given in metric tons of dry organic matter. 1 metric ton (t) = 1000 kg.

Data from Lieth and Whittaker, 1975.

available for cultivation is limited—much of the world's surface is either too hot or too cold, too dry or too salty for agriculture. Cultivation may be extended into the less productive so-called marginal lands, but only at the expense of additional costs for irrigation or other management practices to compensate for water, nutrient, and other environmental deficiencies. Recent introductions of new strains of cereal crops—the "green revolution"—have improved production and yields by a few percent, but mostly on lands already in use for agriculture. Pressures on agricultural productivity will no doubt increase as the world population continues to increase and productive lands are withdrawn from agriculture to accommodate expanding urbanization.

At the time the estimates listed in Table 3.1 were made, 90 percent of the estimated total world biomass of 1.84×10^{12} metric tons was forest. In the time since these estimates were made, conversion of the Amazonian and other tropical rainforests to agricultural and other uses has accelerated. Recent estimates based on data collected by earth satellite indicate that in the Amazon region alone, approximately 12 percent of the land area had been cleared through 1988 and that the rate of deforestation was increasing exponentially (Anderson, 1990). These estimates do not include other tropical forest areas or the cutting of temperate and coastal forests across North America, Europe, and Asia for pulp and timber. As the world's forests are converted to other uses or old-growth forests are replaced by plantations of younger stands to be harvested before they reach maturity, forest biomass and therefore total world biomass will necessarily continue to decline.

SUMMARY

Carbon assimilation by plants creates the plant biomass that supports humans and virtually all other heterotrophic organisms. The study of carbon gain, or productivity, at both the organismal and population level is an important component of agricultural and ecosystem research. Overall carbon gain depends on net primary productivity—the balance between carbon uptake by photosynthesis and carbon loss to respiration. Carbon loss to respiration can be divided into the carbon cost of growth, or growth respiration, and the cost of simply maintaining structure and processes that do not result in a net increase in dry matter. Several studies have shown a negative correlation between respiration and growth rate. The implication is that respiration has a significant impact on biomass accumulation and yield, and that improving yield by manipulating respiration might be feasible.

Productivity is also influenced by a variety of genetic and environmental factors that influence photosynthesis. These include light, available carbon dioxide, temperature, soil water, nutrients, and canopy structure. The rate of photosynthesis increases between the light compensation point and saturation. Because irradiance changes constantly throughout the day, long-term carbon gain depends on the cumulative irradiance over the growing season.

Although the carbon dioxide content of the atmosphere is relatively constant in the short term, there is evidence that plants use the stomata to keep the internal carbon dioxide concentration in balance with electron transport capacity. On the other hand, with adequate light plants will respond to carbon dioxide enrichment by increased productivity.

Photosynthesis and respiration respond differently to temperature. Thus the optimum temperature for net photosynthesis is not the same as the optimum temperature for gross photosynthesis. Plants also require an adequate water and nutrient supply in order to maximize their leaf photosynthetic capacity. Productivity in a stand depends on the pattern of leaf senescence and the structure of the canopy. The ideal canopy maximizes the efficiency of light interception and carbon gain by balancing leaf area, leaf angle, leaf orientation, plant density, and senescence of older leaves.

CHAPTER REVIEW

1. Distinguish between growth respiration and maintenance respiration. Which might best be manipulated in order to improve productivity?
2. Describe how various environmental factors influence plant productivity.
3. Why do large values for leaf area index lead to an overall decline in productivity of a stand?
4. Why are terrestrial ecosystems more productive than marine ecosystems on a unit area basis? What is the current trend toward productivity on a global scale?

FURTHER READING

Beadle, C. L., S. P. Long, S. K. Imbamba, D. O. Hall, R. J. Olembo. 1985. *Photosynthesis in Relation to Plant Production in Terrestrial Environments.* Oxford: Tycooly Publishing.

Coombs, J., D. O. Hall, S. P. Long, J. M. O. Scurlock (eds.). 1985. *Techniques in Bioproductivity and Photosynthesis.* 2nd ed., Oxford: Pergamon Press.

Cooper, J. P. (ed.). 1975. *Photosynthesis and productivity in different environments.* Cambridge: Cambridge University Press.

Gupta, U. S. 1992. *Crop Improvement. Vol. I. Physiological Attributes.* Boulder, CO: Westview Press.

Lieth, H., R. H. Whittaker (eds.). 1975. *Primary Productivity of the Biosphere.* New York, Heidleberg: Springer-Verlag.

Pearcy, R. W., O. Björkman, M. M. Caldwell, J. E. Keeley, R. K. Monson, B. R. Strain. 1987. Carbon gain by plants in natural environments. *Bioscience* 37:21–29.

REFERENCES

Anderson, A. B. 1990. *Alternatives to Deforestation: Steps toward Sustainable Use of the Amazon Rain Forest.* New York: Columbia University Press.

Bolton, J. K., R. H. Brown. 1980. Photosynthesis of grass species differing in carbon dioxide fixation pathways. V. Response of *Panicum maximum*, *Panicum miloides*, and tall fescue (*Fescuca arundinacea*) to nitrogen nutrition. *Plant Physiology* 66:97–100.

Chabot, B. F., D. J. Hicks. 1982. The ecology of leaf life spans. *Annual Review of Ecology and Systematics* 13:229–259.

Duncan, W. G. 1971. Leaf angles, leaf area, and canopy photosynthesis. *Crop Science* 11:482–485.

Ehleringer, J., I. Forseth. 1980. Solar tracking by plants. *Science* 210:1094–1098.

Farquhar, G. D., T. D. Sharkey. 1982. Stomatal conductance and photosynthesis. *Annual Review of Plant Physiology* 33:317–345.

Küppers, M. 1984. Carbon relations and competition between woody species in a central European hedgerow. III. Carbon and water balance at the leaf level. *Oecologia* 65:94–100.

Lambers, H. 1985. Respiration in intact plants and tissues. Its regulation and dependence on environmental factors, metabolism and invaded organisms. In: R. Dource, D. A. Day (eds.), *Higher Plant Cell Respiration. Encyclopedia of Plant Physiology, New Series.* 18:418–473.

Musgrave, M. E., J. Antonovics, J. N. Siedow. 1986. Response of two pea hybrids to CO_2 enrichment: A test of the energy overflow hypothesis for alternative respiration. *Proceedings of the National Academy of Science USA* 83:8157–8161.

Pearcy, R. W., J. Ehlenger, H. A. Mooney, P. W. Rundel. 1989. *Plant Physiological Ecology: Field Methods and Instrumentation.* London, New York: Chapman and Hall.

Wilson, D. 1982. Response to selection for dark respiration rate of mature leaves in *Lolium perenne* and its effects on growth of young plants and simulated swards. *Annals of Botany* (London) 49:303–312.

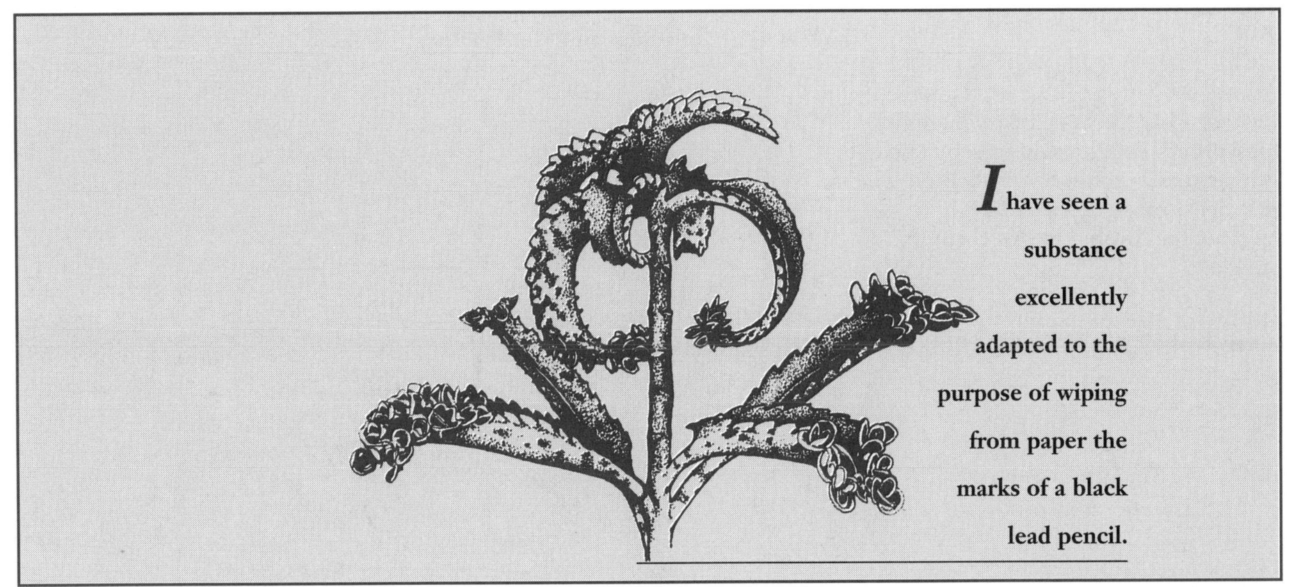

Joseph Priestley (1770)

14

Molecules and Metabolism

The sum of all of the chemical reactions that take place in an organism constitutes **metabolism**. Previous chapters in Part II have focused primarily on photosynthesis and respiration: the metabolism of carbon assimilation and energy conversions. Much of that carbon and energy ends up in proteins, nucleic acids, lipids, and other molecules that are common to all cells and required for the proper functioning of cells and organisms. But in many plants a significant proportion of assimilated carbon and energy is diverted to the synthesis of organic molecules that may have no obvious role in growth and development. These molecules are known as secondary metabolites. Secondary metabolites generally occur in low quantities and their production may be widespread or restricted to particular families, genera, or even species. Many secondary metabolites serve to reduce the impact of insect and animal predation or provide other protective functions. Others have found use in antiquity as folk remedies, soaps, essences, and so forth. They include medicinal products, dyestuffs, feed stocks for chemical industries (gums, resins, rubber), and a variety of substances used to flavor food and drink.

The principal primary metabolites—lipids, proteins, and carbohydrates—have been reviewed in earlier chapters. In this chapter, we examine some of the broader aspects of metabolism. The focus will be on secondary metabolites, with an emphasis on the biosynthesis, physiology, and ecological roles of four major classes:

- terpenes and the mevalonic acid pathway;
- phenolic compounds and the shikimic acid pathway;
- saponins, cardiac glycosides, cyanogenic glycosides, and glucosinolates; and
- alkaloids.

PRIMARY AND SECONDARY METABOLITES

Metabolism *(Gr. = metabole, change) is a very dynamic process. Molecules are constantly turning over; the composition of a cell at any given time represents the balance between synthesis and degradation. Most of the synthesis is directed toward molecules that are essential to the structure and function of the cell. In plants, however, a significant amount of carbon and energy is directed toward molecules whose function is less clear.*

The biosynthetic, or constructive, parts of metabolism, such as photo*synthesis*, protein *synthesis*, lipid *synthesis*, and so forth, are considered **anabolic** (Gr. *ana*, up + *bolein*, to throw) reactions, or **anabolism**. The degradative parts of metabolism—the breakdown of com-

FIGURE 14.1 A schematic to illustrate biosynthetic relationships among some primary and secondary metabolites. The principal groups of secondary metabolites covered in this chapter are circled.

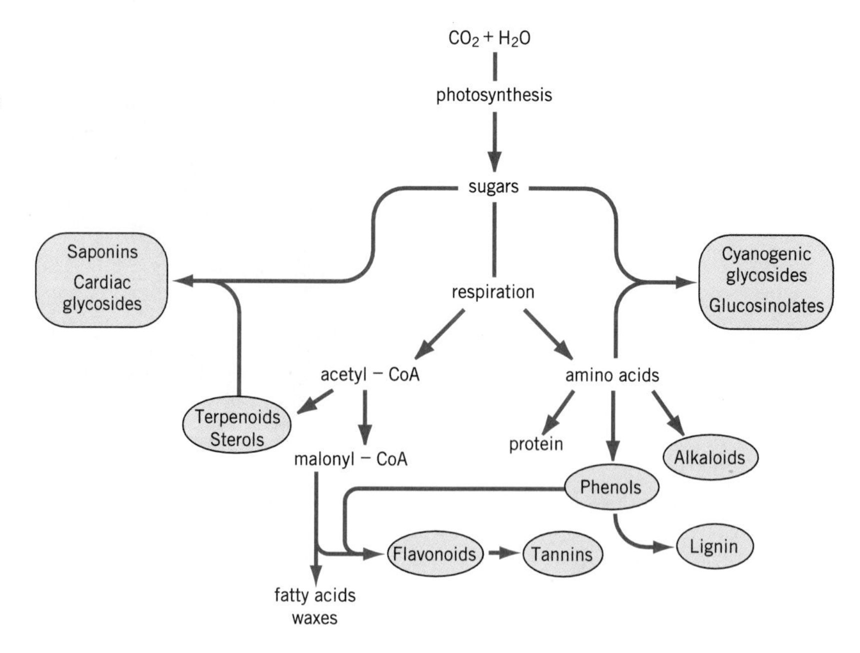

plex molecules into simpler components—are considered **catabolic** (Gr. = *kata*, down + *bolein*, to throw) reactions, or **catabolism.** Central to the metabolism of the cell are glycolysis and the citric acid cycle. Here is where many of the catabolic and anabolic reactions merge (Chap. 12, Fig. 12.14). For this reason, glycolysis and the citric acid cycle are commonly referred to as **intermediary metabolism.**

Metabolism can be subdivided in other ways as well. For example, all cells contain sugar phosphates, amino acids, lipids, proteins, and nucleic acids; these molecules comprise the basic molecular machinery of the cell and are known as **primary metabolites.** However, as noted in the introduction to this chapter, many plants synthesize a variety of other organic molecules that do not appear to qualify as primary metabolites. These molecules are referred to as **secondary metabolites** or **secondary products.**

The distinction between primary and secondary metabolites is not always easily made. In the strictest sense, secondary metabolites are not part of the basic molecular structure of the cell. When present, they are normally found only in particular tissues or organs, or at particular stages of development. By this traditional rule, both chlorophyll and lignin would qualify as secondary products because neither is found universally in all cells and tissues. However, the fundamental importance of both chlorophyll and lignin to growth and development surely qualifies both as primary metabolites. At the biosynthetic level, primary and secondary metabolites share many of the same intermediates and are derived from the same metabolic pathways (Fig. 14.1). The distinction between primary and secondary metabolites is more easily visualized at the organismic level. While primary metabolites are the stuff of life, secondary metabolites have no known role in assimilation or growth and development of the organism. Moreover, although some secondary products are widespread, most are typically limited to particular species or taxonomic groups.

Organic and natural products chemists have studied secondary metabolites for centuries because of their medicinal and economic value. Yet, as far as the plant is concerned, they have, until recently, been considered little more than metabolic waste products. Over the last three decades, it has been shown that many secondary metabolites have significant ecological roles (Harborne, 1988; Rosenthal and Berenbaum, 1991). The important functions of warding off microbial infection, deterring herbivory, and attracting pollinators or fruit-dispersal agents have become a central theme in the study of secondary metabolites.

TERPENOIDS

The terpenoids are a functionally and chemically diverse group of molecules. With nearly 15,000 structures known, terpenoids are probably the largest and most diverse class of organic compounds found in plants (Gershenzon and Croteau, 1991).

Terpenes are generally lipophilic substances derived from a simple five-carbon unit. The large diversity arises from the number of basic units in the chain and the various ways in which they are assembled. Formation of cyclic structures, addition of oxygen-containing functions, and conjugation with sugars or other molecules all add to the possible complexity.

The terpene family includes hormones (gibberellins and abscisic acid): the carotenoid pigments (carotene and xanthophyll); sterols (e.g., ergosterol, sitosterol, cholesterol) and sterol derivatives (e.g., cardiac glycosides); latex (the basis for natural rubber); and many of the essential oils that give plants their distinctive odors and flavors. Cytokinin hormones and chlorophyll, although not terpenoids per se, do contain terpenoid side chains. It is apparent from this list that many terpenoids have significant commercial value as well as important physiological roles.

TERPENE BIOSYNTHESIS

In spite of their diversity, all terpenes and terpene derivatives share a common biosynthetic pathway, called the **mevalonic acid pathway** after a key intermediate. Because of widespread importance of terpenoids, the mevalonic acid pathway is an aspect of plant metabolism that deserves further attention. Terpenoids and terpenoid derivatives may be considered polymers of the 5-carbon 2-methyl-1,3-butadiene, or **isoprene**.

$$CH_2{=}\overset{\displaystyle CH_3}{\overset{|}{C}}{-}CH{=}CH_2$$

Consequently, terpenoids are often referred to as **isoprenoid** compounds. Plants produce large quantities of isoprene and, because it is volitile, also emit substantial amounts into the atmosphere where it contributes to air pollution (see Chap. 22). The actual building blocks for terpenoids, however, are not isoprene itself, but two phosphorylated derivatives known as **Δ^3iPP (isopentenyl pyrophosphate** or "active isoprene") and its isomer **Δ^2iPP (dimethylallyl pyrophosphate)** (Fig. 14.2).

Terpenoid biosynthesis begins with acetyl-Coenzyme A (acetyl-CoA), an intermediate in the respiratory breakdown of carbohydrate and fatty acid metabolism. Three molecules of acetyl-CoA condense in a two-step reaction to form hydroxymethylglutaryl-CoA (HMG-CoA), which is then reduced to mevalonic acid (MVA) (Fig. 14.3). Because the reduction of HMG-CoA to MVA is virtually irreversible (at least *in vitro*), MVA appears to be a very efficient precursor for terpenes. MVA then undergoes a two-step phosphorylation, at the expense of two molecules of ATP, to form C_6 mevalonic acid pyrophosphate (MVA-PP). Removal of a carboxyl group converts MVA-PP to the C_5 compound Δ^3iPP, which is reversibly isomerized to Δ^2iPP.

One molecule of Δ^2iPP then undergoes a head-to-tail condensation with one molecule of Δ^3iPP to form the C_{10} unit, geranylpyrophosphate (GPP), the parent compound for all **monoterpenes**. Two more condensations with Δ^3iPP give rise first to the C_{15} intermediate farnesylpyrophosphate (FPP) and then to the C_{20} geranylgeranylpyrophosphate (GGPP). FPP is a branch point that can give rise to C_{15} **sesquiterpenes**, possibly including abscisic acid, and the C_{30} **triterpenes** (e.g., steroids). GGPP is a major branch, giving rise to linear C_{20} **diterpenes** (e.g., phytol), cyclic diterpenes (e.g., gibberellins), and C_{40} **tetraterpenes** (e.g., carotenoids).

Many terpenoids and terpenoid derivatives may be considered primary metabolites. The hormones abscisic acid and gibberellin, the carotenoid and chlorophyll pigments, and sterols (steroid alcohols), all play significant roles in plant growth and development. The vast majority of terpenoids, however, are secondary metabolites, many of which appear to act as toxins or feeding deterrents to herbivorous insects.

TERPENOIDS AND HERBIVORY

Among the best known insect toxins are the **pyrethroids**, monoterpenoid esters extracted from the flowers of *Chrysanthemum cineraiifolium* and related species. The pyrethroids are neurotoxins that interfere with sodium channels in the insect nerve membrane. They are popular as herbicides because they have a low mammalian toxicity and are biodegradable. The resins of certain conifers accumulate mixtures of terpenoids, including the monoterpenes, α- and β-pinene and myrcene (Fig. 14.4). Boring insects, particularly bark beetles, are repelled by these terpenoids, which they encounter when they sever the resin ducts.

Many plants, such as citrus, mint, *Eucalyptus*, and various herbs (sage, thyme, etc.), produce complex mixtures of alcohols, aldehydes, ketones, and terpenoids, known generally as **essential oils**. Essential oils are responsible for the characteristic odors and flavors of these plants but they are also known to have insect-repellant properties. The terpenoids and terpenoid derivatives found in essential oils are predominantly hemi-, mono-, and sesquiterpenoids, which can be moderately to highly volatile. Several examples are shown in Figure 14.5. In most plants, the essential oils are synthesized in special glandular trichomes (hairs) on the leaf surface (Fig. 14.6).

Many plants respond to insect attack by producing additional quantities of toxic metabolites. Unfortunately, in some cases chemical deterrents can turn against the plant. As insects evolve resistance, they are able to use the same chemical as a host-recognition cue to help locate hosts they can feed on without ill effect.

STEROIDS AND STEROLS

Steroids are known as tetracyclic triterpenoids. They are synthesized from the acyclic triterpene squalene (see Fig. 14.2), although they generally are modified and have fewer than 30 carbon atoms. Steroids with an alcohol group, which is the case with practically all plant steroids, are known as **sterols**. The most abundant sterols in higher plants are stigmasterol and sitosterol (Fig. 14.7), which together often constitute more than 70 percent of

Number of Carbons	Class	Example
5	Hemiterpenoid	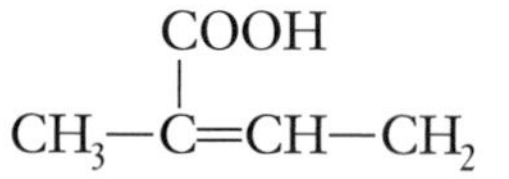 $CH_3{-}C(COOH){=}CH{-}CH_2$ Tigilic Acid (*Geranium* sp.)
10	Monoterpenoid	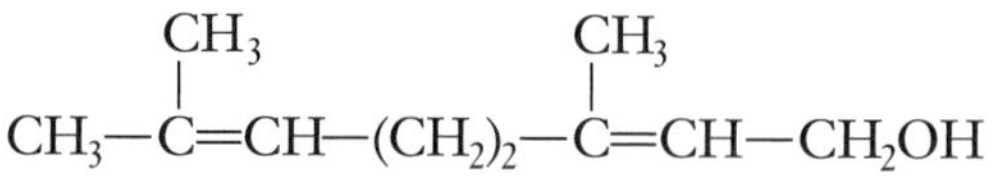 $CH_3{-}C(CH_3){=}CH{-}(CH_2)_2{-}C(CH_3){=}CH{-}CH_2OH$ Geraniol (*Pelargonium*)
10	Cyclic Monoterpenoid	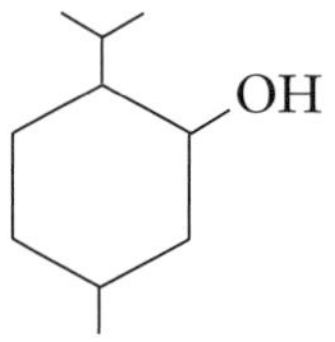 Menthol (Peppermint oil)
15	Sesquiterpenoid	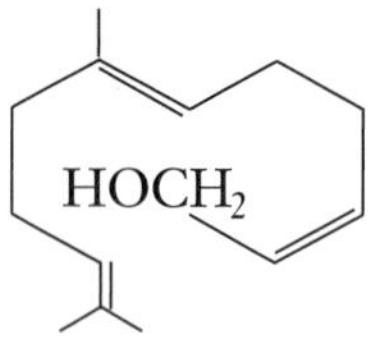 Farnesol (widespread)
20	Diterpenoid	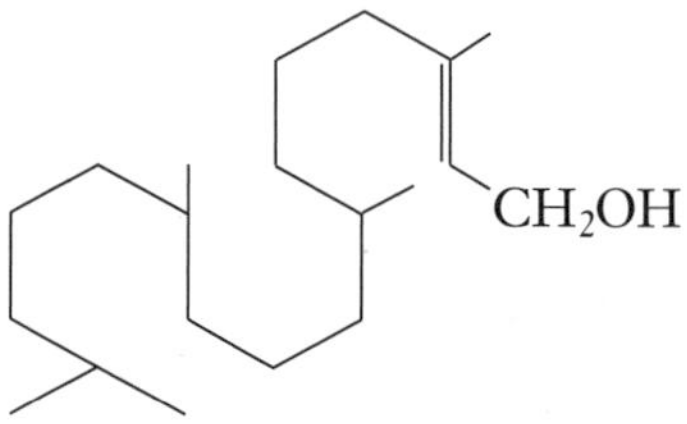 Phytol (Chlorophyll)
30	Triterpenoid	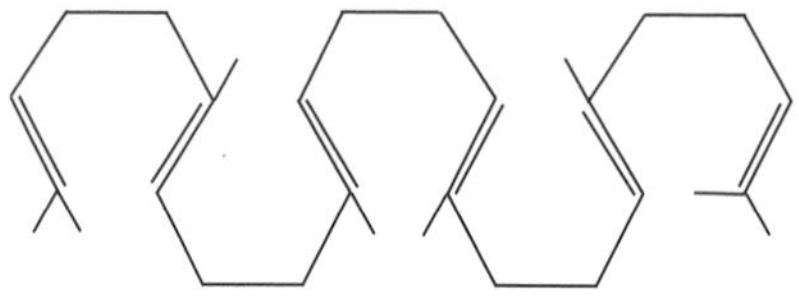 Squalene (A steroid precursor)
30	Triterpenoid	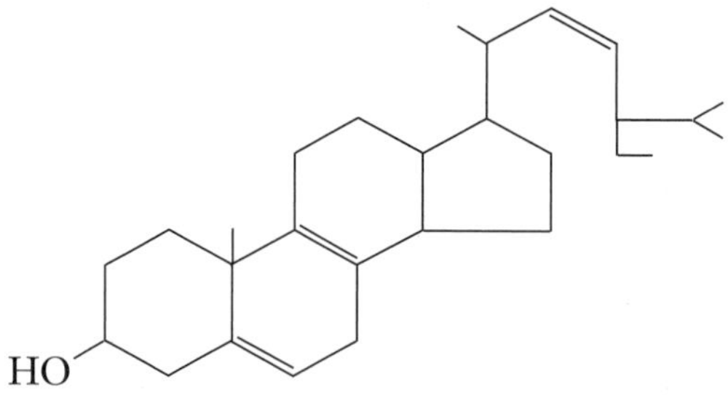 Stigmasterol (glycine max) (A sterol)
40	Tetraterpenoid	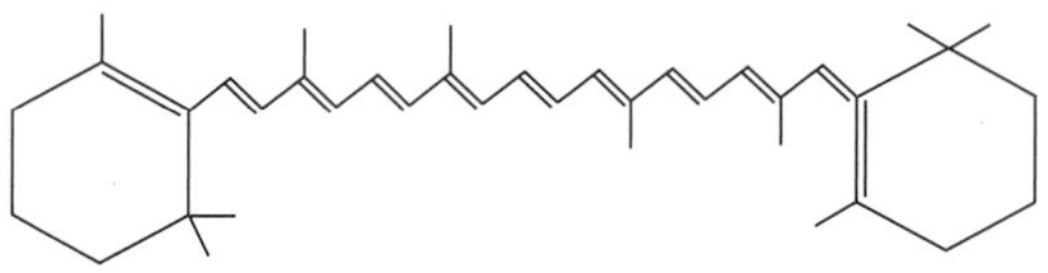 β-carotene

FIGURE 14.2 The principal classes of terpenoids are listed along with examples to illustrate the carbon skeletons in each class.

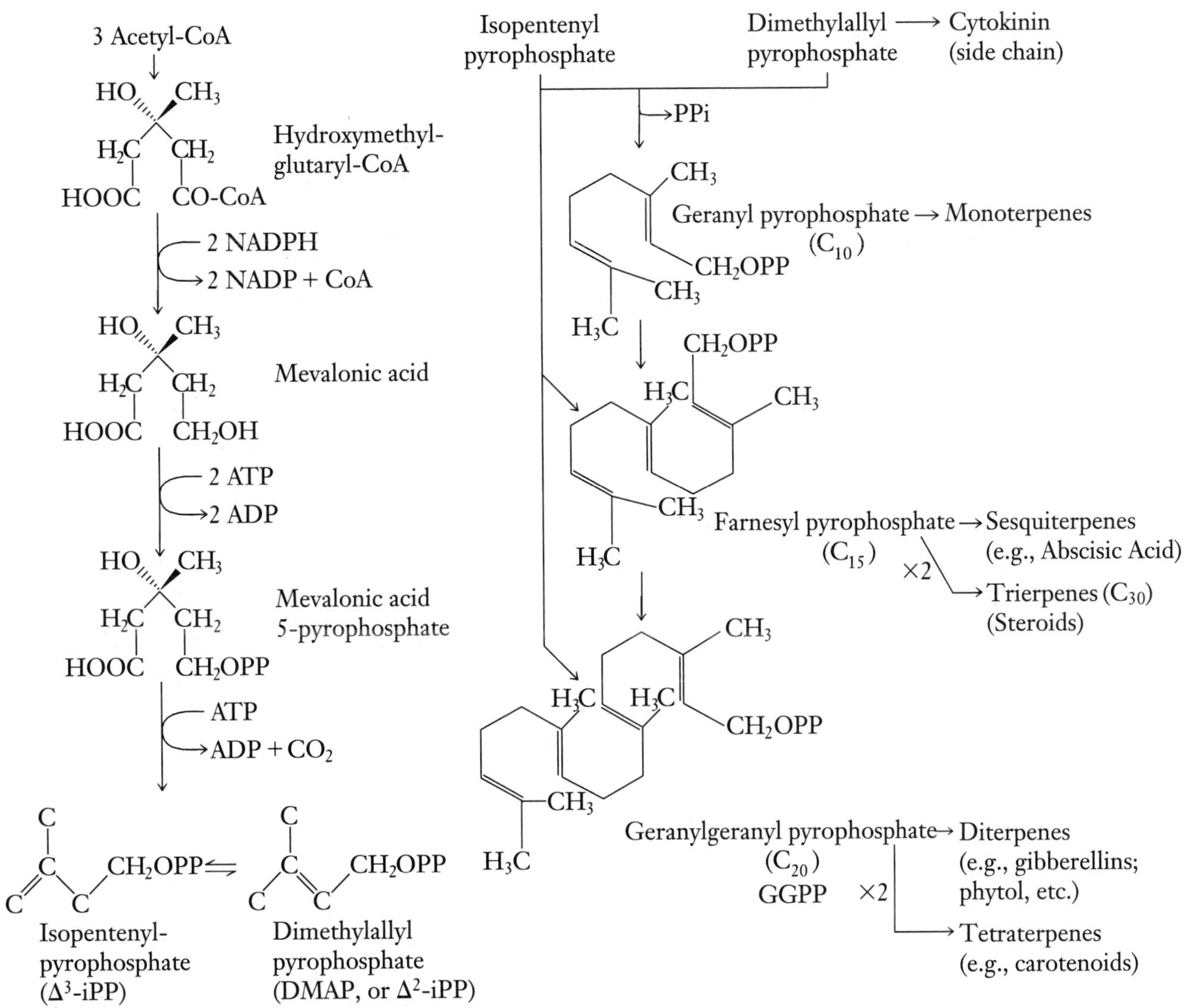

FIGURE 14.3 The mevalonic acid pathway for the biosynthesis of terpenoids. Left: Synthesis of isopentenyl-pyrophosphate from acetyl-CoA. Right: sequential addition of isopentenyl-pyrophosphate units to form terpenes of increasing carbon number. All terpenes are built of multiples of 5-carbon atoms.

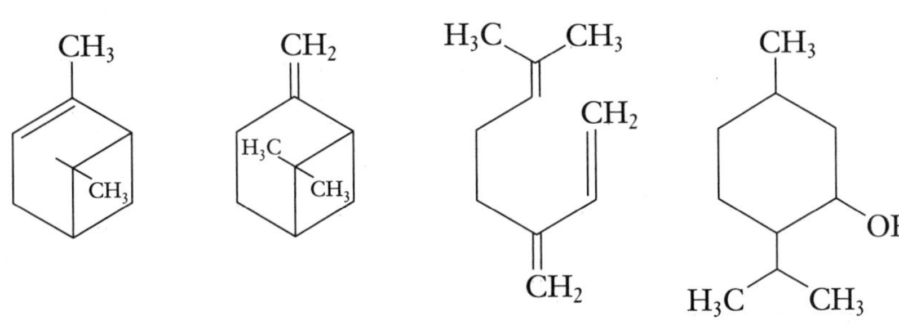

FIGURE 14.4 Monoterpenes contain 10 carbon atoms. Pinine and myrcene are found in the resins of some conifers. Menthol is a principal component of the essential oil of peppermint (*Mentha piperita*).

iso-Amyl Alcohol
(*Mentha, Eucalyptus*)

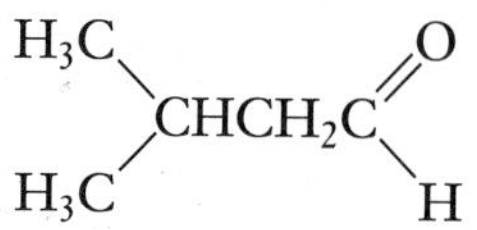

iso-Valeraldehyde
(*Eucalyptus*)

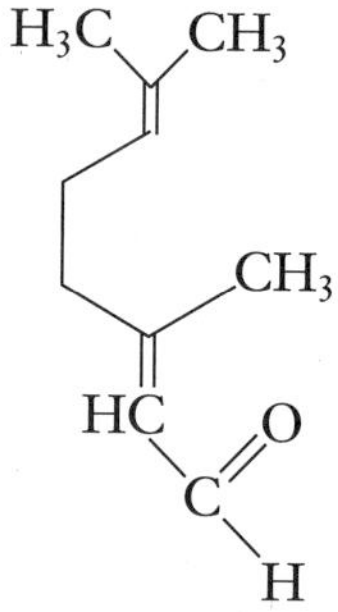

Geranial
(*Ctenium aromaticum*)
(lemon grass)

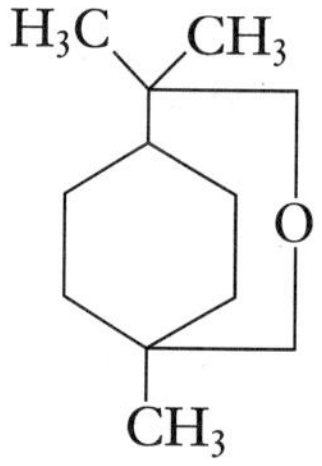

1:8 Cineole
(*Artemesia*)

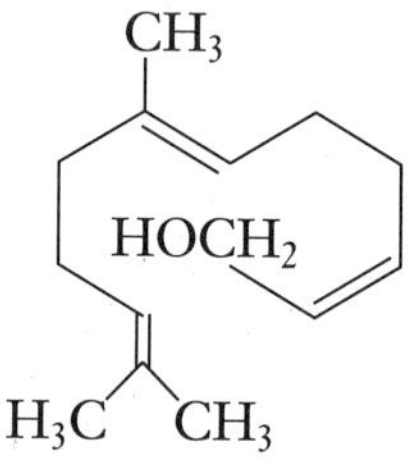

Farnesol
(Widespread)

FIGURE 14.5 Representative terpene constituents of essential oils, including hemiterpenes (*iso*amyl alcohol, *iso*valeraldehyde), monoterpenes (geranial, cineole), and the sesquiterpene farnesol.

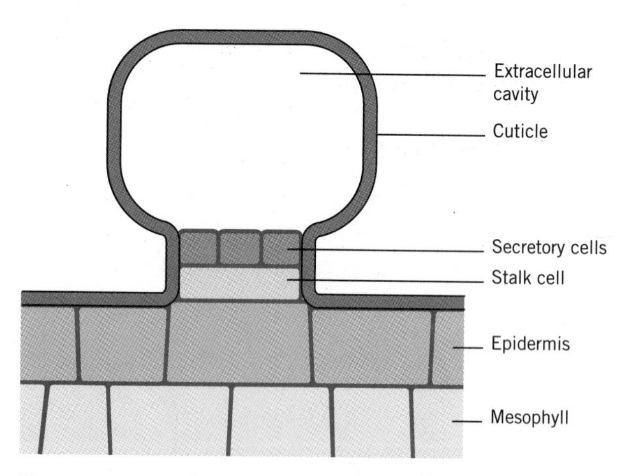

FIGURE 14.6 Schematic diagram of a glandular hair in cross section. Essential oils are produced in the secretory cells and accumulated in a cavity that forms between the secretory cells and the overlying cuticle. Glandular hairs are found on the leaf surface, where it is thought they might serve to deter feeding by herbivores.

the total sterols (Benveniste, 1986). However, plants contain a large number of the more than 150 sterols known to occur in nature, including cholesterol which, although widespread, is present in only trace quantities.

Sterols are a constituent of plant membranes, which is perhaps their most important known function in plants. Sterols are planar molecules and their packing properties are different from phospholipids. Sterols therefore tend to increase the viscosity and enhance the stability of membranes. Otherwise, little is known about the function of sterols in plants. Unlike the steroid hormones in animals, there is no known role for sterols in plant development. Some sterols may have a protective function, such as the **phytoecdysones**, which have a structure similar to the insect molting hormones. When ingested by insect herbivores, phytoecdysones disrupt the insect's molting cycle. Other sterols are present as glycosides, which give rise to a number of interesting and economically significant secondary products. Steroid glycosides are discussed below.

POLYTERPENES

Larger terpenes include the tetraterpenes (40-carbon) and the polyterpenes. The principal tetraterpenes are the carotenoid family of pigments, which were discussed

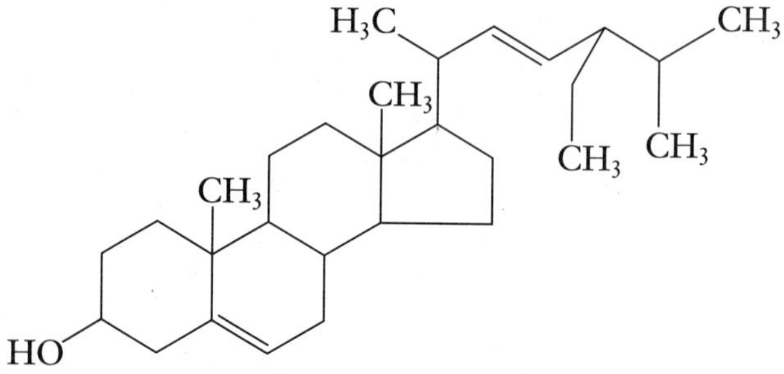

Stigmasterol

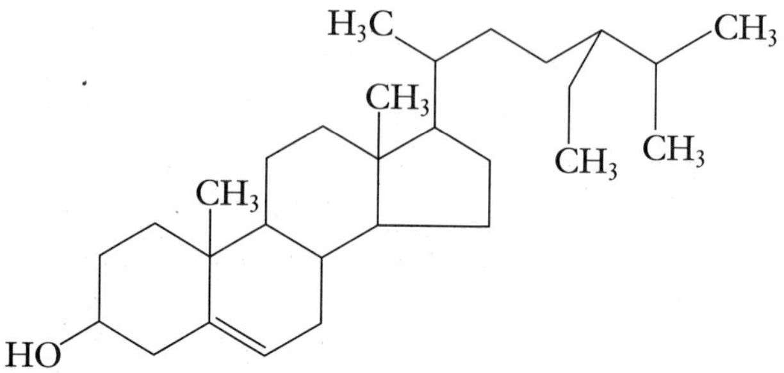

Sitosterol

FIGURE 14.7 Stigmasterol and sitosterol are the two most abundant sterols in plants.

PLATE 1. Nutrient deficiency symptoms in tomato. (See Figure 4.4.)

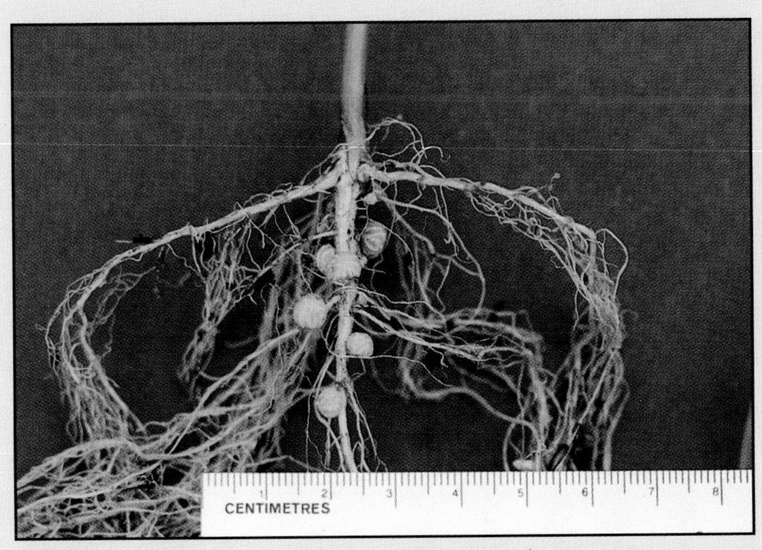

PLATE 2. Soybean root nodules. (See Figure 6.2.)

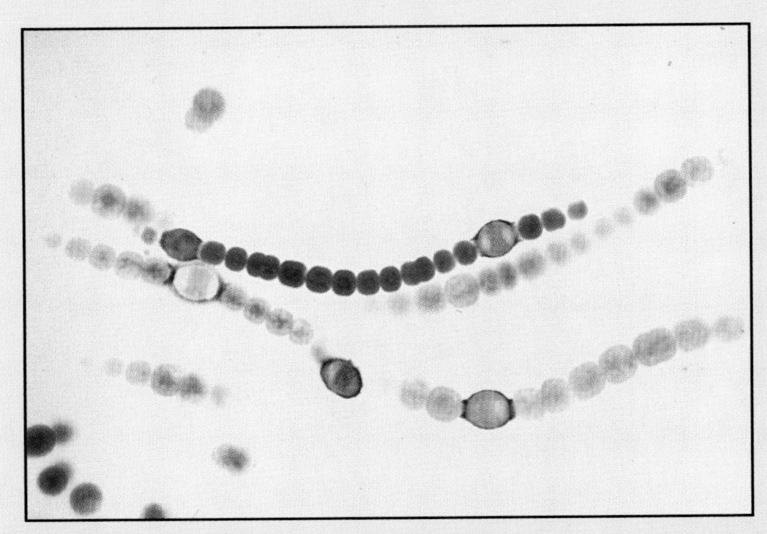

PLATE 3. Heterocysts in the cyanobacterium *Anabaena*. (See Figure 6.8.)

PLATE 4. Chlorophyll fluorescence. (See Figure 7.4.)

PLATE 5. Tobacco stem callus in culture. (See Figure 15.1.)

PLATE 6. Apical dominance in *Vicia faba*. (See Figure 16.4.)

PLATE 7. Gibberellin-stimulated stem elongation in a rosette plant. (See Figure 16.8.)

PLATE 8. Crown gall. (See Figure 16.13.)

PLATE 9. Cytokinin effects in genetically-transformed tobacco callus. (See Figure 16.14.)

PLATE 10. Photomorphogenesis in bean seedlings. (See Figure 18.10.)

PLATE 11. Sulfur dioxide injury on squash leaves. (See Figure 22.17.)

PLATE 12. Photoperiodic control of flowering in *Pharbitis nil* (SD plant) and *Hyoscyamus niger* (LD plant). (See Figure 20.2.)

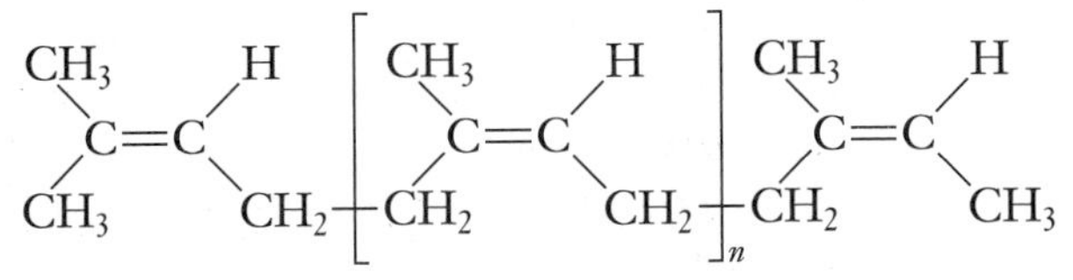

$$\begin{matrix} CH_3 & & H & CH_3 & & H & CH_3 & & H \\ & C{=}C & & & C{=}C & & & C{=}C & \\ CH_3 & & CH_2 & CH_2 & & CH_2 & CH_2 & & CH_3 \end{matrix}$$

$$(CH_3)_2C{=}CH{-}CH_2{-}[CH_2{-}C(CH_3){=}CH{-}CH_2]_n{-}CH_2{-}C(CH_3){=}CH{-}CH_3$$

FIGURE 14.8 The basic structure of rubber. Most rubber is a linear polymer where the value of n may range from a few hundred to several thousand.

earlier (Chap. 7). The only important isoprene derivatives with a molecular mass greater than the tetraterpenes are **rubber** and **gutta**. Rubber is a polymer consisting of up to 15,000 isopentenyl units. The polymer may be linear, as shown in Figure 14.8, or crosslinked into more complex configurations. The only difference between rubber and gutta is the configuration of the double bonds. In rubber the double bonds are all *cis* configuration while in gutta the double bonds are all *trans*.

In the plant, rubber occurs as small particles suspended in a milky white emulsion called **latex**. Latex production is widespread in plants, with estimates ranging from a few hundred to several thousand species that produce latex in some form. The principal commercial source is *Hevea brasiliensis*, a tree native to the Amazon rainforest. Others include the ornamental rubber tree (*Ficus elastica*), milkweed (*Asclepias*), and Russian dandelion (*Taraxacum kok-saghyz*). Latex contains about 30 to 40 percent rubber and 50 percent water. The balance is a complex mixture of resins, terpenes, proteins, and sugars (Klein, 1987). In most plants, latex is produced in the phloem, accumulating in a series of long, interconnected vessels called **laticifers**.

The best known source of gutta is a desert shrub, *Parthenium argentatum*, that grows in northern Mexico and southwestern United States. *Parthenium* (commonly known as guayule) may contain as much as 20 percent latex by weight, which is stored not in laticifers but in the vacuoles of stem and root cells. Guayule was at one time a significant commercial source of gutta for use in rubber products. However, while a single rubber tree, if properly tapped, can continue to produce for up to 30 years, guayule plants must be harvested (and, of course, replanted) annually.

Finally, there is a connection between terpenoids and air pollution. Many of the essential oils (essence, as in perfume), especially hemiterpenes, monoterpenes, and sesquiterpenes, are highly volatile and are given off in large quantities by plants, particularly during warm weather. Known generally as **volatile organic carbon (VOC)**, these natural emissions from plants contribute to the formation of haze and cloud, and are involved in the formation of toxic tropospheric ozone (see Chap. 22).

GLYCOSIDES

Some of the more interesting, if not important, derivatives synthesized by plants are glycosides. Most glycosides are thought to function as deterrents to herbivores.

The term **glycoside** (Gr. = *glykys*, sweet) refers to the bond formed (a glycosidic bond) when a sugar molecule condenses with a hydroxyl group. Sugars may form glycosidic bonds with other sugars, such as when linked together to form polysaccharides, or with hydroxyl groups on noncarbohydrate molecules, such as steroids or amino acids. The sugar most commonly found in glycosides is glucose, although rare sugars are found in specific glycosides.

Three particularly interesting secondary metabolites that occur as glycosides are the saponins, the cardiac glycosides (cardonelides), and the cyanogenic glycosides. A fourth family, the glucosinolates, although technically not glycosides, are similar in structure and so are included here.

SAPONINS

Saponins are terpene glycosides. They may be (1) steroid glycosides, (2) steroid-alkaloid glycosides, or (3) triterpene glycosides (Fig. 14.9). They may also occur as aglycones (the terpenoid without the sugar), which are known as **sapogenins**. The combination of hydrophobic triterpene with a hydrophillic sugar gives saponins the properties of a surfactant or detergent—when agitated in water, saponins form a soapy lather. The name saponin is derived from the soapwort or bouncing Bet (*Saponaria*), which at one time was employed as a soap substitute.

The effect of saponins on animals is somewhat variable. While not significantly toxic to mammals, saponins do have a bitter, acrid taste and will cause severe gastric irritation if ingested. If injected into the bloodstream, however, saponins will hemolyze red blood cells. This action is presumably because of their detergent properties and their ability to disrupt membranes generally. Saponins are highly toxic to fish and have been used as fish poisons. Saponins have also been implicated in reports of livestock poisoning. Alfalfa saponins, for example, can cause digestive problems and bloating in cattle. At the same time, there are reports that saponins in alfalfa sprouts will lower serum cholesterol levels (Olezek, 1996).

Commercially, saponins from the bark of *Quillaja saponaria* have been used as surfactants in photographic film, in shampoos, liquid detergents, toothpastes, and beverages (as emulsifiers). The saponin glycyrrhizin from licorice (*Glyscyrrhiza glabra*) has been used in medicines and as a sweetener and flavor enhancer in foods and cigarettes.

FIGURE 14.9 Saponins. (*A*) Medigenic acid glucoside, a triterpenoid saponin from alfalfa (*Medicago sativa*). (*B*) Diosgenin-glycoside, a steroid saponin isolated from clover (*Meliotus* sp.).

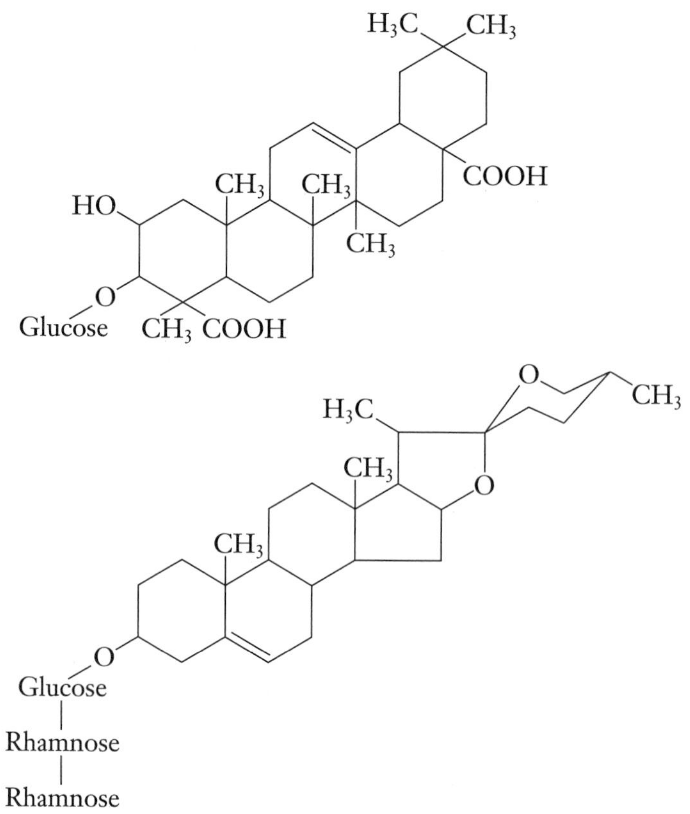

CARDIAC GLYCOSIDES

The *cardiac glycosides*, or *cardenolides*, are structurally similar to the steroid saponins and have similar detergent properties (Malcolm, 1991). They are distinguished from other steroid glycosides by the presence of a lactone ring (attached at C17) and the rare sugars (found almost exclusively in this group of steroids) that form the glycoside (Fig. 14.10). Like the saponins, cardenolides occur naturally as either the glycoside or the aglycone or genin.

The cardenolides have a wide distribution, being recorded in over 200 species representing 55 genera and 12 families. Perhaps the best known is "digitalis," a mixture of cardenolides extracted from the seeds, leaves, and roots of purple foxglove, *Digitalis purpurea* or Grecian foxglove, *D. lanata*. The two principal cardenolides in digitalis are **digitoxin** and its close analog **digoxin**. *Digitalis* is also the source of a saponin, **digitonin**.

Since the late eighteenth century, digitalis has been used for its therapeutic value in treating heart conditions such as atherosclerosis. Because they disrupt the heart muscle Na^{+}/K^{+}-ATPase pump, cardenolides are highly toxic to vertebrates.[1] In therapeutic use, however, carefully regulated doses can both slow and strengthen the heartbeat. Unfortunately, the lethal and therapeutic doses are very close, so the therapy must be carefully monitored.

[1]The extreme toxicity of cardenolides has long been exploited by African hunters, who coated their arrows and spears with cardenolide-rich extracts from plants.

Another common source of cardenolides are the milkweeds *Asclepias* and *Calotropis*. These two species are known as "milkweeds" because they produce a milky-white, cardenolide-rich latex. The milkweeds are particularly interesting because they are the principal host for ovipositing monarch butterflies. The emerging larvae feed on the milkweed and sequester the cardenolides

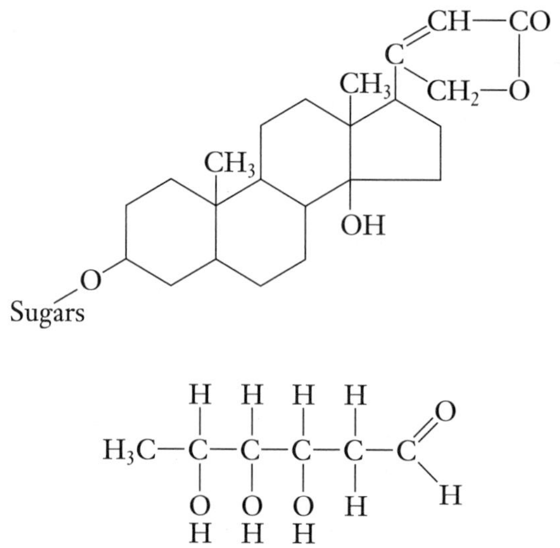

FIGURE 14.10 (*A*) Digitoxin, a cardiac glycoside. The sugar component consists of 2 molecules of digitoxose, 1 acetyl-digitoxose, and 1 glucose. (*B*) The structure of digitoxose, one of the rare sugars found in cardiac glycosides.

without ill effect. The cardenolides are retained through metamorphosis and are present in the adult monarchs. When birds, such as blue jays, attempt to feed on monarchs, the accumulated cardenolides induce an emetic reaction, forcing the bird to vomit. The bird then wisely avoids attempting to feed on monarchs for some time.

CYANOGENIC GLYCOSIDES

It might seem odd that plants synthesize chemicals capable of releasing deadly hydrogen cyanide (HCN), but more than 60 different cyanogenic compounds of plant origin have been described (Seigler, 1991). Predominant among these are the **cyanogenic glycosides** (Fig. 14.11). Most cyanogenic glycosides appear to be derived from four amino acids (phenylalanine, tyrosine, valine, and isoleucine) or from nicotinic acid. Cyanogenic glycosides are not themselves toxic, but when the plant is damaged by an herbivore, the glycoside undergoes an enzymatic breakdown and cyanide is released.

Enzymatic breakdown is a two-step process (Fig. 14.11). First, the sugar is released by the enzyme β-*glycosidase*. The resulting cyanohydrin is moderately unstable and will slowly decompose, liberating HCN in the process. Normally, however, decomposition is accelerated by the enzyme *hydroxynitrile lyase*. Enzymatic release of cyanide does not normally occur in intact plants because the enzymes and the substrate are spatially separated. In some cases, separation is maintained within the cell, but in others, the enzymes are in one cell and the cyanogenic glycosides in another. In *Sorghum*, for example, the cyanogenic glycoside **dhurrin** is synthesized and stored in epidermal cells, while the glycosidase and lyase enzymes are found in the mesophyll cells. Only when the tissue is crushed and the contents of the two cells are mixed will cyanogenesis occur.

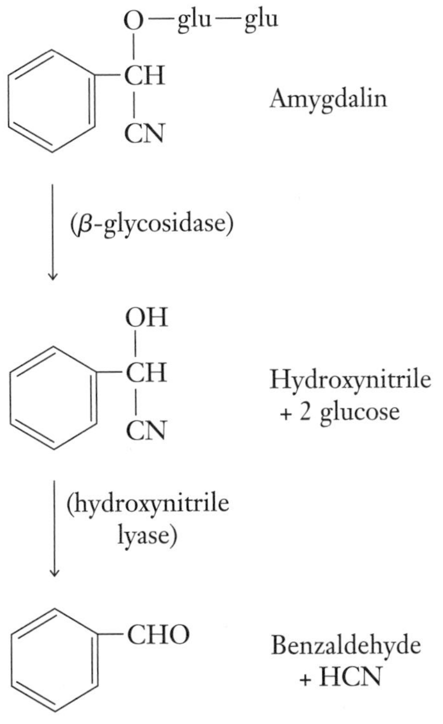

There is some evidence that the presence of cyanogenic glycosides deters feeding by insects and other herbivores, although most animals have the ability to detoxify small quantities of cyanide. Clearly the effectiveness of cyanogenic glycosides as a deterrent depends on many factors, such as the amount present, the rate of release of cyanide, and the ability of the animal to detoxify. The level of cyanogenic glycosides in plants is highly variable, influenced by both genetic control and environmental stress. The latter is a concern when using *Sorghum* for livestock forage. Dhurrin accumulates rapidly and can cause livestock poisoning when *Sorghum* plants are stressed by drought or frost. Another food source containing cyanogenic glycosides is cassava (*Manihot esculenta*), a major source of starch for millions of people in tropical countries. However, poisoning is avoided by careful preparation of the plant.

O—glu—glu
CH
CN
Amygdalin

(β-glycosidase)

OH
CH
CN
Hydroxynitrile
+ 2 glucose

(hydroxynitrile lyase)

CHO
Benzaldehyde
+ HCN

FIGURE 14.11 The hydrolysis of amygdalin, a cyanogenic glycoside found in seeds of apple, cherry, and apricots (*Prunus* sp.).

GLUCOSINOLATES

Glucosinolates are S- and N-containing compounds found primarily in the mustard family (Brassicaceae) and related families in the order Capparales. They are precursors to the mustard oils, an economically important class of flavor constituents that gives the pungent taste to condiments such as mustards and horseradish as well as the distinctive flavor of cabbages, broccoli, and cauliflower.

All glucosinolates are **thioglucosides** with the general structure shown in Fig. 14.12A. The sugar is always glucose. The diversity encountered in structure and properties is due to the R group, which may range from a simple methyl group to large linear or branched chains containing aromatic or heterocyclic structures. The biological activity of glucosinolates depends primarily on their hydrolysis to mustard oils (Fig. 14.12B). Hydrolysis is catalyzed by an enzyme called myrosinase (a thioglucosidase). The hydrolysis product is unstable and immediately undergoes a rearrangement to form a thiocyanate or *iso*thiocyanate. Like the cyanogenic glycosides, glucosinolates are spatially separated from the hydrolytic enzymes so that the mustard oils are normally formed only when the cells are disrupted, allowing the enzyme and substrate to come together. As with other defense compounds, some herbivores are deterred or repelled by the presence of glucosinolates in a plant, while others have adapted to use the glucosinolates or mustard oils as attractants to stimulate feeding and ovipositing.

Glucosinolates, or rather their absence, have had a significant impact on the oil seed industry. Rape seed (*Brassica napus*) is a good source of vegetable oil, but its

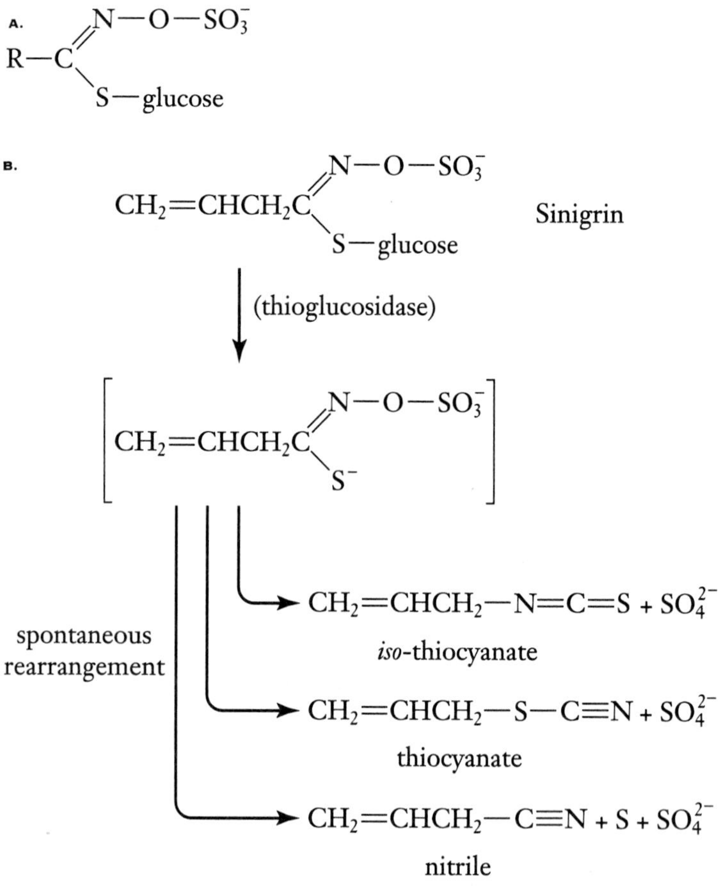

FIGURE 14.12 **Glucosinolates. (*A*) All glucosinolates have the same basic structure. (*B*) Enzymatic hydrolysis of sinigrin, a glucosinolate from black mustard (*Brassica nigra*). Enzymatic removal of glucose creates an unstable product that spontaneously rearranges to form the pungent mustard oils thiocyanate, *iso*thiocyanate, or nitrile.**

A. $R{-}C(=N{-}O{-}SO_3^-){-}S{-}\text{glucose}$

B. $CH_2{=}CHCH_2C(=N{-}O{-}SO_3^-){-}S{-}\text{glucose}$ Sinigrin

↓ (thioglucosidase)

$[CH_2{=}CHCH_2C(=N{-}O{-}SO_3^-){-}S^-]$

spontaneous rearrangement

→ $CH_2{=}CHCH_2{-}N{=}C{=}S + SO_4^{2-}$
iso-thiocyanate

→ $CH_2{=}CHCH_2{-}S{-}C{\equiv}N + SO_4^{2-}$
thiocyanate

→ $CH_2{=}CHCH_2{-}C{\equiv}N + S + SO_4^{2-}$
nitrile

high content of glucosinolate together with high erucic acid (a 22-carbon fatty acid) gives the oil undesirable taste and poor storage properties. New strains have been bred with low glucosinolates and erucic acid. These strains, called **canola** in order to distinguish them from normal rape, are now an economically important oil source.

PHENOLICS

The shikimic acid pathway gives rise to the aromatic amino acids, which, in turn, may be directed toward either primary or secondary metabolism. The family of secondary metabolites derived from the aromatic amino acids are known generally as phenolics, polyphenols, or phenylpropane derivatives.

Phenolics or polyphenols are a large and chemically diverse family of compounds ranging from simple phenolic acids to very large and complex polymers such as tannins and lignin. Also included are the flavonoid pigments described earlier in Chapter 7. The basic structure is phenol, a hydroxylated aromatic ring:

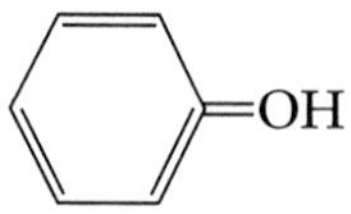

$C_6H_5{-}OH$

As with other secondary products, many phenolics appear to be involved in plant/herbivore interactions. Some (e.g., lignin) are important structural components, while others appear to be simply metabolic end-products with no obvious function.

THE SHIKIMIC ACID PATHWAY

The biosynthesis of most phenolics begins with the aromatic amino acids phenylalanine, tyrosine, and tryptophan. These aromatic amino acids are, in turn, synthesized from phospho*enol*pyruvate and erythrose-4-phosphate by a sequence of reactions

FIGURE 14.13 **The shikimic acid pathway for biosynthesis of aromatic amino acids. Initial precursors are erythrose-4-P from the pentose phosphate pathway and phospho*enol*pyruvate from glycolysis. The enzyme EPSPS, which catalyzes the second reaction in the conversion of shikimate to 3-enolpyruvylshikimate-5-P, is blocked by the herbicide glyphosate. Enzymes indicated as 1, 2, and 3 are subject to regulation by feedback inhibition.**

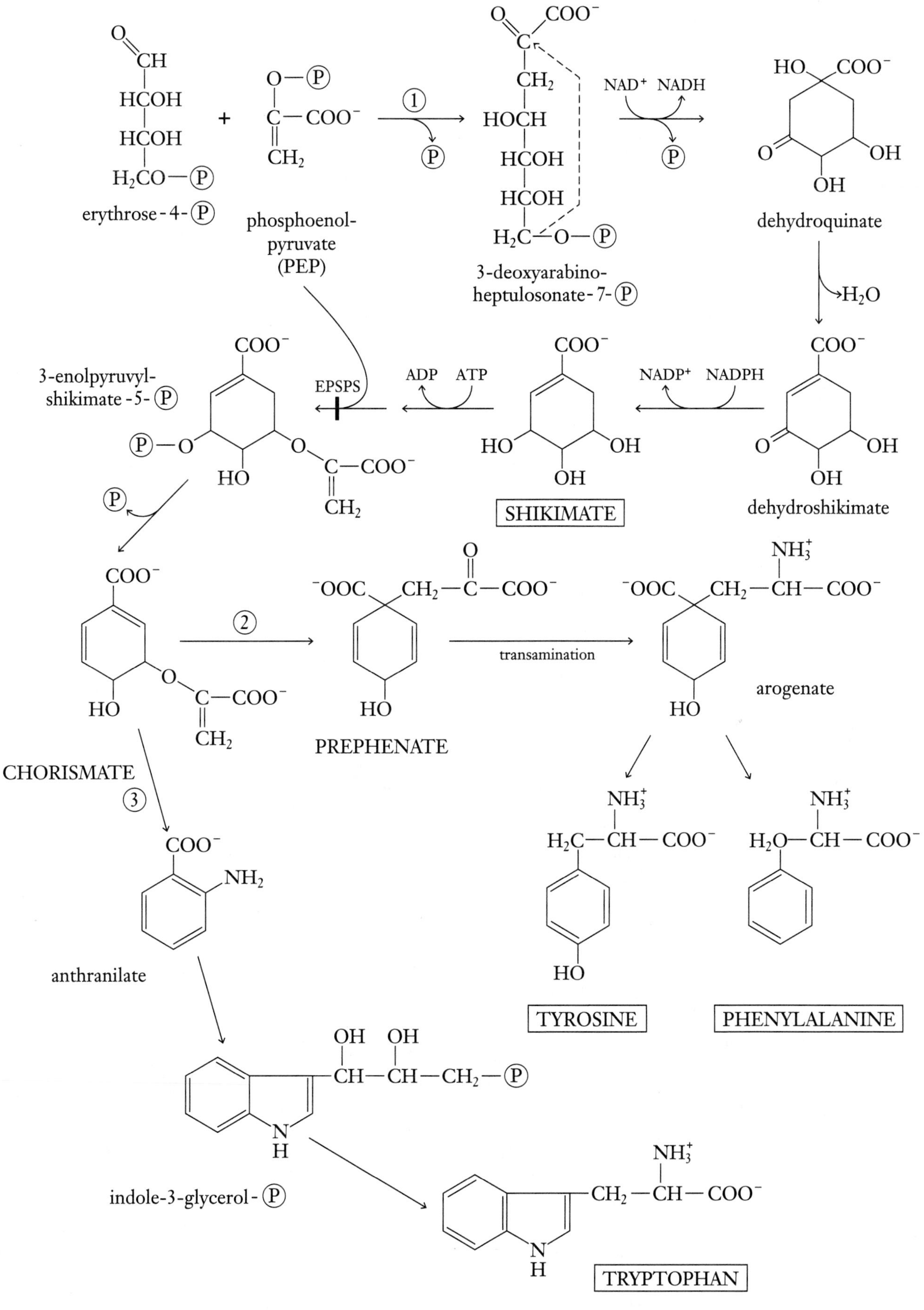

erythrose-4-(P)

phosphoenol-pyruvate (PEP)

(1)

3-deoxyarabino-heptulosonate-7-(P)

NAD^+ NADH

dehydroquinate

H_2O

dehydroshikimate

$NADP^+$ NADPH

SHIKIMATE

ADP ATP

EPSPS

3-enolpyruvyl-shikimate-5-(P)

(2)

CHORISMATE

PREPHENATE

transamination

arogenate

(3)

anthranilate

TYROSINE

PHENYLALANINE

indole-3-glycerol-(P)

TRYPTOPHAN

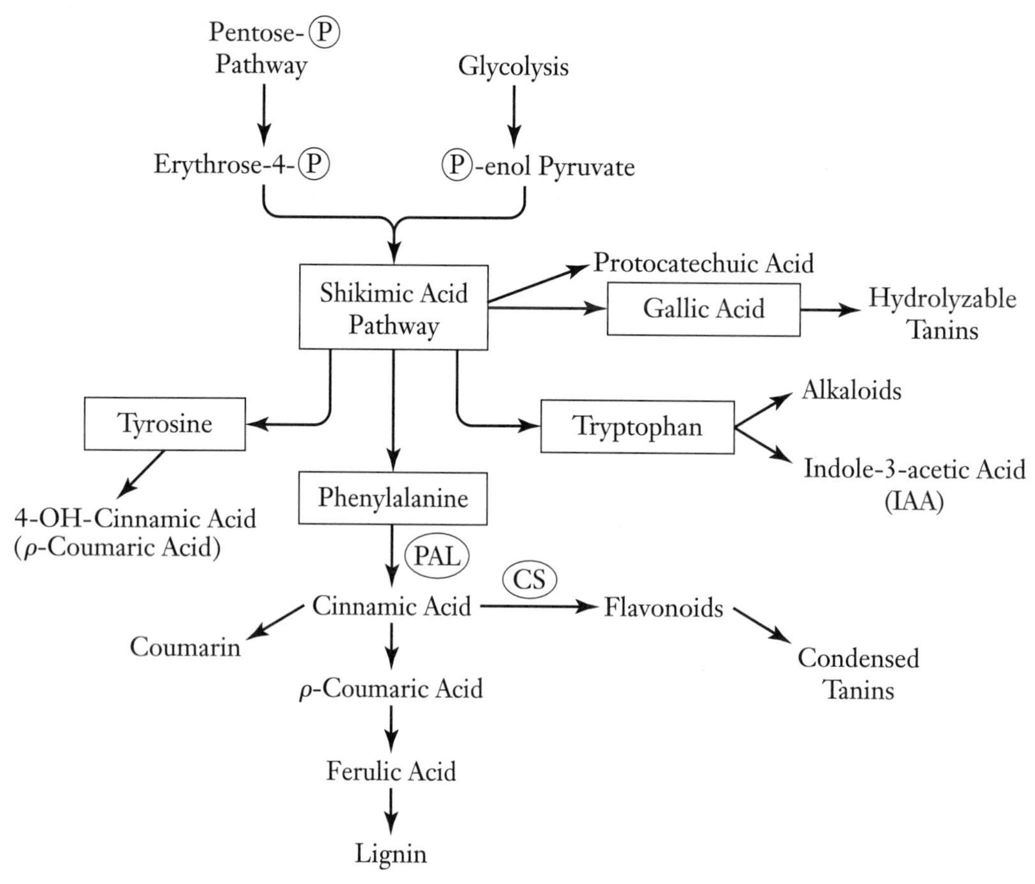

FIGURE 14.14 The central role of the shikimic acid pathway in the synthesis of various primary and secondary metabolites. PAL = phenylalanine ammonia lyase. CS = chalcone synthase.

known as the **shikimic acid pathway** (Fig. 14.13). The shikimic acid pathway is common to bacteria, fungi, and plants, but is not found in animals. Phenylalanine and tryptophan are consequently among the ten amino acids considered essential to animals (including humans) and represent the principal source of all aromatic molecules in animals (Voet and Voet, 1990).[2]

Synthesis of the aromatic amino acids begins with the condensation of one molecule of erythrose-4-P, from the pentose phosphate respiratory pathway, with one molecule of phospho*enol*pyruvate (PEP) from glycolysis. The resulting 7-carbon sugar is then cyclized and then reduced to form shikimate, from which the pathway derives its name. Chlorismate is a branch point that leads either to phenylalanine on the one hand, or to tryptophan on the other. The shikimic acid pathway is an excellent example of feedback regulation (see Box 6.2). Key enzymes in the pathway include the aldolase that catalyzes the initial condensation reaction, which is inhibited by all three end products. In a similar manner, the conversion of chlorismate to prephenate is inhibited by both phenylalanine and tyrosine, while the conversion of chlorismate to anthranilate is inhibited by tryptophan. Another step of interest in the pathway is the second step in the conversion of shikimate to 3-*enol*pyruvylshikimate-5-P to shikimate. The enzyme, **3-*enol*-pyruvylshikimate-5-phosphate synthase (EPSPS)**, is inhibited by the herbicide *glyphosate* (Chaps. 11, 23). Plants treated with glyphosate will die because they are unable to synthesize the aromatic amino acids and their derivatives, especially protein.

[2]Tyrosine, classified as nonessential, can be synthesized by hydroxylation of the essential amino acid phenylalanine. The dietary requirement for phenylalanine thus reflects the need for tyrosine as well.

SIMPLE PHENOLICS

Synthesis of most secondary phenolic products begins with the deamination of phenylalanine to cinnamic acid (Figs. 14.14, 14.15). The enzyme that catalyzes this reaction, **phenylalanine ammonia lyase (PAL)**, is a key enzyme because it effectively controls the diversion of carbon from primary metabolism, such as protein synthesis, into the production of phenolics. An alternate route, the deamination of tyrosine to *p*-coumaric acid, appears to be limited largely, if not exclusively, to grasses. The relative importance of PAL is illustrated by several observations. For example, PAL activity is stim-

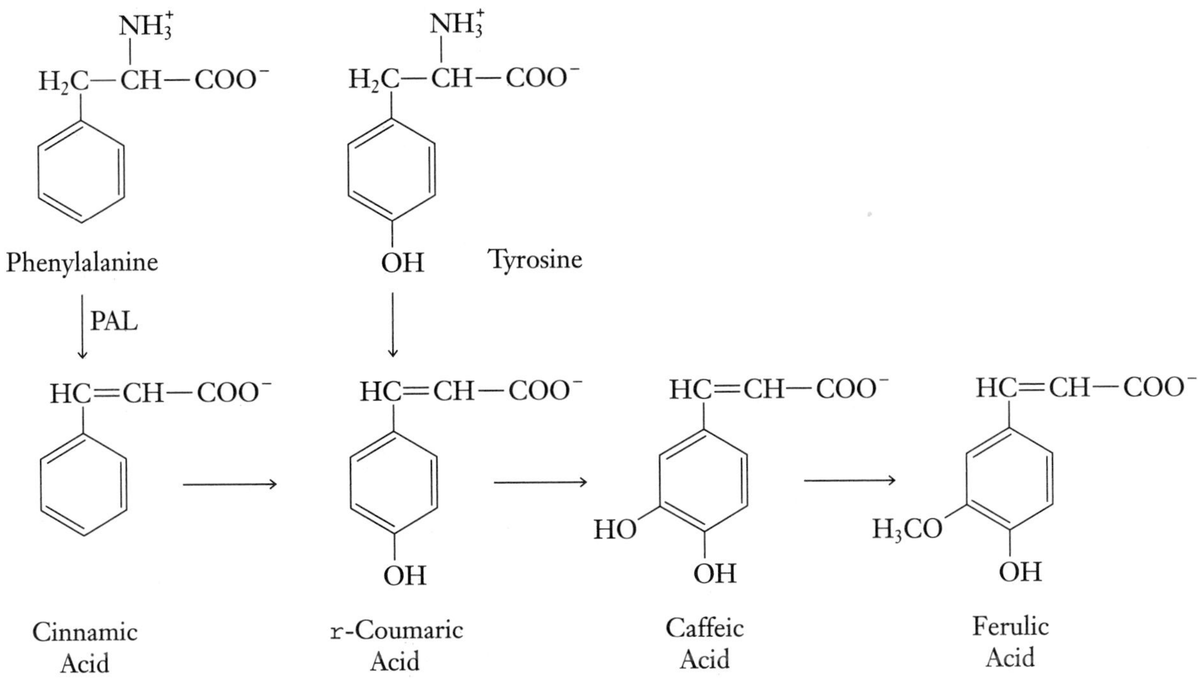

NH_3^+ | $H_2C-CH-COO^-$ Phenylalanine

NH_3^+ | $H_2C-CH-COO^-$ OH Tyrosine

PAL

$HC=CH-COO^-$ Cinnamic Acid

$HC=CH-COO^-$ OH r-Coumaric Acid

$HC=CH-COO^-$ HO OH Caffeic Acid

$HC=CH-COO^-$ H_3CO OH Ferulic Acid

FIGURE 14.15 Phenolic building blocks. Deamination of phenylalanine to cinnamic acid, followed by hydroxylation to *p*-coumaric acid, is the first step in phenolic biosynthesis. In members of the grass family, tyrosine can be deaminated directly to *p*-coumaric acid. PAL = phenylalanine ammonia lyase.

ulated by red and UV radiation, both of which also stimulate an increase in various flavonoids. Similarly, when levels of the pytoalexin glyceollin rise following inoculation of soybean with fungus, there is a corresponding increase in PAL activity. The product of PAL activity, *trans*-cinnamic acid, can inhibit the activity of auxin, a plant hormone, but there is no evidence that it has this function *in vivo*.

Cinnamic acid is readily converted to *p*-coumaric acid by the addition of a hydroxyl group (Fig. 14.15). The sequential addition of hydroyxl and methoxy groups give rise to caffeic acid and ferulic acid respectively. None of these four simple phenols accumulate to any extent. Their principal function appears to be as precursors to more complex derivatives such as coumarins, lignin, tannins, flavonoids, and isoflavonoids.

COUMARINS

The **coumarins** (Fig. 14.16) are a widespread group of lactones formed by ring closure of hydroxycinnamic acid. Coumarin gives new-mown hay its characteristic pleasantly sweet odor. Coumarin is also a constituent of Bergamot oil, that is used to flavor pipe tobacco, tea, and other products. While coumarin itself is not toxic, it can be converted by fungi to a toxic product, **dicoumarol**, that is typically found in moldy hay. Dicoumarol causes fatal hemorrhaging in cattle by inhibiting vitamin K, an essential cofactor in blood clotting. The discovery of dicoumarol in the 1940s led to the development of a coumarin derivative, *warfarin*, as a rat poison. Scopoletin, the most prevalent coumarin in higher plants, is often present in seed coats. It is suspected of being a germination inhibitor that must be leached out of the seed coat before germination can proceed.

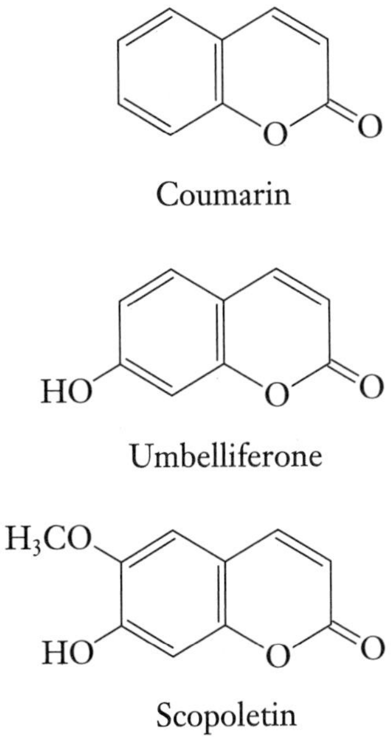

O O Coumarin

HO O O Umbelliferone

H_3CO HO O O Scopoletin

FIGURE 14.16 Coumarins. More than 300 coumarins have been reported from more than 70 plant families. Scopoletin is probably the most widespread coumarin in plants.

LIGNIN

Lignin is a highly branched polymer of three simple phenolic alcohols (Fig. 14.17). Gymnosperm lignin is comprised mainly of coniferyl alcohol subunits while angiosperm lignin is a mixture of coniferyl and sinapyl alcohol subunits. The alcohols are oxidized to free radicals by a ubiquitous plant enzyme called **peroxidase**. The free radicals then react spontaneously and randomly to form lignin.

Lignin is found in cell walls, especially the secondary walls of tracheary elements in the xylem, where it contributes mechanical strength and rigidity to woody stems. Lignin is in fact what makes wood, wood! In spite of its abundance (second only to cellulose), the structure of lignin is not well understood. Lignin is a very large polymer; it is insoluble in water and most organic solvents, and impossible to extract without considerable degradation. Moreover, the three basic monomers may link together in a variety of ways to form a multibranched, three-dimensional structure. The complexity is so great that, like snowflakes, each lignin "molecule" may be unique.

Although the principal function of lignin is structural, it has also been implicated as a defensive chemical. Lignin itself is not readily digested by herbivores and, because it is covalently linked to cellulose and hemicellulose, its presence decreases the digestibility of these polymers as well. Also, when fungal pathogens enter host cells, they do so by enzymatically degrading the host cell wall. Several studies have shown that lignins and other phenolic derivatives accumulate at the site of fungal penetration, presumably slowing the rate of cell wall degradation.

FLAVONOIDS

The **flavonoids** were described earlier in Chapter 7. Here we look briefly at their biosynthesis in order to see how they fit into the phenolic family. Recall that flavonoids consist of three rings, labeled A, B, and C (Chap. 7; Fig. 14.18). Ring B and the three-carbon bridge (ring C) are derived from the shikimic acid pathway via phenylalanine and *p*-coumaric acid. The six carbons that make up the A ring are derived from **malonic acid**, in the form of a malonyl-coenzyme A (malonyl-CoA) complex. (Malonyl-CoA is a principal intermediate in fatty acid synthesis.) The key enzyme is **chalcone synthase (CHS)**. CHS catalyzes the stepwise condensation of three molecules of malonyl-CoA with one molecule of *p*-coumaroyl-CoA to form naringenin chalcone. A chalcone is a C_6-C_3-C_6 pattern wherein ring C has not yet closed.

As noted earlier in Chapter 7, flavonoids include the anthocyanin pigments that serve to attract insect pollinators and the isoflavonoids that function as phytoalexins. The role of flavonoids in symbiont recognition between rhizobia and host legume roots was discussed in Chapter 6. It has also been demonstrated that exposure of plants to UV radiation increases the content of flavonoids, suggesting that flavonoids may offer a measure of protection by screening out harmful UV-B radiation (Caldwell et al., 1983).

TANNINS

The name **tannins** is derived from the historic practice of using plant extracts to "tan" animal hides, that is, to convert hides to leather. These extracts contain phenol derivatives that bind to, and thus denature, proteins. Two categories of tannins are now recognized: condensed tannins and hydrolyzable tannins. Condensed tanins are polymers of flavonoid units linked by strong carbon–carbon bonds. These bonds are not subject to hydrolysis but can be oxidized by strong acid to release anthocyanidins (see Chap. 7). The basic structural unit of hydrolyzable tannins is a sugar, usually glucose, with its hydroxyl groups esterified to gallic acid. Gallic acid residues are in turn joined to form an extensively cross-linked polymer.

Like other phenolics, the role of tannins is not clear. Tannins do appear to deter feeding by many animals when tannin-free alternatives are available. This effect

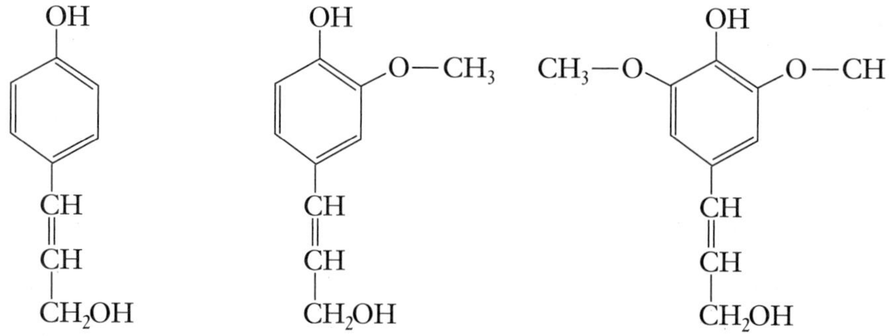

r-Coumaryl Alcohol   Coniferyl Alcohol   Sinapyl Alcohol

FIGURE 14.17 The principal lignin monomers. Cross-linkages most commonly form between the ring alcohol and the double-bonded carbon atoms.

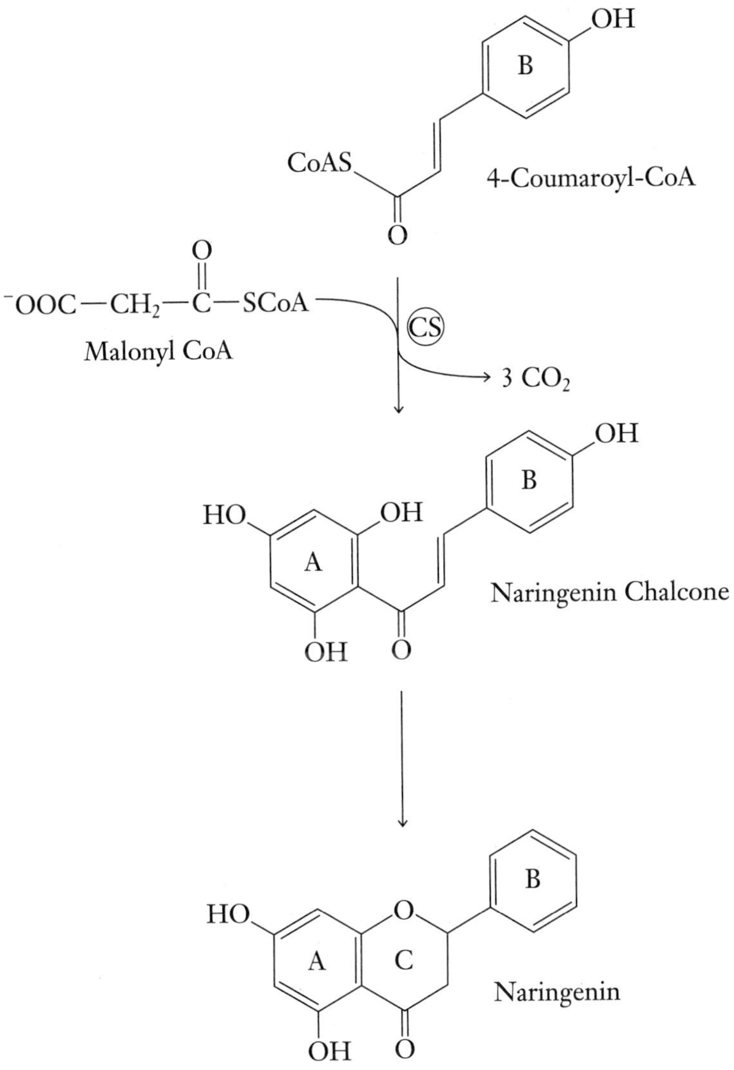

FIGURE 14.18 The biosynthesis of the flavonoid ring structure. The B ring is derived from the shikimic acid pathway. The A ring is derived from three molecules of malonyl-CoA. CS = chalcone synthase.

could be related to the astringency—a sharp, somewhat unpleasant sensation in the mouth—for which tannins are noted.[3] As well, tannins tend to suppress the efficiency of feed utilization, growth rate, and survivorship. The conventional interpretation has been that tannins reduce digestibility of dietary protein, presumably by binding with protein in the gut (Harborne, 1988). Other evidence, however, has cast doubt on this interpretation, suggesting that tannins may be toxic in other, yet unknown, ways (Blytt et al., 1988).

ALKALOIDS

The alkaloids are a very large and heterogeneous family of secondary metabolites that are of interest primarily because of their pharmacological properties and medical applications.

Unlike terpenoids and polyphenols, the **alkaloids** are a large family of chemically unrelated molecules. The principal characters that they share are solubility in water, the presence of at least one nitrogen atom, and high biological activity. Often the nitrogen will accept a proton, which gives it a slightly basic character in solution (hence the name *alkaloid*). Alkaloids are for the most part heterocyclic, although a few aliphatic (noncyclic) nitrogen compounds, such as **mescaline** and **colchicine**, are sometimes considered alkaloids. Altogether, some 10,000 alkaloids have been found to occur in approximately 20 percent of the species of flowering plants, mostly herbaceous dicots (Southon and Buckingham, 1989).

The word "alkaloids" is virtually synonymous with the word "drug"; 10 of the 12 commercially most-important plant-derived drugs are alkaloids (Balandrin et al., 1985). Alkaloids generate varying degrees of physiological and psychological response in humans, largely by interfering with neurotransmitters. In large doses, most alkaloids are highly toxic, but in smaller doses they

[3]The astringent property of tannin is a component in the flavor of many fruits as well as drinks such as tea and red wine.

Alkaloid Class	Example	Other Representatives
Quinoline	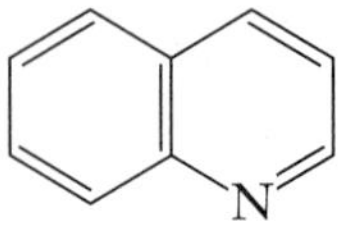 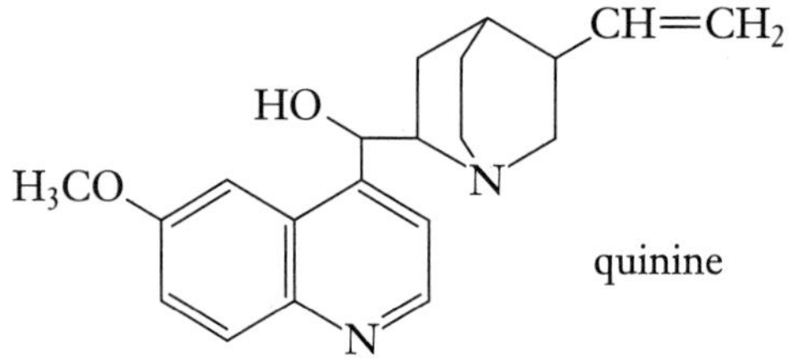 quinine	
Isoquinoline	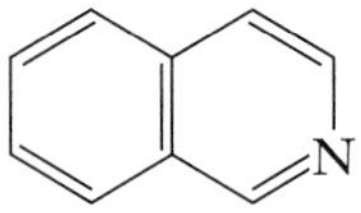 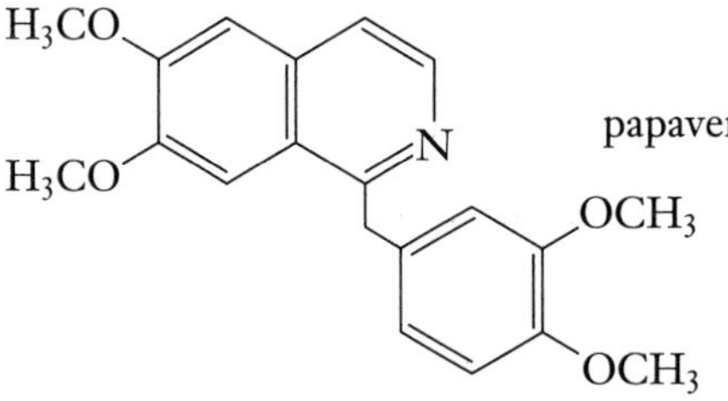 papaverine	morphine codeine berberine
Indole	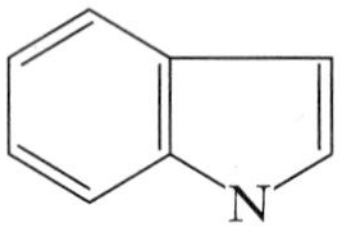 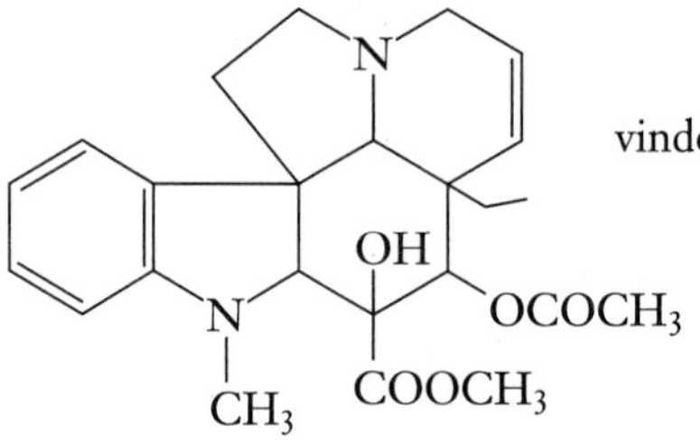 vindoline	vinblastine reserpine strychnine
Pyrrolizidene	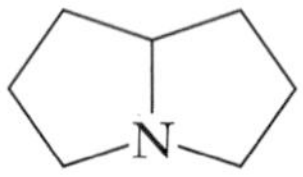 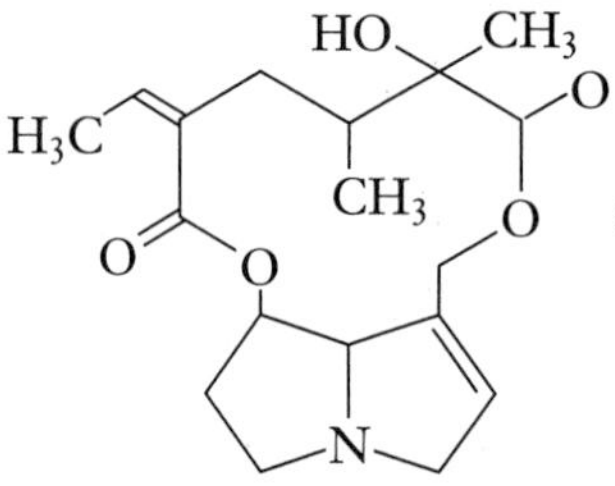 senecionine	retrorsine
Quinolizidine	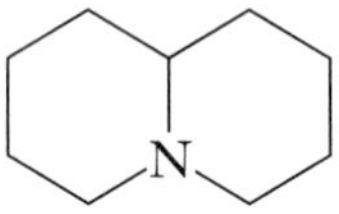 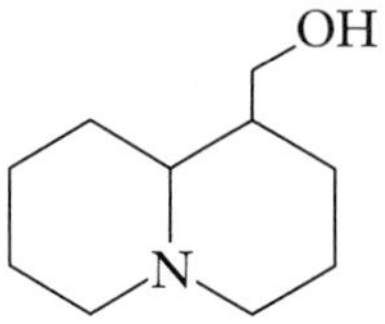 lupinine	cytisine
Tropane	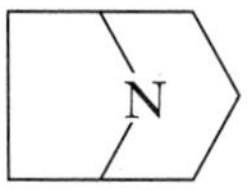 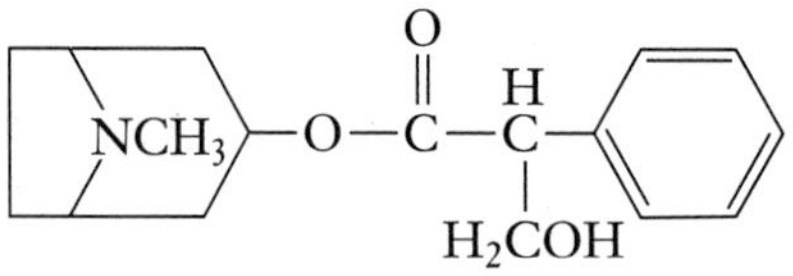 atropine	scopolamine cocaine
Piperidine	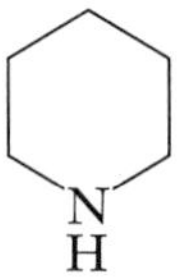 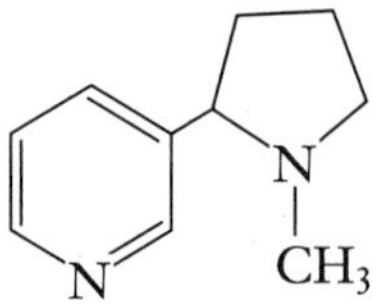 nicotine	coniine
Purine	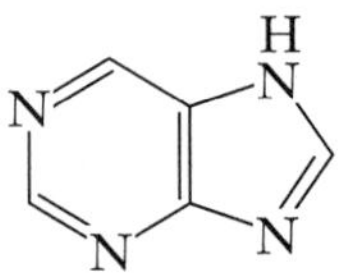 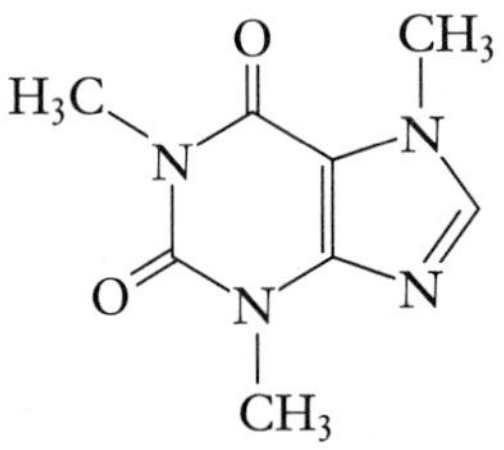 caffeine	

FIGURE 14.19 The principal classes of alkaloids showing the basic ring structure and representative examples of each class.

may have therapeutic value. From prehistory to the present, alkaloids or alkaloid-rich extracts have been used for a variety of pharmacological purposes, such as muscle relaxants, tranquilizers, pain killers, and mind-altering drugs.

Alkaloids are generally classified on the basis of the predominant ring system present in the molecule (Fig. 14.19). In spite of the extensive variation in structure, however, alkaloids are generated from a limited number of simple precursors. Most alkaloids are synthesized from a few common amino acids (tyrosine, tryptophan, ornithine or argenine, and lysine). The tobacco alkaloid **nicotine** is synthesized from nicotinic acid and **caffeine** is a purine derivative.

Although a few alkaloids are found in several genera or even families, most species display their own unique, genetically determined pattern. As with other secondary metabolites, individual alkaloids may be restricted to particular organs, such as roots, leaves, or young fruit. The isoquinoline alkaloids **morphine**, **codeine**, and **papaverine**, for example, are but three of 20 alkaloids produced in the immature seed capsule of the opium poppy, *Papaver somniferum*. **Berberine**, however, is found in several sources, including seeds of barberry (*Caulophyllum thalictroides*). Barberry was at one time a common horticultural hedge but is now out of fashion due to the high incidence of poisoning in young children attracted to its bright-red berries.

The quinolizidine alkaloids, such as **lupinine**, are frequently called lupine alkaloids because of their high abundance in the genus *Lupinus*. Although range animals are deterred from eating lupines because of their bitter taste, grazing on lupines is a common cause of cattle poisonings. The highest concentration of alkaloids occurs in the seed, so livestock losses are generally highest in the fall. Other alkaloids that cause poisoning of livestock include **senecionine** (*Senecio*, groundsel), **lycotonine** (*Delphinium*, larkspur), **scopolamine** (*Datura stramonium*, jimson weed), and **atropine** (also known as hyoscyamine) from black henbane (*Hyoscyamus niger*, see Plate 12).

The indole alkaloids are often referred to as terpenoid-alkaloids because, although the basic indole ring structure is derived from tryptophan, the rest of the molecule is derived from the mevalonic acid pathway. Two well-known terpenoid-indole alkaloids, **vinblastine** and **vincristine**, are produced by the Madagascar periwinkle, *Catharanthus roseus*. In the 1950s it was discovered that vinblastine and vincristine arrest cell division in metaphase by inhibiting microtubule formation. They have since been used in the treatment of Hodgkin's lymphoma, leukemia, and other forms of cancer. Unfortunately, *C. roseus* produces a complex array of indole alkaloids of which vinblastine and vincristine represent only a very small proportion (about 0.00025% of leaf dry weight). This makes their extraction and purification difficult and costly. Recent efforts have led to the development of cell culture systems for the production of higher yields at reasonable cost (Misawa and Goodbody, 1996). Other well-known alkaloids such as **cocaine**, (from leaves of the coca plant, *Erythroxylum coca*), **codeine** (from the opium poppy, *Papaver somniferum*), **nicotine** (tobacco), and **caffeine** (coffee beans and tea leaves) are widely used as stimulants or sedatives.

Whether alkaloids have any specific function in plants is a matter of continuing debate. Like other secondary products, there are a number of general arguments supporting a defensive role (Hartmann, 1991). Most alkaloids have a bitter taste, which is considered a universally repellant character for all animals, including insects. All alkaloids are biologically active; many are significantly toxic. Nicotine, for example, is a potent insect poison and one of the first insecticides used by humans. Many alkaloids have antibiotic properties, suggesting a defense against microbial infection. Interestingly, it has been known for a long time that alkaloid concentrations are very low when measured on a whole plant basis, but, as noted earlier, may be very high in selected organs or tissues. Often the tissues that accumulate alkaloids are those most vulnerable in terms of plant fitness (young influorescences, for example) or peripheral tissues that would be first attacked by herbivores. Thus the morphine content of whole young poppy capsules is less than 2 percent, but 25 percent or more in the latex that exudes when the capsule is wounded. Quinine accumulates in the outer bank of the tree *Cinchona officinalis* and *Rauwolfia* alkaloids are concentrated primarily in the root bark. Although alkaloids have been studied extensively for their pharmacological and medicinal value, there is much yet to be learned with respect to their physiology and chemical ecology.

SUMMARY

Although the products of plant metabolism are often designated as either primary or secondary metabolites, the distinction between the two is not easily made. Primary metabolites such as protein, lipid, carbohydrate, and nucleic acids comprise the basic metabolic machinery of all cells. Others, such as chlorophyll and lignin, are more restricted in occurrence, but are equally essential to the growth and development of the organism. Secondary metabolites, on the other hand, may be found only in specific tissues or at particular stages of development and have no obvious role in the development or survival of the organism.

Two significant metabolic pathways that give rise to both primary and secondary metabolites are the mevalonic acid pathway and the shikimic acid pathway. The mevalonic acid pathway gives rise to two 5-carbon compounds, isopentenyl pyrophosphate and dimethylallyl pyrophosphate, that form the basis for the terpenoid family. Terpenoids include primary metabolites such as phytol (a portion of the chlorophyll molecule), membrane sterols, carotenoid pigments, and the hormones gibberellin and abscisic acid. Secondary products include a range of 5-carbon hemiterpenoids through 40-carbon tetraterpenes and the polyterpenes, rubber and gutta.

The shikimic acid pathway produces the aromatic amino acids phenylalanine, tyrosine, and tryptophan—all primary metabolites required for protein synthesis. Deamination of phenylalanine to cinnamic acid, catalyzed by the enzyme phenylalanine ammonia lyase (PAL), effectively diverts carbon from primary metabolism into the synthesis of a wide range of secondary metabolites—coumarins, lignin, tannins, flavonoids, and isoflavonoids—based on simple phenolic acids.

Other interesting and useful secondary metabolites include the saponins (terpene glycosides), cardiac glycosides, cyanogenic glycosides, and glucosinolates. The alkaloids are a heterogenous group of nitrogenous compounds with significant pharmacological properties. Although the physiological roles of secondary metabolites are poorly understood, most are toxic to some degree and appear to serve primarily in defense against microbial infection and attack by herbivores.

CHAPTER REVIEW

1. Distinguish between anabolism and catabolism; between intermediary, primary and secondary metabolism.
2. What are terpenoids? How do they originate within the plant and what functions do they serve?
3. Explain why sterols may be considered both primary and secondary metabolites.
4. What do saponins, cardenolides, and amygdalin have in common?
5. Why are amino acids such as phenylalanine and tryptophan essential in the human diet?
6. What are EPSPS and PAL? What key metabolic roles do they play?
7. Alkaloids are, for the most part, chemically unrelated. What is the basis for grouping them together?
8. How does the herbicide glyphosate kill plants? When plants are treated with glyphosate, it often takes a week before any signs of injury are observed. Explain this delay.

FURTHER READING

Harborne, J. 1988. *Introduction to Ecological Biochemistry*. 3rd ed. New York: Academic Press.

Klein, R. M. 1987. *The Green World: An Introduction to Plants and People*. New York: Harper & Row.

Robinson, T. 1980. *The Organic Constituents of Higher Plants*. 4th ed. North Amherst: Cordus Press.

Rosenthal, G. A., M. R. Berenbaum. 1991. *Herbivores: Their Interactions with Secondary Metabolites. Vol. 1. The Chemical Participants*. 2nd ed. San Diego: Academic Press.

REFERENCES

Balandrin, M. F., J. A. Klocke, E. S. Wertele, W. H. Bollinger. 1985. Natural plant chemicals: Sources of industrial and medicinal materials. *Science* 228: 1154–1160.

Benveniste, P. 1986. Sterol biosynthesis. *Annual Review of Plant Physiology* 37:275–308.

Blytt, H. J., T. K. Guscar, L. G. Butler. 1988. Antinutritional effects and ecological significance of dietary condensed tannins may not be due to binding and inhibiting digestive enzymes. *Journal of Chemical Ecology* 14:1455–1465

Caldwell, M. M., R. Robberecht, S. D. Flint. 1983. Internal filters: Prospects for UV-acclimation in higher plants. *Physiologia Plantarum* 58:445–450.

Gershenzon, J., R. Croteau. 1991. Terpenoids. In: G. A. Rosenthal, M. R. Berenbaum (eds.), *Herbivores: Their Interactions with Secondary Metabolites. Vol. 1. The Chemical Participants*. 2nd ed. San Diego: Academic Press, pp. 165–219.

Harborne, J. 1988. *Introduction to Ecological Biochemistry*. 3rd ed. New York: Academic Press.

Hartmann, T. 1991. Alkaloids. In: G. A. Rosenthal, M. R. Berenbaum (eds.), *Herbivores: Their Interactions with Secondary Metabolites. Vol. 1. The Chemical Participants*. 2nd ed. San Diego: Academic press, pp. 79–121.

Malcolm, S. B. 1991. Cardenolide-mediated interactions between plants and herbivores. In: G. A. Rosenthal, M. R. Berenbaum (eds.), *Herbivores: Their Interactions with Secondary Metabolites. Vol. 1. The Chemical Participants*. 2nd ed. San Diego: Academic Press, pp. 251–296.

Misawa, M., A. E. Goodbody. 1996. In: F. DiCosmo, M. Misawa (eds.), *Plant Cell Culture Secondary Metabolism: Toward Industrial Application*. Boca Raton: CRC Press, pp. 123–138.

Olezek, W. 1996. Alfalfa saponins: Structure, biological activity, and chemotaxonomy. In: G. R. Waller, K. Yamasaki (eds.), *Saponins Used in Food and Agriculture*. New York: Plenum Press, pp. 155–170.

Seigler, D. S. 1991. Cyanide and cyanogenic glycosides. In: G. A. Rosenthal, M. R. Berenbaum (eds.), *Herbivores: Their Interactions with Secondary Metabolites. Vol. 1. The Chemical Participants*. 2nd ed. San Diego: Academic Press, pp. 35–77.

Southon, I. W., J. Buckingham (eds.), 1989. *Dictionary of Alkaloids*. London: Chapman and Hall.

Voet, D., J. Voet. 1990. *Biochemistry*. New York: Wiley.

Part Three

Regulation of Plant Development

Preview

The development of a multicellular organism is an enormously complex process. From a single cell, the fertilized egg, complicated patterns of cell division, cell enlargement, and cell differentiation give rise to complex tissues, organs, and organisms as different as duckweeds, the smallest flowering plant, and redwood trees, among the largest. Starting with questions as basic as how cells grow, plant physiologists attempt to understand how these intricate patterns of development are regulated.

Regulation begins at the genetic level. The DNA sequences in the chromosomes determine whether that single cell will become a duckweed or a redwood tree. But development is malleable and, within broad limits set by the genes, development may proceed at different rates or take different paths. A duckweed can never become a redwood tree, but many plants may exhibit differences in the rate of growth, the extent of internode elongation, root development, flowering behavior, and a host of other spatial and temporal variations in their developmental patterns.

There are generally two classes of mechanisms that can influence and regulate the developmental pattern of a plant. The first is a system of endogenous chemical messengers, called hormones, that coordinate the development of individual organs as well as the organism as a whole. At least five classes of plant hormones are known along with several other classes of chemicals that interact to influence patterns of development. The second class of mechanisms is comprised of extrinsic factors, such as light, temperature, and gravity, that provide the plant with information about its environment. Environmental factors help the plant to orient in space in order to optimize the use of available resources, such as light for driving photosynthesis or efficient absorption of water and nutrients from the soil. Through a combination of light, photoperiod, temperature, and internal biological clocks, plants also orient in time, maximizing opportunity for flowering and sexual reproduction, or anticipating and thus surviving unfavorable climatic periods.

The final pattern of development and behavior of each individual plant is the result of a complex interplay between these genetic, hormonal, and environmental factors.

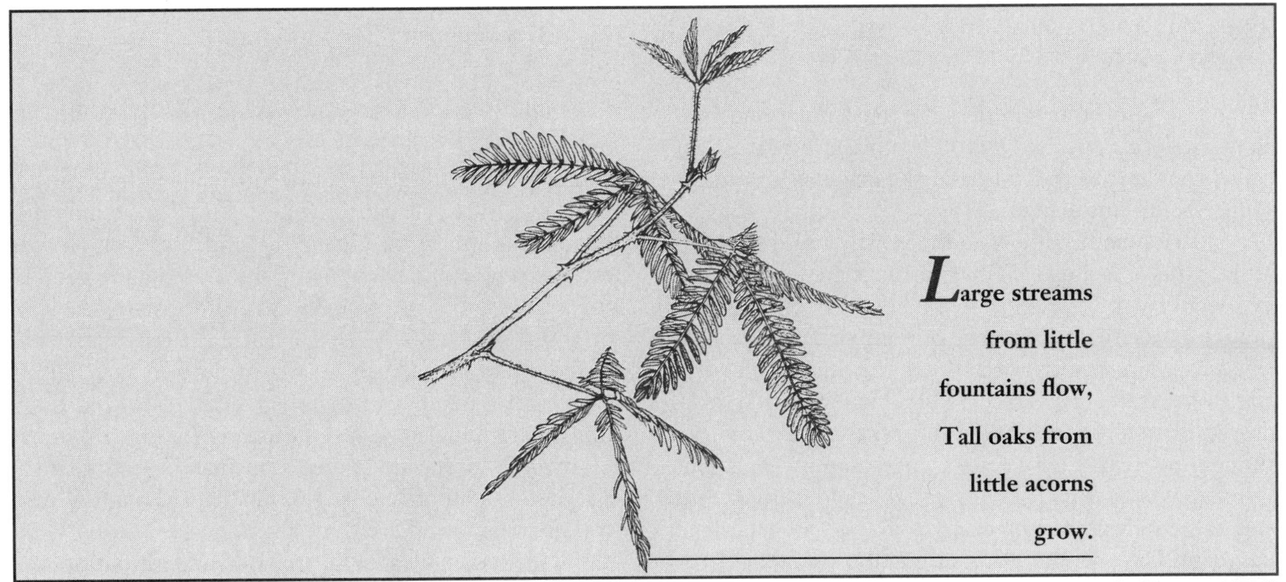

D. Everett (1791)

15

Patterns in Plant Development

The development of a majestic oak tree from a small acorn requires a precise and highly ordered succession of events. Starting as a single fertilized egg, plant cells divide, grow, and differentiate into increasingly complex tissues and organs. In the end, these events give rise to the complex organization of a mature plant that flowers, bears fruit, senesces, and eventually dies. These events, along with their underlying biochemistry and the many factors that either impose or modulate an unfailing and orderly progression through the life cycle, constitute development. Understanding development is one of the major goals of plant physiology.

The purpose of this chapter is to serve as an introduction to plant development. The discussion is generalized and is intended to provide a vocabulary for the chapters on hormones and environmental regulation of development that follow. Included in this chapter are

- a brief discussion of the distinctions between growth, differentiation, and development;
- a brief survey of the principal stages in the development of a higher plant, from seed through shoot and root elongation, flowering, and fruit development;
- an overview of the three principal means for regulating development through gene expression, hormones, and environmental stimuli;
- a discussion of how plant cells grow—how the cell overcomes limitations on cell enlargement that are imposed by the presence of a rigid cell wall; and
- a brief introduction to growth curves and their physiological significance.

For more details, the student should refer to the sources listed under *Further Reading* at the end of the chapter.

GROWTH, DIFFERENTIATION, AND DEVELOPMENT

Three terms used routinely to describe various aspects of the changes that a plant undergoes during its life cycle are **growth, differentiation,** and **development.** In order to understand descriptions of plant development and its regulation, it is necessary to clarify what is meant by these three terms. *Development* is an umbrella term, referring to the sum of *all* of the changes that an organism goes through in its life cycle—from germination of the seed through growth, maturation, flowering, and senescence. Development applies equally well to cells, tissues, and organs, each experiencing similar patterns of change. Development is most readily manifest in changes in form of the organism or organ, such as the transition from the vegetative to flowering condition or from leaf primordium to fully expanded leaf. Development may also be manifest at the subcellular and biochemical levels, such as when chloroplasts appear in leaf cells brought into the light and the enzymes of photosynthesis become active.

Development is the sum of growth and differentiation. *Growth* is a quantitative term, related to changes in size and mass. It can be assessed by a variety of quantitative measures. Growth of cells in culture is sometimes measured as an increase in **cell number** or the **fresh weight** of packed cells. For higher plants, however, fresh weight is not always a reliable measure. Although most plant tissues are approximately 80 percent water, water content is highly variable and fresh weight will fluctuate widely with changes in the water status of the plant. **Dry weight,** a measure of the amount of protoplasm or dry matter, is used more often than fresh weight, but even dry weight can be misleading as a measure of growth. For example, the dry weight of a seedling (including the seed) germinated in darkness will actually decrease compared with the seed alone, although we intuitively sense that considerable growth has occurred. Loss of dry weight occurs in this situation because the growing seedling respires carbon stored in the seed with the attendant loss of carbon dioxide. In darkness, this lost carbon cannot be replaced by photosynthesis and so the seedling experiences a net loss of dry matter. In such a situation, fresh weight or **length** of the seedling axis would be a better measure or growth. Length, and perhaps **width,** would also be suitable measures for an expanding leaf. Length and width would not only provide a measure of the amount of growth, but a length-to-width ratio would also provide information about the pattern of leaf development. It should be obvious that many parameters could be invoked to measure growth, dependent to some extent on the needs of the observer. Whatever the measure, however, all attempts to quantify growth reflect a fundamental understanding that *growth is an irreversible increase in volume or size.*

While cell division and cell enlargement normally go hand-in-hand, it is important to keep in mind that growth can occur without cell division and cell division can occur without growth. For example, cell division is normally completed very early in the development of grass coleoptiles and the substantial enlargement of the organ that follows is due almost entirely to cell enlargement. When seeds of wheat (*Triticum* sp.) are irradiated with gamma rays sufficient to block DNA synthesis and cell division, germination will nevertheless proceed. The result is small seedlings produced by cell enlargement alone. Such seedlings generally do not survive more than two or three weeks, but, except for the giant cells, their morphology is more or less normal. On the other hand, during early stages of embryo development in the flower, the embryo goes through a stage in which cell division continues to produce more but smaller cells, with no overall increase in the size of the embryo.

Differentiation is a qualitative term, referring to differences other than size that arise among cells, tissues and organs. Differentiation occurs when a dividing cell gives rise to two daughter cells destined to assume different anatomical characteristics and functions. In the earliest stages of development, for example, division of the zygote gives rise to cells that will become the root or shoot of the plant. Unspecialized parenchyma cells differentiate into xylem vessels or phloem sieve tubes, each with a distinct morphology and specialized function. Differentiation does not lend itself easily to quantitative interpretation but must normally be described as a series of qualitative changes. Finally, although growth and differentiation are normally concurrent events, examples abound of growth without differentiation and differentiation without growth.

Differentiation is a two-way street. Even though plant cells may appear to be highly differentiated or specialized, they may often be stimulated to revert to a more embryonic form; that is, cells **dedifferentiate.** It is as though the cells have been genetically reprogrammed, allowing them to reverse the process and then to differentiate along new and different paths. Thus, cells taken from the center of the stem of a tobacco plant or a soybean cotyledon may be stimulated to once again undergo cell division, grow as undifferentiated callus and eventually to give rise to a new plant (Fig. 15.1). This ability of differentiated cells to regenerate new plants demonstrates that most living plant cells are **totipotent**; they retain a complete genetic program even though not all of the information is used by the cell at any given time. Thus, development does not reflect a progressive loss of genetic information, only the selective use of that

FIGURE 15.1 Shoot regeneration in callus culture. A piece of pith tissue from the center of a tobacco stem was explanted onto a medium containing mineral salts, vitamins, sucrose, and hormones. The tissue proliferated as an undifferentiated callus (left) for several weeks before regenerating new plantlets (right). These plantlets can eventually planted into soil and will produce a mature tobacco plant. (See color plate 5.)

information in order to achieve particular developmental ends.

Of course, not all cells are totipotent. Highly specialized cells whose development has been locked in, such as by exceptionally thick and rigid cell walls or severely modified protoplasts, are not as likely to be capable of renewed differentiation. On the other hand, it is probable that all tissues contain at least some potentially totipotent cells; cells that have the morphogenetic potential of a zygote. Plant development proceeds in an orderly fashion because that potential is carefully limited. When those limitations are removed, totipotent cells simply revert to the zygotic state and begin the developmental program anew.

In a somewhat simplified way, the genetic program being read in a particular cell depends on the position of that cell with respect to other cells and tissues and the inputs it receives. The position of a cell determines its interaction with its neighbors as well as its place to nutrient and hormone gradients. The importance of position has recently been demonstrated in an elegant study of young *Arabidopsis* roots (van den Berg et al., 1995). *Arabidopsis* roots lend themselves well to this kind of study because of their relatively simple structure. The root tip is comprised of *single layers* of epidermis, cortex, endodermis, and pericycle surrounding a vascular bundle, which makes it easier to track individual cells. Van den Berg and his colleagues labeled root cells with marker genes that were expressed differently in vascular and root cap cells. They then used a laser microscope to surgically remove cells in the quiescent center of the root cap. The dead cells were displaced toward the root tip and replaced by daughters of adjacent vascular cells. In their new position, the former vascular cells expressed the root cap marker. This is a strong indication that the reading of the genetic program in a cell is determined by positional information rather than cell lineage.

Practical applications of totipotency in tissue culture and micropropagation are discussed briefly in Chapter 23. The control of development is the subject of the next section and the several chapters that follow.

CONTROL OF GROWTH AND DEVELOPMENT

The orderly growth and development of complex multicellular organisms require coordination and are subject to controls at three distinct levels. *Intrinsic* controls operate at both the *intracellular* level and the *intercellular* level. Typically, intracellular controls involve changes in gene expression that influence cellular activities through altering the kinds of proteins in the cell. Intercellular controls focus on hormones and their roles in coordinating the activities of groups of cells. *Extracellular* controls are extrinsic; that is, they originate outside the organism and convey information about the environment. Although investigators tend to focus on one level of control at a time in any particular experimental system, it is important to keep in mind that interactions and overlap must inevitably occur. Thus, changes in gene expression may give rise to changes in hormone levels or changes in the sensitivity of cells to hormones. Alternatively, many hormone responses can be traced to effects on gene expression. At some point, even environmental signals must be interpreted in terms of intracellular events. Ultimately, controls at all three levels interact in various ways to determine the overall development of a plant.

GENETIC CONTROL OF DEVELOPMENT

Totipotentcy of plant cells indicates that all of the information required for the development of a complete plant is contained within the genetic complement of each cell, even highly differentiated cells. In other words, cells do not lose genes although many genes are not expressed or may be turned off as differentiation and development progress. The orderly development of a plant requires a programmed sequence of gene activations in order to produce the required gene products, that is, proteins, at the appropriate time. The cells must also have the capacity to respond to those products. As the new techniques of molecular genetics are brought to bear on problems of plant development, it becomes increasingly evident that change in gene expression is a principal factor in regulating development at the intracellular level (see Box 15.1).

Genes consist of specific sequences of nucleotides in the **deoxyribonucleic acid (DNA)** molecule (Fig. 15.2). A sequence of three nucleotides (a **codon**) codes for each amino acid. The nucleotide sequences in the gene thus determine the primary structure or sequence of amino acids of proteins, principally enzymes that determine the course of cellular metabolism. **Gene expression** refers to the synthesis of specific proteins encoded by specific genes. Not all genes are active all the time, but may be turned on or off depending on the requirements of the developmental program or in response to changing environmental conditions. Differential gene expression is thus the principal means for altering the complement of enzymes in the cell and, consequently, the course of metabolism and differentiation of the cell.

Gene expression in eukaryote organisms can be conveniently divided into five principal stages: (1) gene activation, (2) transcription, (3) RNA processing, (4) translation, and (5) protein processing (Murphy and Thompson, 1988). Any one of these stages represents a potential site for regulation.

Gene expression begins with the preparation, or activation, of the DNA that makes up the genes. Plant cells

Box 15.1
Development in a Mutant Weed

Many aspects of plant development have proven difficult to dissect using exclusively physiological and biochemical approaches. This is especially true of the complex metabolic sequences that connect a developmental event with the original signal that initiates that event. Metabolic events associated with a signal transduction pathway often represent a very small portion of all the biochemical processes within a plant and may not be discernible against this background of biochemical "noise."

Because all development can be traced back to the expression of genes, one way around this dilemma is to incorporate genetic mutations into a research program. The use of mutant genes to study metabolism and development has been a constructive approach since the pioneering work of G. Beadle and E. Tatum with *Neurospora* mutants in the 1940s. Over the years, mutants in maize, tomato, pea, and a host of other plants have provided important insights into normal development. More recently, however, recombinant DNA techniques have enabled us to ask questions about genes and developmental events in ways never before possible. The principal strategy is to identify and genetically map a mutation that modifies the physiological response of interest. It is then possible to physically isolate and clone the wild type gene and, based on its nucleotide sequence, deduce the amino acid sequence of the encoded protein that normally operates in the pathway. This provides investigators with useful information about the function of the wild type gene and a means to further probe the step that it controls. Furthermore, by studying interactions between mutant genes and multiple input signals, it is possible to dissect complex interactions between different pathways.

Arabidopsis thaliana

One organism that has come to dominate this new approach to the study of plant development is *Arabidopsis thaliana*, a member of the Brassicaceae, or mustard family. *A. thaliana* (mouse-ear cross) is a small, herbaceous weed that grows in dry fields and along roadsides throughout the temperate regions of the northern hemisphere. Most *Arabidopsis* are winter annuals. Their seeds germinate in the fall, forming an overwintering vegetative rosette. In the spring, as the days grow longer, the stem elongates and flowers.

There are many reasons why *Arabidopsis* has become such a popular experimental tool and model system for molecular genetic experiments on development. *Arabidopsis* is easily grown in the laboratory and its life cycle is complete in 5 to 6 weeks. It is also easily crossed or self-fertilized and produces prodigious numbers of seeds (up to 10,000 per plant). *Arabidopsis* has one of the smallest known plant genomes (approximately 10^8 nucleotides, or 28,000 genes), which makes complete mapping of its genome a reasonable possibility. Finally, mutants are easily induced by treating the seeds with chemical mutagens. The surviving seeds are then germinated and mutant progeny are recovered for analysis. The small size of the plant together with its rapid growth and fecundity make it easy to screen for mutants with reasonable frequency.

One must always be cautious with so-called model systems. What is learned about *Arabidopsis*, for example, may well extrapolate to all brassicas, but not necessarily to all plants. Nonetheless, *Arabidopsis* is an ideal plant for doing molecular genetic and developmental experiments. The isolation and study of mutants in *Arabidopsis* has already made significant contributions to many areas of plant development. Moreover, what is learned about signal transduction in *Arabidopsis* gives investigators important clues about what to look for in other plants.

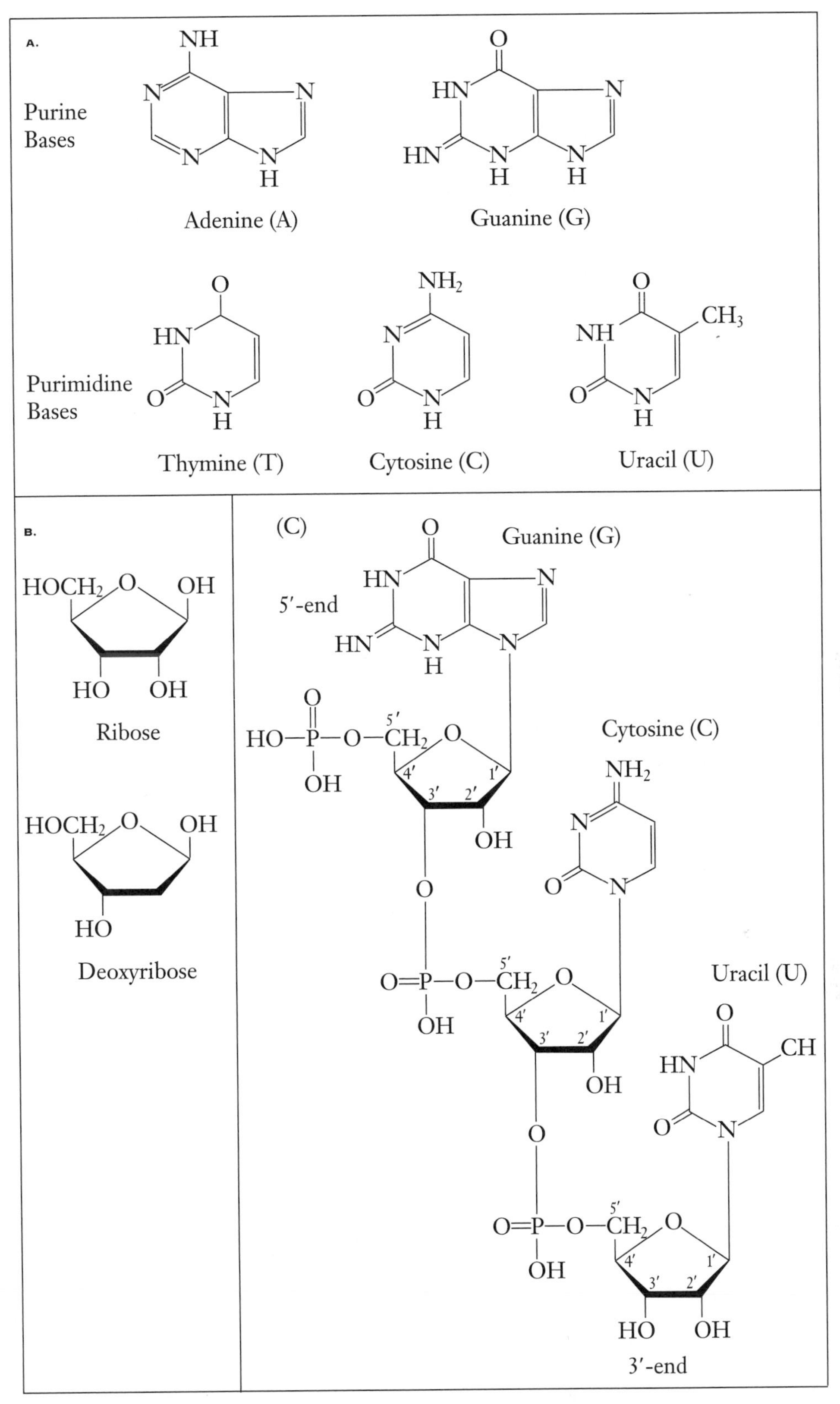

FIGURE 15.2 The building blocks that make up DNA and RNA molecules. (*A*) The purine and pyrimidine bases. A, T, G, and C are found in DNA. U substitutes for T in RNA. (*B*) Nucleotides are composed of a base, a pentose sugar, and a phosphoric acid group. The pentose sugars are ribose and deoxyribose, which lack the —OH group at the 2′ position in the pentose ring. (*C*) Ribonucleotides (RNA) or deoxyribonucleotides (DNA) are linked by an ester bond between the phosphate group attached to the 5′ carbon on one nucleotide with the alcohol on the 3′ carbon of another nucleotide. This is a triribonucleotide; reading GCU from the 5′-end. The triplet GCU in mRNA codes for the amino acid alanine.

contain extensive amounts of DNA that must be contained within a very small space. If the DNA in a single rye plant cell were fully extended, it would reach a length of five meters. Yet this DNA must be packed into a nucleus that measures only five μm in diameter (Murphy and Thompson, 1988). It is possible to do this because the double-stranded DNA that contains the nuclear genes in a eukaryotic cell is wound around a globular aggregate of special protein called **histone.** This arrangement, called **chromatin,** helps to condense the DNA and pack it into a small space. However, it is unlikely the enzymes necessary for the second step of gene expression—transcription—would have access to the DNA while it is in this tightly packed configuration. Although there is no direct evidence for DNA activation, changes in the physical structure of chromatin have been observed. It is reasonable to suspect that, as a first step in gene expression, the DNA must be removed from the histones and either extended or in some other way prepared for transcription.

The next stage in gene expression is the synthesis of a molecule of **ribonucleic acid (RNA),** whose nucleotide sequence is complementary to that of the DNA in the gene. Because the information in the DNA is *transcribed* as the nucleotide sequence in the RNA, this process is known as **transcription.** Transcription is catalyzed by a nuclear enzyme, **RNA polymerase II** (Fig. 15.3A) and, because this RNA molecule carries the information from the nucleus, where the DNA is located, to the cytosol, where protein synthesis takes place, it is called **messenger RNA (mRNA).**

Before the mRNA can be exported from the nucleus and used for translation of protein, it must be processed (Fig. 15.3B). Processing involves the addition of a molecule of 7-methyl-guanosine triphosphate (GTP) to one end of the RNA molecule (called **capping**) and a polyadenylic acid "tail" to the other.[1] In addition, eukaryotic genes contain noncoding regions of DNA called **introns,** which also appear in the complementary strand of mRNA. In the final processing step, these introns are removed and the remaining coding regions, called **exons,** are spliced together. The RNA is then ready to be transported to the cytosol through a pore in the nuclear membrane (Fig. 15.4).

Once in the cytosol, the mRNA attaches to a ribosome, where the message is **translated** into the sequence of amino acids that makes up a protein (Fig. 15.4). Amino acids are delivered to the ribosome one at a time by **amino-acyl transfer RNA (tRNA)** molecules. As each amino acid is attached to the elongating peptide chain, the ribosome advances along the mRNA to the next coded sequence. While one peptide chain is being assembled, additional ribosomes will bind with the mRNA, so that several peptide chains are in the process of being assembled at the same time. A strand of mRNA with several ribosomes attached is called a **polysome.** Most eukaryotic mRNA has a relatively long lifetime (measured in hours) and will synthesize thousands of peptide chains before it is degraded by nucleases in the cytosol. The assembly of peptide chains on polysomes is known as **translation.**

Many proteins are not immediately useful when released from the ribosome, but must undergo some form of post-translational processing before they become active. Proteins destined for membranous compart-

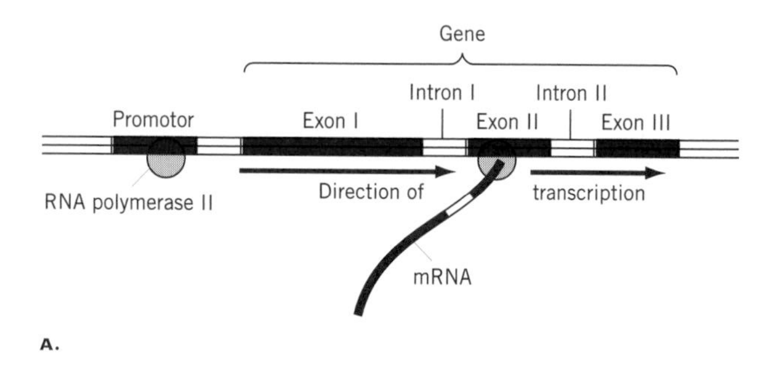

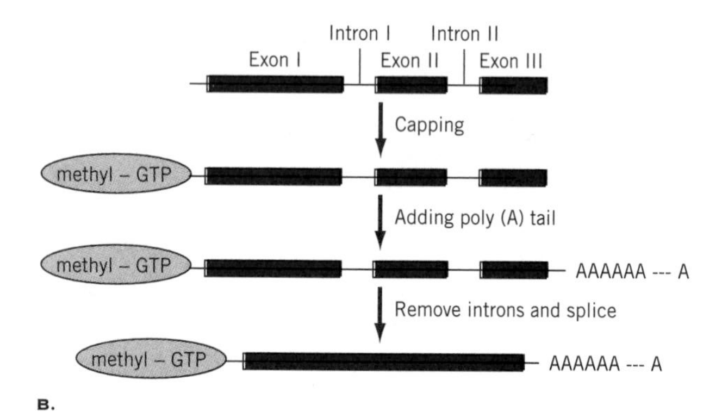

FIGURE 15.3 **(*A*) Transcription of messenger RNA (mRNA) from the gene. The eukaryote gene contains regions that code for protein (exons) alternating with regions that do not code for proteins (introns). Transcription is carried out by the enzyme RNA polymerase II, which moves along the DNA molecule as the mRNA chain elongates. (*B*) mRNA must be processed before it is exported from the nucleus. Processing involves capping the 5′-end with 7-methyl-GTP and adding a polyadenylate tail containing 100 to 200 adenylate molecules to the 3′-end. Finally, the introns are removed and the remaining exons spliced together. The transcript is now ready to be exported through a nuclear pore into the cytosol.**

[1]The cap assists in binding the mRNA to the ribosome and defines the site for starting translation. The function of the poly(A) tail is not clear. It is not required for mRNA translation, but may help to slow the rate of degradation of the mRNA molecule and, consequently, increase the total number of polypeptide chains that may be translated. The poly(A) tail may also play a role in the export of the processed mRNA from the nucleus.

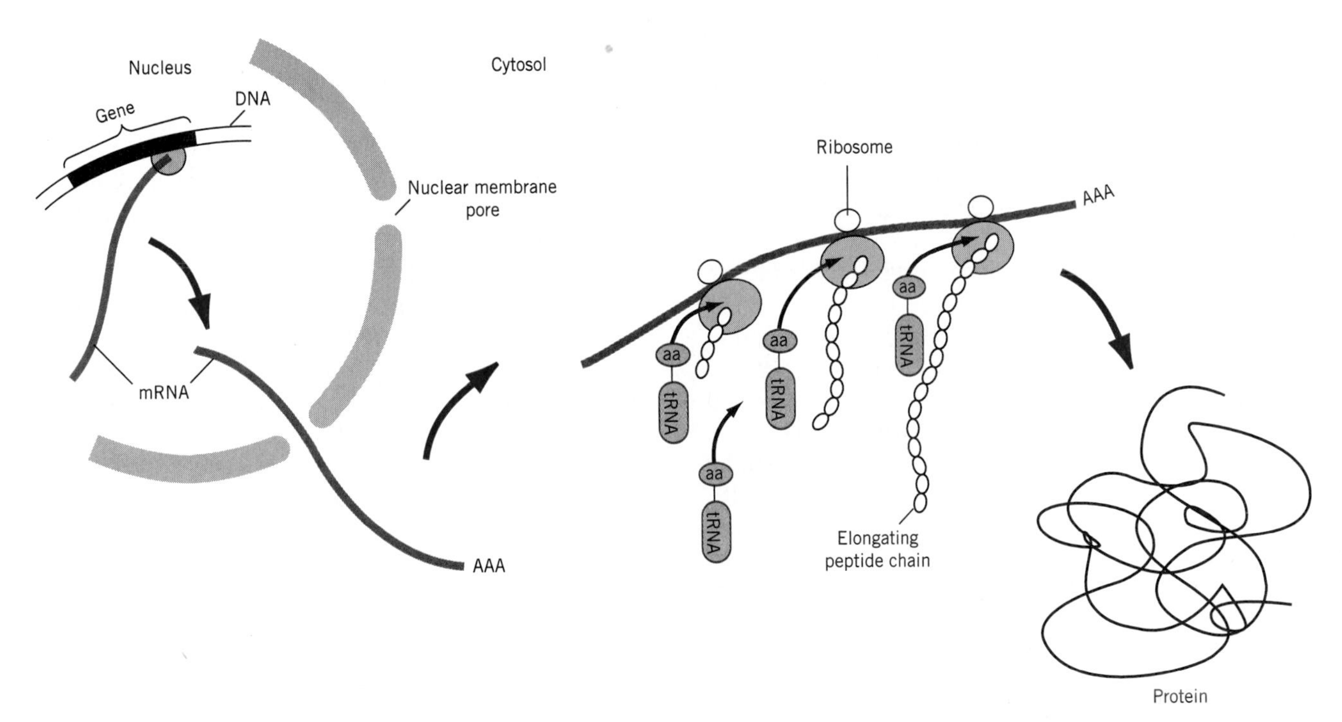

FIGURE 15.4 Translation of protein in eukaryotes. Transcribed and processed mRNA is exported from the nucleus to the cytosol where it joins with ribosomes to initiate the synthesis of protein. Amino acids delivered to the ribosomes by a transfer RNA (tRNA) are added to the elongating polypeptide chain one at a time. The completed polypeptide chain is released from the ribosome and folded into the conformation of the mature protein.

ments—the chloroplast, for example—contain hydrophobic amino acid sequences, called **leader sequences** that facilitate transport of the protein through the membrane. Once the protein is in position, the leader sequence is cleaved off and degraded. Other proteins remain inactive until the peptide chain is shortened or modified by the action of proteolytic enzymes, while still others require the addition of carbohydrate groups. Many enzyme proteins must be activated by phosphorylation, the transfer of a phosphate group from ATP through the action of a protein kinase enzyme.

Many steps are required for the successful expression of a gene and each step represents a *potential* point at which the expression of the gene may be regulated during development. There is evidence for differential transcription as well as control of translation and post-translational processing throughout plant development. Specific examples will be illustrated in the chapters that follow.

HORMONAL REGULATION OF DEVELOPMENT

Interaction between cells is an exceedingly complex affair. A mature pith cell of a tobacco stem, for example, no longer divides or grows and is differentiated primarily as a storage cell. However, freed from the constraints imposed by its position in the stem and cultured on an artificial medium, this same cell will be stimulated to divide and produce a shapeless mass of undifferentiated cells. With an appropriate nutrient balance and the right kinds of hormones in the medium, roots and shoots will form and, eventually, a new, fully competent plant will regenerate (Fig. 15.1). Although there is some evidence for so-called "position effects" and nutrient gradients, much of this kind of behavior can be interpreted in terms of hormones, chemical messengers that carry information between cells. Several classes of hormones are known, which promote or inhibit various developmental responses, either singly or in combination. The details of hormones and their actions in regulating plant development will be discussed in the following two chapters.

ENVIRONMENTAL REGULATION OF DEVELOPMENT

A variety of external or environmental stimuli can at various times be involved in regulating plant development. Most environmental stimuli are physical parameters. Light, temperature, and gravity have the most obvious and dramatic impact. Other parameters such as magnetic field, sound, and wind (a mechanical stimulus) may have more subtle effects, but these have been difficult to establish experimentally. Other environmental factors such as soil moisture, humidity, and nutrition may also influence development in some cases. More recently it has become evident that a variety of air and water pollutants represent a significant environmental challenge to plants and may significantly modify developmental patterns.

Because environmental signals originate outside the plant, plants must have some means of perceiving the signal and converting, or transducing, the information into some permanent metabolic or biochemical change. Understanding the nature of signal perception is one of the first steps in understanding the chain of events that leads to the ultimate response. These processes will be the focus of several of the chapters that follow. It is also becoming increasingly evident that most, if not all, environmental stimuli act at least in part through modifying gene expression or hormonal activities.

A SURVEY OF PLANT DEVELOPMENT

Plant development is a cyclical process. If any cycle can be said to have a beginning, in plants the beginning would be germination of the **seed.** The seed is a convenient place to begin because seeds are **quiescent,** or resting, organs that represent a normal hiatus in the life cycle. Seeds are severely dehydrated—their water content is normally about 5 percent or less—and metabolic reactions take place so slowly they are scarcely detectable. Seeds thus appear to be in a state of suspended animation, capable of surviving adverse conditions for long periods of time without growing.

SEED STRUCTURE AND DEVELOPMENT

Seeds develop in the ovary of the flower. It is here that the rudimentary plant, or **embryo,** takes form. An embryo consists of an **embryonic axis** bearing one or more **cotyledons,** or seedling leaves (see Chap. 1, Fig. 1.13). At one end of the embryonic axis is the **plumule,** which will form the **shoot** (stems and leaves). At the other end is the **radicle,** which gives rise to the roots. Seeds of **monocotyledonous** plants (or **monocots**) have a single cotyledon, called the **scutellum.** The function of the scutellum is primarily to mobilize and absorb nutrients during germination. Seeds of **dicotyledonous** plants (or **dicots**) have two cotyledons that supply the germinating seed with nutrients.

While still in the ovary, the developing embryo is surrounded by a nutritive tissue, or **endosperm.** In some cases, such as the cereal grains and most other monocots, the endosperm is retained until maturity and may comprise the bulk of the seed. These are called **endospermic seeds.** In most dicots, the endosperm is consumed by the embryo as it develops until, at maturity, virtually the entire seed is occupied by the embryo itself (**nonendospermic seeds**). The endosperm of mature endospermic seeds consists of nonliving cells filled with starch along with protein and some small amounts of lipid. In some monocot seeds, most notably the cereal grains such as *Triticum* (wheat), *Hordeum* (barley), and *Avena* (oats), the endosperm is surrounded by one or more layers of living cells, called the **aleurone.** The aleurone consists of living cells distinguished by the presence of numerous protein bodies.

Endospermic dicot seeds, such as *Ricinus communis* (castor bean), have retained a significant amount of endosperm and the cotyledons are thin, leaflike structures. In nonendospermic dicot seeds, such as *Pisum* (pea) and *Phaseolus* (bean), the cotyledons enlarge and may occupy as much as 90 percent of the seed volume at maturity. The endosperm and cotyledons characteristically contain large quantities of stored carbon (in the form of carbohydrates, lipids, and protein), mineral elements, and hormones that support the growth and development of the seedling until it can establish itself as a photosynthetically competent plant.

Surrounding the seed is a hard coat called the seed coat or **testa.** The testa is derived from maternal tissues (the **integuments**), which surrounded the seed during its development in the ovary. Comprised of heavy-walled cells and covered with a thick, waxy cuticle, the testa often presents a significant barrier to the uptake of both water and oxygen by the seed. The testa may therefore have a significant role in regulating the hydration of the seed and the resumption of active metabolism that is required for germination.

SEED GERMINATION

When conditions are appropriate, the embryo will renew its growth and the seed germinates. The initial step in germination of seeds is the uptake of water and rehydration of the seed tissues by the process of **imbibition.** Like osmosis, imbibition involves the movement of water down a water potential gradient (Chap. 2). Imbibition differs from osmosis, however, in that it does not require the presence of a differentially permeable membrane and is driven primarily by surface-acting or **matric forces.** In other words, imbibition involves the chemical and electrostatic attraction of water to cell walls, proteins, and other hydrophillic cellular materials. The contribution of matric forces to water potential can be defined by adding the **matric potential** term (Ψ_M) to Equation 2.10:

$$\Psi = \Psi_P + \Psi_S + \Psi_M \quad (15.1)$$

Matric potential, like osmotic potential, is always negative. Imbibition causes a swelling of the imbibing material, which may generate substantial pressure (called **imbibition pressure**). Imbibition pressure developed by a germinating seed will cause the testa to rupture, thus permitting the embryo to emerge.

Imbibition of water is followed by a general activation of seed metabolism. The specific biochemical events that trigger germination are unknown, but increased respiration is one of the earliest to be detected in moist seeds. This is followed closely by (a) the release of hydrolytic enzymes that digest and mobilize the stored reserves, and (b) renewed cell division and cell enlargement in the embryonic axis. Detailed respiratory pathways have been studied thoroughly in only a few species of seeds, but it is believed that glycolysis and the citric acid cycle are active to varying degrees in most, if not all, seeds. These pathways (discussed earlier in Chap. 12) produce the carbon skeletons and ATP required to support growth and development of the embryo. The pentose phosphate pathway (Chap. 12) is also important in seeds as it produces the reducing potential in the form (NADPH) required for the reductive synthesis of fatty acids and other essential cellular constituents. The pentose phosphate pathway also generates intermediates in the synthesis of aromatic compounds and perhaps nucleic acids. Seeds that store carbon reserves principally in the form of fats and oils will carry out the synthesis of hexose sugars via gluconeogenesis (Chap. 12).

The mobilization of stored carbon in seeds has been studied most extensively in cereals. The endosperm of cereals has long served as a principal source of nutrition for man and domesticated animals, as well as a basic feedstock in the brewing industries. These needs have provided a strong incentive for research into the mobilization of starch reserves in cereal grains. Pathways for the breakdown of starch, starting with the α- and β-amylases, have been described earlier in Chapter 12. The role of these and other enzymes in the mobilization of starch in cereal grains, and how those enzymes are regulated by hormones, will be discussed further in Chapters 16 and 17. In nonendospermic seed such as the legumes (peas, beans), the initial stages of radicle elongation appear to depend on reserves stored in the tissues of the radicle itself. Later, carbon reserves are mobilized from the cotyledons and transported to the elongating axis.

In most species, germination is culminated when the radicle emerges from the seed coat. Radicle emergence occurs through a combination of cell enlargement within the radicle itself and imbibition pressures developed within the seed. Rupture of the seed coat and protrusion of the radicle allows it to make contact with water and nutrient salts required to support further growth of the young seedling.

SHOOT DEVELOPMENT

Emergence of the radicle is followed by elongation of the shoot axis. In some dicot seedlings, the **hypocotyl** (*hypo*, below the cotyledons) is the first to elongate, pulling the cotyledons and the enclosed first foliage leaves up through the soil. This is known as **epigeal** germination (Fig. 15.5A). In other dicots and most monocots, the hypocotyl remains short and compact and the cotyledons remain underground. Instead, the **epicotyl** (*epi*, above the cotyledons) undergoes extensive elongation to bring the first leaves above the soil. This pattern of germination is known as **hypogeal** (Fig. 15.5B).

Once the embryonic tissues have elongated to establish the seedling above the soil, new cells and organs are contributed by regions of active cell division and enlargement, called **meristems.** The **shoot apical meristem,** located at the apex of the stem, is a dome-shaped structure usually surrounded by clasping **leaf primordia** (Fig. 15.6A). The shoot apical meristem contains a small number of dividing cells that give rise to all of the other cells and tissues in the primary shoot, including stem, leaves, branches, and flowers. With each cell division, one daughter cell is left behind to elongate and differentiate while the other daughter cell remains in the meristem to undergo further cell divisions. The shoot apical meristem is generally divided into two zones: an outer **tunica** overlying a central **corpus** (Fig. 15.6B). The tunica may be composed of one or several cell layers and is distinguished by the fact that the cells regularly divide in a plane perpendicular, or **anticlinal,** to the surface of the meristem. The tunica thus gives rise to the outer layers of the plant body, or **epidermis.** The corpus, on the other hand, divides more randomly in both anticlinal and parallel, or **periclinal,** planes, thereby giving rise to the bulk of the internal tissues of the stem and leaves.

Elongation of the shoot axis proceeds through a combination of cell division and enlargement of the cells laid down by the meristem. The rate and extent of elongation is subject to a variety of controls, including nutrition, hormones, and environmental factors such as light and temperature. The final height of a shoot is determined by the rate and extent to which **internodes**—the sections of stem between leaf nodes—elongate. In some plants, such as pea (*Pisum sativum*), elongation occurs primarily near the apical end of the youngest internode (Fig. 15.7) (see Torrey, 1967). The older internodes effectively complete their elongation before the next internode begins. In other plants, elongation may be spread through several internodes, which elongate and mature more or less simultaneously. Still others exhibit changing rates of elongation with successive internodes, usually increasing toward the apex.

In some plants, internodes fail to elongate, thus giving rise to the rosette habit in which all the leaves appear to originate from more or less the same point on the

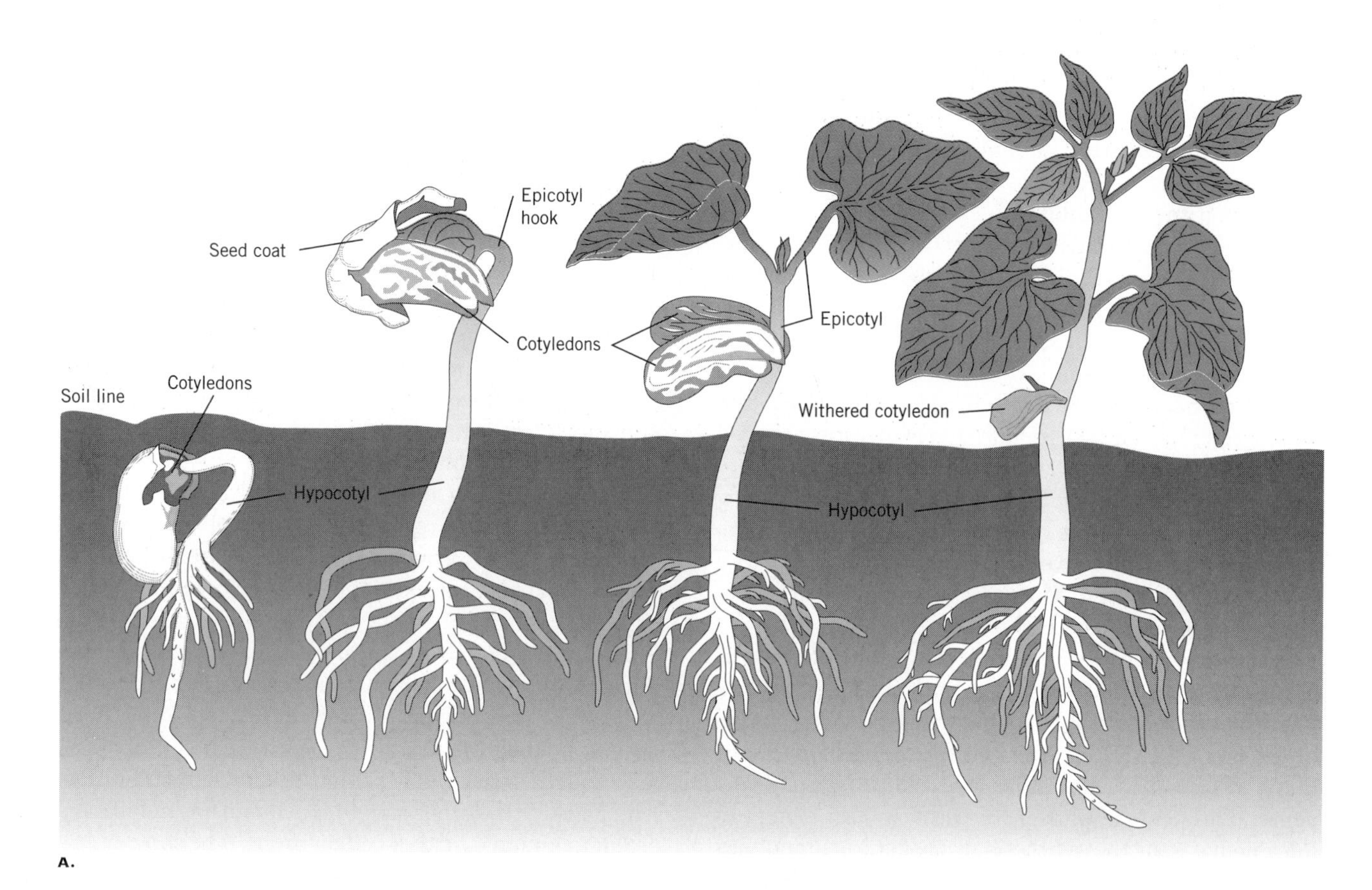

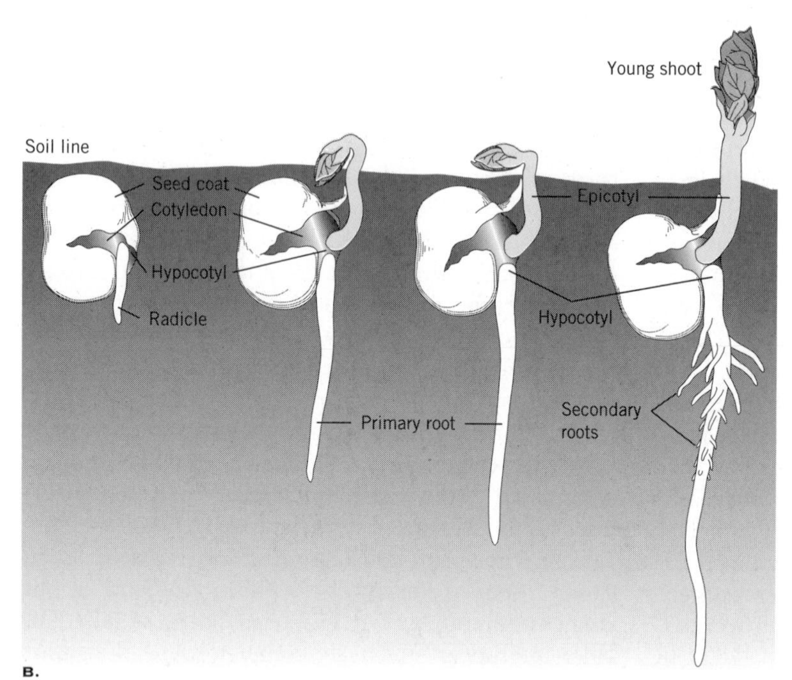

FIGURE 15.5 Germination and seedling development. (*A*) Stages in the germination of bean (*Phaseolus vulgaris*), an example of epigeal germination. (*B*) Hypogeal germination of pea (*Pisum sativum*).

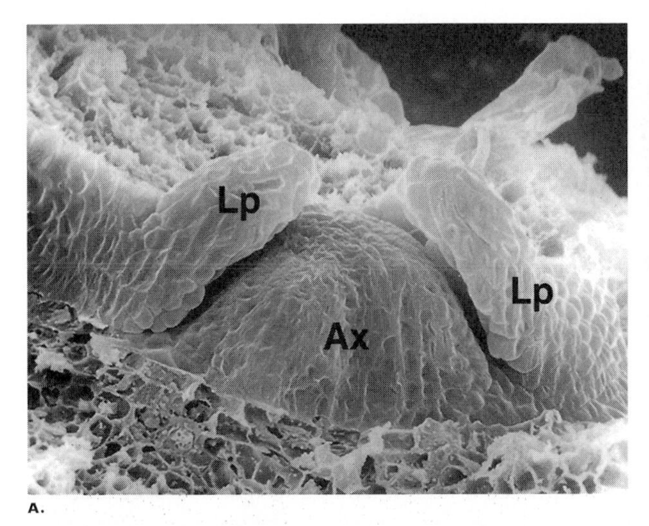

A.

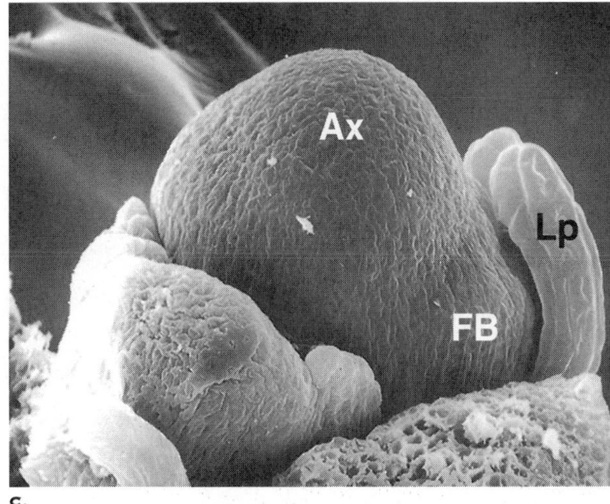

C.

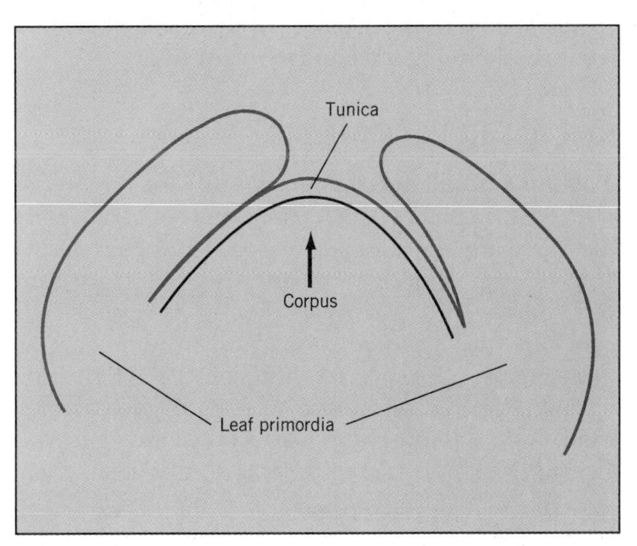

B.

FIGURE 15.6 Meristems. (*A*) Scanning electron micrograph of a vegetative meristem of *Brassica napus*. (*B*) Schematic cross-section of meristem in A, showing position of the tunica, corpus, and leaf primordia. (*C*) Scanning electron micrograph of a flowering meristem of *Brassica napus*. Ax, apex; Lp, leaf primorium; FB floral bud. (*A* and *C* from V. K. Sawhney, P. L. Polowick, *American Journal of Botany* 73:254–263, 1986. With permission of the *American Journal of Botany*. Original photographs kindly provided by V. K. Sawhney and P. L. Polowick.)

FIGURE 15.7 Internode elongation in broad bean *Vicia faba* over 48 hours. The initial spacing between marks was 2mm. This internode has elongated more or less uniformly over its entire length.

stem (Fig. 15.8). The rosette habit is common in biennial plants (those which flower in the second year) and root crops such as carrot (*Daucus carota*) and radish (*Raphanus sativus*) before they reach the flowering stage. Failure of internode elongation is commonly related to low levels of the plant hormone, gibberellin, since application of the hormone usually stimulates internode elongation in rosette plants (Chaps. 16, 17).

Leaf primordia arise as small protrusions on the flanks of the meristem, due to increased cell divisions in both the tunica and corpus. Leaf initiation occurs in a precisely ordered sequence that is species specific and that determines the arrangement, or **phyllotaxy,** of leaves on the mature stem (Fig. 15.9). The pattern is in most cases so predictable that experimenters can surgically damage cells in the region of the meristem where the next primordium would form and thus prevent its

FIGURE 15.8 A rosette leaf arrangement in beach thistle (*Cirsium pitcheri*). The rosette habit results from a failure of internode elongation.

formation. Leaves may be arranged oppositely, in which case each pair of primordia arise at 90° to the previous pair, or in spiral patterns, in which primordia arise at about 137° or 180° to the previous one. What determines such a precise phyllotactic pattern is a matter of considerable speculation. One of the more widely considered theories holds that each leaf primordium occupies a certain amount of space that places physical constraints on the meristem. Subsequent primordia then arise in the next available space not yet occupied. Once the leaf primordium is established as a fingerlike projection, leaf expansion continues through the activity of marginal meristems. Predominantly anticlinal divisions give rise to the normal dorsi-ventral, bladelike leaf morphology.

As the leaf primordium develops, small groups of meristematic cells remain trapped in the juncture of the leaf with the stem (called the **axil**). These cells give rise to an axillary bud, or future branch, which maintains its own apical meristem.

The cells and tissues derived from the apical meristem comprise the **primary plant body** (see Chap. 1). Secondary tissues arise from the activity of secondary meristems, such as the vascular cambium (Fig. 15.10), and account primarily for increase in girth. The vascular cambium is a zone of cells lying between the xylem and phloem tissues in the stem, which has retained the capacity for cell division. By adding new xylem cells to the inside and new phloem cells to the outside, the vascular cambium adds to the diameter of the stem. These new tissues constitute the **secondary plant body.**

ROOT DEVELOPMENT

The **root apical meristem** appears structurally less complex than the shoot apical meristem, in part because the apical meristem is not responsible for producing branch roots. Branch roots arise some distance back from the root tip. The tip of the root is covered by a **root cap,** which provides mechanical protection for the meristem as the root grows through the abrasive soil medium. The root cap also secretes polysaccharides, which form a mucilaginous matrix called **mucigel.** Mucigel lubricates the root tip as it moves through the soil. The root cap along with its coating of mucigel is also involved in perception of gravity by roots (Chap. 19).

A.

B.

FIGURE 15.9 Spiral and opposite phyllotaxy. Both plants are viewed from above to emphasize the phyllotactic pattern. (*A*) Cocklebur (*Xanthium strumarium*), an example of spiral phyllotaxy with a spacing of 137° between successive leaves. (*B*) *Coleus,* an example of opposite phyllotaxy. Leaves arise in opposite pairs with each pair rotated 90° to the previous pair.

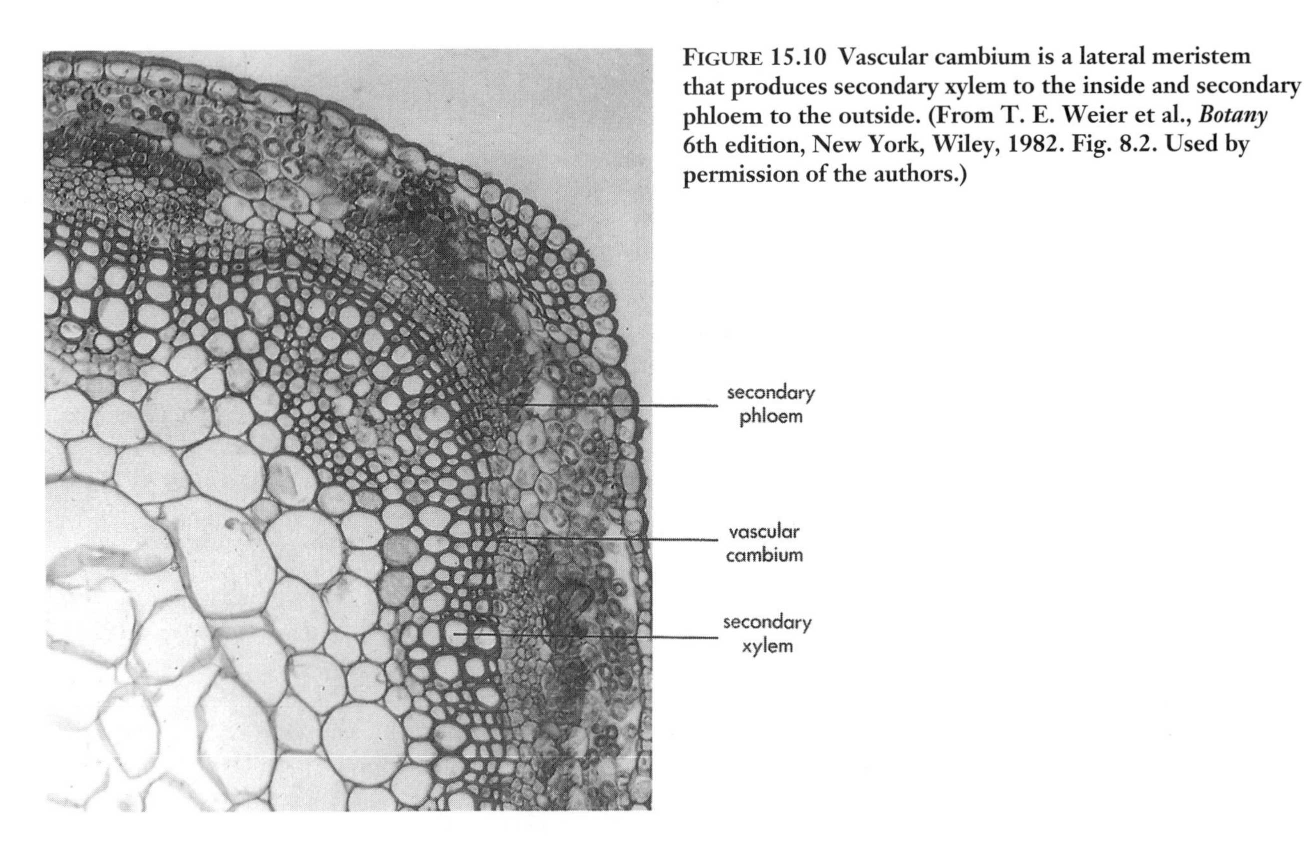

FIGURE 15.10 Vascular cambium is a lateral meristem that produces secondary xylem to the inside and secondary phloem to the outside. (From T. E. Weier et al., *Botany* 6th edition, New York, Wiley, 1982. Fig. 8.2. Used by permission of the authors.)

The root apex has been a favorite for the study of meristematic activity because of its relative simplicity and because mitotic figures (that is, nuclei in stages of chromosome replication and division) are easily rendered visible under the light microscope. Moreover, radioactive thymidine is readily incorporated into the DNA of dividing root tip cells, which allows regions of active DNA synthesis to be detected by autoradiography. The root apical meristem lies immediately below the root cap (Fig. 15.11). In the center of the meristem is a region of slowly dividing cells that incorporate little, if any, radioactive thymidine. This region is called the **quiescent zone.** Cell divisions responsible for new tissues in the elongation root and regeneration of the root cap take place around the periphery of the quiescent zone. Differentiation of vascular tissues begins soon after cells are produced by the root apical meristem. Differentiated xylem elements can be detected within the first 100 μm or so from the region of dividing cells and mature sieve tubes within the first 400 to 500 μm (Fig. 15.12).

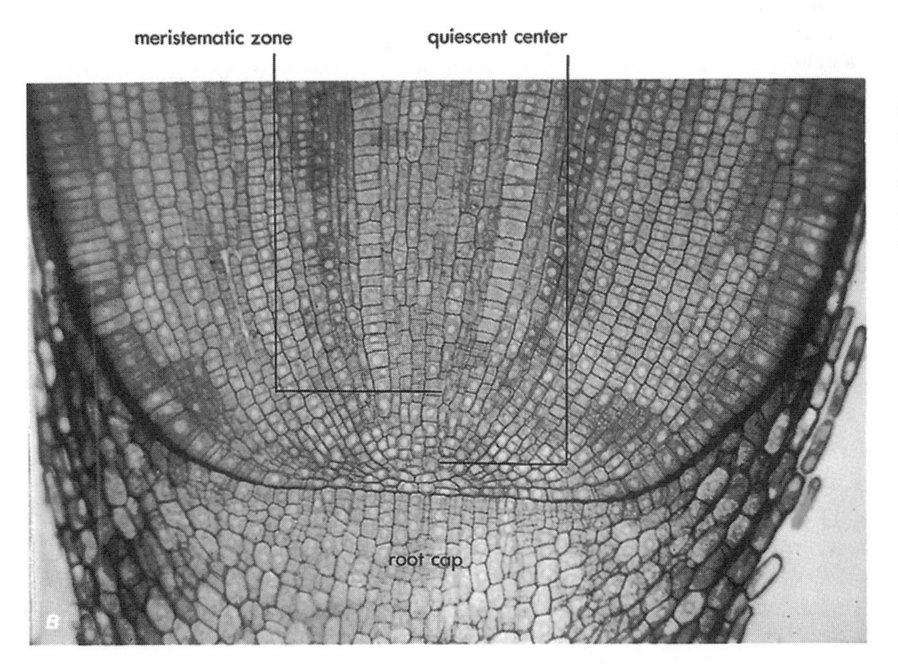

FIGURE 15.11 Long section through a root tip of onion (*Allium* sp.) showing the meristematic region and quiescent center. (From T. E. Weier et al., *Botany* 6th edition, New York, Wiley, 1982. Fig. 9.8. Used by permission of the authors.)

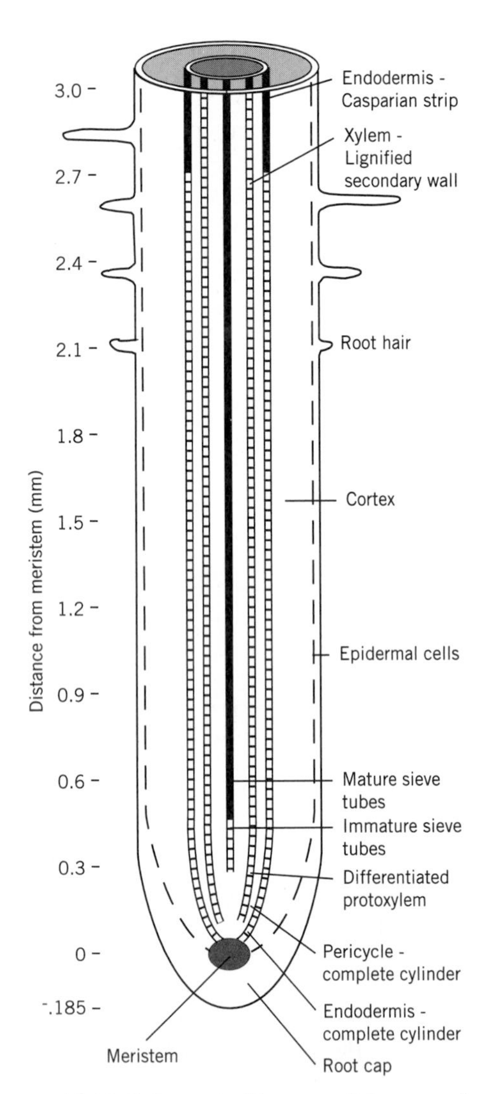

FIGURE 15.12 Schematic diagram of the root tip of a white mustard (*Sinapus alba*) seedling, showing the relative positions of meristems and differentiating tissues. Note that the differentiating phloem and xylem tissues alternate around the periphery of a cylinder. (Redrawn from R. L. Peterson, *Canadian Journal of Botany* 45:319–331, 1967. Reprinted by permission of the National Research Council.)

Lateral or branch root primordia originate in the **pericycle,** a ring of meristematic cells that lies just inside the endodermis of the primary root (Fig. 15.12; see also Fig. 1.15). The growing lateral root works its way through the cortex, either by mechanically forcing its way through or by secreting enzymes that digest the cortical cell walls. Lateral root primordia arise in close proximity to the newly differentiated xylem tissue, which allows vascular elements developing behind the growing tip of the secondary root to maintain connections with the xylem and phloem in the stele of the primary root.

FLOWER EVOCATION AND DEVELOPMENT

Certainly one of the most dramatic and still enigmatic developmental events in the history of a flowering plant is the conversion of the shoot apical meristem from production of vegetative structures to the production of reproductive structures. Flowering is conveniently divided into three stages: induction, evocation, and development. **Induction** refers to events that signal the plant to alter its developmental program. As a consequence, the stem apical meristem reorganizes to produce floral primordia rather than leaf primordia. Induction need not occur at the apex where the flowers will eventually appear but may occur elsewhere in the plant. One well-documented example is the flowering response to the timing of light and dark periods (photoperiod, Chap. 20). Here an external signal, actually the length of a dark period, is perceived by the leaf. In response to this signal, the leaf undergoes a metabolic change resulting in the formation of a floral stimulus. The stimulus, presumably chemical in nature, is then transmitted to the apex where it initiates floral evocation. In spite of the fact that photoperiodic induction and other physiological control processes have been studied extensively over the past 70 years, very little is known of the cascade of events that must certainly be involved linking signal perception to transmission of the floral stimulus from the leaf to the apex (McDaniel, 1991).

Floral evocation refers to events at the shoot apex following induction; events that *commit* the meristem to formation of flower primordia in place of leaves. Floral evocation remains even more of a black box than induction. We know virtually nothing about what causes vegetative leaf primordia to form and we know even less about the cellular and molecular events involved in floral evocation. Although there have been numerous efforts to describe molecular, anatomical, and morphological events associated with evocation, a complete and systematic description of the changes accompanying the conversion of a vegetative apex to a floral apex in a single species has yet to be attempted. Traditional studies have shown that increases in respiratory substrate and respiration rate as well as changes in RNA and protein synthesis in the meristem are all associated with evocation. These events point a finger toward changes in gene expression. Genetic changes appear to precede a general stimulation of cell division in the meristem along with changes in the pattern of cell divisions that give rise to floral primordia. With the advent of gene cloning and other modern molecular genetic techniques, several investigators have confirmed changes in the pattern of gene expression in various regions of the meristem during the transition from a vegetative to floral apex (Meeks-Wagner, 1993). Further application of these

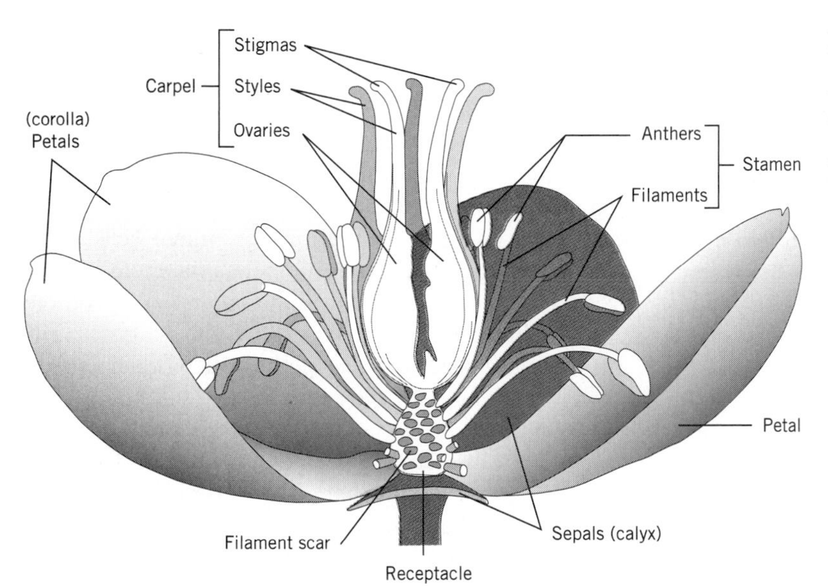

FIGURE 15.13 The anatomy of a flower (*Helleborus*; Christmas rose). (From T. E. Weier et al., *Botany* 6th edition, New York, Wiley, 1982. Fig. 15.1. Used by permission of the authors.)

tools to the study of floral evocation will undoubtedly lead to significant advances in the future.

There is evidence that, in its early stages, floral evocation can be reversed. However, the developmental program eventually reaches a point of no return, beyond which the meristem is irreversibly set upon a path of **floral development.** At this point—when the developmental program continues even though conditions that may cause reversion have been removed—floral development is said to be **determined.** Floral primordia will then arise on the flanks of the apical meristem in place of leaves (Fig. 15.6C). The primordia continue to enlarge, giving rise to primordia for separate floral structures and finally a mature flower.

FLOWER AND FRUIT DEVELOPMENT

The flower is basically a stem with very shortened internodes supporting a series of appendages that are essentially modified leaves. Although the number of variations is almost infinite, the basic structural plan of a flower is relatively simple (Fig 15.13). All of the flower parts illustrated in Figure 15.13 are of the diploid (2N, or sporophyte) generation. Meiosis occurs at two locations within the flower. Within the **ovule,** a single **megaspore mother cell** undergoes meiotic divisions to generate one haploid (N) **egg nucleus,** two **polar nuclei,** and five other haploid nuclei. These eight nuclei represent the female gametophyte generation. In the anther, a **microspore mother** cell undergoes meiotic divisions to produce four **pollen grains.**

When a pollen grain lands on a suitable surface (called **pollination**), such as the stigma of a compatible plant, it will germinate, producing a **pollen tube** (Fig. 15.14). The nucleus of the germinating pollen grain divides to produce a vegetative **tube nucleus** and a **generative nucleus,** which together represent the male gametophyte generation. The pollen tube continues to elongate down through the style, until it eventually reaches the ovule. The control and direction of pollen tube growth is an interesting problem that has yet to be satisfactorily resolved. The interior of the style no doubt provides required nutritional support for tube elongation, but this alone is not sufficient to explain the unerring progress toward the ovule (Owens, 1992). Che-

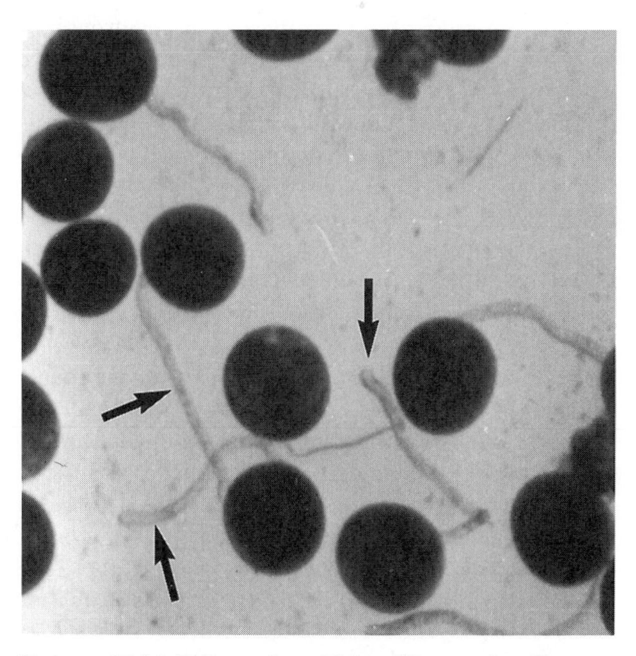

FIGURE 15.14 Pollen tubes. Maize (*Zea mays*) pollen grains were germinated on an agar surface. Elongating pollen tubes are indicated by the arrows. (Photograph courtesy of D. B. Walden.)

motropism could be involved, perhaps in response to a calcium gradient, although evidence for this is not altogether convincing.

By the time the pollen tube reaches the ovule, the generative nucleus has given rise to two sperm nuclei. The two nuclei are delivered into the embryo sac where one fuses with the egg nucleus to produce a diploid **zygote** and the other fuses with the two polar nuclei. The zygote will develop into the embryo while the triploid tissue resulting from fusion of the sperm and polar nuclei gives rise to the endosperm.

The **fruit** is the final stage in the growth of the reproductive organ. A fruit is generally considered to be a mature or ripened ovary, although in some plants other tissues may become involved. In its simplest form, such as peas or beans, the fruit consists of the seed enclosed within an enlarged ovary (the pod). The tomato is another example of an enlarged, fleshy ovary. In corn, the fruit consists of a single seed with its seed coats fused with the dry ovary wall. The strawberry consists of individual fruits (called achenes) borne on the surface of an enlarged, fleshy receptacle. In many cases, it is clear that the fruit undergoes considerable cell division and cell enlargement as well as significant qualitative changes. These changes are due largely to changes in hormone contents.

HOW DO CELLS GROW?

Because the cell is the basic unit of life, the growth of an organism reflects the growth of its individual cells. Therefore, in order to understand how organisms grow and develop, it is first necessary to understand how cells grow. We have previously defined growth as an irreversible increase in volume. Since most of the volume of any cell is water, it follows that for a cell to increase its volume it must take up water. Conversely, if a cell cannot take up water, it will not grow. This dependence of cell growth on water uptake can be demonstrated by bathing cells in an *isotonic* solution of mannitol or similar solute. An isotonic solution will prevent any net uptake of water by cells (Chap. 2). Even though all other requirements for growth may have been met and the cell is fully capable of synthesizing additional protoplasm, cells in isotonic solutions will not enlarge. If the cells are subsequently transferred to a hypotonic solution or pure water, they will take up water and enlarge quite rapidly. From this we can conclude that *the driving force for cell enlargement is water uptake.*

Recall that cells take up water by the process of osmosis (Chap. 2). The high solute concentration of vacuolar sap decreases the water potential in that region so that there is a tendency for water to diffuse into the cell. With no means to compensate, a cell surrounded by water might continue to swell indefinitely or at least until internal pressures exceeded the tensile strength of the membrane. The consequences of such a situation is vividly demonstrated when mammalian red blood cells are placed in water. The cells quickly swell until the plasma membrane bursts, releasing their contents (mostly hemoglobin) into the medium. Most animal cells avoid such osmotic disaster by using metabolic energy to excrete either solute or water and thus maintain a favorable pressure balance. Plant cells have found a different solution—they surround the plasma membrane with a strong, more or less rigid cell wall.

In our earlier discussion of cellular water relations (Chap. 2), we noted that the water potential of a cell is regulated primarily by changes in turgor, the pressure generated by the expanding protoplast against the cell wall. Turgor pressures developed in cells can be quite large—as much as 2 MPa. In order to resist such pressures, cell walls must be very strong indeed. Besides being very strong, cell walls are also rigid. It is this property that causes cells to resist deformation and that allows them to maintain nonspherical shapes. These two properties—the strength and rigidity of the cell wall—impose critical restrictions on the capacity of plant cells to grow. We can then assume that *in order for a cell to increase in size, the strength and rigidity of the cell wall must be modified.* Our question of how cells grow now becomes one of *how the strength and rigidity of the cell wall can be modified to permit water uptake and cell enlargement.* If we can answer this question, we will have a reasonably good understanding of how plant cells grow.

The rate of water uptake by an enlarging cell may be measured by the increase in cell volume. This relationship is described by the fundamental osmotic equation,

$$dV/dt = L\Delta\Psi \quad (15.2)$$

where dV/dt is the change in cell volume over time. L, the hydraulic conductance, is a property of the cell membrane. Hydraulic conductance is a measure of the ability of water to cross the membrane. $\Delta\Psi$ is the water potential gradient between the cell and its surroundings.

We can further simplify our argument by assuming that the cell's surroundings (in this case, water) are at atmospheric pressure. $\Delta\Psi$ can then be defined in terms of the difference in osmotic pressure between the cell and its surroundings and the turgor pressure of the cell:

$$\Delta\Psi = \Delta\pi - P \quad (15.3)$$

Equation 15.3 may then be substituted in Equation 15.2 to give

$$dV/dt = L(\Delta\pi - P) \quad (15.4)$$

Equation 15.4 tells us that the rate of water uptake by a cell, measured by an increase in cell volume, is a function of the water conductance by the plasma membrane,[2] the osmotic pressure gradient between the cell and its surroundings and the turgor pressure of the cell.

When a cell is not growing, there is, by definition, no increase in cell volume and, consequently, no net water uptake. Therefore $\Delta\Psi$ must become zero (Eq. 15.2). This happens because, as the cell takes up water in response to $\Delta\Psi$, the pressure of the protoplast against the rigid cell wall quickly generates a turgor pressure, P, which balances the osmotic potential of the cell. As P increases, the water potential of the cell increases (becomes less negative), and $\Delta\Psi$ approaches zero. *Clearly, in order for a cell to grow, $\Delta\Psi$ must not be allowed to reach zero.*

Since cell expansion requires an increase in volume, it follows that cell expansion also requires an increase in the surface area of the surrounding wall, or **wall extension.** Investigators know that wall extension is driven by turgor pressure—this has been demonstrated empirically. For example, when turgor pressure is experimentally reduced, the rate of cell expansion also declines. Furthermore, wall extension and growth do not occur in cells at very low or zero turgor pressure, even though the cells remain metabolically active and appropriate growth stimuli are present. James Lockhart summarized the interdependence of wall extension and turgor pressure in the following equation (Lockhart, 1965):

$$dV/dt = m\,(P - Y) \qquad (15.5)$$

where Y is the minimum turgor pressure necessary for growth, called the **yield threshold.** The term m is a proportionality constant between growth rate and turgor pressure in excess of the yield threshold. It is known as **wall extensibility** or, simply, **extensibility.** Extensibility is a quantitative measure of the capacity of the wall to irreversibly increase its surface area. Wall extension is evidently under metabolic control of the cell, since it is prevented by metabolic inhibitors such as cyanide or 2,4-dinitrophenol (DNP). Any action of the cell that leads to an increase in wall extensibility is called **wall loosening.**

From the above discussion, it is apparent that the growing cell is faced with conflicting roles of turgor pressure. On the one hand, turgor pressure promotes irreversible wall extension (Eq. 15.5). At the same time turgor pressure opposes the continued uptake of water, which is the driving force for cell expansion (Eq. 15.3). How does the cell resolve this conflict? The answer to this question appears to have been provided by the work of Daniel Cosgrove (Cosgrove, 1987). Cosgrove argues that **stress relaxation** in the wall is central to the process of cell growth. Recall from Chapter 1 that the cell wall is composed of intertwined and cross-linked chains of cellulose and other polysaccharides. Turgor pressure develops because these entanglements resist deformation due to the expanding protoplast. The force of the expanding protoplast pushing against the wall thus generates stress (defined as force/unit area) within the wall. Cell growth appears to be initiated when these stresses are relaxed by wall-loosening events that cause load-bearing elements (perhaps cross-links and entanglements between polymers) in the wall to yield (Fig. 15.15). Stress relaxation, in this sense, would result in a *simultaneous and proportionate reduction in turgor pressure.* A reduction in turgor pressure leads to a decrease in the water potential of the cell ($\Delta\Psi$ becomes more negative), followed by the passive uptake of water. The influx of water in turn increases cell volume, extends the cell wall, and tends to restore both wall stress and turgor pressure. The process of cell growth is thus seen as a continuous adjustment of turgor pressure through stress relaxation in order to balance its conflicting roles in water uptake and cell wall extension.

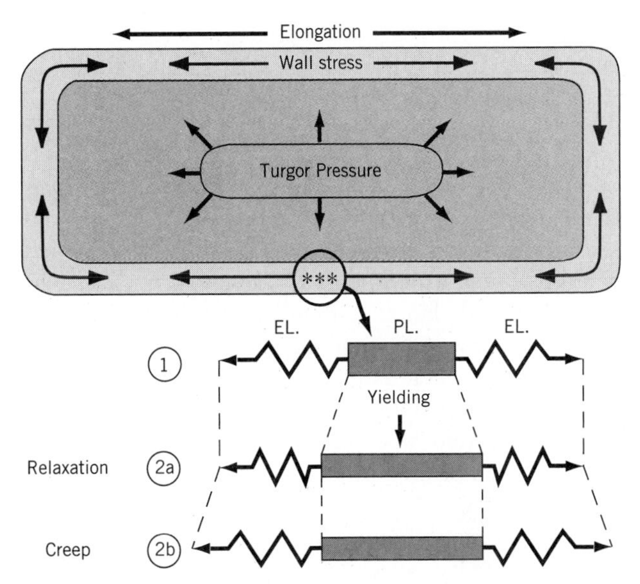

FIGURE 15.15 A model for stress relaxation in the wall of a growing cell. Both elastic (EL) and plastic (PL) components of the wall bear the stress of an expanding protoplast. In 1, turgor causes stress in the wall and extension of the elastic component, represented as springs. In 2a, yielding of the plastic components allows relaxation in the elastic component, illustrated by contraction of the springs. In 2b, turgor reestablishes wall stress and the wall expands to the same extent that the plastic component has lengthened. (From D. I. Cosgrove, *Plant Physiology* 84:561–564, 1987. Copyright the American Society of Plant Physiologists.)

[2]As noted earlier, it is the practice when studying water movement to treat the plasma membrane, cytoplasm and vacuolar membrane as a single differentially permeable barrier.

KINETIC ANALYSIS OF GROWTH

The first half of the twentieth century was marked by numerous attempts to analyze plant growth and development from a mathematical perspective. The idea was that subjecting growth to mathematical analysis would lead to predictions about the nature of the process that could be tested experimentally. Despite extensive analyses of growth kinetics, these efforts have had little impact on our understanding of plant growth or development or their control. The principal difficulty is that while simple models can help to describe growth, they do not readily incorporate qualitative changes; that is, they cannot satisfactorily account for differentiation.

More recently, aided by the advent of computer analysis, there has been a resurgence of interest in mathematical modeling of plant growth, especially in the fields of agriculture and the physiological ecology of plant communities. Starting with simple growth models for individual plants, it is possible to simulate the impact of climatic factors and inputs such as soil structure, nutrient levels, water, temperature, and sunlight on growth, development, and productivity of crop systems or natural vegetation stands. These models have value in breeding programs designed to develop new cultivars especially suited to particular environmental conditions, or in directing the efficient use of fertilizers and water, or in estimating the impact of pollutants on productivity of stands.

Although mathematical analyses have not been particularly useful with respect to understanding the mechanisms of growth and development, they do reveal something of the nature of the growth process and allow comparisons to be made between different conditions. Growth kinetics are a convenient way of summarizing growth data and are therefore commonly used by plant physiologists for describing and interpreting physiological responses.

GROWTH OF MICROORGANISMS IN CULTURE

Before considering the growth of complex, multicellular higher plants, it is instructive to consider the growth of unicellular microorganisms such as bacteria or algae, which reproduce by cell division, or yeast, which reproduce by budding. Consider the growth of a culture of microorganisms that has an adequate nutrient supply and constant environmental conditions. A plot of cell number against time would show a geometric progression or exponential growth as illustrated in Figure 15.16A.

The exponential character of the curve can be explained by assuming that cell division is synchronous, that is, all cells in the culture divide simultaneously. If

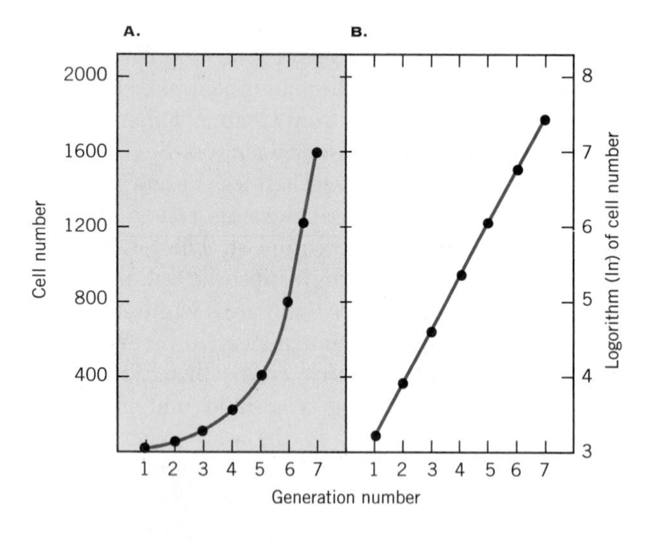

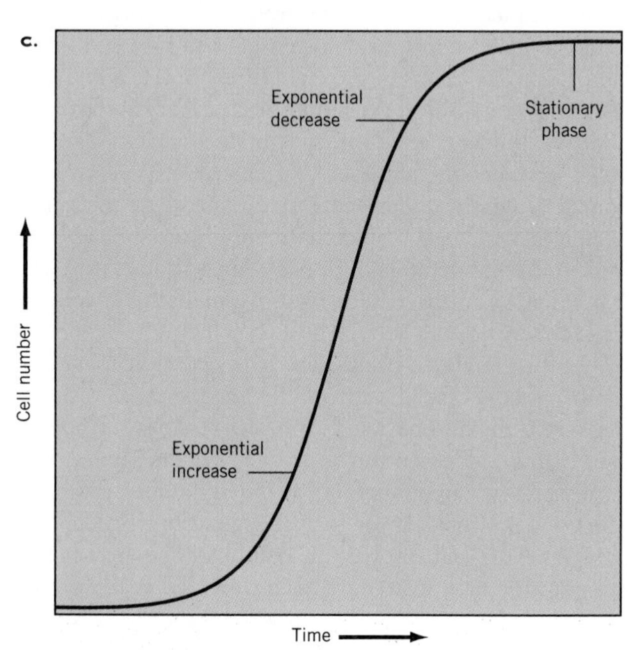

FIGURE 15.16 Idealized growth curves for a population of microorganisms dividing synchronously. The population doubles with each generation. (*A*) Plot of cell number against number of generations. (*B*) A semilog plot. (*C*) An idealized sigmoid growth curve, typical of cells in culture and many higher plants and plant organs.

n_i is the initial number of cells in the culture and n is the number of cells in the culture after a given number of cell divisions, then at the end of the first generation:

$$n = n_i \times 2$$

at the end of the second generation:

$$n = n_i \times 2 \times 2$$

at the end of the third generation:

$$n = n_i \times 2 \times 2 \times 2$$

and at the end of the *x*th generation:

$$n = n_i \times 2^x.$$

Under these conditions the number of cells in the culture will double with each generation and a plot of cell number against time will resemble Figure 15.16A. If the natural logarithm of cell number were plotted against generation number, the result would be a straight line as in Figure 15.16B.

From data of this type, we can calculate two parameters that help to describe growth in the culture. These are identified as the **absolute growth rate (AGR)** and the **relative growth rate (RGR)**. The absolute growth rate at any given time is the slope of the line and may be expressed as the ratio of the change in cell number (dn) over the interval of time (dt):

$$AGR = dn/dt \tag{15.6}$$

Note that although the rate of cell division is constant, the absolute growth rate for a population of single cells increases progressively with time. This is because, at any given time t, the growth rate of the culture is a function of the number of cells present:

$$dn/dt \ (f) \ n \tag{15.7}$$

The absolute growth rate may be divided by the number of cells present to give the relative growth rate:

$$RGR = (dn/dt) * 1/n \tag{15.8}$$

Because in our model culture each cell contributes to the next generation, the relative growth rate remains constant. AGR and RGR are useful because they help to describe the dynamics of cell growth in the culture.

In practice, exponential growth does not continue indefinitely. A depleted nutrient supply, accumulation of toxic products, and other limiting factors eventually lead to a decline in growth rate until the population reaches a stable (or declining) cell number. This gives rise to a sigmoid curve shown in Fig 15.16C.

GROWTH OF MULTICELLULAR ORGANISMS

Measuring or describing the growth of individual cells in culture or small groups of cells such as a coleoptile segment is relatively simple. Growth of multicellular organisms, on the other hand, is another kind of problem.

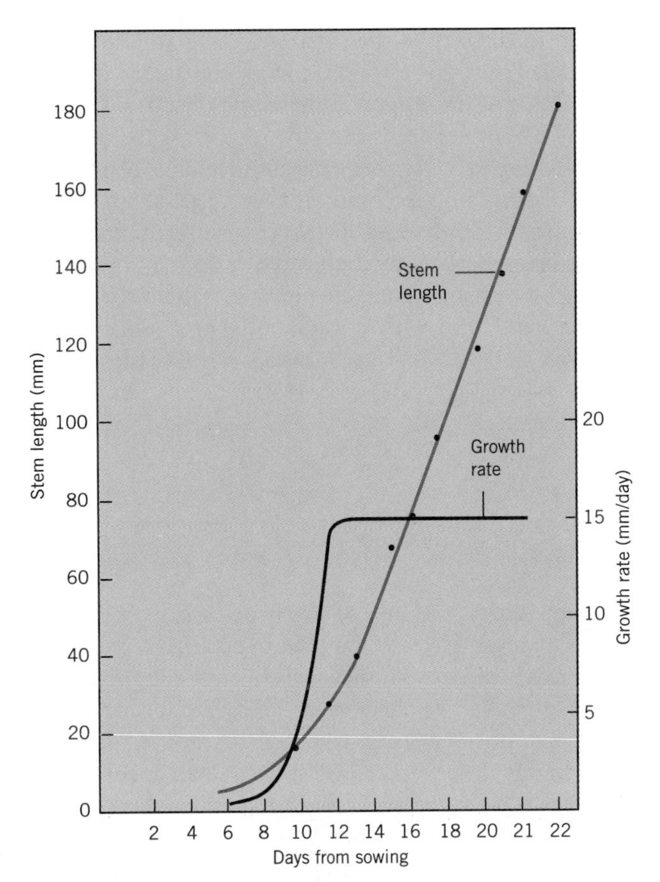

FIGURE 15.17 Growth of a broadbean (*Vicia faba*) seedling over a three-week period. The curve for stem length shows an extended linear phase. Growth rate reaches a maximum that remains constant through the linear phase. In some plants or organs the linear phase may be very short.

The exponential growth curve typical of microorganisms in culture also applies to the growth of some higher plants or plant organs, at least in the early stages. With multicellular organisms, such as an annual plant, AGR often increases during the early phases, then remains constant for a period of time before declining as maturation and senescence sets in (Fig. 15.17). This linear phase is often the dominant character of growth and results largely from the unique meristematic growth plan of higher plants. Unlike cultures of unicellular organisms where the number of dividing cells increases with time, the number of cells in a meristem remains constant. Growth rate is therefore determined principally by the rate of cell division in the meristem and the production of new cells is arithmetic rather than geometric. Another contributing factor is that the subsequent rate of internode elongation is also linear. Increase in size therefore tends to be constant and is not related to the size of the organism or its total number of cells. However, growth curves for bulky organs, which do not have defined meristems, tend to be sigmoidal. Examples are fruits and storage organs.

The growth rate for many multicellular plants and plant organs is quite complex, varying from species to species or with changes in the environment. In many cases, perturbations in the growth rate may be introduced by developmental events. Nevertheless, regardless of whether growth is exponential or linear or multiphasic, all organisms face inherent size limitations and experience a declining growth rate as they approach these limits.

SUMMARY

Development is the sum of two processes: growth and differentiation. Growth is a quantitative term, related to increases in size and mass, while differentiation is a series of qualitative changes in the cell, tissue, or organ. Most plant cells are totipotent: except for the most highly differentiated cells, they retain a complete genetic program and the capacity to renew division, growth, and differentiation. All that is required to change the pattern of differentiation is the right input to select the appropriate genetic information at the right time.

The orderly development of a complex multicellular organism is coordinated by intrinsic and extrinsic controls. Intrinsic controls are expressed at both the intracellular and extracellular levels. Intracellular controls are primarily genetic, requiring a programmed sequence of gene expression. Extracellular controls are primarily hormonal, chemical messengers that allow cells to communicate with one another. Extrinsic controls are environmental cues such as light, temperature, and gravity. Most environmental cues appear to operate at least in part by modifying gene expression or hormonal activities.

A seed is a quiescent structure, containing a food supply (usually an endosperm) and an embryo that contains the rudimentary organs of a young plant—the root and shoot. Seed germination begins with the imbibition of water to hydrate the relatively dry seed tissues. This is followed by the mobilization of stored carbon reserves and the onset of respiration. Elongation of the roots and shoot leads to the establishment of a young seedling, floral evocation, and the development of flowers and fruit.

The driving force for cell enlargement is water uptake. However, in order to resist rupturing in the face of high turgor pressure, plant cells are surrounded by a very strong and relatively rigid wall. Thus, in order for a cell to enlarge, the strength and rigidity of the wall must be modified. In a turgid cell, the force of water pressing against the wall generates stress within the extensively cross-linked wall components. Growth is initiated when these stresses are relieved by wall-loosening events, which causes the load-bearing cross links between wall polymers to yield. Relieved of stress, the wall expands, turgor is reduced, and more water moves in until both turgor and wall stresses are restored.

Growth can be described mathematically, which aids in analyzing and interpreting the growth process.

CHAPTER REVIEW

1. Distinguish between growth, differentiation, and development.
2. How can growth be measured? What are the relative advantages of different parameters for measuring growth?
3. What are the three levels of control over growth and development?
4. Review the principal stages in the development of a higher plant.
5. Describe the conflicting roles of turgor in the growth of plant cells.
6. Distinguish between absolute growth rate and relative growth rate.

FURTHER READING

Bewley, J. D., M. Black. 1985. *Seeds: Physiology of Development and Germination.* New York: Plenum Press.

Esau, K. 1977. *Anatomy of Seed Plants.* New York: Wiley.

Fosket, D. 1994. *Plant Growth and Development: A Molecular Approach.* New York: Academic Press.

Greyson, R. I. 1994. *The Development of Flowers.* New York: Oxford University Press.

Hunt, R. 1978. Plant Growth Analysis. *Studies in Biology No. 96.* London: Edward Arnold.

Raven, P. H., R. F. Evert, S. E. Eichhorn. 1992. *Biology of Plants.* 5th ed. New York: Worth.

Steeves, T. A., I. M. Sussex. 1989. *Patterns in Plant Development,* 2nd ed. Cambridge: Cambridge University Press.

The Plant Cell. 1997. Vol. 9, Number 7. Rockville, MD: American Society of Plant Physiologists. (Special issue containing reviews on plant vegetative development.)

Torrey, J. G. 1967. *Development in Flowering Plants.* New York: Macmillan.

Wareing, P. F., I. D. J. Phillips. 1981. *Growth and Differentiation in Plants.* 3rd ed. Oxford: Pergamon.

REFERENCES

Cosgrove, D. J. 1987. Wall relaxation and the driving forces for cell expansive growth. *Plant Physiology* 84:561–564.

Lockhart, J. 1965. An analysis of irreversible plant cell elongation. *Journal of Theoretical Biology* 8:264–275.

McDaniel, C. N. 1991. Early events in flowering: A black box with a bright future. *The Plant Cell* 3:431–433.

Meeks-Wagner, D. R. 1993. Gene expression in the early floral meristem. *The Plant Cell* 5:1167–1174.

Murphy, T. M., W. E. Thompson. 1988. *Molecular Plant Development.* Englewood Cliffs: Prentice Hall.

Owens, S. J. 1992. Pollination and fertilization in higher plants. In: C. Marshall, J. Grace (eds.), *Fruit and Seed Production: Aspects of Developmental, Environmental Physiology and Ecology.* Cambridge: Cambridge University Press.

van den Berg, C., V. Willemsen, W, Hage, P. Weisbeek, B. Scheres. 1995. Determination of cell fate in the *Arabidopsis* meristem by directional signalling. *Nature* 378:62–65.

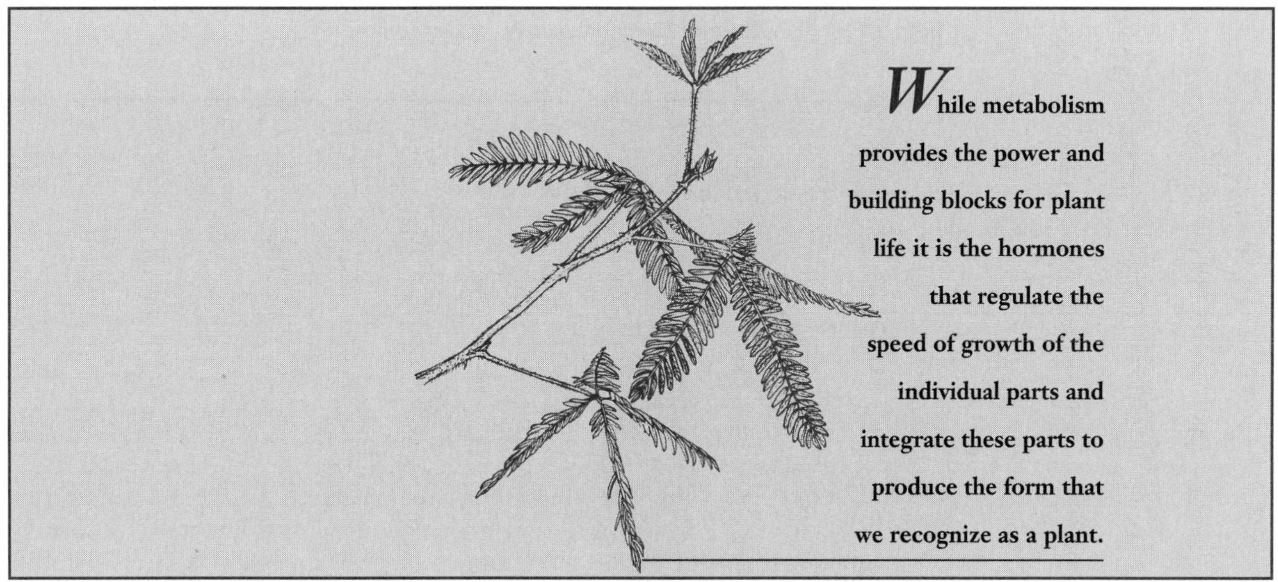

P. J. Davies (1987)

16

The Role of Hormones in Plant Development

Multicellular plants are complex organisms and their orderly development requires an extraordinary measure of coordination between cells. In order to coordinate their activities, cells must be able to communicate with each other, often at some distance. The principal means of intercellular communication are the hormones, chemical messengers that carry information between cells and thus coordinate their growth and development. Plant hormones have been the subject of intensive investigation and frequent controversy since their discovery three-quarters of a century ago.

This first of two chapters on plant hormones serves as our introduction to the concept of hormones and their principal physiological roles in plants. The principal topics to be covered are

- a discussion of the hormone concept in plants and some of the controversy that surrounds it;
- a brief introduction to the five major groups of plant hormones—auxins, gibberellins, cytokinins, abscisic acid, and ethylene—followed by a description of their principal physiological roles; and
- a brief description of two hypothetical hormones, the polyamines, and other biologically active substances that can influence growth and development.

The biochemistry and mode of action of plant hormones will be discussed in Chapter 17.

THE HORMONE CONCEPT IN PLANTS

There are numerous chemical substances—natural and synthetic—that profoundly influence the growth and differentiation of plant cells and organs. Their role in development has been studied for nearly a century, yet the concept of hormones in plants is steeped in controversy.

The latter half of the nineteenth century witnessed exciting advances in mammalian physiology and medicine. By 1850, it was known that blood-borne substances originating in the testis conditioned sexual characteristics. At the same time, physicians pursuing clinical studies had become interested in the effect of glandular extracts and secretions on the course of various diseases. By the turn of the century, a number of substances that elicited specific effects on the growth and physiology of mammals had been demonstrated and the concept that bodily functions were coordinated by the production and circulation of chemical substances was gaining wide acceptance. In 1905, the British physician E. H. Starling introduced the term **hormone** (Gr.; to *excite* or *arouse*) to describe these chemical messengers.

The concept of hormones in plants may be traced back to observations of Duhamel du Monceau in 1758. Du Monceau observed the formation of roots on the

Box 16.1 Historical Perspectives—Discovering Plant Hormones

The discovery of each of the plant hormones represents an important milestone in the progress of plant physiology. But few, if any, investigators actually set out with the intention of "discovering" a hormone. As is often the case in scientific endeavor, these important discoveries arose out of attempts to understand other problems—to explain how various developmental phenomena or abnormal behaviors were regulated. Not infrequently, important scientific discoveries include more than a little serendipity.

The experimental beginnings of plant hormone research in general and auxins in particular can be traced to the work of Charles Darwin. Although Darwin is best known for his work on evolution, later in his career he developed an interest in certain aspects of plant physiology. Some of these studies were summarized in the book *The Power of Movement in Plants*, co-authored by his son, Francis. One of several "movements" studied by the Darwins was the tendency of canary grass (*Phalaris canariensis*) seedlings to bend toward the light coming from a window, a phenomenon we now know as **phototropism** (Chap. 19). The primary leaves of grass seedlings are enclosed in a hollow, sheathlike structure, the **coleoptile,** that encloses and protects the leaves as they grow up through the soil. Darwin observed that coleoptiles, like stems, respond to unilateral illumination by growing toward the light source. However, curvature would not occur if the tip of the coleoptile were either removed or covered in order to exclude light. Since the bending response was observed over the entire coleoptile, Darwin concluded that the phototropic signal was perceived by the tip and "that when the seedlings are freely exposed to lateral light, some influence is transmitted from the upper to the lower part, causing the latter to bend." It was the implications of Darwin's "transmissible influence" that captured the imagination of plant physiologists and set into motion a series of experiments that culminated in the discovery of the plant hormone, auxin—the first plant hormone to be discovered.

Following the publication of Darwin's book, a number of scientists confirmed and extended their observations. In 1910, Boysen-Jensen demonstrated that the stimulus would pass through an agar block and was therefore chemical in nature. In 1918, Paal showed that if the apex were removed and replaced asymmetrically, curvature would occur even in darkness. In the climate of the time—Baylis and Starling's characterization of animal hormones had appeared only a few years earlier—plant physiologists were quick to interpret these observations as strong support for a **phytohormone.**

The active substance was first successfully isolated in 1928 by F. W. Went, then a graduate student working in his father's laboratory in Holland. Following up on the earlier work of Boysen-Jensen and Paal, Went removed the apex of oat (*Avena sativa*) coleoptiles and stood the apical pieces on small blocks of agar. Allowing a period of time for the substance to diffuse from the tissue into the agar block, he then placed each agar block asymmetrically on a freshly decapitated coleoptile (see Fig. 16.2A). The substance then diffused from the block into the coleoptile, preferentially stimulating elongation of the cells on the side of the coleoptile below the agar block. Curvature of the coleoptile was due to differential cell elongation on the two sides. Moreover, the curvature proved to be proportional to the amount of active substance in the agar. Went's work was particularly significant in two respects: First, he confirmed the existence of regulatory substances in the coleoptile apex, and second, he developed a means for isolation and quantitative analysis of the active substance. Because Went used coleoptiles from *Avena* (oat) seedlings, his quantitative test became known as the **Avena curvature test.** Substances active in this test were called auxin, from the Greek *auxin* (to increase).

The results of Went's studies naturally stimulated intensive efforts to isolate and identify the active substance. One particularly active compound, indole-3-

swellings that occur above girdle wounds around the stems of woody plants (see Chap. 11). In order to explain these and similar phenomena, Julius Sachs (ca. 1860) postulated specific organ-forming substances in plants. Root-forming substances, for example, produced in the leaves and migrating down the stem would account for the initiation of roots above the wound. The real beginning of plant hormone research, however, is found in a series of simple but elegant experiments conducted by Charles Darwin (see Box 16.1: Historical Perspectives—Discovering Plant Hormones). It was Darwin's observations and experiments that ultimately led F. W. Went, almost half a century later, to describe a hormonal-like substance as the causative agent when plants grew toward the light. At about the same time, H. Fitting introduced the term *hormone* into the plant physiology literature.

According to current usage, *hormones are naturally*

acetic acid (IAA), was isolated from human urine in 1934. This peculiar source was selected because it was suspected that female sex hormones, secreted in urine, might have some plant growth activity. In a beautiful piece of scientific serendipity, the impure urine preparation initially assayed was highly active, while subsequently purified hormone preparations were inactive. This led the investigators back to the material from which the female sex hormones were initially extracted—the urine of pregnant women—and the identification of IAA. At the same time, IAA was isolated from yeast extracts and the following year, from cultures of *Rhizopus suinus.* IAA was isolated from immature corn kernels in 1946 and since then has been found to be ubiquitous in higher plants. There is general agreement that IAA is the principal auxin in higher plants.

The discovery of gibberellins came about from more practical concerns. During the late nineteenth and early twentieth century, Japanese rice farmers grew concerned about a disease that seriously reduced the yield of their crops. Plants infected with the *bakanae* ("foolish seedling") disease exhibited weak, elongated stems and produced little or no grain. Japanese plant pathologists, interested in developing means for controlling the disease, soon established a connection with the presence of a fungus, *Gibberella fujikuroi.* In 1926, E. Kurosawa reported the appearance of symptoms of the disease in uninfected rice plants that had been treated with sterile filtrates from cultures of this fungus. By 1938, Japanese investigators had isolated and crystallized the active material, which they called *gibberellin* after the genus name for the fungus.

Gibberellin did not come to the attention of western plant physiologists until after the 1939–1945 war, when two groups—one headed by Cross in England and one by Stodola in the United States—isolated and chemically characterized *gibberellic acid* from fungal culture filtrates. At the same time, Japanese workers isolated three gibberellins, which they named gibberellin A_1, gibberellin A_2, and gibberellin A_3. Gibberellin A_3 proved to be identical with gibberellic acid.

The known effect of gibberellins on rice and several other plant systems indicated that similar substances might be present in higher plants as well. The first higher plant gibberellin to be characterized was isolated from immature seeds of runner bean (*Phaseolus coccineus*) and found to be identical with gibberellin A_1. Since then, gibberellins have been shown to be ubiquitous in higher plants.

The discovery of cytokinins came about because plant cells in culture would not divide. The first experimental evidence for chemical control of plant cell division was provided by Haberlandt in 1913, when he demonstrated that phloem sap could cause nondividing, parenchymatous potato tuber tissue to revert to an actively dividing meristematic state. Other cell-division factors were later demonstrated in wounded bean pod tissue, extracts of *Datura* ovules, and the liquid (milky) endosperm of coconut.

In the 1940s and 1950s, plant tissue culture was attracting the attention of physiologists as a tool for study of cell division and development. One group, under the direction of F. Skoog at the University of Wisconsin, was studying the nutritional requirements of tissue cultures derived from tobacco stem segments. Skoog and co-workers found that stem tissue explants containing vascular tissue would proliferate on a defined medium containing auxin. On the same auxin-containing medium, however, tissue explants freed of vascular tissue would exhibit cell enlargement, but failed to divide. They soon found that extracts of vascular tissue, coconut milk, and yeast would all stimulate cell division in the presence of auxin.

C. O. Miller, then working as a postdoctoral student in Skoog's laboratory, began the task of isolating the active principle from yeast extract, which was readily available in large quantity. Based on his initial work, Miller was able to provisionally identify the active material as a purine, one of the nitrogenous bases found in nucleic acids. Earlier, Skoog and co-workers had shown that adenine, also a purine, elicited some cell division in the tobacco pith culture system. This led to a search for active material in nucleic acid preparations, a source high in adenine. In 1956, Miller and his colleagues reported the isolation and crystallization of a highly active substance, identified as the adenine derivative N^6-furfurylaminopurine, from autoclaved

occurring, organic substances that, at low concentration, exert a profound influence on physiological processes. In addition, hormones, at least in animals, are (1) *synthesized in a discrete organ or tissue*, and (2) *transported in the bloodstream to a specific target tissue* where they (3) control a physiological response *in a concentration-dependent manner.* While there are many parallels between animal and plant hormones, there are also some significant differences. Like animal hormones, plant hormones are naturally occurring organic substances that profoundly influence physiological processes at low concentration. The site of synthesis of plant hormones, however, is not so clearly localized. Although some tissues or parts of tissues may be characterized by higher hormone levels than others, synthesis of plant hormones appears to be much more diffuse and cannot always be localized to discrete organs. For example, there is good evidence that auxin, the prototypic plant hormone, is synthesized in herring sperm DNA. Because the compound elicited cell division, or *cytokinesis*, in tissue culture, Miller and his colleagues named the substance *kinetin*. In 1965, Skoog and his colleagues proposed the term *cytokinin*.

Even though kinetin remains one of the most biologically active cytokinins, it is an artifact of isolation from DNA and has not been found in plants. However, the discovery of kinetin and its dramatic effect on cell division stimulated physiologists to look for naturally occurring cytokinins. In the early 1960s, Miller, then at Indiana University, and D. S. Letham, working in Australia, independently reported the isolation of a purine with kinetinlike properties from young, developing maize seed and plum fruitlets. This substance was characterized as 6-(4-hydroxy-3-methyl-*trans*-2-butenylamino)purine, which was given the trivial name *zeatin*. Since the discovery of zeatin, a number of other naturally occurring cytokinins have been isolated and characterized.

As more investigators became interested in plant hormone research, it soon became evident that ether extracts of plant material—used to extract auxins—frequently contained substances that interfered with the auxin response in the *Avena* coleoptile curvature test. Initially, the principal interest of investigators was to rid extracts of these interfering substances. As time went on, however, interest turned toward the possibility that these inhibitors might themselves be growth regulators in their own right. The advent of paper chromatography as an analytical tool made it possible to achieve better separation of the various substances in crude extracts. In 1953, Bennet-Clark and Kefford reported that plant extracts contained, in addition to IAA, a substance that inhibited growth of coleoptile sections, which they called *inhibitor* β. The observation that large amounts of inhibitor β could be isolated from axillary buds and the outer layer of dormant potato tuber led Kefford to suggest that it was involved in apical dominance and maintaining dormancy in potatoes. Meanwhile, other investigators reported the occurrence of inhibitors in buds and leaves that appeared to correlate with the onset of dormancy in woody plants. In 1964, P. F. Waring proposed the term "dormin" for these endogenous, dormancy-inducing substances.

In another line of study, substances that accelerated abscission were isolated from senescing leaves of bean and from cotton and lupin fruits. These substances would accelerate abscission when applied to excised abscission zones and were called "abscission II." These several lines of study came to a head in the mid-1960s when three laboratories independently reported the purification and chemical characterization of abscisin II, inhibitor β, and dormin. All three substances proved to be chemically identical.

It is not unusual in such cases that there was some disagreement over what this substance should be called. Although abscisin II had priority (it was the first to be crystallized and chemically characterized), some felt the term awkward and argued it did not adequately describe its range of effects. Finally, a panel of scientists active in research on abscisin II and dormin was charged with proposing an acceptable name. The name *abscisic acid* and abbreviation *ABA* were recommended by this panel to the 1967 International Conference on Plant Growth Substances, which met in Ottawa. The recommendation was accepted by the Conference and the term abscisic acid is now in universal use.

Those whose business involves the shipping and storing of fruit have long been aware that ripe and rotting fruit could accelerate the ripening of other fruit stored nearby. For example, bananas picked in Cuba and shipped by boat often arrived in New York in an overripe and unmarketable condition. One of the earliest reports that these effects were due to a volatile substance given off by plant tissue was published in 1910 by H. H. Cousins in an annual report of the Jamaican Department of Agriculture. He discovered that ripe oranges released a volatile product that would accelerate ripening of bananas stored with them. A number of similar reports appeared in the early 1930s, showing that volatile emanations from apples caused epinasty in tomato seedlings and respiratory changes associated with the ripening process. In 1934, R. Gane provided indisputable evidence that the volatile substance was ethylene.

the tip of a grass coleoptile but influences the elongation of cells lower down in the same organ. Another plant hormone, cytokinin, is synthesized in the root and transported to the leaves where it influences metabolic activity and delays senescence. Yet there are many other examples where plant hormones appear to act within the same tissue or even the same cell in which they are synthesized. Clearly, action at a distance is not an *essential* property of a plant hormone.

Finally, whether plant hormones act in a concentration-dependent manner is a subject of continuing dispute among students of plant hormones (Trewavas, 1981, 1991; Trewavas and Cleland, 1983). While some argue that plant cells respond to hormone concentration, as they do in animals, others argue that it is not the hormone concentration that is important, but changing sensitivity of the target cells to the hormone. There is evidence to support both views and, as we con-

TABLE 16.1 The influence of plant hormone groups on different categories of development. An x indicates a demonstrated effect of that hormone group on one or more aspects of that developmental category. The absence of an x does not mean that the hormone is ineffective, only that an effect has not been reported in the literature.

	Hormone Group				
	Auxin	Gibberellin	Cytokinin	Abscisic Acid	Ethylene
Dormancy		x	x	x	x
Juvenility	x	x			
Extension Growth	x	x	x	x	x
Root Development	x	x	x		x
Flowering	x	x	x	x	x
Fruit Development	x	x	x	x	x
Senescence	x	x	x		x

Modified from C. Leopold. Ethylene as a plant hormone. In: H. Kaldeway, Y. Varder (eds.) *Hormonal Regulation in Plant Growth and Development*. Weinheim: Verlag Chemie. 1972. Reprinted by permission.

tinue to explore hormone action in plants, we shall one day find, no doubt, that the truth lies somewhere in the middle.

Another perceived difficulty with plant hormones is the multiplicity of their effects. Each group of plant hormones is known to influence a wide variety of developmental events. Moreover, most of these events can be influenced by more than one hormone group (Table 16.1). The complexity of hormonal interactions creates additional problems for experimenters. Typically the effect of a single hormone has been studied in isolation, even though several or even all of the hormones may be present and active in the cell at the same time. It is important in the discussions that follow to remember that the way a plant grows and develops is not the product of any one hormone acting alone. Nor is control of development restricted to hormones, but is the result of control at several levels, including genetic, chemical, and environmental factors. Not incidentally, the multiplicity of hormone action in plants makes it virtually impossible to catalog all of the possible effects of each hormone within the limited coverage of this text. The reader is advised to consult the reference list at the end of this chapter for more extensive coverage of the effects that hormones can have on plants.

The subtle differences between animal and plant hormones have given rise to vigorous debate among some plant physiologists concerning the applicability of the term *hormone* to regulatory chemicals in plants. It has also led to some confusion in terminology. The term *plant growth substances* is preferred by some—there is even an International Plant Growth Substance society. Others argue that "substance" is too vague and growth is only one of many processes influenced by these chemicals. A second term, *plant growth regulator*, is preferred in the agrochemical industry. It is normally used to denote synthetic compounds that exhibit hormonal activity.

The debate over whether plants have "hormones" is largely semantic, however, and is perhaps best resolved by the suggestion of A. C. Leopold that "hormones are what physiologists call hormones" (Leopold, 1987). Plant hormones clearly fulfill most of the criteria set out in Starling's original definition. It is easy enough to accept that plant hormones may differ *in detail* from animal hormones and proceed accordingly. The comparative biology of animal and plant hormones remains a useful and productive approach toward understanding how plants work. Still, plant physiologists should not be blinded by the animal literature but be receptive to new approaches and ideas.

THE PLANT HORMONES

There are currently five recognized groups of plant hormones: **auxins, gibberellins, cytokinins, abscisic acid (ABA)**, and **ethylene.** In addition to the five principal hormones, two other groups sometimes appear to be active in regulating plant growth, the **brassinosteroids** and **polyamines.** Each of these groups will be briefly introduced at this time, followed by a general discussion of its physiological role.

AUXINS

Auxins were the first plant hormones to be discovered. Auxins are synthesized in the stem and roots apices and transported through the plant axis. They are characterized principally by their capacity to stimulate cell elongation in excised stem and coleoptile sections, but also influence a host of other developmental responses, including root initiation, vascular differentiation, tropic responses, and the development of axillary buds, flowers, and fruits.

NATURAL AND SYNTHETIC AUXINS

The principal auxin in plants is indole-3-acetic acid (IAA) (Fig. 16.1). In addition to IAA, several other naturally occurring indole derivatives are known to express auxin activity, including indole-3-ethanol, indole-3-acetaldehyde, and indole-3-acetonitrile. However, these compounds all serve as precursors to IAA and their activity is probably due to conversion to IAA in the tissue (Chap. 17).

The initial discovery of IAA in plants and recognition of its role in growth and development stimulated the search for other chemicals with similar activity. The result has been an array of synthetic chemicals that express auxinlike activity (Fig. 16.1). One of these, **indole butyric acid (IBA)** (IV, Fig. 16.1) was originally thought to be strictly synthetic, but recently IBA has been isolated from seeds and leaves of maize and other species (Epstein et al., 1989). A chlorinated analog of IAA (**4-chloroindoleacetic acid,** or **4-chloroIAA;** II, Fig. 16.1) has also been reported in extracts of legume seeds (Engvild, 1986) and a closely related, naturally occurring aromatic acid, **phenylacetic acid (PAA;** III, Fig. 16.1) has recently been reported to have auxin activity (Leuba and LeToureau, 1990). Because IBA, 4-chloroIAA, and PAA have now been isolated from

FIGURE 16.1 The chemical structures of some naturally occurring and synthetic auxins. IAA (I) is believed to represent the only true auxin. Phenylacetic acid (III) is widespread in plants and two others, 4-chloroindole-3-acetic acid and indole-3-butyric acid, have recently been described in plant extracts. The latter three exhibit auxin responses when applied exogenously, but little is known of their physiology and biochemistry or whether they normally function as hormones in the plant.

Naturally-occurring Auxins

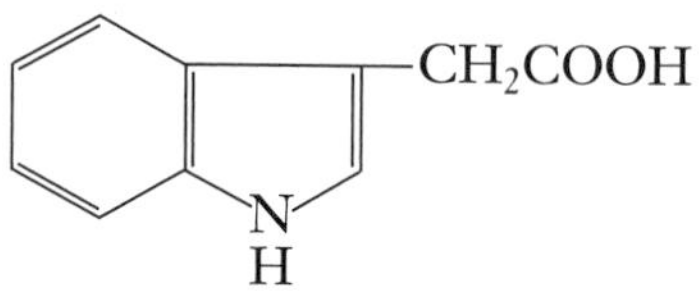

I. Indole-3-acetic Acid (IAA) — CH_2COOH

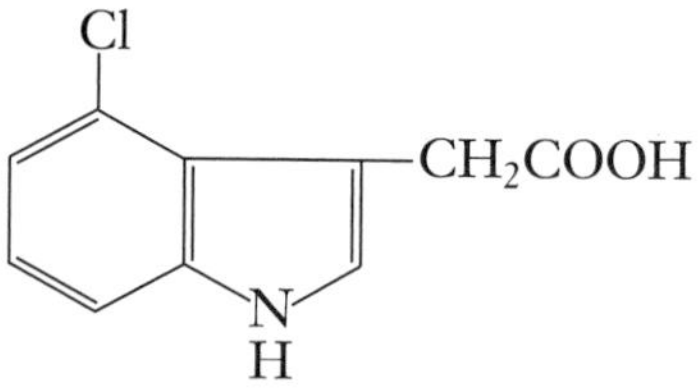

II. 4-Chloroindole-3-acetic Acid — Cl, CH_2COOH

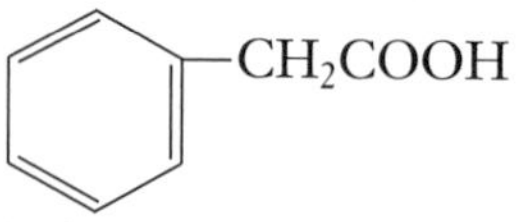

III. Phenylacetic Acid — CH_2COOH

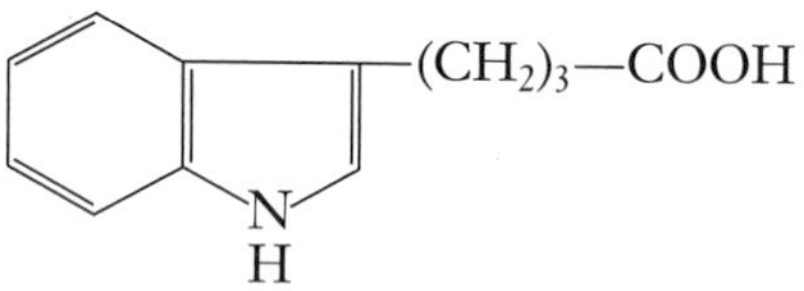

IV. Indole-3-butyric Acid (IBA) — $(CH_2)_3—COOH$

Synthetic Auxins

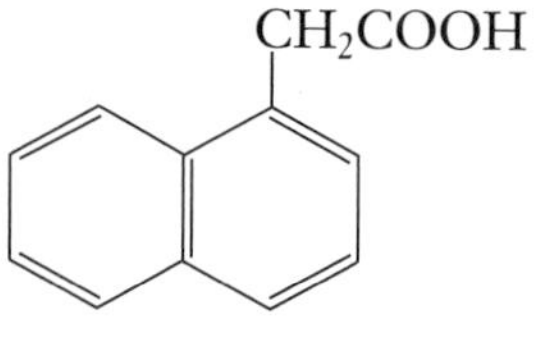

V. Naphthalene acetic Acid — CH_2COOH

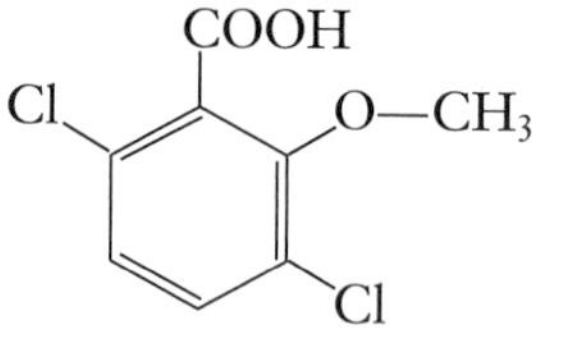

VI. 2-Methoxy-3,6-dichloro-benozic Acid (dicamba) — COOH, Cl, $O—CH_3$, Cl

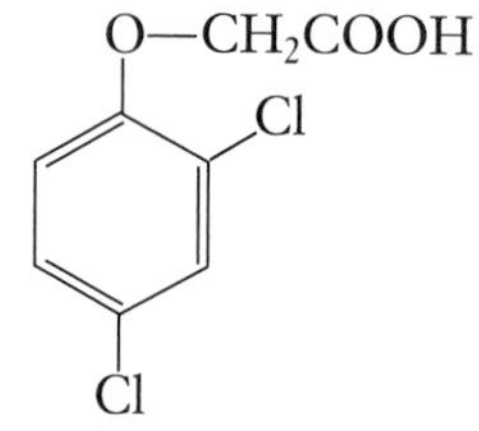

VII. 2,4-Dichlorophenoxyacetic Acid (2,4-D) — $O—CH_2COOH$, Cl, Cl

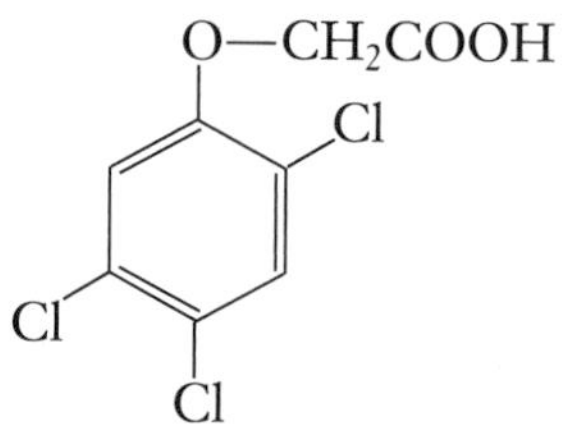

VIII. 2,4,5-Trichlorophenoxy-acetic Acid (2,4,5-T) — $O—CH_2COOH$, Cl, Cl, Cl

plants, are structurally similar to IAA, and elicit many of the same responses as IAA, there is a strong argument for considering them natural hormones. It is not yet established, however, whether or not they are converted to IAA in the tissue before they become active. Chemically, the single unifying character of molecules that express auxin activity appears to be an acidic side chain on an aromatic ring.

PHYSIOLOGICAL ACTION OF AUXIN

Cell Growth and Differentiation It is generally agreed that auxin serves primarily to regulate cell growth and stem elongation. Auxin-regulated cell elongation in *Avena* coleoptiles was the basis for its discovery and this action has been demonstrated repeatedly with *excised* plant tissues such as subapical coleoptile tissues and stem segments cut from dark-grown pea seedlings.

Auxin concentration-response curves typically show an increasing response with increasing concentrations of auxin until an optimum concentration is reached (Fig. 16.2). Concentrations exceeding the optimum characteristically result in reduced growth. If the auxin concentration is high enough, growth may be inhibited compared with controls. Growth responses such as these are often used to assay for unknown hormone concentrations, a technique known as **bioassay.**

Another very characteristic feature of auxin physi-

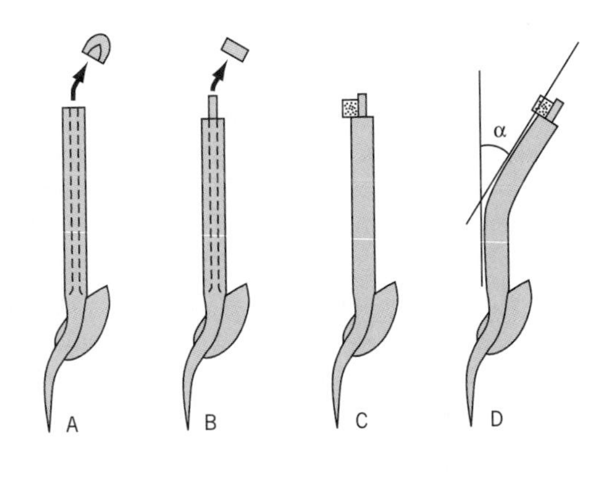

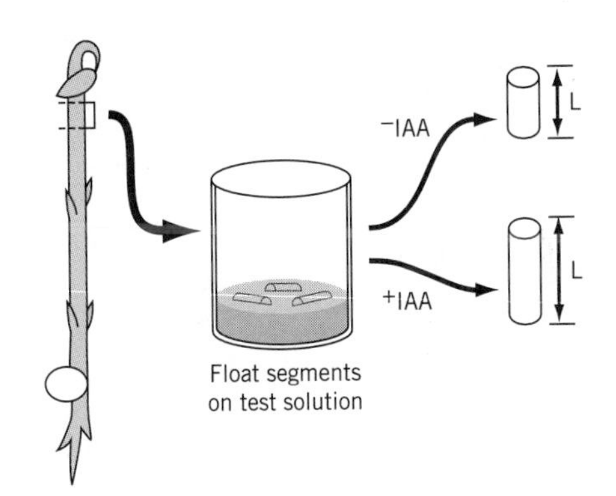

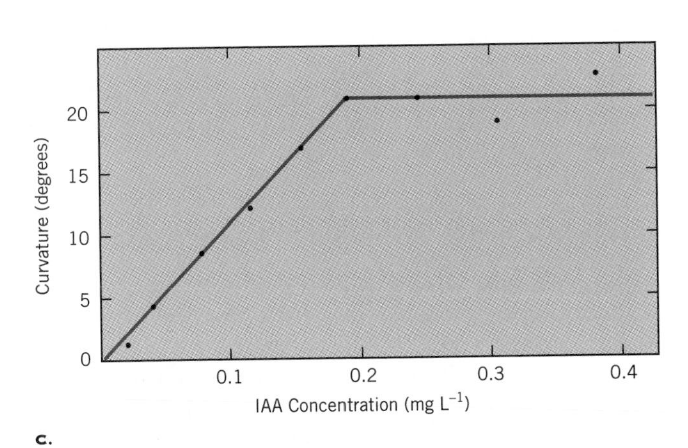

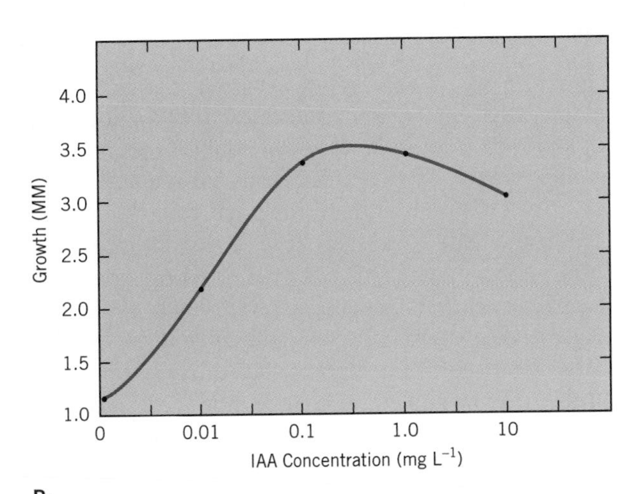

FIGURE 16.2 Concentration response curves for two classic auxin-regulated responses. (*A*) Went's *Avena* curvature test. A small cube of agar containing auxin is placed on the cut surface of a decapitated oat coleoptile. The auxin diffuses into the coleoptile, stimulating growth of the cells below the agar cube. The differential growth causes the coleoptile to curve away from the block. (*B*) Curvature in the *Avena* test is linearly related to auxin concentration. (Redrawn from the data of F. W. Went and K. V. Thimann, *Phytohormones*, 1937. By permission of K. V. Thimann.) (*C*) Pea stem segment test. Stem sections from dark-grown pea seedlings are floated on a medium with or without auxin. (*D*) Typical concentration-response in a pea stem section test. Note auxin concentration is expressed on a logarithmic scale. (Redrawn from the data of Galston and Hand, 1949. With permission of the *American Journal of Botany*.)

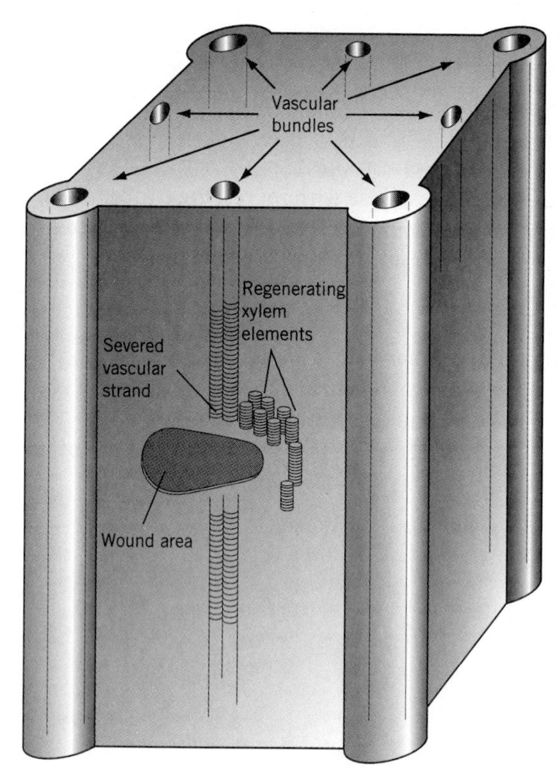

FIGURE 16.3 A diagram of vascular regeneration around a wound in Coleus stimulated by auxin. A vascular bundle was surgically severed and the wounded tissue treated with auxin. After several days, the tissue was cleared and examined microscopically for signs of xylem regeneration around the wound. (Drawn from a student experiment.)

ology is that intact stems and coleoptiles do not show a significant response to *exogenous* application of the hormone. Apparently the endogenous auxin content of intact tissues is high enough to support maximum elongation and added auxin has little or no additional effect. Thus, it is a general rule that the effect of exogenously supplied auxin can be demonstrated only in tissues, such as excised segments of stems and coleoptiles, that have been removed from the normal auxin supply. This "rule" may be subject to change, however, since a sustained growth of intact pea internodes following exogenous applications of IAA has been reported (Yang et al., 1993). Although it is commonly assumed that auxin is essential for cell enlargement and growth of leaves, flowers, and other organs as well as stems, there is little direct evidence available. Auxin-induced cell enlargement is also the basis for the capacity of auxins to initiate and sustain growth of undifferentiated cells when plant tissue are cultured on artificial media.

In addition to stimulating cell enlargement, auxins are also involved in regulating cellular differentiation. For example, the induction of vascular differentiation in shoots is under control of auxin produced in the young, rapidly developing leaves. Some of the most definitive studies in this area have been conducted by W. P. Jacobs and coworkers, working with *Coleus* (Jacobs, 1970). Jacobs found that the production of xylem strands at the base of the petiole is directly proportional to the stream of diffusible IAA moving through the petiole. Defoliation of *Coleus* epicotyls strongly reduces xylem differentiation, but this effect can be reversed by applying equivalent amounts of IAA in lanolin paste.

Jacobs and his coworkers also found that regeneration of vascular tissues around wounds in *Coleus* stems is also under the control of auxin (Fig. 16.3). *Coleus*, like other members of the mint family (Lamiaceae), have characteristic square stems with a vascular bundle at each corner. If a wedge-shaped incision is made that interrupts one of these vascular bundles, parenchyma cells in the region of the wound will differentiate into new vascular elements. These vascular elements will eventually reestablish continuity with the original bundle.

The differentiation of both xylem elements and phloem sieve tubes around the wound is limited and controlled by auxin supply. This can be shown by removal of leaves (a source of auxin) above the wound, for example, which reduces vascular regeneration. On the other hand, because auxin moves preferentially down the stem, removal of leaves below the wound has little or no effect. Furthermore, the extent of vascular regeneration is directly proportional to the auxin supply when exogenous auxin is substituted for the leaves.

Auxin is also required for vascular differentiation in plant tissue cultures (see Chap. 15). When buds, which are a source of auxin, are implanted into clumps of undifferentiated callus tissue in culture, differentiation of callus parenchyma into vascular tissue occurs in regions adjacent to the implant. The same effect is achieved when agar wedges containing IAA and sugars are substituted for the implanted bud.

SHOOT AND ROOT DEVELOPMENT

Axillary Bud Growth As a shoot continues to grow and the apical meristem lays down new leaf primordia, small groups of cells in the **axil** (the angle between the stem and the leaf primordium) of the primordia become isolated from the apical meristem and produce an **axillary bud.** In some cases, such as the bean (*Phaseolus*), the bud continues to grow, although at a much slower rate than the apical bud. In many plants, however, mitosis and cell expansion in the axillary bud is arrested at an early stage and the bud fails to grow. It has been known for some time that removal of the shoot apex, a common horticultural technique for producing bushy plants, stimulates the axillary bud to resume growth (Fig. 16.4). Apparently the apical bud is able to exert a dominant influence that suppresses cell division and enlargement in the axillary bud. For this reason, the phenomenon of coordinated bud development is known as **apical dominance.**

FIGURE 16.4 Apical dominance in broadbean (*Vicia faba*) (*A*) Control plants. (*B*) Removal of the stem apex, a source of auxin, promotes axillary bud growth. (*C*) Dominance can be restored by applying auxin (in lanolin paste) to the cut stem surface. (See color plate 6.)

Shortly after auxin was first discovered, K. V. Thimann and F. Skoog (1934) questioned whether there might be a relationship between the capacity of the shoot tip to release auxin and its capacity to suppress axillary bud development—in other words, is apical dominance controlled by auxin? Thimann and Skoog tested this idea by decapitating broad bean (*Vicia faba*) plants and applying auxin to the cut stump. Axillary bud development remained suppressed in the presence of auxin. Since this initial demonstration, the capacity of auxin to substitute for the shoot tip in maintaining apical dominance has been confirmed repeatedly.

How does auxin from the shoot apex suppress axillary bud development? The most widely accepted theory holds that the optimum auxin concentration for axillary bud growth is much lower than it is for the elongation of stems. The stream of auxin flowing out of the shoot apex toward the base of the plant is thought to maintain an inhibitory concentration of auxin at the axillary bud. Removal of this auxin supply by decapitation reduces the supply of auxin in the region of the axillary and thereby relieves the bud of inhibition. More direct evidence for the role of auxin transport is offered by the observation that triiodobenzoic acid (TIBA) and naphthylphthalamic acid (NPA), both inhibitors of auxin transport (to be discussed in Chap. 17), stimulate release of buds from dominance when applied to the stem between the shoot apex and the bud. In addition, lines of tomato that exhibit prolific branching (that is, the absence of apical dominance) also fail to export radioactively labeled IAA from the shoot apex.

It is now clear that auxin does not operate alone in the control of apical dominance. Cytokinins will antagonize the auxin effect. In many species the application of cytokinins either to the shoot apex or directly to the axillary bud will release the bud from inhibition. Tomato mutants expressing strong apical dominance contain lower amounts of cytokinins than those with normal dominance. Other experiments have shown that there is a correlation between the inhibition of bud growth and abscisic acid (ABA) content *of the bud.* For example, the ABA content of axillary buds in decapitated *Phaseolus* plants is lower than in intact controls. It is not clear whether ABA is imported from other tissues or synthesized locally in the bud, but the lowering of ABA content can be prevented by the application of IAA to the cut stem. Thus, it appears that the ABA content in the axillary bud is under control of IAA moving down from the shoot apex. Application of ABA to the shoot apex also releases axillary buds from inhibition. Ethylene production, known to be stimulated by auxin, has also been implicated in axillary bud inhibition, but the results are inconclusive. At this point, the mechanism for regulating axillary bud growth remains unresolved. It is clear that synthesis and transport of auxin from the apex is a significant component of the process. Other hormones, primarily cytokinins and ABA, may well be involved, but the nature of their interaction is complex and has not been clearly defined.

Leaf Abscission The process of shedding organs such as leaves and fruits is known as **abscission.** Abscission occurs as a result of the development of a special layer of cells, called the **abscission layer,** near the base of the petiole. As the organ ages, the cell walls in the abscission layer weaken and eventually separate. Abscission appears to be dependent on the relative concentrations of auxin on either side of the abscission layer. The auxin content of young, rapidly growing organs is relatively high, declining as the organ ages and approaches senescence. The effect of auxin on abscission can be demonstrated by excising a leaf blade while leaving the petiole attached to the stem. If auxin is applied to the cut petiole surface distal to the abscission layer, abscission of the petiole will be delayed when compared with the controls. These and similar experiments have shown that auxin delays abscission when applied in the early stages of leaf development but hastens abscission when applied in the later stages.

Root Elongation and Development Root elongation is particularly sensitive to auxin. IAA will promote the growth of excised root sections and intact roots, but only at very low concentrations (10^{-8} M or less). Higher concentrations of auxin, in the range that normally stimulates elongation of shoots (10^{-5} to 10^{-6} M), cause a significant inhibition of root growth. Inhibition of root growth by auxin is at least partly due to ethylene production, which is stimulated by high auxin concentrations. Removal of the root tip or application of auxin antagonists often promotes the growth of roots, which suggests that the endogenous auxin production in root tips is normally high enough to be inhibitory.

FIGURE 16.5 Auxin stimulated adventitious root development on American holly (*Ilex opaca*) cuttings. Row A: Treated with a 0.01% solution of indolebutyric acid solution for 17 hours before being rooted in sand. Row B: Untreated controls. (From T. E. Weier et al., *Botany*. 6th edition. New York, Wiley. 1982. Fig. 20.7. Used by permission of the authors.)

Although auxin commonly inhibits root elongation, high auxin concentrations will promote initiation of secondary, or branch, roots. Conversely, removal of young leaves or buds, both sources of auxin, will often reduce the number of secondary roots formed. These results indicate that secondary root initiation is normally controlled by auxin supplied by the shoot. Auxins also promote adventitious root formation on stems (Fig. 16.5). Adventitious roots arise from phloem tissue or, in some woody species such as willows and poplars, from preformed primordia in nodal regions. In stem sections of willow and similar species, adventitious roots will always form at the morphological basal end of the section, regardless of the orientation of the section. This orientation is presumably related to the polar movement of auxin in the stem (Chap. 17).

Flower and Fruit Development Production of flowers and fruit is clearly an important event in the developmental history of a plant. Auxin does not seem to play a major role in the initiation of flowers, since exogenous auxin for the most part tends to inhibit flower formation. Inhibition appears to be a secondary effect, however, resulting largely from auxin-induced ethylene production. In one exceptional case, members of the family Bromeliaceae exhibit a strong stimulation of the flowering response following application of either auxin or ethylene. This phenomenon has enabled commercial growers to control flowering and, consequently, production schedules in bromeliads such as pineapple simply by spraying with either auxin or chemicals that release ethylene.

Once flowers have been initiated, auxin appears to play several important roles in subsequent floral and fruit development. Flowers may be either **perfect,** a bisexual condition having both male and female reproductive structures in the same flower, or **imperfect,** in which a flower has either stamens or pistils but not both. Imperfect flowers may be further subdivided as **monoecious,** in which the staminate and pistillate flowers are borne on the same plant, or **dioecious,** in which the male and female flowers are borne on different plants. Corn (*Zea mays*), cucurbits (*Cucumis* sp.), and the oaks (*Quercus* sp.) are examples of the monoecious condition. In corn, the tassel is the male inflorescence and the ears are produced from the female flowers. Examples of dioecious plants are spinach (*Spinacea oleracea*), American holly, and *Ginkgo*.

The sex of imperfect flowers is genetically determined but it is also subject to modification by nutritional and environmental factors and exogenous hormones. In the cucurbits and many other plants the flower is bisexual in its early stages but one sex organ or the other aborts. Application of auxin during the bisexual stage ensures the formation of female flowers. There is some evidence that this effect may also be mediated by auxin-stimulated ethylene release. Nevertheless, it is a general rule that auxin promotes femaleness in flowers.

Normally, successful pollination and fertilization are required for fruit set (i.e., the initiation of ovary development) to occur. In the mid-1930s, it was found that pollen was a rich source of auxin and that extracts of pollen could stimulate fruit set in unpollinated solanaceous plants (e.g., tomato). The phenomenon of fruit development in the absence of pollination is known as **parthenocarpy.** Because parthenocarpy results in seedless fruits, the phenomenon has significant economic implications (Box 16.2). In California, the synthetic auxin 4-chlorophenoxyacetic acid is used to stimulate fruit set in early tomato crops, when cool night temperatures would otherwise tend to reduce fruit set. Auxin induces parthenocarpy in a small number of plants, particularly in the families Solanaceae (tomato, peppers) and Cucurbitae (cucumber, pumpkin) and in *Citrus*. A classic study of strawberry, conducted by J. P. Nitsch in the 1950s, indicated that the developing seed was the source of auxin for continuing fruit development. Strawberry bears its seeds (achenes) on the surface of the fruit. Nitsch found that removal of the seeds prevented further development of the fruit, but supplying the fruit with auxin restored normal development.

Box 16.2 Commercial Applications of Hormones

Hormones and other regulatory chemicals are now used in a variety of applications where it is desirable for commercial reasons to control some aspect of plant development.

The synthetic auxins are used in commercial applications largely because they are resistant to oxidation by enzymes that degrade IAA. In addition to their greater stability, the synthetic auxins are often more effective than IAA in specific applications. One of the most widespread uses of auxin encountered by the consumer is the use of 2,4-D in weed control. 2,4-D and other synthetic compounds, such as 2,4,5-T and dicamba, express auxin activity at low concentrations, but at higher concentrations are effective herbicides. The introduction of 2,4-D and 4-chlorophenoxyacetic acid (4-CPA) as a herbicide in 1946 revolutionized our approach to agriculture. For reasons that are not clear, chlorinated phenoxyacetic acids are selectively toxic to broadleaf species. 2,4-D remains the principal component of "weed-and-feed" mixtures for home lawn care as well as for control of broadleaf weeds in cereal crops. The synthetic auxins are favored in commercial applications because of their low cost and greater chemical stability.

Indolebutyric acid and naphthaleneacetic acid are both widely used in vegetative propagation—the propagation of plants from stem and leaf cuttings. This application can be traced to the propensity for auxin to stimulate adventitious root formation. Generally marketed as "rooting hormone" preparations, the auxins, usually a synthetic auxin such as NAA or IBA, are mixed with an inert ingredient such as talcum powder. Stem cuttings are dipped in the powder prior to planting in a moist sand bed in order to encourage root formation.

4-CPA may be sprayed on tomatoes to increase flowering and fruit set while NAA is commonly used to induce flowering in pineapples. This latter effect is actually due to auxin-induced ethylene production. NAA is also used both to thin fruit set and prevent preharvest fruit drop in apples and pears. These seemingly opposite effects are dependent on timing the auxin application with the appropriate stage of flower and fruit development. Spraying in early fruitset, shortly after the flowers bloom, enhances abscission of the young fruits (again, due to auxin-induced ethylene production). Thinning is necessary in order to reduce the number of fruit and prevent too many small fruits from developing. Spraying as the fruit matures has the opposite effect, preventing premature fruit drop and keeping the fruit on the tree until it is fully mature and ready for harvest.

The use of synthetic auxins, especially the chlorinated forms, as herbicides has come under close scrutiny by environmental groups because of potential health hazards. 2,4,5-T, for example, has been banned in many jurisdictions because commercial preparations contain significant levels of dioxin, a highly carcinogenic chemical.

The principal commercial use of gibberellins is in the production of table grapes, such as the "Thompson Seedless." A gibberellin spray in the early stages of flowering thins the cluster by stimulating elongation of the floral stems. This spreads the flowers out, allowing for the development of larger fruit. Larger fruit size is encouraged by a second spray at time of pollination and fruit set. Gibberellins have also been used to enhance germination and stimulate early seedling emergence and growth in species such as grape, citrus, apples, peach, and cherry. Treatment of cucumber plants with gibberellin will promote formation of male flowers, which is useful in the production of hybrid seed.

Auxins may also be used to control abscission of fruits. Depending upon the timing and level of application, auxin may cause early fruit drop or prevent premature fruit drop. The early enhancement of fruit drop by auxin, shortly after fruit set, is a way of chemically thinning the fruit to increase the size of those remaining on the tree. Applied later, auxin delays abscission, thus preventing premature fruit drop. This helps to maintain fruit on the tree for a more efficient harvest.

One particular complication that arises with respect to auxin physiology is the interaction between auxin and ethylene. Numerous examples are known where auxin, particularly at high concentrations, can cause inhibitory effects. As already noted in several instances, ethylene production can be promoted by auxin. There is increasing evidence that many of the effects of auxin, especially inhibitory effects, may actually be attributable to auxin-stimulated ethylene production. Many of the known effects of auxin on flower and fruit development can be explained in this way. Because ethylene is a volatile gas and is effective in low concentrations, its presence or role is not necessarily evident unless the investigator specifically looks for it.

GIBBERELLINS

***Gibberellins** are produced by both fungi and higher plants. The exogenous application of gibberellins causes hyperelongation of intact stems. Gibberellins are also prominently involved in seed germination and mobilization of endosperm reserves during early embryo growth, as well as flower and fruit development.*

Gibberellins are the only group of plant hormones that can be defined on the basis of their chemical structure rather than their biological activity. The gibberellins are an extensive chemical family based on the ***ent-***

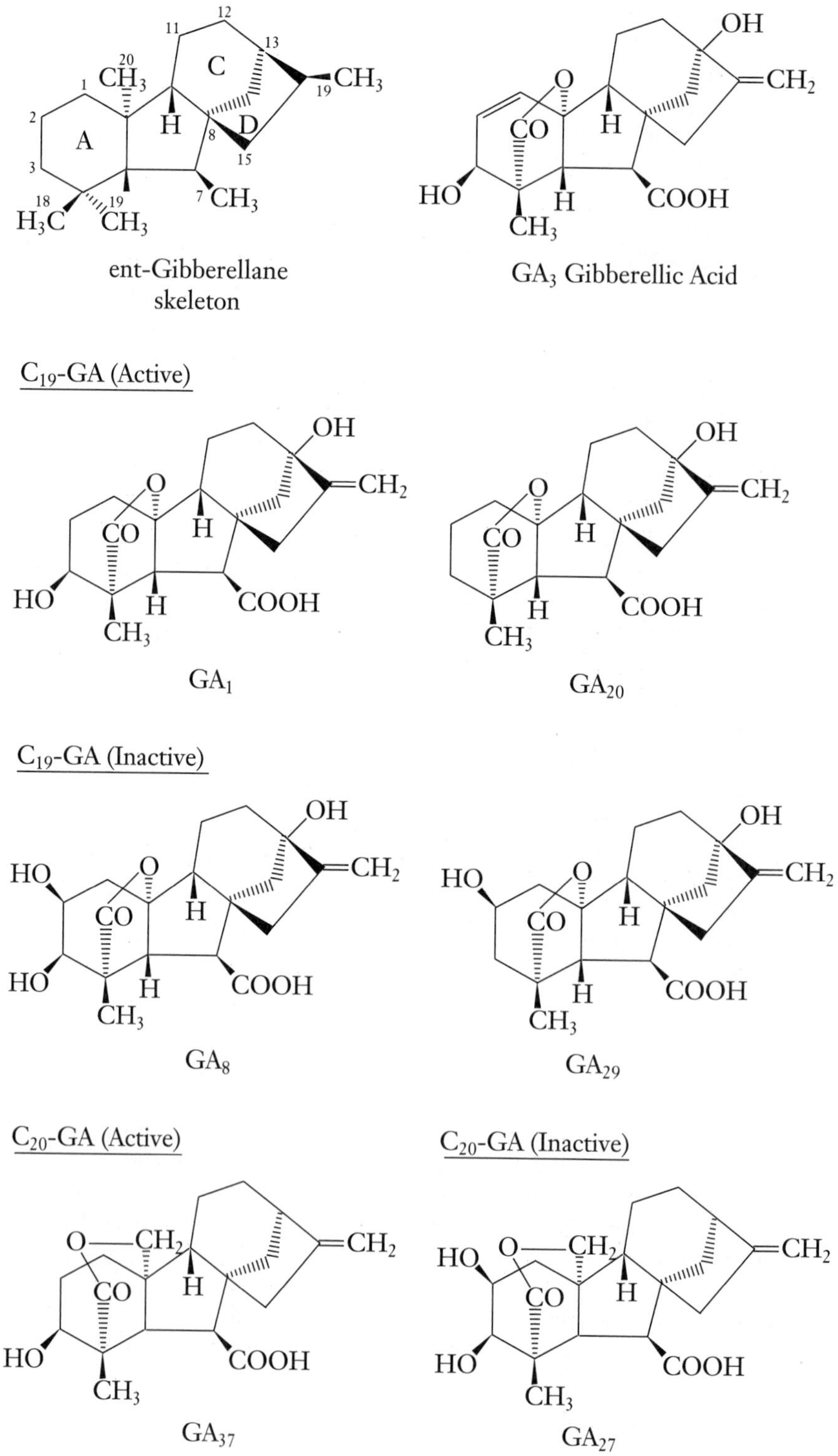

FIGURE 16.6 The *ent*-gibberellane skeleton and chemical structures of selected active and inactive gibberellins.

gibberellane structure (Fig. 16.6). More than 80 are now known and additional members are added almost every year. A little more than one-third of the gibberellins characterized to date have retained the full complement of 20 carbon atoms and are known as C_{20}-gibberellins. The others have lost carbon atom number 20 and are consequently known as C_{19}-gibberellins.

All gibberellins that are demonstrated to be naturally occurring and that have been chemically characterized are assigned an "A" number (MacMillan and Takahashi, 1968). This number does not imply chemical relationships; it is assigned roughly in order of discovery. GA_3, a C_{20} gibberellin also known as **gibberellic acid,** was one of the first to be isolated and characterized. Because GA_3 is readily extracted from fungal cultures it is also the most common commercially available form and, consequently, is perhaps the most studied of the gibberellins. On the other hand, GA_1 and GA_{20}, both C_{19}-GAs, are probably the most active and, consequently, the most important gibberellins in higher plants.

One might ask why there are so many gibberellins. The answer seems to be that a large proportion of the known gibberellins have little or no biological activity. They are either intermediates in the synthesis of active forms or metabolic products that still retain the basic structure but are no longer biologically active (to be discussed in Chap. 17). It is worth noting that the number of gibberellins found in any one species or organ may be very small and the number of active gibberellins smaller yet. It is believed, for example, that GA_1 may be the principal, if not the only, active GA regulating stem elongation in higher plants (Phinney, 1984).

There are certain structural requirements for GA activity. A carboxyl group at carbon-7 is a feature of all GAs and is required for biological activity, and C_{19}-GAs are more biologically active than C_{20}-GAs. In addition, those GAs with 3-β-hydroxylation, 3-β,13-dihydroxylation, or 1,2-unsaturation are generally more active; those with both 3-β-OH and 1,2-unsaturation exhibit the highest activity. Introduction of a hydroxyl (—OH) at the 2 position clearly inactivates the molecule and may be regarded as a mechanism for removing hormones from the active pool (see Chap. 17).

THE PHYSIOLOGICAL ACTION OF GIBBERELLINS

Control of Shoot Elongation

Dwarf Plants It was excessive stem elongation in infected rice plants that led to the discovery of gibberellins and hyperelongation of stem tissue remains one of the more dramatic effects of gibberellins on higher plants. Unlike auxins, gibberellins promote elongation almost exclusively in intact plants rather than excised tissues. Nowhere is this more evident than in the control of internode elongation in genetic dwarfs. The relationship between dwarfing or internode length genes and gibberellins was pioneered by the work of B. O. Phinney on maize (*Zea mays*) and P. W. Brian and coworkers on garden pea (*Pisum sativum*). Since these pioneering studies, experiments have been conducted with dwarf mutants of rice (*Oryza sativa*), bean (*Phaseolus vulgaris*), *Arabidopsis thaliana* and several others. In all cases, application of exogenous gibberellin to the dwarf mutant restores a normal, tall phenotype (Fig. 16.7). Exogenous gibberellin has no appreciable effect on the genetically normal plant.

In maize, more than thirty mutants that influence plant height have been described. Maize plants expressing these mutations have shortened internodes, due to reduced cell division and cell elongation, and at maturity reach only 20 to 25 percent of the height of normal plants. At least five of these mutants (d_1, d_2, d_3, d_5, an_1) exhibit the normal phenotype when treated with GA_3, but show no response to other hormones or growth regulators. Activity assays such as these have been supported by other biochemical and radiotracer experiments, demonstrating conclusively that internode elongation in maize is under control of gibberellins. Specifically, each mutation blocks a different step in the biosynthetic pathway toward GA_1, which is the active form of gibberellin in maize (Chap. 17). Similar experiments with the *Le* allele in garden pea (the same allele studied by Gregor Mendel in his pioneering genetic studies) have demonstrated that this dwarf genotype (the

FIGURE 16.7 The effect of gibberellic acid on dwarf pea. Left: Control, showing reduced internode elongation. Right: Gibberellin treated. Enhanced internode elongation following a foliar-drench with 5 × 10^{-4} M gibberellin.

homozygous recessive *le/le*) also blocks the synthesis of GA_1. In both maize and pea, it has been shown that the dwarf genotype leads to a significant reduction in the gibberellin levels.

While studies with dwarf plants have been instrumental in linking gibberellins with stem elongation, there are other dwarf mutants known that do not respond to application of gibberellin. These mutants may be unrelated to gibberellin-controlled growth and subject to other, as yet unknown, regulating factors.

Rosette Plants Additional support for the role of gibberellins in stem elongation comes from the study of rosette plants. A rosette is essentially an extreme case of dwarfism in which the absence of any signifant internode elongation results in a compact growth habit characterized by closely spaced leaves. The failure of internode to elongate may result from a genetic mutation, or may be environmentally induced. Regardless of the cause, hyperelongation of stems in rosette plants is invariably brought about by the application of small amounts of gibberellin (Fig. 16.8).

Environmentally limited rosette plants such as spinach (*Spinacea oleraceae*) and cabbage (*Brassica* sp.) generally do not flower in the rosette form. Just before flowering, these plants will undergo extensive internode elongation, a phenomenon known as **bolting.** Bolting is normally triggered by an environmental signal, either photoperiod (as in spinach) or a combination of low temperature and photoperiod (as in cabbage). We will return to the phenomena of photoperiod and cold requirement in later chapters. It is sufficient to note here that, under conditions normally conducive to the rosette habit, spinach, cabbage, and many other rosette plants can be induced to bolt by an exogenous application of gibberellic acid.

FIGURE 16.8 Gibberellin-stimulated stem growth in a rosette genotype of *Brassica napus*. Treatments were (from left): 0; 0.5, 1.0, 10.0 ng GA_3 per plant, applied to the meristem. (See color plate 7.)

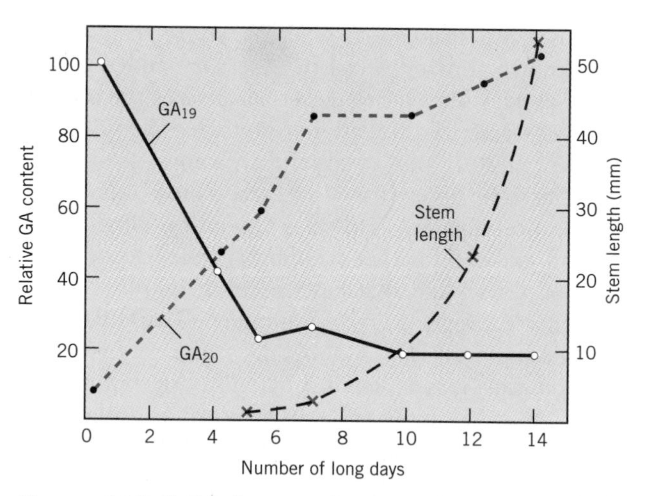

FIGURE 16.9 Following transfer from short to long days, spinach plants (*Spinacea oleraceae*) exhibit extensive stem elongation, accompanied by changes in gibberellin content. A decrease in the level of the inactive GA_{19} is matched by a corresponding increase in the active form G_{20}. (Redrawn from Metzger and Zeevaart, 1980. Copyright American Society of Plant Physiologists.)

The above results suggest that (a) gibberellins are a limiting factor in the stem growth of rosette plants and (b) the effect of long days or cold treatment is to remove that limitation. These possibilities have been confirmed in spinach and *Silene armeria*, both photoperiodic plants requiring long days to flower, by the extensive investigations of J. A. D. Zeevaart and co-workers. Spinach contains six gibberellins, including GA_{19} and GA_{20}. GA_{20} will cause bolting in spinach under short day conditions while GA_{19} is biologically inactive. Zeevaart and coworkers found that rosette plants of spinach contain high levels of the inactive form GA_{19} and low levels of the active GA_{20} (Metzger and Zeevaart, 1980). Upon transfer to long day conditions, however, the level of GA_{19} declined while the level of GA_{20} increased (Fig. 16.9). The reciprocal changes in GA_{19} and GA_{20} levels suggests a precursor-product relationship, which was confirmed in whole plants by feeding deuterium (^2H) labeled precursors (Gianfagna et al., 1983). In other experiments, ^{14}C-labeled GA_{19} was converted to GA_{20} by cell-free extracts from spinach plants maintained under long days, but not in extracts from short day plants (Gilmour et al., 1985). On the basis of these studies, it may be concluded that gibberellins have a significant role in the control of stem elongation in rosette plants.

The relationship between gibberellins and stem elongation in cold-requiring plants has not been studied as thoroughly as it has for photoperiodically sensitive plants. As mentioned above, exogenous application of GA_3 will substitute for the cold requirement in many plants and there is some evidence, based on bioassays, that gibberellin-like activity increases in plants following cold-treatment. It is reasonable to expect on the basis of these studies that changes in gibberellin biosynthesis or metabolism are involved, but a more thorough study is required.

Inhibition of Stem Growth The growth of many stems can be reduced or inhibited by synthetic chemicals that block gibberellin biosynthesis. These so-called growth retardants or antigibberellins include AMO-1618, cycocel (or, CCC), Phosphon-D, ancymidol (known commercially as A-REST), and alar (or, B-nine) (Fig. 16.10). These inhibitors mimic the dwarfing genes by blocking specific steps in gibberellin biosynthesis (see Chap. 17), thus reducing endogenous gibberellin levels and suppressing internode elongation. Commercial flower growers have found these inhibitors useful in producing shorter, more compact poinsettias, lilies, and chrysanthemums. Soil drenches or direct spray of young plants with these chemicals will significantly reduce stem length without affecting flower size.

In some areas of the world, wheat tends to "lodge" near harvest time, that is, it becomes top-heavy with grain and falls over. Spraying the plants with growth retardants produces a shorter, stiffer stem and thus prevents lodging. Growth retardants also have been used to reduce the need for pruning of vegetation under power lines. Alar has been widely used as a spray on cherries and apple; it enhances fruit color and produces a firmer fruit that facilitates harvesting. Use of Alar for such applications has come under severe criticism from environmentalists and consumer groups.

Seed Germination A role for gibberellins in mobilization of reserves during seed germination was first suggested by the work of L. G. Paleg in the late 1950s. Germinating cereal grains secrete a range of hydrolytic enzymes, including α-amylase and proteases, which are involved in the digestion of carbohydrate and protein. Cereal grains such as barley may be transected to produce two half-seeds, with one half-seed containing the embryo and one without. The embryo-containing half-seed will proceed to secrete α-amylase, digest the starchy endosperm, and germinate. The embryoless half-seed cannot, of course, germinate but neither does it produce elevated levels of α-amylase or any of the other hydrolytic enzymes required for germination. Treatment of the embryoless half-seed with gibberellic acid, however, will stimulate the half-seed to produce high levels of a-amylase (Fig. 16.11). GA-stimulated α-amylase secretion requires the presence of the aleurone, a layer of cells surrounding the endosperm in grass seeds. GA-stimulated α-amylase secretion can be blocked by inhibitors such as actinomycin D and cycloheximide, which inhibit RNA and protein synthesis, respectively, which indicates that gibberellin-stimulated *de novo* synthesis of α-amylase by the aleurone layer is an early event in germination. It should be noted that much of this work has been conducted on a single cultivar (cv. "Himalaya") of barley and the responsiveness of seeds to gibberellin can be significantly affected by environment during seed development and maturation. Seed from barley grown at high temperatures, for example, produce high levels of α-amylase in the absence of added

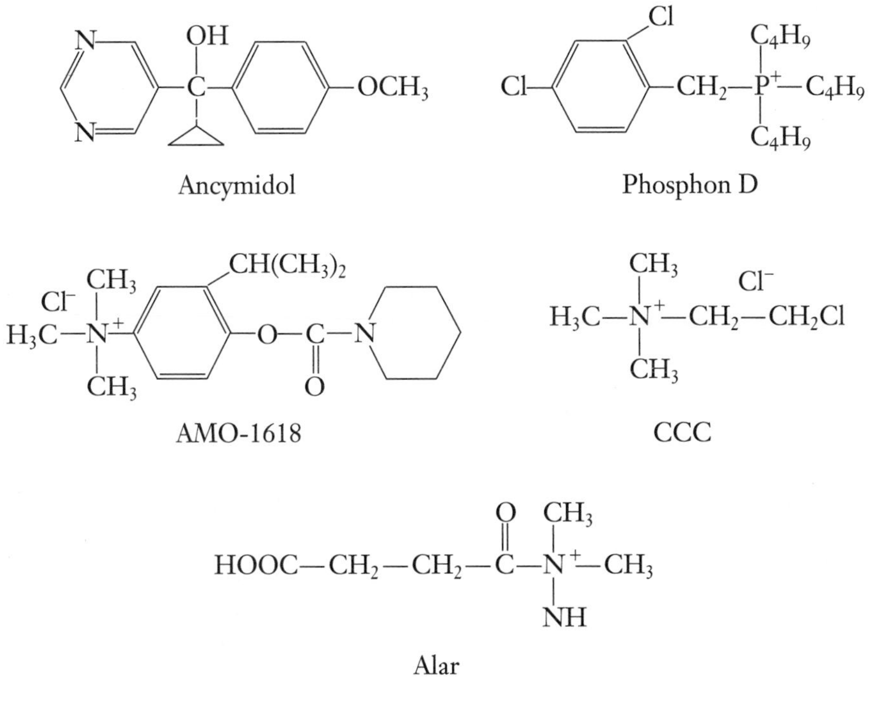

FIGURE 16.10 Chemical structures of five growth retardants, also known as antigibberellins.

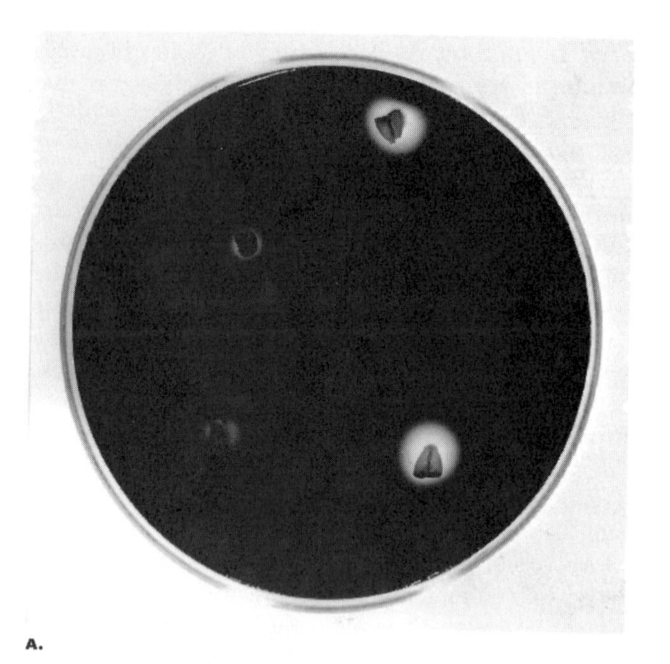

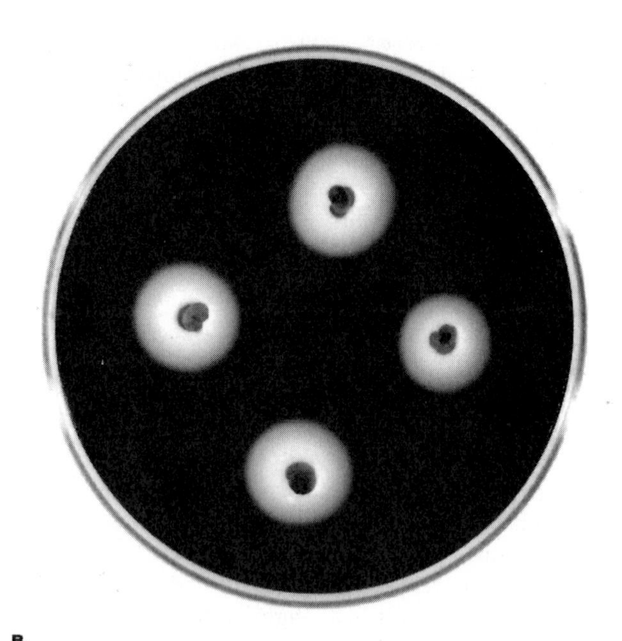

FIGURE 16.11 Gibberellin-stimulated release of α-amylase from barley half seeds. Embryoless half seeds were incubated for 48 hours on the surface of a starch-agar gel. The gel was then washed with IKI reagent, which reacts with starch to form a blue-black color. A clear circle surrounding the half-seed shows that the starch has been digested, indicating α-amylase activity. The plate on the right also contained 10 nanomoles gibberellic acid. The control plate (left) also contains four half-seeds, but two produced no a-amylase. (From a student experiment.)

GA. Nonetheless, the principles that have emerged from the barley system appear to be widely applicable to wheat, rye, oats, and other cereal grains. The mechanism of gibberellin action in enzyme secretion will be discussed further in Chapter 17.

Flowering The role of gibberellins in bolting of rosette plants was introduced earlier. In addition to genetic mutants, bolting will also occur in biennials or in long day flowering plants that require a cold treatment before flowering (see Chaps. 20, 21). In either situation, it is difficult to differentiate between control of stem elongation on the one hand and control of flowering on the other. In the normal course of development, stem elongation often, but not always, appears to be a necessary prelude to flowering. At one time, the substitution of gibberellin for the long day or cold requirement seemed to indicate a role for gibberellin in the flowering process itself. However, inhibitors of GA biosynthesis will suppress photoperiodic-induced stem elongation without interfering with flowering. It thus appears that stem elongation and flowering are separate, albeit interdependent, parts of the bolting phenomenon. The role of gibberellins is in stem elongation, not flowering. Still, gibberellins do influence the capacity of plants to flower as well as sexual characteristics of flowers and subsequent fruit development.

Some plants may be induced to flower almost immediately following germination. Under appropriate conditions, for example, *Chenopodium rubrum* (goosefoot) and *Pharbitis nil* (Japanese morning glory) will produce flowers with only the cotyledonary leaves expanded. Many perennial plants, however, must achieve a minimum stage of development before they are capable of flowering. Such plants are said to pass through a **juvenile phase** (Metzger, 1987). The length of the juvenile phase can range from a few weeks to many years (Table 16.2). The juvenile phase is often difficult to recognize. While in some plants the juvenile phase differs in morphology or leaf shape from that of the mature or adult plant, in most cases it can be distinguished only by the inability to flower. The juvenile leaf of the bean plant (*Phaseolus*), for example, is simple while the adult leaves are compound (trifoliate). The juvenile broad bean (*Vicia faba*) produces four or five bifoliate leaves, but the adult leaf is comprised of three or four leaflet pairs. One plant that has been extensively studied is English ivy (*Hedera helix*), a perennial vine. Juvenile branches of ivy have longer internodes between palmately lobed leaves and do not flower. Adult branches

TABLE 16.2 **Duration of the juvenile phase in selected species. Abridged from Metzger, 1987. (Reprinted by permission of Kluwer Academic Publishers.)**

Plant	Duration (years)
Perilla (*Perilla crispa*)	≤0.25
Bryophyllum (*Bryophyllum daigremontianum*)	1–2
Apple (*Malus pumila*)	6–8
Scotch pine (*Pinus sylvestris*)	5–10
Pear (*Pyrus communis*)	8–12
Douglas fir (*Pseudotsuga menzesii*)	15–20
White fir (*Abies alba*)	25–30
European beech (*Fagus sylvatica*)	30–40

have shortened internodes, bear entire, ovate leaves, and will flower. Axillary buds on the adult branch normally develop into new adult branches, but treatment of the bud with GA_3 will cause the branch to revert from the adult to the nonflowering juvenile form.

In conifers, where the juvenile phase is typically 10 to 20 years, the inability to produce cones in young plants presents an obstacle to commercial breeding programs. This obstacle can be overcome by the application of exogenous gibberellins, which promotes precocious cone formation in many conifers. The mechanism for this response is unknown. However, the gibberellin effects are only temporary. Cessation of the gibberellin treatment leads to abscission of newly formed cones and reversion to juvenile characteristics.

In cases such as *Cucumis* and *Cannabis sativa*, where auxins or ethylene promote femaleness in imperfect flowers, an application of gibberellins will promote formation of male flowers. Growth retardants (which lower gibberellin levels) will feminize the flowers, an effect that can be reversed by gibberellic acid (GA_3). These observations tempt one to conclude that sex expression is normally mediated by the internal balance of auxin/ethylene and gibberellin. Indeed, in those few cases where measurements of endogenous hormone levels in imperfect flowers have been reported, the results are consistent with this postulated role. As always, it seems, there are other results that are not so consistent. Thus, while *exogenous* hormones clearly influence sexual expression in plants, the role of *endogenous* hormones remains obscure.

CYTOKININS

***Cytokinins** are N^6-substituted derivatives of the nitrogenous purine base adenine, characterized by their ability to stimulate cell division in tissue culture.*

Kinetin (N^6-furfurylamino purine) was the first cytokinin to be discovered (Fig. 16.12). Kinetin does not occur naturally but was originally synthesized from her-

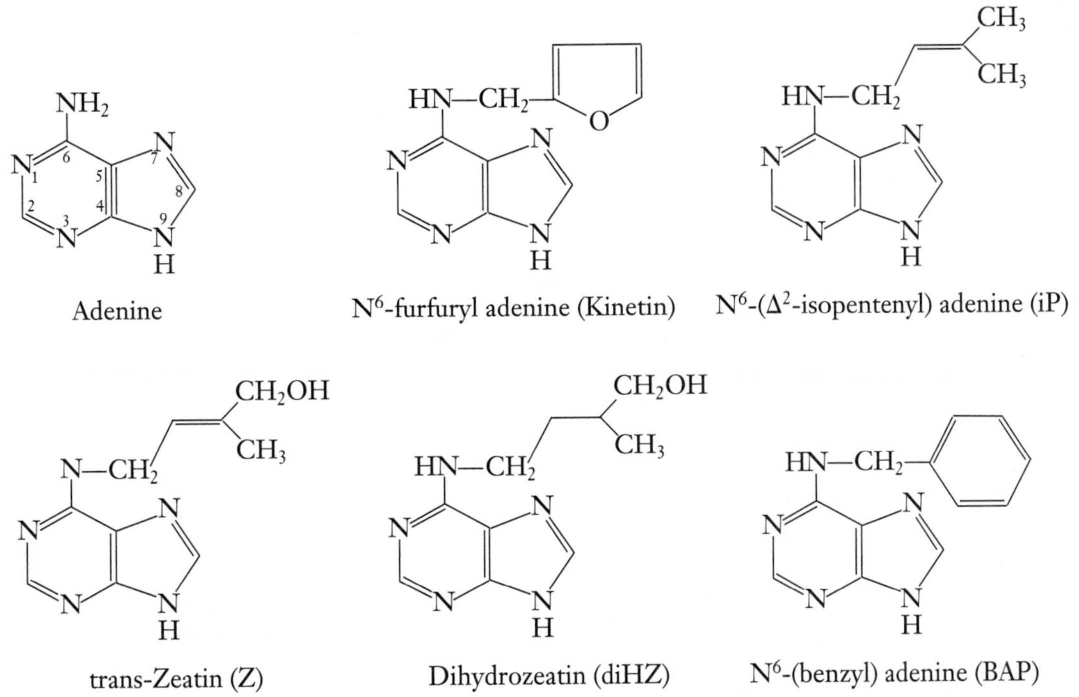

Adenine

N^6-furfuryl adenine (Kinetin)

N^6-(Δ^2-isopentenyl) adenine (iP)

trans-Zeatin (Z)

Dihydrozeatin (diHZ)

N^6-(benzyl) adenine (BAP)

FIGURE 16.12 **The chemical structures of adenine and five adenine derivatives with cytokinin activity.**

ring sperm DNA (see Box 16.1: Historical Perspectives—Discovering Plant Hormones). The most widespread naturally occurring cytokinin in higher plants is **zeatin.** Zeatin and other naturally occurring cytokinins are commonly found with a ribose (the **riboside**) or ribose-phosphate (the **ribotide**) at the 9-position. In addition to stimulating cell division, cytokinins also influence shoot and root differentiation in tissue culture, the growth of lateral buds and leaf expansion, chloroplast development, and leaf senescence.

THE PHYSIOLOGICAL ROLES OF CYTOKININS

Cell Division and Morphogenesis Cytokinins are noted primarily for the ability to induce cell division in plant tissue and cell cultures. Although most mature, differentiated plant cells do not normally divide, many cells may be induced to undergo division when cultured on artificial media containing vitamins, mineral salts, a carbon source, and an optimal concentration of hormones. On a solid agar medium, cells derived from stem pith and cortex, cotyledon, leaf, and other tissues will divide and enlarge to produce a mass of largely undifferentiated cells referred to as **callus tissue** (Chap. 15). Small lumps of callus will also form in liquid culture, but with gentle agitation the cells may be encouraged to separate and form a **free cell culture.** In either case, continued cell division and growth does not normally occur in the absence of cytokinin.

Occasionally, after a prolonged period of continuous subculture, cultured tissues may spontaneously acquire the capacity to synthesize cytokinin and will thus grow on a medium without added hormone. Such tissues are referred to as **habituated.** Once acquired, the habituated state is quite stable; that is, such cultures do not revert to the cytokinin-requiring state. Just how the capacity to synthesize cytokinins is switched on in habituated tissues is not known but it clearly has developmental implications.

Cytokinins also influence morphogenesis in cultured tissues. Maintenance of undifferentiated callus growth is generally achieved with roughly equal molar concentrations of cytokinin and auxins. High molar ratios of cytokinin to auxin tend to induce bud development, while high ratios of auxin to cytokinin will encourage root development. It is thus possible, by manipulating the balance of cytokinin and auxin in the medium, to regenerate complete plantlets from undifferentiated callus tissue in sterile culture. The plantlets can then be transplanted into soil in the greenhouse or field where they grow into fully competent, mature plants. This capacity to regulate morphogenesis in cultured tissues has found considerable use as a propagation method in horticulture and agriculture. Referred to as **micropropagation,** tissue culture can be used to produce large numbers of virtually identical, viral-free clones in a short time (Chap. 23).

FIGURE 16.13 Crown gall, a neoplastic growth on the stem of a *Bryophyllum* plant. Crown gall is the result of infection with the bacterium *Agrobacterium tumefaciens*. The host plant cells are genetically transformed; that is, the bacterial genes that cause an overproduction of auxin and cytokinin are incorporated into the host cell genome. (See color plate 8.)

Another example of cytokinin-mediated cell division with significance to plant physiology is the development of tumorous growths (called **crown gall**) following bacterial infection (Fig. 16.13). *Agrobacterium tumefaciens* is a bacterium capable of invading wounds or lesions, usually near the junction of the stem and the root (the "crown"). Cells in the region of the invasion proliferate to form large, undifferentiated cell masses, called crown gall tumors. There is evidence that these infected cells have been genetically transformed with the capacity to produce cytokinins. Crown gall tissue may be rid of viable bacteria by heating to a temperature of 42 °C. Bacteria-free tissue can then be subcultured indefinitely in the absence of added cytokinin. The tumorous tissue, unlike control stem tissue, has the biosynthetic machinery to produce cytokinins and contains significant amounts of free cytokinins.

Virulent strains of *Agrobacterium* contain a circular piece of extrachromosomal DNA: the Ti or tumor-inducing plasmid. A portion of this plasmid (referred to as the T-DNA) contains genes encoding enzymes for the synthesis of both auxin and cytokinin as well as a class of nitrogenous compounds called **opines.** These genes are not expressed in the bacterium, but upon invasion become incorporated into the nuclear DNA of the host plant cell—that is, the host plant becomes **transformed**—where they are expressed. The transformed cells thus produce an excess of both auxin and cytokinins that stimulate neoplastic growth. The transformed cells also produce the opines, which serve primarily as a nitrogen source for the invading bacterium. Because of the ease with which it transforms plant cells, the Ti plasmid has become widely used for genetic engineering, the introduction of foreign genes into plants (see Chap. 23). For these purposes, the plasmid is first disarmed by removing the genes for auxin and cytokinin biosynthesis. Then other genes of interest may be cloned into the plasmid, which is then used to transform target cells.

Nutrient Mobilization and Senescence When a mature leaf is detached from a plant, it undergoes a process known as **senescence.** Senescence is characterized by the breakdown of protein, nucleic acids and other macromolecules, a loss of chlorophyll, and the accumulation of soluble nitrogen products such as amino acids. Senescence is a normal consequence of the aging process and will occur even when the supply of water and minerals is maintained.

At present, there are three kinds of evidence indicating a role for cytokinins in control of senescence. First is the observation that exogenous application of cytokinin to detached leaves will delay the onset of senescence, maintain protein levels, and prevent chlorophyll breakdown. Application of cytokinins will also delay the natural senescence of leaves on intact plants. The second kind of evidence consists of correlations between endogenous cytokinin content and senescence. For example, detached leaves that have been treated with auxin to induce root formation at the base of the petiole will remain healthy for weeks. In this case, the growing root is a site of cytokinin synthesis and the hormone is transported through the xylem to the leaf blade. If the roots are continually removed as they form, senescence of the leaf will be accelerated. It has also been observed that when a mature plant begins its natural senescence, there is a sharp decrease in the level of cytokinins exported from the root.

A third and particularly convincing line of evidence comes from recent studies employing recombinant DNA techniques. Tobacco plants (*Nicotiana tobacum*) have been transformed with the *Agrobacterium* gene for cytokinin biosynthesis, designated *tmr*, described above (Smart et al., 1991). The *tmr* gene encodes for the enzyme ***iso*-pentenyl transferase**, which catalyzes the rate-limiting step in cytokinin biosynthesis (see Chap. 17). In this case, the *tmr* gene was linked to a **heat shock promoter.** A **promoter** is a sequence of DNA that signals where the transcription of messenger RNA (mRNA) should begin. The heat shock promoter is normally involved in the heat shock response of plants, which is induced by a brief period of high temperature. The heat shock response consists of the synthesis of a new set of proteins called the **heat shock proteins.** The heat shock promoter is thus active only when subjected to a high temperature treatment. By linking the *tmr* gene to the heat shock promoter, cytokinin biosynthesis can be turned on in the transformed plants simply by subjecting the plants to a brief period at high temperature. Catharine Smart and her colleagues noted that a heat shock of 42 °C for 2 hours caused a 17-fold increase in zeatin levels in transformed plants compared with untransformed control plants. When subjected to heat shock on a weekly basis over a twelve-week period, transformed plants exhibited a marked release of lateral buds from apical dominance as well as delayed senescence (Fig. 16.14). Transformed but non-heat-shocked plants also remained green longer than normal plants

FIGURE 16.14 Cytokinin control of senescence and bud growth in tobacco. Tobacco callus cells, genetically transformed such that cytokinin production could be stimulated by heat shock, were allowed to regenerate plantlets. From left to right: transformed heat-shocked plantlets; untransformed heat-shocked plantlets; transformed controls (no heat shock); untransformed controls. Note especially the proliferation of lateral buds and absence of senescence in the transformed, heat-shocked plantlets. (From C. Smart et al., *The Plant Cell* 3:647, 1991. Copyright American Society of Plant Physiologists. Photo courtesy of C. Smart). (See color plate 9.)

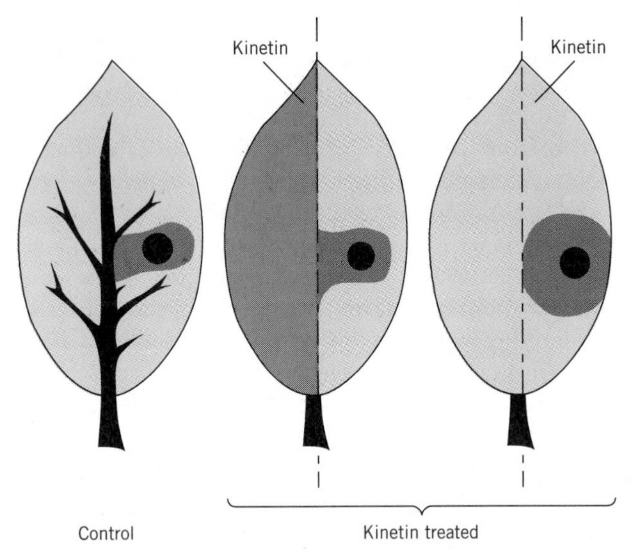

FIGURE 16.15 Diagram of an experiment demonstrating the role of cytokinin in nutrient mobilization. Radioactive ^{14}C-aminobutyric acid was applied to the area indicated by the dark spot. Left: Control. Radioactivity spreads into the vascular tissue for export through the petiole. Center: Radioactivity accumulates in the left half of the leaf, which has been treated with kinetin. Right: Radioactivity is retained near the point of application when the right half of the leaf is treated with kinetin. (Based on experiments of K. Mothes, *Régulateurs Naturels de la Croissance Végetale,* 123, 1963. Centre National de la Recherche Scientifique.)

but did not exhibit release from apical dominance. This is probably due to "leakiness" on the part of the promoter, allowing production of a very small but effective amount of cytokinin even at normal temperature.

The mechanism by which cytokinins are able to delay senescence is not clear, but there is some evidence that cytokinins exert a role in mobilizing nutrients. The classic experiment of K. Mothes and coworkers, summarized in Figure 16.15, illustrates this point. In this experiment, a nutrient labeled with radioactive carbon (e.g., ^{14}C-glycine) is applied to a leaf after a portion of the leaf has been treated with cytokinin. Invariably the radioactivity is transported to and accumulates in the region of cytokinin treatment. A variety of similar experiments have led to the hypothesis that cytokinins direct nutrient mobilization by stimulating metabolism in the area of cytokinin application. This creates a new sink—an area that preferentially attracts metabolites from the region of application (the source). It is unlikely that cytokinins act directly through stimulating protein synthesis since the mobilization of nonmetabolites such as a-aminoisobutyric acid is directed by cytokinins equally well.

Other Cytokinin Effects In addition to the effects described above, exogenously applied cytokinins will influence a number of other developmental phenomena. Cytokinins will stimulate cell enlargement in a limited number of systems, in particular the cotyledons of cucumber and sunflower. It should be noted, however, that cotyledon expansion is not stimulated by either auxins or gibberellins. As described earlier, the application of cytokinins will stimulate release of axillary buds from apical dominance, thus antagonizing the effect of auxins. This cytokinin-auxin antagonism is believed to account for the phenomenon of "witch's broom," an example of extreme axillary bud release (Fig. 16.16). The witch's broom syndrome appears on a wide variety of plants and most often results from parasitism by fungi, bacteria, or mistletoes (*Arceuthobium* sp.), a dwarf flowering shrub that parasitizes predominantly spruce (*Picea*), larch (*Larix*), and pine (*Pinus*). Although the exact nature of the hormonal imbalance is not known, it is believed that the parasitism stimulates an overproduction of cytokinin. The resulting release of apical dominance produces a dense mass of short branches.

FIGURE 16.16 Witch's broom on white pine (*Pinus strobus*). Witch's broom is the result of a fungal infection that stimulates an overproduction of cytokinin and uncontrolled axillary bud development.

ABSCISIC ACID

Abscisic acid is a terpenoid involved primarily in regulating seed germination, inducing storage protein synthesis, and modulating water stress.

Unlike the previous three hormone classes, abscisic acid (ABA) is a single compound (Fig. 16.17). Originally thought to be involved in regulating both abscission and bud dormancy, ABA now appears to have little to do with either of these phenomena. Two major areas of ABA action appear to be in the mobilization of reserves during seed development and germination and in the response of leaves to water stress. ABA is known to in-

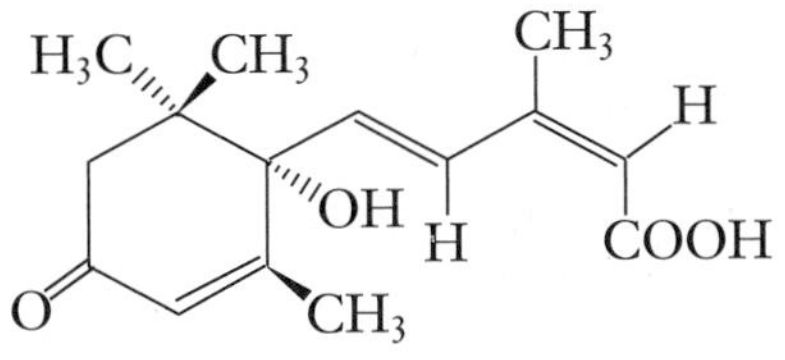

Abscisic Acid

FIGURE 16.17 The chemical structure of abscisic acid.

duce transport of photosynthate toward developing seeds and to promote the synthesis of storage protein. During germination of cereal grains, ABA antagonizes the promotory effect of gibberellin on α-amylase synthesis. Relatively large amounts of ABA are rapidly synthesized in the leaves in response to water stress, where it appears to play a major role in regulating stomatal opening and closure.

THE PHYSIOLOGICAL ROLES OF ABSCISIC ACID

Most temperate zone woody plants experience a cessation of shoot growth during the growing season. This applies to terminal and axillary buds alike. At some point after shoot growth ceases, normally in the late summer or early fall, terminal and axillary buds alike typically enter a period of physiological dormancy. In contrast with the phenomenon of apical dominance imposed by auxins, removal of auxin supply by pruning has no effect on dormant bud growth. The intensity of dormancy reaches a maximum in early- to midwinter and then gradually declines, apparently in response to chilling temperatures, until normal bud regrowth occurs in the spring. Since the original studies on release of bud dormancy in *Acer pseudoplatanus*, which led Waring to the discovery of ABA, numerous studies have attempted to establish a link between ABA, chilling, and dormancy. While some have reported a positive correlation between ABA concentration and the intensity of dormancy, in most cases the results are less than convincing. In general, there is no commanding evidence that ABA is a primary factor in either inducing or maintaining bud dormancy.

What, then, is the role of ABA in the plant? It seems that ABA plays an important regulatory role in two very distinct processes—seed maturation and stomatal function.

Seed development is characterized by often dramatic changes in hormone levels. In most seeds, cytokinin levels are highest during the very early stages of embryo development when the rate of cell division is also highest. As the cytokinin level declines and the seed enters a period of rapid growth, both GA and IAA levels increase. In the early stages of embryogenesis, there is little or no detectable ABA. It is during the latter stages of embryo development, as GA and IAA levels begin to decline, that ABA levels begin to rise. ABA levels generally peak during the maturation stage when seed volume and dry weight also reach a maximum. Maturation of the embryo is characterized by cessation of seed growth, accumulation of nutrient reserves, and the development of tolerance to desiccation.

The timing of ABA accumulation to coincide with embryo maturation suggests that ABA may play a physiological role in the maturation process. One possibility is that ABA serves to prevent **vivipary,** a precocious germination before the embryo reaches maturity or the seed is released from the fruit. Soybean embryos, for example, can be encouraged to germinate precociously by treatments such as washing or slow drying, both of which lower the endogenous ABA level (Ackerson, 1984). Precocious germination will occur when the ABA concentration is reduced to 3 to 4 μg-g fresh weight of seed. This level of ABA is not normally reached until late seed development. ABA also stimulates protein accumulation in the latter stages of soybean embryo development and, as discussed in greater detail below, will prevent GA-induced α-amylase biosynthesis in cereal grains. A number of viviparous mutants in *Arabidopsis* and corn have reduced levels of ABA. Vivipary can also be induced in corn by treatment of the developing ear at the appropriate time with fluridone, a chemical inhibitor of carotenoids. Since carotenoids and ABA share early biosynthetic steps, fluridone also inhibits ABA biosynthesis. Fluridone-induced vivipary can be at least partially alleviated by application of exogenous ABA. All of these results establish a strong connection between ABA and seed maturation and/or prevention of precocious germination.

One of the more dramatic and perhaps best known effects of ABA is the inhibition of stomatal opening in many species (see Chaps. 8, 22). In leaves of plants that have been grown to ensure minimum endogenous levels of ABA, exogenous ABA at concentrations of 10^{-3} to 10^{-4} M will induce complete stomatal closure. This appears to be a means for regulating water balance in the plant since the endogenous level of ABA in leaves is generally very low if the plants are well watered. Subjecting leaves (e.g., wheat) to a water deficit, however, will induce as much as a fortyfold increase in the ABA level within as little as 30 minutes.

ETHYLENE

***Ethylene** is a simple gaseous hydrocarbon with the chemical structure: $H_2C = CH_2$. Ethylene is apparently not required for normal vegetative growth, although it can have a significant impact on the development of roots and shoots.*

Ethylene appears to be synthesized primarily in response to stress and may be produced in large amounts by tissues undergoing senescence or ripening. Ethylene is commonly used to enhance ripening in bananas and other fruits that are picked green for shipment. As noted earlier, ethylene is frequently produced when high concentrations of auxins are supplied to plant tissues. Since the auxin concentrations applied in many of these experiments is higher than the normal physiological concentration, this might also be considered a stress (see Chap. 22). Many of the inhibitory responses to exogenously applied auxin appear to be due to auxin-stimulated ethylene release rather than auxin itself.

THE PHYSIOLOGICAL ROLES OF ETHYLENE

Because ethylene is a simple gaseous hydrocarbon that readily diffuses from its site of synthesis, study of its role as a hormone presents a unique set of problems. Although known primarily for its effects on fruit ripening and its synthesis by many tissues in response to stress, ethylene is known to affect virtually every aspect of plant growth and development. As a byproduct of hydrocarbon combustion, ethylene is also a common environmental pollutant that can play havoc with greenhouse cultures or laboratory experiments. Still, our understanding of ethylene physiology has made tremendous strides over the past two decades, owing largely to development of the gas chromatograph and the availability of ethylene releasing agents. The gas chromatograph has made possible quantitative analysis of ethylene at extremely low concentrations that could not otherwise be measured. **Ethephon** (2-chloroethylphosphonic acid) is a compound that, at physiological pH, readily decomposes to produce ethylene. Use of ethephon is advantageous in the laboratory as its application and concentration is often more easily controlled compared with gaseous ethylene.

Vegetative Development Ethylene has been shown to stimulate elongation of stems, petioles, roots, and floral structures of aquatic and semiaquatic plants. The effect is particularly noted in aquatic plants because submergence reduces gas dispersion and thus maintains higher internal ethylene levels. In rice, ethylene is ineffective in the presence of saturating levels of gibberellins, which also promotes stem elongation. Moreover, ethylene promotes gibberellin synthesis in rice and the elongation effect can be blocked with antigibberellins, which suggests that gibberellin mediates the ethylene effect (Suge, 1985). By contrast, root and shoot elongation in peas is inhibited by ethylene. Ethylene stimulates abnormal growth responses such as swelling of stem tissues and the downward curvature of leaves (called **epinasty;** Chap. 19). Ethylene effects have also been noted for promotion of seed germination, inhibition of bud break, reduced apical dominance, root initiation, and a variety of other developmental phenomena (Abeles et al., 1992).

Fruit Development The stimulation of fruit ripening is one of the earliest and best known effects of ethylene. A variety of fruits, including bananas, apples, kiwi, and avocados, release ethylene gas during the latter stages of the ripening process. In most cases, the release is coincident with a sharp rise in the respiratory rate measured by CO_2 evolution (the **climacteric rise;** see Chap. 12). Ethylene is autocatalytic; that is, ethylene released by ripening fruits will in turn stimulate premature climacteric and ethylene production by other fruits stored nearby. As a consequence of the climacteric and ethylene production, a number of qualitative metabolic changes are initiated in the fruit. These include the hydrolysis of starches to sugars, softening of the tissue through the action of cell wall—degrading enzymes, and the synthesis of pigments and flavor components. Even so-called nonclimacteric fruits such as citrus, in which ripening is not associated with a climacteric rise in respiration or ethylene production, will show enhanced ripening when exposed to ethylene gas. Ethylene-stimulated ripening is of considerable economic importance. Bananas and many other fruits are picked green to facilitate shipping. They are then exposed to ethylene gas to ensure ripening before reaching the supermarket counter. For long-term storage, apples are placed under conditions designed to minimize the production and accumulation of ethylene. These include low temperature and high ambient CO_2 concentration to suppress respiration or continuous exchange of air to prevent a buildup of ethylene.

Flowering Ethylene normally suppresses or delays flowering. However, ethylene stimulates flowering in the family Bromeliaceae (e.g., pineapple). Originally this phenomenon was attributed to auxin, but it is now known to be due to auxin-stimulated ethylene generation. Commercial growers now use ethephon and other ethylene releasing agents to stimulate uniform flowering in pineapple fields.

POLYAMINES

Polyamines are ubiquitous bioactive compounds found in virtually all organisms, including microorganisms, animals, and plants.

The term **polyamine** generally refers to a group of polyvalent compounds containing two or more amine groups (Figure 16.18). They were first observed as crystals in human semen (hence, spermine) by van Leeuwenhoek, aided by his primitive microscope, more than 300 years ago. Of interest primarily to chemists following their chemical characterization in the 1920s, polyamines began to attract the attention of plant physiologists in the early 1970s.

$H_2N—(CH_2)_4—NH_2$

Putrescine

$H_2N—(CH_2)_3—NH—(CH_2)_4—NH_2$

Spermidine

$H_2N—(CH_2)_3—NH—(CH_2)_4—NH—(CH_2)_3—NH_2$

Spermine

FIGURE 16.18 The chemical structures of three common polyamines.

Polyamines are derived biosynthetically from the amino acids arginine and lysine. In addition, spermidine and spermine biosynthesis involves S-adenosylmethionine, an intermediate in ethylene biosynthesis (Chap. 17). In plant cells, polyamines frequently occur conjugated with phenolic compounds such as hydroxycinnamic acid, coumaric acid, or caffeic acid: These forms may be as functionally important as the free polyamines.

A. W. Galston and R. Kaur-Sawhney have proposed that polyamines be considered as active regulators of plant growth (Galston and Kaur-Sawhney, 1987). At normal intracellular pH, polyamines are polycationic; that is, they carry multiple positive charges. Polyamines thus bind readily to nucleic acids, which are polyanionic, and the phospholipids of the plasma membrane. It is possible that this binding characteristic could affect the synthesis and/or activity of macromolecules and membrane permeability.

Thermophilic bacteria apparently produce polyamines as a means of protecting against thermal inactivation of enzymes and spermine stabilizes DNA against thermal denaturation *in vitro.* Polyamines have been shown to be obligate growth factors for both prokaryote and eukaryote microorganisms, and mammalian cells in culture. They also stabilize oat protoplasts, influence cell division and embryogenesis in carrot tissue culture, and delay senescence in some tissues. While there are provocative correlations of polyamines within many areas of plant physiology, their precise physiological role, and consequently their status as a plant hormone, remains to be determined. Whether polyamines can or should qualify as hormones may be an unprofitable question. Their ubiquitous occurrence and interesting effects indicate a role in plants that clearly warrants further study.

HYPOTHETICAL PLANT HORMONES

Flowering and certain low temperature responses are two phenomena for which the existence of a hormone appears to be a logical necessity, yet in neither case has the active substance been isolated or its existence formally proved.

Studies on flowering clearly indicate the transmission of a diffusible chemical signal from the leaf to the apex (Chap. 20). The existence of a flowering hormone, called **florigen,** has been postulated, but it has never been isolated. Similarly, the hormone **vernalin** has been postulated to account for the effect of low temperature on the flowering behavior of winter cereals and biennials (Chap. 21). Vernalin remains as elusive as florigen.

Phenomena as complex and precise as flowering no doubt require participation of some regulatory substances. Failure to isolate the responsible factors could result from several factors. For example, the molecule or molecules could be extremely labile and readily broken down during extraction. On the other hand, we may simply lack an appropriate test for biological activity. "Florigen" might also be a complex interaction between several regulatory molecules, rather than a single hormone.

OTHER BIOLOGICALLY ACTIVE SUBSTANCES

In addition to the plant hormones described earlier, there are a number of other naturally occurring organic compounds that exhibit strong biological activity when applied exogenously in low concentrations, usually to excised tissues.

A complex mixture of lipids that stimulate elongation of bean second internodes has been isolated from pollen of the rape plant (*Brassica napus L.*) The active substances are known collectively as **brassins** or **brassinosteroids.** The structure of one brassinosteroid, **brassinolide,** is shown in Figure 16.19. Brassinolide is present in very low concentrations—in the initial isolation, 500 lb of rape pollen were extracted to yield only 10 mg of crystalline brassinolide! Brassinosteroids have since been isolated from pollen, stems, leaves, and flowers of a variety of species (Mandava, 1988). Like auxins, brassinolide is active in micromolar concentrations and stimulates elongation of hypocotyl and epicotyl tissue from a variety of legume seedlings as well as coleoptile tissue from wheat. In one study, for example, brassinolide at a concentration of 10^{-7} M caused a fourfold increase in the length of soybean epicotyl sections (Clouse et al., 1992).

It is not known how brassinolide stimulates growth. One popular theory is that it operates synergistically with auxin to enhance the natural growth-promoting activity of that hormone. A number of recent studies employing mutations, however, indicate a more fundamental role for brassinosteroids. For example, mutations of the *deetiolated 2* (*DET2*) gene cause defects in the light-regulated development of *Arabidopsis* (Li et al.,

Figure 16.19 Brassinolide: a brassinosteroid.

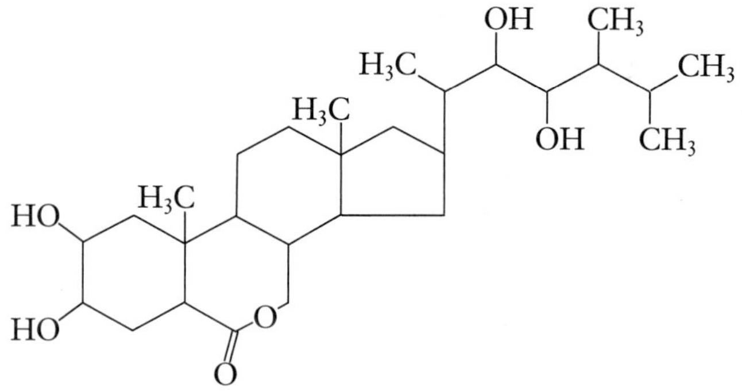

Brassinolide

1996). These defects can be ameliorated by the application of brassinolide. The *DET2* gene encodes a protein with significant homology with a mammalian reductase enzyme necessary for the synthesis of steroid hormones. The implication is that *DET2* may encode a similar reductase necessary for brassinosteroid biosynthesis in plants. Several other mutants have been described that lead to dwarfism in *Arabidopsis*, all of which can be restored to normal growth by exogenous application of brassinosteroids (Somerville and Somerville, 1996). Studies such as these indicate that brassinosteroids have an important role in plant development and may qualify as a distinct class of plant hormones.

Two other biologically active compounds of interest are **coumarin** and ***trans*-cinnamic acid**. Both are phenylpropane derivatives introduced earlier in Chapter 14. Coumarin (Fig. 14.6) is known to inhibit auxin-induced elongation of *Avena* coleoptiles and other tissues *in vitro*, although at sufficiently low concentration elongation may be promoted. Coumarin also inhibits seed germination and has been implicated in natural seed dormancy. It is thought that in some cases germination may proceed only after coumarin levels in the seed coat have been reduced by leaching. *Trans*-cinnamic acid also inhibits auxin activity in stem section assays and has long been considered an antiauxin. There is no hard evidence, however, that substances such as these regulate growth in intact plants. On the other hand, the possibility that many of these biologically active substances may be used to increase yields has attracted attention in the fields of agriculture and horticulture.

SUMMARY

Hormones are naturally occurring, organic substances that exert, at low concentration, profound effects on the function and development of organisms. Developed originally to account for coordination of bodily functions in animals, the hormone concept as it applies to plants has been the subject of much debate. At issue are questions regarding sites of synthesis and action; whether plant hormones act in a concentration-dependent manner or the target sensitivity changes; and the multiplicity of effects.

Auxins were the first plant hormones to be discovered. Auxins are synthesized in the stem and root apices and transported through the plant axis. They are characterized principally by their capacity to stimulate cell elongation in excised stem and coleoptile sections, but also influence a host of other developmental responses, including root initiation, vascular differentiation, tropic responses, and the development of axillary buds, flowers, and fruits.

Gibberellins are produced by both fungi and higher plants. The exogenous application of gibberellins causes hyperelongation of intact stems. Gibberellins are also prominently involved in seed germination and mobilization of endosperm reserves during early embryo growth, as well as flower and fruit development.

Cytokinins are N^6-substituted derivatives of the nitrogenous purine base adenine, characterized by their ability to stimulate cell division in tissue culture. Abscisic acid is a terpenoid derivative involved primarily in regulating seed germination, inducing storage protein synthesis, and modulating water stress. Ethylene is a simple gaseous hydrocarbon. Ethylene is apparently not required for normal vegetative growth but it can have a significant impact on the development of roots and shoots. Its synthesis is often stimulated by auxin, which can account for many auxin effects. Ethylene is especially noted for its role in stimulating fruit development. The brassinosteroids have recently attracted increasing attention, largely through studies of mutations in *Arabidopsis*. Brassinosteroids appear to have an important role in plant development and may soon be considered a distinct class of hormones.

Other possible plant hormones include the polyamines and the hypothetical hormones, florigen and vernalin.

CHAPTER REVIEW

1. In what ways do plant and animal hormones differ? In what ways are they similar?
2. Why does removal of the apical bud stimulate axillary bud development? Can you offer an explanation for why auxins inhibit axillary bud growth but not growth of the apical bud that produces auxin?
3. How would you characterize the five major classes of plant hormones on the basis of their primary action?
4. In Chapter 16 you will learn that an unequal distribution of auxin causes a grass coleoptile to grow toward a light source. What physiological property of auxin would help to explain this response?
5. Auxin and gibberellin both control stem elongation. In what ways do responses of stem to auxin and gibberellin differ?
6. What is a bioassay and how can bioassays be used to distinguish between auxins, gibberellins, and cytokinins? Describe a possible bioassay for each of the three hormones.
7. What are the differences between zeatin, zeatin riboside, and zeatin ribotide?
8. Why does the bacterium *Agrobacterium tumifaciens* cause hyperplastic growth in plants?
9. Which hormones influence flowering and in what ways?
10. What unique problems are related to the study of ethylene as a plant hormone?

FURTHER READING

Abeles, F. B., P. W. Morgan, M. E. Saltveit. 1992. *Ethylene in Plant Biology.* 2nd ed. New York: Academic Press.

Davies, P. J. (ed). 1995. *Plant Hormones.* Dordrecht: Kluwer Academic Publishers.

Evans, P. T, R.L. Malmberg. 1989. Do polyamines have roles in plant development? *Annual Review of Plant Physiology and Plant Molecular Biology* 40:235–269.

Leopold, A. C. 1987. Contemplations on hormones as biological regulators. In: G. V. Hoad, J. R. Lenton, M. B. Jackson, R. K. Atkin, *Hormone Action in Plant Development—A Critical Appraisal.* London: Butterworths. pp. 3–15.

Slocum, R. D., H. E. Flores. 1991. *Biochemistry and Physiology of Polyamines in Plants.* Boca Raton, Fla.: CRC Press.

Zeevaart, J. A. D., R. A. Creelman. 1988. Metabolism and physiology of abscisic acid. *Annual Review of Plant Physiology and Plant Molecular Biology* 39:439–433.

REFERENCES

Ackerson, R. C. 1984. Abscisic acid and precocious germination in soybeans. *Journal of Experimental Botany* 35:414–421.

Clouse, S. D., D. M. Zurek, T. C. McMorris, M. E. Baker. 1992. Effect of brassinolide on gene expression in elongation soybean epicotyls. *Plant Physiology* 100:1377–1383.

Engvild, K. C. 1986. Chlorine-containing natural compounds in higher plants. *Phytochemistry* 25:781–791.

Epstein, E., K-H. Chen, J.D. Cohen. 1989. Identification of indole-3-butyric acid as an endogenous constituent of maize kernals and leaves. *Journal of Plant Growth Regulation* 8:215–223.

Galston, A. W., M. E. Hand. 1949. Studies on the physiology of light action. I. Auxin and the light inhibition of growth. *American Journal of Botany* 36:85–94.

Galston, A. W., R. Kaur-Sawhney. 1987. Polyamines as endogenous growth regulators. In: P. J. Davies, (ed.), *Plant Hormones and their Role in Plant Growth and Development.* Dordrecht: Kluwer Academic Publishers, pp. 280–295.

Gianfagna, T., J. A. D. Zeevaart, W. J. Lusk. 1983. The effect of photoperiod on the metabolism of deuterium-labeled GA_{53} in spinach. *Plant Physiology* 72:86–89.

Gilmour, S. J., J. A. D. Zeevaart, L. Schwenen, J. E. Graebe. 1985. The effect of photoperiod on gibberellin metabolism in cell-free extracts from spinach. *Plant Physiology* S77, 92.

Jacobs, W. P. 1070. Regeneration and differentiation of xylem around a wound. *International Review of Cytology* 28:239–273.

Leuba, V., D. LeTourneau. 1990. Auxin activity of phenylacetic acid in tissue culture. *Journal of Plant Growth Regulation* 9:71–76.

Li, J., P. Nagpal, V. Vitart, T. C. McMorris, J. Chory. 1996. A role for brassinosteroids in light-dependent development of *Arabidopsis. Science* 272:398–401.

MacMillan, J., N. Takahashi. 1968. Proposed procedure for the allocation of trivial names to the gibberellins. *Nature* 217:170–171.

Mandava, N. B. 1988. Plant growth-promoting brassinosteroids. *Annual Review of Plant Physiology and Plant Molecular Biology* 39:23–52.

Metzger, J. D. 1987. Hormones and reproductive development. In: P. J. Davies, P. J. (ed.), *Plant Hormones and their Role in Plant Growth and Development.* Dordrecht: Kluwer Academic Publishers, pp.431–462.

Metzger, J. D., J. A. D. Zeevaart. 1980. Effect of photoperiod on the levels of endogenous gibberellins in spinach as measured by combined gas chromatography-selected ion current monitoring. *Plant Physiology* 66:844–846.

Phinney, B. O. 1984. Gibberellin A_1 dwarfism and shoot elongation in higher plants. In: A. Crozier, J. R. Hillman (eds.), *The Biosynthesis and Metabolism of Plant Hormones.* Cambridge: Cambridge University Press, pp. 17–41.

Smart, C. M., S. R. Scofield, M. W. Bevan, T. A. Dyer. 1991. Delayed leaf senescence in tobacco plants transformed with *tmr* gene for cytokinin production in *Agrobacterium. The Plant Cell* 3:647–656.

Somerville, S., C. Somerville. 1996. Arabidopsis at 7: Still growing like a weed. *Plant Cell* 8:1917–1933.

Suge, H. 1985. Ethylene and gibberellin: Regulation of internodal elongation and nodal root development in floating rice. *Plant and Cell Physiology* 26:607–614.

Thimann, K. V., F. Skoog. 1934. Inhibition of bud development and other functions of growth substances in *Vicia faba. Proceedings of the Royal Society* (London). B114:317.

Trewavas, A. 1981. How do plant growth substances act? *Plant Cell Environment* 4:203–228.

Trewavas, A. 1991. How do plant growth substances work? II. *Plant, Cell and Environment* 14:1–12.

Trewavas, A., R. E. Cleland. 1983. Is plant development regulated by changes in the concentration of growth substances or by changes in the sensitivity to growth substances? *Trends in Biochemical Sciences* 8:354–357.

Yang, T., D. M. Law, P. J. Davies. 1993. Magnitude and kinetics of stem elongation induced by exogenous indole-3-acetic acid in intact light-grown pea seedlings. *Plant Physiology* 102:717–724.

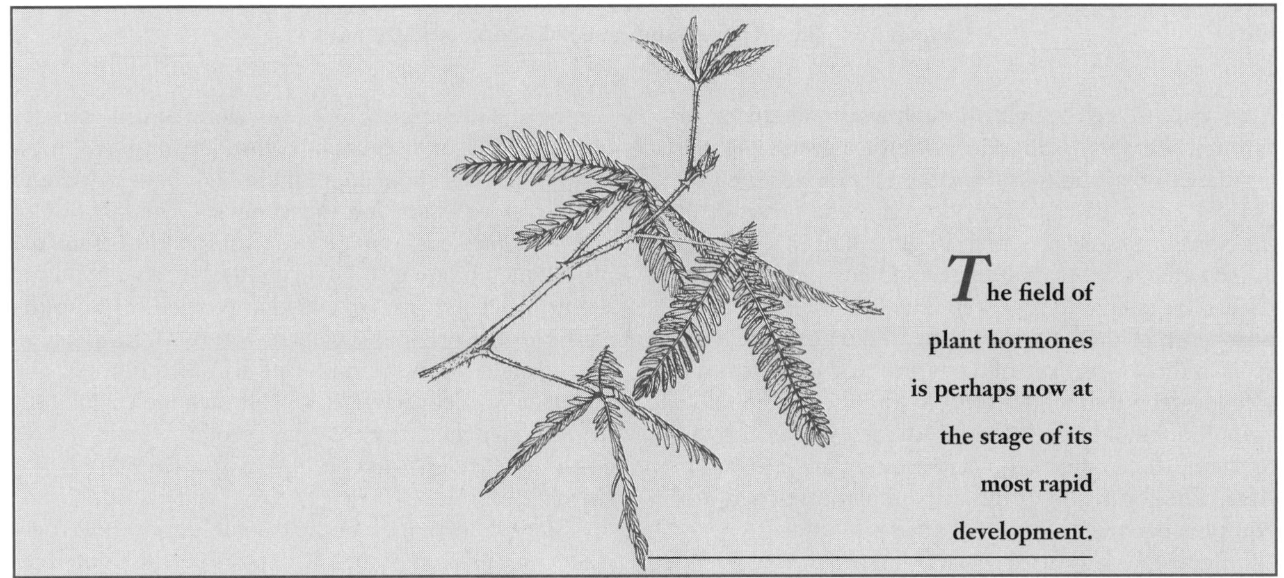

F. W. Went, K. V. Thimann
(Phytohormones, 1937)

17

Biochemistry and Mode of Action of Hormones

How do hormones bring about such profound effects on the physiology of cells? This is perhaps the most challenging question about hormones. Despite years of research into hormone action and the preponderance of descriptive information about the effects hormones have on plants, our understanding of hormone action at the molecular level is only just beginning. That being the case, where does one begin to explore a question such as how plant hormones work? One approach that has stood the test of time is to first establish general principles in studies with other systems. Experiments are then designed to test whether those same principles might also apply to higher plants. Over the past three decades, for example, our understanding of how animal hormones operate has improved markedly and several good models are now available. Since we know that many aspects of biological structure and function are highly conserved, particularly at the cellular level, it is reasonable to expect that plant and animal hormones may share common features. A comparative approach to animal and plant hormone action has proven useful thus far and will no doubt continue to guide plant hormone research for some time to come.

In this chapter, we will

- describe two general models for how hormones function, emphasizing the role of hormone-binding proteins in signal perception and of second messengers in the signal transduction pathway;
- develop the case for hormones in plants by asking whether plants have hormone-binding proteins and whether plant hormone responses involve second messengers;
- show that hormone action depends on rapid turnover of the hormone in the cellular pool, and therefore examine the biosynthesis, transport, and metabolism of auxins, gibberellins, cytokinins, abscisic acid, and ethylene; and
- address the mechanism of hormone action, with particular attention to auxin-regulated cell enlargement and the action of gibberellins in seed germination.

HOW DO HORMONES WORK?

The sequence of events initiated by hormones can generally be resolved into three sequential stages: (1) the initial signal ***perception****, (2) a signal* ***transduction*** *pathway, and (3) the final* ***response****.*

Signal perception involves the reaction of the hormone with a receptor site. Plant hormones may diffuse from cell to cell either through plasmodesmata or through the apoplastic space. In either event, the cell destined to respond to the hormone, known as the **target cell,** must be capable of detecting the presence of the hormone molecule either in the cell or in the fluids immediately surrounding the cell. Detection is accomplished by interactions between the hormone and a cellular receptor that is both specific to the hormone molecule and characteristic of the target cell. Receptors are glycoproteins that bind reversibly with the hormone. As a result of binding the hormone, the receptor is induced to change its conformation and assumes an "activated" state. The formation of this active **hormone-receptor complex** completes the signal perception stage.

Receptors play a particularly important role in the hormone story for two reasons. The first reason is that the presence or absence of a particular receptor determines which cells are able to respond to the hormone. Only those cells that contain a particular receptor at the time a hormone arrives will be able to respond to that hormone. Second, different cell types may possess different receptors that in turn elicit different responses to the same hormone. This could partially explain the multiple effects expressed by many hormones. Even a single cell may contain more than one type of receptor for each hormone, with interacting effects. The smooth muscle cells of blood vessels, for example, contain two epinephrine receptors. Binding of epinephrine to the α-receptor opens a calcium channel that admits calcium into the cell. Binding to the β-receptor activates a metabolic sequence resulting in the active efflux of calcium from the cell.

The second stage of hormone action is the **signal transduction and amplification** stage. In this stage, the activated hormone-receptor complex sets into motion a cascade of biochemical events that ultimately leads to the final, characteristic response. Signal transduction in animals has been studied extensively. Receptors for animal peptide hormones, such as insulin and epinephrine, are commonly located on the extracellular surface of the plasma membrane (Fig. 17.1). The hormone-receptor complex activates a membrane protein called the "**G protein,**" which in turn binds to a third membrane protein—in this example the enzyme **adenylate cyclase**—located at the cytoplasmic surface of the membrane. Binding of the G protein to adenylate cyclase activates the enzyme, thus stimulating the formation of cyclic adenosine monophosphate (**cAMP**) in the cytoplasm. Alternatively, as with certain hormone-receptor complexes, the G protein may interact with an ion channel that controls the flow of calcium into the cell. Once in the cytoplasm, the calcium will bind with one of a number of cytosolic calcium-binding proteins, such as **calmodulin.** The effect of either cAMP or the Ca^{2+}-calmodulin complex is to activate specific **protein kinases** (Chap. 6).

At this stage, it is useful to distinguish between two classes of **messengers.** The hormone is considered a **first messenger** because it brings the original "message" to the cell surface. In Figure 17.1, both cAMP and calcium serve as **second messengers.** The function of a second messenger is to relay information from the plasma membrane to the biochemical machinery inside the cell. Second messengers also provide for **amplification** of the original signal. In Figure 17.1, for example, a single hormone molecule will activate only one molecule of adenylate cyclase. But each molecule of activated adenylate cyclase may produce hundreds of cAMP molecules, thereby initiating a cascade of biochemical events.

Steroid hormones are lipid soluble and readily permeate the cell membranes. Receptors for the animal steroid hormones are thus located inside the cell, commonly in the nucleus (Fig 17.2). Consequently, the hormone-receptor complex also forms in the nucleus where it interacts with specific segments of DNA to either trigger or suppress transcription of mRNA. Note that a second messenger is not necessarily involved in the action of steroid hormones. The effect of the steroid hormones is altered synthesis of either enzyme or structural protein, which in turn alters the biochemistry and physiology of the cell.

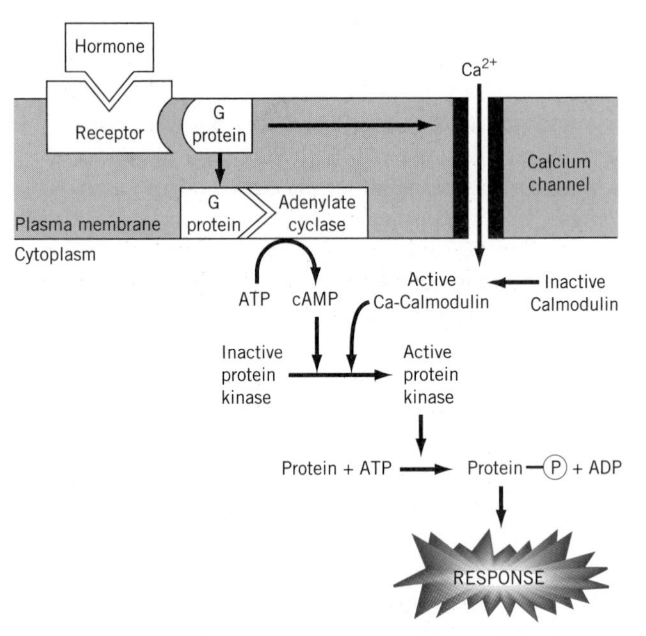

FIGURE 17.1 A model for hormone action involving a plasma membrane-bound receptor. This model is based on mammalian peptide hormones. The G-protein may activate the enzyme adenylate cyclase, stimulating the formation of cyclic-AMP (cAMP). Alternatively, the G-protein may open a membrane calcium channel, thus stimulating the formation of an active Ca^{2+}-calmodulin complex. The cAMP or Ca^{2+}-calmodulin then activate protein kinase enzymes that phosphorylate other proteins, giving rise to the hormone response.

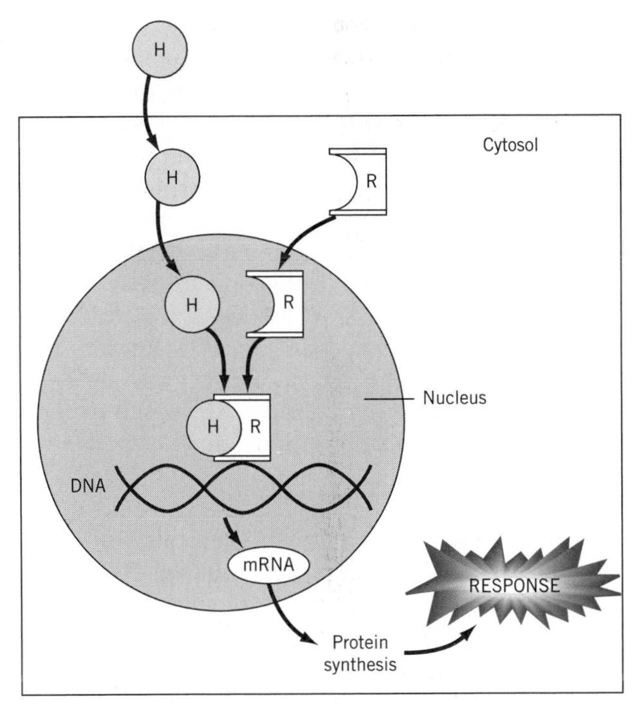

FIGURE 17.2 A model for hormone action involving a nuclear hormone receptor. The hormone-receptor complex regulates the transcription of messenger RNA (mRNA) and protein synthesis. H = hormone; R = receptor.

THE CASE FOR PLANT HORMONES

Based upon the general mechanisms for hormone action outlined above, recent efforts to unravel the mystery of plant hormone action have focused on three key questions:

1. *Are there hormone-binding proteins in plants?*
2. *Do plant hormone responses involve second messengers?*
3. *Do plant hormones alter gene action?*

HORMONE-BINDING PROTEINS IN PLANTS

One interesting property of proteins in general is that they readily bind smaller molecules in a nonspecific manner, especially when tissues are disrupted during isolation protocols. This property is a two-edged sword for the experimentalist. It can be useful, as when serum proteins are included in isolation media to bind small phenolic constituents and control oxidation. It can also cause complications, such as when attempting to identify hormone receptors. For this reason, there are four generally accepted criteria that must be satisfied in order to distinguish between nonspecific binding and hormone-binding properties (Vennis, 1985). First, binding must be specific; that is, a candidate for hormone-binding status must bind only one type of hormone and its structural analogs. Second, the receptor should exhibit a high affinity for the hormone. Third, receptors can be saturated by increasing the concentration of hormone molecules. Fourth, the hormone must bind reversibly with the putative receptor. This last requirement can be demonstrated by showing that molecules of similar structure are able to compete with the hormone for the binding site. It should be emphasized, however, that demonstrating hormone-binding properties for a protein is only the first step. More extensive experimentation is required to establish a true receptor function.

Auxin-Binding Proteins Auxin-binding proteins (ABP) have been sought primarily in two tissues that respond to auxin by cell enlargement: callus cultures of tobacco (*Nicotiana tobacum*) pith tissue (Libbenga and Mennes, 1987) and maize (*Zea mays*) coleoptiles (Venis, 1985). Three classes of IAA-binding proteins have been identified in tobacco callus tissue; two are associated with membrane fractions—probably the plasma membrane—and one is found distributed between the soluble cytoplasmic and nuclear fractions. Of the two membrane-bound binding proteins, one has a relatively low affinity for auxin but a high affinity for **naphthylphthalamic acid (NPA)**, an auxin transport inhibitor. This might be the same auxin-efflux carrier implicated in polar transport of auxins, described later in this chapter. The second membrane-bound receptor has a moderate binding affinity for IAA but does not bind NPA.

The affinity of the cytoplasmic-nuclear binding protein for IAA is several orders of magnitude greater than that of the membrane-localized binding proteins. Its location and its high affinity for IAA suggest that the cytoplasmic-nuclear binding protein is probably capable of detecting relatively low *intra*cellular concentration of auxin. Although a receptor function has not been conclusively demonstrated for either class of binding proteins, Libbenga and Mennes (1987) suggest that the cytoplasmic-nuclear auxin-binding proteins are probably receptors functioning in a manner similar to the steroid hormone receptors in animals.

Only membrane-associated auxin-binding proteins have been detected in maize. One, designated ABP1, is a 43 kDa glycoprotein dimer of 22 kDa subunits that has a high affinity for IAA. ABP1 has been localized primarily in the endoplasmic reticulum, but recent studies have shown that it is also found associated with the plasma membrane and in the cell wall (Jones and Herman, 1993). ABP1 is a prime candidate for the auxin receptor that mediates cell elongation, although the evidence for this role is indirect. Perhaps the most compelling evidence is that antibodies (designated IgG) raised against the auxin-binding protein (designated IgG-antiABP) specifically inhibit both auxin-induced coleoptile elongation and auxin-induced hyperpolarization of the plasma membrane.[1] When IgG-antiABP was used to label coleoptile sections, it was shown that the binding proteins were localized in the outer epidermal cells. The outer epidermal cells are believed to be the most auxin-responsive tissue in the coleoptile. These results suggest very strongly that the binding protein is involved in auxin-induced coleoptile elongation. The role of auxin-binding proteins will be discussed in greater detail later in this chapter.

Cytokinin-Binding Proteins Binding sites for cytokinins have not been so extensively studied as for auxins, but a number of both soluble and particulate binding sites have been reported. The most extensively characterized is the **CBF-1 protein** (CBF = cytokinin binding factor) from wheatgerm (Erion and Fox, 1981; Polya and Davies, 1983). Interestingly, CBF-1 appears to be loosely associated with ribosomes since it is prepared by washing the ribosomal fraction with salt. The association with ribosomes would suggest that the CBF-1-cytokinin complex might have a role in regulating the protein translation process. This is consistent with physiological evidence suggesting that cytokinins modulate protein synthesis, discussed in the previous chapter.

Gibberellin- and Abscisic Acid–Binding Proteins Even though gibberellin-induced enzyme synthesis in cereal grains has been worked out in some detail, as discussed later in this chapter, the mechanism of perception of GA has received little attention. There are no confirmed reports of high-affinity binding proteins for gibberellins. The situation is similar for both abscisic acid and ethylene. There is, however, strong evidence for a high-affinity ABA-binding *site* on guard cell protoplasts. Preliminary results indicate that these sites are proteins located on the apoplastic surface of the plasma membrane.

From the above discussion it is apparent that considerable progress has been made in the study of hormone-binding proteins. Still, direct evidence for hormone receptors in plants is limited. Further study in this important area will be necessary before the action of plant hormones is truly understood.

SECOND MESSENGERS IN PLANTS

Discovery of the biological function of cAMP by E. Sutherland and coworkers in the 1960s was instrumental in unraveling the biochemical mechanism of hormone action in animals. It is only natural that Sutherland's discovery should stimulate plant physiologists to search for a role for cAMP in plants. Although there are scattered reports of the presence of cAMP in plants, there are no plant responses known to be sensitive to cAMP. In addition, neither adenylate cyclase nor cAMP-dependent protein kinase, key enzymes in the cAMP pathway, has been found in plants. At this time, there is no substantive evidence that cAMP functions in signal transduction in plants as it does in animals. This does not, however, discredit the second-messenger concept in plants. There are other candidates for second messengers in plants; the most promising are calcium and the phosphoinositides.

Calcium Calcium reportedly controls numerous physiological processes in plants, including cell elongation and division, protoplasmic streaming, the secretion and activity of various enzymes, hormone action, and tactic and tropic responses. Plants also contain several calcium-binding proteins, although **calmodulin** appears to be the dominant type (Marmé, 1989). Calmodulin is a ubiquitous protein that can be isolated from a variety of higher plants, yeasts, fungi, and green algae. Calmodulin from several plant sources, including spinach, peanut, barley, corn, and zucchini, has been well characterized. Interestingly, many of its properties are similar to calmodulin isolated from bovine brain tissue. Plant and bovine calmodulin have similar molecular mass (17 to 19 kDa), amino acid composition, and calcium-binding properties. Based on spectroscopic analyses, plant and animal calmodulin also undergo similar calcium-induced conformational changes. The highly conserved nature of calmodulin may be considered evidence of a fundamental role in signal transduction.

For calcium to function effectively as a second messenger, the cytosolic Ca^{2+} concentration must be low and under metabolic control. Large amounts of calcium are stored in the endoplasmic reticulum, the mitochondria, and the large central vacuole but the cytosolic Ca^{2+} concentration is kept low through the action of membrane-bound, calcium-dependent ATPases. Activity of the ATPase and, consequently, the cytoplasmic Ca^{2+} concentration, is presumably under control of various stimuli such as light and hormones (Fig. 17.3). In the cytosol, Ca^{2+} reacts with calmodulin and the resulting complex ($CaM{\cdot}Ca^{2+}$) serves to activate certain enzymes. At least two classes of enzymes—NAD kinases and protein kinases—are known to be stimulated by $CaM{\cdot}Ca^{2+}$. NAD kinase catalyzes the phosphorylation of NAD to NADP in the presence of ATP. Because many redox enzymes are specific for one of these cofactors, regulating the balance between NAD and NADP is an effective way to regulate metabolism. Similarly, many other enzymes are activated by protein kinase—catalyzed phos-

[1]**Antibodies** are proteins produced by the immune system of an animal in response to the presence of **antigens.** Antibodies will bind with the antigen, usually a foreign protein, to render it inactive. Antibodies can be raised against plant proteins by injecting purified protein into an animal such as a mouse or rabbit. Antibodies are a useful experimental tool because of the specificity of the antibody-antigen reaction. Antibodies can also be tagged with fluorescent chemicals or other markers so that their location can be readily visualized.

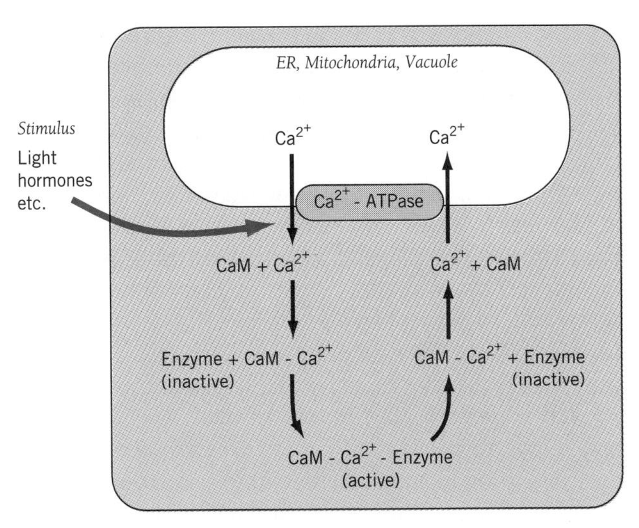

FIGURE 17.3 Calcium as a second messenger. Exchange of calcium between the vacuole and the cytosol may be regulated by hormones or other factors such as light. Cytosolic Ca^{2+} forms an active complex with calmodulin (CaM) or other calcium-binding protein.

Diacyl glycerol (DAG)

Phosphatidylinositol -4, 5 - bisphosphate (PIP_2)

Inositol triphosphate (IP_3)

A.

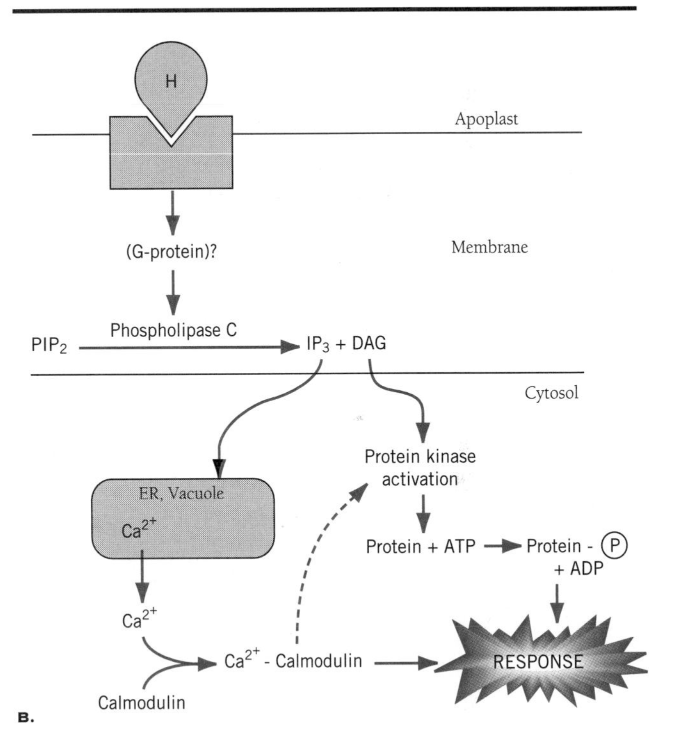

FIGURE 17.4 (*A*) The chemical structures of phosphotidylinositol bisphosphate (PIP_2), inositol triphosphate (IP_3), and diacylglycerol (DAG). (*B*) A model for the inositol triphosphate system. The hormone activates a plasma membrane enzyme phospholipase C, which catalyzes the breakdown of PIP_2 to IP_3 and DAG. IP_3 moves into the cytoplasm where it stimulates the release of calcium and formation of a Ca^{2+}-calmodulin complex. The Ca^2-calmodulin may stimulate the response directly or through activation of a protein kinase. DAG mediates the response through protein kinase activation. R^1 and R^2 are fatty acid-based acyl groups.

phorylation. Several calcium-dependent and calcium/calmodulin-dependent NAD and protein kinases have been isolated from both soluble and membrane fractions from a large number of plants.

Thus, there is a substantial measure of support for the existence of calcium-mediated signal-transduction pathways in plants. However, simply demonstrating the presence of calmodulin and calcium-dependent kinase activity is only a beginning and a number of questions about the stimulus-response coupling still need to be answered. It is necessary, for example, to identify the substrates for these enzymes, and to ask what proteins are phosphorylated and what is their relation to a cellular response. At the other end of the signal chain, it remains to be conclusively demonstrated that plant plasma membranes contain hormone-sensitive calcium channels.

Phosphoinositides Another second messenger that probably does operate in plants is the **inositol triphosphate system** (Boss, 1989). In this system, the hormone receptor-complex activates a plasma membrane enzyme known as **phospholipase C** (Fig 17.4). The hormone-receptor complex may act through a G protein (Millner and Causier, 1996). Phospholipase C catalyzes the breakdown of phosphotidylinositol bisphosphate (PIP_2), a membrane phospholipid, to inositol triphosphate (IP_3) and diacylglycerol (DAG). Both IP_3 and DAG may function as second messengers. IP_3 moves into the cytoplasm where it stimulates the release of calcium, most probably from the vacuole. Note that IP_3 functions as a second messenger to mobilize yet another second messenger, calcium. At the same time DAG activates a particular protein kinase called **protein kinase C.** Most of the essential criteria necessary for involvement of phosphoinositides in signal transduction have been met: hydrolysis of PIP_2 by phospholipase C, an increase in IP_3 and DAG, transient increases in cytosolic calcium, and activation of protein kinase C. Unfortunately, many details of phosphoinositide metabolism in plants remain unclear and the system has not been demonstrated in its entirety in a single plant.

In addition to calcium and the phosphoinositides, there are other potential candidates for second messengers in plants. These include acetylcholine and certain lipids (see Boss and Morré, 1989). The study of secondary messengers and hormone-initiated signal-transduction pathways in plant cells is in its infancy, but promises to be an important and exciting area of research in the future.

HORMONES AND GENE ACTION

The traditional approach to the study of hormone action was centered around efforts to identify primary events at the biochemical level. This approach has been relatively unsuccessful, largely because it is difficult to identify hormone-specific events against a background of complex and concurrent cellular metabolism. A more recent approach has been to identify hormone-induced developmental events that are susceptible to molecular analysis. This approach has been encouraged by recent advances in molecular technology that make it possible to identify, isolate, and reproduce (or, clone) specific genes. The intent is to identify genes whose activities are specifically modified by the hormone. The gene may be turned on or off by the hormone, or its output increased—the gene is *upregulated*—or decreased—the gene is *downregulated.* Although many details have yet to be resolved, it is clear that plant hormones do act, at least in part, by regulating gene action. Interactions between specific hormones and gene action will be described in greater detail below, when we discuss the mechanism of hormone action.

BIOSYNTHESIS, TRANSPORT, AND METABOLISM OF HORMONES

A hormone can serve effectively as a chemical messenger only if the molecule has a limited lifetime in the target cell. Any molecule sufficiently long-lived to be used repeatedly would sacrifice its dynamic, regulatory function. This means that the amount of a hormone in a cellular pool must be closely regulated and exhibit a rate of metabolic turnover that is rapid relative to the response that it controls.

The amount of hormone available to a target cell will be governed primarily by the rates at which active hormone molecules enter and exit the hormone pool. The paths by which hormones enter the pool are called **inputs** and will include: (1) *de novo* synthesis of the hormone; (2) retrieval of active hormone from an inactive storage form, such as a chemical conjugate; and (3) transport of hormone into the pool from a site elsewhere in the plant. Means for removing hormone from the pool once it has acted are called **outputs.** Outputs include: (1) oxidation or some other form of breakdown that renders the molecule inactive; (2) synthesis of inactive storage forms (conjugates); and (3) "consumption" as a result of its activity. The third output is not at all well understood in the case of plant hormones and may be synonymous with oxidative breakdown. Clearly, in order to understand the dynamic regulation of hormone activity in plants, it is essential to know something of these inputs and outputs. No understanding of hormone function can be complete without a working knowledge of hormone biosynthesis, transport, and metabolism.

AUXIN BIOSYNTHESIS AND METABOLISM

Auxin (IAA) is synthesized from the amino acid tryptophan in actively growing regions of the plant. IAA transport is polar, in a predominantly basipetal direction. IAA is readily oxidized to inactive forms.

The auxin IAA is generally ubiquitous in the plant. Highest concentrations of the hormone are detected in meristematic regions and actively growing organs such as coleoptile apices, root tips, the apical buds of growing stems and germinating seeds (Fig. 17.5). Young, rapidly growing leaves, developing inflorescence, and embryos following pollination and fertilization are also major sites of IAA synthesis. Cells in older leaves, stems, and roots do not synthesize appreciable amounts of IAA.

The amount of IAA present will depend upon a number of factors, such as the type and age of tissue and its state of growth. In vegetative tissues, for example, the amount of IAA generally falls in the range between 1 μg and 100 μg (5.7 to 570 nanomoles) kg^{-1} fresh weight, but in seeds it appears to be much higher. In one study, it was estimated that the endosperm of a *single* maize seed four days after germination contains 308 picomoles (pmole = 10^{-12} mole) of IAA (Epstein et al., 1980). At the same time, the maize shoot contained 27 pmoles of IAA and required an estimated input of approximately 10 pmole of IAA hr^{-1} in order to support its growth. The high level of hormone in the seed apparently serves to support the rapid growth of the young seedling when the seed germinates.

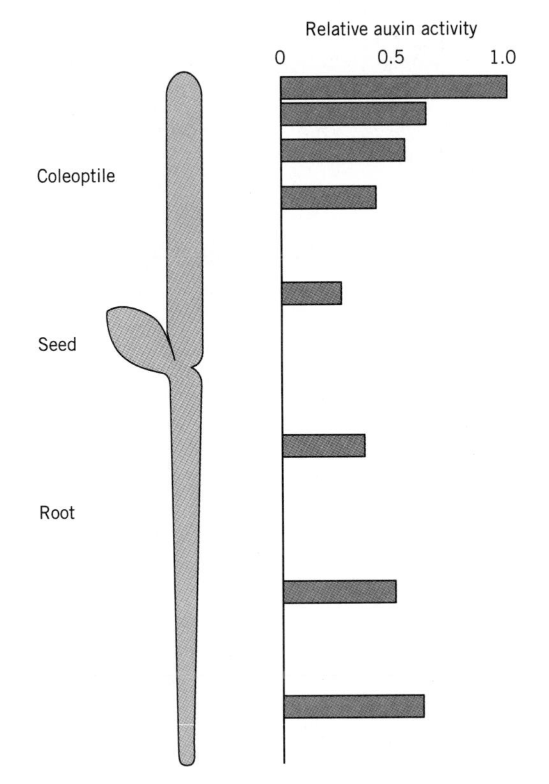

FIGURE 17.5 Auxin distribution in an oat seedling (*Avena sativa*), showing higher concentrations of hormone in the actively growing coleoptile and root apices. (Based on data from K. V. Thimann, *Journal of General Physiology* 18:23–34, 1934.)

BIOSYNTHESIS OF IAA

It is now well established that IAA is synthesized from the aromatic amino acid, tryptophan. Since the 1930s when K.V. Thimann first observed the synthesis of IAA in the mould *Rhizopus suinus*, which had been fed the amino acid tryptophan, the conversion of tryptophan to IAA has since been studied *in vivo* in over 20 different plant species and *in vitro* with at least 10 different cell-free enzyme preparations. The synthesis of IAA is normally studied by feeding plants tryptophan carrying a radioactive label, usually carbon (^{14}C) or tritium (^{3}H), and examining the radioactivity of subsequently isolated IAA or its intermediates. Such feeding experiments are complicated by several factors, however, and the results must always be approached with caution. For example, radiolabeled tryptophan can apparently undergo radiochemical decomposition, thus giving rise to IAA by nonenzymatic reactions. In addition, the pool size of tryptophan (also a precursor for protein synthesis) is very large relative to that of IAA and there is little data on the actual quantity of IAA synthesized. Finally, care must be taken to ensure that experiments are conducted under sterile conditions, since many microorganisms readily convert tryptophan to IAA. While these complications make it difficult to ascertain the exact pathway that functions *in vivo*, the available evidence clearly establishes that plants are able to synthesize IAA from tryptophan.

In most higher plants, synthesis of IAA occurs in three steps, beginning with the conversion of tryptophan to indole-3-pyruvic acid (IPA) (Fig. 17.6). This transamination reaction is catalyzed by a widely distributed **tryptophanamino transferase,** a multispecific enzyme that appears to act as well to remove amino groups from structural analogs of tryptophan such as phenylalanine and tyrosine. The second step is the decarboxylation of IPA to form indole-3-acetaldehyde (IAAld). The enzyme that catalyzes this step, **indole-3-pyruvate decarboxylase,** has been described in several plant tissues and cell-free extracts. Finally, IAAld is oxidized to IAA by a NAD-dependent **indole-3-acetaldehyde dehydrogenase.** The presence of this enzyme has been demonstrated in a number of tissues, including oat coleoptile. IAAld may also be reversibly reduced to indole-3-ethanol. Indole-3-ethanol exhibits auxin activity in bioassays using stem sections, but this is probably due to its conversion to IAA in the tissue.

There is some evidence for alternate biosynthetic pathways involving intermediates other than IPA (Fig. 17.6). In some species, for example, tryptophan may first be decarboxylated to the intermediate tryptamine. Other minor routes have also been postulated, but these are either not clearly established or appear to be restricted to particular species. The burden of biochemical evidence to date, however, seems to indicate that IPA is the dominant intermediate in the synthesis of IAA in higher plants. In most species, the IPA pathway is probably the only pathway.

Apparently not all plants, however, synthesize auxin from tryptophan. IAA may also be formed from **indole-3-acetonitrile (IAN),** an indole derivative found particularly in members of the Brassicaceae (mustard family). Exogenously supplied IAN exhibits auxin activity, probably by conversion to IAA through the action of a nitrilase enzyme. Over the last decade, several tryptophan-requiring mutants have been isolated from the brassica *Arabidopsis* (Leyser, 1997). At least two mutants, known as *trp2* and *trp3*, block the conversion of indole-3-glycerol phosphate to tryptophan (see Fig. 14.3). These mutants contain elevated levels of both IAN and IAA. This observation, plus the evidence that *Arabidopsis* also contains nitrilases capable of converting IAN to IAA, strongly suggests a tryptophan-independent pathway for auxin biosynthesis in *Arabidopsis.* The source of IAN is not known, although its accumulation in tryptophan mutants suggests a tryptophan-independent pathway for the biosynthesis of IAN as well. It is known that IAN can be derived from glucobrassicin, a major indolyl glucosinolate present in crucifers (Chap. 14). Details of the IAN pathway for auxin biosynthesis and whether it is limited to *Arabidopsis* or the brassicas, or is more widespread, remain to be determined.

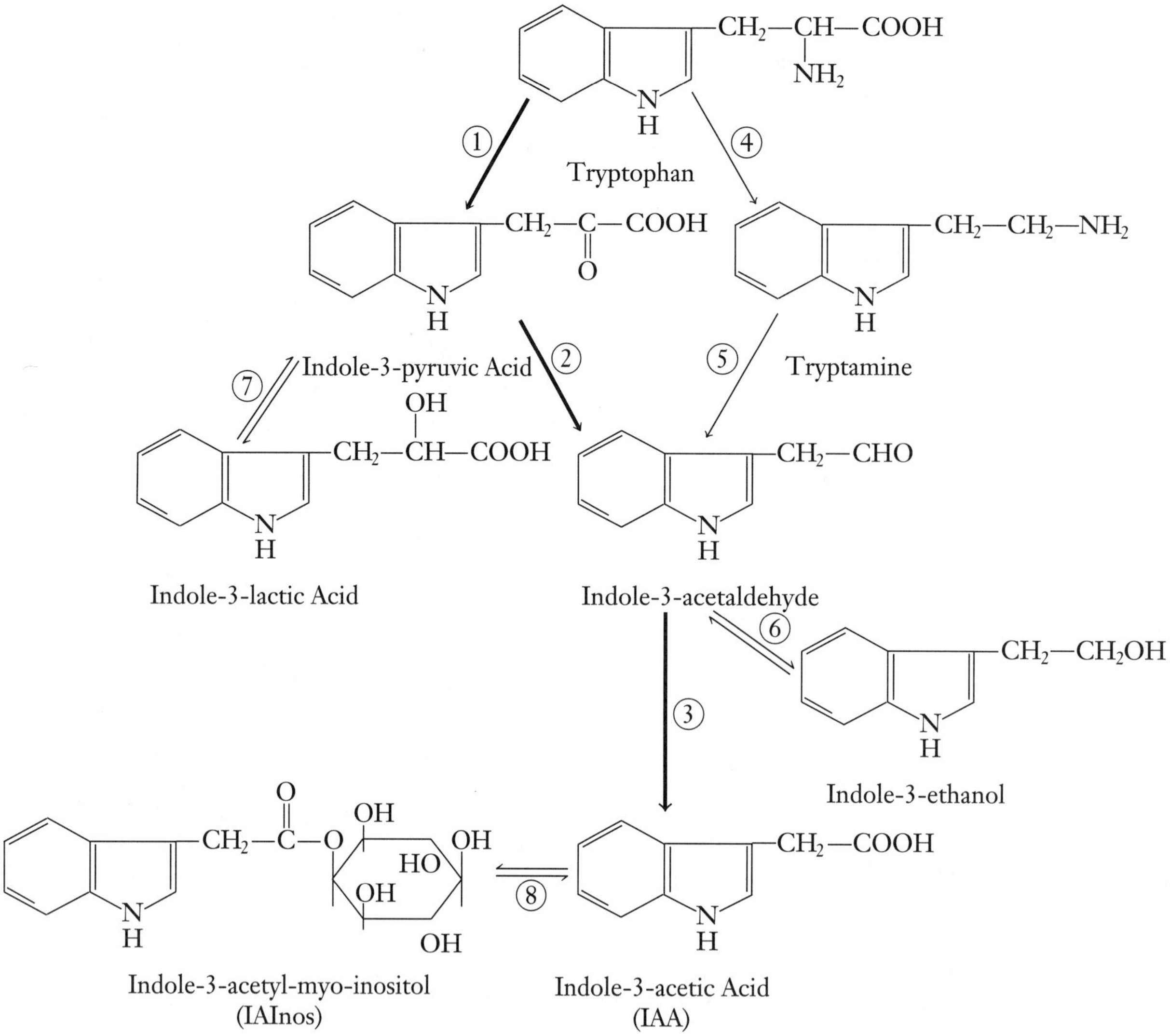

Tryptophan

Indole-3-pyruvic Acid

Tryptamine

Indole-3-lactic Acid

Indole-3-acetaldehyde

Indole-3-ethanol

Indole-3-acetyl-myo-inositol (IAInos)

Indole-3-acetic Acid (IAA)

FIGURE 17.6 The biosynthetic pathway for auxin (indole-3-acetic acid).

IAA CONJUGATES

Very early in the study of auxins, two populations of the hormone were recognized—one was free-moving and could be obtained by diffusion into agar; the other appeared to be bound in the cell and could be isolated only by extraction with solvents or by hydrolysis under alkaline conditions. This latter population, called **bound auxin,** is now recognized as IAA that has formed chemical conjugates, such as as glycosyl esters or peptides (Cohen and Bandurski, 1982). A small portion of the bound auxin may also represent hormone that is complexed with cellular receptor sites. The structures of some of the more common IAA conjugates are shown in Figure 17.7.

IAA-conjugates are themselves inactive but release free, active IAA upon solvent extraction, alkaline hydrolysis, or *in vivo* enzymatic hydrolysis. Although quantitative data are lacking for most plants, large pools of IAA esters have been demonstrated in germinating maize seedlings. At least in germinating seedlings, these pools of IAA conjugates appear to be an important source of active hormone. It has been estimated, for example, that between 20 and 60 percent of the IAA requirement of a germinating maize shoot may be met by hydrolysis of IAA conjugates initially supplied by the endosperm (Epstein et al., 1980). Since most of our knowledge of IAA release by hydrolysis of conjugates comes from studies with germinating seedlings, it is not yet known whether conjugate hydrolysis is equally important in the growth of mature plants.

The sites of synthesis of IAA conjugates are not well understood. The conjugates found in seeds of *Zea mays* are synthesized in the seed itself during later stages of seed development. However, little is known about the synthesis of conjugates in vegetative tissues.

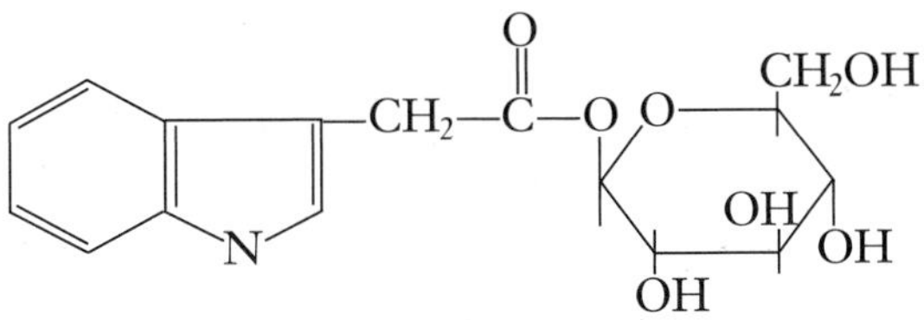

Indole-3-acetyl-β-D-glucose

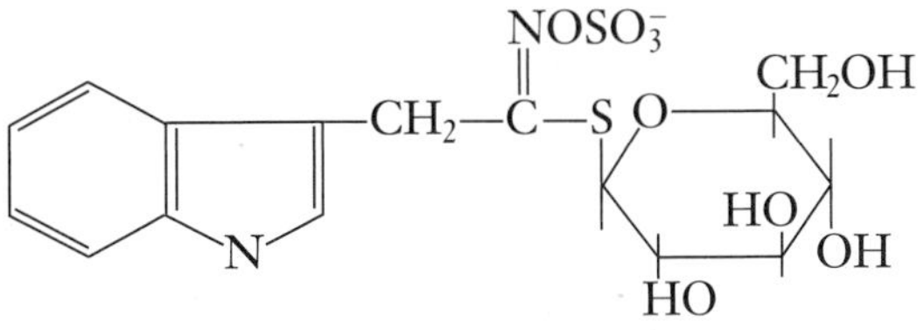

Glucobrassicin

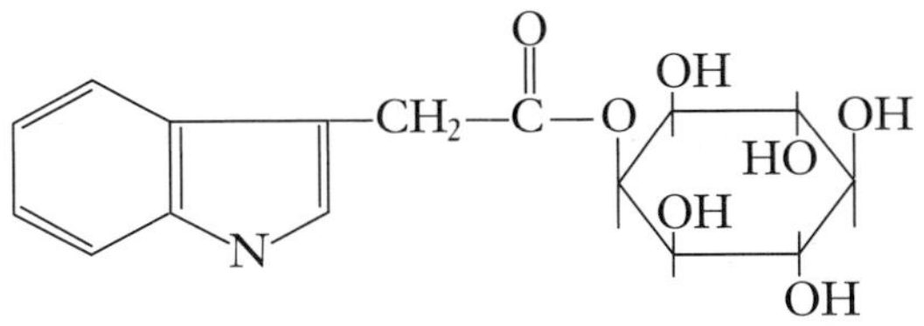

Indole-3-acetyl-*myo*-inositol

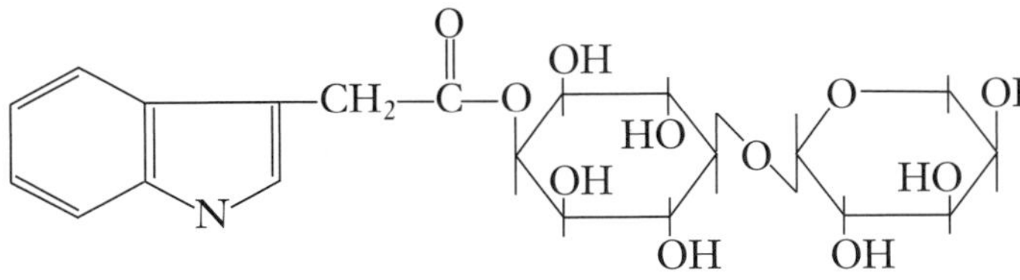

Indole-3-acetyl-*myo*-inositol-arabinose

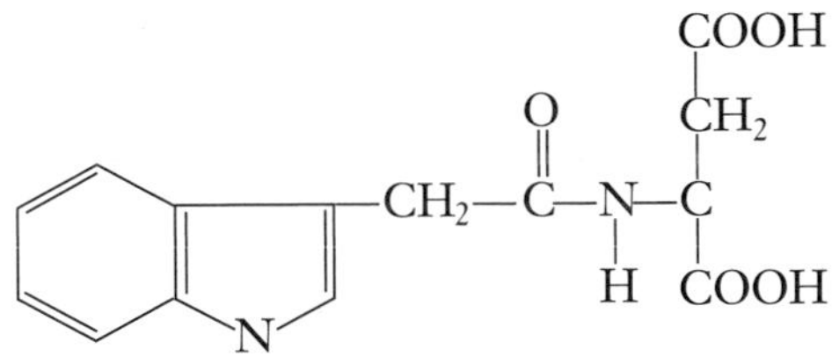

Indole-3-acetyl-L-aspartate

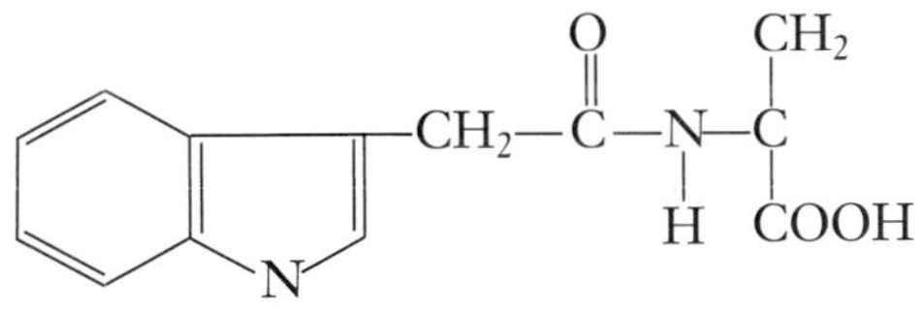

Indole-3-acetyl-L-alanine

FIGURE 17.7 Examples of known conjugates of indole-3-acetic acid.

IAA TRANSPORT

Although transport and action at a distance is not an essential property of plant hormones, hormones do indeed move between tissues and organs within the plant. Transport into or out of a tissue or organ will naturally influence the level of active hormone within that tissue or organ. One example of auxin transport—the mobilization of IAA conjugates from endosperm to shoot in germinated corn seedlings—was described above. Although free IAA is present in the endosperm of the seed, its rate of transport into the shoot is some 400 times slower than is necessary to maintain the measured supply of IAA in the transport stream of the shoot (Nowacki and Bandurski, 1980). In this case, it appears that IAInositol (see Fig. 17.7) is the form that is actually translocated. More complex esters such as IAInositol-galactose or IAInositol-arabinose are first hydrolyzed to IAInos in the scutellum before moving through the vascular tissue into the growing shoot. Once in the shoot, the conjugate is enzymatically hydrolyzed to yield free IAA.

A more perplexing example of auxin transport is that of **polar transport,** originally described in cereal seedlings but later shown to be widespread in shoots and roots. Polarity in auxin transport is expressed as a preferential movement of auxin from top to bottom in a coleoptile or shoot axis (Fig. 17.8). When movement is preferentially away from the morphological apex toward the morphological base of the plant, the direction of movement is described as **basipetal.** Movement in the opposite direction, toward the apex, is referred to as **acropetal.** Roots also exhibit basipetal transport of auxin, from the root tip toward the base of the root.

The phenomenon of polar auxin transport has attracted widespread attention because of the assumption that auxin concentration is an important variable in several developmental responses. Auxin gradients due to polar transport have been invoked to explain, at least in part, polar developmental phenomena such as tropisms, apical dominance, and adventitious root formation. Auxins, in particular IAA along with some of its synthetic analogs, are the only plant hormones known to exhibit polar transport. The path of auxin transport is not clear. Although auxins can be detected in vascular exudates, the concentrations are relatively low. The polar transport path appears to involve parenchyma cells that sheath differentiating vascular tissue, but not the vascular elements themselves.

How is polarity in auxin transport established? Several observations indicate the involvement of some form

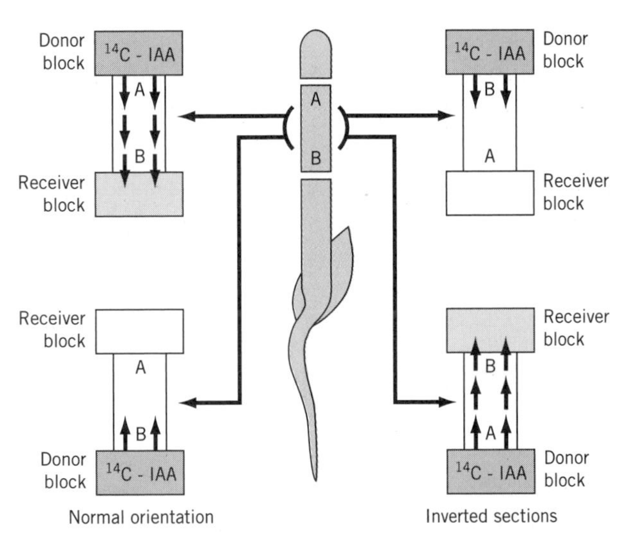

FIGURE 17.8 Polarity in auxin transport in an oat coleoptile segment. The donor block contains ^{14}C-IAA. Regardless of the orientation of the segment, translocation of the radiolabeled IAA is always from the morphologically apical end (*A*) to the morphologically basal end (*B*) of the segment.

of carrier-mediated, active transport mechanism. (Carrier-mediated active transport was discussed in Chap. 5.) First, it can be shown that polar transport is inhibited by respiratory poisons such as cyanide and 2,4-dinitrophenol. It is thus an energy-requiring process dependent on oxidative metabolism in the mitochondrion. Second, certain chemicals, called **phytotropins,** have been known for some time to be specific, noncompetitive inhibitors of polar transport. These include **2,3,5-triiodobenzoic acid (TIBA)**, 9-hydroxyfluorine-9-carboxylic acid (HFCA, also known as **morphactin**), and **N-1-naphthylphthalamic acid (NPA)** (Fig. 17.9). Third, the uptake of radioactive IAA is partially inhibited by nonradioactive IAA. These results suggest that the labeled and unlabeled IAA compete for a limited number of carrier sites.

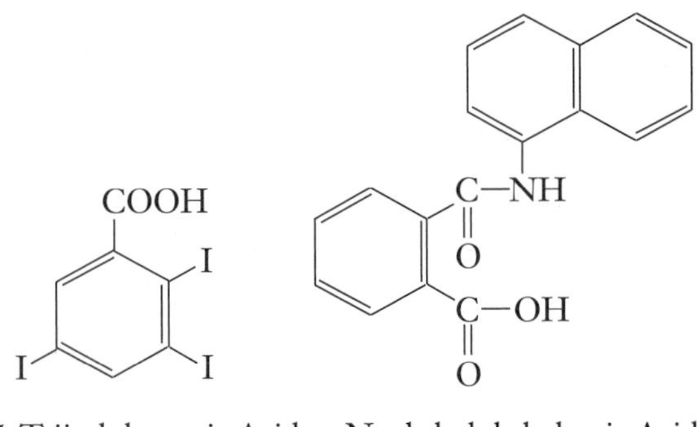

COOH, I, I, I — 2,3,5-Triiodobenzoic Acid (TIBA)

C—NH, O, C—OH, O — Naphthylphthalamic Acid (NPA)

FIGURE 17.9 Inhibitors of polar IAA transport.

Additional support for a carrier-mediated transport is provided by studies of the distribution of a protein believed to be involved in IAA transport (Jacobs and Gilbert, 1983). The presumptive IAA- transport protein binds inhibitors of IAA transport such as NPA, although it does not bind IAA itself. By using antibodies to the NPA-binding protein labeled with the fluorescent dye fluorescein, Jacobs and Gilbert were able to show that, in pea stem sections, *the protein is located primarily at the base of the auxin-transporting cells.*

An asymmetrical distribution of transport properties of cells is not unique. Asymmetric transport has been well documented in animal systems; the asymmetry of glucose and sodium transport systems in intestinal epithelial cells is one good example. While it is not known how the NPA-binding protein is involved in IAA polar transport, its localization at the base of the transporting cells appears to offer a partial explanation. The results of other experiments have indicated there is a diffusion component in IAA transport in addition to the active transport component. The velocity of polar IAA transport, for example, has been estimated in the range of 5 to 20 mm hr^{-1}. Such a rate is consistent with movement by diffusion.

IAA (along with most synthetic auxins) is a weakly acidic, lipophilic molecule. Depending on the pH, IAA may exist either in the protonated (IAAH) or the unprotonated, anionic form (IAA^-). The cell wall space is moderately acidic with a pH of about 5.0. At that pH, IAAH will predominate. Consequently, any IAA^- that finds itself in the cell wall space will rapidly protonate to form IAAH. IAAH has a higher lipid solubility and would be expected to penetrate cell membranes more readily than IAA^-. It has been confirmed that the uptake of IAA into cells increases as the extracellular environment is made more acidic (Rubery and Sheldrake, 1973). Thus cells will take up auxin from the cell wall space as IAAH diffuses down its concentration gradient. Once in the cytoplasm, where the pH is about 7.0, IAAH will dissociate to IAA^- and H^+. The pH difference between the cell wall space and the cytoplasm serves to maintain the IAAH concentration gradient and thus encourage IAAH to continue moving into the cell. It also drives an accumulation of auxin in the cell in the anionic form, which is trapped inside the cell because it does not readily cross the membrane.

The diffusion of IAAH into the cell and the preferential localization of an auxin carrier at the base of the cell may be incorporated into a model for polar auxin transport, which M. H. Goldsmith (1977) has called the "chemiosmotic polar diffusion hypothesis" (Fig. 17.10). In this model, the basally located auxin carrier is viewed as an IAA^- uniport, catalyzing the efflux of IAA^- from the cell. Efflux of the negatively charged IAA- is also favored by the normal membrane potential (outside posi-

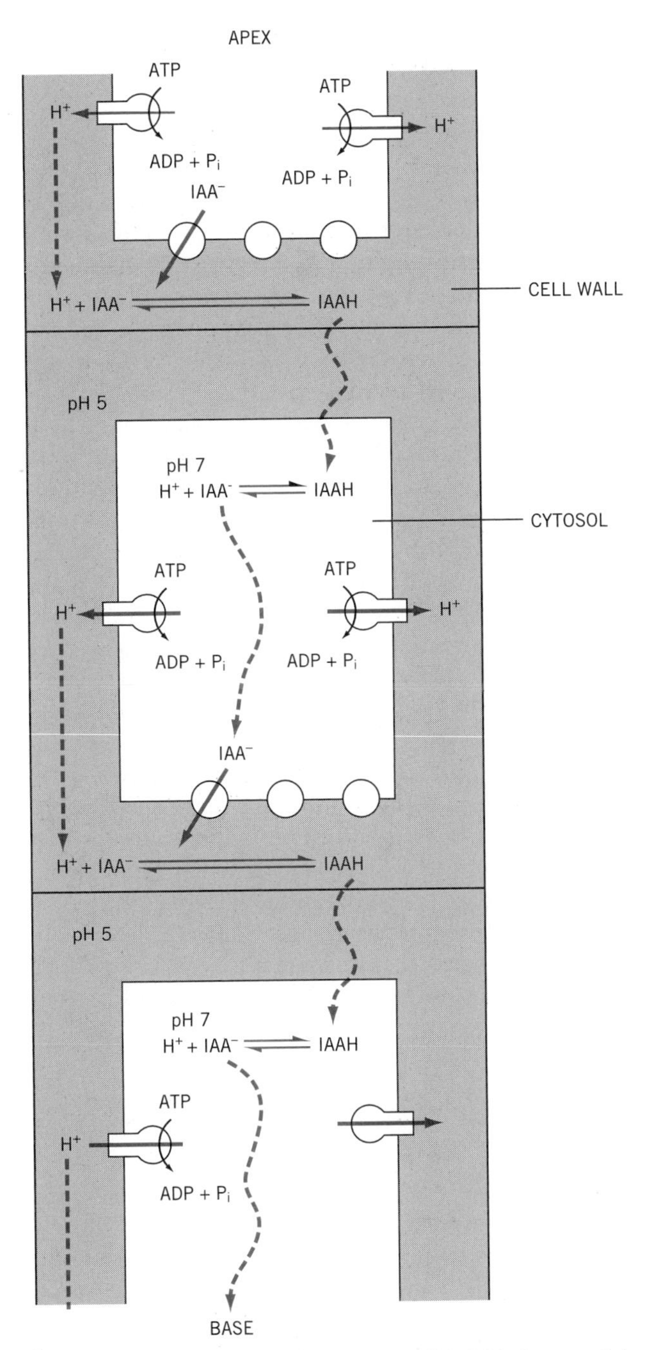

FIGURE 17.10 The chemiosmotic-polar diffusion model for polar transport of IAA in tissues. Protonated IAA, formed in the acidic cell wall space (pH 5.0), is free to diffuse into cells or may be taken up by a proton/IAA symport carrier. In the cell (pH 7.0) IAAH is deprotonated to IAA^-, which can leave the cell only through efflux carriers located at the base of the cell. Membrane-bound ATPase-proton pumps help to maintain the appropriate pH differential across the membrane and provide protons for IAA symport.

tive) across the plasma membrane. Once in the cell wall space, the IAA^- is protonated to IAAH, which then diffuses through the cell walls and is taken up by the subjacent cell. Similarly, the IAA^- formed in the cell either diffuses or is carried by cytoplasmic streaming toward the basal end where it is once again extruded into the cell wall space. In this way, there is a preferential basipetal movement of auxin through a column of cells. ATP-driven proton pumps located in the plasma membrane operate to prevent the accumulation of protons in the cytoplasm and to maintain both the acidity of the cell wall environment and a favorable membrane potential.

Although protonated IAA can diffuse into cells from the cell wall space, recent studies of the *Arabidopsis AUX1* gene suggest a carrier protein may also be involved in IAAH uptake. The *AUX1* gene has been linked to auxin metabolism and transport because mutations at that locus exhibit IAA-resistant root growth, reduced lateral root initiation, and reduced response of the root to gravity. Such a phenotype is consistent with a reduced capacity to take up IAA. Bennett et al. (1996) have cloned the *AUX1* gene and found that the polypeptide sequence of the protein is similar to that of known amino acid permeases, membrane proteins that function as amino acid/proton symport carriers. The protein homologies together with the structural similarities between IAA and tryptophan have led Bennett et al. to suggest that the *Aux1* protein functions as an auxin/proton symporter. Thus both the uptake of IAAH and the basal efflux of IAA^- appear to be mediated by carrier proteins. It is not known whether the uptake carrier is restricted to the upper portion of the cell or is more uniformly distributed. Either way, the involvement of an uptake carrier would not materially alter the polar diffusion hypothesis. The key to polar auxin transport remains the inability of IAA^- to diffuse across the membrane and the restricted location of the basal IAA^- efflux carrier. The uptake permease would only ensure a more rapid IAAH uptake than could be expected by simple diffusion alone.

Polar transport has been studied most extensively in seedling tissues, such as coleoptile sections from *Avena* or *Zea*, bean hypocotyl sections or epicotyls of young pea seedlings. There have been limited studies of auxin movement in mature tissues. In one study, however, W. P. Jacobs found there is some acropetal movement of auxin in *Coleus* stem sections: The ratio of basipetal to acropetal transport is approximately 3 : 1 (Jacobs, 1979). Even in seedling tissue transport may not be strictly polar as has been traditionally believed. There have been reports of limited acropetal movement of IAA in both *Avena* and *Zea* coleoptile sections.

Finally, there are some examples of nonpolar auxin transport. IAA supplied exogenously to mature leaves will be exported along with sugars through the vascular tissue. The movement of inactive IAA conjugates in the vascular tissue was noted above. In this case, the conjugate is probably activated by hydrolysis at the point where it is unloaded from the vascular elements and enters the polar transport stream.

More recently, inhibitors of polar transport are proving to be useful tools for exploring the role of auxins in developmental phenomena that may not be accessible by traditional methods. One example is early stages in the formation of flower buds in *Arabidopsis* (Okada et al., 1991). *Pin1* is a mutant of *Arabidopsis* that results in abnormal floral development. Polar transport of auxin is also affected by the *Pin1* mutation. The polar transport of exogenously supplied ^{14}C-labeled IAA is reduced to approximately 10 percent of normal. What is interesting here is that the mutant phenotype could be generated in wild-type seedlings by applying the auxin transport inhibitors HFCA and NPA. On the other hand, another auxin antagonist (2-[p-chlorophenoxy]-isobutyric acid, or CPIB) is known to inhibit auxin activity but does not interfere with polar transport and does not generate the mutant phenotype. In another study, Liu and coworkers have shown that HFCA stimulates the formation of fused cotyledons in cultured mustard (*Brassica juncea*) embryos (Liu et al., 1993). HFCA thus interferes with the normal initiation of two cotyledons and the transition from the axial symmetry of the early globular-shaped embryo to the bilateral symmetry of the more advanced embryo. Not only does auxin have a role in developmental process, but the polar transport of auxin is itself a critical factor in the early stages of floral bud and embryo development. Both of these studies show how inhibitors of auxin polar transport can be used as developmental probes.

OXIDATION OF IAA

IAA in aqueous solution is readily degraded by a variety of agents, including acids, ultraviolet and ionizing radiation, and visible light, in the presence of sensitizing pigments such as riboflavin. The most prevalent form of IAA degradation, however, appears due to oxygen and peroxide, either separately or in combination, in the presence of a suitable redox system.

Inactivation of the *Avena* growth-promoting substance by aqueous extracts of leaves was first reported in the 1930s, even before the active principle was identified as IAA. An enzyme responsible for inactivating IAA was first isolated from plant extracts by Tang and Bonner (1947) and was called **IAA oxidase.** Later, the enzyme peroxidase, in concert with a flavoprotein, was shown to catalyze the oxidation of IAA while at the same time releasing CO_2 (Galston et al., 1953). Oxidative decarboxylation of IAA is now known to be catalyzed by peroxidases from a variety of plant sources. There are differing requirements for enzymes from different sources and even multiple isozymes within a single species. The oxidative decarboxylation of IAA by peroxidase is now recognized by some physiologists to be synonymous with IAA oxidase (Reinecke and Bandurski, 1987).

IAA oxidation has been most extensively studied in etiolated pea epicotyls and, more recently, with purified horseradish peroxidase. The horseradish peroxidase sys-

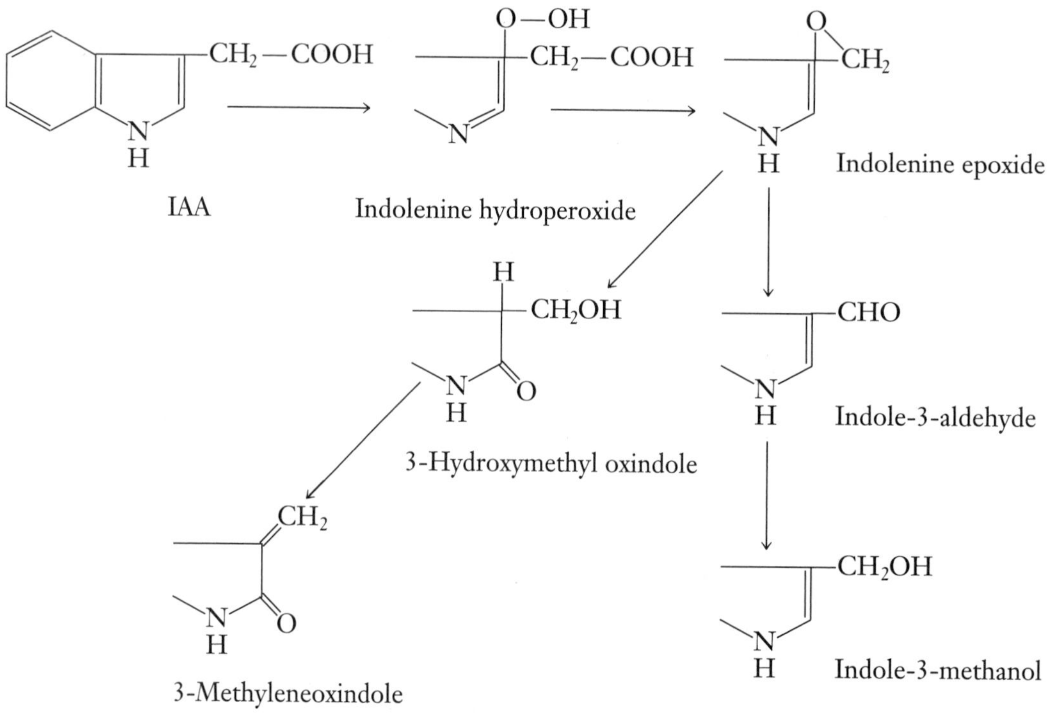

IAA
Indolenine hydroperoxide
Indolenine epoxide
3-Hydroxymethyl oxindole
Indole-3-aldehyde
3-Methyleneoxindole
Indole-3-methanol

FIGURE 17.11 Schematic pathway for oxidative degradation of IAA.

tem will catalyze oxidative decarboxylation without added cofactors, although manganese ion, simple monophenols, and m-diphenols will increase the reaction rate. On the other hand, p-diphenols, o-diphenols and polyphenols will generally inhibit IAA oxidation. These types of compounds encompass a variety of products that occur naturally in plants as precursors to lignin biosynthesis and other pathways. It has been suggested that these compounds may serve a regulatory function in IAA peroxidative oxidation *in vivo*.

The pathway for oxidative decarboxylation of IAA is shown in Figure 17.11. The major end products of oxidation by cell-free enzyme preparations are 3-hydroxymethyl oxindole and 3-methyleneoxindole. There is some evidence suggesting these products may also be detected *in vivo*. An alternate route leading to the formation of indole-3-aldehyde has been demonstrated in some plants, such as stems of tomato, barley, pea, and bean.

Because the end products of IAA oxidation are physiologically inactive, IAA oxidation is an effective way of removing the hormone once it has accomplished its purpose. IAA oxidase activity is generally higher in the older, nongrowing tissues than it is in younger, actively growing tissues, which have a high auxin requirement. Oxidative breakdown is the only known means for irreversibly removing IAA from the active pool and may be very important in regulating IAA-mediated responses. It is worth noting that storage of IAA as conjugates renders them immune to oxidative attack.

GIBBERELLIN BIOSYNTHESIS AND METABOLISM

Gibberellins are diterpenoids, related biosynthetically to carotenoids and other isoprene derivatives. Gibberellins are removed from the active hormone pool by conversion to inactive forms; either by 2-β-hydroxylation or conversion to glucoside conjugates.

It is generally accepted that there are three principal sites of gibberellin biosynthesis: developing seeds and fruits, the young leaves of developing apical buds and elongating shoots, and the apical regions of roots. Immature seeds and fruits are prominent sites of gibberellin biosynthesis. This is based on the observation that young fruits, seeds, and seed parts contain large amounts of gibberellins, particularly during stages of rapid increase in size. In addition, cell-free preparations from many seeds, for example, wild cucumber (*Marah macrocarpus*) and pea (*Pisum sativum*), are able to carry out active gibberellin biosynthesis. The site of gibberellin biosynthesis may be the developing endosperm, as it is in the cucurbits, the young cotyledons of legumes or the scutellum of cereal grains. As the seed matures, metabolism appears to shift in favor of gibberellin-sugar conjugates.

Clear evidence that gibberellin biosynthesis occurs in shoots and roots is not so easily obtained as it is for fruits and seeds. This is partly because gibberellin levels are much lower in vegetative tissues. Vegetative tissues also yield cell-free preparations that are less active, suggesting that enzyme levels for gibberellin metabolism are also lower than for reproductive tissues. Gibberellin synthesis in vegetative tissues is generally supported by the occurrence of gibberellins in tissue exudates and the effects of inhibitors of gibberellin biosynthesis (see Fig.

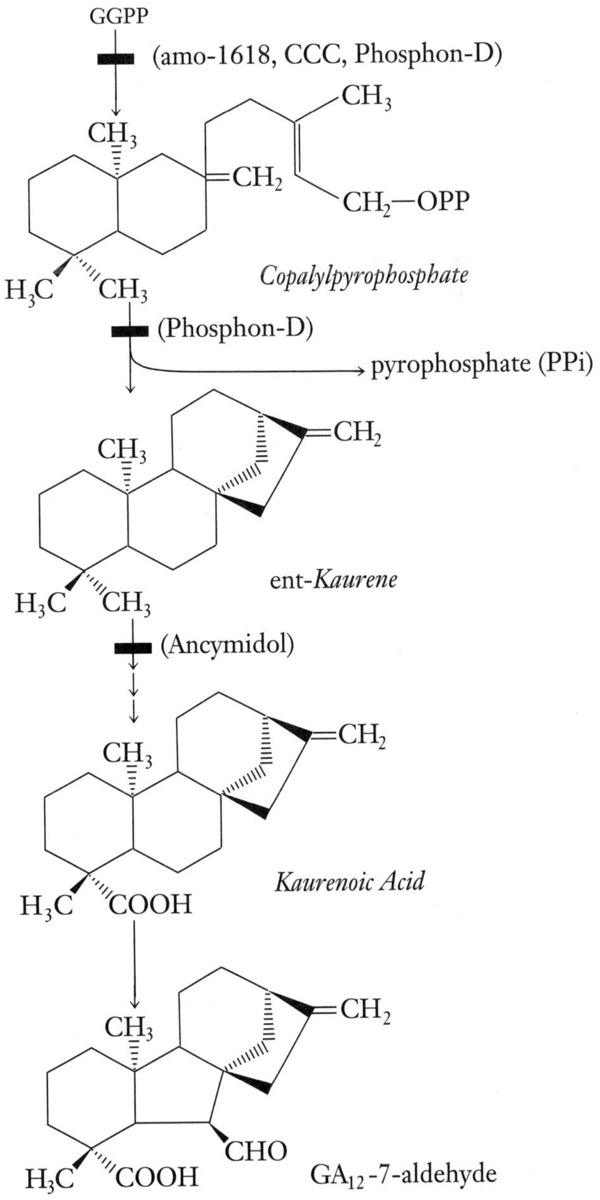

FIGURE 17.12 Gibberellin biosynthesis: the conversion of geranylgeranyl pyrophosphate to GA_{12}-7-aldehyde. GA_{12}-7-aldehyde is inactive, but serves as the precursor to all other gibberellins. The positions at which antigibberellins (dwarfing agents) block gibberellin biosynthesis are indicated.

Box 17.1 Quantitative Analysis of Hormones

Hormone concentrations in most tissues are very low and hormone molecules may undergo complex metabolic interactions. Consequently, in order to study the physiology and biochemistry of these substances in the laboratory, it is essential to have sensitive methods for assessing their presence in plant extracts and for testing the biological activity of purified molecules. The most common hormone assays fall within three broad categories: biological assay or **bioassay, instrumental analysis,** and some form of **immunoassay.**

BIOASSAYS

The biological activity of hormones or plant extracts is commonly tested by applying them to a plant system that is known to respond to that particular class of hormone. The use of living material in this way to test for activity is known as a **bioassay.** Bioassays were introduced earlier in Chapter 16 and the two responses illustrated in Figure 16.2 represent typical bioassays. For decades, bioassays were the principal means, if not the only means, for physiologists to obtain both qualitative and quantitative information about hormones.

In order for a plant response to be useful as a bioassay, there are two principal criteria that must be met. First, the system must respond specifically to that hormone or class of hormones. In other words, the test should not be sensitive to other hormones or metabolites and should not be subject to inference by commonly encountered biochemicals. Second, the response must be sensitive to low concentrations of hormone. This usually requires that the plant material contain no significant amount of any endogenous, active substance. Reduction in the endogenous supply of hormone may be achieved by using excised tissue or mutants that fail to synthesize the hormone in question. Third, the magnitude of the response must bear a quantitative relationship to the concentration of the hormone. To be useful, a bioassay should also be relatively easy to set up, give reproducible results within a reasonable time frame, and be free of inhibition by other compounds commonly encountered in plant extracts. In theory, almost any growth or developmental response could serve as a bioassay, but some have proven more useful than others.

Although the use of bioassays to test for biological activity is still a viable element of hormone research, modern advances in instrumental analysis and immunochemistry have largely replaced the bioassay for routine analysis.

INSTRUMENTAL ANALYSIS

Modern developments in analytical chemistry and instrumental analysis have enabled investigators to achieve major advances in plant hormone research. Although bioassays are generally necessary for detection of plant hormones—and still must be used to test for biological activity—their use for quantitative analysis is limited. Physical techniques such as *high pressure liquid chromatography* (*HPLC*) and *gas chromatography* teamed with *mass spectrometry* (*GC-MS*) have made possible the quantitative analysis of hormones with a speed, sensitivity, and precision never before possible.

Initial purification and concentration of a hormone usually begins with solvent extraction of the plant tissue, followed by some form of column chromatography, which separates groups of molecules on the basis of charge or molecular size. The partially purified sample may then be purified further by HPLC. HPLC is essentially a capillary column with a very small diameter bore and a coating on the inner surface. The sample containing the hormones is injected into the column and the hormones bind to the coating. A solvent is then pumped through at very high pressure (necessary because of the narrow bore). As the solvent moves through the column, the hormones partition between the solvent and the coating, gradu-

17.12). Application of (2-chloroethyl) trimethylammonium chloride (CCC, an "antigibberellin") to roots, for example, rapidly decreases the amount of gibberellin appearing in exudates. Furthermore, removal of the root apices from seedlings of scarlet runner bean (*Phaseolus coccineus*) results in decreased shoot growth and the disappearance of GA_1 (Crozier and Reid, 1971).

GIBBERELLIN BIOSYNTHESIS

Gibberellins are **diterpenoid acids,** chemically related to a large group of naturally occurring compounds called **terpenoids.** Terpenoids may normally be recognized on the basis of their chemical structure, which may be dissected into an appropriate number of five-carbon **isoprene** units (Chap. 14).

ally moving down the column until they exit. Different molecular species will partition to different degrees, and consequently migrate through the column at different rates. Migration rate will depend upon the individual characteristics of the molecule as well as the characteristics of the coating and the solvent.

While substantially increasing purity, HPLC fractions may still contain more than one molecular species (as in the case of gibberellins). These mixtures of similar molecules may be further separated by GC-MS. Like HPLC, GC employs a coated capillary column. However, in GC the molecules are partitioned between a solid coating and a moving gas stream at elevated temperature. With the exception of ethylene, which is already a gas, separation by GC requires that hormone molecules be chemically modified to make them volatile. This can be done by masking carboxyl and hydroxyl groups with methyl and trimethylsilyl groups, respectively. The molecules of interest are then eluted from the column by a programmed, steady increase in temperature.

Molecules separated by GC can be identified and chemically characterized by mass spectrometry. Usually, this is accomplished by passing the sample from the GC directly into the mass spectrometer (MS). In the MS, the molecule is bombarded with electrons, which fragments the molecule and gives each fragment an electrical charge. The charged fragments then pass through an oscillating electrical field that deflects the individual fragments as a function of their mass-to-charge ratio. By varying the field strength, fragments with a particular mass-to-charge ratio may be permitted to enter the detector. Since each molecule produces a unique set of fragments, the mass spectrograph is, in effect, a fingerprint of the molecule that can be used for specific identification.

IMMUNOASSAY

Another technique that has gained considerable importance for hormone analysis is immunoassay, including both **radioimmunoassay (RIA)** and the **enzyme-linked immunosorbent assay (ELISA)**. Immunoassays, available for all four nongaseous hormone groups, employ antibodies raised against the hormone (the antigen). Since hormone molecules are themselves too small to elicit antibody formation, they must first be covalently linked to a large protein before injection into the appropriate animal. The antibodies elicited in the animal are then reactive against the free hormone molecule.

RIA is based on immunoprecipitation of radioactive hormone. In a control reaction, the antibody is used to precipitate a known quantity of radiolabeled hormone. Radiolabeled hormone is then mixed with an unknown quantity of unlabeled hormone from a tissue extract. The unlabeled hormone will compete with the radiolabeled hormone in the precipitation reaction, reducing the quantity of radioactivity in the precipitate. The difference between the precipitated radioactivity in the control and test samples is proportional to the amount of hormone in the unknown sample.

The ELISA test is similar, except that an enzyme (e.g., alkaline phosphatase) is linked to the antibody and the enzyme reaction is used to quantify the immunoprecipitate rather than radioactivity. Immunoassays are relatively rapid and do not require the sophisticated and expensive equipment necessary for instrumental analysis. Yet they are considered at least as sensitive as GC and HPLC. For example, it has been reported that RIA will detect as little as 0.3 picomole (1 pmole = 10^{-12} mole) of IAA and ELISA as little as 20 femtomole (= fmole = 10^{-15} mole). Although relatively new as an analytical tool for hormones, immunoassays, with their combination of intrinsic specificity, sensitivity, and ease of operation, have considerable potential as a method for hormone analysis in the future.

Gibberellin biosynthesis may be conveniently considered in three stages. The first stage in the synthesis of gibberellins is the same mevalonic acid pathway, described in Chapter 14, which leads to other terpenoids. In this pathway, the 5-carbon (C_5) isoprenoid unit isopentenylpyrophosphate (IPP) is synthesized from acetyl coenzyme A (acetyl—CoA) and used to build up the C_{20} geranylgeranyl-pyrophosphate (GGPP). The second stage in the synthesis of gibberellins is the biosynthesis of ***ent*-kaurene** from GGPP and its conversion to **GA_{12}-7-aldehyde** (Fig. 17.12). GA_{12}-7-aldehyde is the first compound in the pathway with the true gibberellane ring system and is believed to be the precursor for all other known gibberellins. The path to GA_{12}-aldehyde has been studied extensively and appears to be the same in the fungus, *G. fujikuroi*, and all higher plants studied so far. The third stage is the biosynthesis of all gibberellins from GA_{12}-7-aldehyde.

The first two steps in the synthesis of gibberellins from GGPP involve the cyclization of GGPP first to copalylpyrophosphate, then to kaurene (Fig. 17.12). It is these two cyclization steps that are inhibited by the antigibberellin dwarfing agents, Amo-1618, CCC, and phophon-D, thus leading to a deficiency of gibberellin in the plant and reduced growth. Following cyclization, the carbon at position 19 on kaurene undergoes three successive oxidations in the sequence $CH_3 \rightarrow CH_2OH \rightarrow CHO \rightarrow COOH$ to form kaurenoic acid. The oxidation of kaurene to kaurenoic acid is inhibited by ancymidol, another dwarfing agent. The final two steps involve a hydroxylation at carbon-7 and contraction of the ring with extrusion of carbon-7 to form GA_{12}-7-aldehyde.

As noted above, GA_{12}-7-aldehyde is the first compound with the true gibberellane skeleton and is the precursor to all other gibberellins. Oxidation of the aldehyde group on carbon-7 to a carboxyl group gives GA_{12}. This carboxyl group is a feature of all GAs and is required for biological activity. The C_{19}-GAs arise by subsequent oxidative elimination of carbon-20. While the biosynthetic pathway up to GA_{12}-7-aldehyde is the same in all plants, subsequent pathways can vary substantially from genus to genus or even in different tissues in the same plant. A brief summary of demonstrated interconversions among gibberellins in pea seed and seedlings is presented in Figure 17.13. The 13-hydroxylation pathway (bold arrows), leading to GA_{20} and GA_1 is probably of widespread occurrence in higher plants.

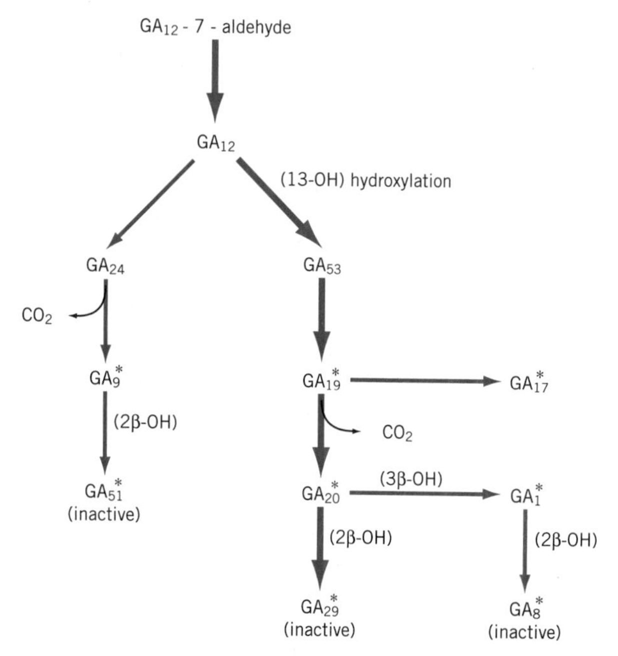

FIGURE 17.13 Proposed pathways for gibberellin biosynthesis in pea (*Pisum sativum*). The major pathway, shown in bold arrows, occurs in seeds and shoots. The pathway shown by dashed arrows occurs only in shoots. The asterisk (*) indicates known endogenous forms. Hydroxylation steps are indicated.

GIBBERELLIN METABOLISM AND TRANSPORT

In immature, actively developing seed of *Phaseolus vulgaris* the principal free gibberellins are GA_1 and its 2-β-hydroxyl analog GA_8, although small amount of GA_4, GA_5, GA_6 (all C_{19} GAs), and GA_{37} and GA_{38} (C_{20} GAs) are also found. Mature seeds, however, contain mainly GA_8-glucoside, with smaller amounts of glucosyl esters of GA_1, GA_4, GA_{37}, and GA_{38} glucosyl esters. As with auxins, the conjugated forms probably represent storage of inactive hormone that can be activated by hydrolysis to free, active hormone upon germination.

Gibberellin transport studies have been conducted largely by application of radioactively labeled GAs to stem or coleoptile sections. Gibberellins have been detected in both the phloem and xylem saps. Transport of gibberellins does not appear to be polar, as it is with auxin, but moves along with other phloem-translocated organic materials according to a source-sink relationship. Whether gibberellins are actually transported in the xylem is not clear; they could end up there simply by lateral translocation from the phloem. On the other hand, it is likely that any gibberellins synthesized in the root tip are distributed to the aerial portions of the plant through the xylem stream. It is not known whether gibberellins are transported as free hormones or in conjugated form.

CYTOKININ BIOSYNTHESIS AND METABOLISM

Cytokinins are synthesized by a condensation of an isopentenyl group with the amino group of adenosine monophosphate. Cytokinins will also form conjugates with sugars and are metabolized by oxidation.

A major site of cytokinin biosynthesis in higher plants is the root. High cytokinin levels have been found in roots, especially the mitotically active root tip, and in the xylem sap of roots from a variety of sources. It is generally concluded that roots are the principal source of cytokinins in most, if not all, plants and that they are transported to the aerial portion of the plant through the xylem. Indirect support for this conclusion is provided by the observation that excised leaves from many species can be maintained in a moist sand bed only if adventitious roots are permitted to form at the base of the petiole. If these roots do not form or are removed as they form, the leaves will quickly undergo senescence. The delayed senescence is apparently due to cytokinins, which are synthesized in the root and transported to the leaf through the vascular tissue.

Immature seeds and developing fruits contain high levels of cytokinins; the first naturally occurring cytokinins were isolated from milky endosperm of maize and developing plum fruits. While there is some evidence that seeds and fruits are capable of synthesizing cytokinins, there is also evidence to the contrary. Thus it remains equally possible that developing seeds, because of their high metabolic activity and rapid growth, may simply function as a sink for cytokinins transported from the roots.

BIOSYNTHESIS OF CYTOKININS

Cytokinins are commonly found in the cell as modified bases in transfer ribonucleic acid (tRNA) molecules. These cytokinins, along with other modified bases such as methylated purines, are not incorporated during transcription of the tRNA, but are synthesized during post-transcriptional processing. For example, enzymes have been isolated from both bacteria (*Escherichia coli*) and maize (*Zea mays*), which catalyze the condensation of Δ^2-isopentenyl pyrophosphate with the appropriate adenosine residue in tRNA to give [9R]iP. It is not yet known whether, *in vivo*, the hydroxylation to form zeatin occurs before or after incorporation of iPP into the tRNA.

Since tRNA contains cytokinins, it is natural to ask whether hydrolysis of tRNA into its constituent nucleotides might account for the occurrence of free cytokinins in plant cells. This does not seem likely for several reasons. In the case of zeatin, which may exist in either the *cis* or *trans* form, the free cytokinin and the tRNA cytokinin are structurally distinct; that is, free zeatin is mainly *trans* while zeatin found in tRNA is in the *cis* form. In addition, cytokinins are found in the tRNA of tissue cultures, even those that require exogenously supplied cytokinins for growth. Finally, the rate of tRNA turnover (i.e., the balance of synthesis and hydrolysis) is generally considered too low to account for the level of free cytokinins encountered in most tissues or the high rate of ^{14}C-labeled adenine incorporation into free cytokinin *in vivo*. While it remains possible that tRNA hydrolysis may contribute some free cytokinin to the cellular pool, the dominant view is that most free cytokinins arise by *de novo* synthesis.

Enzymes that direct *de novo* synthesis of cytokinins from adenosine-5′-monophosphate have been isolated from the slime mould *Dictyostelium discoideum*, tobacco callus tissue, and crown gall tissue (Fig. 17.14). This reaction is specific for the nucleotide; the enzyme will not add the isopentenyl group to either adenine or adenosine. The product, [9R-5′P]iP, is believed to be the precursor to all other naturally occurring cytokinins. However, little [9R-5′P]iP accumulates in tissue; it probably undergoes a rapid hydroxylation of the side chain to give the comparable zeatin ribonucleotide. Reduction of the double bond in the side chain would give the dihydrozeatin derivative, while sequential hydrolysis of the phosphate group and the ribose moiety would give rise to zeatin.

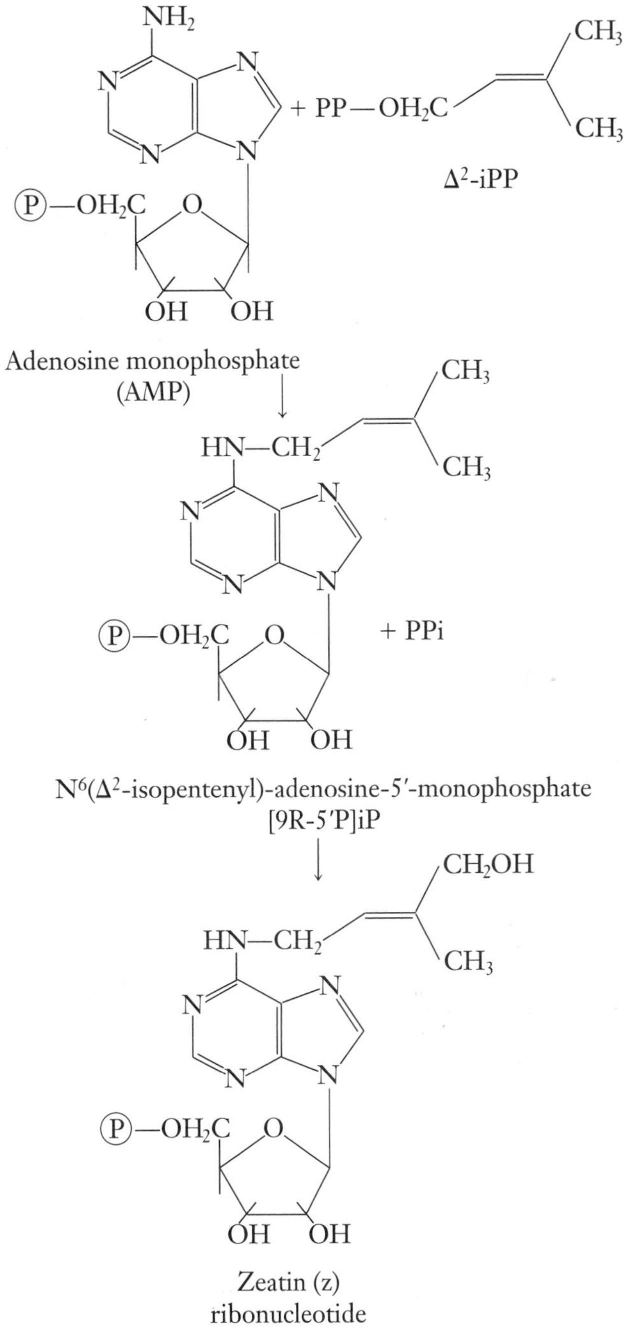

Adenosine monophosphate (AMP) + Δ^2-iPP → N^6(Δ^2-isopentenyl)-adenosine-5′-monophosphate [9R-5′P]iP + PPi → Zeatin (z) ribonucleotide

FIGURE 17.14 The pathway for *de novo* cytokinin biosynthesis. This reaction is catalyzed by the enzyme *iso*-pentenyltransferase.

Cytokinins are known to undergo extensive interconversion between the free base, ribosides, and ribotides when experimentally supplied to tissues. These interconversions have been studied extensively in wheatgerm by C-M. Chen and his colleagues (Chen, 1982). Enzymes have been identified in wheatgerm that catalize the conversion of iP to its riboside ([9R]iP) or to its ribotide ([9R-5′P]iP) as well as enzymes that catalyze the hydrolysis of the ribotides and ribosides to the free base (iP). Naturally, these rapid interconversions make it very difficult to ascertain which is the truly "active" form of the hormone.

CYTOKININ METABOLISM AND TRANSPORT

There are two principle routes for regulating cytokinin activity levels by removal of cytokinins from the active pool: conjugation with either glucose or amino acids, and oxidation. Glucose conjugates are formed primarily at the nitrogen in position 7 or in position 9 (in place of ribose) on the purine ring, or as O-glucosides on the side chain (Fig. 17.15). Both the 7- and 9-glucosides are biologically inactive, while the O-glucosides are biologically very active. The N-glucosyl conjugates are very stable and do not appear to be hydrolyzed readily to give the active free base. Their formation thus appears to be more a means for inactivation of cytokinins than for storage. O-glucosides, on the other hand, appear to be storage forms that are readily hydrolyzed to yield biologically active cytokinins when needed by the plant.

Cytokinins also form conjugates with the amino acid alanine (Fig. 17.15). 9-Alanyl conjugates of zeatin and dihydrozeatin have been identified in tissues of lupine (*Lupinus*) fruit and root nodules, immature apple seeds, and bean (*Phaseolus*) seedlings. These too are very stable conjugates that probably serve to inactivate the cytokinin in the same manner as N-glucosides.

A major route for removal of exogenously supplied cytokinins in many tissues is oxidation by the enzyme **cytokinin oxidase.** Cytokinin oxidase, partially purified from tobacco tissue, maize seed, and crown gall tissue, cleaves the isopentenyl side chain from either zeatin or iP or their ribosyl derivatives (Fig. 17.16).

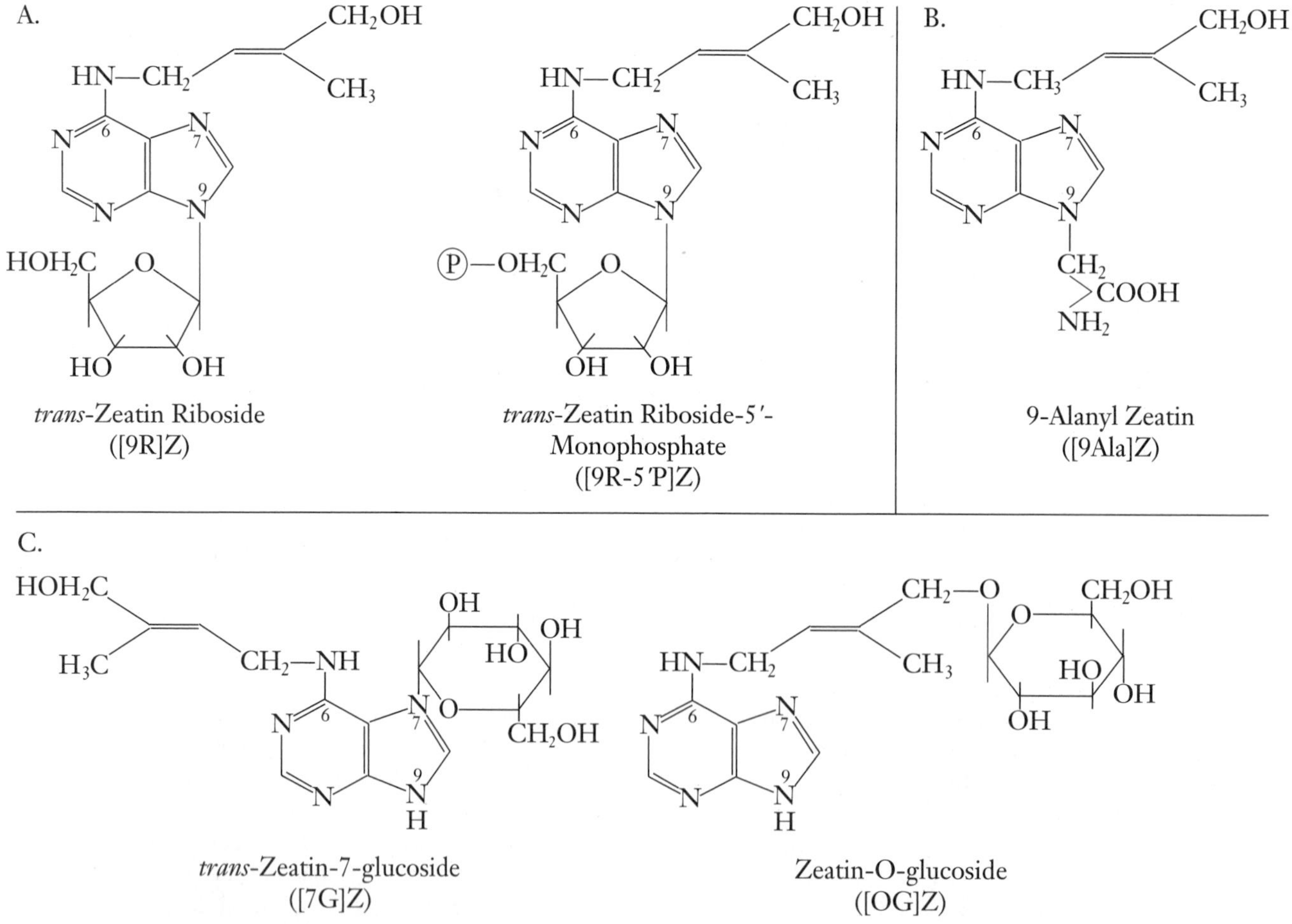

A.

trans-Zeatin Riboside ([9R]Z)

trans-Zeatin Riboside-5′-Monophosphate ([9R-5′P]Z)

B.

9-Alanyl Zeatin ([9Ala]Z)

C.

trans-Zeatin-7-glucoside ([7G]Z)

Zeatin-O-glucoside ([OG]Z)

FIGURE 17.15 Some examples of known zeatin conjugates. (*A*) Ribosides and ribotides. (*B*) Amino acids. (*C*) Glucosides.

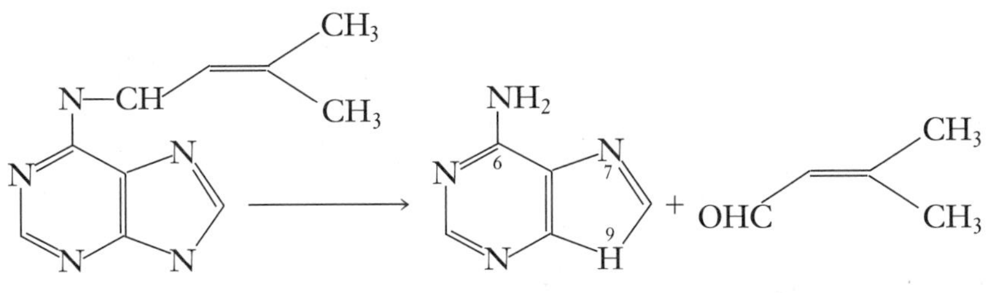

iP Adenine 3-Methyl-2-butenal

FIGURE 17.16 Oxidative degradation of *iso*-pentenyl adenine (iP) by cytokinin oxidase.

ABSCISIC ACID BIOSYNTHESIS AND METABOLISM

Abscisic acid is a 15-carbon isoprene derivative that appears to be synthesized by cleavage from a 40-carbon carotenoid precursor.

There is relatively little detailed knowledge of the localization of abscisic acid biosynthesis. Abscisic acid appears to occur predominantly in mature, green leaves and is probably synthesized in the chloroplast. Numerous reports have shown ABA to be present in chloroplasts and Milborrow (1974) has demonstrated synthesis of ABA from ^{14}C-labeled mevalonate in isolated chloroplasts of ripening bean fruit and avocado leaves. Studies of ABA synthesis and metabolism in stressed and nonstressed plants indicate that stress-induced ABA synthesis also occurs in the chloroplasts and that the ABA rapidly migrates to other regions of the plant. Chloroplasts should probably be considered a major, but not necessarily exclusive, site of ABA synthesis (see Chap. 22).

Abscisic acid is a 15-carbon sesquiterpene; this number and arrangement of carbon atoms suggest that ABA is also derived from mevalonic acid. However, beyond demonstration that ^{14}C-labeled mevalonic acid is incorporated into ABA, details of the pathway of biosynthesis remain obscure. There may be several reasons why progress in this area has been slow. Like other hormones, ABA is normally present in very low concentrations in most plant tissues—on the order of 10 to 50 ngm per gram fresh weight. Only in water-stressed leaves, where the concentration may reach 400 ngm per gram fresh weight, or in young developing seeds do ABA concentrations exceed these values. Moreover, presumed precursors are not readily incorporated into ABA. This could be because of limited solubility and/or uptake into the region of the cell where synthesis occurs. It might also be a result of a feedback loop by which ABA inhibits its own formation when the cell senses that concentrations are adequate. If this were the case and the cell already contains normal levels of ABA, little ABA would be synthesized from added precursors.

There appear to be two possible pathways for the synthesis of ABA: (1) direct synthesis from a 15-carbon precursor or (2) cleavage of a 40-carbon **xanthophyll** (an oxygenated carotenoid (Fig. 17.17). The most likely intermediate for direct synthesis of ABA is the 15-carbon farnesylpyrophosphate, although there is scant evidence to support the direct pathway. On the other

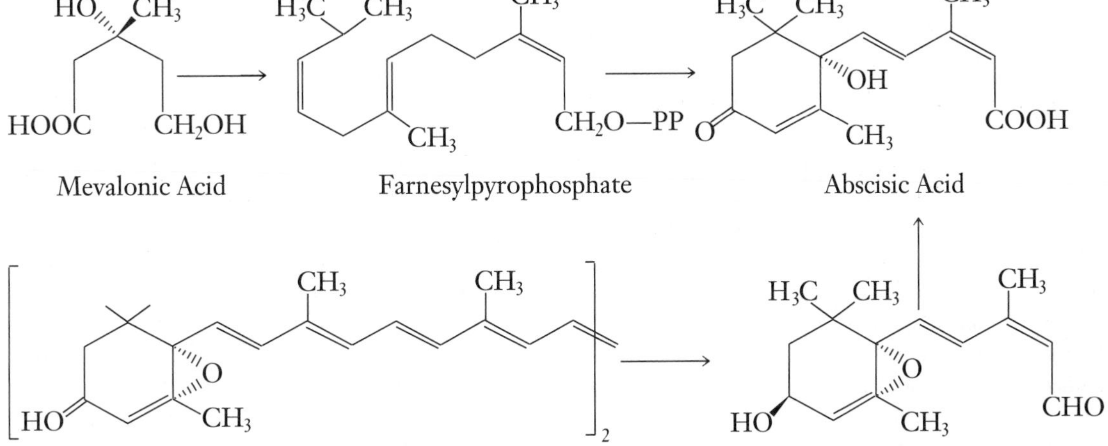

Mevalonic Acid Farnesylpyrophosphate Abscisic Acid

Violaxathin Xanthoxin

FIGURE 17.17 Possible pathways for the biosynthesis of abscisic acid from mevalonic acid or violaxanthin (a diterpene).

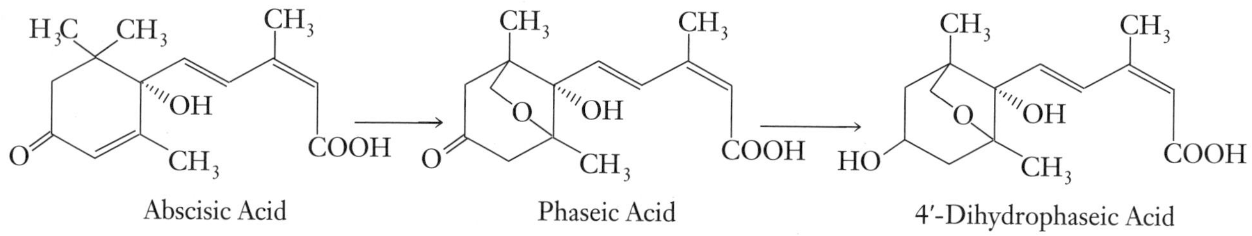

Abscisic Acid → Phaseic Acid → 4′-Dihydrophaseic Acid

FIGURE 17.18 A scheme for the oxidative degradation of abscisic acid.

hand, there is a growing body of supporting evidence for the synthesis of ABA from **violaxanthin.** First, the carbon skeleton of ABA and the position of the oxygen-containing substituents is very similar to that of violaxanthin. J. A. D. Zeevaart and his colleagues compared the incorporation of $^{18}O_2$, a stable isotope of oxygen, into ABA in water-stressed leaves and turgid leaves of several species. The pattern of $^{18}O_2$-enrichment in the carboxyl group of ABA was consistent with the cleavage of a xanthophyll and its rapid conversion to ABA in water-stressed leaves (Zeevaart et al., 1989). Second, it is known that violaxanthin can be degraded in the light *in vitro* to a 15-carbon derivative, xanthoxin, a natural constituent of plants. If radiolabeled xanthoxin is fed to bean or tomato plants, some of the radioactivity appears in ABA. In ABA-deficient tomato mutants, however, conversion of radiolabeled xanthophyll into ABA is reduced relative to wild-type plants (Neil and Horgan, 1985). Third, at least two groups have reported a stoichiometric relationship between losses of violaxanthin and increases in ABA in stressed etiolated bean leaves (Parry and Horgan, 1991). The weight of evidence increasingly leans toward synthesis of ABA from violaxanthin or a closely related xanthophyll in a wide variety of plant systems studied thus far.

Studies of ABA metabolism, usually involving the fate of ^{14}C-labeled ABA, are complicated by the fact that ABA contains an asymmetric carbon atom. An asymmetric carbon atom gives rise to two different forms called **enantiomers.** There are thus two enantiomeric forms of ABA, designated R-ABA and S-ABA. S-ABA is the naturally occurring enantiomer, but most feeding experiments use the racemic mixture containing both the R- and the S-forms. There is evidence that the two enantiomers are metabolized at different rates and that the products of their metabolism are not always the same. Thus it is important to distinguish between those metabolites that can be shown to occur naturally in the tissue and those that may be artifacts of the presence of R-ABA in the experiment.

Abscisic acid is rapidly metabolized when it is applied exogenously to plant tissues. In wilted bean leaves, for example, the half-time for turnover (the time for one-half of the labeled ABA to be destroyed) was estimated to be about three hours. A glucose ester of ABA has been found in low concentration in a variety of plants, but the principal metabolic route seems to be oxidation to phaseic acid (PA) and subsequent reduction of the ketone group on the ring to form dihydrophaseic acid (DPA) (Fig. 17.18). At least some tissues appear to carry the metabolism further to form the 4′-glucoside of DPA. DPA and its glucoside are both metabolically inactive. PA, while inactive or exhibiting reduced activity in most bioassays, is equally effective as a GA_3 antagonist in barley aleurone-α-amylase system.

There is little definitive evidence regarding long-distance transport of ABA. Like gibberellins, ABA is found in both xylem and phloem fluids as well as in parenchyma cells outside the vascular tissue, and there is no evidence for a polarity in transport.

ETHYLENE BIOSYNTHESIS AND METABOLISM

Ethylene is a gaseous hydrocarbon synthesized from the amino acid methionine. Control of ethylene production is exercised by transcriptional regulation of the rate-limiting enzyme, ACC synthase.

Ethylene occurs in all plant organs—roots, stems, leaves, bulbs, tubers, fruits, seeds, and so on—although the rate of production may vary depending on the stage of development. Ethylene production will also vary from tissue to tissue within the organ, but is frequently located in peripheral tissues. In peach and avocado seeds, for example, ethylene production appears to be localized primarily in the seed coats, while in tomato fruit and mung bean hypocotyls it originates from the epidermal regions.

Despite the early discovery of ethylene, its known importance in plant development, and its relatively uncomplicated chemistry, the pathway for ethylene biosynthesis has proven difficult to unravel. This is partly because there are a large number of potential precursors (sugars, organic acids or peptides) that are known to be present in plant tissues. In addition, until recently, the enzymes involved have proven too labile to isolate. Consequently, most of the work has been done *in vivo*, with all the pitfalls inherent in such experiments.

M. Lieberman and L. W. Mapson (1964) first showed that methionine was rapidly converted to ethylene in a cell-free, nonenzymatic model system. In subsequent studies, Lieberman and co-workers confirmed that plant tissues such as apple fruit converted L-[^{14}C]-methionine to [^{14}C]-ethylene and that the ethylene was derived from the third and fourth carbons of methionine. Little progress was made until 1977 when D. Adams and F. Yang demonstrated that S-adenosylmethionine (SAM) was an intermediate in the conversion of methionine to ethylene by apple tissue. In 1979, Adams and Yang further demonstrated the accumulation of 1-aminocyclopropane-1-carboxylic acid (ACC) in apple tissue fed L-[^{13}C]-methionine under anaerobic conditions, which is known to block ethylene production. However, upon reintroduction of oxygen, the labeled ACC was rapidly converted to ethylene. ACC is a nonprotein amino acid that had been isolated from ripe apples in 1957, but its relationship to ethylene was not obvious at that time.

The three-step pathway for ethylene biosynthesis in higher plants is shown in Figure 17.19. In the first step, an adenosine group (i.e., adenine plus ribose) is donated to methionine by a molecule of ATP, thus forming SAM. An ATP requirement is consistent with earlier evidence that ethylene production is inhibited by 2,4-dinitrophenol, an inhibitor of oxidative phosphorylation. Conversion of methionine to SAM is catalyzed by the enzyme methionine adenosyltransferase.

The cleavage of SAM to yield 5′-methylthio-adenosine (MTA) and ACC, mediated by the enzyme **ACC synthase,** is the rate-limiting step. ACC synthase is the only enzyme in the pathway that has been studied in detail. The enzyme has been partially purified from tomato and apple fruit but, because of its instability and low abundance, progress toward its purification and characterization has been slow. More recently, genes for ACC synthase have been isolated from zucchini (*Cucurbita*) fruit (Sato and Theologis, 1989) and tomato pericarp tissue (Van Der Straeten et al., 1990). The cloned genes direct the synthesis of active ACC synthase in the bacterium *E. coli* and yeast, making it now possible to produce the enzyme in sufficient quantity for further study. Having access to the genes will also make it possible to study regulation of ethylene synthesis.

The enzyme or enzyme system that catalyzes the conversion of ACC to ethylene, generally referred to as the **ethylene-forming enzyme (EFE)**, has yet to be isolated. The system has been studied extensively in cells, protoplasts, and intact vacuoles. However, EFE activity is sensitive to detergents and is lost as soon as the plasma or vacuolar membrane is ruptured. This suggests a requirement for an intact membrane and that oxidation of ACC may be linked to an electron transport chain or a transmembrane proton flow. Isolation of the EFE remains one of the key challenges of ethylene research.

Another important aspect of ethylene biosynthesis is the limited amount of free methionine available in plants. In order to sustain normal rates of ethylene production, the sulphur must be recycled back to methionine. Double-labeling experiments have shown that the CH_3S— group is salvaged and recycled as a unit. The remaining four carbon atoms of methionine are supplied by the ribose moiety of the ATP used originally to form SAM (Fig. 17.19). The amino group is provided by a transamination.

Ethylene production is promoted by a number of factors including IAA, wounding, and water stress, principally by the induction of the synthesis of ACC synthase. Induction of this enzyme in plant tissues is blocked by inhibitors of both protein and RNA synthesis, suggesting that induction probably occurs at the transcriptional level. In *E. coli* carrying the cloned ACC

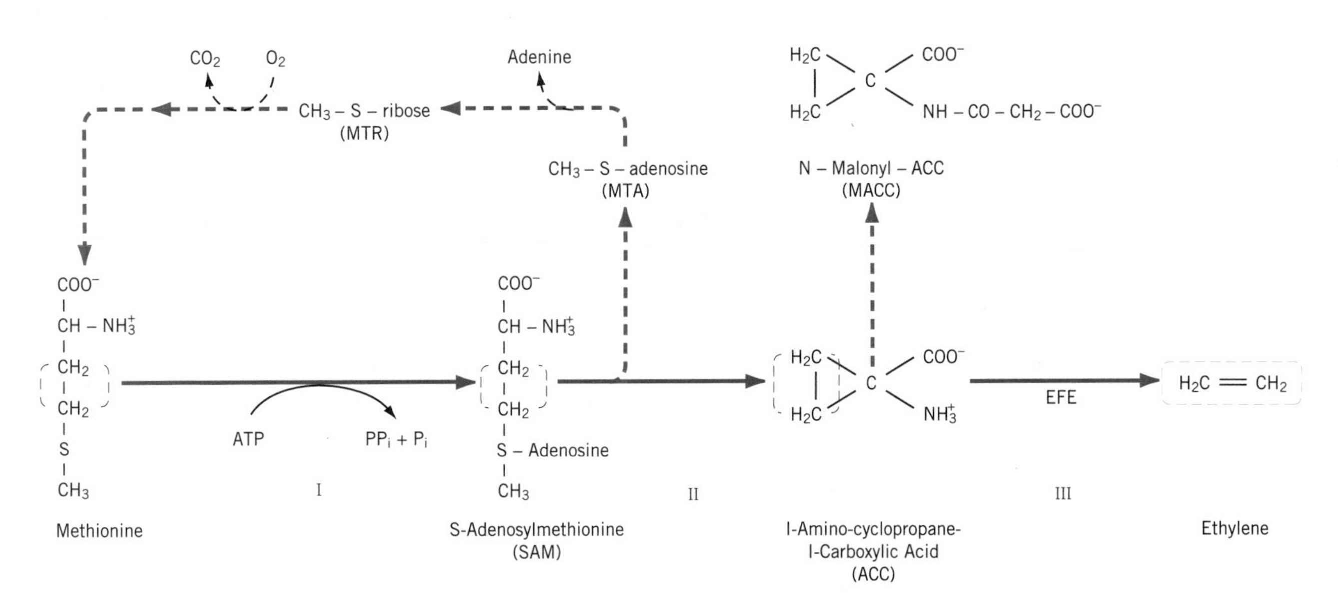

FIGURE 17.19 A scheme for ethylene biosynthesis in higher plants. See text for details.

synthase gene, the physical abundance of ACC synthase messenger RNA also increases in response to IAA and wounding. Control of ethylene production thus appears to be exercised primarily through transcriptional regulation of the ACC synthase gene.

MECHANISMS OF HORMONE ACTION

The primary mechanism of hormone action in plants generally and auxin in particular continues to elude us more than half a century after its discovery. Although auxin appears to be involved in a wide range of growth and developmental responses, efforts to understand how auxin works have focused largely on the fundamental role of auxin in cell expansion.

AUXIN AND CELL EXPANSION

Auxin-induced cell expansion can be quite dramatic. In oat coleoptiles it begins within ten minutes of auxin application and may increase the growth rate as much as tenfold. This short latent period, compared with hours or even days for more complex developmental responses, has made cell expansion an attractive vehicle for the study of primary hormone action. Indeed, cell expansion is probably the most studied hormonal response in plants. For a variety of reasons, it has also become one of the most controversial. Investigators cannot always agree on the validity of experimental proofs or the interpretation of data. Most investigators have been guided by the assumption that a single mechanism will explain cell expansion when, in fact, multiple mechanisms could be involved. To further confound matters, initial growth responses, occurring within the first 30 to 90 minutes of auxin application, may differ from steady-state growth achieved after several hours. Finally, the older literature especially is filled with contradictions that probably result from the choice of experimental tissue. Stem sections, for example, have a large variety of cell types and each type of cell may respond differently to auxin. In dark-grown *Avena* coleoptiles, on the other hand, all of the cell layers seem to respond more or less equally (Cleland, 1987). This, along with the fact that coleoptiles are easy to grow quickly, has made it a popular experimental system for nearly a century. We must be reminded, however, that the coleoptile is a rather specialized organ, found only in the family Gramineae. It often does not respond in the same way as green stems.

Two major theories have been proposed to account for auxin-induced cell expansion. In the 1960s, shortly after Watson and Crick published their now-famous paper on DNA structure and the DNA-RNA-protein dogma was in its ascendency, it was proposed that auxin activated the genes for certain proteins that were necessary for cell growth. There was some evidence in support of this gene activation theory, but techniques available at the time did not allow for sufficiently precise analysis of mRNA and proteins within the very short time periods required to demonstrate primary action. In the meanwhile, Mitchell's chemiosmotic model for oxidative phosphorylation was gaining universal acceptance and interest was turning toward cellular membranes and the control of ion flux across membranes—and ATPase proton pumps in particular. A second theory, the acid growth theory, was proposed, which attributed cell expansion to auxin-induced proton excretion. With recombinant gene technology now available, it has been possible to revisit the gene-regulation hypothesis and there is now evidence that supports both gene activation and acid growth as viable hypotheses.

Whatever its primary action, auxin can alter the rate of cell expansion only by ultimately influencing one or more of the parameters identified previously in Equations 15.4 or 15.5 (Chap. 15). An increase in growth rate, for example, would require an increase in wall extensibility (m), an increase in turgor pressure (P), or a decrease in yield threshold (Y). (Hydraulic conductance, L, of the plasma membrane is not normally a limiting parameter.) Direct measurements of P, using a micropressure probe, indicate that turgor pressure does not change significantly during auxin-stimulated increase in the growth rate of pea stem sections (Cosgrove and Cleland, 1983). Although Y cannot be measured directly, the results of indirect tests indicate that yield threshold does not change either. That leaves extensibility, **m.** Extensibility is difficult to assess. It is on the one hand a rate coefficient, but it is also a measure of the capacity of cell walls to undergo irreversible (plastic) deformation. R. Cleland and his colleagues have devised a number of tests—all of them indirect—to measure extensibility. Whichever the method, the principle is invariably the same—a change in **m** should be reflected as a change in the physical properties of the wall, especially plasticity or its capacity to undergo permanent deformation. On the basis of these tests, there is general agreement that induction of rapid cell enlargement by auxin is accompanied by large increases in **m** (Cleland, 1987). It is reasonable to conclude that auxin stimulates cell expansion by increasing wall extensibility.

Acid Growth Theory In 1970, D. Rayle and R. Cleland proposed a simple but rather provocative theory to explain auxin-stimulated increases in cell wall extensibility. They suggested that auxin causes acidification of the cell wall environment by stimulating cells to excrete protons. There the lower pH activates one or more wall-loosening enzymes, which have an acidic pH optima. The idea that acid pH stimulated elongation was not new—this has been demonstrated repeatedly since 1934—but the linking of acid-stimulated growth to auxin activity has had a major impact on auxin research.

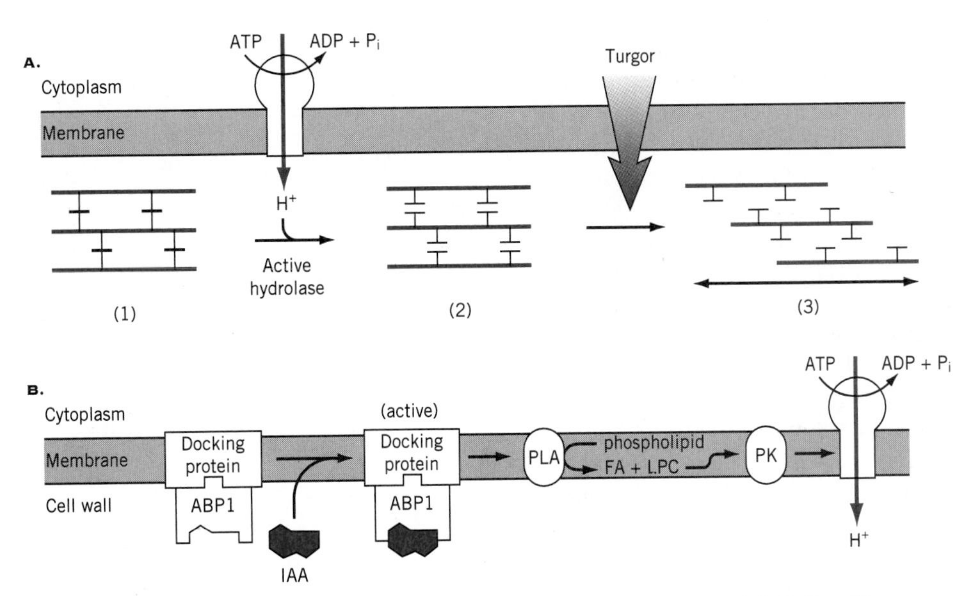

FIGURE 17.20 The acid growth hypothesis for cell enlargement. (*A*) Cell wall polymers are extensively cross-linked with load-bearing bonds (1), which limits the capacity of the cell to expand. An ATPase proton pump located in the plasma membrane acidifies the cell wall space by pumping protons from the cytoplasm. The lower pH activates wall-loosening enzymes that cleave the load-bearing bonds (2). The force of turgor acting on the membrane and cell wall cause the polymers to displace (3) and allow the cell to enlarge. (*B*) The signal transduction chain linking auxin with activation of the ATPase proton pump. See text for details. Abbreviations: ABP1, auxin binding protein 1; PLA, phospholipase A_2; FA, fatty acids; LPC, lysophospholipid; PK, protein kinase. (*B* based on Macdonald, 1997).

At about the same time, A. Hager, working in Germany, published a similar proposal but went further to suggest that auxin stimulated proton excretion by activating a plasma membrane-bound ATPase proton pump. The combined Cleland-Hager proposals, known as the **acid growth theory,** are summarized in Figure 17.20A. An excellent review of the acid growth theory is given by D. L. Rayle and R. E. Cleland (1992). Although the acid growth theory has been tested in relatively few tissues (it has been tested thoroughly only in *Avena* coleoptiles), the evidence is generally supportive.

Plant cell membranes, including the plasma membrane, are known to contain ATPase enzymes that catalyze the electrogenic transport of protons. It is important to note, however, that auxin-binding proteins do not exhibit ATPase activity. It is therefore unlikely that the plasma membrane ATPase is itself the auxin receptor. Still, auxin does cause a hyperpolarization of the cell membrane beginning about 8 to 10 minutes after auxin application. This timing is coincident with the onset of auxin-induced elongation noted earlier. Hyperpolarization of the membrane would result from the activation of an electrogenic ATPase proton pump. It appears that auxin activates a membrane-bound ATPase proton pump, but that it does so indirectly through a separate auxin-receptor complex.

Auxin will also cause growing cells to excrete protons. This is an energy-dependent process and both metabolic inhibitors as well as inhibitors of auxin-induced growth will also inhibit auxin-induced proton excretion. With *Avena* coleoptiles, the pH of the apoplastic solution drops from 5.7 to 4.7 within 8 to 10 minutes of auxin application. Acid solutions at a pH of 3 to 4 are normally required to induce a rate of elongation comparable to optimum auxin concentrations. However, peeled coleoptiles, from which the cuticle has been removed, will respond to pH 5. On the other hand, if the wall space of coleoptiles is infiltrated with neutral buffers to prevent pH change, auxin-induced growth is almost completely prevented. Finally, agents other than auxin that cause proton excretion would be expected to have a similar effect on growth. One such agent is fusicoccin, a phytotoxin from the fungus *Fusicoccum amygdali,* which causes cells to excrete protons at a great rate. It is at least as effective as auxin in stimulating elongation of *Avena* coleoptiles.

If auxin does not directly activate the ATPase-proton pump, two questions arise. First, how is the signal (auxin) perceived by the cell? Second, what is the signal transduction pathway that leads from perception to acidification of the cell wall environment? Although auxin is involved in a range of responses at various levels, studies on auxin perception and the signal transduction pathway have progressed largely with respect to the acid growth response and closely related changes in the polarization of the plasma membrane and properties of ion channels. Much of this work has been carried out with isolated protoplasts, which readily lend themselves to study with microelectrodes and patch-clamp experiments (see Box 5.1) (Napier and Venis, 1995).

As discussed earlier in this chapter, immunological evidence indicates that auxin is perceived at the outer surface of the plasma membrane by the auxin-binding protein ABP1. For two reasons, the significance of ABP1 has attracted some controversy (Venis, 1995). The principal difficulty has to do with the location of ABP1 in the cell. ABP1 is found predominantly in the lumen of the endoplasmic reticulum (ER) and some investigators have been unable to detect any ABP1 at the plasma membrane. ABP1 even contains amino acid sequences at either end of the molecule that are typical of proteins normally retained within the lumen of the ER. However, more sensitive immunolocalization techniques have now confirmed a small population (perhaps 1,000 molecules) on the plasma membrane of maize protoplasts. A second problem is that, based on amino acid sequence, the ABP1 protein appears to have no lipophilic membrane-spanning domain. To reconcile these observations, it has been proposed that ABP1 is complexed with a transmembrane **"docking protein"** (Macdonald, 1997). According to this model, the docking protein provides the necessary lipid solubility to anchor ABP1 to the membrane. The ABP1-docking protein complex is then exported from the ER to the plasma membrane where it is inserted with ABP1 facing the outside (Fig. 17.20B). The ABP1-docking protein complex is itself inactive, but attachment of an auxin molecule activates the complex and initiates the signal transduction pathway. The proposed docking protein has yet to be identified, but there is some evidence that it might be a receptor in the family of G-proteins (see Fig. 17.1).

Auxin also activates **phospholipase A_2 (PLA_2)**, a membrane-bound enzyme that hydrolytically excises the fatty acid from the central (C2) glycerol carbon of a phospholipid (see Box 1.1). The product is a free fatty acid plus a phospholipid with a single fatty acid, called a **lysophospholipid (LPC)**. Several experiments have implicated PLA_2 in the signal transduction chain (Fig. 17.20B). For example, activation of PLA_2 can be blocked by IgG-antiABP and the products of PLA_2, both LPC and fatty acids, stimulate proton secretion and elongation. These effects are inhibited by vanadate, which specifically blocks the plasma membrane proton-ATPase. These data suggest that PLA_2 follows ABP1 in the chain and that lipids, both LPC and fatty acids, function as second messengers. Finally, both the IAA and LPC effects on proton secretion and elongation can be blocked by protein kinase inhibitors, suggesting that the lipids activate the proton-ATPase by a phosphorylation-dependent mechanism.

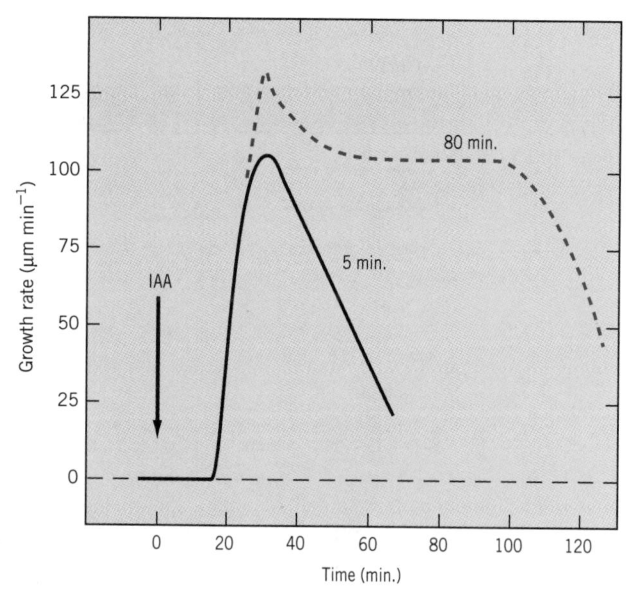

FIGURE 17.21 Kinetics of auxin-induced elongation of maize (*Zea mays*) coleoptiles. The two curves differ in the duration of auxin action. In each case, auxin (10^{-5} M IAA) was added at time = 0 and removed after the indicated period (5 or 80 min.). (From R. K. Dela Feunte and A. C. Leopold, *Plant Physiology* 46:186, 1970. Copyright American Society of Plant Physiologists.)

In spite of the apparent support for acid-induced growth stimulation in *Avena* coleoptiles, the theory does not alone resolve the question of how auxin regulates cell growth, let alone more complex developmental problems. One difficulty is that green stem sections, which respond to auxins, do not respond well (if at all) to acids. Another difficulty is that exogenous acid induces a transitory growth stimulation of coleoptiles. Neither acid nor fusicoccin is effective after the first 30 to 60 minutes. The kinetics for auxin-induced growth show an initial rapid increase in the growth rate, reaching a maximum within 30 to 60 minutes, followed by a steady or gradually declining rate over the next 16 hours (Fig. 17.21). The most plausible explanation for such a two-phase response curve is that the acid-growth response is limited primarily to the rapid initial growth response. Additional auxin-regulated factors might then be required for the maintenance of growth over the longer term. These additional factors could involve the transcription of genes and synthesis of growth-promoting proteins, to be discussed in the following section.

Auxin and Gene Action If auxin action were limited to loosening of the cell wall, it would be expected that turgor-induced cell expansion would stretch the cell wall, causing it to become thinner. The fact that thinning of the wall has not been observed in rapidly elongating cells indicates that new material must be added at a rate consistent with expansion, such that cell wall thickness is maintained. Thus a second effect of auxin appears to be to stimulate the synthesis of cell wall components. This effect could be achieved through auxin regulation of specific gene activity.

Several studies going back as far as the late 1960s indicated that continued RNA and protein synthesis was required to sustain elongation of coleoptile and hypocotyl sections. Most of the early reports were based on electrophoretic analysis of radioactively labeled protein. Sections of maize (*Zea mays*) coleoptile or pea (*Pisum sativum*) hypocotyl, for example, were floated for up to three hours on solutions containing radioactive amino acids with or without auxin. Polypeptides were then isolated from the tissue sections and analyzed by one- or two-dimensional polyacrylamide gel electrophoresis. Following auxin treatment, a limited number of changes in the labeling pattern were observed. Although considered a powerful tool at the time, this technique required extended incubations with label and was not sufficiently sensitive to detect any rapid auxin-induced changes in translation.

More progress was made in the early 1980s, when techniques for *in vitro* translation of isolated messenger RNA (mRNA) became available. Again, the labeled products were analyzed by 2D polyacrylamide gel electrophoresis. Most of this work was carried out with tissue sections, such as pea epicotyl, maize coleoptile, and soybean (*Glycine max*) hypocotyl sections. Following auxin treatment, as many as 30 to 40 translation products were seen to either increase or decrease in labeling intensity. Some of these changes occurred within minutes of auxin application. In soybean and pea, new translation products were evident within 15 to 20 minutes of IAA application while in maize they appeared within 10 minutes. The subsequent use of cDNA probes (DNA sequences complementary to the mRNA) established that auxin stimulates an increase in the physical abundance of mRNA, not simply the efficiency with which the mRNA is translated *in vitro.* The quantity of mRNA may increase as much as 50-fold in response to auxin treatment and some of the changes in mRNA levels were detected within minutes of auxin application.

With the advent of recombinant DNA techniques, it is now possible to isolate and characterize mRNAs and to clone DNA sequences (i.e., genes) that are specifically modulated by auxin. The cDNA clones appear to be induced within minutes of exposure to auxin (short-term responses) or after an hour or more (long-term responses). Of course, the rapidly induced or short-term auxin responses have generated the greatest interest.

One family of auxin-responsive genes that has been identified in both soybean and *Arabidopsis* are the **SAURs,** or **s**mall **a**uxin **u**p-regulated **R**NAs (Gil et al., 1994). SAUR genes encode short, relatively unstable RNA transcripts. In soybean hypocotyls, the expression of SAUR genes appears to be localized in tissues that normally respond to auxin and the RNA transcripts can be detected within 2 to 3 minutes of auxin application—even before auxin-induced elongation can be observed. Furthermore, an asymmetrical distribution of SAUR transcripts has been detected in gravity-stimulated seedlings. The asymmetry correlates with the differential cell elongation in responding seedlings, but can be detected even before any visible signs of curvature. Finally, several auxin-resistant mutants in *Arabidopsis* show low levels of SAUR expression in response to auxin treatment. Although the specific role of the SAUR genes is not yet known, these correlations suggest that they may be an early part of the signal transduction chain for auxin responses (Abel and Theologis, 1996).

Earlier, L. Vanderhoef and C. Stahl (1975) had reported that auxin-induced growth could be separated into two phases on the basis of a differential response to cytokinins. Cytokinins would inhibit the prolonged auxin growth response while the initial growth response was insensitive to cytokinin. This led Vanderhoef and R. Dute (1981) to propose a parallel model for auxin-induced growth. According to this model, proton secretion is responsible for the initial growth response with a lag of 8 to 10 minutes. After about one hour, the prolonged growth response then comes under control of the transcriptional and translational machinery, which produces proteins necessary to sustain growth. With the discovery of rapid gene regulation by auxin, A. Theologis (1986) proposed that the accumulated evidence is more compatible with a series model, represented by the sequence of events: mRNA induction → H^+ secretion → cell elongation. Whether the truth lies with either of these models or a third, as yet undiscovered model, remains to be seen.

This is an exciting period for auxin research, but enthusiasm must be tempered with caution. It is important to note, for example, that very little is known about the mechanism by which auxin regulates gene expression, although the recent work on localization of auxin-binding proteins in the cell appears to offer some direction. Moreover, the identities and functions of most of the early auxin-induced peptides have yet to be determined and auxin is known to induce a number of enzymes that are not directly involved in cell elongation. Nevertheless, for the first time in a century of plant hormone research, we may well have on hand the tools necessary to resolve some of these problems and a solution to the primary mechanism of auxin action could be within our grasp.

GIBBERELLIN CONTROL OF STEM ELONGATION

In spite of the dramatic nature of gibberellin effects on stem elongation, very little is actually known regarding the mechanism. Gibberellin acts to stimulate both cell division and cell elongation in stems. In rosette plants, the rapid elongation that follows gibberellin application is accompanied by a large increase in the number of cell divisions in the region just below the apical meristem and subsequent hyperelongation of the daughter cells. In peas, normal internodes have both more and larger cells than dwarf internodes.

The stimulation of cell and internode elongation by gibberellins naturally raises the question of whether gibberellins operate in a manner similar to auxins. There are indeed some similarities but there are also substantial differences. In contrast with auxins, for example, only a few excised tissues will respond to exogenous gibberellin. One is lettuce hypocotyls, which respond slowly to gibberellins after a 10- to 15-minute lag period. In other systems (e.g., oat stem) the lag period may extend to 3.5 hours. Gibberellin increases wall extensibility in lettuce hypocotyls (Stuart and Jones, 1977) but decreases wall yield threshold in pea (Behringer et al., 1990). More critically, however, there is no significant acidification of the cell wall in response to gibberellin treatment.

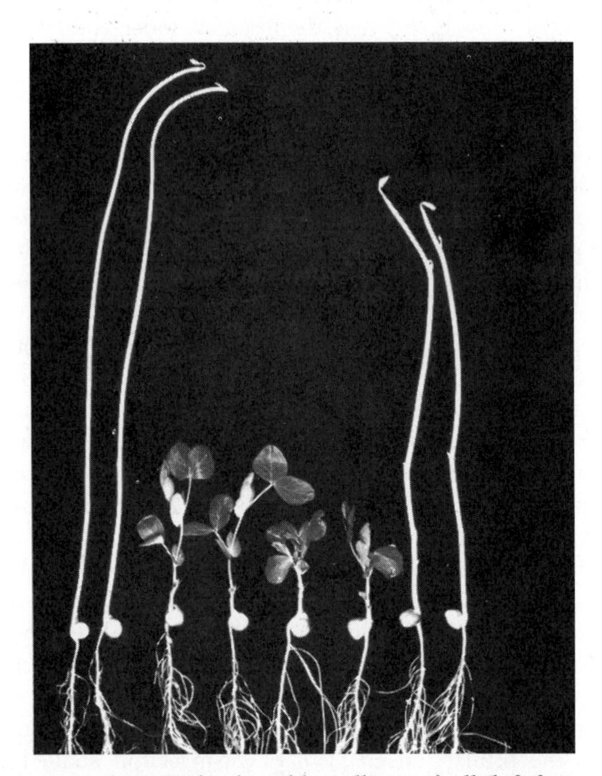

FIGURE 17.22 Eight-day-old seedlings of tall (left four plants) and dwarf (right four plants) pea (*Pisum sativum*) cultivars. The etiolated plants at either end were grown in the dark. The center four plants were grown in the light. Note that the dwarf character is not expressed in the dark. (From T. E. Weier et al., *Botany*, 6th edition, New York, Wiley, 1982. Fig. 20.11. Used by permission of the authors.)

Another complicating factor in the study of gibberellins and stem elongation is the involvement of light. The gibberellin response in lettuce hypocotyls, for example, is a reversal of blue and far-red light inhibition. Dwarf pea and cucumber stems are inhibited by red light (Fig. 17.22); this inhibition is also reversed by gibberellin. Clearly any understanding of the mechanism of gibberellin-induced growth must ultimately involve gibberellin-light interactions.

GIBBERELLIN CONTROL OF SEED GERMINATION

While relatively little is known about the role of gibberellins in stem elongation, the action of gibberellins in the germination of cereal grains is probably the best understood hormonal mechanism in plants. Indeed, the discovery that α-amylase is released in response to GA treatment of barley grains, along with the subsequent isolation of viable aleurone cells as an experimental tool, has led to widespread use of the cereal aleurone as a model for hormone action at the molecular and genetic level.

The α-amylase secreted from barley aleurone consists of multiple isozymes that fall into two major families characterized by their isoelectric points (pI). The low group has pIs in the range of 4.5 to 4.85 and the high group has pIs in the range of 5.9 to 6.3. Within each group the isozymes are quite similar but between the two groups there are major differences with respect to calcium requirement for secretion and sensitivity to EDTA and heavy metals. The two groups of isozymes are translated *in vitro* from two different mRNA populations encoded by two multigene families located on different chromosomes. The situation in wheat is similar except that there are at least three gene families located on two chromosomes. In addition to α-amylase, proteolytic enzymes (proteases), β-amylase, and other starch-degrading enzymes are involved in mobilizing the endosperm reserves (Fig. 17.23).

Earlier physiological studies have established that GA-stimulated α-amylase synthesis is inhibited by inhibitors of transcription and that GA induces significant changes in RNA metabolism, especially mRNA. Later, D. Baulcombe, and D. Buffard (1983) reported an increase in at least seven different mRNAs in GA-treated wheat grains from which the embryo had been removed. If GA acts to regulate gene expression, as these studies suggest, it clearly must regulate a large number of genes from several different families spread throughout the genome.

Does gibberellin regulate transcription of α-amylase mRNA? An unequivocal answer to this question was possible only after techniques for isolation of protoplasts from barley aleurone cells were perfected (Hooley,

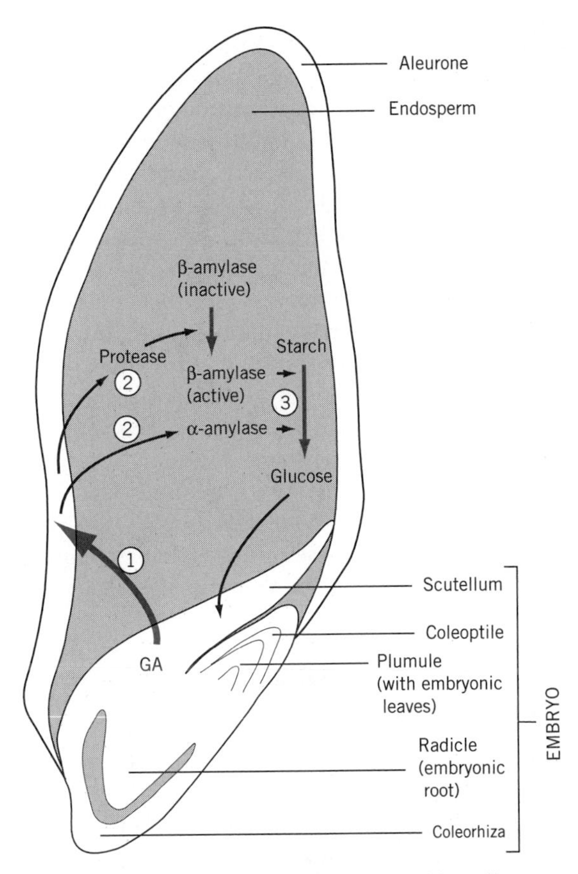

FIGURE 17.23 A schematic illustrating gibberellin-induced release of enzymes and carbohydrate mobilization during germination of barley (*Hordeum vulgare*) seed. Gibberellin moves from the embryo (1) to the aleurone where it stimulates the synthesis of α-amylase and protease enzymes (2). The protease converts an inactive β-amylase to the active form. α- and β-amylase together digest starch to glucose (3), which is mobilized to meet the metabolic demands of the growing embryo.

1982).[2] Aleurone protoplasts provide a useful experimental system. They respond normally to GA by exhibiting the same ultrastructural changes and producing the same isozymes with the same efficiency as intact aleurone cells. Most importantly, high yields of transcriptionally active nuclei can be readily isolated from protoplasts.

Based on evidence from several lines of investigation, it is clear that gibberellin dose regulates transcription of α-amylase mRNA. Both *in vivo* pulse labeling of protein and *in vitro* translation of protein from total aleurone RNA, followed by electrophoretic and autoradiographic analysis, show significant increases in the amount of α-amylase translated following the application of gibberellin (Fig 17.24). α-Amylase may constitute as much as 50 to 60 percent of the total translated protein. Analysis by 2-dimensional gel electrophoresis confirms that the translation products include α-amylase from both gene families. Gibberellin response is apparently not limited to α-amylase alone, as a small number of other peptides either increase or decrease following gibberellin treatment. Northern blot hybridization of α-amylase cDNA clones with RNA from aleurone layers has confirmed that gibberellin causes an increase in the physical abundance of α-amylase mRNA (Fig. 17.25). Following gibberellin treatment, α-amylase mRNA may comprise as much as 20 percent of the total translatable mRNA. Finally, time course studies have shown that the rate of α-amylase synthesis following

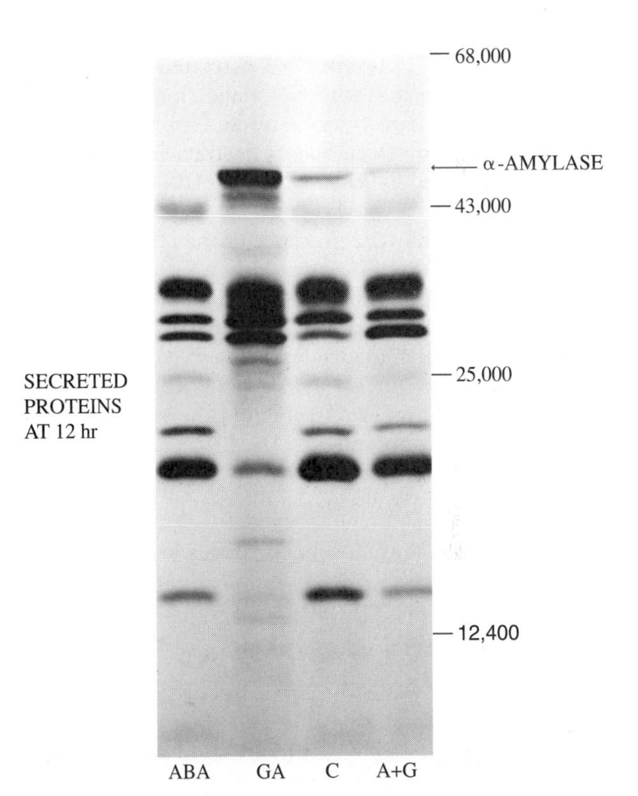

FIGURE 17.24 Hormonal control of a-amylase biosynthesis by barley aleurone layers. Polypeptides synthesized by isolated aleurone layers were labeled with the radioactive amino acid [^{35}S]methionine incubated in the presence of ABA, GA_3, or ABA + GA_3 (A+G). Polypeptides secreted into the incubation medium were separated by electrophoresis on polyacrylamide gels. Polypeptides incorporating the radioactive label were detected by exposing the gel to X-ray film. Note the pronounced stimulation of a-amylase biosynthesis in the presence of GA_3, a stimulation almost completely abolished in the presence of ABA. (*C*) represents untreated controls. Numbers indicate approximate molecular mass. (From Higgins et al., *Plant Molecular Biology* 1:191, 1982. Reprinted by permission of Kluwer Academic Publishers. Original photograph kindly provided by Dr. J. V. Jacobsen.)

[2]Protoplasts are cells from which the cell wall has been removed (see Chap. 1).

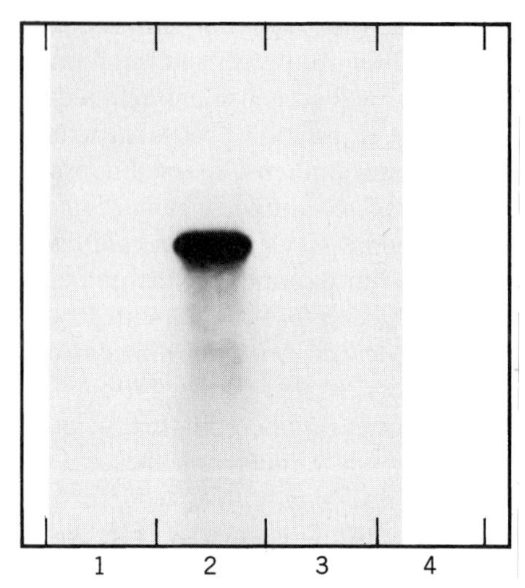

FIGURE 17.25 Influence of gibberellin and abscisic acid on a-amylase mRNA levels. RNA extracted from aleurone layers treated with GA_3 (lane 2), GA_3 + ABA (lane 3), or ABA (lane 4) was extracted from aleurone layers, separated by electrophoresis, transferred to blotting paper, and hybridized with a cDNA clone complementary to α-amylase mRNA. This technique identifies specifically the presence of the mRNA for α-amylase. Note that GA_3 stimulates production of α-amylase mRNA and that this stimulation is largely abolished in the presence of ABA. Lane 1 is control RNA from untreated aleurone. (From P. M. Chandler et al., *Plant Molecular Biology* 3:407, 1984. Reprinted by permission of Kluwer Academic Publishers. Original photograph kindly provided by Dr. J. V. Jacobsen.)

gibberellin treatment closely correlates with the rate of mRNA accumulation. These studies establish with little doubt that a primary action of gibberellin, at least with respect to seed germination, is to regulate gene transcription.

MECHANISM OF CYTOKININ ACTION

The action of cytokinins is poorly understood. It is not yet possible to identify any particular point of biochemical action. There is some evidence that cytokinins have a role in regulating protein synthesis. In cultured soybean cells, for example, cytokinins cause an increase in the overall rate of protein synthesis and change the pattern of proteins that incorporate ^{35}S-methionine. This activity is reflected in an increase in the polyribosome content of cultured cells following cytokinin treatment (Tepfer and Fosket, 1978). An increase in polyribosomes might result from either an increase in the rate of transcription of mRNA or an increase in the stability of the mRNA. There are a few reports of cytokinin-induced mRNAs, but no evidence of a dramatic increase in the overall level of transcription. E. Tobin and her colleagues, on the other hand, have reported evidence for post-transcriptional stabilization of mRNA in duckweed (*Lemna gibba*), an aquatic photoautotroph (Flores and Tobin, 1987). Duckweed can be grown heterotrophically in the dark if supplied with sucrose as a carbon source. In darkness, however, the abundance of mRNA for two nuclear-encoded chloroplast proteins gradually decreases to a low-level. Either brief irradiation with low level red light or the addition of cytokinin to the medium will stimulate an increase in the abundance of mRNA for both the small subunit of ribulose-1,5-bisphosphate carboxylase-oxygenase and the principal chlorophyll a/b-binding polypeptide of the light-harvesting complex. Tobin and her colleagues were able to demonstrate that transcription of the mRNA is largely under control of red light while cytokinin seems to stimulate an increase in the abundance of mRNA. A reasonable interpretation of these results is that cytokinin acts post-transcriptionally to stabilize the mRNA.

MECHANISM OF ABSCISIC ACID ACTION

There is little doubt that ABA exerts its effects on seed maturation and germination largely through regulating gene expression and it does so in direct competition with gibberellins, as described earlier. On the basis of *in vivo* labeling experiments and cell-free translation of barley aleurone mRNA it is clear that ABA suppresses GA-induced synthesis of α-amylase and other hydrolases. It also promotes the synthesis of several ABA-specific polypeptides (Higgins et al., 1982). The effect of ABA can be overcome by providing an excess of gibberellin. ABA apparently controls α-amylase at more than one level. Several studies have shown that ABA operates at the transcriptional level to suppress accumulation of GA-induced α-amylase mRNA (Chandler et al., 1984), but an ABA-induced inhibitor of α-amylase activity has also been identified in mature starchy endosperm (Mundy, 1984). Thus it appears that ABA can prevent germination not only by suppressing transcription of α-amylase but also by inhibiting the activity of any enzyme that might be present in the endosperm.

SUMMARY

The sequence of events initiated by hormones can be resolved into three sequential stages: (1) the initial signal perception, (2) a signal transduction pathway, and (3) the final response. The signal transduction pathway usually involves one or more second messengers that serve to amplify the original signal. Plant hormones appear to qualify on all three counts. It has been known for the past 100 years that plant hormones have significant effect on development. More recently, binding proteins that could serve in the perception stage have been identified for auxin and cytokinin, and possibly ethylene. Lipid-based molecules and calcium appear to be involved in the signal transduction pathway in plants.

In most plants, auxin is synthesized from the amino acid tryptophan or released from storage as conjugates. In some, such as *Arabidopsis*, IAA is probably synthesized from glucosinolates via acetonitrile. Although action at a distance is not an essential property of auxins, IAA is transported in predominantly basipetal direction due to the location of specific auxin efflux carriers near the base of each cell. Once IAA has accomplished its purpose, it can be removed primarily by oxidation to inactive products.

Gibberellins are diterpenoids, related biosynthetically to carotenoids and other isoprene derivatives. Principal intermediates are geranyl-geranyl pyrophosphate and GA_{12}-7-aldehyde. Gibberellins appear to be transported primarily in the phloem in response to source-sink relationships. Gibberellins can be inactivated by hydroxylation at the C2 position or by conversion to inactive conjugates.

Cytokinins are synthesized by a condensation of an isopentenyl group with the amino group of adenosine monophosphate (AMP). Cytokinins will also form conjugates with sugars and are metabolized by oxidation. Abscisic acid (ABA) is a 15-carbon isoprene derivative that appears to be synthesized by cleavage from a 40-carbon xanthophyll violaxanthin. ABA is inactivated by oxidation to phaseic acid and subsequent reduction to dihydrophaseic acid. Ethylene is synthesized from the amino acid methionine. The principal intermediate is 1-aminocyclopropane-1-carboxylic acid (ACC). Because there are limited amounts of methionine in most plants, the sulphur is salvaged and recycled during ethylene biosynthesis. Ethylene biosynthesis is controlled by transcriptional regulation of the rate-limiting enzyme, ACC synthase.

The mechanism of hormone action in plants has been extensively studied in only two cases: auxin control of cell elongation and gibberellin control of cereal grain seed germination. Auxin control of cell elongation is a two-stage process. The first is an acid-stimulated loosening of cell wall polymers followed by turgor-induced expansion. Acid-induced expansion is complete within 30 to 60 minutes. Sustained cell expansion requires continued RNA and protein synthesis, which in turn requires gene activation. Two categories of genes are known to be activated by auxin: short-term responses that are activated within minutes of auxin application and long-term responses that appear after an hour or more.

Auxin is perceived on the outer surface of the plasma membrane by a complex of auxin-binding protein (ABP) and a transmembrane "docking protein." The signal transduction chain for initial acid growth involves G-protein, fatty acids and lysophospholipids, and a protein kinase that activates a membrane-bound proton ATPase. The ATPase pumps protons out of the cell to acidify the cell wall space. The signal transduction chain for gene activation has yet to be determined, although the rapidly induced SAUR genes may be an early step in the pathway toward gene-regulated auxin responses.

Gibberellin stimulates seed germination by regulating gene transcription. Gene products include amylases and other enzymes responsible for degrading storage carbohydrate and mobilizing the sugars for use by the developing embryo. The mechanisms for action of cytokinins and abscisic acid are poorly understood.

CHAPTER REVIEW

1. Describe the two basic models for hormone action. How can low concentrations of hormone molecules have such a profound effect on cellular metabolism?
2. Describe the four requirements for hormone receptors. What is the evidence for hormone receptors in plants?
3. What are second messengers? What are the principal second messengers in plants? Summarize, in general terms, how they operate.
4. Why is it necessary for a hormone to be rapidly turned over? Describe how the size of the active hormone pool is regulated for auxins; for gibberellins.
5. Some seeds appear to accumulate auxin conjugates. Can you suggest a physiological advantage for this?
6. Auxin transport is said to be polar. What does this mean and how is it achieved?
7. In what way are the biosynthesis of gibberellins, cytokinins, and abscisic acid related?
8. How do auxins regulate cell enlargement? What is the evidence that auxin regulates gene action? How can gene regulation by auxin be reconciled with the acid growth theory?
9. Some experiments have shown that gibberellin levels decline as seeds mature. At the same time, ABA levels increase, only to fall just before the seed is shed. What physiological or survival advantage would these changes in hormone levels provide seeds?
10. Describe the process of carbon mobilization in germinating cereal grains.

FURTHER READING

Boss, W. F., D. J. Morré (eds). 1989. *Second Messengers in Plant Growth and Development.* New York: A. R. Liss.

Davies, P. J. (ed). 1995. *Plant Hormones.* Dordrecht: Kluwer Academic Publishers.

Jones, A. M., P. Prasad. 1992. Auxin-binding proteins and their possible roles in auxin-mediated plant cell growth. *Bioessays* 14:43–48.

Parry, A. D., R. Horgan. 1991. Carotenoids and abscisic acid (ABA) synthesis in higher plants. *Physiologia Plantarum* 82:320–326.

Schroeder, J. I., R. Hedrich. 1989. Involvement of ion channels and active transport in osmoregulation and signalling of higher plant cells. *Trends in Biochemical Sciences* 14:187–192.

Theologis, A. 1986. Rapid gene regulation by auxin. *Annual Review of Plant Physiology* 37:407–438.

Venis, M. A., R. M. Napier. 1995. Auxin receptors and auxin-binding proteins. *Critical Reviews in Plant Science* 14:27–47.

Zeevaart, J. A. D., R. A. Creelman. 1988. Metabolism and physiology of abscisic acid. *Annual Review of Plant Physiology and Plant Molecular Biology* 39:439–433.

REFERENCES

Abel, S., A. Theologis. 1996. Early genes and auxin action. *Plant Physiology* 1119–17.

Adams, D. O., S. F. Yang. 1977. Methionine metabolism in apple tissue. Implication of S-adenosylmethionine as an intermediate in the conversion of methionine to ethylene. *Plant Physiology* 60:892–896.

Adams, D. O., S. F. Yang. 1979. Ethylene biosynthesis: Identification of 1-aminocyclopropane-1-carboxylic acid an intermediate in the conversion of methionine to ethylene. *Proceedings of the National Academy of Science USA* 76:170–174.

Baulcombe, D. C., D. Buffard. 1983. Gibberellic-acid-regulated expression of α-amylase and six other genes in wheat aleurone layers. *Planta* 157:493–501.

Behringer, F. J., D. I. Cosgrove, J. B. Reid, P. J. Davies. 1990. Physical basis for altered stem elongation rates in internode length mutants of *Pisum. Plant Physiology* 94:166–173.

Bennett, M. J., A. Marchant, H. G. Green, S. T. May, S. P. Ward, P. A. Millner, A. R. Walker, B. Schulz, K. A. Feldmann. 1996. *Arabidopsis AUX1* gene: A permease-like regulator of root gravitropism. *Science* 273:948–950.

Boss, W. F. 1989. Phosphoinositide metabolism: Its relation to signal transduction in plants. In: Boss, W. F., D. J. Morré (eds.), *Second Messengers in Plant Growth and Development.* New York: A. R. Liss, pp. 29–56.

Chandler, P. M., J. A. Zwar, J. V. Jacobsen, T. J.V. Higgins, A. S.Inglis. 1984. The effects of gibberellic acid and abscisic acid on α-amylase mRNA levels in barley aleurone layers: Studies using an α-amylase cDNA clone. *Plant Molecular Biology* 3:407–418.

Chen, C-M. 1982. Cytokinin biosynthesis in cell-free systems. In: P. F. Waring (ed.), *Plant Growth Substances.* London: Academic Press, pp. 155–164.

Cleland, R. E. 1987. Auxin and cell elongation. In: P. J. Davies (ed.), *Plant Hormones and their Role in Plant Growth and Development.* Boston: Martinus Nijhoff, pp. 132–148.

Cohen, J. D., R.S. Bandurski. 1982. Chemistry and the physiology of the bound auxins. *Annual Review of Plant Physiology* 33:403–454.

Cosgrove, D. J., R. E. Cleland. 1983. Cell wall yield properties of growing tissues: Evaluation by *in vivo* stress relaxation. *Plant Physiology* 78:347–356.

Crozier, A., D. M. Reid. 1971. Do roots synthesise gibberellins? *Canadian Journal of Botany* 49:967–975.

Epstein, E., J. D. Cohen, R. S. Bandurski. 1980. Concentration and metabolic turnover of indoles in germinating kernels of *Zea mays. Plant Physiology* 65:415–421.

Erion, J. L., J. E. Fox. 1981. Purification and properties of a protein which binds cytokinin- active 6-substituted purines. *Plant Physiology* 67:156–162.

Flores, S., E. H. Tobin. 1987. Benzyladenine regulation of the expression of two nuclear genes for chloroplast proteins. In: J. E. Fox, M. Jacobs (eds.), *Molecular Biology of Plant Growth Control.* New York: A. R. Liss, pp. 123–132.

Galston, A. W., J. Bonner, R. S. Baker. 1953. Flavoprotein and peroxidase as components of the indoleacetic acid oxidase system of peas. *Archives of Biochemistry and Biophysics* 49:456

Gil, P., Y. Liu, V. Orbovic, E. Verkamp, K. L. Poff, P. J. Green. 1994. Characterization of the auxin-inducible *SAUR-ACI* gene for use as a molecular genetic tool in *Arabidopsis. Plant Physiology* 104:777–784.

Goldsmith, M. H. M. 1977. The polar transport of auxin. *Annual Review of Plant Physiology* 28:439–478.

Higgins, T. J. V., J. V. Jacobsen, J. A. Zwar. 1982. Gibberellic acid and abscisic acid modulate protein synthesis and mRNA levels in barley aleurone layers. *Plant Molecular Biology* 1:191–215.

Hooley, R. 1982. Protoplasts isolated from aleurone layers of wild oat (*Avena fatua* L.) exhibit the classic response to gibberellic acid. *Planta* 154:29–40.

Jacobs, M., S. F. Gilbert. 1983. Basal localization of the presumptive auxin transport carrier in pea stem cells. *Science* 220:1297–1300.

Jacobs, W. P. 1979. *Plant Hormones and Plant Development.* Cambridge: Cambridge University Press.

Jones, A. M., E. M. Herman. 1993. KDEL-containing auxin-binding protein is secreted to the plasma membrane and cell wall. *Plant Physiology* 101:595–606.

Libbenga, K. R., A. M. Mennes. 1987. Hormone binding and its role in hormone action. In: P. J. Davies (ed.), *Plant Hormones and their Role in Plant Growth and Development.* Boston: Martinus Nijhoff, pp. 194–221.

Lieberman, M., L. W. Mapson. 1964. Genesis and biogenesis of ethylene. *Nature* 204:343–345.

Liu, C., Z. Xu, N-H. Chua. 1993. Auxin polar transport is essential for the establishment of bilateral symmetry during early plant embryogenesis. *The Plant Cell* 5:621–630.

Leyser, O. 1997. Auxin: Lessons from a mutant weed. *Physiologia Plantarum* 100:407–414.

Macdonald, H. 1997. Auxin perception and signal transduction. *Physiologia Plantarum* 100:423–430.

Marmé, D. 1989. The role of calcium and calmodulin in signal transduction. In: W. F. Boss, D. J. Morré (eds.), *Second Messengers in Plant Growth and Development.* New York: A. R. Liss, pp. 57–80.

Milborrow, B. V. 1974. Biosynthesis of abscisic acid by a cell-free system. *Phytochemistry* 13:131–136.

Millner, P. A., B. E. Causier. 1996. G-protein coupled receptors in plant cells. *Journal of Experimental Botany* 47:983–992.

Mundy, J. 1984. Hormonal regulation of α-amylase inhibitor synthesis in germinating barley. *Carlsberg Research Communications* 49:439–444.

Napier, R. M., M. A. Venis. 1995. Auxin and auxin-binding proteins. *New Phytologist* 129:167–201.

Neil, S. J., R. Horgan. 1985. Abscisic acid production and water relations in wilty tomato mutants subjected to water deficiency. *Journal of Experimental Botany* 36:1222–1231.

Nowacki, J., R. S. Bandurski. 1980. *Myo*-inositolesters of indole-3-acetic acid as seed auxin precursors of *Zea mays. Plant Physiology* 65:422–427.

Okada, K., J. Ueda, M. K. Komaki, C. J. Bell, Y. Shimura. 1991. Requirement of the auxin polar transport system in early stages of *Arabidopsis* floral bud formation. *The Plant Cell* 3:677–684.

Polya, G. M., P. J. Davies. 1983. Resolution and properties of a protein kinase catalyzing the phosphorylation of a wheat germ cytokinin-binding protein. *Plant Physiology* 71:482–488.

Rayle, D. L., R. E. Cleland. 1970. Enhancement of wall loosening and elongation by acid solutions. *Plant Physiology* 46:250–253.

Rayle, D. L., R. E. Cleland. 1992. The acid growth theory of auxin-induced cell elongation is alive and well. *Plant Physiology* 99:1271–1274.

Reinecke, D. M., R. S. Bandurski. 1987. Auxin biosynthesis and metabolism. In: P. J. Davies (ed.), *Plant Hormones and their Role in Plant Growth and Development.* Boston: Martinus Nijhoff, pp. 24–42.

Rubery, P. H., P. A. Sheldrake. 1973. Effect of pH and surface charge on cell uptake of auxin. *Nature* (*New Biology*) 244:285–288.

Sato, T., A. Theologis. 1989. Cloning the mRNA encoding 1-aminocyclopropane-1-carboxylate synthase, the key enzyme for ethylene biosynthesis in plants. *Proceedings of the National Academy of Science USA* 86:6621–6625.

Stuart, D. A., R. L. Jones. 1977. The roles of extensibility and turgor in gibberellin- and dark-stimulated growth. *Plant Physiology* 59:61–68.

Tang, Y. W., J. Bonner. 1947. The enzymatic inactivation of indoleacetic acid. *Archives of Biochemistry and Biophysics* 13:11.

Tepfer, D. A., D. E. Fosket. 1978. Hormone mediated translational control of protein synthesis in cultured cells of *Glycine max. Developmental Biology* 62:486–497.

Vanderhoef, L. N., R. R. Dute. 1981. Auxin regulated wall loosening and sustained growth in elongation. *Plant Physiology* 67:146–149.

Vanderhoef, L. N., C. A. Stahl. 1975. Separation of two responses to auxin by means of cytokinin inhibition. *Proceedings of the National Academy of Science USA* 72:1822–1825.

Van der Straeten, D., L. Van Wiemeersch, H. M. Goodman, M. Van Montagu. 1990. Cloning and sequence of two different cDNAs encoding 1-aminocyclopropane-1-carboxylate synthase in tomato. *Proceedings of the National Academy of Science USA* 87:4859–4863.

Venis, M. A. 1985. *Hormone Binding in Plants.* London: Longman.

Venis, M. A. 1995. Auxin binding protein is a red herring? Oh no it isn't! *Journal of Experimental Botany* 46:463–465.

Zeevaart, J. A. D., T. G. Heath, D. A. Gage. 1989. Evidence for a universal pathway of abscisic acid biosynthesis in higher plants from 18O incorporation patterns. *Plant Physiology* 91:1594–1601.

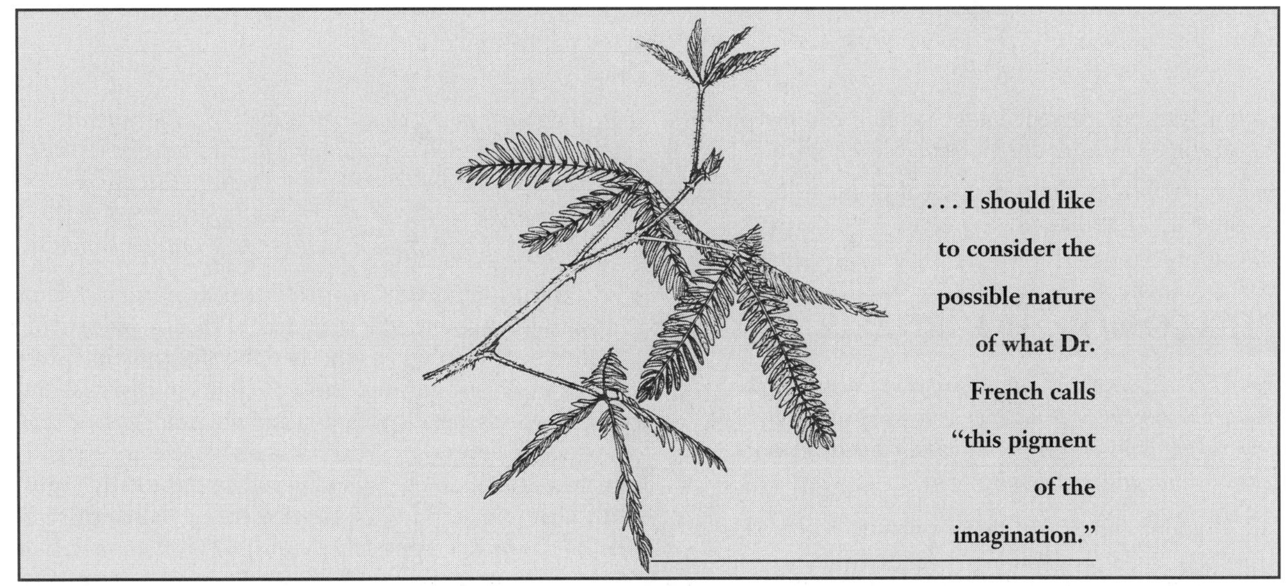

G. Oster (1957)

18

Photomorphogenesis— Responding to Light

The rigid wall that surrounds a plant cell plays an important role in regulating water potential and is a major component in the structural support of plants. Cell walls also limit motility. Plants are literally rooted in place and do not enjoy the luxury of being able to change their environment by changing their location as animals do. Plants cannot move about, seeking shelter from adverse environmental conditions as animals do. Yet plants must avoid adverse conditions if the species is to survive.

The germination of seeds and survival of the seedling that emerges are dictated by conditions in their immediate environment. Many seeds will not germinate if buried too deeply or if they lay in the shade of a forest canopy. Seedlings that emerge beneath a canopy tend to have elongated stems, as if reaching out for the light. Weeds growing in full sun at the edge of a wheat field will be shorter, more compact than plants of the same species competing with the crop in the center of the field. Plants flower at different times, spreading flowering through the season as though each species is awaiting some environmental cue. These and similar patterns of plant behavior clearly have significant survival advantage. They enable plants to make the best use of available resources, compete effectively with other species, or anticipate unfavorable environmental change.

But how can seeds and seedlings know where they are? How do plants measure the passing of the season? As we saw in Chapter 15, the answers to these and many other questions directly related to plant survival may be found in their capacity to detect and interpret a variety of environmental signals. One of those environmental signals is light. We know that photosynthesis will occur only in the light but photosynthesis is only one example of the capacity of plants to discriminate between light and darkness. Plants can also sense light gradients and detect subtle differences in spectral composition. In this way, plants are able to determine whether they are shaded or in full sun, or are able to mark the beginning and end of day. Plants make use of the informational content of light in many different ways to direct their growth, form, and reproduction.

In this chapter we

- introduce the concept of light-regulated plant development or photomorphogenesis;
- describe the discovery of phytochrome and review the basic chemistry of this uniquely photoreversible pigment
- review the physiological effects of phytochrome, showing how it is involved in virtually every aspect of development;
- show how phytochrome can be used to monitor changes in the natural light environment;

- analyze our current understanding of how phytochrome works at the molecular level; and
- review briefly the responses of plants to blue light and UV-B radiation.

PHOTOMORPHOGENESIS

Since light is one of the more significant features of the natural environment, it should not be too surprising that the response of plants to light, called ***photomorphogenesis,*** *is a central theme in plant development.*

While it is true that developmental potential of all organisms is determined by its genes, *realization* of that potential in plants is largely dictated by light. Light-mediated effects on plant development have traditionally been selected for discussion as a topic distinct from normal morphogenesis, but this distinction is more convenient than real. Indeed, the true significance of photomorphogenesis was best summarized by H. Mohr and W. Shropshire when they wrote: "*Normal development in higher plants is photomorphogenesis*" (Mohr and Shropshire, 1983).

How do plant cells acquire and interpret this information so critical to their development? Since light signals, like hormones, originate outside the cell, we can imagine a perception-transduction-response chain somewhat analogous to that found in hormone responses. Perception of light signals requires a pigment that absorbs the light and becomes photochemically active. A pigment acting in this way is commonly referred to as a **photoreceptor**. By selectively absorbing different wavelengths of light, the photoreceptor "reads" the information contained in the light and interprets that information for the cell in the form of a primary action. Primary action may involve simply a conformation change in a protein, a photochemical redox reaction, or some other form of chemical transduction. Whatever the nature of the primary event, absorption of light by the photoreceptor sets into motion a cascade of biochemical events, called the **signal transduction chain,** leading ultimately to a developmental response.

Most photomorphogenic responses in higher plants appear to be under control of one of three signal-transducing photoreceptors: (1) **phytochrome,** which absorbs primarily in the red (R) and far-red (FR) regions of the spectrum; (2) a blue and UV-A-absorbing receptor, **cryptochrome;** and (3) one or more **UV-B receptors.**

PHYTOCHROME

It is now well established that the ubiquitous chromoprotein called phytochrome plays a critical role in almost every stage of plant development. Its existence was predicated on the basis of a simple physiological observation: Seed germination and growth of etiolated seedlings exhibited photoreversible responses to red and far-red light. Because of its uniquely photoreversible character, the proposed pigment system was initially greeted with skepticism in the scientific community.

It has long been recognized that light can influence plant development under conditions that exclude significant levels of photosynthesis. Indeed dramatic differences in the growth and form of plants in darkness and light have fascinated botanists and physiologists for centuries. However, little real progress toward understanding these phenomena was achieved until the early 1950s. At that time, H. A. Borthwick, a botanist, and S. B. Hendricks, a physical chemist, along with several of their colleagues, began a study of **action spectra** (see Chap. 7) for such diverse phenomena as germination of photosensitive lettuce seed, pea stem elongation, and photoperiodic control of flowering. One exciting observation was the similarity of action spectra, with peaks in the red and far-red. Apparently these diverse developmental events all shared a common photoreceptor. More remarkable, however, was the discovery of **photoreversibility**—a response potentiated by red light could be negated if the red light treatment were followed immediately with far-red light. Such clear photoreversibility had never before been described in biology.

The data presented in Table 18.1 illustrate what is meant by photoreversibility. In this study, groups of seeds were allowed to imbibe water in darkness for three hours before being subjected to various brief light treatments. The light treatments were either 1 minute of red light (R; about 660 nm) or 4 minutes of far-red light (FR; $\lambda > 700$ nm) at low fluence rates, or alternating R and FR in rapid succession. Following irradiation, the seeds were returned to darkness for 48 hours, after which the number of germinated seeds in each lot were counted. Note that R promotes germination to about 88 percent but that a R,FR treatment (red followed im-

TABLE 18.1 Photoreversible control of germination. Lettuce seeds were imbibed for 3 hours prior to irradiations. Irradiation times were: red, 1 min; Fr, 3 min. Germination was scored after 48 h in darkness at 20 °C.

Irradiations	Germination (%)
R	88
R, Fr	22
R, Fr, R	84
R, Fr, R, Fr	18
R, Fr, R, Fr, R	72
R, Fr, R, Fr, R, Fr	22

Data from a student experiment.

mediately by far-red) maintains germination at the dark level (22 percent). When the R and FR treatments are alternated, *the percent germination appears to depend solely on whether R or FR was presented last.* It is as though germination were dependent upon a switch that could be thrown one way or the other by red or far-red light.

Borthwick, Hendricks, and their colleagues found similar photoreversibility in the control of flowering behavior and stem elongation by R and FR. These observations led them to propose the existence of a novel pigment system, later called **phytochrome**. This hypothetical pigment would exist in two forms: a red-absorbing form called **Pr** and a far-red-absorbing form called **Pfr**. The pigment would also be *photochromic*, which means that *absorption of light would alter its absorbance properties.* Absorption of red light by Pr would convert the pigment to the far-red-absorbing form while subsequent absorption of far-red light by Pfr would drive the pigment back to the red-absorbing form (Fig. 18.1).

Solely on the basis of simple physiological experiments like that described in Table 18.1, the Beltsville group was able to predict several other features of this hypothetical pigment system. First, because seeds and dark-grown seedling tissues responded *initially* to red light, not far-red, the pigment was probably synthesized as the Pr form, which accumulated in darkness. Moreover, Pr was stable and probably physiologically inactive. Second, because treatment with red light initiated germination and other developmental events, Pfr was probably the active form. On the other hand, Pfr was apparently unstable and was either destroyed or could revert to Pr in darkness by a nonphotochemical, temperature-dependent reaction. Third, because the pigment could not be seen in dark-grown, chlorophyll-free tissue, it was no doubt present at very low concentration. Borthwick and Hendricks further surmised that the pigment must be acting catalytically and was therefore possibly a protein. It is a tribute to the scientific acumen of these investigators and their co-workers that all of these predictions were later proven true.

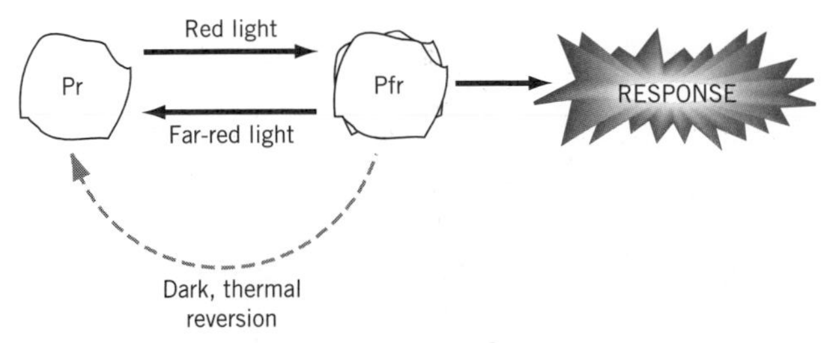

FIGURE 18.1 The photoreversible pigment system as originally postulated on the basis of seed germination and other plant responses to alternating red and far-red light. Absorption of red light by the red-absorbing form of the pigment (Pr) converts it to a form that absorbs far-red light (Pfr). Pfr activates a signal transduction chain that leads to a physiological response.

Based on a peak absorbance in the red region of the spectrum, it was expected that Pr would be a bluish pigment. Yet no blue color was evident in the chlorophyll-free tissues of dark-grown stems and coleoptiles, which nonetheless exhibited photoreversible behavior in their growth. Because of this and its uniquely photoreversible character, the existence of phytochrome was met with some skepticism within the scientific community (Sage, 1992). There was simply no precedent for such a photoreversible pigment in the plant or animal research literature. Most critics expected that two separate pigments were involved.

It was clearly necessary to obtain physical evidence for the existence of phytochrome and, ultimately, to isolate the pigment and characterize it *in vitro*. The strategy that led to a satisfactory resolution of this problem turned on the same unique photoreversible character that generated skepticism in the first place. Since phytochrome was the only known photochromic pigment present in plants, it should be possible to detect, in dark-grown tissue, absorbance changes related to photoconversion of the pigment from one form to the other. Hendricks and his colleagues thus predicted that the conversion of Pr to Pfr in red light would be accompanied by a decrease in absorbance in the red (the maximum absorbance of Pr) and a corresponding increase in absorbance in the far-red. Subsequent irradiation with far-red light should cause an increase in absorbance in the red and a decrease in the far-red. These experiments would require a special kind of spectrophotometer,[1] one capable of measuring very small absorbance changes in dense, light-scattering tissue samples. Fortunately, such an instrument was under development in another laboratory at Beltsville and relatively straightforward modifications were required to adapt its use for phytochrome detection.

The predicted photoreversible absorbance changes were demonstrated in dark-grown maize shoots by Butler et al. in 1959 (Fig. 18.2). This spectral analysis was the first physical evidence that phytochrome actually existed. A short time later, Siegelman and Firer (1964), using protein purification techniques available at the time, successfully isolated and purified the pigment from dark-grown cereal seedlings. They were thus able to demonstrate its photoreversible character *in vitro* as well as obtain absorption spectra for purified Pr and Pfr (Fig. 18.3). In the years that followed, phytochrome was found to be ubiquitous. It is found in algae, bryophytes (mosses and liverworts), and probably all higher plants where it plays a significant role in biochemistry, growth, and development. It is also known now that there are

[1]A spectrophotometer is an instrument for measuring absorption of light by pigments in a solution. Conventional spectrophotometers are by design limited to optically clear solutions, free of light-scattering particles.

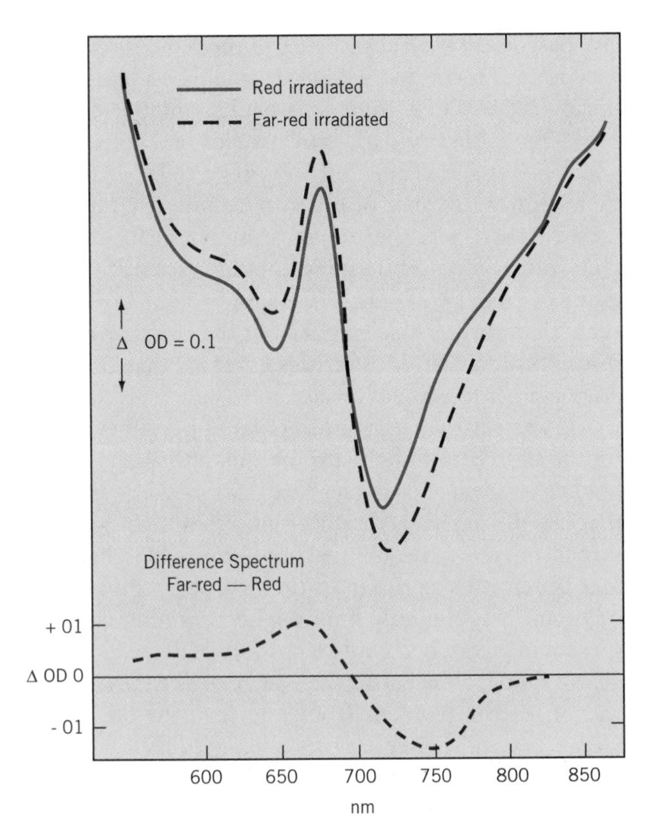

FIGURE 18.2 Absorbance curves for maize shoots following red or far-red irradiations. Note that these curves represent the absorbance of whole tissue, not just the pigment. Note also that conversion of the pigment from Pfr (solid curve) to Pr (dashed curve) causes an increase in absorbance in the red and a decrease in the far-red regions of the spectrum. The difference spectrum effectively represents the absorption spectrum of the Pr form. (From W. Butler et al., *Proceedings of the National Academy of Sciences USA* 45:1703–1708, 1959. Reprinted by permission.)

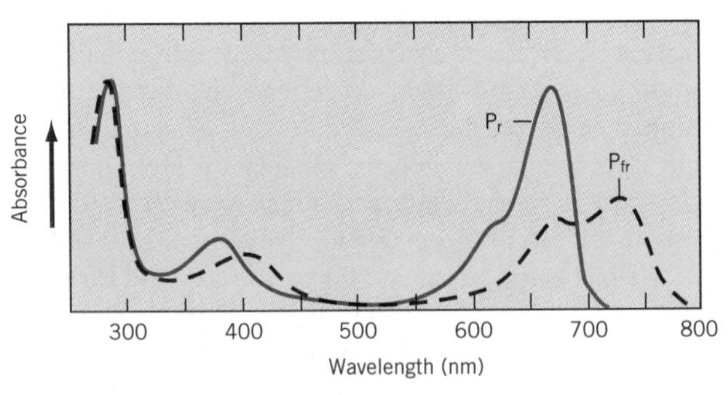

FIGURE 18.3 Absorption spectra of purified phytochrome. Note differential absorption in the blue region of the spectrum as well as the red/far-red region. Some blue light effects are mediated by phytochrome, but photoconversion by red light is 50 to 100 times more effective than in the blue. Because both forms absorb equally in the green region (500 to 550 nm) green light does not change the state of the pigment and can be used as a safe light.

multiple phytochromes, encoded by a family of genes. The phytochrome demonstrated by Butler and his colleagues and first isolated by Siegleman and Firer in 1964 is now known as phytochrome A (PHYA). Because it is also the form that accumulates in etiolated seedlings, the physiology and biochemistry of PHYA has been studied most extensively over the past four decades. The significance of multiple phytochromes will be discussed later. First we will review the physiological and biochemical properties of the phytochrome system, based on studies of phytochrome in dark-grown seedlings.

PHYTOCHROME IN DARK-GROWN SEEDLINGS

Phytochrome accumulates in dark-grown seedlings as Pr, which is stable. Once exposed to light and converted to Pfr, phytochrome undergoes transformations that lead to a loss of both Pfr and total phytochrome.

As predicted, phytochrome is synthesized as the red-absorbing Pr form, which is stable and accumulates in seeds and dark-grown seedlings. Dark-grown (etiolated) seedlings have been a favored source for the study of phytochrome since the beginning for two reasons. First, dark-grown seedlings accumulate relatively large amounts of pigment. Second, the absence of chlorophyll makes it possible to measure phytochrome directly in tissue with spectrophotometers adapted for use with optically dense, light-scattering materials. In this way, changes in the total amount of phytochrome and the relative proportions of Pr and Pfr can be monitored fol-

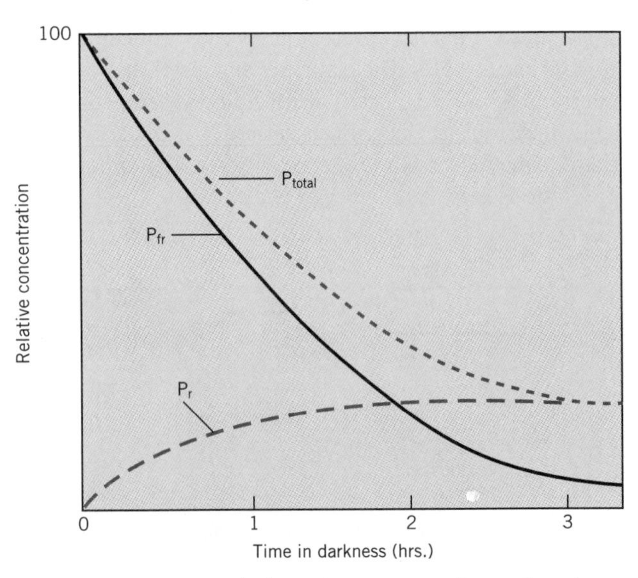

FIGURE 18.4 Typical phytochrome transformations in etiolated seedling tissue. Dark-grown tissue is given a short exposure to low fluence red light at time 0, then monitored spectrophotometrically for total pigment and Pfr in the ensuring dark period. Pr is calculated as the difference.

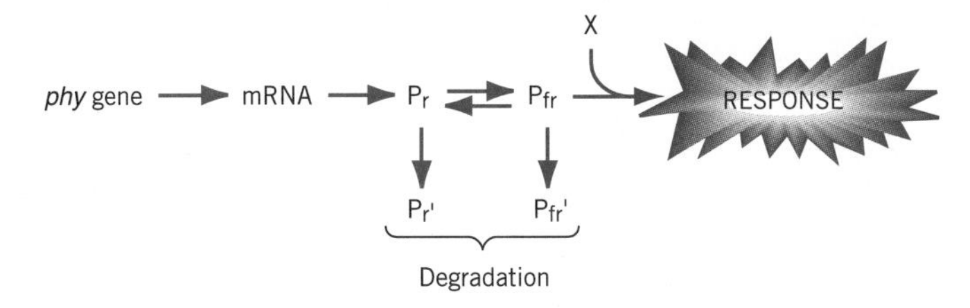

FIGURE 18.5 The phytochrome system. The pigment is synthesized as the physiologically inactive red-absorbing form (Pr), which accumulates in dark-grown seedlings. Red light (660 nm) drives a phototransformation to the far red-absorbing form (Pfr). Absorption of far red light (735nm) returns the pigment to the Pr form. Pfr, the active form, enters into some unknown reaction (X) to give a response. Pr′ and Pfr′ represent inactive degradation products of Pr and Pfr, respectively.

lowing controlled irradiations. These *in vivo* studies have confirmed many of the original predictions.

In dark-grown seedlings, a brief pulse of red light converts Pr to Pfr. A pulse of far-red light immediately following the red treatment converts the pigment back to the Pr form. Far-red light thus negates the effect of red light and cancels any physiological response. It turns out that Pfr, at least in dark-grown seedlings, is relatively unstable, and if not removed photochemically its concentration will decline with time in darkness with a half-life of 1 to 1.5 hours (Fig. 18.4). The loss of Pfr is accompanied by a corresponding decline in the total amount of phytochrome. These kinetics can be explained by the fact that both forms of the pigment are subject to irreversible chemical degradation (Fig. 18.5). Pr accumulates in darkness until its rate of synthesis is matched by the rate of Pr degradation. Following a red light treatment that converts Pr to Pfr, total phytochrome content declines because the rate of Pfr degradation is approximately 100 times greater than the rate of Pr degradation.

Another interesting feature of Figure 18.4 is the apparent increase in Pr. Originally believed to represent dark reversion of the pigment from Pfr to Pr, this feature is now known to be largely an artifact of the spectrophotometric method for measuring phytochrome. Both forms of the pigment have broad overlapping absorption spectra (Fig. 18.3). Because the two spectra overlap, it is not possible to convert 100 percent of the pigment to either form, even when using "pure" red or far-red light. Instead, irradiation with visible light at a sufficiently high fluence rate establishes a **photoequilibrium,** a state in which both Pr and Pfr are present. Photoequilibrium (Φ) is defined as

$$\Phi = \frac{[P_{fr}]}{[P_{TOT}]}$$

in which P_{TOT} is the total phytochrome or the sum of Pr and Pfr.[2] The photoequilibrium established by red light (660 nm) is 0.8 (Table 18.2). That is, even with "pure" red light at 660 nm, it is impossible to convert 100 percent of the pigment to Pfr. This is because Pfr also absorbs, although less efficiently than Pr, light at 660 nm and, thus, about 20 percent of the Pfr is immediately phototransformed back to Pr. In a similar manner, Pr also absorbs a small amount of far-red light (735 nm), so that far-red light phototransforms some Pr to Pfr. The Pr that reappears in Figure 18.4 is simply the gradual uncovering, due to peculiarities of the measuring system, of the 20 percent that was not converted to Pfr by the pulse of red light at time zero. It is important to remember that photoequilibrium, whether under red, far-red, or mixed wavelengths (e.g., white light), is a dynamic process involving a constant cycling between Pr and Pfr.

Note that in the green region of the spectrum (540 nm) there is little or no absorption by either form of phytochrome (Fig. 18.3). Consequently, green light is commonly used as a "safe" light for phytochrome studies on the basis that it causes no *measurable* conversion of phytochrome. Later in this chapter, however, we will see that some phytochrome responses may actually be caused by this traditional "safe light."

Another interesting complication is the presence of several shortlived intermediates in the phototransformations between Pr and Pfr (Fig. 18.6). At ambient temperature, these intermediates have lifetimes measured in the nanosecond-to-millisecond range. At low temperature, however, they can be stabilized and identified on the basis of absorption spectra. Photoexcitation of Pr (to Pr*) is followed by dark relaxation through three intermediates to Pfr. The role of these intermediates, beyond

TABLE 18.2 Typical phytochrome photoequilibria (Φ) established by blue, red, and far-red light.

Light	λ (nm)	Φ
Blue	450	0.4
Red	660	0.8
Far-red	720	0.03
Far-red	756	0.01

[2]The reader should be careful not to confuse the use of the symbol Φ to designate phytochrome photoequilibrium with its use introduced earlier to designate quantum yield (Chap. 9). The same symbols are frequently used by workers in different fields to designate different concepts. The context of use is your guide to its meaning.

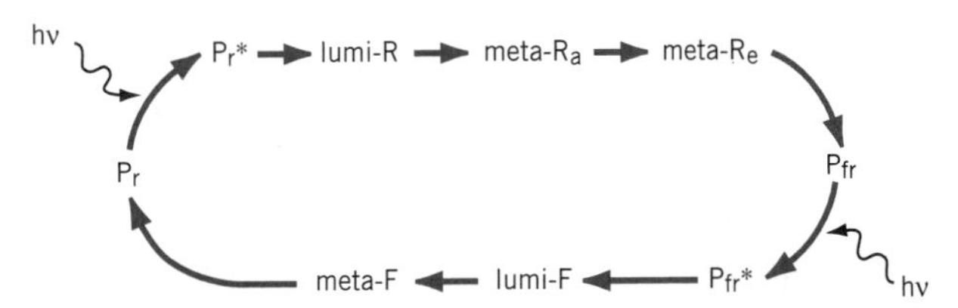

FIGURE 18.6 Photochemical intermediates in the photoconversion of phytochrome. The lifetimes of photochemical intermediates are measured in milliseconds, but in a steady state that there will be a finite concentration of each intermediate over time.

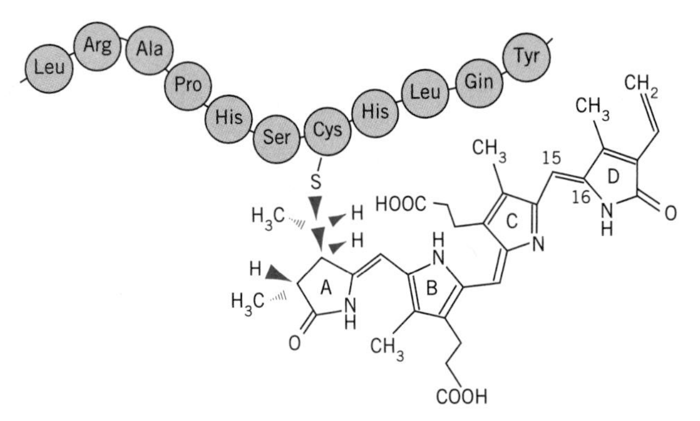

FIGURE 18.7 Probable structure of the phytochrome chromophore and its binding to the apoprotein. The chromophore is covalently linked to the protein at cysteine-321 via a thioether bond. Pr is shown. Photoconversion between Pr and Pfr involves a *cis-trans* isomerization in the C-15, C-16 methine bridge. (After W. Rüdiger, The chromophore, in R. E. Kendrick, G. H. M. Kronenberg, (eds.), *Photomorphogenesis in Plants*, Dordrecht, Nijhoff, 1986, pp. 17–33; R. D. Vierstra, P. H. Quail, The protein, same volume, pp. 35–60. With permission of Kluwer Academic Publishers.)

contributing to understanding the photochemistry, is not clear. However, as the pigment cycles under continuous irradiation there will exist a small but finite steady-state pool of each intermediate. Moreover, relaxation between intermediates is a thermal process. As thermal relaxations are bound to be slower than photochemical excitation of Pr, it is possible that some intermediates may accumulate to physiologically significant levels. Some of these intermediates are also photoreactive and it has been suggested that they may play a significant role in certain phytochrome-mediated phenomena—especially under prolonged irradiation.

THE CHEMISTRY OF PHYTOCHROME

Phytochrome is a chromoprotein, consisting of a chromophore and an apoprotein. Both the chromophore and the protein moiety undergo conformational changes during conversion between Pr and Pfr.

Phytochrome is a bluish **chromoprotein** with absorption maxima of 667 nm for Pr and 730 nm for Pfr. Its unique photochemical properties result from a complex interaction between the chromophore and the apoprotein. The chromophore is an open chain tetrapyrrole similar in structure to phycocyanin (Fig. 18.7). It differs from phycocyanin only in the substitution of a —$CH{=}CH_2$ group for —CH_2—CH_3 on the D ring. The A ring of the chromophore is covalently linked to the apoprotein through a thioether bond to a cysteine residue.

Phytochrome is believed to exist *in vivo* as a dimer with one chromophore per monomer. Estimates of monomer molecular mass have varied considerably, owing largely to the difficulty in isolating the protein in undegraded form. It appears that crude extracts of all plants contain a Pr-specific protease that cleaves the protein into several fragments. Isolation of intact, or native, protein can be optimized by first converting the pigment to the Pfr form, adding protease inhibitors and working rapidly at ice temperature (Vierstra and Quail, 1983). Molecular mass estimates for native monomer range from 120 kDa (zucchini) to 127 kDa (maize).

The molecular mass of oat phytochrome, which has been studied most extensively, is 124 kDa. A polypeptide map containing 1128 amino acids has been deduced from DNA nucleotide sequence analysis. The chromophore is attached at cysteine-321, part of a unique 11-amino acid sequence at the NH_2 terminal end of the protein (Fig. 18.7). Chemical data indicate that the chromophore is housed within a cavity in the folded protein, which shields the chromophore from the external aqueous environment. Of particular interest are the results of a study of the distribution of hydrophobic and hydrophillic amino acid in the protein, called a *hydropathy profile*. Some experimental results have indicated that the pigment is strongly associated with cellular membranes. However, the hydropathy profile for oat phytochrome indicates that it is a relatively hydrophillic protein, which is more consistent with a soluble globular protein rather than an intrinsic membrane protein.

As noted earlier, it is generally agreed that Pr is biologically inactive and that formation of Pfr initiates an active physiological response. The question that naturally arises then concerns the structural differences between the Pr and Pfr forms and whether these differences provide any clues as to the biological activity. Unfortunately, the exact nature of the phototransformation between the two forms is not clear, although both the chromophore and the apoprotein are believed to undergo conformational changes. The principal difference between the Pr chromophore and the Pfr chromophore appears to be a *cis-trans* isomerization of the methine bridge between rings C and D (Rüdiger, 1986). The absorption of red light provides the energy required to overcome a high activation energy for rotation around the double bond, which is not normally achieved at ambient temperature.

Several lines of evidence indicate that the protein also undergoes a conformational change. On the one hand, enzymatic removal of the 6 kDa and 4 kDa fragments from the NH_2-terminal end of the polypeptide (Fig. 18.8) induces several changes in the characteristics of the molecule. Among these changes are a shift in the absorption maximum of Pfr by approximately 8 nm, a tenfold increase in the rate of reversion to Pr, and a decrease in the quantum yield for photoconversion of Pr to Pfr. Apparently the NH_2-terminal fragment is a structurally important domain that influences the association of the Pfr chromophore with the protein. This domain is more susceptible to proteolytic cleavage and more accessible to antibody binding when the pigment is in the Pr form than when it is in the Pfr form. On the other hand, a 55 kDa fragment at the carboxy-terminal end of the protein is more susceptible to attack by proteolytic enzymes when the pigment is in the Pfr form. These observations can best be explained by assuming there are photochemically induced changes in the conformation of the protein. Whatever the differences in conformation between the Pr and Pfr, they cannot be large since there are no detectable differences in the two forms of the pigment by other measures normally used to distinguish between proteins. These measures include the rate at which Pr and Pfr sediment in a centrifugal field (called sedimentation coefficient) or migrate in an electrical field (called electrophoretic mobility). How these changes in the conformation of the protein are brought about or their significance for the biological action of phytochrome remain to be elucidated.

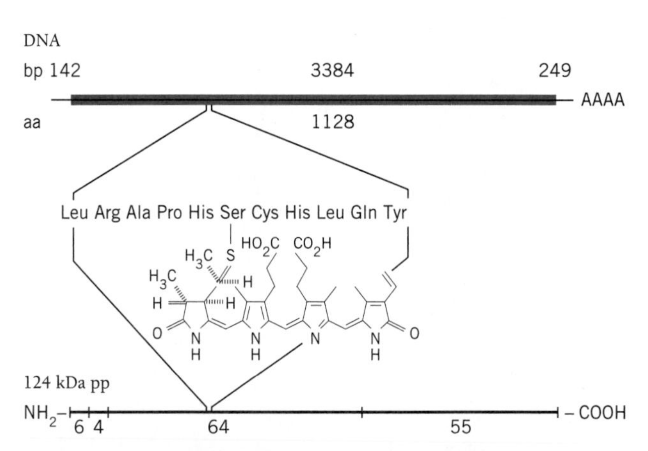

FIGURE 18.8 A schematic showing the alignment of the *Avena* 124-kDa phytochrome polypeptide map with phytochrome DNA sequence. DNA represents the phytochrome gene. The phytochrome apoprotein is shown as the 124 kDa polypeptide (pp). The amino acid sequence that binds the chromophore is located near the middle of a 64 amino acid domain. The short N-terminal domains and the 55 amino acid carboxy-terminal domains are shown. (From R. D. Vierstra, P. H. Quail, in R. E. Kendrick, G. H. M. Kronenberg (eds.), *Photomorphogenesis in Plants*, Kluwer, 1986, pp. 35–60. Reprinted by permission of Kluwer Academic Publishers.)

PHYSIOLOGICAL EFFECTS OF PHYTOCHROME

At least three kinds of phytochrome responses are known, depending on the fluence requirements. Classical low fluence responses include a wide range of physiological and developmental phenomena from membrane potential through seedling development and flowering. Nonphotoreversible high irradiance reactions are also mediated by phytochrome.

Phytochrome mediated effects are conveniently grouped into three categories on the basis of their energy requirements (Fig. 18.9). The classical red, far-red photoreversible responses discovered by Hendricks and Borthwick and their colleagues are known as **low fluence responses (LFRs)**. Photon-fluence requirements for LRFs are in the range of 10^{-1} to 10^{2} μmol m^{-2} of red light. **Very low fluence responses (VLFRs),** as the

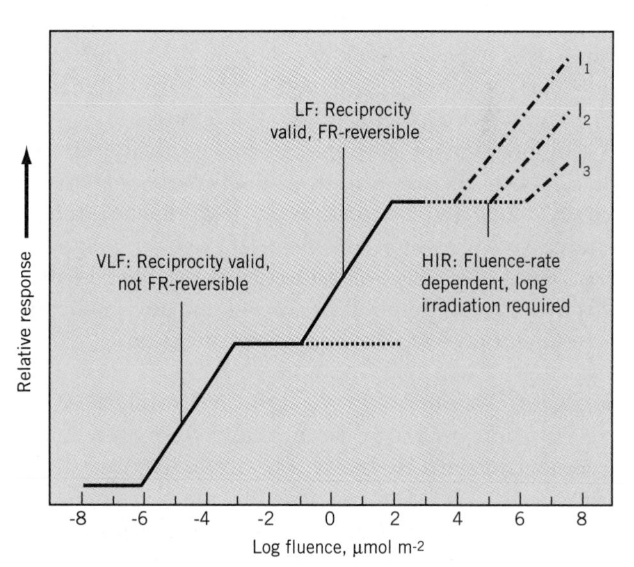

FIGURE 18.9 A schematic illustration of Very Low Fluence (VLF), Low Fluence (LF), and High Irradiance Response (HIR) phytochrome responses as a function of increasing red light fluence. VLF and LF responses may be potentiated by short pulses (1s). The HIR is both fluence and time dependent, as shown. Fluence rates $I_1 < I_2 < I_3$. (From W. R. Briggs et al., in G. Colombetti et al. (eds.), *Sensory Perception and Transduction in Aneural Organisms*, Plenum Publishing, 1985. Reprinted by permission.)

name implies, are induced by much lower light levels, on the order of 10^{-6} to 10^{-3} μmol m^{-2} red light. **High irradiance reactions (HIRs)** require continuous irradiation, usually with FR or blue light, over several hours or longer and are dependent on the actual fluence rate.

LOW FLUENCE PHYTOCHROME RESPONSES

Seed Germination The germination of most seeds is influenced by light as evident in the flush of germination in areas of cultivation or natural disturbance. Seeds that are stimulated to germinate by light, which probably includes the majority of nonagricultural species, are known as **positively photoblastic**. Those in which germination is inhibited by light are **negatively photoblastic**. Some seeds, mostly agriculturally important species that have been selected for high germinability, are not affected by light.

Seeds, such as lettuce (*Lactuca sativa*), may require only brief exposure to light, measured in seconds or minutes, while others may require as much as several hours or even days of constant or intermittent light (*Lythrum salicaria, Epilobium cephalostigma*). In all cases, the responsible pigment appears to be phytochrome. Soil attenuates light very quickly; a 1 mm thickness of fine soil, for example, passes less than 1 percent of the light and then only at wavelengths longer than 700 nm. Thus most light-requiring seeds need not be buried very deeply for germination to be held in check. However some seeds (e.g., *Sinapis arvensis*) require very little Pfr ($\Phi = 0.05$) to stimulate gemination and may exhibit germination when covered with up to 8 mm of soil.

Suppression of germination in negatively photoblastic seeds, such as wild oats (*Avena fatua*), generally requires long-term exposures at high fluence rates. Far-red and blue light are most effective, although in some cases (e.g., *Phacelia tanacetifolia*) red light is also effective. Photoinhibition of seed germination appears to be an example of a high irradiance reaction.

Seedling Development As virtually any school child knows, plants grown in darkness take on a rather unusual appearance (Fig. 18.10). The details may vary from one species to another, but generally the stems of dicot seedlings are very long and spindly, usually with a pronounced recurve just below the leaves. The leaves themselves undergo limited development and remain small and clasping, almost as though they were still in the embryo. Chlorophyll is absent and the seedlings appear white or yellow in color. The leaves of grass seedlings continue to elongate but remain tightly rolled up. The first internode, or mesocotyl, of grass seedlings elongates excessively in the dark and the coleoptile, which is a modified leaf, grows slowly. This general condition exhibited by dark-grown plants is called **etiolation**. Other characteristics of the etiolated condition include arrested chloroplast development and low activities of many enzymes.

In dicotyledonous seedlings, hypocotyl elongation, plumular hook opening, and leaf expansion have received the most attention. Upon irradiation with white light, the growth rate of the hypocotyl slows, the hypocotyl hook gradually straightens, and elongation of the epicotyl accelerates. Light also stimulates the leaves to unfold, enlarge, and complete their development. Chloroplast development proceeds and the leaves green up as chlorophyll accumulates.

The ecological significance of these responses is not difficult to construct. Remember that a plant is fundamentally a photosynthetic organism. A seed carries a limited amount of nutritive tissue that must suffice to support development of the seedling until such time as the seedling is established in the light and photosynthesis can take over the supply of energy and carbon. In the dark, the limited reserves of a seed are committed to extension growth of the axis in order to maximize the possibility that the plumule, composed of the young

FIGURE 18.10 Photomorphogenesis in seedlings of bean (*Phaseolus vulgaris*). The seedling at left was grown under normal light conditions. The seedling on the right was grown in darkness. The center seedling was exposed to 5 minutes of weak red light daily. (See color plate 10.)

TABLE 18.3 Selected examples of phytochrome-mediated responses.

Photoperiodic floral induction
Nyctinastic leaf movements
Phototropic sensitivity
Seed germination
Stem elongation
Plumular hook opening
Leaf and cotyledon expansion
Chloroplast development
Chlorophyll and carotenoid synthesis
Anthocyanin synthesis
Enzyme activation
Protein synthesis
mRNA transcription
Chloroplast phototactic movement
Surface potential (root tips)
Transmembrane potential

leaves, will reach the light and be able to carry out photosynthesis before the reserves are exhausted. Once established in the light, the remaining reserves may be invested in development of photosynthetic tissue, such as leaf expansion, chloroplast development, and so forth. Therefore, the role of phytochrome in seedling development appears to be one of conveying information to the seedling about its position relative to the soil surface.

In addition to the more obvious morphological changes in etiolated seedlings, phytochrome also modulates a variety of other changes at the morphological, biochemical, and biophysical levels (Table 18.3).

It is important to understand that LFRs are studied under carefully controlled laboratory conditions where they are most evident. All manipulations are carried out under a low fluence rate green safelight, which is presumed to be photochemically inactive. Irradiations are restricted to relatively pure filtered red and far-red light, as opposed to white light. Finally, the irradiations themselves are very short (typically one to three minutes) and the plants are returned to darkness for development of the response. LFRs are thus induced by poising the system with a maximum level of Pfr for a very brief period of time. The active Pfr then leads to a physiological response. Red, far-red photoreversibility with short-term irradiations, so characteristic of LFRs, is the key test to be met for phytochrome involvement in any physiological action. Phytochrome is the only photobiological system known that exhibits this unique characteristic. LFRs also exhibit reciprocity between duration of irradiation and fluence rate, but only up to fluence levels that establish a photoequilibrium between Pr and Pfr. LFRs are not dependent on fluence rate once photoequilibrium between Pr and Pfr is established.

In most LFR experiments, the level of the response with FR, either alone or as the last irradiation in sequence with R, is typically higher than dark controls. That is to say, FR never establishes complete photoreversibility. This can be explained by the overlap of the absorption spectra for Pr and Pfr (Fig. 18.3). Because the absorption spectra of Pr and Pfr overlap in the far-red region, Pr will absorb a small amount of FR. Thus even the purest FR source will convert a small amount, typically 1 to 3 percent, of Pr to Pfr. This amount of Pfr is usually sufficient to initiate a low level of response.

Bioelectric Potentials and Ion Distribution The response time for most phytochrome-mediated developmental effects is measured in hours or even days, but there are some responses with response times measured in minutes or seconds. Most, but not all, of these rapid responses appear to related to membrane-based activities such as bioelectric potential or ion flux.

One of the earliest indications that phytochrome influenced the electrical properties of tissues was a curious effect on the surface charge of root tips reported by T. Tanada (1968). He observed that dark-grown barley root tips would float freely in a glass beaker with a specially prepared negatively charged surface. Within 30 seconds following a brief red irradiation the root tips would adhere to the surface. A subsequent far-red treatment would release the root tips from the glass. M. Jaffe demonstrated that adhesion and release was correlated with phytochrome-induced changes in the **surface potential** of the root tips (Jaffe, 1968). A brief red treatment generated a positive surface potential, attracting the tips to the negatively charged surface. A far-red treatment generated a negative surface potential, thereby causing the tips to detach. Similar effects of red and far-red light on surface potential of *Avena* coleoptiles have been demonstrated by I. Newman (1981).

Phytochrome-modulated **transmembrane potentials** have been reported for a variety of tissues from several laboratories. The results are not completely consistent, but in most cases red light induces a depolarization of the membrane within 5 to 10 second following a red light treatment (Fig. 18.11). A subsequent far-red treatment causes a slow return to normal polarity or small hyperpolarization. At this point, it is not known whether such effects are due to a direct action of phytochrome on the membrane or whether a second messenger system is involved.

One of the oldest and most detailed studies of membrane-based phytochrome effects is chloroplast rotation in *Mougeotia*. Each cell of this filamentous green alga contains a single flat chloroplast that is capable of ro-

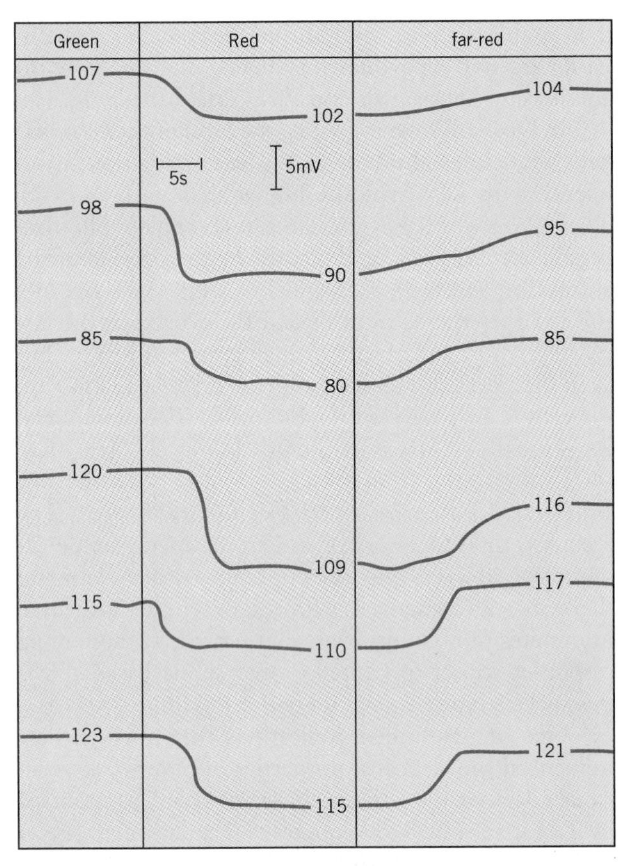

FIGURE 18.11 Chart recordings of membrane potentials for six oat (*Avena*) coleoptile parenchyma cells during successive exposures to green, red, and far-red light. Numbers indicate the mean potential value for each treatment. (From R. H. Racusen, Phytochrome control of electrical potentials and intercellular coupling in oat-coleoptile tissue, *Planta* 132:25–29, Fig. 2, 1976. Copyright Springer-Verlag.)

tating around its long axis so that either its face or its edge is oriented toward incident light (Fig. 18.12A, B). Reorientation of the chloroplast is mediated by phytochrome: Red light is most effective and it is far-red reversible. In a series of elegant experiments, W. Haupt employed plane-polarized light and microbeams of red and far-red light to irradiate specific locations in the cell. He found that the phytochrome responsible for chloroplast rotation was located not in the chloroplast itself but in the cortical cytoplasm; that is, the region of cytoplasm lying just inside the plasma membrane. Haupt also found that reorientation of the chloroplast in plane-polarized light was dependent on the direction of polarization relative to the long axis of the cell. Polarized red light, for example, was most effective when its electrical vector was parallel to the long axis of the cell (Fig. 18.12C). Far-red was most effective at reversing the red effect when polarized in a plane at right angles to the plane of red light (Fig. 18.12D). These results suggest that the phytochrome molecule assumes a particular orientation in the cytoplasm and that the orientation of Pfr is normal to the orientation of Pr.

Unless phytochrome itself migrates from the plasma membrane to the chloroplast, which is unlikely, there must be a signal chain that links the two. One possibility is that calcium may function as a second messenger for phytochrome, perhaps interacting with the cytoskeleton to control chloroplast orientation. The uptake of Ca^{2+} into *Mougeotia* cells is stimulated by red light. In addition, application of the calcium ionophore A23187 to specific sites on the cell will stimulate the chloroplast to reorient, but only if calcium is available in the suspension medium (Serlin and Roux, 1984). Similar application of valinomycin, a K^+ ionophore, has no effect. It appears that Pfr stimulates the uptake of calcium by the cell, which in turn causes reorientation of the chloroplast.

Changes in bioelectric potential are electrochemical phenomena, related to ion movements across the plasma membrane. In the case of mung bean root tips, for example, Jaffe found that the red light-induced positive increase in surface potential was accompanied by an efflux of hydrogen ions. A correlation between phytochrome and ion movements has also been demonstrated in **nyctinastic** or sleep movements of leaves. These movements will be discussed further in Chapter 19, but because of their interaction with phytochrome and ion movements a brief discussion here is appropriate. Paired leaves or leaflets of some species are generally horizontal during the day but fold together when darkened. Plants that show this behavior have a bulbous zone, called the **pulvinus,** at the base of the leaf or leaflet. The pulvinus drives leaf movement by altering its shape as a result of differential changes in the volume of cells on the upper and lower side of the organ. Changes in the volume and shape of these motor cells is osmotically driven in response to rapid redistribution of solutes, principally K^+, Cl^+ and malate. Ruth Satter and her colleagues have documented phytochrome control over K^+ flux in these organs (see Chap. 19). It appears, however, that phytochrome does not directly control K^+ flux. K^+ moves through electrically gated K^+ channels which are opened by phytochrome-driven H^+ efflux. The role of phytochrome may be to activate a plasma membrane-bound ATPase proton pump that in turn depolarizes the membrane to open the K^+ channels. There is some indication that Ca^{2+} may also be involved as a second messenger in nyctinastic movements as well. Indeed, Ca^{2+}/calmodulin has been implicated as a mediator of several phytochrome responses. Those phytochrome responses for which Ca^{2+} has been implicated as a mediator are usually accompanied by light-induced uptake of Ca^{2+} and are generally prevented by calmodulin inhibitors. It remains to be demonstrated, however, that phytochrome promotes an increase of free Ca^{2+} in the cytoplasm of plant cells.

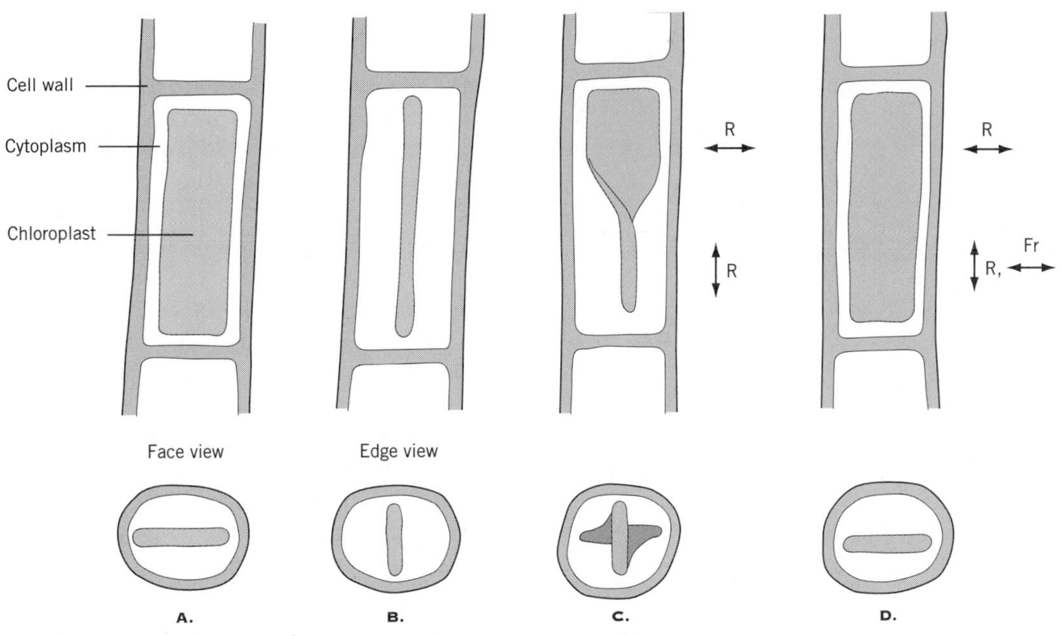

FIGURE 18.12 **Diagram of Haupt's experiments with *Mougeotia* chloroplasts. *Mougeotia* is a filamentous green alga with a single, platelike chloroplast in each cell. (*A*) and (*B*) illustrate the chloroplast in face and edge views. Arrows (*C* and *D*) indicate the plane of polarization of polarized red (R) or far-red (Fr) light.**

VERY LOW FLUENCE RESPONSES

Several studies have indicated that dark-grown seedlings are capable of responding to very low levels of light. Red light, for example, promotes an increase in sensitivity of cereal grain seedlings to a subsequent phototropic stimulus. But the red light fluence required to saturate the response was found to be at least 100 times less than that required to induce a *measurable* conversion of Pr to Pfr! A low far-red fluence also promotes phototropic sensitivity just as red light does, indicating that less than 1 percent of the pigment need be converted to Pfr in order to saturate the response. D. F. Mandoli and W. R. Briggs (1981) found that exposure to even the traditional dim green safelights could elicit or even saturate elongation responses in dark-grown *Avena* seedlings. They have estimated that as little as 0.01 percent Pfr is required to elicit inhibition of mesocotyl elongation. This extreme sensitivity to light makes the study of VLFRs technically difficult. VLFRs, for example, are not photoreversible. The principal evidence that a VLF response is mediated by phytochrome is the similarity of its action spectrum to the absorption spectrum of Pr (Mandoli and Briggs, 1981). The phenomenon, however, raises perplexing yet intriguing questions about experimental photocontrol of plant development.

HIGH IRRADIANCE REACTIONS

In the natural environment, plants are exposed to long periods of sunlight at relatively high fluence rates. Under such conditions, characterized by relatively high energy over long periods of time, the photomorphogenic program achieves maximum expression and responses such as leaf expansion and stem elongation are far more striking. Such light-dependent responses are known as high irradiance responses (HIRs). High irradiance responses generally share the following characteristics (Mancinelli, 1980): (a) full expression of the response requires prolonged exposure to high irradiance; (b) the magnitude of the response is a function of the fluence rate and duration (i.e., the Bunsen-Roscoe reciprocity law does not apply); (c) in contrast with LFRs, HIRs are not fully red, far-red photoreversible.

HIRs were first revealed in studies of anthocyanin synthesis in red cabbage seedlings (Siegelman and Hendricks, 1957). Like other responses of etiolated seedlings, the initiation of anthocyanin accumulation is a classic phytochrome-dependent LFR. The red, far-red photoreversibility, however, is limited to brief irradiations. When longer-term irradiations are applied, the action peak for anthocyanin accumulation is shifted to the far-red, with reduced effectiveness in the red. This effect of prolonged far-red irradiation has been interpreted as maintaining a low level of Pfr over time—long enough to avoid rapid depletion of the Pfr pool by degradation.

The HIR has been implicated in a wide range of responses that also qualify as LFRs, including stem growth, leaf expansion, and seed germination. However, HIRs may exhibit strikingly different action spectra de-

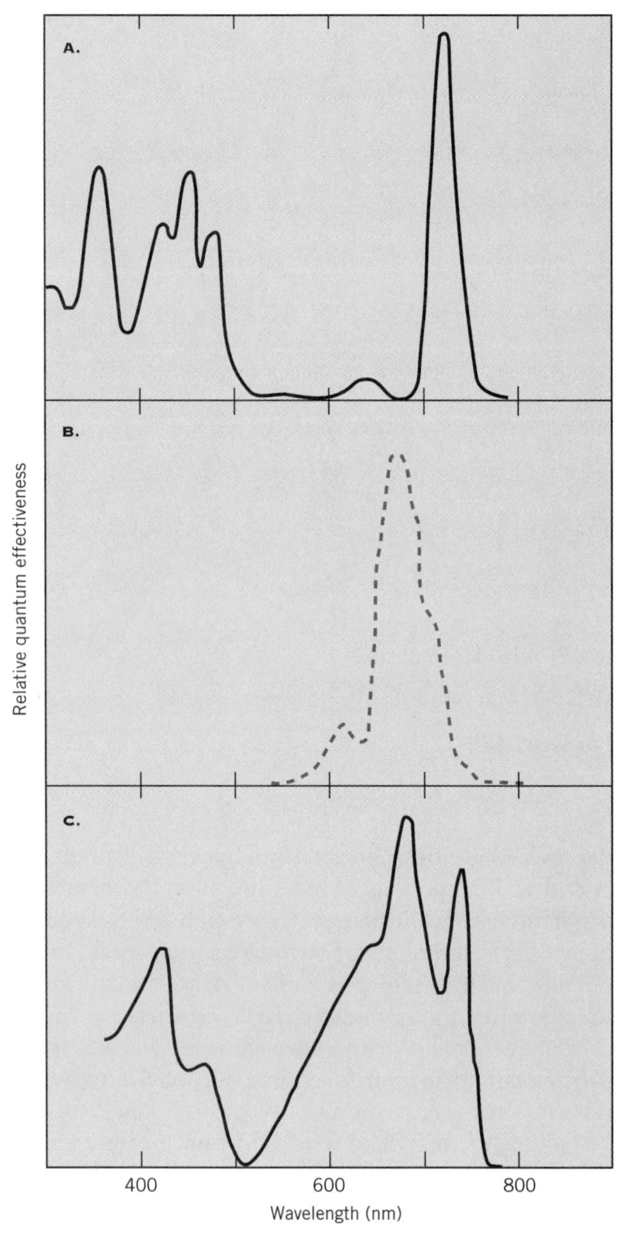

FIGURE 18.13 Action spectra for inhibition of stem elongation by continuous irradiation. (*A*) Etiolated lettuce seedlings (*Lactuca sativa*). (*B*) Light grown and (*C*) etiolated white mustard (*Sinapis Alba*). (*A* From K. M. Hartman, *Z. Naturforsch* 1967, 22b:1172, copyright Verlag der Zeitschrift der Naturforschung; *B* and *C* from C. J. Beggs et al., 1980, *Plant Physiology* 66:615, copyright American Society of Plant Physiologists.)

pending on the species or growth conditions (Fig. 18.13). Etiolated seedlings, for example, respond to blue, red, and far-red light. As de-etiolation progresses, there is a shift from a far-red-sensitive HIR to a red-sensitive HIR. Not surprisingly then, light-grown, green tissues are more responsive to red light rather than far-red. Some systems, such as anthocyanin synthesis in *Sorghum* seedlings, respond only to blue-UV-A light.

Is phytochrome the photoreceptor for HIRs? Although the effectiveness of red and far-red light argues in favor of phytochrome as a photoreceptor, the unique characteristic of phytochrome reactions—photoreversibility—is conspicuously absent from high irradiance reactions. Nevertheless, several theoretical models have been developed in order to explain HIR in terms of phytochrome concentration, the rate of cycling between Pr and Pfr, steady-state levels of photochemical intermediates, and other known features of the phytochrome system. None of the models, however, can satisfactorily explain all of the characteristics of the HIR, especially action in the blue-UV-A region of the spectrum. Based on variations in action spectra, at least three categories of HIRs can be recognized: (a) action in the blue-UV-A, red, and far-red; (b) action in the blue-UV-A and red; (c) action in the blue-UV-A only (Mancinelli, 1980). In order to explain these differences, it is generally conceded that at least two photoreceptors must be involved: phytochrome, and a blue-UV-A receptor.

K. M. Hartmann has presented a very strong argument that the 716 nm peak in the action spectrum for inhibition of hypocotyl growth in lettuce seedlings under continuous light (Fig. 18.13A) is due to phytochrome. In a truly elegant experiment, Hartmann presented seedlings with light of either 658 nm or 766 nm (Hartmann, 1967; see also Cosgrove, 1986). These wavelengths were chosen because they are absorbed by Pr and Pfr, respectively, but avoid as much as possible regions of overlapping absorption between the two forms of the pigment. Presented separately, 658 and 766 nm light were ineffective at inhibiting hypocotyl elongation. Presented simultaneously, however, 658 and 766 nm light could inhibit elongation as effectively as 716 nm light. The trick was to ensure that fluence rates at 658 nm and 766 nm were adjusted to give the same intermediate level of Pfr (Pfr/PTOT) as that established by irradiation at 716. Hartmann's thesis is that 766 nm light is ineffective because it converts phytochrome predominantly to the inactive Pr form. On the other hand, 658 nm light converts the pigment predominantly to the Pfr form that, although initially inhibitory, is rapidly lost by degradation reactions. Light at 716 nm is effective because it establishes an intermediate level of Pfr, balancing the competing reactions of Pfr action and Pfr degradation. Light at 716 nm thus maintains the most effective concentration of Pfr over the duration of the prolonged light treatment.

There is also evidence that supports a separate blue-UV-A receptor that may, in some systems, interact with phytochrome in controlling the high irradiance reaction. A blue light-dependent inhibition of hypocotyl elongation in light-grown cucumber (*Cucurbita*), lettuce (*Lactuca*), and tomato (*Lycopersicum*) seedlings can be demonstrated by simultaneous irradiation with blue and white light. Such conditions do not appreciably alter the ratio of Pfr to total phytochrome, but do cause an inhibition of stem growth compared with controls receiving white light alone. In another system, phytochrome (i.e., red, far-red photoreversible) control of anthocyanin biosynthesis in *Sorghum* seedlings becomes effective only after prolonged exposure to blue or white light.

In still another system, hypocotyl elongation in wild-type *Arabidopsis* is inhibited by continuous Fr, blue, UV-A, and red light. There are two mutants of *Arabidopsis*, known as *hy-1* and *hy-2*, which have reduced amounts of phytochrome. In both mutants there is virtually complete loss of Fr and red inhibition of hypocotyl elongation, with no change in the response to blue and UV-A. Finally, phytochrome does not induce anthocyanin biosynthesis in totally dark-grown seedlings. However, red, far-red photoreversibile control is apparent following a prolonged blue light treatment. These results strongly support the hypothesis that a separate blue-UV-A photoreceptor may be operative in some HIRs and that it may act cooperatively with phytochrome.

PHYTOCHROME IN GREEN PLANTS

Phytochrome in green plants is different from phytochrome in etiolated tissues. The form of phytochrome expressed in etiolated tissues is only one of five gene products. Four other gene products are expressed at low levels in both dark- and light-grown tissues.

The assumption that phytochrome takes an active role in directing morphogenesis under natural radiation conditions raises an interesting question. If Pfr is degraded as rapidly in green plants as it is in etiolated tissues, how do plants under continuous illumination maintain adequate levels of active phytochrome? It has long been argued, on the basis of both physiological and *in vivo* spectrophotometric studies, that some properties of phytochrome in light-grown green plants differed from those of etiolated seedlings. Only recently, though, has direct molecular evidence in support of this argument become available.

It has traditionally been difficult to study phytochrome in green plants owing to its low concentration and intense interference by chlorophyll absorption and fluorescence in the red and far-red regions. Chlorophyll interference can now be partially circumvented by procedures aimed at rapid elimination of chlorophyll from extracts, thereby permitting *in vitro* spectral analysis. Alternatively, light-grown seedlings may be treated with Norflurazon, a herbicide that inhibits carotenoid biosynthesis and, as a result, induces photobleaching of chlorophyll. The resulting chlorophyll-free tissue is then amenable to *in vivo* spectrophotometric analysis.

Several lines of evidence now confirm the earlier suspicions that green-tissue phytochrome is distinct from that in dark-grown tissue. Phytochrome from light-grown *Avena* tissue is smaller (apparent molecular mass = 118 kDa) and has a shorter Pr absorption maximum (652 nm) compared with the pigment from etiolated tissue (124 kDa; 666 nm). In addition, phytochrome from light-grown *Avena* seedlings is neither immunoprecipitated nor recognized on immunoblots by antibodies raised against phytochrome from etiolated seedlings. The kinetics of photoconversion (Pr to Pfr and back to Pr) appear to be similar, but Pfr in light-grown tissue has a considerably longer half-life. For example, M. Jabben has measured phytochrome transfomations in seedlings of maize *Zea mays* treated with the herbicide SAN 9789, which inhibits chlorophyll accumulation in the light. The half-life of Pfr in treated seedlings grown under continuous light is about 8 hours (Fig 18.14) compared with 1.0 to 1.5 hours in etiolated seedlings (Fig. 18.4). Similarly long half-lives for Pfr have been observed in light-grown *Avena* and several dicot seedlings as well.

The labile form of phytochrome that accumulates in dark-grown seedlings is referred to as Type I phytochrome, while the more stable form found in green seedlings is known as Type II phytochrome (Furuya, 1989; Vince-Prue, 1991). Recent studies using recombinant DNA techniques, however, have shown there are multiple forms of Type II phytochrome, encoded by a small family of differentially regulated genes (Vierstra, 1993; Chory, 1997). The best characterized family of genes has been isolated from *Arabidopsis thaliana*, in

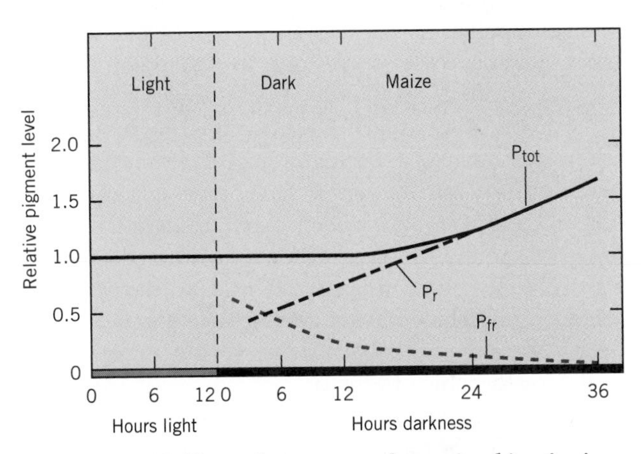

FIGURE 18.14 Phytochrome transformation kinetics in seedlings of maize (*Zea mays*) treated with a herbicide that prevents chlorophyll accumulation. Phytochrome levels were monitored in darkness following 9 days of growth under continuous white light. (From M. Jabben, *Planta* 149:91–96, Fig. 3, 1980. Copyright Springer-Verlag.)

which there are five phytochrome genes (designated *phyA*, *phyB*, *phyC*, *phyD*, *phyE*). The gene *phyA* is expressed in dark-grown tissue and encodes the labile Type I form of phytochrome (PHYA), which accumulates in the dark. PHYA does not accumulate in the light because, as described below, transcription of *phyA* is inhibited by PfrA (the far-red absorbing form of PHYA). In addition, PHYA protein is rapidly degraded. The remaining Type II phytochrome genes (*phyB*, *phyC*, *phyD*, *phyE*) are expressed at low levels in both light and darkness. Their products (PHYB-E) are light stable.

The existence of different forms of phytochrome in green plants with a more stable Pfr would seem to offer at least a partial answer to the question asked earlier. It is highly probable that Pfr degradation in green tissues is slow enough to be balanced by Pr synthesis, thus maintaining a stable pool of phytochrome in continuous light. But why, then, do plants require multiple forms of phytochrome? Put another way, why do etiolated plants accumulate an excess of labile PhytA? There is no answer to this question, but H. Smith has advanced an interesting argument (Vince-Prue, 1991). Smith points out that PhytA accumulates in two situations: (1) seeds that require red light to germinate and consequently do not germinate when buried deep in the soil and (2) germinated seedlings in which phytochrome is used to detect light as the seedling approaches the soil surface. Thus, the large amount of PhytA that accumulates under these conditions appears to function as a sensitive antenna, detecting primarily the *presence* of light rather than light quality. Once the seed or seedling is exposed to adequate light, the excessive quantity of labile phytochrome disappears. This allows the more stable Type II phytochrome to monitor the R-FR ratio and direct development accordingly, as described above. It may be difficult to obtain direct evidence in support of such a scenario, but it is an important first step in taking phytochrome studies out of the laboratory into the real world.

Finally, as noted earlier, the accepted dogma of phytochrome is that Pr is biologically inactive and that formation of Pfr initiates active developmental responses. There is a large body of evidence supporting this dogma. However, some investigators have questioned whether Pr might also play a more active role, at least with respect to stem elongation in light-grown plants (Smith, 1983). Recently, it has been shown that the normal vertical growth habit of *Arabidopsis* seedlings is markedly reduced when grown in red light (Liscum and Hangarter, 1993). Seedlings grown in red light assume a much more random orientation. (The normal, erect growth habit is an example of negative gravitropism—see Chap. 19.) On the other hand, mutants that lack photochemically functional phytochrome and, consequently, are unable to produce Pfr, exhibit a normally erect habit regardless of light treatment. Based on a genetic analysis, Liscum and Hangarter have concluded that the normal erect growth habit occurs when phytochrome B is in the Pr form. It is assumed, of course, that the erect habit and not randomized growth is the "active response." Whether control of other plant responses may be attributed to Pr in the same way remains to be seen.

PHYTOCHROME UNDER NATURAL CONDITIONS

One of the more challenging questions about phytochrome is how it functions in a natural environment, under prolonged exposure to varying fluence rates for varying times. Is there information in the natural light environment that can be interpreted by phytochrome?

The largest body of information about phytochrome is derived from studies of etiolated seedlings subject to brief irradiations, that is, LFRs. But of course, except in the laboratory, plants do not grow in dark boxes with occasional flashes of red and far-red light. Light initially perceived by germinated seedlings as they reach the soil surface no doubt operates as a classical LFR, serving to convert a large proportion of the Pr present to Pfr and initiate the deetiolation process. But it would also serve to stimulate the loss of Pfr and a lowering of the total phytochrome content. This raises the question of how phytochrome functions, if indeed it has any role, in green plants that are exposed to continuous daylight at high fluence rates.

The responses of green plants are at times strikingly different from those of etiolated seedlings. Quantitatively, green plants are more sensitive to red light. Inhibition of hypocotyl elongation in light-grown mustard seedlings, for example, requires lower fluence rates than do dark-grown seedlings. In our earlier discussion of the high irradiance responses it was pointed out that deetiolation was accompanied by a shift from far-red sensitivity to red-sensitivity. Indeed, a high level of far-red light actually stimulates extension growth in most green plants.

Many studies on light-grown plants indicate that extension growth is determined at least in part by the concentration of Pfr or Pfr/P ratio (symbol = Φ). One of the earliest indications was provided by studies manipulating "end of day" treatments (Downs et al., 1957). Several varieties of beans (*Phaseolus vulgaris L*), sunflower (*Helianthus*), and morning glory were subjected to either FR or FR followed by R at the end of 8-hour photoperiods of cool white fluorescent light (Fig 18.15). Red light given at the end of the photoperiod, which established a high proportion of Pfr at the beginning of the dark period, resulted in a strong inhibition of stem growth. Light that reduced the level of Pfr led to a cor-

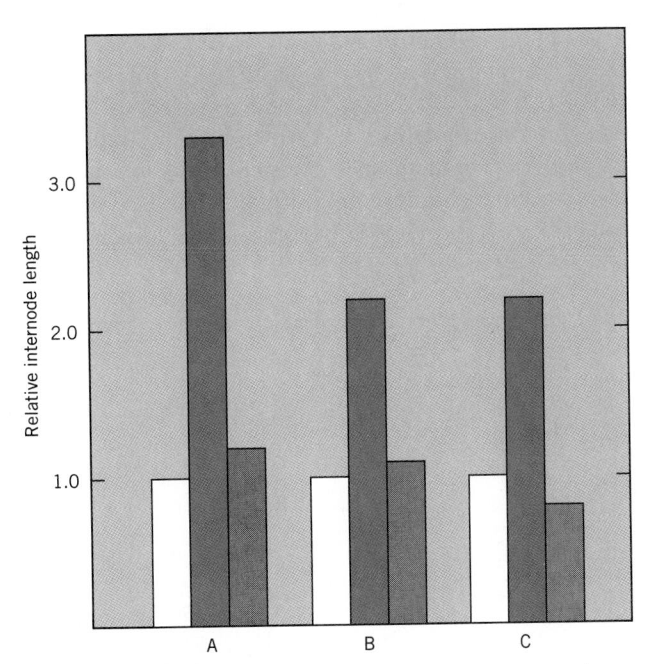

FIGURE 18.15 Effect of an end-of-day treatment with either far-red (green bar) or far-red, red (grey bar) on internode lengths in *Helianthus* (*A*), *Phaseolus* (*B*), or *Pharbitus* (*C*). Treatments were 5 minutes duration at the end of an 8-hour photoperiod of white fluorescent light. Internode lengths are plotted relative to controls (open bars) that received no end-of-day treatment. (Data from Downs et al., *Botanical Gazette* 118:199–208, 1957.)

responding reduction in the inhibition effect. In other words, the extent of second internode extension growth was determined by the proportion of Pfr present at the beginning of the dark period. Control of extension growth was R, FR reversible. Other experiments with *Synningia* and *Fuchsia* have demonstrated a similar dependence of extension growth on conditions that establish a low level of Pfr at the end of a normal photoperiod.

Experiments such as these have helped to focus attention on the behavior of phytochrome in green plants and whether phytochrome has a significant role to play in the survival strategy of plants. Most attention has focused on plant growth under canopies, since it is here that alterations to the spectral distribution of natural light are most dramatic. Radiation within a canopy is markedly deficient of red and blue light, which are largely absorbed by the chlorophyll in the overlying leaves. By contrast, chlorophyll is transparent to far-red light; attenuation of far-red is limited almost solely to reflection. The effect of canopy shading can thus be described by taking the ratio of red to far-red fluence rates (R/FR, or ζ; Gr. zeta). The value of ζ in unfiltered daylight is typically in the range of 1.05 to 1.25. The value in shadelight beneath a canopy will, of course, vary with the nature of the vegetation and the density of the canopy. Some representative values are listed in Table 18.4.

TABLE 18.4 Approximate values of R/FR (ζ) for canopy filtered light.

Canopy	R/FR
wheat	0.5
maize	0.20
oak woodland	0.12–0.17
maple woodland	0.14–0.28
spruce forest	0.15–0.33
tropical rainforest	0.22–0.30

These values fall well within the range where a small change in ζ would cause a relatively large change in the proportion of Pfr (Fig. 18.16). This being the case, phytochrome should make an excellent sensor of shadelight quality.

There is now good evidence that plants *can* indeed detect these characteristic differences between shadelight and unfiltered daylight. Shadelight can be mimicked in the laboratory or growth chamber by supplementing white fluorescent light (ζ = 2.28) with various

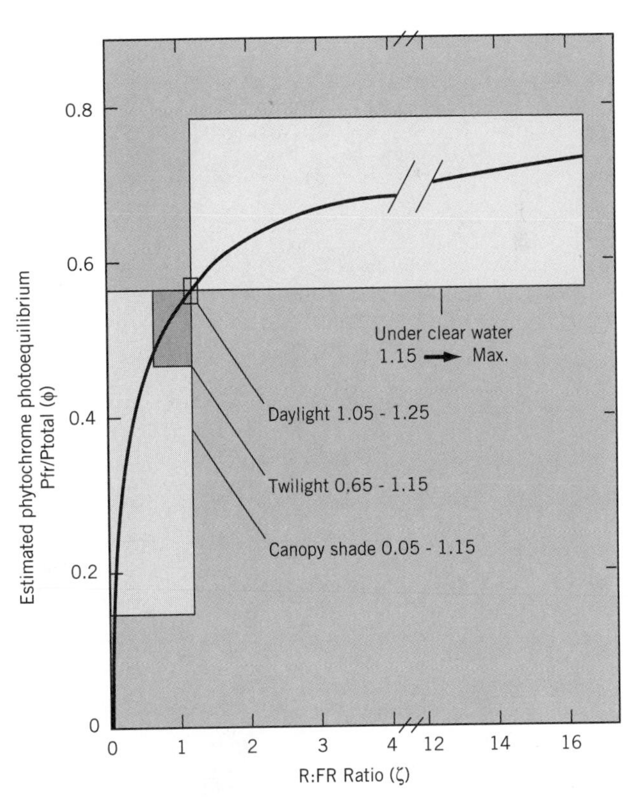

FIGURE 18.16 The relationship between R:Fr ratio (ζ) and phytochrome photoequilibrium. Blocked areas indicate range of values for ζ observed under indicated ecological conditions. (Reproduced with permission from the *Annual Review of Plant Physiology*, Vol. 33. Copyright 1982 by Annual Reviews, Inc.)

amounts of far-red light through the entire photoperiod. This is done in such a way that the fluence rate of photosynthetically active radiation (PAR) remains constant from one treatment regime to the next. Any differences in growth and morphogenesis are thus attributable to the phytochrome photoequilibrium value (Φ), which can then be estimated from the measured R/FR ratio for each regimen. When young plants of *Chenopodium album* were grown this way, the logarithmic stem extension rate was found to be linearly related to Φ (Fig. 18.17). The response to light quality may be quite rapid—in *Chenopodium* an increase in the stem extension rate can be observed within seven minutes of adding FR light to the background fluorescent source.

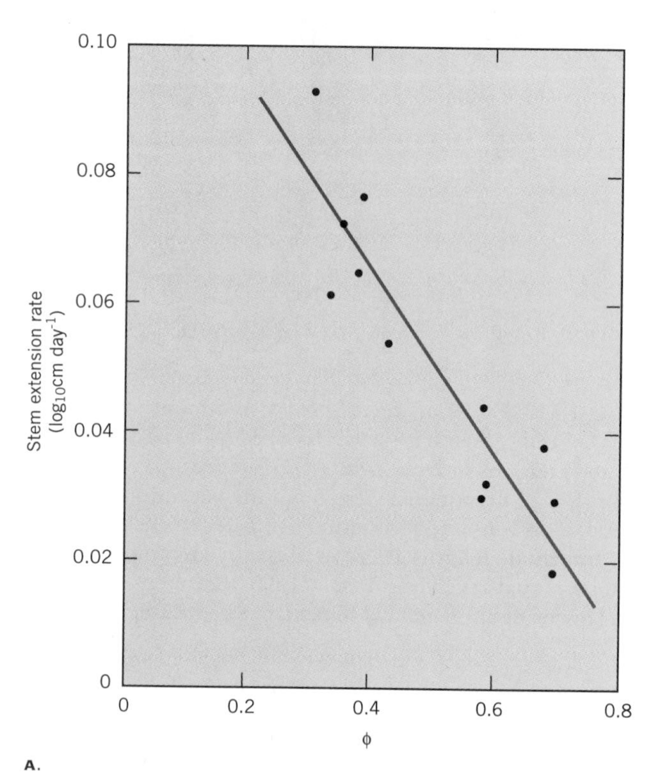

FIGURE 18.17 Photomorphogenesis in light-grown seedlings. (*A*) The relationship between the proportion of Pfr (Φ) and stem extension rate of *Chenopodium album* grown for 9 days in simulated shadelight. (*B*) Seedlings of *Chenopodium album* grown for 14 days in simulated shadelight. Shadelight was simulated by providing supplementary far-red light sufficient to provide the indicated R:FR ratios (ζ). (*A* from D. C. Morgan and H. Smith, The relationship between phytochrome photoequilibrium and development in light-grown *Chenopodium album* L., *Planta* 142:187–193, Fig. 4, 1978, copyright Springer-Verlag; *B* reproduced by permission of Dr. David Morgan and Professor Harry Smith, University of Leicester, UK. Photograph courtesy of Prof. Smith.)

There are also substantial changes in the spectral energy distribution of natural daylight on a daily basis (Holmes and Smith, 1977; Hughes et al., 1984). At both dawn and dusk, when the sun sits low on the horizon, there are significant relative decreases in the value of ζ compared with the main part of the day. In one study, for example, ζ values at dusk were reduced by 14 to 44 percent of those at midday (Holmes and Smith, 1977). A detailed examination of the response of *Cucurbita pepo* L. to end-of-day red or far-red light reveals that a reduction in the proportion of phytochrome maintained as Pfr at the end of the photoperiod is associated with drastic changes in the developmental pattern. Lowering of Φ from high (ca. 0.65 to 0.75) to very low (ca. 0.03) values accentuates stem and petiole extension, reduces leaf expansion and branching, and lowers the chlorophyll content. These experiments do not, of course, prove a causal link between phytochrome photoequilibria and morphogenic responses to changes in the radiation environment. They do, however, demonstrate that plants have the capacity to respond to changes in spectral energy distribution similar to those that occur naturally in the environment. It seems highly likely at this point that phytochrome is the photoreceptor that detects both end-of-day signals and shading.

MECHANISM OF PHYTOCHROME ACTION

Phytochrome appears to operate at more than one level. Those best documented are related to membrane properties or regulation of mRNA transcription.

Deciphering causal relationships at the cellular or subcellular level is a difficult task at best. In a situation where the stimulus induces such a complex array of effects, as is the case with phytochrome, the task of identifying the primary action becomes even more daunting. Although Pfr is a common initiator for so many developmental events, the subsequent chain of metabolic and biochemical events could vary markedly between responses or from plant to plant or even with developmental age. Is, for example, the primary action of Pfr the same for LFRs as it is for HIRs or can phytochrome induce similar responses in different ways?

Additional complications are introduced by variations in response times. Some effects can be observed within seconds while others develop over hours or days. In spite of these complexities, however, two consistent themes have emerged. One is that many of the effects of Pfr can be explained as changes in the properties of cellular membranes. The early studies on phytochrome-induced leaflet movement in *Mimosa* and *Albizzia* and Tanada's clever experiment demonstrating modulation of surface charge on root tips both demonstrate rapid changes in ion distribution across cellular membranes. Given the established role of membranes in mediating the regulatory action of hormones, sometimes through controlling ion channels, it would perhaps be surprising if phytochrome did not also have some impact on membrane-based activities.

The second dominant theme is that Pfr regulates gene expression. Given the wide diversity of developmental change induced by phytochrome, it is to be expected that at some point the transcription of new genetic information must come into play. First proposed by H. Mohr in the mid-1960s, it has been only with the more recent development of modern molecular techniques that the gene activation hypothesis could be tested. Between 1993 and 1997, mutations have been identified in at least three of the five *Arabidopsis* genes encoding phytochrome protein (*phyA*, *phyB*, *phyD*) and two involved in chromophore biosynthesis (*hy1*, *hy2*) (see Chory, 1997). Unraveling the effects of these and other mutants on various phytochrome-mediated phenomena will be an important first step toward elucidating the phytochrome signal transduction pathway(s).

PHYTOCHROME AND MEMBRANES

Since most of the rapid phytochrome-induced phenomena are most easily interpreted as effects on membrane properties, the first obvious question to ask is whether the phytochrome molecule is associated with membranes in the cell. There are two strategies for answering this question. The first strategy to be attempted was to fractionate cells and, using dual-wavelength difference spectrophotometry, assay for phytochrome in the various fractions. If phytochrome were localized in one or more membrane systems, one would expect a high specific activity in those fractions. Using this approach, phytochrome has been reported in association with virtually every fraction of the cell, including plastids (both etioplast and chloroplast envelopes), mitochondria, endoplasmic reticulum, nuclei and plasma membranes, as well as the soluble fraction.

The results of fractionation studies are problematic. For example, most of the phytochrome extracted from dark-grown plants is soluble while more than half of the phytochrome from red-irradiated plants can be pelleted with particulate fractions. Still there appears to be no preferential association with any one membrane fraction. In most cases, the specific activity of phytochrome in the particulate fraction was about the same or less than in the crude homogenate from which the fraction was derived. These results would be expected if organelle-bound phytochrome were simply a nonspecific surface association due to disruption of the cells. They are also consistent with the hydropathic profile of phytochrome, which favors a soluble protein. On the other hand, at least 75 percent of mitochondrial-associated phytochrome is inaccessible to digestion by proteolytic enzymes, indicating a more intimate relationship with the membrane.

The subcellular distribution of phytochrome has also been studied by immunocytochemistry. In this technique, material fixed for examination by either light or electron microscopy is probed with antibodies to phytochrome. The antibodies are in turn labeled with the enzyme peroxidase or some other marker that makes possible direct visualization of the antigen-antibody complex. In dark-grown oat coleoptile parenchyma cells, phytochrome appears to be uniformly distributed throughout the cytosol, although some of the pigment appears to be associated with the plasma membrane, endoplasmic reticulum, and the nuclear envelope. While there are technical difficulties with this approach, not the least of which is the assumption that phytochrome localization within the cell was not affected by the fixation process, these results are nonetheless generally consistent with the fractionation studies.

One particularly interesting result of the immunocytochemical studies is the observation that phytochrome distribution is at least in part determined by its form. The Pr of nonirradiated coleoptile cells is uniformly distributed but Pfr is not! Within seconds of its formation by irradiation, Pfr becomes **sequestered**. That is, Pfr aggregates at discrete loci within the cell. A similar event has been observed in pea epicotyl tissue. Attempts to identify the location and function of this sequestered phytochrome have been unsuccessful—it is not, for example, associated with any identifiable membrane or organelle. Sequestering of Pfr in the cell, however, does appear to correlate with enhanced pelletability of Pfr observed *in vitro*. Both are photoreversible and may be manifestations of the same biochemical event.

Since isolated organelles have phytochrome associated with them, it should be a simple matter to test purified organelles for R/Fr photoreversible functions. Such experiments have been conducted successfully with isolated mitochondria, plastids, peroxisomes and nuclei. Isolated mitochondria have been most thoroughly documented and exhibit photoreversible NADP reduction, calcium fluxes, and ATPase activity. These functions have been demonstrated using either endogenous phytochrome, that is, phytochrome extracted along with the organelle, or exogenously added phytochrome.

On balance, the evidence would seem to indicate that phytochrome is not an intrinsic membrane protein. To the extent that it may induce changes in membrane properties, it must do so either by a loose association with the membrane or through an intermediate, but yet unidentified, signal chain.

PHYTOCHROME AND GENE ACTION

The prevailing expectation in biology is that changes in gene expression underlie overt developmental events. Demonstrated participation of phytochrome in major developmental events such as seed germination, de-etiolation, and conversion of a vegetative apex to a flowering apex naturally suggests that changes in gene expression are involved. This gives rise to the expectation that changes in the level of specific gene products (i.e., proteins) and ultimately mRNA levels are subject to regulation by phytochrome.

Phytochrome regulation at the level of protein was first reported in 1960 by A. Marcus, who reported red-far-red reversible control of glyceraldehyde-3-phosphate dehydrogenase activity in bean seedlings. Since then, the list of enzymes and other proteins whose activities are known to be light-regulated, in most cases by phytochrome, has grown to more than 60. Nine years later, M. Jaffe reported phytochrome-dependent increases in the RNA content of pea buds 24 hours after a red light treatment. Other workers reported changes in the level of various RNAs or inhibition of light-mediated effects by actinomycin-D, an inhibitor of transcription. However, direct evidence for photoregulation of genes in higher plants awaited the development of technology for isolation and *in vitro* translation of mRNA in the late 1970s. By 1985, E. M. Tobin and J. Silverthorne were able to list nine identified and multiple unidentified proteins whose genes were expressed differently in light-grown and dark-grown plants. Seven of these genes have been shown to be regulated by phytochrome.

It is curious that the majority of phytochrome-regulated genes studied thus far are *nuclear* genes encoding for mRNAs of *chloroplast* proteins. Two have been studied most extensively: the small subunit of ribulose-1,5-bisphosphate carboxylase/oxygenase (Rubisco) and the light-harvesting chlorophyll a/b binding proteins (LHCP). Rubisco is an abundant soluble chloroplast enzyme that catalyzes the carboxylation step in photosynthesis (Chap. 10). It is a multimeric protein, consisting of eight identical large subunits (LSU) and eight identical small subunits (SSU). The LSU is encoded by the chloroplast genome and synthesized in the chloroplast. The SSU is encoded in the nuclear genes and synthesized in the cytoplasm as a precursor polypeptide. LHCP is the major component of an antenna complex that gathers light for photosynthesis (Chap. 9). Like SSU, the polypeptides of the LHCP apoprotein (also known as chlorophyll a/b binding protein) are encoded in the nucleus, synthesized on cytoplasmic ribosomes, and imported into the chloroplast.

Early studies demonstrated that light stimulated an increase in Rubisco protein in barley seedlings. Using *in vitro* translation of poly(A) RNA isolated from light-grown and dark-grown duckweed (*Lemna gibba*), E. M. Tobin later demonstrated that such changes were a consequence of phytochrome-dependent increases in the level of translatable SSU mRNA. Similar results have been reported by Tobin in the United States and K. Apel and his colleagues in Germany for the chlorophyll a/b binding protein in *Lemna* and barley. Nuclear runoff experiments have confirmed that phytochrome regulates the genes for both of these proteins at the transcriptional level. In nuclear runoff experiments, nuclei isolated immediately following the light treatment are incubated in the presence of a radiolabeled RNA precursor (^{32}P uridine-triphosphate). Conditions are chosen such that only transcripts that were initiated *prior* to isolation of the nuclei will incorporate the label. The amount of RNA transcribed from a particular gene can then be measured by hybridizing the labeled transcripts to cDNA known to contain that gene. Results of these experiments confirm that Pfr acts to increase the rate of transcription of these two genes.

Not all genes are stimulated by phytochrome. There are at least two examples of important genes whose transcription is negatively regulated by phytochrome. For these genes Pfr causes a *decrease* in transcription. One negatively regulated gene encodes for NADPH-protochlorophyllide oxidoreductase, the enzyme that catalyzes reduction of protochlorophyllide to chlorophyllide. In this case, the level of translatable mRNA decreases within an hour following a brief red pulse and remains low in continuous light. The effect of a red pulse is reversible with far-red. A decrease in mRNA would at least partially explain the observed decrease in activity of this enzyme in the light.

The second example of negatively regulated transcription is the phytochrome gene itself (Colbert et al., 1983). As little as five seconds of red light causes a rapid decline in translatable phytochrome mRNA in etiolated seedlings. After a 15-minute lag period, the level of mRNA drops by 50 percent within the first hour and by more than 95 percent in two hours (Fig. 18.18A). The decline in mRNA is far-red reversible although the level of Pfr established by far-red light alone is sufficient to induce a significant loss of mRNA. Thus it appears that phytochrome autoregulates transcription of its own mRNA by some form of feedback inhibition (Fig. 18.18B).

With the aid of modern molecular techniques it has been clearly established that phytochrome does control gene expression and that it *can* do so at the transcription level. How Pfr regulates gene transcription, however, is not clear. Some studies have suggested that Pfr may activate another protein (a second messenger?) that binds to certain DNA sequences called light-regulated elements (LREs) (Fig. 18.19). LREs are short DNA sequences located in the promoter region downstream from the gene itself. Binding of the regulatory protein to the LRE apparently stimulates transcription of the gene; transcription will not proceed in its absence (see Moses and Chua, 1988). LREs have been identified for two genes in particular—those encoding the small subunit of Rubisco (rbcS) and the apoprotein of LHCP (cab). It is interesting to note that the rbcS LRE may also regulate other genes placed under its control. The LRE for the rbcS gene will, for example, impose light sensitivity on reporter genes that are not normally sensitive to light if the reporter gene is inserted into the chromosome in the region of the LRE.

In another interesting study, D. Ernst and D. Oesterheldt (1984) reported an increased *transcription rate in vitro* when phytochrome was added to *isolated* rye nuclei. These observations suggest the possibility of a direct interaction between phytochrome and the nucleus, obviating the need for an intervening protein or second messenger.

In spite of these exciting recent discoveries, unraveling the molecular action of phytochrome in gene regulation is only the first step in a long journey. There is still much to be learned about the molecular biology of

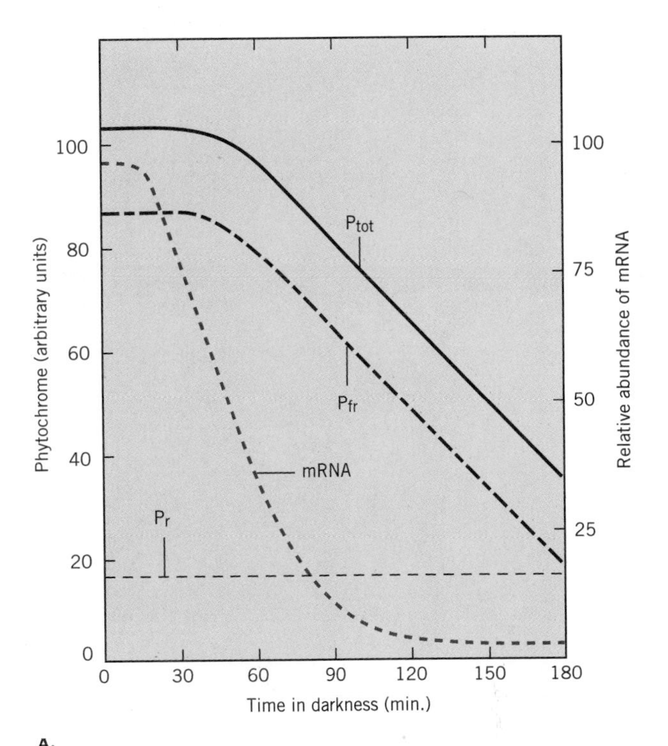

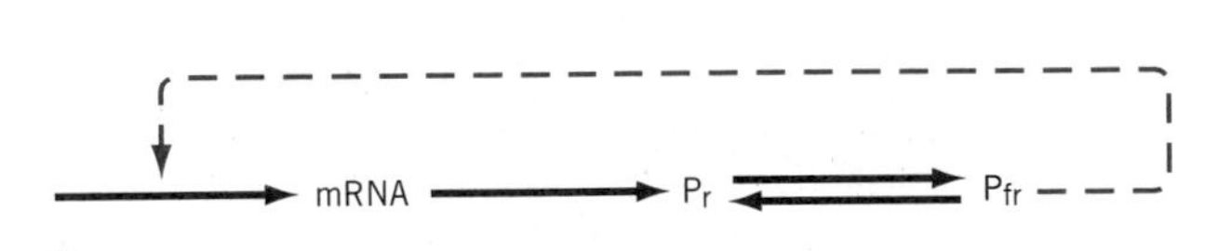

FIGURE 18.18 Phytochrome-induced decline in phytochrome mRNA. (*A*) The kinetics of phytochrome mRNA in darkness following 5 seconds of red light are compared with the loss of Pfr and total phytochrome. (*B*) A schematic interpretation of *A*, showing the feedback inhibition of phytochrome mRNA by Pfr. (From J. T. Colbert et al., *Proceedings of the National Academy of Sciences USA* 80:2248, 1983. Reprinted by permission.)

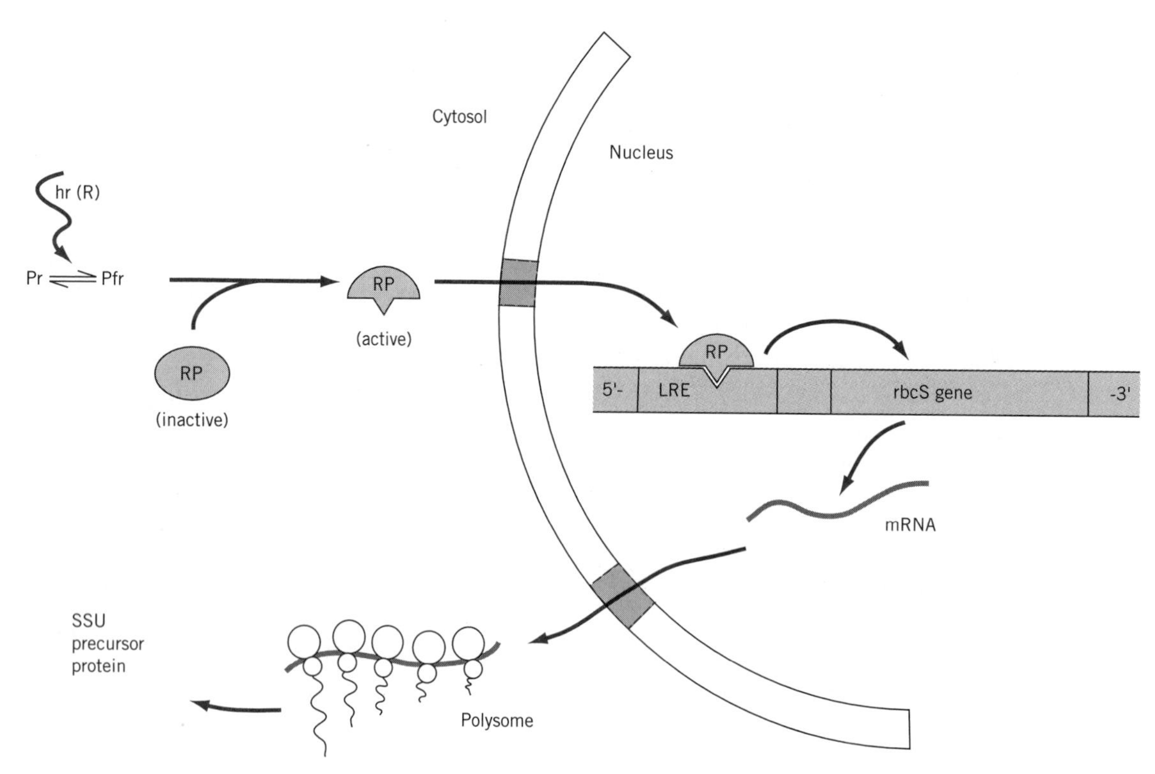

FIGURE 18.19 Schematic model for phytochrome regulation of the Rubisco small subunit gene (rbcS). Pfr activates a regulatory protein (RP) that moves into the nucleus and binds with the light regulatory element (LRE). LREs are located in the uncoded promoter region downstream (i.e., toward the 5′-end) of the rbcS gene. The LRE stimulates transcription of the gene and the resulting mRNA is exported to the cytosol where it directs the translation of the small subunit (SSU) precursor protein. The SSU precursor protein is then transported into the chloroplast, processed, and combined with the large subunits to form active Rubisco.

development. There is no single gene for leaf expansion or stem elongation. These and other complex developmental events require the coordinated input of many gene products that must be expressed in the right tissues at the right time. Only when we understand something of these complex spatial and temporal interactions will we begin to understand and appreciate the true role of phytochrome in regulating plant responses to the natural radiation environment.

BLUE-LIGHT RESPONSES

In addition to the blue-UV-A regulated HIRs described above, there are other responses that are specifically activated by blue/UV-A light (see Senger, 1987).

Sensitivity to blue and UV-A light is particularly characteristic of the lower plants and fungi, which suggests that blue light photoreceptors might be evolutionarily more primitive. It has been retained in higher plants in control of a number of responses in higher plants including phototropism (also a prominent response in the lower plants and fungi as well), stem elongation, stomatal opening, and activation of flavoenzymes.

In the past it was argued that, because Pr and Pfr absorb differentially in the blue, control of responses such as stem elongation by blue light might also be explained by phytochrome. However, as in the case of the high energy responses, several lines of evidence point to a separate photoreceptor. (1) In some seedlings red and blue light are perceived by different organs while in others sensitivity to blue light is lost during development but sensitivity to red is not. (2) There are also kinetic differences. Using sensitive transducers to measure rapid growth responses, Cosgrove (1981) has shown that blue light inhibition of stem growth can be detected within 60 seconds whereas the red response does not begin for 15 to 90 minutes. (3) Elongation of hypocotyl growth can be inhibited by low fluence rate blue light against a background of high fluence rate yellow light (Φ = 0.74) which overrides any influence of the blue light on phytochrome photoequilibrium (Thomas and Dickinson, 1979).

TABLE 18.5 Coaction of phytochrome and a blue light receptor in photocontrol of anthocyanin biosynthesis in milo *(Sorghum vulgare)* seedlings. Treatments were begun five days after sowing. Anthocyanin content is expressed as absorbance at 510 nm.

Treatment	Anthocyanin content
27 hrs R	0.0
27 hrs Fr	0.0
3 hrs B + 24 hrs dark	0.19
3 hrs B + 5 min R + 24 hrs dark	0.19
3 hrs B + 5 min Fr + 24 hrs dark	0.05
3 hrs B + 5 min Fr + 5 min R + 24 hrs dark	0.19

Data from Drumm and Mohr, 1978.

Finally, the control of anthocyanin biosynthesis, a prominent blue light effect, indicates a **coaction** between the blue receptor and phytochrome. Typically, control by phytochrome does not become effective unless preceded by a relatively long treatment with blue light (Table 18.5). The identity of the blue light photoreceptor that interacts with phytochrome is not yet known. It could be the flavoprotein cryptochrome (Chap. 7) or the blue-UVA photoreceptor involved in phototropism (Chap. 19).

UV-B RESPONSES

A number of plant responses are attributed to radiation in the UV-B region of the spectrum.

A positive effect of ultraviolet light on anthocyanin accumulation has been known since the mid-1930s. Later it was recognized that sunlight filtered through window glass, which absorbs ultraviolet rays, was less effective than unfiltered sunlight. This effect was finally characterized by Wellmann in 1971 when he showed that flavonoid biosynthesis in parsley (*Petroselinum crispum*) cell suspension cultures and seedlings was induced by UV-B radiation (280–320 nm) (Beggs et al., 1986). Maximum effectiveness was at 290 to 300 nm with little or no activity beyond 320 nm. By 1986, Beggs et al. listed eleven species of higher plants for which UV-B induced anthocyanin and flavonoid biosynthesis in coleoptiles, hypocotyls, seedling roots, and cell culture.

In *Sorghum bicolor* the action spectrum shows three peaks: 290, 385, and 650 nm. Action at 385 and 650 nm could be reversed by a subsequent exposure to far-red, but the peak at 290 could not. Yatsuhashi et al. (1982) attributed the 385 and 650 peaks to phytochrome, leaving the 290 peak due to a UV-B receptor. In parsley it appears that flavonoid biosynthesis results from the coaction of three pigments: phytochrome, a separate blue receptor, and a UV-B receptor. The UV-B system is a necessary prerequisite to flavonoid biosynthesis since neither the blue receptor nor phytochrome is effective unless preceded by a UV-B light treatment.

The identity of the UV-B receptor is unknown. Phytochrome has been suggested—the protein moiety does absorb UV-B light—but results such as those described above would argue against it. In members of the Leguminoseae family, ultraviolet-induced flavonoid synthesis can be reversed with blue light in a manner reminiscent of photoreactivation of UV damage in microorganisms. This could implicate DNA itself as a UV photoreceptor, but the action peak is shifted to wavelengths somewhat shorter than those normally characteristic of UV-B action. The identity of the UV-B receptor remains to be resolved.

SUMMARY

The phytochromes are a unique family of chromoproteins that play a critical role in almost every stage of plant development from seed germination to flowering. The existence of phytochrome was predicted on the basis of physiological experiments that demonstrated photoreversibility with red (660 nm) and far-red (730 nm) light. The pigment exists in two forms: Pr absorbs maximally at 660 nm and Pfr absorbs at 730 nm. When Pr absorbs red light, it is converted to Pfr. When Pfr absorbs far-red light, it is converted back to Pr. The physical presence of phytochrome was established by demonstrating the predicted photoreversible absorbency changes *in vivo*. In *Arabidopsis* there are five phytochrome genes encoding five species of phytochrome (PHYA-E).

Phytochrome A (PHYA) accumulates in dark-grown seedlings as PrA, which is stable. PfrA is unstable and is destroyed with a half-life of 1 to 1.5 hours. PHYB is expressed at low levels in both light and dark. PfrB is stable, with a half-life of 8 hours or more. A mixture of red and far-red light (or white light) will establish a photoequilibrium mixture of Pr and Pfr. Pfr is the physiologically active form. Phytochrome is a bluish chromoprotein with a molecular mass of about 124 kDa. The chromophore is an open chain tetrapyrrole similar in structure to phycocyanin. The phototransformation of phytochrome involves conformational changes in both the chromophore and the protein.

Phytochrome-mediated effects are conveniently grouped into three categories on the basis of their energy requirements: very low fluence responses (VLFR), low fluence responses (LFR), and high irradiance reactions (HIR). LFRs include the classically photoreversible phytochrome responses such as seed germination and deetiolation. LFRs convey information to the seed about its position relative to the soil surface and maximize the potential for a seedling to become established in light and initiate photosynthesis before the nutrient reserves of the seedling are exhausted. VLFRs are not photoreversible and are difficult to study because they saturate at light levels below those that cause a measurable conversion of Pr to Pfr. HIRs require prolonged exposure to high irradiance, are time dependent, and are not photoreversible.

Under natural conditions, the phytochrome photoequilibrium value (Pfr/P) is related to the red to far-red fluence rates. It is likely that phytochrome (probably PHYB) is the sensor that detects changes in red/far-red fluence ratio that occur under canopies and as end-of-day signals. If so, phytochrome gives a plant information about the proximity of its neighbors and contributes to the time-sensing mechanism (Chap. 20).

Phytochrome appears to operate at more than one level. The best documented are related to membrane properties or gene expression (i.e., regulation of mRNA transcription). The signal transduction pathway for phytochrome action is unknown, but the identification of mutants deficient of one or more phytochromes is an important first step in deciphering this important regulatory system.

CHAPTER REVIEW

1. What is meant by the statement "normal development in higher plates is photomorphogenesis?"
2. What unique character distinguishes phytochrome from all other pigments?
3. Distinguish between low fluence responses and high irradiance responses.
4. Phytochrome appears to influence a wide range of plant responses at all levels. Why might a ubiquitous control such as this be advantageous to the plant?
5. In what way is phytochrome particularly suited to monitor the natural light environment?
6. Most of the phytochrome in dark-grown tissue is lost following a brief exposure to low fluence light. How is this observation reconciled with the assumption that phytochrome controls development in plants that normally grow in the light?

FURTHER READING

Furuya, M. (ed.). 1987. *Phytochrome and Photoregulation in Plants*. Tokyo: Academic Press.

Furuya, M. 1993. Phytochromes: Their molecular species, gene families, and functions. *Annual Review of Plant Physiology and Plant Molecular Biology* 44:617–645.

Kendrick, R. E. and G. H. M. Kronenberg. 1994. *Photomorphogenesis in Plants*. 2nd ed. Dordrecht: Kluwer Academic Publishers.

Plant, Cell and Environment. Vol. 20, No. 6. June 1997. London: Blackwell Science Ltd. (A special issue containing reviews on various aspects of photomorphogenesis.)

Sage, L. C. 1992. *Pigment of the Imagination. A History of Phytochrome Research*. New York: Academic Press.

Senger, H. (ed.). 1987. *Blue Light Responses: Phenomena and Occurrence in Plants and Microorganisms*. Boca Raton: CRC Press.

Smith, H. (ed.). 1981. *Plants and the Daylight Spectrum*. London: Academic Press.

Smith, H. 1995. Physiological and ecological function within the phytochrome family. *Annual Review of Plant Physiology and Plant Molecular Biology* 46:289–315.

Vierstra, R. D. 1993. Illuminating phytochrome functions. *Plant Physiology* 103:679–684.

REFERENCES

Beggs, C. J., E. Wellmann, H. Grisebach. 1986. Photocontrol of flavonoid biosynthesis. In: R. E. Kendrick, G. H. M. Kronenberg, (eds.), *Photomorphogenesis in Plants*. Dordrecht: Martinus Nijhoff, pp. 467–499.

Butler, W. L., K. H. Norris, H. W. Siegelman, S. B. Hendricks. 1959. Detection, assay, and preliminary purification of the pigment controlling photoresponsive development of plants. *Proceedings of the National Academy of Sciences USA* 45:1703–1708.

Chory, J. 1997. Light modulation of vegetative development. *The Plant Cell* 9:1225–1234.

Colbert, J. T., H. P. Hershey, P. H. Quail. 1983. Autoregulatory control of translatable phytochrome mRNA levels. *Proceedings of the National Academy of Sciences USA* 80:2248–2252.

Cosgrove, D. J. 1981. Rapid suppression of growth by blue light: Occurrence, time course, and general characteristics. *Plant Physiology* 67:584–590.

Cosgrove, D. J. 1986. Photomodulation of growth. In: R. E. Kendrick, G. H. M. Kronenberg, (eds.), *Photomorphogenesis in Plants*. Dordrecht: Martinus Nijhoff, pp. 341–366.

Downs, R. J., S. B. Hendricks, H. A. Bothwick. 1957. Photoreversible control of elongation of pinto beans and other plants under normal conditions of growth. *Botanical Gazette* 118:199–208.

Drumm, H., H. Mohr. 1978. The mode of interaction between blue(UV) light photoreceptor and phytochrome in anthocyanin formation in the *Sorghum* seedling. *Photochemistry and Photobiology* 40:261–266.

Ernst, D., D. Oesterheldt. 1984. Purified phytochrome influences *in vitro* transcription in rye nuclei. *EMBO Journal* 3:3075–3078.

Furuya, M. 1989. Molecular properties and biogenesis of phytochrome I and II. *Advances in Biophysics* 25:133–167.

Hartmann, K. M. 1967. Ein Wirkungsspectrum der Photomorphogenese unter Hochenergiebedingungen und seine Interpretation auf der Basis des Phytochrome (Hypokotylwachstumshemmung bei *Lactuca sativa* L.). *Zeitschrift für Naturforschung* 22b:1172–1175.

Holmes, M. G., H. Smith. 1977. The function of phytochrome in the natural environment. I. Characterisation of daylight for studies in photomorphogenesis and photoperiodism. *Photochemistry and Photobiology* 25:533–538.

Hughes, J. E., D. C. Morgan, P. A. Lambton, C. R. Black, H. Smith. 1984. Photoperiodic time signals during twilight. *Plant, Cell and Environment* 7:269–277.

Jaffe, M. J. 1968. Phytochrome-mediated bioelectric potentials in mung bean seedlings. *Science* 162:1016–1017.

Liscum, E., R. P. Hangarter. 1993. Evidence that the red-absorbing form of phytochrome B modulates gravitropism in *Arabadopsis thaliana. Plant Physiology* 103:15–19.

Mancinelli, A. 1980. The photoreceptors of the high irradiance responses of plant photomorphogenesis. *Photochemistry and Photobiology* 32:853–857.

Mandoli, D. F., W. R. Briggs. 1981. Phytochrome control of two low-irradiance responses in etiolated oat seedlings. *Plant Physiology* 67:733–739.

Mohr, H., W. Shropshire. 1983. An introduction to photomorphogenesis for the general reader. In: W. Shropshire, H. Mohr. *Photomorphogenesis. Encyclopedia of Plant Physiology*, NS. Vol. 16A, pp. 24–38.

Moses, P. B., N.-H. Chua. 1988. Light switches for plant genes. *Scientific American* 258 (April):88–93.

Newman, I. A. 1981. Rapid responses of oats to phytochrome show membrane processes unrelated to pelletability. *Plant Physiology* 68:1494–1499.

Quail, P. H., H. P. Hershey, K. B. Idler, R. A. Sharrock, A. H. Cristensen, B. M. Parks, D. Sommers, J. Tepperman, W. B. Bruce, K. Dehesh. 1991. In: B. Thomas, C. B. Johnson (eds.), *Phytochrome Properties and Biological Action*: Berlin, Heidelberg: Springer-Verlag, pp. 13–38.

Rüdiger, W. 1986. The chromophore. In: R. E. Kendrick, G. H. M. Kronenberg (eds.), *Photomorphogenesis in Plants*. Dordrecht: Martinus Nijhoff, pp. 17–33.

Serlin, B. S., S. J. Roux. 1984. Modulation of chloroplast movement in the green alga *Mougeotia* by the Ca^{2+} ionophore A23187 and by calmodulin antagonists. *Proceedings of the National Academy of Science USA* 81:6368–6372.

Siegelman, H. W., E. M. Firer. 1964. Purification of phytochrome from oat seedlings. *Biochemistry* 3:418–423.

Siegelman, H. W., S. B. Hendricks. 1957. Photocontrol of anthocyanin formation in turnip and red cabbage seedlings. *Plant Physiology* 32:393–398.

Smith, H. 1983. Is Pfr the active form of phytochrome? *Philosophical Transactions, Royal Society of London*, Series B, 303:443–452.

Tanada, T. 1968. A rapid photoreversible response of barley root tips in the presence of 3-indoleacetic acid. *Proceedings of the National Academy of Science USA* 50:376–380.

Thomas, B., H. Dickinson, 1979. Evidence for two photoreceptors controlling growth in etiolated seedlings. *Planta* 146:545–550.

Tobin, E. M., J. Silverthorne. 1985. Light regulation of gene expression in higher plants. *Annual Review of Plant Physiology* 36:569–593.

Vierstra, R. D. 1993. Illuminating phytochrome functions. There is light at the end of the tunnel. *Plant Physiology* 103:679–684.

Vierstra, R. D., P. H. Quail. 1983. Purification and initial characterization of 124-kilodalton phytochrome from *Avena. Biochemistry* 22:2498–2505.

Vince-Prue, D. 1991. Phytochrome action under natural conditions. In: B. Thomas, C. B. Johnson (eds.), *Phytochrome Properties and Biological Action*. Berlin, Heidelberg: Springer-Verlag, pp. 313–319.

Yatsuhashi, H., T. Hashimoto, S. Shimizu. 1982. Ultraviolet action spectrum for anthocyanin formation in Broom sorghum first internodes. *Plant Physiology* 70:735–741.

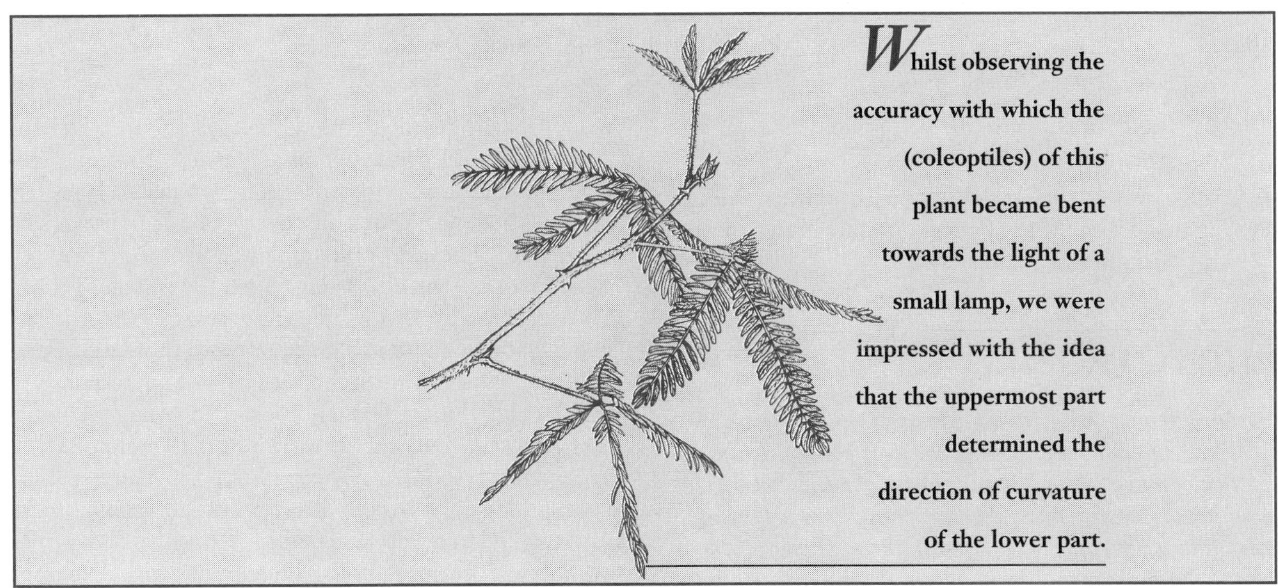

C. and F. Darwin
(1881)

19

Plant Movements—Orientation in Space

The power of movement is generally perceived as an animal trait not normally associated with plants. Yet, to the surprise of many, movement pervades the life of the green plant. Movement in higher plants does not involve locomotion as it does in animals, nor is it so dramatic. Plant movement is mostly slow and deliberate, but it is a key factor in determining the orientation of plants in space. Plants that have been inadvertently placed in the horizontal position will reorient their root and shoot to the vertical. House plants will bend, appearing to seek light coming through a window. Leaves may periodically rise and fall throughout the day while others track the sun as it moves across the sky. Leaves of the venus flytrap snap closed on a hapless insect. While most plant movements are relatively slow, they nonetheless serve important functions by positioning organs for the uptake of nutrients and water, optimal interception of sunlight, or (in the case of the fly trap) obtaining nitrogen.

We recognize two principal categories of movement in plants, based on the distinctiveness of their mechanism. **Growth movements** are *irreversible.* They arise as the result of differential growth within an organ or between two different organs. **Turgor movements** are *reversible,* resulting from simple volume changes in certain cells—most often in a special organ called the pulvinus. Within each group, we can further distinguish between **nutations, tropisms,** and **nastic** responses. The term nutation (or circumnutations) denotes a regular rotary or helical movement of plant organs, most typically the stem apex, in space (Brown, 1993). Nutations are normally evident only with patient observation or time-lapse photography. Tropic responses are directionally related to the stimulus—the response may be in the same direction, opposite, or at some specific angle to it. Examples are phototropism (a response to light), gravitropism (gravity), hydrotropism (water), or thigmotropism (touch). Nastic responses are not obviously related to any vector in the stimulus. Directionality of nastic responses is inherent in the tissue and includes epinasty (bending down), hyponasty (bending up), nyctinasty (the rhythmic sleep movements of leaves), seismonasty (response to mechanical shock), thermonasty (temperature), and thigmonasty (touch).

This chapter will focus on the three plant movements that have been explored most thoroughly. These are

- phototropism, particularly the nature of the photoreceptor and the role of auxin in the signal transduction chain;
- gravitropism, including a brief discussion of the nature of the gravitational stimulus and the mechanism of graviperception, the particular character of gravitropism in shoots and roots, and the role of

auxin and calcium in the differential growth response; and

- nastic movements, including the structure of motor organs and the role of potassium flux in nyctinastic and seismonastic movements.

PHOTOTROPISM

Phototropism has been studied most extensively in grass coleoptiles. Coleoptiles bend toward the light because the translocation of auxin out of a shoot apex is light-sensitive. A gradient in fluence rate established across a coleoptile causes a lateral redistribution of auxin such that a higher proportion of the auxin is translocated down the side experiencing the lower fluence. The resulting differential growth rate causes the shoot axis to curve toward the light.

Most people are familiar with the sight of house plants bending toward the light from an open window, an everyday example of the phenomenon called **phototropism** (Fig 19.1). Phototropism is a classic plant physiology problem. It has attracted the interest of botanists since the middle of the nineteenth century, especially in Germany. Darwin's study of phototropism, published in his book *The Power of Movement in Plants* (1881), is credited with overcoming the preoccupation of English-speaking botanists with descriptive and taxonomic biology and stimulating an interest in the more dynamic aspects of plant function. Phototropism has played an important role in the early history of plant physiology and its study and debate about the nature of the process continues even today. Phototropism occurs in lower plants and fungi as well as in higher plants, where presentation of photosynthetic structures to the light is essential to survival.

Aside from its intrinsic intellectual appeal, phototropism provides a useful model system in which to study some very fundamental aspects of plant function. Cell elongation in phototropically stimulated grass coleoptiles led to Went's discovery of plant hormones (Chap. 16). The coleoptile remains one of the more useful systems in which to study control of cell elongation. Phototropism is also a blue light response and it has many features that make it attractive to those interested in studying the blue light receptor. The response is rapid, readily quantified, and occurs in young seedlings that are easily and quickly grown in the laboratory.

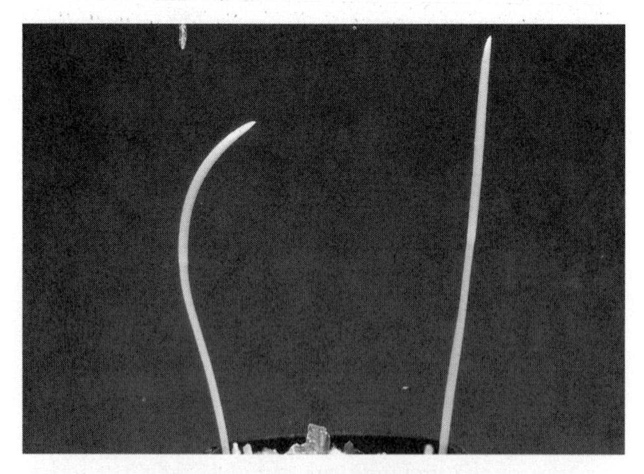

FIGURE 19.1 Phototropic response in oat (*Avena sativa*) coleoptile. Left: Dark-grown seedling placed in unilateral blue light (from right) for 90 minutes. Right: Unlighted control.

Tropic responses may be either positive or negative. If a plant responds in the direction of the stimulus (e.g., toward a light source) it is said to be positive. If it grows away from the stimulus it is said to be negative. Whether the phototropic response is positive or negative depends largely on the nature of the organ or its age. For example, stems and other aerial organs are for the most part positively phototropic while the tendrils of most climbing plants are negatively phototropic. Leaves are normally **plagiotropic,** which means they orient at angles intermediate to the light. Roots, on the other hand, are largely nonphototropic, although some may exhibit a weakly negative response. The stems of ivy (*Hedera helix*) are negatively phototropic during the shade-loving juvenile stage, but older branches become positively phototropic. The stems of ivy-leafed toad flax (*Cymbalaria muralis*) become negatively phototropic following fertilization. This interesting behavior helps to place ripening seed pods into crevices in the walls on which the plant is found.

Phototropism is often defined as a response to unilateral light and so it is in the laboratory. Under normal circumstances, however, the bending response will occur in plants that are receiving light from all sides. It is only required that the fluence rate be unequally distributed. In experiments with bilaterally illuminated grass coleoptiles, for example, as little as 20 percent difference in fluence rate on the two sides of the organ will induce a bending response (Fig. 19.2). Thus light can be presented unilaterally (as it is in most laboratory experiments), bilaterally, from all sides, and even from above, providing only that a gradient is created across the organ. Thus, *phototropism is a growth response to a light gradient.*

The magnitude of a light gradient across an organ such as a coleoptile is dependent on optical properties of the tissue as well as differences in incident light. A light gradient across an organ can be intensified by **screening** within the organ; that is, pigments, including but not limited to the photoreceptor itself, will attenuate the light as it passes through the organ. Light can also be attenuated by scattering, reflection, or diffraction in the cells. Thus gradients across individual cells, measured by using microfiberoptic probes, may vary from 5 : 1 to 50 : 1. To further complicate matters, organs such

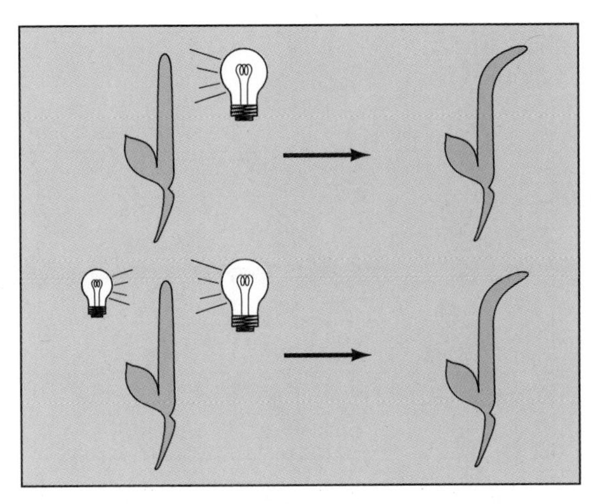

FIGURE 19.2 The phototropic response will occur whenever there is a light gradient established across the stem or coleoptile axis.

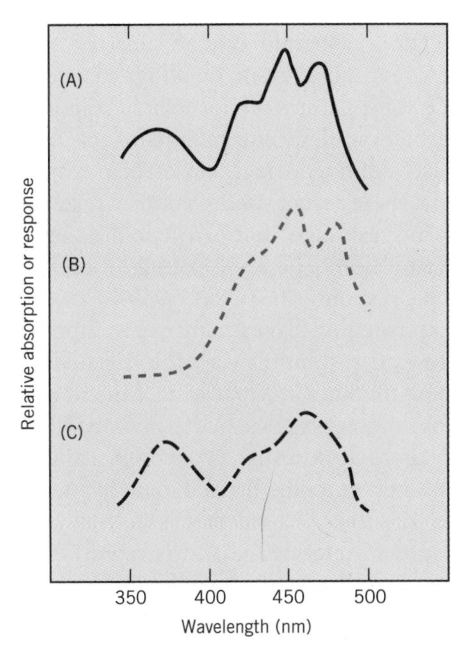

FIGURE 19.3 A comparison of the action spectrum for *Avena* coleoptile phototropism (*A*) with the absorption spectra for (*B*) carotene and (*C*) riboflavin.

as coleoptiles appear to function as light pipes. Light applied to the tip, for example, will be transmitted through the coleoptile to cells further down the organ. Thus the phototropic stimulus is not necessarily a simple matter. These complex interactions between light and the optical properties of tissue have led to significant difficulties in experimental design as well as in interpretation of the resulting experimental data.

PHOTOTROPIC SIGNAL PERCEPTION

The Photoreceptor Phototropism, as with all other responses to an external stimulus, can be considered in three stages: signal perception, signal transduction, and response. Since the phototropic stimulus is light, we can conclude that a photoreceptor is involved. However, even though phototropism in grass coleoptiles is the classical physiological blue light effect, the identity of the photoreceptor has been the subject of vigorous debate for many years. Since the 1930s, action spectra have been repeatedly determined for a number of organisms but have been most thoroughly documented for coleoptiles of *Avena* and *Zea* and for sporangiophores of the fungus *Phycomyces*. The action spectra for *Avena* and *Phycomyces* are virtually identical, indicating that they share a common photoreceptor. All other phototropic action spectra are similar and consistently show two peaks in the blue near 475 nm and 450 nm and a small peak or shoulder at 420 nm (Fig. 19.3A). In addition there is a large action peak in the UV-A region near 370 nm. The action spectra for both *Avena* and *Phycomyces* phototropism show an additional peak in the region of 280 nm, suggesting that the photoreceptor might be a chromoprotein, analogous to phytochrome. These action spectra have suggested two possible candidates—carotenes and flavins or flavoprotein—as the phototropic photoreceptor.

Originally it seemed likely for several reasons that the photoreceptor was a carotenoid. Carotenoids are a ubiquitous pigment, occurring almost universally in biological materials. They are localized in tissues and regions of high phototropic sensitivity and the absorption spectra of carotenoids have a characteristic triple peak structure that matches well the phototropic action peaks in the blue region of the spectrum (compare Fig. 19.3B with 19.3A). Carotenoids do not, however, absorb significantly in the 300 to 400 nm range and so cannot account for the action peak near 370 nm.

In the 1940s, it was suggested that the photoreceptor could be a flavin molecule such as riboflavin. Flavins are also ubiquitous in organisms, occurring most frequently as prosthetic groups in the electron carriers flavin mononucleotide (FMN) and flavin-adenine dinucleotide (FAD). Riboflavin absorbs blue light and has a peak in the region of 370 nm as well (Fig. 19.3C). However, in aqueous solution at room temperature, the blue absorbance is rather broad and lacks the characteristic three-peak structure of the action spectrum. Since a composite of the carotene and flavin absorption spectra provided a good match with the phototropic action spectrum, the possibility of coaction between the two pigments has long been considered an attractive alternative hypothesis.

Recent experiments, however, have more or less ruled out a role for carotenoids as the photoreceptor (Palmer et al., 1996). When the enzyme phytoene desaturase is blocked, either by mutation or by treatment of maize seedlings with the herbicide SAN 9789 (norflurazon), no carotenoid can be detected. Yet carotenoid-deficient mutants and seedlings treated with norflurazon exhibit a normal phototropic response to blue light. Carotenoid deficient mutants of the fungus *Phycomyces* also exhibit normal phototropic responses. Interestingly, these recent studies confirm results reported nearly forty years ago that carotenoid-deficient maize mutants and albino barley seedlings exhibited strong phototropic responses (Galston, 1959)!

The carotenoid-flavin controversy appears to be headed toward resolution with the recent isolation of phototropic mutants in *Arabidopsis* (Liscum and Briggs, 1995). Four mutations, designated *nph1*, *nph2*, *nph3*, and *nph4* (for *n*on-*p*hototropic *h*ypocotyl), exhibit no response to unilateral blue light. The *nph1* mutant is also missing a 120 kDa protein that is normally associated with the plasma membrane and is rapidly phosphorylated when irradiated with blue light. This protein and the phosphorylation response have been detected in the wild type *(NPH1) Arabidopsis* as well as a variety of other species, both monocotyledonous and dicotyledonous. It has been suggested that this protein is the apoprotein for the phototropic photoreceptor and that activation of the protein by phosphorylation may be the first step in the signal transduction chain (Briggs and Liscum, 1997).

The nature of the NPH1 chromophore has not yet been established, but it is likely a flavin or a pterin, or both (Chap. 7). One thing is clear. The putative phototropic photoreceptor is biochemically and genetically distinct from cryptochrome (CRY1) (Chap. 7). First, CRY1 is a 75 kDa cytoplasmic protein while the NPH1 protein is a plasma membrane protein with a mass of 120 kDa. *Arabidopsis* seedlings that carry a cryptochrome-deficient mutation *(hy4)* do not show normal suppression of hypocotyl elongation in blue light but do exhibit normal phototropism. Conversely, the nonphotoropic mutant *nph1* shows normal suppression of hypocotyl elongation. Seedlings that carry both the *hy4* and the *nph1* mutations exhibit neither blue light–suppressed hypocotyl elongation nor phototropism (Briggs and Liscum, 1997). Clearly, cryptochrome and the putative phototropic photoreceptor are two different, genetically distinct, blue light photoreceptors.

Fluence Response Curves Perhaps no aspect of phototropism has indicated the complexity of the process so much as attempts to define relationships between fluence and response. Phototropism is characterized by a rather curious fluence-response curve, quite unlike most photobiological responses. Fluence response curves are generally obtained by maintaining a constant fluence rate but varying the presentation time. Figure 19.4 shows fluence-response curves determined for *Avena* phototropism at three different fluence rates. The lowermost curve in Figure 19.4, curve c, exhibits the characteristic features of the classic response to increasing fluence. There is an initial rise to a first peak, which is called **first positive curvature.** With increasing fluence, curvature declines, to the point that this may even result in a bending *away* from the light source. This decline and negative response is called **first negative curvature.** Note that first negative curvature is not necessarily "negative" in the sense of bending away from the light. It may be simply a reduced positive response. Following the region of first negative curvature, the response curve again rises into what is called **second positive curvature.** In some cases, a second negative and even third positive curvature have been reported.

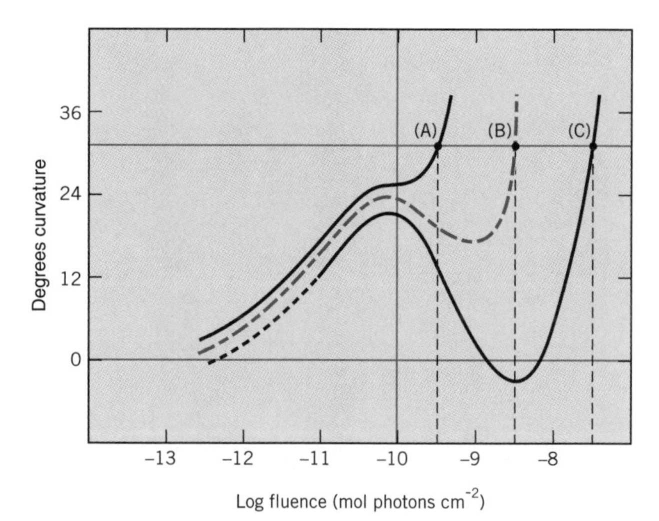

FIGURE 19.4 Phototropic photon fluence-response curves for *Avena* coleoptiles at three different fluence rates of blue light. (*A*) 1.4 × 10^{-13} mol cm^{-2} s^{-1}. (*B*) 1.4 × 10^{-12} mol cm^{-2} s^{-1}. (*C*) 1.4 × 10^{-11} mol cm^{-2} s^{-1}. The solid vertical line indicates the peak of first positive curvature. The three dashed vertical lines indicate the fluence required to generate 30° of second positive curvature. Curves (*A*) and (*B*) are offset slightly in the vertical direction in order to emphasize their similarity in the first positive region. (Data replotted from B. K. Zimmerman and W. R. Briggs, *Plant Physiology* 38:248, 1963.)

Note that the first positive region is constant for all three fluence rates—the curves would superimpose if not offset as they are in Figure 19.4—but first negative and second positive curvature are not. This indicates that the Bunsen-Roscoe **reciprocity** law applies to first positive curvature; that is, the response is a product of the fluence rate times the presentation time. However, reciprocity does not apply to first negative and second positive curvatures. Indeed, examination of the three different curves leads to the conclusion that second positive curvature is more a function of presentation time than fluence. Note, for example, that approximately 1000 s was required to produce 30° of second positive curvature, regardless of the fluence rate. Since reciprocity generally indicates participation of a single photoreceptor, failure of reciprocity for second positive curvature suggests the possibility that more than one photoreceptor might be involved. On the other hand, action spectra have been determined for both first and second positive curvature and they are identical. It must therefore be concluded that both are mediated by the same, single photoreceptor. Thus, the complexities of second positive curvature must be due to subsequent events in the signal transduction chain rather than the primary photoact.

SIGNAL TRANSDUCTION IN PHOTOTROPISM

At the same time F. W. Went and his contemporaries had chosen to study the influence of the apex on coleoptile elongation, parallel studies on the role of the root apex were being conducted in Germany by N. Cholodny. The result was independent proposals by Cholodny, in 1924 and Went, in 1926, that the apex was able to influence cell elongation lower down in the extension region of the organ (see Weevers, 1949). These ideas were drawn together in the 1920s in an attempt to explain phototropism. The **Cholodny-Went hypothesis** states that unilateral illumination induces a *lateral redistribution* of endogenous auxin near the apex of the organ. This asymmetry in auxin distribution is maintained as the auxin is transported longitudinally toward the base of the organ. The cells on the shaded side of the organ, which receives the higher amount of auxin, are thereby stimulated to elongate more than those on the lighted side. It is this differential growth that causes curvature toward the light source.

The experimental basis for the Cholodny-Went hypothesis is derived largely from agar diffusion experiments originally conducted by Went. In Went's experiments, coleoptiles were first stimulated with unilateral light. The coleoptile apices were then excised, split longitudinally and the two halves placed on agar blocks in order to collect the auxin that diffused out of the base. The amount of auxin collected in the agar blocks was then assayed by the *Avena* curvature test (Chap. 16). Went reported a significantly higher quantity of auxin was collected from the shaded half of the coleoptile apex than from the lighted half, indicating that unilateral lighting caused a greater proportion of the auxin to be transported down the shaded side of the coleoptile.

In the decades following Went's original experiments, a sizable literature accumulated offering evidence both for and against the Cholodny-Went hypothesis and at least two alternative hypotheses have been proposed to account for differential growth during phototropism. One, primarily of historical interest, proposed that an asymmetric distribution of auxin could be explained on the basis of an inactivation of auxin on the lighted side. This was supported by evidence that light catalyzed the destruction of IAA *in vitro* in the presence of riboflavin and the observation that in Went's original split-tip experiments there was a slightly lower total yield of auxin from phototropically stimulated apices compared with unstimulated controls. However, results of later experiments, described below, show that phototropic stimulation does not significantly decrease total auxin yield *in vivo.* A second alternative hypothesis, proposed by A. H. Blaauw in the early part of the century, does not require an asymmetric distribution of auxin but involves a direct inhibition of cell growth by blue light—the so-called **light growth response.** Because the hormonal model was more fashionable at the time, Blaauw's light-growth model received relatively little attention at the time. However, more recent studies have revived interest in Blaauw's model and we will return to this point later.

Doubts as to the validity of the Cholodny-Went hypothesis also arose from numerous unsuccessful attempts to verify asymmetric auxin distribution by applying ^{14}C-IAA to tropically stimulated coleoptiles. (Similar problems were encountered in experiments with gravitropic stimulation, which is also attributed to a lateral redistribution of auxin.) These problems, however, may be largely attributed to poor experimental technique. It is now evident that a large proportion of the radioactive auxin taken up by the tissue does not enter the auxin transport stream. When care is taken to discount this nondiffusible auxin, a clear differential in auxin transport can be detected (Briggs, 1963; Gillespie and Thimann, 1961). For example, when maize coleoptile tips were supplied with ^{14}C-IAA, approximately 65 percent of the radioactivity was recovered from the shaded side. There was no significant asymmetry when subapical sections were used.

The Cholodny-Went hypothesis has been systematically reevaluated by one of its strongest advocates, W. R. Briggs and his colleagues. Briggs has confirmed that auxin production in coleoptiles of *Zea mays* is confined to the apical 1 to 2 mm. Lateral redistribution during phototropic stimulation probably occurs within the most apical one-half mm. Furthermore, phototropic stimuli (including fluences well beyond that required to stimulate second positive curvature) do not reduce the yield of diffusible auxin. Briggs repeated Went's original split-tip experiments but, unlike Went, he excised the tips and placed them on agar blocks *before* presenting the phototropic stimulus. The results (Fig. 19.5) clearly demonstrate that when the tip is partially split, leaving tissue continuity only at the very apex of the coleoptile, exposure to unilateral light causes an increase in the amount of diffusible auxin on the shaded side and a decrease on the lighted side. The total amount of auxin recovered, however, remains effectively constant. When lateral diffusion of auxin is prevented throughout the entire length of the tip, no such asymmetric auxin dis-

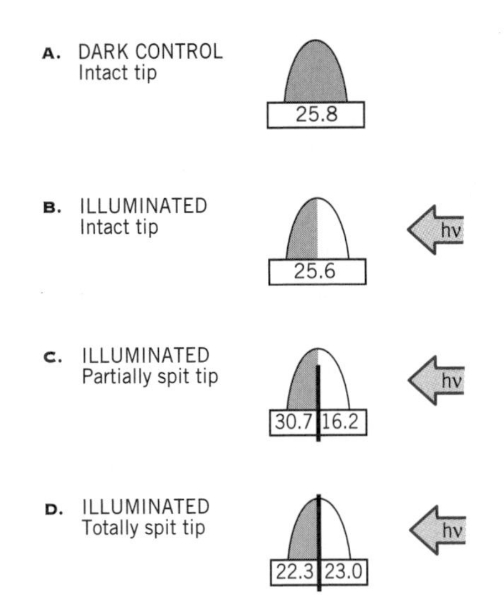

FIGURE 19.5 Asymmetric distribution of diffusible auxin in excised *Zea mays* coleoptile apices following phototropic stimulation. (*A*, *B*) Intact control apices. (*C*) Tips were partially split, leaving tissue continuity only at the very apex. A microscope cover slip was inserted to provide a barrier to lateral diffusion. (*D*) Tips were totally split and the diffusion barrier passed through the apex. Numbers indicate the amount of auxin collected in the agar blocks over a 3-hour period, based on degrees of curvature in the *Avena* curvature bioassay. Values are for auxin collected from 3 tips (*A*, *B*) or 6 half-tips (*C*, *D*). (Data from W. R. Briggs, *Plant Physiology* 38:237, 1963.)

tribution is observed. These results clearly support the principal tenet of the Cholodny-Went hypothesis, namely that unilateral light induces a preferential migration of auxin down the shaded side of the coleoptile.

But is auxin asymmetry actually responsible for phototropic curvature in coleoptiles? To answer this question, it is necessary to establish (1) that endogenous auxin concentrations are limiting for growth of the coleoptile, (2) that auxin asymmetry can be sustained as it is transported down the coleoptile, and (3) that the auxin differential is sufficient to cause the observed differences in growth. Briggs and his colleagues addressed these questions in a study of the effects of exogenous auxin application on the growth of maize coleoptiles (Baskin et al., 1986). Various concentrations of IAA were mixed with lanolin paste and applied as a small spot on one or both sides of the coleoptile tip. Control experiments established that lanolin paste itself has no effect on the growth of coleoptiles, but a curvature similar to first positive curvature can be induced by spot application of the IAA-lanolin mixture to just one side of the tip. Careful measurements of elongation on both sides of the coleoptile during IAA-induced curvature showed that only the side receiving the application was stimulated above controls. Furthermore, blue light-induced phototropic curvature could be completely eliminated by a spot application of 125 ng of IAA to the *lighted* side immediately following stimulation. The results of this study appear to answer the first two questions; endogenous auxin is limiting for growth and there is no lateral diffusion of auxin below the point of IAA-lanolin application.

One question remains—is the differential in auxin concentration induced by a phototropic stimulus sufficient to cause the required changes in growth rate on the two sides of the coleoptile? Straight-growth of isolated coleoptile segments incubated in IAA solutions is a function of the logarithm of IAA concentration. This would seem to indicate that rather large differences in hormone concentration would be required to effect a significant difference in growth rate (Firn and Digby, 1980, Trewavas, 1981). Agar-diffusion experiments with tropically stimulated coleoptiles show a differential of, at best, 2 : 1 between the shaded and lighted sides. However, Baskin et al. (1986) point out that agar diffusion experiments average the auxin gradients on both the lighted and shaded halves of the coleoptile. This, they argue, would underestimate the true magnitude of the gradient. Moreover, both the *Avena* curvature test, in which IAA is applied directly to the cut surface of decapitated plants, and IAA-induced curvature of intact maize coleoptiles exhibit a near-linear dependence on hormone concentration (Chap. 16, Fig. 16.2).

Another approach to the study of phototropism makes use of marker beads and time-lapse photography to measure growth kinetics of responding coleoptiles. By placing small glass or styrene beads along the coleoptile, it is possible to measure simultaneously elongation rates on the lighted and shaded sides of bending coleoptiles (Fig. 19.6). The Cholodny-Went model predicts increased elongation rates on the shaded side and reduced elongation rates on the lighted side. A rapid cessation of elongation on the lighted side is a consistent feature of all kinetic experiments. For the shaded side, the results are less consistent—elongation rates may increase, remain unchanged, or increase following a brief

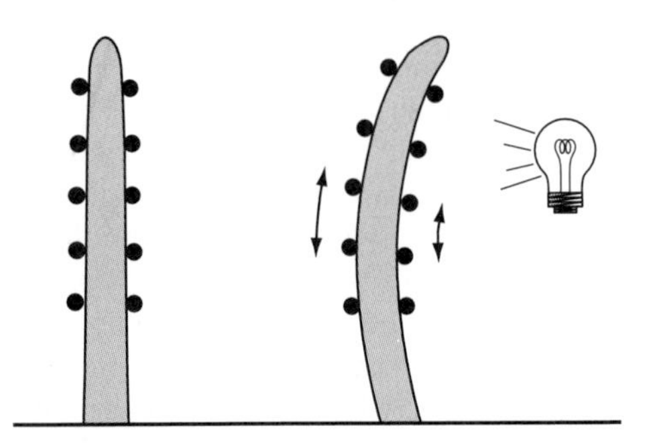

FIGURE 19.6 Using glass or styrene beads to measure coleoptile growth rate. The beads adhere to the coleoptile and are displaced as the cells elongate. The growth rate can be calculated from measurements made on photographic images.

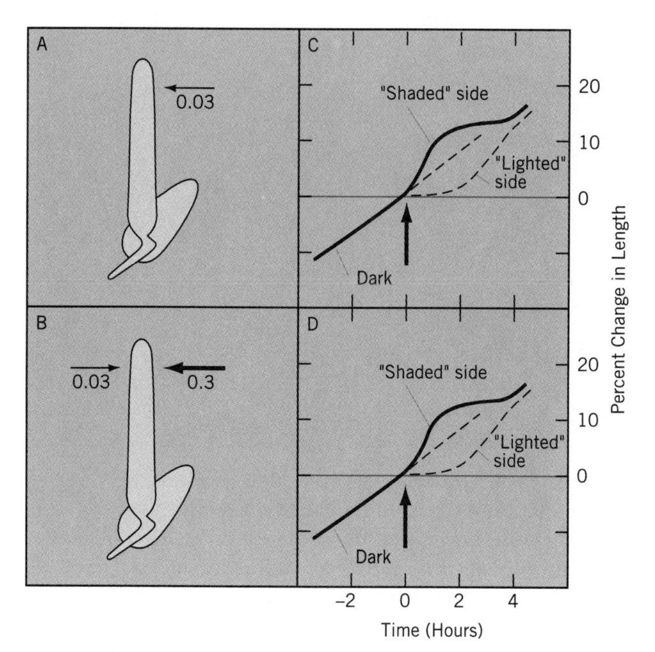

FIGURE 19.7 Growth kinetics giving rise to phototropism. ***Avena*** **coleoptiles were subjected to either unilateral (*A* and *C*) or bilateral (*B* and *D*) blue light at the indicated fluence rates. Both treatments stimulate phototropic curvature. Note that a fluence rate of 0.03 μmol m^{-2} s^{-1} causes a significant inhibition of growth on the lighted side when presented unilaterally but shows no such response when presented as the lower fluence rate during bilateral irradiation. Vertical arrows indicate light on. (Data from K. Macleod et al., *Journal of Experimental Botany* 36:312, 1985.)**

period of decrease. Since it is known that both blue and red light will partially inhibit coleoptile elongation, experiments in which the rate of elongation decreases on the lighted side and remains unchanged on the shaded side appear to offer support for Blaauw's light-growth model. Blaauw's model equates blue-light inhibition of elongation with blue-light induction of phototropism and argues that a light gradient across the coleoptile would establish a gradient of photoinhibition. There are two lines of evidence, however, that argue against Blaauw's model. Blue light at 0.003 μmol m^{-2} s^{-1} significantly inhibits cell growth when presented unilaterally but has no inhibitory effect when presented bilaterally with light of a higher fluence rate (Macleod et al., 1985) (Fig. 19.7). Results such as these show that cells are not independent as predicted by the Blaauw model. Cell elongation in a phototropically stimulated organ depends not only on the fluence rate at each cell, but on the *gradient* in fluence rate across the organ. Other experiments have shown the two phenomena—light-dependent growth inhibition and the tropic response—can be separated. If coleoptiles are uniformly irradiated with red or blue light sufficient to saturate the growth inhibition response before the phototropic stimulus is presented, the reduction of growth on the lighted side of the coleoptile is compensated by an increased rate of growth on the shaded side.

All in all, the evidence in support of the Cholodny-Went hypothesis seems rather compelling, at least with respect to coleoptiles stimulated at low fluence rates presented for short duration—that is, first positive curvature in grass seedlings. The mechanism by which blue light establishes lateral redistribution of auxin is not yet known, although phototropic mutants, such as those now available in *Arabidopsis*, may help further our understanding of the signal transduction chain. As described earlier, phosphorylation of the NPH1 protein may be a first step in the chain. The other *nph* mutants (*nph2–4*) may provide additional clues. The *nph4* mutant, for example, differs from the other three in that both gravitropism and phototropism are impaired. Since both are auxin-mediated differential growth responses, the *NPH4* gene product is probably involved close to the response end of the transduction chain. In *nph2* and *nph3* only phototropism is impaired, but both mutants contain normal amounts of the putative receptor protein and undergo normal phosphorylation in blue light. These mutants thus appear to be involved somewhere in the middle of the transduction chain (Briggs and Liscum, 1997).

A final area of concern is phototropism in light-grown plants, where relatively little is known about the phototropic process. As with phytochrome, discussed in Chapter 18, most of what we know about phototropism is derived from laboratory studies with etiolated seedlings. However, in light-grown cucumber (*Cucumis sativus*) and sunflower (*Helianthus annuus*) seedlings subjected to uniform lighting, curvature of the stem can be induced by simply shading one of the cotyledons (Fig. 19.8) and the phototropic response of sunflower seed-

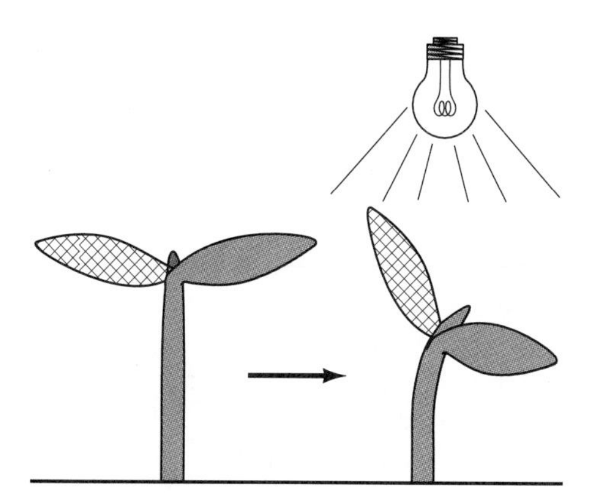

FIGURE 19.8 Curvature in cucumber (*Cucumis sativus*) seedlings induced by shading cotyledons. The left-hand cotyledon was covered with aluminum foil and the seedling uniformly irradiated with white light for 8 hours. (After J. E. Shuttleworth and M. Black, *Planta* 135:51, 1977.)

lings is markedly decreased if the leaves are removed. The cucumber response, at least, differs from the classical phototropic response in that it is induced by red light rather than blue light. This appears to be a phytochrome-mediated response and is related to inhibition of hypocotyl elongation below the irradiated cotyledon. Both the cucumber and sunflower responses may be attributed to the fact that the leaves are a prime source of auxin required for the growth response. Both cucumber and white mustard (*Sinapis alba*) also exhibit a classical phototropic response induced by irradiating the hypocotyls directly with blue light. Clearly the control of stem growth in deetiolated plants is an area where there is still much to learn.

GRAVITROPISM

Gravity is perceived in roots by starch-containing cells in the root cap. Gravitropism, like phototropism, is a differential growth response caused by a redistribution of auxin in the translocation stream.

Gravitropism[1] is probably one of the most unfailingly obvious and familiar plant phenomena to most people (Fig. 19.9). As every school child knows, shoots always grow "up" and roots always grow "down." But do they? A casual walk through the wood or garden should reveal how over simplified this view is. The lateral branches of most trees and shrubs do not grow up; they grow outward in a more or less horizontal position. Stolons (or runners) of strawberry (*Fragaria*) plants and buttercups (*Ranunculus*) also grow horizontally along the soil surface. Dig into the soil and you will find rhizomes (underground stems) and many roots growing horizontally. Many pendulous inflorescences show no directional preference for growth, but hang down simply of their own weight.

It is true that the root and shoot of the primary plant axis do align themselves parallel with the direction of gravitational pull. Such an alignment is said to be **orthogravitropic.** The primary root, which grows toward the center of the earth, exhibits **positive gravitropism.** The shoot, which grows away from the center of the earth, exhibits **negative gravitropism.** Organs such as stolons, rhizomes, and some lateral branches, which grow at right angles to the pull of gravity, are said to be **diagravitropic.** Organs oriented at some intermediate angle (between 0° and 90° to the vertical) are said to be **plagiogravitropic.** Lateral stems and lateral roots are commonly plagiogravitropic. Organs that exhibit little or no sensitivity to gravity are said to be **agravitropic.**

[1]Except in the more recent literature, the reader will encounter the term **geotropism,** which was introduced in 1868 by A. B. Frank to describe this phenomenon. Since the response is to the force of gravity, the term gravitropism is now considered more appropriate.

FIGURE 19.9 Gravitropism in maize (*Zea mays*) seedlings. Four-day-old dark-grown seedlings were placed in the horizontal position for three hours. Note the shoot exhibits negative gravitropism and the root exhibits positive gravitropism.

The advantages to the plant of positive and negative gravitropic growth responses are fairly obvious. Seeds may assume a random orientation in the soil, but in order to ensure survival, the shoot with its photosynthetic structures must be above ground in order to take advantage of sunlight. The root system must penetrate the soil in order to secure anchorage and a reliable supply of nutrients and water. The primary root most often exhibits a strongly negative orthogravitropic response. Secondary roots (that is, first-level branch roots), however, tend to grow more horizontally while tertiary roots are generally agravitropic. This hierarchy of gravitational responses ensures that the root system more effectively fills the available space and thus more efficiently mines the soil of water and nutrients (Fig. 19.10). In a similar fashion, a hierarchy of positive orthogravitropic, diagravitropic, and plagiotropic responses in the shoot system ensures a more efficient capture of sunlight to drive photosynthesis.

GRAVIPERCEPTION

Like phototropism and other responses induced by external stimuli, gravitropism can be divided into partial processes of perception, transduction, and response. Unlike other stimuli, however, the force of gravity is omnipresent and nonvarying. It does not vary in magnitude as temperature does, for example. Gravity cannot be turned on and off, such as light at dawn and dusk. Moreover, gravity is not a unilateral stimulus; there is no gradient component in gravity. Cells on the lower side of a stem or root are subjected to the same gravitational force as those on the upper side. Consequently, it is likely that

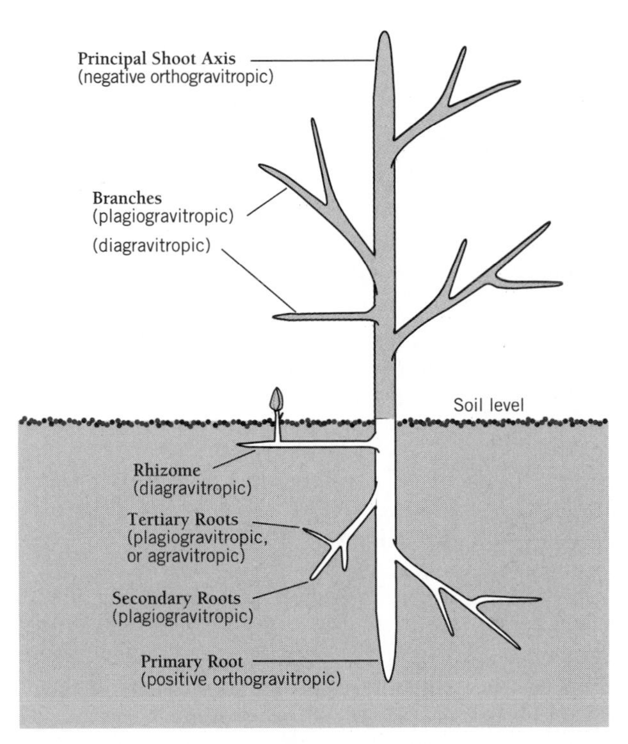

FIGURE 19.10 Diagram illustrating the range of gravitropic responses in shoots and roots.

gravity can be detected only by the movement of some structure or structures within the cell—a movement that establishes an initial asymmetry in the cell and is translated in terms of pressure. The mass and movement of whatever structure is involved must be consistent with the sensitivity and speed of the gravitational response and there must be a mechanism for transducing the pressure signal into a differential growth response.

Sensitivity of Gravistimulation Gravitational stimulation (stimulus quantity or dose) is the product of the intensity of the stimulus and the time over which the stimulus is applied:

$$d = t \cdot a$$

where a is the acceleration of mass due to gravity (in *g*), t is the time (in seconds) over which the stimulus is applied, and d is the dose (in *g* seconds) (see Box 19.1: Methods in the Study of Gravitropism). The minimum dose required to induce gravitropic curvature is called the **threshold dose.** Threshold dose will vary depending on the organism or experimental conditions. Values of d in the range of 240 g·s (at 22.5 °C) to 120 g·s (at 27.5 °C) have been reported for *Avena* coleoptiles, but more careful mathematical analyses suggest that less than 30 g·s (i.e., an acceleration of 1 *g* for less than 30 seconds) is sufficient to induce gravitropic curvature in roots. Three other parameters are of interest when defining gravistimulation: presentation time, reaction time, and threshold intensity.

The minimum duration of stimulation required to induce a curvature that is just detectable is known as the **presentation time.** The intensity of stimulation should also be defined, although a stimulus of 1 *g* at 90° is more or less standard. A force of 1 *g* is easily obtained by simply placing the stem or root in a horizontal position. Presentation times of 12 seconds for cress roots and 30 seconds for *Avena* coleoptiles have been determined (Volkmann and Sievers, 1979). Pickard (1973) reported that a brief 1-second stimulus will induce curvature in *Avena* coleoptiles if the stimulus is repeated every 5 seconds. This suggests that "some cumulative receptive process begins the instant the plant is turned on its side" (Pickard, 1985).

Presentation time should not be confused with **reaction time,** which is the interval between the presentation of the stimulus and the actual development of curvature. Reaction times involve the complete signal transduction sequence that leads to the asymmetric growth response. Typically, 10 minutes is required before curvature can be visually detected, although reaction times may vary from a few minutes to hours, depending upon the species and conditions. In experiments employing sensitive electronic position-sensing transducers, curvature of maize coleoptiles could be detected within about 1.5 minutes of horizontal placement, while bending of the mesocotyl could not be detected before 3.5 minutes (Bandurski et al., 1984).

The minimum stimulus intensity required to induce a response is known as the **threshold intensity.** Threshold intensities have been determined for a variety of plant organs under different experimental conditions. The results are remarkably consistent and indicate that roots are perhaps an order of magnitude more sensitive than shoots. In land-based clinostat experiments (see Box 19.1), values for threshold intensity for *Avena* coleoptiles and roots were found to be 1.4×10^{-3} *g* and 1.4×10^{-4} *g* (Shen-Miller et al., 1968). Values calculated for lettuce seedling hypocotyl and roots in experiments aboard the Salyut 7 spacecraft were 2.9×10^{-3} *g* and 1.5×10^{-4} *g*, respectively (Halstead and Dutcher, 1987). It is apparent that many plants are very sensitive to gravitational stimulus.

Tissue Sensitivity Gravitropic sensitivity in the root is localized in the root cap, a thimblelike mass of cells that covers the tip of the root (Fig 19.11). Traditionally the function of the root cap is believed to be twofold; it provides physical protection for the root apical meristem and its outer cells secrete a mucilaginous polysaccharide that lubricates the path of the growing root. A third function, that of graviperception, has more recently been established by experiments in which the root cap is surgically removed. Removal of the root cap does not interfere with the elongation of the root but completely abolishes any gravitropic response. Decapped roots will recover gravisensitivity after about 24 hours,

Box 19.1
Methods in the Study of Gravitropism

A fundamental requirement for any form of scientific experimentation is that of controlling the application of the stimulus in the form of intensity (or concentration) and duration. Since the gravitational field of earth cannot be extinguished (except in the microgravity conditions of space), experimentation on gravitational effects on plants and other organisms has required some unique approaches.

Most experiments require a mass acceleration in the range of 1 g or less, which can easily be achieved by simply orienting the organ (e.g., a coleoptile or primary root) away from the vertical. The force (at least for short-term stimulation) is generally proportional to the sine of the angular deviation from the vertical. Thus the force is greatest in the horizontal position since sine 90° = 1 (sine of angles less than 90° is less than 1). Seedling shoots generally must be oriented between 0.5 to 10° from the vertical in order to induce curvature.

Forces greater than 1 g can be achieved by the use of specially designed centrifuges. Similar centrifuges have been used earlier in the century for studying the properties of animal (egg) membranes, and so forth. Centrifugation has not been used extensively in the study of gravitropisn, but those experiments in which it has been used have provided some useful insights.

The problem of extinguishing gravitational forces has been approached in two ways: **clinostats** and **space flight.** A clinostat is a device that holds the plant axis in a horizontal position while continuously rotating it about the horizontal axis. This does not actually extinguish the gravitational field, of course, but the summated effect is a nondirected constant stimulation. With the clinostat, plants can first be subjected to a brief stimulus and then rotated to, in effect, remove any further stimulus. As might be expected, continuous multilateral stimulation has been found to influence a variety of physiological parameters. *Avena* seedlings, for example, respond with increased growth rate and increased respiration. Some of these changes may be incidental, but others may influence the gravitropic response. Results must always be interpreted with caution.

SPACE—THE FINAL FRONTIER

The advent of space flight in the 1950s has provided plant scientists with unique opportunities to study responses to microgravity conditions. Since 1960, when the first wheat and maize seeds were carried aloft on Sputnik 4, experiments with plants have been conducted on manned and unmanned spacecraft from both the United States and the Soviet Union (now Russia). Perhaps not unexpectedly, physiological effects of microgravity are not limited to gravitropism but embrace a variety of other cellular events. Many of the effects are deleterious, including reduced growth, chromosomal aberrations, and other cytological abnormalities. Death at the flowering stage was common until, in 1982, *Arabidopsis thaliana* were successfully carried through a complete life cycle and produced viable seed. Many of the difficulties could be attributed simply to the logistics of trying to maintain plants in space, but even with improved methods, difficulties are still being encountered. As yet the returns may be modest, but the use of microgravity as an experimental tool is ripe for exploitation.

---

which correlates with the regeneration of a new cap. It is interesting to note, however, that decapped maize roots failed to regenerate root caps over a 5- to 7-day time period in the microgravity of space, while ground-based controls regenerated caps well within the same time frame. This raises an interesting conundrum: Since regeneration of the root cap is itself an indication of graviresponsiveness, does this mean that decapped roots must also sense gravity (Evans et al., 1986)?

The site of gravisensitivity in shoots is not so readily apparent as it is in roots. A variety of experimental approaches over the past century have indicated that, as with phototropism, gravisensitivity is greatest in the extreme apical region of grass coleoptiles, although sensitivity is not limited to the apex. Indeed most shoots appear to be most sensitive in the apex but retain some residual sensitivity if the tip is removed. Higher sensitivity in the apical regions, particularly in dicotyledonous plants, could simply reflect their higher capacity for growth. Since gravitropism is a differential growth response, sensitivity to gravity would not be expected in regions where growth is largely complete. There are reports that hypocotyls can be stripped of their epidermal layers without a change in growth rate, but that "peeled" hypocotyls do not respond to gravity (Firn and Digby, 1980). This suggests that the site of graviperception is in the peripheral cells layers of the hypocotyl. Firn and Digby also reported no evidence for longitudinal trans-

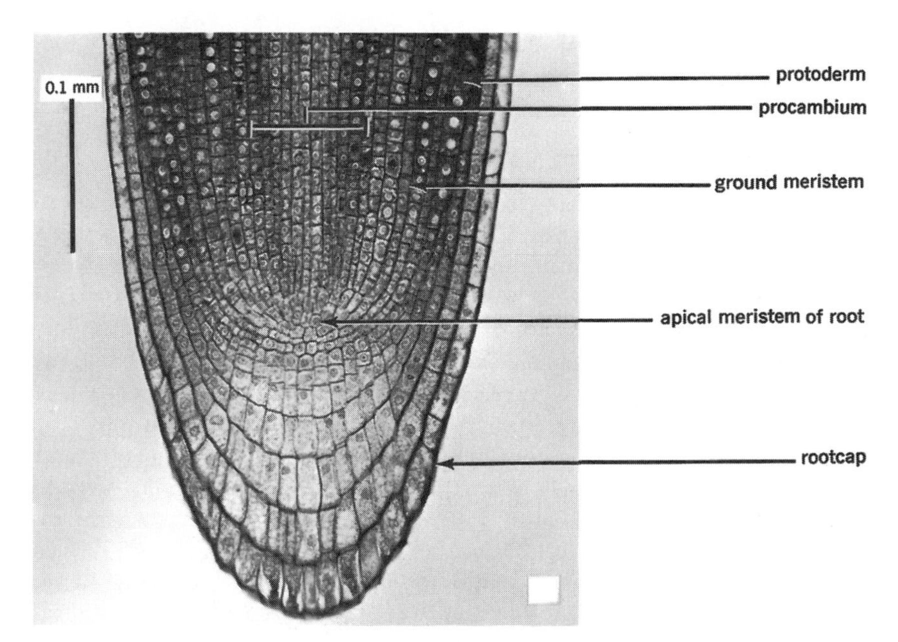

FIGURE 19.11 Longitudinal section of an onion (*Allium* sp.) root tip showing the root cap. (From K. Esau, *Anatomy of Seed Plants*, 1977, New York, Wiley. Reprinted by permission.)

mission of the signal, indicating that the sites of graviperception and response are not separate. This conclusion does not agree with the general conclusion arrived at on the basis of coleoptile and root studies described below.

The Mechanism of Graviperception As pointed out earlier, a distinction between "up" and "down" by plant organs must almost certainly involve sedimentation of some structure within the cell such that a physiological asymmetry is established. Here again, animal-based models were introduced at the turn of the century and have dominated the thinking of plant physiologists ever since. F. Noll was the first to suggest, in 1892, that plants might sense gravity in a manner similar to some animals (Weevers, 1949). Crustaceans, molluscs, and many other invertebrates have gravity-sensing organs called **statocysts,** small innervated cavities lined with sensory hairs. Within the cavity are one or more **statoliths,** tiny granules of sand or calcium carbonate that are pulled downward by gravity. When the statocyst changes position the statoliths also shift position, bending the sensory hairs and sending an action potential to inform the central nervous system of the change.

In 1900, G. Haberlandt and E. Nemec independently adapted the statolith theory to account for plant responses to gravity (Weevers, 1949). Based on careful cytological studies, they proposed that starch grains found in specialized tissues function as statoliths. According to Haberlandt and Nemec, statocytes are cells containing sedimentable starch grains. Tissues that contain statocytes are known as **statenchyma.** Support for the statolith hypothesis was found in earlier reports by Darwin and others that removal of the root tip, where most of the starch grains are found, resulted in a loss of gravitropic response. Nonetheless, the hypothesis was not universally accepted and over the decades a number of investigators have attempted to prove or disprove it.

In its current form, we understand that the statolith is not simply a starch grain, but an **amyloplast** (Fig. 19.12). An amyloplast is a group of starch grains con-

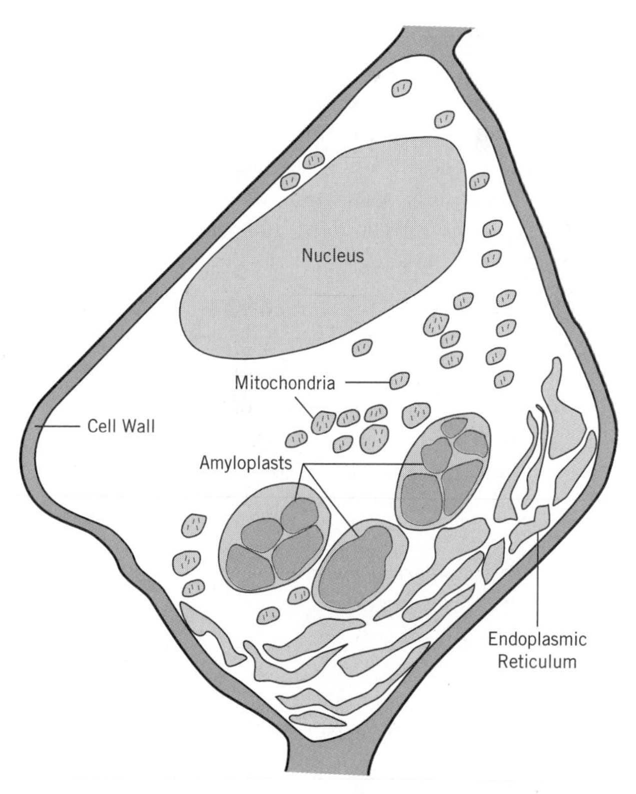

FIGURE 19.12 A statocyte containing three statoliths. (Based on an electron micrograph of *Lepidium* root, Volkmann and Sievers, 1979.)

tained within a membrane. There may be 1 to 8 individual grains within an amyloplast and as many as a dozen amyloplasts in each statocyte (Wilkins, 1984). This compares with the single large grains characteristic of starch storage organs. Not all amyloplasts in all cells are readily mobile. In fact, detection of putative statoliths appears to be largely confined to regions of high gravitropic sensitivity. These include the mass of nonvacuolated cells in the central core of the root cap, also known as the **columella,** and a zone of endodermal cells that sheath the vascular tissues (the **starch sheath**) in many shoots. (There is a bit of circular argument here—mobile amyloplasts are presumed to function as statoliths in part because they are found in tissues that are believed to be most sensitive to gravity, in part because they contain putative statoliths.) Mobile amyloplasts may also be found in the inner cortical cells of aerial organs and the **pulvini,** or motor organs in the nodes of grass stems that are responsive to gravity.

It is understood that any gravity-sensing mechanism involving particle sedimentation would have to operate with a speed and sensitivity consistent with the known speed and sensitivity of the response. In the 1960s, L. J. Audus undertook a careful examination of various subcellular particles (Audus, 1969). Audus concluded that, of all the cellular organelles, only starch grains have the mass and density to move through the viscous cytoplasm within known presentation times. Ultrastructural examination has shown that other cellular organelles, such as the endoplasmic reticulum, may become shifted in gravistimulated cells, but these movements are thought to be a consequence of starch grain sedimentation.

Although there is still no direct proof for the starch-statolith hypothesis, there is a large body of evidence that is more *consistent* with that idea than any other that has been put forward to date. This evidence is summarized below:

1. There are no well-documented cases of plant species that respond to gravity but have no starch grains or amyloplasts. In lower plants such as algae and fungi, excess carbohydrate may not be stored as starch. In these cases, some other substance may function as statoliths. In the alga *Chara*, for example, the role of starch grains is replaced by granules of barium sulphate.
2. There is a strong correlation between the rate of starch sedimentation and presentation time. In sweet pea (*Lathyrus odoratus*), for example, there is a parallel increase in both sedimentation time and presentation time as the temperature is lowered from 30°C to 10°C. The decline in both is presumably related to an increase in protoplasmic viscosity.
3. Loss of starch by hormone treatment or mutation is accompanied by a loss of graviresponse. For example, roots of cress seedlings (*Lepidium sativum*) treated with cytokinin or gibberellin at 35°C become starch-free in 29 hours. The growth rate of treated roots is reduced only slightly (0.48 mm h^{-1} vs. 0.64 mm h^{-1}) but any response to gravity is completely eliminated (Iversen, 1969). Transfer of the roots to water in the light results in a parallel recovery of both amyloplasts and gravitropic responsiveness after 20 to 24 hours.

   The maize mutant *amylomaize* produces smaller amyloplasts than the wild type. In studies of the percentage and speed of amyloplast sedimentation the degree of coleoptile curvature was strictly correlated with the size of the amyloplast (Hertal et al., 1969). Another mutant of maize, hcf-3, is unable to carry out photosynthesis and thus can form no starch in the leaf base statocytes when the endosperm reserves have been exhausted (Miles, 1981). Such seedlings do not respond to gravity unless fed sucrose, in which case recovery of both amyloplasts and gravisensitivity was noted.

As noted above, the available evidence offers only correlative support for the starch-statolith hypothesis, and not all of the data is consistent. A starchless mutant of *Arabidopsis* has no starch in the root cap or hypocotyl, but its roots and hypocotyls still exhibit a gravitropic response (Casper and Pickard, 1989). In another study, J. Kiss et al. (1997) compared the graviresponses of *Arabidopsis* roots and hypocotyls with differing starch contents. Mutants with an intermediate starch content (50 percent of wild-type) showed reduced curvature (80 percent of wild-type). However, even hypocotyls of starch-deficient mutants (0 percent of wild-type) showed some curvature (38 percent of wild-type). On the basis of results such as these, it has been suggested that starch per se is not required for gravitropism, but it is required for full sensitivity (see Kaufman et al., 1995).

SIGNAL TRANSDUCTION IN GRAVITROPISM

It is not known how the sedimentation of statoliths creates physiological asymmetry in the cell or tissue, although a number of models have been proposed. Most models agree that it is not the change in position of the statolith or the process of movement per se that is important. The preferred view is that the statolith exerts pressure on one or more membranes or other cellular components. Although there is no direct evidence for pressure-sensitive membranes in plants, both the plasma membrane or the endoplasmic reticulum (ER) have been suggested. Interaction with the plasma membrane could result in altered transport through plasmodesmata. Some statocytes, such as the columella cells of root caps, are characterized by a larger-than-normal number of plasmodesmata. Alternatively, interaction with the plasma membrane could lead to an activation of certain enzymes that modulate hormone metabolism. It is the ER, however, that has attracted the most attention. Ultrastructural studies have indicated that the ER is displaced by sedimenting statoliths when roots of maize and *Vicia faba* are oriented horizontally. The ER appears to be randomly distributed in cells of vertically oriented maize root, but in the horizontal position is pushed away from the lower regions of the cell by the amyloplasts.

Unlike maize and *Vicia*, the ER is normally located in the vicinity of the apical (lower) wall in vertically oriented cress (*Lepidium*) and pea (*Pisum sativum*) root caps. When normally vertical roots are reoriented to the horizontal position, the amyloplasts sediment toward the lateral (now lower) wall but the ER remains in place.

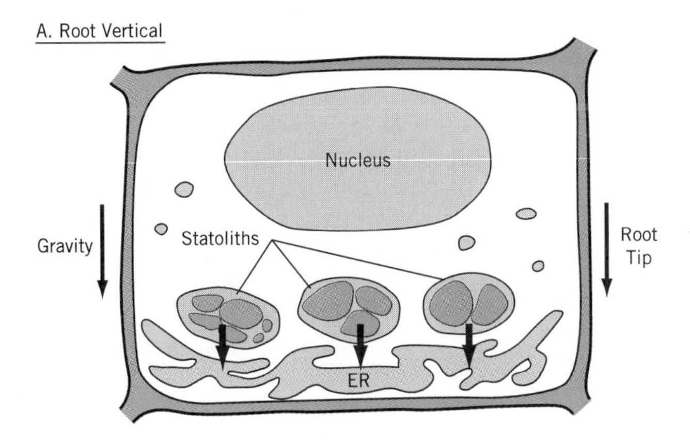

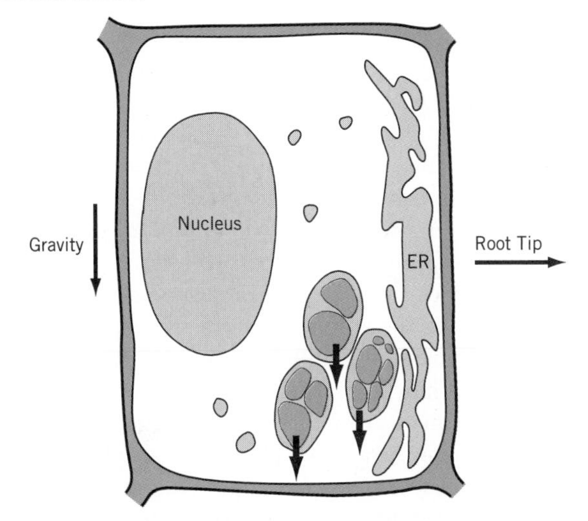

FIGURE 19.13 Schematic representation of the Volkmann-Sievers model for graviperception. (*A*) When the root is oriented vertically, starch-laden statoliths exert pressure uniformly on the endoplasmic reticulum (ER). (*B*) In the horizontal position, the statoliths are redistributed in the cell in response to gravity, exerting unequal pressure on the ER.

Volkmann and Sievers (1979) have constructed a model for graviperception that takes into account the relative positions of amyloplasts and ER in *Lepidium* roots. According to this model (Fig. 19.13), the pressure of amyloplasts on the ER is the primary transducer. In the vertical position, the pressure of the statoliths on the ER is equal in all statocytes (Fig. 19.13A). When reoriented in the horizontal position (Fig. 19.13B), the amyloplasts fall away from the ER in the upper part of the statocyte but continue to exert pressure on the ER in the lower part. The Volkmann-Sievers model requires relatively little movement of the amyloplasts and thus is consistent with both rapid presentation times and the small angle of deviation from vertical (say, 10°) necessary to induce a response. There are difficulties regarding the general applicability of the Volkmann-Sievers model. In particular, the distribution of ER found in *Lepidium* has been found in root caps of some species, but not, as mentioned above, in others.

An interesting variation on the pressure transduction theme is the proposal that the gravireceptor lies at the junction between the plasma membrane and the cell wall. In this scenario, the receptor, probably a protein, is stimulated when the protoplast as a whole settles in response to gravity (Wayne and Staves, 1997). This proposal is based on experiments with internodal cells of the alga *Chara*. *Chara* cells have no starch or statoliths, yet gravity induces a polarity in cytoplasmic streaming. It was found that hydrostatic pressure could mimic the force of gravity and induce a similar polarity. With the application of hydrostatic pressure under a variety of conditions, Wayne and his co-workers concluded that *Chara* cells sense gravity by sensing compression rather than "up" versus "down." Wayne and Staves (1997) argue that not only could their "gravitational pressure model" apply to higher plants, it would explain why starchless mutants are able to respond to gravity, although less efficiently than wild-type cells. The presence of statoliths would simply increase the overall density of the cytoplasm and thus increase the efficiency of gravisensing in wild-type cells. These provocative ideas will no doubt stimulate further interest and research into the mechanics of gravisensing and the initial stages in signal transduction.

THE GROWTH RESPONSE IN GRAVITROPISM

While little is certain of the initial signal transduction in gravitropism, there is no doubt that development of curvature ultimately involves a differential growth response. As with phototropism, the Cholodny-Went hypothesis of asymmetric auxin distribution has dominated thinking and research into gravitropism for more than 60 years. Accordingly, the hypothesis states that horizontal orientation of the shoot or roots induces a lateral translocation of auxin toward the *lower* side of the organ. In a manner similar to phototropism, auxin redistribution would bias the growth rate in favor of the lower side such that negatively gravitropic organs (e.g., coleoptiles and shoots) would turn upward. In positively gravitropic organs such as roots, the higher concentration of auxin is thought to *inhibit* elongation on the lower side relative to the upper. The responses of shoots and root will be discussed separately.

Coleoptiles and Shoots The study of asymmetric auxin distribution in gravistimulated coleoptiles is based largely on split-tip agar-diffusion experiments similar to those previously described for phototropism. In the late 1920s H. Dolk demonstrated that roughly two-thirds of the diffusible auxin yield from horizontal coleoptile apices could be collected from the lower half. This has been confirmed by many other investigators. Perhaps the clearest evidence for lateral redistribution of auxin is provided by the experiments by M. Wilkins and his colleagues (Wilkins, 1984). In these experiments, a microdrop of tritium-labeled IAA (5-^{3}H-IAA) was applied asymmetrically to the apex of a maize coleoptile. After a period of time, the coleoptiles were bisected longitudinally and the amount of radioactivity in the two halves was measured. When the coleoptiles were maintained in the vertical position (and in the absence of any phototropic stimulus), 90 percent of the activity was retained in the coleoptile half below the point of application. In the horizontal position, however, only two-thirds was retained and the remaining one-third was translocated into the lower half. This pattern of IAA redistribution has been confirmed using radioimmunoassay techniques (Mertens and Weiler, 1983) and is entirely dependent on metabolic energy.

In spite of results like those described above, the validity of the Cholodny-Went hypothesis for gravitropism has been questioned just as it has for phototropism. Digby and Firn (1979) have analyzed changes in the growth rate for the upper and lower sides of gravistimulated maize coleoptiles. During the initial stages, the growth rate on the lower side almost doubles while growth on the upper side quickly falls off to zero. Such a disproportionate change in growth rate is not predicted by the Cholodny-Went hypothesis and has led others to question whether the magnitude of auxin asymmetry is sufficient to account for such large differences in growth rate. This same question has been raised in regard to phototropism (see earlier discussion). Brigg's counterargument on the basis of curvature response to a *linear* increase in auxin concentration might apply equally well here.

Asymmetric distribution of apically synthesized auxin appears to play a significant role in the gravitropic response of maize and oat coleoptiles, although it may not be the only factor. The gravitropic response of dicotyledonous shoots, on the other hand, is quite dramatic but poorly understood. Lateral transport of radiolabeled IAA and 2,4-D has been observed in *Helianthus* and *Phaseolus* hypocotyls. but could not be detected in green epicotyls (Phillips and Hartung, 1976). Using radioimmunoassay techniques, Mertens and Weiler (1983) were unable to detect significant redistribution of endogenous IAA in horizontal sunflower hypocotyls. F. B. Salisbury and co-workers have observed that the upper and lower sides of sunflower and soybean hypocotyls continue to react differently when immersed in a solution of auxin (Salisbury et al., 1988). They have argued that, since all cells would be exposed to the same auxin concentration, hypocotyls respond not to differential auxin concentration but to changes in the sensitivity of cells to auxin.

Still, auxin is required for graviresponse in dicots. S. Iwami and Y. Masuda (1974) found that removal of the uppermost region of *Cucumis* hypocotyls reduced the gravitropic response, but that it could be replaced with IAA. *Helianthus* hypocotyls that have been depleted of their endogenous auxin supply by decapitation do not respond gravitropically when held in the horizontal position for up to three hours. If, however, auxin is supplied at the time of restoring the seedling to the vertical position, a tropic curvature will develop (Brauner and Hager, 1958). It appears that induction of a physiological difference does not require the presence of auxin but completion of the growth response does. Moreover, it appears that the tropic stimulus is in some way "remembered" by the hypocotyls.

Grass Stems A particularly interesting and agriculturally significant manifestation of gravitropism is seen in the stems of cereals and other grasses that have been **lodged,** that is, laid prostrate by heavy wind or rain. The lower portion of grass leaves form a sheath that is inserted into the node and clasps the stem. The swollen base of the leaf sheath, just above the internode, is called a psuedopulvinus or **false pulvinus** (pl. = *pulvini*). The false pulvinus is a self-contained organ that both senses and responds to gravity. Unlike the classical pulvinus, a reversible turgor-driven motor organ to be described in the next section, the false pulvinus of grass stems is dedicated to graviresponse and undergoes irreversible cell elongation. When the grass stem is laid prostrate, the younger pulvini nearest the apex undergo differential cell elongation—the cells on the lower side elongate extensively while those on the upper side elongate very little or may even be compressed (Fig. 19.14).

Pulvini cells to the inside of the vascular bundles have prominent statoliths that fall rapidly whenever the stem is tilted toward the horizontal. Most of the growth response occurs, however, in collenchyma cells that are found to the outside of the statocytes on the lower side of the pulvinus. Just how differential growth is estab-

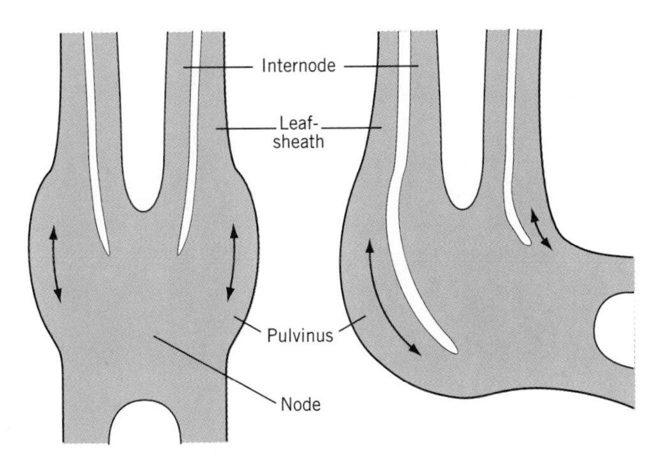

FIGURE 19.14 Diagram of a false-pulvinus in a grass leaf-sheath. (*A*) A vertically oriented stem. (*B*) A gravistimulated stem. Arrows indicate the relative growth of cells on the two sides of the pulvinus.

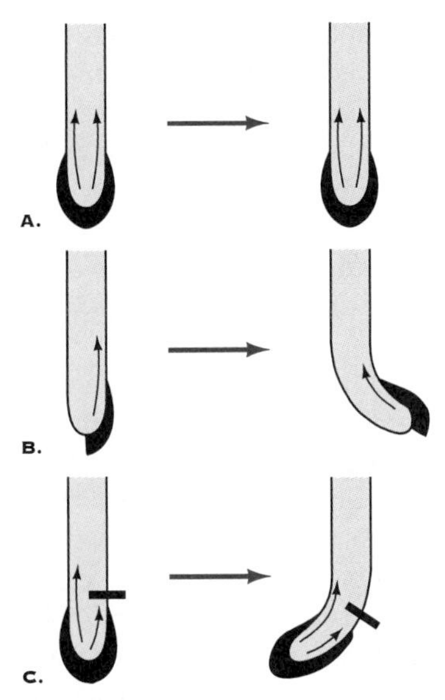

FIGURE 19.15 The role of root cap-secreted inhibitors in curvature of vertically oriented roots. Arrows indicate the flow of inhibitors from the root cap to the elongation region of the root. (*A*) Control root; the root cap is left intact. Inhibitor flow and root growth are uniform. (*B*) The root cap is surgically removed from one-half of the root. The root grows toward the side with the cap remaining. (*C*) A piece of microscope cover slip is inserted into the root on one side to block the flow of inhibitor. The root grows toward the side without the block.

lished is not well understood, although auxins are believed to be involved. The free IAA content of *Avena* pulvini (from 45-day-old vertical shoots) is approximately 60 to 70 ng per gram dry weight of tissue (Kaufman and Song, 1987). Following 24 hours of gravistimulation, the free IAA content increases to 420 ng, two-thirds of which is found in the lower half of the pulvinus. The auxin differential is not due to lateral translocation since radiotracer experiments with ^{14}C-IAA failed to detect any downward movement across gravistimulated pulvini. The excess free IAA on the lower side may be attributed to increased IAA synthesis or released from stored conjugates. Changes have also been found in the amounts of free gibberellins and GA conjugates, raising speculation that gibberellins may also be implicated in grass shoot gravitropism.

Roots Involvement of hormones in gravitropic responses of roots has been the subject of controversy for many years. This is in part because very little is known about the role of hormones in normal root growth. Auxin has been prominent in most theories of root gravitropism, generally on the basis that even moderate concentrations of exogenous auxin will inhibit root growth. There is, in fact, a considerable body of evidence for involvement of an inhibitor in root gravitropism. Wilkins and his colleagues have shown that when the cap is removed from only one side of a maize root, the root will curve toward the side where the cap remains (Fig. 19.15). Similarly, when an impermeable barrier is imposed between the cap and the elongation zone, but only on one side of the root, the root will curve away from the barrier. This growth pattern indicates that the root cap is the source of a diffusible substance that is transported to the elongation region where it inhibits root growth.

What is the nature of the inhibitor? The Cholodny-Went hypothesis would say that the inhibitor is auxin and there is a growing body of evidence that would support this view (Moore and Evans, 1986). It is known, for example, that auxins very effectively inhibit root growth when exogenously supplied at concentrations of 10^{-6} M or greater, concentrations that normally stimulate elongation of coleoptiles and shoots. However, although there is some evidence that auxin may be synthesized in the root tip, the primary source of auxin for roots appears to be the shoot and the auxin is transported toward the root tip in a highly polarized manner.

One of the early events in gravitropism is a rapid but asymmetric efflux of protons in the elongation zone. This is easily demonstrated by placing roots on agar containing the pH indicator dye bromcreosol purple. The change of color from red to yellow, indicating an increase in acidity, is seen *only* on the *upper side* of gravistimulated roots. Such observations are consistent with the acid-growth theory for auxin action and with kinetic measurements indicating an inhibition of elongation on the lower side. In addition, both gravitropic curvature and asymmetric proton efflux are prevented by inhibitors of auxin transport (TIBA, NPA).

Asymmetrical exogenous application of IAA to roots will induce curvature toward the side of application. An asymmetric distribution of exogenously supplied auxin has also been detected; three times more auxin moves toward the lower side than in the reverse direction. The key question is, of course, whether gravitropic stimulation induces asymmetries in the *endogenous* auxin content of responding roots. This question has yet to be conclusively answered, but no gradient was detected in horizontal roots of maize and *Vicia faba* (Mertens and Weiler, 1983).

Inhibitors other than auxin have been suggested. One possibility is ethylene. If auxin does accumulate on the lower side of the root, it could stimulate the synthesis of ethylene. Concentrations of IAA that normally inhibit root growth will promote growth instead in the presence of inhibitors of ethylene synthesis. Another possible inhibitor is abscisic acid. ABA is present in root caps, will inhibit elongation if applied exogenously, and reportedly accumulates on the lower side of horizontally oriented roots. Wilkins (1984) presents a strong argument in favor of the view that gravitropism depends on ABA. However, this view has been challenged on several fronts (Moore and Evans, 1986). At physiological concentrations (0.1 μM), the initial effect of ABA is to *stimulate* root elongation. Inhibition is seen only at relatively high concentrations and prolonged exposure. In addition, roots of both mutant and fluridone-treated maize seedlings, which have no detectable ABA, exhibit normal gravitropic responses. At present, the role of hormones in root gravitropism is unclear although the case for auxin involvement is still viable.

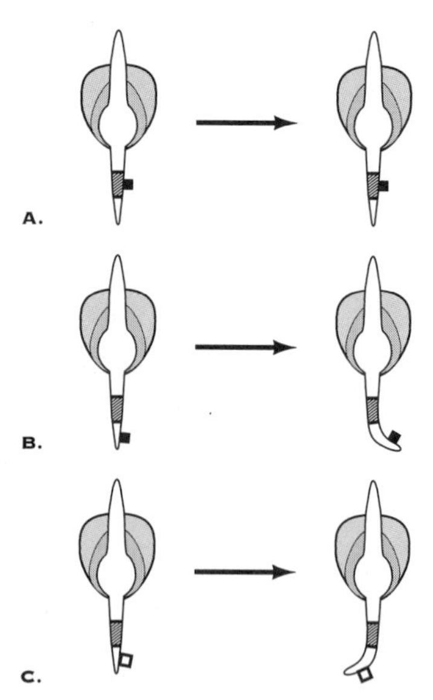

FIGURE 19.16 The response of maize primary root growth to calcium. (*A*) An agar block containing calcium placed on the side of the root in the elongation zone has no effect. (*B*) When the agar block containing calcium is applied to the root tip, the root grows toward the source of calcium. (*C*) An agar block containing EGTA, a calcium chelator, causes the root to grow in the opposite direction.

THE ROLE OF CALCIUM IN GRAVITROPISM

Several investigators have highlighted a role for calcium as a primary effector of both coleoptile and root gravicurvature (Moore and Evans, 1986). K. Goswami and L.J. Audus (1976) demonstrated a differential accumulation of radiolabeled calcium (^{45}Ca) in the *upper* half of gravistimulated sunflower hypocotyls and maize coleoptiles. Calcium redistribution, also observed in phototropically stimulated organs, occurred within one hour. A link between calcium and auxin was indicated as well. Calcium redistribution also occurred following asymmetric application of exogenous IAA and could be prevented in horizontal organs treated with the auxin transport inhibitor NPA. These results were confirmed for *Avena* coleoptiles by R. Slocum and S. Roux (1983) using a histochemical technique. By fixing tissue in a mixture containing potassium antimonate ($KSb(OH)_6$), unbound calcium rapidly forms calcium antimonate, which precipitates in place and can be visualized in tissue sections. Slocum and Roux found evidence of calcium localization in cells of the upper epidermis and underlying parenchyma within ten minutes of gravistimulation. Both calcium redistribution and gravitropism are prevented by prior treatment with EGTA, a chelator that ties up free calcium. In addition, the graviresponse of coleoptiles is prevented by treatment with chloropromazine, an inhibitor of the calcium-binding protein calmodulin. This suggests that the response to gravity might at one stage involve a Ca^{2+}/calmodulin complex.

Experiments with root gravitropism have provided equally convincing evidence of a role for calcium. Work from M. L. Evans' laboratory has shown that asymmetrically applied agar blocks containing 10 mM $CaCl_2$ will induce curvature of maize roots, but only if the block is applied to the root tip (Fig. 19.16). Migration of calcium in roots appears to be restricted to the root cap—migration is prevented if the cap is removed—and is directed toward the *lower* side of the root (Moore and Evans, 1986). Moreover, the calcium appears to move not through root cap cells but through the thick muscilaginous layer that coats the root cap. The importance of this mucilaginous coating is reflected in the observation that its continual removal by washing renders the root insensitive to gravity (Evans et al., 1986; Moore and Evans, 1986). Note that the direction of calcium asymmetry relative to auxin is opposite in gravistimulated coleoptiles and roots. In both organs *calcium migrates toward the potentially concave side.* Thus in a horizontally oriented root, calcium moves downward to accumulate on the lower side of the cap but it moves toward the upper side of the coleoptile.

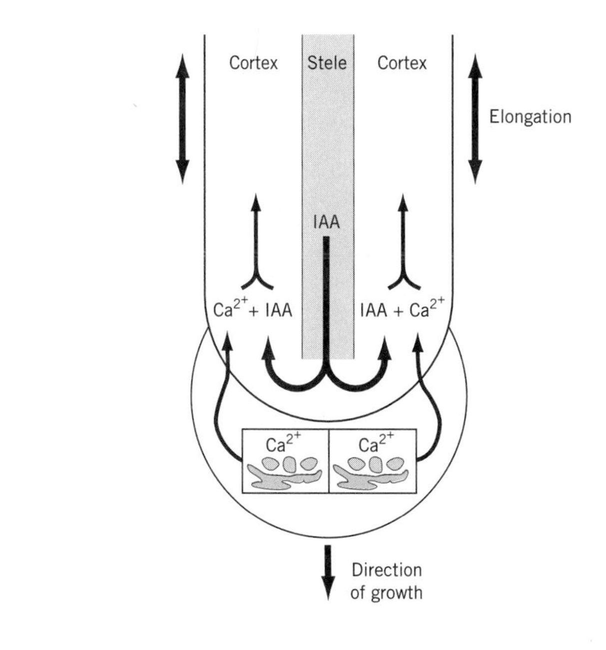

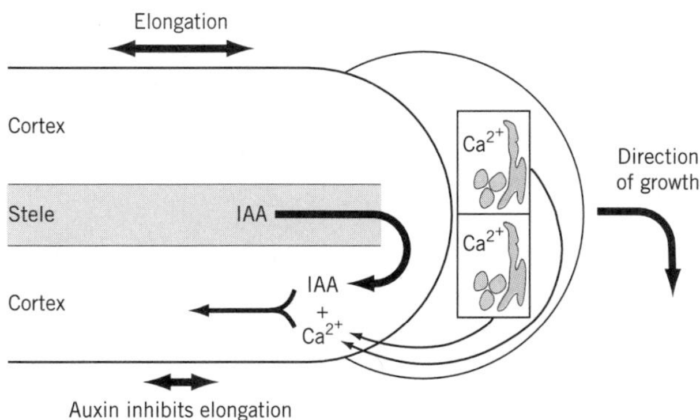

FIGURE 19.17 The Evans-Moore model for root gravitropism. Orientation of the root in a horizontal position causes a displacement of statoliths, which in turn stimulates a preferential flow of calcium toward the lower side of the root. High calcium concentration leads to a high auxin concentration and inhibition of cell elongation on the lower side.

Primarily on the basis of a gravity-induced redistribution of calcium, Moore and Evans have proposed a model to explain root gravitropism (Fig. 19.17). In vertically oriented roots, calcium is secreted by peripheral root cap cells into the surrounding mucilage. Secretion is uniform; there is no calcium gradient across the roots tips and the roots continue to grow downward. As well, there is a symmetrical flow of auxin from the cap into the root tip. When placed horizontally, an electrochemical gradient is established across the root. This electrochemical gradient is due at least in part to the secretion of protons (H^+) on the now upper side of the root. Horizontal orientation also induces an acropetal movement of calcium in the columella tissue. The source of this calcium is unknown, but it could be released from the ER due to a statolith-ER interaction. The electrochemical gradient is thought to induce an "electrophoresis" of calcium downward through the mucilage and backward along the lower side of the root. Thus calcium will accumulate along the lower side of the root cap and the elongation zone of the root.

How the accumulation of calcium on the lower side of the root is translated into a differential growth response is not clear, but there are two possibilities. The first is that *calcium acts as a sink for auxin*, thereby attracting auxin to the lower side of the root where the excessively high concentration inhibits elongation. This is in effect a refinement of the Cholodny-Went hypothesis; the auxin asymmetry is induced by the calcium asymmetry. As there is no evidence in support of an endogenous auxin asymmetry in roots, however, this possibility seems unlikely.

The second possibility is that *calcium sensitizes the tissue to auxin*. It is known that auxin does not inhibit elongation of calcium-depleted roots. Given this observation, it is not necessary to invoke a gravity-induced auxin asymmetry. Instead, the calcium asymmetry alone would give rise to differential auxin *activity*, which would in turn lead to the observed growth response.

Although partly speculative, the Moore-Evans model for root gravitropism is consistent with the reported absence of measurable auxin gradients in gravi-responding roots as well as a variety of other observations. On the other hand, a link between calcium and tissue sensitivity to auxin remains to be demonstrated by experiment. The Moore-Evans model will no doubt stimulate further research into the effector status of calcium gradients and the mechanism of gravitropism in roots.

PLANTS IN ZERO GRAVITY

Is gravitropism expressed in the microgravity environment of space flight? How do plants orient themselves when the gravitational stimulus is removed? Questions such as these are of particular interest to scientists concerned with developing plans for prolonged spaceflights of the future, where astronauts might be required to grow plants as part of the life support system or for a supply of fresh food.

In the microgravity of space, seedlings of jackpine (*Pinus sylvestris*) and wheat (*Triticum compactum*) maintain a normal, linear growth form—both the root and shoot grow as linear extensions of the embryonic radicle and coleoptile (Halstead and Dutcher, 1984, 1987). Tomato seedlings, however, grow in the shape of a horseshoe—maintaining the characteristic shape of their embryo. The growth of pea (*Pisum sativum*) seedlings in microgravity also reflects the arc shape of the pea embryo. In the absence of external gravitational stimulus, the orientation of seedlings appears to be determined primarily by the shape and position of the embryo in the seed. In other words, given no influence to the contrary, the seedling parts simply continue as they were. The only other determinant appears to be the substrate on which the seed is germinated. The seedling axis is maintained, regardless of its shape, on the plane of the substrate.

NASTIC MOVEMENTS

In addition to the directed movements of tropisms, many plants and plant parts, especially leaves, exhibit nastic movements, in which the direction of movement is not related to any vectorial component of the stimulus. Nastic responses may involve differential growth, in which case the movement is permanent. Alternatively the movement may be reversible, caused by turgor changes in a specialized motor organ.

Epinasty and thermonasty are examples of nastic responses involving differential growth. **Epinasty** is the downward bending of an organ, commonly petioles and leaves whose tips are inclined toward the ground. It is not a response to gravity, however, but appears to depend on an unequal flow of auxin through the upper and lower sides of the petiole. Epinasty is also a common response to ethylene or excessive amounts of auxin. The reverse response, called **hyponasty,** is less common but can be induced by gibberellins. A typical example of **thermonasty** is the repeated opening and closure of some flower petals, such as tulip and crocuses. In spite of their repeated nature, however, thermonastic movements are permanent and result from alternating differential growth on the two surfaces of the petals.

The most dramatic nastic movements are all turgor movements, which may be broadly separated into three categories: (1) the leisurely rhythmic sleep leaf movements in **nyctinastic** plants, (2) the very rapid **seismonastic** movements in a limited number of species, and (3) the **thigmonastic** or **thigmotropic** curling of threadlike appendages in climbing plants and vines. Nyctinastic and seismonastic responses depend on differential turgor movements in specialized motor organs, called the **pulvinus** (pl. = *pulvini*). The pulvinus is a bulbous structure most often encountered in plant families characterized by compound leaves, such as the Leguminoseae and Oxalidaceae (Figs. 19.18; 19.19). It occurs at the base of the petiole (*primary pulvinus*), the pinna (*secondary pulvinus*), or the pinnule (*tertiary pulvinus*). The pulvinus contains a number of large, thin-walled **motor cells,** which alter the position of the leaf by undergoing reversible changes in turgor.

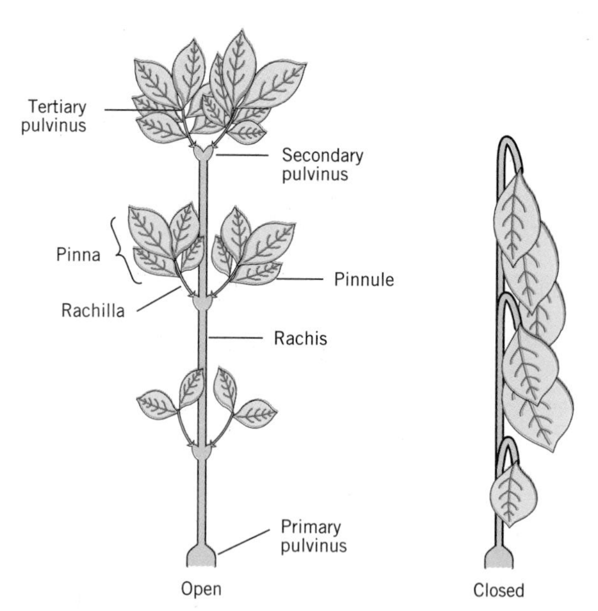

FIGURE 19.18 A leaf of *Samanea samanan* illustrating the location of the pulvini. Closed leaflets fold upward, parallel to the rachis. (Reproduced from the *Journal of General Physiology* 40:413–430, 1974, by copyright permission of the Rockefeller University Press.)

NYCTINASTY

Nyctinastic movements (Gr. *nyctos* = night + *nastos* = closure) are most evident in leaves that take up a different position in the night from that taken during the day. Typically leaves or leaflets are in the horizontal, or open, position during the day and assume a more vertical, or closed, orientation at night. The primary leaf of common bean plants exhibits particularly strong nyctinastic movements but this can also be seen in *Coleus*, prayer plants, and other common garden and house plants. Observations of nyctinastic movements can be traced back as far as the writings of Pliny in ancient Greece. The Swedish botanist C. Linnaeus (in 1775) coined the term "plant sleep" to describe nyctinastic movements and they are commonly referred to as sleep movements today (Sweeny, 1987).

Sleep movements have been studied by several eighteenth- and nineteenth-century botanists, including Darwin. The process, however, has been studied most extensively by Ruth Satter and her colleagues in *Samanea saman*, a member of the Leguminoseae with doubly compound leaves (Fig. 19.18). In *Samanea* the paired pinnae and pinnules are separated while in closing they fold toward each other. Also in *Samanea,* the paired organs fold basipetally (i.e., downward), but in other species, such as *Mimosa pudica* and *Albizzia julibrissin*, closure of paired pinnules is acropetal. The doubly compound leaves of *Mimosa*, *Albizzia*, and *Samanea* all have three pulvini, but the simple leaf of *Phaseolus* (bean) has only two. It is the secondary pulvinus that generally exhibits the more rapid or dramatic change and has consequently been studied most extensively. They are also relatively large (2 to 3 mm diameter, 4 to 7 mm long in

A.

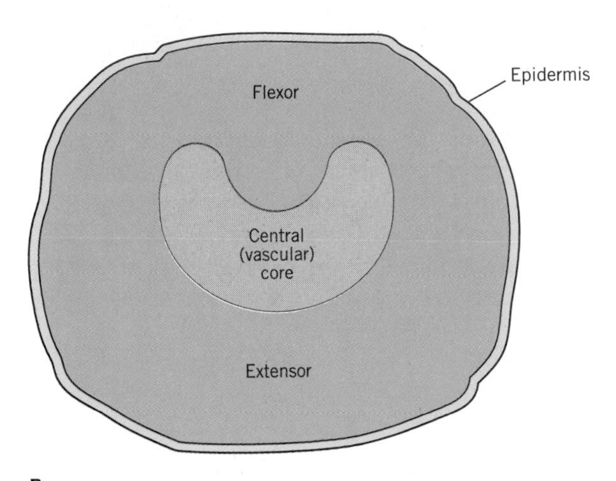

B.

FIGURE 19.19 The pulvinus. (*A*) Secondary pulvini of bean (*Phaseolus vulgaris*). The pulvini are the swollen areas at the juncture of the petiole with the stem. (B) Schematic diagram of a pulvinus in cross-section.

Samanea) and the changes in curvature are readily visible to the naked eye.

All nyctinastic responses depend on reversible turgor changes in the pulvinus. The pulvinus is typically cylindrical in shape with prominent transverse furrows, which facilitate bending, on the adaxial and abaxial sides (Fig. 19.19A). It contains a central vascular core with both xylem and phloem surrounded by sclerenchyma tissue. The vascular tissue assumes a linear arrangement as it passes through the pulvinus, apparently enhancing the flexibility of the pulvinar region. Outside the vascular core is a cortex comprised of 10 to 20 layers of parenchyma cells. The cells of the outer cortex have thin, elastic walls and exhibit large changes in size and shape during movement. These are called **motor cells.** It is the change in size and shape of the motor cells that is responsible for leaf movement.

The adaxial and abaxial sides of the pulvinus are also known as the **extensor** and **flexor** regions, respectively (Fig. 19.19B). In the extensor region, cells gain turgor during opening and lose turgor during closure. In the flexor region, turgor decreases during opening and increases during closure. The relative positions of extensor and flexor regions in the pulvinus will be reversed, depending on whether closure is basipetal or acropetal. Thus, the swelling of extensor motor cells and shrinkage of flexor motor cells straighten the pulvinus and open or spread apart leaflets.

Nyctinastic movements are sensitive to blue light, the physiological status of phytochrome, and endogenous rhythms. Although the mechanism of signal perception and how these three stimuli interact is not known, it is clear that both the receptors and the responding system (the motor cells) are located in the pulvinus—at most a few cells apart. It is known that phytochrome can "reset" the endogenous clock that regulates nyctinastic leaf oscillations. The subject of phytochrome and endogenous rhythms will be discussed in the next chapter.

Regardless of the nature of the stimulus, motor cell volume changes are due to osmotic water uptake (or loss) as a result of ion accumulation (or loss) across the cell membrane. What ions are exchanged, how are they transported, and what is the driving force for ion movement? These questions have been approached with a variety of techniques, including histochemical and radiochemical methods, and scanning electron microscopy coupled with X-ray analysis. Leaf movement in all nyctinastic plants studied thus far is associated with a massive redistribution of potassium ion (K^+) between the extensor and flexor regions of the pulvinus (Fig. 19.20). Swollen extensor cells are characterized by high protoplasmic K^+ and low apoplastic K^+. In *Phaseolus vulgaris*, fully 30 percent of the osmotic potential change can be accounted for by K^+ movement. The charge carried by K^+ is compensated primarily by chloride and possibly small organic anions such as malate and citrate.

K^+ exchange between the motor cells and the apoplast occurs through channels in the plasma membrane. Patch-clamp experiments on isolated *Samanea* motor cell protoplasts have shown that membrane polarity regulates the influx and efflux of K^+. There is also a pH gradient across the plasma membrane of motor cells that changes during movement. During upward movement of the bean leaf, for example, swelling of the extensor motor cells is accompanied by a decrease in pH of the

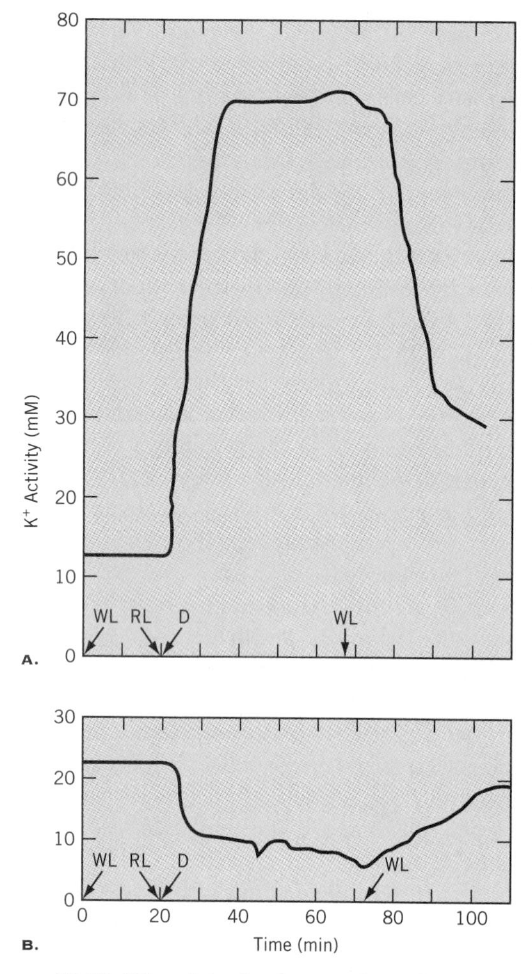

FIGURE 19.20 K^+ activity in the apoplast of extensor (upper curve) and flexor (lower curve) cells during closure of *Samanea* leaflets. (From C. Z. Lowen, R. L. Satter, Light promoted changes in the apoplastic K^+ activity in the *Samanea samanan* pulvinus, *Planta* 179:412–427, Fig. 1, 1989. Copyright Springer-Verlag.)

extensor apoplast by almost a full unit. Proton extrusion is almost certainly driven by a plasma membrane-based ATPase proton pump. Although quantitative relationships between H^+ and K^+ have not yet been tested, proton extrusion could establish the electrochemical gradient necessary to drive K^+ uptake.

The prominent roles of K^+ channels and H^+ pumps have been incorporated into the current model for motor cell movement (Satter and Galston, 1981; Morse et al., 1990). A simplified version of the model is shown in Figure 19.21. In this model, the light signal activates phytochrome (or cryptochrome), which accelerates **inositol phospholipid** turnover. Recent experiments have shown that light that stimulates opening of *Samanea* pulvini also decreases the level of phosphotidylinositol 4,5-bisphosphate (PIP_2) and increases the level of the second messenger inositol 1,4,5-triphosphate (IP_3) (Morse et al., 1990). There is a transient stimulation of **diacylglycerol (DAG)**. These changes are qualitatively similar but quantitatively smaller than those normally detected in animal systems. The assays, however, involved whole pulvini, which contain vascular, collenchyma, and epidermal tissues as well as the motor cells. The changes could be appropriately greater if restricted to the smaller population of motor cells. If inositol phospholipid metabolism functions in plants as it does in animals, DAG would be expected to activate a protein-kinase C (or its plant equivalent) to phosphorylate certain proteins. IP_3 would be expected to release free calcium—exogenous IP_3 *does* liberate calcium—although from which compartment is not known. Both the phosphorylated protein and/or transient increases in free calcium stimulate proton extrusion by activating the proton pump. The resulting electrochemical gradient energizes the uptake of K^+ and other ions, which in turn stimulates the osmotic uptake of water and motor cell

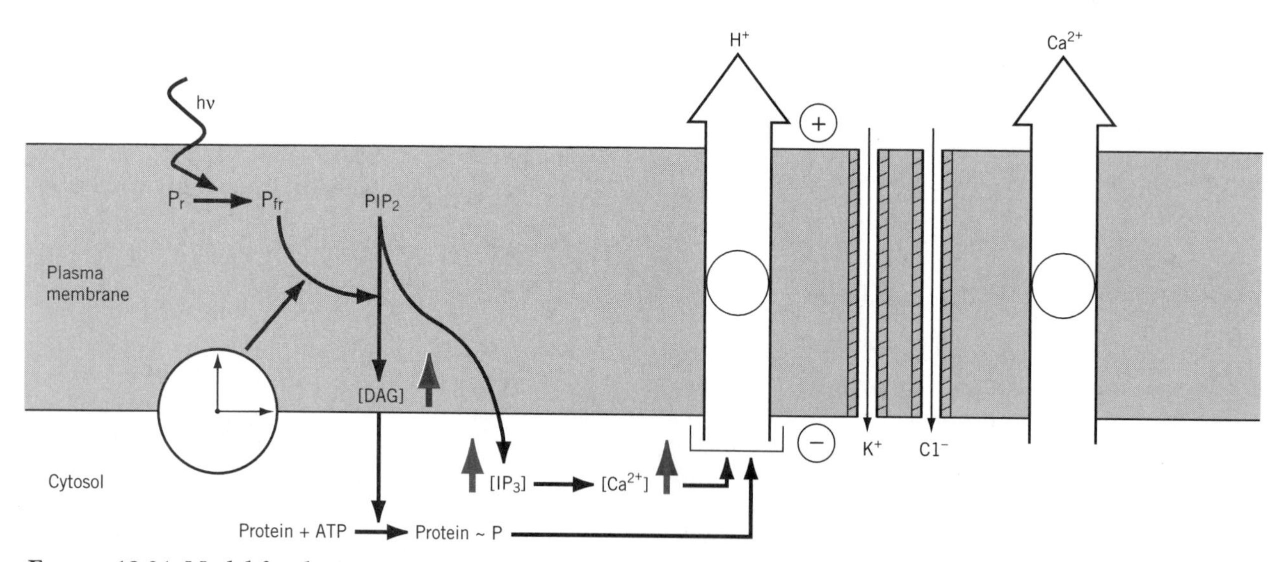

FIGURE 19.21 Model for the interaction of phytochrome, biological clocks and the inositol triphosphate system in leaf movements of nyctinastic plants. (See text for details.)

swelling. The presence of a calcium pump which extrudes Ca^{2+} would help to ensure the restoration of Ca^{2+} homeostasis.

Many details of this model remain to be described, especially the function of the inositol phospholipid cycle in plants. Still, plant cells are known to contain virtually all the required components and the model is consistent with what has been observed in pulvini thus far. Significant advances are to be expected in the future, especially now that patch-clamp techniques—long a mainstay of electrophysiology research for animal cells—can now be applied to plant cell protoplasts (see Box 5.1, Chap. 5). This state-of-the-art technique has been in use for plant cells only since about 1984, but has already proven invaluable for the study of ion channels.

SEISMONASTY

A limited number of leguminous plants that possess pulvini and exhibit nyctinastic movements also exhibit a response to mechanical stimulation. This phenomenon is known as **seismonasty.** Since seismonastic plants respond to touch, they are sometimes considered thigmonastic. However, seismonastic plants respond to a wider variety of stimuli including shaking or wind, falling raindrops, wounding by cutting, and intense heat or burning.

The best known example of seismonastic plants is the tropical shrub *Mimosa pudica* (Fig 19.22). The survival advantage of such a response is not certain. Some have suggested that since these plants grow in arid, exposed areas where they are exposed to drying winds, folding of the leaves may be a means of reducing water loss. Others suggest that it is a means of protection from large herbivores or insects. However, one thing is clear—the response is very rapid. When the pulvinus is stimulated directly, bending begins in less than one second!

The ultimate response, leaf movement, of course involves movement of pulvini motor cells just as in nyctinastic movements. However, there are three essential characteristics of the seismonastic response that have served to focus attention on the early steps of signal transduction. The first of these is the rapidity of the response. Second, seismonasty follows the "all or none principle." There is no obvious relationship between the intensity of the stimulus and the extent of the response. Third, excitation is propagated from the place of stimulation. The similarity of these characteristics to animal nerve transmission has given rise to the expectation that plants may also be capable of transmitting stimuli in the form of potential changes. Indeed it has now been well established that virtually any part of the *Mimosa* plant can perceive stimuli and transmit them as electric pulses

A.

B.

FIGURE 19.22 The seismonastic (sensitive) plant *Mimosa pudica* in the open (left) and closed (right) positions. The plant on the right was photographed about ten seconds after closure was stimulated by a sharp tap to the stem with a pencil.

to the pulvini (Fromm and Eschrich, 1990). Although plants do not have discrete nerve tissue, it appears that phloem sieve tubes can and do function as conduits for signal transmission. Stimulation of the petiole results in a rapid depolarization that is propagated basipetally along the sieve tube at a rate of about 2 cm s^{-1}. The unique structure of the sieve tube with its protoplasmic continuity through the sieve plates appears to be well suited for transmission of electrical signals. The appearance of the action potential is correlated with a rapid uptake of protons, suggesting that protons are responsible for the depolarization. Fromm and Eschrich believe that when the action potential reaches the pulvinus, it stimulates a rapid unloading of both K^+ and sugars into the apoplast. Water would follow and cause collapse of the motor cells.

Other investigators have found that substances isolated from phloem sap of Mimosa and other species will stimulate closure of Mimosa pulvini when applied to the cut end of the stem. The active substance has been identified as a glycosylated derivative of gallic acid (4-0-(β-D-gluco-pyranosyl-6′-sulphate)). Called "turgorin," this substance has been isolated from 14 higher plants that exhibit nyctinastic movements. Schildknecht and Meir-Augenstein (1990) suggest that turgorin may give rise to action potentials in a manner similar to the animal neurotransmitter, acetylcholine.

SUMMARY

The power of movement is generally perceived to be an animal trait, yet movement pervades the lives of green plants. Plant movements are important because they serve to orient the plant body in space. Thus roots exhibit positive gravitropism, growing down in order to mine the soil for mineral nutrients and water. Shoots exhibit negative gravitropism and positive phototropism in order to optimize the interception of sunlight for photosynthesis.

There are several categories of plant movements. Growth movements involve cell division and elongation and are consequently irreversible. Turgor movements involve changes in turgor pressure and cell volume, and are reversible. Tropisms are directionally related to the stimulus whereas the directionality of nastic movements are inherent in the tissue and are not related to any vector in the stimulus. Nutations are rotary or helical movements best observed with time-lapse photography. Since plant movements are in response to some form of external signal, they are conveniently dissected into three stages: signal perception, signal transduction, and response.

Under natural conditions, phototropism is a growth response to a light gradient, although in the laboratory it is usually studied by subjecting organs to unilateral light. Organs may either grow toward (positive phototropism) or away from (negative phototropism) the higher irradiance. Phototropism is a response to blue and UV-A light; signal perception therefore requires a pigment that absorbs those wavelengths. A comparison of action spectra with the absorption spectra of known pigments present in grass coleoptiles has indicated either carotenoids or flavins might serve as photoreceptors for phototropism. However, on the basis of experiments with carotenoid-deficient plants and phototropic mutants of *Arabidopsis*, it appears likely that the photoreceptor for phototropism is a flavoprotein located in the plasma membrane.

The phototropic response is characterized by differential growth on the lighted and shaded sides of the responding organ. The most generally accepted theory to account for differential growth in coleoptiles and stems is the Choladny-Went theory. This theory proposes a lateral redistribution of auxin as it flows basipetally from the apex where it is synthesized. In the case of positive phototropism, the higher concentration of auxin flows down the shaded side of the organ, causing cells on the shaded side to elongate more rapidly than those on the lighted side.

Unlike most stimuli to which plants are exposed, gravity is omnipresent and nonvarying. There is no gravitational gradient. Gravity can be sensed only by movement of cellular structures (statoliths), which then establishes an asymmetry that is translated in terms of pressure.

Sensitivity to gravity in the root is localized in the root cap, particularly in the columella, a mass of nonvacuolated cells in the central core of the root cap. Although there is no direct proof, the weight of evidence indicates that statoliths are the starch-containing plastids, amyloplasts. The primary transducer that senses the pressure of the statoliths and initiates the signal transduction chain remains unknown. Some evidence suggests it might be the endoplasmic reticulum. Another theory proposes that the pressure-sensing transducer lies at the plasma membrane-cell wall interface and responds to the weight of the protoplast as a whole. As in phototropism, the gravitropic response involves differential growth that can be explained by redistribution of auxin transport. The steps between pressure sensing and auxin redistribution are unknown, but several experiments indicate that calcium ion plays a major role.

Plants exhibit a variety of nastic responses. One of the most prominent is the periodic movement of leaves known as sleep movements, or nyctinasty. Leaf movement is mediated by turgor changes in specialized motor cells located in structures called pulvini, found at the distal end of the petiole. Turgor changes are mediated by a flux of potassium ion induced by an interaction between phytochrome, biological clocks and the inositol triphosphate system. Another nastic response is illustrated by seismonasty in the sensitive plant *Mimosa*. Seismonasty involves similar turgor changes in response to physical disturbance.

CHAPTER REVIEW

1. Define phototropism. Is phototropism restricted to unilateral lighting? In what way(s) can a light gradient be established across a plant organ such as a stem?
2. What pigment(s) function as the photoreceptor for phototropism? List the evidence that supports your conclusion.
3. Review the experimental basis for the Cholodny-Went hypothesis. What arguments can be raised against this hypothesis?
4. Shoots and roots express various levels of gravitropic response. What are the physiological advantages to be gained by such variation in response?
5. Review the statolith theory for gravitropism as it applies to roots. How is the gravitational stimulus perceived by a root and how does it respond? What is the evidence that calcium is involved in root gravitropism?
6. Describe nyctinasty. What might be the physiological significance or survival value of nyctinasty? In what ways is the seismonastic response similar to nyctinasty? In what ways is it different?

FURTHER READING

Briggs, W. R., T. I. Baskin. 1988. Phototropism in higher plants—controversies and caveats. *Botanica Acta* 101:133–139.

Brock, T. G., P. B. Kaufman. 1990. Movement in grass shoots. In: R. L. Satter, H. I. Gorton, T. C. Vogelmann, (eds.). *The Pulvinus: Motor Organ for Leaf Movement.* Rockville, Md: American Society of Plant Physiologists, pp. 59–71.

Brown, A. H. 1993. Circumnutations: From Darwin to space flights. *Plant Physiology* 101:345–348.

Darwin, C. 1881. *The Power of Movement in Plants.* New York: Appleton-Century-Crofts.

Firn, R. D. 1986. Phototropism. In: R. E. Kendrick, G. H. M. Kronnenberg (eds.), *Photomorphogenesis in Plants.* Dordrecht: Martinus Nijhoff Publishers, pp. 369–390.

Fukaki, H., H. Fujisawa, M. Tasaka. 1996. How do plant shoots grow up? The initial step to elucidate the molecular mechanisms of shoot gravitropism using *Arabidopsis thaliana*. *Journal of Plant Research* 109:129–137.

Kaufman, P. B., L.-L. Wu, T. G. Brock, D. Kim. 1995. Hormones and the orientation of growth. In: P. J. Davies (ed.), *Plant Hormones.* Dordrecht: Kluwer Academic Publishers, pp. 547–571.

Satter, R. L., H. I. Gorton, T. C. Vogelmann. 1990. *The Pulvinus: Motor Organ for Leaf Movement.* Rockville, Md.: American Society of Plant Physiologists.

REFERENCES

Audus, L. J. 1969. Geotropism. In: M. B. Wilkins (ed.), *Physiology of Plant Growth and Development.* London: McGraw-Hill, pp. 205–242.

Bandurski, R. S., A. Schulze, P. Dayanandan, P. B. Kaufman. 1984. Response to gravity by *Zea mays* seedlings. I. Time course of the response. *Plant Physiology* 74:284–288.

Baskin, T. I., Briggs, W. R., Iino, M. 1986. Can lateral redistribution of auxin account for phototropism of maize coleoptiles? *Plant Physiology* 81:306–309.

Briggs, W. R. 1963. Mediation of phototropic responses of corn coleoptiles by lateral transport of auxin. *Plant Physiology* 38:237–247.

Briggs, W. R., E. Liscum. 1997. The role of mutants in the search for the photoreceptor for phototropism in higher plants. *Plant, Cell and Environment.* 20:768–772.

Brauner, L., A. Hager. 1958. Versuche zur Analyse der geotropischen perzeption. I. *Planta* 51:115–147.

Casper, T., B. G. Pickard. 1989. Gravitropism by a starchless mutant of *Arabidopsis*: implications for the starch-statolith theory of gravity sensing. *Planta* 177:187–197.

Digby, J., R. D. Firn. 1979. An analysis of the changes in growth rate occurring during the initial stages of geocurvature in shoots. *Plant, Cell and Environment* 2:145–148.

Evans, M. L., R. Moore, K. H. Hasenstein. 1986. How roots respond to gravity. *Scientific American* 255(6):112–119.

Firn, R. D., J. Digby. 1980. The establishment of tropic curvatures in plants. *Annual Review of Plant Physiology* 31:131–148.

Fromm, J., W. Eschrich. 1990. Seismonastic movements in Mimosa. In: R. L. Satter, H. L. Gorton, T. C. Vogelmann, (eds.) *The Pulvinus: Motor Organ for Leaf Movement.* Rockville, Md: American Society of Plant Physiologists, pp. 25–43.

Galston, A. W. 1959. Phototropism of stems, roots, and coleoptiles. In: W. Ruhland (ed.), *Handbuch der Pflanzenphysiologie*, pp. 492–529. Berlin: Springer-Verlag.

Gillespie, B., K. V. Thimann. 1961. The lateral transport of indoleacetic acid-C^{14} in geotropism. *Experimentia* 17:126–129.

Goswami, K. K. A., L. J. Audus. 1976. Distribution of calcium, potassium and phosphorous in *Helianthus annuus* hypocotyls and *Zea mays* coleoptiles in relation to tropic stimulation and curvature. *Ann. Bot.* 40:49–64.

Halstead, T. W., F. R. Dutcher. 1984. Experiments on plants grown in space: Status and prospects. *Ann. Bot.* 54(Suppl. 3):3–18.

Halstead, T. W., F. R. Dutcher. 1987. Plants in space. *Ann. Rev. Plant Physiol.* 38:317–345.

Hertal, R., R. K. de la Feunte, A. C. Leopold. 1969. Geotropism and the lateral transport of auxin in the corn mutant amylomaize. *Planta* 88:204–214.

Iversen, T. H. 1969. Elimination of geotropic responsiveness in roots of cress (*Lepidium sativum*) by removal of statolith starch. *Physiologia Plantarum* 22:1251–1262.

Iwami, S., Y. Masuda. 1974. Geotropic response of cucumber hypocotyls. *Plant Cell Physiol.* 15:121–129.

Kaufman, P. B., I. Song. 1987. Hormones and the orientation of growth. In: P. J. Davies, *Plant Hormones and Their Role in Plant Growth and Development.* Dordrecht: Kluwer Academic Publishers, pp. 375–392.

Kiss, J. Z., M. M. Guisinger, A. J. Miller, K. S. Stackhouse. 1997. Reduced gravitropism in hypocotyls of starch-deficient mutants of *Arabidopsis. Plant and Cell Physiology* 38:518–525.

Liscum, E., W. R. Briggs. 1995. Mutations in the *NPH1* locus disrupt the perception of phototropic stimuli. *Plant Cell* 7:473–485.

Macleod, K., J. Digby, D. B. Firn. 1985. Evidence inconsistent with the Blaauw model of phototropism. *Journal of Experimental Botany* 36:312–319.

Mertens, R., E. W. Weiler. 1983. Kinetic studies on the redistribution of endogenous growth regulators in gravireacting plant organs. *Planta.* 158:339–348.

Miles, C. D. 1981. The relationship between geotropic response and statolith in a nuclear mutant maize. In D. J. Carr (ed.), *Abstracts of the XIII International Botanical Congress,* Sydney, Australia, p. 250.

Moore, R., M. L. Evans. 1986. How roots perceive and respond to gravity. *American Journal of Botany* 73:574–587.

Morse, M. J., R. C. Crain, G. G. Cote, R. L. Satter. 1990. Light-signal transduction via accelerated inositol phospholipid turnover in *Samanea* pulvini. In: D. J. Morre, W. F. Boss, F. A. Loewus (eds.), *Inositol Metabolism in Plants.* New York: Wiley-Liss, pp. 201–215.

Palmer, J. M., K. M. Warpeha, W. R. Briggs. 1996. Evidence that zeaxanthin is not the photoreceptor for phototropism in maize coleoptiles. *Plant Physiology* 110:1323–1338.

Phillips, I. D. J., W. Hartung, 1976. Longitudinal and lateral transport of [3,4-^{3}H]gibberellin A_1 and 3-indole(acetic acid-2-^{14}C) in upright and geotropically responding green internode segments from *Helianthus annus. New Phytol.* 76:1–9.

Pickard, B. 1973. Geotropic response patterns of the *Avena* coleoptile. I. Dependence on angle and duration of stimulation. *Can. J. Bot.* 51:1003–1021.

Pickard, B. 1985. Early events in geotropism of seedling shoots. *Ann. Rev. Plant Physiol.* 36:55–75.

Salisbury, F. B., L. Gillespie, P. Rorabaugh. 1988. Gravitropism in higher plant shoots. V. Changing sensitivity to auxin. *Plant Physiology* 88:1186–1194.

Satter, R. L., A. W. Galston. 1981. Mechanisms of control of leaf movement. *Ann. Rev. Plant Physiology* 32:83–110.

Schildknecht, H., W. Meir-Augenstein. 1990. Role of turgorins in leaf movement. In: R. L. Satter, H. L. Gorton, T. C. Vogelmann (eds.), *The Pulvinus: Motor Organ for Leaf Movement.* Rockville, Md: American Society of Plant Physiologists, pp. 205–213.

Shen-Miller, J., R. Hinchman, S. A. Gordon. 1968. Thresholds for georesponse to acceleration in gravity-compensated *Avena* seedlings. *Plant Physiology* 43:338–344.

Slocum, R. D., S. J. Roux. 1983. Cellular and subcellular localization of calcium in gravistimulated oat coleoptiles and its possible significance in the establishment of tropic curvature. *Planta* 157:481–492.

Sweeny, B. 1987. *Rhythmic Phenomena in Plants,* 2nd ed. New York: Academic Press.

Trewavas, A. 1981. How do plant growth substances work? *Plant Cell Environ.* 4:203–228.

Volkmann, D., A. Sievers. 1979. Graviperception in multicellular organs. In: W. Haupt and M. E. Feinleib (eds.), *Physiology of Movements, Encyclopedia of Plant Physiology,* NS. Vol. 7, pp. 573–600. Berlin: Springer-Verlag.

Wayne, R., M. P. Staves. 1997. A down to earth model of gravisensing or Newton's Law of Gravitation from the apple's perspective. *Physiologia Plantarum* 98:917–921.

Weevers, T. 1949. *Fifty Years of Plant Physiology.* Waltham, Mass: Chronica Botanica.

Wilkins, M. B. 1984. Gravitropism. In: M. B. Wilkins (ed.), *Advanced Plant Physiology.* London: Pitman, pp. 163–185.

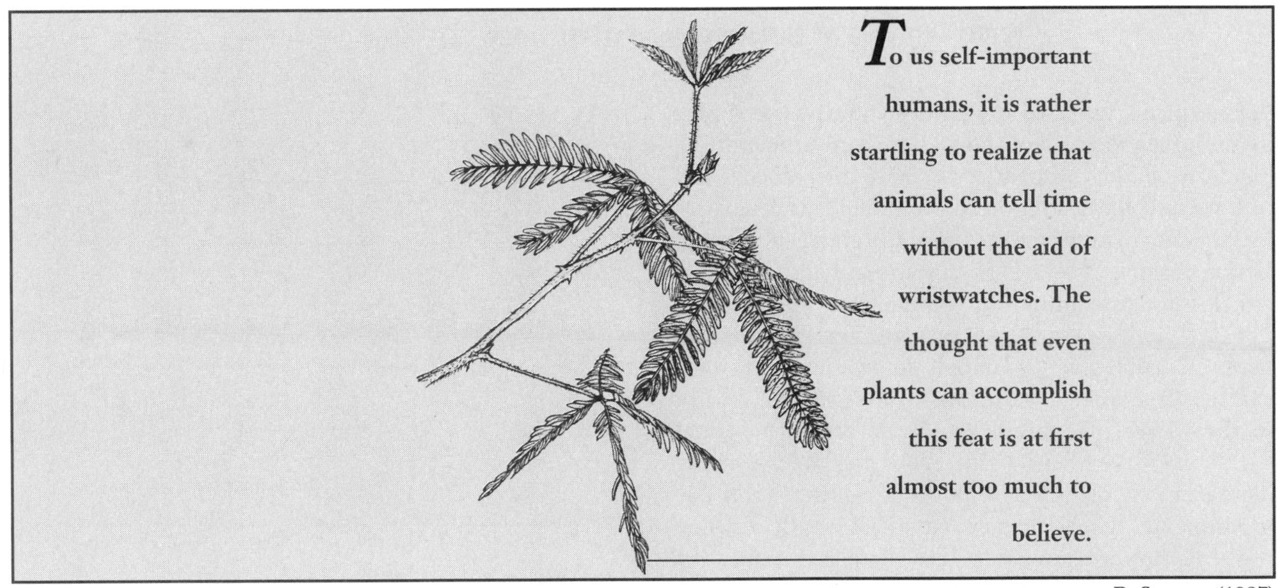

B. Sweeny (1987)
Rhythmic Phenomena in Plants

20

Measuring Time: Photoperiodism and Rhythmic Phenomena

It may indeed be difficult for the layman to believe that plants can tell time, but many aspects of plant behavior can be interpreted in no other way. One example is the consistent flowering of various species at particular times of the year. Roses always bloom in the summer and chrysanthemums in the fall. Indeed, the flowering of some plants is so predictable from one year to the next that gardeners have for centuries incorporated them into their gardens as floral calendars, unerringly marking the progress of the seasons. In the northern latitudes, perennial plants sense the short days of autumn as a signal to induce bud dormancy, thus anticipating the unfavorable conditions of winter. It is the length of day that gives the most reliable indication of the advancing season and an organism's capacity to measure daylength is known as **photoperiodism.** But photoperiodism is only one of the more outward manifestations of a far more fundamental timekeeping mechanism, known as the **biological clock,** that is reflected in the daily rhythmicity of nyctinastic leaf movements.

In this chapter we will examine

- photoperiodism; including the distinction between short day plants, long day plants, and other response types; the central role of the dark period; the nature of photoperiodic perception; and a discussion of the elusive floral hormone;
- timing of biological processes by the internal biological clock;
- time measurement in photoperiodism, including the hour-glass hypothesis and the role of the biological clock
- vernalization—the low temperature requirement for flowering in winter annuals and biennial plants; and
- a brief discussion of the significance of photoperiodism in nature.

PHOTOPERIODISM

The study of photoperiodism has focused above all on the switch from the vegetative state to the flowering state—arguably one of the most dramatic and mysterious events in the life of a flowering plant. Photoperiodism influences many other aspects of plant development such as tuber development, leaf fall and dormancy, but it is flowering that has attracted the major share of interest.

Although it had earlier been suggested that latitudinal variations in daylength contributed to plant distribution, the first efforts at controlled experimentation were conducted by a French scientist in 1912. J. Tournois found that both *Humulus* (hops) and *Cannabis* (hemp) plants flowered precociously during the winter in the greenhouse. Tournois eliminated temperature, humidity, and light intensity as environmental cues and in 1914 concluded that the shortening of daylength or lengthening of night was responsible for early flowering. Unfortunately, World War I intervened and Tournois did not live to continue his experiments. At the same time, H. Klebs was studying *Sempervivum funkii* of the family Crassulaceae (commonly known as "hens-and-chickens"). *Sempervivum* grows as a vegetative rosette in the winter greenhouse. By supplementing normal daylight with artificial light, Klebs was able to stimulate stem elongation and induce flowering. From his experiments, Klebs concluded that length of day triggered flowering in nature. However, it remained for W. W. Garner and H. A. Allard to demonstrate the full impact of daylength on flowering and coin the term **photoperiodism.**

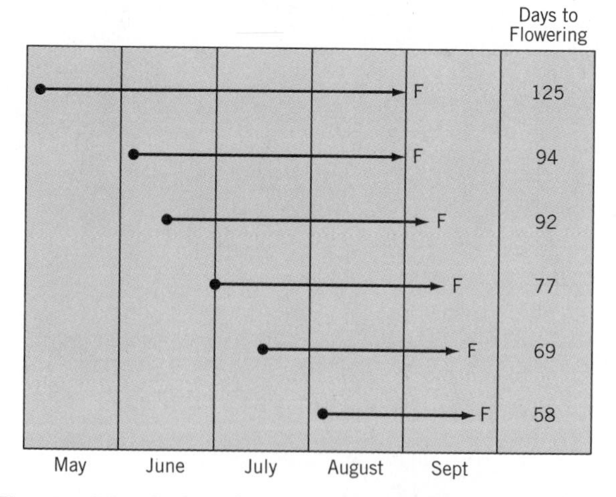

FIGURE 20.1 September soybeans. Soybeans (*Glycine max*, cv. Biloxi) sown over a three-month period all flower within a three-week period in September.

GIANT TOBACCO AND SEPTEMBER SOYBEANS

W. W. Garner and H. A. Allard were scientists with the United States Department of Agriculture near Washington, D.C. The initial focus of their work was a mutant cultivar of tobacco (*Nicotiana tabacum*), called Maryland Mammoth. In the field, Maryland Mammoth plants grew to be very tall with large leaves. Such characteristics would obviously be advantageous to the tobacco industry at the time (in the early 1920s), but breeding efforts were frustrated by the fact that the plants would not flower in the field during the normal growing season at that latitude. In the greenhouse, however, even very small plants flowered in the winter and early spring. Clearly, flowering was not simply a matter of the age of the plants. Another problem of interest to Garner and Allard concerned flowering in soybean (*Glycine max*). When the cultivar Biloxi was sown over a three-*month* period from May to August, all of the plants flowered within a three-*week* period in September (Fig. 20.1). The earliest seeded plants thus took 125 days to flower while those seeded last required only 58 days. Again it appeared that all plants, regardless of age, were simply awaiting some signal to initiate flowering.

Like Tournois, Garner and Allard eliminated a variety of environmental conditions (such as nutrition, temperature, and light intensity) as the "signal," coming finally (and with some reluctance) to the conclusion that flowering was controlled by the relative length of day and night. Using a crude but effective system of rolling plant benches in and out of darkened garagelike buildings at predetermined times, Garner and Allard proceeded to describe the flowering characteristics of scores of different species with respect to daylength (Garner and Allard 1920, 1923). They went on to suggest that bird migration might also be keyed to daylength—a phenomenon that is now well documented. We now know that photoperiodic control is not limited to flowering, but is a basic regulatory component in many aspects of plant and animal behavior.

PHOTOPERIODIC RESPONSE TYPES

Photoperiodic responses generally fall into three fundamental categories, of which the first two are the most commonly studied. They are: **short-day plants** (SD plants), **long-day plants** (LD plants), and daylength-indifferent or **day-neutral plants** (DNP) (Table 20.1 and Fig. 20.2). Short-day plants are those that flower only, or flower earlier, in response to daylengths that are *shorter* than a certain value *within a 24-hour cycle.* Long-day plants correspondingly respond to daylengths that are *longer* than a certain value while day-neutral plants flower irrespective of daylength. Within the long- and short-day categories, we also recognize **qualitative** (or obligate) and **quantitative** (or facultative) requirements. Plants that have an absolute requirement for a particular photoperiod before they will flower are considered qualitative photoperiodic types. The common cocklebur (*Xanthium strumarium*), for example, is a *qualitative short-day plant.* This means that *Xanthium* will not flower unless it receives an appropriate short photoperiod. On the other hand, most spring cereals such as wheat (*Triticum* sp.) and rye (*Secale cereale*) are *quantitative long-day plants.* Although spring cereals will eventually flower even if maintained under continuous short days, flowering is dramatically accelerated under long days. However, the distinction between qualitative and

TABLE 20.1 Representative plants exhibiting the principal photoperiodic response types.

Short-Day Plants	
Chenopodium rubrum	red goosefoot
Chrysanthemum sp.	chrysanthemum
Cosmos sulphureus	yellow cosmos
Euphorbia pulcherrima	poinsettia
Glycine max	soybean
Nicotiana tabacum	tobacco (Maryland Mammoth)
Perilla crispa	purple perilla
Pharbitis nil	Japanese morning glory
Xanthium strumarium	cocklebur
Long-Day Plants	
Anethum graveolens	dill
Beta vulgaris	Swiss chard
Hyoscyamus niger	black henbane
Lolium sp.	rye grass
Raphanus sativus	radish
Secale cereale	spring rye
Sinapis alba	white mustard
Spinacea oleracea	spinach
Triticum aestivum	spring wheat
Day-Neutral Plants	
Cucumis sativus	cucumber
Gomphrena globosa	globe amaranth
Helianthus annuus	sunflower
Phaseolus vulgarus	common bean
Pisum sativum	garden pea
Zea mays	corn

quantitative response is not always hard and fast for a particular species or cultivar. Photoperiod requirement is often modified by external conditions such as temperature. A plant may, for example, have a qualitative requirement at one temperature but respond quantitatively at another temperature.

In addition to these three basic categories, there are a number of other response types that respond to long and short days in some combination or in which the response is modified by some environmental condition, usually temperature (Salisbury, 1963; Vince-Prue, 1975). Various species of the genus *Bryophyllum* are, for example, **long-short-day plants** (LSD plant)—they will flower only if a certain number of short days are preceded by a certain number of long days. The reverse is true of the **short-long-day plant** (SLD plant) *Trifolium repens* (white clover). Some plants, such as winter cereals, require a low temperature treatment before they become responsive to photoperiod while others may have a qualitative photoperiodic requirement at one temperature but only a quantitative requirement at another temperature. A few plants have highly specialized requirements. **Intermediate-daylength plants,** for example, flower only in response to daylengths of intermediate length but remain vegetative when the day is either too long or too short. Another type of behavior is **amphophotoperiodism,** illustrated by *Madia elegans* (tarweed). In this case, flowering is delayed under intermediate daylength (12 to 14 hours) but occurs rapidly under daylengths of 8 hours or 18 hours.

There are many, often subtle, variations to the three basic response types, encompassing a large number of flowering plants. A complete listing is beyond the scope of this book, but both Salisbury (1963) and Thomas and Vince-Prue (1997) provide extensive lists of plants with their known photoperiodic characteristics. However, most of what is known about the physiology of photoperiodism in plants has been learned from a relatively small number of short-day and long-day plants.

CRITICAL DAYLENGTH

It is important to understand that the distinction between SD plant and LD plant is not based on the absolute length of day. Consider, for example, that both *Xanthium* and *Hyoscyamus niger* (black henbane) will flower with 12 to 15 hours of light per day (Fig. 20.3). Yet the former is properly classified as a SD plant and the latter as a LD plant. Whether a plant is classified as a SD plant or LD plant depends on its behavior relative to a **critical daylength.** Plants that flower when the daylength is *shorter* than some critical maximum are classified as SD plant. Those that flower in response to daylengths longer than a critical minimum are classified as LD plant. Thus the critical daylength for the SD plant *Xanthium* is 15.5 hours, meaning that it will flower whenever the daylength is *less than* 15.5 hours out of every 24. The critical daylength for the LD plant *Hyoscyamus* is 11 hours and it will flower when the daylength exceeds that value. Clearly the actual daylength under which a plant will flower is, in the absence of further information, no indication of its response type. A corollary to this observation is that, although SD plants tend to flower in the spring and fall and LD plants tend to flower in midsummer, the classification as a SD plant or LD plant is not necessarily an indication of the time of year that species will flower.

Pharbitis nil

SHORT DAYS

A.

B.

LONG DAYS

C.

D.

FIGURE 20.2 Flowering response of a SD plant (*Pharbitis nil*) and a LD plant (*Hyoscyamus niger*) to short days and long days. Note the prominent flowers (arrows) in *Pharbitis* under short days and *Hyoscyamus* under long days. Note also that *Hyoscyamus* remains as a rosette under short days. Plants of each species under both photoperiod regimes are of the same age. (See color plate 12.)

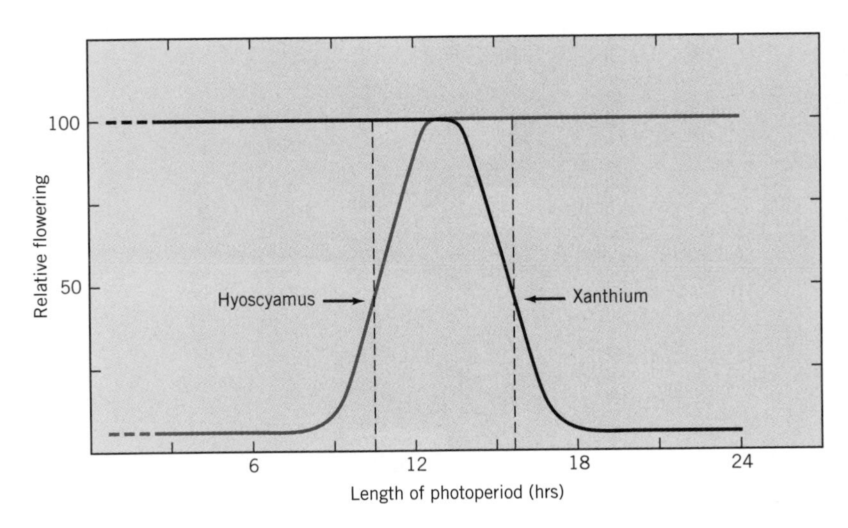

FIGURE 20.3 A diagram to illustrate the concept of critical daylength in populations of *Xanthium strumarium* (cocklebar), a short day plant, and in *Hyoscyamus niger* (black henbane), a long day plant. Critical day lengths are indicated by the vertical dotted lines. Note that *Xanthium* flowers when the daylength is *shorter* than its critical daylength and *Hyoscyamus* flowers when the daylength is *longer* than its critical daylength.

PHOTOPERIODIC INDUCTION

Many plants require more or less continuous exposure to the appropriate photoperiod, at least until floral primordia have been developed, in order to flower successfully. Others will proceed to flower even if, once exposed to even a single proper photoperiod, the plant is returned to unfavorable photoperiods. Such plants are said to be **induced** and the appropriate photoperiod is referred to as an **inductive** treatment. The phenomenon of induction raises intriguing, though unresolved, questions about the physiological properties of the induced state. Clearly a physiological change has taken place in induced plants and this change persists, even though no anatomical or morphological change is evident at the apex where flowers will appear. Induction can also be experimentally useful. One of the reasons *Xanthium* has been so widely used for studies of photoperiodism is that a single short-day cycle will irreversibly lead to flowering, even in plants that are returned to long days. Such an extreme sensitivity to induction is not widespread, but it has been demonstrated in other SD plants such as Japanese morning glory (*Pharbitis nil*), duckweed (*Lemna purpusilla*), and pigweed (*Chenopodium rubrum*), and in the LD plants dill (*Anethum graveolens*) and rye grass (*Lolim temulentum*).

Induction is not an all-or-none process, but can be achieved in degrees. Although *Xanthium* will respond to a single inductive cycle, the initiation of floral primordia is more rapid and more prolific if multiple cycles are given. Other plants may exhibit fractional induction—a summation of inductive photoperiods despite interruption with noninductive cycles. *Plantago lanceolata* (plantain), for example, normally requires a threshold of about 25 long days to induce flowering. Plants given only ten long days will remain vegetative, but a schedule of 10 long days followed by 10 short days and then another 15 long days will induce flowering. The plants are able to sum the long days in spite of the intervening short days. Examples of summation with up to thirty intervening noninductive periods have been reported (Vince-Prue, 1975).

THE CENTRAL ROLE OF THE DARK PERIOD

In their original publications, Garner and Allard suggested that plants responded to the relative lengths of day and night. The term photoperiodism, which combines the Greek roots for *light* and *duration*, turns out to be misleading because it implies that plants measure the duration of daylight. In fact, plants measure neither the relative length of day and night nor the length of the photoperiod—they measure the length of the dark period. This was elegantly demonstrated by the experiments of K. C. Hamner and J. Bonner in 1938 (Fig. 20.4). Under 24-hour cycles of light and dark, *Xanthium* flowered with dark periods longer than 8.5 hours but remained vegetative on schedules of 16 hours light and 8 hours darkness (Fig 20.4A, B). On schedules of 4 hours light–8 hours darkness, plants remained vegetative even though the 4-hour photoperiod is much shorter than the 15.5-hour critical photoperiod (Fig 20.4C). On the other hand, schedules of 16 hours light—32 hours darkness induced rapid flowering even though the photoperiod exceeded the critical daylength (Fig. 20.4D).

The above results allow two conclusions. First, the relative length of day and night is not the determining factor in photoperiodism, since the ratio of light to dark is the same in schedules B, C, and D (Fig. 20.4) but with different results. Second, it is the length of the dark period that is important. The consistent feature of these experiments is that *Xanthium* will flower whenever the dark period exceeds 8.5 hours and will remain vegetative whenever the dark period is less than 8.5 hours. Hamner and Bonner confirmed the critical role of the dark pe-

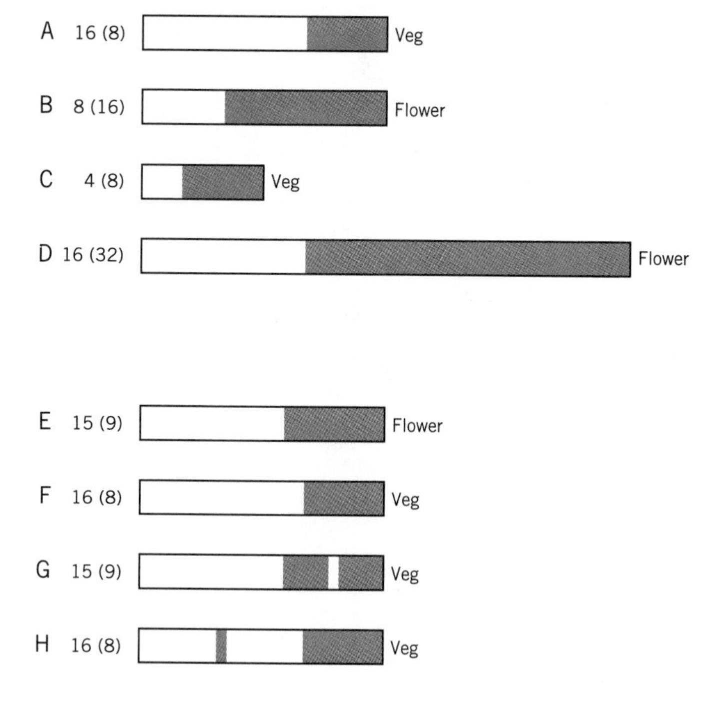

FIGURE 20.4 The central role of dark period in *Xanthium strumarium*, a SD plant. The photoperiod regime is shown to the left. The number enclosed in brackets indicates the length of the dark period. Note that the plants flower whenever the dark period is uninterrupted for nine hours or more. (The diagram represents the experiments of K. C. Hamner and J. Bonner, 1938.)

riod by interrupting the dark period with brief light exposures (Fig. 20.4 E–H). The flowering effect of an inductive 9-hour dark period can be nullified by interrupting the dark period with a brief light-break (Fig. 20.4G), but a "dark interruption" of a long light period has no effect (Fig. 20.4H). Experiments with LD plant give similar results; LD plants require a dark period shorter than some critical maximum. With LD plants, a light-break in the middle of an otherwise noninductive long dark period will shorten the dark period to less than the maximum and permit flowering to occur. At this point it is clear that photoperiodism has relatively little to do with daylength per se. Rather it is a *response to the duration and timing of light and dark periods* (see Hillman, 1962). Thus, the critical daylength for a SD plant actually represents the maximum length of day in a normal 24-hour regime that will allow a dark period of sufficient length. In the case of LD plants, long dark periods are inhibitory and the critical daylength is the minimum in a 24-hour regime that will keep the dark period short enough to allow flowering.

The fluence given during a light-break need not be very high to be effective. As little as one minute of incandescent light at a low fluence will prevent flowering in *Xanthium*. Even bright moonlight is sufficient to delay flowering in some SD plants. This raises an interesting possible relationship between nyctinastic leaf movements, discussed in the previous chapter, and control of flowering. It is possible, at least in some cases, that nyctinastic leaf movements could serve to reorient the leaves parallel to incident moonlight and thus reduce its impact on the time-sensing mechanism.

PERCEPTION OF THE PHOTOPERIODIC SIGNAL

What part of the plant perceives the photoperiodic signal? The actual change from the vegetative to reproductive growth occurs, of course, in meristematic areas—usually beginning at the shoot apex and appearing later in the axillary buds. Contrary to expectations, however, the photoperiodic signal is perceived not by the stem apex but by the leaf. This has been demonstrated in a variety of ways. Some of the earliest experiments were conducted by the Russian physiologist M. Chailakhyan. He reported flowering in *Chrysanthemum morifolium*, a SD plant, in which the apical, defoliated portion was kept on long days but the leafy portion was subjected to short days. When conditions were reversed—the upper, defoliated portion kept on short days and the leafy portion on long days—the plants remained vegetative (Fig. 20.5). Although the plants in this experiment still contained axillary buds, flowering has been successfully induced in plants from which the axillary buds have been removed. In later experiments it was shown that plants such as *Perilla* and *Xanthium* stripped of all but one leaf could be induced to flower if the remaining leaf were provided the appropriate photoperiod. J. A. D. Zeevaart has shown that leaves may also be removed from induced plants and grafted to noninduced receptors where they will induce a flowering response (Zeevaart, 1958). Finally, Zeevaart has also demonstrated that leaves need not even be attached to the plant in order to be induced. When excised leaves of *Perilla* (SD plant) were exposed to short days and

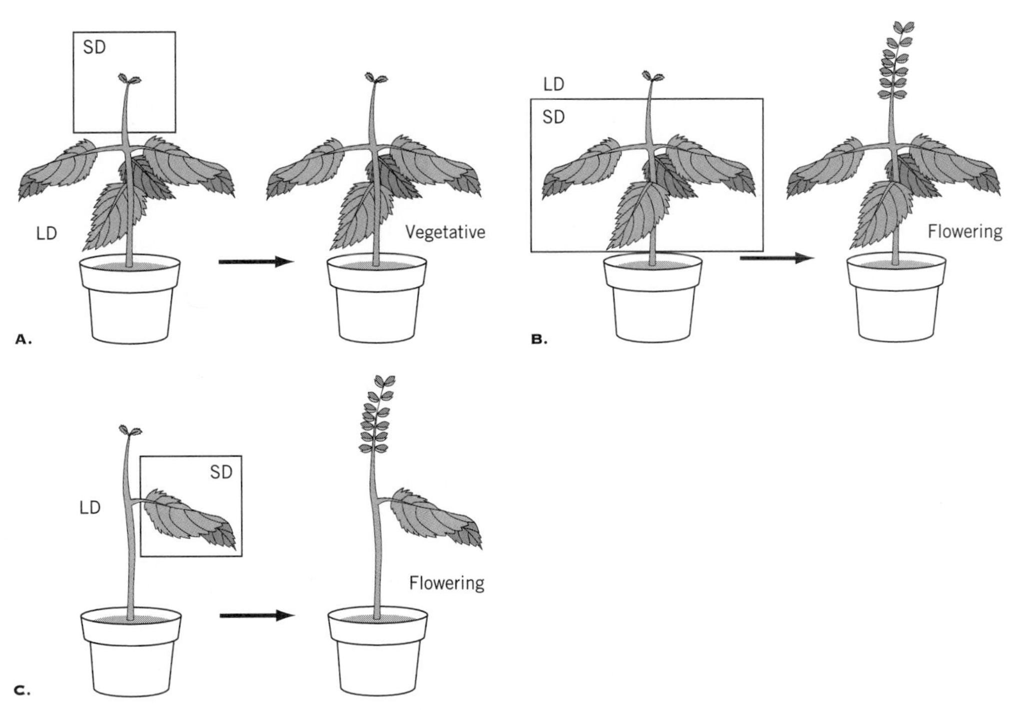

FIGURE 20.5 The role of the leaf in perception of the photoperiodic stimulus in the short-day plant *Perilla*. (*A*) Plants remain vegetative when the shoot apex is covered to provide short days and the leaves are maintained under long days. (*B*) Plants flower when the leaves are given short days but the meristem is maintained under long days. (*C*) Flowering will occur when only a single leaf is provided short days. (Based on the work of M. Chailakhyan, *Canadian Journal of Botany* 39:1817, 1961. Reprinted by permission.)

grafted back to noninduced plants, the plants flowered even when maintained under long days.

The sensitivity of the leaf may vary with age. In *Chrysanthemum*, *Perilla*, and *Glycine* the youngest fully expanded leaf was found to be most sensitive. In experiments with *Xanthium* in which the plants are stripped of all but the most sensitive leaf, it has been shown that peak sensitivity is reached during the period of most rapid expansion, when the leaf is about half its final size. These observations lead to two conclusions: first, that the leaf is independently responsible for perceiving the phototropic signal, and second, that the leaf initiates a signal chain that communicates this information to the apex.

THE ROLE OF PHYTOCHROME

If the leaf perceives the photoperiodic stimulus, then the leaf must be capable of measuring time. Just how leaves measure time is not clear, but we do know that phytochrome is involved. This is another area where the light-break phenomenon, described above, has proven particularly useful. We learned above that photoperiodism is a response to the length of a dark period, but the length of the dark period is defined by the timing between photoperiods. In other words, the length of the dark period is determined by the timing of light-off and light-on signals. In the case of a SD plant such as *Xanthium*, a light-break given in the middle of an otherwise inductive long dark period may be construed as a premature light-on signal that interferes with the timing process. Light-breaks are useful because they are effective with short exposures of low-fluence-rate light and can be applied to a single induced leaf. The light-break thus provides an opportunity to explore the nature of the pigment involved in photoperiodism by determining an action spectrum.

Early action spectra on several SD plants and LD plants in the late 1940s indicated that red light was most effective as a light-break, with a maximum near 660 nm. Then, at the same time that photoreversibility of seed germination was demonstrated, H. A. Borthwick and his colleagues also reported that red light inhibition of flowering in *Xanthium* was reversible with far-red (Borthwick et al., 1952). Red, far-red photoreversibility of the light-break clearly implicates phytochrome in the photoperiodic timing process. The role of phytochrome is far from clear at this point, but based on recent work with phytochrome mutants in *Arabidopsis*, it has been suggested that PHYA is required to promote flowering of an LDP under certain conditions (Coupland, 1997). PHYB, on the other hand, seems to inhibit flowering. Phytochrome is not the entire answer, however, since other data indicates involvement of an internal biological clock as well. For this reason, we will return to a discussion of time measurement in photoperiodism later in this chapter—after we have discussed biological clocks more thoroughly.

LIGHT REQUIREMENTS AND FLORAL HORMONES

The simple observation that photoperiodic perception occurs in the leaves and flowering occurs in the meristem suggests the logical necessity of a floral stimulus to be transmitted between the two often-distant organs. Because of its profound developmental impact at the meristem, this floral stimulus is generally assumed to be hormonal in nature.

Although a SD plant will flower in response to a single long dark period, to be most effective the inductive dark period must be preceded by a period of light. The function of the pre-dark light requirement is not clear since the requirements vary markedly depending on the species and conditions of the experiment. For maximum flowering with a single inductive dark period, *Xanthium* requires 8 to 12 hours of light. *Pharbitis* requires at least 6 hours. In others such as *Kalanchoe*, a few seconds of light per day were sufficient to induce flowering. Where longer light periods are necessary, it may be because photosynthetic products are required for the processes initiated in the dark period. Where very brief periods of light are effective, clearly photosynthesis cannot be involved and some other explanation for the light requirement must be sought.

In many cases it has been demonstrated that high-fluence light following the inductive dark period is also important. Again the experimental details are sketchy and the requirements seem to vary, but two explanations have been offered. Carr (1957) has proposed that the postinductive light period provides a stream of photosynthate that enhances translocation of the floral stimulus out of the leaf. However, some of the results are not readily explained as translocation effects. There is also evidence suggesting that the floral stimulus is subject to destruction or inactivation in the leaf if the dark period is too long. Consequently, Salisbury (1963) suggested that light may be required to stabilize or prevent inactivation of the floral stimulus in the leaf. Actually, most of the data fit a combination of these two ideas and, as suggested by Vince-Prue (1975), the effect of the postinduction light requirement may be to quickly move the stimulus away from a site of inactivation in the leaf.

The Russian plant physiologist M. Chailakhyan, in 1936, was the first to suggest that the floral stimulus might be a hormone, for which he proposed the name **florigen** (Chailakhyan, 1968). Unfortunately, since numerous attempts to isolate and identify the florigen have met with limited success, most of the evidence for the existence of a floral hormone is indirect, resting on the results of physiological experiments.

Chailakhyan and numerous investigators since have shown that the floral stimulus can be transmitted through a graft union. When several *Xanthium* plants are approach-grafted in sequence, all can be brought to flower if only the first is induced by short days (Fig. 20.6). Members of the same family, such as the SD plant tobacco (*Nicotiana tabacum*) and the LD plant black henbane (*Hyoscyamus niger*), both in the family Solanaceae, can be grafted with relative ease. In such a partnership, *Hyoscyamus* will flower under short days if the tobacco is also maintained under short days, but not if the tobacco is maintained under long days. Conversely, tobacco will flower under long days when grafted to *Hyoscyamus* maintained under long days.

In some of the early experiments, it was suggested that florigen, like auxin, could be transmitted through a nonliving connection, but this proved not to be true. More careful experiments, involving anatomical studies, proved that transmission of the stimulus occurred only when a tissue union had been established. Consequently, a number of successful interspecific and intergeneric grafts have yielded results similar to that described above for tobacco and *Hyoscyamus*. These results have led to the conclusion that the final product of photoperiodic induction appears to be physiologically equivalent in plants of different photoperiodic classes and is probably identical to the constitutive floral stimulus in day-neutral plants.

Given the universal nature of other plant hormones,

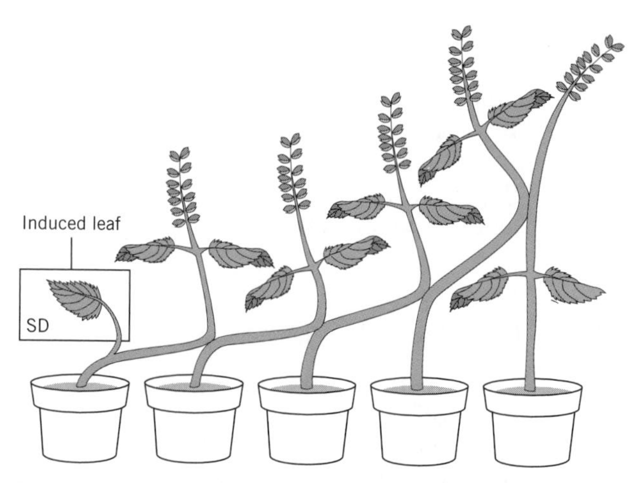

FIGURE 20.6 Transmission of the floral stimulus in grafted plants. Several plants are "approach" grafted and the terminal plant is induced to flower. All plants will flower, indicating that the floral stimulus has been transmitted from the single induced leaf through all of the plants.

it should not be too surprising that the same floral stimulus operates in all photoperiodic classes. The major unanswered question that remains, however, is with regard to the chemical identity of the floral hormone. One approach toward answering this question is to prepare from flowering plants an extract that will evoke flowering in noninduced plants. Subsequent fractionation of the extract should lead to identification of the active substance. Unfortunately, although several attempts have been made, they have met with limited success. In one of the most successful attempts to date, Lincoln et al. (1961) prepared a methanol extract from freeze-dried *Xanthium* plants that evoked a weak flowering response when applied to leaves of other *Xanthium* plants kept under long days. Few other attempts have been successful and none has been consistently repeated.

Other efforts to identify florigen have focused on the known plant hormones, several of which are known to modify flowering behavior in a wide variety of plants. Auxins, cytokinins, and ethylene have been reported to either enhance or suppress flowering in various species. As a rule, high concentrations are required, and the results are often not very spectacular and have not always stood the test of critical evaluation. One must keep in mind that exogenous application of hormones, particularly in high concentrations, frequently causes abnormalities. Many of the effects of these hormones on flowering could be extremely indirect.

Of the several classes of plant hormones, only gibberellins have been shown consistently to evoke flowering in a wide variety of species. Lang (1957) showed that repeated applications of dilute gibberellin solutions (containing principally GA_3) to the apex of annual *Hyoscyamus, Samolus parviflorus* and *Silene armeria* (all LD plants) elicited a flowering response under short days. Does this mean that gibberellin is equivalent to florigen? The answer to this question is clearly negative. To begin with, evocation of flowering in response to gibberellin application is almost entirely restricted to LD plants, which normally grow as rosettes under short days. This includes annual LD plants and biennial species which require an overwintering cold treatment before flowering as LD plants (vernalization, Chap. 21). For example, carrot (*Daucus carota*), chinese cabbage (*Brassica pekinensis*), and biennial strains of black henbane (*Hyoscyamus niger*) grow as rosettes and remain vegetative during the first growing season. The meristems are then subjected to an overwinter cold treatment. The following spring, the stems undergo rapid internode elongation and the plants will flower in response to long days. In the absence of any cold treatment but under long days, exogenously applied gibberellin will stimulate stem elongation and flowering. It thus appears that gibberellin will substitute for the cold requirement of biennial species or the long day requirement of annual LD plants. But gibberellin will not evoke flowering in most SD plants (such as *Xanthium* or Biloxi soybean) kept under long days. Some LD plants also do not appear to respond to gibberellins but this may be largely due to varying sensitivity to gibberellins. Grand Rapids lettuce (*Lactuca sativa*), for example, will respond to GA_1 and GA_3 but not to GA_2 and GA_4. Most of the earlier experiments were attempted with a narrow selection of gibberellins, principally GA_3. Were some of these experiments to be repeated with the wider selection of gibberellins now available, it is highly probable that positive responses would be detected.

The principal effect of gibberellins when applied to rosette plants under short days is to stimulate cell division in the subapical region just below the apical meristem, much as it does in stems of dwarf cultivars and other elongating stems. It has also been found that the endogenous gibberellin content is usually higher in plants under long days regardless of the photoperiodic type. It must be noted that stem elongation and flowering, even in biennials, are independent developmental events—even though one seems to be a necessary prelude to the other. It appears likely that the action of gibberellin is related more to its role in stem elongation, rather than specifically in floral evocation itself (Zeevaart, 1984).

One interesting hypothesis, which has yet to be critically tested, was put forward by Chailakhyan. Chailakhyan proposed that flowering is under control of two hormones: gibberellin and a hypothetical substance, **anthesin.** Since gibberellin does not promote flowering in SD plants, he proposes that SD plants have sufficient gibberellin but lack anthesin. Synthesis of anthesin would be stimulated by short days. LD plants, on the other hand, would have sufficient anthesin but lack adequate levels of gibberellin. Gibberellin synthesis is then stimulated by long days. Accordingly, day-neutral plants would flower irrespective of daylength because they are able to synthesize adequate levels of both "hormones." While certain experiments—principally the grafting experiments described earlier—are consistent with this hypothesis, it does little but substitute one hypothetical substance for another.

TEMPERATURE AND PHOTOPERIODISM

Temperature can influence photoperiodic time measurement in a variety of ways. In some cases the flowering response is simply enhanced at certain temperatures while others may exhibit one response type at high temperature and another at low temperature. One of the more dramatic and thoroughly studied interactions between temperature and photoperiod is the low temperature requirement of winter annuals and biennial flowering plants. In both cases, the absence of a cold treatment will prevent or delay flowering.

The term **vernalization** describes the use of low temperature treatments in order to hasten flowering. Originally applied to low temperature treatment of germinated seedlings, especially winter cereals, the term is now more broadly interpreted to include low temperature requirements of developed biennial and woody perennial plants as well. In this section, we will examine vernalization specifically as it relates to flowering behavior. Vernalization as a low temperature phenomenon will be discussed more fully in Chapter 21.

Winter cereals will not normally flower during a single growing season, but must be planted in the fall in order to flower and produce a crop the following year. Spring strains will flower and produce grain in the same year they are planted. One of the more thorough studies of vernalization in winter cereals has been conducted by F. G. Gregory and O. N. Purvis and their collaborators in England. The spring strain of rye (*Secale cereale*, var. Petkus) is a typical quantitative LD plant—under long days flowers are initiated after seven leaves have been produced. Under short days, however, flowers will not appear until at least 22 leaves have been produced. The winter strain, germinated and grown at normal temperatures (say, 18 °C), is not a LD plant and flowers only after 22 leaves have been produced, regardless of photoperiod. However, if germinated seedlings of the winter strain are subjected to a low temperature treatment (around 1 °C) for several weeks, it will subsequently flower early in response to long days just as the spring strain does. A similar pattern is found in the biennial strain of *Hyoscyamus niger*. Annual strains of *Hyoscyamus* are a typical LD plant, growing in a rosette habit under short days, but undergoing extensive stem elongation (bolting) and flowering under long days. The biennial strain differs from the annual strain in that it has a cold requirement that must be satisfied before it will bolt and flower as a LD plant. Note that the winter cereals have a quantitative vernalization requirement while *Hyoscyamus* has a qualitative, or absolute, requirement.

Although vernalization is most commonly linked to a long-day photoperiodic requirement, other combinations are known. Some day-neutral plants have a cold requirement, but once the cold requirement is satisfied flowering will proceed regardless of photoperiod. Several varieties of *Chrysanthemum*, a SD plant, require vernalization. *Campanula medium* is also a SD plant that grows as a rosette under long days. Stem elongation and flowering may be induced by either short days or cold treatment. Gibberellins will normally substitute for the cold requirement in vernalizable LD plants (see Chap. 21) and for long days in many other LD plants. Gibberellins do not usually substitute for short days in SD plant, although there are exceptions in both cases. One exception is *Campanula*, where gibberellins will induce stem elongation but not flowering in plants maintained under long days. The common theme here is the need to break the rosette habit. Overall, available evidence seems to favor the view that stem elongation and flowering are separate processes under control of separate but closely linked genes. In any event, stem elongation must precede flowering and it is stem elongation that is controlled by gibberellins, not flowering.

THE BIOLOGICAL CLOCK

Many aspects of plant behavior, including nyctinasty and photoperiodism, exhibit periodic oscillations that appear to be controlled by an internal time-measuring system, or biological clock.

One long-observed and dramatic manifestation of the biological clock is the diurnal rise and fall of leaves (nyctinastic, or sleep, movements) described in the previous chapter. Superficially, nyctinastic movements appear to be subject to external, or **exogenous,** control—namely the daily pattern of light and dark periods. The possibility that an internal, or **endogenous,** timekeeper might be involved was first raised by the French astronomer M. De Mairan in 1729. De Mairan found that leaf movements in the sensitive plant (*Mimosa*) persisted even when the plants were placed in darkness for several days. Subsequently studies by J. G. Zinn (in 1759) and A. P. De Candolle (in 1825) confirmed De Mairan's findings (see Sweeny, 1987). Curiously, De Candolle found that under continuous light the time between maximum opening of the leaves was closer to 22 or 23 hours rather than the 24 hours under natural conditions. It would be a full century before the significance of this finding was fully appreciated!

The study of leaf movements continued to interest botanists and plant physiologists through the latter half of the eighteenth and the early nineteenth centuries. Much of the work simply confirmed the widespread occurrence of leaf movements and that they persisted under either continuous light or continuous darkness. In 1863, J. Sachs reported no correlation between leaf movements and temperature fluctuations, thus eliminating temperature as a cause. One difficulty, from an experimental perspective, was that studies of periodic phenomena required around-the-clock monitoring. In 1875, W. Pfeffer devised an apparatus for automatic and continuous recording of leaf position. Pfeffer attached the leaf, via a fine thread, to a stylus, which in turn recorded the position of the leaf on a rotating drum coated with carbon (lampblack). With some improvements, the same apparatus is occasionally used even today (Fig. 20.7). However, more commonly a form of time-lapse photography is now used to record leaf position.

Over a period of 40 years, from 1875 to 1915, Pfeffer contributed several papers devoted to leaf movements in *Phaseolus vulgaris*, the common garden bean

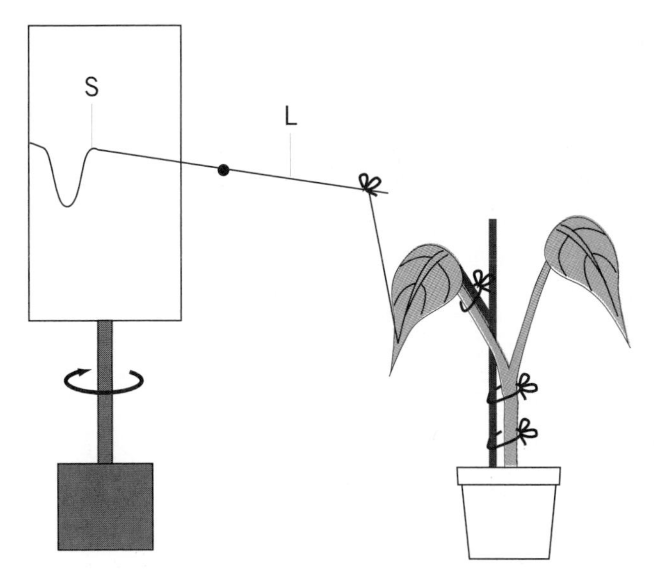

FIGURE 20.7 A diagram illustrating the principle of the drum recording apparatus used by Bünning and others for recording leaf movements. The recording stylus (S) is attached to a finely balanced lever (L), which is in turn tied to the midvein of the leaf. As the leaf changes position, the stylus describes a tracing on the rotating drum.

A.

B.

FIGURE 20.8 Sleep movements. Seedlings of bean (*Phaseolus vulgaris*) are shown with the primary leaf in the horizontal day position (*A*) and vertical night position (*B*).

(Fig. 20.8). At one point, he showed that plants that had lost their rhythmic leaf movements (as they will under prolonged continuous light or darkness) will regain them if exposed to a new light-dark cycle. If the new cycle is inverted with respect to natural day and night, leaf movement will also be reversed. Pfeffer concluded that persistent leaf movements under continuous light or darkness were a "learned" behavior. Others showed that regardless of any previous light-dark cycle, under continuous illumination sleep movements clearly reverted to a 24 hour oscillation. In the end, Pfeffer was forced to conclude that leaf movements were an endogenous, and probably inherited, behavior.

During the 1920s, improvements in the technology for maintaining constant environments, especially with respect to light and temperature, set the stage for significant advances in understanding leaf movements. One key observation was made by Rose Stoppel in Germany. Maintaining bean plants in a dark room at constant temperature, Stoppel observed that maximum night position of the leaves (i.e., near vertical orientation) occurred at the same time (between 3:00 and 4:00 a.m.) every day, exactly 24 hours apart. She reasoned that an endogenous, biological timer could be that accurate only if some environmental factor acted to, in effect, reset the clock on a daily basis. Stoppel referred to this factor as *factor X*.

The endogenous nature of the biological clock finally became evident through the work of two young botanists who had been given the task of determining whether subtle atmospheric factors might influence plants. E. Bünning and K. Stern became interested in Stoppel's factor X and, in order to achieve satisfactory constant temperature conditions, set up their bean plants and recording devices in Stern's potato cellar. Like Stoppel, Bünning and Stern found that the maximum night position came every 24 hours, but, surprisingly, some 7 to 8 hours later than in Stoppel's experiments! Bünning and Stern quickly recognized that the key was the very weak red light used when watering the plants and tending the recording equipment. Stoppel visited her experiments early in the morning while Bünning and Stern, because Stern's potato cellar was some distance from the laboratory, made their visits in the late afternoon. Interestingly, at that time the textbook

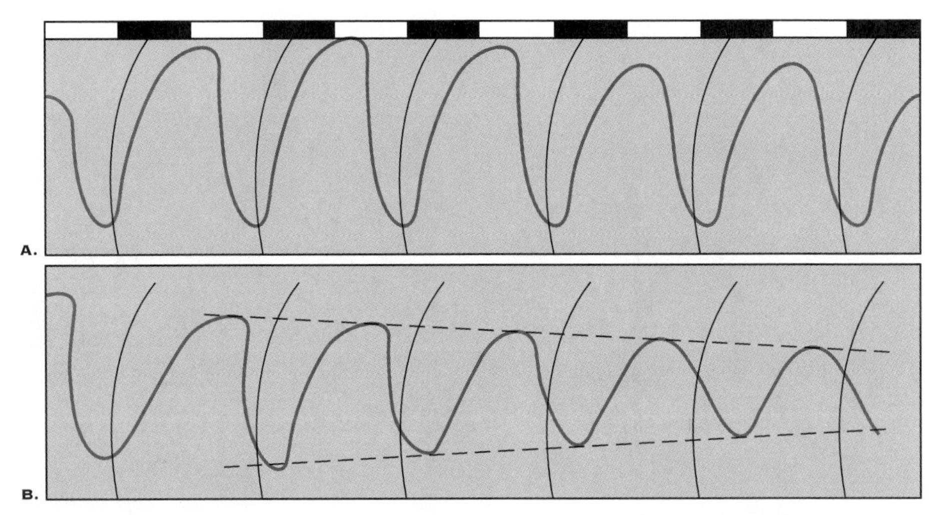

FIGURE 20.9 Leaf movement in bean (*Phaseolus vulgaris*) is a manifestation of an endogenous rhythm. (*A*) A 6-day record of primary leaf position with alternating light and dark periods. Light and dark bars represent light and dark periods, respectively. The vertical lines indicate midnight solar time. Period length is 24 hours. (*B*) Free-running rhythmicity under continuous light. After the first period, the period length is extended to 25.7 hours. Note also that the amplitude (dashed lines) diminishes with time under continuous light. (From the data of E. Bünning and M. Tazawa, *Planta* 50:107, 1957.)

dogma was that red light had no effect on plant morphogenesis. Bünning and Stern concluded that the dogma was wrong—even weak red light apparently synchronized leaf movement so that the maximum night position always occurred about 16 hours later. Indeed, when the red light was eliminated, the period between maximum night positions was no longer 24 hours but 25.4 hours (Fig. 20.9). Red light was factor X!

THE LANGUAGE OF BIOLOGICAL CLOCKS

A graphic plot of a biological rhythm against time describes a repeating pattern that resembles physical wave phenomena. Consequently, much of the terminology that describes physical oscillations has been adopted to describe biological cycles. At the beginning, however, it is necessary to distinguish between simple periodic phenomena and endogenous rhythms. Photosynthetic carbon uptake, for example, describes a periodicity because it is light-driven and daylight is periodic over time. Photosynthesis is thus diurnal, in that it is active only during daylight hours and is controlled by fluctuations in an external factor (light). The key to an **endogenous rhythm** is that it persists, for at least several cycles, under constant conditions (usually constant light or constant darkness). The rhythmicity expressed under constant conditions is described as **free-running.**

The time required to complete a cycle is known as the **period** (τ; tau) (Fig. 20.10A). Period is conveniently described as the time from peak to peak, but it applies equally well to any two comparable points in the repeating cycle. Constant rhythms are traditionally classified according to the length of their **free-running period.** Thus a **circadian** rhythm has a period of approximately 24 hours (*circa* = about + *diem* = day). Bean leaf movement is a circadian rhythm because its period length is about 25.4 hours. A period of about 28 days, the time between one full moon and the next, describes a **lunar** rhythm and a period of one year is an **annual** rhythm. Also of interest are rhythms in metabolic activity with periods substantially less than 24 hours (measured in minutes or hours). These are known as **ultradian** rhythms.

The difference between the maximum and minimum, or peak and trough, of a rhythm is known as the **amplitude** (A) (Fig. 20.10A). The amplitude of a free-running circadian rhythm usually diminishes with time until it eventually disappears altogether. In some cases this may be due to declining energy reserves in prolonged darkness since the amplitude can be maintained, at least for a while, by feeding sucrose. More often, however, the clock seems to run down and an external signal is required to start up the rhythm again. The term **phase** has two slightly different but related usages. Any point of the cycle that can be identified by its relationship to the rest of the cycle can be considered a phase. The position of the peak (maximum night position of leaves, maximum flowering, etc.) is the most common reference point for phase relationships because it is usually most readily identified. In Figure 20.10A, for example, the two rhythms are displayed out of phase by approximately 6 hours. Phase may

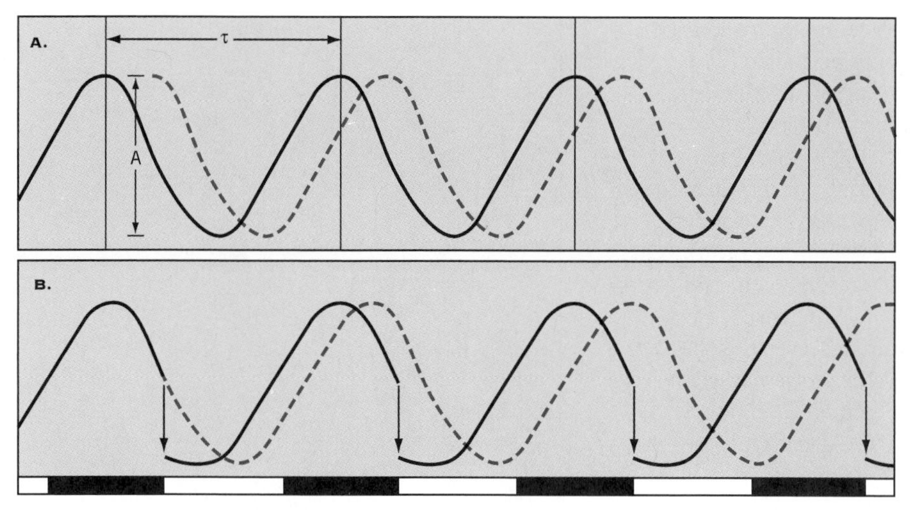

FIGURE 20.10 Examples of circadian oscillators. (*A*) Two rhythms with period (τ) and amplitude (A). Although the period and amplitude for both rhythms are the same, the two are slightly out of phase. (*B*) Free-running rhythm and entrainment. A free-running rhythm with a period of 28 hours (dashed line) is entrained to 24 hours by a daily light-on signal. The open and closed bars represent light and dark conditions.

also be used to describe an arbitrary part of the cycle, such as night phase or day phase.

Discussions of endogenous circadian rhythms are sometimes complicated by the fact that two time frames are involved: **solar time** and **circadian time** (**CT**). Solar time is based on a normal 24-hour day. Circadian time, on the other hand, is based on the free-running period. One cycle is considered to be 24 hours long, regardless of its actual length in solar time. Each hour of circadian time is therefore 1/24 of the free-running period. Thus if the free-running period is 30 hours, events that occur at 0, 15, and 30 hours of darkness will have occurred at circadian times CT: 0, CT: 12, and CT: 24 respectively. The circadian time scale is useful in assessing phase relationships within experiments or between rhythms with different periods. Finally, the phase of the free-running cycle that corresponds to day in a normal light-dark environment is known as **subjective day** and that which corresponds to normal night is the **subjective night.**

The rhythmic movements of bean leaves are normally coupled, or synchronized, to the solar day-night cycle (Fig. 20.10B). The same coupling was evident in both Stoppel's and Bünning's experiments when the rhythm was coupled to the red light signal. Such a coupling of a circadian rhythm to a regular external environmental signal is known as **entrainment.** The signal that synchronizes the rhythms is often referred to as a *zeitgeber* (Ger. *zeit* = time + *geben* = to give).

CIRCADIAN RHYTHMS, TEMPERATURE, AND LIGHT

A large number of circadian rhythms have now been described in a wide range of organisms. The list includes single-celled flagellates, algae, fungi, crustacea, insects, birds, and mammals (including humans), in addition to flowering plants. A full list of known rhythms in flowering plants alone would cover several pages. An abbreviated list is provided in Table 20.2. A more extensive list, with references, is provided by Sweeny (1987).

If plants do have some kind of internal clock, what and where is it and how does it operate? Unfortunately, in spite of massive volumes of literature related to biological clocks, the answers to these questions are simply not available, although there is some information bearing on the third part. Difficulties in unraveling the mystery of the biological clock arise from several sources. One is that the clock appears to be almost exclusively internal and is not generally subject to manipulation from the outside. This means the clock is not amenable to most experimental strategies, since these normally require that the investigator be able to control or manipulate the system in some way. Another difficulty is our inability to distinguish between oscillations that are part of the timekeeping mechanism and those that are simply the "hands" responding to the output of the basic oscillator.

Most chemical reactions, and thus growth and other biological responses, respond to temperature with a Q_{10} near 2. This means that a 10°C increase in temperature will approximately double the rate of the process. A decrease in temperature leads to a decrease in the rate by

TABLE 20.2 Examples of circadian rhythmic phenomena in higher plants.

Rhythm	Organism
Sleep movements	Many species
Stomatal opening	Banana, tobacco, *Vicia*
Stem growth	Tomato, *Chenopodium*
CO_2 production	Orchid flowers
Gas uptake	Dry onion seeds
Membrane potential	Spinach leaves
mRNA expression	Pea

the same amount. In most cases, such temperature sensitivity is no doubt advantageous to the organism. The accuracy of a biological clock, however, would be severely compromised if it were sensitive to often random temperature fluctuations brought about by local environments. As it turns out, the period length of circadian rhythms is relatively insensitive to temperature. A classic example is again bean leaf movement (Bünning, 1956). When seedlings are raised in the dark from seed, leaf movements tend to be small and unsynchronized. A single flash of light (the zeitgeber) initiates larger, synchronized movements. In Bünning's experiments, the synchronized movements had a periodicity of 28.3 hours at a constant 15°C and 28 hours at 25°C. Although these data would seem to suggest that the circadian rhythm is insensitive to temperature, this is not strictly true. When seedlings were shifted from 20°C to 15°C, the initial period was 29.7 hours. Seedlings shifted from 20°C to 25°C had a period of 23.7 hours. These periods, however, lasted only for the first cycle or two—later cycles returned to periods of approximately 28 hours. Clearly the circadian rhythm is temperature-sensitive, but some mechanism quickly compensates for variations in temperature. Consequently, the Q_{10} for most circadian rhythms is near 1 (Sweeny, 1987). Amplitude may be affected by temperature, but **temperature-compensation** is clearly a characteristic of the period for most circadian rhythms.

The action of the biological clock or endogenous rhythms is to ensure that certain functions occur at a particular time of day. For example, the oscillations of the clock in beans determines that the leaves rise during the day and fall at night. The period of the endogenous rhythm is fixed but, it may be "fast" or "slow" relative to the 24-hour solar period. Moreover, the daily duration of light and dark within the 24-hour solar period changes steadily throughout the season. How does the organism reconcile a nonvarying endogenous periodicity with these seasonal changes in daylight? How are these rhythms kept in phase? The answer is found in *entrainment.* Entrainment is, in effect, a means for moving the oscillations of the clock forward or back in time on a daily basis, just as you might reset the time of an alarm clock every night before retiring. Entrainment is also useful to the experimentalist because it is the one significant way in which circadian rhythms can be manipulated in the laboratory.

Entrainment is not limited to solar periodicity. Within limits, rhythms can be entrained to light-dark cycles either shorter (18 to 20 hours) or longer (up to 30 hours or more) than 24 hours. Entrainment to extremely short or long cycles is rare. More useful information, however, can be obtained by studying whether the rhythm is moved forward or back in time and by how much. The experiment normally entails giving brief light pulses (usually 1 hour or less) at various times during an established free-running rhythm in constant darkness.

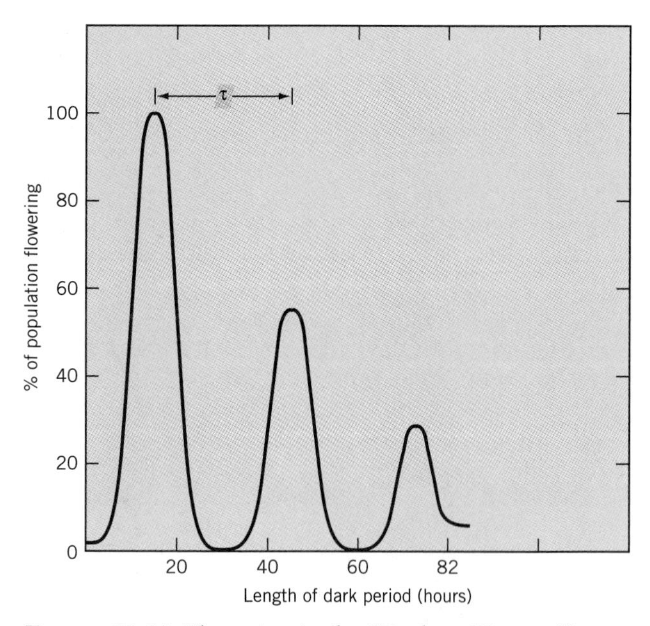

FIGURE 20.11 Flowering in the SD plant *Chenopodium rubrum* responds rhythmically to the length of a single dark period. Populations of *C. rubrum* seedlings were exposed to a single dark period of varied length as indicated. The free-running periodicity (τ) is approximately 30 hours. The amplitude diminishes because the young seedlings are depleting their carbon supply over the extended dark period. Amplitude can be maintained by supplying the seedlings with glucose. (From R. W. King, Time measurement in photoperiodic control of flowering, Ph.D. thesis, University of Western Ontario, 1971. With permission of R. W. King.)

The timing of the next peak is then compared with controls that have not been given a light pulse. For example, when populations of *Chenopodium* seedlings are exposed to single dark periods of various lengths, their capacity to flower fluctuates rhythmically for at least three cycles (Fig. 20.11). In this case, a common light-off signal sets the rhythm in motion and the timing of the light-on signal (i.e., the length of dark) determines whether the plants will flower. If, however, relatively brief pulses of light are given at various times during the dark period, the pulses will reset the clock, or shift the phase of the rhythm. The result is a **phase response curve** such as that shown in Figure 20.12. This curve demonstrates that a light pulse given early in the subjective night causes a delay of the first and subsequent peaks relative to the control. Somewhere near the middle of the subjective night there is a *phase jump* such that pulses given in the latter half of the subjective night cause subsequent peaks to be advanced relative to the controls. Note that light pulses given during the subjective day have very little effect on phase relationships. Similar phase response curves have been demonstrated for a variety of circadian rhythms and help to explain entrainment. It seems to be a character of the system that rephasing is accomplished in such a way as to require the least net displacement of the rhythm. Thus, during the early part of the subjective night, the light pulse is apparently interpreted as a delayed light-off, or *dusk*, signal and the phase of the endogenous rhythm is adjusted accordingly. As the pulse arrives later, the delay is increased until at some point the pulse is now interpreted as an early light-on, or *dawn*, signal. This causes the rhythm to be advanced. Phase-shifting in this way constantly adjusts or entrains the rhythm to local solar time.

The observation that similar phase response curves can be described for such different phenomena as insect pupal eclosion, bioluminescence in the dinoflagellate *Gonyaulax*, and CO_2 evolution and flowering in higher plants indicates similar properties, if not mechanisms, for the circadian clock in a variety of organisms.

Although light plays an obvious role in resetting circadian rhythms, the photoreceptor involved is clearly not the same in all cases. Action spectra for the rhythms in *Gonyaulax*, the protozoan *Paramecium*, fungi and insects all share a large peak in the blue region of the spectrum, suggesting that a flavoprotein blue-receptor might be involved. In several higher plants, however, phytochrome appears to be involved. For example, establishment of rhythms in dark-grown bean plants and phase-setting of leaf movement in *Samanea* both show a classic phytochrome photoreversibility with brief red and far-red light treatments. Others, however, such as the CO_2-evolution rhythm in *Bryophyllum* leaves, can be reset with red light but the effect is not reversible with far-red. The effect of far-red light in the *Bryophyllum* system is to abolish the rhythm altogether! This might also be a phytochrome effect, but these and other experiments make it clear that photocontrol of phase-setting is not a straightforward process.

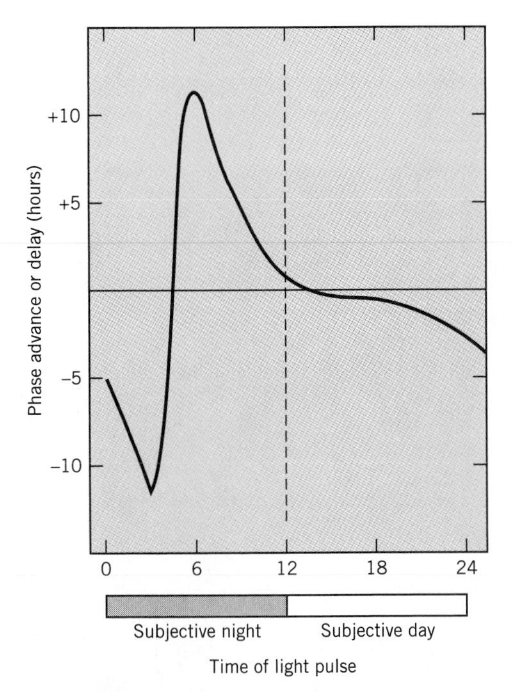

FIGURE 20.12 Phase response curve. Light-on signals during the early part of the subjective night cause a delay in the timing of the next peak; given late in the subjective night, the result is an advance of the next peak. Light-on signals given during the subjective day have little or no effect on the phase of the rhythm. (Redrawn from R. W. King, Time measurement in photoperiodic control of flowering, Ph.D. thesis, University of Western Ontario, 1971. With permission of R. W. King.)

TIME MEASUREMENT IN PHOTOPERIODISM

We can now turn our attention to the question of how plants measure photoperiodic time. Two theories have been advanced to account for photoperiodic time measurement. One theory invokes the operation of a biochemical interval timer while the other, originally proposed by Bünning, ties photoperiodic time measurement with circadian rhythms.

With the discovery that flowering in SD plants was determined by the length of a critical dark period, it seemed possible that the length of the critical dark period could be measured rather simply by the appearance or disappearance of some metabolite. Such an approach to timing was likened to an hourglass (Fig 20.13) which, when the sand runs out, must be inverted to restart the timing process. Thus this hypothesis has come to be

FIGURE 20.13 An hourglass timer. According to the hourglass hypothesis, the critical dark period represents the time required for Pfr to "decay" out of the system, much as the sand runs out of the upper chamber in an hourglass.

known as the "hourglass" hypothesis. When action spectra showed that phytochrome was the photoreceptor responsible for the light-break phenomenon, it seemed reasonable to conclude that Pfr, present at the end of the photoperiod, was inhibitory to flowering. The length of the critical dark period thus represented the time required for Pfr to fall below some critical value (at least in SD plants) long enough to allow for synthesis of florigen. The effectiveness of a red light break in the middle of an otherwise inductive long dark period was seen simply as raising the level of Pfr, thus restarting the timing process before sufficient florigen could be produced.

While the hourglass hypothesis has its attractions, experiments bearing on it appear to raise more problems than solutions. For example, spectrophotometric measurements indicated a half-life for Pfr of 1.0 to 1.5 hours, which is far too short to account for the lengths of most critical dark periods. Furthermore, no characteristic differences in phytochrome decay could be detected in etiolated seedlings of plants from several photoperiodic classes. The rate of Pfr loss is also highly temperature dependent yet, in at least two SD plants (*Xanthium* and *Chenopodium*), temperature was found to have little influence on the length of the critical dark period. Interestingly, a decrease in temperature from 25°C to 18°C almost doubled the length of the critical dark period for *Pharbitis*, also a SD plant, A similar temperature change significantly shortened the critical dark period for *Hyoscyamus*, a LD plant. With the more recent finding that phytochrome in green tissues behaves differently from that in etiolated tissue, the idea of phytochrome reversion and timing may have to be reevaluated. In the meantime, although there is some data consistent with the hourglass hypothesis, the bulk of the data clearly favors an interaction of photoperiodism with circadian clocks.

Several experimental strategies have been employed to demonstrate rhythmicity in flowering. The difficulty, of course, is that, unlike leaf movement or carbon dioxide evolution, flowering is not a continuous process. One example is the flowering of the SD plant *Chenopodium rubrum*, introduced in the previous section. Seedlings of *Chenopodium*, can be induced to flower with a single dark period when the seedlings are only 4½ days old (King and Cumming, 1972). Before and after the dark period, the seedlings are maintained under continuous light. When the length of the dark period is varied, light during the first 8 to 10 hours inhibits flowering as would be expected with a SD plant. Thereafter, the capacity of the seedlings to flower expresses a rhythmic pattern for at least three cycles, with a free-running period of 30 hours (Fig. 20.11). With minor variations in the experimental strategy, essentially the same results have been demonstrated with a variety of plants including cocklebur, soybean, Japanese morning glory, and duckweed.

The above results suggest that the light-on signal at the end of the dark period will either promote or inhibit flowering depending on how it interacts with the phasing of an endogenous circadian rhythm. Bünning, who originally proposed in 1936 that photoperiodism was tied to circadian rhythms, suggested just such an interaction. He proposed that the rhythm was comprised of two phases—the **photophile,** or light-loving phase, and the **scotophile,** or dark-loving phase—which alternated about every 12 hours. According to Bünning's hypothesis, light falling on the plant during the photophile phase would promote flowering while light during the scotophile phase would inhibit flowering. In most experimental situations, when the plant is placed under continuous conditions, the photophile phase is probably equivalent to subjective day and the scotophile phase equivalent to subjective night.

Although the results of many (but not all) experiments can be interpreted in terms of Bünning's hypothesis, perhaps the strongest support for a close dependency of photoperiodic time measurement on rhythmic phases comes from experiments with the aquatic SD plant *Lemna purpusilla* (Hillman, 1964) (Fig. 20.14). As an experimental organism *Lemna* offers the unique advantage of growing heterotrophically in darkness when supplied with glucose. This means that long light periods can be eliminated and timing can be controlled by skeleton photoperiods—short pulses of light that serve to mark the beginning and end of dark periods. All that

FIGURE 20.14 Duckweed (*Lemna* sp.). The genus *Lemna* is the smallest flowering plant. Each leaflike frond is about 1 to 2mm across. Flowers arise at the point of the frond and consist of male and female parts only (no petals or sepals). There are short day and long day species of *Lemna*.

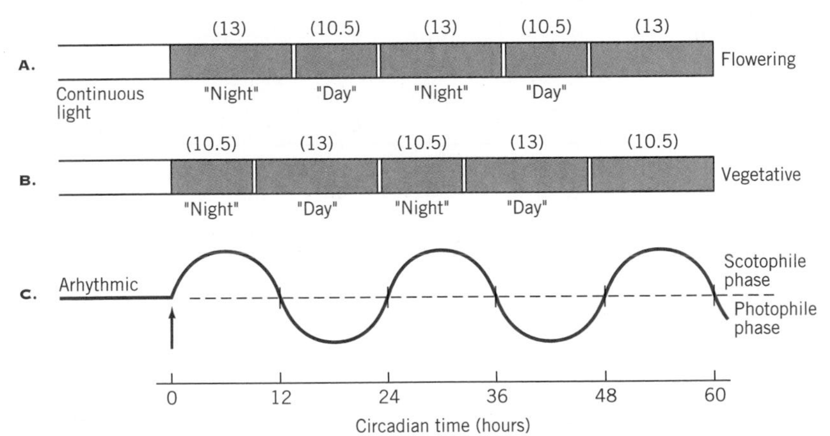

FIGURE 20.15 Skeletal photoperiods and the control of flowering in *Lemna purpusillia*, a SD plant. *Lemna* is arrhythmic under continuous light. The first light-off signal (arrow) starts up the rhythm. Because the first phase is the scotophile phase, *Lemna* interprets the first dark period, regardless of its length, as night. (*A*) If the "night" dark period exceeds the critical value, the plants will flower; if not (*B*), the plants remain vegetative. (From the data of W. S. Hillman, 1964.)

is required to induce flowering in *Lemna* are two 15-minute (0.25 h) light periods every 24 hours (Fig. 20.15). Thus, when taken from *continuous light* to a schedule of 13 hours dark: 0.25 hours light: 10.5 hours dark: 0.25 hours light, *Lemna* will flower. On a schedule of 10.5 hours dark: 0.25 hours light: 13 hours dark: 0.25 hours light, however, the plants remain vegetative. Note that the only difference between the two schedules is the length of the *first* dark period following continuous light. Even though both schedules contain a 13-hour inductive dark period, flowering is induced only when the long dark period comes first.

Clearly *Lemna* recognizes the brief light pulses as the beginning and end of two different dark periods, but how does the plant know which is which? As with other plants under continuous light, *Lemna* becomes arrhythmic; that is, the amplitude of the rhythm damps out until the rhythm disappears. Transfer of the plants to darkness starts up the rhythm. Since flowering occurred only when the first dark period exceeded the critical night length, the first dark period (and every alternate period thereafter) must have been interpreted as *night*. This appears to be a general rule—the light-off signal following a period of continuous light starts up the rhythm and the first dark period (CT 0 to 12) is always the scotophile or night phase. The second phase (CT 12 to 24) is the photophile or day phase. In subsequent experiments, Hillman confirmed that flowering occurs only when a dark period longer than the critical night length coincides with the night or scotophile phase of its circadian rhythm. It is interesting to note that this interpretation is not limited to flowering plants—skeleton photoperiods elicit similar responses with respect to photoperiodic effects in insects and birds.

The circadian clock is clearly an important component of photoperiodic time measurement, but experiments to date do not explain the differences between SD plants and LD plants, nor do they explain how the measurement of critical night length and circadian clock time are connected. As well, the location and exact nature of the clock in plants remains an enigma yet to be unraveled.

GENETIC APPROACHES TO PHOTOPERIODISM AND RHYTHMS

The physiology and biochemistry of photoperiodism and biological rhythms have been extensively studied for the past 70 years. While each phenomenon has been thoroughly described in numerous model systems, little has been revealed of the underlying mechanisms. Both photoperiodism and endogenous rhythms are fundamentally questions of time measurement, a concept that is not easily imagined within the framework of conventional biochemistry. Only recently has a solution to this impasse been sought in genetic approaches to the study of photoperiodism and rhythms.

The rationale of the genetic approach is that if a mutant can be identified that influences timing at any level, the wild-type gene can be isolated and its gene product analyzed for clues to its role in the timing mechanism (Thomas and Vince-Prue, 1997). Genes that influence photoperiodic control of development are not novel. Flowering genes in particular have been a part of plant breeding programs for years. Because early flowering and insensitivity to photoperiod are most desired in crop species, however, most major crop cultivars have been bred to exclude sensitivity to photoperiod. In only two crop species—pea (*Pisum sativum*) and wheat (*Triticum aestivum*)—have the genetics of photoperiodic processes received much attention (Thomas and Vince-Prue, 1997). Interestingly, both peas and wheat are quantitative long-day plants. In peas, several genes that affect photoperiodic timing and the onset of flowering have been identified. One in particular, *fsd* (*flowering short days*), is a recessive mutant that causes the plant to behave as a qualitative short-day plant. When the mutant is grafted to a wild-type stock under long days, the mutant will flower. These results reaffirm the grafting experiments described earlier, indicating that the wild type provides a transmissible floral promoter.

Recently there has been a surge of interest in flowering genes in *Arabidopsis* in which a number of flowering genes have been identified. *Arabidopsis* is a quantitative long-day plant with a critical photoperiod of 8 to 10 hours. Under long days, it flowers with 4 to 7 leaves in the rosette (about 3 weeks). Under short days, flowering is delayed until 20 leaves have formed (7 to 10 weeks). Flowering is also promoted by exposure to blue or far-red light, reflecting the role of phytochrome in photoperiodic phenomena. The search in *Arabidopsis* has focused on flowering time mutants—mutants that cause early or late flowering (Coupland, 1997). Here flowering time refers not to elapsed time (for example, days to flowering) but to the number of rosette leaves produced before the flowering stem appears.

As might be expected, mutants that affect phytochrome also influence flowering. Thus the *hy1* mutant is defective in the synthesis of the phytochrome chromophore. In the absence of functional photoreceptor, *hy1* mutants show an elongated hypocotyl, but they also flower earlier than wild type under both long and short days. However, because they flower early under *both* conditions, the mutant still shows a response to photoperiod.

Several photoperiod mutants have also been identified. These are mutants that are essentially insensitive to photoperiod and flower either earlier or later than wild-type plants. The *early flowering 3* (*elf3*) is an example of the early flowering type. *Constans* (*co*) and *gigantea* (*gi*) are late flowering, daylength-insensitive mutants. Both *co* and *gi* delay flowering under long days, but have no effect on flowering time under short days. Both wild-type genes *CO* and *GI* have been cloned and studied in some detail (Coupland, 1997). At this stage it appears that the *GI* gene operates before the *CO* gene in the same pathway and that floral promotion under long days depends on the amount of *CO* mRNA transcribed.

The *elf3* mutant is particularly interesting because it appears to interact with the endogenous clock. This conclusion is based on the effect of *elf3* on two other phenomena that are regulated by circadian rhythms in *Arabidopsis:* leaf movement and transcription of the chlorophyll *a/b* binding protein (CAB). The *elf3* gene not only advances flowering, but also renders both leaf movement and *CAB* gene expression arhythmic when entrained plants are shifted to continuous light. Curiously, the rhythm for *CAB* expression is not lost if the entrained plants are shifted to continuous dark. This indicates that *ELF3* does not encode a component of the endogenous clock itself, but may in some way link the clock to the initial photoreception or light-on signal.

The photoperiodic signal transduction chain and endogenous clock have proven particularly difficult to study by traditional physiology and biochemistry. Although still in its infancy, the molecular genetic approach has already provided insights into mechanisms that have eluded investigators for more than 70 years. Prospects for future resolution of these classical problems are quite exciting.

PHOTOPERIODISM IN NATURE

Time is nature's way of ensuring that everything doesn't happen at once. (Anonymous)

Photoperiodism almost certainly reflects the need for plants to synchronize their life cycles to the time of year. Outside of the tropics, daylength is the most reliable predictor of seasonal change (Fig. 20.16). Not surprisingly, photoperiodism is more important to plants in the subtropical and temperate latitudes where seasonal variations in daylength are more pronounced. But even many tropical plants respond to the small changes in daylength that occur within 5 or 10 degrees of the equator. Does this mean that the photoperiod response ties a species to a particular latitude? Probably not, since the critical photoperiod only sets the upper (for SD plant) or lower (for LD plant) limits for daylength. Beyond that, flowering and other responses to photoperiod can usually occur within fairly broad limits. Moreover, there is evidence that populations of plants are able to genetically adapt to latitude, thus giving rise to **physiological ecotypes.** In a variety of species, including *Betula* (birch), *Chenopodium* (pigweed), *Oxyria digyna* (mountain sorrel), and *Xanthium*, there are known ecotypes or photoperiodic races characterized by different critical daylengths. As a rule, the length of the critical day is longer as the individuals are collected at more northerly latitudes. In most cases, the critical daylength seems to key flowering to a consistent time interval before the arrival of damaging autumn frosts at that latitude.

Photoperiodism helps to ensure that plants flower in their temporal niche, reducing competition with others or ensuring that reproductive development is completed before the onslaught of unfavorable winter conditions. In many species, germination, for a variety of reasons, may not be uniform. If flowering relied solely on plant size, nonuniform germination would be expected to spread flowering out in time as well. To the extent that cross-pollination is required or advantageous, flowering synchronized by photoperiod would serve to ensure the maximum pollinating population or to coordinate flowering with the appearance of a particular pollinating insect.

Photoperiod and its effects on geographical distribution of plants can have a direct impact on humans, as illustrated by the case of common ragweed (*Ambrosia artemisifolia*). Ragweed is an annual SD plant, with a critical daylength of about 14.5 hours. The further north one goes, the longer the summer daylength. At the latitude of Winnepeg, Canada, (50 °N), for example, the daylength exceeds 16 hours through most of June and

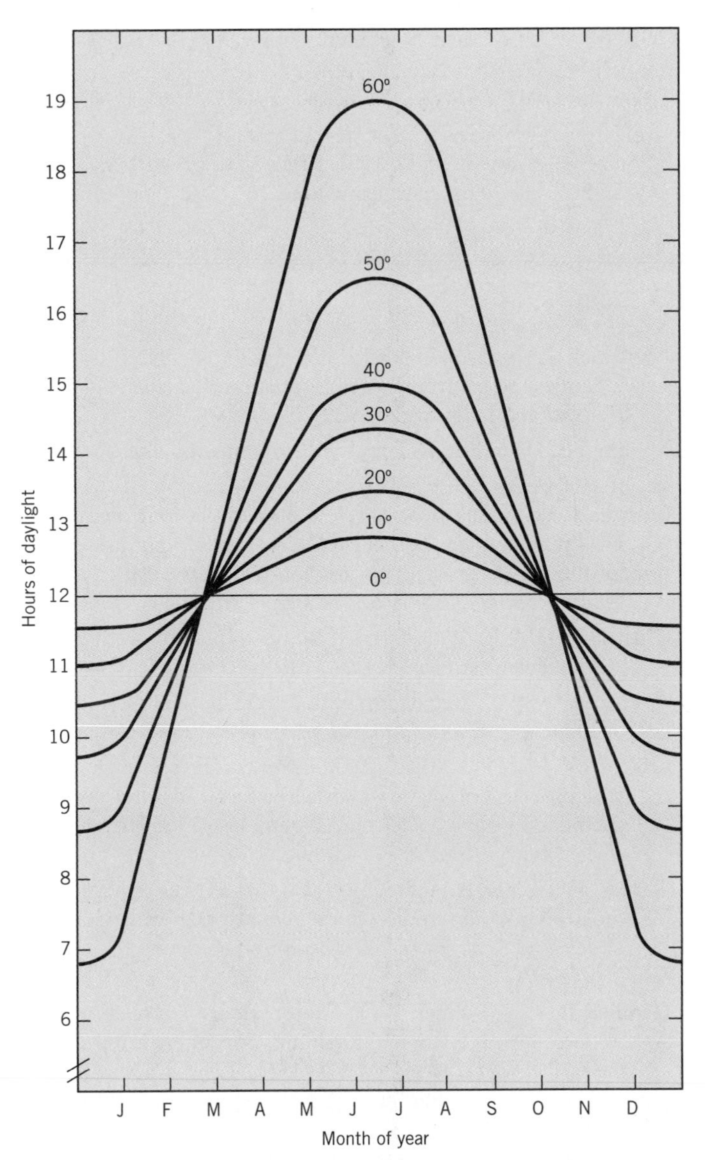

FIGURE 20.16 Daylength as function of latitude and month of year. Daylength is plotted as the time between sunrise and sunset on the 20th of each month. (Data from *The American Ephemeris and Nautical Almanac*, U.S. Naval Observatory, 1969.)

July and doesn't drop below 14.5 hours until mid-August (Fig. 20.16). Ragweed induced to flower at that time of the year would have insufficient time to flower and produce mature seed before the arrival of killing frosts in early fall. Since common ragweed can reproduce only by seed, it is abundant only in the more southerly regions of Ontario, Quebec and the maritime provinces. It is rarely found, and then only on scattered patches of agricultural land, throughout most of western and central Canada. Hay-fever sufferers in these regions are at least spared the inconvenience of highly allergenic ragweed pollen, which plagues their neighbors to the south. This is not to suggest that photoperiod is the only factor that limits the distribution of plants such as ragweed, but it is clearly a significant part of the equation.

SUMMARY

Photoperiodism is a response to the duration and timing of light and dark periods. There are three basic photoperiodic response types: short-day plants (SDP), long-day plants (LDP), and day-neutral plants (DNP). Other response types are variations on the three basic types and may be modified by environmental conditions such as temperature. A photoperiod requirement may be qualitative, in which case the requirement is absolute, or quantitative, in which case the favorable photoperiod merely hastens the response. The distinction between LDP and SDP is based on their response to daylengths greater than or shorter than the critical daylength. The absolute critical daylength varies from one species to another and the critical daylength for a LDP may be shorter than the critical daylength for a SDP.

Photoperiodism is somewhat of a misnomer, in that the plant actually senses the length of a dark interval between the light-off and light-on signals. Action spectra of the light break, which interrupts an otherwise inductive dark period, indicate that phytochrome is involved in the light signals.

Photoperiodic light signals are perceived in the leaf but the response ultimately occurs elsewhere in the plant. This separation of perception and response suggests the logical necessity for a transmissible stimulus. In the case of flowering, the stimulus is believed to be a hormone that has been called florigen. All attempts to isolate and characterize florigen have been unsuccessful.

Temperature can influence photoperiodic timing in many ways. One of the more dramatic, thoroughly studied temperature-photoperiod interactions is vernalization—the low-temperature requirement of winter annuals and biennial flowering plants. Typically, vernalization of winter annuals converts a daylength-insensitive plant into a quantitative LDP, although some SDP requires a low-temperature treatment as well. In biennials, vernalization induces bolting, or stem elongation, which is a necessary prelude to flowering.

Many aspects of plant development, including photoperiodism and nyctinasty, are controlled by an internal biological clock. The biological clock is difficult to study by traditional methods because, other than shifting the phase ("setting the clock"), the clock is not readily manipulated by external influence. Time measurement in photoperiodism involves an interaction between phytochrome and the biological clock. The nature of the clock and of phytochrome/clock interactions remains an enigma, although recent studies of flowering time mutants in *Arabidopsis* promise to shed some light on these complex problems.

Because changing daylength is the most reliable predictor of seasonable change, photoperiodism almost certainly reflects the need for plants to synchronize their life cycle to the time of year. Photoperiodism helps ensure that plants flower in their temporal niche, reducing competition with others, or that reproduction is complete before the onslaught of unfavorable winter conditions.

CHAPTER REVIEW

1. You have discovered a new plant whose photoperiod characteristics are not yet described. How would you go about determining whether this plant were a SD plant, a LD plant, or a day neutral plant?
2. Many metabolic processes appear to vary throughout a normal day. How would you determine whether these processes were regulated by an endogenous, circadian clock?
3. What is the physiological significance of physiological ecotypes, or photoperiodic races within a species that are characterized by different critical day lengths?
4. How does the response of *Lemna* to skeletal photoperiods lend support to Bünning's notion of a photophile and scotophile phase?
5. Distinguish between diurnal, circadian, and free-running rhythms.
6. Endogenous circadian rhythms have been described extensively, but virtually nothing is known of their underlying mechanism. Why is it so difficult to experimentally unlock the secrets of the biological clock?

FURTHER READING

Atherton, J. G. (ed.). 1987. *Manipulation of Flowering*. London: Butterworths.

Halevy, A. H. (ed.). 1985. *CRC Handbook of Flowering*. Boca Raton: CRC Press.

Hillman, W. S. 1962. *The Physiology of Flowering*. New York: Prentice-Hall.

Pharis, R. P., R. W. King. 1985. Gibberellins and reproductive development in seed plants. *Annual Review of Plant Physiology* 36:517–568.

Salisbury, F. B. 1963. *The Flowering Process*. New York: Pergamon Press.

Sweeny, B. 1987. *Rhythmic Phenomena in Plants*. 2nd ed. New York: Academic Press.

Thomas, B., Vince-Prue, D. 1997. *Photoperiodism in Plants*. 2nd ed. San Diego: Academic Press.

REFERENCES

Borthwick, H. A., S. B. Hendricks, M. W. Parker. 1952. The reaction controlling floral initiation. *Proceedings of the National Academy of Sciences USA* 38:929–934.

Bünning, E. 1956. Endogenous rhythms in plants. *Annual Review of Plant Physiology* 7:71–90.

Carr, D. J. 1957. On the nature of photoperiodic induction. IV. Preliminary experiments on the effect of light following the inductive long dark period in *Xanthium pennsylvanicum*. *Physiologia Plantarum* 10:249–265.

Chailakhyan, M. K. 1968. Internal factors of plant flowering. *Annual Review of Plant Physiology* 19:1–36.

Coupland, G. 1997. Regulation of flowering by photoperiod in *Arabidopsis*. *Plant, Cell and Environment* 20:785–789.

Garner, W. W., H. A. Allard. 1920. Effect of the relative length of day and night and other factors of the environment on growth and reproduction in plants. *Journal of Agricultural Research* 18:553–606.

Garner, W. W., H. A. Allard. 1923. Further studies in photoperiodism, the response of the plant to relative length of day and night. *Journal of Agricultural Research* 23:871–920.

Hamner, K. C., J. Bonner. 1938. Photoperiodism in relation to hormones as factors in floral initiation and development. *Botanical Gazette* 100:388–431.

Hillman, W. S. 1964. Endogenous circadian rhythms and the response of *Lemna purpusilla* to skeleton photoperiods. *American Naturalist* 98:323–328.

King, R. W., B. Cumming. 1972. Rhythms as photoperiodic timers in the control of flowering in *Chenopodium rubrum*. *Planta* 103:281–301.

Lang, A. 1957. The effect of gibberellin upon flower formation. *Proceedings of the National Academy of Sciences USA* 43:709–717.

Lincoln, R. G., D. L. Mayfield, A. Cunningham. 1961. Preparation of a floral initiating extract from *Xanthium*. *Science* 133:756.

Vince-Prue, D. 1975. *Photoperiodism in Plants*. New York: McGraw-Hill.

Zeevaart, J. A. D., 1958. Flower formation as studied by grafting. *Mede-delingen Landbouwhogeschool Wageningen*. 58:1–88.

Zeevaart, J. A. D. 1984. Photoperiodic induction, the floral stimulus and flower-promoting substances. In: D. Vince-Prue, B. Thomas, K. E. Cockshull (eds.), *Light and the Flowering Process*. London: Academic Press, pp. 137–142.

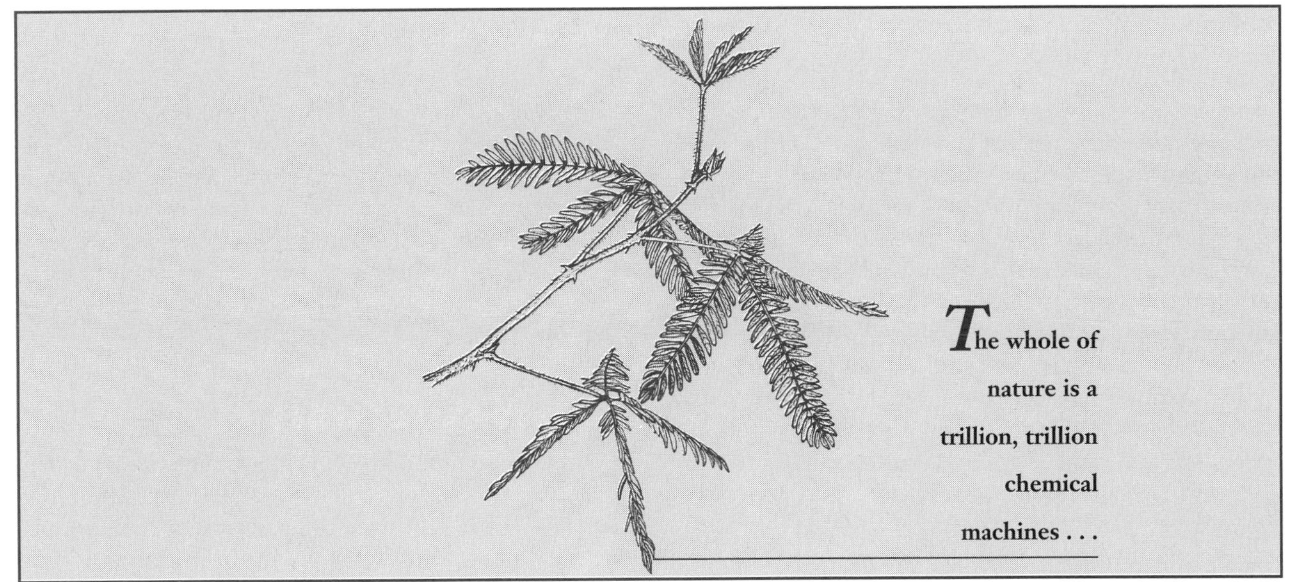

H. H. Seliger and W. D. McElroy (1965)

21

Temperature and Plant Development

One universal characteristic of chemical machines is their sensitivity to temperature. Along with light and water, temperature is one of the most critical factors in the physical environment of plants. It exerts a profound influence on plant growth and development, the geographic distribution of plants, and even their survival. All of the chemical machinery of nature—every individual enzymic reaction, every metabolic function, every physiological process—has temperature limits above and below which it cannot function and an optimum temperature range where it proceeds at a maximum rate. Temperature also affects the integrity of cell structure (especially the structure and properties of membranes), limits the distribution of species in space and time, and influences the direction of specific developmental events.

In this chapter we will introduce some of the more significant ways in which temperature is known to influence plant growth, development, and distribution. Specific topics include:

- a brief summary of temperature as a factor in the plant's environment;
- the influence of temperature on the metabolism and growth of plants, with particular reference to the influence of temperature in the natural environment; and
- the influence of temperature on development, in particular the phenomenon of vernalization as it affects flowering behavior, and temperature influences on bud and seed dormancy.

Many developmental responses involve complex interactions between temperature and other factors, which will be examined in the course of our discussion. In addition to the role of temperature in the "normal" growth and development of plants, extremes of temperature often represent a major stress for plants. This topic will be developed further in the next chapter (Chap. 22).

TEMPERATURE IN THE PLANT ENVIRONMENT

The range of temperatures encountered over most of the earth's surface is generally considered favorable for life as we know it for a variety of reasons. Most importantly, the temperature at which biological processes can occur is generally limited by the freezing point of water on the low side and the irreversible denaturation of proteins on the high side (see Chap. 13, Fig. 13.6). Between these two extremes, individual organisms and species exhibit for their growth and survival the same **cardinal temperatures** (minimum, optimum, maximum) that apply to individual enzyme reaction or multiple enzyme-catalyzed metabolic sequences.

Green plants probably first evolved in the tropical

regions, not so much because of warmer temperatures (although that may have been a factor), but because the temperatures there were relatively stable. With time, plants gradually migrated into the temperate and polar regions as they adapted to wider variations in temperature on a daily and seasonal basis. Green plants are now found in regions as extreme as the antarctic continent and northern tundras, where temperatures over much of the year are near or below freezing, and in the warmest places on earth such as Death Valley (California), where summer temperatures commonly approach or even exceed 50 °C.

Living organisms may be broadly classified according to their ability to withstand temperature. Those that grow optimally at low temperatures, that is, between 0 °C and 10 °C, are called **psychrophiles.** The psychrophiles include primarily algae, fungi, and bacteria. Most higher plants fall into the category of **mesophiles,** whose optimum temperatures lie roughly between 10 °C and 30 °C. **Thermophiles** will grow unhindered at temperatures between 30 °C and 65 °C, although there are reports of cyanobacteria growing at temperatures as high as 85 °C. These temperature ranges apply to hydrated, actively growing organisms. Most dry seeds, with moisture contents as low as 5 percent, are able to withstand a much broader range of temperatures for extended periods of time.

INFLUENCE OF TEMPERATURE ON GROWTH AND PLANT DISTRIBUTION

Temperature is a principal factor in limiting the distribution of plants. Often the distribution limits reflect temperature characteristics of major metabolic processes.

The temperature range compatible with growth of higher plants lies generally between 0 °C and 45 °C. Within those limits, temperature compatibility is very much species dependent. Various cultivars of wheat (*Triticum vulgare*), for example, will grow at temperatures from near zero to over 40 °C, although the temperature optimum for growth falls in the range of 20 °C to 25 °C. The optimum for maize (*Zea mays*), a plant of tropical origin, is in the range 30 °C to 35 °C and it will not grow below 12 °C to 15 °C. Garden-cress (*Lepidium sativum*) will grow at temperatures as low as 2 °C but its maximum temperature for growth is 28 °C. As a general rule, temperatures optimum for growth reflect the geographical region in which the species originated. Thus, plants native to warm regions either require or perform better at higher temperatures than those that originated in cooler areas of the world. The effects of temperature on physiology and metabolism in turn influence plant distribution, called **biogeography.** At times, temperature-related metabolic effects not only limit distribution, but have significant economic implications as well. Cotton (*Gossypium*), for example, is a southern crop in part because cool night temperatures in northern latitudes adversely affect fiber cell wall thickening (Roberts et al., 1992). The northern limit for maize production is very much limited by its inability to grow at lower temperatures.

COASTS AND DESERTS

A striking example of the impact of temperature on plant growth and distribution is seen in experiments conducted by O. Björkman and his colleagues at the Carnegie Institution of Washington in California (Björkman et al., 1973a). This study is included here because it illustrates the kinds of problems plants face in environments with temperature extremes and some of the ways they might deal with these problems. A variety of species native to either the cool coastal regions of northern California or to the hot, dry desert of Death Valley were transplanted into experimental gardens in both locations. The Death Valley site was located at (appropriately named!) Furnace Creek, where summer air temperatures commonly reached 50 °C. A temperature this high is lethal for many organisms. By contrast, average daily maximum temperatures at the coastal site were less than 20 °C.

Plants at both sites were irrigated and fertilized so that water supply and nutrients were not limiting factors and their performance with respect to growth and survival was assessed on a regular basis. On the basis of their growth responses, the plants could be grouped into three main categories: (1) those that were unable to survive the summer months; (2) those that survived but grew slowly during the summer; and (3) those that grew most rapidly during the summer months (Table 21.1).

Virtually all of the species native to the cool coastal climate were unable to survive the high desert temperatures. Of the plants tested, only *Tidestromia oblongifolia*, a deciduous C4 perennial native to Death Valley, was able to thrive in the summer desert heat. Strikingly, *T. oblongifolia* was unable to survive the cool coastal temperatures. At the other extreme, *Atriplex glabriuscula*, a C3 annual native to the coastal region, thrived in the coastal garden but died in the desert in spite of ample irrigation. Two clones of the C4 species *Atriplex lentiformis* were also tested; one native to the coastal regions of southern California and one that occurs naturally in Death Valley. In terms of biomass production, the desert clone outperformed the coastal clone in the Death Valley garden; their relative performance was reversed in the coastal garden.

In order to assess the performance of these plants with respect to temperature under more carefully controlled conditions, additional experiments were con-

TABLE 21.1 Growth responses of selected *Atriplex* and *Tidestromia* species planted in hot desert and cool coastal climates.

Summer Growth	Death Valley Garden	Coastal Garden
1. No survival	*A. glabriuscula*	*T. oblongifolia*
2. Slow summer growth	*A. lentiformis* (Coastal clone)	
	A. lentiformis (Desert clone)	
3. Rapid summer growth	*T. oblongifolia*	*A. glabriuscula*
		A. lentiformis (Coastal clone)
		A. lentiformis (Desert clone)

Data from Björkman et al., 1973a.

ducted using controlled environment facilities (Björkman et al., 1973b). The relative growth responses of *A. glabriuscula* and *T. oblongifolia* to temperature are shown in Figure 21.1. Maximum growth rate of the coastal species *A. glabriuscula* was achieved with day and night temperatures of 25 °C and 17 °C, respectively. Growth of *A. glabriuscula* fell off sharply with daily temperatures above 35 °C. *T. oblongifolia*, on the other hand, grew maximally under the hottest temperature regime tested (44 °C day/30 °C night) and showed little or no growth under a 16 °C day/10 °C night regime. In all these experiments, water and nutrient supplies were nonlimiting and other potentially stressful factors (such as wind, salt spray, and predators) were eliminated. Therefore, the differing abilities of *A. glabriuscula* and *T. oblongifolia* to survive and grow in such contrasting habitats as the cool coastal garden and the hot desert floor must be due primarily to intrinsic differences in the temperature-dependence of their growth responses (Björkman et al., 1973b).

What accounts for the success of *T. oblongifolia* under the extraordinarily high temperatures of Death Valley? It appears to be related to some extent to a superior photosynthetic performance of *T. oblongifolia* at high temperature. The temperature-dependence of photosynthesis for both *T. oblongifolia* and *A. glabriuscula* (Fig. 21.2) appears to parallel very closely the effect of temperature on relative growth rate (Fig. 21.1). The coastal C3 species *A. glabriuscula* is capable of high rates of photosynthesis at low and moderate temperatures but, as with growth, photosynthesis falls off sharply near 40 °C. *T. oblongifolia*, on the other hand, photosynthesizes poorly at low and moderate temperatures and reaches a maximum at 45 °C to 50 °C. Because *T. oblongifolia* is a C4 species, it might be expected to outperform *A. glabriuscula*, a C3 species, under desert conditions. Some of the metabolic reasons for this generalization were discussed earlier (Chap. 10). While improved water use efficiency might be a factor under natural circumstances, it should not be so important when the plants are well irrigated as they were in transplant experiments. Moreover, *T. oblongifolia* also outperforms other species of

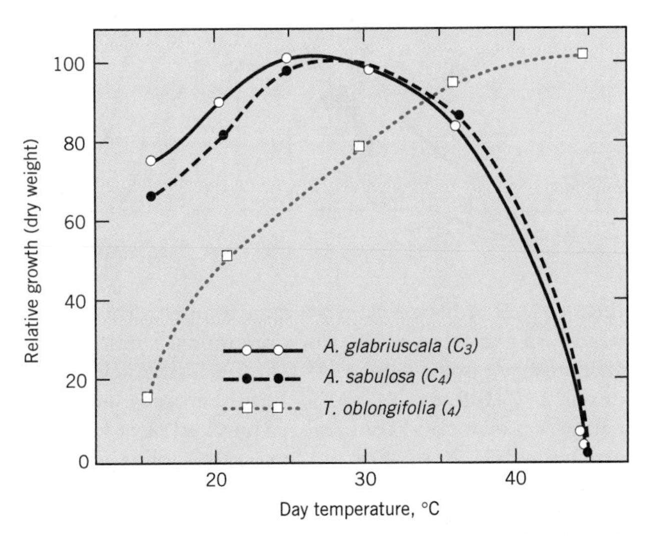

FIGURE 21.1 Relative daily growth rate of *A. glabriuscula* and *T. oblongifolia* as a function of growth temperature. (From O. Björkman et al., 1973a. Reprinted by permission.)

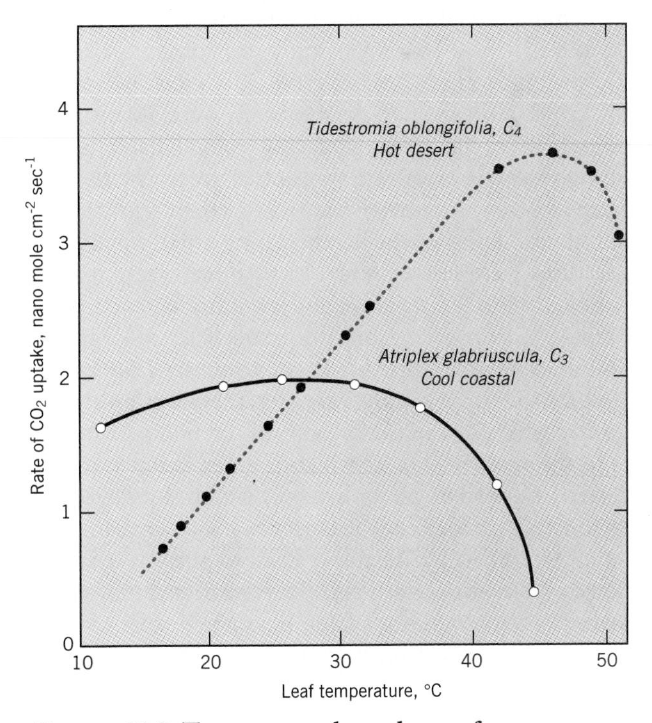

FIGURE 21.2 Temperature dependence of photosynthesis in *A. glabriuscula* and *T. oblongifolia* under high light conditions. CO_2 concentrations and stomatal conductance were equivalent. (From O. Björkman et al., 1973b. Reprinted by permission.)

Atriplex which, even though they are C4 species, also fail to survive the desert summers. Finally, differences between the two genera are not limited to photosynthetic metabolism—the response of respiration to temperature is much the same as photosynthesis.

The superior performance of *T. oblongifolia* is unrelated to differences in stomatal conductance, the diffusive transport of CO_2 to fixation sites inside the cells, or other factors that might be expected to limit photosynthetic capacity. It appears that the photosynthetic apparatus (and probably the respiratory apparatus as well) of *T. oblongifolia* is intrinsically more stable to high temperature. This is illustrated by comparing photosynthesis in plants of each species grown at both high and low temperatures. When *A. glabriuscula* was grown at 17 °C, the temperature optimum for photosynthesis was approximately 25 °C. When grown at temperatures near 40 °C there was a slight shift in the optimum toward higher temperatures, but the maximum rate of photosynthesis was reduced to about one-half that of the cool-grown plants. The response of *T. oblongifolia* to growth temperature was similar. When grown at cool temperatures, the optimum temperature for photosynthesis was shifted to 30 °C from the normal 42 °C. However, the maximum rate of photosynthesis in the cool-grown plants was reduced by a factor of six compared with plants grown at 42 °C. Clearly the capacity to assemble a competent photosynthetic apparatus is a major factor in determining where these two species can survive.

Other species are more flexible with regard to temperature. In the Björkman study, for example, several species, including *A. hymenelytra*, *Nerium oleander*, and the creosote bush (*Larrea divaricata*), were able to survive in both the desert and cool coastal habitats, although their growth rate was not as great as either *A. glabriuscula* or *T. oblongifolia*. Most of their growth was in fact accomplished during the spring or fall when temperatures were less extreme. In all three cases, growth at low or high temperature under controlled conditions caused an appropriate shift (by as much as 15 °C) in the optimum temperature for photosynthesis (Björkman, 1980). More importantly, however, there was no significant change in the maximum rate of photosynthesis, only the temperature at which the maximum rate occurred. Thus some plants exhibit a significant degree of phenotypic plasticity with respect to photosynthesis and temperature, which enables them to survive a wider range of climatic conditions. Plants restricted to one climate or another do not exhibit the same degree of plasticity in their metabolic reactions.

MOUNTAIN SLOPES

Another example of the relationship between temperature and environment is the distribution of plants along an elevational gradient, such as up a mountainside where temperature decreases with increasing altitude. This decrease in temperature with elevation is called the **adiabatic lapse rate.** As air rises, it expands and cools. The term *adiabatic* refers to the fact that cooling occurs without an exchange of heat; cooling as the air rises is entirely due to expansion of the air mass as the pressure decreases. It is because of adiabatic lapse that the temperature gradient remains stable and the cooler air does not descend from high in the mountains to displace the warmer air in the valleys. The adiabatic lapse rate for dry air is constant at about 1 °C/100 m elevation. The lapse rate for moist air (wet lapse rate) is more variable and lower than the dry lapse rate because condensation of water vapor releases heat (Fig. 21.3).

P. W. Rundel has studied the distribution of C3 and C4 grasses along an elevational gradient in Hawaii (Rundel, 1980). Rundel found a sharp transition in the distribution of the two photosynthetic types at about 1400 m. C4 grasses were predominant at warmer, drier elevations below 1400 m, while in the cooler, moist environment above 1400m the C3 grasses were predominant. The midpoint of the transition zone is the elevation where the maximum daily temperature for the warmest month of the year is approximately 21 °C. Similar distributions of C4 and C3 grasses have been reported in other elevational studies carried out in Africa and Costa Rica, and in latitudinal gradients in North America. In the latter case, the transition temperatures are slightly lower, but the principle is still valid.

The exact physiological basis for differences in plant

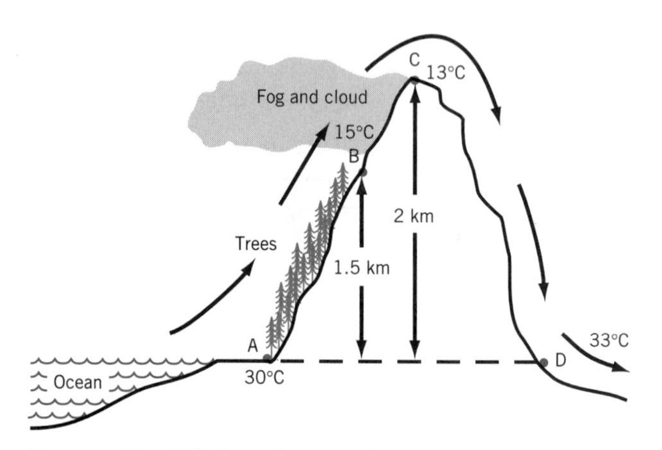

FIGURE 21.3 Adiabatic lapse and air temperature on the windward and leeward sides of a mountain. Unsaturated air rises from point A to point B at the adiabatic lapse rate of 1 °C/100 m. At point B the air becomes saturated with water vapor and cools more slowly at a wet lapse rate of 0.4 °C/100 m. The dry lapse rate applies as the air warms on its descent to D and is no longer saturated. This warming trend is responsible for the Chinook winds that blow out of the Rocky Mountains or the Foehn winds of alpine Europe. (From N. J. Rosenberg et al., *Microclimate. The Biological Environment*, New York, Wiley, 1983. With permission.)

distribution is far from fully understood, but these kinds of studies provide some measure of support for the generalization that plants with the C4 photosynthetic pathway enjoy an advantage in hot, arid environments. At the other extreme, low temperature is a major factor regulating the northerly limits of most agriculturally significant species. Although it would no doubt be overly simplistic to expect a clear-cut distribution based on a single environmental parameter such as temperature, it is clear that temperature does play a major role.

INFLUENCE OF TEMPERATURE ON DEVELOPMENT

While the temperature in tropical climes is relatively stable, plants growing in temperate regions and closer to the poles are subject to more or less predictable variations in temperature on a daily and seasonal basis. It is perhaps not too surprising that plants have evolved ways to incorporate this information in their developmental and survival strategies. Plants use this information to ensure dormancy of buds, tubers, and seeds, and to modify their flowering behavior, all of which appear to be keys to survival over periods unfavorable to normal growth and development.

TEMPERATURE AND FLOWERING RESPONSE

There are many examples of interactions between temperature and photoperiod, particularly with respect to flowering behavior (Salisbury, 1963; see also the discussion by Hillman, 1962). In most cases the interaction results in relatively subtle changes in the length of the critical photoperiod or a tendency toward day-length neutrality or an inability to flower altogether at high or low temperature extremes. There are other plants, however, for which flowering is either quantitatively or qualitatively dependent on exposure to low temperature. This phenomenon, known as **vernalization,** was introduced briefly in Chapter 20.

Vernalization refers specifically to the promotion of flowering by a period of low temperature and should not be confused with other miscellaneous effects of low temperature on plant development. The term itself is a translation of the Russian *yarovizatsya:* both words combining the root for spring (Russian, *yarov;* Latin, *ver*) with a suffix meaning "to make" or "become" (Hillman, 1962). Coined by the Russian T. D. Lysenko in the 1920s, *vernalization* reflects the ability of a cold treatment to make a winter cereal behave like a spring cereal with respect to its flowering behavior. The response had actually been observed many years earlier by agriculturalists, but didn't receive critical attention of the scientific community until J. G. Gassner showed in 1918 that the cold requirement of winter cereals could be satisfied during seed germination. For his part, Lysenko received considerable notoriety for his conviction that the effect was an inheritable conversion of the winter strain to a spring strain. His position—a form of the thoroughly discredited Lamarkian doctrine of inheritance of acquired characteristics—was adopted as Soviet dogma in biology and remained so until the 1950s. The adoption of Lysenko's views as official dogma had a significant impact on Soviet biology and placed agriculture in the USSR at a severe disadvantage for decades.[1]

Occurrence of Vernalization Vernalization occurs most commonly in winter annuals and biennials. Typical winter annuals are the so-called "winter" cereals (wheat, barley, rye). "Spring" cereals are normally planted in the spring, come to flower and produce harvestable grain before the end of the growing season. Winter strains planted in the spring would normally fail to flower or produce mature grain within the span of a normal growing season. Winter cereals are instead planted in the fall. They germinate and overwinter as small seedlings, resume growth in the spring, and are harvested usually about midsummer.

One of the most thorough studies of vernalization and photoperiodism has been carried out on the Petkus strain of rye (*Secale cereale*) by F. G. Gregory and O. N. Purvis beginning in the 1930s. The spring strain of Petkus rye is a quantitative, long-day plant (LDP) (see Chap. 20). Under short days, floral initiation does not occur until after about 22 leaves have been produced, typically requiring a season of about 4.5 months. Under the appropriate long-day regime, however, flowering in the spring strain is initiated after only about seven leaves have been produced (requiring about two months). The winter strain, on the other hand, is not a LDP. Germinated at normal temperature, the winter strain flowers equally slowly—requiring four to five months—regardless of daylength. Planted in the fall, however, the winter strain receives an overwintering low temperature treatment. When it resumes growth in the spring, the winter strain responds to photoperiod in exactly the same way as the spring strain. The effect of the overwintering cold treatment can also be achieved by vernalizing the seed; that is, by holding the *germinated seed* near 1 °C for several weeks (Purvis, 1934). Note that the low temperature treatment, at least in the case of winter annuals, does not alone promote early flower initiation.

[1]The story of vernalization is a classic example of what can happen when science becomes enmeshed in political ideology. For the interested student, this unfortunate episode in the history of science has been artfully documented by D. Joravsky in his book The *Lysenko Affair* (Chicago: University of Chicago Press, 1970).

Rather, the effect of vernalization is to *render the seedling sensitive to photoperiod.*

Biennials are monocarpic plants that normally flower (and die) in the second season, again following an overwintering cold treatment. Typical biennials include many varieties of sugar- and table beet (*Beta vulgaris*), cabbages and related plants (*Brassica oleraceae*), carrots (*Daucus carota*) and other members of the family Umbellifereae, foxglove (*Digitalis purpurea*), and some strains of black henbane (*Hyoscyamus niger*). Biennials share with the winter annuals the property that subjecting the growing plant to a cold treatment stimulates a subsequent photoperiodic response.

Biennials typically grow as a rosette, characterized by shortened internodes, in the first season (Fig. 21.4). Over winter, the leaves die back but the crown, including the apical meristem, remains protected. New growth the following spring is characterized by extensive stem elongation, called **bolting,** followed by flowering. The cold requirement in biennials is qualitative (i.e., absolute). In the absence of a cold treatment many biennials can be maintained in the nonflowering rosette habit indefinitely. As a rule, vernalizable plants, whether winter annuals or biennials, tend to respond as long-day flowering plants, although some biennials are day-indifferent following vernalization. One exception to the rule is the perennial *Chrysanthemum morifolium*, a SDP. Some varieties of *Chrysanthemum* require vernalization before responding as a quantitative SDP. As a perennial, *Chrysanthemum* normally requires vernalization on an annual basis (Schwabe, 1950). Many other plants such as pea (*Pisum sativum*) and spinach (*Spinacea oleracea*) can be induced to flower earlier with a cold treatment but it is not an absolute requirement.

Effective Temperature The range of temperatures effective in vernalization varies widely depending on the species and duration of exposure. In Petkus rye, the effective range is −5 °C to +15 °C, with a broad optimum between +1 °C and +7 °C. Within these limits, vernalization is proportional to the duration of treatment. Flowering advances sharply after as little as one to two weeks treatment at 1 °C to 2 °C and is maximally effective after about seven weeks at that temperature (Fig. 21.5). Within the effective range, the temperature optimum is generally higher for shorter treatment periods. Presumably, a longer exposure to lower temperatures within the effective range is required because the metabolic reactions leading to the vernalized state progress more slowly.

FIGURE 21.4 Vernalization and stem elongation in cabbage (*Brassica* sp.). Left: Cabbage plants were vernalized for six weeks at 5 °C. Center: Plants were sprayed weekly with a solution containing 5 × 10^{-4}M gibberellic acid. Right: Control plants remain in a rosette habit. Except for the vernalization treatment, all plants were maintained in the greenhouse under a 16-hour photoperiod.

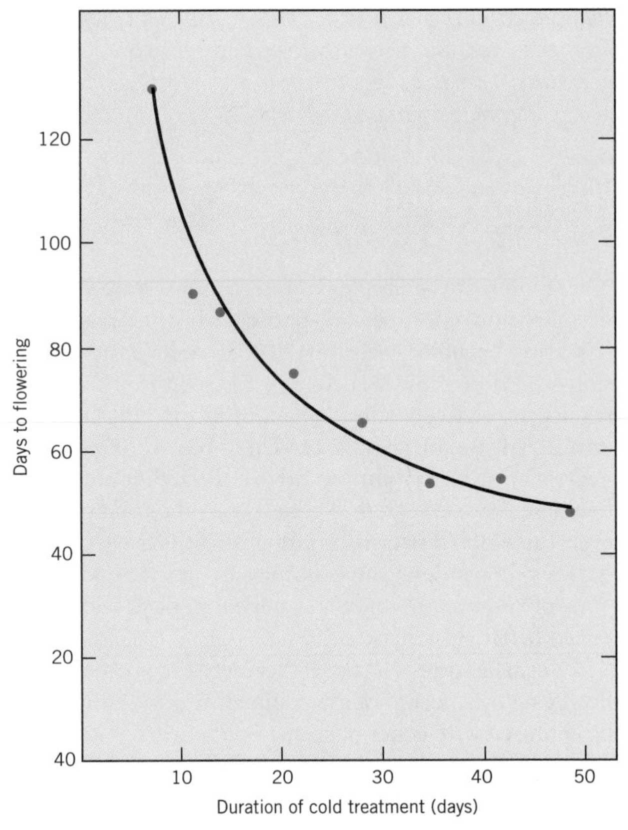

FIGURE 21.5 Vernalization in Petkus rye (*Secale cereale*). Seeds were germinated in moist sand at 1 °C for the time indicated. Cold treatments were scheduled so that all seeds were returned to the greenhouse at the same time. The number of days to flowering progressively decreased with increasing length of the cold treatment. (From O. N. Purvis, F. G. Gregory, *Annals of Botany*, N.S. 1:569–591, 1937. Copyright The Annals of Botany Company.)

Like flowering, the vernalized state is more or less permanent in most species, giving rise to the concept of an **induced** state. Vernalized *Hyoscyamus*, a LDP, can be held under short days for up to ten months before losing the capacity to respond to long day treatment. On the other hand, all cold-requiring plants that have been studied are capable of being **devernalized;** vernalization can be reversed if followed immediately by a high temperature treatment. Flowering in vernalized winter wheat, for example, can be fully nullified if the seedlings are held near 30 °C for three to five days. For most plants, then, there is a "neutral" temperature where neither vernalization nor devernalization occurs. For Petkus rye the neutral temperature is about 15 °C. Vernalized seeds of Petkus rye can also be devernalized by drying them for several weeks or by maintaining the seeds under anaerobic conditions for a period of time following the cold treatment.

Perception of the Stimulus A vernalization treatment is effective only on actively growing plants—cold treatment of dry seeds will not suffice. Thus winter cereals may be vernalized as soon as the embryo has imbibed water and the germination process has been initiated. Other plants, in particular the biennials, must reach a certain minimum size before they can be vernalized. *Hyoscyamus*, for example, is not sensitive before 10 days of age and does not reach maximum sensitivity until 30 days of age. In either case, the cold treatment appears to be effective only in the meristematic zones of the shoot apex. This can be shown by localized cooling treatments or vernalization of moistened embryos. Early studies by Gregory and Purvis showed that even the cultured apex of isolated rye embryos was susceptible to vernalization. Thus the induced state established in a relatively few meristematic cells can be maintained throughout the development of the plant. Most biennials, however, cannot be induced as seed. In these plants it is the overwintering stem apex that perceives the stimulus, although there are some reports suggesting that leaves and even isolated roots may be susceptible in some cases.

The Nature of the Vernalization Stimulus Experiments with isolated embryos have also shown that vernalization treatments are effective only when the embryo is supplied with carbohydrate and oxygen is present, indicating that it is an energy-dependent metabolic process. Still, the nature of the induced state has eluded researchers for many years. To the extent that the meristem itself is the site of perception, the necessity for a transmissible hormone appears to be ruled out. A cold-induced, permanent change in the physiological or genetic state of the meristematic cells would be self-propagating, that is, it could be passed on to daughter cells by cell division. There is some support for this argument. In plants such as Petkus rye and *Chrysanthemum*, only tissue produced in a direct cell line from the induced meristem is vernalized. If the cold treatment is localized to a single apex, it will flower, but all the buds that did not receive the cold treatment will remain vegetative. In other experiments, especially with *Hyoscyamus*, transmission of the vernalized state across a graft union has been demonstrated. A list of successful experiments has been tabulated by Lang in his 1965 review (Lang, 1965). These experiments are comparable to the transmission of "florigen" across a graft union (Chap. 20) and result in flowering in nonvernalized receptor plants. If a vernalized *Hyoscyamus* plant is grafted to an unvernalized plant, both will flower under long days. Transmission requires a successful (i.e., living) graft union and appears to be coordinated with the flow of photoassimilate between the donor and receptor.

Experiments such as these led G. Melchers to propose the existence of transmissible vernalization stimulus called **vernalin.** Like florigen, vernalin has resisted all attempts at isolation and remains a hypothetical substance. Unfortunately, the vernalin story is to some extent clouded by interpretation. The grafting experiments all require vernalization followed by long days. They do not unequivocally distinguish between the transmission of "vernalin" and the possibility that the unvernalized partner is responding to the floral stimulus itself, which would be transmitted from the vernalized donor under long days.

Adding to the complexity of vernalization is the apparent involvement of gibberellins in the response to low temperature (Fig. 21.4). This was dramatically demonstrated by A. Lang in 1957 when he showed that repeated application of 10 μg of GA_3 to the apex would stimulate flowering in nonvernalized plants of *Hyoscyamus* and several other biennials maintained under short days (Lang, 1957). No such promotion occurred in *Xanthium* and other SDP treated with gibberellin under noninductive long days. Subsequently it has been shown that gibberellin levels tend to increase in response to low temperature treatments in several cold-requiring species.

Results such as these have raised the question: Are vernalin and gibberellin equivalent? The answer is most likely no, for several reasons. It is true that gibberellin appears to substitute for the cold requirement of some vernalizable plants and for the long-day requirement in some LDP, or, in the case of vernalization, both. But it is important to note that virtually every situation in which gibberellin has successfully substituted for low temperature or long days in promoting flowering involves bolting—the rapid elongation of stems from the rosette vegetative state. Far less success has been achieved with gibberellins in **caulescent** LDP—those whose stems are already elongated in the vegetative state. Moreover, the developmental pattern in responsive plants differs significantly depending on whether stem elongation is stimulated by low temperature or gibberellin treatment. Following low temperature treatment, flower buds are evident at the time stem elongation begins. Following gibberellin treatment, on the other hand, the stem first elongates to produce a vegetative shoot. Flower buds do not appear until later. One resolution to the problem has been suggested by M. Chailakhyan, who has modified his anthesin hypothesis (described in the previous chapter) to suggest that vernalin is simply a gibberellin precursor. Vernalin would be accumulated in response to cold treatment in those plants requiring vernalization, but long days are required to complete its conversion to gibberellin. Unfortunately, these hypotheses are not very satisfying since both anthesin and vernalin remain hypothetical. Clearly, the regulation of floral induction is one of the most challenging problems facing plant physiologists today.

BUD DORMANCY

The growth of deciduous perennial plants typical of temperate regions is cyclical—such plants normally undergo a cessation of active growth on an annual basis. Buds formed in the latter part of the summer or early fall overwinter in a dormant condition and do not normally renew their growth until more favorable temperature conditions return in the spring. Bud dormancy is thus a defense mechanism, helping to ensure that the plant is able to survive the adverse conditions of winter. To ensure survival, dormancy mechanisms must be in place *before* the arrival of unfavorable conditions. This means that the plant must be able to anticipate climatic change. Mechanisms must also be in place to ensure that the buds do not break dormancy until such time that conditions in the environment are appropriate to sustain normal growth and development. Premature breaking of buds during an unseasonably warm period in the winter, for example, could have serious consequences. In other words, dormancy is more than simply a period of inactivity. Dormancy is a precisely regulated phenomenon, cued by factors in the environment and maintained, and ultimately broken, by specific metabolic changes in the organism.

What is **dormancy**? The term is generally used to indicate tissues such as buds, seeds, tubers, and corms that fail to grow even though they are provided with adequate moisture and oxygen at an appropriate (i.e., physiological) temperature. Unfortunately, dormancy terminology has been quite inconsistent and can be confusing. Terms such as quiescence, rest, true dormancy, and imposed or enforced dormancy have been used by different authorities to describe various states and/or conditions. Part of the problem is that dormancy is a progressive process, often occurring in degrees. Buds that are just entering dormancy, for example, may be stimulated to renew growth rather easily. On the other hand, buds that have developed full dormancy may require extensive or severe treatment to break dormancy and renew growth. Also, failure to grow may be "imposed" by factors intrinsic to the tissue or may result from environmental limitations. In most cases, it appears that both intrinsic and extrinsic factors interact to induce, maintain, and release the dormant state.

Dormancy studies have focused on three principal questions: (1) What are the environmental signals that stimulate the onset of dormancy and how are they perceived? (2) What metabolic changes are responsible for the reduced activity?, and (3) What signals the startup of renewed growth at the appropriate time?

The onset of bud dormancy is coincident with leaf fall, decreased cambial activity, and increased cold hardiness (to be discussed further in Chap. 22). Bud dormancy appears to be a typical short-day response (Chap. 20), initiated by the short days of late summer. In most cases, phytochrome is involved and the leaf perceives the photoperiodic stimulus. However, in at least one case, that of birch (*Betula pubescens*), the response appears to be localized in the apical bud.

Changes in hormone levels coincident with dormancy have been detected, but it is not always clear whether these changes are causative or simply a result of reduced growth. In some species, a decline in gibberellin and auxin levels can be detected prior to the cessation of bud growth and the onset of dormancy. In the late 1940s and into the 1950s, a number of investigators reported large increases in growth-inhibiting activities of crude extracts from tissues entering dormancy. Indeed, it was these experiments that eventually led to the discovery of the hormone abscisic acid (ABA) (Chap. 16). Originally called "dormin" (i.e., giving rise to dormancy), an increase in ABA level was thought to be a principal causative factor in the onset of bud dormancy. In some experiments, application of ABA has suppressed or delayed breaking dormancy on buds of woody branches. More recent evidence, however, based on more sensitive analytical techniques such as gas-liquid-chromatography (GLC) have cast doubt on any such role for endogenous ABA. In a number of species, the ABA content of leaves and buds actually decreases with the onset of dormancy. There is, in fact, a considerable amount of conflicting evidence relating ABA levels to dormancy (Phillips et al., 1980). Other inhibitory factors may be involved but, if so, their identity remains unknown.

Relatively little is known about the physiological state of buds during this period of inactive growth. It appears to be characterized by low respiratory activity and an inability to grow even if temperature, oxygen, and water supply are adequate. A bud is actually a meristem in which the internodes have failed to elongate

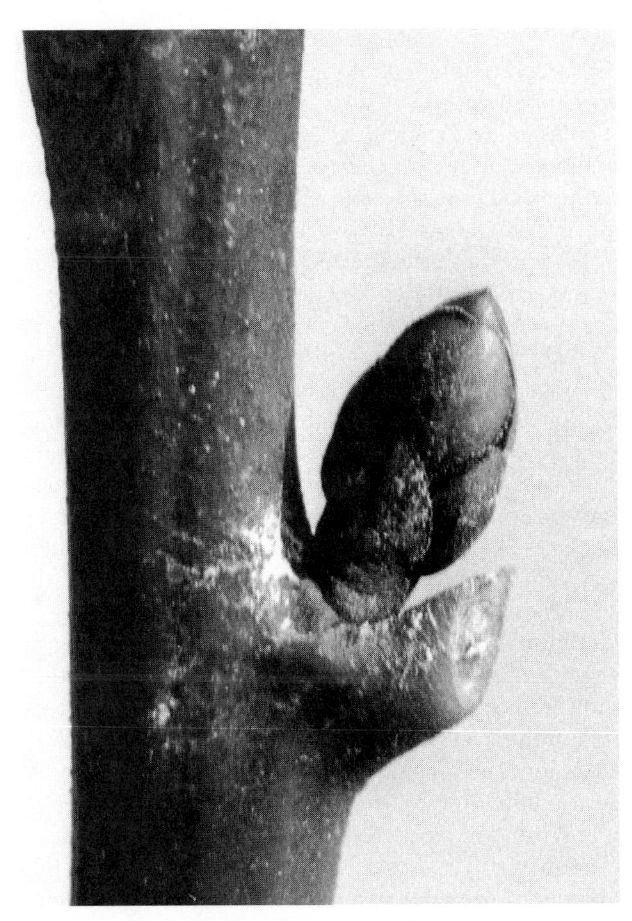

FIGURE 21.6 An axillary bud of sweet bay (*Laurus noblis*). Note the prominent bud scales that enclose and protect the bud.

and leaves have not enlarged. The whole is enclosed in a set of modified leaves, called **bud scales** (Fig. 21.6). Bud scales serve a protective function, insulating the bud and preventing desiccation. Bud scales also contribute to reduced respiration since their removal usually results in a significant increase in respiratory rate. There is some older evidence (dating from the 1960s) that inhibitors of nucleic acid and protein synthesis can prevent the onset of dormancy in some buds, indicating that dormancy may involve transcription of new genetic information. Modern recombinant DNA techniques might be usefully applied to the study of bud dormancy.

Dormancy can be induced by low temperature in concert with short days, but the principal role of temperature appears to be in breaking dormancy. In many cases a period of low temperature is required before active growth resumes. In other words, many dormant buds have a chilling requirement that must be met before the cells are capable of cell division and enlargement. Most studies on the chilling requirement for breaking dormancy have concentrated on commercial fruit species and deciduous ornamentals. This is because in the northern hemisphere, it is the chilling requirement that largely determines their southerly limits of cultivation. The process is especially critical in fruit trees since the flower bud that bears fruit is initiated in the previous summer. It then overwinters and, following satisfaction of the chilling requirement, continues its development the following spring.

Temperatures near or just above freezing appear to be most effective at breaking dormancy. The amount of chilling required varies with species, cultivar, and even location of the buds on the trees. Species such as apple (*Malus pumila*), pear (*Pyrus communis*), and cherry (*Prunus sps.*) require approximately 7 to 9 weeks of exposure to temperatures below 7 °C in order to overcome dormancy. Others may require up to 22 weeks (American plum, *Prunus americana*) or as few as 4 to 6 weeks (apricot, *Prunus armeniaca*). Persimmons (*Diospyros kaki*) require only 4 days of low temperature and so can be grown successfully much further south than other fruit trees. The temperature in temperate regions often varies widely throughout the winter, but this generally poses no problem for dormant tissues. In most cases, buds and other dormant tissues are able to sum the periods of cold and will not renew growth until the appropriate amount of cold treatment has been accumulated.

There is wide variation in the chilling requirement of different species and **ecotypes** or genetic races of maple. More than 12 weeks of chilling are required to break dormancy in sugar maple (*Acer saccharum*) collected in southern Canada while those collected from the warmer regions near the southern limits of its distribution required only a few weeks of low temperature. Similar results were obtained for seedlings of red maple (*Acer rubrum*) (Perry and Wang, 1960; Taylor and Dumbroff, 1975).

SEED DORMANCY AND GERMINATION

Seeds are in many respects similar to buds—they consist of a small embryonic axis (along with some storage tissues) enclosed by a series of membranes, collectively called the **seed coat.** The seed coat serves a protective function much as bud scales do. Its presence often suppresses germination by restricting the uptake of water and exchange of oxygen, it mechanically limits the expansion of the embryo and, in some cases, it contains inhibitors that prevent growth of the embryo. These limitations can be removed and the germination of many seeds accelerated by mechanically disrupting or removing the seed coat, a process called **scarification.** In the laboratory, scarification may be accomplished with files or sandpaper. In nature, abrasion by sand, microbial action, or passage of the seed through animal gut will accomplish the same end. Seed coats can be very tough. Uniformity and rate of germination of morning glory (*Pharbitis nil*), cotton, and some tropical legume seed, for example, can be improved by soaking in concentrated sulphuric acid for up to an hour. Scarification by passage through animal gut no doubt occurs as a result of the acidic conditions in the gut.

As with buds, dormancy in seeds refers to the situation wherein the embryo fails to grow because of physiological or environmental limitations. As noted above, limitations commonly include the inability of water or oxygen to penetrate the seed coat. Seeds of some plants, particularly in the family Leguminoseae, have specialized structures that control seed moisture content. E. Hyde described a structure in seeds of lupine (*Lupinus arboreus*) that functions as a hygroscopically operated check valve and limits imbibition of water by the seed (Hyde, 1954). Because water cannot pass through the unscarified seed coat, the only possible route of entry is through a small pore, called the **hilum** (Fig. 21.7). When the water content of the seed is higher than ambient, the hilum is open to permit the exit of water and allow the seed to dry. But when the moisture content outside the seed is higher than inside, cells surrounding the hilum swell, thus closing off the pore and preventing the uptake of water. In addition, as the seed dries out the permeability of the seed coat to water also decreases and the dormancy of the seed increases. Other seeds have pores that are blocked with a plug, called the **strophiolar plug,** which must be mechanically dislodged before water and oxygen can enter.

There is a considerable body of evidence to suggest that seed coats also interfere with gas exchange; oxygen uptake in particular. As noted earlier, removal of the

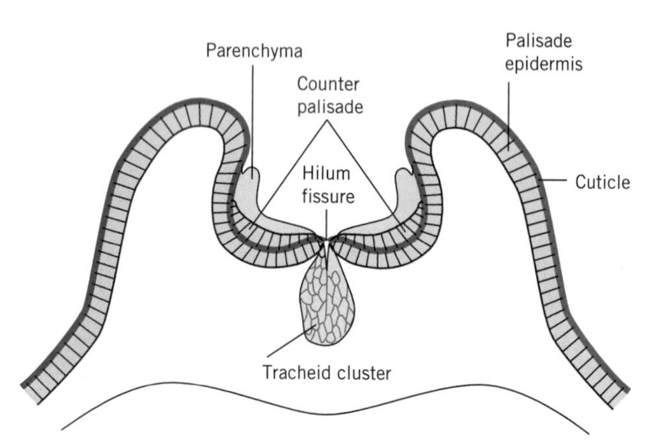

FIGURE 21.7 A cross-section through the seed of *Lupinus arboreus* (tree lupine) showing the hilum, a hygroscopic valve that regulates water loss from the seed. The counter palisade is a group of thick-walled cells lying on the *outer* surface of the cuticle. When the seed moisture content is higher than the moisture level in the ambient air, the hilum fissure is open and water escapes from the tracheid cluster. When the water ambient moisture level is higher than inside the seed, the counter palisade cells swell and close off the hilum fissure in order to prevent the seed from rehydrating. (From E. O. C. Hyde, *Annals of Botany* 18:241, 1954. With permission of The Annals of Botany Company.)

seed coat often leads to a significant increase in respiratory consumption of oxygen. Measurements of the oxygen permeability of seed coats have been made and there is general agreement that permeability is very low in those seeds tested. However, it is not always clear that limited oxygen permeability is the primary cause of dormancy. The complexity of the situation and problems of interpretation are well illustrated by studies of the genus *Xanthium*, or cocklebur (Bewley and Black, 1982).

A cocklebur contains two seeds: an upper, dormant seed and a lower, nondormant seed. Dormancy of the upper seed can be overcome either by removing the seed coat or by subjecting the intact seed to high oxygen tension. The inference is that seed coat permeability in the dormant seed limits the supply of oxygen to the embryo and thus prevents germination. However, several other observations have cast doubt on this conclusion. There are, for example, no measurable differences between the dormant and nondormant seed with respect to the permeability of the seed coat to oxygen. Moreover, the rate of oxygen diffusion through the seed coats is more than sufficient to support measured rates of oxygen consumption by the embryos inside. Clearly, dormancy of the upper seed in *Xanthium* cannot be due to limited permeability of the seed coat to oxygen.

Why then, do the upper, dormant seeds require a higher oxygen level to elicit germination? It appears that the seed coat is a barrier, not to the uptake of oxygen but to the removal of an inhibitor from the embryo. Aqueous extracts of *Xanthium* seeds have revealed the presence of two unidentified inhibitors, based on tests of the extracts in a wheat coleoptile elongation assay. The same two inhibitors are found in diffusate collected from isolated embryos placed on a moist medium, but not in diffusates from seeds surrounded by an intact seed coat. Thus germination in the dormant seed appears to be prevented by the presence of these inhibitors and the seed coat serves as a barrier that prevents those inhibitors from being leached out. The oxygen requirement can be explained by the observation that high oxygen tension reduces the amount of extractable inhibitor, presumably by some oxidative degradation.

Even the role of inhibitors in seed dormancy is not clear. Along with hormones such as auxins and gibberellins, a large number of inhibitors have been identified in seeds, fruits, and other dispersal units. These include hormones (ABA), unsaturated lactones (coumarin), phenolic compounds (ferulic acid), various amino acids, and cyanogenic compounds (i.e., compounds that release cyanide) characteristic of apple and other seeds in the family Rosaceae. The simple presence of an inhibitor does not, however, prove its role in dormancy. The inhibitors could be localized in tissues not directly involved in growth of the embryo or otherwise sequestered so as to preclude any role in preventing germination. Evidence in support of a role for inhibitors is generally limited to leaching experiments such as that described above for *Xanthium*. In some cases, dormancy can then be restored by exposing the leached seed to the inhibitor. In order to clearly establish whether an inhibitor has an active role in regulating germination, it is necessary to establish whether inhibitor levels in the seed correlate with the onset and termination of dormancy (Wareing, 1965). In spite of the voluminous literature relating inhibitors to dormancy, there is very little critical support for a direct role. For the present, evidence for the imposition and maintenance of dormancy by inhibitors remains largely circumstantial.

The dormancy and germination of seeds is also influenced by light and hormones. The role of light was described in Chapter 18. The roles of the hormones gibberellin and ABA were discussed in Chapter 17 and need no further comment here.

Temperature and Seed Dormancy Temperature has a significant impact on termination of dormancy in many seeds. In fully imbibed seeds, both alternating and low (chilling) temperatures are known to terminate dormancy.

Alternating Temperature. Many seeds, even though fully hydrated, will not germinate when maintained under constant temperature. They require instead a diurnal cycle of fluctuating temperature. The required temperature differential between the high and low temperature is often not great, ranging from a few degrees to perhaps 5 °C or 10 °C, depending on the species. Germination of broad-leaved dock (*Rumex obtusifolia*) seeds, for example, exceeds 90 percent when the temperature differential is about 10 °C and when the high temperature is given for 16 hours each day (Totterdell and Roberts, 1980).

The reaction to alternating temperature is complex and poorly understood. In *Rumex*, alternating treatments are effective only when the high temperature is greater than 15 °C. Also, when the high temperature is given for only 8 hours each day, a differential of only 5 °C is required to induce 90 percent germination. Although in some cases the effect of alternating temperature appears to be localized in the embryo itself, there are many well-documented cases where the effect of alternating temperature is mechanical. It is, in effect, a form of scarification, releasing the seed from some kind of seed coat—imposed dormancy.

Low Temperature. It has long been known that freshly shed seeds of many herbaceous and woody species have dormant embryos that can be induced to growth only by a prolonged low temperature treatment. These include maples (*Acer*), hazel (*Corylus*), and many genera in the family Rosaceae (pear, *Pyrus;* apple, *Malus;* hawthorne, *Crateagus*). Normally, following the required

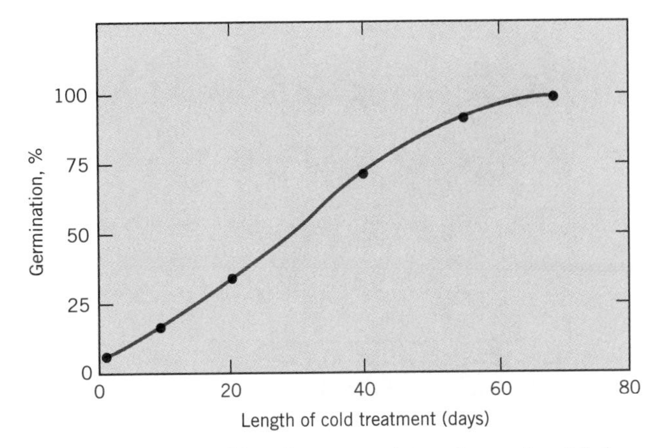

FIGURE 21.8 Breaking dormancy in apple seeds with low temperature. Moist seeds were held at 4 °C for the time indicated. (Redrawn from L. C. Luckwell, *Journal of Horticultural Science* 27:53, 1952.)

period of low temperature, the seeds will not germinate until temperatures are more favorable for embryo emergence and seedling development. In most cases this requirement ensures that the seed shed in late summer or fall will not germinate until spring.

The exposure to low temperature that satisfies this germination requirement is known as either **pre-chilling** or **stratification.** The latter term has its origin in the horticultural practice of layering seeds in moist sand or peat moss and exposing them to low temperature for several weeks or months to induce germination. It is important that the prechilling requirement for release of seed dormancy not be confused with vernalization, which is a cold treatment to an already germinated seedling, as discussed earlier in this chapter.

As with breaking of bud dormancy, temperatures near freezing but below 10 °C are most effective for terminating seed dormancy. The optimum for most species is near 5 °C. In a population of seeds, the effectiveness is also a function of the length of the cold treatment (Fig. 21.8).

It is presumed that seeds undergo some metabolic changes during the period of low temperature, but the exact nature of these changes is unclear. There is some evidence for redistribution of carbon and nitrogen from the endosperm to the embryo, a decline in the inhibitor content, and a rise in gibberellin and cytokinin content. Gibberellin treatment will substitute at least partially for the cold requirement in many seeds, just as they do in other cold-requiring systems.

RESPONSE TO CHANGES IN TEMPERATURE

Growers have long recognized the beneficial effect on plant growth of lowering greenhouse temperatures during the night. This effect has been particularly well doc-

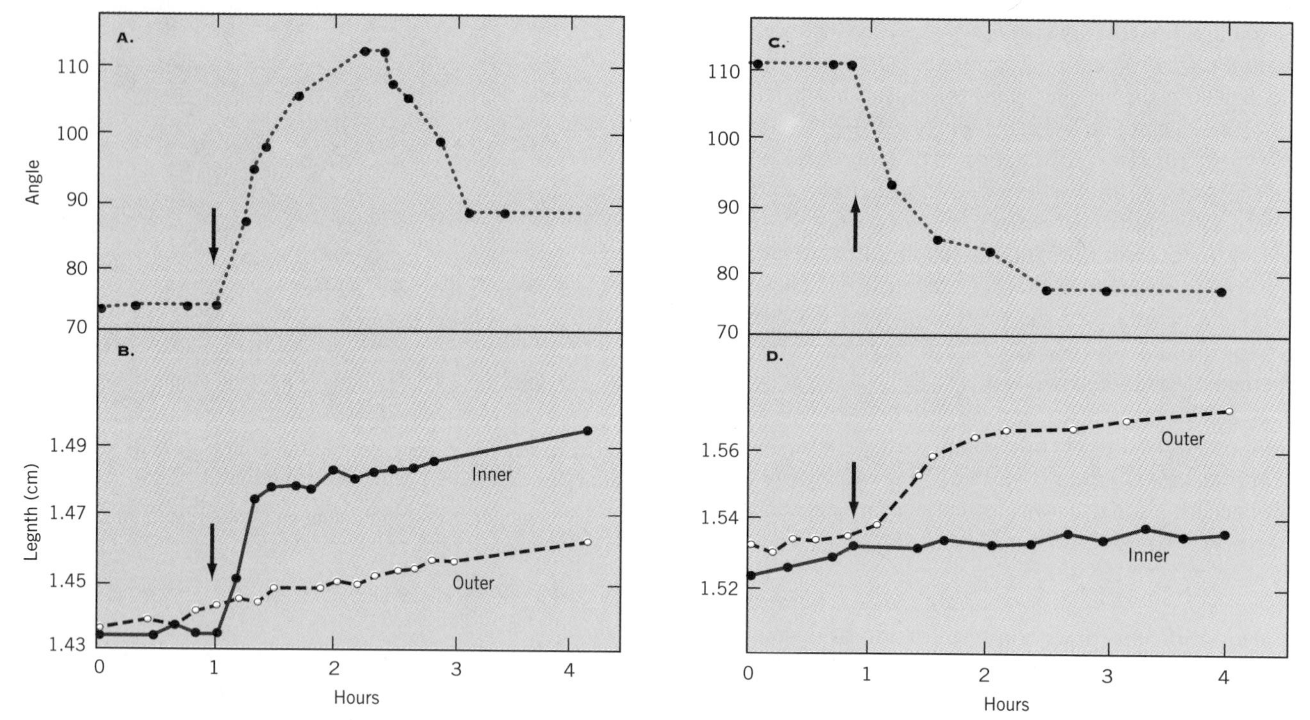

FIGURE 21.9 The effect of 10 °C temperature shifts on flower opening and differential growth in *Tulipa* flowers. (*A* and *B*) The effect of raising temperature from 7 °C to 17 °C. (*C* and *D*) The effect of lowering temperature from 20 °C to 10 °C. The temperature shift is indicated by the arrows. (*A*) and (*C*) show floral opening and closure, respectively. (*B*) and (*D*) show the growth of cells on the inner and outer surfaces of the tepals. (From W. M. Wood, *Journal of Experimental Botany* 4:65–77, 1953. Reprinted by permission of The Company of Biologists, Ltd.)

umented by the work of F. Went and his colleagues in the 1940s (Went, 1961). Went found that tomatoes (*Lycopersicum esculentum*) grown at constant temperatures of 26 °C and 18 °C grew poorly and (at 26 °C) failed to produce fruit. Plants maintained under alternating conditions of 26 °C during the day and 18 °C at night grew vigorously and produced a maximum number of fruit. In order to be effective, the day-night differential had to be synchronized with the light-dark cycle. If the temperature cycle was inverted, with the high temperature falling during the dark period, growth was even poorer than at a constant 26 °C. To describe this phenomenon, Went coined the term **thermoperiodism.**

It is now recognized that many, but certainly not all, plants perform better under regimes with a similar temperature differential. For those that do, the effect is primarily on vegetative development, in contrast to photoperiodism where the influence is primarily on floral production. In some plants, such as potato (*Solanum*) and tobacco (*Nicotiana*), low night temperature leads to a decline in shoot-to-root ratio, due to preferential root growth.

Another example of the effect of temperature differentials is illustrated by floral movements in members of the Liliaceae, such as tulip (*Tulipa*) and *Crocus*. Flowers in these plants normally open during the day and close at night, but these movements are only slightly affected by light. Instead, the perianth segments or tepals[2] respond to changing temperature. This is a form of **thermonasty,** involving a differential growth response of cells on the inner (or adaxial) and outer (or abaxial) surfaces of the tepals.

According to W. M. Wood, the optimum temperature for growth differs by approximately 10 °C between the two surfaces (Fig. 21.9) (Wood, 1953). The opening of the flower in response to an increase in temperature (Fig. 21.9A) corresponds to a sharp but transient increase in the growth rate of cells on the inner surface (Fig. 21.9B). Conversely, closure following a drop in temperature (Fig. 21.9C) appears to be caused by a similar change in the growth rate of cells on the outer surface (Fig. 21.9D). Other investigators have reported opening following a rise of as little as 0.2 °C for *Crocus* and 1 °C to 2 °C for tulip. There are lower limits—*Crocus*, for example, will not open at temperatures below

[2]**Tepal** is the collective term for sepals and petals when the two share a common morphology and are indistinguishable one from the other.

8 °C. Thus if the spring days are very cold, the flowers may not open at all.

SUMMARY

All living organisms can be broadly classified according to their ability to withstand temperature. Psychrophiles grow optimally at temperatures of 0 °C to 10 °C; mesophiles, 10 °C to 30 °C; and thermophiles, 30 °C to 65 °C. Most higher plants are mesophiles, although plants will generally survive temperatures between 0 °C and 45 °C. Temperature limits generally reflect the freezing point of water on the low side and denaturation of protein on the high side.

Temperature is a principal factor in the distribution of plants, or biogeography. In one study, several species were planted into a cool, coastal environment and a hot, dry desert environment. Survival in the extreme environments appeared to be due to intrinsic differences between species in the temperature-dependence of their growth responses. In those that survived the desert environment, both photosynthesis and respiration appeared to be more stable at high temperature. In a study of elevational gradients up a mountainside, a sharp transition was found between C4 species (in the warmer, drier, lower elevations) and C3 species (in the cooler, moister, higher elevations). It is clear that temperature stability of principal metabolic pathways is a significant determinant in plant distribution.

Plants also use temperature as a cue in their developmental and survival strategies. Vernalization is the promotion of flowering by a period of low temperature. In the case of winter annuals, such as cereals, vernalization changes the photoperiodic behavior from daylength indifference to a quantitative long-day response. Biennials typically grow as a rosette until vernalized. The flowering stem then bolts (elongates) and responds as a long-day plant. A temperature of approximately 0 °C to 5 °C, applied to the actively growing apex of the plant for several weeks, is required for vernalization to be effective.

Photoperiod and low temperature also induce dormancy in buds, characterized by low respiratory rate and an inability to grow even if temperature, oxygen, and water supply are adequate. Many dormant buds have a chilling requirement that must be met before dormancy can be broken and growth renewed.

Dormancy is also a property of many seeds, a situation in which the seed fails to germinate because of environmental and physiological limitations. Seed coats may interfere with water uptake or oxygen uptake, or may contain inhibitors that must be broken down or leached out before germination can proceed. Many seeds require alternating temperatures or a period of low temperature (pre-chilling or stratification) to break dormancy.

Some plants, such as tomato, grow poorly at constant temperature, but require alternating day/night temperatures (thermoperiodism) for optimum growth. In others, such as tulip and crocus, changing temperature regulates the opening and closure of floral petals.

CHAPTER REVIEW

1. In what ways does temperature influence physiological processes? Does temperature interact with other environmental factors? If so, which ones?
2. How does temperature influence the geographical distribution of plants? What modifications might you expect to find in plants adapted to high-temperature habitats? To plants in arctic or alpine habitats?
3. Review the distinction between vernalization and stratification. How does the concept of induction apply to vernalization?
4. Dormant buds and seeds normally require an extended treatment of some sort in order to break dormancy. What is the survival value of such a requirement?
5. In what ways do the daily movements of floral petals, such as tulip and *Crocus*, differ from the movements of bean leaves?

FURTHER READING

Bewley, J. D., M. Black. 1985. *Seeds: Physiology of Development and Germination.* New York: Plenum.

Hillman, W. S. 1962. *The Physiology of Flowering.* New York: Holt, Rinehart and Winston.

Lang, A. 1965. Physiology of flower initiation. In: W. Ruhland (ed.), *Handbuch der Pflanzenphysiologie. Encyclopedia of Plant Physiology.* Berlin: Springer-Verlag, XV (1):1380–1536.

Salisbury, F. B. 1963. *The Flowering Process.* Oxford: Pergamon Press.

REFERENCES

Bewley, J. D., M. Black. 1982. *Physiology and Biochemistry of Seeds. Vol. 2. Viability, Dormancy and Environmental Control.* Berlin: Springer-Verlag.

Björkman, O., 1980. The response of photosynthesis to temperature. In: J. Grace, E. D. Ford, P. G. Jarvis (eds.), *Plants and Their Atmospheric Environment.* Oxford: Blackwell Scientific Publications, pp. 273–301.

Björkman, O., M. Nobs, H. Mooney, J. Troughton, J. Berry, F. Nicholson, W. Ward. 1973a. Growth responses of plants from habitats with contrasting thermal environments. Transplant studies in the Death Valley and the Bodega Head Experimental Gardens. *Carnegie Institution of Washington Yearbook* 73:748–757.

Björkman, O., B. Mahall, M. Nobs, W. Ward, F. Nicholson, H. Mooney. 1973b. Growth responses of plants from habitats with contrasting thermal environments. An analysis of the temperature dependence of growth under controlled conditions. *Carnegie Institution of Washington Yearbook* 73:757–767.

Hyde, E. O. C. 1954. The function of the hilum in some Papilionaceae in relation to the ripening of the seed and permeability of the testa. *Annals of Botany* 18:241–256.

Lang, A. 1957. The effects of gibberellin upon flower formation. *Proceedings of the National Academy of Science USA* 43:709–717.

Perry, T. O., C. W. Wang. 1960. Genetic variation in the winter chilling requirement for date of dormancy break for *Acer rubrum*. *Ecology* 41:790–794.

Phillips, I. D. J., J. Miners, J. R. Roddick. 1980. Effects of light and photoperiodic conditions on abscisic acid in leaves and roots of *Acer pseudoplatanus*. *Planta* 149:118–122.

Purvis, O. N. 1934. An analysis of the influence of temperature during germination on the subsequent development of certain winter cereals and its relation to the length of day. *Annals of Botany* 48:917–955.

Roberts, E. M., N. R. Rao, J-Y. Huang, N. L. Trolinder, C. H. Haigler. 1992. Effects of cycling temperatures on fiber metabolism in cultured cotton ovules. *Plant Physiology* 100:979–986.

Rundel, P. W. 1980. The ecological distribution of C4 and C3 grasses in the Hawaiian Islands. *Oecologia* 45:354–359.

Schwabe, W. W. 1950. Factors controlling flowering of the chrysanthemum. I. The effects of photoperiod and temporary chilling. *Journal of Experimental Botany* 1:329–343.

Taylor, J. S., E. B. Dumbroff. 1975. Bud, root, and growth regulator activity in *Acer saccharum* during the dormant season. *Canadian Journal of Botany* 53:321–331.

Totterdell, S., E. H. Roberts. 1980. Characteristics of alternating temperatures which stimulate loss of dormancy in seeds of *Rumex obtusifolius* L. and *Rumex crispus* L. *Plant, Cell and Environment* 3:3–12.

Wareing, P. F. 1965. Endogenous inhibitors in seed germination and dormancy. In: W. Ruhland (ed.), *Handbuch der Pflanzenphysiologie. Encyclopedia of Plant Physiology.* Berlin: Springer-Verlag, XV (2):909–924.

Went, F. W. 1961. Temperature. In: W. Ruhland (ed.), *Handbuch der Pflanzenphysiologie. Encyclopedia of Plant Physiology.* Berlin: Springer-Verlag, XVI:1–23.

Wood, W. M. L. 1953. Thermonasty in tulip and crocus flowers. *Journal of Experimental Botany* 4:65–77.

Part Four

Stress Physiology and Biotechnology

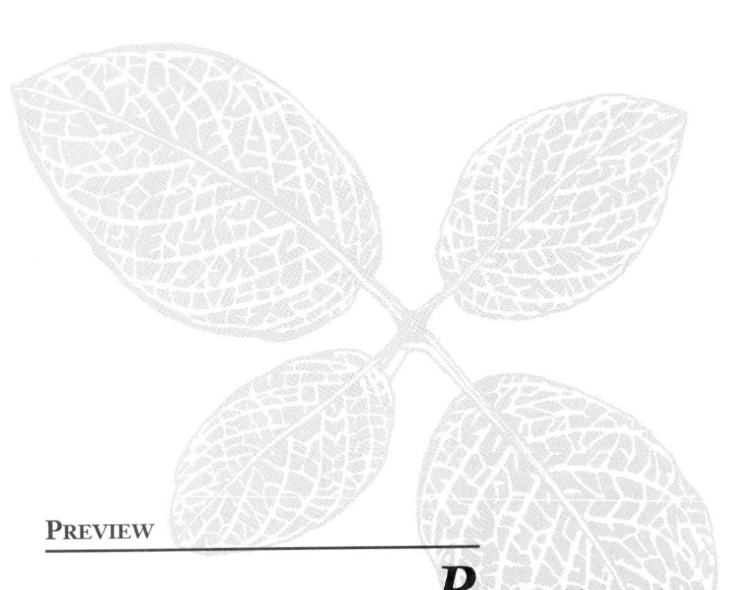

Preview

Plants seldom enjoy optimal environmental conditions, but are often subject to extremes of water potential, temperature, salinity, and other factors that may push an organism to their limits for survival. The study of plants under such conditions, known as stress physiology, is an important aspect of physiological plant ecology for three reasons. First, plants often respond to stressful conditions by altering their physiology and metabolism. The study of plants under stress may thus provide insights into normal physiological mechanisms. Second, the study of stress physiology contributes to our understanding of those factors that limit plant distribution. Finally, in agriculture, the ability of crops to withstand stress is a major factor in determining yield.

This part ends with a discussion of the revolutionary, yet ancient, science of biotechnology. Our newly acquired ability to manipulate genes provides powerful tools for altering the genetic makeup of plants. Plant scientists can now selectively increase tolerance to herbicides, enhance pesticide resistance, and improve the quality of seed storage protein. The prospects of higher yields and nitrogen fixation in genetically engineered plants, growing plants for renewable fuels, and new ways to provide drugs and other useful products from tissue culture portend an exciting future for practical applications of plant physiology and related fields of study.

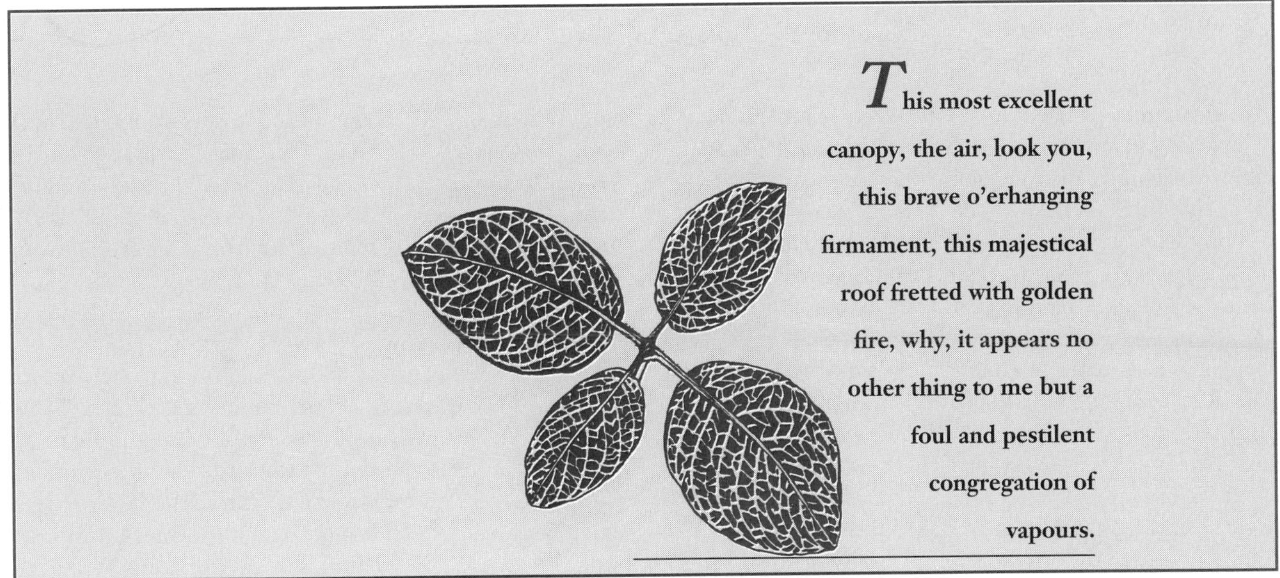

*T*his most excellent canopy, the air, look you, this brave o'erhanging firmament, this majestical roof fretted with golden fire, why, it appears no other thing to me but a foul and pestilent congregation of vapours.

W. Shakespeare
Hamlet, Act 2, scene 2.

22

The Physiology of Plants Under Stress

The previous chapters have focused on plant growth and development under normal environmental conditions. But plants often encounter unusual or extreme conditions: Trees and shrubs in the northern temperate latitudes experience the extreme low temperatures of winter; alpine plants experience cold, drying winds and high levels of UV radiation; and agricultural crops may experience periods of extended drought or their roots may be subjected to high concentrations of salt in the soil. More recently, soil, water, and air pollutants generated through human activities have added to the list of factors that plants must cope with in their environment. Extremes in environmental parameters create stressful conditions for plants, which may have a significant impact on their physiology, development, and survival.

The study of plant responses to environmental stress has long been a central theme for plant environmental physiologists and physiological ecologists. How plants respond to stress helps to explain their geographic distribution and their performance along environmental gradients. Because stress invariably leads to reduced productivity, **stress responses** are also important to agricultural scientists. Understanding stress responses is essential in attempts to breed stress-resistant cultivars that can withstand drought, salinity, and other yield-limiting conditions. Finally, because stressful conditions cause perturbations in the way a plant functions, they provide the plant physiologist with another very useful tool for the study of basic physiology and biochemistry.

This chapter will examine some of the stresses that plants encounter in their environment. The principal topics to be addressed include

- the concept of biological stress and how plants generally respond to stressful conditions;
- the effects of water deficits or drought on shoot and root growth, the nature of injury under conditions of water stress, and how plants respond to water deficits by decreasing stomatal conductance and adjusting the osmotic properties of cells;
- the challenge of low temperature stress and how herbaceous and woody plants are able to avoid low-temperature injury by acclimation;
- the effects of high-temperature stress on membranes and metabolism, including the synthesis of heat shock proteins; and
- the response of plants to high-salt environments, insects and disease, and environmental pollution.

WHAT IS STRESS?

Biological stress is not easily defined, but it implies adverse effects on an organism.

Terminology used to describe stressful environments and plant responses to those stresses has been the subject of controversy for some years. **Stress** is fundamentally a mechanical concept, defined by engineers and physical scientists as a force per unit area applied to an object. In response to a stress, an object develops a **strain,** or change in dimension. This stress–strain relationship is readily described in mechanical examples, such as the stretching of an elastic band or the bending of a metal bar subjected to a load. In mechanical systems, both stress, as the force applied, and strain, as dimensional change, are readily quantified.

It is difficult to define stress so precisely in a biological sense. J. Levitt has argued that the physical terminology can and should be applied to living organisms (see Levitt, 1972; Turner and Kramer, 1980), but, in practice, the biological concept of stress carries more general connotations. At the ecosystem level, for example, any external constraint that limits productivity (that is, carbon gain) below the genetic potential of the plant may be considered a stress (Grime, 1979). This approach may have utility in particular situations, such as agriculture, where mathematical models can be used to estimate genetic potential under optimal environments. However, without a practical means for estimating genetic potential, it becomes far more difficult to judge the impact of stress according to this criterion.

When evaluating environmental stress, another difficulty arises from the question of when, along a continuous environmental gradient, a condition becomes a stress. Plant species are highly variable with respect to their optimum environments and their susceptibility to extremes of, for example, temperature, water potential, or salinity (Table 22.1). At what point in the continuum does temperature or soil water potential become a stress? There are no simple answers and no system of terminology has received the support of all scientists working in the field of stress physiology (Jones and Jones, 1989).

Perhaps the most useful definition of biological stress is an "adverse force or influence that tends to inhibit normal systems from functioning" (Jones and Jones, 1989). Even this definition has its problems, since an interpretation of what is "normal" is highly subjective and will vary among species or even ecotypes. This definition of stress also carries with it the implication that the effects of a stressful environment are harmful or potentially harmful to the plant. Some stress physiologists feel this concept is too restrictive because it raises questions about adaptive mechanisms that enable plants to grow in environments that might otherwise be considered stressful. Is stress a function of the environment or the organism? For example, are the extreme environments encountered in deserts or arctic tundra stressful for plants that normally thrive there? Are these environments stressful only to some species but not to others? Perhaps the examples of environmental stress to be discussed in this chapter will provide some insights into these questions.

TABLE 22.1 **Principal environmental stresses to which plants may be subjected.**

High temperature (heat)
Low temperature (chilling, freezing)
Excess water (flooding, anoxia)
Water deficit (drought, low water potential)
Salinity
Radiation (visible, ultraviolet)
Chemical (pesticides, heavy metals, air pollutants)
Biotic (pathogens, competition)

PLANT RESPONSES TO STRESS

Plants can respond to stress in several ways. Plants may escape the effects of stress by completing their growth during less stressful periods or they may suffer injury if the stress is present and they cannot cope. Alternatively, specific alterations in metabolism may enable the plant to either avoid or tolerate the impact of stress.

Some plants may be *injured* by a stress, which means that they exhibit one or more metabolic dysfunctions. If the stress is moderate and short term, the injury may be temporary and the plant may recover when the stress is removed. If the stress is severe enough, it may prevent flowering and seed formation or otherwise impair the survival of the plant. Some plants escape the stress altogether, such as ephemeral, or short lived, desert plants. **Ephemeral plants** germinate, grow, and flower very quickly following seasonal rains. They thus complete their life cycle during a period of adequate moisture and form dormant seeds before the onset of the dry season. In a similar manner, many arctic annuals rapidly complete their life cycle during the short arctic summer and survive over winter in the form of seeds. Because ephemeral plants never really experience the stress of drought or low temperature, they are known as **stress escapers.**

Many plants have the capacity to resist stress, through either **stress avoidance** or **stress tolerance.** Avoidance mechanisms reduce the impact of a stress, even though the stress is present in the environment. Established plants of alfalfa (*Medicago sativa*), for example, survive dry habitats as adult plants by sending down deep root systems that penetrate the water table. Alfalfa is thereby ensured an adequate water supply under conditions in which more shallow-rooted plants would experience drought. Other plants develop fleshy leaves that store water, thick cuticles or pubescence (leaf hairs) to help reduce evaporation, or other modifications that help to either conserve water or reduce water loss. Cacti with their fleshy photosynthetic stems and leaves reduced to simple thorns are another example of drought avoiders. Most drought avoiders would be severely injured should they ever actually experience desiccation.

Tolerance requires that the organism come to thermodynamic equilibrium with the stress; that is, internal conditions are in equilibrium with conditions outside the plant. Drought tolerance, for example, requires that the organism survive desiccation of its protoplasm without injury, retaining the capacity for normal growth and development when the protoplasm is rehydrated. An extreme example of tolerance is a heterogenous group of ferns and flowering plants (more than 100 species are known) collectively known as resurrection plants (Gaff, 1977). While the vegetative parts of most plants are intolerant of even moderate dehydration, the foliage of resurrection plants will survive air-drying to as little as 7 percent water without injury.

Two other terms that require explanation are **adaptation** and **acclimation.** Both are means of achieving tolerance to a particular stress. Adaptation refers to *heritable* modifications in structure or function which increase the fitness of the organism in the stressful environment. The morphological and physiological modifications associated with crassulacean acid metabolism (CAM) plants (Chap. 10) are examples of adaptation. Acclimation, on the other hand, refers to *nonheritable* physiological modifications that occur over the life of an individual. These modifications are induced by gradual exposure to the stress, such as chilling temperatures or slow drying, and enable that individual to live and reproduce in the stressful environment. The *capacity* to acclimate is, of course, a genetic trait, but the specific changes brought about in response to stress are not themselves passed on to the next generation. The ability of biennial plants and winter strains of cereal grains to survive over winter, discussed in the previous chapter, is an example of acclimation to low temperature. The process of acclimation to a stress is known as **hardening** and plants that have acclimated are commonly referred to as **hardy.** Thus, frost-hardy plants are those that have acclimated to low temperature and are able to survive the freezing stress of winter and drought-hardy plants are able to survive water stress.

Finally, another controversy over terminology is concerned with use of the term **strategy.** Strategy is often used to describe the manner in which a plant responds successfully to a particular stress. Some physiologists object to use of the term for the reason that strategy implies a conscious plan; that is, it is teleological.[1] As H. G. Jones and M. B. Jones point out, however, strategy can validly describe a genetically programmed sequence of responses that enable an organism to survive in a particular environment (Jones and Jones, 1989).

[1]Teleology is the doctrine of final causes, assigning purpose to natural processes. Teleological arguments are considered inappropriate in natural science.

WATER STRESS

The stress of water deficits is a persistent threat to plant survival, yet many plants develop morphological and physiological modifications that enable them to survive in regions of inadequate rainfall and low soil moisture content.

Water stress may arise through either an excess of water or a water deficit. An example of excess water is flooding. Flooding stress is most commonly an oxygen stress, due primarily to reduced oxygen supply to the roots. Reduced oxygen in turn limits respiration, nutrient uptake, and other critical root functions. Stress due to water deficit is far more common, so much so that the correct term **water deficit stress** is usually shortened to simply **water stress.** Because water stress in natural environments usually arises due to lack of rainfall, a condition known as drought, this stress is often referred to as **drought stress** (Fig. 22.1). In the laboratory, water stress can be simulated by allowing transpirational loss from leaves, a condition commonly referred to as **desiccation stress.** Water stress is also a component of both **salt stress** and **osmotic stress,** to be discussed later in this chapter. The unifying feature through all of these variations on water stress is low water potential (Ψ).

MEMBRANES AND WATER STRESS

Damage resulting from water stress is related to the detrimental effects of desiccation on protoplasm. Removal of water, for example, leads to an increase in solute concentration as the protoplast volume shrinks, which may itself have serious structural and metabolic consequences. The integrity of membranes and proteins is also affected by desiccation, which in turn leads to metabolic dysfunctions.

Although the full extent of injury to membranes is not yet resolved, it is understood that removal of water from membranes disrupts the normal bilayer structure and introduces water-filled channels lined with the polar phospholipid head groups. In other words, membranes become exceptionally porous when desiccated. When membranes are rehydrated, these channels permit large amounts of solute to be leaked between compartments or from the cell into the extracellular space. Stresses within the lipid bilayer may also displace membrane proteins, which, together with solute leakage, contributes to a loss of membrane selectivity, a general disruption of cellular compartmentation, and a loss of activity of membrane-based enzymes.

FIGURE 22.1 A desert scrub in southwestern United States. Plants in desert regions are often subject to water stress due to lack of rainfall and drying winds. (Courtesy of the National Park Service. Photograph by Woodbridge Williams.)

In addition to membrane damage, numerous studies have shown that cytosolic and organellar proteins may undergo substantial loss of activity or even complete denaturation when dehydrated. Loss of membrane integrity and protein stability may both be exacerbated by high concentrations of cellular electrolytes that accompany dehydration of protoplasm. The consequence of all these events is a general disruption of metabolism in the cell upon rehydration.

PHOTOSYNTHESIS AND WATER STRESS

Photosynthesis is particularly sensitive to water stress. Photosynthesis can be affected by water stress in two ways. First, closure of the stomata normally cuts off access of the chloroplasts to the atmospheric supply of carbon dioxide. Second, there are direct effects of low cellular water potential on the structural integrity of the photosynthetic machinery. The role of water stress in stomatal closure will be discussed in the following section.

Direct effects of low water potential on photosynthesis have been studied extensively in chloroplasts isolated from sunflower (*Helianthus annuus*) leaves subjected to desiccation (Rao et al., 1987). Sunflower has proven useful for these studies because stomatal closure has only a minor effect on photosynthesis. This is because direct effects on the photosynthetic activity of chloroplasts decrease the demand for CO_2 and the CO_2 level inside the leaf remains relatively high. Both electron transport activity and photophosphorylation are reduced in chloroplasts isolated from sunflower leaves with leaf water potentials below about -1.0 MPa. These effects reflect damage to the thylakoid membranes and ATP synthetase protein (CF^0–CF^1 complex; see Chapter 9). Furthermore, the inhibitory effects of low water potential are enhanced by high concentrations of magnesium—concentrations that are likely to occur in dehydrated leaves (Rao et al., 1987).

STOMATAL RESPONSES TO WATER DEFICIT

Plants are often subjected to acute water deficits due to a rapid drop in humidity or increase in temperature when a warm, dry air mass moves into their environment. The result can be a dramatic increase in the vapor pressure gradient between the leaf and the surrounding air. Consequently, the rate of transpiration increases (Chap. 3). An increase in the vapor pressure gradient will also enhance drying of the soil. Because evaporation occurs at the soil surface, the arrival of a dry air mass has particular consequences for the uptake of water by shallow-rooted plants.

Plants generally respond to acute water deficits by

closing their stomata in order to match transpirational water loss through the leaf surfaces with the rate at which water can be resupplied by the roots. It has been shown in virtually all plants studied thus far, including plants from desert, temperate, and tropical habitats, that stomatal opening and closure is responsive to ambient humidity (Mansfield and Atkinson, 1990). Unlike the surrounding epidermal cells, the surfaces of the guard cells are not protected with a heavy cuticle. Consequently, guard cells lose water directly to the atmosphere. If the rate of evaporative water loss from the guard cells exceeds the rate of water regain from underlying mesophyll cells, the guard cells will become flaccid and the stomatal aperture will close. The guard cells may thus respond directly to the vapor pressure gradient between the leaf and the atmosphere. Closure of the stomata by direct evaporation of water from the guard cells is sometimes referred to as **hydropassive closure.** Hydropassive closure requires no metabolic involvement on the part of the guard cells; guard cells respond to loss of water as a simple osmometer (Chap. 2).

Stomatal closure is also regulated by *hydroactive* processes. **Hydroactive closure** is metabolically dependent and involves essentially a reversal of the ion fluxes that cause opening (Chap. 8). Hydroactive closure is triggered by decreasing water potential in the leaf mesophyll cells and appears to involve abscisic acid (ABA) and other hormones. Since the discovery of ABA in the late 1960s, it has been known to have a prominent role in stomatal closure due to water stress. ABA accumulates in water-stressed (that is, wilted) leaves and external application of ABA is a powerful inhibitor of stomatal opening. Furthermore, two tomato mutants, known as *flacca* and *sitiens*, fail to accumulate normal levels of ABA and both will wilt very readily. The precise role of ABA in stomatal closure in water-stressed whole plants has, however, been difficult to decipher with certainty. This is because ABA is ubiquitous, often occurring in high concentrations in nonstressed tissue. Also, some early studies indicated that stomata would begin to close before increases in ABA content could be detected.

In most well-watered plants, ABA appears to be synthesized in the cytoplasm of leaf mesophyll cells but, because of intracellular pH gradients, ABA accumulates in the chloroplasts (Fig. 22.2). At low pH, ABA exists in the protonated form ABAH, which freely permeates most cell membranes. The dissociated form ABA^- is impermeant; because it is a charged molecule it does not readily cross membranes. Thus, ABAH tends to diffuse from cellular compartments with a low pH into compartments with a higher pH. There some of it dissociates to ABA^- and becomes trapped. It is well established that in actively photosynthesizing mesophyll cells the cytosol will be moderately acidic (pH 6.0 to 6.5) while the chloroplast stroma is alkaline (pH 7.5 to 8.0) (see Chap. 9). It has been calculated that if the stroma pH is 7.5 and cytosolic pH is 6.5, the concentration of ABA in the chloroplasts will be about tenfold higher than in the cytosol (Milborrow, 1984).

According to the current model, the initial detection of water stress in leaves is related to its effects on photosynthesis, described earlier in this chapter. Inhibition of electron transport and photophosphorylation in the chloroplasts would disrupt proton accumulation in the thylakoid lumen and lower the stroma pH (Cowen et al., 1982). At the same time, there is an increase in the pH of the apoplast surrounding the mesophyll cells (Hartung et al., 1988). The resulting pH gradient stimulates a release of ABA from the mesophyll cells into the apoplast, where it can be carried in the transpiration stream to the guard cells (Fig. 22.3).

Just how ABA controls turgor in the guard cells remains to be determined. Evidence indicates that ABA does not need to enter the guard cell, but acts instead

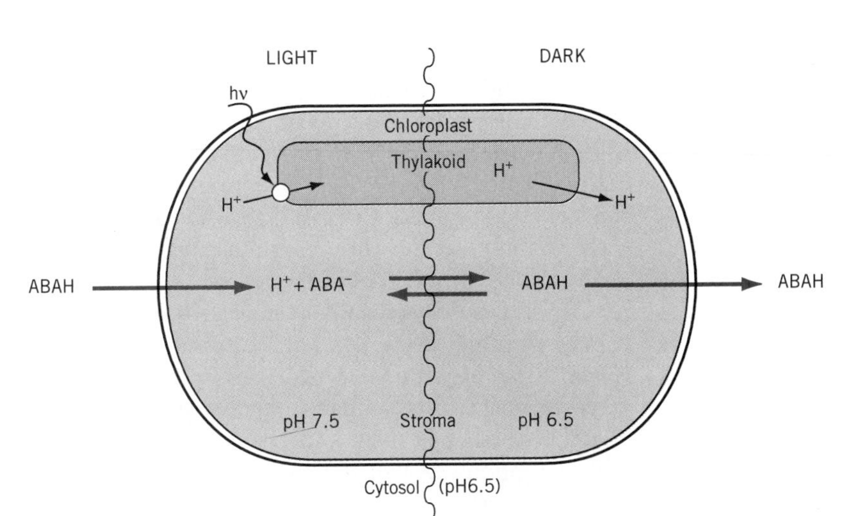

FIGURE 22.2 ABA storage in chloroplasts. In the light, photosynthesis drives protons into the interior of the thylakoid, creating a pH gradient between the stroma and the cytosol. The pH gradient favors movement of ABAH into the chloroplast, where it dissociates to ABA^-. The membrane is less permeable to ABA^-. In the dark, protons leak back into the stroma, the pH gradient collapses, and ABAH moves back into the cytosol.

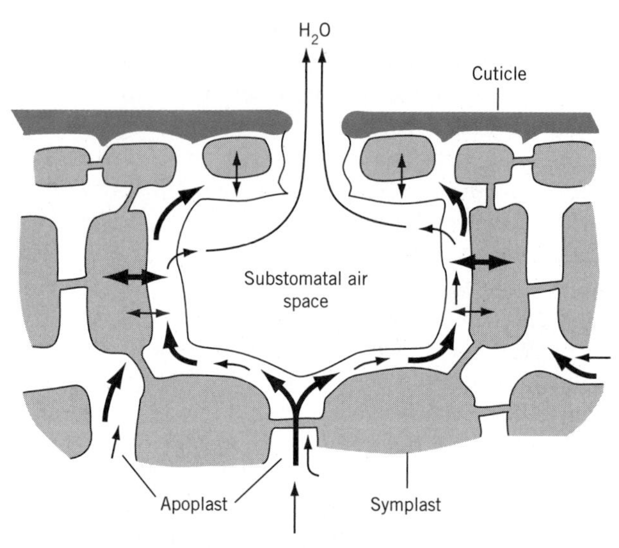

FIGURE 22.3 ABA movement in the apoplast. ABA synthesized in the roots is carried to the leaf mesophyll cells (heavy arrows) in the transpiration stream (light arrows). ABA equilibrates with the chloroplasts of the photosynthetic mesophyll cells or is carried to the stomatal guard cells in the apoplast.

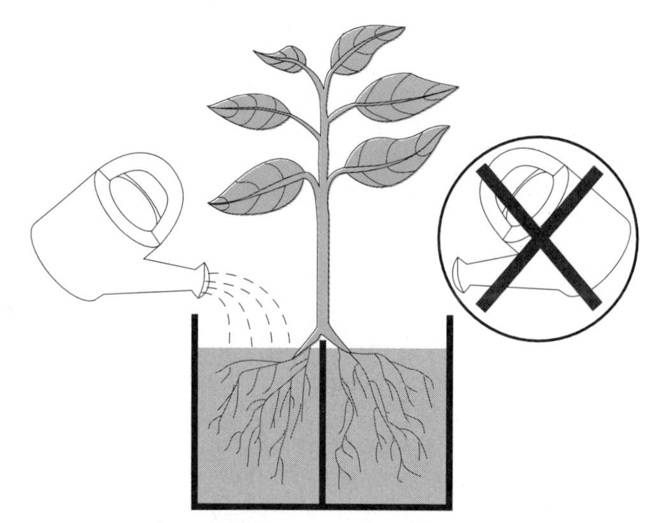

FIGURE 22.4 An experimental setup for testing the effects of desiccated roots on ABA synthesis and stomatal closure. Roots of a single plant are divided equally between two containers. Water supplied to one container maintains the leaves in a fully turgid state while water is withheld from the second container. Withholding water from the roots leads to stomatal closure, even though the leaves are not stressed.

on the outer surface of the plasma membrane. Note that guard cells are symplastically isolated from the surrounding mesophyll cells; that is, there are no plasmodesmata connecting the guard cells with the surrounding mesophyll cells. Thus, ABA can reach the guard cells only through the apoplastic space, not through the symplast. Presumably ABA interacts with the high-affinity binding sites on the plasma membrane (see Chap. 17), although the existence of such sites has yet to be confirmed. Nonetheless, there are strong indications that ABA interferes with plasma membrane proton pumps and, consequently, the uptake of K^+, or that it stimulates K^+ efflux from the guard cells. Either way, the guard cells will lose turgor, leading to closure of the stomata.

As noted above, wilted leaves accumulate large quantities of ABA. In most cases, however, stomatal closure begins before there is any significant increase in the ABA concentration. This can be explained by the release of stored ABA into the apoplast, which occurs early enough and in sufficient quantity—the apoplast concentration will at least double—to account for initial closure. Increased ABA synthesis follows and serves to prolong the closing effect.

Stomatal closure does not always rely on the perception of water deficits and signals arising within the leaves. In some cases it appears that the stomata close in response to soil desiccation *before* there is any measurable reduction of turgor in the leaf mesophyll cells (Mansfield and Atkinson, 1990). Several studies have indicated a feed-forward control system that originates in the roots and transmits information to the stomata (Blackman and Davies, 1985; Zhang and Davies, 1987). In these experiments, plants are grown such that the roots are equally divided between two containers of soil (Fig. 22.4). Water deficits can then be introduced by withholding water from one container while the other is watered regularly. Control plants receive regular watering of both containers. Stomatal opening along with factors such as ABA levels, water potential, and turgor are compared between half-watered plants and fully watered controls. Typically, stomatal conductance, a measure of stomatal opening, declines within a few days of withholding water from the roots (Fig. 22.5), yet there is no measurable change in water potential or loss of turgor in the leaves. In experiments with day flower (*Commelina communis*), there was a significant increase in ABA content of the roots in the dry container and in the leaf epidermis (Fig. 22.6). Furthermore, ABA is readily translocated from roots to the leaves in the transpiration stream, even when roots are exposed to dry air (Zhang and Davies, 1987). These results provide reasonably good evidence that ABA is involved in a kind of early warning system that communicates information about soil water potential to the leaves.

Hormones other than ABA may also be involved in communication between water-stressed roots and leaves. In an experiment with half-watered maize (*Zea mays*) plants, results similar to those with *Commelina* were obtained (Blackman and Davies, 1985). One notable exception was that the ABA content of the leaves did not increase, and the application of cytokinins to the

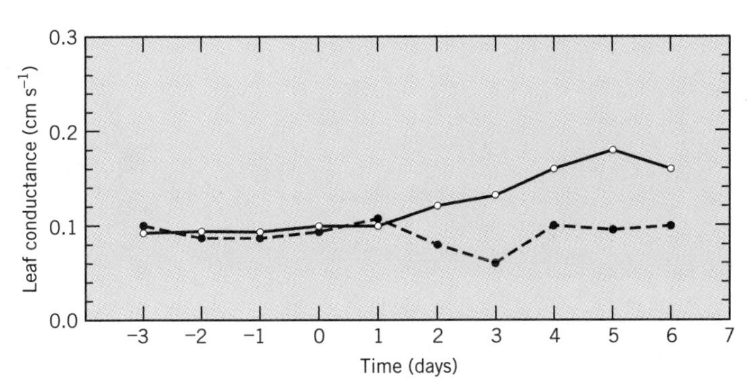

FIGURE 22.5 Stomatal closure in a split-root experiment. Maize (*Zea mays*) plants were grown as shown in Fig. 22.4. Control plants (open circles) had both halves of the root system well-watered. Water was withheld from half the roots of the experimental plants (closed circles) on day zero. Stomatal opening, measured as leaf conductance, declined in the plants with water-stressed roots. (From P. G. Blackman, W. J. Davies, *Journal of Experimental Botany* 36:39–48, 1985. Reprinted by permission of The Company of Biologists, Ltd.)

leaves prevented stomatal closure. At least in *Commelina*, it appears that closure is brought about by decreased movement of cytokinins out of the drying roots.

Both hydropassive and hydroactive closure of stomata represent mechanisms that enable plants to anticipate potential problems of water availability through either excessive transpirational loss from the leaves or chronic, but nonlethal, soil water deficit. Although considerable progress has been achieved in this field over the last decade, there is clearly much yet to be learned about stomatal behavior and the response of plants to water stress.

OSMOTIC ADJUSTMENT

Another pronounced response to water stress in many plants is a decrease in osmotic potential resulting from an accumulation of solutes. This process is known as **osmotic adjustment.** While some increase in solute concentration is expected as a result of dehydration and decreasing cell volume, osmotic adjustment refers specifically to a net increase in solute concentration due to metabolic processes triggered by stress. Osmotic adjustment generates a more negative leaf water potential, thereby helping to maintain water movement into the leaf and, consequently, leaf turgor.

Solutes accumulate slowly during osmotic adjustment and the decreases in osmotic potential due to osmotic adjustment are relatively small, less than 1.0 MPa. Nevertheless, the role of solutes in maintaining turgor at relatively low water potentials represents a significant form of acclimation to water stress. Osmotic adjustment may also play an important role in helping partially wilted leaves to regain turgor once the water supply recovers. By helping to maintain leaf turgor, osmotic adjustment also enables plants to keep their stomata open and continue taking up CO_2 for photosynthesis under conditions of moderate water stress.

Solutes implicated in osmotic adjustment include a range of inorganic ions (especially K^+), sugars, and

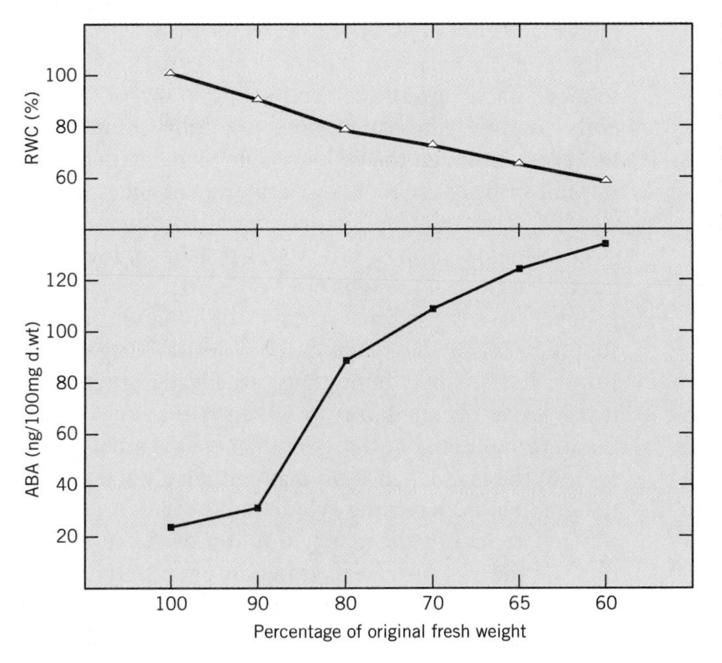

FIGURE 22.6 Effect of air drying on the ABA content of *Commelina communis* root tips. Root tips were air dried to the relative water contents shown in the upper curve. Lower curve shows the dramatic increase in ABA content as the fresh weight decreases. (From J. Zhang, W. J. Davies, *Journal of Experimental Botany* 38:2015–2023, 1987. Reprinted by permission of The Company of Biologists, Ltd.)

FIGURE 22.7 Three solutes typically involved in osmotic adjustment.

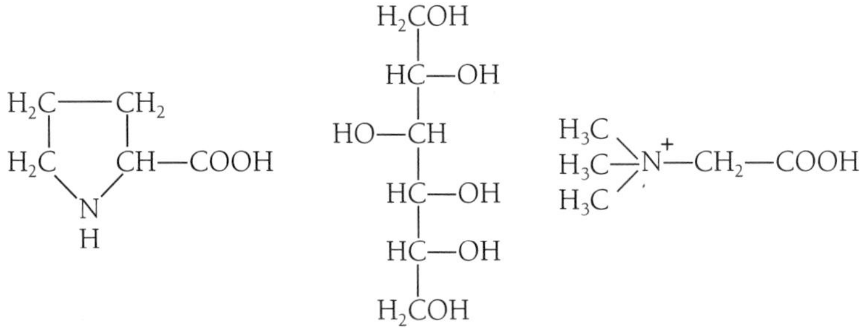

Proline Sorbitol Glycine betaine

amino acids (Fig. 22.7). One amino acid that appears to be particularly sensitive to stress is **proline.** A large number of plants synthesize proline from glutamine in the leaves. The role of proline is demonstrated by experiments with tomato cells in culture. Cells subjected to water (osmotic) stress by exposure to hyperosmotic concentrations of polyethylene glycol (PEG) responded with an initial loss of turgor and rapid accumulation of proline. As proline accumulation continued, however, turgor gradually recovered (Handa et al., 1986). **Sorbitol,** a sugar alcohol, and **betaine** (N,N,N-trimethyl glycine), are other common accumulated solutes. Most chemicals associated with osmotic adjustment share the property that they do not significantly interfere with normal metabolic processes.

Although osmotic adjustment appears to be a general response to water stress, not all species are capable of adjusting their solute concentrations. Sugarbeet (*Beta vulgaris*, on the one hand, synthesizes large quantities of betaine and is known as an **osmotic adjuster.** Osmotic adjustment in cowpea (*Vigna unguiculata*), on the other hand, is minimal and cowpea is known as an **osmotic nonadjuster.** Cowpea instead has very sensitive stomata and avoids desiccation by closing the stomata and maintaining a relatively high water potential. It is interesting to note that there is no long-term advantage of osmotic adjustment over stomatal closure, at least with regard to net carbon gain. While sugarbeet is able to continue photosynthesis at lower water potentials, the advantage over cowpea is shortlived. After one or two days, excessive water loss overrides osmotic adjustment and over the long term carbon assimilation in sugarbeet declines.

EFFECTS OF WATER DEFICIT ON SHOOT AND ROOT GROWTH

One of the early effects of water deficit is a reduction in vegetative growth. Shoot growth, and especially the growth of leaves is generally more sensitive than root growth. In a study in which water was withheld from maize (*Zea mays*) plants, for example, there was a significant reduction of leaf expansion when tissue water potentials reached −0.45 MPa and growth was completely inhibited at −1.00 MPa. At the same time, normal root growth was maintained until the water potential of the root tissues reached −0.85 MPa and was not completely inhibited until the water potential dropped to −1.4 MPa (Westgate and Boyer, 1985). Reduced leaf expansion is beneficial to a plant under conditions of water stress because it leads to a smaller leaf area and reduced transpiration.

Traditionally, the effect of low water potential on cell enlargement has been attributed to a loss of turgor in the cells in the growing region. Recall that in Chapter 15 it was shown that cell enlargement occurs when water moves in to establish full turgor following stress relaxation in the cell wall. It should not be too surprising, then, that an early consequence of limited water supply would be reduced growth. The occurrence of water deficits in young growing plants would be expected to cause a reduction in turgor which would slow leaf expansion. Unfortunately, the explanation is apparently not so straightforward—it is not always possible to observe a direct relationship between cell growth and turgor in stressed plants. Often there is little or no measurable decrease in turgor in response to water stress, even though enlargement is inhibited and the soil water content decreases. This observation appears to be explained by an accumulation of solutes, or osmotic adjustment, in the growing cells as the water potential decreases. The resulting decrease in osmotic potential would help to move water into the cell and maintain turgor. Apparently osmotic adjustment does not fully compensate, however. Although the cells are able to maintain turgor, it is not sufficient to maintain a full rate of enlargement.

If turgor is not the controlling factor, then the answer to the question of how water deficits limit cell enlargement must be sought elsewhere. M. Westgate and J. Boyer have argued that it might be a matter of water supply between the xylem and the leaf (Westgate and Boyer, 1985). They found that as the water potential in maize leaves declined in response to water stress, so did the water potential of the xylem. The movement of water into the leaf depends on maintaining a water potential gradient between the xylem and the leaf. A lowered water potential in the xylem would reduce the magnitude of the xylem-to-leaf gradient, effectively cutting the leaf off from its water supply. While sufficient water is available to maintain leaf turgor, there is not sufficient water to drive cell enlargement.

Other studies have shown that measurable increases in solute concentrations in tissues experiencing water stress cannot be detected until *after* the growth rate has begun to decline and that water potential may have a more direct effect on cell wall properties (Matthews et al., 1984; Van Volkenburgh and Boyer, 1985). In Equation 15.5, it was shown that growth is dependent on cell wall extensibility (m) and the difference between turgor (P) and the yield threshold (Y); that is, $dV/dt = m(P - Y)$. Water stress apparently leads to both a decrease in m and an increase in Y. Either of these changes would have the same effect—a decreased rate of cell enlargement. The decrease in m can be explained by an observed decrease in proton excretion into the cell walls of stressed tissues. Changes in Y would no doubt require modifications to the complex cell wall structure, but the nature of those changes remains open to speculation.

LEAF AREA ADJUSTMENT

The preceding discussion applies primarily to shoots and leaves that are actively growing. Many mature plants, such as cotton (*Gossypium hirsutum*), subjected to prolonged water stress will respond by accelerated senescence and abscission of the older leaves. In the case of cotton, only the youngest leaves at the apex of the stem will remain in cases of severe water stress. This process, sometimes referred to as **leaf area adjustment,** is another mechanism for reducing leaf area and transpiration during times of limited water availability. So long as the buds remain viable, new leaves will be produced when the stress is relieved.

As noted above, roots are generally less sensitive than shoots to water stress. Apparently, osmotic adjustment in roots is sufficient to maintain water uptake and growth down to much lower water potentials than is possible in leaves. Relative root growth may actually be enhanced by low water potentials, such that the root/shoot ratio will change in favor of the proportion of roots. An increase in the root/shoot ratio as the water supply becomes depleted is clearly advantageous, as it improves the capacity of the root system to extract more water by exploring larger volumes of soil.

A changing root/shoot ratio is accompanied by a change in source/sink relationships with the result that a larger proportion of photosynthate is partitioned to the roots. Delivery of carbon to the roots can continue, however, only to the extent that carbon supply can be maintained by photosynthesis or mobilization of reserves stored in the leaves.

TEMPERATURE STRESS

Plants exhibit a wide range of sensitivities to extremes of temperature. Some are killed or injured by moderate chilling temperatures while others, properly acclimated, can survive freezing temperatures tens of degrees below zero celsius. High temperature stress induces synthesis of a unique set of stress proteins.

Each plant has its unique set of temperature requirements for growth and development (Chap. 21). There is an optimum temperature at which each plant grows and develops most efficiently, and upper and lower limits. As the temperature approaches these limits, growth diminishes, and beyond those limits there is no growth at all. Except in the relatively stable climates of tropical forests, temperatures frequently exceed these limits on a daily or seasonal basis, depending on the environment. Deserts, for example, are characteristically hot and dry during the day but experience low night temperatures because, in the absence of a moist atmosphere, much of this heat is reradiated into space. Plants at high altitudes, where much of the daily heat gain is radiated into the thin atmosphere every night, experience similar temperature excursions. Plants native to the northern temperate and boreal forests must survive temperatures as low as −70 °C every winter.

How plants respond to temperature extremes has long captivated plant biologists. In this section, we will consider three temperature extremes that can cause injury to plants: chilling, freezing, and high temperatures.

CHILLING STRESS

Many plants, especially those native to warm habitats, are injured when exposed to low nonfreezing temperatures (Lynch, 1990). Plants such as maize (*Zea mays*), tomato (*Lycopersicon esculentum*), cucumber (*Cucurbita* sp.), soybean (*Glycine max*), cotton (*Gossypium hirsutum*), and banana (*Musa* sp.) are particularly sensitive and will exhibit signs of injury when exposed to temperatures below 10 to 15 °C. Even some temperate plants such as apple (*Malus* sp.), potato (*Solanum tuberosum*), and asparagus (*Asparagus* sp.) experience injury at temperatures above freezing (0 to 5 °C).

Outward signs of chilling injury can take a variety of forms, depending on the species and age of the plant, and the duration of the low-temperature exposure. Young seedlings typically show signs of reduced leaf expansion, wilting, and chlorosis. In extreme cases, browning and the appearance of dead tissue (necrosis) and/or death of the plant will result. In some plants, reproductive development is especially sensitive to chilling temperature. Exposure of rice plants, for example, to chilling temperatures at the time of anthesis (floral opening) results in sterile flowers.

Symptoms of chilling injury reflect a wide range of metabolic dysfunctions in chilling-sensitive tissues, including: impaired protoplasmic streaming; reduced respiration, photosynthesis and protein synthesis; and altered patterns of protein synthesis. Indeed, there appear to be few aspects of cellular biochemistry that are not impaired in chilling-sensitive tissues following exposure to low temperature. There seem to be two possible explanations for such a diversity of responses. Either there are multiple low-temperature-sensitive sites in chilling-sensitive plants, or there is a single primary site that influences a range of metabolic processes (Lynch, 1990). The favored explanation is that low temperature causes reversible changes in the physical state of cellular membranes. Consequently, most research into chilling sensitivity has focused on changes in the physical properties of lipids and the effects of those changes on membrane function.

As discussed in Chapter 1, membrane lipids consist primarily of diacylglycerides containing two fatty acids of either 16 or 18 carbon atoms (see Box 1.1). Some fatty acids are unsaturated, which means that they have one or more carbon—carbon double bonds (—CH═CH—) while others are fully saturated with hydrogen (—CH_2—CH_2—). Because saturated fatty acids—and lipids which contain them—solidify at higher temperatures than unsaturated fatty acids, the relative proportions of unsaturated and saturated fatty acids in membrane lipids have a strong influence on the fluidity of membranes. A change in the membrane from the fluid state to a gel (or semicrystalline) state, is marked by an abrupt transition that can be monitored by a variety of physical methods. The temperature at which this transition occurs is known as the **transition temperature.**

Chilling-sensitive plants tend to have a higher proportion of saturated fatty acids (Table 22.2) and a correspondingly higher transition temperature. For mitochondrial membranes of the chilling-sensitive plant mung bean (*Vigna radiata*), for example, the transition temperature is 14 °C (Raison and Orr, 1986). Mung bean seedlings grow poorly below 15 °C. Chilling-resistant species, on the other hand, tend to have lower proportions of saturated fatty acids and, therefore, lower transition temperatures. During acclimation to low temperature, the proportion of unsaturated fatty acids increases and transition temperature decreases (Raison and Orr, 1986; Williams et al., 1988).

The net effect of the transition from a liquid membrane to a semicrystalline state at low temperature is similar to the effects of water stress described above. The integrity of membrane channels is disrupted, resulting in loss of compartmentation and solute leakage, and the operation of integral proteins that make up respiratory assemblies, photosystems, and other membrane-based metabolic processes is impaired. Membranes of chilling-resistant or acclimated plants are able to maintain membrane fluidity to much lower temperatures and thereby protect these critical cellular functions against damage.

TABLE 22.2 **Ratio of unsaturated/saturated fatty acids of membrane lipids of mitochondria isolated from chilling-sensitive and chilling-resistant tissues.**

Chilling-sensitive tissues		
Phaseolus vulgaris (bean)	shoot	2.8
Ipomoea batatas (sweet potato)	tuber	1.7
Zea mays (maize)	shoot	2.1
Lycopersicon esculentum (tomato)	green fruit	2.8
Chilling-resistant tissues		
Brassica oleracea (cauliflower)	buds	3.2
Brassica campestris (turnip)	root	3.9
Pisum sativum (pea)	shoot	3.8

From data of J. M. Lyons et al., *Plant Physiology* 39:262, 1964.

FREEZING STRESS

Freezing stress is commonly encountered by trees and shrubs overwintering in north temperate, subarctic, and alpine regions. One species of larch (*Laryx dahurica*), for example, survives in the most northerly forests of Siberia where temperatures commonly reach −65 to −70 °C. Yet even the hardiest of plants may experience significant injury or death if exposed to temperatures at or just below the freezing point during periods of active growth. Normally, seedlings of winter rye (*Secale cereale*), for example, will survive temperatures no lower than −4 or −5 °C. However, when acclimated by growth at temperatures near 5 °C, rye seedlings will survive temperatures as low as −28 to −30 °C.

Tolerance to freezing temperatures is important to agriculture, where the ability to withstand late spring or early autumn frosts may determine the success of a crop. In order to fully understand how plants are able to acclimate to such low temperatures, it is helpful to first understand how and where freezing occurs in plants and how it causes damage.

It is ice formation, not low temperature per se that causes freezing injury to plant cells. Dehydrated tissues such as seeds and fungal spores, for example, can withstand prolonged storage at temperatures close to absolute zero (0K) without injury. In addition, even fully hydrated, nonhardy cells can survive extremely rapid freezing, such as by immersion in liquid nitrogen (−196 °C). This is because the water vitrifies; that is, the water essentially solidifies without the formation of ice crystals, or any crystals that do form are so small that they create no mechanical damage to cell structure. To ensure survival, it is necessary that the cells are rewarmed equally rapidly in order to avoid the formation or growth of ice crystals as they are returned to normal temperatures. Storage in liquid nitrogen, called **cryo-**

genic storage, is commonly used to store fungal cells, or sperm for purposes of artificial insemination.

If the freezing rate of cells or tissues is sufficiently slow, say less than 10 °C min^{-1}, ice will form outside the protoplasts where water is purest. Protoplasmic water will migrate out of the cells and add to the extracellular ice crystals. Extracellular ice does not kill plant cells and the tissues can normally be rewarmed without injury. At intermediate freezing rates (10 °C to 100 °C), ice crystals will form within the protoplasts. Intracellular ice formation disrupts the fine structure of the cells and invariably results in death.

Thermal Analysis of Freezing One convenient way to monitor freezing in woody tissues is by **thermal analysis;** that is, by measuring the heat released (the latent heat of fusion) when ice forms. Thermal analysis is carried out by embedding a small thermocouple probe (Appendix) under the bark of a moist woody twig. A second thermocouple is commonly embedded in a *dry* sample of woody tissue as a control. The temperature of the sample tissue is then recorded as the surrounding air temperature is steadily lowered. The result of thermal analysis is a plot of tissue temperature against ambient temperature as shown in Figure 22.8. Note that initially, the sample tissue cools well below the freezing point of water with no apparent ice formation. The failure to form ice at below-freezing temperature, called **supercooling,** is discussed further below. At about −6 to −8 °C there is an abrupt rise in the temperature of the sample tissue. This rise is called an **exotherm.** The first exotherm represents ice formation in the apoplastic space—that is, *extracellular* freezing—of xylem tracheary elements and intercellular spaces in the cortex, bark, phloem, and so forth. Ice forms first in the apoplastic space because the tissue water there contains relatively little dissolved solute and its freezing point is normally only a few tenths of a degree below that of pure water.

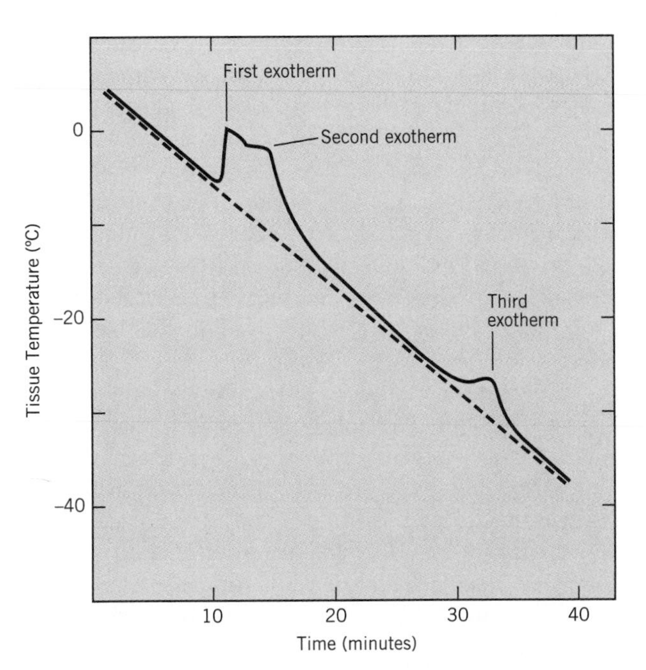

FIGURE 22.8 Tissue temperature during controlled cooling of woody stems. Exotherms are the points at which tissue temperature increases significantly above ambient due to heat released (heat of fusion) as water crystallizes into ice.

Because the vapor pressure of ice is much lower than that of liquid water at the same temperature, ice formation in the apoplast establishes a vapor pressure gradient between the apoplast and the surrounding cells. Unfrozen cytoplasmic water will then migrate down this gradient from the cell into the apoplast, where it contributes to the enlargement of the ice crystals already there. The result is a second extracellular exotherm, seen as a shoulder rather than a distinct peak in Figure 22.8. The continued migration of cellular water into the apoplastic space has two significant consequences. First, loss of water effectively increases the solute concentration of the cytoplasm, thereby lowering its freezing point by 1 or 2 °C. Second, migration of water into the apoplast leads to progressive dehydration of the protoplasm. In the absence of intracellular ice formation, therefore, the effects of freezing stress become very similar to the effects of water stress.

Deep Supercooling in Woody Tissues Supercooling occurs because the tissue water lacks nucleating substances necessary to initiate ice formation. In addition, ice nucleation does not occur readily in very small volumes of water, especially if the spread of ice crystals from adjacent tissues can be prevented. The minimum temperature to which pure water, free of nucleating substances, can be supercooled is −38 °C. At that temperature, spontaneous ice nucleation occurs regardless of whether nucleating substances are present. During controlled freezing, woody tissues frequently supercool to −15 °C.

Some tissues, in particular the vegetative buds of coniferous species and flower buds of apple and other deciduous fruit trees, will **deep supercool** to temperatures as low as −40 °C, despite the formation of ice crystals in bud scales and adjacent tissues (Burke et al., 1976). Supercooling to −47 °C in xylem tissues is the lowest on record. Although deep supercooling does not occur in all plants, it has been observed in overwintering stems of a large number of temperate zone and timberline woody species (George et al., 1982).

Deep supercooling is an effective freezing avoidance mechanism, but only up to a point. Deep supercooling appears to limit the geographical distribution of those plants in which it does occur. This is because temperatures in the range of −40 to −50 °C are very near the limits for spontaneous nucleation for tissue water. When nucleation does occur under such conditions, a third, low-temperature exotherm is seen in the thermal analysis curve (Fig. 22.8). The third exotherm represents *intracellular* ice formation which, as noted above, is invariably lethal to the cells in which it does occur. Note that the northern limits of the deciduous forest in northeastern United States and eastern Canada fall in the region where a minimum temperature of −40 °C may be expected to occur at least once each year. Killing of the xylem ray parenchyma cells is often followed by invasion of microorganisms, a common cause of black heart injury of fruit trees.

Acclimation in Woody Tissues Boreal deciduous trees and shrubs such as paper birch (*Betula papyrifera*), trembling aspen (*Populus tremuloides*), and willow (*Salix* sp.)—all found as far north as the arctic circle—survive because they are able to acclimate to the below-freezing winter temperatures. During their normal growing season, these plants will suffer injury or death if exposed to freezing temperatures. Even a light frost during the spring or summer may be lethal to plants that are actively growing. Yet, acclimated stems of these species may survive temperatures as low as −196 °C (liquid nitrogen) without apparent injury. Acclimated stems of these species do not undergo deep supercooling and do not exhibit a third, low-temperature exotherm.

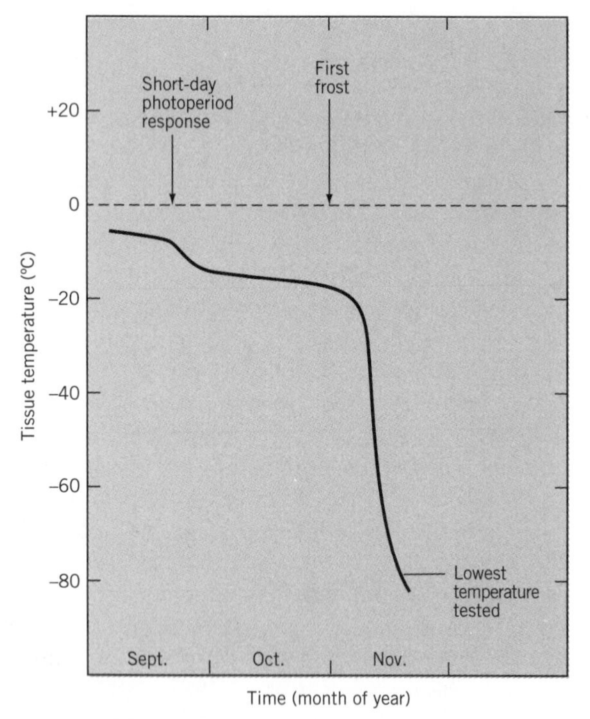

FIGURE 22.9 Acclimation to low temperature in woody stems. The curve depicts the lowest survival temperature as a function of time of year. Note that significant decreases in survival temperature correspond to shortening daylength and the time of the first frost.

Acclimation of woody tissues to freezing stress is a common phenomenon in nature, but the precise mechanism by which acclimation is achieved is not well understood. It is known that acclimation in woody tissues occurs in two distinct stages (Fig. 22.9). It begins in the autumn when growth has slowed but before the leaves have begun to senesce and fall. This first stage of acclimation is induced by short days and is thought to be under control of phytochrome. Acclimation at this stage can be inhibited by long days and early frost. The first stage presumably involves the synthesis of a hardiness-promoting factor in the leaves, which is then exported to the overwintering stems. The identity of this factor is unknown, although, largely on the basis of studies with herbaceous plants, it is thought abscisic acid may be involved.

The second stage of acclimation is triggered by exposure of the overwintering tissue to low temperature, corresponding to the first frost (Fig. 22.9). At this stage, numerous metabolic changes have been observed. There are increases in the level of organic phosphates and the conversion of starch to sugars. Glycoproteins accumulate and the protoplasm becomes generally more resistant to dehydration. Fully acclimated cells can withstand temperatures far below those normally experienced in nature.

Cold Acclimation in Herbaceous Tissues In recent years, research into cold acclimation has focused on herbaceous plants, callus cultures, cell suspensions, and even isolated protoplasts. Exposure of herbaceous cells and tissues to low but nonfreezing temperature for a few days to weeks will markedly increase their freezing tolerance. As noted earlier, herbaceous seedlings such as rye (*Secale cereale*) are able to acclimate and withstand temperatures as low as −30 °C. The study of herbaceous plants such as rye, alfalfa (*Medicago*), and spinach (*Spinacea oleracea*) offers the advantage of being able to follow metabolic and molecular events before, during, and after acclimation, thereby assisting in the search for metabolic and genetic factors involved in the hardening process.

Acclimation to low temperature requires energy, provided by light and photosynthesis, and apparently involves changes in gene expression. During acclimation, there are increases in protein synthesis and qualitative changes in the pattern of proteins synthesized. Eight new proteins were observed during acclimation of alfalfa and three in spinach and all are attributable to transcription of new mRNA. The identity and role of these proteins remains to be determined.

There is good reason to believe that ABA might be involved in cold acclimation of herbaceous tissues. An increase in endogenous ABA levels has been observed during cold acclimation in several species and the amount of increase is greater in cold-tolerant varieties than in cold-sensitive varieties. In addition, significant levels of cold tolerance can be induced by the application of ABA to intact plants, callus, and suspension cultures (Mohapatra et al., 1988). In both intact alfalfa seedlings and suspension-cultured cells of winter rapeseed (*Brassica napus*), exogenous ABA can induce up to 50 to 60 percent survival compared with a normal cold-acclimation treatment. ABA also induces the synthesis of new proteins. In alfalfa, some of the induced proteins are unique to ABA treatment, but some are common to both low-temperature and ABA treatment (Mohapatra et al., 1988). In the case of *Brassica* cells in suspension culture, both cold-acclimation and ABA application induce transcription of the mRNA for a 20 kDa protein (Johnson-Flannagan and Singh, 1987).

HIGH TEMPERATURE STRESS

Plants growing in desert regions and semiarid[2] croplands routinely encounter high temperatures accompanied by high radiation levels, low soil moisture, and high potential transpiration rates. In such an environment it is often difficult to distinguish between the individual effects of multiple stresses, but high temperature is certainly a major factor in plant productivity, if not survival, in these regions (Chap. 21).

One problem faced by plants under conditions of high solar radiation and temperature is energy absorption by leaves, which can easily raise the temperature of a leaf by 5 °C or more above ambient. Many plants avoid overheating by assuming a more vertical orientation of the leaves or, as is the case with grasses, rolling their leaves along the long axis. Other morphological adaptations include reflective leaf hairs (pubescence) and waxy surfaces to reflect light and thereby reduce energy absorption, or smaller, deeply dissected leaves that minimize the thickness of the boundary layer and maximize heat loss by convection and conduction. Note that these same modifications also help reduce transpirational water loss which normally plays a significant role in dissipation of heat from leaves.

Upper temperature limits for plant growth were discussed briefly in Chapter 21. Although plants such as *Tidestromia oblongifolia* and some desert grasses thrive at temperatures up to 50 °C or even higher (Chap. 21; Larcher et al., 1989), few vascular plants are able to survive leaf temperatures in excess of 50 °C to 55 °C. The highest known temperature tolerance for vascular plants is found among the agaves and cacti, both families of succulent plants native to arid and semiarid environments. Large numbers of species in these two families are able to tolerate temperatures of 60 °C or more, while some can survive short-term (1hr) treatments as high as 74 °C (Nobel, 1988).

High Temperature Effects on Membranes and Metabolism The traditional view is that the high-temperature limit for most plants is determined by irreversible denaturation of enzymes. Although enzyme function must certainly play a critical role, more recently attention has turned to changes in the properties of membranes as a principal cause of high-temperature damage. One of the principal differences between agave and cacti, for example, and C3 plants grown at moderate temperatures is a higher proportion of saturated fatty acids in the membrane lipids of the high-temperature-tolerant species. The fluidity of membranes increases at high temperature, which could cause problems with respect to permeability and the catalytic functions of membrane proteins. More saturation would lead to a less fluid membrane, which would help maintain the strength of hydrophobic interactions at higher temperatures and thereby preserve stability of both the membrane and interactions between the lipids and integral membrane proteins (Raison et al., 1980; Quinn, 1988).

Reactions in the thylakoid membranes of higher plant chloroplasts are most sensitive to high-temperature damage, with consequent effects on the efficiency of photosynthesis. Three lines of evidence indicate that photosystem II and its associated oxygen-evolving complex (Chap. 9) are particularly susceptible to injury. First, the oxygen-evolving complex is directly inactivated by heat, thereby disrupting electron donation to PSII. Second, high-temperature effects on fluidity of the thylakoid membrane disturb the lateral distribution of pigment-protein complexes. Such changes would be expected to interfere with the efficiency of electron transport and photophosphorylation. Third, the high-temperature limit for photosynthesis is generally marked by an abrupt increase in chlorophyll fluorescence, which is easily monitored in intact leaves. This fluorescence rise reflects irreversible damage to the energy transfer mechanism of PSII. The result is that an increasing portion of the absorbed energy cannot be used photochemically by PSII and must be reradiated as light. In *Agave*, *Atriplex*, and *Phaseolus vulgaris* seedlings, short-term acclimation at high but sublethal temperatures leads to an increase by as much as 5 °C in the temperature at which this abrupt rise in chlorophyll fluorescence occurs. Other studies have indicated that the activities of Rubisco and other carbon-fixation enzymes may also be severely compromised at high temperatures.

Heat Shock Proteins Exposure of most organisms to supraoptimal temperatures for brief periods suppresses

[2]The term *desert* is an anthropocentric term referring to dry places that are usually hot for at least a portion of the year. Arid and semiarid are classifications based on annual rainfall. Arid regions receive less than 250 mm of rain annually and semiarid regions between 250 and 450 mm.

the synthesis of most proteins and induces the synthesis of a new family of low molecular mass proteins known as **heat shock proteins (HSPs)**. This interesting class of proteins was originally discovered in *Drosophila melanogaster* (fruit fly) but they have since been discovered in a variety of animals, plants, and microorganisms. Exposures in the range of 15 minutes to a few hours at temperatures 5 °C to 15 °C above the normal growing temperature are usually sufficient to cause full induction of HSPs. HSPs are either not present or present at very low levels in nonstressed tissues. Initially, interest in HSPs centered on their potential for the study of gene regulation. There are, however, several aspects of HSPs that are of physiological interest.

There are three distinct classes of HSPs in higher plants, based on their approximate molecular mass: HSP90, HSP70, and a heterogeneous group with a molecular mass in the range of 17 to 28 kDa (Table 22.3) (Vierling, 1990). One in particular, HSP70, has a high degree of structural similarity—about 70 percent identical—in both plants and animals. Another protein, **ubiquitin,** is also found in all eukaryote organisms subjected to heat stress and is considered a HSP. Ubiquitin has an important role in marking proteins for proteolytic degradation. HSPs are found throughout the cytoplasm as well as in nuclei, chloroplasts, and mitochondria. As well, induction of HSPs does not require a sudden temperature shift: They have been detected in field-grown plants following more gradual temperature rises of the sort that might be expected under normal growing conditions.

Although the genetics of HSPs and regulation of their expression is slowly being unraveled, relatively little is known about their precise role in the cell (Table 22.3). HSPs are synthesized very rapidly following an abrupt increase in temperature; new mRNA transcripts can be detected within 3 to 5 minutes and HSPs form the bulk of newly synthesized protein within 30 minutes. Within a few hours of return to normal temperature, HSPs are no longer produced and the pattern of protein synthesis returns to normal. The speed of their appearance suggests that HSPs might have a critical role in protecting the cell against deleterious effects of rapid temperature shifts. HSP70, for example, appears to function as a molecular chaperone, or **chaperonin.** Chaperonins are a class of proteins normally present in the cell that direct the assembly of multimeric protein aggregates. There is in the chloroplast, for example, a Rubisco-binding protein that helps to assemble the large and small subunits of Rubisco into a functional enzyme. It has been suggested that HSP70 functions to prevent the disassembly and denaturation of multimeric aggregates during heat stress. At the same time, increased ubiquitin levels reflect an increased demand for removal of proteins damaged by the heat shock (Vierling, 1990).

The induction of new proteins is not limited to heat shock. Other stresses, including low temperature, water deficit, ABA treatment, salinity, anoxia, and osmotic stress induced artificially by polyethylene glycol solutions all result in the synthesis of new families of proteins. Although some of these **stress proteins** are similar to HSPs, the type of protein formed varies according to tissue type, growth conditions and plant species. Thus, the synthesis of new proteins appears to be a common response to stress, but there is no universal set of stress proteins.

TABLE 22.3 **Principal heat shock proteins (HSP) found in plants and their probable functions. Families are designated by their typical molecular mass. The number and exact molecular mass of proteins in each family vary depending on plant species.**

HSP Family	Probable Function
HSP 110	Unknown.
HSP 90	Protecting receptor proteins.
HSP 70	ATP-dependent protein assembly or disassembly reactions; preventing protein from denaturation or aggregation (molecular chaperone). Found in cytoplasm, mitochondria, and chloroplasts.
HSP 60	Molecular chaperone, directing the proper assembly of multi-subunit proteins. Found in cytoplasm, mitochondria, and chloroplasts.
LMW HSPs (17–28 kDa)	Function largely unknown. LMW (low molecular weight) HSPs reversibly form aggregates called "heat shock granules." Found in cytoplasm and chloroplasts.
Ubiquitin	An 8 kDa protein involved in targeting other proteins for proteolytic degradation.

Based on Vierling, 1990.

SALT STRESS

High salt concentrations in the rhizosphere generate stress through water deficits and ion toxicity. Exclusion of salt and osmotic adjustment both play major roles in tolerance of high salt environments.

The concentration of inorganic ions in a plant's environment may vary widely between deficient and excess. Although technically a form of low salt stress, ion deficiency is generally manifest as a nutrient problem and, as such, was discussed earlier in Chapter 4. In practice, the term **salt stress** refers only to an excess of ions, particularly, but certainly not limited to, Na^+ and Cl^-.

There are large areas of the earth in which high salinity is a natural part of the environment (Fitter and Hay, 1987). Coastal salt-marshes are characteristic of low-lying regions, frequently in estuaries, that are subject to tidal inundation. Seawater, which is approximately 3 percent sodium chloride, contains 460 mM Na^+, 50 mM Mg^{2+}, and 540 mM Cl^-, with smaller quantities of other ions. Seawater has a solute potential on the order of −2.7 MPa. The actual salinity of salt-marshes, however, will depend on a number of factors, such as its elevation and distance from the sea, the extent of mixing between seawater and fresh water, the amount of evaporation (which causes an increase in salt concentration), or the amount of precipitation (which causes a decrease in the salt concentration). High salinity is also characteristic of inland deserts. Here evaporation exceeds precipitation; there is little if any leaching and salts accumulate in the soil. Desert soils typically have high concentrations of Na^+, Cl^-, Ca^{2+}, SO_4^{2-}, and carbonates. High salinity is also found near the shores of inland lakes, such as the Great Salt Lake and the Dead Sea, that have no outlets and so accumulate salts as water evaporates.

FIGURE 22.10 Extensive irrigation leads to significant increases in the salt content of agricultural soils. (Copyright California Department of Water Resources.)

A third category of highly saline soils is agricultural land that has been heavily irrigated (Fig. 22.10) (Flowers et al., 1977). Since irrigation is particularly intense in drier regions, there is extensive water loss through a combination of evaporation and transpiration, called **evapotranspiration.** The result is that the salts delivered along with the irrigation water are concentrated in the soil. The salinization of agricultural lands has serious consequences, as much of the land must ultimately be withdrawn from production. In China, for example, more than seven million hectares are classified as saline, much of this resulting from centuries of irrigation (Sun, 1987). Unfortunately, most economically important crop species are very sensitive to saline soil conditions. This has led to intensive water management practices in an effort to reduce salinization and efforts to breed varieties with higher salt tolerance.

Plants that grow in high-salt soils are known as **halophytes** (Fig. 22.11). These include plants such as *Suaeda maritima* (fam. Chenopodiaceae) and marshgrass (*Spartina* sp.) that grow in salt-marshes across Europe and North America. The most tolerant halophytes will continue to grow at concentrations of NaCl in the 200 to 500 mM range. Some halophytes are known as **salt regulators.** Salt regulators, such as mangrove (*Rhizophora mangle*), do not absorb salt, but actively exclude it from their roots. Other salt regulators take up the salt but excrete large quantities through specialized **salt glands** in the leaves. Excreted salt crystallizes on the leaf surface where it is no longer harmful. Excretion through

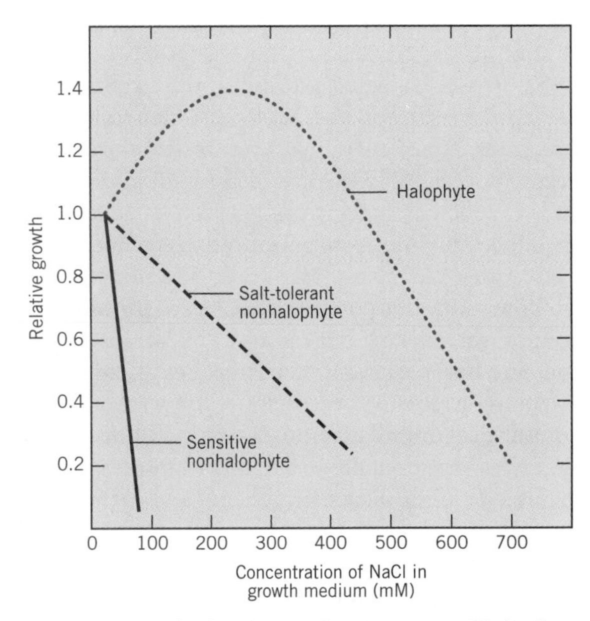

FIGURE 22.11 Idealized growth responses of halophytes, salt-tolerant nonhalophytes, and sensitive nonhalophytes to salt concentration.

salt glands is highly specific; primarily Na^+, Cl^-, and HCO_3^- are excreted against a concentration gradient while ions such as Ca^{2+}, NO_3^-, SO_4^{2-}, and HPO_4^- are retained against a concentration gradient. In one study, it was estimated that *Aeluropus litoralis* excreted salt at the rate of 8970 μmol g^{-1} dry weight of leaf d^{-1} (Pollak and Waisel, 1970). **Salt accumulators,** on the other hand, rely on high ion uptake to maintain cell turgor under conditions of low soil water potential. Some species of *Atriplex*, for example, have leaf water potentials as low as −2 MPa, compared with −0.2 to −0.3 mPa for a nonhalophyte. The excess ions accumulate in the vacuole, while cytoplasmic concentrations of Na^+ and Cl^- ions are kept low. It is interesting to note that even the most tolerant halophytes are not obligate halophytes—virtually all species studied thus far will grow well in low salt environments.

At the other extreme are sensitive nonhalophytes (Fig. 22.11), also called **glycophytes.** Many agriculturally significant species such as beans (*Phaseolus vulgaris*), soybeans (*Glycine max*), rice (*Oryza sativa*), and maize (*Zea mays*) are glycophytes. Glycophytes can tolerate very little salt and may suffer irreparable damage at concentrations of NaCl less than 50 mM. Other nonhalophyte crop species such as tomato (*Lycopersicon esculentum*), cotton (*Gossypium hirsutum*), sugarbeet (*Beta vulgaris*), barley (*Hordeum vulgare*), and wheat (*Triticum aestivum*) will tolerate higher levels of salt. Barley is the most salt-tolerant of the agriculturally important cereals and has been grown successfully in fields that irrigation has rendered unsuitable for other crops. There are cultivars of both barley and wheat, however, that exhibit varying degrees of sensitivity to salt stress. This raises the possibility that increased salt tolerance may be achieved by selective breeding programs (Yeo and Flowers, 1989).

Salt stress can damage plants at three different levels. First, high salt concentrations, and high sodium concentrations in particular, will alter the structure of soils. Because porosity of the soil is decreased, both aeration and hydraulic conductance of soils can be adversely affected. Second, high salt concentrations are inextricably linked with water stress. High salt concentrations generate low soil water potentials, a form of **physiological drought,** that make it increasingly difficult for the plant to acquire both water and nutrients. Because they share the common element of osmotic stress, drought and high salt evoke similar responses as well. For example, halophytes that exclude excess salts from the roots will acclimate to high salt environments and maintain low internal water potentials by osmotic adjustment. The solutes contributing to osmotic adjustment in halophytes are the same solutes—including proline, betaine, and sorbitol—that accumulate in response to water stress. It is often difficult to experimentally distinguish between osmotic impact of high salt and more direct effects of excess ions on plant metabolism, growth, and survival.

A third form of injury in glycophytes involves toxicity effects of specific ions, especially Na^+ and Cl^-. Although the precise mechanism for injury is not yet understood, excess Na^+ might cause problems with membranes, enzyme inhibition, or general metabolic dysfunction. It is known that high salt suppresses growth and reduces carbon assimilation in glycophytes. The reduction in carbon assimilation is due to a combination of reduced photosynthesis and enhanced maintenance respiration. Na^+ leaks passively into the root cells of *glycophytes*, moving down a concentration gradient. The internal concentration is kept low with the aid of sodium pumps that actively extrude sodium. A part of the increased maintenance respiration may represent the increased energy costs of pumping sodium out of the roots under high-salt conditions.

Ions, of course, do not act independently and there are especially strong interactions between Na^+ and Ca^{2+}. Toxicity of NaCl in young barley, maize, and other plants can be reduced by increased Ca^{2+} levels. Apparently Na^+ displaces membrane-associated Ca^{2+} from the roots and this effect can be overcome by providing excess Ca^{2+}. The result is reduced Na^+ uptake.

The physiology and molecular biology of salt tolerance have begun to attract attention only recently. As with other stresses, salt stress evokes changes in the pattern of protein synthesis, suggesting that new genes may be transcribed or, at least, the products of some genes are increased and others decreased. Prior treatment of salt-tolerant cells with ABA also induces new proteins and improves the ability to acclimate to NaCl. Some, but not all, of the proteins induced by both salt and ABA are the same.

Salt-induced gene expression has been studied in cell suspension cultures and intact root systems. Cultured cells are studied in an effort to provide a more homogeneous system and avoid some of the difficulties that might be encountered in more highly organized tissues. Roots, on the other hand, are the organ that must first cope with increased salinity. It has been found that several polypeptides accumulate to high levels in cultured tobacco cells challenged with high concentrations of NaCl, or with polyethylene glycol (PEG) (Singh et al., 1985). One in particular, a 26 kDa polypeptide, is induced by both NaCl and PEG. This polypeptide has been called **osmotin,** in recognition of the fact that it commonly appears under conditions of low water potential, or osmotic shock, induced by a variety of factors. In acclimated tobacco cell cultures, osmotin accumulates in the vacuole and may constitute as much as 12 percent of total cellular protein (Singh et al., 1987). Osmotin is a normal constituent of tomato plants, particularly in the roots, where it occurs in very low amounts in unstressed plants. The amount of protein in the roots increases significantly, however, within eight hours of addition of 171 mM NaCl. The mRNA for osmotin increases at least 100-fold in cultured tomato cells challenged with NaCl, indicating that the response to salt stress is at the level of gene transcription (King et al., 1988).

An increase in the level of osmotin is one of the more striking biochemical responses to salt and osmotic stress. However, there is no obvious relationship between the level of osmotin and salt tolerance in several varieties of tomato, which makes it difficult to assign any specific role in acclimation or tolerance. It is clear that salt stress, tolerance, and high-salt damage is a multifaceted problem. It is often difficult to separate the effects of high salt from those of low water potential, or complex interactions with other ions. Understanding mechanisms of salt tolerance will be an important first step in the search for physiological characteristics that contribute to resistance in salt-sensitive agricultural crops.

INSECTS AND DISEASE

Plants are constantly exposed to herbivorous insects and potentially pathogenic microorganisms. Although farmers and gardeners may not agree, most plants do exhibit a surprising resistance to disease. A number of strategies have been developed by plants to deter herbivory or limit the spread of invading microbes.

In Chapters 7 and 14, it was noted that isoflavonoids and other secondary metabolites, known collectively as **phytoalexins,** were synthesized by plants to help ward off insects and disease. Phytoalexins are, however, only a small part of the intricate defense strategies mounted by plants under attack.

THE HYPERSENSITIVE REACTION

Typically, a plant challenged by insects or potentially pathogenic microorganisms responds with changes in the composition and physical properties of cell walls, the biosynthesis of secondary metabolites that serve to isolate and limit the spread of the invading pathogen and necrotic lesions at the site of invasion. These responses are collectively known as a **hypersensitive reaction**.

The hypersensitive reaction is commonly activated by viruses, bacteria, fungi, and nematodes and occurs principally in plants outside the pathogen's normal specificity range. Although the hypersensitive reaction is complex and can vary depending upon the nature of the causal agent, there are common features which generally apply. An early event in any case is the activation of defense-related genes and synthesis of their products, **pathogenesis-related (PR) proteins**. PR proteins include proteinase inhibitors that disarm proteolytic enzymes secreted by the pathogen and lytic enzymes such as β-1,3-glucanase and chitinase that degrade microbial cell walls. Also activated are genes that encode enzymes for the biosynthesis of isoflavonoids and other phytoalexins that limit the growth of pathogens. Lignin, callose, and suberin are accumulated in cell walls along with hydroxyproline-rich glycoproteins that are believed to provide structural support to the wall. These deposits strengthen the cell wall and render it less susceptible to attack by the invading pathogen. Finally, the invaded cells initiate **programmed death,** a process that results in the formation of necrotic lesions at the infection site. Cell necrosis isolates the pathogen, slowing both its development and its spread throughout the plant. It is not clear at this time to what extent these components of the hypersensitive reaction are sequential or parallel events.

AVIRULENCE GENES

How do plants recognize potential pathogens and initiate defense responses? Because the hypersensitive reaction is a form of developmental response, we must assume that a signal detection and transduction chain is involved. Attempts to explain the susceptibility of plants to infection have shown that disease has an underlying genetic basis. Both pathogens and plants carry genes that determine the nature of their interaction—whether disease will occur (virulent) or not occur (avirulent). According to the **gene-for-gene hypothesis,** pathogenic microorganisms carry **avirulence *(avr)* genes** and host plants carry corresponding **resistance *(R)* genes**. Disease occurs only when the pathogen lacks the *avr* genes or the plant carries recessive alleles at the *R* locus. A matching pair of pathogen *avr* genes and *dominant* plant *R* genes initiates a hypersensitive reaction (Yang et al. 1997).

Although a number of *avr* genes have been isolated from both bacteria and fungi, the specific function of their products is not known. One possibility is that *avr* genes encode enzymes for the production of substances called **elicitors** and that *R* genes encode receptors that recognize elicitors. Elicitors (Latin, *elicere;* to entice) are metabolites isolated from pathogens that evoke a hypersensitive response in host plants. A variety of elicitors have been identified, most of them extracellular microbial products commonly associated with cell walls of bacteria and fungi. For example, fungal elicitors include β-glucans, chitosan (a chitin subunit),[3] and arachidonic acid (an unsaturated lipid). Other elicitors include various polysaccharides, glycoproteins, and small peptides.

[3]The principal carbohydrate in most fungal cell walls is chitin rather than cellulose. Chitin is a polymer of N-acetylglucosamine that forms microfibrils similar to cellulose.

Even pectic fragments, resulting from initial degradation of the plant cell wall pectins, or mechanical damage are capable of eliciting a hypersensitive reaction.

Recognition of elicitors by the plant cell no doubt takes place at the plasma membrane. Because many components of the hypersensitive reaction involve differential gene activation, it is expected that some form of signal transduction pathway is required to relay this information to the nucleus. A variety of common signalling agents have been suggested, including changes in pH, and ion fluxes (especially potassium and calcium). For example, a transient uptake of Ca^{2+} (and efflux of K^{+}) was observed when cultured cells were challenged with a fungal elicitor (Hahlbrock et al., 1995). Moreover, expression of defense response genes can be regulated by regulating intracellular Ca^{2+} levels. Thus, defense responses can be activated by stimulating Ca^{2+} uptake with Ca^{2+} ionophores or inhibited by blocking Ca^{2+} channels. Other early events in elicitor-treated cells include protein phosphorylation and the production of active oxygen species ($O_2^{\cdot-}$ and H_2O_2), known as the oxidative burst. The precise role of these various signals and how they interact in the signal cascade is unknown. It is a topic that is under active investigation in numerous laboratories.

SYSTEMIC ACQUIRED RESISTANCE

Some secondary metabolites associated with the hypersensitive reaction appear to constitute an early warning system, sending signals to other cells and tissues that prepare them to resist secondary infections. Initially the hypersensitive reaction is limited to the few cells at the point of invasion, but over a period of time, ranging from hours to days, the capacity to resist pathogens gradually becomes distributed throughout the entire plant. In effect, the plant reacts to the initial infection by slowly developing a general immune capacity. This phenomenon is known as **systemic acquired resistance (SAR)** (Ryals et al., 1996).

The development of SAR is still poorly understood, but one component of the signalling pathway appears to be **salicylic acid** (Fig. 22.12; Durner et al., 1997). Salicylic acid (2-hydroxybenzoic acid) is a naturally occurring secondary metabolite with analgesic properties. Native North Americans and Eurasians have long used willow bark (*Salix* sps.), a source of the salicylic acid glycoside, **salicin,** to obtain generalized relief from aches and pains.

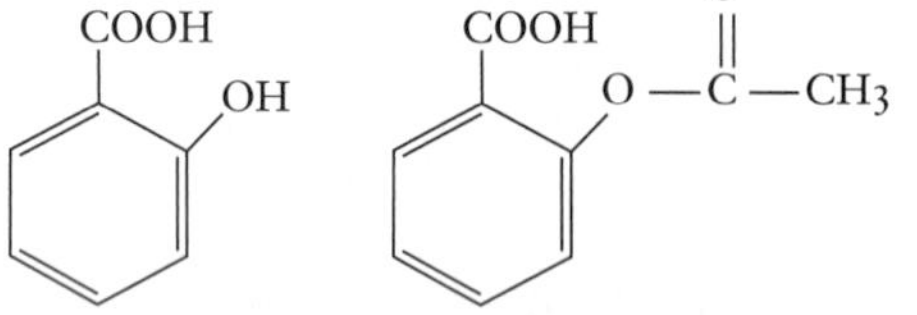

COOH OH   COOH O—C(=O)—CH_3

Salicylic acid   Aspirin (acetylsalicylic acid)

FIGURE 22.12 The chemical structure of salicylic acid and its commercial derivative acetylsalicylic acid. Salicylic acid has been implicated in the immune strategies of plants.

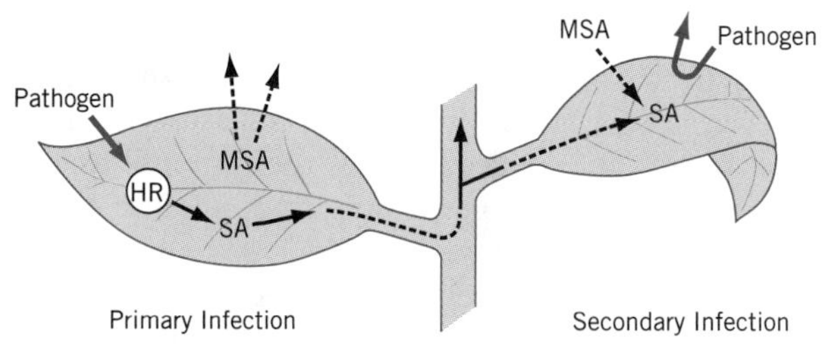

FIGURE 22.13 The possible role of salicylic acid in systemic acquired resistance (SAR). The first pathogens to infect the plant (primary infection) stimulate a localized hypersensitive reaction (HR) and the synthesis of salicylic acid (SA). Salicylic acid is translocated through the phloem to other regions of the plant where it prevents secondary infection by other pathogens. Alternatively, salicylic acid may be converted to methylsalycylic acid (MSA). MSA is moderately volatile and may function as an airborne signal.

The relationship between salicylic acid and resistance to pathogens did not become apparent until the early 1990s, when it was observed that both salicylic acid and its acetyl derivative (aspirin), when applied to tobacco plants, induced PR gene expression and enhanced resistance to tobacco mosaic virus (TMV). Since then, it has been shown in a variety of plants that infection is followed by increased levels of salicylic acid both locally and in distal regions of the plant (Fig. 22.13). For example, when tobacco plants are inoculated with TMV, the salicylic acid level rises as much as 20-fold in the inoculated leaves and 5-fold in the noninfected leaves. Furthermore, the appearance of PR proteins rises in parallel with salicylic acid. The rise in salicylic acid levels usually precedes the development of SAR. There are also a number of *Arabidopsis* mutants and transgenic plants that are characterized by constitutively high levels of both salicylic acid and SAR and, consequently, enhanced resistance to pathogens.

On the other side of the coin, plants with artificially low levels of salicylic acid generally fail to establish SAR. For example, bacteria have a gene designated *nahG* that encodes the enzyme salicylate hydroxylase. *Arabidopsis* plants transformed with the *nahG* gene thus contain little or no salicylic acid. Plants transformed with the *nahG* gene also fail to establish SAR and are compromised in their ability to ward off pathogen attack. Salicylic acid levels can also be reduced by direct inhibition of the enzyme phenylalanine-ammonia lyase (PAL), which catalyzes the first step in the biosynthesis of salicylic acid. PAL-limited *Arabidopsis* plants lose their resistance to disease, but resistance can be restored by applying salicylic acid (Durner et al., 1997). Based on results such as these, it is clear that salicylic acid has a significant role in plant defense responses. However, the mechanism whereby salicylic acid establishes and maintains SAR is yet to be determined.

JASMONATES

On the basis of recent experiments, it appears that **jasmonates,** especially **jasmonic acid** and its methyl ester (**methyl jasmonate**) (Fig. 22.14), also mediate insect and disease resistance (Creelman and Mullet, 1997; Wasternack and Parthier, 1997). Jasmonates have been found to occur throughout plants, with highest concentrations in young, actively growing tissues. Methyl jasmonate is the principal constituent of the essential oil of *Jasminium* and high concentrations of jasmonic acid have been isolated from fungal culture filtrates.

Jasmonates were first recognized for their ability to promote senescence of detached barley leaf segments, but a role in disease resistance was suggested when phytoalexin biosynthesis in cell cultures was linked to jasmonic acid content. It is now known that jasmonic acid accumulates in wounded plants and in plants treated with elicitors. Jasmonic acid also activates a number of genes encoding proteins with antifungal properties.

There are some similarities in the action of salicylic acid and jasmonates with respect to insect and disease resistance, but there are also some important distinctions. In a study of two fungal resistance genes in *Arabidopsis*, for example, it was found that expression of one gene was induced by salicylic acid, but not jasmonic acid, while the second gene was induced by jasmonic acid but not salicylic acid. Apparently there are at least two defensive pathways, one mediated by salicylic acid and one mediated by jasmonates.

Jasmonates have been studied extensively only in the last decade and little is yet known about how they regulate gene expression. However, jasmonic acid is synthesized from the unsaturated fatty acid, linolenic acid,

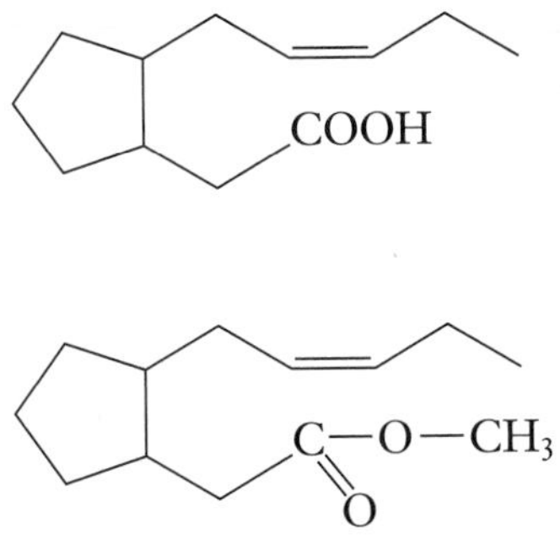

COOH

C—O—CH_3
O

FIGURE 22.14 **The chemical structures of jasmonic acid (above) and methyjasmonate (below). Jasmonic acid is synthesized from linolenic acid (18:3).**

which has led to the proposal that jasmonic acid functions as a type of second messenger (see Creelman and Mullet, 1997). Plant membranes are a rich source of linolenic acid in the form of phospholipids. It is thought that elicitors might bind with a receptor in the plasma membrane. Then, in a manner similar to that proposed for auxins (see Fig. 17.20), the elicitor–receptor complex activates a membrane-bound phospholipase that releases linolenic acid. The linolenic acid is oxidized to jasmonic acid, which in turn acts to modulate gene expression.

Another very interesting but somewhat complicating aspect of jasmonates is that their action is not limited to insect and disease resistance. Through their effect on gene expression, jasmonates modulate a number of other physiological processes. These include seed and pollen germination, vegetative protein storage, root development, and tendril coiling. In most of these effects, the jasmonates appear to work in concert with ethylene. This breadth of jasmonate effects has led some to suggest that jasmonates should be elevated to the status of plant hormones.

ENVIRONMENTAL POLLUTANTS

Although not historically a natural hazard, environmental pollution represents a relatively new set of stresses for plants. Pollution stress is primarily chemical in nature and includes toxic effects of heavy metals, airborne oxides of carbon, nitrogen, and sulfur, and photochemical products.

We hear a lot about environmental pollution these days—soil pollution, water pollution, and air pollution—but what exactly is a pollutant? A pollutant may be most easily described in terms of air. Air is defined as the atmosphere; the colorless, odorless, and tasteless mixture of gases that surrounds the earth. Air consists primarily of nitrogen and oxygen, with smaller amounts of carbon dioxide. But air also contains other gasses, including oxides of carbon, sulfur, and nitrogen. Many of these gasses, along with large quantities of ash, originate from natural sources such as volcanic eruptions and forest fires. Air also contains dust (from deserts and wastelands) and **volatile organic hydrocarbons (VOC)**. VOC include numerous by-products of industrial activity and fossil fuel combustion. Another significant source of VOC is vegetation. Plants emit large quantities of ethylene, alcohols, isoprene, isoprene-based monoterpenes, and a variety of other hydrocarbons into the atmosphere (Sharkey, 1996). These natural contaminants are continuously present in the atmosphere, but normally at concentrations that do not adversely affect plants and other organisms. Air contaminants become pollutants when their concentrations reach levels that *do* have adverse effects on organisms.

Effects may range from a simple unpleasant odor to conditions that present a measurable threat to the organism's wellbeing or survival. *A pollutant, therefore, is a contaminant of air, water, or soil that has an adverse effect on an organism.* Pollutants may be of natural origin but most are **anthropogenic;** that is, the result of human activity.

Environmental pollution has reached crisis proportions over the past several decades and is now having a significant impact on the health of forests and agricultural production. Two principal categories of pollutants that plants must cope with include heavy metals, found primarily in soil and water, and toxic gasses formed in the atmosphere by photochemical reactions.

HEAVY METALS

As discussed in Chapter 4, plants require a relatively small number of elements for their growth and survival. Natural soils, however, contain many other elements that, in spite of the selectivity of root cell membranes, may be detected in plant tissues in trace amounts. Many of these elements, especially heavy metals such as cadmium, lead, and arsenic, can be highly toxic. In addition, as discussed in Chapter 4, required micronutrients such as copper, nickel, and zinc can also be toxic in excess.

There are localized regions where natural geochemical processes have caused unusually high levels of heavy metals in soils and there are species that thrive in these regions. Unfortunately, the widespread accumulation of heavy metals in soils is increasingly becoming a problem as a consequence of twentieth-century industrial activity. Mining wastes, paper mills, and deposits from atmospheric emissions all contribute to increasing levels and more widespread distribution of heavy metals in the environment.

As with other stresses, plant species differ markedly in their sensitivity to heavy metals. Many species will thrive on soils rich with arsenic, selenium, nickel, chromium, gold cyanide, cadmium, and other metals or metalloids.[4] Often the heavy metals are excluded from uptake by roots due to the normal selectivity of root cell membranes (Chap. 5), a form of *avoidance.* Other species take up metals and accumulate them to levels that would be lethal to nontolerant species. Such plants are called **accumulator species** and represent a form of true *tolerance.* Examples of accumulator species include some species of *Astragalus* (Leguminoseae) and *Stanleya pinnata* (Brassicaceae) that may accumulate selenium at levels up to 10 percent of the dry weight of seeds. A number of species endemic to nickel-rich soils will contain in excess of 1000 μg Ni g^{-1} dry weight, compared with normal concentrations on the order of 0.05 μg Ni g^{-1} dry weight. It has been reported that *Viola calaminaria* will accumulate zinc (Zn) and lead (Pb) in excess of 10,000 μg g^{-1} dry weight (Peterson, 1993).

How do plants tolerate such high levels of toxic elements? There appear to be two principal strategies, both involving detoxification by combining the toxic element with an organic molecule. Many plants sequester the metals in the form of low-molecular-weight organometallic compounds, primarily by combining the metal with sulfur-containing amino acids such as cysteine or methionine or organic acids such as acetate, malate, or citrate. Many other plants synthesize small sulfur-rich polypeptides called **phytochelatins** (Grill et al., 1985). Phytochelatins are unusual peptides with the general formula (γ-glutamic acid-cysteine)$_n$-glycine, where n = 2-8 (Fig. 22.15). The unusual structure arises from the fact that the peptide bond between glutamate and cysteine utilizes the side chain, or γ, carboxyl group of glutamate rather than the α-carboxyl group characteristic of proteins. This bonding arrangement suggests that phytochelatins are not synthesized on ribosomes and are thus not a direct gene product but are the product of some biosynthetic pathway. The structure of phytochelatin is similar to that of the tri-peptide **glutathione** (**GSH** = γ-glutamyl-cysteinyl-glycine), the most abundant thiol (—SH compound) in plants. Glutathione plays an important role in detoxification of peroxides that are generated in the presence of active oxygen species. Glutathione also accumulates when plants are fed sulfur compounds, suggesting that glutathione may also serve as a storage pool for cysteine. When sulfur is in short supply, glutathione is slowly degraded and used as a source of sulfur (Schmidt and Jäger, 1992).

Significant amounts of phytochelatins are found only when toxic levels of cadmium, copper, mercury, lead, zinc, and other metal or metalloid elements are present. Although there is little doubt that phytochelatins readily sequester metallic elements to the cysteine thiol groups, their exact role is not clear. They may serve as a shuttle, binding metals in the cytosol and carrying them into the vacuole. Once in the vacuole, the acidic pH would displace the metal, allowing the peptide to return to the cytosol. The metal would then be sequestered by organic acids that are usually present at high concentrations in the vacuole (Steffens, 1990).

AIR POLLUTION

Major airborne pollutants that have an impact on plants include CO_2, CO, SO_2, NO_x (nitrogen oxides; mostly NO and NO_2), fluorides, and a variety of photochemical derivatives including peroxyacetyl nitrate (PAN) and ozone. Although some of these gasses are produced by volcanic eruptions and other natural phenomena, the vast proportion of these pollutants presently in the at-

[4]Metalloids are elements, such as arsenic, that have certain properties intermediate between those characteristic of metals and those of nonmetals.

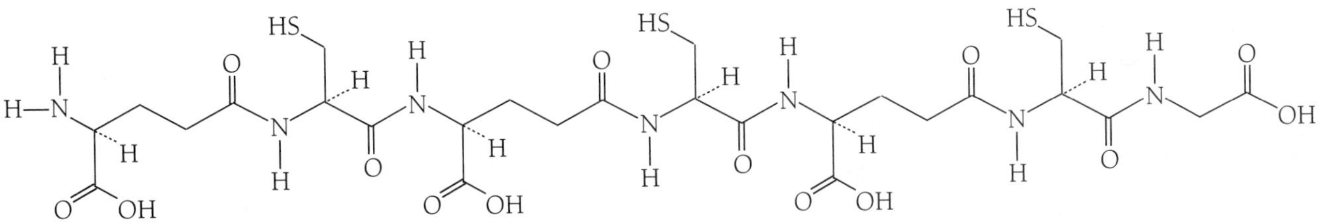

FIGURE 22.15 The structure of (γ-glutamylcysteinyl)$_3$-glycine, a metal-binding polypeptide.

mosphere are anthropogenic in origin. Oxides of carbon, sulfur, and nitrogen are generated primarily through combustion of fossil fuels in motor vehicles, but also from smelters and coal-fired power plants. Fluoride is a common constituent of minerals and is given off as HF during the reduction of aluminum ores and phosphate production.

The effects of most pollutants are highly variable depending on species sensitivity, the intensity and duration of exposure to the pollutant, wind, rainfall, and other meteorological factors. Some regions, such as the Los Angeles basin in California and the Ruhr Valley in Europe, are particularly susceptible to episodes of pollution because geographical features confine the air mass and prevent dilution through dispersal over larger areas.

Gaseous pollutants such as SO_2 and NO_x gain entry into the leaves through open stomata, following the same diffusion pathway as CO_2. Accordingly, plants are most susceptible to these pollutants during daylight hours under conditions conducive to stomatal opening. Once inside the leaf, both SO_2 and NO_x readily dissolve in the apoplastic water to produce mainly sulfite and bisulfite ions (SO_3^{2-}, HSO_3^-) or nitrite and nitrate (NO_2^-, NO_3^-), respectively. Sulfite, bisulfite, and nitrite are toxic at high concentrations. Even nitrate can be toxic at sufficiently high concentrations, but, at lesser concentrations, both nitrate and nitrite can be metabolized through nitrogen reduction. Most leaves have the capacity to detoxify sulfite and bisulfite, if the concentrations are not excessively high, by oxidizing them to the less toxic sulfate ion (Fig. 22.16). At low atmospheric concentrations, SO_2 may actually satisfy a plant's essential nutrient need for sulfur. For this reason, it is often difficult to demonstrate sulfur deficiency in heavily industrialized regions.

The primary site of SO_2 injury appears to be the chloroplast and photosynthesis. Concentrations of SO_2 as low as 0.035 μL L^{-1} of air will cause disruption of chloroplast membranes. Higher concentrations will damage the plasmalemma and other membranes, inhibit enzymes (including Rubisco and PEPcarboxylase), and generally disrupt metabolism. Disruption of chloroplast membranes and breakdown of chlorophyll give rise to visible symptoms of injury such as bleaching and necrosis at the leaf margins and intercostal (between the leaf veins) regions (Fig. 22.17). Yields of sensitive crop species such as current (*Ribes* sp.) and winter cereals may be impaired when the average concentration of SO_2 exceeds 0.01 to 0.02 μL L^{-1}. These numbers are only approximations, however, because the effects are highly dependent on the intensity and duration of peak exposures.

The effects of **peroxyacetyl nitrate (PAN)** on plants were first observed in 1945 in the Los Angeles area as a silvery-bronze glazing on leaf crops such as lettuce (*Lactuca sativa*) and spinach (*Spinacea oleraceae*). PAN is a constituent of smog, the acidic, eye-irritating haze which periodically fills the air over certain large cities and industrial regions. PAN is formed photochemically by the action of UV and visible radiation on a complex mixture of oxygen, NO_x, and volatile unsaturated hydrocarbons generated in automobile exhaust. There is relatively little data on critical concentrations of PAN. Concentrations of 0.01 to 0.02 μL L^{-1} are

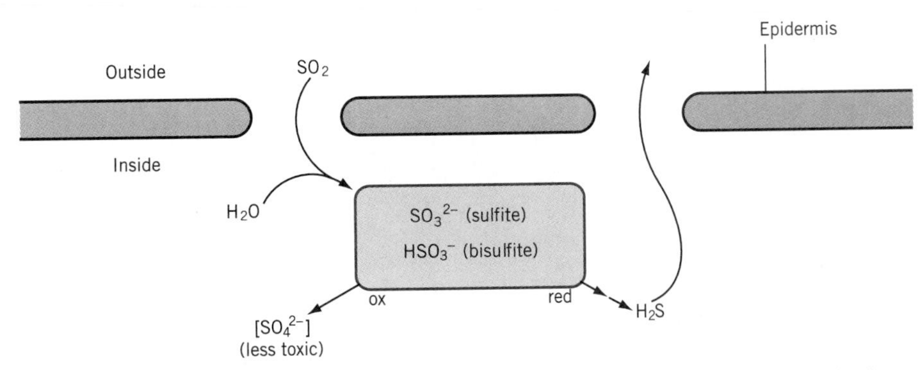

FIGURE 22.16 Entry of pollutant sulfur dioxide and its detoxification in a leaf.

FIGURE 22.17 Sulfur dioxide causes marginal and intercostal necrosis on squash leaves. With modern emission control technology, sulfur dioxide injury is rare in developed countries. (Photograph courtesy of M. Treshow.). (See color plate 11.)

common in urban areas and concentrations as high as 0.21 $\mu L\ L^{-1}$ have been reported. Concentrations in the range of 0.02 to 0.1 $\mu L\ L^{-1}$ for periods of 1 to 2 hours will cause injury to sensitive species and concentrations above 0.1 $\mu L\ L^{-1}$ cause collapse of cells and introduction of large air spaces into the leaves. This cellular collapse is the cause of the visible bronzing and glazing of leaves. Even less is known about the effects of PAN on metabolism, growth, and reproduction. PAN is difficult to generate and monitor in the laboratory and, because it is highly explosive, has proven to be dangerous to work with.

Ozone (O_3) is another photochemical pollutant that presents an even more serious problem than PAN. In Chapter 7, we briefly discussed the beneficial role of stratospheric ozone in filtering out harmful UV radiation. At ground level, however, ozone is another problem entirely. Ozone is highly reactive and tropospheric (ground level) ozone is now recognized as a major cause of injury to plants. Tropospheric ozone is formed by the action of sunlight on nitrogen dioxide (Fig. 22.18). Ozone accumulates because volatile organic hydrocarbons react with nitric oxide to prevent the back reaction. The first visible symptoms of ozone injury are generally minute chlorotic or white flecks or lesions on the leaves.

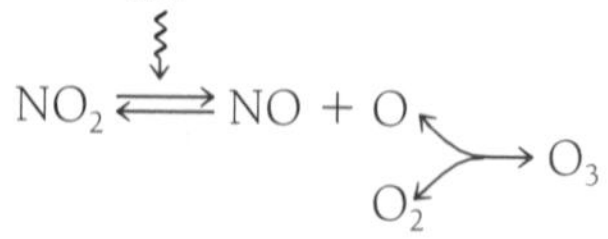

$$NO_2 \overset{h\nu}{\rightleftharpoons} NO + O, \quad O + O_2 \rightarrow O_3$$

FIGURE 22.18 Formation of ozone. Ozone is formed when NO_2 is converted to NO by the action of sunlight. The released oxygen atom reacts with a molecule of oxygen to form ozone. The reactions are reversible and normally an equilibrium is established. When volatile organic carbon is present, it reacts with NO to prevent the back reaction and ozone accumulates.

Ozone, like other gasses, enters the leaf through open stomata. Once inside the leaf, ozone dissolves in the apoplastic water where it rapidly decomposes, giving rise to several highly toxic oxygen-free radicals (HO_2^-, O_2^-, O^-, OH^-). Free radicals are toxic because they are highly reactive, attacking unsaturated fatty acids and sulfhydryl groups. Oxidation of fatty acids will disrupt membranes and alter their permeability. Oxidation of sulfhydryl groups on proteins will result in a loss of enzyme activity, including Rubisco. Ozone therefore disrupts the ultrastructure of the cell and its organelles and inhibits critical metabolic pathways such as photosynthesis and respiration. A principal effect of ozone damage is reduced carbon assimilation, even at ozone concentrations well below those that evoke visible injury, leading to a decline of forest trees and production losses in agricultural crops. Losses due to ozone across North America alone are measured in the tens of millions of dollars annually. Colored beans (such as kidney beans), for example, are particularly sensitive to ozone. Once a dominant crop in southwestern Ontario, in the region between Lake St. Clair and Lake Erie, few colored beans are now grown in this region because of increasing ozone levels.

SUMMARY

Biological stress is not easily defined, but it usually refers to a physical or biotic factor that has an adverse effect on an organism. Physical stresses include extremes of temperature, water (flooding and drought), salinity, radiation, and various chemicals (pesticides, heavy metals, and air pollutants). The principal biotic stresses faced by plants are insects and disease.

Plants may respond to stress in several ways. Stress escapers avoid stress by completing their life cycle during periods of relatively low stress. Stress avoiders have mechanisms that isolate their cells from the stressful condition, while others tolerate stress by altering their metabolism, which allows them to come to thermodynamic equilibrium with the stress but not suffer injury.

Damage resulting from water stress is related to physical changes in the protoplasm and the integrity of membranes. Desiccated protoplasm is metabolically dysfunctional and membrane damage leads to solute leakage and a general disruption of critical cell compartmentation. Most plants are protected against excessive desiccation by stomatal closure due to low turgor in the guard cells. Stomatal closure is triggered by decreasing water potential in the leaf mesophyll. The hormone abscisic acid (ABA) appears to have a significant role in stomatal closure, although its mechanism of action is unknown. ABA also appears to be involved as part of a root-to-leaf signal chain that initiates stomatal closure when soil begins to dry out. Many plants cope with water stress by osmotic adjustment—an accumulation of osmotically active solute. The resulting more negative water potential helps to maintain water movement into the leaf and, consequently, leaf turgor. Yet other plants respond to prolonged water stress by adjusting leaf area. Senescence and abscission of older leaves reduce leaf area and transpiration.

Plants respond differently to chilling stress (temperatures near but above the freezing point) and freezing stress. The membranes of chilling-sensitive plants tend to have a higher proportion of unsaturated fatty acids and, consequently, change from a fluid to semicrystalline gel state at higher temperatures than chilling-resistant plants. The ability to withstand freezing stress is most commonly seen in overwintering trees and shrubs.

Most woody tissues, although not inherently tolerant of freezing conditions during a period of active growth, will acclimate with a proper sequence of short days and low temperature. Water freezes first in the apoplastic spaces where there is relatively little dissolved solute. Protoplasmic water will then migrate out of the cells and add to the extracellular crystals. Both of these freezing events are accompanied by a measurable increase in tissue temperature, or exotherm, due to the heat of fusion released when water crystallizes. A third exotherm occurs when ice forms intracellularly, which kills the tissue.

The upper temperature limit for most plants is determined by a combination of irreversible denaturation of enzymes and problems with membrane fluidity. The membranes of plants adapted to high temperatures, such as cacti, have a higher proportion of saturated fatty acids, which helps to stabilize membranes at high temperatures. Most organisms, including plants, subjected briefly to high temperatures, respond by changes in gene expression. The synthesis of most proteins is suppressed and the synthesis of a new family of low molecular weight, heat shock proteins is induced. Little is known about the precise function of most heat shock proteins, although some appear to operate as molecular chaperones. Chaperones direct the assembly and help to maintain the stability of other, multisubunit proteins.

High salt concentrations in the rhizosphere generate stress by creating low water potential and salt toxicity. Exclusion of salt by roots and osmotic adjustment both play major roles in tolerance of high salt environments.

Plants respond to insect damage and microbial pathogen infection with a hypersensitive reaction. The hypersensitive reaction includes changes in the composition and increased strength of the cell wall, biosynthesis of phytoalexins, and necrotic lesions at the site of infection. These responses serve to isolate the potential pathogen and prevent its development and spread through the plant. According to the gene-for-gene hypothesis, plant disease responses require complementary resistance genes and avirulence genes in the host plant and invading pathogen, respectively. Disease occurs only if the pathogen lacks the avirulence genes or the plant carries recessive genes at the *R* locus.

Avirulence genes may encode enzymes for the production of elicitors, metabolites that evoke a hypersensitive response. A large, heterogeneous group of substances may function as elicitors, although most common are extracellular microbial products associated with microbial cell walls. Recognition of elicitors takes place at the plant cell membrane, probably involving a receptor encoded by the plant resistance gene.

Also associated with the hypersensitive reaction is the production of salicylic acid. Salicylic acid or its methyl ester may serve as a mobile signal, participating in systemic acquired resistance, a form of generalized immune response. Another possible signalling agent is jasmonic acid, a derivative of the fatty acid linolenic acid. The mechanisms of action of both salicylic and jasmonic acids are unknown at this time.

Although not historically a natural hazard, environmental pollutants represent a relatively new set of stresses for plants. Pollution stress is primarily chemical in nature and includes toxic effects of heavy metals, airborne oxides of carbon, nitrogen, and sulfur, and photochemical products. Many plants have the ability to detoxify heavy metals by binding them with small sulfur-rich polypeptides called phytochelatins.

CHAPTER REVIEW

1. Stress and strain are fundamentally physical concepts. Why is it more difficult to define biological stress and strain?
2. Describe how plants may be injured by water stress. In what ways are freezing stress and salt stress similar to water stress?
3. List the major environmental stresses faced by plants. Can you think of any other stresses not on this list?
4. Stomatal closure is an important factor in protecting plants against water stress. How does stomatal closure come about? What evidence is there that a plant can initiate stomatal closure in anticipation of water stress?
5. Salinity is an increasing problem for agriculture. Why is this so? Farmers do not normally fertilize

crops during periods of water stress because it might damage the plants. How might this happen?

6. What is the difference between "good" ozone and "bad" ozone?
7. Distinguish between adaptation and acclimation. Give some examples of each.
8. One early effect of water deficit is reduced shoot growth and leaf expansion. What is the cellular basis for this observation?

FURTHER READING

Alscher, R. G., J. R. Cumming (eds.). 1990. *Stress Responses in Plants: Adaptation and Acclimation Mechanisms.* New York: Wiley-Liss, pp. 377–394.

Crawford, R. M. M. 1989. *Studies in Plant Survival. Ecological Case Histories of Plant Adaptation to Diversity.* Oxford: Blackwell Scientific Publications.

George, M. F., M. J. Burke. 1984. Supercooling of tissue water to extreme low temperature in overwintering plants. *Trends in Biochemical Sciences* 9:211–214.

Jones, H. G., T. J. Flowers, M. B. Jones (eds.). 1989. *Plants Under Stress. Biochemistry, Physiology and Ecology and Their Application to Plant Improvement.* Cambridge: Cambridge University Press.

Katterman, F. (ed.). 1990. *Environmental Injury to Plants.* New York: Academic Press.

Levitt, J. 1980. *Responses of Plants to Environmental Stresses.* New York: Academic Press.

Marchand, P. J. 1987. *Life in the Cold. An Introduction to Winter Ecology.* Hanover, N.H.: University Press of New England.

Smith, J. A. C., H. Griffiths (eds.). 1993. *Water Deficits. Plant Responses from Cell to Community.* Cambridge: BIOS Scientific Publishers.

Treshow, M. (ed.). 1984. *Air Pollution and Plant Life.* New York: Wiley.

Treshow, M., F. K. Anderson. 1989. *Plant Stress from Air Pollution.* New York: Wiley.

REFERENCES

Blackman, P. G., W. J. Davies. 1985. Root to shoot communication in maize plants of the effects of soil drying. *Journal of Experimental Botany* 36:39–48.

Burke, M. J., L. V. Gusta, H. A. Quamme, C. J. Weiser, P. H. Li. 1976. Freezing and injury in plants. *Annual Review of Plant Physiology* 27:507–528.

Cowan, I. R., J. A. Raven, W. Hartung, G. D. Farquhar. 1982. A possible role for abscisic acid in coupling stomatal conductance and photosynthetic carbon metabolism in leaves. *Australian Journal of Plant Physiology* 9:489–498.

Creelman, R. A., J. E. Mullet. 1997. Biosynthesis and action of jasmonates in plants. *Annual Review of Plant Physiology and Plant Molecular Biology* 48:355–381.

Durner, J., J. Shah, D. F. Klessig. 1997. Salicylic acid and disease resistance in plants. *Trends in Plant Science* 2:266–274.

Fitter, A. H., R. K. M. Hay. 1987. *Environmental Physiology of Plants.* 2nd ed. New York: Academic Press.

Flowers, T. J., P. F. Troke, A. R. Yeo. 1977. The mechanism of salt tolerance in halophytes. *Annual Review of Plant Physiology* 28:89–121.

Fowden, L., T. Mansfield, J. Stoddart. 1993. *Plant Adaptation to Environmental Stress.* London: Chapman and Hall.

Gaff, D. F. 1977. Desiccation tolerant vascular plants of southern Africa. *Oecologia* 31:95–109.

George, M. F., M. R. Becwar, M. J. Burke. 1982. Freezing avoidance by deep undercooling of tissue water in winter-hardy plants. *Cryobiology* 19:628–639.

Grill, E., E. L. Winnacker, M. H. Zenk. 1985. Phytochelatins: The principal heavy-metal complexing peptides of higher plants. *Science* 230:674–676.

Grime, J. P. 1979. *Plant Strategies and Vegetation Processes.* Chichester: Wiley.

Hahlbrock, K. D., D. Scheel, E. Logemenn, T. Nürnberger, M. Papniske, S. Reinold, W. R. Sacks, E. Schmelzer. 1995. Oligopeptide elicited defense gene activation in cultured parsley cells. *Proceedings of the National Academy of Sciences US* 92:4150–4157.

Handa, S., A. T. Handa, P. M. Hasegawa, R. A. Bressen. 1986. Proline accumulation and the adaptation of cultured plant cells to stress. *Plant Physiology* 80:938–945.

Hartung, W., J. W. Radin, D. L. Hendrix. 1988. Abscisic acid movement into the apoplastic solution of water-stressed cotton leaves. *Plant Physiology* 86:908–913.

Johnson-Flannagan, A. M., J. Singh. 1987. Alteration of gene expression during the induction of freezing tolerance in *Brassica napus* suspension cultures. *Plant Physiology* 85:6909–6705.

Jones, H. G., M. B. Jones. 1989. Introduction: Some terminology and common mechanisms. In: H. G. Jones, T. J. Flowers, M. B. Jones (eds.), *Plants Under Stress.* Cambridge: Cambridge University Press, pp. 1–10.

King, G. J., V. A. Turner, C. E. Hussy Jr., E. S. Wurtle, S. M. Lee. 1988. Isolation and characterization of a tomato cDNA clone which codes for a salt-induced protein. *Plant Molecular Biology* 10:401–412.

Larcher, W., M. Holzner, J. Pichler. 1989. Temperature resistance of graminoids from a dry valley of the central alps. *Flora (Jena)* 118:115–131.

Levitt, J. 1972. *Responses of Plants to Environmental Stress.* New York: Academic Press.

Lynch, D. V. 1990. Chilling injury in plants: the relevance of membrane lipids. In: F. Katterman (ed.), *Environmental Injury to Plants.* New York: Academic Press, pp. 17–34.

Mansfield, T. A., C. J. Atkinson. 1990. Stomatal behavior in water stressed plants. In: R. G. Alscher, J. R. Cumming (eds.), *Stress Responses in Plants: Adaptation and Acclimation Mechanisms.* New York: Wiley-Liss, pp. 241–264.

Matthews, M. A., E. Van Volkenburgh, J. S. Boyer. 1984. Acclimation of leaf growth to low water potentials in sunflower. *Plant, Cell and Environment* 7:199–206.

Milborrow, B. V. 1984. Inhibitors. In: M. B. Wilkins (ed.), *Advanced Plant Physiology.* London: Pitman, pp. 76–110.

Mohapatra, S. S., R. J. Poole, R. S. Dhindsa. 1988. Abscisic acid-regulated gene expression in relation to freezing tolerance in alfalfa. *Plant Physiology* 87:468–473.

Nobel, P. S. 1988. *Environmental Biology of Agaves and Cacti.* Cambridge: Cambridge University Press.

Peterson, P. J. 1993. Plant adaptation to environmental stress: metal pollutant tolerance. In: L. Fowden, T. Mansfield, J. Stoddart (eds.), *Plant Adaptation to Environmental Stress.* London: Chapman and Hall, pp. 171–188.

Pollak, G., Y. Waisel. 1970. Salt secretion in *Aeluropus litoralis* (Willd.) Parl. *Annals of Botany* 34:879–888.

Quinn, P. J. 1988. Effects of temperature on cell membranes. In: S. P. Long, F. I. Woodward (eds.), *Plants and Temperature. Symposia of the Society for Experimental Biology* 42:237–258.

Raison, J. K., G. R. Orr. 1986. Phase transitions in liposomes formed from polar lipids of mitochondria from chilling-sensitive plants. *Plant Physiology* 81:807–811.

Raison, J. K., J. A. Berry, P. A. Armond, C. S. Pike. 1980. Membrane properties in relation to the adaptation of plants to temperature stress. In: N. C. Turner, P. J. Kramer (eds.), *Adaptation of Plants to Water and High Temperature Stress.* New York: Wiley, pp. 261–273.

Rao, I. M., R. E. Sharp, J. S. Boyer. 1987. Leaf magnesium alters photosynthetic response to low water potentials in sunflower. *Plant Physiology* 84:1214–1219.

Ryals, J. A., U. H. Neuenschwander, M. G. Willits, A. Molina, H.-Y. Steiner, M. D. Hunt. 1996. Systemic acquired resistance. *The Plant Cell* 8:1809–1819.

Schmidt, A., K. Jäger. 1992. Open questions about sulfur metabolism in plants. *Annual Review of Plant Physiology and Plant Molecular Biology* 43:325–49.

Sharkey, T. D. 1996. Emission of low molecular mass hydrocarbons from plants. *Trends in Plant Science* 1:78–82.

Singh, N. K., A. K. Handa, P. M. Hasegawa, R. A. Bressan. 1985. Proteins associated with adaptation of cultured tobacco cells to NaCl. *Plant Physiology* 79:126–137.

Singh, N. K., C. A. Bracker, P. M. Hasegawa, A. K. Handa, S. Bruckel, M. A. Hermodsson, E. Pfankoch, F.E. Regnier, R. A. Bressan. 1987. Characterization of osmotin. *Plant Physiology* 85:529–536.

Steffens, J. C. 1990. Heavy metal stress and the phytochelatin response. In: R. G. Alscher, J. R. Cumming (eds.), *Stress Responses in Plants: Adaptation and Acclimation Mechanisms.* New York: Wiley-Liss, pp. 377–394.

Sun, H. 1987. Good crops from salty soils. In: S. Wittwer, Y. Yu, H. Sun, L. Wang (eds.), *Feeding a Billion. Frontiers of Chinese Agriculture.* East Lansing: Michigan State University Press, pp. 83–92.

Turner, N, C., P. J. Kramer (eds.). 1980. *Adaptation of Plants to Water and High Temperature Stress.* New York: Wiley.

Van Volkenburgh, E., J. S. Boyer. 1985. Inhibitory effects of water deficit on maize leaf elongation. *Plant Physiology* 77:190–194.

Vierling, E. 1990. Heat shock protein function and expression in plants. In: R. G. Alscher, J. R. Cumming (eds.), *Stress Responses in Plants: Adaptation and Acclimation Mechanisms.* New York: Wiley-Liss, pp. 357–375.

Wasternack, C., B. Parthier. 1997. Jasmonate-signalled plant gene expression. *Trends in Plant Science* 2:302–307.

Westgate, M. E., J. S. Boyer. 1985. Osmotic adjustment and the inhibition of leaf, root, stem and silk growth at low water potentials in maize. *Planta* 164:540–549.

Williams, J. P., M. U. Kahn, K. Mitchell, G. Johnson. 1988. The effect of temperature on the level and biosynthesis of unsaturated fatty acids in diacyglycerols of *Brassica napus* leaves. *Plant Physiology* 87:904–910.

Yang, Y., J. Shah, D. F. Klessig. 1997. Signal perception and transduction in plant defense responses. *Genes & Development* 11:1621–1639.

Yeo, A. R., T. J. Flowers. 1989. Selection for physiological characters—examples from breeding for salt tolerance. In: H. G. Jones, T. J. Flowers, M. B. Jones (eds.), *Plants Under Stress. Biochemistry, Physiology and Ecology and Their Application to Plant Improvement.* Cambridge: Cambridge University Press, pp. 218–234.

Zhang, J., W. J. Davies. 1987. Increased synthesis of ABA in partially dehydrated root tips and ABA transport from roots to leaves. *Journal of Experimental Botany* 38:2015–2023.

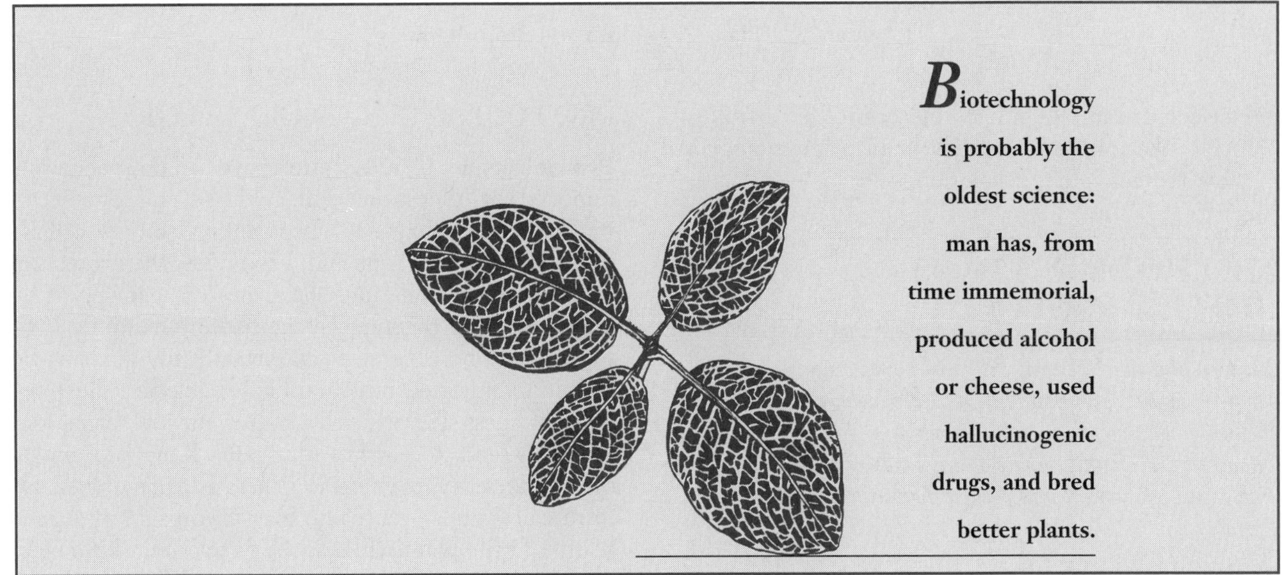

K. W. Fuller, J. R. Gallon
(1985)

23

Plant Physiology and Biotechnology

In the broadest sense, **biotechnology** employs a range of biological, chemical, and engineering disciplines to develop the use of biological organisms in agriculture and industrial processes. The use of living organisms to produce goods for human consumption is not new; its roots were established thousands of years ago. The production of beer, wine, cheese, and bread through the activities of microorganisms all have ancient origins. By the end of the nineteenth century, chemical feedstocks such as ethanol, acetone, and acetic acid were being produced by microbial fermentation. By the middle of the twentieth century, fermentation by microorganisms was also being used for sewage and solid waste treatment. Mass culture of microorganisms is now routinely used for production of antibiotics and a host of other commercial products.

The most obvious use of plant products in traditional biotechnology was as food. Over the centuries, man has learned to select edible crops for yield and quality. Plant breeding was introduced to produce new and better varieties. Early man also exploited plants in other ways. In the beginning it was for fuel and clothing, but as time went on plants provided vegetable dyes for coloring, herbs, spices, drugs, perfumes, rubber, flavorings, oils, waxes and a host of other products. One crop alone, maize (*Zea mays*), is grown as much to provide raw materials for industry as it is to provide food for livestock and human consumption. The starches and oils from maize find their way into literally thousands of industrial products as diverse as ice cream, glues, plastics, pharmaceuticals, paints, and cosmetics.

If biotechnology has been with us for so long, why is it only now receiving public attention? The answer lies in the revolution in molecular biology. Advances in DNA technology have made it possible to move genes between widely divergent organisms, creating new combinations of characters never before possible in nature. These advances were originally developed in the laboratory where they have revolutionized our ability to ask questions—and obtain meaningful answers—about fundamental biological problems. This new technology is now being exploited by industry, where the economic potential is immense.

The role of the plant physiologist in the new technology is rather straightforward. Whether to improve the yield of crops, increase resistance of plants to herbicides and pathogens, enhance vegetative propagation, or produce pharmaceuticals and other useful secondary products from plant cell cultures, plant biotechnology involves the manipulation of biochemistry, physiology, and development.

In this chapter, we describe

- tissue culture, protoplast fusion, and recombinant DNA as techniques fundamental to the revolution in plant biotechnology; and

- selected examples of plant biotechnology that illustrate the role and potential for plant physiology in the biotechnology revolution.

METHODS IN PLANT BIOTECHNOLOGY

Modern plant biotechnology is based on advances in molecular genetics and recombinant DNA technology, but has been facilitated by two other developments in methodology that have had a profound impact on the study of plant physiology: tissue and cell culture and the development of plant cell protoplasts.

TISSUE AND CELL CULTURE

Tissue culture, discussed previously in Chapters 15 and 16, is the technique of maintaining plant tissues indefinitely on an artificial medium. As early as the 1930s, Phillip White showed it was possible to isolate small groups of plant cells and maintain them indefinitely as dedifferentiated callus in artificial culture. Since White's initial experiments, it has become possible to cultivate almost any kind of plant tissue in culture. Callus derived from a single initial group of cells can be subcultured, multiplied, and, in many cases, induced to reinitiate differentiation of roots and shoots and eventually form fully competent plants.

Although plant tissue culture is most often carried out on solid medium, containing agar, tissues can also be cultured in liquid medium. If the cultures are mechanically agitated by shaking or rotating, some cells are dislodged and grow as small clumps or individual cells. Under appropriate conditions of light, nutrition, and hormones, some cells in culture give rise to small clusters of cells morphologically analogous to normal plant embryos. Called **somatic embryos**,[1] these clusters proceed through an apparently normal developmental program to produce plants that are indistinguishable from normal plants. Somatic embryos were first demonstrated by F. C. Stewart in 1958, during his studies of carrot cells in liquid culture. The study of somatic embryos in the laboratory has provided valuable insights into various aspects of plant development as well as practical experience in plant regeneration. The availability of somatic embryos has also given rise to the idea of **artificial seeds**. An artificial seed consists of a somatic embryo encased in a gelatinous matrix. Although not yet commercially viable, artificial seeds offer the possibility for low-cost, large-scale propagation of high-value hybrid vegetable crops and ornamental plants.

[1]The term *somatic* refers to vegetative, or nonsexual, cells. Thus, somatic embryos are formed without going through a sexual reproductive stage.

PROTOPLASTS AND CELL FUSION

Protoplasts are plant cells that have had their cell walls removed by digestion with cellulase and other cell wall-degrading enzymes. The resulting wall-less cell assumes a spherical shape and, having lost the protection of the rigid cell wall, must be protected against lysis by a high osmotic potential in the bathing medium. Protoplasts can be isolated from virtually any plant tissue and, in many cases, remain capable of sustained cell division. Unless steps are taken to prevent wall formation, cell walls tend to reform within a few hours of removal of the digestive enzymes. Regenerated cells divide and form callus and, eventually, may be capable of regenerating fertile plants (Fig. 23.1). Species that have been fully regenerated from protoplasts include potato, tobacco, pepper and tomato. Unfortunately, monocots do not lend themselves to protoplast culture as well as dicots and full regeneration of important cereal crops from protoplasts has not yet been achieved.

Removal of the constraints imposed by the cell wall has stimulated many fundamental studies on plant cell biology. Perhaps most significant is the capacity of protoplast to fuse, giving rise to somatic hybrids; that is, hybrids involving vegetative rather than reproductive or germ cells. This allows hybrids to be formed between species that are genetically incompatible. Moreover, plants have cytoplasmic (that is, chloroplast and mitochondrial) genomes that are maternally inherited. Some of these genes, such as those for atrazine resistance in the chloroplast and cytoplasmic male sterility in the mi-

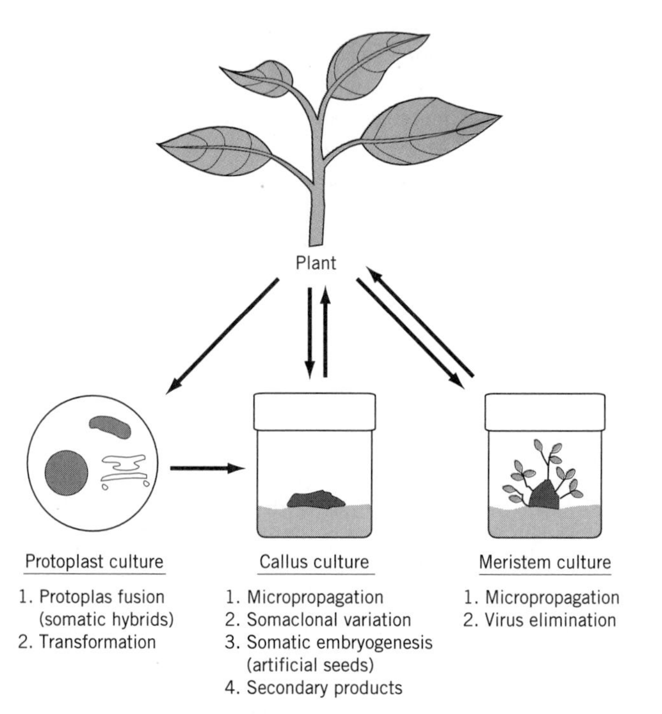

FIGURE 23.1 Micropropagation and other uses of plant cultures.

tochondrion, have practical as well as theoretical importance. Protoplast fusion makes it possible to produce new cytoplasmic-nuclear combinations without the timely and expensive backcrossing required by conventional breeding. Finally, as discussed in the following section, the absence of a cell wall facilitates the genetic transformation of plants by the direct transfer of genes into plant cells.

RECOMBINANT DNA

DNA recombination refers to the insertion of DNA from one genome into the DNA of another genome. DNA recombination is not a new invention—it is probably as old as DNA itself. In nature, however, recombination occurs only between strongly homologous DNA molecules and only when those molecules happen to get inside the same cell. Recombination in nature is thus likely to occur only between closely related species and has a minimal impact on the diversification of genetic information.

By contrast, the new recombinant DNA technology is carried out *in vitro*, allowing scientists to overcome natural genetic barriers and recombine DNA from completely different species. Moreover, the selection of specific DNA sequences, or genes, for insertion into the host DNA is both more rapid and potentially more efficient than conventional methods of plant breeding. Because a foreign gene has been introduced into a genome by artificial means, the resulting plant is said to be **genetically engineered**.

In principle, the technique of genetic engineering is relatively straightforward (Fig. 23.2). There are three essential components: (1) a source of "foreign" DNA containing the desired gene, (2) a **vector** that carries the gene, and (3) a means for introducing the vector into the host plant.

There are two principal methods for isolating genes. In the first method, DNA containing the gene of interest is first cut into small fragments (**restriction fragments**) by use of a **restriction endonuclease**. Restriction endonucleases are enzymes isolated from bacteria that recognize and cleave the doublestranded DNA at specific sequences of four to eight nucleotides. Because the enzymes cleave the entire genome of the cell from which the DNA was isolated, this method has the potential for producing large numbers of DNA fragments. Many of these fragments will contain only a portion of a gene or regions that are not normally transcribed. However, different species of bacteria produce many restriction enzymes with different specificities, so it is not difficult to find an enzyme that will produce a DNA fragment containing a particular gene of interest. In the second method, messenger RNA (mRNA) isolated from cells can be used to direct the synthesis of **complimentary DNA (cDNA)**. cDNA is synthesized from mRNA by the enzyme **reverse transcriptase**, which uses the mRNA molecule as a template. This method has the advantage that each piece of DNA represents a gene actively transcribed in that cell.

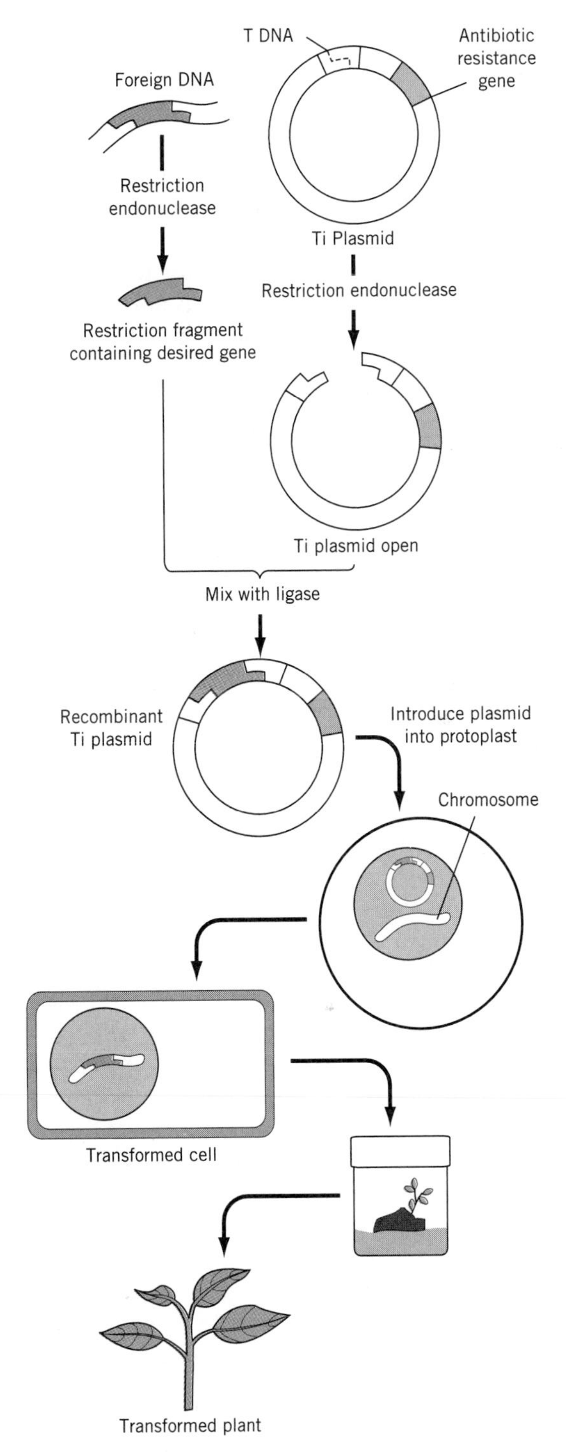

FIGURE 23.2 Using genetic engineering to transform plants with a foreign gene.

The next step is to insert the desired DNA into a vector, most commonly a small, circular piece of double-stranded bacterial DNA known as a plasmid. The plasmid is opened up with the same endonuclease used to isolate the desired gene. This leaves the ends of the DNA to be inserted complementary to the open ends of the plasmid DNA, allowing the two DNA molecules to be mixed together and their ends joined with the enzyme **DNA ligase**. The resulting recombinant plasmid, containing the "foreign" gene, can then be introduced into the host cell.

The most widely used vector for introducing foreign genes into plants is the Ti plasmid of the crown-gall bacterium *Agrobacterium tumifaciens*, described briefly in Chapter 16. *A. tumifaciens* is a natural genetic engineer. Its tumor-inducing Ti plasmid contains a region, known as T-DNA, that is normally integrated into the host cell DNA during infection. T-DNA contains the tumor-inducing genes that are responsible for transformation of the infected tissue into a tumorous gall. For use as a DNA vector, the plasmid is normally "disarmed" by removing the tumor-inducing genes and replacing them with the gene to be cloned. This renders the bacterium avirulent without compromising its ability to infect and transform host plant cells. The vector plasmid also contains a gene for antibiotic resistance (typically kanamycin resistance) that allows transformed cells to be selected.

Plants may be transformed with *A. tumifaciens* either by cocultivation with protoplasts or by infection of leaf disks. In the first case, cultured protoplasts undergoing active cell division are inoculated with bacteria carrying the recombinant plasmid. After two or three days, the bacteria are killed with antibiotics and the protoplasts are allowed to develop as small clumps of cells called **microcalli**. The microcalli are then transferred to a medium containing antibiotic (for example, kanamycin). Only calli derived from transformed cells will carry the resistance gene and survive on the antibiotic-containing medium. Alternatively, small (100 mm^2) disks cut from surface-sterilized leaves (or other suitable tissue) can be incubated with a culture of *A. tumifaciens*, allowing the bacterium to infect the wound cells at the cut surface. The bacteria are then killed and transformed cells selected with antibiotic treatment. In either case, whether protoplasts or leaf disks, the final step is to manipulate the nutritional and hormonal balance of the growth medium in order to induce root and shoot formation. The result is a transformed plant that carries and expresses the new gene and will pass that gene to its progeny through normal sexual reproduction.

A principal shortcoming of *A. tumifaciens* as a transformation vector is its limited host range. Although there are tentative reports of success with monocots, at present only certain families of dicots can be efficiently transformed by this method. Fortunately, there are other methods available for the direct transfer of genes into host cells and tissues. Cells may be induced to take up DNA fragments by treatment with polyethylene glycol (PEG) or electrical fields (electroporation). Both of these treatments render the cell temporarily permeable. In other cases, cells have been successfully transformed by microinjection of DNA directly into protoplasts or by bombardment with DNA particle guns (Klein et al., 1987). In the latter case, microscopic (approximately 1 μm) particles, coated with plasmid DNA, are accelerated to 400 ms^{-1} with a modified 0.22-caliber gun. At this velocity, the particles penetrate the cell walls and plasmic membranes, carrying the DNA into the cells.

PROGRESS AND POTENTIAL IN PLANT BIOTECHNOLOGY

Biotechnology has made possible novel ways to propagate plants, design efficient weed control, and protect crop species against pathogens and insect pests.

MICROPROPAGATION

The success of plant tissue culture techniques has made possible large-scale cloning of plant species. With a relatively small investment in space, technical support, and materials, it is possible to produce literally millions of high-quality, genetically uniform plants. The process is known as **micropropagation**. The most common technique is to place excised meristematic tissue on a medium that reduces apical dominance and encourages axillary bud development. The new shoots can be separated and subcultured to produce more axillary shoots, or placed on a medium that encourages rooting and planted out. Alternatively, tissues can be used to establish callus cultures, which may then be induced to form roots and shoots.

Micropropagation can also be an effective way to eliminate viruses and other pathogens and produce commercial quantities of pathogen-free propagules. The first plants to be mass-produced by tissue culture were virus-free *Cymbidiums* (orchids) but the technique has also been found useful for potato, lilies, tulips, and other species that are normally propagated vegetatively. Potato, for example, is vegetatively propagated through buds on the tubers, a system that readily transmits viruses to the next generation. Micropropagation of potato from meristem cultures has proven to be an effective way to isolate virus-free lines.

Micropropagation is also now being used extensively in the production of forest tree species. Here the propagules are generated primarily from cultures of axillary and adventitious buds; callusing and differentiation of new buds is rarely used. A similar approach has been applied successfully to cultivars of apple (*Malus*), peach (*Pyrus*) and pear (*Pyrus*). Because most temperate fruits are highly heterozygous, they do not breed true from seed but are propagated by vegetative cuttings. Rooting of microcuttings in culture is now a routine procedure in many commercial laboratories. By 1983, growers in the Netherlands were producing more than 21 million plants by micropropagation (Jones, 1985).

In spite of the fact that plantlets derived from tissue culture are cloned from presumably identical somatic (nonsexual) cells, the regenerated plants can exhibit significant variation in their morphology and physiology. This is known as **somaclonal variation**. The cause of somaclonal variation is not clear, but it involves spontaneous genetic variation as a result of the culture conditions. The value of somaclonal variation is that occasionally the variants exhibit disease resistance or some other agronomically useful trait.

PLANT PROTECTION

High productivity levels associated with modern agriculture depend heavily on protection of the crop species against competition from weeds and attack by viruses, fungi, and predatory insects. Since the late 1940s, weeds, pathogens, and predatory insects have been controlled by extensive use of herbicides and insecticides. Serious questions are being raised about the carryover of these agricultural chemicals along with the food product and their potentially harmful effects on consumers and the ecosystem. The advent of genetic engineering presents opportunities for novel methods of plant protection against such pests with decreased reliance on potentially dangerous chemical controls.

Herbicide Resistance Weeds are a major factor in agricultural economics. Yield reductions due to weeds exceed $12 billion annually. Because farmers have become increasingly dependent on herbicides for effective control of weeds, an additional $8 billion is spent every year on weed control. The heavy use of herbicides has had some undesirable side effects, not the least of which is a dramatic rise in the number of species exhibiting herbicide resistance (Holt et al., 1993). This is because repeated applications of herbicide increase the selection pressure in favor of the resistant weed population. Since the early 1960s, more than 120 species have been identified with resistance to a range of herbicides (Table 23.1). Fifty-seven species have been identified with biotypes resistant to triazine herbicides alone. The number is undoubtedly larger since these data were collected.

The triazine herbicides (atrazine, simazine) inhibit electron transport by binding to the 32-kDa D1 protein in photosystem II (Chap. 9). Two forms of resistance to triazine herbicides are known. Triazine herbicides are taken up primarily through the roots, and the roots of some crop species, such as maize (*Zea mays*) and *Sorghum*, have an enzyme, **glutathione-S-transferase**, that rapidly detoxifies the herbicide by conjugating the herbicide with glutathione. For this reason, triazines are an important selective herbicide for maize and sorghum crops. However, many widely distributed weeds such as species of *Amaranthus* and *Chenopodium* have developed triazine-resistant biotypes. Resistance in virtually all cases can be traced to a mutation in the *psbA* gene that codes for the D1 protein in PSII. The change of a single amino acid in the D1 protein from serine to glycine lowers the affinity of the protein for triazines by 1000-fold.

TABLE 23.1 Herbicide resistance in weed biotypes. Resistance means that a population of plants has the ability to withstand a herbicide dosage substantially greater than that which will kill a wild-type plant.

Class of Herbicide	Year First Detected	Number of Species with Resistance
Inhibitors of photosynthesis		
Bypyridiliums	1976	18
Substituted ureas	1983	7
Triazines	1968	57
Uracils	1988	2
Inhibitors of amino acid synthesis		
ALS inhibitors	1986	8
Auxin-type inhibitors		
Phenoxyacetic acids	1962	6
Other herbicides		23
Total		121

Triazine-resistant weeds will continue to be a problem that has to be dealt with. On the other hand, some advantage could be gained if the gene for triazine resistance could be transferred into other useful crop species. Unfortunately, the D1 protein is encoded in the chloroplast genome and efforts to engineer triazine-resistant crops have been hampered by difficulties encountered in transforming chloroplast DNA (Mazur and Falco, 1989). In one case, traditional plant breeding has been used to transfer resistance from a weedy species, *Brassica campestris*, into the economically important *Brassica napus* (canola, or oil seed rape). Although the resulting canola biotype is triazine resistant, the yield of the resistant biotype is about 20 percent less than the sensitive biotype. Attempts to engineer resistant biotypes by introducing the gene for glutathione-S-transferase have also been attempted, but with limited success.

Glyphosate is a broad spectrum herbicide with little residual soil activity. Glyphosate is readily translocated in the phloem and is thus particularly effective against perennial weeds (Chap. 11). Glyphosate inhibits the activity of **5-enolpyruvyl-shikimate-3-phosphate**

synthase (EPSPS), a chloroplast enzyme that catalyzes a step in the synthesis of aromatic amino acids (tryptophan, phenylalanine, tyrosine) (see Chap. 14). This pathway is found only in plants and microorganisms, so glyphosate is not acutely toxic to humans. Because of these properties, glyphosate-resistant cultivars of important crop species would be commercially important.

Two strategies have been exploited to engineer glyphosate resistance. Tobacco (*Nicotiana*), tomato (*Lycopersicum*), and *Petunia* have been transformed with a mutant gene from the bacterium *Salmonella typhimurium*. The gene, which codes for an EPSPS with decreased affinity for glyphosate, conferred increased, but incomplete, glyphosate tolerance to the transgenic plants. In other experiments, overexpression of EPSPS was achieved by transforming *Petunia* plants with an EPSPS gene linked with an active promoter. The resulting transgenic plants were able to tolerate glyphosate dosage approximately four times that which killed untransformed controls.

Glyphosate-tolerant soybeans (*Glycine max*) have been engineered using a resistant EPSPS enzyme isolated from *Agrobacterium*. Resistant crops, marketed as "Roundup-ready," are now being grown across the United States and Canada. In 1996, approximately one million acres of Roundup-ready soybeans were grown in the state of Iowa alone and by 1998 it was expected that more than half that state's soybean acreage would be sown to genetically engineered crop.

Other herbicides that have been the target of genetic engineering include a chemically heterogeneous group of herbicides (sulfonylureas, imidazolinone, and triazolopyrimidine) that share a common target site—the enzyme **acetohydroxyacid synthase (ALS)**. This enzyme, found only in plants and microorganisms, catalyzes an early step in the synthesis of branched chain amino acids (isoleucine, leucine, and valine). The *Hra* line of tobacco carries a mutant ALS gene that is 1000-fold more resistant to sulfonylureas than the wild-type. When wild-type plants were transformed with this gene, it conferred herbicide resistance to the transgenic plants (Fig. 23.3). Other species, including tomato, sugarbeet, cotton, and alfalfa, have been successfully transformed with the mutant gene resulting in varying levels of herbicide resistance.

Insect and Disease Resistance Insects and plant diseases are responsible for annual crop losses in excess of 10 to 15 percent worldwide. This represents not only an economic loss of more than $100 billion, but a significant loss in food production as well (Brears and Ryals, 1994). Clearly, control of insect herbivory and disease is a high priority for world agriculture. Traditional insecticides are not a satisfactory method of control be-

FIGURE 23.3 Field trial of herbicide-resistant transgenic canola (*Brassica napus*) plants. Canola plants were transformed with a gene encoding for a sulfonourea-insensitive form of the enzyme acetohydroxyacid synthase (ALS). The performance of transformed and genetically normal plants was tested in field trials. Transgenic plants (right) were resistant to sulfonourea which stunted the growth of normal plants (left). (Photograph courtesy of B. J. Mazur.)

cause they kill beneficial insects as well as pathogens. Insecticides thus interfere with natural insect controls and, perhaps more importantly, with pollinating vectors. For example, many orchards across North America now find it necessary to import hives of bees during the pollinating season because natural populations of bees have been decimated by long-term insecticide use.

Several strategies have been developed to engineer disease resistance into plants. One strategy relies on the natural defense system. Recall that the hypersensitive response (Chap. 22) is marked by the transcription of a series of mRNAs that are responsible for the synthesis of pathogenesis related (PR) proteins. One of these PR proteins is **chitinase**, an enzyme that catalyzes the hydrolysis of chitin, a principal constituent of fungal cell walls. Chitinase levels, especially in the vacuole and apoplast, increase during fungal infection. The chitinase gene isolated from bean (*Phaseolus vulgaris*) has been successfully cloned into tobacco plants (*Nicotiana tobacum*) and canola (*Brassica napus*) (Broglie et al., 1991). In both cases, the transgenic plants exhibited increased resistance to a soil-borne fungal pathogen. Seedling mortality was reduced and the onset of symptoms was delayed in the transformed plants (Fig. 23.4).

Another approach to engineering insect resistance exploits naturally occurring bacterial enterotoxins. *Bacillus thuringiensis* (Bt) is a bacterium that, when ingested by insect larvae, produces a protein that binds to receptors in the insect gut and interferes with normal digestion. Various strains of Bt are effective against various insects, such as the European corn borer, the Colorado potato beetle, or the cotton boll weevil. However, because mammals do not have the required receptors in the gut lining, Bt proteins are considered nontoxic to humans. Consequently, for several years Bt sprays have been available to home gardeners and for limited agricultural use. However, the Bt gene (*Cry1Ab*) has now been cloned into economically important crops such as maize, potato, and cotton. The gene expresses in the plant and produces the Cry1Ab protein along with the normal complement of plant proteins. Farmers growing the genetically engineered varieties no longer need to spray insecticides in order to control major insect predators because the insect ingests the Cry1A protein when attacking the plant.

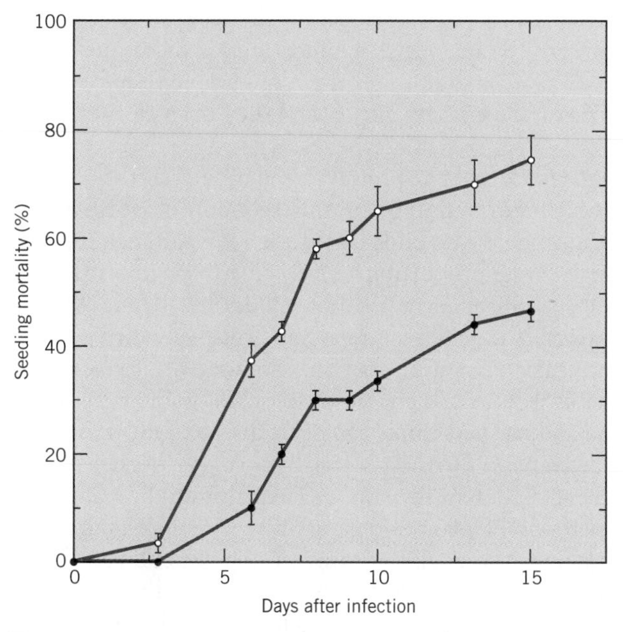

FIGURE 23.4 Disease resistance in transformed canola. Canola (*Brassica napus*) plants were transformed with the gene for chitinase from bean (*Phaseolus vulgaris*). Transformed plants (closed circles) exhibited lower seedling mortality than control plants (open circles) when the soil was inoculated with a fungus. (From K. Broglie et al., *Science* 254:1194–1197, Copyright 1991 by the AAAS.)

Cultivation of Bt engineered crops on a commercial scale does raise some concerns about insect resistance. Because the target insects are continuously exposed to the Cry1A protein as they graze, there is a high probability of selecting for resistance in the insect population. To counter this concern, commercial growers are encouraged to grow nonengineered strains of the same crop adjacent to the engineered strain. It is expected that insects feeding on the nonengineered strains will help to maintain a supply of susceptible genes in the insect population.

Another possible strategy for increasing disease resistance in plants involves a form of "vaccination" (Abel et al., 1986). Cultured tobacco cells were transformed with a cDNA clone for TMV coat protein, using the Ti plasmid of *Agrobacterium* as a vector. Plants regenerated from the transformed cells expressed both the mRNA and the TMV coat protein. The transformed plants were also less susceptible to TMV infection. Up to 60 percent of the transformed plants failed to develop symptoms when inoculated with TMV, and in those that did develop symptoms the appearance of the symptoms was delayed. Although it is not known how the presence of viral coat protein protects the plant against infection, this strategy has since been used to protect other crops, including tomato, alfalfa, and rice, from viral infections.

PLANT METABOLISM

One of the early promises held out by the new biotechnology is improved yield. Over the past three to four decades, intensive plant breeding programs have brought about marked improvements in both quality and yield of food crops. The benefits of this so-called "green revolution" were expected to accrue especially to underdeveloped nations. Further progress in food production is now expected through the genetic engineering of plants. Indeed, some progress has been made, to the extent that engineering built-in resistance to pests and diseases will improve yields.

One of the challenges facing biotechnologists is to modify photosynthesis so as to achieve increases in net carbon gain. One route that has been examined closely is to reduce the oxygenase capacity of Rubisco without unacceptable losses of carboxylase activity. Current biochemical evidence, however, suggests this is not likely (Somerville, 1990). Another possibility would be to reduce carbon drain through maintenance respiration. The problem is that photosynthesis and respiration are complex pathways, interacting not only with each other but with nitrogen and sulfur assimilation and other critical metabolic pathways. No single gene is involved in determining net carbon gain and our basic understanding of the biochemistry and physiology of these processes is inadequate. Any improvements to carbon gain through biotechnology will have to await a more thorough understanding of the metabolic processes involved.

Another area promising potential for biotechnology is to reduce reliance on nitrogen fertilizers by improving nitrogen fixation by bacteria, or even by engineering symbiotic nitrogen fixation in plants presently incapable of forming such associations. Again this is a complex problem, involving a large number of genes spread through two different organisms (Chap. 6). In recent years, however, the application of molecular techniques has revealed in increasing detail how nitrogen-fixing systems operate. Improvement in nitrogen fixation remains an exciting prospect for the future as we understand more about the complex interactions between rhizobia and host plants.

PLANT PRODUCTS

Fuels It is apparent that the world supply of fossil fuel is rapidly diminishing. Even before the supplies are exhausted, the point will be reached where the energy cost of retrieving the remaining supplies exceeds the energy content of the fuel brought to the surface. It is clear that alternative sources must be found. The obvious source is solar energy—it is abundant and free—and the best solar energy converters are green plants.

There are several ways to obtain energy from plant material, or **biomass**. The most efficient, when carried out under controlled conditions, is direct combustion of biomass to produce steam or electricity. Other processes include pyrolysis and gasification, which convert the biomass to oils and gasses, respectively. However, the usefulness of these processes is limited by the need for high temperatures to drive the reactions. Gasification, for example, is carried out at low oxygen levels and temperatures above 1500°C.

Biomass can be also converted to methane by bacterial decomposition under anaerobic conditions. The result is a mixture of methane and CO_2 at a ratio of roughly 2 : 1 by volume. Alternatively, cereal grains and other sources of starch can be readily converted to ethanol by fermentation with yeast (*Saccharomyces cerevisiae*). The best plant for converting solar energy and thus the best feedstock for ethanol fermentation is sugarcane (*Saccharum officinarum*). The world leader in ethanol fermentation is Brazil. By the mid-1980s, Brazil was producing more than 20 percent of its total liquid energy needs from ethanol (Calvin, 1985). There are other benefits as well. Sufficient steam is produced in the refining process to generate excess electricity, and the waste products of sugar refining, known as **bagasse**, can be compressed into pellets and used as a fuel in generating stations to produce still more electricity. Some of this electricity can be used to produce fertilizer for the sugarcane crop. The energy content of ethanol is somewhat less than crude oil (29 MJ kg^{-1} compared with 42 MJ kg^{-1}), but the fact that sugarcane produces fuel on a renewable basis is more than sufficient to compensate.

Another alternative source of fuel is the oils that many plants store in their seeds. The oil content of sunflower seed, for example, is about 40 to 45 percent by weight and it can be used directly in farm machinery in place of diesel oil (Calvin, 1985). Plant oils are much more highly reduced than ethanol, so their energy content is much closer to that of fuel oil. The energy content of sunflower oil, for example is 37 MJ kg^{-1}. Most other plant oils, because they are triglycerides (see Chap. 1, Box 1.1), cannot be used as fuel directly. These must first be processed to separate the fatty acids from the glycerol. Plant oils as a substitute for diesel fuels have several advantages; they are a renewable resource, have a favorable energy output, and they burn cleaner.

Edible Oils The production of oilseed crops is a major contributor to the agricultural economy of North America. Much of the oil is used for cooking, but an increasingly larger proportion is being diverted to industrial use. One popular source of edible oils is oilseed rape (*Brassica napus*). Native oilseed rape has a relatively high content of erucic acid, a 22-carbon unsaturated fatty acid that tends to impart a characteristic undesirable taste to the oil and has been linked to heart disease. One line of oilseed rape, known commercially as canola, has had its erucic acid content lowered to virtually zero by traditional methods of plant breeding. Biotechnologists are now looking to reengineer canola using recombinant DNA techniques. Objectives include increasing the oil content by 15 to 25 percent or increasing the stearic acid content. Stearic acid is an 18-carbon saturated fatty acid and could be used to solidify margarine at room temperature, bypassing the normal (and costly) hydrogenation process. Another variety of canola, already in commercial trials, has a high lauric acid content. Lauric acid is a 12-carbon saturated fatty acid used in soaps, detergents, and specialty confectionery products. Yet another group is attempting to put the erucic acid back into oilseed rape, because erucic acid has potential for industrial uses such as marine lubricants. It is clear that in the future farmers may be tailoring their crops to meet the demands of specific but nontraditional markets.

Biodegradable Plastics The bacterium *Alcaligenes eutrophus* is one of about 90 species of bacteria that synthesize **polyhydroxyalkanoates (PHA)** (Poirier et al., 1995). PHAs are a family of thermoplastic polyesters that accumulate as inclusions inside the bacterial cell, typically when excess carbon is available but growth is otherwise limited. The principal function of PHAs appears to be as a storage carbon that can be catabolized to provide carbon and energy when the growth-limiting condition is removed.

Of the 40 or more PHAs characterized, one of the most widespread and thoroughly characterized is **polyhydroxybutyrate (PHB)**. PHB is synthesized from acetyl-CoA by a series of three enzymes (Fig. 23.5). Recently, the bacterial genes that encode the three enzymes have been cloned into *Arabidoposis thaliana*, with the result that *Arabidopsis* accumulated PHB inclusions that were similar in size and appearance to those found in bacterial cells (Poirier et al., 1995). Interestingly, when the genes were expressed in the plastids, the amount of PHB (14 percent of dry weight) was 100 times that found when the genes were expressed in the cytoplasm. This difference was attributed to the fact that, in plants, the biosynthesis of fatty acids from acetyl-CoA occurs in the plastids. The flux of carbon through acetyl-CoA is therefore much greater in plastids than in the cytoplasm.

Why engineer plants to product PHA plastics? Plastics are, of course, noted for their durability and resistance to degradation. These properties make them ideal for many industrial applications and consumer products, but at the same time they generate resource and environmental problems. Plastics represent about one-fifth of municipal waste across North America, most of which ends up in landfill sites. As noted above, a wide range of bacteria use PHAs as a carbon source and contain the enzymes necessary to catabolize PHAs—that is to say, PHAs are 100 percent biodegradable. A PHA copolymer in which hydroxybutyrate is combined with hydroxyvalerate (a 5-carbon monomer) is currently in commercial production by bacterial fermentation. Fermentation is costly, as it requires a large industrial production facility and considerable energy. Plants, on the other hand, are known for producing large amounts of other products, such as starches and oils, at relatively low cost. It is possible that oil seed crops such as rape-

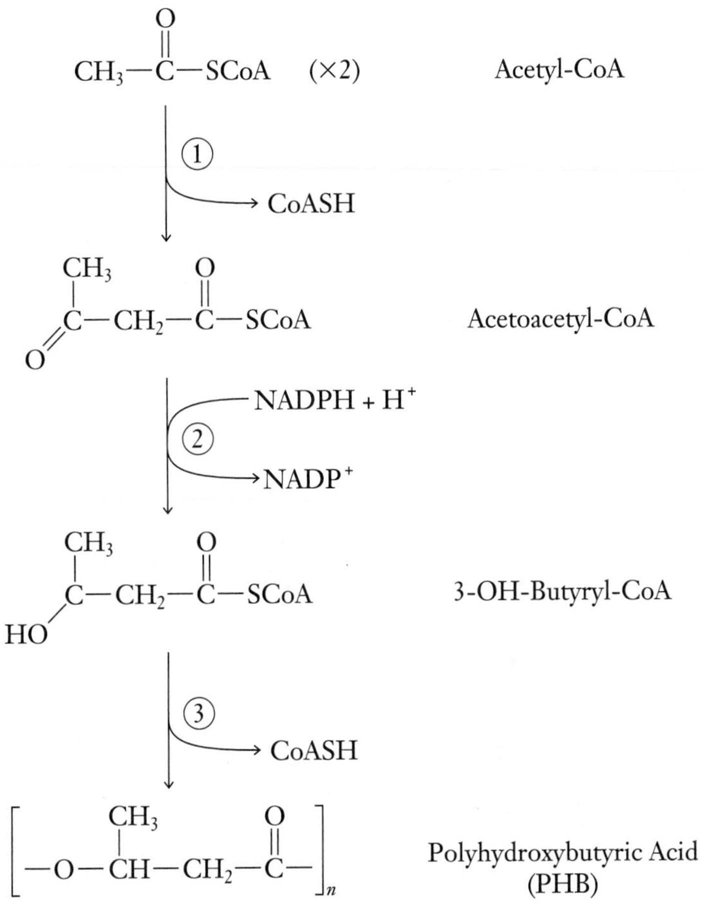

$$CH_3-\overset{\overset{O}{\|}}{C}-SCoA \quad (\times 2) \qquad \text{Acetyl-CoA}$$

$$\downarrow \text{(1)} \rightarrow CoASH$$

$$\overset{CH_3}{\underset{O\!\!\diagup\!\!\diagup}{\overset{|}{C}}}-CH_2-\overset{\overset{O}{\|}}{C}-SCoA \qquad \text{Acetoacetyl-CoA}$$

$$\downarrow \text{(2)} \quad NADPH + H^+ \rightarrow NADP^+$$

$$\overset{CH_3}{\underset{HO\diagup}{\overset{|}{C}}}-CH_2-\overset{\overset{O}{\|}}{C}-SCoA \qquad \text{3-OH-Butyryl-CoA}$$

$$\downarrow \text{(3)} \rightarrow CoASH$$

$$\left[-O-\overset{\overset{CH_3}{|}}{CH}-CH_2-\overset{\overset{O}{\|}}{C}-\right]_n \qquad \text{Polyhydroxybutyric Acid (PHB)}$$

FIGURE 23.5 Biosynthesis of polyhydroxybutyrate (PHB). The enzyme 3-ketothiolase (1) catalyzes the condensation of two acetyl-CoA units to form acetoacetyl-CoA, which is then reduced to 3-hydroxybutyryl-CoA by acetoacetyl-CoA reductase (2). 3-Hydroxybutyryl-CoA is then polymerized by PHA synthase (3). The genes that encode the three enzymes are *phaA*, *phaB*, and *phaC*, respectively.

seed (*Brassica napus*, a close relative of *Arabidopsis*), sunflower (*Helianthus*), and soybean (*Glycine max*) could also produce PHAs on a commercial scale.

Another potential for PHA synthesis by plants is illustrated by the transformation of cotton plants (*Gossypium hirsutum*) with the *phaB* and *phaC* genes (John and Keller, 1996). (Endogenous 3-ketothiolase is present in the cytoplasm of most plants.) In this case the genes were expressed in the fiber or seed hair cells. The amount of PHB that accumulated in the lumen of the fiber cells was not large (0.34 percent of fiber weight), but was sufficient to alter the thermal properties of the fiber. The transgenic fibers conducted less heat than normal fibers, suggesting enhanced insulation characteristics. Further development of this engineered "polyester-cotton blend" could have a significant impact on the textile industry.

Other Products In addition to carbohydrates, amino acids, fatty acids and other compounds involved in primary metabolism, plants produce large numbers of secondary products (Chap. 14). These include alkaloids and other drugs, gums, resins, essential oils, and a variety of other chemicals whose role in the plant are poorly understood. Many of these have been exploited by humans for centuries. Plants were used in China for medicinal purposes as far back as 2700 B.C. By 1820, 70 percent of the drugs listed in *U.S. Pharmacopoeia* originated from plants. Even though today many of our drugs are chemically manufactured, probably 25 percent are derived from plants. There may be many drugs yet to be discovered and exploited. One area of high potential pharmaceutical interest relates to cancer chemotherapeutic drugs from higher plants. This interest has been stimulated by the recent discovery that **taxol**, a diterpenoid harvested from the inner bark of the Pacific yew (*Taxus brevifolia*), shows promise as an antitumor agent in breast and ovarian cancers (Stierle et al., 1993). Natural drugs are sometimes preferred over manufactured drugs because they can be produced in greater yield, at lower cost, and with higher purity than synthetic alternatives. Another exciting possibility is that drugs and other secondary products might be mass-produced at even lower cost in cell or tissue culture.

In addition to drugs, there are a wide variety of other plant chemicals that have potential as chemical feedstocks for industry. Many of these are only just beginning to attract the interest of industry and, like drugs, the potential for efficient, large-scale production of these chemicals in tissue culture is given serious consideration.

Finally, as an alternative to cell and tissue culture, engineered plants could serve as low-cost delivery vehicles for vaccines against common diseases in humans (Mason and Arntzen, 1995). Although recombinant vaccines are being produced in yeasts and other microorganisms, these vaccines are expensive because, as noted earlier, they require fermentation and subsequent purification of the antigens. Oral delivery of vaccines through edible plant tissue could be a very cost-effective alternative that would benefit especially developing countries where vaccines are urgently needed.

As a first step, the gene encoding hepatitis B surface antigen has been successfully cloned into tobacco (*Nicotiana tobacum*). (Tobacco was chosen for its ease of genetic manipulation and its extensive use as a model laboratory plant.) Not only was the gene expressed, but the antigenic protein was synthesized and assembled in an immunologically active form. More recently, the gene encoding Norwalk virus capsid protein (NVCP) has been successfully cloned into potato (*Solanum tuberosum*) (Mason et al., 1996). The Norwalk virus is a common agent of acute gastroenteritis in humans. The cloned gene was expressed in the tubers and the resulting NVCP self-assembled into viral-like particles. When engineered potato tubers were fed to mice, the mice responded by producing antibodies against NVCP. Oral vaccinations are typically less effective and required higher doses than injections and the amount of antibody produced by the mice was low. Nevertheless, these results indicate both the feasibility of producing vaccines in plants and the potential for effective immunization through oral delivery. The system has yet to be tested on humans and there are still questions to be answered. For example, which food product will be the most effective delivery vehicle (bananas are the favored candidate), or can oral vaccines protect against diseases not normally spread through food and water? This work is in its infancy, but it is representative of the exciting advances we can look forward to in the future for plant physiology and biotechnology.

SUMMARY

Biotechnology is as ancient as the production of alcoholic beverages, cheese, and bread. Modern plant biotechnology is based on advances in molecular genetics and recombinant DNA technology, but has been facilitated by developments in cell and tissue culture and plant cell protoplasts. Cell, tissue, and protoplast culture allows plant cells to be modified and then regenerated into mature, sexually competent plants that can pass the modifications on to their progeny.

Recombinant DNA technology provides methods for isolating and modifying genes and inserting new genes (DNA fragments) into plants. Common vectors for inserting DNA into plants include the crown-gall bacterium *Agrobacterium tumifaciens*. Other treatments that render the cell membrane temporarily permeable to large DNA fragments or direct microinjection are also effective.

Micropropagation is a technique for large-scale cloning of high-quality, genetically uniform plants. It is also used to propagate virus-free lines of orchids, potatoes, and forest tree species.

One area where biotechnology is having an impact is on insect and disease resistance. Many crop species genetically engineered to exhibit resistance to specific herbicides or insects are now in commercial production in North America. Another area of importance is in engineering plant metabolism. Efforts are being made to increase edible oil production by plants or to change the fatty acid composition to meet specific needs. The capacity of plants to produce biodegradable plastics, drugs, and oral vaccines is also being developed.

CHAPTER REVIEW

1. What is biotechnology?
2. Describe how genetic engineering can be used to improve weed control.
3. What is systemic acquired immunity and how does it help to protect plants against secondary infection?
4. Describe tissue and protoplast culture and list the various ways these techniques can be used to practical advantage.
5. What role do you see for the plant physiologist in the future?

FURTHER READING

Fuller, K. W., J. R. Gallon (eds.). 1985. *Plant Products and the New Technology.* Oxford: Clarendon Press.

Green, C. E., D. A. Somers, W. P. Hackett, D. D. Biesboer (eds.). 1987. *Plant Tissue and Cell Culture.* New York: Liss.

Gresshoff, P. M. (ed.) 1992. *Plant Biotechnology and Development.* Boca Raton: CRC Press.

Grierson, D. (ed.) 1991. *Plant Genetic Engineering.* London: Blackie.

Grierson, D., S. M. Covey. 1988. *Plant Molecular Biology.* New York: Methuen.

Kung, S., C. J. Arntzen (eds.). 1989. *Plant Biotechnology.* Boston: Butterworth.

Smith, J. E. 1988. *Biotechnology.* 2nd ed. London: Edward Arnold.

REFERENCES

Abel, P. P., R. S. Nelson, B. De, N. Hoffman, S. G. Rogers, R. T. Fraley, R. N. Beachy. 1986. Delay of disease development in transgenic plants that express the tobacco mosaic virus coat protein. *Science* 232:738–741.

Brears, T., J. Ryals. 1994. Genetic engineering for disease resistance in plants. *Agro-Food Industry Hi-Technology* July/Aug:10–13.

Broglie, K., I. Chet, M. Holliday, R. Cressman, P. Biddle, S. Knowlton, C. J. Mauvais, R. Broglie. 1991. Transgenic plants with enhanced resistance to the fungal pathogen *Rhizoctonia solani. Science* 254:1194–1197.

Calvin, M. 1985. Fuel oils from higher plants. In: K. W. Fuller, J. R. Gallon (eds.), *Plant Products and the New Technology.* Oxford: Clarendon Press, pp. 147–160.

Holt, J. S., S. B. Powles, J. A. M. Holtum. 1993. Mechanisms and agronomic aspects of herbicide resistance. *Annual Review of Plant Physiology and Plant Molecular Biology* 44:203–229.

John, M. E., G. Keller. 1996. Metabolic engineering in cotton: Biosynthesis of polyhydroxybutyrate in fiber cells. *Proceedings of the National Academy of Science, USA* 93:12768–12773.

Jones, M. G. K. 1985. New plants through tissue culture. In: K. W. Fuller, J. R. Gallon, (eds.), *Plant Products and the New Technology.* Oxford: Clarendon Press, pp. 215–228.

Klein, T. M., E. D. Wolf, R. Wu, J. C. Stanford. 1987. High velocity microprojectiles for delivering nucleic acids into living cells. *Nature (London)* 327:70–72.

Mason, H. S., C. J. Arntzen. 1995. Transgenic plants as vaccine production systems. *Trends in Biotechnology* 13:388–392.

Mason, H. S., J. M. Ball, J.-J. Shi, X. Jiang, M. K. Estes, C. J. Arntzen. 1996. Expression of Norwalk virus capsid protein in transgenic tobacco and potato and its oral immunogenicity in mice. *Proceedings of the National Academy of Science, USA* 93:5335–5340.

Mazur, B. J., S. C. Falco. 1989. The development of herbicide resistant crops. *Annual Review of Plant Physiology and Plant Molecular Biology* 40:441–470.

Poirier, Y., C. Nawrath, C. Somerville. 1995. Production of polyhydroxyalkanoates, a family of biodegradable plastics and elastomers, in bacteria and plants. *Bio/Technology* 13:142–150.

Somerville, C. R. 1990. The biochemical basis for plant improvement. In: D. T. Dennis, D. H. Turpin (eds.), *Plant Physiology and Plant Molecular Biology.* Essex: Longman Group, pp. 490–501.

Stierle, A., G. Strobel, D. Stierle. 1993. Taxol and taxane production by *Taxomyces andreanae*, an endrophytic fungus of the Pacific yew. *Science* 260:214–216.

Appendix

Measuring Water Potential and Its Components

Because the growth, metabolic activities, and, ultimately, the productivity of plants can be markedly influenced by its water status, it is important to be able to measure water potential and its components both conveniently and accurately. Over the years, physiologists have developed a variety of methods for assessing water potential and its components. By way of example, several of the more commonly employed methods are described in principle here. The student who wishes to pursue methodology in more detail is encouraged to consult the reference list at the end of the chapter.

WATER POTENTIAL

TISSUE WEIGHT-CHANGE METHOD

Water potential of some tissues can be estimated rather simply by equilibrating preweighed samples of tissue in solutions of known osmotic potential. *The objective is to determine which solution has an osmotic potential equivalent to the water potential of the tissue.* If the osmotic potential of the bathing solution is more negative than the water potential of the tissue, water will leave the tissue and the tissue will *lose weight;* if less negative, the tissue will take up water and *gain weight.* That solution in which the tissue neither gains nor loses weight is deemed to have an osmotic potential equivalent to the water potential of the tissue.

In practice, samples of uniform size are prepared, weighed, and placed in solutions of known *molality* (Fig. A.1A). Because it is important that concentration (and, hence, the osmotic potential) of the solution not change significantly during the test, the solute used to prepare the solutions should be one that is not readily taken up by the tissue. Sorbitol, mannitol, and polyethylene glycol (PEG) are commonly used for this purpose. After allowing sufficient time for the tissue and bathing solution to come to equilibrium, the tissue is blotted to remove excess solution and weighed once again. The weight gain or loss is then calculated as a percentage of the original weight and plotted against the concentration of the solution (Fig A.1B).

As noted earlier, osmotic potential is one of the colligative properties of solutions that is solely dependent on mole fraction of dissolved solute. The osmotic potential of a solution can be calculated from the van't Hoff equation:

$$\Psi_S = -C\gamma RT.$$

Van't Hoff determined this relationship empirically by plotting measured osmotic pressures (with an osmometer) against solute concentration. C is the *molal* concentration (moles per kilogram of water). Note that *molality* is independent of temperature and is a more suitable unit for these purposes than *molarity.* Below about 0.2 M, molarity and molality are reasonably close, but at high concentrations molarity can be significantly less than molality for the same solution. Other terms in the above equation are γ, the activity coefficient; R, the universal gas constant (0.00831 kg MPa mol^{-1} $°K^{-1}$); and T, the absolute temperature (°K = °C + 273). If C, γ, and T are known, then the water potential can readily be calculated. Recall that the value of γ is invariably less than one. However, for neutral solutes (such as sorbitol or mannitol) in dilute solution, a value of 1.0 can be assumed without significant error. Hence, for a 1.0 m solution of sorbitol at 20 °C, the osmotic potential is calculated as $-(1.0 \text{ mol kg}^{-1})(1.0)(0.00831 \text{ kg MPa mol}^{-1} \text{°K}^{-1})(293 \text{°K}) = -2.43$ MPa.

As with most methods, water potential estimates based on the tissue weight-change method are an average for all the cells in the tissue. However, because of its simplicity, this method is commonly used in the classroom to demonstrate principles of water potential.

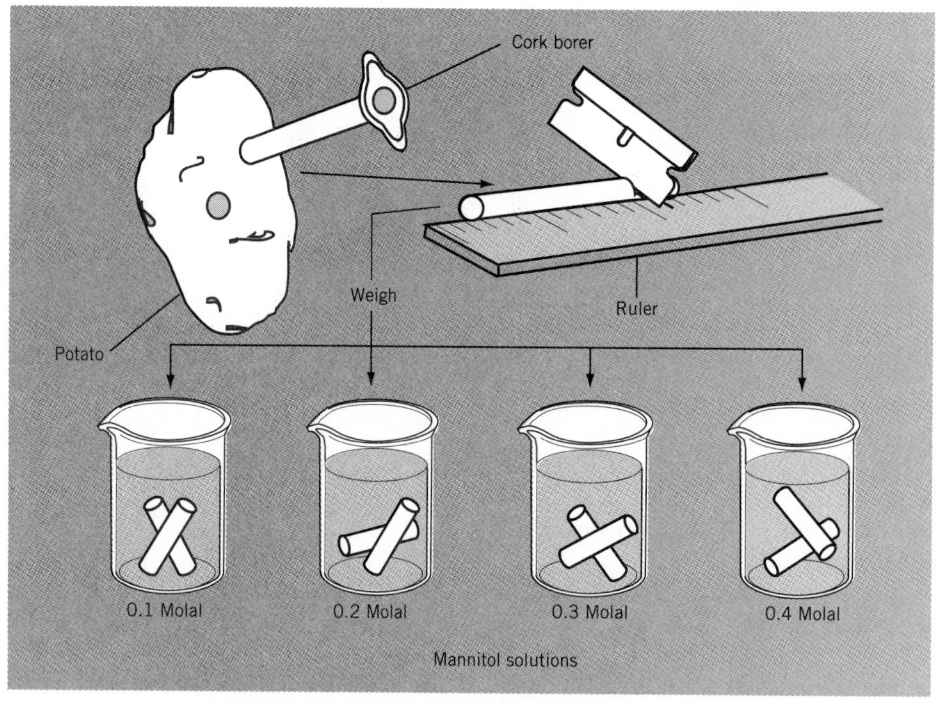

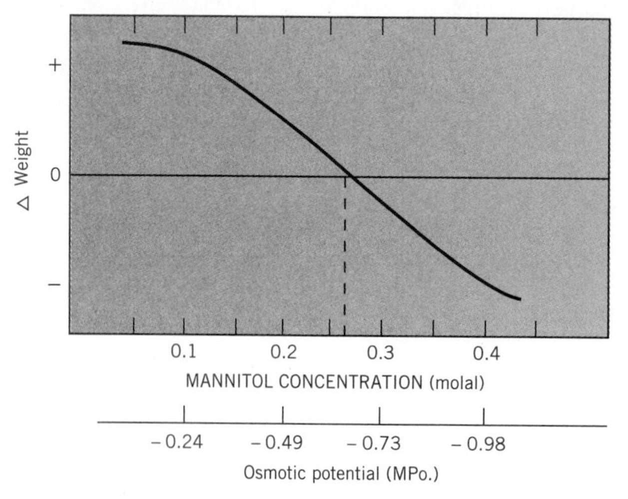

FIGURE A.1 Measuring water potential and its components. (*A*) A laboratory method for determining the water potential of potato tuber tissue by change in mass. (*B*) Typical results for an experiment conducted as shown in (*A*). Water potential of the tissue sample is estimated as equivalent to the osmotic potential of the solution in which there is no net exchange of water between the tissue and the solution.

THERMOCOUPLE PSYCHROMETRY

The favored method of measuring water potential of plant tissues in the laboratory is based on the technique of **thermocouple psychrometry.** Psychrometry refers to measurements of the difference in temperature between an atmosphere and a freely evaporating moist surface in that atmosphere. The rate at which molecules evaporate and, therefore, the rate of cooling depend primarily on the vapor pressure of water in the atmosphere. The lower the vapor pressure, the steeper the vapor pressure gradient and the more rapid the evaporation. The psychrometric technique is thus used to determine small differences in vapor pressure. This is the principle behind use of a wet-and-dry bulb thermometer to measure the relative humidity of an atmosphere. A thermocouple is formed where wires of two dissimilar metals are joined, such as copper and constantan (a copper-nickel alloy). If both ends of the wires are joined to form a closed circuit with two thermocouple junctions and if the two junctions are held at different temperatures, an electrical current will flow through the circuit. This is known as the *Seebeck effect.* The magnitude of the current is a measure of the difference in temperature between the two junctions.

Thermocouple psychrometry depends on the principle that water vapor at equilibrium with plant tissue will have the same water potential as the tissue. The water potential of the tissue can thus be measured by measuring the vapor pressure in the chamber. There are many variations on the theme, but the basic method involves sealing the plant material to be measured in a small chamber (Fig A.2). The chamber also contains a thermocouple junction with a silver ring on which a drop of water has been placed (silver is used because it has a high thermal conductivity). The chamber is very small so that the atmosphere quickly saturates with water vapor. However, because the water potential of the tissue is more negative water than that of the water drop, water from the atmosphere will diffuse into the tissue. As water continues to evaporate from the drop in order to replace the vapor taken up by the tissue, the resulting evaporative cooling depresses the temperature of the thermocouple, which can be measured against a reference. It is assumed that the rate of diffusive transfer is proportional to the difference in water potential between the thermocouple and the tissue. Thus, knowing both the rate of cooling at the junction and the ambient temperature, it is possible to calculate the vapor pressure of the atmosphere and, hence, water potential of the tissue.

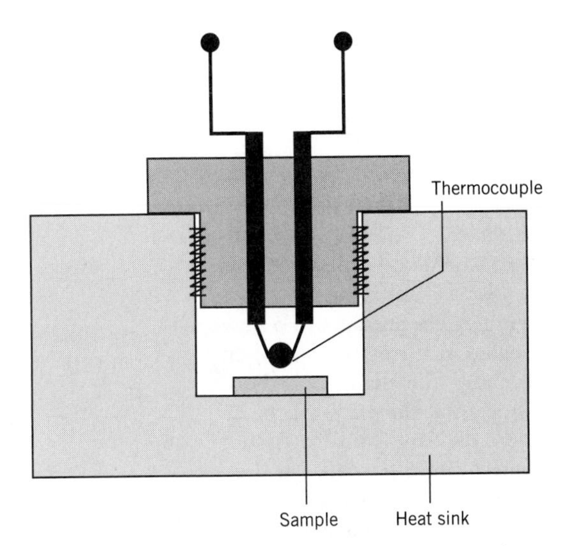

FIGURE A.2 Diagram of the thermocouple psychrometer sample chamber.

There is a certain amount of error inherent in such measurements, due largely to the diffusive resistance, or resistance to the transfer of water vapor, on the part of the tissue. This and certain other inaccuracies inherent in the basic technique were largely eliminated by a variation called the *isopiestic* technique, introduced by Boyer and Knipling (1965). Instead of water, a drop of solution with known solute concentration, and hence known water potential, is placed on the thermocouple junction. If the water potential of this solution ($\Psi_{solution}$) is greater than that of the tissue, water will be transferred from the solution to the tissue and evaporative cooling will occur as above. On the other hand, if $\Psi_{solution}$ is lower than that of the tissue, water vapor will be transferred from the tissue to the solution. The vapor will condense on the thermocouple and its temperature will increase. However, when the water potential of the solution on the thermocouple is the same as that of the tissue, the vapor pressure at the thermocouple and the tissue will be equal. There will be no diffusive transfer of water between the tissue and the thermocouple and, consequently, no evaporative cooling or diffusion error. The temperature of the thermocouple will be the same as the ambient temperature. This technique is particularly useful because it can be used to measure the water potential of solutions and soils as well as plant tissues. It can also be quite sensitive, with a resolution in the range of 0.01 MPa, and is adaptable to field applications.

OSMOTIC POTENTIAL

CRYOSCOPIC METHODS

One of the colligative properties of solutions that can be used to determine osmotic potential of a solution is its **freezing point depression.** A 1.0 molal solution, which has an osmotic potential of −2.43 MPa, will freeze at −1.86 °C. The osmotic potential therefore changes at the rate of −2.43 MPa/1.86 °C, or −1.30 MPa $°C^{-1}$. A variety of instruments are available for measuring freezing point depression. Some instruments monitor temperature as the sample first supercools and then, as the water crystallizes, warms to the actual freezing point. Others involve freezing the sample first. The sample is then gradually warmed, causing the ice to melt. The investigator monitors the sample with a microscope and records the temperature at which the last crystal melts. Results are commonly expressed as **osmolality,** a measure of the total concentration of dissolved particles in a solution without regard to molecular species.

Cell sap is, of course, a complex solution and its osmotic potential represents the sum of the individual osmotic potentials contributed by each solute, many of which dissociate to some extent. A variety of methods for obtaining plant sap have been used. Most commonly the plant tissue is subjected to a freeze–thaw cycle in order to disrupt the membranes and the sap is then squeezed out with a press. Regardless of the method used, the sap is subjected to some degree of mixing with other cytoplasmic contents. Another source of error is the dilution of the cell contents with apoplastic water, which leads to an underestimation of osmotic values. Some instruments, however, will accept nanoliter volumes so that freezing points can be determined on sap extracted from single cells with a micropipet.

INCIPIENT PLASMOLYSIS

Osmotic potentials may also be estimated by observing **incipient plasmolysis.** This is especially useful for leaf epidermal cells and other cells that lend themselves to microscopic observation. The method is based on the assumption that at incipient plasmolysis—when the protoplasts are just beginning to pull away from the cell wall—turgor pressure is equal to zero and the water potential of the cell is equivalent to the osmotic potential alone (refer to Figure 2.6). Tissue samples are allowed to equilibrate (about 30 min) with a series of solutions of known osmotic potentials. The tissue is then examined under a microscope for incipient plasmolysis. The solution in which 50 percent of the cells exhibit incipient plasmolysis is considered to have an osmotic potential equivalent to the osmotic potential of the cells.

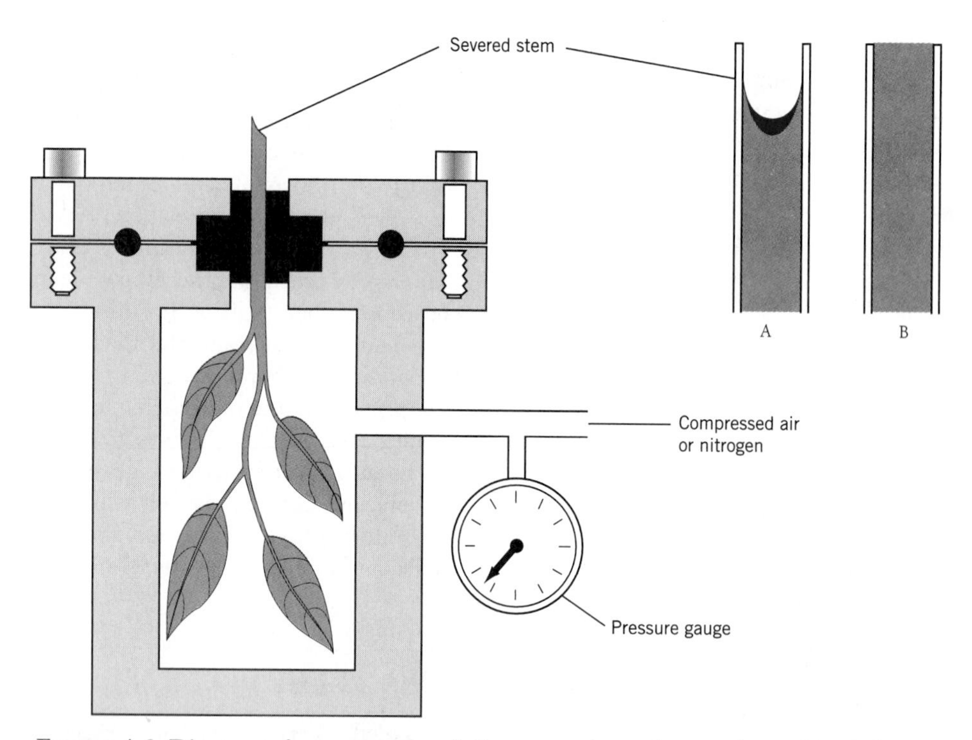

FIGURE A.3 Diagram of a pressure bomb for measuring xylem pressure. To make a measurement, a severed shoot or branch is quickly sealed in a pressure chamber. Only the severed end is left protruding from the chamber. The chamber is then pressurized with nitrogen gas until xylem sap just emerges at the cut surface. At this point, the positive pressure required to force xylem exudate from the tissue (displayed by the pressure gauge) is equal to the negative water potential of the xylem. If it is assumed that the osmotic potential of the xylem is small and can be ignored, the pressure in the xylem is approximately equal to its water potential. The position of the water column in a xylem vessel before and after pressurizing the chamber is illustrated in (*A*) and (*B*), respectively.

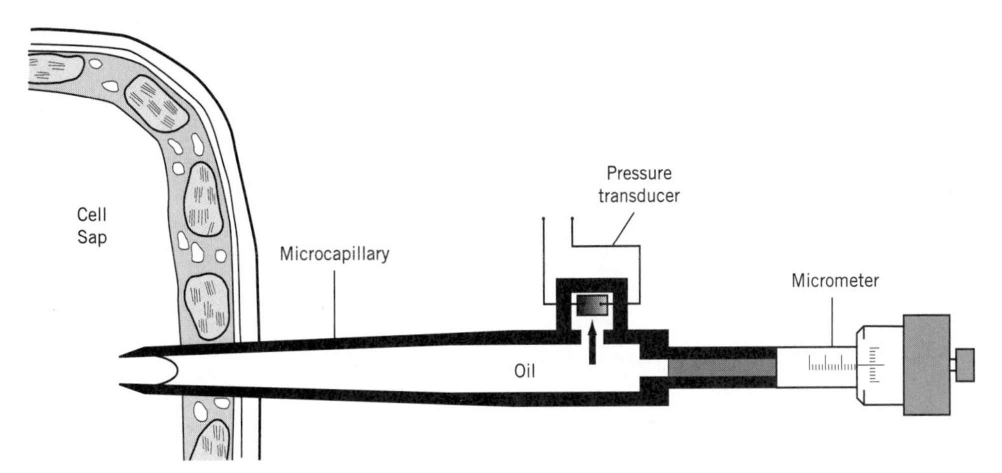

FIGURE A.4 A micromanometer for measuring pressure changes in small cells.

PRESSURE MEASUREMENTS

The **pressure bomb** (Fig. A.3) is a useful device for measuring the negative hydrostatic pressures (tensions) normally present in xylem vessels (Scholander et al., 1965). The shoot is excised and sealed in a pressure chamber with only the cut surface of the stem exposed. If the water column in the xylem is under tension, as it normally is in a transpiring shoot, it will withdraw from the surface when the stem is cut. Pressuring the chamber will force the water back to the surface. The pressure required to bring the water back to the surface is known as the *balance pressure.* It is considered of equal magnitude but of opposite sign to the tension than existed in the xylem prior to excision.

Methods for direct measurement of turgor pressure in cells were pioneered by P. B. Green (Green and Stanton, 1967). Green devised a micromanometer (a manometer is a pressure measuring device) by closing off one end of a microcapillary tube and drawing out the other end to a fine point. When inserted into the vacuole of a giant cell of *Nitella*, a filamentous alga, the pressure of the cell serves to compress the gas volume in the tube. By measuring the volume change (using a microscope, of course) and applying the ideal gas laws (the product of pressure and volume is constant), the turgor pressure of the cell can be calculated. A more sophisticated version was subsequently developed by Zimmermann and Steudle (1974) for measurements in the smaller cells of higher plants (Fig. A.4). The glass microcapillary tube in this case is oil-filled and attached to a micrometer fitted with a sensitive pressure-sensing transducer. A transducer is a device that converts one form of energy to another—in this case, pressure is converted to electrical energy, which can be amplified and displayed on a meter. As with Green's micromanometer, when the tip of the manometer penetrates the cell, the cell fluids will flow into the capillary. A turn of the micrometer screw will return the cytoplasm to the tip of the capillary and restore the original cell volume. At this point the pressure inside the cell is balanced exactly by the pressure in the manometer, which is detected by the pressure-sensing transducer. This device also makes it possible to introduce small changes in volume and to measure the pressure changes that result, which is the basis for calculating the volumetric modulus of elasticity for cell walls.

REFERENCES

Boyer, J. S., E. B. Knipling. 1965. Isopiestic technique for measuring leaf water potentials with a thermocouple psychrometer. *Proceedings of the National Academy of Sciences USA* 54:1044–1051.

Green, P. B., R. W. Stanton. 1967. Turgor pressure: Direct manometric measurement in cells of *Nitella. Science* 155:1675–1676.

Scholander, P. F., H. T. Hammel, E. D. Bradstreet. 1965. Sap pressure in vascular plants. *Science* 148:339–346.

Zimmermann, U., E. Steudle. 1974. The pressure-dependence of the hydraulic conductivity, the membrane resistance and membrane potential during turgor regulation in *Valonia utricularis. Journal of Membrane Biology* 16:331–352.

Index

H